A TREATISE
on the Theory of
Determinants

A TREATISE

on the Theory of Determinants

BY

THOMAS MUIR, C.M.G., M.A., F.R.S., F.R.S.E.

REVISED AND ENLARGED BY

WILLIAM H. METZLER, Ph.D., F.R.S.E., F.R.S.C.

DOVER PUBLICATIONS, INC.
NEW YORK NEW YORK

Published in the United Kingdom by Constable and Company Limited, 10 Orange Street, London, W.C. 2.

This new Dover edition, first published in 1960, is an unabridged and corrected republication of the edition originally published by Longmans, Green and Company in 1933.

Manufactured in the United States of America

Dover Publications, Inc.
180 Varick Street
New York 14, N. Y.

PREFACE

When I started some years ago to make a revised edition of this text I did not anticipate that it would be such a large task and take so long a time, but the more I got into it, the more I found to do and as the object was to include pretty much all the known theorems to date, the work of covering the literature was great and indeed for one who was carrying a heavy load of teaching and administrative work it would have proven nigh impossible had not Dr. Muir's *Theory of Determinants in the Historical Order of Development* been available and down so near to date.

I have tried to preserve the wording of the original test in so far as possible but it is now forty-five years since it was published and much has been done in the field since then so there has of necessity been some change and much additional material. A number of important results are imbodied among the problems and for the most part references have been intentionally omitted since the History is so available and gives every detail. The book will undoubtedly have some imperfections but I trust not over many, and for any such the blame should fall upon the reviser and not upon Dr. Muir, who has left me with a free hand.

The reviser wishes to record his thanks to many former students for their assistance and especially to Mr. Ralph Beaver. He also wishes to express his indebtedness to Professor J. J. Nassau who practically put into form the Chapter on Alternants and to Professor L. H. Rice who wrote the chapter on Determinants of Higher Class.

W. H. METZLER

New York State College for Teachers.
Albany, N. Y.
July, 1928.

TABLE OF CONTENTS

A TREATISE
on the Theory of
Determinants

CHAPTER I

Section 1

PERMUTATIONS

1. With the n quantities $a_1, a_2, \cdots, a_n$, which we may suppose arranged in order of magnitude, we may form $n!$ permutations. The elements of a permutation are said to be arranged in their natural order when the suffixes are arranged in the order of magnitude. The element a_r is said to be greater than the element a_s when $r>s$. Any permutation in which an element precedes another with a less suffix is said to contain an *inversion*. The number of inversions in any permutation is the same as the number of interchanges of consecutive elements *necessary* to arrange them in their natural order.

Thus the elements in the permutation $a_4a_7a_8a_9$ are arranged in their natural order and the permutation $a_6a_5a_7a_3a_8$ contains the following inversions: a_7a_3, a_5a_3, a_6a_3, a_6a_5.

Permutations naturally divide themselves into two classes according as they contain an even or an odd number of inversions.

Since an element of any permutation may be changed from one position to any other by repeated interchanges with adjacent elements it is easily seen that we can from any one obtain all the other $n!$ permutations by the successive interchange of two elements.

2. *If from any permutation we form another by the interchange of any two elements then the difference between the number of inversions in the two will be an odd number.*

Suppose the number of elements between the two interchanged is m and let the given permutation be represented by

$$A\,a_rBa_sC$$

where a_r and a_s are the elements to be interchanged, A is the group of elements preceding a_r, B is the group of elements between a_r and a_s, and C is the group of elements following a_s.

The resulting permutation would be

$$A\,a_sBa_rC\,.$$

Let α denote the number of elements in $B<a_r$ and let β denote the number of elements in $B<a_s$, then $m-\alpha$ is the number of elements in $B>a_r$ and $m-\beta$ is the number of elements in $B>a_s$.

We have then to consider the change in the number of inversions due, first to carrying a_r over the elements B, second to the interchange of a_r and a_s, third to carrying a_s forward over the elements B. In the first case the number of inversions will be diminished by α and increased by $m-\alpha$, in the second it will be either increased or diminished by one according as $a_r \lesseqgtr a_s$, and in the third it will be increased by β and be diminished by $m-\beta$. The change in the number of inversions due to the interchange of two elements is therefore

$$-\alpha + m - \alpha \pm 1 + \beta - m + \beta = 2(\beta - \alpha) \pm 1$$

which is an odd number.

If a_r and a_s are adjacent elements then the change in the number of inversions is ± 1 according as $a_r \lessgtr a_s$.

EXAMPLES.

1. Find the number of inversions of order in the series

$$7,\ 8,\ 4,\ 1,\ 3,\ 2,\ 9,\ 6,\ 5.$$

Taking 7 along with each of the numbers which follow it, we have the couplets

$$(7, 8),\ (7, 4),\ (7, 1),\ (7, 3),\ (7, 2),\ (7, 9),\ (7, 6),\ (7, 5),$$

and of these it is evident that *six* are instances of inversion of order; taking 8 along with each of the numbers which follow it, we obtain other *six* inversions; and proceeding in like manner with 4 and the other numbers of the series, we find that the total number of inversions

$$= 6 + 6 + 3 + 0 + 1 + 0 + 2 + 1,$$
$$= 19.$$

2. Count the number of inversions of order in

$$3, 6, 4, 1, 5, 2;\ \ 7, 1, 6, 5, 3, 4, 2;\ \ 3, 2, 9, 4, 1, 6, 7;$$
$$4, 8, 6, 7, 2, 5, 3;\ \ 7, 8, 3, 2, 1, 4, 6, 5;\ \ 3, 1, 9, 8, 2, 6, 5, 7, 4.$$

3. If one element of a permutation be made to take up a new place, by being, as it were, passed over m other elements, the number of inversions in the new permutation will differ from the number in the original by m.

Let a be the element made to pass over m other elements, and suppose h of these m elements are less than a and consequently $m-h$ of them will be greater than a. The number of changes of inversions will therefore be $m-h+h$ or m.

4. If, in the natural series of integers $1, 2, 3, \cdots, n$, a group of m consecutive numbers taken in any order be passed backward over r

groups of s consecutive numbers each, the numbers of each group being arranged in any order, the number of additional inversions in the permutations thus obtained is $m \cdot r \cdot s$.

This is self-evident, for the number of numbers passed over is clearly $r \cdot s$, and each of the m numbers to which precedence has been given will thus give rise to $r \cdot s$ inversions, making $m \cdot r \cdot s$ inversions in all.

If the number of numbers in the group moved is the same as the number in each of the groups passed over,—that is, when $s = m$,—then the number of additional inversions is, of course, $r \cdot m^2$. Consequently, when m is even, the number of additional inversions will be always even whatever r may be; and when m is odd, the number of additional inversions will be odd or even according as r is odd or even. From these we get the following theorem:

If the first n m integers be separated into n groups of m each, say the groups $A_1, A_2, \cdots, A_n$, *so that*

A_1 *stands for any permutation of* $1, 2, \cdots, m$

A_2 *stands for any permutation of* $m+1, m+2, \cdots, 2m$

. . . .

A_n *stands for any permutation of* $mn-m+1, mn-m+2, \cdots, nm$ *then, when m is even, any one of the n! permutations of* $A_1A_2, \cdots, A_n$, *has an even number of inversions of the* $n \cdot m$ *integers more than the standard permutation* $A_1, A_2, \cdots, A_n$ *has; and, when m is odd, has an odd or an even number of inversions more, according as the suffixes of the A's have an odd or an even number of inversions.*

5. *Of the n! permutations of the elements* $a_1, a_2, \cdots, a_n$, *there are as many that have an even number of inversions as there are that have an odd number; that is there are* $n!/2$ *in each of the two classes.*

This is easily seen to be true when n has the values 2 or 3 and suppose it is true when n has the value $m-1$; that is, suppose that of the $(m-1)!$ permutations of the elements $a_1, a_2, \cdots, a_{m-1}$ there are $(m-1)!/2$ in each class.

Starting now with the element a_m at the end of any one of these permutations and moving it successively over each of the $m-1$ elements on its left we get m resulting permutations of m elements of which the first, third, fifth, $\cdots$, belong to the same class as the given permutation and the second, fourth, $\cdots$, belong to the opposite class, that is, $m/2$ or $(m+1)/2$ belong to the same class and $m/2$ or $(m-1)/2$ belong to the opposite class according as m is even or odd.

It follows therefore that if half of the permutations of the first $m-1$ elements belong to each class half of the permutations of the first m

elements will belong to each class and since this is true for the first two or three elements it is true in general.

6. If in the permutation $a_1a_2 \cdots a_n$ we move each element one place to the right the last element being placed first we get the permutation $a_na_1a_2 \cdots a_{n-1}$ which is said to be obtained from the first by a cyclic interchange.

This is evidently equivalent to $n-1$ interchanges and therefore the resulting permutation will belong to the same class as the original or to the opposite according as n is odd or even.

Performing $n-1$ successive cyclic interchanges will give $n-1$ cyclic permutations. The next cyclic interchange would give the original.

7. Taking any one of the $n!$ permutations of the elements a_1, a_2, $\cdots$, a_n we can form $(n-r)!$ others by permuting in all possible ways the last $(n-r)$ elements leaving the first r the same in each. It is evident that no two of these are alike and that no other of the $n!$ permutations contains the same r elements in the same position. It follows therefore that any $(n-r)!$ of the $n!$ permutations of the n things cannot have more than r elements in the same position.

8. Two permutations of the numbers 1, 2, 3, $\cdots$, n are called *conjugate* when each number and the number of the place which it occupies in one permutation are interchanged in the case of the other. For example, the permutations

(A) 3 8 5 10 9 4 6 1 7 2

(B) 8 10 1 6 3 7 9 2 5 4

are conjugate, because 3 is in the 1st place of (A) and 1 is in the third place of (B) and so on in every case. A permutation may be conjugate with itself. Thus the permutation

6 3 2 4 5 1

which has 6 in the 1st place, has also 1 in the 6th place, and so on. A *self-conjugate* permutation may consequently be defined independently, as one in which each element is either in its original position or has taken part in one interchange.

Conjugate permutations are identical with those which Jacobi calls *reciprocal*, his definition being that two permutations are so called when the performance of the one upon the other gives rise to the primitive permutation. For example 3 5 2 1 4 and 4 3 1 5 2 are reciprocal permutations, because $(3\ 5\ 2\ 1\ 4)(4\ 3\ 1\ 5\ 2) = 1\ 2\ 3\ 4\ 5$.

9. The number of self-conjugate permutations of the elements 1, 2, 3, $\cdots$, n is

$$1 + {}_nC_2 + \tfrac{1}{2}{}_nC_2\,{}_{n-2}C_2 + \tfrac{1}{2}\cdot\tfrac{1}{3}{}_nC_2\,{}_{n-2}C_2\,{}_{n-4}C_2 + \cdots$$

where ${}_nC_r$ stands for

$$\frac{n(n-1)\cdots(n-r+1)}{1\cdot 2\cdot 3\cdots r} \quad\text{or}\quad \frac{n!}{(n-r)!r!}\,.$$

This is best established as it was first obtained, namely, by classifying the instances of self-conjugateness, and then making a census of the classes. A basis of classification exists in the varying number of elements retaining their primitive positions in the permutations. Taking this basis we see at once that we have to consider in order the classes

1. where no element is changed in position,
2. where two are changed
3. where four are changed
4. where six are changed,

. . . .

In the first class there is manifestly only 1, namely, the primitive permutation. In the second class there are as many as there are different pairs of elements, viz., ${}_nC_2$, for example,

$$2\ 1\ 3\ 4\ 5 \cdots n, \qquad 3\ 2\ 1\ 4\ 5 \cdots n.$$

In the case of the third class we have to find how many pairs of pairs are possible, each having the four elements involved all different. For the first of the two pairs we have, as has just been seen, ${}_nC_2$ to choose from; for the second there are only ${}_{n-2}C_2$ to choose from, because there are two fewer elements to make the pairs of; consequently the number required is $\frac{1}{2}\ {}_nC_2\ {}_{n-2}C_2$, the $\frac{1}{2}$ being due to the fact that the order in which the two pairs are taken is immaterial. In the case of the fourth class, we have to ascertain how many triads of pairs are possible, each having the six elements involved all different; and the result is similarly found to be $1/3!\,{}_nC_2\,{}_{n-2}C_2\,{}_{n-4}C_2$. The remaining classes manifestly follow the same law, consequently the proposition is established.

10. If U_n stands for the number of self-conjugate permutations of the elements $1, 2, 3, \cdots, n$, then

$$U_n = U_{n-1} + (n-1)U_{n-2}$$

If, having examined the case of the elements $1, 2, 3, \cdots, (n-1)$ we bring our nth element n to join them, it is clear that we have two possibilities to consider, namely. (1) the element n remaining in its place, (2) the said element suffering interchange with one of the others. Now, when it remains in its own proper position, the number of self-conjugate permutations is unaltered from the previous case, that is to say, is U_{n-1}; and when it is taken with one of the $(n-1)$ other elements to form a pair for interchange, there will arise, by reason of the remaining $n-2$ elements U_{n-2} self-conjugate permutations, that is to say, in all $(n-1)U_{n-2}$. We have, consequently, as was to be proved

$$U_n = U_{n-1} + (n-1)U_{n-2}.$$

As there is evidently 1 self-conjugate permutation when $n=1$, and 2 when $n=2$, the first ten values U_n are

$$\begin{array}{lllll} U_1 = 1, & U_3 = 4, & U_5 = 26, & U_7 = 232, & U_9 = 2620, \\ U_2 = 2, & U_4 = 10, & U_6 = 76, & U_8 = 764, & U_{10} = 9496. \end{array}$$

The expression first got for the number of self-conjugate permutations ought, of course, and may be shown to satisfy this difference equation.* To show directly that it does, it is best in the first place to simplify it a little. The typical term

$$= \frac{1}{r!}\,{}_nC_2\cdot{}_{n-2}C_2\cdot{}_{n-4}C_2 \cdots {}_{n-2r+2}C_2,$$

$$= \frac{1}{r!}\,\frac{n(n-1)}{1\cdot 2}\cdot\frac{(n-2)(n-3)}{1\cdot 2}\cdots\frac{(n-2r+2)(n-2r+1)}{1\cdot 2}$$

$$= \frac{n(n-1)\cdots(n-2r+1)}{r!2^r}$$

$$= \frac{n(n-1)\cdots(n-2r+1)}{(2r)!}\cdot\frac{(2r)!}{r!2^r}$$

$$= 1\cdot 3\cdot 5\cdots(2r-1)_nC_{2r}.$$

The whole expression consequently becomes

$$1 + 1\cdot{}_nC_2 + 1\cdot 3_nC_4 + 1\cdot 3\cdot 5_nC_6 + \cdots.$$

Taking this for U_n, we have

$$\begin{aligned} U_{n-1} + (n-1)U_{n-2} &= 1 + 1\cdot{}_{n-1}C_2 + 1\cdot 3\cdot{}_{n-1}C_4 + 1\cdot 3\cdot 5\cdot{}_{n-1}C_6 \\ &\quad + (n-1)\{1 + 1\cdot{}_{n-2}C_2 + 1\cdot 3\cdot{}_{n-2}C_4 + \cdots\} \\ &= 1 + 1\{{}_{n-1}C_2 + (n-1)\} \\ &\quad + 1\cdot 3\left\{{}_{n-1}C_4 + \frac{n-1}{3}\,{}_{n-2}C_2\right\} + \cdots. \end{aligned}$$

* See Muir, Proc. Royal Soc. Edin. Dec. 1889.

But the general term on the right here

$$= 1\cdot 3\cdot 5 \cdots (2r-1)\left\{{}_{n-1}C_{2r} + \frac{n-1}{2r-1}{}_{n-2}C_{2r-2}\right\}$$

$$= 1\cdot 3\cdot 5 \cdots (2r-1)\{{}_{n-1}C_{2r} + {}_{n-1}C_{2r-1}\}$$

$$= 1\cdot 3\cdot 5 \cdots (2r-1){}_nC_{2r}.$$

It therefore follows that

$$U_{n-1} + (n-1)U_{n-2} = 1 + 1\cdot {}_nC_2 + 1\cdot 3{}_nC_4 + 1\cdot 3\cdot 5{}_nC_6 + \cdots$$
$$= U_n.$$

11. Let A denote any permutation of n elements and let A' denote its conjugate. If B denote the permutation formed from A by interchanging any two of its elements as for instance a and b, then B' the conjugate of B is evidently obtained by interchanging the elements in the ath and bth places in A'. Since this is true and since all permutations may be obtained from any one by the successive interchange of two elements, it follows that conjugate permutations belong either to the same or to opposite classes. But the two permutations 3 1 2 4 5 $\cdots$ n and its conjugate 2 3 1 4 5 $\cdots$ n belong to the same class; therefore conjugate permutations belong to the same class. Instead of these two conjugate permutations we might have taken any self-conjugate permutation.

Section II

Combinations

12. If we are given any combination of n numbers m at a time, the combination of the remaining $(n-m)$ numbers is said to be the complementary with respect to n of the given combination.

Let $(n|m_1), (n|m_2), \cdots, (n|m_\mu)$ denote the $(n)_m \equiv {}_nC_m = \mu$ combinations of the numbers $1, 2, \cdots, n$ taken m at a time, and let $(\bar{n}|m_\alpha)$ denote the complementary of the combination $(n|m_\alpha)$. Let $(n|m_\alpha|l_1), (n|m_\alpha|l_2), \cdots, (n|m_\alpha|l_\lambda)$ denote the $(m)_l = \lambda$ combinations of the numbers in the combination $(n|m_\alpha)$ taken l at a time, and let $(n|\overline{m}_\alpha|l_\beta)$ denote the combination which is the complementary with respect to m of the combination $(n|m_\alpha|l_\beta)$, that is, the combination of the $m-l$ numbers remaining after the numbers in the combination $(n|m_\alpha|l_\beta)$ are taken out of the combination $(n|m_\alpha)$.

This principle of notation may be extended indefinitely, thus, $(n|m_\alpha|l_\beta|k_\gamma)$ will denote the γth selection k at a time of the l numbers in the βth selection of the m numbers l at a time in the αth selection of the n numbers 1, 2, 3, $\cdots$, n, m at a time; $(\bar{n}\,|\bar{m}_\alpha\,|l_\beta\,|k_\gamma)$ will denote the γth selection, k at a time, of the $(n-m-l)$ numbers remaining after the βth selection, l at time, of the (n-m) numbers remaining after the αth selection, m at a time, is taken from the n numbers 1, 2, 3, $\cdots$, n, and so on.

Let $(n\,|\bar{m}_\alpha\,|l_\beta)(n\,|m_\alpha\,|l_\beta)$ denote the combination made up of the numbers in $(n\,|\bar{m}_\alpha|l_\beta)$ in loco followed by the numbers in $(n\,|m_\alpha\,|l_\beta)$ in loco and let $(n\,|\bar{m}_\alpha\,|l_\beta)(n\,|m_\alpha\,|l_\beta)$ denote the combination of the numbers in $(n\,|\bar{m}_\alpha\,|l_\beta)$ and $(n\,|m_\alpha\,|l_\beta)$ in their natural order. Thus (1237)(4568) is the combination 12374568 while (1237)(4568) is the combination 12345678.

Let s followed by any combination denote the sum of the numbers in the combination. Thus $s(n\,|m_\alpha\,|l_\beta)$ denotes the sum of the numbers in the combination $(n\,|m_\alpha\,|l_\beta)$.

If from the n numbers 1, 2, 3, $\cdots$, n we form ρ combinations of m, l, k, $\cdots$, h numbers respectively, where $m+l+k+\cdots+h=n$ then we will extend the idea of complementary combinations by calling this a system of complementary combinations. There would be

$$\left\{\frac{n!}{m!l!k!\cdots h!}\right\}$$

ways of forming such a system. The system would be denoted by

$$(n\,|\,m_\alpha),\quad (\bar{n}\,|\,m_\alpha\,|\,l_\beta),\quad (\bar{n}\,|\,\bar{m}_\alpha\,|\,l_\beta\,|\,k_\gamma),\ \cdots,\ (\bar{n}\,|\,\bar{m}_\alpha\,|\,\bar{l}_\beta\,|\,\bar{k}_\gamma\,|\,\cdots\,|\,h_\delta).$$

13. The total number of inversions in any combination $(n\,|\bar{m}_\alpha\,|l_\beta)(n\,|m_\alpha\,|l_\beta)$ may obviously be divided into three parts, first the number of inversions in $(n\,|\bar{m}_\alpha\,|l_\beta)$; second the number in $(n\,|m_\alpha\,|l_\beta)$; and third the number in the combination formed by the numbers in $(n\,|\bar{m}_\alpha\,|l_\beta)$ arranged in their natural order followed by the numbers $(n\,|m_\alpha\,|l_\beta)$ arranged in their natural order. It is easily seen that if all the numbers in $(n\,|m_\alpha\,|l_\beta)$ are greater than those in $(n\,|\bar{m}_\alpha\,|l_\beta)$ the number of inversions in the third division is zero. We shall hereafter consider only the third of the three parts of the total number of inversions in $(n\,|\bar{m}_\alpha\,|l_\beta)(n\,|m_\alpha\,|l_\beta)$ and therefore let it always be understood that the numbers in any of the simple combinations $(n\,|m_\alpha\,|l_\beta\,|\,\cdots)$ are arranged in their natural order.

14. The number of inversions in the combinations $(\bar{n}\,|m_\alpha)(n\,|m_\alpha)$ is $s(\bar{n}\,|m_\alpha)-(n-m)(1+n-m)/2$. For if the numbers in $(\bar{n}\,|m_\alpha)$ are $a_1a_2\cdots a_{n-m}$ and the numbers in $(n\,|m_\alpha)$ are $b_1b_2\cdots b_m$, then

since the a's and b's together make up the first n integers it follows that there are

$$\begin{array}{ll} a_1 - 1 & b\text{'s less than } a_1 \\ a_2 - 2 & b\text{'s less than } a_2 \\ a_3 - 3 & b\text{'s less than } a_3 \\ \cdots\cdots & \cdots\cdots\cdots \\ a_{n-m} - (n-m) & b\text{'s less than } a_{n-m}. \end{array}$$

The number of inversions is therefore

$$a_1 + a_2 + \cdots + a_{n-m} - \{1 + 2 + \cdots + (n-m)\},$$

or

$$s(\bar{n} \mid m_\alpha) - \tfrac{1}{2}(n-m)(1+n-m).$$

15. The combination $(n \mid \bar{m}_\alpha \mid l_\beta)(n \mid m_\alpha \mid l_\beta) = \pm (n \mid m_\alpha)$ according as the number of inversions is even or odd. Let

$$\begin{aligned} &(n \mid \bar{m}_\alpha \mid l_1)(n \mid m_\alpha \mid l_1) + (n \mid \bar{m}_\alpha \mid l_2)(n \mid m_\alpha \mid l_2) + \cdots \\ &\qquad + (n \mid \bar{m}_\alpha \mid l_\lambda)(n \mid m_\alpha \mid l_\lambda) \end{aligned}$$

(1) $$= \phi(m, l) \cdot (n \mid m_\alpha)$$

then

$$\begin{aligned} &(n \mid m_\alpha \mid l_1)(n \mid \bar{m}_\alpha \mid l_1) + (n \mid m_\alpha \mid l_2)(n \mid \bar{m}_\alpha \mid l_2) + \cdots \\ &\qquad + (n \mid m_\alpha \mid l_\lambda)(n \mid \bar{m}_\alpha \mid l_\lambda) \\ &\qquad = \phi(m, m-l) \cdot (n \mid m_\alpha) \end{aligned}$$

But it is readily seen that $(n \mid \bar{m}_\alpha \mid l_\beta)(n \mid m_\alpha \mid l_\beta) = (-1)^{l(m-l)}$ $(n \mid m_\alpha \mid l_\beta)(n \mid \bar{m} \mid l_\beta)$. Therefore $\phi(m, l) = \pm \phi(m, m-l)$ according as $l(m-l)$ is even or odd. If $l=1$ the signs of the left-hand member of equation (1) are evidently alternately positive and negative and therefore

$$\phi(m, 1) = 1 \text{ or } 0$$

according as m is odd or even.

It is also apparent that $\phi(m, m) = 1$.

16. The set of combinations

$$(n \mid \bar{m}_\alpha \mid l_1)(n \mid m_\alpha \mid l_1),\ (n \mid \bar{m}_\alpha \mid l_2)(n \mid m_\alpha \mid l_2),\ \cdots\ (n \mid \bar{m}_\alpha \mid l_\lambda)(n \mid m_\alpha \mid l_\lambda)$$

may be divided up into groups as follows:

the 1st group containing the first $(m-1)_{l-1}$ combinations
the 2nd group containing the next $(m-2)_{l-1}$ combinations
. . . .
the $(m-l+1)$st group containing the last $(l-1)_{l-1} = 1$ combination.

The first number of the second part of each combination of the rth group is the same and is the rth of the selection of m numbers, that is, the rth of the numbers in $(n|m_\alpha)$. The first $r-1$ numbers in the first part of each combination of the rth group are the same and are the first $r-1$ of the numbers in $(n|m_\alpha)$.

It follows from this that the signs of the combinations of the rth group are the same as or the opposite to (according as $m-l-r+1$ is even or odd, there being $m-l-r+1$ numbers in the first part greater than the first number in the second part) the signs of the corresponding members of the set obtained by striking out the $r-1$ numbers common to the first part and the one number common to the second part of each combination of the group.

We have therefore

$$\begin{aligned}\phi(m, l) &= \phi(l-1, l-1) - \phi(l, l-1) + \phi(l+1, l-1)\\ &\quad - \cdots + (-1)^{m-l-r+1}\phi(m-r, l-1)\\ &\quad + \cdots + (-1)^{m-l}\phi(m-1, l-1)\end{aligned}\tag{2}$$

a reduction formula for $\phi(m, l)$. As an immediate consequence of this we have

$$\begin{aligned}\phi(m, l) &= \phi(m-r, l) + (-1)^{m-r-l+1}\phi(m-r, l-1)\\ &\quad + \cdots + (-1)^{m-l}\phi(m-1, l-1)\\ &= \phi(m-1, l) + (-1)^{m-l}\phi(m-1, l-1)\end{aligned}\tag{3}$$

17. We have by successive applications of (2)

$$\begin{aligned}\phi(2m, 2l+1) &= \phi(2l, 2l) - \phi(2l+1, 2l) + \cdots - \phi(2m-1, 2l)\\ &= \phi(2l-1, 2l-1) - \phi(2l-1, 2l-1) + \phi(2l, 2l-1)\\ &\quad + \phi(2l-1, 2l-1) - \phi(2l, 2l-1) + \phi(2l+1, 2l-1)\\ &\quad \cdots\cdots\cdots\cdots\cdots\cdots\\ &\quad - \phi(2l-1, 2l-1) + \phi(2l, 2l-1)\\ &\quad \cdots + \phi(2m-2, 2l-1) = \phi(2l, 2l-1)\\ &\quad + \phi(2l+2, 2l-1) + \cdots + \phi(2m-2, 2l-1)\end{aligned}\tag{4}$$

Put $l=1$ then

$$\phi(2m, 3) = \phi(2, 1) + \phi(4, 1) + \cdots + \phi(2m-2, 1) = 0. \qquad (\S 15)$$

Put $l=2$ then

$$\phi(2m, 5) = \phi(4, 3) + \phi(6, 3) + \cdots + \phi(2m-2, 3) = 0.$$

In this way it may be shown that

$$(5) \qquad \phi(2m, 2l+1) = 0 \quad (l = 1, 2, \cdots, m-1).$$

From equations (3) and (5) we have

$$\phi(2m+1, 2l) = \phi(2m, 2l) - \phi(2m, 2l-1) = \phi(2m, 2l)$$

and

$$\begin{aligned}\phi(2m+1, 2l+1) &= \phi(2m, 2l+1) + \phi(2m, 2l)\\ &= \phi(2m, 2l).\end{aligned}$$

Therefore

$$(6) \qquad \phi(2m+1, 2l+1) = \phi(2m+1, 2l) = \phi(2m, 2l).$$

It follows from this and §15 that

$$\phi(m, l) = \phi(m, m-l).$$

If $l=m$ then $\phi(m, m)=\phi(m, 0)=1$.

18. From equations (2), (5), and (6) we have

$$\begin{aligned}(7)\ \phi(2m, 2l) &= \phi(2m-1, 2l-1) + \phi(2m-3, 2l-1)\\ &\qquad + \cdots + \phi(2l-1, 2l-1)\\ &= \phi(2m-2, 2l-2) + \phi(2m-4, 2l-2)\\ &\qquad + \cdots + \phi(2l-2, 2l-2).\end{aligned}$$

These properties at once suggest that

$$\phi(2m, 2l) = {}_mC_l \text{ or } m_l$$

and it may be easily proved that this is true.

If in equation (7) we put 1st $l=1$, then

$$\begin{aligned}\phi(2m, 2) &= \phi(2m-2, 0) + \phi(2m-4, 0) + \cdots + \phi(0, 0)\\ &= m \text{ or } m_1;\end{aligned}$$

2nd $l=2$, then

$$\begin{aligned}\phi(2m, 4) &= \phi(2m-2, 2) + \phi(2m-4, 2) + \cdots + \phi(2, 2)\\ &= (m-1)_1 + (m-2)_1 + \cdots + 1\\ &= m_2;\end{aligned}$$

3rd $l=3$ then

$$\begin{aligned}\phi(2m, 6) &= \phi(2m-2, 4) + \phi(2m-4, 4) + \cdots + \phi(4, 4)\\ &= (m-1)_2 + (m-2)_2 + \cdots + 2_2\\ &= m_3.\end{aligned}$$

In this way we see from equation (7) itself that if it is true for any value of l it is true for a value one greater and is therefore true for all values; hence

$$\phi(2m, 2l) = m_l (l = 1, 2, \cdots m).$$

Problem. Show that if the permutations of any group be separated into sub-groups, (1) those which begin with α, (2) those which begin with β, and so on, then the series of signs of the 3rd, 5th, and other odd sub-groups is identical with the series of signs of the 1st sub-group, and the signs of any one of the even sub-groups is got by changing each sign of the first sub-group into the opposite sign.

Note. This is obvious on observing that the 2nd sub-group is obtained by interchanging α and β in each member of the 1st sub-group.

CHAPTER II

Definitions and Notations

19. If a particular form of algebraical expression be lengthy and of frequent occurrence, it becomes desirable to introduce a suggestive *name* and *symbol* for it; and if the form be one of a family, it is also desirable that the names and symbols of all of them should indicate this relationship. Such a nomenclature and notation are advantageous, not merely from the convenience thereby afforded in speaking and writing, but as helps to the actual discovery of the properties of the forms in question.

Note. The following—with which the learner is probably already familiar—may be taken as an instance of this. Early in the history of the science, it being a common requirement to make use of the product resulting from the multiplication of a number by itself, this product was *named* the ***Power*** of the number, and was *symbolized* in various ways; also the product resulting from the multiplication of the "power" (as thus understood) of a number by the number itself, was called the ***Cube*** of the number, and was variously expressed by means of a symbol; and similarly with other such products. Then, in the latter half of the sixteenth century, the fact that these products were members of a family was recognized in the nomenclature by calling them all ***Powers***, and distinguishing them as ***second power, third power,*** etc. and in the next century there came into use a like improvement in notation, the second power of a, third power of a, etc., being denoted by a^2, a^3, $\cdots$. Thence arose the subsidiary terms *exponent* and *base*, and thus the known truths regarding powers became easily expressible either in words or symbols, and the way was opened to the generalization of these truths, and to the suggestion and discovery of others.

20. The expressions

$$a_1b_2 - a_2b_1,\ a_1b_2c_3 + a_3b_1c_2 + a_2b_3c_1 - a_3b_2c_1 - a_2b_1c_3 - a_1b_3c_2, \text{ etc.}$$

are instances of important forms which often occur in analysis. They may arise as the result of eliminating the variables from a system of linear homogeneous equations. Thus, eliminating x and y from the equations

$$a_1x + a_2y = 0$$
$$b_1x + b_2y = 0$$

we get as a result

$$a_1b_2 - a_2b_1 = 0.$$

Eliminating x, y, and z from the equations

$$a_1x + a_2y + a_3z = 0$$
$$b_1x + b_2y + b_3z = 0$$
$$c_1x + c_2y + c_3z = 0$$

we get as a result

$$a_1b_2c_3 - a_1b_3c_2 + a_2b_3c_1 - a_2b_1c_3 + a_3b_1c_2 - a_3b_2c_1 = 0$$

and in general if we eliminate the variables from a system of n linear homogeneous equations, we get a result of the form

$$\sum \pm a_1b_2c_3 \cdots l_n = 0.$$

These forms are called *determinants*, and since they are functions of the coefficients only and may be considered without reference to their origin—the preceding being but one way in which they may arise—the definition of a determinant should obviously contain no reference either to its origin or to the variables.

21. Definition. If we have n^2 quantities arranged in a square of n rows and n columns, then the sum of all the terms that can be formed by taking the product of n quantities, one from each column and one from each row, the sign preceding any term being determined by writing in succession the numbers of the rows from which the quantities composing it have come, and in a separate series the numbers of the columns, and taking $+$ or $-$ according as the total number of inversions of order in these two series* is even or odd, is called the determinant of these quantities and is said to be of the nth order.

* It is manifest that the quantities in any term may be so arranged that either the numbers indicating the rows or the numbers indicating the columns from which the quantities composing it have come are in their natural order. When thus arranged it is only necessary in determining the sign to consider the number of inversions in the series where the numbers are not in their natural order.

E. H. Moore has suggested the following:

A determinant of order n is uniquely defined by the unique definition of its n^2 elements a_{rs}, where the suffixes rs run independently over any (the same) set of n distinct marks, and by the law of signs for the terms.

For most purposes the marks $1, 2, \cdots, n$ are convenient, but for the investigation of some determinants of special form it is more convenient to use some other set of n marks.

The ordinary notation for determinants is

$$\begin{vmatrix} a_1 & a_2 \\ b_1 & b_2 \end{vmatrix}, \quad \begin{vmatrix} a_1 & a_2 & a_3 \\ b_1 & b_2 & b_3 \\ c_1 & c_2 & c_3 \end{vmatrix}, \quad \cdots, \quad \begin{vmatrix} a_1 & a_2 \cdots & a_n \\ b_1 & b_2 \cdots & b_n \\ \cdots & \cdots & \cdots \\ l_1 & l_2 \cdots & l_n \end{vmatrix}.$$

The quantities a_1, a_2, etc., are called the *constituents* or *elements* and the products a_1b_2 etc. are called the *terms* of the determinant.

In the square array representing a determinant the diagonal from the left-hand top corner to the right-hand bottom corner is called the *principal diagonal* and the diagonal from the right-hand top corner to the left-hand bottom corner is called the *secondary diagonal* of the determinant. The term formed by taking the product of all the constituents along the principal diagonal is called the *principal* or *leading term* of the determinant. Constituents are said to be *conjugate* to each other, when the place which either occupies in the rows is the same as that which the other occupies in the columns. Constituents along the principal diagonal are termed *self-conjugate.* Two *terms* of a determinant are said to be *conjugate* when the constituents of the one are the conjugate of the constituents of the other. A *self-conjugate term* is one into which there might enter self-conjugate constituents and pairs of conjugate constituents.

EXAMPLES

1. Tell the signs of the terms

$$ingd, \quad lgbm, \quad gdjm, \quad nkah$$

of the determinant

$$\begin{vmatrix} a & e & i & m \\ b & f & j & n \\ c & g & k & o \\ d & h & l & p \end{vmatrix}.$$

2. Find the full expression for the preceding determinant in the ordinary notation.

3. Tell the signs of the terms

$$a_5b_4c_3d_2e_1, \quad a_3e_2c_4d_1b_5, \quad b_2a_1e_3c_4d_5, \quad b_3c_2a_1d_5e_4$$

of the determinant

$$\begin{vmatrix} a_1 & a_2 & a_3 & a_4 & a_5 \\ b_1 & b_2 & b_3 & b_4 & b_5 \\ c_1 & c_2 & c_3 & c_4 & c_5 \\ d_1 & d_2 & d_3 & d_4 & d_5 \\ e_1 & e_2 & e_3 & e_4 & e_5 \end{vmatrix}.$$

22. The notation indicated in §21 where all the letters in each row are the same and all the suffixes in each column are the same, the letters being in alphabetical order from top to bottom and the suffixes in order of magnitude from left to right is very convenient for finding the expansion of the determinant. For in each symbol denoting a constituent the letter will indicate the row to which the constituent belongs, and the suffix will indicate the column, so that to form a term we have only to write all the letters in alphabetical order and attach one of the suffixes to each, and to determine the sign of the term we have only to count the number of inversions of order in the suffixes as there written.

Thus, in the case of the determinant of the third order

$$\begin{vmatrix} a_1 & a_2 & a_3 \\ b_1 & b_2 & b_3 \\ c_1 & c_2 & c_3 \end{vmatrix},$$

we write all the permutations of the suffixes, namely,

$$1\,2\,3,\quad 1\,3\,2,\quad 2\,3\,1,\quad 2\,1\,3,\quad 3\,1\,2,\quad 3\,2\,1,$$

and taking each permutation along with the letters a, b, c, we have at once the expansion

$$a_1b_2c_3-a_1b_3c_2+a_2b_3c_1-a_2b_1c_3+a_3b_1c_2-a_3b_2c_1.$$

The same result will obviously be obtained if we write all the permutations of the letters and take each permutation along with the suffixes 1, 2, 3.

23. It is also possible in the case of such determinants to use an abridged notation. One form of this consists in writing one of the terms, namely, the leading term, as a type of them all, prefixing to it the symbol $\pm$ to indicate the variability of the signs, enclosing this in brackets, and before all placing the symbol of summation, $\sum$. Thus the determinant

$$\begin{vmatrix} a_1 & a_2 & a_3 \\ b_1 & b_2 & b_3 \\ c_1 & c_2 & c_3 \end{vmatrix},$$

is denoted by

$$\sum(\pm\, a_1b_2c_3)\,.$$

Equally efficient notations are

$$D(a_1b_2c_3)\,,\quad |\, a_1b_2c_3 \,|$$

which do not, like the other, aim at a partial definition, but are meant merely as suggestive contractions for the longer form, or for *the determinant of which the first row consists of a's, the second row of b's the third row of c's, and of which the elements in the first column have the suffix* 1, *those in the second column the suffix* 2, *and those in the third column the suffix* 3.

The learner should accustom himself to the use of these shorter forms, and expecially to pass mentally from them with ease to the standard notation. Thus, to take another instance,

$$D(x_0y_2z_4w_6)\,,\quad \sum(\pm\, x_0y_2z_4w_6)\,,\ \text{or}\ |\, x_0y_2z_4w_6 \,|$$

should readily suggest

$$\begin{vmatrix} x_0 & x_2 & x_4 & x_6 \\ y_0 & y_2 & y_4 & y_6 \\ z_0 & z_2 & z_4 & z_6 \\ w_0 & w_2 & w_4 & w_6 \end{vmatrix}.$$

EXAMPLE.

In the determinant

$$\begin{vmatrix} x_1 & y_1 & z_1 & v_1 & w_1 \\ x_2 & y_2 & z_2 & v_2 & w_2 \\ x_3 & y_3 & z_3 & v_3 & w_3 \\ x_4 & y_4 & z_4 & v_4 & w_4 \\ x_5 & y_5 & z_5 & v_5 & w_5 \end{vmatrix},$$

find

1. the terms containing x_5w_1;
2. the terms containing y_2w_3;
3. the terms containing x_4v_1;

24. Another convenient notation for determinants is to write each constituent with a double suffix, the first suffix indicating the row

and the second the column to which the constituent belongs. Thus the determinant of the nth order would, in this notation, be written

$$\begin{vmatrix} a_{11} & a_{12} & a_{13} \cdots a_{1n} \\ a_{21} & a_{22} & a_{23} \cdots a_{2n} \\ a_{31} & a_{32} & a_{33} \cdots a_{3n} \\ \cdots & \cdots & \cdots \\ a_{n1} & a_{n2} & a_{n3} \cdots a_{nn} \end{vmatrix} .$$

Since only one letter is employed we may, if we choose, denote such a determinant by a shorter symbol even than

$$| a_{11}\, a_{22} \cdots a_{nn} |$$

namely, by

$$| a_{1n} | \text{ or } | a_{nn} |$$

25. The notation of the preceding article may be modified by the omission of the a's in writing the constituents of $|a_{1n}|$; so that rs is put for a_{rs}, and

$$\begin{vmatrix} 11 & 12 & 13 \cdots 1n \\ 21 & 22 & 23 \cdots 2n \\ \cdots & \cdots & \cdots \\ n1 & n2 & n3 \cdots nn \end{vmatrix} \text{ for } | a_{1n} | .$$

Sylvester who calls this his *umbral* notation, writes also

$$\begin{vmatrix} 1 & 2 & 3 \\ 1 & 2 & 3 \end{vmatrix} \text{ for } |a_{13}|$$

and generally

$$\begin{vmatrix} a & b & c \cdots \\ \alpha & \beta & \gamma \cdots \end{vmatrix} \text{ for } | x_{a\alpha}\, x_{b\beta}\, x_{c\gamma} \cdots | .$$

26. Symmetry. In a determinant there obviously may be three kinds of symmetry: (1) symmetry with respect to the principal diagonal, (2) symmetry with respect to the secondary diagonal,* and (3) symmetry with respect to the centre (the point of intersection of the two diagonals).

* By reversing the order of the rows and columns it is apparent that symmetry with respect to the secondary diagonal may be changed to symmetry with respect to the principal diagonal.

These three imply that: (1) $a_{rs}=a_{sr}$, that is conjugate elements are equal, (2) $a_{rs}=a_{n+1-s,n+1-r}$, (3) $a_{rs}=a_{n+1-r,n+1-s}$, respectively.

When (1) is true the determinant is said to be *axisymmetric*, when (3) is true it is said to be *centrosymmetric*. When two of the three types exist at once the determinant is said to be *bisymmetric*.

27. The function obtained by making all the negative signs in a determinants positive is called a permanent and is usually denoted by the symbol

$$\overset{+}{\Big|} \qquad\qquad \overset{+}{\Big|}$$

28. Another very convenient notation which we shall use is the following: $(a, b, c)(x, y, z)$ or $\frac{a\,b\,c}{x\,y\,z}$ for $ax+by+cz$ and the matrix form

$$\left(\begin{vmatrix} a_{11} & a_{12} & a_{13} \\ a_{21} & a_{22} & a_{23} \\ a_{31} & a_{32} & a_{33} \end{vmatrix}\right)(x_1x_2x_3)(y_1y_2y_3),$$

or the bipartite form

$$\begin{array}{ccc|c} x_1 & x_2 & x_3 & \\ \hline a_{11} & a_{12} & a_{13} & y_1 \\ a_{21} & a_{22} & a_{23} & y_2 \\ a_{31} & a_{32} & a_{33} & y_3 \end{array}$$

for the lineo-linear expression

$$(a_{11}x_1 + a_{12}x_2 + a_{13}x_3)y_1 + (a_{21}x_1 + a_{22}x_2 + a_{23}x_3)y_2 + (a_{31}x_1 + a_{32}x_2 + a_{33}x_3)y_3.$$

Bipartites are treated in chapter VIII.

29. *If* $\iota_1\iota_2\iota_3 \cdots \iota_n=1$ *and* $\iota_1, \iota_2, \cdots,$ *be symbols subject to the laws of ordinary algebra except that* $\iota_r\iota_s=-\iota_s\iota_r$ *and* $\iota_r^2=0$, *then*

$$|a_{1n}| = (a_{11}\iota_1 + a_{12}\iota_2 + \cdots + a_{1n}\iota_n)(a_{21}\iota_1 + a_{22}\iota_2 + \cdots + a_{2n}\iota_n) \cdots (a_{n1}\iota_1 + a_{n2}\iota_2 + \cdots + a_{nn}\iota_n).$$

Writing in a column the factors

$$a_{11}\iota_1 + a_{12}\iota_2 + a_{13}\iota_3 + \cdots + a_{1n}\iota_n,$$
$$a_{21}\iota_1 + a_{22}\iota_2 + a_{23}\iota_3 + \cdots + a_{2n}\iota_n,$$
$$\cdots\cdots\cdots\cdots\cdots\cdots\cdots$$
$$a_{n1}\iota_1 + a_{n2}\iota_2 + a_{n3}\iota_3 + \cdots + a_{nn}\iota_n,$$

it becomes evident that the identity to be established is but a symbolical statement of the definition (§21) of a determinant. For, firstly, owing to the constitution of the factors, the product must consist of all terms of the form

$$a_{1r}a_{2s}a_{3u}\cdots a_{nz}\times\iota_r\iota_s\iota_u\cdots\iota_z,$$

which can be got by taking one and only one element from each row of the determinant: secondly, the condition $\iota_r^2=0$ necessitates the disappearance from this of every term containing two or more elements from the same column; and thirdly, the conditions $\iota_1\iota_2\iota_3\cdots\iota_n=1$, $\iota_r\iota_s=-\iota_s\iota_r$ ensure that the sign-factor of any term shall be $+1$ or -1 according as the number of inversions of order in the suffixes of its ι's, that is, in the second suffixes of its a's, is even or odd.

Example.

$$(a\iota_1 + b\iota_2 + c\iota_3)(d\iota_1 + e\iota_2 + f\iota_3)(g\iota_1 + h\iota_2 + k\iota_3)$$
$$= (bd\iota_2\iota_1 + cd\iota_3\iota_1 + ae\iota_1\iota_2 + ce\iota_3\iota_2 + af\iota_1\iota_3 + bf\iota_2\iota_3)(g\iota_1 + h\iota_2 + k\iota_3),$$
$$= \{(ae - bd)\iota_1\iota_2 + (af - cd)\iota_1\iota_3 + (bf - ec)\iota_2\iota_3\}(g\iota_1 + h\iota_2 + k\iota_3),$$
$$= g(bf - ce)\iota_2\iota_3\iota_1 + h(af - cd)\iota_1\iota_3\iota_2 + k(ae - bd)\iota_1\iota_2\iota_3,$$
$$= \{g(bf - ce) - h(af - cd) + k(ae - bd)\}\iota_1\iota_2\iota_3,$$
$$= \begin{vmatrix} a & b & c \\ d & e & f \\ g & h & k \end{vmatrix}.$$

Exercises

Set I

Write the following algebraical expressions in the usual notation.

1. $\begin{vmatrix} c & d & e \\ f & g & h \\ k & l & m \end{vmatrix}$.
2. $\begin{vmatrix} x & y & z \\ v & w & u \\ t & r & s \end{vmatrix}$.
3. $\begin{vmatrix} a_1 & a_2 & a_3 \\ b_1 & b_2 & b_3 \\ c_1 & c_2 & c_3 \end{vmatrix}$.

4. $\begin{vmatrix} a & f & -c \\ d & -e & f \\ g & h & -k \end{vmatrix}$. 5. $\begin{vmatrix} a & -2a & b \\ 3b & -c & 4d \\ 2c & 3d & -4b \end{vmatrix}$. 6. $\begin{vmatrix} a & b & c \\ 0 & e & f \\ -g & 0 & h \end{vmatrix}$.

7. $\begin{vmatrix} x & 0 & y \\ 0 & x & y \\ -x & -y & 0 \end{vmatrix}$. 8. $\begin{vmatrix} x & -2z & -y^2 \\ -y & -2x & z^2 \\ -z & 2y & -x^2 \end{vmatrix}$. 9. $\begin{vmatrix} -a & -b & -c \\ -b & -c & -a \\ -c & -a & -b \end{vmatrix}$.

Find the single numbers to which the following determinants are equivalent.

10. $\begin{vmatrix} 1 & 2 & 3 \\ 2 & 3 & 4 \\ 3 & 4 & 5 \end{vmatrix}$. 11. $\begin{vmatrix} 1 & \frac{1}{2} & \frac{1}{3} \\ \frac{1}{2} & \frac{1}{3} & \frac{1}{4} \\ \frac{1}{3} & \frac{1}{4} & \frac{1}{5} \end{vmatrix}$. 12. $\begin{vmatrix} 4 & 5 & 2 \\ -1 & 2 & -3 \\ 6 & -4 & 5 \end{vmatrix}$.

13. $\begin{vmatrix} 1 & 1 & 1 \\ 4 & -3 & 0 \\ 3 & 2 & -5 \end{vmatrix}$. 14. $\begin{vmatrix} 4 & -1 & -2 \\ 0 & 3 & 0 \\ 3 & -7 & 4 \end{vmatrix}$. 15. $\begin{vmatrix} -1 & -1 & 1 \\ -3 & 1 & -4 \\ 2 & -3 & -5 \end{vmatrix}$.

Write the following in determinant form.

16. $bfj+eid+hcg-hfd-ecj-big$

17. $m_1n_2r_3-m_1n_3r_2+m_2n_3r_1-m_2n_1r_3+m_3n_1r_2-m_3n_2r_1$.

Find the values of x in the following equations.

18. $\begin{vmatrix} x & -4 & 1 \\ -6 & 3 & -2 \\ x & 2 & 1 \end{vmatrix} = 0.$ 19. $\begin{vmatrix} 1 & 1 & 1 \\ a & x & c \\ b & b & x \end{vmatrix} = 0.$

20. $\begin{vmatrix} x & a & a \\ a & x & a \\ a & a & x \end{vmatrix} + \begin{vmatrix} b & b & x \\ b & x & b \\ x & b & b \end{vmatrix} = 0.$

21. Find the expression in the ordinary notation which is the equivalent of

$$\begin{vmatrix} 2c & a+b+c & a+b+c \\ a+b+c & 2a & a+b+c \\ a+b+c & a+b+c & 2b \end{vmatrix}.$$

22. $\begin{vmatrix} 0 & 0 & 0 & a_1 \\ 0 & 0 & b_1 & a_2 \\ 0 & c_1 & b_2 & a_3 \\ d_1 & c_2 & b_3 & a_4 \end{vmatrix}$. 23. $\begin{vmatrix} a_1 & b_1 & c_1 & 0 \\ a_2 & b_2 & c_2 & 0 \\ 0 & 0 & 0 & d_2 \\ 0 & 0 & 0 & d_3 \end{vmatrix}$. 24. $\begin{vmatrix} b_2 & b_3 & b_4 & b_5 \\ 0 & c_3 & 0 & 0 \\ 0 & d_3 & d_4 & d_5 \\ 0 & e_3 & 0 & e_5 \end{vmatrix}$.

25. $\begin{vmatrix} x & 0 & 0 & 0 & y \\ y & x & 0 & 0 & 0 \\ 0 & y & x & 0 & 0 \\ 0 & 0 & y & x & 0 \\ 0 & 0 & 0 & y & x \end{vmatrix}$. 26. $\begin{vmatrix} a_4 & 0 & c_2 & 0 & x \\ a_3 & 0 & c_1 & x & d_1 \\ a_2 & 0 & x & 0 & 0 \\ a_1 & x & b_1 & b_2 & b_3 \\ x & 0 & 0 & 0 & 0 \end{vmatrix}$.

27. Write the terms of $|x_0y_1z_2w_3|$ which contain y_2w_3.

28. What other term of $|b_0c_1d_2e_3f_4|$, besides the secondary diagonal term, contains the elements $f_0c_3e_1$?

29. Find the number of inversions of order in

$$n,\ n-1,\ n-2,\ \cdots,\ 3,\ 2,\ 1.$$

30. What other term of $|a_mb_nc_pd_qe_r|$ besides $a_nb_pc_md_re_q$ contains the elements $b_pc_me_q$?

31. What is the sign-factor of the secondary diagonal term in a determinant of the nth order?

32. Write the terms of $|a_1b_3c_5d_6|$ which contain the element b_5.

33. If the integral numbers d, e, f, g, h, i, j, k, be "cyclically" transposed so as to become k, d, e, f, g, h, i, j, how many more or fewer inversions of order will there be?

34. Find how many terms of a determinant of the nth order contain any particular element.

35. If the sign of a term be determined from one arrangement of the elements composing it, show that the same sign would be got from a different arrangement.

CHAPTER III

General Properties

30. If we combine any permutation of the numbers 1, 2, 3, · · · , n, used as column numbers, with the row numbers 1, 2, 3, · · · , n of a determinant we get, in umbral notation, a term of the determinant. We proceed to the investigation of the properties in general.

31. *The full number of terms in a determinant of the nth order is* $n!$ This follows at once from the fact that each term contains one constituent from each row and one from each column and there are therefore as many terms as there are ways of arranging n things all together which is $n!$

It is apparent from this, since the factors in any term $a_{rs}a_{tu}a_{vw} \cdots$ may be arranged in any order, that we may suppose the elements of one of the lines of suffixes $r, t, v, \cdots$, or $s, u, w, \cdots$, are arranged in their natural order. We shall in general suppose that the elements in the first line of suffixes are arranged in their natural order, and therefore the terms of a determinant are the $n!$ products having the $n!$ permutations of the elements 1, 2, 3, · · · , n, for the suffixes in the second line.

32. If we combine any self-conjugate permutation with the row numbers 1, 2, 3, · · · , n of a determinant we evidently get a self-conjugate term of the determinant. Thus, if we combine the self-conjugate permutation 361452 with the row numbers 123456 we obtain

$$a_{13}a_{26}a_{31}a_{44}a_{55}a_{62}$$

a self-conjugate term of the determinant

$$\left| a_{11}a_{22}a_{33} \cdots a_{66} \right| .$$

There are therefore as many self-conjugate terms in a determinant of the nth order as there are self-conjugate permutations of the elements 1, 2, 3, · · · , n. (§9.)

33. From §11, we see that *conjugate terms of a determinant have the same sign.*

It should be observed that if we write any term of a determinant using the double suffix notation and keeping the row numbers in their natural order and then rewrite it keeping the numbers indicating the columns in their natural order, the numbers indicating the columns in the first and those indicating the rows in the second are conjugate permutations.

34. *In a determinant of the nth order not more than* $(n-r)!$ *terms can have r elements in common.*

This is an immediate consequent of the definition of a determinant and §7.

35. It follows from §5 that of *the full number of terms of a determinant exactly as many are positive as are negative.*

36. *If all the constituents of a row or column of a determinant be zero, so also is the determinant itself.*

For by definition every term must contain a constituent from this row or column and will therefore vanish.

37. *Two determinants which differ only in that the rows of the one are in order the columns of the other are equal.*

Every term of the first determinant must contain one and only one element from each row and each column of that determinant; therefore it must contain one and only one element from each column and each row of the second determinant, and therefore it must be a term of that determinant also. Similarly we can show that every term of the second determinant is a term of the first; therefore the terms of the two determinants are alike in magnitude. Any constituent in the first is the conjugate of the corresponding constituent of the second and since conjugate terms have the same sign the terms of the two determinants are alike in sign.

From this it is evident that any theorem in the statement of which the word *row* or the words *row* and *column* occur would also be true if the word *column* or the words *column* and *row* respectively be substituted. For in proving the former theorem in regard to the determinant

$$\Delta \equiv \begin{vmatrix} a_1 & a_2 & a_3 \cdots a_n \\ b_1 & b_2 & b_3 \cdots b_n \\ c_1 & c_2 & c_3 \cdots c_n \\ \cdots & \cdots & \cdots \\ l_1 & l_2 & l_3 \quad l_n \end{vmatrix},$$

we are proving the latter theorem in regard to the determinant

$$\begin{vmatrix} a_1 & b_1 & c_1 \cdots l_1 \\ a_2 & b_2 & c_2 \cdots l_2 \\ a_3 & b_3 & c_3 \cdots l_3 \\ \cdots & \cdots & \cdots \\ a_n & b_n & c_n \cdots l_n \end{vmatrix}.$$

38. Since by definition every term of a determinant contains one, and only one, constituent from each row and from each column, it follows that *a determinant is a linear homogeneous function of the constituents in any row or any column.* Thus Δ of the last section can be written in the form

$$a_1A_1 + a_2A_2 + a_3A_3 + \cdots + a_nA_n$$

or

$$a_1A_1 + b_1B_1 + c_1C_1 + \cdots + l_1L_1$$

where $A_1, A_2, \cdots, A_n$ contain no constituents from the first row and $A_1, B_1, \cdots, L_1$ contain no constituents from the first column.

39. *If all the constituents of a row of a determinant be multiplied by the same quantity, the resulting determinant equals the product of the original determinant and the said quantity.*

For we may write the determinant as a linear function of the constituents in this row, in which form the constant multiplier will appear as a factor.

EXAMPLES.

$$m\begin{vmatrix} a & b & c \\ d & e & f \\ g & h & k \end{vmatrix} = \begin{vmatrix} ma & mb & mc \\ d & e & f \\ g & h & k \end{vmatrix} = \begin{vmatrix} a & mb & c \\ d & me & f \\ g & mh & k \end{vmatrix},$$

$$= (1/m)\begin{vmatrix} a & mb & c \\ md & m^2e & mf \\ g & mh & k \end{vmatrix} = m\begin{vmatrix} a & mb & c \\ d/m & e & f/m \\ g & mh & k \end{vmatrix} = \cdots .$$

$$\begin{vmatrix} 1 & 7 & -1 \\ 3 & 21 & 0 \\ 4 & -28 & -6 \end{vmatrix} = 7\begin{vmatrix} 1 & 1 & -1 \\ 3 & 3 & 0 \\ 4 & -4 & -6 \end{vmatrix} = 7 \times 2\begin{vmatrix} 1 & 1 & -1 \\ 3 & 3 & 0 \\ 2 & -2 & -3 \end{vmatrix},$$

$$= 7 \times 2 \times 3\begin{vmatrix} 1 & 1 & -1 \\ 1 & 1 & 0 \\ 2 & -2 & -3 \end{vmatrix}$$

$$= 7 \times 2 \times 3 \times (-3 + 2 + 2 + 3),$$

$$= 168.$$

40. *If in any determinant* $|a_{1n}|$ *we change the sign of every element, the sum of whose suffixes is odd, the determinant is unaltered.*

This amounts to multiplying every odd numbered row and column by -1, which is equivalent to multiplying the determinant by $(-1)^{2k}$ or 1.

This theorem may be more generally stated as follows:

The determinant remains unaltered if we multiply each a_{ij} *by* p^{i-j} *where* p *is any number*, for each term would be multiplied by $p^{\alpha-\alpha}$ or 1, where $\alpha=1+2+3+4+\cdots+n$.

41. *If the same* m *elements of* $n-m+k$ *rows of a determinant of the* n*th order contain* u *as a common factor, then* u^k *is a factor of the determinant.*

Move to the right the columns, and downward the rows, that have elements which do not contain the given factor, until the elements which do contain this factor form a rectangular array in the upper left hand corner of the determinant—an array having m columns and $n-m+k$ rows. Now it will be impossible to find a term of the determinant that does not contain k of these elements, for when we have taken an element from each row and column outside this rectangular array we have but $n-k$ elements for our term and must take the remaining k elements from within the array. Hence every term contains u to a power $\geqq k$.

The particular case where $n=m$ and $k=1$ gives the theorem of §39.

Exercises. Set II

1. Find what relation exists between the two algebraical expressions

$$\begin{vmatrix} a & b & c \\ d & e & f \\ g & h & k \end{vmatrix}, \quad \begin{vmatrix} a & b & c \\ g & h & k \\ d & e & f \end{vmatrix}.$$

2. Find how the expressions

$$\begin{vmatrix} a & b & c \\ d & e & f \\ g & h & k \end{vmatrix}, \quad \begin{vmatrix} d & e & f \\ a & b & c \\ g & h & k \end{vmatrix}, \quad \begin{vmatrix} d & e & f \\ g & h & k \\ a & b & c \end{vmatrix}$$

are related to each other.

3. State the probable theorem regarding determinants, to which the results of Ex. 1 and 2 point.

4. Show that

$$\begin{vmatrix} a & b & c \\ a & b & c \\ d & e & f \end{vmatrix} = 0.$$

5. Without changing from the determinant notation, show that

$$\begin{vmatrix} a & b & c \\ d & e & f \\ a & b & c \end{vmatrix} = 0$$

by using what has already been proved.

6. Show that

$$\begin{vmatrix} x_1 & y_1 & z_1 \\ x_2 & y_2 & z_2 \\ x_3 & y_3 & z_3 \end{vmatrix} = x_1 \begin{vmatrix} y_2 & z_2 \\ y_3 & z_3 \end{vmatrix} - y_1 \begin{vmatrix} x_2 & z_2 \\ x_3 & z_3 \end{vmatrix} + z_1 \begin{vmatrix} x_2 & y_2 \\ x_3 & y_3 \end{vmatrix},$$

$$= x_1 \begin{vmatrix} y_2 & z_2 \\ y_3 & z_3 \end{vmatrix} - x_2 \begin{vmatrix} y_1 & z_1 \\ x_3 & z_3 \end{vmatrix} + x_3 \begin{vmatrix} y_1 & z_1 \\ y_2 & z_2 \end{vmatrix}.$$

7. Supply the elements in the following blank determinant-forms of the second order:

$$\begin{vmatrix} x_1 & y_1 & z_1 \\ x_2 & y_2 & z_2 \\ x_3 & y_3 & z_3 \end{vmatrix} = z_1 \begin{vmatrix} \quad & \quad \\ \quad & \quad \end{vmatrix} - z_2 \begin{vmatrix} \quad & \quad \\ \quad & \quad \end{vmatrix} + z_3 \begin{vmatrix} \quad & \quad \\ \quad & \quad \end{vmatrix},$$

$$= -x_2 \begin{vmatrix} \quad & \quad \\ \quad & \quad \end{vmatrix} + y_2 \begin{vmatrix} \quad & \quad \\ \quad & \quad \end{vmatrix} - z_2 \begin{vmatrix} \quad & \quad \\ \quad & \quad \end{vmatrix}.$$

8. Show that

$$\begin{vmatrix} a_1 & b_1 \\ a_2 & b_2 \end{vmatrix} \cdot \begin{vmatrix} x_1 & y_1 \\ x_2 & y_2 \end{vmatrix} = \begin{vmatrix} a_1 & b_1 \\ x_1 & y_1 \end{vmatrix} \cdot \begin{vmatrix} a_2 & b_2 \\ x_2 & y_2 \end{vmatrix} + \begin{vmatrix} a_1 & b_1 \\ x_2 & y_2 \end{vmatrix} \cdot \begin{vmatrix} x_1 & y_1 \\ a_2 & b_2 \end{vmatrix},$$

$$= \begin{vmatrix} x_1 & b_1 \\ x_2 & b_2 \end{vmatrix} \cdot \begin{vmatrix} a_1 & y_1 \\ a_2 & y_2 \end{vmatrix} + \begin{vmatrix} y_1 & b_1 \\ y_2 & b_2 \end{vmatrix} \cdot \begin{vmatrix} x_1 & a_1 \\ x_2 & a_2 \end{vmatrix}.$$

9. Show that

$$\begin{vmatrix} a+x & b & c \\ d+y & e & f \\ g+z & h & k \end{vmatrix} = \begin{vmatrix} a & b & c \\ d & e & f \\ g & h & k \end{vmatrix} + \begin{vmatrix} x & b & c \\ y & e & f \\ z & h & k \end{vmatrix}.$$

10. Show that

$$\begin{vmatrix} a+mc & b & c \\ d+mf & e & f \\ g+mk & h & k \end{vmatrix} = \begin{vmatrix} a & b & c \\ d & e & f \\ g & h & k \end{vmatrix} = \begin{vmatrix} a-mb & b & c \\ d-me & e & f \\ g-mh & h & k \end{vmatrix}.$$

Write the following expressions in determinant form:

11. $ayp - myc + xnc - anz + mbz - bxp.$

12. $x_1y_2 - x_2y_1 + x_2y_3 - x_3y_2 + x_3y_1 - x_1y_3.$

13. $3abc - a^3 - b^3 - c^3.$

14. $acf + 2bed - cd^2 - b^2f - ae^2.$

Find single determinants of the third order equivalent to the expressions

15. $$x\begin{vmatrix} x & c \\ c & x \end{vmatrix} - a\begin{vmatrix} a & b \\ c & x \end{vmatrix} + b\begin{vmatrix} a & b \\ x & c \end{vmatrix}.$$

16. $$\begin{vmatrix} 0 & a_2 & a_4 \\ b_1 & 0 & b_4 \\ c_1 & c_2 & 0 \end{vmatrix} + a_1\begin{vmatrix} 0 & b_4 \\ c_2 & 0 \end{vmatrix} + b_2\begin{vmatrix} 0 & a_4 \\ c_1 & 0 \end{vmatrix} + c_4\begin{vmatrix} 0 & a_2 \\ b_1 & 0 \end{vmatrix} + a_1b_2c_4.$$

17. $$\left\{\begin{vmatrix} a & b \\ d & e \end{vmatrix}\begin{vmatrix} e & f \\ y & z \end{vmatrix} - \begin{vmatrix} d & e \\ x & y \end{vmatrix}\begin{vmatrix} b & c \\ e & f \end{vmatrix}\right\} \div e.$$

42. As an instance of how determinants come into use in algebra, there may be taken the case of the solution of a set of simultaneous equations of the first degree.

If

(1) $$a_1x + b_1y = c_1,$$

and

(2) $$a_2x + b_2y = c_2,$$

then we have

$$a_1b_2x + b_1b_2y = b_2c_1$$

$$- a_2b_1x - b_1b_2y = - b_1c_2.$$

Therefore

$$x = \frac{\begin{vmatrix} c_1 & b_1 \\ c_2 & b_2 \end{vmatrix}}{\begin{vmatrix} a_1 & b_1 \\ a_2 & b_2 \end{vmatrix}},$$

and, similarly,

$$y = \frac{\begin{vmatrix} a_1 & c_1 \\ a_2 & c_2 \end{vmatrix}}{\begin{vmatrix} a_1 & b_1 \\ a_2 & b_2 \end{vmatrix}}.$$

Again, if

(1) $$a_1x + b_1y + c_1z = d_1$$

(2) $$a_2x + b_2y + c_2z = d_2$$

(3) $$a_3x + b_3y + c_3z = d_3$$

then, using

$$\begin{vmatrix} b_2 & c_2 \\ b_3 & c_3 \end{vmatrix}, \quad -\begin{vmatrix} b_1 & c_1 \\ b_3 & c_3 \end{vmatrix}, \quad \begin{vmatrix} b_1 & c_1 \\ b_2 & c_2 \end{vmatrix},$$

as multipliers, we have

$$a_1\begin{vmatrix} b_2 & c_2 \\ b_3 & c_3 \end{vmatrix}x + b_1\begin{vmatrix} b_2 & c_2 \\ b_3 & c_3 \end{vmatrix}y + c_1\begin{vmatrix} b_2 & c_2 \\ b_3 & c_3 \end{vmatrix}z = d_1\begin{vmatrix} b_2 & c_2 \\ b_3 & c_3 \end{vmatrix},$$

$$-a_2\begin{vmatrix} b_1 & c_1 \\ b_3 & c_3 \end{vmatrix}x - b_2\begin{vmatrix} b_1 & c_1 \\ b_3 & c_3 \end{vmatrix}y - c_2\begin{vmatrix} b_1 & c_1 \\ b_3 & c_3 \end{vmatrix}z = -d_2\begin{vmatrix} b_1 & c_1 \\ b_3 & c_3 \end{vmatrix},$$

$$a_3\begin{vmatrix} b_1 & c_1 \\ b_2 & c_2 \end{vmatrix}x + b_3\begin{vmatrix} b_1 & c_1 \\ b_2 & c_2 \end{vmatrix}y + c_3\begin{vmatrix} b_1 & c_1 \\ b_2 & c_2 \end{vmatrix}z = d_3\begin{vmatrix} b_1 & c_1 \\ b_2 & c_2 \end{vmatrix}.$$

Hence, by addition, we have (Exercises, Set II. 6)

$$\begin{vmatrix} a_1 & b_1 & c_1 \\ a_2 & b_2 & c_2 \\ a_3 & b_3 & c_3 \end{vmatrix}x + \begin{vmatrix} b_1 & b_1 & c_1 \\ b_2 & b_2 & c_2 \\ b_3 & b_3 & c_3 \end{vmatrix}y + \begin{vmatrix} c_1 & b_1 & c_1 \\ c_2 & b_2 & c_2 \\ c_3 & b_3 & c_3 \end{vmatrix}z = \begin{vmatrix} d_1 & b_1 & c_1 \\ d_2 & b_2 & c_2 \\ d_3 & b_3 & c_3 \end{vmatrix}.$$

Now, the second and third determinants in this equation are each equal to 0; therefore

$$x = \begin{vmatrix} d_1 & b_1 & c_1 \\ d_2 & b_2 & c_2 \\ d_3 & b_3 & c_3 \end{vmatrix} \div \begin{vmatrix} a_1 & b_1 & c_1 \\ a_2 & b_2 & c_2 \\ a_3 & b_3 & c_3 \end{vmatrix}.$$

In a similar way, or, more shortly, by using the result just obtained, we may show that

$$y = \begin{vmatrix} a_1 & d_1 & c_1 \\ a_2 & d_2 & c_2 \\ a_3 & d_3 & c_3 \end{vmatrix} \div \begin{vmatrix} a_1 & b_1 & c_1 \\ a_2 & b_2 & c_2 \\ a_3 & b_3 & c_3 \end{vmatrix},$$

and

$$z = \begin{vmatrix} a_1 & b_1 & d_1 \\ a_2 & b_2 & d_2 \\ a_3 & b_3 & c_3 \end{vmatrix} \div \begin{vmatrix} a_1 & b_1 & c_1 \\ a_2 & b_2 & c_2 \\ a_3 & b_3 & c_3 \end{vmatrix}.$$

The learner should compare the values of x, y, z, just found with those of x, y, in the case of the preceding set of equations, noting that the denominator is always the determinant whose elements are in order the coefficients of the unknown quantities in the given equations, and that the numerator of the value of any of the unknown quantities differs from the denominator simply in having the right-hand members of the equations occupying in order the places of the coefficients of the unknown quantity in question.

43. One advantage of these solutions lies in the fact that the results obtained are such as can be exceedingly easily remembered, so that we are thus enabled to derive the benefit usually attached to remembered results. namely, of being able to utilize them in the solution of similar problems. Thus, if the given set of equations be

$$\begin{aligned} 3x + 2y - 4z &= -5, \\ 2x - 3y + z &= -1, \\ 4x + y - 2z &= 0; \end{aligned}$$

and it be required to find the value of y, we have at once

$$\begin{aligned} y &= \begin{vmatrix} 3 & -5 & -4 \\ 2 & -1 & 1 \\ 4 & 0 & -2 \end{vmatrix} \div \begin{vmatrix} 3 & 2 & -4 \\ 2 & -3 & 1 \\ 4 & 1 & -2 \end{vmatrix}, \\ &= \frac{6 + 0 - 20 - 16 - 20 - 0}{18 - 8 + 8 - 48 + 8 - 3}, \\ &= \frac{-50}{-25}, \\ &= 2. \end{aligned}$$

Exercises. Set III

Tell immediately the values of x and y which satisfy the following pairs of equations:

1. $\begin{cases} 4x + 3y = 24, \\ 5x + 2y = 23. \end{cases}$

2. $\begin{cases} 3x + 5y = 17, \\ 2x + 3y = 11. \end{cases}$

3. $\begin{cases} 6x - 4y = 6, \\ 7x - 3y = 12. \end{cases}$

4. $\begin{cases} 4x - 5y = 15, \\ -3x + 17y = 2. \end{cases}$

5. $\begin{cases} -ax + by = a^2, \\ bx - ay = -b^2. \end{cases}$

6. $\begin{cases} -4x + 7y - 10 = 0, \\ 7x - 4y + 1 = 0. \end{cases}$

Find, by means of determinants, the values of x, y, and z which satisfy the following sets of equations:

7. $\begin{cases} 3x - 4y + 2z = 1, \\ 2x + 3y - 3z = -1, \\ 5x - 5y + 4z = 7. \end{cases}$ 8. $\begin{cases} 3x + 4y - 5z = -2, \\ 4x + 5y - 3z = 11, \\ 5x + 3y - 4z = 3. \end{cases}$

9. $\begin{cases} 4x - 7y + z = 16, \\ 3x + y - 2z = 10, \\ 5x - 6y - 3z = 10. \end{cases}$ 10. $\begin{cases} 6x + 8y + 3z = 6, \\ 5x + 6y - 9z = 1, \\ 7x - 10y - 3z = 0. \end{cases}$

11. $\begin{cases} 5x - 4z = 42, \\ 3z + 5y = 1, \\ 4y - 3x = -10. \end{cases}$ 12. $\begin{cases} 6/x - 2/y + 1/z = 4, \\ 2/x + 5/y - 2/z = 3/4, \\ 5/x - 1/y + 3/z = 63/4. \end{cases}$

13. Having given

(1) $$a_1x + b_1y + c_1 = 0$$

(2) $$a_2x + b_2y + c_2 = 0$$

(3) $$a_3x + b_3y + c_3 = 0$$

show, by solving for x and y in (2) and (3) and substituting the results in (1), that

$$\begin{vmatrix} a_1 & b_1 & c_1 \\ a_2 & b_2 & c_2 \\ a_3 & b_3 & c_3 \end{vmatrix} = 0.$$

44. Addition Theorem. *If each of the constituents of a row of a determinant consists of two terms, the determinant may be expressed as the sum of two determinants, the first of which is got from the original determinant by excluding one term of each of the constituents in question, and the second by replacing these and excluding the other terms.*

For if the determinant be

$$\Delta \equiv \begin{vmatrix} a_1 + \alpha_1 & a_2 + \alpha_2 & \cdots & a_n + \alpha_n \\ b_1 & b_2 & \cdots & b_n \\ \cdots & \cdots & \cdots & \cdots \\ l_1 & l_2 & \cdots & l_n \end{vmatrix}.$$

Then we have seen that we may write

$$\Delta = (a_1 + \alpha_1)A_1 + (a_2 + \alpha_2)A_2 + \cdots + (a_n + \alpha_n)A_n$$

$$= a_1A_1 + a_2A_2 + \cdots a_nA_n + \alpha_1A_1 + \alpha_2A_2 + \cdots \alpha_nA_n$$

$$\begin{vmatrix} a_1 & a_2 & \cdots & a_n \\ b_1 & b_2 & \cdots & b_n \\ \cdot & \cdot & \cdots & \cdot \\ l_1 & l_2 & \cdots & l_n \end{vmatrix} + \begin{vmatrix} \alpha_1 & \alpha_2 & \cdots & \alpha_n \\ b_1 & b_2 & \cdots & b_n \\ \cdot & \cdot & \cdots & \cdot \\ l_1 & l_2 & \cdots & l_n \end{vmatrix},$$

which proves the proposition.

EXAMPLES.

$$\begin{vmatrix} a & A + B & h \\ b & C + D & k \\ c & E - F & l \end{vmatrix} = \begin{vmatrix} a & A & h \\ b & C & k \\ c & E & l \end{vmatrix} + \begin{vmatrix} a & B & h \\ b & D & k \\ c & -F & l \end{vmatrix},$$

and

$$\begin{vmatrix} a & b & c \\ d & e & f \\ g + h & k & l + m \end{vmatrix} = \begin{vmatrix} a & b & c \\ d & e & f \\ g & k & l \end{vmatrix} + \begin{vmatrix} a & b & c \\ d & e & f \\ h & 0 & m \end{vmatrix}.$$

45. *If each of the constituents of a row consists of n terms, it is evident that the given determinant can be partitioned in similar fashion into n determinants.*

And generally *if the constituents of the first row consist each of m_1 terms, those of the second row of m_2 terms, and so on, the determinant can be partitioned into $m_1\, m_2 \cdots$ determinants.*

EXAMPLE.

$$\begin{vmatrix} a_1 + b_1 + c_1 & d_1 & e_1 - f_1 \\ a_2 + b_2 + c_2 & d_2 & e_2 - f_2 \\ a_3 + b_3 + c_3 & d_3 & e_3 - f_3 \end{vmatrix} = \begin{vmatrix} a_1 & d_1 & e_1 - f_1 \\ a_2 & d_2 & e_2 - f_2 \\ a_3 & d_3 & e_3 - f_3 \end{vmatrix}$$

$$+ \begin{vmatrix} b_1 & d_1 & e_1 - f_1 \\ b_2 & d_2 & e_2 - f_2 \\ b_3 & d_3 & e_3 - f_3 \end{vmatrix} + \begin{vmatrix} c_1 & d_1 & e_1 - f_1 \\ c_2 & d_2 & e_2 - f_2 \\ c_3 & c_3 & e_3 - f_3 \end{vmatrix}$$

$$= \begin{vmatrix} a_1 & d_1 & e_1 \\ a_2 & d_2 & e_2 \\ a_3 & d_3 & e_3 \end{vmatrix} - \begin{vmatrix} a_1 & d_1 & f_1 \\ a_2 & d_2 & f_2 \\ a_3 & d_3 & f_3 \end{vmatrix} + \begin{vmatrix} b_1 & d_1 & e_1 \\ b_2 & d_2 & e_2 \\ b_3 & d_3 & e_3 \end{vmatrix},$$

$$- \begin{vmatrix} b_1 & d_1 & f_1 \\ b_2 & d_2 & f_2 \\ b_3 & d_3 & f_3 \end{vmatrix} + \begin{vmatrix} c_1 & d_1 & e_1 \\ c_2 & d_2 & e_2 \\ c_3 & d_3 & e_3 \end{vmatrix} - \begin{vmatrix} c_1 & d_1 & f_1 \\ c_2 & d_2 & f_2 \\ c_3 & d_3 & f_3 \end{vmatrix}.$$

46. If $m_1=m_2=\cdots=m_n=p$ and if

$$\Delta \equiv \begin{vmatrix} \sum a_{11} & \sum a_{12}\cdots\sum a_{1n} \\ \sum a_{21} & \sum a_{22}\cdots\sum a_{2n} \\ \cdots & \cdots \\ \sum a_{n1} & \sum a_{n2}\cdots\sum a_{nn} \end{vmatrix}$$

where

$$\sum a_{11}=a_{11}+b_{11}+\cdots+p_{11} \text{ etc., that is,}$$

where each constituent is the sum of p terms it may be written as the sum of p^n determinants.

Let $A=|a_{1n}|$, $B=|b_{1n}|\cdots P=|p_{1n}|$ and let $\Delta a_\alpha b_\beta\cdots p_\pi$ denote the determinant formed as follows: the first α columns are taken from A, the next β columns are taken from B, the next γ columns are taken from C, and so on, the last π columns being taken from P, with the proviso that no two columns thus taken have the same column number, that is, come from corresponding positions, and where $\alpha+\beta+\cdots+\pi=n$.

We may then write

$$\Delta = \sum_0^n \alpha \sum_0^n \beta \cdots \sum_0^n \pi \Delta a_\alpha b_\beta \cdots p_\pi .$$

47. The identity established in §44 may be otherwise viewed as a theorem for the *addition* of two or more determinants which are related to each other in a particular way; that is to say, beginning with the second member of the identity, the theorem is: *The sum of any number of determinants which are alike except as regards a particular row, the rth, say, is equal to a determinant which is like each of the given determinants except that any constituent of its rth row is the sum of the corresponding constituents of all the given determinants.*

EXAMPLES.

$$\begin{vmatrix} 4 & 3 & 4 \\ 7 & -1 & 3 \\ 2 & 8 & 6 \end{vmatrix} + \begin{vmatrix} 4 & 3 & 4 \\ 5 & 1 & -3 \\ 2 & 8 & 6 \end{vmatrix} + \begin{vmatrix} 5 & -12 & 3 \\ 3 & 0 & 8 \\ 4 & 0 & 6 \end{vmatrix} = \begin{vmatrix} 4 & 3 & 4 \\ 12 & 0 & 0 \\ 2 & 8 & 6 \end{vmatrix}$$

$$+ \begin{vmatrix} 5 & 3 & 4 \\ -12 & 0 & 0 \\ 3 & 8 & 6 \end{vmatrix} = \begin{vmatrix} 9 & 3 & 4 \\ 0 & 0 & 0 \\ 5 & 8 & 6 \end{vmatrix}$$

$$= 0,$$

$$m\,|a_1 b_2 c_3 d_4| + n\,|a_1 b_2 c_3 e_4| = |a_1 b_2 c_3 \overline{md_4 + ne_4}| .$$

48. If we multiply a determinant by a polynomial we get different forms, thus

$$(x_1 + x_2 + \cdots + x_k)\begin{vmatrix} a_{11} & a_{12} \cdots a_{1n} \\ \cdots & \cdots \\ \cdots & \cdots \\ a_{n1} & a_{n2} \cdots a_{nn} \end{vmatrix}$$

$$= \begin{vmatrix} a_{11}x_1 + a_{11}x_2 + \cdots + a_{11}x_k & a_{12} \cdots a_{1n} \\ a_{21}x_1 + a_{21}x_2 + \cdots + a_{21}x_k & a_{22} \cdots a_{2n} \\ \cdots\cdots & \cdots \\ a_{n1}x_1 + a_{n1}x_2 + \cdots + a_{n1}x_k & a_{n2} \cdots a_{nn} \end{vmatrix}.$$

This is the sum of k determinants

$$\begin{vmatrix} a_{11}x_1 & a_{12} \cdots a_{1n} \\ a_{21}x_1 & a_{22} \cdots a_{2n} \\ \cdots & \cdots \\ a_{n1}x_1 & a_{n2} \cdots a_{nn} \end{vmatrix} + \cdots + \begin{vmatrix} a_{11}x_k & a_{12} \cdots a_{1n} \\ a_{21}x_k & a_{22} \cdots a_{2n} \\ \cdots & \cdots \\ a_{n1}x_k & a_{n2} \cdots a_{nn} \end{vmatrix}.$$

This is but the simplest case of a whole set of forms obtained by taking various arrangements of the x's before multiplying the constituents of the columns, or what amounts to the same thing, by partitioning the product determinant up into parts in all possible ways thus

$$(x_1 + x_2 + x_3)\begin{vmatrix} a_{11} & a_{12} \\ a_{21} & a_{22} \end{vmatrix}$$

$$= \begin{vmatrix} a_{11}x_1 & a_{12} \\ a_{21}x_1 & a_{22} \end{vmatrix} + \begin{vmatrix} a_{11}x_2 & a_{12} \\ a_{21}x_2 & a_{22} \end{vmatrix} + \begin{vmatrix} a_{11}x_3 & a_{12} \\ a_{21}x_3 & a_{22} \end{vmatrix},$$

or

$$= \begin{vmatrix} a_{11}x_1 & a_{12} \\ a_{21}x_2 & a_{22} \end{vmatrix} + \begin{vmatrix} a_{11}x_2 & a_{12} \\ a_{21}x_3 & a_{22} \end{vmatrix} + \begin{vmatrix} a_{11}x_3 & a_{12} \\ a_{21}x_1 & a_{22} \end{vmatrix},$$

or

$$= \begin{vmatrix} a_{11}x_1 & a_{12} \\ a_{21}x_3 & a_{22} \end{vmatrix} + \begin{vmatrix} a_{11}x_2 & a_{12} \\ a_{21}x_1 & a_{22} \end{vmatrix} + \begin{vmatrix} a_{11}x_3 & a_{12} \\ a_{21}x_2 & a_{22} \end{vmatrix}.$$

49. *If two rows of a determinant be interchanged the resulting determinant differs only in sign from the original determinant.* As this amounts to the interchange of two suffixes its truth follows from §2.

50. *If from a determinant* Δ *another determinant* Δ' *be got as if by making one of the rows of the former pass from its place over* p *rows, then* $\Delta=(-1)^p\Delta'$.

The transference may be affected by the transposition of the row in question with the p rows in succession, beginning with the nearest of them. This would occasion p changes of sign; hence the truth of the theorem.

EXAMPLES.

$$\begin{vmatrix} a & b & c & d \\ e & f & g & h \\ k & l & m & n \\ o & p & q & r \end{vmatrix} = (-1)^3 \begin{vmatrix} b & c & d & a \\ f & g & h & e \\ l & m & n & k \\ p & q & r & o \end{vmatrix} = (-1)^5 \begin{vmatrix} b & c & d & a \\ p & q & r & o \\ f & g & h & e \\ l & m & n & k \end{vmatrix} = \cdots \; ;$$

$$|a_1b_2c_3d_4e_5| = (-1)^3|a_1b_5c_2d_3e_4| = (-1)^7|a_4b_1c_5d_2e_3| = \cdots \; ;$$

and

$$|a_1b_2c_3d_4e_5| + |a_1e_2b_3c_4d_5| = |a_1b_2c_3d_4e_5| + (-1)^3|a_1b_2c_3d_4e_5| ,$$
$$= 0.$$

$$\begin{vmatrix} a_1 & a_2 & a_3 & a_4 \\ b_1 & b_2 & b_3 & b_4 \\ c_1 & c_2 & c_3 & c_4 \\ d_1 & d_2 & d_3 & d_4 \end{vmatrix} = (-1)^{4(4-1)/2} \begin{vmatrix} d_1 & d_2 & d_3 & d_4 \\ c_1 & c_2 & c_3 & c_4 \\ b_1 & b_2 & b_3 & b_4 \\ a_1 & a_2 & a_3 & a_4 \end{vmatrix} = \begin{vmatrix} d_1 & d_2 & d_3 & d_4 \\ c_1 & c_2 & c_3 & c_4 \\ b_1 & b_2 & b_3 & b_4 \\ a_1 & a_2 & a_3 & a_4 \end{vmatrix}$$

$$|a_1\,b_2\,c_3\,d_4\,e_5| = (-1)^{5(5-1)/2}|e_1\,d_2\,c_3\,b_4\,a_5| = |e_1\,d_2\,c_3\,b_4\,a_5| \; ;$$

and

$$|a_1\,b_2\,c_3\,d_4\,e_5\,f_6| = (-1)^{6(6-1)/2}|f_1e_2d_3c_4b_5a_6| = -|f_1e_2d_3c_4b_5a_6| .$$

51. *If two rows of a determinant be identical the determinant is equal to zero.*

Let the determinant be Δ. Then, transposing the two rows referred to, we get a determinant which by §49 is equal to $-\Delta$. But this new determinant is exactly the same as the original, on account of the identity of the two rows transposed. Hence we have

$$\Delta = -\Delta$$

so that

$$2\Delta = 0$$

and therefore

$$\Delta = 0$$

52. *If the constituents in any row differ only by the same factor from the constituents in any other row, the determinant vanishes.*

For taking out the common factor we have left a determinant with two rows identical which vanishes.

If in a determinant the elements of the rows are connected by the same linear relation, the determinant may be reduced to one having a column of zeros and is therefore zero.

53. If we subtract corresponding elements of the two determinants

$$A \equiv \begin{vmatrix} a_1 & a_1 \cdots a_1 \\ a_2 & a_2 \cdots a_2 \\ \cdots & \cdots \\ a_n & a_n \cdots a_n \end{vmatrix}, \quad B \equiv \begin{vmatrix} b_1 & b_2 \cdots b_n \\ b_1 & b_2 \cdots b_n \\ \cdots & \cdots \\ b_1 & b_2 \cdots b_n \end{vmatrix}$$

to form a new determinant D, then $D=0$ for all values of $n>2$.

Thus when $n=3$

$$\begin{vmatrix} a_1 - b_1 & a_1 - b_2 & a_1 - b_3 \\ a_2 - b_1 & a_2 - b_2 & a_2 - b_3 \\ a_3 - b_1 & a_3 - b_3 & a_3 - b_3 \end{vmatrix}$$

is seen to be equal to

$$\begin{vmatrix} b_2 - b_1 & b_3 - b_2 & a_1 - b_3 \\ b_2 - b_1 & b_3 - b_2 & a_2 - b_3 \\ b_2 - b_1 & b_3 - b_2 & a_3 - b_3 \end{vmatrix} = 0$$

by subtracting column 2 from column 1, and column 3 from column 2.

The same would be true for any higher order.

54. *If two determinants Δ and Δ' of the nth order be such that the first row of the one is the same as the last row of the other, the second row of the one the same as the $(n-1)$th row of the other, the third row of the one the same as the $(n-2)$th row of the other, and so on, then*

$$\Delta = (-1)^{(1/2)n(n-1)}\Delta'.$$

The transformation of the one determinant into the other may be effected by making the first row of Δ pass over the $n-1$ other rows, then making what was before this the second row pass over $(n-2)$ rows, next making what was originally the third row pass over $(n-3)$ rows, and so on until what was originally the $(n-1)$th row is made to pass over the one next it, namely, that which originally was the nth. The number of changes of sign consequent upon these alterations is $(n-1)+(n-2)+(n-3)+\cdots+2+1=\frac{1}{2}n(n-1)$ hence the truth of the theorem.

55. *If two determinants Δ, Δ' of the nth order be such that the first row of the one is, when reversed, the last row of the other, the second row, when reversed, the $(n-1)$st row of the other, and so on, then $\Delta=\Delta'$.*

The transformation of one determinant into the other may be effected by reversing the order of the rows, and then in the result reversing the order of the columns. The number of changes of sign occasioned by this is

$$\tfrac{1}{2}n(n-1)+\tfrac{1}{2}n(n-1)$$

that is,

$$n(n-1).$$

Therefore

$$\begin{aligned}\Delta &= (-1)^{n(n-1)}\Delta' \\ &= \Delta'\end{aligned}$$

since $n(n-1)$ is an even number.

EXAMPLES.

$$\begin{vmatrix} a & b & c \\ d & e & f \\ g & h & k \end{vmatrix} = - \begin{vmatrix} g & h & k \\ d & e & f \\ a & b & c \end{vmatrix} = \begin{vmatrix} k & h & g \\ f & e & d \\ c & b & a \end{vmatrix}.$$

and

$$|a_1b_2c_3d_4| = |d_4c_3b_2a_1|.$$

56. *A determinant being given, it is possible to transfer any element to the place occupied by any other, and yet have the resulting determinant equal in magnitude to the original one.*

If the two elements be in the same row it is at once seen that the transposition of the columns to which they belong effects the change referred to, and, contrariwise, if the two elements be in the same column.

If the two be neither in the same row nor in the same column, what is necessary is the transposition of the rows they belong to, followed, in the form which results, by the transposition of the columns.

For example, if we wish the element b_2 of the determinant $|a_1\, b_2\, c_3\, d_4\, e_5|$ to be in the fourth row and fifth column, we proceed as follows:

$$\begin{vmatrix} a_1 & a_2 & a_3 & a_4 & a_5 \\ b_1 & b_2 & b_3 & b_4 & b_5 \\ c_1 & c_2 & c_3 & c_4 & c_5 \\ d_1 & d_2 & d_3 & d_4 & d_5 \\ e_1 & e_2 & e_3 & e_4 & e_5 \end{vmatrix} = - \begin{vmatrix} a_1 & a_2 & a_3 & a_4 & a_5 \\ d_1 & d_2 & d_3 & d_4 & d_5 \\ c_1 & c_2 & c_3 & c_4 & c_5 \\ b_1 & b_2 & b_3 & b_4 & b_5 \\ e_1 & e_2 & e_3 & e_4 & e_5 \end{vmatrix} = \begin{vmatrix} a_1 & a_5 & a_3 & a_4 & a_2 \\ d_1 & d_5 & d_3 & d_4 & d_2 \\ c_1 & c_5 & c_3 & c_4 & c_2 \\ b_1 & b_5 & b_3 & b_4 & b_2 \\ e_1 & e_5 & e_3 & e_4 & e_2 \end{vmatrix} .$$

The attainment of the end desired is dependent upon the transference of one row to the place of another, and, as this may be accomplished in other ways than by the transposition of the two rows, there is a corresponding possible variety in the form of the results. Thus

$$|a_1\, b_2\, c_3\, d_4\, e_5| = |a_1\, c_2\, d_3\, b_4\, e_5| = -|a_1\, c_3\, d_4\, b_5\, e_2| ,$$

or, transposing cyclically,

$$|a_1\, b_2\, c_3\, d_4\, e_5| = |d_1\, e_2\, a_3\, b_4\, c_5| = |d_3\, e_4\, a_5\, b_1\, c_2| .$$

57. *If the first constituent of a row of a determinant be increased by any multiple of the first constituent of any other row, the second constituent of the former be increased by the same multiple of the second constituent of the latter, and so on with all the other constituents of the two rows, the new determinant thus obtained is equal to the original one.*

Let the given determinant be

$$\begin{vmatrix} \cdots & \cdots & \cdots\cdots\cdots \\ a & b & c\cdots\cdots\cdots \\ \cdots & \cdots & \cdots\cdots\cdots \\ x & y & z\cdots\cdots\cdots \\ \cdots & \cdots & \cdots\cdots\cdots \end{vmatrix} , \quad \text{or} \quad \Delta,$$

and let the common multiplier be m, so that the new determinant is

$$\begin{vmatrix} \cdots & \cdots & \cdots & \cdots \\ a+mx & b+my & c+mz & \cdots \\ \cdots & \cdots & \cdots & \cdots \\ x & y & z & \cdots \\ \cdots & \cdots & \cdots & \cdots \end{vmatrix} \quad \text{or} \quad \Delta'$$

On account of the row of binomial elements in Δ', we have (§44),

$$\Delta' = \begin{vmatrix} \cdots & \cdots & \cdots \\ a & b & c \cdots \\ \cdots & \cdots & \cdots \\ x & y & z \cdots \\ \cdots & \cdots & \cdots \end{vmatrix} + \begin{vmatrix} \cdots & \cdots & \cdots \\ mx & my & mz \cdots \\ \cdots & \cdots & \cdots \\ x & y & z \cdots \\ \cdots & \cdots & \cdots \end{vmatrix}$$

$$= \begin{vmatrix} \cdots & \cdots & \cdots \\ a & b & c \cdots \\ \cdots & \cdots & \cdots \\ x & y & z \cdots \\ \cdots & \cdots & \cdots \end{vmatrix} + m \begin{vmatrix} \cdots & \cdots & \cdots \\ x & y & z \cdots \\ \cdots & \cdots & \cdots \\ x & y & z \cdots \\ \cdots & \cdots & \cdots \end{vmatrix}$$

$$= \Delta + m(0) \qquad (\S 51)$$

$$= \Delta.$$

EXAMPLES.

Without changing the first determinant into the ordinary notation, show that

$$\begin{vmatrix} aA + bB & aC + bD \\ cA + dB & cC + dD \end{vmatrix} = \begin{vmatrix} a & b \\ c & d \end{vmatrix} \cdot \begin{vmatrix} A & B \\ C & D \end{vmatrix}.$$

Without passing from the determinant notation, show that

$$\begin{vmatrix} 1 & a & b + c \\ 1 & b & c + a \\ 1 & c & a + b \end{vmatrix} = 0.$$

58. If another row of the determinants Δ, Δ' in the preceding be

$$p \quad q \quad r \cdots$$

it follows that

$$\begin{vmatrix} \cdots & \cdots & \cdots & \cdots \\ a + mx + np & b + my + nq & c + mz + nr & \cdots \\ \cdots & \cdots & \cdots & \cdots \\ x & y & z & \cdots \\ \cdots & \cdots & \cdots & \cdots \end{vmatrix} = \Delta'$$

and therefore also$=\Delta$. So that from a continued application of the theorem of §57, we have a more general result, namely: *If the constituents of any row of a determinant be increased by any equimultiples of the corresponding constituents of a second row and by any equimultiples of the corresponding constituents of a third row, and so on, the resulting determinant is equal to the original one.*

59. We have seen from §52 that if the constituents in one row differ only by a constant multiplier from those in another row the determinant vanishes. As a more general case of this we have

$$\begin{vmatrix} a_{11} & a_{12} & a_{13} & \cdots a_{1n} \\ a_{21} & a_{22} & a_{23} & \cdots a_{2n} \\ \cdots & \cdots & \cdots & \cdots \\ a_{n-1,1} & a_{n-1,2} & a_{n-1,3} & \cdots a_{n-1,n} \\ T_1 & T_2 & T_3 & \cdots T_n \end{vmatrix} = 0$$

where

$$T_r = m_1 a_{1r} + m_2 a_{2r} + \cdots + m_{n-1} a_{n-1,r}, \qquad (r = 1, 2, \cdots, n).$$

60. These theorems may be advantageously employed in the simplification of determinants, more especially those whose constituents are expressed in figures.

For example consider the determinant

$$\begin{vmatrix} 14 & 15 & 11 \\ 21 & 22 & 16 \\ 23 & 29 & 17 \end{vmatrix}$$

Subtracting each constituent of the third column from the corresponding constituents of the first and second columns we have the equivalent determinant

$$\begin{vmatrix} 3 & 4 & 11 \\ 5 & 6 & 16 \\ 6 & 12 & 17 \end{vmatrix},$$

and this we know (§39) is equal to

$$2\begin{vmatrix} 3 & 2 & 11 \\ 5 & 3 & 16 \\ 6 & 6 & 17 \end{vmatrix}.$$

Now subtracting each element of the second column from the corresponding element of the first column, and multiplying each element of the second column by 5 and subtracting the result from the third column we have

$$2\begin{vmatrix} 1 & 2 & 1 \\ 2 & 3 & 1 \\ 0 & 6 & -13 \end{vmatrix},$$

which if we please we may alter similarly into

$$2\begin{vmatrix} 1 & 2 & 1 \\ 0 & -1 & -1 \\ 0 & 6 & -13 \end{vmatrix}$$

the given determinant being thus equal to 2(13+6), that is, to 38.

Operating on the rows instead of the columns, we might have proceeded thus:

$$\begin{vmatrix} 14 & 15 & 11 \\ 21 & 22 & 16 \\ 23 & 29 & 17 \end{vmatrix} = \begin{vmatrix} 14 & 15 & 11 \\ 7 & 7 & 5 \\ 2 & 7 & 1 \end{vmatrix} = \begin{vmatrix} 0 & 1 & 1 \\ 5 & 0 & 4 \\ 2 & 7 & 1 \end{vmatrix} = 35 + 8 - 5 = 38,$$

as in the preceding case.

EXAMPLE 1. Show that

$$\begin{vmatrix} a & b & c & d \\ b & c & d & a \\ c & d & a & b \\ d & a & b & c \end{vmatrix} = (a+b+c+d)(-a+b-c+d)\begin{vmatrix} 0 & 1 & -1 & 1 \\ 1 & c & d & a \\ 1 & d & a & b \\ 1 & a & b & c \end{vmatrix}.$$

The given determinant

$$= \begin{vmatrix} a+b+c+d & b & c & d \\ b+c+d+a & c & d & a \\ c+d+a+b & d & a & b \\ d+a+b+c & a & b & c \end{vmatrix}, \qquad (\S 57)$$

$$= \begin{vmatrix} 1 & b & c & d \\ 1 & c & d & a \\ 1 & d & a & b \\ 1 & a & b & c \end{vmatrix}(a+b+c+d), \qquad (\S 39)$$

$$= \begin{vmatrix} 0 & b-c+d-a & c-d+a-b & d-a+b-c \\ 1 & c & d & a \\ 1 & d & a & b \\ 1 & a & b & c \end{vmatrix} (a+b+c+d),$$

$$= \begin{vmatrix} 0 & 1 & -1 & 1 \\ 1 & c & d & a \\ 1 & d & a & b \\ 1 & a & b & c \end{vmatrix} (a+b+c+d)(-a+b-c+d).$$

EXAMPLE 2. Show that

$$\begin{vmatrix} 1 & a & a^2 & a^3 \\ 1 & b & b^2 & b^3 \\ 1 & c & c^2 & c^3 \\ 1 & d & d^2 & d^3 \end{vmatrix} = (b-a)(c-a)(c-b)(d-a)(d-b)(d-c),$$

without finding the expansion of the determinant.

If each element of the first row be subtracted from the corresponding element of the second row, it is at once clear that $b-a$ is a factor of the determinant, and that $c-a$, $c-b$, $d-a$, $d-b$, $d-c$ are factors follows equally readily in the same way. Further we see that if these factors were multiplied together, the first term of the product would be bc^2d^3, and that all the other terms would be unlike this, so that the coefficient of bc^2d^3 in the product is $+1$. On the other hand, looking at the principal diagonal, we see that the coefficient of bc^2d^3 in the expansion of the determinant is also $+1$. Thus the identity is established.

EXAMPLE 3. Show that

$$\begin{vmatrix} 0 & a^2 & b^2 & c^2 \\ a^2 & 0 & \gamma^2 & \beta^2 \\ b^2 & \gamma^2 & 0 & \alpha^2 \\ c^2 & \beta^2 & \alpha^2 & 0 \end{vmatrix} = \begin{vmatrix} 0 & a\alpha & b\beta & c\gamma \\ a\alpha & 0 & c\gamma & b\beta \\ b\beta & c\gamma & 0 & a\alpha \\ c\gamma & b\beta & a\alpha & 0 \end{vmatrix}.$$

Taking the first determinant and multiplying the elements of its first and second rows by α, and the elements of its third and fourth rows by a, we obtain as its equivalent

$$\frac{1}{a^2\alpha^2}\begin{vmatrix} 0 & a^2\alpha & b^2\alpha & c^2\alpha \\ a^2\alpha & 0 & \gamma^2\alpha & \beta^2\alpha \\ b^2a & \gamma^2a & 0 & \alpha^2a \\ c^2a & \beta^2a & \alpha^2a & 0 \end{vmatrix},$$

and dividing the elements of the first and second *columns* of this new determinant by a, and the elements of the third and fourth by α, we thus have the original determinant equal to

$$\begin{vmatrix} 0 & a\alpha & b^2 & c^2 \\ a\alpha & 0 & \gamma^2 & \beta^2 \\ b^2 & \gamma^2 & 0 & a\alpha \\ c^2 & \beta^2 & a\alpha & 0 \end{vmatrix}.$$

Continuing in exactly similar fashion we find it

$$= \frac{1}{b^2\beta^2}\begin{vmatrix} 0 & a\alpha\beta & b^2\beta & c^2\beta \\ a\alpha\beta & 0 & \gamma^2b & \beta^2b \\ b^2\beta & \gamma^2\beta & 0 & a\alpha\beta \\ c^2b & \beta^2b & a\alpha b & 0 \end{vmatrix} = \begin{vmatrix} 0 & a\alpha & b\beta & c^2 \\ a\alpha & 0 & \gamma^2 & b\beta \\ b\beta & \gamma^2 & 0 & a\alpha \\ c^2 & b\beta & a\alpha & 0 \end{vmatrix},$$

$$= \frac{1}{c^2\gamma^2}\begin{vmatrix} 0 & a\alpha\gamma & b\beta\gamma & c^2\gamma \\ a\alpha c & 0 & \gamma^2c & b\beta c \\ b\beta c & \gamma^2c & 0 & a\alpha c \\ c^2\gamma & b\beta\gamma & a\alpha\gamma & 0 \end{vmatrix} = \begin{vmatrix} 0 & a\alpha & b\beta & c\gamma \\ a\alpha & 0 & c\gamma & b\beta \\ b\beta & c\gamma & 0 & a\alpha \\ c\gamma & b\beta & a\alpha & 0 \end{vmatrix},$$

as was required.

By a further combination of multiplications and divisions the process assumes a neater form, thus:—Taking for the rows the multipliers

$$\alpha\beta\gamma,\quad \alpha bc,\quad a\beta c,\quad ab\gamma,$$

respectively, that is, multiplying in all by $(abc\alpha\beta\gamma)^2$, we have

$$\begin{vmatrix} 0 & a^2\alpha\beta\gamma & b^2\alpha\beta\gamma & c^2\alpha\beta\gamma \\ a^2\alpha bc & 0 & \gamma^2\alpha bc & \beta^2\alpha bc \\ b^2a\beta c & \gamma^2a\beta c & 0 & \alpha^2a\beta c \\ c^2ab\gamma & \beta^2ab\gamma & \alpha^2ab\gamma & 0 \end{vmatrix},$$

and then all that is required is to divide by $(abc\alpha\beta\gamma)^2$ by operating on the *columns* with the divisors

$$abc, \quad a\beta\gamma, \quad \alpha b\gamma, \quad \alpha\beta c,$$

respectively.

EXERCISES. SET IV

Find the simplest forms of the following numerical expressions:

1. $\begin{vmatrix} 15 & 17 & 16 \\ 12 & 18 & 14 \\ 19 & 17 & 13 \end{vmatrix}$. 2. $\begin{vmatrix} 15 & 13 & 10 \\ 12 & 17 & 10 \\ 16 & 11 & 19 \end{vmatrix}$. 3. $\begin{vmatrix} 20 & 15 & 25 \\ 17 & 12 & 22 \\ 19 & 20 & 16 \end{vmatrix}$.

4. $\begin{vmatrix} 27 & 37 & 47 \\ 33 & 23 & 29 \\ 25 & 28 & 24 \end{vmatrix}$. 5. $\begin{vmatrix} 22 & 29 & 27 \\ 25 & 23 & 30 \\ 28 & 26 & 24 \end{vmatrix}$. 6. $\begin{vmatrix} 30 & 36 & 35 \\ 33 & 31 & 37 \\ 38 & 34 & 32 \end{vmatrix}$.

7. What effect is produced on a determinant of the nth degree by multiplying all its elements by -1?

8. Find the simplified expansion of the determinant

$$\begin{vmatrix} a & a+3 & a+6 \\ a+1 & a+4 & a+7 \\ a+2 & a+5 & a+8 \end{vmatrix}.$$

9. Show that

$$\begin{vmatrix} 1 & 1 & 1 & 1 \\ 1 & 1+x & 1 & 1 \\ 1 & 1 & 1+y & 1 \\ 1 & 1 & 1 & 1+z \end{vmatrix} = xyz.$$

10. Find the simplified expansion of

$$\begin{vmatrix} a_1-b_1 & a_2-b_2 & a_3-b_3 \\ b_1-c_1 & b_2-c_2 & b_3-c_3 \\ c_1-a_1 & c_2-a_2 & c_3-a_3 \end{vmatrix}, \text{ and of } \begin{vmatrix} a+c & 2a-b & b+2c \\ b+a & 2b-c & c+2a \\ c+b & 2c-a & a+2b \end{vmatrix}.$$

11. Prove that, if the sum or difference of every pair of corresponding elements of two rows of a determinant be a constant multiple of the corresponding element of another row, the determinant is equal to zero.

12. Express

$$\begin{vmatrix} a_1 + h_1 + k_1 & a_2 + h_1 + k_2 & a_3 + h_1 + k_3 & 1 \\ b_1 + h_2 + k_1 & b_2 + h_2 + k_2 & b_3 + h_2 + k_3 & 1 \\ c_1 + h_3 + k_1 & c_2 + h_3 + k_2 & c_3 + h_3 + k_3 & 1 \\ 1 & 1 & 1 & 1 \end{vmatrix}$$

in a simpler form as a determinant of the fourth order.

13. Show that

$$\begin{vmatrix} a_1 & a_2 & a_3 \\ b_1 & b_2 & b_3 \\ c_1 & c_2 & c_3 \end{vmatrix} = \frac{1}{a_1a_2a_3} \begin{vmatrix} 1 & 1 & 1 \\ a_2a_3b_1 & a_1a_3b_2 & a_1a_2b_3 \\ a_2a_3c_1 & a_1a_3c_2 & a_1a_2c_3 \end{vmatrix}.$$

14. Show that

$$\begin{vmatrix} 0 & a_2 & a_3 \\ b_1 & b_2 & b_3 \\ c_1 & c_2 & c_3 \end{vmatrix} = \begin{vmatrix} 0 & 1 & 1 \\ 1 & a_3b_2c_1 & a_2b_3c_1 \\ 1 & a_3b_1c_2 & a_2b_1c_3 \end{vmatrix}.$$

15. Solve the equation

$$\begin{vmatrix} a_1 + b_1x & c_1 & d_1 \\ a_2 + b_2x & c_2 & d_2 \\ a_3 + b_3x & c_3 & d_3 \end{vmatrix} = 0$$

16. Prove that

$$\begin{vmatrix} a+c & b+d & a+c & b+d \\ b+d & a+c & b+d & a+c \\ a+b & b+c & c+d & d+a \\ c+d & d+a & a+b & b+c \end{vmatrix} = 0$$

17. Find the simplified expansion of

$$\begin{vmatrix} x & a & b & c+d \\ x & b & c & d+a \\ x & c & d & a+b \\ x & d & a & b+c \end{vmatrix}, \text{ and of } \begin{vmatrix} a & b+c+d & a+b & c+d \\ b & c+d+a & b+c & d+a \\ c & d+a+b & c+d & a+b \\ d & a+b+c & d+a & b+c \end{vmatrix}$$

18. Express

$$\begin{vmatrix} a_2x^2 + a_1x + a_0 & a_3 & a_4 \\ b_2x^2 + b_1x + b_0 & b_3 & b_4 \\ c_2x^2 + c_1x + c_0 & c_3 & c_4 \end{vmatrix}$$

in terms arranged according to ascending powers of x.

19. Prove that, if the sum or difference of every pair of corresponding elements of two rows of a determinant be a constant multiple of the sum or difference of the corresponding pair of elements of two other rows, the determinant is equal to zero.

20. Show that

$$\begin{vmatrix} 0 & a & b & c \\ a & 0 & c & b \\ b & c & 0 & a \\ c & b & a & 0 \end{vmatrix} = \begin{vmatrix} 0 & 1 & 1 & 1 \\ 1 & 0 & c^2 & b^2 \\ 1 & c^2 & 0 & a^2 \\ 1 & b^2 & a^2 & 0 \end{vmatrix}.$$

21. Express either determinant of the preceding exercise as the product of four linear factors.

22. Prove that a determinant remains substantially the same if the signs of the elements be changed in every alternate member of the set of lines consisting of either diagonal and the lines parallel to it, the diagonal itself being one of the lines left unaltered.

23. Prove that

$$\begin{vmatrix} a & b & c & 0 \\ b & a & 0 & c \\ c & 0 & a & b \\ 0 & c & b & a \end{vmatrix} = \begin{vmatrix} -a & b & c & 0 \\ b & -a & 0 & c \\ c & 0 & -a & b \\ 0 & c & b & -a \end{vmatrix}.$$

Resolve the following determinants into simple factors:

24. $\begin{vmatrix} a & b & c & d \\ b & a & d & c \\ c & d & a & b \\ d & c & b & a \end{vmatrix}.$ 25. $\begin{vmatrix} x & 1 & 1 & 1 \\ 1 & x & 1 & 1 \\ 1 & 1 & x & 1 \\ 1 & 1 & 1 & x \end{vmatrix}.$

26. $\begin{vmatrix} a & a & a & a \\ a & b & b & b \\ a & b & c & c \\ a & b & c & d \end{vmatrix}.$ 27. $\begin{vmatrix} x^3 & ax^3 & a^2x & a^3 \\ y^3 & by^2 & b^2y & b^3 \\ z^3 & cz^2 & c^2z & c^3 \\ w^3 & dw^2 & d^2w & d^3 \end{vmatrix}.$

28. Show that

$$\begin{vmatrix} bcd & a & a^2 & a^3 \\ cda & b & b^2 & b^3 \\ dab & c & c^2 & c^3 \\ abc & d & d^2 & d^3 \end{vmatrix} = \begin{vmatrix} 1 & a^2 & a^3 & a^4 \\ 1 & b^2 & b^3 & b^4 \\ 1 & c^2 & c^3 & c^4 \\ 1 & d^2 & d^3 & d^4 \end{vmatrix}.$$

Find a single determinant equivalent to

29. $|a_1\, b_4\, c_5| + |a_2\, b_4\, c_5| - |a_3\, b_4\, c_5|$.

30. $|a_0\, b_2\, c_5| - |a_0\, b_3\, c_5| - |a_1\, b_3\, c_5| + |a_1\, b_2\, c_5|$.

61. Let $A = |a_{1n}|$, then if $\Delta a_{rs} = a_{rs+1} - a_{rs}$ we have

$$A = \begin{vmatrix} a_{11} & \Delta a_{11} & \Delta a_{11} \cdots \Delta a_{1,n-2} \\ a_{21} & \Delta a_{21} & \Delta a_{21} \cdots \Delta a_{2,n-2} \\ \cdots & \cdots & \cdots \\ a_{n1} & \Delta a_{n1} & \Delta a_{n1} \cdots \Delta a_{n,n-2} \end{vmatrix}$$

as an immediate consequent of §57.

Repeat the same operation beginning with the third column, and we have

$$A = \begin{vmatrix} a_{11} & \Delta a_{11} & \Delta^2 a_{11} \cdots \Delta^2 a_{1,n-2} \\ a_{21} & \Delta a_{21} & \Delta^2 a_{21} \cdots \Delta^2 a_{2,n-2} \\ \cdots & \cdots & \cdots \\ a_{n1} & \Delta a_{n1} & \Delta^2 a_{n1} \cdots \Delta^2 a_{n,n-2} \end{vmatrix}$$

where the second difference, $\Delta^2 a_{rs} = \Delta a_{rs+1} - \Delta a_{rs}$.

Continuing this process, leaving out a column each time, we have

$$A = \begin{vmatrix} a_{11} & \Delta a_{11} & \Delta^2 a_{11} \cdots \Delta^{n-1} a_{11} \\ a_{21} & \Delta a_{21} & \Delta^2 a_{21} \cdots \Delta^{n-1} a_{21} \\ \cdots & \cdots & \cdots \\ a_{n1} & \Delta a_{n1} & \Delta^2 a_{n1} \cdots \Delta^{n-1} a_{n1} \end{vmatrix}$$

where generally the kth difference $\Delta^k a_{rs} = \Delta^{k-1} a_{rs+1} - \Delta^{k-1} a_{rs}$.

An arithmetical series is said to be of the *kth order* if the kth differences are constant.

62. *If the n elements in each of the rows of A form an arithmetical series of order not greater than* $(n-2)$, *it follows from* §61 *that A is zero*, for the elements in the last column will be zeros.

63. The following more general theorem is readily seen:

A determinant is zero if the elements in the same h columns of each row form an arithmetical series of order not greater than $(h-2)$.

EXAMPLE. If

$$(a)_k = \frac{a(a-1)\cdots(a-k+1)}{1\cdot 2\cdots k},$$

$$\begin{vmatrix} (a)_0 & (a)_1 & \cdots & (a)_k \\ (a+d)_0 & (a+d)_1 & \cdots & (a+d)_k \\ \cdots & \cdots & \cdots & \cdots \\ (a+kd)_0 & (a+kd)_1 & \cdots & (a+kd)_k \end{vmatrix} = 1\cdot d\cdot d^2\cdots d^k,$$

$$= d^{r(r+1)/2}$$

since

$$\Delta^t(a+td)_t = d^t.$$

EXERCISES. SET V

1. Without finding the expansions of the determinants show that

$$\begin{vmatrix} a & b & c \\ d & e & f \\ g & h & k \end{vmatrix} = \begin{vmatrix} h & g & k \\ e & d & f \\ b & a & c \end{vmatrix}.$$

2. Show that

$$\begin{vmatrix} a & b & c & d \\ e & f & g & h \\ i & j & k & l \\ m & n & o & p \end{vmatrix} = \begin{vmatrix} a & c & b & d \\ i & k & j & l \\ e & g & f & h \\ m & o & n & p \end{vmatrix}.$$

3. Show that

$$\begin{vmatrix} a_1 & a_2 & a_3 & a_4 \\ b_1 & b_2 & b_3 & b_4 \\ c_1 & c_2 & c_3 & c_4 \\ d_1 & d_2 & d_3 & d_4 \end{vmatrix} = \begin{vmatrix} c_1 & b_1 & a_1 & d_1 \\ c_4 & b_4 & a_4 & d_4 \\ c_3 & b_3 & a_3 & d_3 \\ c_2 & b_2 & a_2 & d_2 \end{vmatrix}.$$

4. Show that

$$\begin{vmatrix} a & b & c & d \\ e & f & g & h \\ 0 & j & 0 & 0 \\ k & l & m & n \end{vmatrix} = \begin{vmatrix} j & 0 & 0 & 0 \\ f & e & g & h \\ b & a & c & d \\ l & k & m & n \end{vmatrix}.$$

5. Show that

$$|a_1\, b_2\, c_3\, d_4| = |d_1\, b_4\, c_3\, a_2|,$$

and

$$|a_1\, b_2\, c_3\, d_4\, e_5| = |a_3\, b_2\, e_1\, d_4\, c_5|.$$

Find by cyclical transposition of the rows and columns of $|b_0 c_1 d_2 e_3 f_4|$ a determinant equal to the said determinant and having

6. d_3 in the first row and first column;
7. e_1 in the first row and first column;
8. c_2 in the fifth row and second column;
9. f_4 in the second row and third column.
10. Show that

$$\begin{vmatrix} 0 & b_1 & b_2 & 0 \\ a_1 & a_2 & a_3 & a_4 \\ d_1 & d_2 & d_3 & d_4 \\ 0 & c_1 & c_2 & 0 \end{vmatrix} = \begin{vmatrix} a_2 & a_1 & a_4 & a_3 \\ b_1 & 0 & 0 & b_2 \\ c_1 & 0 & 0 & c_2 \\ d_2 & d_1 & d_4 & d_3 \end{vmatrix}.$$

11. Show that

$$\begin{vmatrix} l & 0 & 0 & 0 \\ d & a & c & b \\ k & 0 & h & 0 \\ g & 0 & f & e \end{vmatrix} = \begin{vmatrix} a & b & c & d \\ 0 & e & f & g \\ 0 & 0 & h & k \\ 0 & 0 & 0 & l \end{vmatrix}.$$

12. Show that

$$\begin{vmatrix} c_3 & c_6 & c_4 & c_2 & c_5 \\ 0 & a_6 & 0 & a_2 & 0 \\ d_3 & d_6 & d_4 & d_2 & d_5 \\ 0 & b_6 & 0 & b_2 & 0 \\ e_3 & e_6 & e_4 & e_2 & e_5 \end{vmatrix} = \begin{vmatrix} a_6 & a_2 & 0 & 0 & 0 \\ b_6 & b_2 & 0 & 0 & 0 \\ c_6 & c_2 & c_3 & c_4 & c_5 \\ d_6 & d_2 & d_3 & d_4 & d_5 \\ e_6 & e_2 & e_3 & e_4 & e_5 \end{vmatrix}.$$

13. Transform

$$\begin{vmatrix} a & b & c & d \\ b & c & d & a \\ c & d & a & b \\ d & a & b & c \end{vmatrix}$$

so as to have the principal diagonal composed (1) of the four a's, (2) of the four b's, (3) of the four c's, (4) of the four d's.

14. Show that

$$\begin{vmatrix} a_1 + a_2 & a_2 + a_3 & a_3 + a_1 \\ b_1 + b_2 & b_2 + b_3 & b_3 + b_1 \\ c_1 + c_2 & c_2 + c_3 & c_3 + c_1 \end{vmatrix} = 2 \begin{vmatrix} a_1 & a_2 & a_3 \\ b_1 & b_2 & b_3 \\ c_1 & c_2 & c_3 \end{vmatrix}.$$

15. Show that

$$\begin{vmatrix} a_1 + a_2 + a_3 & a_2 + a_3 + a_4 & a_3 + a_4 + a_1 & a_4 + a_1 + a_2 \\ b_1 + b_2 + b_3 & b_2 + b_3 + b_4 & b_3 + b_4 + b_1 & b_4 + b_1 + b_2 \\ c_1 + c_2 + c_3 & c_2 + c_3 + c_4 & c_3 + c_4 + c_1 & c_4 + c_1 + c_2 \\ d_1 + d_2 + d_3 & d_2 + d_3 + d_4 & d_3 + d_4 + d_1 & d_4 + d_1 + d_2 \end{vmatrix}$$

$$= 3 \begin{vmatrix} a_1 & a_2 & a_3 & a_4 \\ b_1 & b_2 & b_3 & b_4 \\ c_1 & c_2 & c_3 & c_4 \\ d_1 & d_2 & d_3 & d_4 \end{vmatrix}.$$

16. Use the principles employed in §§ 37 and 39 to show that

$$\begin{vmatrix} 0 & a & b & c & d \\ -a & 0 & e & f & g \\ -b & -e & 0 & h & i \\ -c & -f & -h & 0 & j \\ -d & -g & -i & -j & 0 \end{vmatrix} = 0.$$

17. Similarly show that

$$\begin{vmatrix} a & b & c & d & 0 \\ e & f & g & 0 & d \\ h & i & 0 & g & -c \\ j & 0 & i & -f & b \\ 0 & j & -h & e & -a \end{vmatrix} = 0.$$

18. If m rows, namely, the h_1th, h_2th, $\cdots$, h_mth be transferred so as to become the 1st, 2nd, $\cdots$, mth, without altering the relative positions of the remaining rows, and then n columns, namely, the k_1th, k_2th, $\cdots$, k_nth, be similarly transferred, the determinant thus obtained is the same as the original or differs from it only in sign according as

$$h_1+h_2+\cdots+h_m-\tfrac{1}{2}m(m+1)+k_1+k_2+\cdots+k_n-\tfrac{1}{2}n(n+1)$$

is even or odd.

19. Without finding the expansion of the determinant show that $ab+bc+ca$ is a factor of

$$\begin{vmatrix} ab & c^2 & c^2 \\ a^2 & bc & a^2 \\ b^2 & b^2 & ac \end{vmatrix}.$$

Establish the following identities:

20. $$\begin{vmatrix} a+b & c & c \\ a & b+c & a \\ b & b & c+a \end{vmatrix} = 4abc.$$

21. $$\begin{vmatrix} (a^2+b^2)/c & c & c \\ a & (b^2+c^2)/a & a \\ b & b & (c^2+a^2)/b \end{vmatrix} = 4abc.$$

22. $$\begin{vmatrix} a+b-c & c & c \\ a & b+c-a & a \\ b & b & c+a-b \end{vmatrix} = \begin{vmatrix} a & b & c \\ b & c & a \\ c & a & b \end{vmatrix}.$$

23. $$\begin{vmatrix} (a+b)^2 & ca & bc \\ ca & (b+c)^2 & ab \\ bc & ab & (c+a)^2 \end{vmatrix} = 2abc(a+b+c)^3.$$

24. $$\begin{vmatrix} a+b+nc & (n-1)a & (n-1)b \\ (n-1)c & b+c+na & (n-1)b \\ (n-1)c & (n-1)a & c+a+nb \end{vmatrix} = n(a+b+c)^3.$$

25. $$\begin{vmatrix} (a+b)^3 & -c^3 & -c^3 \\ -a^3 & (b+c)^3 & -a^3 \\ -b^3 & -b^3 & (c+a)^3 \end{vmatrix} = 3abc(a+b+c)^3 \sum a^2b.$$

26. $$\begin{vmatrix} a+b+c & d & d & d \\ a & b+c+d & a & a \\ b & b & c+d+a & b \\ c & c & c & d+a+b \end{vmatrix} = 4 \sum a^2bc.$$

27. $$\begin{vmatrix} a+x & a-x & a-y & a+y \\ a-x & a-y & a+y & a+x \\ a-y & a+y & a+x & a-x \\ a+y & a+x & a-x & a-y \end{vmatrix} = -16a(x-y)(x+y)^2.$$

28. $$\begin{vmatrix} 1 & a & a & a^2 \\ 1 & b & b & b^2 \\ 1 & c & c' & cc' \\ 1 & d & d' & dd' \end{vmatrix} = (a-b)\begin{vmatrix} 1 & ab & a+b \\ 1 & cd' & c+d' \\ 1 & c'd & c'+d \end{vmatrix}.$$

CHAPTER IV

MINORS AND EXPANSION

64. *Definition of Minor.* If in the determinant $\Delta \equiv |a_{1n}|$ we delete any r rows and any r columns the determinant whose elements are in order the elements thus left is called a *minor* of order $n-\text{r}$ of Δ.

Minors obtained by the deletion of one row and one column are called *first minors*, those obtained by the deletion of two rows and two columns are called *second minors*, and generally those obtained by the deletion of r rows and r columns are called *rth minors*.

A minor formed by the deletion of the λth row and λth column is called a *principal first minor*, a minor formed by the deletion of the λ_1th and λ_2th rows and the λ_1th and the λ_2th columns is called a *principal second minor*, and generally a minor formed by the deletion of the λ_1th, λ_2th, $\cdots$, λ_rth rows and the λ_1th, λ_2th, $\cdots$, λ_rth columns is called a *principal rth minor*.

Principal minors are sometimes called *coaxial minors*. Two minors which are such that the rows and columns deleted to obtain the one are exactly those not deleted to obtain the other are called *complementary minors*. If the original determinant is of the nth order then the complementary of a minor of the rth order is of the $(n-r)$th order.

65. The minor formed by the deletion of the hth row and the rth column will be denoted by A_{hr}, the minor formed by deleting the hth and kth rows and the rth and the sth columns will be denoted by $A_{hk,rs}$ and in general the minor formed by deleting the hth, kth, lth, $\cdots$, rows and the rth, sth, tth, $\cdots$ columns will be denoted by $A_{hkl\cdots,rst\cdots}$.

The minor formed by deleting all except the hth, kth, lth, $\cdots$ rows and all except the rth, sth, tth, $\cdots$ columns will be denoted by $A'_{hkl\cdots,rst}$. Both of these notations will be found convenient. It is to be understood that in both cases $h, k, l, \cdots$ and $r, s, t, \cdots$ are in ascending order of magnitude. Then in $A'_{hkl\cdots,rst}$ since the numbers in the first line of the suffix denote rows and the numbers in the second line of the suffix denote columns, it follows from §49 that an odd number of inversions will change the sign of the minor. Thus

$$A'_{hkl\ldots,rst} = A'_{klh\ldots rst} = -A'_{khl\ldots,rst} = -A'_{hkl\ldots,srt}, \text{ etc.}$$

66. Though the numbers in the suffix of $A_{hkl\cdots,rst}$ do not denote the rows and columns which are contained by the minor as in the

other case, but denote the rows and columns which are not contained by it we shall for convenience sake suppose the same law to hold and that $A_{hk,rs} = -A_{kh,rs} = \text{etc.}$

67. We shall also use $A'_{(n|m_\alpha),(n|m_\beta)}$, and $A'_{(\bar{n}|m_\alpha)(\bar{n}|m_\beta)}$ or when no ambiguity can arise $|(n|m_\alpha),(n|m_\beta)|$, and $|(\bar{n}|m_\alpha),(\bar{n}|m_\beta)|$ or $\vdots\ (n|m_\alpha), (n|m_\beta)\ \vdots$ to denote the minor formed by the elements in the intersection of the αth selection of m rows and the βth selection of m columns and its complementary respectively.

Minors in which the rows and columns taken to form the one are the same as the columns and rows taken to form the other are called *conjugate minors.* Thus $A_{(n|m_\alpha),(n|m_\beta)}$ and $A_{(n|m_\beta),(n|m_\alpha)}$ are conjugate minors.

From §12, it is evident that $A'_{(n|m_\alpha),(n|m_\beta)} \equiv A_{(\bar{n}|m_\alpha),(\bar{n}|m_\beta)}$.

68. We have seen (§38) that a determinant

$$\Delta = \begin{vmatrix} a_{11} & a_{12} & \cdots & a_{1n} \\ a_{21} & a_{22} & \cdots & a_{2n} \\ \cdot & \cdot & \cdots & \cdot \\ a_{n1} & a_{n2} & \cdots & a_{nn} \end{vmatrix}$$

may be written

$$a_{11}\mathcal{A}_1 + a_{12}\mathcal{A}_2 + a_{13}\mathcal{A}_3 + \cdots + a_{1n}\mathcal{A}_n$$

where $\mathcal{A}_r$ contains no constituent from either the first row or the rth column and is the complete coefficient of a_{1r} in Δ. Now since every term of Δ which contains a_{11} can contain no other constituents from the first row and first column, and by definition a_{11} must be multiplied by all possible combinations of products of $n-1$ constituents, taken one from each of the other rows and columns, it is evident that $\mathcal{A}_1$ is the minor of Δ complementary to a_{11}, that is, A_{11}. To find the value of $\mathcal{A}_p$, we bring a_{1p} into the first column by carrying the pth column over the $p-1$ columns which precede it. This changes the sign of the determinant or not according as p is even or odd and therefore $\mathcal{A}_p \doteq (-1)^{p-1}A_{1p}$. $\mathcal{A}_p$ is called the *cofactor* of a_{1p} in Δ and is the aggregate of all the terms in Δ containing a_{1p}. We shall generally use $\mathcal{A}_{rs}$ or $[rs]$ to denote the cofactor of a_{rs}.

Therefore

$$\Delta = a_{11}A_{11} - a_{12}A_{12} + a_{13}A_{13} - \cdots + (-1)^{n-1}a_{1n}A_{1n},$$

or

$$a_{11}\mathcal{A}_{11} + a_{12}\mathcal{A}_{12} + a_{13}\mathcal{A}_{13} + \cdots + a_{1n}\mathcal{A}_{1n}.$$

69. In like manner we can write Δ in terms of the constituents of any other row.

Thus

$$\Delta = (-)^{p-1}a_{p1}A_{p1} + (-)^{p}a_{p2}A_{p2} + \cdots + (-)^{n+p-2}a_{pn}A_{pn}.$$

EXAMPLE.

$$\begin{vmatrix} a_{11} & a_{12} & a_{13} & a_{14} \\ a_{21} & a_{22} & a_{23} & a_{24} \\ a_{31} & a_{32} & a_{33} & a_{34} \\ a_{41} & a_{42} & a_{43} & a_{44} \end{vmatrix} = -a_{21}\begin{vmatrix} a_{12} & a_{13} & a_{14} \\ a_{32} & a_{33} & a_{34} \\ a_{42} & a_{43} & a_{44} \end{vmatrix} + a_{22}\begin{vmatrix} a_{11} & a_{13} & a_{14} \\ a_{31} & a_{33} & a_{34} \\ a_{41} & a_{43} & a_{44} \end{vmatrix}.$$

$$-a_{23}\begin{vmatrix} a_{11} & a_{12} & a_{14} \\ a_{31} & a_{32} & a_{34} \\ a_{41} & a_{42} & a_{44} \end{vmatrix} + a_{24}\begin{vmatrix} a_{11} & a_{12} & a_{13} \\ a_{31} & a_{32} & a_{33} \\ a_{41} & a_{42} & a_{43} \end{vmatrix}.$$

70. It follows from this that, *if all the elements in the pth row of a determinant Δ be zero except the one in the qth column*

$$\Delta = (-)^{p+q}a_{pq}A_{pq}$$

For

$$\begin{aligned} \Delta &= (-)^{p-1}a_{p1}A_{p1} + \cdots + (-)^{p+q-2}a_{pq}A_{pq} \\ &\qquad + \cdots + (-)^{n+p-2}a_{pn}A_{pn} \\ &= (-)^{p+q-2}a_{pq}A_{pq} = (-)^{p+q}a_{pq}A_{pq}. \end{aligned}$$

If $p=q=1$, then $\Delta=a_{11}A_{11}$ and if $a_{11}=1$, then $\Delta=A_{11}$.

EXAMPLE. In the expansion of the determinant

$$\begin{vmatrix} a_{11} & a_{12} & a_{13} & a_{14} \\ a_{21} & a_{22} & a_{23} & a_{24} \\ a_{31} & a_{32} & a_{33} & a_{34} \\ a_{41} & a_{42} & a_{43} & a_{44} \end{vmatrix}$$

the portion whose terms contain the element a_{33} is equal to

$$\begin{vmatrix} a_{11} & a_{12} & a_{13} & a_{14} \\ a_{21} & a_{22} & a_{23} & a_{24} \\ 0 & 0 & a_{33} & 0 \\ a_{41} & a_{42} & a_{43} & a_{44} \end{vmatrix} \text{ or } \begin{vmatrix} a_{11} & a_{12} & 0 & a_{14} \\ a_{21} & a_{22} & 0 & a_{24} \\ a_{31} & a_{32} & a_{33} & a_{34} \\ a_{41} & a_{42} & 0 & a_{44} \end{vmatrix},$$

and therefore is equal to

$$a_{33}\begin{vmatrix} a_{11} & a_{12} & a_{13} & a_{14} \\ a_{21} & a_{22} & a_{23} & a_{24} \\ 0 & 0 & 1 & 0 \\ a_{41} & a_{42} & a_{43} & a_{44} \end{vmatrix} \text{ or } a_{33}\begin{vmatrix} a_{11} & a_{12} & 0 & a_{14} \\ a_{21} & a_{22} & 0 & a_{24} \\ a_{31} & a_{32} & 1 & a_{34} \\ a_{41} & a_{42} & 0 & a_{44} \end{vmatrix},$$

and consequently to

$$a_{33}\begin{vmatrix} a_{11} & a_{12} & a_{14} \\ a_{21} & a_{22} & a_{24} \\ a_{41} & a_{42} & a_{44} \end{vmatrix}.$$

71. Reversing the order of the members of the last equation of the preceding article we may view the theorem as affirming that *without altering the value of a determinant its order may be raised by superposing a zero on every column and prefixing unity to the row of zeros thus formed and a constituent of any finite magnitude to each of the other rows, thus*

$$\begin{vmatrix} a & b & c \\ d & e & f \\ g & h & k \end{vmatrix} = \begin{vmatrix} 1 & 0 & 0 & 0 \\ \alpha & a & b & c \\ \beta & d & e & f \\ \gamma & g & h & k \end{vmatrix} = \begin{vmatrix} 1 & A & B & C \\ 0 & a & b & c \\ 0 & d & e & f \\ 0 & g & h & k \end{vmatrix}.$$

By repeating this process we may evidently raise the order of a determinant to any desired extent.

72. *A determinant of the nth order is expressible as the sum of n determinants, the first of which is obtained by changing into zero all the constituents of any row or any column of the original determinant except the first constituent, the second by changing into zero all the constituents of the same row or column except the second constituent, and so on.*

Let $a_{p1}\ \ a_{p2}\ \ \cdots\ \ a_{pn}$ be the specified row of the original determinant so that

$$\begin{matrix} a_{p1} & 0 & 0 & 0 & \cdots 0 \\ 0 & a_{p2} & 0 & 0 & \cdots 0 \\ 0 & 0 & a_{p3} & 0 & \cdots 0 \\ 0 & 0 & 0 & a_{p4} & \cdots 0 \\ \cdots & \cdots & \cdots & \cdots & \cdots \\ 0 & 0 & 0 & 0 & \cdots a_{pn} \end{matrix}$$

are the corresponding rows of the n determinants.

Then the first of the n determinants equals $(-)^{p-1}a_{p1}A_{p1}$, the second equals $(-)^{p}a_{p2}A_{p2}$, and so on, the qth being equal to $(-)^{p+q-2}$ $a_{pq}A_{pq}$. The sum of these make up the original determinant. Hence the truth of the theorem.

This theorem also is evident from the *addition theorem*. (§44).

EXAMPLE.

$$\begin{vmatrix} a_1 & a_2 & a_3 & a_4 \\ b_1 & b_2 & b_3 & b_4 \\ c_1 & c_2 & c_3 & c_4 \\ d_1 & d_2 & d_3 & d_4 \end{vmatrix} = \begin{vmatrix} a_1 & a_2 & a_3 & a_4 \\ b_1 & b_2 & 0 & b_4 \\ c_1 & c_2 & 0 & c_4 \\ d_1 & d_2 & 0 & d_4 \end{vmatrix} + \begin{vmatrix} a_1 & a_2 & 0 & a_4 \\ b_1 & b_2 & b_3 & b_4 \\ c_1 & c_2 & 0 & c_4 \\ d_1 & d_2 & 0 & d_4 \end{vmatrix}$$

$$+ \begin{vmatrix} a_1 & a_2 & 0 & a_4 \\ b_1 & b_2 & 0 & b_4 \\ c_1 & c_2 & c_3 & c_4 \\ d_1 & d_2 & 0 & d_4 \end{vmatrix} + \begin{vmatrix} a_1 & a_2 & 0 & a_4 \\ b_1 & b_2 & 0 & b_4 \\ c_1 & c_2 & 0 & c_4 \\ d_1 & d_2 & d_3 & d_4 \end{vmatrix}.$$

73. The number of terms in a determinant of the second order being two, it follows from §72 that the number of terms in a determinant of the third order is 2×3, that therefore the number of terms in a determinant of the fourth order is $2\times 3\times 4$, and so generally as in §31 the number of terms in a determinant of the nth order is $2\times 3\times 4\cdots\times n$ or $n!$ Similarly we have another proof of the theorem of §35.

74. *If the constituents on one side of either diagonal of a determinant be all zero, the determinant consists of only one term, namely the term composed of the constituents of the said diagonal.*

EXAMPLE.

$$\begin{vmatrix} 0 & 0 & 0 & d \\ 0 & 0 & c & g \\ 0 & b & f & i \\ a & e & h & j \end{vmatrix} = -a \begin{vmatrix} 0 & 0 & d \\ 0 & c & g \\ b & f & i \end{vmatrix},$$

$$= -ab \begin{vmatrix} 0 & d \\ c & g \end{vmatrix} = abcd.$$

75. The result of §69 will be easily seen to be of paramount importance in reference to the work of simplifying and expanding determinants, the finding of the expansion being made dependent upon the finding of the expansions of a number of determinants of

the next lower order. Evidently also the advantage thus obtained will be augmented if by the use of previously established theorems we can succeed in changing several of the elements of a row of the original determinant into 0, for then the number of the said determinants of the next lower order will be correspondingly lessened. For example, consider the determinant

$$\begin{vmatrix} 10 & 4 & 17 & 13 \\ 4 & 2 & 8 & 6 \\ 3 & -1 & 8 & 1 \\ 7 & 5 & 20 & 17 \end{vmatrix},$$

and denote it by C. Then

$$C = 10\begin{vmatrix} 2 & 8 & 6 \\ -1 & 8 & 1 \\ 2 & 20 & 17 \end{vmatrix} - 4\begin{vmatrix} 4 & 17 & 13 \\ -1 & 8 & 1 \\ 5 & 20 & 17 \end{vmatrix} + 3\begin{vmatrix} 4 & 17 & 13 \\ 2 & 8 & 6 \\ 5 & 20 & 17 \end{vmatrix} - 7\begin{vmatrix} 4 & 17 & 13 \\ 2 & 8 & 6 \\ -1 & 8 & 1 \end{vmatrix}$$

$$= 10(48) \qquad - 4(58) \qquad + 3(-4) \qquad - 7(16)$$

$$= 124.$$

Here we have at once applied the theorem of §69.

Again

$$C = \begin{vmatrix} 2 & 4 & 1 & 1 \\ 0 & 2 & 0 & 0 \\ 5 & -1 & 12 & 4 \\ -3 & 5 & 0 & 2 \end{vmatrix} = 2\begin{vmatrix} 2 & 1 & 1 \\ 5 & 12 & 4 \\ -3 & 0 & 2 \end{vmatrix},$$

$$= 2\begin{vmatrix} 0 & 1 & 0 \\ -19 & 12 & -8 \\ -3 & 0 & 2 \end{vmatrix} = -2\begin{vmatrix} -19 & -8 \\ -3 & 2 \end{vmatrix},$$

$$= -2\begin{vmatrix} -31 & 0 \\ -3 & 2 \end{vmatrix} = 124.$$

Here we diminish each element of the first column by twice the corresponding element of the second column, each element of the third column by four times the corresponding element of the second column, and each element of the fourth column by three times the corresponding element of the second column, the result being a determinant with a row containing three zero elements, from which by means of the theorem of §69 or of §70 we pass to a single determinant of the next lower order; then this determinant is treated in similar fashion; and so on.

It will be observed that one of the elements of the second row of C is a measure of each of the other elements of the row, and that to this peculiarity is due the possibility of transforming C, as above, into a determinant with a row containing three zero elements. The second column possesses the same peculiarity, so that we might also proceed as follows:

$$C = \begin{vmatrix} 22 & 0 & 49 & 17 \\ 10 & 0 & 24 & 8 \\ 3 & -1 & 8 & 1 \\ 22 & 0 & 60 & 22 \end{vmatrix} = \begin{vmatrix} 22 & 49 & 17 \\ 10 & 24 & 8 \\ 22 & 60 & 22 \end{vmatrix},$$

$$= \begin{vmatrix} 5 & -2 & 17 \\ 2 & 0 & 8 \\ 0 & -6 & 22 \end{vmatrix} = -204 + 88 + 240,$$

$$= 124.$$

Had such not been the case we could first have transformed C into a determinant having the peculiarity referred to, for example a determinant having one of its elements 1; and in this way the second mode of procedure can be seen to be always possible.

EXAMPLE. Show that

$$\begin{vmatrix} b_1 & a & a & a \\ a & b_2 & a & a \\ a & a & b_3 & a \\ a & a & a & b_4 \end{vmatrix}$$

$$= (b_1 - a)(b_2 - a)(b_3 - a)(b_4 - a)\left\{1 + \frac{a}{b_1 - a} + \frac{a}{b_2 - a} + \frac{a}{b_3 - a} + \frac{a}{b_4 - a}\right\}.$$

It is clear that (§71) the given determinant is equal to

$$\begin{vmatrix} 1 & 1 & 1 & 1 & 1 \\ 0 & b_1 & a & a & a \\ 0 & a & b_2 & a & a \\ 0 & a & a & b_3 & a \\ 0 & a & a & a & b_4 \end{vmatrix}.$$

Multiplying each element of the first row of this determinant by a, and subtracting the result from the corresponding element of each of the other rows, we have

$$\begin{vmatrix} 1 & 1 & 1 & 1 & 1 \\ -a & b_1 - a & 0 & 0 & 0 \\ -a & 0 & b_2 - a & 0 & 0 \\ -a & 0 & 0 & b_3 - a & 0 \\ -a & 0 & 0 & 0 & b_4 - a \end{vmatrix}.$$

This we know (§45) is equal to

$$\begin{vmatrix} b_1 - a & 0 & 0 & 0 \\ 0 & b_2 - a & 0 & 0 \\ 0 & 0 & b_3 - a & 0 \\ 0 & 0 & 0 & b_4 - a \end{vmatrix} + a \begin{vmatrix} 1 & 1 & 1 & 1 \\ 0 & b_2 - a & 0 & 0 \\ 0 & 0 & b_3 - a & 0 \\ 0 & 0 & 0 & b_4 - a \end{vmatrix}$$

$$-a \begin{vmatrix} 1 & 1 & 1 & 1 \\ b_1 - a & 0 & 0 & 0 \\ 0 & 0 & b_3 - a & 0 \\ 0 & 0 & 0 & b_4 - a \end{vmatrix} + a \begin{vmatrix} 1 & 1 & 1 & 1 \\ b_1 - a & 0 & 0 & 0 \\ 0 & b_2 - a & 0 & 0 \\ 0 & 0 & 0 & b_4 - a \end{vmatrix}$$

$$-a \begin{vmatrix} 1 & 1 & 1 & 1 \\ b_1 - a & 0 & 0 & 0 \\ 0 & b_2 - a & 0 & 0 \\ 0 & 0 & b_3 - a & 0 \end{vmatrix}.$$

Here the first two determinants have all their elements 0 on one side of the principal diagonal, and the others have each a column with all the elements 0 except one, hence we transform the expression into

$$(b_1 - a)(b_2 - a)(b_3 - a)(b_4 - a) + a(b_2 - a)(b_3 - a)(b_4 - a)$$

$$+ a \begin{vmatrix} b_1 - a & 0 & 0 \\ 0 & b_3 - a & 0 \\ 0 & 0 & b_4 - a \end{vmatrix}$$

$$+ a \begin{vmatrix} b_1 - a & 0 & 0 \\ 0 & b_2 - a & 0 \\ 0 & 0 & b_4 - a \end{vmatrix} + a \begin{vmatrix} b_1 - a & 0 & 0 \\ 0 & b_2 - a & 0 \\ 0 & 0 & b_3 - a \end{vmatrix}$$

whence readily comes the result desired.

In the preceding process, the first step, namely, taking an equivalent determinant of higher order, is worthy of the learner's attention as being conducive to symmetry in obtaining the result in the particular form wanted.

EXERCISES. SET VI

Find the single numbers to which the following determinants are equal:

1. $\begin{vmatrix} 3 & 1 & 4 & 1 \\ 2 & 2 & 8 & 5 \\ 1 & 6 & 4 & 2 \\ 3 & 2 & 5 & 3 \end{vmatrix}$. 2. $\begin{vmatrix} 3 & 7 & 4 & 3 \\ 7 & 4 & 3 & 5 \\ 2 & 1 & 9 & 4 \\ 8 & 6 & 4 & 7 \end{vmatrix}$. 3. $\begin{vmatrix} 2 & 1 & 3 & 4 \\ 7 & 4 & 5 & 9 \\ 3 & 3 & 6 & 2 \\ 1 & 7 & 7 & 5 \end{vmatrix}$.

4. $\begin{vmatrix} 8 & 7 & 5 & 10 \\ 4 & 3 & 9 & 2 \\ 8 & 9 & 6 & 12 \\ 3 & 1 & 2 & 4 \end{vmatrix}$. 5. $\begin{vmatrix} 10 & 8 & 9 & 14 \\ 17 & 15 & 18 & 11 \\ 15 & 19 & 10 & 13 \\ 16 & 17 & 18 & 10 \end{vmatrix}$. 6. $\begin{vmatrix} 21 & -22 & 6 & 14 \\ 12 & 13 & -7 & 18 \\ 25 & 14 & 18 & -26 \\ -7 & 17 & -12 & 4 \end{vmatrix}$.

7. $\begin{vmatrix} 3 & 1 & 5 & 4 & 2 \\ 7 & 6 & 4 & 1 & 3 \\ 1 & 3 & 2 & 9 & 4 \\ 2 & 2 & 9 & 2 & 1 \\ 8 & 6 & 1 & 3 & 4 \end{vmatrix}$. 8. $\begin{vmatrix} 5 & -1 & 4 & 6 & -2 \\ -1 & 4 & 6 & -2 & 5 \\ 4 & 6 & -2 & 5 & -1 \\ 6 & -2 & 5 & -1 & 4 \\ -2 & 5 & -1 & 4 & 6 \end{vmatrix}$.

9. $\begin{vmatrix} 2 & 4 & 3 & 1 & 4 & 3 \\ -4 & 2 & -3 & 2 & -1 & 2 \\ 5 & -1 & 6 & 2 & -1 & 5 \\ 1 & 1 & 1 & -2 & -2 & -2 \\ 7 & -3 & -5 & 1 & 4 & 2 \\ 3 & 1 & 2 & -1 & 2 & 3 \end{vmatrix}$. 10. $\begin{vmatrix} 12 & 22 & 14 & 17 & 20 & 10 \\ 16 & 4 & 7 & 1 & -2 & 15 \\ 10 & -3 & -2 & 3 & -2 & 8 \\ 7 & 12 & 8 & 9 & 11 & 6 \\ 11 & 2 & 4 & -8 & 1 & 9 \\ 24 & 6 & 6 & 3 & 4 & 22 \end{vmatrix}$

Find the ordinary expansion of the following determinants:

11. $\begin{vmatrix} 0 & 0 & k & 1 & x \\ 0 & 0 & h & x & 0 \\ 0 & 0 & x & 0 & 0 \\ 0 & x & e & f & g \\ x & a & b & c & d \end{vmatrix}$. 12. $\begin{vmatrix} a_1 & b_1 & c_1 & d_1 & e_1 \\ 0 & b_2 & 0 & 0 & 0 \\ 0 & b_3 & c_3 & d_3 & e_3 \\ 0 & b_4 & 0 & d_4 & 0 \\ 0 & b_5 & 0 & d_5 & e_5 \end{vmatrix}$.

13. $\begin{vmatrix} a & 1 & 0 & 0 \\ -1 & b & 1 & 0 \\ 0 & -1 & c & 1 \\ 0 & 0 & -1 & d \end{vmatrix}$. 14. $\begin{vmatrix} 0 & d & d & d \\ a & 0 & a & a \\ b & b & 0 & b \\ c & c & c & 0 \end{vmatrix}$. 15. $\begin{vmatrix} 0 & a & b & c \\ a & 0 & \gamma & \beta \\ b & \gamma & 0 & \alpha \\ c & \beta & \alpha & 0 \end{vmatrix}$.

16. $\begin{vmatrix} 1 & a & a & a \\ 1 & x & a & a \\ 1 & a & x & a \\ 1 & a & a & x \end{vmatrix}$. 17. $\begin{vmatrix} 1 & y & 0 & 0 \\ 1 & x & y & 0 \\ 1 & 0 & x & y \\ 1 & 0 & 0 & x \end{vmatrix}$.

18. $\begin{vmatrix} 0 & 1 & 1 & 1 \\ 1 & b+c & a & a \\ 1 & b & c+a & b \\ 1 & c & c & a+b \end{vmatrix}$. 19. $\begin{vmatrix} a & b & c & d \\ -a & b & x & y \\ -a & -b & c & z \\ -a & -b & -c & d \end{vmatrix}$.

20. $\begin{vmatrix} 1+a & 1 & 1 & 1 \\ 1 & 1+b & 1 & 1 \\ 1 & 1 & 1+c & 1 \\ 1 & 1 & 1 & 1+d \end{vmatrix}$. 21. $\begin{vmatrix} a+b & b & c & d \\ a & b+c & c & d \\ a & b & c+d & d \\ a & b & c & d+e \end{vmatrix}$.

22. $$\begin{vmatrix} x & 0 & 0 & 0 & \cdots & 0 & a_n \\ -1 & x & x^2 & x^3 & \cdots & x^{n-1} & a_{n-1} \\ 0 & -1 & 0 & 0 & \cdots & 0 & a_{n-2} \\ 0 & 0 & -1 & 0 & \cdots & 0 & a_{n-3} \\ \cdots & \cdots & \cdots & \cdots & \cdots & \cdots & \cdots \\ 0 & 0 & 0 & 0 & \cdots & 0 & a_1 \\ 0 & 0 & 0 & 0 & \cdots & -1 & a_0 \end{vmatrix}.$$

23. $$\begin{vmatrix} a_1 & b_2 & 0 & 0 & \cdots & 0 & 0 \\ a_2 & -b_1 & b_3 & 0 & \cdots & 0 & 0 \\ a_3 & 0 & -b_2 & b_4 & \cdots & 0 & 0 \\ a_4 & 0 & 0 & -b_3 & \cdots & 0 & 0 \\ \cdots & \cdots & \cdots & \cdots & \cdots & \cdots & \cdots \\ a_n & 0 & 0 & 0 & \cdots & -b_{n-1} & b_{n+1} \\ a_{n+1} & 0 & 0 & 0 & \cdots & 0 & -b_n \end{vmatrix}.$$

Establish the following identities:

24. $$\begin{vmatrix} a^2+1 & ab & ac & ad \\ ab & b^2+1 & bc & bd \\ ac & bc & c^2+1 & cd \\ ad & bd & cd & d^2+1 \end{vmatrix} = a^2+b^2+c^2+d^2+1.$$

25. Resolve into simple factors

$$\begin{vmatrix} x & a_1 & a_2 & a_3 & 1 \\ a_1 & x & a_2 & a_3 & 1 \\ a_1 & a_2 & x & a_3 & 1 \\ a_1 & a_2 & a_3 & x & 1 \\ a_1 & a_2 & a_3 & a_4 & 1 \end{vmatrix}.$$

26. Without finding the ordinary expansion of the determinants, show that

$$\begin{vmatrix} a_1 & a_2 & a_3 & a_6 \\ 1 & a_1 & a_2 & a_5 \\ 0 & 1 & a_1 & a_4 \\ 0 & 0 & 1 & a_3 \end{vmatrix} + \begin{vmatrix} a_1 & a_2 & a_3 & a_5 \\ 1 & a_1 & a_3 & a_4 \\ 0 & 1 & a_2 & a_3 \\ 0 & 0 & a_1 & a_2 \end{vmatrix} = \begin{vmatrix} a_2 & a_3 & a_4 & a_6 \\ 1 & a_1 & a_2 & a_4 \\ 0 & 1 & a_1 & a_3 \\ 0 & 0 & 1 & a_2 \end{vmatrix}.$$

76. *If the first constituent of a row of a determinant be multiplied by the complimentary of the first constituent in another row, the second constituent in the former row be multiplied by the complimentary of the second constituent in the latter, and so on, the sum of the products with alternately positive and negative signs is equal to zero.*

For if in the identity,

$$(-)^{p-1}a_{p1}A_{p1} + (-)^p a_{p2}A_{p2} + \cdots (-)^{n+p-2}a_{pn}A_{pn} \equiv \Delta$$

we write a_{qr} for a_{pr} $(r=1, 2, \cdots, n)$ $q \neq p$. The right hand member will vanish having two rows identical. Therefore

$$a_{q1}A_{p1} - a_{q2}A_{p2} + \cdots (-)^{n-1}a_{qn}A_{pn} \equiv 0 \quad q \neq p.$$

If $\Delta = 0$, then this is true for $p = 1, 2, \cdots, n$.

77. *If the first pair of elements in the first row of a determinant be taken in succession with every pair below it, and the determinants of the second order which have these pairs for rows be placed in order as the elements of the first column of a new determinant, and if the like be*

done in the case of the second and following pairs of consecutive elements in the row, then the new determinant thus obtained divided by the product of all the elements of the first row of the original determinant except the first and last is equal to the original determinant.

Let the given determinant be

$$\begin{vmatrix} a_1 & b_1 & c_1 & d_1 \cdots k_1 & l_1 \\ a_2 & b_2 & c_2 & d_2 \cdots k_2 & l_2 \\ a_3 & b_3 & c_3 & d_3 \cdots k_3 & l_3 \\ \cdots & \cdots & \cdots & \cdots & \cdots \\ a_n & b_n & c_n & d_n \cdots k_n & l_n \end{vmatrix} \quad \text{or} \quad \Delta.$$

Multiplying each element of the first column by $-b_1$ and adding to the result a_1 times the corresponding element of the second column, we have (§58)

$$-b_1\Delta = \begin{vmatrix} 0 & b_1 & c_1 & d_1 \cdots k_1 & l_1 \\ -a_2b_1 + b_2a_1 & b_2 & c_2 & d_2 \cdots k_2 & l_2 \\ -a_3b_1 + b_3a_1 & b_3 & c_3 & d_3 \cdots k_3 & l_3 \\ \cdots & \cdots & \cdots & \cdots & \cdots \\ -a_nb_1 + b_na_1 & b_n & c_n & d_n \cdots k_n & l_n \end{vmatrix}.$$

Again, multiplying each element of the second column of this determinant by $-c_1$ and adding to the result b_1 times the corresponding element of the third column, we have

$$(-1)^2b_1c_1\Delta = \begin{vmatrix} 0 & 0 & c_1 & d_1 \cdots k_1 & l_1 \\ -a_2b_1+b_2a_1 & -b_2c_1+c_2b_1 & c_2 & d_2 \cdots k_2 & l_2 \\ -a_3b_1+b_3a_1 & -b_3c_1+c_3b_1 & c_3 & d_3 \cdots k_3 & l_3 \\ \cdots & \cdots & \cdots & \cdots & \cdots \\ -a_nb_1+b_na_1 & -b_nc_1+c_nb_1 & c_n & d_n \cdots k_n & l_n \end{vmatrix}.$$

This process being continued, the final result is $(-1)^{n-1}b_1c_1d_1 \cdots e_1l_1\Delta$

$$= \begin{vmatrix} 0 & 0 & 0 & 0 & l_1 \\ -a_2b_1+b_2a_1 & -b_2c_1+c_2b_1 & -c_2d_1+d_2c_1 \cdots & -k_2l_1+l_2k_1 & l_2 \\ -a_3b_1+b_3a_1 & -b_3c_1+c_3b_1 & -c_3d_1+d_3c_1 \cdots & -k_3l_1+l_3k_1 & l_3 \\ \cdots & \cdots & \cdots & \cdots & \cdots \\ -a_nb_1+b_na_1 & -b_nc_1+c_nb_1 & -c_nd_1+d_nc_1 \cdots & -k_nl_1+l_nk_1 & l_n \end{vmatrix},$$

$$= (-1)^{n-1} l_1 \begin{vmatrix} -a_2b_1+b_2a_1 & -b_2c_1+c_2b_1 & -c_2d_1+d_2c_1 & \cdots & -k_2l_1+l_2k_1 \\ -a_3b_1+b_3a_1 & -b_3c_1+c_3b_1 & -c_3d_1+d_3c_1 & \cdots & -k_3l_1+l_3k_1 \\ \cdots & \cdots & \cdots & \cdots & \cdots \\ -a_nb_1+b_na_1 & -b_nc_1+c_nb_1 & -c_nd_1+d_nc_1 & \cdots & -k_nl_1+l_nk_1 \end{vmatrix}$$

and dividing by $(-1)^{n-1}b_1c_1d_1 \cdots l_1$, we obtain

$$\begin{vmatrix} a_1 & b_1 & c_1 & d_1 \cdots l_1 \\ a_2 & b_2 & c_2 & d_2 \cdots l_2 \\ a_3 & b_2 & c_3 & d_3 \cdots l_3 \\ \cdots & \cdots & \cdots & \cdots \\ a_n & b_n & c_n & d_n \cdots l_n \end{vmatrix}$$

$$= \frac{1}{b_1c_1d_1 \cdots k_1} \begin{vmatrix} \begin{vmatrix} a_1 & b_1 \\ a_2 & b_2 \end{vmatrix}, & \begin{vmatrix} b_1 & c_1 \\ b_2 & c_2 \end{vmatrix}, & \ldots, & \begin{vmatrix} k_1 & l_1 \\ k_2 & l_2 \end{vmatrix} \\ \begin{vmatrix} a_1 & b_1 \\ a_3 & b_3 \end{vmatrix}, & \begin{vmatrix} b_1 & c_1 \\ b_3 & c_3 \end{vmatrix}, & \ldots, & \begin{vmatrix} k_1 & l_1 \\ k_3 & l_3 \end{vmatrix} \\ \cdots & \cdots & \cdots & \cdots \\ \begin{vmatrix} a_1 & b_1 \\ a_n & b_n \end{vmatrix}, & \begin{vmatrix} b_1 & c_1 \\ b_n & c_n \end{vmatrix}, & \ldots, & \begin{vmatrix} k_1 & l_1 \\ k_n & l_n \end{vmatrix} \end{vmatrix},$$

as was to be proved.

78. The new determinant found in the preceding paragraph being one degree lower than the original, the theorem is important as affording an easy means of evaluating a determinant whose elements are expressed in figures. Thus, taking the example already dealt with (§75), we have

$$\begin{vmatrix} 10 & 4 & 3 & 7 \\ 4 & 2 & -1 & 5 \\ 17 & 8 & 8 & 20 \\ 13 & 6 & 1 & 17 \end{vmatrix} = 1/12 \begin{vmatrix} 4 & -10 & 22 \\ 12 & 8 & 4 \\ 8 & -14 & 44 \end{vmatrix},$$

$$= 4/3 \begin{vmatrix} 2 & -5 & 44 \\ 3 & 2 & 1 \\ 4 & -7 & 22 \end{vmatrix},$$

$$= -4/15 \begin{vmatrix} 19 & -27 \\ 6 & -33 \end{vmatrix},$$

$$= -4/5(-209 + 54),$$

$$= 124.$$

Here, looking at the first two rows of the given determinant, we at once mentally evaluate

$$\begin{vmatrix} 10 & 4 \\ 4 & 2 \end{vmatrix}, \quad \begin{vmatrix} 4 & 3 \\ 2 & -1 \end{vmatrix}, \quad \begin{vmatrix} 3 & 7 \\ -1 & 5 \end{vmatrix},$$

and place the results 4, -10, 22 for the first row of the new determinant; similarly we proceed with the first and third rows, and with the first and fourth rows. This gives us a determinant of the third order, from which we remove the factors 2, 4, 2; then we treat the resulting determinant as the first was treated, and thus have at last only to deal with a determinant of the second order.

If one of the elements included between the first and last of the first row be zero, the theorem is clearly inapplicable; but then it has to be remembered in such a case that any row or any column may be made the first row.

Examples. Use the process of §77 to find the simplest equivalent of each of the determinants

1. $$\begin{vmatrix} 3 & 4 & 1 & 2 \\ 6 & 9 & 7 & 5 \\ 7 & 10 & 5 & 8 \\ 4 & 2 & 9 & 3 \end{vmatrix}.$$

2. $$\begin{vmatrix} 7 & 3 & 9 & 4 \\ 3 & 9 & 4 & 7 \\ 2 & 6 & 1 & 5 \\ 10 & 7 & 14 & 10 \end{vmatrix}.$$

3. $$\begin{vmatrix} 3 & 4 & 7 & 2 & 5 \\ -3 & 1 & 2 & 5 & -1 \\ 6 & -2 & 3 & -1 & 4 \\ 5 & 9 & -2 & 3 & 2 \\ 1 & -3 & 5 & 3 & 7 \end{vmatrix}.$$

4. $$\begin{vmatrix} 16 & 14 & 12 & 9 & 15 \\ 5 & 4 & 3 & 1 & 6 \\ 2 & 8 & 4 & 9 & 7 \\ 3 & -2 & -2 & -3 & -6 \\ 1 & 3 & 2 & 3 & 2 \end{vmatrix}.$$

Prove that

$$\begin{vmatrix} a_1 & a_2 & a_3 & a_4 \cdots a_n \\ b_1 & b_2 & b_3 & b_4 \cdots b_n \\ c_1 & c_2 & c_3 & c_4 \cdots c_n \\ d_1 & d_2 & d_3 & d_4 \cdots d_n \\ \cdots & \cdots & \cdots & \cdots \\ l_1 & l_2 & l_3 & l_4 \cdots l_n \end{vmatrix}$$

$$= \frac{1}{a_1^{n-2}} \begin{vmatrix} \begin{vmatrix} a_1 & b_2 \end{vmatrix} & \begin{vmatrix} a_1 & b_3 \end{vmatrix} & \begin{vmatrix} a_1 & b_4 \end{vmatrix} & \cdots & \begin{vmatrix} a_1 & b_n \end{vmatrix} \\ \begin{vmatrix} a_1 & c_2 \end{vmatrix} & \begin{vmatrix} a_1 & c_3 \end{vmatrix} & \begin{vmatrix} a_1 & c_4 \end{vmatrix} & \cdots & \begin{vmatrix} a_1 & c_n \end{vmatrix} \\ \begin{vmatrix} a_1 & d_2 \end{vmatrix} & \begin{vmatrix} a_1 & d_3 \end{vmatrix} & \begin{vmatrix} a_1 & d_4 \end{vmatrix} & \cdots & \begin{vmatrix} a_1 & d_n \end{vmatrix} \\ \cdots & \cdots & \cdots & \cdots & \cdots \\ \begin{vmatrix} a_1 & l_2 \end{vmatrix} & \begin{vmatrix} a_1 & l_3 \end{vmatrix} & \begin{vmatrix} a_1 & l_4 \end{vmatrix} & \cdots & \begin{vmatrix} a_1 & l_n \end{vmatrix} \end{vmatrix}.$$

79. If a determinant is equal to zero it is possible to so transform it as to have all the elements in a row or column equal to zero.

Let $A \equiv |a_{nn}|$ be the determinant which is given equal to zero. Add to the nth column multiplied by $\mathcal{A}_{1n}$, the first times $\mathcal{A}_{11}$, the second times $\mathcal{A}_{12}$, the third times $\mathcal{A}_{13}$, and so on, and it is readily seen the elements of the last column become zeros. Similarly we may make the elements in any column or row zeros.

Since in a determinant which has two of its columns proportional we can readily reduce the elements in one of the two to zeros, it follows that this is a necessary and sufficient condition that the determinant is zero.

80. Given the set of n equations in n variables

$$(1) \qquad \begin{aligned} a_{11}x_1 + a_{12}x_2 + \cdots + a_{1n}x_n &= y_1 \\ a_{21}x_1 + a_{22}x_2 + \cdots + a_{2n}x_n &= y_2 \\ \cdots\cdots\cdots\cdots\cdots\cdots\cdots\cdots \\ a_{n1}x_1 + a_{n2}x_2 + \cdots + a_{nn}x_n &= y_n \end{aligned}$$

let us denote the determinant of the coefficients $|a_{1n}|$ by A and by $\mathcal{A}_{rs}$ the cofactor of a_{rs} in A.

In §42 we saw that the problem was to find multipliers which would, after multiplication and addition cause all terms to vanish except one.

From §§68, 76 these multipliers are obviously cofactors and if we multiply these equations by $\mathcal{A}_{1k}, \mathcal{A}_{2k}, \cdots, \mathcal{A}_{nk}$ respectively and add, we have on the left x_kA, for the coefficients of all the other x's vanish by §76. On the right we have $y_1\mathcal{A}_{1k}+y_2\mathcal{A}_{2k}+ \cdots +y_n\mathcal{A}_{nk}$ or as it may be written

$$\begin{vmatrix} a_{11} \cdots a_{1,k-1} & y_1 & a_{1,k+1} \cdots a_{1n} \\ a_{21} \cdots a_{2,k-1} & y_2 & a_{2,k+1} \cdots a_{2n} \\ \cdots\cdots & \cdots & \cdots\cdots \\ a_{n1} \cdots a_{n,k-1} & y_n & a_{n\ k+1} \cdots a_{nn} \end{vmatrix} \equiv A_k \quad \text{say},$$

Therefore $x_k A = A_k$, or if $A \neq 0$

$$x_k = A_k/A. \tag{2}$$

Giving k the values from 1 to n we have the values for the n x's.

81. If now we add to the above set of equations another, namely

$$a_{n+1,1}x_1 + a_{n+1,2}x_2 + \cdots + a_{n+1,n}x_n = y_{n+1}$$

and substitute in it the values of the x's from (2) §80, then we get

$$a_{n+1,1}A_1 + a_{n+1,2}A_2 + \cdots + a_{n+1,n}A_n = y_{n+1}A$$

or as it may be written

$$\begin{vmatrix} a_{11} & a_{12} & \cdots a_{1n} & y_1 \\ a_{21} & a_{22} & \cdots a_{2n} & y_2 \\ \cdots & \cdots & \cdots & \cdots \\ a_{n1} & a_{n2} & \cdots a_{nn} & y_n \\ a_{n+1,1} & a_{n+1,2} & \cdots a_{n+1,n} & y_{n+1} \end{vmatrix} = 0.$$

This is then the relation which must exist between the coefficients of the $n+1$ equations in n variables if they are to be satisfied by the same set of values of the x's. It is called the *Resultant* or *Eliminant* of the set of equations.

If in equations (1) §80 the y's are all zero then we see that the x's must all be zero unless the determinant of the set is zero.

82. If the equations are to be satisfied by values of the x's not all zero we may suppose x_n not zero and divide them all by it, then we have n non-homogeneous equations in the $n-1$ ratios x_k/x_n $(k=1, \cdots, n-1)$. Then by §81 we must have $A=0$.

We see therefore that if $A \neq 0$, the only solution is that the x's are all zero, but if $A=0$ then there may be a solution other than all zeros for the x's.

Exercises. Set VII

With the help of determinants solve the following sets of equations:

1. $$\begin{aligned} 4x + 7y + 3z - 2w &= 9 \\ 2x - y - 4z + 3w &= 13 \\ 3x + 2y - 7z - 4w &= 2 \\ 5x - 3y + z + 5w &= 13. \end{aligned}$$

2. $$\begin{aligned} 3x + 2y + 4z - w &= 13 \\ 5x + y - z + 2w &= 9 \\ 2x + 3y - 7z + 3w &= 14 \\ 4x - 4y + 3z - 5w &= 4. \end{aligned}$$

3.
$$\begin{aligned} v + w - y &= a \\ w + x - z &= b \\ x + y - v &= c \\ y + z - w &= d \\ z + v - x &= e. \end{aligned}$$

4.
$$\begin{aligned} v + w + x - y &= a \\ w + x + y - z &= b \\ x + y + z - v &= c \\ y + z + v - w &= d \\ z + v + w - x &= e. \end{aligned}$$

5.
$$\begin{aligned} w + x + y + z &= 1 \\ aw + bx + cy + dz &= e \\ a^2w + b^2x + c^2y + d^2z &= e^2 \\ a^3w + b^3x + c^3y + d^3z &= e^3. \end{aligned}$$

6.
$$\begin{aligned} w + x + y + z &= 1 \\ w + ax + by + cz &= 0 \\ w + a^2x + b^2y + c^2z &= 0 \\ w + a^3x + b^3y + c^3z &= 0. \end{aligned}$$

7.
$$\begin{aligned} v - 2w - 2x + y + 3z &= a \\ w - 2x - 2y + z + 3v &= b \\ x - 2y - 2z + v + 3x &= c \\ y - 2z - 2v + w + 3x &= d \\ z - 2v - 2w + x + 3y &= e. \end{aligned}$$

8. What relation must exist between a, b, c, d if the equations

$$\begin{aligned} ax + by + cz + d &= 0, \\ bx + ay + dz + c &= 0, \\ ax + cy + bz + d &= 0, \\ cx + ay + dz + b &= 0, \end{aligned}$$

be simultaneously true?

9. If the equations

$$\begin{aligned} a_1x^3 + b_1x^2 + c_1x + d_1 &= 0, \\ a_1x^4 + b_1x^3 + c_1x^2 + d_1x &= 0, \\ b_2x^2 + c_2x + d_2 &= 0, \\ b_2x^3 + c_2x^2 + d_2x &= 0, \\ b_2x^4 + c_2x^3 + d_2x^2 &= 0, \end{aligned}$$

be simultaneously true (which evidently will be the case if the first and third be simultaneously true, that is, have a common root), find the relation which must exist between a_1, b_1, c_1, d_1, b_2, c_2, d_2.

Similarly find the resultant in the case of each of the following pairs of equations:

10.
$$a_1x^2 + b_1x + c_1 = 0$$
$$a_2x^2 + b_2x + c_2 = 0.$$

11.
$$a_1x^4 + b_1x^3 + c_1x^2 + d_1x + e_1 = 0$$
$$c_2x^2 + d_2x + e_2 = 0.$$

12.
$$a_1x^3 + b_1x^2 + c_1x + d_1 = 0$$
$$a_2x^3 + b_2x^2 + c_2x + d_2 = 0.$$

13.
$$a_1x^4 + b_1x^3 + c_1x^2 + d_1x + e_1 = 0$$
$$b_2x^3 + c_2x^2 + d_2x + e_2 = 0.$$

14.
$$a_1x^4 + c_1x^2 + d_1x + e_1 = 0$$
$$a_2x^4 + b_2x^3 + d_2x + e_2 = 0.$$

15.
$$a_1x^6 + b_1x^2 + c_1 = 0$$
$$d_1x^4 + e_1x^2 + f_1 = 0.$$

83. Looking upon the determinant $|a_{1n}|$, or Δ as a function of n^2 independent variables, namely, the constituents, and since all terms of Δ which contain a_{11} are included in $a_{11}A_{11}$ and A_{11} is independent of a_{11}, we have by differentiation of

$$\Delta = a_{11}A_{11} - a_{12}A_{12} + a_{13}A_{13} - \cdots + (-)^{n-1}a_{1n}A_{1n},$$

the relation

$$\frac{\partial\Delta}{\partial a_{11}} = A_{11}.$$

Similarly

$$\frac{\partial\Delta}{\partial a_{12}} = -A_{12}, \qquad \frac{\partial\Delta}{\partial a_{13}} = A_{13}, \cdots.$$

$$\frac{\partial\Delta}{\partial a_{21}} = -A_{21}, \qquad \frac{\partial\Delta}{\partial a_{22}} = A_{22}, \cdots;$$

and generally

$$\frac{\partial\Delta}{\partial a_{pq}} = (-)^{p+q}A_{pq}.$$

We have thus a less arbitrary notation for the cofactor of the constituents than that of §68, and we may write

$$\Delta = a_{11}\frac{\partial\Delta}{\partial a_{11}} + a_{12}\frac{\partial\Delta}{\partial a_{12}} + \cdots + a_{1n}\frac{\partial\Delta}{\partial a_{1n}}, \quad \text{etc.}$$

84. If all the constituents are functions of the same variable x then

$$\frac{d\Delta}{dx} = \sum \frac{\partial\Delta}{\partial a_{pq}} \frac{da_{pq}}{dx}.$$

If

$$a_{ki} = ea_{ik},$$

then

$$\frac{d\Delta}{da_{ik}} = \frac{\partial\Delta}{\partial a_{ik}} + \frac{\partial\Delta}{\partial a_{ki}} \cdot \frac{da_{ki}}{da_{ik}}$$
$$= A_{ik} + eA_{ki}.$$

If

$$a_{ki} = a_{ik},$$

then

$$A_{ik} = A_{ki}.$$

and

$$\frac{d\Delta}{da_{ik}} = 2A_{ik}.$$

If

$$a_{ki} = -a_{ik} \quad \text{and} \quad a_{ii} = 0$$

then when n is even $A_{ik} = -A_{ki}$,

and

$$\frac{d\Delta}{da_{ik}} = 2A_{ik}$$

but when n is odd $A_{ik} = A_{ki}$

and

$$\Delta = 0, \qquad \frac{d\Delta}{da_{ik}} = 0.$$

85. For the differential we have

$$d\Delta = \sum \frac{\partial\Delta}{\partial a_{pq}} da_{pq} \qquad (p, q = 1, 2, \cdots, n)$$
$$= \sum (-)^{p+q} A_{pq} da_{pq}$$
$$= \sum (-)^{p+1} A_{p1} da_{p1} + \sum (-)^{p+2} A_{p2} da_{p2} + \cdots,$$
$$(p = 1, 2, \cdots, n),$$

Therefore

$$d\Delta = \begin{vmatrix} da_{11} & a_{12} \cdots a_{1n} \\ da_{21} & a_{22} \cdots a_{2n} \\ \cdots & \cdots \\ da_{n1} & a_{n2} \cdots a_{nn} \end{vmatrix} + \begin{vmatrix} a_{11} & da_{12} \cdots a_{1n} \\ a_{21} & da_{22} \cdots a_{2n} \\ \cdots & \cdots \\ a_{n1} & da_{n2} \cdots a_{nn} \end{vmatrix}$$

$$+ \cdots + \begin{vmatrix} a_{11} & a_{12} \cdots da_{1n} \\ a_{21} & a_{22} \cdots da_{2n} \\ \cdots & \cdots \\ a_{n1} & a_{n2} \cdots da_{nn} \end{vmatrix}$$

so that the differential of a determinant is the sum of n determinants formed by substituting for the constituents of each column of Δ in succession their differentials. The rth derivative of Δ may be written $(d_1+d_2+ \cdots +d_n)^r\Delta$ where d_h denotes differentiation of the hth column.

86. Making use of this principle to find the derivative with respect to x of the determinant

$$L_n(x) \equiv \begin{vmatrix} a_{11}-x & a_{12} & \cdots a_{1n} \\ a_{21} & a_{22}-x & \cdots a_{2n} \\ \cdots & \cdots & \cdots \\ a_{n1} & a_{n2} & \cdots a_{nn}-x \end{vmatrix},$$

we have

$$\frac{d}{dx}L_n(x) = (-1)^1 \sum L_{n-1,ii}(x)$$

$$\frac{d^2}{dx^2}L_n(x) = (-1)^2 2 \sum L_{n-2,ii}(x)$$

$$\cdots\cdots\cdots\cdots$$

$$\frac{d^r}{dx^r}L_n(x) = (-1)^r r! \sum L_{n-r,ii}(x)$$

Where $L_{n-r,ii}(x)$ represents a coaxial minor of order $n-r$.

EXAMPLE 1. Let

$$\Delta = \begin{vmatrix} y_1 & y_1' & y_1'' \cdots y_1^{(n-1)} \\ y_2 & y_2' & y_2'' \cdots y_2^{(n-1)} \\ \cdots & \cdots & \cdots \\ y_n & y_n' & y_n'' \cdots y_n^{(n-1)} \end{vmatrix}$$

where $y_1, y_2, \cdots, y_n$ are functions of x and the accents denote differential coefficients with respect to x. Then

$$\frac{d\Delta}{dx} = \begin{vmatrix} y_1 & y_1' & \cdots & y_1^{(n-2)} & y_1^{(n)} \\ y_2 & y_2' & \cdots & y_2^{(n-2)} & y_2^{(n)} \\ \cdots & \cdots & \cdots & \cdots & \cdots \\ y_n & y_n' & \cdots & y_n^{(n-2)} & y_n^{(n)} \end{vmatrix}.$$

Since each of the other $n-1$ determinants vanish having two columns identical.

If the functions $y_1, y_2, y_3, \cdots, y_n$ are each multiplied by any arbitrary function u, then the whole determinant Δ will be multiplied by u^n.

EXAMPLE 2.

Let

$$\Delta_n = \begin{vmatrix} 1 & 1 & \cdots 1 \\ y_1 & y_2 & \cdots y_n \\ y_1^2 & y_2^2 & \cdots y_n^2 \\ \cdots & \cdots & \cdots \\ y_1^{(n-1)} & y_2^{(n-1)} & \cdots y_n^{(n-1)} \end{vmatrix}.$$

Then

$$\frac{\partial \Delta_n}{\partial y_1} = \begin{vmatrix} 0 & 1 & \cdots 1 \\ 1 & y_2 & \cdots y_n \\ 2y_1 & y_2^2 & \cdots y_n^2 \\ \cdots & \cdots & \cdots \\ (n-1)y_1{}^{n-2)} & y_2^{(n-1)} & \cdots y_n^{(-1)} \end{vmatrix}$$

and

$$\frac{\partial^{n-1}\Delta_n}{\partial y_1 \partial y_2 \cdots \partial y_{n-1}} = \begin{vmatrix} 0 & 0 & \cdots 1 \\ 1 & 1 & \cdots y \\ 2y_1 & 2y_2 & \cdots y_n^2 \\ \cdots & \cdots & \cdots \\ (n-1)y_1^{(n-2)} & (n-1)y_2^{(n-2)} & \cdots y_n^{(n-1)} \end{vmatrix}$$

$$= (-)^{n-1}(n-1)!\,\Delta_{n-1}.$$

EXAMPLE 3. Find the complete differential of

$$\begin{vmatrix} a_1 & a_2 & a_3 & a_4 \\ b_1 & b_2 & b_3 & b_4 \\ c_1 & c_2 & c_3 & c_4 \\ d_1 & d_2 & d_3 & d_4 \end{vmatrix}$$

due to the independent variation of all the elements.

87. *If a determinant be differentiated with respect to the element common to the first row and column deleted to form any minor, this differential coefficient be differentiated with respect to the element common to the next row and column deleted, and so on, the result equals the said minor or differs from it only in sign, according as the sum of the numbers indicating the deleted rows and columns is even or odd.*

Let Δ be the determinant, and let the rows which are deleted to form the minor be in order the hth, kth, lth, $\cdots$, and the columns the rth, sth, uth, $\cdots$, so that the minor would be denoted by $A_{hkl\cdots,rsu\cdots}$. Then (§83) we have

$$A_{hr} = (-1)^{h+r}\frac{\partial\Delta}{\partial a_{hr}},$$

and, since the element of the kth row and sth column of Δ is the element of the $(k-1)$th row and $(s-1)$th column of A_{hr}, we have further

$$\begin{aligned} A_{hk,rs} &= (-1)^{k-1+s-1}\frac{\partial}{\partial a_{ks}}\left\{(-1)^{h+r}\frac{\partial\Delta}{\partial a_{hr}}\right\}, \\ &= (-1)^{k+s+h+r}\frac{\partial^2\Delta}{\partial a_{ks}\partial a_{hr}}; \end{aligned}$$

so that if the minor be of the mth order, we have generally

$$A_{hkl\cdots,rsu\cdots} = (-1)^{h+k+l+\cdots+r+s+u+\cdots}\frac{\partial^m\Delta}{\partial a_{hr}\partial a_{ks}\partial a_{lu}\cdots}$$

It has to be carefully noted that in the identity here proved $h<k<l<\cdots$ and $r<s<u<\cdots$. If this were not the case, there would still be equality as to magnitude, but the sign would be otherwise determined than from the sum of $h, k, l, \cdots, r, s, u, \cdots$.

If this proviso with regard to $h, k, l, \cdots$ and $r, s, u, \cdots$ be not made, the sign factor would evidently be

$$(-)^{\Sigma h+\Sigma r+\nu_1+\nu_2}$$

where $\Sigma h = h+k+\cdots$, $\Sigma r = r+s+\cdots$,

ν_1 = number of inversions in $h, k, l, \cdots$,

ν_2 = number of inversions in $r, s, t, \cdots$.

88. Since

$$A_{hk,rs} = (-)^{h+k+r+s+\nu_1+\nu_2} \frac{\partial^2\Delta}{\partial a_{ks}\partial a_{hr}},$$

$$A_{hk,sr} = (-)^{h+k+r+s+\nu_1+\nu_2} \frac{\partial^2\Delta}{\partial a_{kr}\partial a_{hs}}$$

and

$$A_{hk,rs} = -A_{hk,sr}.$$

Therefore

$$\frac{\partial^2\Delta}{\partial a_{ks}\partial a_{hr}} = -\frac{\partial^2\Delta}{\partial a_{kr}\partial a_{hs}}.$$

89. *In a determinant the cofactor of the product of any number of elements which may come together in one term is equal to the result of differentiating the determinant in succession with respect to the said elements.*

Just as by differentiating $|a_{1n}|$, or Δ say, with respect to a_{hr} we obtain the cofactor of a_{hr} in Δ, so by differentiating the result with respect to an element of another row and column, a_{ks} say, we obtain the cofactor of a_{ks} in $\partial\Delta/\partial a_{hr}$; and thus we see that the cofactor of $a_{hr}a_{ks}$ in

$$\Delta = \frac{\partial^2\Delta}{\partial a_{hr}\partial a_{ks}}.$$

Similarly, the cofactor of $a_{hr}a_{ks}a_{lu}$ in

$$\Delta = \frac{\partial^3\Delta}{\partial a_{hr}\partial a_{ks}\partial a_{lu}},$$

$$= \frac{\partial^3\Delta}{\partial a_{kr}\partial a_{ls}\partial a_{hu}}$$

$$= \text{etc.}$$

and so generally.

90. If, in a determinant of the nth order, the minor formed of the elements common to the first m rows and the first m columns be multiplied by its complementary, the product gives $m!(n-m)!$ terms of the original determinant.

Let the determinant be denoted by Δ. The minors would be $A'_{12\cdots m,12\cdots m}$ and $A'_{m+1\cdots n,m+1\cdots n}$ respectively, which for convenience we will denote by M and N.

If any term of M be taken and any term of N, the product of the two terms must contain one and only one constituent from each row and each column of Δ, and must therefore, setting aside the question of sign, be a term of Δ.

Further, suppose that the numbers whose inversions of order determine the sign of the term taken from M are

$$\beta,\ \delta,\ \gamma,\ \cdots,\ \kappa.$$

The number of said inversions being r; and suppose that the corresponding numbers in the case of the term taken from N are

$$\xi,\ \pi,\ \rho,\ \tau,\ \cdots$$

the number of inversions being s. Then the sign factor of the product of the two terms would be

$$(-)^r \times (-)^s.$$

But the series of numbers for determining the sign of it viewed as a term of Δ would be $\beta, \delta, \gamma, \cdots, \kappa, m+\xi, m+\pi, \cdots, m+\tau$; and as each of the numbers $m+\xi, m+\pi, \cdots, m+\tau$ is greater than any one of the numbers $\beta, \delta, \cdots, \kappa$, and the number of inversions in $m+\xi, \cdots, m+\tau$, is the same as in $\xi, \pi, \cdots, \tau$, the total number of inversions in the series must therefore be $r+s$. Consequently the sign-factor of the product viewed as a term of Δ would be

$$(-)^{r+s},$$

which, as we have just seen, is the sign it actually bears.

Thus the product of any term of M and any term of N is a term of Δ; therefore if M which consists of $m!$ terms be multiplied by N which consists of $(n-m)!$ terms, there will result $m!(n-m)!$ terms of Δ.

This theorem is also evident from the fact that the combination of the numbers in the first line of the suffixes of the elements in the term of the one minor is the complementary with respect to n of the combination of the numbers in the first line of the elements in the term of the other minor and that the numbers in the one combination are all greater than those of the other.

91. *If, in a determinant of the nth order, the minor of the constituents common to m rows, namely, the hth, kth, lth, · · · , and m columns, namely, the rth, sth, uth, · · · be multiplied by its complementary, the product taken with the sign* $(-)^{h+k+l+\cdots+r+s+u+\cdots}$ *gives* $m!(n-m)!$ *terms of the original determinant.*

Let the determinant be Δ, the minor M and its complementary N. Making the hth row pass over the $h-1$ rows which precede it, then the kth row over $k-2$ preceding rows, then the lth row over $l-3$ preceding rows, and so on, we have a determinant Δ' whose first m rows are the hth, kth, lth, · · · of Δ and such that

$$\Delta' = (-)^{h-1+k-2+l-3+\cdots}\Delta.$$

Again by treating the rth, sth, uth, · · · , columns of this determinant in like fashion, we obtain a third determinant Δ'' whose first m rows and first m columns are the m rows and m columns of Δ out of which M is formed, and such that

$$\begin{aligned}\Delta'' &= (-)^{h-1+k-2+l-3+\cdots}(-)^{r-1+s-2+u-3+\cdots}\Delta\\ &= (-)^{h+k+l+\cdots+r+s+u+\cdots}\Delta\end{aligned}$$

since $-1-2-3\cdots-1-2-3\cdots$ is even.

Now (§90) the product $M\times N$ gives $m!(n-m)!$ terms of Δ''; therefore, taken with the sign-factor $(-)^{h+k+l+\cdots+r+s+u\cdots}$ it will give $m!(n-m)!$ terms of Δ.

Example. Taking the first, second, fifth rows, and the third, fourth, fifth columns of

$$\begin{vmatrix} a_1 & a_2 & a_3 & a_4 & a_5\\ b_1 & b_2 & b_3 & b_4 & b_5\\ c_1 & c_2 & c_3 & c_4 & c_5\\ d_1 & d_2 & d_3 & d_4 & d_5\\ e_1 & e_2 & e_3 & e_4 & e_5 \end{vmatrix}$$

we have the minor

$$\begin{vmatrix} a_3 & a_4 & a_5\\ b_3 & b_4 & b_5\\ e_3 & e_4 & e_5 \end{vmatrix}$$

whose complementary is

$$\begin{vmatrix} c_1 & c_2 \\ d_1 & d_2 \end{vmatrix}$$

hence $(-1)^{1+2+5+3+4+5}\,|a_3b_4e_5|\,|c_1d_2|$ gives $3!2!$ terms of $|a_1b_2c_3d_4e_5|$. Similarly, $|c_1d_2e_3|\,|a_4b_5|$ and $-|b_2c_3d_5|\,|a_1e_4|$ each give twelve terms; and, as we already know (§31), $-b_3|a_1c_2d_4e_5|$ and $e_1|a_2b_3c_4d_5|$ each give twenty-four.

92. The proposition of (§91) is equivalent to the statement that the cofactor of a minor of a determinant is the complementary minor, the sign to be taken with the product being $+$ or $-$ according as the sum of the numbers indicating the deleted rows and columns is even or odd. As, however, the sign of the product of the two minors must be the same as the sign of the product of their principal terms, it is evident that we have another and perhaps easier mode of fixing it, namely, by determining the sign of the latter product as a term of the original determinant. Thus, taking the first example above, the sign to be prefixed to $|a_3b_4e_5|\,|c_1d_2|$ is the sign of $a_3b_4e_5c_1d_2$ or $a_3b_4c_1d_2e_5$ as a term of $|a_1b_2c_3d_4e_5|$, and therefore (§21) is $(-1)^4$.

Exercises. Set VIII

Write down the complementaries of the following minors of $|a_0b_1c_2d_3e_4f_5|$:

1. b_3. 2. $|c_2\,e_4|$. 3. $|c_0\,f_5|$. 4. $|b_0\,c_5|$.

5. $|d_2\,\epsilon_3\,f_4|$. 6. $|b_1\,c_3\,d_4|$. 7. $|b_0\,e_2\,f_4|$. 8. $|c_0\,d_4\,e_2|$.

9. What are the complementary minors of $|a_{13}a_{35}|$ and $|a_{12}a_{35}a_{56}|$ in $|a_{01}a_{12}a_{23}a_{34}a_{45}a_{56}|$?

10. Express as determinants.

$$\frac{\partial^2}{\partial a_{25}\,\partial a_{53}}|a_{16}| \quad\text{and}\quad \frac{\partial^3}{\partial a_{13}\,\partial a_{42}\,\partial a_{35}}|a_{01}\,a_{12}\,a_{23}\,a_{34}\,a_{45}\,a_{56}|\,.$$

11. What is the cofactor of $|a_{22}a_{43}|$ in $|a_{21}a_{32}a_{43}a_{54}a_{65}|$ and in $|a_{01}a_{12}a_{23}a_{34}a_{45}a_{56}|$?

12. Express in symbols the theorem regarding the effect of transposing two rows of a determinant.

13. Find the single number which is the equivalent of

$$\begin{vmatrix} 1 & a & b & c \\ a & 1 & 0 & 0 \\ b & 0 & 1 & 0 \\ c & 0 & 0 & 1 \end{vmatrix} + \begin{vmatrix} 1 & a & b \\ -a & 1 & c \\ -b & -c & 1 \end{vmatrix}.$$

14. Find the final expansion of

$$\begin{vmatrix} 0 & -b & -c & -d \\ b & 0 & -e & -f \\ c & e & 0 & -g \\ d & f & g & 0 \end{vmatrix}.$$

15. Express in terms arranged according to descending powers of x

$$\begin{vmatrix} x & -1 & -1 & -1 \\ 1 & x^3 & -1 & -1 \\ 1 & 1 & x^5 & -1 \\ 1 & 1 & 1 & x^7 \end{vmatrix}.$$

16. Find the quadratic in x, y, z equivalent to

$$\begin{vmatrix} 0 & x & y & z \\ x & a & h & g \\ y & h & b & f \\ z & g & f & c \end{vmatrix}.$$

17. Find the quadratic in x, y, z, w equivalent to

$$\begin{vmatrix} 0 & x & y & z & w \\ x & 1 & -1 & -1 & -1 \\ y & -1 & 1 & -1 & -1 \\ z & -1 & -1 & 1 & -1 \\ w & -1 & -1 & -1 & 1 \end{vmatrix}.$$

18. Find the quadratic in x, y, z equivalent to

$$\begin{vmatrix} 1 & 0 & 0 & ax+hy+gz \\ 0 & 1 & 0 & hx+by+fz \\ 0 & 0 & 1 & gx+fy+cz \\ x & y & z & 0 \end{vmatrix}.$$

19. Prove that

$$\begin{vmatrix} 1 & 0 & 0 & 0 & ax+hy+gz \\ 0 & 1 & 0 & 0 & hx+by+fz \\ 0 & 0 & 1 & 0 & gx+fy+cz \\ 0 & 0 & 0 & 1 & lx+my+nz \\ x & y & z & 1 & k \end{vmatrix} = \begin{vmatrix} 1 & 0 & 0 & ax+hy+gz+l \\ 0 & 1 & 0 & hx+by+fz+m \\ 0 & 0 & 1 & gx+fy+cz+n \\ x & y & z & k \end{vmatrix}$$

$$= -(ax^2+by^2+cz^2+2fyz+2gzx+2hxy+lx+my+nz-k).$$

93. Laplace's Theorem.* *If any m rows of a determinant be selected and every possible minor of the mth order be formed from them, and if each be multiplied by its complementary and the sign + or − be affixed to the product according as the sum of the numbers indicating the rows and columns from which the minor is formed be even or odd, the aggregate of the products thus obtained is equal to the original determinant.*

Let the given determinant be of the nth order.

The number of different minors of the mth order that can be formed from these m selected rows is clearly the number of sets of m columns that can be formed out of n, and therefore is

$$\frac{n!}{m!(n-m)!}.$$

Now the product of each of these minors and its complementary gives, when its sign is fixed in the manner stated, $m!(n-m)!$ terms of A. Using all the different minors, therefore, we obtain

$$\frac{m!(n-m)!n!}{m!(n-m)!}, \qquad \text{or} \qquad n!$$

different terms of A, that is, the full expansion of A.

* See Hist. I p. 115.

EXAMPLES. Taking the first two rows of

$$\begin{vmatrix} a_1 & a_2 & a_3 & a_4 \\ b_1 & b_2 & b_3 & b_4 \\ c_1 & c_2 & c_3 & c_4 \\ d_1 & d_2 & d_3 & d_4 \end{vmatrix}$$

we have in all six minors, namely

$$\begin{vmatrix} a_1 & a_1 \\ b_1 & b_2 \end{vmatrix}, \quad \begin{vmatrix} a_1 & a_3 \\ b_1 & b_3 \end{vmatrix}, \quad \begin{vmatrix} a_1 & a_4 \\ b_1 & b_4 \end{vmatrix}, \quad \begin{vmatrix} a_2 & a_3 \\ b_2 & b_3 \end{vmatrix}, \quad \begin{vmatrix} a_2 & a_4 \\ b_2 & b_4 \end{vmatrix}, \quad \begin{vmatrix} a_3 & a_4 \\ b_3 & b_4 \end{vmatrix};$$

hence

$$\begin{aligned} |a_1 b_2 c_3 d_4| = {} & |a_1 b_2|\,|c_3 d_4| - |a_1 b_3|\,|c_2 d_4| + |a_1 b_4|\,|c_2 d_3| \\ & + |a_2 b_3|\,|c_1 d_4| - |a_2 b_4|\,|c_1 d_3| + |a_3 b_4|\,|c_1 d_2|\,. \end{aligned}$$

By selecting any other pair of rows except the last pair or any pair of columns we should obtain a like development: by selecting one row or three the development is that of §69.

Similarly, we find

$$\begin{aligned} |a_1 b_2 c_3 d_4 e_5| = {} & |a_1 b_2 c_3|\,|d_4 e_5| - |a_1 b_2 c_4|\,|d_3 e_5| \\ & + |a_1 b_2 c_5|\,|d_3 e_4| + |a_1 b_3 c_4|\,|d_2 e_5| \\ & - |a_1 b_3 c_5|\,|d_2 e_4| + |a_1 b_4 c_5|\,|d_2 e_3| \\ & - |a_2 b_3 c_4|\,|d_1 e_5| + |a_2 b_3 c_5|\,|d_1 e_4| \\ & - |a_2 b_4 c_5|\,|d_1 e_3| + |a_3 b_4 c_5|\,|d_1 e_2|\,. \end{aligned}$$

The theorem of this article is an extension of that in §69 and may be symbolically expressed thus:

$$\Delta = \sum_1^{\mu} \alpha(-)^{\nu} A_{(n\,|\,m\alpha),\,(n\,|\,m\beta)} \cdot A_{(\bar{n}\,|\,m\alpha),\,(\bar{n}\,|\,m\beta)}$$

where $\mu = n_m$ and ν is the number of inversions in $(n\,|m_\alpha)(\bar{n}\,|m_\alpha)+$the number in $(\dot{n}\,|m_\beta)(\bar{n}\,|m_\beta)$, that is,

$$s(n\,|\,m_\alpha) - \frac{m(1+m)}{2} + s(n\,|m_\beta) - \frac{(m+1)m}{2}$$

or $s(n\,|m_\alpha)+s(n\,|m_\beta)-m(1+m)$ (§14).

94. Since either of the determinants in each of the products referred to in the theorem of the preceding article may itself be expressed as an aggregate of products of complementary minors, we see that by repeated use of the theorem we can develop a determinant as an aggregate of products of more than two minors. Thus

$$
\begin{aligned}
|a_1 b_2 c_3 d_4 e_5 f_6| &= |a_1 b_2 c_3|\,|d_4 e_5 f_6| - |a_1 b_2 c_4|\,|d_3 e_5 f_6| + \cdots \\
&= a_1|b_2 c_3|\,|d_4 e_5 f_6| - a_2|b_1 c_3|\,|d_4 e_5 f_6| \\
&\quad + a_3|b_1 c_2|\,|d_4 e_5 f_6| - a_1|b_2 c_4|\,|d_3 e_5 f_6| \\
&\quad + a_2|b_1 c_4|\,|d_3 e_5 f_6| - a_4|b_1 c_2|\,|d_3 e_5 f_6| + \cdots
\end{aligned}
$$

where each term is a product of three minors, one of the first, one of the second and one of the third order.

The symbolic expression for this general theorem would be

$$
\Delta = \sum_1^{nh} \alpha \sum_1^{(n-h)_i} \beta \cdots \sum_1^{(n-h-i-\cdots-k)_l} \delta \; (-)^\nu A_{(n|h\alpha),(n|h\alpha')} A_{(\bar{n}|h\alpha|i\beta),(\bar{n}|h\alpha'|i\beta')}
$$
$$
\cdots A_{(\bar{n}|\bar{h}\alpha|\bar{i}\beta|\cdots|m_1),(\bar{n}|\bar{h}\alpha'|\bar{i}\beta'|\cdots|m_1)}
$$

where ν is the number of inversions in

$$
+\text{ the number in } \begin{matrix} (n|\,h_\alpha)(\bar{n}|\,h_\alpha|\,i_\beta)(\cdots)(\bar{n}|\,\bar{h}_\alpha|\,\bar{i}_\beta|\cdots|\,m_1) \\ (n|\,h_{\alpha'})(\bar{n}|\,h_{\alpha'}|\,i_{\beta'})(\cdots)(\bar{n}|\,\bar{h}_{\alpha'}|\,\bar{i}_{\beta'}\cdots|\,m_1) \end{matrix}
$$

and where $n=h+i+\cdots+l+m$.

We might extend the meaning of complementary minors and call a set whose row and column numbers form a set of complementary combinations (§12), as in any term of this expansion, a *System of Complementary Minors.*

Making use of the theorem of the preceding article to expand the determinant of §46 we get:

$$
\Delta a_h b_i \cdots k_m = \sum_1^{nh} \alpha' \sum_1^{(n-h)_i} \beta'
$$
$$
\cdots \sum_1^{(n-h-\cdots-k)_l} \delta' \quad (-)^\nu A_{(n|h\alpha),(n|h\alpha')} B_{(\bar{n}|h\alpha|i\beta),(\bar{n}|h\alpha'|i\beta')}
$$
$$
\cdots P_{(\bar{n}|\bar{h}\alpha|\bar{i}\beta|\cdots|m_1),(\bar{n}|\bar{h}\alpha'|\bar{i}\beta'|\cdots|m_1)}
$$

where ν is the number of inversions in

$$(n \mid h_\alpha \between \bar{n} \mid h_\alpha \mid i_\beta \between \cdots \between \bar{n} \mid \bar{h}_\alpha \mid \bar{i}_\beta \mid \cdots \mid m_1)$$

+ the number of inversions in

$$(n \mid h_{\alpha'} \between \bar{n} \mid h_{\alpha'} \mid i_{\beta'} \between \cdots \between \bar{n} \mid \bar{h}_{\alpha'} \mid \bar{i}_{\beta'} \mid \cdots \mid m_1).$$

95. Making use of the theorem of the preceding article, together with the expansion of §46 we get:

$$\Delta = \sum_{0}^{n} h \sum_{0}^{n} i \cdots \sum_{0}^{n} m \sum_{1}^{n_h} \alpha' \sum_{1}^{(n-h)_i} \beta'$$
$$\cdots \sum_{1}^{(n-h-i\cdots)_l} \delta' \; (-)^{\nu} A_{(n|h\alpha),(n|h\alpha')} B_{(\bar{n}|h\alpha|i\beta),(\bar{n}|h\alpha'|i\beta')}$$
$$\cdots \; P_{(\bar{n}|\bar{h}\alpha|\bar{i}\beta|\cdots|m_1),(\bar{n}|\bar{h}\alpha'|\bar{i}\beta'|\cdots|m_1)}$$

where the value of ν is found as in the above case.

96. For any fixed value of $\alpha, \beta, \gamma, \cdots, 1$, there are $n!/h!i!j! \cdots m!$ ways of forming a system of complementary minors which is the same as the coefficient of the general term in the expansion of the multinomial $(x_1+x_2+x_3+\cdots+x_n)^n$.

We may write symbollically for the expansion of Δ

$$\Delta = \Big\{ A_{(n|h\alpha),(\bar{n}|h\alpha')} + A_{(\bar{n}|h\alpha|i\beta),(\bar{n}|h\alpha'|i\beta')}$$
$$+ \cdots + A_{(\bar{n}|\bar{h}\alpha|\bar{i}\beta|\cdots|m_1),(\bar{n}|\bar{h}\alpha'|\bar{i}\beta'|\cdots|m_1)} \Big\}^n$$

where the exponents are to be dropped having no effect but simply indicating the order of the determinant factors, and where the coefficient of every term is to be replaced by a summation sign, the number of terms in the sum being given by the coefficient. This is *Albeggiani's theorem.*

EXAMPLE. If

$$B = \begin{vmatrix} \cdot & \cdot\;\cdot\;\cdot & \cdot & y_{11} & y_{12} & \cdots & y_{1n} \\ \cdot & \cdot\;\cdot\;\cdot & \cdot & \cdot & \cdot & \cdots & \cdot \\ \cdot & \cdot\;\cdot\;\cdot & \cdot & y_{r1} & y_{r2} & \cdots & y_{rn} \\ x_{11} & \cdots & x_{1r} & a_{11} & a_{12} & \cdots & a_{1n} \\ \cdot & \cdot\;\cdot\;\cdot & \cdot & \cdot & \cdot & \cdots & \cdot \\ x_{21} & \cdots & x_{2r} & a_{21} & a_{22} & \cdots & a_{2n} \\ x_{n1} & \cdots & x_{nr} & a_{n1} & a_{n2} & \cdots & a_{nn} \end{vmatrix}$$

and if X be any r-line minor of the x array, Y any r-line minor of the y array, and A the minor got by deleting the rows and columns of $|a_{1n}|$ which are the continuations of the rows and columns of X and Y respectively then

$$B = \sum(-1)^{r+\sigma}X\cdot Y\cdot A\,.,$$

where σ is the sum of the numbers of the columns taken from the y array and the number of the rows taken from the x array.

97. Starting with a determinant of the 4th order, we have by Laplace's expansion-theorem

$$\begin{vmatrix} a_1 & a_2 & a_3 & a_4 \\ b_1 & b_2 & b_3 & b_4 \\ c_1 & c_2 & c_3 & c_4 \\ d_1 & d_2 & d_3 & d_4 \end{vmatrix} \begin{aligned} &= |a_1 b_2|\,|c_3 d_4| - |a_1 c_2|\,|b_3 d_4| \\ &\quad + |a_1 d_2|\,|b_3 c_4| + |b_1 c_2|\,|a_3 d_4| \\ &\quad - |b_1 d_2|\,|a_3 c_4| + |c_1 d_2|\,|a_3 b_4|\,. \end{aligned}$$

Now the sum of the first and last terms here is clearly equal to the permanent

$$\overset{+}{}\!\begin{vmatrix} |a_1 b_2| & |a_3 b_4| \\ |c_1 d_2| & |c_3 d_4| \end{vmatrix}\!\overset{+}{},$$

the sum of the second and fifth equal to

$$-\overset{+}{}\!\begin{vmatrix} |a_1 c_2| & |a_3 c_4| \\ |b_1 d_2| & |b_3 d_4| \end{vmatrix}\!\overset{+}{},$$

and the sum of the third and fourth equal to

$$\overset{+}{}\!\begin{vmatrix} |a_1 d_2| & |a_3 d_4| \\ |b_1 c_2| & |b_3 c_4| \end{vmatrix}\!\overset{+}{};$$

so that we have

$$\begin{aligned} |a_1 b_2 c_3 d_4| &= \overset{+}{}\!\big|\,|a_1 b_2|\;|c_3 d_4|\,\big|\!\overset{+}{} - \overset{+}{}\!\big|\,|a_1 c_2|\;|b_3 d_4|\,\big|\!\overset{+}{} \\ &\quad + \overset{+}{}\!\big|\,|a_1 d_2|\;|b_3 c_4|\,\big|\!\overset{+}{}. \end{aligned}$$

98. Taking next the determinant of the 6th order we have, as before

$$
\begin{aligned}
|a_1 b_2 c_3 d_4 e_5 f_6| = {} & |a_1 b_2 c_3|\,|d_4 e_5 f_6| - |a_1 b_2 d_3|\,|c_4 e_5 f_6| \\
& + |a_1 b_2 e_3|\,|c_4 d_5 f_6| - |a_1 b_2 f_3|\,|c_4 d_5 e_6| \\
& + |a_1 c_2 d_3|\,|b_4 e_5 f_6| - |a_1 c_2 e_3|\,|b_4 d_5 f_6| \\
& + |a_1 c_2 f_3|\,|b_4 d_5 e_6| + |a_1 d_2 e_3|\,|b_4 c_5 f_6| \\
& - |a_1 d_2 f_3|\,|b_4 c_5 e_6| + |a_1 e_2 f_3|\,|b_4 c_5 d_6| \\
& - |b_1 c_2 d_3|\,|a_4 e_5 f_6| + |b_1 c_2 e_3|\,|a_4 d_5 f_6| \\
& - |b_1 c_2 f_3|\,|a_4 d_5 e_6| - |b_1 d_2 e_3|\,|a_4 c_5 f_6| \\
& + |b_1 d_2 f_3|\,|a_4 c_5 e_6| - |b_1 e_2 f_3|\,|a_4 c_5 d_6| \\
& + |c_1 d_2 e_3|\,|a_4 b_5 f_6| - |c_1 d_2 f_3|\,|a_4 b_5 e_6| \\
& + |c_1 e_2 f_3|\,|a_4 b_5 d_6| - |d_1 e_2 f_3|\,|a_4 b_5 c_6| \;;
\end{aligned}
$$

and as the sum of the first and last terms of this development is equal to

$$
\begin{vmatrix} |a_1 b_2 c_3| & |a_4 b_5 c_6| \\ |d_1 e_2 f_3| & |d_4 e_5 f_6| \end{vmatrix}
$$

the sum of the second from the beginning and the second from the end equal to

$$
-\begin{vmatrix} |a_1 b_2 d_3| & |a_4 b_5 d_6| \\ |c_1 e_2 f_3| & |c_4 e_5 f_6| \end{vmatrix},
$$

and so on, there results the identity

$$
|a_1 b_2 c_3 d_4 e_5 f_6| = \sum \big\|\, |a_1 b_2 c_3| \;\; |d_4 e_5 f_6| \,\big\|,
$$

there being ten terms on the right included under the sign of summation, and the sign preceding each being the same as the sign of that particular term of the original determinant which is brought into prominence by the notation employed.

99. Taking next the expansion of $|a_1 b_2 c_3 d_4 e_5 f_6|$ in terms of minors of the 2nd order, we should obtain 90 terms of the form

$$
|a_1 b_2| \cdot |c_3 d_4| \cdot |e_5 f_6| \;;
$$

and these we should find capable of being collected into 15 sets of 6, with each set expressible as a permanent of the 3rd order, the identity reached being

$$|a_1\, b_2\, c_3\, a_4\, e_5\, f_6| = \sum \overset{+}{\left|\begin{matrix} |a_1\, b_2| & |a_3\, b_4| & |a_5\, b_6| \\ |c_1\, d_2| & |c_3\, d_4| & |c_5\, d_6| \\ |e_1\, f_2| & |e_3\, f_4| & |e_5\, f_6| \end{matrix}\right|}^{+} .$$

100. Towards the establishment of the theorem in all its generality the first step necessary is to prove the following:

If the rows of a determinant of the (mn)th order be separated by horizontal lines into n sets of m rows each, and the columns be similarly divided, the result may be viewed as roughly representing a determinant of the nth order (called a compound determinant) each of whose elements is a determinant of the mth order and a minor of the original determinant, and each term of the compound determinant thus arising will produce $(m!)^n$ terms, differing at most only in sign from terms of the original determinant.*

The original determinant being denoted by

$$|a_{11}\, a_{22}\, a_{33} \cdots a_{mn,mn}|$$

the compound determinant referred to will be

$$\left|\begin{matrix} \begin{vmatrix} a_{11} & a_{12} & \cdots & a_{1m} \\ a_{21} & a_{22} & \cdots & a_{2m} \\ \cdots & \cdots & \cdots & \cdots \\ a_{m1} & a_{m2} & \cdots & a_{mm} \end{vmatrix}, & \begin{vmatrix} a_{1,m+1} & a_{1,m+2} & \cdots & a_{1,2m} \\ a_{2,m+1} & a_{2,m+2} & \cdots & a_{2,2m} \\ \cdots & \cdots & \cdots & \cdots \\ a_{m,m+1} & a_{m,m+2} & \cdots & a_{m,2m} \end{vmatrix}, & \cdots \\ \begin{vmatrix} a_{m+1,1} & a_{m+1,2} & \cdots & a_{m+1,m} \\ a_{m+2,1} & a_{m+2,2} & \cdots & a_{m+2,m} \\ \cdots & \cdots & \cdots & \cdots \\ a_{2m,1} & a_{2m,2} & \cdots & a_{2m,m} \end{vmatrix}, & \begin{vmatrix} a_{m+1,m+1} & a_{m+1,m+2} & \cdots & a_{m+1,2m} \\ a_{m+2,m+1} & a_{m+2,m+2} & \cdots & a_{m+2,2m} \\ \cdots & \cdots & \cdots & \cdots \\ a_{2m,m+1} & a_{2m,m+2} & \cdots & a_{2m,2m} \end{vmatrix}, & \cdots \\ \cdots & \cdots & \cdots \end{matrix}\right| ,$$

or say, for shortness sake,

$$\begin{vmatrix} M_{11} & M_{12} & \cdots & M_{1n} \\ M_{21} & M_{22} & \cdots & M_{2n} \\ \cdots & \cdots & \cdots & \cdots \\ M_{n1} & M_{n2} & \cdots & M_{nn} \end{vmatrix} ,$$

* Such determinants are defined and dealt with in Chapter VI.

any term of which will, of course, be of the form

$$M_{hk}\, M_{pq}\, M_{rs} \cdots$$

where no two of the M's, by the definition of a determinant, belong to the same set of m rows or m columns of the original. Now, each term of the development of the minor M_{hk} being in this way the product of m elements taken from m rows and m columns of the original determinant, and each term of the development of the minor M_{pq} being the product of m elements taken from m other rows and m other columns, and each term of the development of the minor M_{rs} being the product of m elements taken from a third set of m rows and a third set of m columns, and so on, it follows that each term got by multiplying together a term of each of the n minors M_{hk}, M_{pq}, M_{rs}, $\cdots$ will contain an element from each of the mn rows and mn columns, and therefore by definition be a term of that determinant. Further, since the number of terms in each minor is $m!$, the number of terms of the original determinant which arise from the term $M_{hk}M_{pq}M_{rs} \cdots$ of the compound determinant is $(m!)^n$, and therefore the number of those which arise from all the terms of the compound determinant is $n!(m!)^n$.

101. In the preceding the mn rows of the original determinant were separated in the simplest way into n sets of m rows each, but it is clear that if they had been separated in a different way into n sets of m rows each, the same mode of reasoning would have led to the same result. Now the number of different ways of breaking up mn things into n sets of m each is

$$\frac{C_{mn,m}\, C_{mn-m,m} \cdot C_{mn-2m,m} \cdots C_{m,m}}{n!},$$

or(what is the same thing, since $C_{rs} = r/sC_{r-1,s-1}$),

$$C_{mn-1,m-1} \cdot C_{mn-m-1,m-1} \cdot C_{mn-2m-1,m-1} \cdots$$

that is

$$\frac{\dfrac{(mn)!}{m!(mn-m)!} \cdot \dfrac{(mn-m)!}{m!(mn-2m)!} \cdot \dfrac{(mn-2m)!}{m!(mn-3m)!} \cdots \dfrac{m!}{m!0!}}{n!}$$

$$= \frac{(mn)!}{(m!)^n n!}\,.$$

If, therefore, we form this number of different compound determinants, the total number of terms of the original determinant which we shall thus obtain is

$$\frac{(mn)!}{(m!)^n n!}\cdot n!(m!)^n$$

that is $(mn)!$ which is exactly the full number of terms in the original determinant. It is consequently manifest that the terms of the original determinant can be represented, so far as magnitude is concerned, by a sum of compound determinants obtainable in the manner indicated.

102. From §4 we see that when m is even the sign in every case for $M_{hk}M_{pq}\cdots$ is positive and when m is odd the sign must be the sign which it bears in the compound determinant to which it belongs. This is equivalent to saying that when m is even the compound determinants with which we started should all be viewed as permanents and that this change is not necessary when m is odd.

103. When the given determinant of the (mn)th order is of the form $|a^0b^1c^2d^3|$ which is called an *alternant*, (Ex. 2, §60) interesting changes are possible in connection with each term of the development.

Thus in the case of the simplest form of alternant of the 4th order we have

$$|a^0\,b^1\,c^2\,d^3| = \overset{+}{\left|\begin{matrix} |a^0\,b^1| & |a^2\,b^3| \\ |c^0\,d^1| & |c^2\,d^3| \end{matrix}\right|}^{+} - \overset{+}{\left|\begin{matrix} |a^0\,c^1| & |a^2\,c^3| \\ |b^0\,d^1| & |b^2\,d^3| \end{matrix}\right|}^{+}$$

$$+ \overset{+}{\left|\begin{matrix} |a^0\,d^1| & |a^2\,d^3| \\ |b^0\,c^1| & |b^2\,c^3| \end{matrix}\right|}^{+}$$

$$= \overset{+}{\left|\begin{matrix} b-a & a^2b^3-a^3b^2 \\ d-c & c^2d^3-c^3d^2 \end{matrix}\right|}^{+} - \cdots,$$

$$= (b-a)(d-c)\overset{+}{\left|\begin{matrix} 1 & a^2b^2 \\ 1 & c^2d^2 \end{matrix}\right|}^{+} - \cdots,$$

In the case of the simplest alternant of the 6th order we have

$$|a^0\,b^1\,c^2\,d^3\,e^4\,f^5| = \sum\left|\begin{matrix} |a^0\,b^1\,c^2| & |a^3\,b^4\,c^5| \\ |d^0\,e^1\,f^2| & |d^3\,e^4\,f^5| \end{matrix}\right|,$$

$$= \sum|a^0\,b^1\,c^2|\cdot|d^0\,e^1\,f^2|\left|\begin{matrix} 1 & a^3b^3c^3 \\ 1 & d^3e^3f^3 \end{matrix}\right|,$$

$$= \sum|a^0\,b^1\,c^2|\cdot|d^0\,e^1\,f^2|\cdot|(abc)^0\,(def)^3|,$$

where it should be noted that each term of the development consists of three factors, each of which is a simple alternant, and each of which is, therefore, expressible as a product of binomial factors. Thus the specimen term here given after Σ is equal to

$$(c-b)(c-a)(b-a)(f-e)(f-d)(e-d)(d^3e^3f^3-a^3b^3c^3).$$

Similarly we have

$$|a^0b^1c^2d^3e^4f^5g^6h^7i^8| = \sum |a^0b^1c^2|\cdot|d^0e^1f^2|\cdot|g^0h^1i^2|\cdot|(abc)^0(def)^3(ghi)^6|,$$

where each term under the sign of summation is expressible as a product of twelve binomial factors, the first, for example, being equal to

$$(c-b)(c-a)(b-a)(f-e)(f-d)(e-d)(i-h)(i-g)(h-g)$$
$$\times(g^3h^3i^3-d^3e^3f^3)(g^3h^3i^3-a^3b^3c^3)(d^3e^3f^3-a^3b^3c^3).$$

104. *Let*

$$D_{n,p}\equiv\Delta\equiv\begin{vmatrix} a_{11}+b_{11}+\cdots+p_{11} & \cdots & a_{1n}+b_{1n}+\cdots+p_{1n} \\ a_{21}+b_{21}+\cdots+p_{21} & \cdots & a_{2n}+b_{2n}+\cdots+p_{2n} \\ \cdots\cdots & \cdots & \cdots\cdots \\ \cdots\cdots & \cdots & \cdots\cdots \\ a_{n1}+b_{n1}+\cdots+p_{n1} & \cdots & a_{nn}+b_{nn}+\cdots+p_{nn} \end{vmatrix}$$

be a determinant of the nth order, each of whose elements consists of p terms; let $\sum D_{n,p-1}$ denote the sum of p determinants formed from $D_{n,p}$ by omitting, first, all the first terms of the elements, secondly, all the second terms and so on; let $\sum D_{n,p-2}$ denote the $\frac{1}{2}p(p-1)$ determinants formed by omitting, firstly, all the first and all the second terms of the elements, secondly, all the first and all the third terms, and so on; and let $\sum D_{n,p-3}$, $\sum D_{n,p-4}$, etc., bear similar interpretations, then

$$(1)\quad \begin{aligned} D_{n,p} - \sum D_{n,p-1} + \sum D_{n,p-2} - \cdots + (-)^{p-1}\sum D_{n,1} \\ = \sum \Delta a_{n-p+1-\alpha} b_{\alpha+1-\beta} c_{\beta+1-\gamma} \cdots o_{\nu+1-\pi} p_{\pi+1} \end{aligned}$$

where $n-p\geqq\alpha\geqq\beta\geqq\gamma\geqq\cdots\geqq\nu\geqq\pi\geqq0$, and where, since the subscripts of $a, b, \cdots, p$ are by definition essentially positive, we are to interpret the righthand side of (1) as zero in case of a negative subscript which occurs when $p>n$.

To prove this we have but to partition all the determinants in (1) having polynomial elements into determinants with monomial elements according to §46 and then it may be seen that the complete coefficient of any one of these determinants with monomial elements is zero.

Take for instance one of these δ_1, whose elements are in order the first n terms of the n^2 polynomials. It is clear that the number of times δ_1, occurs in $D_{n,p}$ is 1 time; in $D_{n,p-1}$ is $(p-1)$ times; in $D_{n,p-2}$ is $\frac{1}{2}(p-1)(p-2)$ times; $\cdots$.

Hence the coefficient of δ_1 is

$$1-\binom{p-1}{1}+\binom{p-1}{2}-\cdots=(1-1)^{p-1}=0.$$

The same holds for any other of the p^n determinants with monomial elements. If $p=n$, the right-hand side of (1) becomes $\sum\Delta a_{11}b_{11}\cdots p_{n}$.

If we make all the terms vanish in all the elements of Δ except those in the principal diagonal, it reduces to the product of n p-termed polynomials.

If $\pi_{n,p}$ denote the product of the n p-termed expressions; if $\sum\pi_{n,p-1}$ denote the sum of the p products formed from $\pi_{n,p}$ by omitting firstly all the first terms of the expressions, secondly all the second terms, and so on; if $\sum\pi_{n,p-2}$, $\sum\pi_{n,p-3}\cdots$ bear similar interpretations; and if $\pi a_\alpha b_\beta c_\gamma \cdots p_\pi$ denote the products of α a's, β b's, γ c's $\cdots$, πp's then (1) *becomes*

$$(2)\quad \pi_{n,p}-\sum\pi_{n,p-1}+\cdots+(-)^{p-1}\sum\pi_{n,1} = \sum\pi a_{n-p+1-\alpha}b_{\alpha+1-\beta}\cdots p_{\pi+1}.$$

If further we make all the polynomials in $\pi_{n,p}$ identical there results the relation

$$(3)\quad \begin{aligned}&(a_1+a_2+\cdots+a_p)^n-\sum(a_1+a_2+\cdots+a_{p-1})^n\\ &\qquad+\cdots+(-)^{p-2}\sum(a_1+a_2)^n+(-)^{p-1}\sum a_1{}^n\\ &\qquad=\sum a_1{}^{n-p+1-\alpha}a_2{}^{\alpha+1-\beta}\cdots a_p{}^{\pi+1}\end{aligned}$$

If $a_1=a_2=\cdots=a_p=1$, we have from (3)

$$p^n-p(p-1)^n+\frac{p(p-1)(p-2)^n}{2!}-\cdots$$

$$+(-)^{p-2}\frac{p(p-1)2^n}{2!}+(-)^{p-1}p$$

$$=\sum\binom{n}{n-p+1-\alpha}\binom{p-1+\alpha}{\alpha+1-\beta}\binom{p-2+\beta}{\beta+1-\gamma}\cdots\binom{\pi+1}{\pi-1}$$

or

$$(4) \qquad = \sum \frac{n!}{(n-p+1-\alpha)!(\alpha+1-\beta)!\cdots(\pi+1)!}.$$

If $p>n$ the right-hand sides of (1), (2), (3), (4) become zero.

If $p=n$ the right-hand side of (1) is $\sum \Delta a_{11}b_{11}c_{11}\cdots p_{11}$, of (2) is $\sum \pi a_{11}b_{11}c_{11}\cdots p_{11}$ of (3) is $n!a_1a_2\cdots a_n$, and of (4) is $n!$.

105. As illustrations of these identities we have for $n=3$, $p=3$

$$\begin{vmatrix} a_{11}+b_{11}+c_{11} & a_{12}+b_{12}+c_{12} & a_{13}+b_{13}+c_{13} \\ a_{21}+b_{21}+c_{21} & a_{22}+b_{22}+c_{22} & a_{23}+b_{23}+c_{23} \\ a_{31}+b_{31}+c_{31} & a_{32}+b_{32}+c_{32} & a_{33}+b_{33}+c_{33} \end{vmatrix}$$

$$-\sum \begin{vmatrix} a_{11}+b_{11} & a_{12}+b_{12} & a_{13}+b_{13} \\ a_{21}+b_{21} & a_{22}+b_{22} & a_{23}+b_{23} \\ a_{31}+b_{31} & a_{32}+b_{32} & a_{33}+b_{33} \end{vmatrix} + \sum \begin{vmatrix} a_{11} & a_{12} & a_{13} \\ a_{21} & a_{22} & a_{23} \\ a_{31} & a_{32} & a_{33} \end{vmatrix}$$

$$= \sum \begin{vmatrix} a_{11} & b_{12} & c_{13} \\ a_{21} & b_{22} & c_{23} \\ a_{31} & b_{32} & c_{33} \end{vmatrix}.$$

If

$$a_{11}=a_{22}=a_{33}=a, \qquad b_{11}=b_{22}=b_{33}=b,$$
$$c_{11}=c_{22}=c_{33}=c, \qquad a_{ij}=b_{ij}=c_{ij}=0 \quad (i\neq j)$$

this becomes:

$$(a+b+c)^3-(a+b)^3-(a+c)^3-(b+c)^3+a^3+b^3+c^3=6abc$$

for $p=2$ and $n=3$, we have

$$\begin{vmatrix} a_{11}+b_{11} & a_{12}+b_{12} & a_{13}+b_{13} \\ \cdots\cdots & \cdots\cdots & \cdots\cdots \\ a_{31}+b_{31} & \cdots\cdots & a_{33}+b_{33} \end{vmatrix} - \begin{vmatrix} a_{11} & a_{12} & a_{13} \\ \cdots & \cdots & \cdots \\ a_{31} & a_{32} & a_{33} \end{vmatrix} - \begin{vmatrix} b_{11} & b_{12} & b_{13} \\ \cdots & \cdots & \cdots \\ b_{31} & b_{32} & b_{33} \end{vmatrix}$$

$$= \begin{vmatrix} a_{11} & a_{12} & b_{13} \\ \cdots & \cdots & \cdots \\ a_{31} & a_{32} & b_{33} \end{vmatrix} + \begin{vmatrix} a_{11} & b_{12} & a_{13} \\ \cdots & \cdots & \cdots \\ a_{31} & b_{32} & a_{33} \end{vmatrix} + \begin{vmatrix} b_{11} & a_{12} & a_{13} \\ \cdots & \cdots & \cdots \\ b_{31} & a_{32} & a_{33} \end{vmatrix}$$

$$+ \begin{vmatrix} a_{11} & b_{12} & b_{13} \\ \cdots & \cdots & \cdots \\ a_{31} & b_{32} & b_{33} \end{vmatrix} + \begin{vmatrix} b_{11} & a_{12} & b_{13} \\ \cdots & \cdots & \cdots \\ b_{31} & a_{32} & b_{33} \end{vmatrix} + \begin{vmatrix} b_{11} & b_{12} & a_{13} \\ \cdots & \cdots & \cdots \\ b_{31} & b_{32} & a_{33} \end{vmatrix}.$$

For $n=2$ and $p=4$, we have,

$$\begin{vmatrix} a_{11}+b_{11}+c_{11}+d_{11} & a_{12}+b_{12}+c_{12}+d_{12} \\ a_{21}+b_{21}+c_{21}+d_{21} & a_{22}+b_{22}+c_{22}+d_{22} \end{vmatrix}$$

$$-\begin{vmatrix} a_{11}+b_{11}+c_{11} & a_{12}+b_{12}+c_{12} \\ a_{21}+b_{21}+c_{21} & a_{22}+b_{22}+c_{22} \end{vmatrix} - \begin{vmatrix} a_{11}+b_{11}+d_{11} & a_{12}+b_{12}+d_{12} \\ a_{21}+b_{21}+d_{21} & a_{22}+b_{22}+c_{22} \end{vmatrix}$$

$$-\begin{vmatrix} a_{11}+c_{11}+d_{11} & a_{12}+c_{12}+d_{12} \\ a_{21}+c_{21}+d_{21} & a_{22}+c_{22}+d_{22} \end{vmatrix} - \begin{vmatrix} b_{11}+c_{11}+d_{11} & b_{12}+c_{12}+d_{12} \\ b_{21}+c_{21}+d_{21} & b_{22}+c_{22}+d_{22} \end{vmatrix}$$

$$+\begin{vmatrix} a_{11}+b_{11} & a_{12}+b_{12} \\ a_{21}+b_{21} & a_{22}+b_{22} \end{vmatrix} + \begin{vmatrix} a_{11}+c_{11} & a_{12}+c_{12} \\ a_{21}+c_{21} & a_{22}+c_{22} \end{vmatrix} + \begin{vmatrix} a_{11}+d_{11} & a_{12}+d_{12} \\ a_{21}+d_{21} & a_{22}+d_{22} \end{vmatrix}$$

$$\begin{vmatrix} b_{11}+c_{11} & b_{12}+c_{12} \\ b_{21}+c_{21} & b_{22}+c_{22} \end{vmatrix} + \begin{vmatrix} b_{11}+d_{11} & d_{12}+b_{12} \\ b_{21}+d_{21} & b_{22}+d_{22} \end{vmatrix} + \begin{vmatrix} c_{11}+d_{11} & c_{12}+d_{12} \\ c_{21}+d_{21} & c_{22}+d_{22} \end{vmatrix}$$

$$-\begin{vmatrix} a_{11} & a_{12} \\ a_{21} & a_{22} \end{vmatrix} - \begin{vmatrix} b_{11} & b_{12} \\ b_{21} & b_{22} \end{vmatrix} - \begin{vmatrix} c_{11} & c_{12} \\ c_{21} & c_{22} \end{vmatrix} - \begin{vmatrix} d_{11} & d_{12} \\ d_{21} & d_{22} \end{vmatrix} = 0 ;$$

and putting $a_{ij}=b_{ij}=c_{ij}=d_{ij}=0$, $(i \neq j)$, we have

$$\begin{aligned} &(a_{11}+b_{11}+c_{11}+d_{11})(a_{22}+b_{22}+c_{22}+d_{22}) \\ &- (a_{11}+b_{11}+c_{11})(a_{22}+b_{22}+c_{22}) - (a_{11}+b_{11}+d_{11})(a_{22}+b_{22}+d_{22}) \\ &- (a_{11}+c_{11}+d_{11})(a_{22}+c_{22}+d_{22}) - (b_{11}+c_{11}+d_{11})(b_{22}+c_{22}+d_{22}) \\ &+ (a_{11}+b_{11})(a_{22}+b_{22}) + (a_{11}+c_{11})(a_{22}+c_{22}) + (a_{11}+d_{11})(a_{22}+d_{22}) \\ &+ (b_{11}+c_{11})(b_{22}+c_{22}) + (b_{11}+d_{11})(b_{22}+d_{22}) + (c_{11}+d_{11})(c_{22}+d_{22}) \\ &- a_{11}a_{22} - b_{11}b_{22} - c_{11}c_{22} - d_{11}d_{22} = 0, \end{aligned}$$

or

$$\begin{aligned} &- (a_{11}+b_{11}+c_{11}+d_{11})(a_{22}+b_{22}+c_{22}+d_{22}) \\ &- (a_{11}+b_{11}+c_{11},\ a_{11}+b_{11}+d_{11},\ a_{11}+c_{11}+d_{11}, \\ &\quad b_{11}+c_{11}+d_{11})(a_{22}+b_{22}+c_{22},\ a_{22}+b_{22}+d_{22},\ a_{22}+c_{22}+d_{22}, \\ &\quad b_{22}+c_{22}+d_{22}) - (a_{11}+b_{11},\ a_{11}+c_{11},\ a_{11}+d_{11},\ b_{11}+c_{11}, \\ &\quad b_{11}+d_{11},\ c_{11}+d_{11})(a_{22}+b_{22},\ a_{22}+c_{22},\ a_{22}+d_{22},\ b_{22}+c_{22}, \\ &\quad b_{22}+d_{22},\ c_{22}+d_{22}) - (a_{11},\ b_{11},\ c_{11},\ d_{11})(a_{22},\ b_{22},\ c_{22},\ d_{22} = 0) \end{aligned}$$

and finally by making $a_{11}=a_{22}=a$, $b_{11}=b_{22}=b$, $c_{11}=c_{22}=c$, $d_{11}=d_{22}=d$ there results the identity expressing the sum of seven squares as the sum of eight squares, namely,

$$(a+b+c+d)^2+(a+b)^2+(a+c)^2+(a+d)^2+(b+c)^2$$
$$+(b+d)^2+(c+d)^2=(a+b+c)^2+(a+b+d)^2+(a+c+d)^2$$
$$+(b+c+d)^2+a^2+b^2+c^2+d^2.$$

Example. If θ is an imaginary nth root of unity, show that for $n=2s+1$

$$(-)^{s+1}\sum(1+\theta+\theta^2+\cdots+\theta^{s-1})^{2s+1}+(-)^{s+2}\sum(1+\theta+\cdots+\theta^{s-2})^{2s+1}+\cdots+(-)^{2s-1}\sum(1+\theta)^{2s+1}+(-)^{2s}(2s+1)=\tfrac{1}{2}(2s+1)!$$

and for $n=2s$

$$(-)^{s}\sum(1+\theta+\theta^2+\cdots+\theta^{s-1})^{2s}+(-)^{s+1}\,2\sum(1+\theta+\cdots+\theta^{s-2})^{2s}+\cdots+(-)^{2s-2}\,2\sum(1+\theta)^{2s}+(-)^{2s-1}\,2.2s=(2s)!$$

106. *If any m rows of a determinant be selected, and every possible minor of the mth order be formed from them, and each minor be multiplied by the complementary of the corresponding minor formed from other m rows, and the sign + or − be affixed to the product according as the sum of the numbers indicating the rows and columns from which the complementary is formed be even or odd, the aggregate of the products thus obtained is equal to zero.*

Let $|a_{1n}|$ be the determinant, then the aggregate of products referred to is equal to a determinant of the nth order having for m of its rows the m rows from which the first factors are found, and for its other rows the $n-m$ rows from which the second factors are found. But the rows of the latter set cannot be all different from those of the former; for, if from n things a set of m be taken, and from the same n things another set of m, the $n-m$ left the second time must include one or more of those taken the first time. Hence, (§52) the aggregate of products is equal to zero.

Expressed symbolically this theorem is:

$$\sum_{1}^{\mu}\alpha(-1)^{\nu}A_{(n\,|\,m\alpha),(n\,|\,m\beta)}A_{(\bar{n}\,|\,m\alpha),(\bar{n}\,|\,m\gamma)}=0$$

where $\beta\neq\gamma$

Example. Taking the first and second rows of $|a_1b_2c_3d_4|$, we have the minors

$$|a_1\,b_2|\;,\quad |a_1\,b_3|\;,\quad |a_1\,b_4|\;,\quad |a_2\,b_3|\;,\quad |a_2\,b_4|\;,\quad |a_3\,b_4|\;;$$

and the corresponding minors formed from other two rows, the second and third, being

$$|b_1\,c_2|\;,\quad |b_1\,c_3|\;,\quad |b_1\,c_4|\;,\quad |b_2\,c_3|\;,\quad |b_2\,c_4|\;,\quad |b_3\,c_4|\;,$$

we have as their complementaries

$$|a_3\,d_4|\;,\quad |a_2\,d_4|\;,\quad |a_2\,d_3|\;,\quad |a_1\,d_4|\;,\quad |a_1\,d_3|\;,\quad |a_1\,d_2|\;:$$

then

$$-|a_1b_2||a_3d_4| + |a_1b_3||a_2d_4| + |a_1b_4||a_2d_3| + |a_2b_3||a_1d_4|$$
$$-|a_2b_4||a_1d_3| + |a_3b_4||a_1d_2| = 0,$$

being in fact equal to

$$|a_1\,b_2\,a_3\,d_4|\;.$$

The theorem here exemplified is seen to include that of §76 as the theorem of §93 includes that of §68.

107. *If in any determinant of the nth order there be m rows all having in the same places n − m zero constituents, the determinant is expressible as the product of two of its minors, namely the minor whose constituents are the remaining constituents of the m rows, and its complementary: the sign of the product being + or − according as the sum of the numbers indicating the rows and columns from which the minor is found is even or odd.*

Seeking to find a development of the determinant as an aggregate of products of complementary minors by Laplace's theorem, we see that there are in the m rows only m vertical lines of non-zero constituents, and that consequently there can be formed only one non-zero minor of the mth order. The products therefore all vanish except that arising from this minor and its complementary.

Example.

$$\begin{vmatrix} a_1 & a_2 & a_3 & a_4 & a_5 \\ b_1 & 0 & b_3 & 0 & b_5 \\ c_1 & 0 & c_3 & 0 & c_5 \\ d_1 & d_2 & d_3 & d_4 & d_5 \\ e_1 & 0 & e_3 & 0 & e_5 \end{vmatrix} = \begin{vmatrix} a_1 & a_2 & a_3 & a_4 & a_5 \\ d_1 & d_2 & d_3 & d_4 & d_5 \\ b_1 & 0 & b_3 & 0 & b_5 \\ c_1 & 0 & c_3 & 0 & c_5 \\ e_1 & 0 & e_3 & 0 & e_5 \end{vmatrix} = -\begin{vmatrix} a_2 & a_4 & a_1 & a_3 & a_5 \\ d_2 & d_4 & d_1 & d_3 & d_5 \\ 0 & 0 & b_1 & b_3 & b_5 \\ 0 & 0 & c_1 & c_3 & c_5 \\ 0 & 0 & e_1 & e_3 & e_5 \end{vmatrix},$$

$$= -\begin{vmatrix} a_2 & a_4 \\ d_2 & d_4 \end{vmatrix}\begin{vmatrix} b_1 & b_3 & b_5 \\ c_1 & c_3 & c_5 \\ e_1 & e_3 & e_5 \end{vmatrix}.$$

108. *If, in any determinant of the nth order there be m rows all having in the same places more than $n-m$ zero-elements, the determinant vanishes.*

109. In like manner we see that conversely, *the product of two determinants of the rth and sth orders may be expressed as a determinant of the $(r+s)$th order whose elements are* (1) *the r^2+s^2 elements of the two determinants, so placed that the said determinants may be complementary minors of the new determinant, and that the sum of the numbers of the rows and columns they occupy may be even,* (2) *rs zeros completing the rows in which the elements of one of these minors stand and* (3) *any rs finite elements whatever for the remaining places.* Thus

$$\begin{vmatrix} a_1 & a_2 \\ b_1 & b_2 \end{vmatrix} \begin{vmatrix} x_1 & x_2 \\ y_1 & y_2 \end{vmatrix} = \begin{vmatrix} a_1 & a_2 & w_1 & w_2 \\ b_1 & b_2 & w_3 & w_4 \\ 0 & 0 & x_1 & x_2 \\ 0 & 0 & y_1 & y_2 \end{vmatrix} = \begin{vmatrix} a_1 & 0 & a_2 & 0 \\ \pi_1 & x_1 & \pi_2 & x_2 \\ b_1 & 0 & b_2 & 0 \\ \pi_3 & y_1 & \pi_4 & y_2 \end{vmatrix}.$$

110. Cauchy's Theorem: *Any row and column of a determinant being selected, if the element common to them be multiplied by its cofactor in the determinant, and every product of another element of the row by another element of the column be multiplied by its cofactor, the sum of the results is equal to the given determinant.*

Let the selected row and column be the rth and sth respectively, then the theorem may be symbolically stated thus:

$$\Delta = a_{rs}\frac{\partial\Delta}{\partial a_{rs}} + \sum a_{ri}a_{ks}\frac{\partial^2\Delta}{\partial a_{ri}\,\partial a_{ks}} \qquad \begin{matrix}(i,k = 1,2,\cdots,n)\\ i \neq s\\ k \neq r\end{matrix}$$

$$= (-1)^{r+s}a_{rs}A_{rs} + \sum \pm\, a_{ri}a_{ks}A_{rk,is}.$$

The sign before $a_{ri}a_{ks}A_{rk,is}$ is determined by the formula $(-1)^{\nu_1+\nu_2}$ where ν_1 is the total number of inversions in the suffix and ν_2 denotes the sum of the numbers $r+i+k+s$.

The multiplication of a_{rs} by its cofactor gives all the terms of the determinant which contain a_{rs}. As each of the terms containing $a_{r,s-1}$ must also contain one of the constituents of the column and cannot contain a_{rs}, we see that by multiplying $a_{r,s-1}a_{1s}$, $a_{r,s-1}a_{2s}$, $\cdots$, $a_{r,s-1}a_{r-1,s}$, $a_{r,s-1}a_{r+1,s}$, $\cdots$, $a_{r,s-1}a_{ns}$, by their respective cofactor, we obtain all the terms containing $a_{r,s-1}$. Similarly by multiplying $a_{r,s-2}a_{1s}$, $a_{r,s-2}a_{2s}$, $\cdots$, $a_{r,s-2}a_{r-1,s}$, $a_{r,s-2}a_{r+1,s}$, $\cdots$, $a_{r,s-2}$, a_{ns}, by their respective cofactors, we obtain all the terms containing $a_{r,s-2}$. Consequently if we continue this process we shall finally have every term in which one of the constituents of the row occurs, that is to say, we shall have the full development of the determinant—and this was the theorem to be proved.

If $a_{rs}=0$, then $\Delta=\sum(-1)^{\nu_1+\nu_2}a_{ri}a_{ks}A_{rk,is}$.

EXAMPLE.

$$\begin{vmatrix} a_{11} & a_{12} & a_{13} & a_{14} \\ a_{21} & a_{22} & a_{23} & a_{24} \\ a_{31} & a_{32} & a_{33} & a_{34} \\ a_{41} & a_{42} & a_{43} & a_{44} \end{vmatrix} = a_{22}\begin{vmatrix} a_{11} & a_{13} & a_{14} \\ a_{31} & a_{33} & a_{34} \\ a_{41} & a_{43} & a_{44} \end{vmatrix} - a_{21}a_{12}\begin{vmatrix} a_{33} & a_{34} \\ a_{43} & a_{44} \end{vmatrix} + a_{21}a_{32}\begin{vmatrix} a_{13} & a_{14} \\ a_{43} & a_{44} \end{vmatrix}$$

$$- a_{21}a_{42}\begin{vmatrix} a_{13} & a_{14} \\ a_{33} & a_{34} \end{vmatrix} + a_{23}a_{12}\begin{vmatrix} a_{31} & a_{34} \\ a_{41} & a_{44} \end{vmatrix} - a_{23}a_{32}\begin{vmatrix} a_{11} & a_{14} \\ a_{41} & a_{44} \end{vmatrix}$$

$$+ a_{23}a_{42}\begin{vmatrix} a_{11} & a_{14} \\ a_{31} & a_{34} \end{vmatrix} - a_{24}a_{12}\begin{vmatrix} a_{31} & a_{33} \\ a_{41} & a_{43} \end{vmatrix} + a_{24}a_{32}\begin{vmatrix} a_{11} & a_{13} \\ a_{41} & a_{43} \end{vmatrix}$$

$$- a_{24}a_{42}\begin{vmatrix} a_{11} & a_{13} \\ a_{31} & a_{33} \end{vmatrix}.$$

111. The theorem of the preceding article is useful for expanding a determinant which has been bordered. For example let the determinant $\Delta = |a_{1n}|$ be bordered thus

$$\begin{vmatrix} \alpha_{00} & \alpha_{01} & \alpha_{02} & \alpha_{03} & \cdots & \alpha_{0n} \\ \alpha_{10} & a_{11} & a_{12} & a_{13} & \cdots & a_{1n} \\ \alpha_{20} & a_{21} & a_{22} & a_{23} & \cdots & a_{2n} \\ \cdots & \cdots & \cdots & \cdots & \cdots & \cdots \\ \alpha_{n0} & a_{n1} & a_{n2} & a_{n3} & \cdots & a_{nn} \end{vmatrix} = \Delta' \quad \text{say}$$

then

$$\Delta' = \alpha_{00}\Delta + \sum \alpha_{i0}\alpha_{0k}\frac{\partial\Delta}{\partial a_{ik}}$$

$$= \alpha_{00}\Delta + (-1)^{i+k+\nu}\sum \alpha_{i0}\alpha_{0k}A_{ik}$$

where ν = number of inversions in ik.

Again let the determinant

$$\Delta = \begin{vmatrix} x_1 & a_2 & a_3 \cdots a_n \\ a_1 & x_2 & a_3 \cdots a_n \\ a_1 & a_2 & x_3 \cdots a_n \\ a_1 & a_2 & a_3 \cdots a_n \\ \cdots & \cdots & \cdots \\ a_1 & a_2 & a_5 \cdots x_n \end{vmatrix}$$

be bordered thus

$$\begin{vmatrix} 1 & 0 & 0 & 0 \cdots 0 \\ 1 & x_1 & a_2 & a_3 \cdots a_n \\ 1 & a_1 & x_2 & a_3 \cdots a_n \\ 1 & a_1 & a_2 & x_3 \cdots a_n \\ \cdots & \cdots & \cdots & \cdots \\ 1 & a_1 & a_2 & a_3 \cdots x_n \end{vmatrix}$$

and before expanding let us subtract a_1 times the first column of the bordered determinant from the second column, a_2 times the first from the third, a_3 times the first from the fourth and so on. This will not alter the value of Δ and we have

$$\begin{vmatrix} 1 & -a_1 & -a_2 & -a_3 \cdots & -a_n \\ 1 & x_1 - a_1 & 0 & 0 \cdots & 0 \\ 1 & 0 & x_2 - a_2 & 0 \cdots & 0 \\ 1 & 0 & 0 & x_3 - a_3 \cdots & 0 \\ \cdots & \cdots & \cdots & \cdots & \cdots \\ 1 & 0 & 0 & 0 \cdots & x_n - a_n \end{vmatrix}$$

which when expanded by *Cauchy's* theorem gives

$$\Delta = \prod_1^n i(x_i - a_i) + \sum_1^n i\, a_i \frac{d}{dx_i} \prod_1^n i(x_i - a_i),$$

a form due to Sardi.

112. That the sum of the signed primary minors of the determinant $|a_{1n}|$ is

$$-\begin{vmatrix} 0 & 1 & \cdots 1 \\ 1 & a_{11} & \cdots a_{1n} \\ 1 & a_{21} & \cdots a_{2n} \\ \cdots & \cdots & \cdots \\ 1 & a_{n1} & \cdots a_{nn} \end{vmatrix} \equiv \Delta' \text{ say},$$

is readily seen on expanding this latter determinant by Cauchy's theorem in terms of products of the elements of the first row and first column.

If on Δ' we perform the following operations: col 2−col 3; col 3−col 4; col 4−col 5; · · · then row 2−row 3; row 3−row 4; row 4−row 5; · · · we see that the order reduces by two and we have

$$\Delta' = -\begin{vmatrix} a_{11}-a_{12}-a_{21}+a_{22} & a_{12}-a_{13}-a_{22}+a_{23} & a_{13}-a_{14}-a_{23}+a_{24} & \cdots \\ a_{21}-a_{22}-a_{31}+a_{32} & a_{22}-a_{23}-a_{32}+a_{33} & a_{23}-a_{24}-a_{33}+a_{34} & \cdots \\ a_{31}-a_{32}-a_{41}+a_{42} & a_{32}-a_{33}-a_{42}+a_{43} & a_{33}-a_{34}-a_{43}+a_{44} & \cdots \\ \cdots & \cdots & \cdots & \cdots \end{vmatrix}$$

The element in the position (r, s) being

$$a_{r,s} - a_{r,s+1} - a_{r+1,s} + a_{r+1,s+1}.$$

It is obvious that we might obtain different forms of order $n-1$ for Δ', the law of formation for the element in the position $(r+1, s+1)$ being

$$a_{r+1,s+1} - a_{k,s+1} - a_{r+1,h} + a_{k,h}$$

where $k \neq r+1$ and $h \neq s+1$.

113. *If the elements of a determinant Δ are all increased by the same quantity w, the determinant is thereby increased by w times the sum of the signed primary minors.*

$$\Delta' = \begin{vmatrix} 1 & 1 & 1 & 1 & 1 \\ 0 & a_{11}+w & a_{12}+w & a_{13}+w & a_{14}+w \\ 0 & a_{21}+w & a_{22}+w & a_{23}+w & a_{24}+w \\ 0 & a_{31}+w & a_{32}+w & a_{33}+w & a_{34}+w \\ 0 & a_{41}+w & a_{42}+w & a_{43}+w & a_{44}+w \end{vmatrix}$$

$$= \begin{vmatrix} 1 & 1 & 1 & 1 & 1 \\ -w & a_{11} & a_{12} & a_{13} & a_{14} \\ -w & a_{21} & a_{22} & a_{23} & a_{24} \\ -w & a_{31} & a_{32} & a_{33} & a_{34} \\ -w & a_{41} & a_{42} & a_{43} & a_{44} \end{vmatrix} = \Delta - w \begin{vmatrix} 0 & 1 & \cdots & 1 \\ 1 & a_{11} & \cdots & a_{14} \\ 1 & a_{21} & \cdots & a_{24} \\ 1 & a_{31} & \cdots & a_{34} \\ 1 & a_{41} & \cdots & a_{44} \end{vmatrix}.$$

114. If from a determinant $A = |a_{nn}|$ a new determinant be formed by altering the order of p of the elements of the first row and then

multiplying these altered elements by $e_1, e_2, \cdots, e_p$ respectively, a second determinant by making the same alteration of order in the corresponding elements of the second row and multiplying by $e_1, e_2, \cdots, e_p$ as in the first row, and so on, then the sum of the n determinants thus derived is $(n-p)$ times the original determinant.

Without loss of generality we may assume that the p elements which have their order changed are the first and let $\alpha_1, \alpha_2, \cdots, \alpha_p$ be the numbers 1, 2, 3, $\cdots$, p in their changed order, then the 1st, 2nd, $\cdots$, nth rows of the 1st, 2nd, $\cdots$, nth determinants respectively would be

$$\begin{array}{cccccc} e_1 a_{1\alpha_1} & e_2 a_{1\alpha_2} \cdots & e_p a_{1\alpha p} & a_{1,p+1} \cdots & a_{1n}, \\ e_1 a_{2\alpha_1} & e_2 a_{2\alpha_2} \cdots & e_p a_{2\alpha p} & a_{2,p+1} \cdots & a_{2n}, \\ \cdots & \cdots & \cdots & \cdots & \cdots \\ e_1 a_{n\alpha_1} & e_2 a_{n\alpha_2} \cdots & e_p a_{n\alpha p} & a_{n,p+1} \cdots & a_{nn}. \end{array}$$

Expanding these n determinants in terms of the elements in the altered rows and their complementaries we have

$$\begin{array}{l} e_1 a_{1\alpha_1}\mathcal{A}_{11}+e_2 a_{1\alpha_2}\mathcal{A}_{12}+\cdots+e_p a_{1\alpha p}\mathcal{A}_{1p}+a_{1,p+1}\mathcal{A}_{1,p+1}\cdots+a_{1n}\mathcal{A}_{1n} \\ e_1 a_{2\alpha_1}\mathcal{A}_{21}+e_2 a_{2\alpha_2}\mathcal{A}_{12}+\cdots+e_p a_{2\alpha p}\mathcal{A}_{2p}+a_{2,p+1}\mathcal{A}_{2,p+1}\cdots+a_{2n}\mathcal{A}_{2n} \\ \cdots\cdots\cdots\cdots\cdots\cdots\cdots\cdots\cdots\cdots\cdots\cdots \\ e_1 a_{n\alpha_1}\mathcal{A}_{n1}+e_2 a_{n\alpha_2}\mathcal{A}_{n2}+\cdots+e_p a_{n\alpha p}\mathcal{A}_{np}+a_{n,p+1}\mathcal{A}_{n,p+1}\cdots+a_{nn}\mathcal{A}_{nn} \end{array}$$

Summing these column-wise we readily see that each of the first p columns is zero and each of the last $n-p$ columns gives A, so that the sum of the n determinants is $(n-p)A$.

When $p=n$, then the sum is zero.

We may apply the theorem to each of the n determinants while keeping one row exempt from further change.

Similar equalities may be obtained by operating on columns instead of rows.

115. If from any n-line determinant Δ we form another Δ' such that the rth row of Δ' is the sum of all the rows of Δ except the rth, then

$$\Delta' = (-)^{n-1}(n-1)\Delta$$

For example

$$\begin{vmatrix} x'+x'' & y'+y'' & z'+z'' \\ x''+x & y''+y & z''+z \\ x+x' & y+y' & z+z' \end{vmatrix} = 2\begin{vmatrix} x & y & z \\ x' & y' & z' \\ x'' & y'' & z'' \end{vmatrix}$$

The theorem is readily established by performing the following operations in order

$$\begin{aligned}&\text{row } 1 + \text{row } 2 + \cdots + \text{row } n,\\ &\text{row } 1 \div (n-1),\\ &\text{row } 2 - \text{row } 1, \text{row } 3 - \text{row } 1, \cdots,\\ &\text{row } 1 + \text{row } 2 + \cdots + \text{row } n.\end{aligned}$$

116. If from a determinant Δ of the nth order, we form another Δ' such that the first row of Δ' is the sum of the first p rows of Δ and every other row of Δ' is got by subtracting the corresponding row of Δ from the row preceding it in Δ then

$$\Delta' = (-1)^{n-1} p\Delta.$$

The proof of this may be brought about by performing the following operations upon Δ'. Row $1+(p-1)$ row $2+(p-2)$ row $3+\cdots$ $+$row p, which will give p times row 1. Take out the factor p and then add the first row to the second, then the second to the third, and so on. The elements of Δ' will now be the same as those of Δ except that those of the last $n-1$ rows are negative.

117. The determinant Δ' got from $\Delta \equiv |a_{1n}|$ by diminishing every row by the sum of all the other rows is equal to

$$-(n-2)2^{n-1}|a_{1n}|.$$

The truth of this may be seen on performing the following operations: row $n-$row $(n-1)$, row $(n-1)-$row $(n-2)$, $\cdots$, row 3 $-$row 2, row $2-$row 1; then take out the 2 which will be a factor of all rows except the first. Next perform the operations row $1+(n-1)$ row $2+(n-2)$ row $3+\cdots+$row n and $-(n-2)$ times the first row of $|a_{1n}|$ will be the result. Take out this factor $-(n-2)$ and perform the operations row $2+$row 1, row $3+$row 2, $\cdots$, row $n+$row $(n-1)$ in order.

118. *If any n-line determinant has all the elements in any p by q array multiplied by x and all the elements in the complementary array divided by x, then the determinant is multiplied by x^{p+q-n}.*

If the array has p rows and q columns we may multiply the remaining $n-q$ columns by x and then the result will be p rows with x a factor of all the elements in those rows and hence the truth of the theorem.

It is readily seen that x^{p+q-n} will be a factor even though the elements of the complementary array are not divided by x.

PROBLEM. Use this theorem to show that

$$|a_1 b_2 c_3 d_4 e_5| = \frac{\begin{vmatrix} 12345 & 21345 & 32145 & 42315 & 52341 \\ 21345 & 21345 & 23145 & 24315 & 25341 \\ 32145 & 31245 & 32145 & 32415 & 32541 \\ 42315 & 41325 & 42135 & 42315 & 42351 \\ 52341 & 51342 & 52134 & 52314 & 52341 \end{vmatrix}}{21345 \cdot 32145 \cdot 42315 \cdot 52341}$$

where $pqrst$ stands for the term $a_p b_q c_r d_s e_t$ of the determinant $|a_1 b_2 c_3 d_4 e_5|$.

In general, *any n-line determinant can be expressed by means of* $1+(n-1)^2$ *of its terms.*

If we use $\overline{rs}$ to denote the result of performing on 12345 the interchange of r and s, and $\overline{rs}\cdot\overline{uv}$ to denote the result of performing on $\overline{rs}$ the interchange of u and v, then the foregoing may be written as follows:

$$|a_1 b_2 c_3 d_4 e_5| = \frac{\begin{vmatrix} \overline{11} & \overline{12} & \overline{13} & \overline{14} & \overline{15} \\ \overline{21} & \overline{21} \text{ or } \overline{12} & \overline{21}\cdot\overline{13} & \overline{21}\cdot\overline{14} & \overline{21}\cdot\overline{15} \\ \overline{31} & \overline{31}\cdot\overline{12} & \overline{31} \text{ or } \overline{13} & \overline{31}\cdot\overline{14} & \overline{31}\cdot\overline{15} \\ \overline{41} & \overline{41}\cdot\overline{12} & \overline{41}\cdot\overline{13} & \overline{41} \text{ or } \overline{14} & \overline{41}\cdot\overline{15} \\ \overline{51} & \overline{51}\cdot\overline{12} & \overline{51}\cdot\overline{13} & \overline{51}\cdot\overline{14} & \overline{51} \text{ or } \overline{15} \end{vmatrix}}{\overline{12}\cdot\overline{13}\cdot\overline{14}\cdot\overline{15}}$$

119. *If the successive diagonals on the upper side of and parallel to the main diagonal of any determinant be multiplied by* $x, x^2, x^3, \cdots$ *respectively and those on the underside by* $x^{-1}, x^{-2}, x^{-3}, \cdots$ *respectively, the value of the determinant remains unaltered.*

This is equivalent to multiplying the element in the (r, s)th place by x^{s-r} and therefore equivalent to multiplying any term by $x^{\Sigma s - \Sigma r} = x^0$.

When $x = -1$, this is the same as saying that a determinant is not altered by changing the sign of every element whose place-indexes have an odd sum.

120. The product of the two determinants $|a_{11}a_{22}a_{33}a_{44}|$ and $|b_{11}b_{22}|$ may be written as a determinant of the 6th order, thus

$$\begin{vmatrix} a_{11} & a_{12} & a_{13} & a_{14} & \cdot & \cdot \\ a_{21} & a_{22} & a_{23} & a_{34} & \cdot & \cdot \\ a_{31} & a_{32} & a_{33} & a_{34} & \cdot & \cdot \\ a_{41} & a_{42} & a_{43} & a_{44} & \cdot & \cdot \\ b_{11} & b_{12} & b_{13} & \cdot & b_{11} & b_{12} \\ b_{21} & b_{22} & b_{23} & \cdot & b_{21} & b_{22} \end{vmatrix},$$

where dots are used instead of zeros.*

This becomes, after performing the operations, $\text{col}_6 - \text{col}_2$; $\text{col}_5 - \text{col}_1$,

$$\begin{vmatrix} a_{11} & a_{12} & a_{13} & a_{14} & -a_{11} & -a_{12} \\ a_{21} & a_{22} & a_{23} & a_{24} & -a_{21} & -a_{22} \\ a_{31} & a_{32} & a_{33} & a_{34} & -a_{31} & -a_{32} \\ a_{41} & a_{42} & a_{43} & a_{44} & -a_{41} & -a_{42} \\ b_{11} & b_{12} & b_{13} & \cdot & \cdot & \cdot \\ b_{21} & b_{22} & b_{23} & \cdot & \cdot & \cdot \end{vmatrix}$$

Expanding in terms of minors from the last 3 columns and their complementaries we get

$$-\left| a_{11}\, a_{22}\, a_{34} \right| \begin{vmatrix} a_{41} & a_{42} & a_{43} \\ b_{11} & b_{12} & b_{13} \\ b_{21} & b_{22} & b_{23} \end{vmatrix} + \left| a_{11}\, a_{22}\, a_{44} \right| \begin{vmatrix} a_{31} & a_{32} & a_{33} \\ b_{11} & b_{12} & b_{13} \\ b_{21} & b_{22} & b_{23} \end{vmatrix}$$

$$-\left| a_{11}\, a_{32}\, a_{44} \right| \begin{vmatrix} a_{21} & a_{22} & a_{23} \\ b_{11} & b_{12} & b_{13} \\ b_{21} & b_{22} & b_{23} \end{vmatrix} + \left| a_{21}\, a_{32}\, a_{44} \right| \begin{vmatrix} a_{11} & a_{12} & a_{13} \\ b_{11} & b_{12} & b_{13} \\ b_{21} & b_{22} & b_{23} \end{vmatrix}.$$

If before expansion we had inserted b_{14}, b_{24} in their places in the last two lines the original product would not be changed but the expansion would be

$$\begin{aligned} & \left| a_{31}\, a_{42}\, b_{13}\, b_{24} \right| \left| a_{11}\, a_{22} \right| - \left| a_{21}\, a_{42}\, b_{13}\, b_{24} \right| \left| a_{11}\, a_{32} \right| \\ + & \left| a_{11}\, a_{42}\, b_{13}\, b_{24} \right| \left| a_{21}\, a_{32} \right| - \left| a_{11}\, a_{32}\, b_{13}\, b_{24} \right| \left| a_{21}\, a_{42} \right| \\ + & \left| a_{11}\, a_{22}\, b_{13}\, b_{24} \right| \left| a_{31}\, a_{42} \right| + \left| a_{21}\, a_{32}\, b_{13}\, b_{24} \right| \left| a_{11}\, a_{42} \right|. \end{aligned}$$

* This method of indicating zeros by dots will hereafter be generally used, where no confusion can thereby arise.

So in general, writing $|a_{1p}|\,|b_{1q}|$ in the form

$$\begin{vmatrix} a_{11} & \cdots & a_{1p} & 0 & \cdots & 0 \\ a_{21} & \cdots & a_{2p} & 0 & \cdots & 0 \\ \cdot & \cdot & \cdot & \cdot & \cdot & \cdot \\ a_{p1} & \cdots & a_{pp} & 0 & \cdots & 0 \\ b_{11} \cdots b_{1k} & \cdots & 0 & b_{11} & \cdots & b_{1q} \\ \cdot & \cdot & \cdot & \cdot & \cdot & \cdot \\ b_{q1} \cdots b_{qk} & \cdots & 0 & b_{q1} & \cdots & b_{qq} \end{vmatrix}$$

where $q \leqq k \leqq p$ and subtracting the

1st col. from the $(p+1)$th
2nd col. from the $(p+2)$nd
. . . .
qth col. from the $(p+q)$th

we get

$$(-1)^q \begin{vmatrix} a_{11} & \cdots & a_{1p} & a_{11} & \cdots & a_{1q} \\ a_{21} & \cdots & a_{2p} & a_{21} & \cdots & a_{2q} \\ \cdot & \cdot & \cdot & \cdot & \cdot & \cdot \\ a_{p1} & \cdots & a_{pp} & a_{p1} & \cdots & a_{pq} \\ b_{11} \cdots b_{1k} & \cdots & 0 & 0 & \cdots & 0 \\ \cdot & \cdot & \cdot & \cdot & \cdot & \cdot \\ b_{q1} \cdots b_{qk} & \cdots & 0 & 0 & \cdots & 0 \end{vmatrix}$$

Expanding this by Laplace's theorem in terms of minors of order $p+q-k$ formed from the last $p+q-k$ columns and their complementaries gives us the theorem that this expansion is equal to the product of the two determinants $|a_{1p}|\,|b_{1q}|$. Since k can take $p-q$ different values we would have $p-q$ different forms in which this product might be expressed and since all of these forms are equal to the same product any two may be placed equal to each other.

121. It is readily seen that

$$(1)\quad |a_1\, b_2\, c_3\, d_4\, e_5| = \begin{vmatrix} a_1 & a_2 & a_3 & \cdot & \cdot \\ b_1 & b_2 & b_3 & b_4 & b_5 \\ c_1 & c_2 & c_3 & c_4 & c_5 \\ d_1 & d_2 & d_3 & d_4 & d_5 \\ e_1 & e_2 & e_3 & e_4 & e_5 \end{vmatrix} + \begin{vmatrix} \cdot & \cdot & \cdot & a_4 & a_5 \\ b_1 & b_2 & b_3 & b_4 & b_5 \\ c_1 & c_2 & c_3 & c_4 & c_5 \\ d_1 & d_2 & d_3 & d_4 & d_5 \\ e_1 & e_2 & e_3 & e_4 & e_5 \end{vmatrix}$$

$$= \begin{vmatrix} a_1 & a_2 & a_3 & \cdot & \cdot \\ b_1 & b_2 & b_3 & b_4 & b_5 \\ c_1 & c_2 & c_3 & c_4 & c_5 \\ d_1 & d_2 & d_3 & d_4 & d_5 \\ e_1 & e_2 & e_3 & e_4 & e_5 \end{vmatrix} + \begin{vmatrix} a_1 & a_2 & a_3 & a_4 & a_5 \\ b_1 & b_2 & b_3 & \cdot & \cdot \\ c_1 & c_2 & c_3 & c_4 & c_5 \\ d_1 & d_2 & d_3 & d_4 & d_5 \\ e_1 & e_2 & e_3 & e_4 & e_5 \end{vmatrix}$$

$$(2) \qquad - \begin{vmatrix} a_1 & a_2 & a_3 & \cdot & \cdot \\ b_1 & b_2 & b_3 & \cdot & \cdot \\ c_1 & c_2 & c_3 & c_4 & c_5 \\ d_1 & d_2 & d_3 & d_4 & d_5 \\ e_1 & e_2 & e_3 & e_4 & e_5 \end{vmatrix} + \begin{vmatrix} \cdot & \cdot & \cdot & a_4 & a_5 \\ \cdot & \cdot & \cdot & b_4 & b_5 \\ c_1 & c_2 & c_3 & c_4 & c_5 \\ d_1 & d_2 & d_3 & d_4 & d_5 \\ e_1 & e_2 & e_3 & e_4 & e_5 \end{vmatrix}$$

$$\text{or} = \begin{vmatrix} \cdot & \cdot & \cdot & a_4 & a_5 \\ b_1 & b_2 & b_3 & b_4 & b_5 \\ c_1 & c_2 & c_3 & c_4 & c_5 \\ d_1 & d_2 & d_3 & d_4 & d_5 \\ e_1 & e_2 & e_3 & e_4 & e_5 \end{vmatrix} + \begin{vmatrix} a_1 & a_2 & a_3 & a_4 & a_5 \\ \cdot & \cdot & \cdot & b_4 & b_5 \\ c_1 & c_2 & c_3 & c_4 & c_5 \\ d_1 & d_2 & d_3 & d_4 & d_5 \\ e_1 & e_2 & e_3 & e_4 & e_5 \end{vmatrix}$$

$$(3) \qquad - \begin{vmatrix} \cdot & \cdot & \cdot & a_4 & a_5 \\ \cdot & \cdot & \cdot & b_4 & b_5 \\ c_1 & c_2 & c_3 & c_4 & c_5 \\ d_1 & d_2 & d_3 & d_4 & d_5 \\ e_1 & e_2 & e_3 & e_4 & e_5 \end{vmatrix} + \begin{vmatrix} a_1 & a_2 & a_3 & \cdot & \cdot \\ b_1 & b_2 & b_3 & \cdot & \cdot \\ c_1 & c_2 & c_3 & c_4 & c_5 \\ d_1 & d_2 & d_3 & d_4 & d_5 \\ e_1 & e_2 & e_3 & e_4 & e_5 \end{vmatrix}.$$

The general theorem exemplified in these examples may be stated as follows:

If Δ denote any n-line determinant with its first r rows conceived as separated into two arrays, namely, an r-by-s array ($r \leqq s$) called P and an r-by-($n-s$) array called Q, and if $\sum \Delta_{p1}$ be the sum of all the determinants got from Δ by zero-ising in it a single row of P, $\sum \Delta_{p2}$ the similar sum got by zero-ising in Δ two rows of P, and so on, then

$$\Delta = \sum \Delta_{p1} - \sum \Delta_{p2} + \sum \Delta_{p3} = \cdots + (-1)^{r+1}\Delta_{pr} - (-1)^{r+1}\Delta_{qr},$$

where of course Δ_{qr} is the determinant obtained from Δ by zero-ising all r rows of Q.

The truth of this general theorem is seen on expanding the terms on the right in terms of minors of order s formed from the first s columns and their complementaries.

There is of course an alternative statement for this theorem corresponding to (2) and (3). Thus

$$\Delta = \sum\Delta_{q1} - \sum\Delta_{q2} + \cdots + (-1)^{r+1}\Delta_{qr} - (-1)^{r+1}\Delta_{pr}.$$

122. If the sum of the elements in every row of a determinant $A \equiv |a_{1n}|$ have the same value then

$$A_{1r} - A_{1s} = (-1)^{r+s-1}\, S.Q,$$

where A_{1k} is the signed complementary minor of a_{1k} in the determinant, S is the sum of the elements in each row, and Q is the determinant formed from A by deleting the first row and the rth and sth columns and inserting a column of units in the first place.

If we bring the column of A_{1r} headed by a_{2s} and the column of A_{1s} headed by a_{2r} to the first place the two differ by the first column only and may therefore be added and if in the result we add all the other columns to the first then S will be the values of all the elements in this column. After S is removed the determinant Q remains.

Exercises. Set IX

1. Show that if when x is put equal to a in the determinant Δ the elements of p columns become identical or proportional, then Δ is divisible by $(x-a)^{p-1}$.

2. If $s=a_1+a_2+\cdots+a_u$, and $A_k=s-a_k$ show that:

(a)
$$\begin{vmatrix} x-A_1 & a_2 & \cdots & a_u \\ a_1 & x-A_2 & \cdots & a_u \\ \cdot & \cdot & \cdot & \cdot \\ a_1 & a_2 & \cdots & x-A_u \end{vmatrix} = x(x-s)^{u-1}$$

(b)
$$\begin{vmatrix} x-a_1 & A_2 & \cdots & A_u \\ A_1 & x-a_2 & \cdots & A_u \\ \cdot & \cdot & \cdot & \cdot \\ A_1 & A_2 & \cdots & x-a_u \end{vmatrix} = \{x-(u-2)s\}(x-s)^{u-1}$$

3. Show that

$$\begin{vmatrix} a_1 & \cdot & \cdot & b_1 & \cdot & \cdot \\ \cdot & a_2 & \cdot & \cdot & b_2 & \cdot \\ \cdot & \cdot & a_3 & \cdot & \cdot & b_3 \\ c_1 & c_2 & c_3 & f_1 & f_2 & f_3 \\ d_1 & d_2 & d_3 & g_1 & g_2 & g_3 \\ e_1 & e_2 & e_3 & h_1 & h_2 & h_3 \end{vmatrix} = \begin{vmatrix} a_1f_1-b_1c_1 & a_2f_2-b_2c_2 & a_3f_3-b_3c_3 \\ a_1g_1-b_1d_1 & a_2g_2-b_2d_2 & a_3g_3-b_3d_3 \\ a_1h_1-b_1e_1 & a_2h_2-b_2e_2 & a_3h_3-b_3e_3 \end{vmatrix}$$

4. If

$$Q_n = \begin{vmatrix} 1 & a & a(a+d) & a(a+d)(a+2d) \cdots\cdots \\ 1 & a+d & (a+d)\ (a+2d) & (a+d)\ (a+2d)(a+3d) \cdots \\ 1 & a+2d & (a+2d)(a+3d) & (a+2d)(a+3d)(a+4d) \cdots \\ \cdot & \cdot & \cdot & \cdot \end{vmatrix}$$

show that

$$Q_n = d^{n-1}(n-1)! \cdot d^{n-2}(n-2)! \cdots d^2 2! \cdot d \cdot 1!$$

123. As in §110 we expanded a determinant by products in pairs of the constituents of a row and column so now we may expand it by products in pairs of the minors formed from the constituents in two or more rows and two or more columns.

Expanding the determinant Δ, by §93 we have

$$\Delta = A_{12,12}A'_{12,12} - A_{12,13}A'_{12,13} + A_{12,14}A'_{12,14} - \cdots$$
$$+ A_{12,23}A'_{12,23} - A_{12,24}A'_{12,24} + \cdots$$
$$+ A_{12,34}A_{12,34} - A_{12,35}A'_{12,35} + \cdots + \text{etc.}$$

$$= A_{12,12}A'_{12,12} + \sum_3^n i(-1)^{\nu_1}A_{12,1i}A'_{12,1i}$$
$$+ \sum_3^n i(-1)^{\nu_2}A_{12,2i}A'_{12,2i}$$
$$+ \sum_3^{n-1} i \sum_4^n k(-1)^{\nu_3}A_{12,ik}A'_{12,ik}$$

where $k > i$ and $\nu_1 = i+4$, $\nu_2 = i+5$, $\nu_3 = i+k+3$, and since

$$A_{12,ik} = \sum_{\sigma=4,\rho=3}^{\sigma=n,\rho=n-1} (-1)^{\nu_4}A'_{\rho\sigma,12}A_{12\rho\sigma,12ik}$$

where $\nu_4 = \rho+\sigma+3$, and $\sigma > \rho$, we have

$$\Delta = A_{12,12}A'_{12,12} + \sum_3^n i(-1)^{\nu_1}A_{12,1i}A'_{12,1i} + \sum_3^n i(-1)^{\nu_2}A_{12,2i}A'_{12,2i}$$
$$+ \sum_{k=4,i=3}^{k=n,i=n-1} \sum_{\sigma=4,\rho=3}^{\sigma=n,\rho=n-1} (-1)^{\nu_3+\nu_4}A'_{12,ik}A'_{\rho\sigma,12}A_{12\rho\sigma,12ik};$$

which is the expansion by products in pairs of the constituents of the first two rows and the first two columns.

If $a_{11}=a_{12}=a_{21}=a_{22}=0$, then

$$\Delta = \sum_{k=4,\,i=3}^{k=n,\,i=n-1} \sum_{\sigma=4,\,\rho=3}^{\sigma=n,\,\rho=n-1} (-1)^{\nu_3+\nu_4} A'_{12,ik} A'_{\rho\sigma,12} A_{12\rho\sigma,12ik}.$$

124. In a determinant of the nth order it is evident that the number of coaxial minors of the rth order is the number of ways of taking n things r at a time or n_r. The total number of coaxial minors of all orders is therefore

$$n_1 + n_2 + n_3 + \cdots + n_{n-1} + n_n = 2^n - 1.$$

125. CAYLEY'S THEOREM. *Any determinant of the nth order may be developed in a series of terms, the first of which is got from the given determinant by changing all the elements of the principal diagonal into zero, the next n by multiplying each element of the principal diagonal by its cofactor in the determinant and altering the said cofactors as the given determinant was altered, the next* $\frac{1}{2}n(n-1)$ *by multiplying the product of each pair of elements of the principal diagonal by its cofactor altered as before, and so on, the last term being simply the product of the elements of the principal diagonal.*

By taking the given determinant and changing all the elements of the principal diagonal into zero, we delete all the terms containing any of these elements; also no other terms are thereby deleted. Thus the altered determinant represents the sum of the terms which are independent of the elements in question.

By multiplying any element of the principal diagonal by its cofactor we obtain exactly all the terms containing that element. This cofactor, however, has its principal diagonal composed of elements from the original principal diagonal. If, therefore, we change these into zero, the altered product will represent the sum of the terms which contain the element in question and none of its fellows in the principal diagonal.

In this way it is seen that the expansion specified in the theorem gives, first, all the terms of the determinant involving *no* element of the principal diagonal; secondly, all those involving only *one element;* thirdly, all those involving only *two* elements, and so on; so that the full number of terms is in the end obtained.

It is worthy of notice that there is a break in the series: we pass from those terms involving only $n-2$ elements of the diagonal to the term involving them all, there being no term involving only $n-1$ of them.

EXAMPLE:

$$\begin{vmatrix} a_{11} & a_{12} & a_{13} \\ a_{21} & a_{22} & a_{23} \\ a_{31} & a_{32} & a_{33} \end{vmatrix} = \begin{vmatrix} 0 & a_{12} & a_{13} \\ a_{21} & 0 & a_{23} \\ a_{31} & a_{32} & 0 \end{vmatrix} + a_{11}\begin{vmatrix} 0 & a_{23} \\ a_{32} & 0 \end{vmatrix} + a_{22}\begin{vmatrix} 0 & a_{13} \\ a_{31} & 0 \end{vmatrix}$$

$$+ a_{33}\begin{vmatrix} 0 & a_{12} \\ a_{21} & 0 \end{vmatrix} + a_{11}a_{22}a_{33}.$$

A determinant having the elements along the principal diagonal all zero has been called by Sylvester an *invertebrate* or *zero-axial determinant.*

126. The theorem of the last article may be described as giving an expression for a determinant in terms of its own *devertebrated* (if we may so term them) coaxial minors and its principal diagonal elements. Exclusive of the elements of the principal diagonal there are 2^n-1-n coaxial minors and if we use Cayley's expansion theorem in connection with each of these we get 2^n-1-n equations, linear with respect to the devertebrated minors. So that on solving for the latter, there must result an expression for each devertebrated coaxial minor in terms of the vertebrated coaxial minors and the principal diagonal elements.

The general theorem thus obtained is

$$\begin{vmatrix} 0 & a_{12} & a_{13} & \cdots & a_{1n} \\ a_{21} & 0 & a_{23} & \cdots & a_{2n} \\ a_{31} & a_{32} & 0 & \cdots & a_{3n} \\ \cdot & \cdot & \cdot & \cdots & \cdot \\ a_{u1} & a_{n2} & a_{n3} & \cdots & 0 \end{vmatrix} = (a_{11}\,a_{22}\cdots a_{nn}) - \sum a_{11}(a_{22}\,a_{33}\cdots a_{nn}) + \sum a_{11}\,a_{22}(a_{33}\cdots a_{nn}) + \cdots$$

It may be viewed as a sort of converse of Cayley's, which in outward form it very closely resembles.

127. The determinant

$$\Delta \equiv \begin{vmatrix} a_{11}-x_{11} & a_{12} & \cdots & a_{1n} \\ a_{21} & a_{22}-x_{22} & \cdots & a_{2n} \\ \cdot & \cdot & \cdots & \cdot \\ a_{n1} & a_{n2} & \cdots, & a_{nn}-x_{nn} \end{vmatrix}$$

may be looked upon as a special case of the determinant

$$\begin{vmatrix} a_{11}+x_{11} & a_{12}+x_{12} \cdots, a_{1n}+x_{1n} \\ \cdot & \cdot \quad \cdots \quad \cdot \\ a_{n1}+x_{n1} & a_{n2}+x_{n2} \cdots a_{nn}+x_{nn} \end{vmatrix}$$

in which every x_{ij} $(i\neq j)$ is zero.

It may therefore, by §44, be partitioned into 2^n determinants with monomial elements. The term independent of the x's is $|a_{1n}|$ or A, the coefficient of x_{ii} is A_{ii}, the coefficient of $x_{ii}x_{jj}$, is $A_{ij,\,ij}$ and so in general the coefficient of $x_{ii}x_{jj}\cdots x_{kk}$ is $(-1)^k A_{ij\cdots k,\, ij\cdots k}$, and we have

$$\begin{aligned} \Delta = A - \sum x_{11}A_{11} + \sum x_{11}x_{22}A_{12,12} - \sum x_{11}x_{22}x_{33}A_{123,123} \\ (1) \qquad + (-1)^k \sum x_{11}x_{22}\cdots x_{kk}A_{12\cdots k,12\cdots k} \\ + \cdots + (-1)^n x_{11}\cdots x_{nn}. \end{aligned}$$

If $x_{11}=x_{22}=\cdots=x_{nn}=x$, then

$$\begin{aligned} \Delta = A - x\sum A_{11} + x^2\sum A_{12,12} + \cdots + (-1)^k x^k \sum A_{1\cdots k,1\cdots k} \\ (2) \qquad + \cdots + (-1)^n x^n. \end{aligned}$$

If in Δ we put $x_{ii}=a_{ii}$ $(i=1, 2, \cdots, n)$ then Δ becomes an invertebrate or zero-axial determinant which we may denote by $\Delta_{0,n}$ and (1) becomes

$$\begin{aligned} (3) \qquad \Delta_{0,n} = A - \sum a_{11}A_{11} + \sum a_{11}a_{22}A_{12,12} + \cdots + (-1)^k \\ \sum a_{11}\cdots a_{kk}A_{1\cdots k,1\cdots k} \\ + \cdots + (-1)^{n-1}a_{11}a_{22}\cdots a_{nn}. \end{aligned}$$

This suggests a simple way of obtaining Cayley's expansion theorem as follows:

Expanding as a determinant with binomial elements

$$A \equiv \begin{vmatrix} (a_{11}-x_{11})+x_{11} & a_{12} & \cdots & a_{1n} \\ a_{11} & (a_{22}-x_{22})+x_{22} & \cdots & a_{2n} \\ \cdots & \cdots & \cdots & \cdots \\ a_{n1} & a_{n2} & \cdots & (a_{nn}-x_{nn})+x_{nn} \end{vmatrix}$$

we have

$$\begin{aligned} (4) \qquad A \equiv \Delta + \sum x_{11}\Delta_{11} + \sum x_{11}x_{22}\Delta_{12,12} \\ + \cdots + \sum x_{11}x_{22}\cdots x_{kk}\Delta_{12\cdots k,12\cdots k} + \cdots + x_{11}x_{22}\cdots x_{nn}. \end{aligned}$$

If now we put $x_{ii}=a_{ii}$ $(i=1, 2, \cdots, n)$ we have Cayley's theorem, for the determinants on the right of (4) all become invertebrate coaxial minors of A.

128. If in Δ we put $x_{ii}=-a_{ii}$ $(i=r_1, r_2, \cdots, r_k)$ and $x_{ii}=0$ for $(i=r_{k+1}, \cdots, r_n)$ where $r_1, r_2, \cdots, r_n$ are the numbers $1, 2, \cdots, n$ in some order and denote by $\Delta_{0,k}$ the resulting determinant, then equation (1) §127 becomes

$$\Delta_{0,k} = A - \sum a_{r_1r_1}A_{r_1r_1} + \sum a_{r_1r_1}a_{r_2r_2}A_{r_1r_2,r_1r_2} + \cdots + (-1)^k a_{r_1r_1}\cdots a_{r_kr_k}A_{r_1r_2\cdots r_k,r_1r_2\cdots r_k}.$$

If $k=n$ this becomes (3) §127.

PROBLEMS. Prove the truth of the general theorem as exemplified in the example

$$|a_1\,b_2\,c_3| = \sum\begin{vmatrix} x & a_2 & a_3 \\ b_1 & b_2 & b_3 \\ c_1 & c_2 & c_3 \end{vmatrix} - \sum\begin{vmatrix} x & a_2 & a_3 \\ b_1 & y & b_3 \\ c_1 & c_2 & c_3 \end{vmatrix} + \begin{vmatrix} x & a_2 & a_3 \\ b_1 & y & b_3 \\ c_1 & c_2 & z \end{vmatrix} + (a_1-x)(b_2-y)(c_3-z).$$

When $x=y=z=0$ we have the theorem that $\Delta-\sum\Delta_1+\sum\Delta_2-\cdots$ *=the diagonal term, where Δ_r is got from Δ by deleting r diagonal elements.*

Prove that

$$|a_1\,b_2\,c_3\,d_4| = a_1\,|b_2\,c_3\,d_4| + b_2\begin{vmatrix} \cdot & a_3 & a_4 \\ c_1 & c_3 & c_4 \\ d_1 & d_3 & d_4 \end{vmatrix} + c_3\begin{vmatrix} \cdot & a_2 & a_4 \\ b_1 & \cdot & b_4 \\ d_1 & d_2 & d_4 \end{vmatrix} + d_4\begin{vmatrix} \cdot & a_2 & a_3 \\ b_1 & \cdot & b_3 \\ c_1 & c_2 & \cdot \end{vmatrix} + \begin{vmatrix} \cdot & a_2 & a_3 & a_4 \\ b_1 & \cdot & b_3 & b_4 \\ c_1 & c_2 & \cdot & c_4 \\ d_1 & d_2 & d_3 & \cdot \end{vmatrix}$$

Prove that

$$|a_1\,b_2\,c_3\,d_4\,e_5| = \begin{vmatrix} a_1 & a_2 & a_3 & a_4 & a_5 \\ b_1 & b_2 & b_3 & b_4 & b_5 \\ c_1 & c_2 & \cdot & c_4 & c_5 \\ d_1 & d_2 & d_3 & \cdot & d_5 \\ e_1 & e_2 & e_3 & e_4 & \cdot \end{vmatrix} + \sum\left(c_3\begin{vmatrix} a_1 & a_2 & a_4 & a_5 \\ b_1 & b_2 & b_4 & b_5 \\ d_1 & d_2 & \cdot & d_5 \\ e_1 & e_2 & e_4 & \cdot \end{vmatrix}\right)$$

$$+\sum\left(c_3d_4\begin{vmatrix} a_1, & a_2 & a_5 \\ b_1 & b_2 & b_5 \\ e_1 & e_2 & \cdot \end{vmatrix}\right)+c_3d_4e_5\begin{vmatrix} a_1 & a_2 \\ b_1 & b_2 \end{vmatrix}$$

$$=\begin{vmatrix} a_1 & a_2 & a_3 & a_4 & a_5 \\ b_1 & \cdot & b_3 & b_4 & b_5 \\ c_1 & c_2 & \cdot & c_4 & c_5 \\ d_1 & d_2 & d_3 & \cdot & d_5 \\ e_1 & e_2 & e_3 & e_4 & \cdot \end{vmatrix}+\sum\left(b_2\begin{vmatrix} a_1 & a_3 & a_4 & a_5 \\ c_1 & \cdot & c_4 & c_5 \\ d_1 & d_2, & \cdot & d_5 \\ e_1 & e_2 & e_4 & \cdot \end{vmatrix}\right)$$

$$+\sum\left(b_2c_3\begin{vmatrix} a_1 & a_4 & a_5 \\ d_1 & \cdot & d_5 \\ e_1 & e_4 & \cdot \end{vmatrix}\right)$$

$$+\sum\left(b_2c_3d_4\begin{vmatrix} a_1 & a_5 \\ e_1 & \cdot \end{vmatrix}\right)+b_2c_3d_4e_5a_1.$$

129. *In a determinant of the nth order the full number of terms which are independent of the elements of the principal diagonal is*

$$1\cdot 2\cdot 3\cdots, n\left\{\frac{1}{1\cdot 2}-\frac{1}{1\cdot 2\cdot 3}+\frac{1}{1\cdot 2\cdot 3\cdot 4}-\cdots+(-1)^n\frac{1}{1\cdot 2\cdot 3\cdots n}\right\}.$$

Let $\Delta_{0,n}$ be the determinant, and $\phi(n)$, as yet unknown, the number of terms in it of the kind referred to; so that $\phi(3)=2$ and $\phi(2)=1$. Then the general theorem (3) of §127 readily gives

$$\phi(n) = n! - n\cdot(n-1)! + n_2\cdot(n-2)! - n_3\cdot(n-3)! + \cdots.$$

The first two terms on the right cancel each other and taking $n!$ out of the remaining terms we get

$$\phi(n) = n!\left(\frac{1}{1\cdot 2}-\frac{1}{1\cdot 2\cdot 3}+\frac{1}{1\cdot 2\cdot 3\cdot 4}-\cdots+(-1)^n\frac{1}{1\cdot 2\cdot 3\cdots n}\right)$$

as required.

130. Let $u_{n,r}$ represent a determinant with a diagonal of r zeros, then by expanding, it may readily be found that

(1) $$u_{n,r} = (n-r)u_{n-1,r-1} + (r-1)u_{n-1,r-2}$$

and

(2) $$u_{n,r} = u_{n,r+1} + u_{n-1,r}$$

If E and F are two operators operating on u, and such that E refers to n alone and F refers to r alone then from (2) we have

(3) $$E = EF + 1$$

or

$$1 = F + \frac{1}{E}\cdot$$

Using this last form on (1) we get

(4) $$u_{n,r} = (n-1)u_{n-1,r-1} + (r-1)u_{n-2,r-2}$$

Giving r in this, the value n we get

$$u_{n,n} = (n-1)u_{n-1,n-1} + (n-1)u_{n-2,n-2}$$

or

(5) $$u_{n,n} + u_{n-1,n-1} = nu_{n-1,n-1} + (n-1)u_{n-2,n-2}$$
$$u_{n,n} = nu_{n-1,n-1} + (-1)^n$$

131. If $V_{n,r}$ represents a determinant of the nth order having two adjacent diagonals of zeros, the one having r and the other $r-1$, then it may by expansion be found that

(1) $$V_{n,r} = (n-1)V_{n-1,r-1} - (n-2r+1)V_{n-2,r-2} + V_{n-3,r-3.}$$

which when r is n becomes

(2) $$V_{n,n} = (n-1)(V_{n-1,n-1} + V_{n-2,n-2}) + V_{n-3,n-3}$$

and hence

(3) $$(n-1)V_{n,n} = (n^2-n-1)V_{n-1,n-1} + nV_{n-2,n-2} - (-1)^n 2$$

132. If W_n denotes the number of terms in a determinant having zeros in the main and the lower adjacent diagonal and also zero in the position $(1, n)$, then it has been found that

$$W_n = nW_{n-1} + \frac{n}{n-2}W_{n-2} + (-1)^{n-1}\frac{4}{n-2}$$

which implies that $(n-2)W_n + (-1)^{n+1}4$ is a multiple of n.

Exercises. Set X

1. Find the number of possible arrangements of n things a_1, a_2, a_3, $\cdots$, a_n, subject to the condition that no one shall be in its original place.

We may have a_1 in any place except the first, a_2 in any place except the second, and so on—data which we may present to the eye in the form

$$\begin{matrix} (\ ) & a_1 & a_1 & a_1 \cdots & a_1 \\ a_2 & (\ ) & a_2 & a_2 \cdots & a_2 \\ a_3 & a_3 & (\ ) & a_3 \cdots & a_3 \\ a_4 & a_4 & a_4 & (\ ) \cdots & a_4 \\ \cdot & \cdot & \cdot & \cdot & \cdots \cdot \\ a_n & a_n & a_n & a_n \cdots & (\ ), \end{matrix}$$

an a_1 being written in the places which it is allowable for a_1 to occupy, and ( ) signifying that the suffixed letter found in the same line with it may not occupy its place. Looking to this table we see that for the first place in any of the arrangements we may take any letter that is in the first column; for the second place any letter that is in the second column, provided it be not in the same line with the letter taken from the first column; for the third place any letter that is in the third column, provided it be not in the same line with either of the letters previously taken, and so on. This law of formation, however, is identical with that in accordance with which the terms of a determinant are got from the elements; so that the problem we are concerned with is transformed into finding the full number of terms of a zero-axial determinant of the nth order. Hence (§129) the required number of arrangements is

$$1 \cdot 2 \cdot 3 \cdots n \left\{ \frac{1}{1 \cdot 2} - \frac{1}{1 \cdot 2 \cdot 3} + \frac{1}{1 \cdot 2 \cdot 3 \cdot 4} - \cdots + (-1)^n \frac{1}{1 \cdot 2 \cdot 3 \cdot 4 \cdots n} \right\}.$$

2. If $\phi(n)$ denote as in §129 the number of terms of a zero-axial determinant of the nth order, prove with the help of §124 that

$$\phi(n) + \frac{n}{1}\phi(n-1) + \frac{n(n-1)}{1 \cdot 2}\phi(n-2) + \cdots + \frac{n(n-1)}{1 \cdot 2}\phi(2) + n\phi(1) + 1 = 1 \cdot 2 \cdot 3 \cdots n.$$

3. Find the number of minors of the $(n-2)$th order in a determinant of the nth order.

4. If k particular elements of the principal diagonal of a determinant of the nth order be fixed upon, find the number of terms which contain these elements and no other elements of the same diagonal.

5. Find the difference between the number of positive and the number of negative terms in a zero-axial determinant of the nth order.

6. Prove that

$$\begin{vmatrix} 0 & 1 & 1 & 1 & 1 \\ 1 & 0 & 1+a & 1+b & 1+c \\ 1 & a+1 & 0 & a+b & a+c \\ 1 & b+1 & b+a & 0 & b+c \\ 1 & c+1 & c+a & c+b & 0 \end{vmatrix} = 2^3 \begin{vmatrix} 1+a & 1 & 1 \\ 1 & 1+b & 1 \\ 1 & 1 & 1+c \end{vmatrix}.$$

7. Prove that

$$\begin{vmatrix} a_1 a_2 & a_1 & a_2 & 1 \\ b_1 b_2 & b_2 & b_1 & 1 \\ c_1 c_2 & c_1 & c_4 & 1 \\ d_1 d_2 & d_2, & d_1, & 1 \end{vmatrix} = \begin{cases} \quad (a_1 - b_2)(b_1 - c_2)(c_1 - d_2)(d_1 - a_2) \\ -(a_2 - b_1)(b_2 - c_1)(c_2 - d_1)(d_2 - a_1). \end{cases}$$

8. From the theorem of example 2 above, prove that

$$\phi(n) = \begin{vmatrix} n! & C_{n,1} & C_{n,2} & \cdots & C_{n,n-1} & C_{n,n} \\ (n-1)! & 1 & C_{n-1,1} & \cdots & C_{n-1,n-2} & C_{n-1,n-1} \\ (n-2)! & 0 & 1 & \cdots & C_{n-2,n-3} & C_{n-2,n-2} \\ \cdot & \cdot & \cdot & \cdots & \cdot & \cdot \\ 1! & 0 & 0 & \cdots & 1 & C_{1,1} \\ 1 & 0 & 0 & \cdots & 0 & 1 \end{vmatrix},$$

9. $$\begin{vmatrix} \delta_{11} & \delta_{12} & \xi_{11} & \xi_{12} & \xi_{13} \\ \delta_{21} & \delta_{22} & \xi_{21} & \xi_{22} & \xi_{23} \\ \eta_{11} & \eta_{12} & a_{11} & a_{12} & a_{13} \\ \eta_{21} & \eta_{22} & a_{21} & a_{22} & a_{23} \\ \eta_{31} & a_{3.} & a_{31} & a_{32} & a_{33} \end{vmatrix}$$

$$= \left|\delta_{11}\,\delta_{22}\right| \cdot \left|a_{11}\,a_{22}\,a_{33}\right| + \delta_{11}\begin{vmatrix} \cdot & \xi_{21} & \xi_{22} & \xi_{23} \\ \eta_{12} & a_{11} & a_{12}, & a_{13} \\ \eta_{22} & a_{21} & a_{22} & a_{23} \\ \eta_{32} & a_{31} & a_{32} & a_{33} \end{vmatrix}$$

$$- \delta_{12}\begin{vmatrix} \cdot & \xi_{21} & \xi_{22} & \xi_{23} \\ \eta_{11} & a_{11} & a_{12} & a_{13} \\ \eta_{21} & a_{21} & a_{22} & a_{23} \\ \eta_{31} & a_{31} & a_{32} & a_{33} \end{vmatrix} - \delta_{21}\begin{vmatrix} \cdot & \xi_{11} & \xi_{12} & \xi_{13} \\ \eta_{12} & a_{11} & a_{12} & a_{13} \\ \eta_{22} & a_{21} & a_{22} & a_{23} \\ \eta_{32} & a_{31} & a_{32} & a_{33} \end{vmatrix}$$

$$+ \delta_{22}\begin{vmatrix} \cdot & \xi_{11} & \xi_{12} & \xi_{13} \\ \eta_{11} & a_{11} & a_{12} & a_{13} \\ \eta_{21} & a_{21} & a_{22} & a_{23} \\ \eta_{31} & a_{31} & a_{32} & a_{33} \end{vmatrix} + \begin{vmatrix} \cdot & \cdot & \xi_{11} & \xi_{12} & \xi_{13} \\ \cdot & \cdot & \xi_{11} & \xi_{22} & \xi_{23} \\ \eta_{11} & \eta_{12} & a_{11} & a_{12} & a_{13} \\ \eta_{21} & \eta_{22} & a_{21} & a_{22} & a_{23} \\ \eta_{31} & \eta_{32} & a_{31} & a_{32} & a_{33} \end{vmatrix}$$

10. Show that the number of positive terms in the ordinary development of a determinant having negative elements in the diagonal and positive elements elsewhere is

$$\frac{n!}{2} - (-2)^{n-2}(n-2).$$

11. Show that the number of terms containing a particular set of m diagonal elements is $\phi(n-m)$; that the number containing m diagonal elements without restriction is $(n)_m\ \phi(n-m)$; that the number containing not more than m is $\sum_{\mu=0}^{m} (n)_\mu\phi(n-\mu)$; that the number containing not less than m is $1+\sum_{\mu=m}^{n-2} (n)_\mu\phi(n-\mu)$.

12. If $\psi(n)$ denote the number of terms in a determinant with zeros in both diagonals, then

$$\psi(n) = (n-1)\psi(n-1) + \begin{cases} 2(n-2)\psi(n-3) & \text{if } n \text{ is even} \\ 2(n-1)\psi(n-2) & \text{if } n \text{ is odd} \end{cases}$$

133. It is evident that if the rows of a determinant of the nth order be multiplied by $x_1, x_2, \cdots, x_n$ respectively, and the columns be then divided by $x_1, x_2, \cdots, x_n$ respectively, the determinant is unaltered in value, and each of the minors of the transformed determinant is, to a factor *pres*, equal to the corresponding minor of the original determinant, the connecting multiplier being $x_h x_k x_l \cdots$

$/x_r x_s x_t \cdots$ if the minor belong to the hth, kth, lth, $\cdots$ rows, and the rth, sth, tth, $\cdots$ columns of the original. The connecting multiplier in the case of the coaxial minors is manifestly 1: in other words, the principal minors remain unaltered by the transformation.

Next it is clear that $x_1, x_2, \cdots, x_n$ may be so chosen that all the elements of any one of the rows or columns, except the diagonal element, shall be 1.

For example, the first row,

$$a_{11},\ a_{12}\frac{x_1}{x_2},\ a_{13}\frac{x_1}{x_3},\ \cdots,\ a_{1n}\frac{x_1}{x_n}$$

may be made to take the form

$$a_{11},\ 1,\ 1,\ \cdots,\ 1,$$

by giving $x_1, x_2, x_3, \cdots, x_n$ the values $1, a_{12}, a_{13}, \cdots, a_{1n}$ respectively.

It follows that

Any determinant of the nth order may be transformed so as to have 1 *for* $n-1$ *of the elements, and yet the determinant itself and all its coaxial minors remain unaltered in value.*

134. This is the same as saying that 2^n-1 quantities, namely, the determinant and its coaxial minors, can be expressed in terms of $n^2-(n-1)$ others, namely, the modified elements, which are not equal to unity. Eliminating the latter, we have 2^n-n^2+n-2 relations connecting the former, which gives us the theorem:

There are 2^n-n^2+n-2 *relations between the coaxial minors of any determinant of the nth order.*

135. In the case of a determinant of order 4 the actual expressions for these special relations have been found but are too ungainly in their explicit form to be reproduced here.†

† For their consideration and for problems 1, 2, 3, 4 vide. Nanson: Phil. Mag. 5th series, xliv, pp. 362–367. Muir: Trans. Royal Soc. Edinb. vol. XXXIX, part II, pp. 323–339. Cf. Stouffer: Trans. Am. Math. Soc., Vol. 26, pp. 356+

PROBLEMS.

1. Show that the rationalizant of the expression

$$x + h(bc)^{1/2} + k(ca)^{1/2} + l(ab)^{1/2} \text{ is}$$

(a)
$$\begin{vmatrix} x & h(bc)^{1/2} & k(ca)^{1/2} & l(ab)^{1/2} \\ h(bc)^{1/2} & x & l(ab)^{1/2} & k(ac)^{1/2} \\ k(ac)^{1/2} & l(ab)^{1/2} & x & h(bc)^{1/2} \\ l(ab)^{1/2} & k(ca)^{1/2} & h(bc)^{1/2} & x \end{vmatrix}$$

(b)
$$\begin{vmatrix} x & h & k & l \\ hbc & x & lb & kc \\ kca & la & x & hc \\ lab & ka & hb & x \end{vmatrix}.$$

2. Use example 1(a) to obtain a determinant expression for

(a) $\cos(\alpha+\beta+\gamma)$ in terms of $\cos\alpha$, $\cos\beta$, $\cos\gamma$.

(b) $\cos(\alpha+\beta+\gamma+\delta)$ in terms of $\cos\alpha$, $\cos\beta$, $\cos\gamma$, $\cos\delta$.

3. Express $\cos^{-1}x+\cos^{-1}y+\cos^{-1}z+\cos^{-1}w=0$ in purely algebraic form.

4. Express the resultant of

$$Bx^2 - 2C'xy + Ay^2 = 0$$
$$Cy^2 - 2A'yz + Bz^2 = 0$$
$$Ax^2 - 2B'xz + Cx^2 = 0$$

in the form

$$\begin{vmatrix} 1 & A'/(BC)^{1/2} & C'/(AB)^{1/2} \\ A'/(BC)^{1/2} & 1 & B'/(CA)^{1/2} \\ C'/(AB)^{1/2} & B'/(CA)^{1/2} & 1 \end{vmatrix} = 0.$$

136. *The product of two determinants of the same order is equal to the sum of like products obtained from the original by interchanging a particular column of the one determinant with the columns of the other in succession.*

Let $|a_{1n}|$, $|b_{1n}|$ be the two determinants, and, first, let the particular column fixed upon be the first column of $|a_{1n}|$.

If a determinant of the $2n$th order be formed, having $|a_{1n}|$ for the minor situated in the first n rows and first n columns, a determinant of n^2 zero elements for the minor situated in the last n rows and first n columns, and $|b_{1n}|$ for the complementary of both of these; and if the first column of $|a_{1n}|$ be interchanged with the column of zeros below it, we have the determinant

$$\begin{vmatrix} 0 & a_{12} \cdots a_{1n} & b_{11} & b_{12} \cdots b_{1n} \\ 0 & a_{22} \cdots a_{2n} & b_{21} & b_{22} \cdots b_{2n} \\ \cdots & \cdots & \cdots & \cdots \\ 0 & a_{n2} \cdots a_{nn} & b_{n1} & b_{n2} \cdots b_{nn} \\ a_{11} & 0 \cdots 0 & b_{11} & b_{12} \cdots b_{1n} \\ a_{21} & 0 \cdots 0 & b_{21} & b_{22} \cdots b_{2n} \\ \cdots & \cdots & \cdots & \cdots \\ a_{n1} & 0 \cdots 0 & b_{n1} & b_{n2} \cdots b_{nn} \end{vmatrix}, \quad \text{or } \Delta, \text{ say.}$$

Looking at the first n rows of this and seeking to form from them all the minors of the nth order, with a view to obtain the expansion of the determinant as an aggregate of products of complementary minors, we see that we need only take those minors into which enter the $n-1$ columns surmounting the zeros, for every other minor has a complementary which vanishes. Consequently all the minors worth attending to are got by taking along with these $n-1$ columns the columns of $|b_{1n}|$ in succession; and thus the expansion referred to is

$$(-1)^n \begin{vmatrix} a_{12} \cdots a_{1n} & b_{11} \\ a_{22} \cdots a_{2n} & b_{21} \\ \cdots & \cdots \\ a_{n2} \cdots a_{nn} & b_{n1} \end{vmatrix} \begin{vmatrix} a_{11} & b_{12} \cdots b_{1n} \\ a_{21} & b_{22} \cdots b_{2n} \\ \cdots & \cdots \\ a_{n1} & b_{n2} \cdots b_{nn} \end{vmatrix}$$

$$+ (-1)^{n+1} \begin{vmatrix} a_{12} \cdots a_{1n} & b_{12} \\ a_{22} \cdots a_{2n} & b_{22} \\ \cdots & \cdots \\ a_{n2} \cdots a_{nn} & b_{n2} \end{vmatrix} \begin{vmatrix} a_{11} & b_{11} & b_{13} \cdots \\ a_{21} & b_{21} & b_{23} \cdots \\ \cdots & \cdots & \cdots \\ a_{n1} & b_{n1} & b_{n3} \cdots \end{vmatrix}$$

$$+ \cdots + (-1)^{2n-1} \begin{vmatrix} a_{12} \cdots a_{1n} & b_{1n} \\ a_{22} \cdots a_{2n} & b_{2n} \\ \cdots & \cdots \\ a_{n2} \cdots a_{nn} & b_{nn} \end{vmatrix} \begin{vmatrix} a_{11}, & b_{11} \cdots b_{1,n-1} \\ a_{21} & b_{21} \cdots b_{2,n-1} \\ \cdots & \cdots \\ a_{n1} & b_{n1} \cdots b_{n,n-1} \end{vmatrix},$$

the index of the sign-factor of the first product being n because n interchanges make its first factor the leading nth minor of Δ. Making the b column of each of the first factors pass over the columns before it, and making the a column of each of the second factors pass over to the place of the missing b column, we have

$$-\begin{vmatrix} b_{11} & a_{12} & \cdots, a_{1n} \\ b_{21} & a_{22} & \cdots, a_{2n} \\ \cdots & \cdots & \cdots \\ b_{n1} & a_{n2} & \cdots, a_{nn} \end{vmatrix} \cdot \begin{vmatrix} a_{11} & b_{12} & \cdots b_{1n} \\ a_{21} & b_{22} & \cdots b_{2n} \\ \cdots & \cdots & \cdots \\ a_{n1} & b_{n2} & \cdots, b_{nn} \end{vmatrix}$$

$$-\begin{vmatrix} b_{12} & a_{12} & \cdots a_{1n} \\ b_{22} & a_{22} & \cdots a_{2n} \\ \cdots & \cdots & \cdots \\ b_{n2}, & a_{n2} & \cdots a_{nn} \end{vmatrix} \cdot \begin{vmatrix} b_{11} & a_{11} & b_{13} & \cdots b_{1n} \\ b_{21} & a_{21} & b_{23} & \cdots b_{2n} \\ \cdots & \cdots & \cdots & \cdots \\ b_{n1} & a_{n1} & b_{n3} & \cdots b_{nn} \end{vmatrix}$$

$$- \cdots - \begin{vmatrix} b_{1n} & a_{12} & \cdots a_{1n} \\ b_{2n} & a_{22} & \cdots a_{2n} \\ \cdots & \cdots & \cdots \\ b_{nn} & a_{n2} & \cdots, a_{nn} \end{vmatrix} \cdot \begin{vmatrix} b_{11} & \cdots b_{1,n-1} & a_{11} \\ b_{21} & \cdots b_{2,n-1} & a_{21} \\ \cdots & \cdots & \cdots \\ b_{n1} & \cdots b_{n,n-1} & a_{n1} \end{vmatrix}$$

the sign of every product being now $-$, since the number of changes of sign in the first is $n-1$, in the second n, in the third $n+1$, and so on.

Again, returning to Δ, and subtracting each element of the $(n+1)$th row from the corresponding element of the first row, each element of the $(n+2)$th row from the corresponding element of the second row, and so on, we have a determinant from whose first n rows it is possible to form but one non-zero minor of the nth order, namely, $-|a_{1n}|$; hence it is seen (§108) that

$$\Delta = -|a_{1n}| \cdot |b_{1n}| .$$

This and the former expression obtained for Δ at once give the required identity in the special case under consideration.

Secondly, let the particular column fixed upon be not the first of $|a_{1n}|$, but some other, the kth say. Then it is readily seen that to establish the theorem we have only to make this kth column pass over the $k-1$ which precede it, and apply to the product of the resulting determinant and $|b_{1n}|$ the already established case, and in every first factor of the result make the first column pass over the next $k-1$ columns.

As another way to write the theorem of this article we have

$$|a_1 b_2 c_3| \cdot |a_4 b_5 c_6| - |a_1 b_2 c_4| \cdot |a_3 b_5 c_6| + |a_1 b_3 c_4| \cdot |a_2 b_5 c_6|$$
$$- |a_2 b_3 c_4| \cdot |a_1 b_5 c_6|$$

$$= \begin{vmatrix} a_1, & b_1 & c_1 & |a_1 b_5 c_6| \\ a_2, & b_2 & c_2 & |a_2 b_5 c_6| \\ a_3 & b_3 & c_3 & |a_3 b_5 c_6| \\ a_4 & b_4 & c_4 & |a_4 b_5 c_6| \end{vmatrix} = - \begin{vmatrix} a_1 & a_2 & a_3 & a_4 & a_5 & a_6 \\ b_1 & b_2 & b_3 & b_4 & b_5 & b_6 \\ c_1 & c_2 & c_3 & c_4 & c_5 & c_6 \\ a_1 & a_2 & a_3 & a_4 & 0 & 0 \\ b_1 & b_2 & b_3 & b_4 & 0 & 0 \\ c_1 & c_2 & c_3 & c_4 & 0 & 0 \end{vmatrix}.$$

137. SYLVESTER'S THEOREM. *The product of two determinants of the same order is equal to the sum of like products obtained from the original by interchanging k chosen columns of the one determinant with every set of k columns of the other in succession; the interchange of k columns with k columns being effected by interchanging the first column of the one set with the first column of the other, the second of the one with the second of the other, and so on.*

Let $|a_{1n}|$, $|b_{1n}|$ be the two determinants, and, first, let the k columns fixed upon be the first k columns of $|a_{1n}|$.

If a determinant of the $2n$th order be formed exactly as in §136, and the first k columns of its minor $|a_{1n}|$ be then interchanged in order with the k columns of zeros below them, we have the determinant

$$\begin{vmatrix} 0 & \cdots & 0 & a_{1,k+1} & \cdots & a_{1n} & b_{11} & b_{12} & \cdots & b_{1n} \\ 0 & \cdots & 0 & a_{2,k+1} & \cdots & a_{2n} & b_{21} & b_{22} & \cdots & b_{2n} \\ \cdots & \cdots & \cdots & \cdots & \cdots & \cdots & \cdots & \cdots & \cdots & \cdots \\ 0 & \cdots & 0 & a_{n,k+1} & \cdots & a_{nn} & b_{n1}, & b_{n2} & \cdots & b_{nn} \\ a_{11} & \cdots & a_{1k} & 0 & \cdots & 0 & b_{11} & b_{12} & \cdots & b_{1n} \\ a_{21} & \cdots & a_{2k} & 0 & \cdots & 0 & b_{21} & b_{22} & \cdots & b_{2n} \\ \cdots & \cdots & \cdots & \cdots & \cdots & \cdots & \cdots & \cdots & \cdots & \cdots \\ a_{n1} & \cdots & a_{nk} & 0 & \cdots & 0 & b_{n1} & b_{n2} & \cdots & b_{nn} \end{vmatrix} \quad \text{or} \quad \Delta, \quad \text{say}.$$

Preparing to express this as an aggregate of products of complementary minors of the nth order formed from the first and last n

rows, we see that the first factor of any non-zero product of this kind must contain the last $n-k$ columns of $|a_{1n}|$ and a set of k columns from $|b_{1n}|$, and that the cofactor must contain the remaining columns from $|a_{1n}|$ and $|b_{1n}|$; in other words, that this product is derivable from $|a_{1n}|\,|b_{1n}|$ by transferring the first k columns of $|a_{1n}|$ to $|b_{1n}|$ and a set of k columns of $|b_{1n}|$ to $|a_{1n}|$,—the transferred columns occupying the last k places in the first factor and the first k places in the second. If the columns taken from $|b_{1n}|$ be the θ_1th, θ_2th, $\cdots$, θ_kth, the index of the sign-factor of the product is

$$(1+2+\cdots+n)+(\overline{k+1}+\overline{k+2}+\cdots+n+\overline{n+\theta_1}$$
$$+\overline{n+\theta_2}+\cdots+\overline{n+\theta_k})\ ;$$

that is

$$\tfrac{1}{2}n(n+1)+\left\{\tfrac{1}{2}(n-k)(n+k+1)+kn+\theta_1+\theta_2+\cdots+\theta_k\right\}\ ;$$

that is

$$\tfrac{1}{2}n(n+1)+\tfrac{1}{2}n(n+1)-\tfrac{1}{2}k(k+1)+kn+\theta_1+\theta_2+\cdots+\theta_k.$$

Now, were it not for the positions of the transferred columns, the series of products thus obtained would be that referred to in the statement of the theorem; and if, in each case, we pass the b columns of the first factor in order over the $n-k$ preceding a columns, and in the second factor transfer the kth column to occupy the θ_kth place, the $(k-1)$th column to occupy the θ_{k-1}th place, and so on, we obtain the said series of products exactly. The number of changes of sign caused by these transferences of column is seen to be

$$k(n-k)+(\overline{\theta_r-k}+\cdots+\overline{\theta_2-2}+\overline{\theta_1-1})\ ;$$

that is $k(n-k)-\frac{1}{2}k(k+1)+\theta_1+\theta_2+\cdots+\theta_k$. Consequently the index of the sign-factor of the product will now be the sum of this number and the former index; so that, those parts of the sum being neglected which are even, the sign-factor is found to be

$$(-1)^{-k^2} \text{ that is } (-1)^{k^2} \text{ that is } (-1)^k.$$

Hence Δ is equal to the aggregate of the products referred to in the theorem each taken with the sign-factor $(-1)^k$.

Again, subtracting each element of the $(n+1)$th row of Δ from the corresponding element of the first row, each element of the $(n+2)$th row from the corresponding element of the second row, and so on, we have (§ 107) also

$$\Delta = (-1)^k |a_{1n}|\, |b_{1n}| .$$

Hence, by equating these two expressions for Δ, we have the required identity established for the particular case under consideration.

When the k columns fixed upon for interchange are not the first k columns of $|a_{1n}|$, we may prove the theorem by making them the first k columns and then using the already proved case, exactly after the manner of §136.

EXAMPLE. Taking the determinants $|a_1b_2c_3d_4|$ and $|x_1y_2z_3w_4|$, and selecting the first two columns of the former as the columns for interchange, we have

$$\begin{aligned} |a_1b_2c_3d_4|\,|x_1y_2z_3w_4| &= |x_1y_2c_3d_4|\,|a_1b_2z_3w_4| + |x_1z_2c_3d_4|\,|a_1y_2b_3w_4| \\ &+ |x_1w_2c_3d_4|\,|a_1y_2z_3b_4| + |y_1z_2c_3d_4|\,|x_1a_2b_3w_4| \\ &+ |y_1w_2c_3d_4|\,|x_1a_2z_3b_4| + |z_1w_2c_3d_4|\,|x_1y_2a_3b_4|. \end{aligned}$$

138. From the elements of two determinants $|a_{1n}|$, $|b_{1n}|$ a number of different zero-determinants of the $2n$th order can be formed resembling Δ in §137, and thus a corresponding number of identities similar to the one there established can be at once obtained by expanding in terms of products of complementary minors of the nth order. Any one of the first n columns of such a zero-determinant is formed by taking for the one half of it the corresponding column of $|a_{1n}|$ and for the other either a repetition of this or n zeros; the last n columns are formed in like manner, but from $|b_{1n}|$; and the number of columns independent of zeros is not less than $n+1$, this being necessary in order that the determinant may vanish in accordance with the theorem of §108. If, in the determinant, $|a_{1n}|$ and $|b_{1n}|$ are complementary minors, the resulting identity may be expressed like that of §137, namely, so as to give an equivalent for the product of $|a_{1n}|$ and $|b_{1n}|$.

EXAMPLE. Taking the determinants $|a_1b_2c_3|$, $|x_1y_2z_3|$, and seeking to express their product as an aggregate of like products, in which the first factor shall contain the column of b's and the second factor the column of x's, we form the determinant

$$\begin{vmatrix} a_1 & b_1 & c_1 & 0 & y_1 & z_1 \\ a_2 & b_2 & c_2 & 0 & y_2 & z_2 \\ a_3 & b_3 & c_3 & 0 & y_3 & z_3 \\ a_1 & 0 & c_1 & x_1 & y_1 & z_1 \\ a_2 & 0 & c_2 & x_2 & y_2 & z_2 \\ a_3 & 0 & c_3 & x_3 & y_3 & z_3 \end{vmatrix},$$

which (§106) is equal to

$$\begin{aligned} |a_1b_2c_3|\,|x_1y_2z_3| + |a_1b_2y_3|\,|c_1x_2z_3| - |a_1b_2z_3|\,|c_1x_2y_3| \\ + |b_1c_2y_3|\,|a_1x_2z_3| - |b_1c_2z_3|\,|a_1x_2y_3| \\ - |b_1y_2z_3|\,|a_1c_2x_3|\,. \end{aligned}$$

But by subtracting each element of the last three rows from the corresponding element of the first three rows, it is at once seen to be also equal to 0; hence

$$\begin{aligned} |a_1b_2c_3|\,|x_1y_2z_3| = |a_1b_2y_3|\,|x_1c_2z_3| + |a_1b_2z_3|\,|x_1y_2c_3| \\ + |y_1b_2c_3|\,|x_1a_2z_3| + |z_1b_2c_3|\,|x_1y_2a_3| \\ - |y_1b_2z_3|\,|x_1a_2c_3|\,. \end{aligned}$$

139. If the two given determinants in the immediately preceding theorems have one or more columns in common, the number of products in the resulting identity is less (§51) than it would otherwise be. Special cases of this kind are of sufficiently frequent occurrence to merit the student's attention.

140. It is readily seen that identities, similar to those of §138, but having for the factors of each product two determinants of *different* orders, might also be established. They do not admit of very simple statement, but are often of use.

EXAMPLE. Since

$$\begin{vmatrix} b_1 & b_2 & 0 & b_3 & b_4 \\ c_1 & c_2 & 0 & c_3 & c_4 \\ b_1 & 0 & b_2 & b_3 & b_4 \\ c_1 & 0 & c_2 & c_3 & c_4 \\ d_1 & 0 & d_2 & d_3 & d_4 \end{vmatrix} = 0$$

it at once follows that

$$|b_1c_2|\,|b_2c_3d_4| = |b_2c_3|\,|b_1c_2d_4| - |b_2c_4|\,|b_1c_2d_3|\ .$$

141. SCHWEINS'S THEOREM. *If we expand the determinant A by Laplace's theorem, first in terms of the minors of order r formed from any r rows, with their complementaries, and second in terms of the minors of order m, formed from any m columns* ($r<m$), *with their complementaries; then the sum of the* $(n-r)_{m-r}$ *terms of the second expansion which have in common the elements in the intersection of the selected r rows and m columns is equal to the sum of the* m_r *terms of the first expansion which have for one factor the minors of the rth order formed from the elements in the intersection of the selected r rows and m columns.*

Stated symbolically the theorem is

$$\sum_{1}^{(n-r)_{m-r}} i(-1)^{\nu_1} A_{(\bar{n}|n-r\beta)(n|n-r\beta|m-r_i),(n|m\alpha)} A_{(n|\overline{(n-r\beta}|m-r_i),(\bar{n}|m\alpha)}$$

$$= \sum_{1}^{m_r} j(-1)^{\nu_2} A_{(\bar{n}|n-r\beta),(n|m\alpha|r_j)} A_{(n|n-r\beta),(\bar{n}|m\alpha)(n|\bar{m}\alpha|r_j)}$$

where ν_1=no. of inversions in $(\bar{n}|n-r_\beta)(n|n-r_\beta|m-r_i)(n|\overline{n-r_\beta}|m-r_i)$+no. of inversions in $(n|m_\alpha)(\bar{n}|m_\alpha)$,

ν_2=no. of inversions in $(\bar{n}|n-r_\beta)(n|n-r_\beta)$+no. of inversions in $(n|m_\alpha|r_j)((\bar{n}|m_\alpha)(n|\bar{m}_\alpha|r_j))$, and where $(n|n-r_\beta)$ is used to denote the βth combination of the n numbers $1, 2, \cdots, n$, taken $n-r$ at a time.

Expanding the determinant

$$A_{(\bar{n}|n-r\beta)(n|n-r\beta|m-r_i),(n|m\alpha)}$$

in terms of its minors of the rth and $(m-r)$th orders by Laplace's theorem we have

$$A_{(\bar{n}|n-r\beta)(n|n-r\beta|m-r_i),(n|m\alpha)}$$

$$= \sum_1^{m_r} j(-1)^{\nu_3} A_{(\bar{n}|n-r\beta),(n|m\alpha|r_j)} A_{(n|n-r\beta|m-r_i),(n|\bar{m}\alpha|r_j)}.$$

But

$$\sum_1^{(n-r)_{m-r}} i(-1)^{\nu_4} A_{(n|n-r\beta|m-r_i),(n|\bar{m}\alpha|r_j)} A_{(n|(\overline{n-\beta r})|m-r_i),(\bar{n}|m\alpha)}$$

$$= A_{(n|n-r\beta),(\bar{n}|m\alpha)(n|\bar{m}\alpha|r_j)}$$

therefore since evidently $\nu_1+\nu_3=\nu_2+\nu_4$ the truth of the theorem appears.

This theorem is due to Schweins.

If $m=n$ it becomes

$$A = \sum_1^{n_r} j(-1)^{\nu} A_{(n|r\beta),(n|r_j)} A_{(\bar{n}|r\beta),(\bar{n}|r_j)}$$

which is Laplace's theorem.

142. Perhaps the simplest proof of Schweins's theorem is obtained on equating the two expansions by Laplace's theorem of the determinant

$$\begin{vmatrix} a_{11} & a_{12} & \cdots & a_{1m} & 0 & \cdots & 0 \\ a_{21} & a_{22} & \cdots & a_{2m} & 0 & \cdots & 0 \\ \cdots & \cdots & \cdots & \cdots & \cdots & \cdots & \cdots \\ a_{r1} & a_{r2} & \cdots & a_{rm} & 0 & \cdots & 0 \\ a_{r+1,1} & a_{r+1,2} & & a_{r+1,m} & a_{r+1,m+1} & \cdots & a_{r+1,n} \\ \cdots & \cdots & \cdots & \cdots & \cdots & \cdots & \cdots \\ a_{n1} & a_{n2} & \cdots & a_{nm} & a_{n,m+1} & \cdots & a_{nn} \end{vmatrix}$$

first in terms of minors of the mth order got from the first m columns and second in terms of minors of the rth order got from the first r rows.

143. As an immediate consequence of Schweins's theorem we have

$$\sum_1^{(n-1)_r} i(-1)^{\nu} \Big[A_{(n|n-1\alpha|r\gamma)(\bar{n}|n-1\alpha),(n|n-1\beta|r_i)(\bar{n}|n-1\beta)}$$

$$\times A_{(n|(\overline{n-1\alpha})|r\gamma),(n|(\overline{n-1\beta})|r_i)}$$

$$+ A_{(n|n-1\alpha|r\gamma),(n|n-1\beta|r_i)} A_{(n|(\overline{n-1\alpha})|r\gamma)(\bar{n}|\overline{n-1\alpha}),(n|(\overline{n-1\beta})|r_i)(\bar{n}|n-1\beta)}\Big]$$

$$= \sum \Big[A_{(n|r+1\alpha|1_i),(\bar{n}|n-1\beta)} A_{(n|(\overline{n-1\alpha})|r\gamma)(n|(\overline{r+1\alpha})|1_j),(n|n-1\beta)}$$

$$+ A_{(n|n-1\alpha|r\gamma)(n|(\overline{n-r\alpha})|1_i),(n|n-1\beta)} A_{(n|n-r\alpha|1_j),(\bar{n}|n-1\beta)}\Big]$$

where $(n|r+1_\alpha)$ represents the $r+1$ numbers in $(n|n-1_\alpha|r_\gamma)$ $(\bar{n}|n-1_\alpha)$ and $(n|n-r_\alpha)$ represents the $n-r$ numbers in $(n|\overline{n-1_\alpha}|r_\gamma)$ $(\bar{n}|n-1_\alpha)$.

$$= A + A_{(n|n-1\alpha),(n|n-1\beta)}A_{(\bar{n}|n-1\alpha),(\bar{n}|n-1\beta)}$$

$$= A_{(\bar{n}|n-1\alpha),(\bar{n}|n-1\beta)}A_{(n|n-1\alpha),(n|n-1\beta)}$$

$$+ \sum_1^n iA_{(n|n-1_i),(n|n-1\beta)}A_{(\bar{n}|n-1_i),(\bar{n}|n-1\beta)}$$

$$= 2A_{(\bar{n}|n-1\alpha),(\bar{n}|n-1\beta)}A_{(n|n-1\alpha),(n|n-1\beta)}$$

$$+ \sum_1^{(n-1)_1} i\, A_{(n|n-1\alpha|1_i),(\bar{n}|n-1\beta)}A_{(n|\overline{(n-1\alpha)}|1_i)(\bar{n}|n-1\alpha),(n|n-1\beta)}\,.$$

144. In §137 we have seen that the theorem comes out of expanding a certain determinant in two different ways. The determinant in question may be pictorially represented by

k	$n-k$	n	
(0)	(y)	(z)	n
(x)	(0)	(z)	n

where (0) denotes that the elements in the rectangle are all zeros.

In §141 we have another theorem (Schweins's) arising from the twofold expansion of the determinant

m	$n-m$	
(x)	(0)	r
(y)	(z)	$n-r$

There are certain types of special cases of this determinant which are worthy of consideration. The first would be the determinant represented by

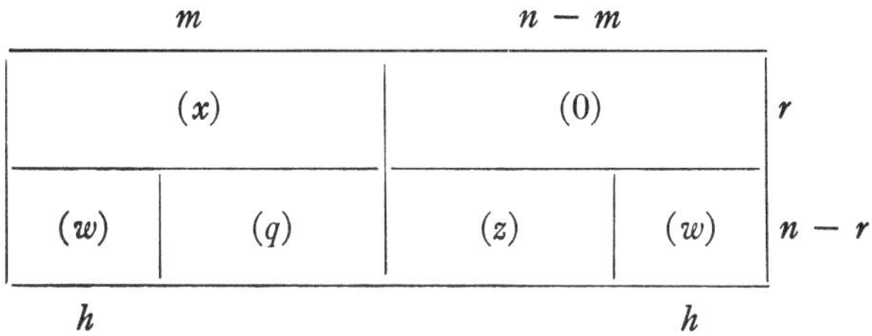

where the elements in h of the m columns of the $n-r$ rows are the same as in h of the $n-m$ columns of the $n-r$ rows; or what amounts to the same thing

(a)

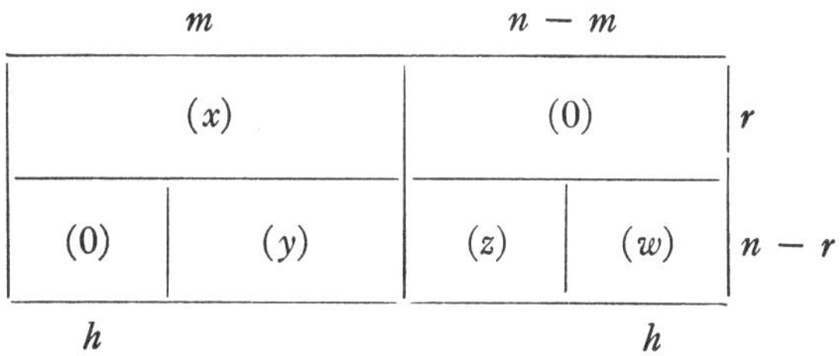

since we may subtract the last h columns from the first h.

Another type would be represented by

(b)

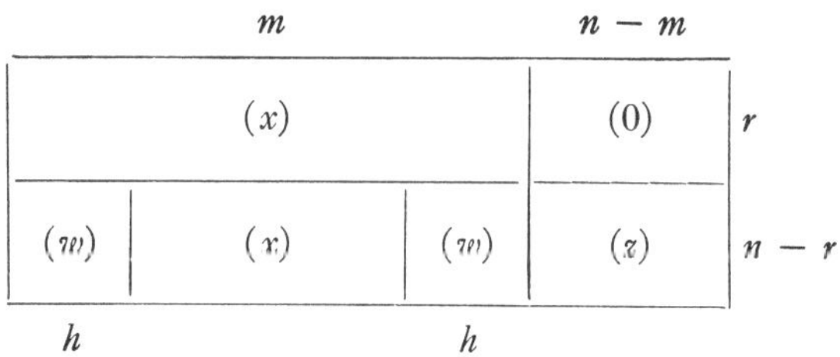

Still another type is the determinant represented by

(c)

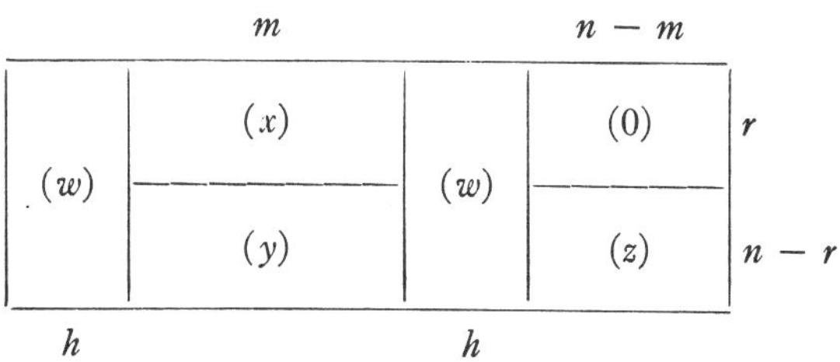

which is obviously zero.

145. The symbolic relation* obtained from the expansion of (a) is

$$(1)\quad \sum_{1}^{(n-r)_{m-r}} i(-1)^{\nu_1} \left| \begin{matrix} (\bar n \mid n - r_{\alpha_1})(n \mid n - r_{\alpha_1} \mid m - r_i) \\ (n \mid m_{\beta_1} \mid h_{\beta_2})(n \mid \bar m_{\beta_1} \mid h_{\beta_2}) \end{matrix} \right| \cdot \left| \begin{matrix} (n \mid \overline{n - r_{\alpha_1}} \mid m - r_i) \\ (\bar n \mid m_{\beta_1} \mid h_{\beta_3})(\bar n \mid \bar m_{\beta_1} \mid h_{\beta_2}) \end{matrix} \right|$$

$$= \sum_{1}^{(m-h)_{r-h}} j(-1)^{\nu_2} \left| \begin{matrix} (\bar n \mid n - r_{\alpha_1}) \\ (n \mid m_{\beta_1} \mid h_{\beta_2})(n \mid \bar m_{\beta_1} \mid h_{\beta_2} \mid r - h_j) \end{matrix} \right| \cdot \left| \begin{matrix} (n \mid n - r_{\alpha_1}) \\ (\bar n \mid m_{\beta_1} \mid h_{\beta_3})(\bar n \mid \bar m_{\beta_1} \mid h_{\beta_3})(n \mid \bar m_{\beta_1} \mid \bar h_{\beta_2} \mid r - h_j) \end{matrix} \right|$$

where the numbers in $(n \mid m_{\beta_1} \mid h_{\beta_2})$ and $(\bar n \mid m_{\beta_1} \mid h_{\beta_3})$ are the same by hypothesis. In the place of $(n \mid m_{\beta_1} \mid h_{\beta_2})(n \mid \bar m_{\beta_1} \mid h_{\beta_2})$ and $(\bar n \mid m_{\beta_1} \mid h_{\beta_3})(\bar n \mid \bar m_{\beta_1} \mid h_{\beta_3})$ we may of course write $(n \mid m_{\beta_1})$ and $(\bar n \mid m_{\beta_2})$ respectively if we remember that they have a set of h numbers common to the two.

The symbolic relation obtained from the expansion of (b) is

$$\sum_{1}^{(n-r)_{m-r-h}} i(-1)^{\nu_1} \left| \begin{matrix} (\bar n \mid n - r_{\alpha_1})(n \mid n - r_{\alpha_1} \mid m - r - h_i) \\ (n \mid m_{\beta_1} \mid 2h_{\beta_2} \mid h_{\beta_3})(n \mid \bar m_{\beta_1} \mid 2h_{\beta_2}) \end{matrix} \right| \cdot \left| \begin{matrix} (n \mid \overline{n - r_{\alpha_1}} \mid m - r - h_i) \\ (\bar n \mid m_{\beta_1})(n \mid m_{\beta_1} \mid \overline{2h_{\beta_2}} \mid h_{\beta_3}) \end{matrix} \right|$$

$$(2)\quad = \sum_{1}^{(m-2r)_{r-h}} j(-1)^{v_2} \left| \begin{matrix} (\bar n \mid n - r_{\alpha_1}) \\ (n \mid m_{\beta_1} \mid 2h_{\beta_2} \mid h_{\beta_3})(n \mid \bar m_{\beta_1} \mid 2h_{\beta_2} \mid r - h_j) \end{matrix} \right| \cdot \left| \begin{matrix} (n \mid n - r_{\alpha_1}) \\ (\bar n \mid m_{\beta_1})(n \mid m_{\beta_1} \mid \overline{2h_{\beta_2}} \mid h_{\beta_3})(n \mid \bar m_{\beta_1} \mid \overline{2h_{\beta_2}} \mid r - h_j) \end{matrix} \right|$$

where the numbers in $(n \mid m_{\beta_1} \mid 2h_{\beta_2} \mid h_{\beta_3})$ and $(n \mid m_{\beta_1} \mid \overline{2h_{\beta_2}} \mid h_{\beta_3})$ are the same.

If we proceed to write the symbolic relation obtained from the expansion of (c) we will find that it is exactly (2). This is due to the fact that the elements in (x) which are in the columns $(n \mid m_{\beta_1} \mid \overline{2h_{\beta_2}} \mid h_{\beta_3})$ are not involved in relation (2), and therefore the relation is independent of these elements.

146. We may of course have similar sets of rows with identical elements at the same time that they occur in the columns. Thus if k

* Whenever, as in this formula, it presents no printing difficulties we shall use $\left| \begin{matrix} x_1\, x_2 \cdots \\ y_1\, y_2 \cdots \end{matrix} \right|$ to indicate a determinant whose row numbers are $x_1\, x_2 \cdots$ and whose column numbers are $y_1\, y_2 \cdots$ instead of the notation gives in §67. For convenience in printing we shall also use $(n \mid m - r_i)$ to denote the i th selection of n numbers taken $m-r$ at a time.

is the number in each of two sets of rows with identical elements and we combine (1) and (2) the result is

$$\sum_{1}^{(n-r-2k)(m-r-k)} i \; (-1)^{\nu_1}$$

$$\left|\begin{matrix}(\bar{n}\,|\,n-r_{\alpha_1})(n\,|\,n-r_{\alpha_1}\,|\,2k_{\alpha_2}\,|\,k_{\alpha_3})(n\,|\,\overline{n-r_{\alpha_1}}\,|\,2k_{\alpha_2}\,|\,m-r-k_i)\\(n\,|\,m_{\beta_1})\end{matrix}\right|$$

$$\cdot\left|\begin{matrix}(n\,|\,n-r_{\alpha_1}\,|\,\overline{2k_{\alpha_2}}\,|\,k_{\alpha_3})(n\,|\,\overline{n-r_{\alpha_1}}\,|\,\overline{2k_{\alpha_2}}\,|\,m-r-k_i)\\(\bar{n}\,|\,m_{\beta_1})\end{matrix}\right|$$

$$(3)\qquad = \sum_{1}^{(m-h)_{r+k-h}} j(-1)^{\nu_2}\left|\begin{matrix}(\bar{n}\,|\,n-r_{\alpha_1})(n\,|\,n-r_{\alpha_1}\,|\,2k_{\alpha_2}\,|\,k_{\alpha_3})\\(n\,|\,m_{\beta_1}\,|\,h_{\beta_2})(n\,|\,\bar{m}_{\beta_1}\,|\,h_{\beta_2}\,|\,r+k-h_j)\end{matrix}\right|$$

$$\cdot\left|\begin{matrix}(n\,|\,n-r_{\alpha_1}\,|\,\overline{2k_{\alpha_2}}\,|\,k_{\alpha_3})(n\,|\,\overline{n-r_{\alpha_1}}\,|\,2k_{\alpha_2})\\(\bar{n}\,|\,m_{\beta_1})(n\,|\,\bar{m}_{\beta_1}\,|\,\bar{h}_{\beta_2}\,|\,r+k-h_j)\end{matrix}\right|,$$

where the numbers in $(n\,|n-r_{\alpha_1}|2k_{\alpha_2}|k_{\alpha_3})$ and $(n\,|n-r_{\alpha_1}|\overline{2k_{\alpha_2}}|k_{\alpha_3})$ are the same; and where the h numbers in $(n\,|m_{\beta_1}|h_{\beta_2})$ are found in $(\bar{n}\,|m_{\beta_1})$.

147. If in (3) §146 we put $k=0$ it reduces to (1).

If in (1) we put $n=2n$, $h=k$, $m=n+k$ and $r=n$, then it becomes the theorem of §137.

If in (1) $h=r$, then it becomes

$$\sum_{1}^{(n-r)_{m-r}} i(-1)^{\nu_1}\left|\begin{matrix}(\bar{n}\,|\,n-r_{\alpha_1})(n\,|\,n-r_{\alpha_1}\,|\,m-r_i)\\(n\,|\,m_{\beta_1})\end{matrix}\right|$$

$$\cdot\left|\begin{matrix}(n\,|\,\overline{n-r_{\alpha_1}}\,|\,m-r_i)\\(\bar{n}\,|\,m_{\beta_1})\end{matrix}\right|$$

$$(1')\qquad = (-1)^{\nu_2}\left|\begin{matrix}(\bar{n}\,|\,n-r_{\alpha_1})\\(n\,|\,m_{\beta_1}\,|\,r_{\beta_2})\end{matrix}\right|\cdot\left|\begin{matrix}(n\,|\,n-r_{\alpha_2})\\(\bar{n}\,|\,m_{\beta_1})(n\,|\,\bar{m}_{\beta_1}\,|\,r_{\beta_2})\end{matrix}\right|.$$

If in (1) $h>r$ then it becomes

$$(1'')\qquad \sum_{1}^{(n-r)_{m-r}} i(-1)^{\nu_1}\left|\begin{matrix}(\bar{n}\,|\,n-r_{\alpha_1})(n\,|\,n-r_{\alpha_1}\,|\,m-r_i)\\(n\,|\,m_{\beta_1})\end{matrix}\right| \cdot\left|\begin{matrix}(n\,|\,\overline{n-r_{\alpha_1}}\,|\,m-r_i)\\(\bar{n}\,|\,m_{\beta_1})\end{matrix}\right| = 0$$

Similar results would be obtained from (2) by putting first $h=r$, and then $h>r$.

It should be observed that in (1), h, the number of numbers common to the two factors in the columns, is less than r, the number of numbers constant in the rows, in (1′) these numbers are just equal, and in (1″) the first is greater than the second.

Example of (1). Let $n=8$, $m=5$, $r=4$, $h=2$,

$$(n\,|\,m_{\beta_1}) \equiv 12345, \qquad (\bar{n}\,|\,m_{\beta_1}) \equiv 678, \qquad (n\,|\,m_{\beta_1}\,|\,h_{\beta_2}) \equiv 12$$

$$(\bar{n}\,|\,n - r_{\alpha_1}) \equiv 1234, \qquad (n\,|\,n - r) \equiv 5678$$

$$\begin{vmatrix}(1234)\ (5)\\(12)\ (345)\end{vmatrix}\cdot\begin{vmatrix}(678)\\(12)\ (6)\end{vmatrix} - \begin{vmatrix}(1234)\ (6)\\(12)\ (345)\end{vmatrix}\cdot\begin{vmatrix}(578)\\(12)\ (6)\end{vmatrix}$$

$$+\begin{vmatrix}(1234)\ (7)\\(12)\ (345)\end{vmatrix}\cdot\begin{vmatrix}(568)\\(12)\ (6)\end{vmatrix} - \begin{vmatrix}(1234)\ (8)\\(12)\ (345)\end{vmatrix}\cdot\begin{vmatrix}(567)\\(12)\ (6)\end{vmatrix}$$

$$=\begin{vmatrix}(1234)\\(12)\ (34)\end{vmatrix}\cdot\begin{vmatrix}(5678)\\(12).(5).(6)\end{vmatrix} - \begin{vmatrix}(1234)\\(12)\ (35)\end{vmatrix}\cdot\begin{vmatrix}(5678)\\(12)\ (4)\ (6)\end{vmatrix}$$

$$+\begin{vmatrix}(1234)\\(12)\ (45)\end{vmatrix}\cdot\begin{vmatrix}(5678)\\(12).(3).(6)\end{vmatrix}.$$

This may be verified by expanding the second factor of each term on the right. Thus

$$\begin{vmatrix}(5678)\\(12)\ (5)\ (6)\end{vmatrix} = \begin{matrix}5\\5\end{matrix}\begin{vmatrix}678\\(12)\ (6)\end{vmatrix} - \begin{matrix}6\\5\end{matrix}\begin{vmatrix}578\\(12)\ (6)\end{vmatrix} + \begin{matrix}7\\5\end{matrix}\begin{vmatrix}568\\(12)\ (6)\end{vmatrix} - \begin{matrix}8\\5\end{matrix}\begin{vmatrix}567\\(12)\ (6)\end{vmatrix}$$

$$\begin{vmatrix}(5678)\\(12)\ (4)\ (6)\end{vmatrix} = \begin{matrix}5\\4\end{matrix}\begin{vmatrix}678\\(12)\ (6)\end{vmatrix} - \begin{matrix}6\\4\end{matrix}\begin{vmatrix}578\\(12)\ (6)\end{vmatrix} + \begin{matrix}7\\4\end{matrix}\begin{vmatrix}568\\(12)\ (6)\end{vmatrix} - \begin{matrix}8\\4\end{matrix}\begin{vmatrix}567\\(12)\ (6)\end{vmatrix}$$

$$\begin{vmatrix}(5678)\\(12)\ (3)\ (6)\end{vmatrix} = \begin{matrix}5\\3\end{matrix}\begin{vmatrix}678\\(12)\ (6)\end{vmatrix} - \begin{matrix}6\\3\end{matrix}\begin{vmatrix}578\\(12)\ (6)\end{vmatrix} + \begin{matrix}7\\3\end{matrix}\begin{vmatrix}568\\(12)\ (6)\end{vmatrix} - \begin{matrix}8\\3\end{matrix}\begin{vmatrix}567\\(12)\ (6)\end{vmatrix}$$

Then recombine them as follows

$$\begin{vmatrix}678\\(12)\ (6)\end{vmatrix}\left\{\begin{vmatrix}1234\\(12)\ (34)\end{vmatrix}\begin{matrix}5\\5\end{matrix} - \begin{vmatrix}1234\\(12)\ (35)\end{vmatrix}\begin{matrix}5\\4\end{matrix} + \begin{vmatrix}1234\\(12)\ (45)\end{vmatrix}\begin{matrix}5\\3\end{matrix}\right\}$$

$$-\begin{vmatrix}578\\(12)\ (6)\end{vmatrix}\left\{\begin{vmatrix}1234\\(12)\ (34)\end{vmatrix}\begin{matrix}6\\5\end{matrix} - \begin{vmatrix}1234\\(12)\ (35)\end{vmatrix}\begin{matrix}6\\4\end{matrix} + \begin{vmatrix}1234\\(12)\ (45)\end{vmatrix}\begin{matrix}5\\3\end{matrix}\right\}$$

$$+\begin{vmatrix}568\\(12)\ (6)\end{vmatrix}\left\{\begin{vmatrix}1234\\(12)\ (34)\end{vmatrix}\begin{matrix}7\\5\end{matrix} - \begin{vmatrix}1234\\(12)\ (35)\end{vmatrix}\begin{matrix}7\\4\end{matrix} + \begin{vmatrix}1234\\(12)\ (45)\end{vmatrix}\begin{matrix}7\\3\end{matrix}\right\}$$

$$-\begin{vmatrix}567\\(12)\ (6)\end{vmatrix}\left\{\begin{vmatrix}1234\\(12)\ (34)\end{vmatrix}\begin{matrix}8\\5\end{matrix} - \begin{vmatrix}1234\\(12)\ (35)\end{vmatrix}\begin{matrix}8\\4\end{matrix} + \begin{vmatrix}1234\\(12)\ (45)\end{vmatrix}\begin{matrix}8\\3\end{matrix}\right\}$$

and by addition get the four terms on the left.

Example of (1′). Let $n=7$, $m=5$, $r=2$, $h=2$,

$$(n \mid m_{\beta_1}) = 12345, \quad (\bar{n} \mid m_{\beta_1}) = 67,$$

$$(n \mid m_{\beta_1} \mid h_{\beta_2}) = 12, \quad (\bar{n} \mid n - r_{\alpha_1}) = 12, \quad (n \mid n - r_{\alpha_1}) = 34567$$

$$\begin{vmatrix} (12)\,345 \\ 12\ \ 345 \end{vmatrix} \cdot \begin{vmatrix} 67 \\ 12 \end{vmatrix} - \begin{vmatrix} 12346 \\ 12345 \end{vmatrix} \cdot \begin{vmatrix} 57 \\ 12 \end{vmatrix} + \begin{vmatrix} 12347 \\ 12345 \end{vmatrix} \cdot \begin{vmatrix} 56 \\ 12 \end{vmatrix} + \begin{vmatrix} 12356 \\ 12345 \end{vmatrix} \cdot \begin{vmatrix} 47 \\ 12 \end{vmatrix}$$

$$- \begin{vmatrix} 12357 \\ 12345 \end{vmatrix} \cdot \begin{vmatrix} 46 \\ 12 \end{vmatrix} + \begin{vmatrix} 12367 \\ 12345 \end{vmatrix} \cdot \begin{vmatrix} 45 \\ 12 \end{vmatrix} - \begin{vmatrix} 12456 \\ 12345 \end{vmatrix} \cdot \begin{vmatrix} 37 \\ 12 \end{vmatrix} + \begin{vmatrix} 12457 \\ 12345 \end{vmatrix} \cdot \begin{vmatrix} 36 \\ 12 \end{vmatrix}$$

$$- \begin{vmatrix} 12467 \\ 12345 \end{vmatrix} \cdot \begin{vmatrix} 35 \\ 12 \end{vmatrix} + \begin{vmatrix} 12567 \\ 12345 \end{vmatrix} \cdot \begin{vmatrix} 34 \\ 12 \end{vmatrix} = \begin{vmatrix} 12 \\ 12 \end{vmatrix} \cdot \begin{vmatrix} 34567 \\ 12345 \end{vmatrix}.$$

Example of (1″). Let $n=5$, $m=3$, $r=1$, $h=2$,

$$(n \mid m_{\beta_1}) = 123, \qquad (\bar{n} \mid m_{\beta_1}) = 45$$

$$(n \mid m_{\beta_1} \mid h_{\beta_2}) = 12, \quad (\bar{n} \mid n - r_{\alpha_1}) = 1, \quad (n \mid n - r_{\alpha_1}) = 2345$$

$$\begin{vmatrix} (1)\ \ (23) \\ (12)\ \ (3) \end{vmatrix} \cdot \begin{vmatrix} 45 \\ 12 \end{vmatrix} - \begin{vmatrix} 124 \\ 123 \end{vmatrix} \cdot \begin{vmatrix} 35 \\ 12 \end{vmatrix} + \begin{vmatrix} 125 \\ 123 \end{vmatrix} \cdot \begin{vmatrix} 34 \\ 12 \end{vmatrix} - \begin{vmatrix} 134 \\ 123 \end{vmatrix} \cdot \begin{vmatrix} 25 \\ 12 \end{vmatrix}$$

$$+ \begin{vmatrix} 135 \\ 123 \end{vmatrix} \cdot \begin{vmatrix} 24 \\ 12 \end{vmatrix} - \begin{vmatrix} 145 \\ 123 \end{vmatrix} \cdot \begin{vmatrix} 23 \\ 12 \end{vmatrix} = 0$$

Example of (3). Let $n=8$, $m=6$, $r=3$, $h=1$, $k=1$,

$$(\bar{n} \mid n - r_{\alpha_1}) \equiv 123, \qquad (n \mid n - r_{\alpha_1}) \equiv 45678$$

$$(n \mid m_{\beta_1}) \equiv 123456, \qquad (\bar{n} \mid m_{\beta_1}) \equiv 78,$$

$$(n \mid m_{\beta_1} \mid h_{\beta_2}) \equiv 1, \qquad (n \mid n - r_{\alpha_1} \mid 2k_{\alpha_2}) \equiv 48,$$

$$(n \mid n - r_{\alpha_1} \mid 2k_{\alpha_2} \mid k_{\alpha_3}) \equiv 4, \qquad (n \mid n - r_{\alpha_1} \mid \overline{2k_{\alpha_2}} \mid k_{\alpha_3}) \equiv 8,$$

$8 \equiv 4$ in rows, $8 \equiv 1$ in columns.

$$\begin{vmatrix} (123)\ (4)\ (56) \\ 123456 \end{vmatrix} \cdot \begin{vmatrix} 47 \\ 17 \end{vmatrix} - \begin{vmatrix} (123)\ (4)\ (57) \\ 123456 \end{vmatrix} \cdot \begin{vmatrix} 46 \\ 17 \end{vmatrix} + \begin{vmatrix} (123)\ (4)\ (67) \\ 123456 \end{vmatrix} \cdot \begin{vmatrix} 45 \\ 17 \end{vmatrix}$$

$$= \begin{vmatrix} 1234 \\ (1)\,234 \end{vmatrix} \cdot \begin{vmatrix} 4567 \\ (17)\,56 \end{vmatrix} - \begin{vmatrix} 1234 \\ (1)\,235 \end{vmatrix} \cdot \begin{vmatrix} 4567 \\ (17)\,46 \end{vmatrix} + \begin{vmatrix} 1234 \\ (1)\,236 \end{vmatrix} \cdot \begin{vmatrix} 4567 \\ (17)\,45 \end{vmatrix}$$

$$+ \begin{vmatrix} 1234 \\ (1)\,245 \end{vmatrix} \cdot \begin{vmatrix} 4567 \\ (17)\,36 \end{vmatrix} - \begin{vmatrix} 1234 \\ (1)\,246 \end{vmatrix} \cdot \begin{vmatrix} 4567 \\ (17)\,35 \end{vmatrix} + \begin{vmatrix} 1234 \\ (1)\,256 \end{vmatrix} \cdot \begin{vmatrix} 4567 \\ (17)\,34 \end{vmatrix}$$

$$- \begin{vmatrix} 1234 \\ (1)\,345 \end{vmatrix} \cdot \begin{vmatrix} 4567 \\ (17)\,26 \end{vmatrix} + \begin{vmatrix} 1234 \\ (1)\,346 \end{vmatrix} \cdot \begin{vmatrix} 4567 \\ (17)\,25 \end{vmatrix} - \begin{vmatrix} 1234 \\ (1)\,356 \end{vmatrix} \cdot \begin{vmatrix} 4567 \\ (17)\,24 \end{vmatrix}$$

$$+ \begin{vmatrix} 1234 \\ (1)\,456 \end{vmatrix} \cdot \begin{vmatrix} 4567 \\ (17)\,23 \end{vmatrix}.$$

148. *The product of a determinant and any one of its minors M is expressible as an aggregate of products of pairs of minors: the first factors of the products being obtained by taking q rows in which the rows of M are included and forming from them every minor of the qth order which contains M, the second factor of any product being that minor which includes M and the complementary of the first factor, and the sign of any product being fixed by transforming the second factor so as to have its principal diagonal coincident with those of the two minors which it was formed to include, and then taking + or − according as the sum of the numbers indicating the rows and columns from which the first factor was formed is even or odd.*

Let $|a_{1n}|$ be the given determinant, $|a_{pp}a_{p+1,p+1} \cdots a_{qq}|$ the chosen minor M, and let the q rows of $|a_{1n}|$ taken to include the rows of M be the *first* q rows.

We have at once an equivalent for the product of $|a_{1n}|$ and its minor M, by forming a new determinant in which $|a_{1n}|$ and M are complementary minors, the former being situated in the first n rows and first n columns with nothing but zero-elements below it. Such a determinant is

$$\begin{vmatrix}
a_{11} & a_{12} & \cdots & a_{1,p-1} & a_{1p} & \cdots & a_{1q} & a_{1,q+1} & \cdots & a_{1n} & 0 & \cdots & 0 \\
a_{21} & a_{22} & \cdots & a_{2,p-1} & a_{2p} & \cdots & a_{2q} & a_{2,q+1} & \cdots & a_{2n} & 0 & \cdots & 0 \\
\cdots & \cdots & \cdots & \cdots & \cdots & \cdots & \cdots & \cdots & \cdots & \cdots & \cdots & \cdots & \cdots \\
a_{p-1,1} & a_{p-1,2} & \cdots & a_{p-1,p-1} & a_{p-1,p} & \cdots & a_{p-1,q} & a_{p-1,q+1} & \cdots & a_{p-1,n} & 0 & \cdots & 0 \\
a_{p1} & a_{p2} & \cdots & a_{p,p-1} & a_{pp} & \cdots & a_{pq} & a_{p,q+1} & \cdots & a_{pn} & 0 & \cdots & 0 \\
\cdots & \cdots & \cdots & \cdots & \cdots & \cdots & \cdots & \cdots & \cdots & \cdots & \cdots & \cdots & \cdots \\
a_{q1} & a_{q2} & \cdots & a_{q,p-1} & a_{qp} & \cdots & a_{qq} & a_{q,q+1} & \cdots & a_{qn} & 0 & \cdots & 0 \\
a_{q+1,1} & a_{q+1,2} & \cdots & a_{q+1,p-1} & a_{q+1,p} & \cdots & a_{q+1,q} & a_{q+1,q+1} & \cdots & a_{q+1,n} & a_{q+1,p} & \cdots & a_{q+1,q} \\
\cdots & \cdots & \cdots & \cdots & \cdots & \cdots & \cdots & \cdots & \cdots & \cdots & \cdots & \cdots & \cdots \\
a_{n1} & a_{n2} & \cdots & a_{n,p-1} & a_{np} & \cdots & a_{nq} & a_{n,q+1} & \cdots & a_{nn} & a_{np} & \cdots & a_{nq} \\
0 & 0 & \cdots & 0 & 0 & \cdots & 0 & 0 & \cdots & 0 & a_{pp} & \cdots & a_{pq} \\
\cdots & \cdots & \cdots & \cdots & \cdots & \cdots & \cdots & \cdots & \cdots & \cdots & \cdots & \cdots & \cdots \\
0 & 0 & \cdots & 0 & 0 & \cdots & 0 & 0 & \cdots & 0 & a_{qp} & \cdots & a_{qq}
\end{vmatrix}$$

or Δ, say, where it has to be specially noticed that the q chosen rows of $|a_{1n}|$ are prolonged with zeros, and that each of the other rows is prolonged by repeating in order the elements of it which are in the same column with any of the elements of M.

In Δ the minor M occurs twice. Adding each element of the first set of rows to which M belongs to the corresponding element of the second set, and then subtracting each element of the second set of columns to which M belongs from the corresponding element of the first set, we find

$$\Delta=\left|\begin{array}{ccccccccccccc}
a_{11} & a_{12} & \cdots & a_{1,p-1} & a_{1p} & \cdots & a_{1q} & a_{1,q+1} & \cdots & a_{1n} & 0 & \cdots & 0 \\
a_{21} & a_{22} & \cdots & a_{2,p-1} & a_{2p} & \cdots & a_{2q} & a_{2,q+1} & \cdots & a_{2n} & 0 & \cdots & 0 \\
\cdots & \cdots & \cdots & \cdots & \cdots & \cdots & \cdots & \cdots & \cdots & \cdots & \cdots & \cdots & \cdots \\
a_{p-1,1} & a_{p-1,2} & \cdots & a_{p-1,p-1} & a_{p-1,p} & \cdots & a_{p-1,q} & a_{p-1,q+1} & \cdots & a_{p-1,n} & 0 & \cdots & 0 \\
a_{p1} & a_{p2} & \cdots & a_{p,p-1} & a_{pp} & \cdots & a_{pq} & a_{p,q+1} & \cdots & a_{pn} & 0 & \cdots & 0 \\
\cdots & \cdots & \cdots & \cdots & \cdots & \cdots & \cdots & \cdots & \cdots & \cdots & \cdots & \cdots & \cdots \\
a_{q1} & a_{q2} & \cdots & a_{q,p-1} & a_{pq} & \cdots & a_{qq} & a_{q,q+1} & \cdots & a_{qn} & 0 & \cdots & 0 \\
a_{q+1,1} & a_{q+1,2} & \cdots & a_{q+1,p-1} & 0 & \cdots & 0 & a_{q+1,q+1} & \cdots & a_{q+1,n} & a_{q+1,p} & \cdots & a_{q+1,q} \\
\cdots & \cdots & \cdots & \cdots & \cdots & \cdots & \cdots & \cdots & \cdots & \cdots & \cdots & \cdots & \cdots \\
a_{n1} & a_{n2} & \cdots & a_{n,p-1} & 0 & \cdots & 0 & a_{n,q+1} & \cdots & a_{nn} & a_{np} & \cdots & a_{nq} \\
a_{p1} & a_{p2} & \cdots & a_{p,p-1} & 0 & \cdots & 0 & a_{p,q+1} & \cdots & a_{pn} & a_{pp} & \cdots & a_{pq} \\
\cdots & \cdots & \cdots & \cdots & \cdots & \cdots & \cdots & \cdots & \cdots & \cdots & \cdots & \cdots & \cdots \\
a_{q1} & a_{q2} & \cdots & a_{q,p-1} & 0 & \cdots & 0 & a_{q,q+1} & \cdots & a_{qn} & a_{qp} & \cdots & a_{qq}
\end{array}\right|$$

If now we take the first q rows of this determinant, and form every minor of the qth order preparatory to finding the expansion of the determinant as an aggregate of products of complementary minors, we see that, although the full list of minors would be exactly the same as if we had been dealing with $|a_{1n}|$, we need take only those which include the selected minor M, because all the others have here complementaries which vanish; also, we see that each of the complementaries of those thus taken includes the complementary of the same minor in $|a_{1n}|$ and the selected minor besides, and that each is itself a minor of $|a_{1n}|$, being formed from those $n-p+1$ rows of $|a_{1n}|$ which are made up of the $n-q$ rows not included in the chosen q rows and the $q-p+1$ rows in which M is situated.

But this aggregate of products is exactly the aggregate of products specified in the enunciation of the theorem, and as it is the equivalent of Δ and therefore of $|a_{1n}|\times M$, the theorem has been established.

Examples. Taking the determinant

$$\left|\begin{array}{ccccc}
a_1 & a_2 & a_3 & a_4 & a_5 \\
b_1 & b_2 & b_3 & b_4 & b_5 \\
c_1 & c_2 & c_3 & c_4 & c_5 \\
d_1 & d_2 & d_3 & d_4 & d_5 \\
e_1 & e_2 & e_3 & e_4 & e_5
\end{array}\right|$$

and its minor

$$\left|\begin{array}{cc}
b_3 & b_4 \\
c_3 & c_4
\end{array}\right|,$$

we have

$$\begin{aligned} |b_3c_4|\,|a_1b_2c_3d_4e_5| &= |a_1b_2c_3d_4|\,|b_3c_4e_5| - |a_1b_3c_4d_5|\,|b_3c_4e_2| \\ &\quad - |a_2b_3c_4{,}d_5| \cdot |b_3c_4{,}e_1| , \\ &= -|a_1b_2c_3e_4|\,|b_3c_4d_5| \\ &\quad + |a_1b_3c_4e_5|\,|b_3c_4d_2| \\ &\quad - |a_2b_3c_4e_5|\,|b_3c_4d_1| , \\ &= \cdots , \end{aligned}$$

the factors of the fourth order being formed in the first case from the first, second, third, fourth rows, and in the second from the first, second, third, fifth rows.

Taking a minor of the next lower order for the cofactor of $|a_1b_2c_3d_4e_5|$, we have

$$\begin{aligned} d_2|a_1b_2c_3d_4e_5| &= |c_1d_2e_3|\,|a_4b_5d_2| - |c_1d_2e_4|\,|a_3b_5d_2| \\ &\quad + |c_1d_2e_5|\,|a_3b_4d_2| + |c_2d_3e_4|\,|a_1b_5d_2| \\ &\quad - |c_2d_3e_5|\,|a_1b_4d_2| \\ &\quad + |c_2d_4e_5|\,|a_1b_3d_2| \\ &= \cdots , \\ &= -|d_1e_2|\,|a_3b_4c_5d_2| - |d_2e_3|\,|a_1b_4c_5d_2| \\ &\quad - |d_2e_4|\,|a_1b_3c_5d_2| \\ &\quad + |d_2e_5|\,|a_1b_3c_4d_2| \\ &= \cdots . \end{aligned}$$

If the cofactor of $|a_1b_2c_3d_4e_5|$ be of a lower order still, namely, the order 0, and as such be taken equal to unity, we have the theorem of §93, which in this way we may view as being here generalized.

149. At the opposite extreme from the theorem of §93 viewed as a case of the foregoing, we have another theorem of sufficient importance to be specially noticed. This is the case in which, the original determinant being of the nth order, its cofactor M is of the $(n-2)$th. Here the rows from which the first factors of the development are formed must be $n-1$ in number, and as the said first factors must include M, there can be only two of them, so that the development must consist of two terms which are each the product of two determinants of the $(n-1)$th order.

EXAMPLES.

$$|b_2 c_3 d_4|\,|a_1 b_2 c_3 d_4 e_5| = |b_2 c_3 d_4 e_5|\,|a_1 b_2 c_3 d_4| - |b_1 c_2 d_3 e_4|\,|a_2 b_3 c_4 d_5| \;;$$

$$|a_1 b_2 c_3|\,|a_1 b_2 c_3 d_4 e_5| = |a_1 b_2 c_3 d_4|\,|a_1 b_2 c_3 e_5| - |a_1 b_2 c_3 d_5|\,|a_1 b_2 c_3 e_4| \;.$$

Denoting $|a_1b_2c_3d_4e_5|$ by D we may (§87) write the first of these in the form

$$D\frac{\partial^2 D}{\partial a_1 \partial e_5} = \frac{\partial D}{\partial a_1}\,\frac{\partial D}{\partial a_5} - \frac{\partial D}{\partial a_5}\,\frac{\partial D}{\partial e_1}\;;$$

similarly, the second; and, quite generally, we have

$$D(a_{1n})\frac{\partial^2 D}{\partial a_{hr}\partial a_{ks}} = \frac{\partial D}{\partial a_{hr}}\,\frac{\partial D}{\partial a_{ks}} - \frac{\partial D}{\partial a_{hs}}\,\frac{\partial D}{\partial a_{kr}}\,,$$

the form in which the theorem is commonly quoted.

The theorems of §§93, 148 are connected in another way, which it is of still greater importance to observe. As an instance of the latter theorem we have (§148),

$$|b_3 c_4|\,|a_1 b_2 c_3 d_4 e_5| = |a_1 b_2 c_3 d_4|\,|b_3 c_4 e_5| - |a_1 b_3 c_4 d_5|\,|b_3 c_4 e_2| + |a_2 b_3 c_4 d_5|\,|b_3 c_4 e_1|$$

If now, in place of each determinant here, we substitute the cofactor which it has in $|a_1b_2c_3d_4e_5|$, we obtain the statement

$$|a_1 d_2 e_5| = e_5|a_1 d_2| - e_2|a_1 d_5| + e_1|a_2 d_5|\,,$$

which is at once recognized as an instance of the theorem of §93. The identities

$$\begin{aligned} d_2|a_1 b_2 c_3 d_4 e_5| = {} & |c_1 d_2 e_3|\,|a_4 b_5 d_2| - |c_1 d_2 e_4|\,|a_3 b_5 d_2| \\ & + |c_1 d_2 e_5|\,|a_3 b_4 d_2| + |c_2 d_3 e_4|\,|a_1 b_5 d_2| \\ & - |c_2 d_3 e_5|\,|a_1 b_4 d_2| \\ & + |c_2 d_4 e_5|\,|a_1 b_3 d_2|\,, \end{aligned}$$

$$\begin{aligned} |a_1 b_3 c_4 e_5| = {} & |a_4 b_5|\,|c_1 e_3| - |a_3 b_5|\,|c_1 e_4| + |a_3 b_4|\,|c_1 e_5| \\ & + |a_1 b_5|\,|c_3 e_4| - |a_1 b_4|\,|c_3 e_5| + |a_1 b_3|\,|c_4 e_5|\,, \end{aligned}$$

are similarly related; and generally it is found that to every instance of the one theorem there corresponds in this way an instance of the other, so that the two may be spoken of as *complementary theorems.*

150. *If the first k elements in the first $k-1$ rows of a determinant Δ_n be taken with the first k elements of the other rows in succession to form as determinants of the kth order the elements of the first column of a new determinant Δ_{n-k+1}, and if the determinants formed in like manner from the second k consecutive columns of Δ_n be made the elements of the second column of Δ_{n-k+1}, and so on, then Δ_n is equal to Δ_{n-k+1} divided by the product of all the determinants of the $(k-1)$th order, except the first and last, formed from the first $k-1$ rows and any $k-1$ consecutive columns of Δ_n.*

The case of this in which $k=2$ has been already established (§77); so that if $\Delta_n = |a_{1n}|$, we know that

$$(1)\quad \begin{vmatrix} a_{11} & a_{12} & a_{13} & \cdots & a_{1n} \\ a_{21} & a_{22} & a_{23} & \cdots & a_{2n} \\ a_{31} & a_{32} & a_{33} & \cdots & a_{3n} \\ \cdots & \cdots & \cdots & \cdots & \cdots \\ a_{n1} & a_{n2} & a_{n3} & \cdots & a_{nn} \end{vmatrix}$$

$$= \frac{1}{a_{12}\,a_{13}\cdots a_{1,n-1}} \begin{vmatrix} |a_{11}\,a_{22}| & |a_{12}\,a_{23}| & \cdots & |a_{1,n-1}\,a_{2n}| \\ |a_{11}\,a_{32}| & |a_{12}\,a_{33}| & \cdots & |a_{1,n-1}\,a_{3n}| \\ \cdots & \cdots & \cdots & \cdots \\ |a_{11}\,a_{n2}| & |a_{12}\,a_{n3}| & \cdots & |a_{1,n-1}\,a_{nn}| \end{vmatrix}.$$

Transforming the second determinant here by means of the identity in which it occurs, and putting

$$\begin{vmatrix} |a_{11}\,a_{22}| & |a_{12}\,a_{23}| \\ |a_{11}\,a_{32}| & |a_{12}\,a_{33}| \end{vmatrix} = a_{12}\,|a_{11}\,a_{22}\,a_{33}|,$$

$$\begin{vmatrix} |a_{12}\,a_{23}| & |a_{13}\,a_{24}| \\ |a_{12}\,a_{33}| & |a_{13}\,a_{34}| \end{vmatrix} = a_{13}\,|a_{12}\,a_{23}\,a_{34}|,$$

$$\cdots\cdots\cdots\cdots\cdots\cdots$$

as §149 entitles us to do, we obtain the result

$$\frac{\begin{vmatrix} a_{12}|a_{11}\,a_{22}\,a_{33}| & a_{13}|a_{12}\,a_{23}\,a_{34}| & \cdots & a_{1,n-1}|a_{1,n-2}\,a_{2,n-1}\,a_{3n}| \\ a_{12}|a_{11}\,a_{22}\,a_{43}| & a_{13}|a_{12}\,a_{23}\,a_{44}| & \cdots & a_{1,n-1}|a_{1,n-2}\,a_{2,n-1}\,a_{4n}| \\ \cdots & \cdots & \cdots & \cdots \\ a_{12}|a_{11}\,a_{22}\,a_{n3}| & a_{13}|a_{12}\,a_{23}\,a_{n4}| & \cdots & a_{1,n-1}|a_{1,n-2}\,a_{2,n-1}\,a_{nn}| \end{vmatrix}}{|a_{12}\,a_{23}|\;|a_{13}\,a_{24}|\cdots|a_{1,n-2}\,a_{2,n-1}|},$$

so that by substitution in (1) we have

$$
(2)\quad |a_{1n}| = \frac{\begin{vmatrix} |a_{11}\,a_{22}\,a_{33}| & |a_{12}\,a_{23}\,a_{34}| & \cdots & |a_{1,n-2}\,a_{2,n-1}\,a_{3n}| \\ |a_{11}\,a_{22}\,a_{43}| & |a_{12}\,a_{23}\,a_{44}| & \cdots & |a_{1,n-2}\,a_{2,n-1}\,a_{4n}| \\ \cdots & \cdots & \cdots & \cdots \\ |a_{11}\,a_{22}\,a_{n3}| & |a_{12}\,a_{23}\,a_{n4}| & \cdots & |a_{1,n-2}\,a_{2,n-1}\,a_{nn}| \end{vmatrix}}{|a_{12}\,a_{23}|\;|a_{13}\,a_{24}|\cdots|a_{1,n-2}\,a_{2,n-1}|},
$$

which is the next case of the theorem.

In exactly similar fashion it follows from this that

$$
(3)\quad |a_{1n}| = \frac{\begin{vmatrix} |a_{11}\,a_{22}\,a_{33}\,a_{44}| & \cdots & |a_{1,n-3}\,a_{2,n-2}\,a_{3,n-1}\,a_{4n}| \\ |a_{11}\,a_{22}\,a_{33}\,a_{54}| & \cdots & |a_{1,n-3}\,a_{2,n-2}\,a_{3,n-1}\,a_{5n}| \\ \cdots & \cdots & \cdots \\ |a_{11}\,a_{22}\,a_{33}\,a_{n4}| & \cdots & |a_{1,n-3}\,a_{2,n-2}\,a_{3,n-1}\,a_{nn}| \end{vmatrix}}{|a_{12}\,a_{23}\,a_{34}|\cdots|a_{1,n-3}\,a_{2,n-2}\,a_{3,n-1}|},
$$

and so on; the extreme case being the identity repeatedly used in the demonstration, namely, that of §149.

Similarly an extension of the identity of Ex. 5, §78, might be established, giving

$$
|a_{1n}| = \begin{vmatrix} |a_{11}\,a_{22}\,a_{33}| & |a_{11}\,a_{22}\,a_{34}| & \cdots & |a_{11}\,a_{22}\,a_{3n}| \\ |a_{11}\,a_{22}\,a_{43}| & |a_{11}\,a_{22}\,a_{44}| & \cdots & |a_{11}\,a_{22}\,a_{4n}| \\ \cdots & \cdots & \cdots & \cdots \\ |a_{11}\,a_{22}\,a_{n3}| & |a_{11}\,a_{22}\,a_{n4}| & \cdots & |a_{11}\,a_{22}\,a_{nn}| \end{vmatrix} \div |a_{11}\,a_{22}|^{n-3},
$$

and so forth.

CHAPTER V

MULTIPLICATION

151. *The product of a determinant of the nth order by an expression of n terms is equal to the sum of n determinants, the first of which is got from the given determinant by multiplying each element of the first row by the corresponding term of the given expression, the second by multiplying similarly each element of the second row, the third by multiplying similarly each element of the third row, and so on.*

Let the given determinant be

$$A \equiv (a_{11}\, a_{22} \cdots a_{nn}),$$

and the given expression

$$u_1 + u_2 + \cdots + u_n,$$

and then the n determinants referred to are

$$\begin{vmatrix} u_1a_{11} & u_2a_{12} & \cdots & u_na_{1n} \\ a_{21} & a_{22} & \cdots & a_{2n} \\ \cdot & \cdot & \cdot & \cdot \\ a_{n1} & aa_{n2} & \cdots & a_{nn} \end{vmatrix}, \quad \begin{vmatrix} a_{11} & a_{12} & \cdots & a_{1n} \\ u_1a_{21} & u_2a_{22} & \cdots & u_na_{2n} \\ \cdot & \cdot & \cdot & \cdot \\ a_{n1} & a_{n2} & \cdots & a_{nn} \end{vmatrix} \quad \cdots .$$

Now the coefficient of u_1 in the first of them is evidently $a_{11}A_{11}$, in the second $-a_{21}A_{21}$, in the third $a_{31}A_{31}$, and so on: therefore in the sum of the n determinants the coefficient of u_1 is $a_{11}A_{11} - a_{21}A_{21} + a_{31}A_{31} - \cdots a_{n1}A_{n1}$ or A.

Similarly the coefficient of u_2 is seen to be $-a_{12}A_{12} + a_{22}A_{22} - \cdots a_{n2}A_{n2}$ or A, and so on.

Hence the sum of the n determinants is

$$(u_1 + u_2 + \cdots + u_n)A.$$

Let $F(0, n_1, n_2, n_3, \cdots, n_k)$ denote the determinant

$$\begin{vmatrix} 1 & 1 & \cdots & 1 \\ a_1^{n_1} & a_2^{n_1} & \cdots & a_{k+1}^{n_1} \\ a_1^{n_2} & a_2^{n_2} & \cdots & a_{k+1}^{n_2} \\ \cdot & \cdot & \cdot & \cdot \\ a_1^{n_k} & a_2^{n_k} & \cdots & a_{k+1}^{n_k} \end{vmatrix} \quad \text{of order } (k+1)$$

then as an immediate application of the foregoing theorem we have

$$(a_1 + a_2 + \cdots + a_{k+1})F(0,n_1,n_2, \cdots ,n_k) = F(1,n_1,n_2, \cdots ,n_k)$$
$$+F(0,n_1 + 1,n_2, \cdots , n_k) + F(0,n_1,n_2 + 1, \cdots , n_k)$$
$$+ \cdots + F(0,n_1,n_2, \cdots , n_k + 1),$$

a theorem due to Malet.

152. More generally we have

$$(a_1^\alpha + a_2^\alpha + a_3^\alpha + \cdots + a_{k+1}^\alpha)F(0,n_1,n_2, \cdots , n_k)$$
$$= F(\alpha,n_1,n_2, \cdots , n_k) + F(0,n_1 + \alpha,n_2, \cdots , n_k)$$
$$+ F(0,n_1,n_2 + \alpha, \cdots , n_k) + \cdots + F(0,n_1,n_2, \cdots , n_k + \alpha).$$

153. *The product of two determinants of the third order is itself a determinant of the third order.*

Let

$$A = | a_{11}\, a_{22}\, a_{33} |, \quad B = | b_{11}\, b_{22}\, b_{33} |,$$

$$C = \begin{vmatrix} a_{11}b_{11}+a_{12}b_{12}+a_{13}b_{13} & a_{11}b_{21}+a_{12}b_{22}+a_{13}b_{23} & a_{11}b_{31}+a_{12}b_{32}+a_{13}b_{33} \\ a_{21}b_{11}+a_{22}b_{12}+a_{23}b_{13} & a_{21}b_{21}+a_{22}b_{22}+a_{23}b_{23} & a_{21}b_{31}+a_{22}b_{32}+a_{23}b_{33} \\ a_{31}b_{11}+a_{32}b_{12}+a_{33}b_{13} & a_{31}b_{21}+a_{32}b_{22}+a_{33}b_{23} & a_{31}b_{31}+a_{32}b_{32}+a_{33}b_{33} \end{vmatrix}$$

where we observe the element in the first row and first column of C, namely, $a_{11}b_{11}+a_{12}b_{12}+a_{13}b_{13}$, is formed by multiplying each element of the first row of A by the corresponding element of the first row of B, and adding the products thus formed, that the element of the first row and second column of C, namely, $a_{11}b_{21}+a_{12}b_{22}+a_{13}b_{23}$, is formed in like manner from the first row of A and the second row of B, and that, generally, the element in the pth row and qth column of C is formed in this manner from the pth row of A and the qth row of B. Now the elements of C being all trinomial, the determinant may be partitioned into twenty-seven determinants having all their elements monomial. Of these, however, twenty-one will be found to vanish on account of the existence in them of identical columns; thus

$$\begin{vmatrix} a_{11}b_{11} & a_{11}b_{12} & a_{13}b_{13} \\ a_{21}b_{11} & a_{21}b_{12} & a_{23}b_{13} \\ a_{31}b_{11} & a_{31}b_{12} & a_{33}b_{13} \end{vmatrix} = b_{11}b_{12}b_{13} \begin{vmatrix} a_{11} & a_{11} & a_{13} \\ a_{21} & a_{21} & a_{23} \\ a_{31} & a_{31} & a_{33} \end{vmatrix} = 0.$$

Indeed it is clear that in forming the twenty-seven determinants we need not take the set of first terms in the first column of C along with the set of first terms in either of the other two columns, nor the set of second terms in the first column along with the set of second terms in either of the other two columns, nor the set of third terms in the first column along with the set of third terms in either of the other two columns. The only determinants which do not vanish will therefore be composed of a set of first terms taken from one column, a set of second terms from another column, and a set of third terms taken from the remaining column; and the number of them will consequently be the number of permutations of the numbers 1, 2, 3, that is, 6. In agreement with this the result will be found to be

$$C = \begin{vmatrix} a_{11}b_{11} & a_{12}b_{22} & a_{13}b_{33} \\ a_{21}b_{11} & a_{22}b_{22} & a_{23}b_{33} \\ a_{31}b_{11} & a_{32}b_{22} & a_{33}b_{33} \end{vmatrix} - \begin{vmatrix} a_{11}b_{11} & a_{12}b_{32} & a_{13}b_{23} \\ a_{21}b_{11} & a_{22}b_{32} & a_{23}b_{23} \\ a_{31}b_{11} & a_{32}b_{32} & a_{33}b_{23} \end{vmatrix}$$

$$- \begin{vmatrix} a_{11}b_{21} & a_{12}b_{12} & a_{13}b_{33} \\ a_{21}b_{21} & a_{22}b_{12} & a_{23}b_{33} \\ a_{31}b_{21} & a_{32}b_{12} & a_{33}b_{33} \end{vmatrix} + \begin{vmatrix} a_{11}b_{12} & a_{12}b_{32} & a_{13}b_{13} \\ a_{21}b_{21} & a_{22}b_{32} & a_{23}b_{13} \\ a_{31}b_{21} & a_{32}b_{32} & a_{33}b_{13} \end{vmatrix}$$

$$- \begin{vmatrix} a_{11}b_{31} & a_{12}b_{22} & a_{13}b_{13} \\ a_{21}b_{31} & a_{22}b_{22} & a_{23}b_{13} \\ a_{31}b_{31} & a_{32}b_{22} & a_{33}b_{13} \end{vmatrix} + \begin{vmatrix} a_{11}b_{31} & a_{12}b_{12} & a_{13}b_{23} \\ a_{21}b_{31} & a_{22}b_{12} & a_{23}b_{23} \\ a_{31}b_{31} & a_{32}b_{12} & a_{33}b_{23} \end{vmatrix}$$

$$= b_{11}b_{22}b_{33}A - b_{11}b_{32}b_{23}A - b_{21}b_{12}b_{33}A + b_{21}b_{32}b_{13}A$$
$$- b_{31}b_{22}b_{13}A + b_{31}b_{12}b_{23}A = AB.$$

154. There are obviously four ways in which multiplication may take place, namely, (1) row by row, (2) column by column, (3) row by column, (4) column by row, and since by §37 a determinant is not altered by interchanging rows and columns it is seen that all four give results having the same value.

155. THE MULTIPLICATION THEOREM. *If two determinants A and B of the same order be given, and a new determinant C be formed such that in every case the element of its pth row and qth column is obtained by multiplying each element of the pth row of A by the corresponding element of the qth row of B, and adding the products thus formed, then $C=AB$.*

Let the two determinants A and B be $|a_{11}a_{22} \cdots a_{nn}|$ and $|b_{11}b_{22} \cdots b_{nn}|$ respectively. The element in the pth row and qth column of C is $a_{p1}b_{q1}+a_{p2}b_{q2}+ \cdots +a_{pn}b_{qn}$. Allowing q to take successively the values $1, 2, \cdots, n$ we get the n elements in the pth row of C and allowing p to take successively the same values we get the n elements in the qth column of C. If p and q both change we get all the elements.

The determinant C can be partitioned into n^n determinants, the columns of each of which are got by taking from each of the columns of C a set of terms in the same vertical line. Of these determinants, however, those containing two columns taken from corresponding places in the columns of C may be neglected, since, when the common factor of the elements of each column of such a determinant is separated from them, the determinant must have two columns identical, and therefore be equal to zero. There are evidently $n!$ of those which do not vanish. Taking any one of them, say the one having for its first column the set of r_1th terms of the first column of C, for its second column the set of r_2th terms of the second column of C, and, generally, for its kth column the set of r_kth terms of the kth column of C, no two values of r_k being alike, the result is the determinant

$$\begin{vmatrix} a_{1r_1}b_{1r_1} & a_{1r_2}b_{2r_2} & \cdots & a_{1r_n}b_{nr_n} \\ a_{2r_1}b_{1r_1} & a_{2r_2}b_{2r_2} & \cdots & a_{2r_n}b_{nr_n} \\ \cdots & \cdots & \cdots & \cdots \\ a_{nr_1}b_{1r_1} & a_{nr_2}b_{2r_2} & \cdots & a_{nr_n}b_{nr_n} \end{vmatrix}$$

Now this is equal to

$$b_{1r_1}b_{2r_2}\cdots b_{nr_n}\begin{vmatrix} a_{1r_1} & a_{1r_2} & \cdots & a_{1r_n} \\ a_{2r_1} & a_{2r_2} & \cdots & a_{2r_n} \\ \cdots & \cdots & \cdots & \cdots \\ a_{nr_1} & a_{nr_2} & \cdots & a_{nr_n} \end{vmatrix}$$

and since $r_1, r_2, r_3, \cdots, r_n$ are the numbers $1, 2, 3, \cdots, n$ in some order this in turn is equal to

$$(-1)^\nu b_{1r_1}b_{2r_2}\cdots b_{nr_n}A\,,$$

Where ν is the number of inversions in the combination $(r_1r_2\cdots r_n)$. But $(-1)^\nu\, b_{1r_1}b_{2r_2}\cdots b_{nr_n}$ is a term, with proper sign, of B.

The $n!$ determinants whose sum is equal to C are thus seen to be each expressible as the product of two factors, one factor being A in every case, and the other being in succession all the terms of B. Hence $C=AB$.

EXAMPLE.

$$\begin{vmatrix} 1 & 2 & 1 & 1 \\ 3 & 0 & 1 & 4 \\ 0 & 2 & 1 & -1 \\ 2 & 3 & 0 & -4 \end{vmatrix} \times \begin{vmatrix} -1 & 4 & 2 & -1 \\ 2 & -1 & 3 & -2 \\ 0 & 2 & -1 & 1 \\ 3 & 0 & 4 & -1 \end{vmatrix}$$

$$
= \begin{vmatrix} -1+8+2-1, & 2-2+3-2, & 0+4-1+1, & 3+0+4-1 \\ -3+0+2-4, & 6-0+3-8, & 0+0-1+4, & 9+0+4-4 \\ 0+8+2+1, & 0-2+3+2, & 0+4-1-1, & 0+0+4+1 \\ -2+2+0+4, & 4-3+0+8, & 0+6-0-4, & 6+0+0+4 \end{vmatrix}
$$

$$
= \begin{vmatrix} 8 & 1 & 4 & 6 \\ -5 & 1 & 3 & 9 \\ 11 & 3 & 2 & 5 \\ 14 & 9 & 2 & 10 \end{vmatrix}.
$$

156. Since it is possible to alter, in accordance with previously established theorems, the form of either or both of the two given determinants before going through the process of forming the determinant which is to be their product, it is clear that in this way there may be obtained more than one form of result. Thus, using the theorem (§37) regarding the transformation of rows into columns, we have

$$
\begin{vmatrix} a_{11} & a_{12} \\ a_{21} & a_{22} \end{vmatrix} \begin{vmatrix} b_{11} & b_{12} \\ b_{21} & b_{22} \end{vmatrix} = \begin{vmatrix} a_{11}b_{11}+a_{12}b_{12} & a_{11}b_{21}+a_{12}b_{22} \\ a_{21}b_{11}+a_{22}b_{22} & a_{21}b_{21}+a_{22}b_{22} \end{vmatrix}
$$

$$
= \begin{vmatrix} a_{11} & a_{12} \\ a_{21} & a_{22} \end{vmatrix} \begin{vmatrix} b_{11} & b_{21} \\ b_{12} & b_{22} \end{vmatrix} = \begin{vmatrix} a_{11}b_{11}+a_{12}b_{21} & a_{11}b_{12}+a_{12}b_{22} \\ a_{21}b_{11}+a_{22}b_{21} & a_{21}b_{12}+a_{22}b_{22} \end{vmatrix}
$$

$$
= \begin{vmatrix} a_{11} & a_{21} \\ a_{12} & a_{22} \end{vmatrix} \begin{vmatrix} b_{11} & b_{12} \\ b_{21} & b_{22} \end{vmatrix} = \begin{vmatrix} a_{11}b_{11}+a_{21}b_{12} & a_{11}b_{21}+a_{21}b_{22} \\ a_{12}b_{11}+a_{22}b_{12} & a_{12}b_{21}+a_{22}b_{22} \end{vmatrix}
$$

$$
= \begin{vmatrix} a_{11} & a_{21} \\ a_{12} & a_{22} \end{vmatrix} \begin{vmatrix} b_{11} & b_{21} \\ b_{12} & b_{22} \end{vmatrix} = \begin{vmatrix} a_{11}b_{11}+a_{21}b_{21} & a_{11}b_{12}+a_{21}b_{22} \\ a_{12}b_{11}+a_{22}b_{21} & a_{12}b_{12}+a_{22}b_{22} \end{vmatrix}.
$$

157. The form of the result in the case in which the two determinants to be multiplied are identical is worthy of notice, each of the elements situated on one side of the principal diagonal being identical with that similarly situated on the other side, that is, conjugate elements are alike and the product is therefore a symmetrical determinant. Thus taking $|a_{11}a_{22}a_{33}|^2$ as an instance, we have

$$
|a_{11}\ a_{22}\ a_{33}|^2
$$

$$
= \begin{vmatrix} a_{11}^2+a_{12}^2+a_{13}^2 & a_{11}a_{21}+a_{12}a_{22}+a_{13}a_{23} & a_{11}a_{31}+a_{12}a_{32}+a_{13}a_{33} \\ a_{21}a_{11}+a_{22}a_{12}+a_{23}a_{13} & a_{21}^2+a_{22}^2+a_{23}^2 & a_{21}a_{31}+a_{22}a_{32}+a_{23}a_{33} \\ a_{31}a_{11}+a_{32}a_{12}+a_{33}a_{13} & a_{31}a_{21}+a_{32}a_{22}+a_{33}a_{23} & a_{31}^2+a_{32}^2+a_{33}^2 \end{vmatrix}
$$

158. If one of the two determinants, whose product in determinant form is wished, be of lower order than the other, we can raise its order to that of the other in the manner already shown (§71), and then proceed as before. As, however, this preliminary change may be accomplished in a variety of ways, there thus arises an increased variety in the possible forms of the result. For example, if the product wanted be that of $|a_{11}a_{22}a_{33}a_{44}|$ and $|b_{11}b_{22}|$, we have

$$\begin{vmatrix} a_{11} & a_{22} & a_{33} & a_{44} \end{vmatrix} \cdot \begin{vmatrix} b_{11} & b_{22} \end{vmatrix}$$

$$= \begin{vmatrix} a_{11} & a_{12} & a_{13} & a_{14} \\ a_{21} & a_{22} & a_{23} & a_{24} \\ a_{31} & a_{32} & a_{33} & a_{34} \\ a_{41} & a_{42} & a_{43} & a_{44} \end{vmatrix} \cdot \begin{vmatrix} b_{11} & b_{12} & 0 & 0 \\ b_{21} & b_{22} & 0 & 0 \\ 0 & 0 & 1 & 0 \\ 0 & 0 & 0 & 1 \end{vmatrix}$$

$$= \begin{vmatrix} a_{11}b_{11} + a_{12}b_{12} & a_{11}b_{21} + a_{12}b_{22} & a_{13} & a_{14} \\ a_{21}b_{11} + a_{22}b_{12} & a_{21}b_{21} + a_{22}b_{22} & a_{23} & a_{24} \\ a_{31}b_{11} + a_{32}b_{12} & a_{31}b_{21} + a_{32}b_{22} & a_{33} & a_{34} \\ a_{41}b_{11} + a_{42}b_{12} & a_{41}b_{21} + a_{42}b_{22} & a_{43} & a_{44} \end{vmatrix}; \quad \text{or}$$

$$= \begin{vmatrix} a_{11} & a_{12} & a_{13} & a_{14} \\ a_{21} & a_{22} & a_{23} & a_{24} \\ a_{31} & a_{32} & a_{33} & a_{34} \\ a_{41} & a_{42} & a_{43} & a_{44} \end{vmatrix} \cdot \begin{vmatrix} b_{11} & b_{12} & b_{13} & 0 \\ b_{21} & b_{22} & b_{23} & 0 \\ 0 & 0 & 1 & 0 \\ b_{41} & b_{42} & b_{43} & 1 \end{vmatrix} = \Delta,$$

where the rth ($r=1, 2, 3, 4$) row of Δ is

$$a_{r1}b_{11} + a_{r2}b_{12} + a_{r3}b_{13}, \; a_{r1}b_{21} + a_{r2}b_{22} + a_{r3}b_{23},$$

$$a_{r3}, \; a_{r1}b_{41} + a_{r2}b_{42} + a_{r3}b_{43}$$

159. If the process of §156 be used to find the product of two determinants, one or both of which contain one or more zero columns, there results a determinant whose value might not otherwise readily appear, but which, from viewing it as arising in the manner stated, we know must equal zero. Thus the determinant

$$\begin{vmatrix} a_{11}b_{11} + a_{12}b_{12} & a_{11}b_{21} + a_{12}b_{22} & a_{11}b_{31} + a_{12}b_{32} \\ a_{21}b_{11} + a_{22}b_{12} & a_{21}b_{21} + a_{22}b_{22} & a_{21}b_{31} + a_{22}b_{32} \\ a_{31}b_{11} + a_{32}b_{12} & a_{31}b_{21} + a_{32}b_{22} & a_{31}b_{31} + a_{32}b_{32} \end{vmatrix} = 0,$$

since (§153) it is equal to

$$\begin{vmatrix} a_{11} & a_{12} & 0 \\ a_{21} & a_{22} & 0 \\ a_{31} & a_{32} & 0 \end{vmatrix} \times \begin{vmatrix} b_{11} & b_{12} & b_{13} \\ b_{21} & b_{22} & b_{23} \\ b_{31} & b_{32} & b_{33} \end{vmatrix}$$

160. Having found as in §155 the product of two determinants, the product of the result and another determinant may be similarly found, and thus we see generally that *the product of any number of determinants of the same or different orders is obtainable as a determinant of the order which is highest among the factors.*

EXAMPLE 1. Show that

$$\begin{vmatrix} y^3 & -y^2x & yx^2 & -x^3 \\ a & b & c & d \\ b & c & d & e \\ c & d & e & f \end{vmatrix} = \begin{vmatrix} ax+by & bx+cy & cx+dy \\ bx+cy & cx+dy & dx+ey \\ cx+dy & dx+ey & ex+fy \end{vmatrix}$$

By §156 we have

$$\begin{vmatrix} y^3 & -y^2x & yx^2 & -x^3 \\ a & b & c & d \\ b & c & d & e \\ c & d & e & f \end{vmatrix} \times \begin{vmatrix} 1 & 0 & 0 & 0 \\ x & y & 0 & 0 \\ 0 & x & y & 0 \\ 0 & 0 & x & y \end{vmatrix}$$

$$= \begin{vmatrix} y^3 & 0 & 0 & 0 \\ a & ax+by & bx+cy & cx+dy \\ b & bx+cy & cx+dy & dx+ey \\ c & cx+dy & dx+ey & ex+fy \end{vmatrix},$$

$$= y^3 \begin{vmatrix} ax+by & bx+cy & cx+dy \\ bx+cy & cx+dy & dx+ey \\ cx+dy & dx+cy & ex+fy \end{vmatrix}$$

whence by division the identity is established.

EXAMPLE 2. Prove that if the expression

$$ax^2 + by^2 + cz^2 + dxy + eyz + fzx + gx + hy + kz + l \equiv F$$

be the product of two linear factors, $\alpha_1x+\beta_1y+\gamma_1z+\delta_1$ and $\alpha_2x+\beta_2y+\gamma_2z+\delta_2$ say, then

$$D \equiv \begin{vmatrix} 2a & d & f & g \\ d & 2b & e & h \\ f & e & 2c & k \\ g & h & k & 2l \end{vmatrix} = 0.$$

Multiplying the factors together and comparing the result with the given expression, we have

$$\begin{aligned} a &= \alpha_1\alpha_2 & f &= \alpha_1\gamma_2 + \alpha_2\gamma_1 \\ b &= \beta_1\beta_2 & g &= \alpha_1\delta_2 + \alpha_2\delta_1 \\ c &= \gamma_1\gamma_2 & h &= \beta_1\delta_2 + \beta_2\delta_1 \\ d &= \alpha_1\beta_2 + \alpha_2\beta_1 & k &= \gamma_1\delta_2 + \gamma_2\delta_1 \\ e &= \beta_1\gamma_2 + \beta_2\gamma_1 & l &= \delta_1\delta_2. \end{aligned}$$

Hence

$$D = \begin{vmatrix} \alpha_1\alpha_2 + \alpha_2\alpha_1 & \alpha_1\beta_2 + \alpha_2\beta_1 & \alpha_1\gamma_2 + \alpha_2\gamma_1 & \alpha_1\delta_2 + \alpha_2\delta_1 \\ \alpha_1\beta_2 + \alpha_2\beta_1 & \beta_1\beta_2 + \beta_2\beta_1 & \gamma_1\beta_2 + \gamma_2\beta_1 & \delta_1\beta_2 + \delta_2\beta_1 \\ \alpha_1\gamma_2 + \alpha_2\gamma_1 & \beta_1\gamma_2 + \beta_2\gamma_1 & \gamma_1\gamma_2 + \gamma_2\gamma_1 & \delta_1\gamma_2 + \delta_2\gamma_1 \\ \alpha_1\delta_2 + \alpha_2\delta_1 & \beta_1\delta_2 + \beta_2\delta_1 & \gamma_1\delta_2 + \gamma_2\delta_1 & \delta_1\delta_2 + \delta_2\delta_1 \end{vmatrix}$$

That is

$$D = \begin{vmatrix} \alpha_1 & \alpha_2 & 0 & 0 \\ \beta_1 & \beta_2 & 0 & 0 \\ \gamma_1 & \gamma_2 & 0 & 0 \\ \delta_1 & \delta_2 & 0 & 0 \end{vmatrix} \times \begin{vmatrix} \alpha_2 & \alpha_1 & 0 & 0 \\ \beta_2 & \beta_1 & 0 & 0 \\ \gamma_2 & \gamma_1 & 0 & 0 \\ \delta_2 & \delta_1 & 0 & 0 \end{vmatrix}, \qquad (\S 153)$$

Not only does D (known as the discriminant of F) itself vanish in the circumstances mentioned, but so also do all its principal minors. Thus, taking the minor obtained by deleting the second row and third column, we find

$$\begin{vmatrix} 2a & d & g \\ f & e & k \\ g & h & 2l \end{vmatrix} = \begin{vmatrix} \alpha_1 & \alpha_2 & 0 \\ \gamma_1 & \gamma_2 & 0 \\ \delta_1 & \delta_2 & 0 \end{vmatrix} \cdot \begin{vmatrix} \alpha_2 & \alpha_1 & 0 \\ \beta_2 & \beta_1 & 0 \\ \delta_2 & \delta_1 & 0 \end{vmatrix} = 0;$$

the first factor here being obtainable from the first factor above by deleting the second row and fourth column, and the second factor from the second factor above by deleting the third row and fourth column.

Example 3. Show that if in the F of Example 2 we put

$$x = \alpha_1 X + \beta_1 Y + \gamma_1 Z,$$
$$y = \alpha_2 X + \beta_2 Y + \gamma_2 Z,$$
$$z = \alpha_2 X + \beta_3 Y + \gamma_3 Z,$$

arrange the result with regard to X, Y, Z as the given expression is arranged with regard to x, y, z, and denote the coefficients in order by $A, B, C, \cdots$, then

$$\Delta \equiv \begin{vmatrix} 2A & D & F & G \\ D & 2B & E & H \\ F & E & 2C & K \\ G & H & K & 2L \end{vmatrix} = \begin{vmatrix} \alpha_1 & \beta_1 & \gamma_1 \\ \alpha_2 & \beta_2 & \gamma_2 \\ \alpha_3 & \beta_3 & \gamma_3 \end{vmatrix} \cdot \begin{vmatrix} 2a & d & f & g \\ d & 2b & e & h \\ f & e & 2c & k \\ g & h & k & 2l \end{vmatrix}.$$

Making the substitutions and arranging, we find

$$A = a\alpha_1^2 + b\alpha_2^2 + c\alpha_3^2 + d\alpha_1\alpha_2 + e\alpha_2\alpha_3 + f\alpha_1\alpha_3,$$
$$G = g\alpha_1 + h\alpha_2 + k\alpha_3$$
$$B = a\beta_1^2 + b\beta_2^2 + c\beta_3^2 + d\beta_1\beta_2 + e\beta_2\beta_3 + f\beta_1\beta_3,$$
$$H = g\beta_1 + h\beta_2 + k\beta_3,$$
$$C = a\gamma_1^2 + b\gamma_2^2 + c\gamma_3^2 + d\gamma_1\gamma_2 + e\gamma_1\gamma_3 + f\gamma_1\gamma_3,$$
$$K = g\gamma_1 + h\gamma_2 + k\gamma_3,$$
$$D = 2a\alpha_1\beta_1 + 2b\alpha_2\beta_2 + 2c\alpha_3\beta_3 + d\alpha_2\beta_1 + d\alpha_1\beta_2 + e\alpha_3\beta_2 + e\alpha_2\beta_3 + f\alpha_3\beta_1 + f\alpha_1\beta_3,$$
$$E = 2a\beta_1\gamma_1 + 2b\beta_2\gamma_2 + 2c\beta_3\gamma_3 + d\beta_2\gamma_1 + d\beta_1\gamma_2 + e\beta_3\gamma_2 + e\beta_2\gamma_3 + f\beta_3\gamma_1 + f\beta_1\gamma_3,$$
$$F = 2a\gamma_1\alpha_1 + 2b\gamma_2\alpha_2 + 2c\gamma_3\alpha_3 + d\gamma_2\alpha_1 + d\gamma_1\alpha_2 + e\gamma_3\alpha_2 + e\gamma_2\alpha_3 + f\gamma_3\alpha_1 + f\gamma_1\alpha_3.$$

Now

$$\begin{vmatrix} \alpha_1 & \beta_1 & \gamma_1 \\ \alpha_2 & \beta_2 & \gamma_2 \\ \alpha_3 & \beta_3 & \gamma_3 \end{vmatrix}^2 \times D = \begin{vmatrix} \alpha_1 & \alpha_2 & \alpha_3 & 0 \\ \beta_1 & \beta_2 & \beta_3 & 0 \\ \gamma_1 & \gamma_2 & \gamma_3 & 0 \\ 0 & 0 & 0 & 1 \end{vmatrix}^2 \cdot D,$$
$$= \left| \alpha_1 \; \beta_2 \; \gamma_3 \right|$$

$$\cdot\begin{vmatrix} 2a\alpha_1+d\alpha_2+f\alpha_3 & d\alpha_1+2b\alpha_2+e\alpha_3 & f\alpha_1+e\alpha_2+2c\alpha_3 & g\alpha_1+h\alpha_2+k\alpha_3 \\ 2a\beta_1+d\beta_2+f\beta_3 & d\beta_1+2b\beta_2+e\beta_3 & f\beta_1+e\beta_2+2c\beta_3 & g\beta_1+h\beta_2+k\beta_3 \\ 2a\gamma_1+d\gamma_2+f\gamma_3 & d\gamma_1+2b\gamma_2+e\gamma_3 & f\gamma_1+e\gamma_2+2c\gamma_3 & g\gamma_1+h\gamma_2+k\gamma_3 \\ g & h & k & 2l \end{vmatrix},$$

$$= \begin{vmatrix} 2A & D & F & G \\ D & 2B & E & H \\ F & E & 2C & K \\ G & H & K & 2L \end{vmatrix},$$

if in multiplying we use $|\alpha_1\beta_2\gamma_3|$ again in its altered form. It is also apparent from this that the complementary minors of $2L$ and $2l$ are connected in the same way by the multiplier $|\alpha_1\beta_2\gamma_3|^2$.

Exercises. Set XI

Perform the following multiplications, giving the results as determinants:

1. $\begin{vmatrix} x & y & 1 \\ x_1 & y_1 & 1 \\ x_2 & y_2 & 1 \end{vmatrix} \, |a_1 b_2 c_3|$.

2. $\begin{vmatrix} a & b & 0 \\ c & 0 & d \\ 0 & e & f \end{vmatrix} \begin{vmatrix} 0 & a & b \\ c & 0 & d \\ e & f & 0 \end{vmatrix}$.

3. $\begin{vmatrix} 1 & a & a^2 \\ 1 & b & b^2 \\ 1 & c & c^2 \end{vmatrix} \begin{vmatrix} a^2 & -a & 1 \\ b^2 & -b & 1 \\ c^2 & -c & 1 \end{vmatrix}$.

4. $\begin{vmatrix} a+b & c & c \\ a & b+c & a \\ b & b & c+a \end{vmatrix} \begin{vmatrix} a+b+\frac{1}{2}c & -\frac{1}{2}a & -\frac{1}{2}b \\ -\frac{1}{2}c & b+c+\frac{1}{2}a & -\frac{1}{2}b \\ -\frac{1}{2}c & -\frac{1}{2}a & c+a+\frac{1}{2}b \end{vmatrix}$

5. By changing x^2+y^2, y^2+z^2, z^2+x^2 into determinant form and multiplying, find an expression for their product as the sum of two squares.

6. Find the product of

$$\begin{vmatrix} a & a & a & a \\ a & b & b & b \\ a & b & c & c \\ a & b & c & d \end{vmatrix} \quad \text{and} \quad \begin{vmatrix} -1 & 1 & 0 & 0 \\ 0 & -1 & 1 & 0 \\ 0 & 0 & -1 & 1 \\ 1 & 1 & 1 & -1 \end{vmatrix},$$

and thence resolve the former determinant into simple factors.

7. Prove the identity of Ex. 14, Set V, by using

$$\begin{vmatrix} 1 & -1 & 1 \\ 1 & 1 & -1 \\ -1 & 1 & 1 \end{vmatrix}$$

as a multiplier. Write down the corresponding multiplier in the case of Ex. 15, Set V.

Give the quotients in the following cases as determinants:

8. $$\begin{vmatrix} 3\alpha^2 & \beta^2+\beta\alpha+\alpha^2 & \gamma^2+\gamma\alpha+\alpha^2 \\ \beta^2+\beta\alpha+\alpha^2 & 3\beta^2 & \gamma^2+\gamma\beta+\beta^2 \\ \gamma^2+\gamma\alpha+\alpha^2 & \gamma^2+\gamma\beta+\beta^2 & 3\gamma^2 \end{vmatrix} \div \begin{vmatrix} 1 & \alpha & \alpha^2 \\ 1 & \beta & -\beta^2 \\ 1 & \gamma & \gamma^2 \end{vmatrix}.$$

9. $$\begin{vmatrix} 2xy & y^2+x & x^2+y \\ x^2+y & 2xy & x+y^2 \\ x+y^2 & y+x^2 & 2xy \end{vmatrix} \div \begin{vmatrix} x & y & 0 \\ y & 0 & x \\ 0 & x & y \end{vmatrix}.$$

10. Use the multiplication theorem to find the simple factors of the determinant of Ex. 24, Set IV.

11. Find the product of

$$\begin{vmatrix} a+bi & -c+di \\ c+di & a-bi \end{vmatrix} \text{ and } \begin{vmatrix} \alpha+\beta i & -\gamma+\delta i \\ \gamma+\delta i & \alpha-\beta i \end{vmatrix},$$

where $i=(-1)^{\frac{1}{2}}$;
and thence show how the product of two sums of four squares is itself expressible as a sum of four squares.

12. Show that

$$2(a+b+c)\begin{vmatrix} a & b & c \\ b & c & a \\ 1 & 1 & 1 \end{vmatrix} = \begin{vmatrix} a^2 & b^2 & c^2 \\ b & c & a \\ 1 & 1 & 1 \end{vmatrix} + \begin{vmatrix} a & b & c \\ b^2 & c^2 & a^2 \\ 1 & 1 & 1 \end{vmatrix} + \begin{vmatrix} ab & bc & ca \\ b & c & a \\ 1 & 1 & 1 \end{vmatrix} + \begin{vmatrix} a & b & c \\ ab & bc & ca \\ 1 & 1 & 1 \end{vmatrix}.$$

13. Show that the identity of Ex. 12, Set IV, follows from finding the product of

$$\begin{vmatrix} a_1 & b_1 & c_1 & 1 \\ a_2 & b_2 & c_2 & 1 \\ a_3 & b_3 & c_3 & 1 \\ 1 & 1 & 1 & 0 \end{vmatrix}, \quad \begin{vmatrix} 1 & 0 & 0 & k_1 \\ 0 & 1 & 0 & k_2 \\ 0 & 0 & 1 & k_3 \\ 0 & 0 & 0 & 1 \end{vmatrix}, \quad \begin{vmatrix} 1 & 0 & 0 & 0 \\ 0 & 1 & 0 & 0 \\ 0 & 0 & 1 & 0 \\ h_1 & h_2 & h_3 & 1 \end{vmatrix}.$$

14. Find the expansion of

$$\begin{vmatrix} a-\lambda & h & g \\ h & b-\lambda & f \\ g & f & c-\lambda \end{vmatrix} \times \begin{vmatrix} a+\lambda & h & g \\ h & b+\lambda & f \\ g & f & c+\lambda \end{vmatrix}$$

according to descending powers of λ, showing that the coefficients are alternately negative and positive.

15. Find the product of

$$\begin{vmatrix} a_1 & a_2 & a_3 & 0 \\ 0 & a_1 & a_2 & a_3 \\ 0 & b_1 & b_2 & b_3 \\ b_1 & b_2 & b_3 & 0 \end{vmatrix} \quad \text{and} \quad \begin{vmatrix} b_2 & b_3 & 0 & 0 \\ b_3 & 0 & 0 & 0 \\ -a_3 & 0 & -1 & 0 \\ -a_2 & -a_3 & 0 & -1 \end{vmatrix},$$

and thence show that the former determinant is equal to

$$\begin{vmatrix} \begin{vmatrix} a_1 \, b_2 \end{vmatrix} & \begin{vmatrix} a_1 \, b_3 \end{vmatrix} \\ \begin{vmatrix} a_1 \, b_3 \end{vmatrix} & \begin{vmatrix} a_2 \, b_3 \end{vmatrix} \end{vmatrix}.$$

Resolve into determinant factors

16. $$\begin{vmatrix} a^2+bc & ab & bd \\ ac & bc+de & df \\ ce & ef & de+f^2 \end{vmatrix}.$$

17. $$\begin{vmatrix} a_1 & b_1x_1+c_1y_1 & b_1x_2+c_1y_2 \\ a_2 & b_2x_1+c_2y_1 & b_2x_2+c_2y_2 \\ a_3 & b_3x_1+c_3y_1 & b_3x_2+c_3y_2 \end{vmatrix}$$

18. $$\begin{vmatrix} ax_1+cz_1 & 0 & fx_1+gz_1 \\ ax_2+by_2+cz_2 & dy_2 & fx_2+gz_2 \\ by_3+cz_3 & dy_3 & gz_3 \end{vmatrix}.$$

19. Prove that

$$-4\begin{vmatrix} 1 & x_1 & y_1 & x_1^2+y_1^2 \\ 1 & x_2 & y_2 & x_2^2+y_2^2 \\ 1 & x_3 & y_3 & x_3^2+y_3^2 \\ 1 & x_4 & y_4 & x_4^2+y_4^2 \end{vmatrix}^2$$

= a determinant of the fourth order, Δ say, where the element in the position (ij) of Δ is $(x_i-x_j)^2+(y_i-y_j)^2$, and therefore 0 when $i=j$.

20. Prove, as in Ex. 15, that

$$\begin{vmatrix} a_1 & a_2 & a_3 & a_4 & 0 & 0 \\ 0 & a_1 & a_2 & a_3 & a_4 & 0 \\ 0 & 0 & a_1 & a_2 & a_3 & a_4 \\ 0 & 0 & b_1 & b_2 & b_3 & b_4 \\ 0 & b_1 & b_2 & b_3 & b_4 & 0 \\ b_1 & b_2 & b_3 & b_4 & 0 & 0 \end{vmatrix} = \begin{vmatrix} \lvert a_1\, b_2\rvert & \lvert a_1\, b_3\rvert & \lvert a_1\, b_4\rvert \\ \lvert a_1\, b_3\rvert & \lvert a_1\, b_4\rvert + \lvert a_2\, b_3\rvert & \lvert a_2\, b_4\rvert \\ \lvert a_1\, b_4\rvert & \lvert a_2\, b_4\rvert & \lvert a_3\, b_4\rvert \end{vmatrix}.$$

21. Prove, in the same way, that

$$\begin{vmatrix} a_1 & a_2 & a_3 & a_4 & 0 \\ 0 & a_1 & a_2 & a_3 & a_4 \\ 0 & 0 & b_1 & b_2 & b_3 \\ 0 & b_1 & b_2 & b_3 & 0 \\ b_1 & b_2 & b_3 & 0 & 0 \end{vmatrix}$$

$$= \begin{vmatrix} \lvert a_1\, b_2\rvert & a_1b_3 & -a_4b_1 \\ \lvert a_1\, b_3\rvert & -a_4b_1 + \lvert a_2\, b_3\rvert & -a_4b_2 \\ -a_4b_1 & -a_4b_2 & -a_4b_3 \end{vmatrix}$$

$$\div(-a_4) = \begin{vmatrix} a_1b_1 & a_1b_2 & a_1b_3 \\ a_1b_2 & \lvert a_1\, b_3\rvert + a_2b_2 & a_2b_3 - a_4b_1 \\ a_1b_3 & a_2\, b_3 - a_4b_1 & a_3b_3 - a_4b_2 \end{vmatrix} \div a_1$$

22. Show that the proposition of §77 may be established by using the multiplication theorem.

23. Find the quotient of

$$\begin{vmatrix} (s-a_1)^2 & a_1^2 & a_1^2 & \cdots & a_1^2 \\ a_2^2 & (s-a_2)^2 & a_2^2 & \cdots & a_2^2 \\ a_3^2 & a_3^2 & (s-a_3)^2 & \cdots & a_3^2 \\ \cdots & \cdots & \cdots & \cdots & \cdots \\ a_n^2 & a_n^2 & a_n^2 & \cdots & (s-a_n)^2 \end{vmatrix}$$

$$\div \begin{vmatrix} s-a_1 & a_1 & a_1 & \cdots & a_1 \\ a_2 & s-a_2 & a_2 & \cdots & a_2 \\ a_3 & a_3 & s-a_3 & \cdots & a_3 \\ \cdots & \cdots & \cdots & \cdots & \cdots \\ a_n & a_n & a_n & \cdots & s-a_n \end{vmatrix},$$

where $s = a_1 + a_2 + \cdots a_n$.

24. Prove that

$$\begin{vmatrix} 0 & a_2(a_1b+ab_1)-2baa_1 & a_2(a_1c+ac_1)-2c_2aa_1 \\ b_2(ba_1+b_1a)-2a_2bb_1 & 0 & b_2(b_1c+bc_1)-2c_2bb_1 \\ c_2(c_1a+ca_1)-2a_2cc_1 & c_2(c_1b+cb_1)-2b_2cc_1 & 0 \end{vmatrix}$$

$$= \begin{vmatrix} 0 & a\,b_1c_1+a_1b_1c\,-2a_1b\,c_1 & a\,b_2c_2+a_2b_2c-2a_2b\,c_2 \\ a_1bc+abc_1-2ab_1c & 0 & a_1b_2c_2+a_2b_2c_1-2a_2b_1c_2 \\ a_2bc+abc_2-2ab_2c & a_2b_1c_1+a_1b_1c_2-2a_1b_2c_1 & 0 \end{vmatrix}.$$

Establish the three following identities:

25. $$\begin{vmatrix} a+b+c+\tfrac{1}{2}d & -\tfrac{1}{2}a & -\tfrac{1}{2}b & -\tfrac{1}{2}c \\ -\tfrac{1}{2}d & b+c+d+\tfrac{1}{2}a & -\tfrac{1}{2}b & -\tfrac{1}{2}c \\ -\tfrac{1}{2}d & -\tfrac{1}{2}a & c+d+a+\tfrac{1}{2}b & -\tfrac{1}{2}c \\ -\tfrac{1}{2}d & -\tfrac{1}{2}a & -\tfrac{1}{2}b & d+a+b+\tfrac{1}{2}c \end{vmatrix}$$

$=\tfrac{1}{2}(a+b+c+d)^4.$

26. $$\begin{vmatrix} (a+b+c)^2 & d^2 & d^2 & d^2 \\ a^2 & (b+c+d)^2 & a^2 & a^2 \\ b^2 & b^2 & (c+d+a)^2 & b^2 \\ c^2 & c^2 & c^2 & (d+a+b)^2 \end{vmatrix}$$

$=2(a+b+c+d)^4 \sum a^2bc.$

27. $$\begin{vmatrix} 3d & s+d & s+d & s+d \\ s+a & 3a & s+a & s+a \\ s+b & s+b & 3b & s+b \\ s+c & s+c & s+c & 3c \end{vmatrix} = -12 \sum a^2bc,$$

if $s=a+b+c+d$.

161. The two different modes which have thus been found for expressing the product of two determinants as a determinant suggest the possibility of deriving the result obtained in the one case (§155) from that obtained in the other (§109). This can really be done, and the process of transformation is sufficiently instructive to merit attention. Taking the particular case in which the two determinants to be multiplied are of the third order, namely:

$$|A_1\,B_2\,C_3| \text{ or } \Delta \text{ and } |a_1\,b_2\,c_3| \text{ or } \Delta',$$

we have (§109)

$$\Delta\Delta' = \begin{vmatrix} a_1 & b_1 & c_1 & -1 & 0 & 0 \\ a_2 & b_2 & c_2 & 0 & -1 & 0 \\ a_3 & b_3 & c_3 & 0 & 0 & -1 \\ 0 & 0 & 0 & A_1 & A_2 & A_3 \\ 0 & 0 & 0 & B_1 & B_2 & B_3 \\ 0 & 0 & 0 & C_1 & C_2 & C_3 \end{vmatrix},$$

where there is specially to be noticed the nine particular elements chosen for the places which may be filled by any nine finite elements whatever. Then, increasing each element of the first column by a_1 times the corresponding element of the fourth column, a_2 times the corresponding element of the fifth column, and a_3 times the corresponding element of the sixth column, and increasing each element of the second column and each element of the third column in a similar fashion, but with the multipliers b_1, b_2, b_3 in the one case, and c_1, c_2, c_3 in the other, the product takes the form

$$\begin{vmatrix} 0 & 0 & 0 & -1 & 0 & 0 \\ 0 & 0 & 0 & 0 & -1 & 0 \\ 0 & 0 & 0 & 0 & 0 & -1 \\ a_1A_1+a_2A_2+a_3A_3 & b_1A_1+b_2A_2+b_3A_3 & c_1A_1+c_2A_2+c_3A_3 & A_1 & A_2 & A_3 \\ a_1B_1+a_2B_2+a_3B_3 & b_1B_1+b_2B_2+b_3B_3 & c_1B_1+c_2B_2+c_3B_3 & B_1 & B_2 & B_3 \\ a_1C_1+a_2C_2+a_3C_3 & b_1C_1+b_2C_2+b_3C_3 & c_1C_1+c_2C_2+c_3C_3 & C_1 & C_2 & C_3 \end{vmatrix}$$

Hence (§93) it is equal to

$$-\begin{vmatrix} a_1A_1+a_2A_2+a_3A_3 & b_1A_1+b_2A_2+b_3A_3 & c_1A_1+c_2A_2+c_3A_3 \\ a_1B_1+a_2B_2+a_3B_3 & b_1B_1+b_2B_2+b_3B_3 & c_1B_1+c_2B_2+c_3B_3 \\ a_1C_1+a_2C_2+a_3C_3 & b_1C_1+b_2C_2+b_3C_3 & c_1C_1+c_2C_2+c_3C_3 \end{vmatrix} \times$$

$$\begin{vmatrix} -1 & 0 & 0 \\ 0 & -1 & 0 \\ 0 & 0 & -1 \end{vmatrix},$$

which becomes at once the result of §153.

162. In §155 the product of two determinants of the nth order is given as a determinant of the nth order; in §109 it is given as a determinant of the $2n$th order. We can, however, further express it

as a determinant of each of the intermediate orders, so that the forms of §§109, 155 may be viewed as the extremes of a series. Thus we have, firstly,

$$(1)\qquad |A_1 B_2 C_3|\,|a_1 b_2 c_3| = \begin{vmatrix} a_1 & b_1 & c_1 & 0 & 0 & 0 \\ a_2 & b_2 & c_2 & 0 & 0 & 0 \\ a_3 & b_3 & c_3 & 0 & 0 & 0 \\ 0 & 0 & 0 & A_1 & A_2 & A_3 \\ 0 & 0 & 0 & B_1 & B_2 & B_3 \\ 0 & 0 & 0 & C_1 & C_2 & C_3 \end{vmatrix};$$

secondly, we have

$$|A_1 B_2 C_3|\,|a_1 b_2 c_3| = \begin{vmatrix} a_1 & b_1 & c_1 & -1 & 0 & 0 \\ a_2 & b_2 & c_2 & 0 & 0 & 0 \\ a_3 & b_3 & c_3 & 0 & 0 & 0 \\ 0 & 0 & 0 & A_1 & A_2 & A_3 \\ 0 & 0 & 0 & B_1 & B_2 & B_3 \\ 0 & 0 & 0 & C_1 & C_2 & C_3 \end{vmatrix},$$

$$= \begin{vmatrix} 0 & 0 & 0 & -1 & 0 & 0 \\ a_2 & b_2 & c_2 & 0 & 0 & 0 \\ a_3 & b_3 & c_3 & 0 & 0 & 0 \\ a_1A_1 & b_1A_1 & c_1A_1 & A_1 & A_2 & A_3 \\ a_1B_1 & b_1B_1 & c_1B_1 & B_1 & B_2 & B_3 \\ a_1C_1 & b_1C_1 & c_1C_1 & C_1 & C_2 & C_3 \end{vmatrix},$$

$$(2)\qquad = \begin{vmatrix} a_2 & b_2 & c_2 & 0 & 0 \\ a_3 & b_3 & c_3 & 0 & 0 \\ a_1A_1 & b_1A_1 & c_1A_1 & A_2 & A_3 \\ a_1B_1 & b_1B_1 & c_1B_1 & B_2 & B_3 \\ a_1C_1 & b_1C_1 & c_1C_1 & C_2 & C_3 \end{vmatrix};$$

thirdly, we have

$$|A_1 B_2 C_3|\,|a_1 b_2 c_3| = \begin{vmatrix} a_1 & b_1 & c_1 & -1 & 0 & 0 \\ a_2 & b_2 & c_2 & 0 & -1 & 0 \\ a_3 & b_3 & c_3 & 0 & 0 & 0 \\ 0 & 0 & 0 & A_1 & A_2 & A_3 \\ 0 & 0 & 0 & B_1 & B_2 & B_3 \\ 0 & 0 & 0 & C_1 & C_2 & C_3 \end{vmatrix},$$

$$= \begin{vmatrix} 0 & 0 & 0 & -1 & 0 & 0 \\ 0 & 0 & 0 & 0 & -1 & 0 \\ a_3 & b_3 & c_3 & 0 & 0 & 0 \\ a_1A_1+a_2A_2 & b_1A_1+b_2A_2 & c_1A_1+c_2A_2 & A_1 & A_2 & A_3 \\ a_1B_1+a_2B_2 & b_1B_1+b_2B_2 & c_1B_1+c_2B_2 & B_1 & B_2 & B_3 \\ a_1C_1+a_2C_2 & b_1C_1+b_2C_2 & c_1C_1+c_2C_2 & C_1 & C_2 & C_3 \end{vmatrix},$$

$$(3) \qquad = \begin{vmatrix} a_3 & b_3 & c_3 & 0 \\ a_1A_1+a_2A_2 & b_1A_1+b_2A_2 & c_1A_1+c_2A_2 & A_3 \\ a_1B_1+a_2B_2 & b_1B_1+b_2B_2 & c_1B_1+c_2B_2 & B_3 \\ a_1C_1+a_2C_2 & b_1C_1+b_2C_2 & c_1C_1+c_2C_2 & C_3 \end{vmatrix};$$

and, fourthly, we have the natural conclusion to these, namely, the procedure and result of §161.

The general theorem, to which we are in this way led, and which can be proved in the manner indicated, is

The product of two determinants of the nth order may be found by substituting zero-columns for m columns in the one and for the corresponding m columns in the other, multiplying the two determinants thus obtained, increasing the number of the columns in the result by appending in order the deleted columns of the first determinant, increasing the number of rows by superposing the deleted columns of the second determinant after changing them in order into rows, putting zeros in the places above the added columns and to the right of the added rows, and prefixing to the determinant thus formed the sign factor $(-1)^{m(n+1)}$.

The multiplication theorem of §155 is the case of this when $m=0$ although strictly to include it the word "multiplying" here would have to give place to a phrase descriptive of the process intended.

163. If we differentiate the product $A\cdot B=C$ with respect to a_{ij}, where $c_{ij}=a_{i1}b_{j1}+\cdots+a_{in}b_{jn}$, we have

$$B\cdot\frac{\partial A}{\partial a_{ij}}=\frac{\partial C}{\partial c_{i1}}b_{1j}+\frac{\partial C}{\partial c_{i2}}b_{2j}\cdots+\frac{\partial C}{\partial c_{in}}b_{nj}.$$

Multiply this by $\partial B/\partial b_{kj}=B_{kj}$ and add the resulting equations obtained by giving j all values from 1 to n; and we have the result

$$B\sum\frac{\partial A}{\partial a_{ij}}\cdot\frac{\partial B}{\partial b_{kj}}=\frac{\partial C}{\partial c_{i1}}\sum B_{kj}b_{1j}+\cdots+\frac{\partial C}{\partial c_{in}}\sum B_{kj}b_{nj}$$

$$=\frac{\partial C}{\partial c_{ik}}\sum B_{kj}b_{kj},$$

since all the other terms vanish. Therefore

$$\sum \frac{\partial A}{\partial a_{ij}} \cdot \frac{\partial B}{\partial b_{kj}} = \frac{\partial C}{\partial c_{ik}}, \qquad (j = 1, 2, \cdots, n).$$

Similarly

$$\frac{\partial^2 C}{\partial c_{ik} \partial c_{rs}} = \frac{1}{1 \cdot 2} \sum \frac{\partial^2 A}{\partial a_{ij} \partial a_{rh}} \cdot \frac{\partial^2 B}{\partial b_{kj} \partial b_{sh}}, \qquad (j, h = 1, 2, \cdots, n)$$

and

$$\frac{\partial^3 C}{\partial c_{ik} \partial c_{rs} \partial c_{pq}} = \frac{1}{1 \cdot 2 \cdot 3} \sum \frac{\partial^3 A}{\partial a_{iu} \partial a_{rv} \partial a_{pw}} \cdot \frac{\partial^3 B}{\partial b_{ku} \partial b_{sv} \partial b_{qw}},$$

$$(u, v, w = 1, 2, \cdots, n)$$

and so in general.

164. The product

$$|\, a\, b'\, c'' \,| \cdot |\, \alpha\, \beta'\, \gamma'' \,| = - \begin{vmatrix} a & b & c & \cdot \\ a' & b' & c' & \cdot \\ a'' & b'' & c'' & \cdot \\ \cdot & \cdot & \cdot & 1 \end{vmatrix} \cdot \begin{vmatrix} \alpha & \beta & \cdot & \gamma \\ \alpha' & \beta' & \cdot & \gamma' \\ \alpha'' & \beta'' & \cdot & \gamma'' \\ \cdot & \cdot & 1 & \cdot \end{vmatrix}$$

$$= - \begin{vmatrix} a\alpha + b\beta & a\alpha' + b\beta' & a\alpha'' + b\beta'' & c \\ a'\alpha + b'\beta & a'\alpha' + b'\beta' & a'\alpha'' + b'b'' & c' \\ a''\alpha + b''\beta & a''\alpha' + b''\beta' & a''\alpha'' + b''\beta'' & c'' \\ \gamma & \gamma' & \gamma'' & 0 \end{vmatrix}$$

$$= -A, \quad \text{say}.$$

It is also

$$= \begin{vmatrix} a & b & c & \cdot & \cdot \\ a' & b' & c' & \cdot & \cdot \\ a'' & b'' & c'' & \cdot & \cdot \\ \cdot & \cdot & \cdot & 1 & \cdot \\ \cdot & \cdot & \cdot & \cdot & 1 \end{vmatrix} \cdot \begin{vmatrix} \alpha & \cdot & \cdot & \beta & \gamma \\ \alpha' & \cdot & \cdot & \beta' & \gamma' \\ \alpha'' & \cdot & \cdot & \beta'' & \gamma'' \\ \cdot & 1 & \cdot & \cdot & \cdot \\ \cdot & \cdot & 1 & \cdot & \cdot \end{vmatrix}$$

$$= \begin{vmatrix} a\alpha & a\alpha' & a\alpha'' & b & c \\ a'\alpha & a'\alpha' & a'\alpha'' & b' & c' \\ a''\alpha & a''\alpha' & a''\alpha'' & b'' & c'' \\ \beta & \beta' & \beta'' & \cdot & \cdot \\ \gamma & \gamma' & \gamma'' & \cdot & \cdot \end{vmatrix} = B \text{ say}.$$

Therefore

$$|\, a\, b'\, c''\,| \cdot |\, \alpha\, \beta'\, \gamma''\,| = -A = B$$

165. The product $|a_1 b_2 c_3 d_4| \cdot |\alpha_1 \beta_2 \gamma_3 \delta_4| \cdot |A_1 B_2 C_3 D_4|$.

$$= \begin{vmatrix}
a_1 & a_2 & a_3 & a_4 & \cdot & \cdot & \cdot & \cdot & \cdot & \cdot & \cdot & \cdot \\
b_1 & b_2 & b_3 & b_4 & \cdot & \cdot & \cdot & \cdot & \cdot & \cdot & \cdot & \cdot \\
c_1 & c_2 & c_3 & c_4 & \cdot & \cdot & \cdot & \cdot & \cdot & \cdot & \cdot & \cdot \\
a_1 & a_2 & a_3 & a_4 & \alpha_1 & \alpha_2 & \alpha_3 & \alpha_4 & \cdot & \cdot & \cdot & \cdot \\
b_1 & b_2 & b_3 & b_4 & \beta_1 & \beta_2 & \beta_3 & \beta_4 & \cdot & \cdot & \cdot & \cdot \\
d_1 & d_2 & d_3 & d_4 & \gamma_1 & \gamma_2 & \gamma_3 & \gamma_4 & \cdot & \cdot & \cdot & \cdot \\
\cdot & \cdot & \cdot & \cdot & \alpha_1 & \alpha_2 & \alpha_3 & \alpha_4 & A_1 & A_2 & A_3 & A_4 \\
\cdot & \cdot & \cdot & \cdot & \gamma_1 & \gamma_2 & \gamma_3 & \gamma_4 & C_1 & C_2 & C_3 & C_4 \\
\cdot & \cdot & \cdot & \cdot & \delta_1 & \delta_2 & \delta_3 & \delta_4 & D_1 & D_2 & D_3 & D_4 \\
\cdot & \cdot & \cdot & \cdot & \cdot & \cdot & \cdot & \cdot & B_1 & B_2 & B_3 & B_4 \\
\cdot & \cdot & \cdot & \cdot & \cdot & \cdot & \cdot & \cdot & C_1 & C_2 & C_3 & C_4 \\
\cdot & \cdot & \cdot & \cdot & \cdot & \cdot & \cdot & \cdot & D_1 & D_2 & D_3 & D_4
\end{vmatrix}$$

$$= \sum |\, a_1\, b_2\, c_3\,| \cdot |\, a_4\, \beta_1\, \gamma_2\,| \cdot |\, \alpha_3\, \gamma_4\, D_1\,| \cdot |\, B_2\, C_3\, D_4\,|.$$

To see the truth of the first equation, perform the following operations: row 4 − row 1, row 5 − row 2, then row 7 − row 4.

The second equation is seen on expanding the determinant of the twelfth order in terms of minors of order three formed from the first, second, third, and fourth sets of three rows.

166. It is readily seen that the determinant

$$\begin{vmatrix}
h_1 m_1 & k_1 x_1 & h_1 m_2 & k_1 x_2 \\
h_2 m_1 & k_2 y_1 & h_2 m_2 & k_2 x_2 \\
p_1 n_1 & q_1 z_1 & p_1 n_2 & q_1 z_2 \\
p_2 n_1 & q_2 z_1 & p_2 n_2 & q_2 z_2
\end{vmatrix}
=
\begin{vmatrix}
h_1 & k_1 & \cdot & \cdot \\
h_2 & k_2 & \cdot & \cdot \\
\cdot & \cdot & p_1 & q_1 \\
\cdot & \cdot & p_2 & q_2
\end{vmatrix}
\cdot
\begin{vmatrix}
m_1 & \cdot & m_2 & \cdot \\
\cdot & x_1 & \cdot & x_2 \\
n_1 & \cdot & n_2 & \cdot \\
\cdot & z_1 & \cdot & z_2
\end{vmatrix}$$

$$= B \cdot C = \begin{vmatrix} h_1 & k_1 \\ h_2 & k_2 \end{vmatrix} \cdot \begin{vmatrix} p_1 & q_1 \\ p_2 & q_2 \end{vmatrix} \cdot \begin{vmatrix} m_1 & n_1 \\ m_2 & n_2 \end{vmatrix} \cdot \begin{vmatrix} x_1 & z_1 \\ x_2 & z_2 \end{vmatrix}.$$

From this it is also readily seen that if we take two determinants each of order three and three determinants each of order two, we in a similar manner obtain as their product a determinant of order six each of whose elements is the product of two factors.

167. More generally we might take n_1 determinants of order n_2 and n_2 determinants of order n_1 and (by row-by-column multiplication) obtain as their product a determinant of order $n=n_1\cdot n_2$, each of whose elements is the product of two factors.

Let

$$B^{(i_2)} \equiv \left| b^{(i_2)}_{i_1j_1} \right| , \quad \left\{ \begin{matrix} i_1,j_1 = 1,2, \cdots , n_1 \\ i_2 = 1,2, \cdots , n_2 \end{matrix} \right\}$$

represent n_2 determinants each of order n_1, let

$$C^{(j_1)} \equiv \left| c^{(j_1)}_{i_2j_2} \right| (j_2 = 1,2, \cdots , n_2)$$

represent n_1 determinants each of order n_2, and let

$$A \equiv \left| a_{i_1i_2j_1j_2} \right| ,$$

where throughout

$$a_{i_1i_2j_1j_2} = b^{(i_2)}_{i_1j_1} c^{(j_1)}_{i_2j_2}$$

then the theorem is $A=B^{(1)}\cdot B^{(2)} \cdots B^{(n_2)}\cdot C^{(1)}\cdot C^{(2)} \cdots C^{(n_1)}$

Form as in §166 the determinant B of order $n=n_1\cdot n_2$ using the n_2 determinants $B^{(i_2)}$. If in B we consider the n rows and n columns divided up into n_2 sets of n_1 in each, then i_2 indicates the set to which any element belongs and i_1j_1 indicates the row and column in the set. The arrangement is such that if we number the sets from left to right and from top to bottom 1, 2, $\cdots$, n_2 then the elements of $B^{(\alpha)}$ are found in order at the intersection of the αth set of rows and the αth set of columns. Also form as in §166 a determinant C of order n using the n_1 determinants $C^{(j_1)}$. If in this we consider the n rows and n columns divided up into n_1 sets of n_2 in each, then j_1 indicates the set to which any element belongs, and i_2j_2 indicates the row and column in the set. The arrangement in this case is such that the elements of $|C^{(1)}{}_{i_2j_2}|$ occupy the intersections of the rows and columns indicated by the numbers 1, (n_1+1), $(2n_1+1)$, $\cdots$ $(\overline{n_2-1}\cdot n_1+1)$; the elements of $|C^{(2)}{}_{i_2j_2}|$ occupy the intersections of the rows and columns indicated by the numbers, 2, (n_1+2), $(2n_1+2)$, $\cdots$ $(\overline{n_2-1}\cdot n_1+2)$, and in general the elements of $|C^{(\alpha)}{}_{i_2j_2}|$ occupy the intersections of the rows and columns indicated by the numbers α, $(n_1+\alpha)$, $(2n_1+\alpha)$, $\cdots$ $(\overline{n_2-1}\cdot n_1+\alpha)$ where $\alpha=1, 2, \cdots, n_1$.

From this it is apparent that the product row-by-column of B and C will give A in which each element is the product of two factors, and since B is the product of the n_2 $B^{(i_2)}$'s, and C is the product of the n_1 $C^{(j_1)}$'s it follows that $A=\pi B^{(i_2)} C^{(j_1)}$ as was to be proved.

168. We may now form two determinants of the nth order $(n=n_1\cdot n_2\cdot n_3)$ precisely as in §166 first by taking n_3 determinants $A^{(g)},(g=1, 2, \cdots, n_3)$ each of whose elements is the product of two factors such as A there found, and second by taking $n_1\cdot n_2$ determinants $D^{(h_3)}\equiv \left|d^{(h_3)}_{i_3j_3}\right|, \left\{\begin{smallmatrix} i_3,j_3=1,2,\cdots n_3 \\ h_3=1,2,\cdots n/n_3\end{smallmatrix}\right\}$ of order n_3. From the first of these we form a determinant B of order n arranged as in B of §166 and from the second we form a determinant C arranged as in C of §166. Then the product of B and C will give a determinant A each of whose elements is the product of three factors, one a b, one a c, and one a d. We have then

$$A = \left| a_{i_1i_2i_3j_1j_2j_3}\right| = \prod_{h_1h_2h_3} \left| b^{(h_1)}_{i_1j_1}\right| \cdot \left| c^{(h_2)}_{i_2j_2}\right| \cdot \left| d^{(h_3)}_{i_3j_3}\right|$$

where throughout

$$a_{i_1i_2i_3j_1j_2j_3} = b^{(h_1)}_{i_1j_1} c^{(h_2)}_{i_2j_2} d^{(h_3)}_{i_3j_3}$$

and where now

$$\begin{cases} h_1 = 1,2, \cdots, n/n_1 \\ h_2 = 1,2, \cdots, n/n_2 \end{cases}$$

169. To put the results of the last article in a form better suited for further generalization let us write for $B^{(h_1)}$, $C^{(h_2)}$, $D^{(h_3)}$,

$$A^{(1,i_2i_3)} = \left| a^{(1,i_2i_3)}_{i_1j_1} \right|, \quad A^{(2,i_3j_1)} = \left| a^{(2,i_3j_1)}_{i_2j_2} \right|, \quad A^{(3,j_1j_2)} = \left| a^{(3,j_1j_2)}_{i_3j_3} \right|,$$

respectively.

The theorem may now be stated as follows:

$$A = \left| a_{i_1i_2i_3j_1j_2j_3}\right| = \prod A^{(1,i_2i_3)}\cdot A^{(2,i_3i_1)}\cdot A^{(3,j_1j_2)}$$

where throughout

$$a_{i_1i_2i_3j_1j_2j_3} = a^{(1,i_2i_3)}_{i_1j_1}\cdot a^{(2,i_3j_1)}_{i_2j_2}\cdot a^{(3,j_1j_2)}_{i_3j_3},$$

where

$$\begin{cases} i_1,j_1 = 1,2, \cdots, n_1 \\ i_2,j_2 = 1,2, \cdots, n_2 \\ i_3,j_3 = 1,2, \cdots, n_3 \end{cases}$$

and where the number of determinants in the product is $n/n_1+n/n_2+n/n_3$.

170. The most general theorem of this type may now be stated as follows: The determinant

$$A = \left| a_{i_1 i_2 \cdots i_k j_1 j_2 \cdots j_k} \right|, \quad \left\{ \begin{matrix} i_\beta, j_\beta = 1, 2, \cdots, n_\beta \\ \beta = 1, 2, \cdots, k \end{matrix} \right\},$$

of order $n = n_1 \cdot n_2 \cdots n_k$, where throughout

$$a_{i_1 i_2 \cdots i_k j_1 j_2 \cdots j_k} = \prod_{\beta=1}^{\beta=k} a_{i_\beta j_\beta}^{(\beta, i_{\beta+1} \cdots j_{\beta-1})},$$

and where $i_{\beta+1} \cdots j_{\beta-1}$ are the $k-1$ indices lying between i_β and j_β in $i_1 i_2 \cdots i_k j_1 j_2 \cdots j_k$, is the product of $\nu = n/n_1 + n/n_2 + \cdots + n/n_k$ determinants

$$A^{(\beta, i_{\beta+1} \cdots j_{\beta-1})} = \left| a_{i_\beta j_\beta}^{(\beta, i_{\beta+1} \cdots j_{\beta-1})} \right|$$

of order n_β.

The proof of this most general theorem is readily obtained by induction, assuming the truth for $k-1$ and then using the theorem of §167 to prove it true for k.

Assume $A^{(g_1 i_k)} = \left| a_{i_1 \cdots i_{k-1} j_1 \cdots j_{k-1}}^{(g_1 i_k)} \right|$, or as we may temporarily denote it $\left| a_{ij}^{(g_1 i_k)} \right|$, of order $n' = n_1 \cdot n_2 \cdots n_{k-1}$, with elements the product of $k-1$ factors, and where throughout

$$a_{i_1 \cdots i_{k-1} j_1 \cdots j_{k-1}}^{(g_1 i_k)} = a_{ij}^{(g_1 i_k)} = \prod_{\beta=1}^{\beta=k-1} a_{i_\beta j_\beta}^{(\beta, i_{\beta+1} \cdots j_{\beta-1})}.$$

Then $A^{(g_1 i_k)}$ is the product of $\nu' = n/n_1 + n/n_2 + \cdots + n/n_{k-1}$ determinants,

$$A^{(\beta, i_{\beta+1} \cdots j_{\beta-1})} = \left| a_{i_\beta j_\beta}^{(\beta, i_{\beta+1} \cdots j_{\beta-1})} \right| \left\{ \begin{matrix} i_\beta, j_\beta = 1, 2, \cdots, n_\beta \\ \beta = 1, 2, \cdots, k-1 \end{matrix} \right\}$$

Take now n_k determinants $A^{(g_1 i_k)} (g_1 = 1, 2, \cdots, n_k)$ and form a determinant B of order $n = n_1 \cdot n_2 \cdots n_k$ just as B of §166 was formed, and also take n' determinants $A^{(g_2 j)} = \left| a_{ij}^{(g_2 j)} \right| \left\{ \begin{matrix} g_2 & = 1, 2, \cdots n' \\ i_k, j_k & = 1, 2, \cdots n_k \end{matrix} \right\}$ of order n_k, with single elements, and form a determinant C of order n in the same manner as C of §166 was formed, then by the theorem of §167, the product $B \cdot C = A = \left| a_{i i_k j j_k} \right| = \left| a_{i_1 \cdots i_k j_1 \cdots j_k} \right|$, where throughout $a_{i_1 i_2 \cdots i_k j_1 j_2 \cdots j_k} = a_{ij}^{(g_1 i_k)} \cdot a_{i_k j_k}^{(g_2, j)}$, is a determinant of order n whose elements are the products of k factors, and is the product of $\nu = n/n_1 + n/n_2 + \cdots + n/n_k$ determinants

$$A^{(\beta, i_{\beta+1} \cdots j_{\beta-1})} = \left| a_{i_\beta j_\beta}^{(\beta, i_{\beta+1} \cdots j_{\beta-1})} \right|, \quad \left\{ \begin{matrix} i_\beta, j_\beta = 1, 2, \cdots, n_\beta \\ \beta = 1, 2, \cdots, k \end{matrix} \right\}$$

as was to be proved.

Corollaries

1. If all the determinants of the same order $n_g (g=1, 2, \cdots, k)$ are equal then

$$A = \prod_{g=1}^{g=k} \left\{ A^{(\beta, i_{\beta+1} \cdots i_{\beta-1})} \right\}^{n/n_g}$$

2. If $n_1 = n_2 = \cdots = n_k = m$ then $n = m^k$ and

$$A = \Pi \left\{ A^{(\beta, i_{\beta+1} \cdots i_{\beta-1})} \right\} = \prod_{g=1}^{g=k \cdot m^{k-1}} \left| a_{1m}^{(\iota)} \right|$$

3. If all determinants are of the same order and all alike then

$$A = \left| a_{1m} \right|^{k m^{k-1}}$$

The case of the theorem where $k=2$ was considered by Zehfuss in 1858 and such determinants have come to be known as Zehfuss determinants.

Determinants of this form occur frequently and their factors may readily be found by means of the theorem. Thus

$$\begin{vmatrix} a_{11}^2 & a_{12}^2 & a_{11}a_{12} & a_{11}a_{12} \\ a_{11}a_{21} & a_{12}a_{22} & a_{12}a_{21} & a_{11}a_{22} \\ a_{11}a_{21} & a_{12}a_{22} & a_{11}a_{22} & a_{12}a_{21} \\ a_{21}^2 & a_{22}^2 & a_{21}a_{22} & a_{22}a_{21} \end{vmatrix} = \begin{vmatrix} a_{11}a_{12} & \cdot & \cdot \\ a_{21}a_{22} & \cdot & \cdot \\ \cdot & \cdot & a_{11}a_{12} \\ \cdot & \cdot & a_{21}a_{22} \end{vmatrix} \cdot \begin{vmatrix} a_{11} & 0 & a_{21} & 0 \\ 0 & a_{12} & 0 & a_{22} \\ a_{12} & 0 & a_{22} & 0 \\ 0 & a_{11} & 0 & a_{21} \end{vmatrix}$$

$$= \begin{vmatrix} a_{11}^2 & a_{12}^2 & 2a_{11}a_{12} & a_{11}a_{12} \\ a_{11}a_{21} & a_{12}a_{22} & a_{11}a_{22} + a_{12}a_{21} & a_{11}a_{22} \\ 0 & 0 & 0 & -\left| a_{11}a_{22} \right| \\ a_{21}^2 & a_{22}^2 & 2a_{21}a_{22} & a_{22}a_{21} \end{vmatrix}$$

$$= - \left| a_{11}a_{22} \right| \begin{vmatrix} a_{11}^2 & a_{12}^2 & 2a_{11}a_{12} \\ a_{21}^2 & a_{22}^2 & 2a_{21}a_{22} \\ a_{11}a_{21} & a_{12}a_{22} & a_{11}a_{22} + a_{12}a_{21} \end{vmatrix} .$$

From which we see that

$$\begin{vmatrix} a_{11}^2 & a_{12}^2 & 2a_{11}a_{12} \\ a_{21}^2 & a_{22}^2 & 2a_{21}a_{22} \\ a_{11}a_{21} & a_{12}a_{22} & a_{11}a_{22}+a_{12}a_{12} \end{vmatrix} = \begin{vmatrix} a_{11} & a_{12} \\ a_{21} & a_{22} \end{vmatrix}^3$$

EXERCISE. Show that

$$\begin{vmatrix} a_1 & a_2 & -1 & \cdot & \cdot \\ b_1 & b_2 & \cdot & -1 & \cdot \\ c_1 & c_2 & \cdot & \cdot & -1 \\ \cdot & \cdot & l_1 & l_2 & l_3 \\ \cdot & \cdot & m_1 & m_2 & m_3 \end{vmatrix} = \begin{vmatrix} a_1 & b_1 \\ a_2 & b_2 \end{vmatrix} \cdot \begin{vmatrix} l_1 & l_2 \\ m_1 & m_2 \end{vmatrix} + \begin{vmatrix} a_1 & c_1 \\ a_2 & c_2 \end{vmatrix} \cdot \begin{vmatrix} l_1 & l_3 \\ m_1 & m_3 \end{vmatrix} + \begin{vmatrix} b_1 & c_1 \\ b_2 & c_2 \end{vmatrix} \cdot \begin{vmatrix} l_2 & l_3 \\ m_2 & m_3 \end{vmatrix}$$

Show that

$$cdh \begin{vmatrix} a & b & c \\ d & e & f \\ g & h & k \end{vmatrix} = -BFC + bB\cdot fF + fF\cdot gG + gG\, bB - (bB + fF + gG)(b + g - cdh) + (bfg - cdh)^2$$

and thence if $bfg = cdh$

$$\begin{vmatrix} a & b & c \\ d & e & f \\ g & h & k \end{vmatrix} = \frac{BF}{g} + \frac{FG}{b} + \frac{GB}{f} - \frac{BFG}{bfg},$$

where B, F, G are the minors complementary to b, f, g, respectively in the determinant.

171. If $(x_iy_iz_i)$ and $(x_i'\,y_i'\,z_i')(i=1, 2, 3, 4)$ represent the coordinates of the vertices of two tetrahedrons we know from Geometry that six times the volumes of these tetrahedrons are given by the determinants

$$\Delta \equiv \begin{vmatrix} x_1 & y_1 & z_1 & 1 \\ x_2 & y_2 & z_2 & 1 \\ x_3 & y_3 & z_3 & 1 \\ x_4 & y_4 & z_4 & 1 \end{vmatrix}, \quad \Delta' = \begin{vmatrix} x_1' & y_1' & z_1' & 1 \\ x_2' & y_2' & z_2' & 1 \\ x_3' & y_3' & z_3' & 1 \\ x_4' & y_4' & z_4' & 1 \end{vmatrix}$$

respectively.

Taking the product of these two in the form

$$\begin{vmatrix} x_1 & y_1 & z_1 & 1 & 0 \\ x_2 & y_2 & z_2 & 1 & 0 \\ x_3 & y_3 & z_3 & 1 & 0 \\ x_4 & y_4 & z_4 & 1 & 0 \\ 0 & 0 & 0 & 0 & 1 \end{vmatrix} \cdot \begin{vmatrix} x_1' & y_1' & z_1' & 0 & 1 \\ x_2' & y_2' & z_2' & 0 & 1 \\ x_3' & y_3' & z_3' & 0 & 1 \\ x_4' & y_4' & z_4' & 0 & 1 \\ 0 & 0 & 0 & -1 & 1 \end{vmatrix}$$

we get

$$\Delta\cdot\Delta' = \begin{vmatrix} \sum x_1x_1' & \sum x_1x_2' & \sum x_1x_3' & \sum x_1x_4' & -1 \\ \sum x_2x_1' & \sum x_2x_2' & \sum x_2x_3' & \sum x_2x_4' & -1 \\ \sum x_3x_1' & \sum x_3x_2' & \sum x_3x_3' & \sum x_3x_4' & -1 \\ \sum x_4x_1' & \sum x_4x_2' & \sum x_4x_3' & \sum x_4x_4' & -1 \\ 1 & 1 & 1 & 1 & . \end{vmatrix}$$

where $\sum x_rx_s' = x_rx_s' + y_ry_s' + z_rz_s'$.

If now we multiply each of the first four columns by 2 and then divide the last row by 2 we will have multiplied the determinant by 8.

Next perform the following operations

$$\text{row}_1 - \sum x_1^2 \times \text{row}_5,\ \text{row}_2 - \sum x_2^2 \times \text{row}_5,\ \cdots$$
$$\text{col}_1 - \sum x_1^2 \times \text{col}_5,\ \text{col}_2 - \sum x_2^2 \times \text{col}_5,\ \cdots$$

and we obtain

$$\Delta\cdot\Delta' = \frac{1}{8}\begin{vmatrix} \sum(x_1-x_1')^2 & \sum(x_1-x_2')^2 & \sum(x_1-x_3')^2 & \sum(x_1-x_4')^2 & 1 \\ \sum(x_2-x_1')^2 & \sum(x_2-x_2')^2 & \sum(x_2-x_3')^2 & \sum(x_2-x_4')^2 & 1 \\ \sum(x_3-x_1')^2 & \sum(x_3-x_2')^2 & \sum(x_3-x_3')^2 & \sum(x_3-x_4')^2 & 1 \\ \sum(x_4-x_1')^2 & \sum(x_4-x_2')^2 & \sum(x_4-x_3')^2 & \sum(x_4-x_4')^2 & 1 \\ 1 & 1 & 1 & 1 & . \end{vmatrix} = \Delta''$$

We have on the right an expression for thirty-six times the product of the two volumes in terms of the distances from the vertices of the one to the vertices of the other. If the two tetrahedrons coincide the accented letters are the same as the unaccented and the principal diagonal terms become zero and we have an expression for the square of the volume. If all four points lie in the same plane the volume is zero and we have a relation satisfied by the coordinates of four points in a plane.

EXERCISE 1. Show that the determinant Δ'' when the accents are dropped may be obtained from the product

$$\begin{vmatrix} x_1^2 & -2x_1 & \cdot & \cdot & 1 \\ x_2^2 & -2x_2 & \cdot & \cdot & 1 \\ x_3^2 & -2x_3 & \cdot & \cdot & 1 \\ x_4^2 & -2x_4 & \cdot & \cdot & 1 \\ 1 & \cdot & \cdot & \cdot & \cdot \end{vmatrix} \times \begin{vmatrix} 1 & x_1 & \cdot & \cdot & x_1^2 \\ 1 & x_2 & \cdot & \cdot & x_2^2 \\ 1 & x_3 & \cdot & \cdot & x_3^2 \\ 1 & x_4 & \cdot & \cdot & x_4^2 \\ \cdot & \cdot & \cdot & \cdot & 1 \end{vmatrix}$$

where the multiplication is row-by-row. Similarly for any order.

2. In the case where the tetrahedrons coincide let $2W$ represent the determinant on the right and let the minors complementary to the elements in the positions (11), (22), (33), (44), be represented by F, G, H, K, respectively.

Show that the result of rationalizing $F^{\frac{1}{2}}+G^{\frac{1}{2}}+H^{\frac{1}{2}}+K^{\frac{1}{2}}$ contains W as a factor.

3. Express Δ in the form

$$\begin{vmatrix} x_1 - x_2 & y_1 - y_2 & z_1 - z_2 \\ x_1 - x_3 & y_1 - y_3 & z_1 - z_3 \\ x_1 - x_4 & y_1 - y_4 & z_1 - z_4 \end{vmatrix}$$

and by squaring express the relation between the mutual distances of four points in a plane in the form

$$\begin{vmatrix} 2d^2_{12} & d^2_{12} + d^2_{13} - d^2_{23} & d^2_{12} + d^2_{14} - d^2_{24} \\ d^2_{12} + d^2_{13} - d^2_{23} & 2d^2_{13} & d^2_{13} + d^2_{14} - d^2_{34} \\ d_{12} + d^2_{14} - d_{24} & d^2_{13} + d^2_{14} - d^2_{34} & 2d^2_{14} \end{vmatrix} = 0$$

where $d^2_{rs} = \sum (x_r - x_s)^2$.

172. If $S_{rs} \equiv (a_r + b_s)$ then the determinant of order $(n+1)$

$$\begin{vmatrix} S^n_{11} & S^n_{12} & \cdots & S^n_{1n} & a^n_1 \\ S^n_{21} & S^n_{22} & \cdots & S^n_{2n} & a^n_2 \\ \cdot & \cdot & \cdots & \cdot & \cdot \\ S^n_{n1} & S^n_{n2} & \cdots & S^n_{nn} & a^n_n \\ b^n_1 & b^n_2 & & b^n_n & 0 \end{vmatrix} = a_1 a_2 \cdots a_n b_1 b_2 \cdots b_n \begin{vmatrix} S^n_{11} & S^n_{12} & \cdots & S^n_{1n} & 1 \\ S^n_{21} & S^n_{22} & \cdots & S^n_{2n} & 1 \\ \cdot & \cdot & \cdots & \cdot & \cdot \\ S^n_{n1} & S^n_{n2} & \cdots & S^n_{nn} & 1 \\ 1 & 1 & \cdots & 1 & 0 \end{vmatrix},$$

for the determinant on the left is the product of

$$\begin{vmatrix} a_1^n & n_1 a_1^{n-1} & \cdots & n a_1 & 1 \\ a_2^n & n_1 a_2^{n-1} & \cdots & n a_2 & 1 \\ \cdots & \cdots & \cdots & \cdots & \cdots \\ a_n^n & n_1 a_n^{n-1} & \cdots & n a_n & 1 \\ 0 & 0 & \cdots & 0 & 1 \end{vmatrix} \cdot \begin{vmatrix} 1 & b_1 & b_1^2 & \cdots & b_1^n \\ 1 & b_2 & b_2^2 & \cdots & b_2^n \\ \cdots & \cdots & \cdots & \cdots & \cdots \\ 1 & b_n & b_n^2 & \cdots & b_n^n \\ 1 & 0 & & \cdots & 0 \end{vmatrix} \equiv A_1 \; B_1,$$

say, where $n_r = {}_nC_r$ and the determinant on the right is the product of

$$\begin{vmatrix} a_1^n & n_1 a_1^{n-1} & \cdots & n a_1 & 1 \\ \cdots & \cdots & \cdots & \cdots & \cdots \\ a_n^n & n_1 a_n^{n-1} & \cdots & n a_n & 1 \\ 1 & 0 & \cdots & 0 & 0 \end{vmatrix} \cdot \begin{vmatrix} 1 & b_1 & b_1^2 & \cdots & b_1^n \\ \cdots & \cdots & \cdots & \cdots & \cdots \\ 1 & b_n & b_n^2 & \cdots & b_n^n \\ 0 & 0 & 0 & \cdots & 1 \end{vmatrix} \equiv A_2 \cdot B_2$$

say. By easy transformations it may be seen that $A_1 = a_1 a_2 \cdots a_n A_2$ and $B_1 = b_1 b_2 \cdots b_2 B_2$ and hence the theorem.

In the foregoing if we make the b's all negative we have

$$\begin{vmatrix} d_{11}^n & d_{12}^n & \cdots & d_{1n}^n & a_1^n \\ \cdots & \cdots & \cdots & \cdots & \cdots \\ d_{n1}^n & d_{n2}^n & \cdots & d_{nn}^n & a_n^n \\ b_1^n & b_2^n & \cdots & b_n^n & 0 \end{vmatrix} = a_1 a_2 \cdots a_n b_1 b_2 \cdots b_n \begin{vmatrix} d_{11}^n & \cdots & d_{1n}^n & 1 \\ \cdots & \cdots & \cdots & \cdots \\ d_{n1}^n & \cdots & d_{nn}^n & 1 \\ 1 & \cdots & 1 & 0 \end{vmatrix}$$

where $d_{rs} = a_r - b_s$.

CHAPTER VI

COMPOUND DETERMINANTS

173. A determinant with elements which are themselves determinants is called a *compound* determinant.

The determinant each of whose elements is the cofactor of the corresponding elements in another determinant is called the determinant *adjugate* to that other. Thus

$$\begin{vmatrix} \begin{vmatrix} b_2 & c_3 \end{vmatrix} & -\begin{vmatrix} a_2 & c_3 \end{vmatrix} & \begin{vmatrix} a_2 & b_3 \end{vmatrix} \\ -\begin{vmatrix} b_1 & c_3 \end{vmatrix} & \begin{vmatrix} a_1 & c_3 \end{vmatrix} & -\begin{vmatrix} a_1 & b_3 \end{vmatrix} \\ \begin{vmatrix} b_1 & c_2 \end{vmatrix} & -\begin{vmatrix} a_1 & c_2 \end{vmatrix} & \begin{vmatrix} a_1 & b_2 \end{vmatrix} \end{vmatrix}$$

is the determinant adjugate to

$$\begin{vmatrix} a_1 & b_1 & c_1 \\ a_2 & b_2 & c_2 \\ a_3 & b_3 & c_3 \end{vmatrix};$$

and using the notation of §68,

$$\left| \mathcal{A}_{11} \mathcal{A}_{22} \cdots \mathcal{A}_{nn} \right| \quad \text{or} \quad \left| \mathcal{A}_{1n} \right|$$

is the determinant adjugate to

$$\left| a_{11}\, a_{22} \cdots a_{nn} \right| \quad \text{or} \quad \left| a_{1n} \right|$$

When the elements of the adjugate determinant are specified as above by means of the complementary minors of the corresponding elements in the original determinant, negative signs must appear in the places whose row-number and column-number have a sum which is odd. These signs may however (Ex. 22, Set IV) be deleted without altering the value of the determinant; hence, in the definition which has been given we might substitute "complementary minor" for "co-factor."

174. *The determinant adjugate to a determinant of the nth order is equal to the $(n-1)$th power of the latter.*

Let the given determinant be

$$A \equiv \begin{vmatrix} a_{11} & a_{12} & a_{13} & \cdots & a_{1n} \\ a_{21} & a_{22} & a_{23} & \cdots & a_{2n} \\ a_{31} & a_{32} & a_{33} & \cdots & a_{3n} \\ \cdot & \cdot & \cdot & \cdots & \cdot \\ a_{n1} & a_{n2} & a_{n3} & \cdots & a_{nn} \end{vmatrix}, \quad \text{or} \quad \left| a_{1n} \right|.$$

Multiply it by its adjugate

$$\begin{vmatrix} \mathcal{A}_{11} & \mathcal{A}_{12} & \mathcal{A}_{13} \cdots \mathcal{A}_{1n} \\ \mathcal{A}_{21} & \mathcal{A}_{22} & \mathcal{A}_{23} \cdots \mathcal{A}_{2n} \\ \mathcal{A}_{31} & \mathcal{A}_{32} & \mathcal{A}_{33} \cdots \mathcal{A}_{3n} \\ \cdots & \cdots & \cdots \\ \mathcal{A}_{n1} & \mathcal{A}_{n2} & \mathcal{A}_{n3} \cdots \mathcal{A}_{nn} \end{vmatrix}, \quad \text{or} \quad |\mathcal{A}_{1n}|,$$

after the manner of §154, the first column of the new determinant is

$$\begin{array}{l} a_{11}\mathcal{A}_{11} + a_{12}\mathcal{A}_{12} + a_{13}\mathcal{A}_{13} + \cdots + a_{1n}\mathcal{A}_{1n} \\ a_{21}\mathcal{A}_{11} + a_{22}\mathcal{A}_{12} + a_{23}\mathcal{A}_{13} + \cdots + a_{2n}\mathcal{A}_{1n} \\ a_{31}\mathcal{A}_{11} + a_{32}\mathcal{A}_{12} + a_{33}\mathcal{A}_{13} + \cdots + a_{3n}\mathcal{A}_{1n} \\ \cdots\cdots\cdots\cdots\cdots\cdots\cdots\cdots \\ a_{n1}\mathcal{A}_{11} + a_{n2}\mathcal{A}_{12} + a_{n3}\mathcal{A}_{13} + \cdots + a_{nn}\mathcal{A}_{1n} \end{array}$$

the first expression of which is (§68) equal to $|a_{1n}|$, and each of the others (§76) equal to zero.

In like manner the elements of the second column are all zero, except the second element, which is $|a_{1n}|$; and so on. Thus

$$A \cdot |\mathcal{A}_{1n}| = \begin{vmatrix} A & 0 & 0 \cdots 0 \\ 0 & A & 0 \cdots 0 \\ 0 & 0 & A \cdots 0 \\ \cdots & \cdots & \cdots \\ 0 & 0 & 0 \cdots A \end{vmatrix} = A^n;$$

therefore

$$|\mathcal{A}_{1n}| = A^{n-1},$$

as was to be proved.

175. *If the determinant adjugate to a given determinant be formed, any minor of it of the mth order is equal to the product obtained by multiplying the cofactor of the corresponding minor in the original determinant by the* $(m-1)$*th power of the original determinant.*

I. When the minor of the adjugate is $|\mathcal{A}_{11}\mathcal{A}_{22}\cdots\mathcal{A}_{mm}|$, the original determinant being

$$\begin{vmatrix} a_{11} & a_{12} & a_{13} & \cdots a_{1m} & a_{1,m+1} & \cdots a_{1n} \\ a_{21} & a_{22} & a_{23} & \cdots a_{2m} & a_{2,m+1} & \cdots a_{2n} \\ a_{31} & a_{32} & a_{33} & \cdots a_{3m} & a_{3,m+1} & \cdots a_{3n} \\ \cdots & \cdots & \cdots & \cdots & \cdots & \cdots \\ a_{m1} & a_{m2} & a_{m3} & \cdots a_{mm} & a_{m,m+1} & \cdots a_{mn} \\ a_{m+1,1} & a_{m+1,2} & a_{m+1,3} & \cdots a_{m+1,m} & a_{m+1,m+1} & \cdots a_{m+1,n} \\ \cdots & \cdots & \cdots & \cdots & \cdots & \cdots \\ a_{n1} & a_{n2} & a_{n3} & \cdots a_{nm} & a_{n,m+1} & \cdots a_{nn} \end{vmatrix}.$$

Taking the adjugate determinant and changing the elements of all the rows after the mth, those occupying the principal diagonal into 1, and all the others into 0, we have

$$\begin{vmatrix} \mathcal{A}_{11} & \mathcal{A}_{12} & \cdots \mathcal{A}_{1m} & \mathcal{A}_{1,m+1} & \mathcal{A}_{1,m+2} & \cdots \mathcal{A}_{1n} \\ \mathcal{A}_{21} & \mathcal{A}_{22} & \cdots \mathcal{A}_{2m} & \mathcal{A}_{2,m+1} & \mathcal{A}_{2,m+2} & \cdots \mathcal{A}_{2n} \\ \cdots & \cdots & \cdots & \cdots & \cdots & \cdots \\ \mathcal{A}_{m1} & \mathcal{A}_{m2} & \cdots \mathcal{A}_{mm} & \mathcal{A}_{m,m+1} & \mathcal{A}_{m,m+2} & \cdots \mathcal{A}_{mn} \\ 0 & 0 & \cdots 0 & 1 & 0 & \cdots 0 \\ 0 & 0 & \cdots 0 & 0 & 1 & \cdots 0 \\ \cdots & \cdots & \cdots & \cdots & \cdots & \cdots \\ 0 & 0 & \cdots 0 & 0 & 0 & \cdots 1 \end{vmatrix},$$

which is clearly equal to the chosen minor $|\mathcal{A}_{11} \cdots \mathcal{A}_{mm}|$. Multiplying the original determinant by this, there results

$$A \cdot |\mathcal{A}_{1m}| = \begin{vmatrix} A & 0 & 0 & \cdots 0 & 0 & \cdots 0 \\ 0 & A & 0 & \cdots 0 & 0 & \cdots 0 \\ 0 & 0 & A & \cdots 0 & 0 & \cdots 0 \\ \cdots & \cdots & \cdots & \cdots & \cdots & \cdots \\ 0 & 0 & 0 & \cdots A & 0 & \cdots 0 \\ a_{1,m+1} & a_{2,m+1} & a_{3,m+1} & \cdots a_{m,m+1} & a_{m+1,m+1} & \cdots a_{n,m+1} \\ a_{1,m+2} & a_{2,m+2} & a_{3,m+2} & \cdots a_{m,m+2} & a_{m+1,m+2} & \cdots a_{n,m+2} \\ \cdots & \cdots & \cdots & \cdots & \cdots & \cdots \\ a_{1n} & a_{2n} & a_{3n} & \cdots a_{mn} & a_{m+1,n} & \cdots a_{nn} \end{vmatrix}$$

$$= A^{m} \cdot \left| a_{m+1,m+1} \cdots a_{nn} \right| ;$$

$$\left| \mathcal{A}_{1m} \right| = A^{m-1} \left| a_{m+1,m+1} \cdots a_{nn} \right| ;$$

and $|a_{m+1,m+1} \cdots a_{nn}|$ being, in the original determinant, the cofactor of the minor corresponding to the chosen minor, one case of the theorem is established.

II. When the minor of the adjugate is any other than

$$\left| \mathcal{A}_{11} \mathcal{A}_{22} \cdots \mathcal{A}_{mm} \right| .$$

Let the rows from which the elements of the minor are taken be the hth, kth, lth, $\cdots$, and the columns the rth, sth, uth, $\cdots$, so that the minor may be denoted by $|\mathcal{A}_{hr} \mathcal{A}_{ks} \mathcal{A}_{lu} \cdots|$, and let $h+k+l+ \cdots +r+s+u+ \cdots =\sigma$.

Translating the said rows in order upwards and the said columns in order towards the left, the chosen minor will occupy the place at first occupied by $|\mathcal{A}_{11} \mathcal{A}_{22} \cdots \mathcal{A}_{mm}|$; and if we change the elements in the rows after the mth, making those 1 which occupy places in the principal diagonal and all the others 0, we have as before a determinant of the nth order, which equals the chosen minor

$$\left| \mathcal{A}_{hr} \mathcal{A}_{ks} \mathcal{A}_{lu} \cdots \right| .$$

Also, translating in the same way the corresponding rows and columns in the original determinant, we have a determinant which (see §91) is equal to

$$(-1)^{\sigma} \left| a_{1n} \right| .$$

In the former of these two resulting determinants each $\mathcal{A}$ of the first m rows occupies the place which the corresponding a occupies in the latter determinant; consequently on multiplying the two together we have as before

$$\left| a_{1n} \right|^{m} \times \text{cofactor of } \left| a_{hr} a_{ks} a_{lu} \cdots \right| \text{ in } (-1)^{\sigma} \left| a_{1n} \right| ,$$

and on division by $(-1)^{\sigma} |a_{1n}|$ there results

$$\left| \mathcal{A}_{hr} \mathcal{A}_{ks} \mathcal{A}_{lu} \cdots \right| = \left| a_{1n} \right|^{m-1} \times \text{cofactor of } \left| a_{hr}\, a_{ks} \cdots \right| \text{ in} \left| a_{1n} \right| ,$$

as was to be proved.

If we write A_{hr} for the complementary minor of a_{hr} in $|a_{1n}|$, then since $\mathcal{A}_{hr}=(-1)^{h+r}A_{hr}$, etc. and the cofactor of $|a_{hr}\, a_{ks}\, a_{lu} \cdots|$ in $|a_{1n}|$ is equal to $(-1)^{\sigma}$ multiplied by the complementary minor of $|a_{hr}\, a_{ks}\, a_{lu} \cdots|$ in $|a_{1n}|$, the result just obtained becomes $|(-1)^{h+r}A_{hr}(-1)^{k+s}A_{ks} \cdots| = |a_{1n}|^{m-1}(-1)^{\sigma}\times$complementary of

$|a_{hr}\ a_{ks} \cdots|$; so that, multiplying the rows of the left-hand member of this by $(-1)^h$, $(-1)^k$, $\cdots$, respectively, and the columns by $(-1)^r$, $(-1)^s$, $\cdots$, respectively, and multiplying the right-hand member by the same, viz. by $(-1)^\sigma$, we have

$$|A_{hr}\ A_{ks}\ A_{lu} \cdots| = |a_{1n}|^{m-1}(-1)^\sigma \times$$

complementary of $|a_{hr}\ a_{ks}\ a_{lu} \cdots|$.

EXAMPLE. The adjugate of

$$\begin{vmatrix} a_1 & a_2 & a_3 & a_4 \\ b_1 & b_2 & b_3 & b_4 \\ c_1 & c_2 & c_3 & c_4 \\ d_1 & d_2 & d_3 & d_4 \end{vmatrix}$$

is

$$\begin{vmatrix} |b_2\, c_3\, d_4| & -|b_1\, c_3\, d_4| & |b_1\, c_2\, d_4| & -|b_1\, c_2\, d_3| \\ -|a_2\, c_3\, d_4| & |a_1\, c_3\, d_4| & -|a_1\, c_2\, d_4| & |a_1\, c_2\, d_3| \\ |a_2\, b_3\, d_4| & -|a_1\, b_3\, d_4| & |a_1\, b_2\, d_4| & -|a_1\, b_2\, d_3| \\ -|a_2\, b_3\, c_4| & |a_1\, b_3\, c_4| & -|a_1\, b_2\, c_4| & |a_1\, b_2\, c_3| \end{vmatrix}$$

and if the chosen minor be

$$\begin{vmatrix} |a_1\, c_3\, d_4| & -|a_1\, c_2\, d_4| \\ |a_1\, b_3\, c_4| & -|a_1\, b_2\, c_4| \end{vmatrix},$$

we take the adjugate and by transposition of rows and columns obtain

$$\begin{vmatrix} |a_1\, c_3\, d_4| & -|a_1\, c_2\, d_4| & -|a_2\, c_3\, d_4| & |a_1\, c_2\, d_3| \\ |a_1\, b_3\, c_4| & -|a_1\, b_2\, c_4| & -|a_2\, b_3\, c_4| & |a_1\, b_1\, c_3| \\ -|b_1\, c_3\, d_4| & |b_1\, c_2\, d_4| & |b_2\, c_3\, d_4| & -|b_1\, c_2\, d_3| \\ -|a_1\, b_3\, d_4| & |a_1\, b_2\, d_4| & |a_2\, b_3\, d_4| & -|a_1\, b_2\, d_3| \end{vmatrix}$$

then altering the elements of the last two rows we have

$$\begin{vmatrix} |a_1\, c_3\, d_4| & -|a_1\, c_2\, d_4| & -|a_2\, c_3\, d_4| & |a_1\, c_2\, d_3| \\ |a_1\, b_3\, c_4| & -|a_1\, b_2\, c_4| & -|a_2\, b_3\, c_4| & |a_1\, b_2\, c_3| \\ 0 & 0 & 1 & 0 \\ 0 & 0 & 0 & 1 \end{vmatrix}$$

Multiplying this by

$$\begin{vmatrix} b_2 & b_3 & b_1 & b_4 \\ d_2 & d_3 & d_1 & d_4 \\ a_2 & a_3 & a_1 & a_4 \\ c_2 & c_3 & c_1 & c_4 \end{vmatrix}$$

there results

$$|b_2\, d_3\, a_1\, c_4| \begin{vmatrix} |a_1\, c_3\, d_4| & -|a_1\, c_2\, d_4| \\ |a_1\, b_3\, c_4| & -|a_1\, b_2\, c_4| \end{vmatrix}$$

$$= \begin{vmatrix} |a_1\, b_2\, c_3\, d_4| & 0 & 0 & 0 \\ 0 & |a_1\, b_2\, c_3\, d_4| & 0 & 0 \\ b_1 & d_1 & a_1 & c_1 \\ b_4 & d_4 & a_4 & c_4 \end{vmatrix},$$

$$= |a_1\, b_2\, c_3\, d_4|^2 \cdot |a_1\, c_4| \;;$$

Therefore

$$\begin{vmatrix} |a_1\, c_3\, d_4| & |a_1\, c_2\, d_4| \\ |a_1\, b_3\, c_4| & |a_1\, b_2\, c_4| \end{vmatrix} = |a_1\, b_2\, c_3\, d_4|\; |a_1\, c_4| .$$

This example serves also to illustrate the fact that the special case of the theorem of §148, noticed in §149, is at the same time a special case ($m=2$) of the present theorem.

176. *If the minor A_{11} complementary to the element a_{11} of a determinant A is zero then the determinant is expressible as the product of two linear functions: one linear in the elements $a_{21}, a_{31}, \cdots, a_{n1}$, and the other linear in the elements $a_{12}, a_{12}, \cdots, a_{1n}$, each with fractional coefficients.* Thus

$$\begin{vmatrix} A_{11} & A_{12} \\ A_{21} & A_{22} \end{vmatrix} = A_{11}A_{22} - A_{12}A_{21} = A \cdot A_{12,12}$$

$$\text{or } A = -\frac{A_{12}A_{21}}{A_{12,12}}, \quad \text{if} \quad A_{12,12} \neq 0.$$

But A_{12} may be expressed as a linear function of $a_{21}, a_{31}, \cdots, a_{n1}$, and A_{21} as a linear function of $a_{12}, a_{13}, \cdots, a_{1n}$ and hence the truth of the theorem is seen.

Instead of A_{11} we might have taken any minor A_{ij} as the vanishing minor and obtained a corresponding result.

EXERCISE. Show that the above theorem may be written

$$-A\cdot\frac{\partial B}{\partial a_{n-1,n-1}}=\left\{\frac{\partial B}{\partial a_{1,n-1}}a_{1n}+\frac{\partial B}{\partial a_{2,n-1}}a_{2n}+\cdots\right.$$
$$\left.+\frac{\partial B}{\partial a_{n-1,n-1}}a_{n-1,n}\right\}$$
$$\times\left\{\frac{\partial B}{\partial a_{n-1,1}}a_{n1}+\frac{\partial B}{\partial a_{n-1,2}}a_{n2}+\cdots\right.$$
$$\left.+\frac{\partial B}{\partial a_{n-1,n-1}}a_{n,n-1}\right\}$$

where $$B=\frac{\partial A}{\partial a_{nn}}\cdot$$

177. Two matrices are said to be the *reciprocal* of each other when row-by-column multiplication gives

$$\begin{vmatrix} 1 & \cdot & \cdot & \cdot & \cdot \\ \cdot & 1 & \cdot & \cdot & \cdot \\ \cdot & \cdot & 1 & \cdot & \cdot \\ \cdot & \cdot & \cdot & \cdot & \cdot \\ \cdot & \cdot & \cdot & \cdot & \cdot \end{vmatrix}=1$$

We may define *reciprocal determinants* in the same way though, for determinants, "row-by-row" may replace "row-by-column" in the definition.

178. It immediately follows from §175 that, in the case of a determinant which is equal to zero, all the minors of the adjugate which are of a higher order than the first must also be equal to zero. Thus, taking minors of the second order,

if $$|a_{1n}|=0,\quad\begin{vmatrix}\mathcal{A}_{h1} & \mathcal{A}_{h2}\\ \mathcal{A}_{k1} & \mathcal{A}_{k2}\end{vmatrix}=\begin{vmatrix}\mathcal{A}_{h1} & \mathcal{A}_{h3}\\ \mathcal{A}_{k1} & \mathcal{A}_{k3}\end{vmatrix}=\cdots=0,$$

and therefore $\mathcal{A}_{h1}:\mathcal{A}_{k1}::\mathcal{A}_{h2}:\mathcal{A}_{k2}::\cdots$ or $\mathcal{A}_{h1}:\mathcal{A}_{h2}:\mathcal{A}_{h3}:\cdots::\mathcal{A}_{k1}:\mathcal{A}_{k2}:\mathcal{A}_{k3}:\cdots$ provided $\mathcal{A}_{ki}\neq 0$ that is to say, *in the case of a zero-determinant the cofactors* of the elements of any one row are in order proportional to the cofactors of the elements of any other row.*

In the same way it may be proved that *if all minors of order* $(m+1)$ *vanish then all m-line minors formed from any set of m rows are proportional to the corresponding minors formed from any other set of m rows.*

* True also for complementary minors; see §175.

179. *To every general theorem which takes the form of an identical relation between a number of the minors of a determinant or between the determinant itself and a number of its minors, there corresponds another theorem derivable from the former by merely subtituting for every minor its cofactor* in the determinant, and then multiplying any term by such a power of the determinant as will make all the terms of the same degree.*

Let the established identity in regard to $|a_{1n}|$ be

$$M_rM_sM_t\cdots+\cdots=M_\rho M_\sigma M_\tau\cdots+\cdots$$

where M_r is used to denote some minor of $|a_{1n}|$ of the order r, and where consequently $r+s+t+\cdots=\rho+\sigma+\tau+\cdots$.

Now since the identity holds in regard to every determinant, it holds in regard to $|A_{1n}|$ the adjugate of $|a_{1n}|$, hence if N_r stand for the minor of $|A_{1n}|$ corresponding to the minor M_r of $|a_{1n}|$, it follows that

$$N_rN_sN_t\cdots+\cdots=N_\rho N_\sigma N_\tau\cdots+\cdots$$

Substituting for every N its equivalent as given by the theorem of §175, and, in order to do so, denoting the cofactor of M_r in $|a_{1n}|$ by M'_{n-r} we have

$$|a_{1n}|^{r-1}M'_{n-r}\cdot|a_{1n}|^{s-1}M'_{n-s}\cdot|a_{1n}|^{t-1}M'_{n-t}\cdots+\cdots$$
$$=|a_{1n}|^{\rho-1}M'_{n-\rho}\cdot|a_{1n}|^{\sigma-1}M'_{n-\sigma}\cdot|a_{1n}|^{\tau-1}M'_{n-\tau}\cdots+\cdots$$

whence, on division by the lowest power of $|a_{1n}|$ contained in any term, there results the identity which was to be established.

This is the *Law of Complementaries* incidentally exemplified in §149.

180. By the application of the Law of Complementaries, some of the already established theorems furnish new theorems of considerable interest. As an example the identity of §77 may be taken, a particular case of which is

$$|a_1b_2c_3d_4|\,b_1c_1=\begin{vmatrix}|a_1b_2| & |b_1c_2| & |c_1d_2|\\ |a_1b_3| & |b_1c_3| & |c_1d_3|\\ |a_1b_4| & |b_1c_4| & |c_1d_4|\end{vmatrix}$$

The complementary of this with respect to $|a_1b_2c_3d_4|$ is

$$(A)\qquad |a_2c_3d_4|\;|a_2b_3d_4|=\begin{vmatrix}|c_3d_4| & |a_3d_4| & |a_3b_4|\\ |c_2d_4| & |a_2d_4| & |a_2b_4|\\ |c_2d_3| & |a_2d_3| & |a_2b_3|\end{vmatrix},$$

* True also for complementary minors; see §175.

an identity not hitherto noticed, but which when known can be established otherwise. The complementary with respect to $|a_1b_2c_3d_4e_5|$ is

$$(B)\quad e_5 \begin{vmatrix} a_2\,c_3\,d_4\,e_5 \end{vmatrix} \begin{vmatrix} a_2\,b_3\,d_4\,e_5 \end{vmatrix} = \begin{vmatrix} |c_3\,d_4\,e_5| & |a_3\,d_4\,e_5| & |a_3\,b_4\,e_5| \\ |c_2\,d_4\,e_5| & |a_2\,d_4\,e_5| & |a_2\,b_4\,e_5| \\ |c_2\,d_3\,e_5| & |a_2\,d_3\,e_5| & |a_2\,b_3\,e_5| \end{vmatrix}$$

and by inserting "$f_6g_7 \cdots$" after every e_5 in this we have a result which includes (A) and (B), viz. the complementary with respect to $|a_1\,b_2 \cdots g_7 \cdots|$. A still more general theorem will be got by taking the complementary of the theorem of §150, of which that of §77 is a particular case.

The student will find it instructive to take every theorem to which the law is applicable and find the complementary theorem. Even where no new result is obtained, some new tie of relationship may be made apparent.

EXAMPLE. Prove that

$$\begin{vmatrix} a & b & g \\ h & b & f \\ g & f & c \end{vmatrix} (ax^2 + by^2 + cz^2 + 2fyz + 2gzx + 2hxy) = \begin{vmatrix} 0 & x & y & z \\ x & A & H & G \\ y & H & B & F \\ z & G & F & C \end{vmatrix}$$

where $A, H, \cdots$ stand for the complmentary minors of $a, h, \cdots$ in the first determinant.

Developing the right-hand member by §110 as a quadratic in $x, y, z, \cdots$, we have the cofactor of x^2 in it

$$= \begin{vmatrix} B & F \\ F & C \end{vmatrix} = a \begin{vmatrix} a & h & g \\ h & b & f \\ g & f & c \end{vmatrix}, \qquad (\S\,175)$$

as it should be. Similarly it is seen that the cofactors of $y^2, z^2, \cdots$ are the same in both members.

PROBLEM. By using the Law of Complementaries on the identity

$$|a_1\,b_2\,c_3\,d_4\,e_5| = \sum\{\,|a_1\,b_2| \cdot |c_3\,d_4|\,e_5\}$$

show that

$$|a_1\,b_2\,c_3\,d_4\,e_5|^2 = \sum\{\,|a_1\,b_2\,c_3\,d_4| \cdot |a_1\,b_2\,e_5| \cdot |c_3\,d_4\,e_5|\,\}.$$

Exercises. Set XII

1. Show that the sum of the numbers indicating the rows and columns from which the elements of a minor are taken and the sum of the corresponding numbers in the case of the complementary minor are either both even or both odd.

Resolve the following into determinant factors:

2. $$\begin{vmatrix} z_1 & 0 & 0 & 0 & 0 & 0 \\ 0 & z_2 & 0 & 0 & y_2 & x_2 \\ 0 & 0 & z_3 & y_3 & 0 & 0 \\ z_3 & z_3 & z_2 & y_2 & y_3 & y_3 \\ 0 & 0 & z_1 & y_1 & 0 & 0 \\ z_2 & z_1 & 0 & 0 & y_1 & y_1 \end{vmatrix}.$$

3. $$\begin{vmatrix} z_1 & 0 & 0 & 0 & 0 & 0 \\ 0 & z_2 & 0 & 0 & y_2 & 0 \\ 0 & 0 & z_3 & y_3 & 0 & x_3 \\ 0 & z_3 & z_2 & y_2 & y_3 & x_2 \\ z_3 & 0 & z_1 & y_1 & 0 & x_1 \\ z_2 & z_1 & 0 & 0 & y_1 & 0 \end{vmatrix}.$$

4. $$\begin{vmatrix} a_1 & 0 & 0 & a_4 & 0 & 0 \\ 0 & a_2 & 0 & 0 & a_5 & 0 \\ 0 & 0 & a_3 & 0 & 0 & a_6 \\ b_1 & 0 & 0 & b_4 & 0 & 0 \\ 0 & b_2 & 0 & 0 & b_5 & 0 \\ 0 & 0 & b_3 & 0 & 0 & b_6 \end{vmatrix}.$$

5. Use §93 to show that

$$\begin{vmatrix} 0 & a_2 & 0 & a_4 & 0 \\ b_1 & 0 & b_3 & 0 & b_5 \\ 0 & c_2 & 0 & c_4 & 0 \\ d_1 & 0 & d_3 & 0 & d_5 \\ 0 & e_2 & 0 & e_4 & 0 \end{vmatrix} = 0.$$

6. Show that if m elements of one row of a determinant of the nth order contain a common factor, which is also contained in the corresponding elements of other $n-m$ rows, this factor is a factor of the determinant.

7. Use §93 to show that

$$\begin{vmatrix} a_2 & b_2 & c_2 & 0 & 0 \\ a_3 & b_3 & c_3 & 0 & 0 \\ a_1A_1 & b_1A_1 & c_1A_1 & A_2 & A_3 \\ a_1B_1 & b_1B_1 & c_1B_1 & B_2 & B_3 \\ a_1C_1 & b_1C_1 & c_1C_1 & C_2 & C_3 \end{vmatrix} = |\,A_1\,B_2\,C_3\,|\;|\,a_1\,b_2\,c_3\,|\,.$$

8. Expand in a series of terms of the form $(p_1-p_2)(a_1-a_2)x_1x_2$ the determinant

$$\begin{vmatrix} x_4 & 0 & 0 & 0 & 1 & a_1 \\ 0 & x_3 & 0 & 0 & 1 & a_2 \\ 0 & 0 & x_2 & 0 & 1 & a_3 \\ 0 & 0 & 0 & x_1 & 1 & a_4 \\ 1 & 1 & 1 & 1 & 0 & 0 \\ p_1 & p_2 & p_3 & p_4 & 0 & 0 \end{vmatrix}.$$

9. Show that

$$\begin{vmatrix} a_1 & a_2 & a_3 & a_4 \\ b_1 & b_2 & b_3 & b_4 \\ b_4 & b_3 & b_2 & b_1 \\ a_4 & a_3 & a_2 & a_1 \end{vmatrix} = \begin{vmatrix} a_1+a_4 & a_2+a_3 \\ b_1+b_4 & b_2+b_3 \end{vmatrix} \begin{vmatrix} a_1-a_4 & a_2-a_3 \\ b_1-b_4 & b_2-b_3 \end{vmatrix}.$$

10. If a determinant of the nth order be developed as a sum of products, the factors of which are a minor of the αth order, a minor of the βth order, and a minor of the γth order, how many terms will there be in the development?

11. Prove that

$$|a_3 b_4| \begin{vmatrix} 0 & a_2 & a_3 \\ b_1 & b_2 & b_3 \\ c_1 & c_2 & c_3 \end{vmatrix} - |a_2 b_3| \begin{vmatrix} 0 & a_3 & a_4 \\ b_1 & b_3 & b_4 \\ c_1 & c_3 & c_4 \end{vmatrix} = a_3 b_1 |a_2 b_3 c_4|.$$

12. Prove that the product of the adjugates of two determinants of the same order is equal to the adjugate of the product of the said determinants.

13. Prove the identities (2), (3) §162 by altering the forms of the factors and using the ordinary multiplication theorem (§153).

14. Prove that if $|a_{1n}|=0$, then

$$A_{r1}A_{1r}:A_{r2}A_{2r}:A_{r3}A_{3r}: \cdots ::A_{11}:A_{22}:A_{33}: \cdots$$

15. Prove that

$$a_1|a_2 b_3 c_4| - a_2|a_1 b_3 c_4| = |a_1 c_2||a_3 b_4| - |a_1 b_2||a_3 c_4|.$$

16. If

$$|a_0 d_1| + |b_0 c_1| = |a_0 d_2| + |b_0 c_2| = |a_1 d_3| + |b_1 c_3| = |a_2 d_3| + |b_2 c_3| = 0,$$

then

$$|a_0 d_3|\,|a_1 d_2| = |b_0 c_3|\,|b_1 c_2|\,.$$

17. Prove that if $|a_{1n}|=1$, then $|a_{1n}|$ and $|A_{1n}|$ are mutually adjugate.

18. Prove that

$$|a_2 b_3|\begin{vmatrix} 0 & 0 & a_3 & a_4 \\ b_1 & b_2 & b_3 & b_4 \\ c_1 & c_2 & c_3 & c_4 \\ d_1 & d_2 & d_3 & d_4 \end{vmatrix} - a_2|a_3 b_4|\,|b_1 c_2 d_3| = a_3\begin{vmatrix} a_2 & a_3 & 0 & 0 & a_4 \\ b_2 & b_3 & 0 & 0 & b_4 \\ b_2 & b_3 & b_1 & b_2 & b_4 \\ c_2 & c_3 & c_1 & c_2 & c_4 \\ d_2 & d_3 & d_1 & d_2 & d_4 \end{vmatrix}$$

19. Using the notation of §179, prove that

$$|A_{1n}| = \frac{N_r N_{n-r+1}}{M'_{n-r}M'_{r-1}}\,.$$

20. Prove that if in $|a_1 b_2 c_3 d_4|$ the cofactor of b_3 be equal to zero, then $|a_1 b_2 c_3 d_4|\,|b_1 d_4| = |a_1 c_3 d_4|\,|a_1 b_2 d_4|$. State the same theorem in regard to $|a_{1n}|$.

21. If

$$|a_0 d_1| + |b_0 c_1| = 0 = |a_0 d_2| + |b_0 c_2|\,,$$

then

$$|a_0 b_1 c_2 d_3| = \{|a_0 d_3| + |b_0 c_3|\}\{|a_1 d_2| + |b_1 c_2|\}.$$

22. Prove that

$$|a_1 c_2 e_3|\,|b_1 d_2 e_3| = |a_1 d_2 e_3|\,|b_1 c_2 e_3| + |a_1 b_2 e_3|\,|c_1 d_2 e_3|\,.$$

23. Prove that if $|a_1 b_2 c_3 d_4|=0$, then

$$|a_1 b_3|\,|a_1 c_2 d_4| = |a_1 b_2|\,|a_1 c_3 d_4| + |a_1 b_4|\,|a_1 c_2 d_3|\,.$$

24. Resolve into determinant-factors of the second order the determinant

$$\begin{vmatrix} a^2 & ab & ab & b^2 \\ ac & ad & bc & bd \\ ac & bc & ad & bd \\ c^2 & cd & cd & d^2 \end{vmatrix}.$$

25. From a determinant D_1 the minors consisting of four adjacent elements are taken in order to be the elements of a new determinant D_2: in D_2 every minor consisting of four adjacent elements is divided by the corresponding element in the minor of D_1 obtained by deleting the first and last rows and columns, and the quotients are placed in order to form D_3: in like manner D_4 is obtained from the elements of D_1 and D_2 and so on. Prove that the final result is equal to D_1.

26. Use the process of Ex. 25 to perform Exs. 1, 2, 3, 4 §78.

181. In the case of a determinant of the nth order it is possible to delete m rows in ${}_nC_m$ different ways, and to delete m columns in ${}_nC_m$ different ways; and thus the number of minors of the $(n-m)$th order belonging to a determinant of the nth order is $({}_nC_m)^2$. Arranging these minors as the elements of a new determinant, and giving precedence in any row to that minor of two which first has a column number less than the corresponding column number of the other, and precedence in any column to that minor of two which first has a row number less than the corresponding row number of the other we obtain what is known as the *determinant of the mth-ary minors of* the original determinant, or, more conveniently and shortly, the *$(n-m)$th Compound* of the original determinant. Just as the $(n-1)$th or highest compound of a determinant is called the adjugate of the determinant, so, generally, the $(n-m)$th compound is called the adjugate of the mth compound.

EXAMPLE. The determinant of the secondary minors—the third compound of

$$\begin{vmatrix} a_1 & a_2 & a_3 & a_4 & a_5 \\ b_1 & b_2 & b_3 & b_4 & b_5 \\ c_1 & c_2 & c_3 & c_4 & c_5 \\ d_1 & d_2 & d_3 & d_4 & d_5 \\ e_1 & e_2 & e_3 & e_4 & e_5 \end{vmatrix}$$

is

$$\begin{vmatrix} |a_1 b_2 c_3| & |a_1 b_2 c_4| & |a_1 b_2 c_5| & \cdots & |a_3 b_4 c_5| \\ |a_1 b_2 d_3| & |a_1 b_2 d_4| & |a_1 b_2 d_5| & \cdots & |a_3 b_4 d_5| \\ |a_1 b_2 e_3| & |a_1 b_2 e_4| & |a_1 b_2 e_5| & \cdots & |a_3 b_4 e_5| \\ \cdots & \cdots & \cdots & \cdots & \cdots \\ |c_1 d_2 e_3| & |c_1 d_2 e_4| & |c_1 d_2 e_5| & \cdots & |c_3 d_4 e_5| \end{vmatrix},$$

or

$$\big|\, |a_1 b_2 c_3| \quad |c_1 b_2 d_4| \quad |a_1 b_2 e_5| \cdots |c_3 d_4 e_5| \,\big|.$$

In the umbral notation (§25), which makes more evident the mode of arranging the minors, it is

$$\left\|\begin{array}{ccccc} \left|\begin{matrix}1&2&3\\1&2&3\end{matrix}\right| & \left|\begin{matrix}1&2&3\\1&2&4\end{matrix}\right| & \left|\begin{matrix}1&2&3\\1&2&5\end{matrix}\right| & \cdots & \left|\begin{matrix}1&2&3\\3&4&5\end{matrix}\right| \\ \left|\begin{matrix}1&2&4\\1&2&3\end{matrix}\right| & \left|\begin{matrix}1&2&4\\1&2&4\end{matrix}\right| & \left|\begin{matrix}1&2&4\\1&2&5\end{matrix}\right| & \cdots & \left|\begin{matrix}1&2&4\\3&4&5\end{matrix}\right| \\ \left|\begin{matrix}1&2&5\\1&2&3\end{matrix}\right| & \left|\begin{matrix}1&2&5\\1&2&4\end{matrix}\right| & \left|\begin{matrix}1&2&5\\1&2&5\end{matrix}\right| & \cdots & \left|\begin{matrix}1&2&5\\3&4&5\end{matrix}\right| \\ \cdots & \cdots & \cdots & \cdots & \cdots \\ \left|\begin{matrix}3&4&5\\1&2&3\end{matrix}\right| & \left|\begin{matrix}3&4&5\\1&2&4\end{matrix}\right| & \left|\begin{matrix}3&4&5\\1&2&5\end{matrix}\right| & \cdots & \left|\begin{matrix}3&4&5\\3&4&5\end{matrix}\right| \end{array}\right\|,$$

or

$$\left\|\begin{matrix} 1&2&3 \\ 1&2&3 \end{matrix}\right| \quad \left|\begin{matrix}1&2&4\\1&2&4\end{matrix}\right| \quad \left|\begin{matrix}1&2&5\\1&2&5\end{matrix}\right| \cdots \left|\begin{matrix}3&4&5\\3&4&5\end{matrix}\right\|;$$

where in any row or column the order of precedence is decided by the order of magnitude of the numbers 123, 124, 125, 134, 135, 145, 234, 235, 245, 345, whose digits are the column or row numbers of the minors.

182. *The mth compound of* $|a_{1n}|$ *is of the same order as the* $(n-m)$*th compound,* (2) *conjugate elements in it are conjugate minors of* $|a_{1n}|$, (3) *if each element of it be replaced by the complementary minor in* $|a_{1n}|$ *the result is equal to the* $(n-m)$*th compound,* and we shall always understand that the corresponding elements in two adjugate determinants are so related.

Here (1) depends on the fact that ${}_nC_m = {}_nC_{n-m}$, and (3) upon §55.

183. *The mth compound of a minor of* $|a_{1n}|$ *is a minor of the mth compound of* $|a_{1n}|$.

184. *The mth compound of* $|a_{1n}|$ *is equal to* $|a_{1n}|^{{}_{n-1}C_{m-1}}$.

Let Δ_m denote the mth compound of Δ, so that Δ_1 and Δ are the same. If in Δ_m we prefix to each element the sign $+$ or $-$ according as the sum of its row and column numbers is even or odd, the signs in any row will be in order all the same as those in any other row, or all opposite, and the signs of conjugate elements will (§182) be alike; the change made is thus equivalent to altering the signs of all the elements in certain rows and afterwards all the elements in the corresponding columns, hence, the value of the determinant itself will remain unaltered. Multiplying Δ_m as thus changed by Δ_{n-m} in the form of §182, we have a determinant whose principal diagonal elements are (§93) all Δ, and other elements (§106) all 0: hence

$$\Delta_m \Delta_{n-m} = (\Delta)^{{}_nC_m}$$

But Δ has in general no factors, therefore Δ_m and Δ_{n-m} must both be powers of Δ. Now Δ_m is of the order ${}_nC_m$ and each element of it is of the mth degree in the elements of Δ: consequently the power which Δ_m is of Δ has for its index $m \cdot {}_nC_m \div n$, that is, ${}_{n-1}C_{m-1}$.

185. *If the mth compound of* $|a_{1n}|$ *be formed, any minor of it of the kth order is equal to the product obtained by multiplying the cofactor of the corresponding minor in the adjugate compound by* $|a_{1n}|^{{}_{n-1}C_m}$.

This is related to the theorem which procedes it exactly as the theorem of §175 is related to that of §174. The proof of §175 with evident modifications need not therefore be repeated.

186. For certain of the minors (§183) of the mth compound an expression of quite different form from the foregoing may be obtained by means of §184: new identities thus arise.

Another special minor is that dealt with in §150, namely

$$\left\| \begin{vmatrix} 1, 2, \cdots, m-1, m \\ 1, 2, \cdots, m-1, m \end{vmatrix}, \begin{vmatrix} 1, 2, \cdots, m-1, m+1 \\ 2, 3, \cdots, \quad m \quad, m+1 \end{vmatrix}, \right.$$

$$\left. \begin{vmatrix} 1, 2, \cdots, m-1, m+2 \\ 3, 4, \cdots, m-1, m+2 \end{vmatrix}, \cdots, \begin{vmatrix} 1, 2 \quad, \cdots, m-1, n \\ n-m+1, \cdots, \qquad n \end{vmatrix} \right\|,$$

which is there shown to be equal to

$$\begin{vmatrix} 1, 2, \cdots, n \\ 1, 2, \cdots, n \end{vmatrix} \times \begin{vmatrix} 1, 2, \cdots, m-1 \\ 2, 3, \cdots, \quad m \end{vmatrix} \times \begin{vmatrix} 1, 2, \cdots, m-1 \\ 3, 4, \cdots, m+1 \end{vmatrix}$$

$$\times \cdots \times \begin{vmatrix} 1, 2 \quad, \cdots, m-1 \\ n-m+1, \cdots, n-1 \end{vmatrix}.$$

Still another is that with which the Complementary of this theorem is concerned (§180); and there are many more having the like property, viz., expressibility as a product of powers of the original determinant and a number of its minors.

187. *If any identical relation be established between a number of the minors of a determinant or between the determinant itself and a number of its minors, the elements of the determinant being letters with single suffixes and the determinants denoted by means of their principal diagonals, then a new theorem is always obtainable by merely taking a line of new letters with new suffixes and annexing it to the end of the*

diagonal of every determinant, including those of order 0, *occurring in the identity.*

Let (A) be the established identity, and $|a_1\, b_2\, c_3 \cdots l_n|$ the determinant whose minors are involved in it. Taking the Complementary (§178) of (A) with respect to $|a_1\, b_2\, c_3 \cdots l_n|$ we obtain an identity, (B) say, likewise involving minors of $|a_1\, b_2\, c_3 \cdots l_n|$. But these minors are also minors of $|a_1\, b_2\, c_3 \cdots l_n\, r_\alpha\, s_\beta \cdots z_\gamma|$, and therefore it is allowable to take the complementary of (B) with respect to this extended determinant. Doing this we pass, not back to (A), but to a new theorem (A') which is seen to be derivable from (A) by annexing to the end of the diagonal of every determinant in it the line of letters $r_\alpha\, s_\beta \cdots z_\gamma$. The clause "including those of order 0" is necessitated by the last clause in the enunciation of the Law of Complementaries. This is the *Law of Extensible Minors.*

188. By means of the Law of Extensible Minors every identity which we have given in Compound Determinants may be made more general. Thus, taking the identity (§186).

$$(1)\qquad \begin{vmatrix} |a_1\, b_2| & |a_2\, b_3| & |a_3\, b_4| \\ |a_1\, c_2| & |a_2\, c_3| & |a_3\, c_4| \\ |a_1\, d_2| & |a_2\, d_3| & |a_3\, d_4| \end{vmatrix} = |a_1\, b_2\, c_3\, d_4|\, a_2\, a_3,$$

and adopting the extension $e_5 f_6$, we have

$$\begin{vmatrix} |a_1\, b_2\, e_5\, f_6| & |a_2\, b_3\, e_5\, f_6| & |a_3\, b_4\, e_5\, f_6| \\ |a_1\, c_2\, e_5\, f_6| & |a_2\, c_3\, e_5\, f_6| & |a_3\, c_4\, e_5\, f_6| \\ |a_1\, d_2\, e_5\, f_6| & |a_2\, d_3\, e_5\, f_6| & |a_3\, d_4\, e_5\, f_6| \end{vmatrix}$$

$$= |a_1\, b_2\, c_3\, d_4\, e_5\, f_6| \quad |a_2\, e_5\, f_6| \quad |a_3\, e_5\, f_6| \;;$$

or, if we view the elements of $|a_1 b_2 c_3 d_4|$ in (1) as themselves determinants of order 1, we have

$$|e_5\, f_6|^3 \begin{vmatrix} |a_1\, b_2\, e_5\, f_6| & |a_2\, b_3\, e_5\, f_6| & |a_3\, b_4\, e_5\, f_6| \\ |a_1\, c_2\, e_5\, f_6| & |a_2\, c_3\, e_5\, f_6| & |a_3\, c_4\, e_5\, f_6| \\ |a_1\, d_2\, e_5\, f_6| & |a_2\, d_3\, e_5\, f_6| & |a_3\, d_4\, e_5\, f_6| \end{vmatrix}$$

$$= \begin{vmatrix} |a_1\, e_5\, f_6| & |a_2\, e_5\, f_6| & |a_3\, e_5\, f_6| & |a_4\, e_5\, f_6| \\ |b_1\, e_5\, f_6| & |b_2\, e_5\, f_6| & |b_3\, e_5\, f_6| & |b_4\, e_5\, f_6| \\ |c_1\, e_5\, f_6| & |c_2\, e_5\, f_6| & |c_3\, e_5\, f_6| & |c_4\, e_5\, f_6| \\ |d_1\, e_5\, f_6| & |d_2\, e_5\, f_6| & |d_3\, e_5\, f_6| & |d_4\, e_5\, f_6| \end{vmatrix}$$

$$|a_2\, e_5\, f_6| \quad |a_3\, e_5\, f_6|$$

In corroboration of these new identities we deduce from them

$$\begin{vmatrix} |a_1\,e_5\,f_6| & |a_2\,e_5\,f_6| & |a_3\,e_5\,f_6| & |a_4\,e_5\,f_6| \\ |b_1\,e_5\,f_6| & |b_2\,e_5\,f_6| & |b_3\,e_5\,f_6| & |b_4\,e_5\,f_6| \\ |c_1\,e_5\,f_6| & |c_2\,e_5\,f_6| & |c_2\,e_5\,f_6| & |c_4\,e_5\,f_6| \\ |d_1\,e_5\,f_6| & |d_2\,e_5\,f_6| & |d_3\,e_5\,f_6| & |d_4\,e_5\,f_6| \end{vmatrix}$$

$$= |a_1\,b_2\,c_3\,d_4\,e_5\,f_6| \cdot |e_5\,f_6|^3,$$

which is the Extensional of the manifest identity

$$\begin{vmatrix} a_1 & a_2 & a_3 & a_4 \\ b_1 & b_2 & b_3 & b_4 \\ c_1 & c_2 & c_3 & c_4 \\ d_1 & d_2 & d_3 & d_4 \end{vmatrix} = |a_1\,b_2\,c_3\,d_4|\ ,$$

and has been already proved (§175).

Other identities may be derived from these extensionals. Thus having the relation

$$\begin{vmatrix} |a_1\,b_2\,c_3\,d_4\,e_5| & |a_1\,b_2\,c_3\,d_4\,e_6| & |a_1\,b_2\,c_3\,d_4\,e_7| \\ |a_1\,b_2\,c_3\,d_4\,f_5| & |a_1\,b_2\,c_3\,d_4\,f_6| & |a_1\,b_2\,c_3\,d_4\,f_9| \\ |a_1\,b_2\,c_3\,d_4\,g_5| & |a_1\,b_2\,c_3\,d_4\,g_6| & |a_1\,b_2\,c_3\,d_4\,g_7| \end{vmatrix}$$

$$= |a_1\,b_2\,c_3\,d_4|^2 \cdot |a_1\,b_2\,c_3\,d_4\,e_5\,f_6\,g_7|\ .$$

We observe that on the left e_1 is contained as an element in each of the determinants of the first row, f_2 is contained as an element of each of the determinants in the second row, and g_3 is contained as an element of each determinant of the third row, while on the right these occur only in the second factor. Consequently we have

$$\begin{vmatrix} |a_2\,b_3\,c_4\,d_5| & |a_2\,b_3\,c_4\,d_6| & |a_2\,b_3\,c_4\,d_7| \\ -|a_1\,b_3\,c_4\,d_5| & -|a_1\,b_3\,c_4\,d_6| & -|a_1\,b_3\,c_4\,d_7| \\ |a_1\,b_2\,c_4\,d_5| & |a_1\,b_2\,c_4\,d_6| & |a_1\,b_2\,c_4\,d_7| \end{vmatrix}$$

$$= |a_1\,b_2\,c_3\,d_4|^2 \cdot |a_4\,b_5\,c_6\,d_7|$$

or

$$\begin{vmatrix} |a_5\,b_2\,c_3\,d_4| & |a_6\,b_2\,c_3\,d_4| & |a_7\,b_2\,c_3\,d_4| \\ |a_1\,b_5\,c_3\,d_4| & |a_1\,b_6\,c_3\,d_4| & |a_1\,b_7\,c_3\,d_4| \\ |a_1\,b_2\,c_5\,d_4| & |a_1\,b_2\,c_6\,d_4| & |a_1\,b_2\,c_7\,d_4| \end{vmatrix}$$

$$= -\,|a_1\,b_2\,c_3\,d_4|^2\,|a_4\,b_5\,c_6\,d_7|$$

—a theorem which has important applications.

Another theorem is obtained by equating the cofactors of f_1g_2 instead of the cofactors of $e_1f_2g_3$ to obtain the last theorem.

Doing this gives

$$\begin{vmatrix} |a_1\,b_2\,c_3\,d_4\,e_5| & |a_1\,b_2\,c_3\,d_4\,e_6| & |a_1\,b_2\,c_3\,d_4\,e_7| \\ |a_2\,b_3\,c_4\,d_5| & |a_2\,b_3\,c_4\,d_6| & |a_2\,b_3\,c_4\,d_7| \\ |a_1\,b_3\,c_4\,d_5| & |a_1\,b_3\,c_4\,d_6| & |a_1\,b_3\,c_4\,d_7| \end{vmatrix}$$
$$= -\,|a_1\,b_2\,c_3\,d_4|^2\,|a_3\,b_4\,c_5\,d_6\,e_7|\,.$$

The next theorem in the series would be obtained by equating the cofactors of g_1.

189. *The difference between any two terms of the adjugate of a determinant* $A \equiv |a_{1n}|$ *is divisible by* A.

Let the two terms be

$$T_n \equiv A_{1i_1}A_{2i_2}\cdots A_{ni_n}$$

and

$$T_n' \equiv A_{1j_1}A_{2j_2}\cdots A_{nj_n}$$

where $i_1i_2\cdots i_n$ and $j_1j_2\cdots j_n$ are any two permutations of the numbers $1, 2, \cdots, n$. Then we are to prove that $T_n - T_n'$ is divisible by A.

By successive interchanges of the i's we can arrive at the permutation of the j's and by adding and subtracting terms corresponding to each of these interchanges we get a series of pairs of terms which may be combined. Thus if i_k is the same as j_1 the first step in the series of interchanges and terms is

$$A_{1i_1}A_{2i_2}\cdots A_{ki_k}\cdots A_{ni_n} - A_{1i_k}A_{2i_2}\cdots A_{ki_1}\cdots A_{ni_n}$$
$$+ A_{1i_k}A_{2i_2}\cdots A_{ki_1}\cdots A_{ni_n} - A_{1j_1}A_{2j2}\cdots A_{nj_n}$$
$$= (A_{1i_1}A_{ki_k} - A_{1i_k}A_{ki_1})A_{2i_2}\cdots A_{ni_n} + \text{3rd and 4th terms}.$$

We get in this way a series of terms on the right each of which has a factor of the form $(A_{ri}A_{sj} - A_{rj}A_{si})$ and since this is a minor of the second order of the adjugate of A it is divisible by A and therefore $T_n - T_n'$ is divisible by A.

It is also true that the difference between any two terms of any minor of the adjugate of A is divisible by A; for if we multiply this difference, which we will denote by $T_k - T_k'$, by any single term S of the complementary minor we obtain $S(T_k - T_k')$ which is the difference between two terms of the full adjugate and therefore is divisible by A and since S is in general not divisible by A it follows that $T_k - T_k'$ must be.

190. A_0 is any n-line determinant and A_1 is its adjugate, and in general A_r is the adjugate of A_{r-1}. Let M_k and M'_{n-k} in their respective determinants be complementary minors so far as position is concerned, the one being of the kth order and the other of the $(n-k)$th. Then

(1) $$M'_{n-k} \text{ in } A_{2m} = A_0^{\nu_1} \cdot M'_{n-k} \text{ in } A_0$$

(2) $$M_k \text{ in } A_{2m+1} = A_0^{\nu_2} \cdot M'_{n-k} \text{ in } A_0$$

where

$$\nu_1 = \{(n-1)^{2m} - 1\}(n-k)/n$$
$$\nu_2 = \{(n-1)^{2m+1} + 1\}(k-n)/n$$

To prove this we have

$$M_k \text{ in } A_1 = A_0^{k-1} M'_{n-k} \text{ in } A_0$$

(3) $$M'_{n-k} \text{ in } A_2 = A_1^{n-k-1} M_k \text{ in } A_1$$

$$M_k \text{ in } A_3 = A_2^{k-1} M'_{n-k} \text{ in } A_2$$

. . . .

By multiplication and division we have

$$M'_{n-k} \text{ in } A_{2m} = (A_0 A_2 \cdots A_{2m-2})^{k-1} (A_1 A_3 \cdots A_{2m-1})^{n-k-1} M'_{n-k} \text{ in } A_0$$

But since $A_r = A_0^{(n-1)^r}$ it follows that

$$(A_0 A_2 \cdots A_{2m-2})^{k-1} = (A_0^{\nu_3})^{k-1},$$

and

$$(A_1 A_3 \cdots A_{2m-1})^{n-k-1} = (A_0^{\nu_4})^{n-k-1},$$

where

$$\nu_3 = \{(n-1)^{2m} - 1\}/n(n-2),$$
$$\nu_4 = (n-1)\nu_3;$$

so that M'_{n-k} in $A_{2m} = A_0^{\nu_1} M'_{n-k}$ in A_0.

For the other we have

$$M_k \text{ in } A_{2m+1} = A_{2m}^{k-1} M'_{n-k} \text{ in } A_{2m}$$

which on substituting gives the desired result.

The results for $k=1$ and $n-1$ are of interest. Using $k=n-1$ in (1) we have

The element in the (r, s) position in A_{2m} is

$$a_{rs} A_0^{\{(n-1)^{2m}-1\}/n}.$$

Using $k=1$ in (2) we have

The element in the (r, s) position in A_{2m+1} is

$$A_{rs}A_0^{\{(n-1)^{2m}-1\}(n-1)/n}.$$

These give a ready rule for writing the elements in any adjugate of a determinant.

From (3) we have

$$M'_{n-k} \text{ in } A_2 = A_1^{n-k-1}A_0^{k-1}M'_{n-k} \text{ in } A_0$$

in which, if we put $k=n-1$, we get

$$\frac{M_1' \text{ in } A_0}{M_1' \text{ in } A_2} = \frac{1}{A_0^{n-2}} = \frac{A_0}{A_1}.$$

In words this says that *any element of a determinant is to the corresponding element of its second adjugate as the determinant is to its first adjugate.*

191. The minor

$$\left\|\begin{array}{c}1\,2\\1\,2\end{array}\right|\left|\begin{array}{c}3\,4\\3\,4\end{array}\right\| = \begin{array}{c|c|}1&2\,3\,4\\1&2\,3\,4\end{array} + \begin{array}{c|c|}2&1\,3\,4\\2&1\,3\,4\end{array} + \begin{array}{c|c|}1&2\,3\,4\\2&1\,3\,4\end{array} + \begin{array}{c|c|}2&1\,3\,4\\1&2\,3\,4\end{array} - \left|\begin{array}{c}1234\\1234\end{array}\right|$$

where

$$\begin{array}{c|c|}1&2\,3\,4\\1&2\,3\,4\end{array}$$

stands for $a_{11}A_{234,234}$, etc.

Again the minor

$$\left\|\begin{array}{c}1\,2\,3\\1\,2\,3\end{array}\right|\left|\begin{array}{c}4\,5\,6\\4\,5\,6\end{array}\right\| = \left|\begin{array}{c}1\,2\,3\\1\,2\,3\end{array}\right|\left|\begin{array}{c}4\,5\,6\\4\,5\,6\end{array}\right| - \left|\begin{array}{c}1\,2\,3\\4\,5\,6\end{array}\right|\left|\begin{array}{c}4\,5\,6\\1\,2\,3\end{array}\right|$$

$$= \Bigg[\left|\begin{array}{c}1234\\1234\end{array}\right|\left|\begin{array}{c}56\\56\end{array}\right| + \left|\begin{array}{c}1235\\1235\end{array}\right|\left|\begin{array}{c}46\\46\end{array}\right| + \left|\begin{array}{c}1236\\1236\end{array}\right|\left|\begin{array}{c}45\\45\end{array}\right| + \left|\begin{array}{c}1234\\1236\end{array}\right|\left|\begin{array}{c}5\,6\\4\,5\end{array}\right|$$

$$+ \left|\begin{array}{c}1236\\1234\end{array}\right|\left|\begin{array}{c}45\\56\end{array}\right| - \left|\begin{array}{c}1235\\1234\end{array}\right|\left|\begin{array}{c}46\\56\end{array}\right| - \left|\begin{array}{c}1234\\1235\end{array}\right|\left|\begin{array}{c}56\\46\end{array}\right| - \left|\begin{array}{c}1235\\1236\end{array}\right|\left|\begin{array}{c}4\,6\\4\,5\end{array}\right|$$

$$- \left|\begin{array}{c}1236\\1235\end{array}\right|\left|\begin{array}{c}45\\46\end{array}\right|\Bigg] - \Bigg[\begin{array}{|c|c}1\,2\,3\,4\,5&6\\1\,2\,3\,4\,5&6\end{array} + \begin{array}{|c|c}1\,2\,3\,4\,6&5\\1\,2\,3\,4\,6&5\end{array} + \begin{array}{|c|c}1\,2\,3\,5\,6&4\\1\,2\,3\,5\,6&4\end{array}$$

$$- \begin{array}{|c|c}1\,2\,3\,4\,5&6\\1\,2\,3\,4\,6&5\end{array} - \begin{array}{|c|c}1\,2\,3\,4\,6&5\\1\,2\,3\,4\,5&6\end{array} + \begin{array}{|c|c}1\,2\,3\,4\,5&6\\1\,2\,3\,5\,6&4\end{array} + \begin{array}{|c|c}1\,2\,3\,5\,6&4\\1\,2\,3\,4\,5&6\end{array}$$

$$- \begin{array}{|c|c}1\,2\,3\,4\,6&5\\1\,2\,3\,5\,6&4\end{array} - \begin{array}{|c|c}1\,2\,3\,5\,6&4\\1\,2\,3\,5\,6&5\end{array}\Bigg] + \left|\begin{array}{c}123456\\123456\end{array}\right| = S_1 - S_2 + S_3,$$

where S_1 represents the terms in the first square brackets, S_2 represents those in the second, and S_3 represents $\begin{vmatrix} 123456 \\ 123456 \end{vmatrix}$

If we represent the left-hand side by S_0 we may write the relation

$$- S_0 + S_1 - S_2 + S_3 = 0.$$

To prove its truth we have but to show that the total coefficient of any product $\begin{vmatrix} \alpha_1\alpha_2\alpha_3 \\ \beta_1\beta_2\beta_3 \end{vmatrix} \cdot \begin{vmatrix} \alpha_4\alpha_5\alpha_6 \\ \beta_4\beta_5\beta_6 \end{vmatrix}$ is zero. Thus taking the various types of products we readily see that the total coefficient of

$$\begin{vmatrix} 1\,2\,3 \\ 1\,2\,3 \end{vmatrix}\begin{vmatrix} 4\,5\,6 \\ 4\,5\,6 \end{vmatrix} \text{ is } [-1+3-3+1]=0,$$

$$\begin{vmatrix} 1\,2\,3 \\ 1\,2\,4 \end{vmatrix}\begin{vmatrix} 4\,5\,6 \\ 3\,5\,6 \end{vmatrix} \text{ is } [\ \ 0-1+2-1]=0,$$

$$\begin{vmatrix} 1\,2\,3 \\ 1\,4\,5 \end{vmatrix}\begin{vmatrix} 4\,5\,6 \\ 2\,3\,6 \end{vmatrix} \text{ is } [\ \ 0+0-1+1]=0,$$

$$\begin{vmatrix} 1\,2\,3 \\ 4\,5\,6 \end{vmatrix}\begin{vmatrix} 4\,5\,6 \\ 1\,2\,3 \end{vmatrix} \text{ is } [\ \ 1+0+0-1]=0,$$

where for instance the coefficient of $\begin{vmatrix} 123 \\ 123 \end{vmatrix} \cdot \begin{vmatrix} 456 \\ 456 \end{vmatrix}$ in S_0 is 1, in S_1 is 3, in S_2 is 3, and in S_3 is 1 and similarly for the others.

The general case may be stated as follows:

$$\left\| \begin{matrix} (n|\,m_\alpha) \\ (n|\,m_\alpha) \end{matrix} \right\| \left\| \begin{matrix} (n|\,m_\beta) \\ (n|\,m_\beta) \end{matrix} \right\| \equiv \begin{vmatrix} (n|\,m_\alpha) \\ (n|\,m_\alpha) \end{vmatrix}\begin{vmatrix} (n|\,m_\beta) \\ (n|\,m_\beta) \end{vmatrix} - \begin{vmatrix} (n|\,m_\alpha) \\ (n|\,m_\beta) \end{vmatrix}\begin{vmatrix} (n|\,m_\beta) \\ (n|\,m_\alpha) \end{vmatrix}$$

$$= \sum_1^m k(-1)^{k+1} \sum_1^{(m)k} i \sum_1^{(m)k} j(-1)^{\nu_k} \begin{vmatrix} (n\,|\,m_\alpha)\ (n\,|\,m_\beta\,|\,k_i) \\ (n\,|\,m_\alpha)\ (n\,|\,m_\beta\,|\,k_j) \end{vmatrix}\begin{vmatrix} (n\,|\,\bar{m}_\beta\,|\,k_i) \\ (n\,|\,\bar{m}_\beta\,|\,k_j) \end{vmatrix}$$

where $(n\,|m_\beta\,|k_i)$ and $(n\,|\bar{m}_\beta\,|k_i)$ are complementary combinations of the numbers in $(n\,|m_\beta)$, and where ν_k is the sum of the inversions in $(n\,|m_\beta\,|k_i)(n\,|\bar{m}_\beta\,|k_i)$ plus the number in $(n\,|m_\beta\,|k_j)(n\,|\bar{m}_\beta\,|k_j)$.

Here the m values of k give the m groups of terms $S_1, S_2, \cdots S_m$. If we write the relation as before

$$- S_0 + S_1 - S_2 + \cdots + (-1)^{m-1} S_m = 0,$$

then the proof proceeds precisely as in the case where m is 3.

It may be observed here that there is no loss of generality in taking a coaxial minor of the mth compound since by a proper shifting of the numbers indicating the rows and columns it may be turned over into any other minor of the second order. It is also to be noted that this relation gives the value of all minors of order two of the compound in terms of minors of higher order. Thus in the first relation if we put 2 for 4 in the column numbers it becomes

$$\left\| \begin{vmatrix} 1\,2 \\ 1\,2 \end{vmatrix} \begin{vmatrix} 3\,4 \\ 3\,2 \end{vmatrix} \right\| = 0 + {}^{2}_{2}\begin{vmatrix} 1\,3\,4 \\ 1\,3\,2 \end{vmatrix} + {}^{1}_{2}\begin{vmatrix} 2\,3\,4 \\ 1\,3\,2 \end{vmatrix} + 0 - 0$$

or

$$\left\| \begin{vmatrix} 1\,2 \\ 1\,2 \end{vmatrix} \begin{vmatrix} 3\,4 \\ 2\,3 \end{vmatrix} \right\| = {}^{2}_{2}\begin{vmatrix} 1\,3\,4 \\ 1\,2\,3 \end{vmatrix} + {}^{1}_{2}\begin{vmatrix} 2\,3\,4 \\ 1\,2\,3 \end{vmatrix} .$$

If in addition we put 2 for 4 in the row numbers we get

$$\left\| \begin{vmatrix} 1\,2 \\ 1\,2 \end{vmatrix} \begin{vmatrix} 2\,3 \\ 2\,3 \end{vmatrix} \right\| = {}^{2}_{2}\begin{vmatrix} 1\,2\,3 \\ 1\,2\,3 \end{vmatrix} .$$

EXAMPLE. If all minors of order $k+1$ of a determinant $A \equiv |a_{1n}|$ of order n be zero but not all minors of order k are zero, show that the elements in any line of the kth compound of A are proportional to the corresponding elements in any other parallel line.

192. *If in the mth compound of a determinant A of order n there is an element α_{jk}, say, in the kth column and also one, α_{hi}, say, in the hth row which do not vanish, then α_{ji} and α_{hk} cannot possibly vanish if all minors of A of order $(m+1)$ vanish.* For

$$\begin{vmatrix} \alpha_{hi} & \alpha_{hk} \\ \alpha_{ji} & \alpha_{jk} \end{vmatrix} = 0,$$

since it is a minor of order two of the mth compound of a vanishing minor of order $m+1$.

We have then

$$\alpha_{hi}\alpha_{jk} - \alpha_{ji}\alpha_{hk} = 0,$$

but since α_{hi} and α_{jk} are not zero it follows that α_{ji} and α_{hk} cannot be zero.

193. *If a determinant of order n has all minors of order $m+1$ equal to zero, but all the principal diagonal minors of the mth compound are different from zero, then all elements of the mth compound are different from zero.* For we would have

$$\alpha_{ii}\alpha_{jj} - \alpha_{ij}\alpha_{ji} = 0$$

for all values of i and j, and since neither α_{ii} nor α_{jj} vanish, it follows that neither α_{ij} nor α_{ji} can vanish.

194. *If all the principal diagonal elements of the mth compound of a determinant of order n vanish as well as all minors of order $m+1$, then not more than $(\lambda-1)$ of the elements of the mth compound can be different from zero, where $\lambda=(n)_m$.*

This follows from the equations

$$\alpha_{ii}\alpha_{jj} - \alpha_{ij}\alpha_{ji} = 0,$$

$$\alpha_{ij}\alpha_{hi} - \alpha_{ii}\alpha_{hj} = 0.$$

195. *If A denotes a determinant of order six, then*

$$\left(\begin{vmatrix}12\\12\end{vmatrix}+\begin{vmatrix}12\\13\end{vmatrix}+\cdots+\begin{vmatrix}12\\56\end{vmatrix}\right)\left(\begin{vmatrix}34\\12\end{vmatrix}+\begin{vmatrix}34\\13\end{vmatrix}+\cdots+\begin{vmatrix}34\\56\end{vmatrix}\right)$$
$$-\left(\begin{vmatrix}13\\12\end{vmatrix}+\begin{vmatrix}13\\13\end{vmatrix}+\cdots+\begin{vmatrix}13\\56\end{vmatrix}\right)\left(\begin{vmatrix}24\\12\end{vmatrix}+\begin{vmatrix}24\\13\end{vmatrix}+\cdots+\begin{vmatrix}24\\56\end{vmatrix}\right)$$
$$+\left(\begin{vmatrix}14\\12\end{vmatrix}+\begin{vmatrix}14\\13\end{vmatrix}+\cdots+\begin{vmatrix}14\\56\end{vmatrix}\right)\left(\begin{vmatrix}23\\12\end{vmatrix}+\begin{vmatrix}23\\13\end{vmatrix}+\cdots+\begin{vmatrix}23\\56\end{vmatrix}\right)$$
$$+\left(\begin{vmatrix}23\\12\end{vmatrix}+\begin{vmatrix}23\\13\end{vmatrix}+\cdots+\begin{vmatrix}23\\56\end{vmatrix}\right)\left(\begin{vmatrix}14\\12\end{vmatrix}+\begin{vmatrix}14\\13\end{vmatrix}+\cdots+\begin{vmatrix}14\\56\end{vmatrix}\right)$$
$$-\left(\begin{vmatrix}24\\12\end{vmatrix}+\begin{vmatrix}24\\13\end{vmatrix}+\cdots+\begin{vmatrix}24\\56\end{vmatrix}\right)\left(\begin{vmatrix}13\\12\end{vmatrix}+\begin{vmatrix}13\\13\end{vmatrix}+\cdots+\begin{vmatrix}13\\56\end{vmatrix}\right)$$
$$+\left(\begin{vmatrix}34\\12\end{vmatrix}+\begin{vmatrix}34\\13\end{vmatrix}+\cdots+\begin{vmatrix}34\\56\end{vmatrix}\right)\left(\begin{vmatrix}12\\12\end{vmatrix}+\begin{vmatrix}12\\13\end{vmatrix}+\cdots+\begin{vmatrix}12\\56\end{vmatrix}\right)$$
$$=2\left(\begin{vmatrix}1234\\1234\end{vmatrix}+\begin{vmatrix}1234\\1235\end{vmatrix}+\begin{vmatrix}1234\\1236\end{vmatrix}+\begin{vmatrix}1234\\1245\end{vmatrix}+\begin{vmatrix}1234\\1246\end{vmatrix}\right.$$
$$+\begin{vmatrix}1234\\1256\end{vmatrix}+\begin{vmatrix}1234\\1345\end{vmatrix}+\begin{vmatrix}1234\\1346\end{vmatrix}+\begin{vmatrix}1234\\1356\end{vmatrix}+\begin{vmatrix}1234\\1456\end{vmatrix}$$
$$\left.+\begin{vmatrix}1234\\2345\end{vmatrix}+\begin{vmatrix}1234\\2346\end{vmatrix}+\begin{vmatrix}1234\\2356\end{vmatrix}+\begin{vmatrix}1234\\2456\end{vmatrix}+\begin{vmatrix}1234\\3456\end{vmatrix}\right).$$

For if we expand the left-hand member, the sum of corresponding terms in the products vanishes or not according as the factors of the terms have or have not a column in common, that is, according as there is or is not a number common to the lower lines of the suffixes of both factors. Thus

$$\begin{vmatrix}1\;2\\1\;2\end{vmatrix}\cdot\begin{vmatrix}3\;4\\1\;2\end{vmatrix}-\begin{vmatrix}1\;3\\1\;2\end{vmatrix}\cdot\begin{vmatrix}2\;4\\1\;2\end{vmatrix}+\begin{vmatrix}1\;4\\1\;2\end{vmatrix}\cdot\begin{vmatrix}2\;3\\1\;2\end{vmatrix}+\begin{vmatrix}2\;3\\1\;2\end{vmatrix}\cdot\begin{vmatrix}1\;4\\1\;2\end{vmatrix}$$
$$-\begin{vmatrix}2\;4\\1\;2\end{vmatrix}\cdot\begin{vmatrix}1\;3\\1\;2\end{vmatrix}+\begin{vmatrix}3\;4\\1\;2\end{vmatrix}\cdot\begin{vmatrix}1\;2\\1\;2\end{vmatrix}=0,$$

$$\begin{vmatrix}1\;2\\1\;2\end{vmatrix}\cdot\begin{vmatrix}3\;4\\1\;3\end{vmatrix}-\begin{vmatrix}1\;3\\1\;2\end{vmatrix}\cdot\begin{vmatrix}2\;4\\1\;3\end{vmatrix}+\begin{vmatrix}1\;4\\1\;2\end{vmatrix}\cdot\begin{vmatrix}2\;3\\1\;3\end{vmatrix}+\begin{vmatrix}2\;3\\1\;2\end{vmatrix}\cdot\begin{vmatrix}1\;4\\1\;3\end{vmatrix}$$
$$-\begin{vmatrix}2\;4\\1\;2\end{vmatrix}\cdot\begin{vmatrix}1\;3\\1\;3\end{vmatrix}+\begin{vmatrix}3\;4\\1\;2\end{vmatrix}\cdot\begin{vmatrix}1\;2\\1\;3\end{vmatrix}=0,$$

$$\begin{vmatrix}1\;2\\1\;2\end{vmatrix}\cdot\begin{vmatrix}3\;4\\3\;4\end{vmatrix}-\begin{vmatrix}1\;3\\1\;2\end{vmatrix}\cdot\begin{vmatrix}2\;4\\3\;4\end{vmatrix}+\begin{vmatrix}1\;4\\1\;2\end{vmatrix}\cdot\begin{vmatrix}2\;3\\3\;4\end{vmatrix}+\begin{vmatrix}2\;3\\1\;2\end{vmatrix}\cdot\begin{vmatrix}1\;4\\3\;4\end{vmatrix}$$
$$-\begin{vmatrix}2\;4\\1\;2\end{vmatrix}\cdot\begin{vmatrix}1\;3\\3\;4\end{vmatrix}+\begin{vmatrix}3\;4\\1\;2\end{vmatrix}\cdot\begin{vmatrix}1\;2\\3\;4\end{vmatrix}=\begin{vmatrix}1234\\1234\end{vmatrix},$$

$$\begin{vmatrix}1\;2\\1\;3\end{vmatrix}\cdot\begin{vmatrix}3\;4\\2\;4\end{vmatrix}-\begin{vmatrix}1\;3\\1\;3\end{vmatrix}\cdot\begin{vmatrix}2\;4\\2\;4\end{vmatrix}+\begin{vmatrix}1\;4\\1\;3\end{vmatrix}\cdot\begin{vmatrix}2\;3\\2\;4\end{vmatrix}+\begin{vmatrix}2\;3\\1\;3\end{vmatrix}\cdot\begin{vmatrix}1\;4\\2\;4\end{vmatrix}$$
$$-\begin{vmatrix}2\;4\\1\;3\end{vmatrix}\cdot\begin{vmatrix}1\;3\\2\;4\end{vmatrix}+\begin{vmatrix}3\;4\\1\;3\end{vmatrix}\cdot\begin{vmatrix}1\;2\\2\;4\end{vmatrix}=-\begin{vmatrix}1234\\1234\end{vmatrix},$$

etc.

It is easily seen that $\begin{vmatrix}1234\\1234\end{vmatrix}$ as well as each of the other terms on the right will be obtained in $4\cdot 3/1\cdot 2=6$ different ways, four of which will give a positive and two a negative sign, since in the set of combinations 1234, 1324, 1423, 2314, 2413, 3412 there are four having an even number of inversions and two having an odd number of inversions. Hence the truth of the equation.

196. *In general, if* $A=|a_{1n}|$, *then*

$$(-1)^{\nu}\sum_{i=1}^{(n)l}A_{(n|m\alpha|l_1),(n|l_i)}\sum_{j=1}^{(n)m-l}A_{(n|\overline{m}\alpha|l_1),(n|m-lj)}$$
$$+(-1)^{\nu}\sum_{i=1}^{(n)l}A_{(n|m\alpha|l_2),(n|l_i)}\sum_{i=1}^{(n)m-l}A_{(n|m\alpha|l_2),(n|m-lj)}+\cdots$$
$$=\phi(m,l)\sum_{h=1}^{(n)m}A_{(n|m\alpha),(n|mh)}$$

or
$$\sum_{k=1}^{(m)_l}(-1)^{\nu}\left\{\sum_{i=1}^{(n)_l} A_{(n|m_\alpha|l_k),(n|l_i)} \sum_{j=1}^{(n)_{m-l}} A_{(n|\bar{m}_\alpha|l_k),(n|m-l_j)}\right\}$$
$$= \phi(m,l) \sum_{h=1}^{(n)_m} A_{(n|m_\alpha),(n|m_h)}$$

where ν denotes the number of inversions in $(n|m_\alpha|l_k)(n|\bar{m}_\alpha|l_k)$, and $\phi(m, l)$ denotes, as in §18, the excess of the number of combinations in the set

$$(n|m_\alpha|l_1)(n|\bar{m}_\alpha|l_1),\ (n|m_\alpha|l_2)(n|\bar{m}_\alpha|l_2),\ \cdots,\ (n|m_\alpha|l_\delta)(n|\bar{m}_\alpha|l_\delta)$$
$$(\delta = (m)_l.)$$

which have an even number of inversions over those which have an odd number.

For if we expand the left-hand member as before, it is easily seen that the sum of corresponding terms of the products will vanish or not according as $(n|l_i)$ and $(n|m-l_j)$ have or have not any numbers in common. Any term $A_{(n|m_\alpha),(n|m_h)}$ which does not vanish will be obtained in $(m)_l$ ways from those particular sums of corresponding terms of products which are given by

$$\sum_{i=1}^{(m)_l}(-1)^{\nu} A_{(n|m_\alpha|l_i),(n|m_h|l_j)} A_{(n|\bar{m}_\alpha|l_i),(n|m_h|l_j)}, \quad (j = 1, 2, \cdots, (m)_l).$$

The coefficient of $A_{(n|m_\alpha),(n|m_h)}$ is therefore $\phi(m, l)$. The theorem is thus established.

197. Sylvester's Theorem. Using $\left|\left|\begin{smallmatrix}(n|m_\alpha)\\(n|m_{\alpha'})\end{smallmatrix}\right|\right|_k$ to denote the kth compound of the minor of order m consisting of the αth selection of rows and the α'th selection of columns of $A \equiv |a_{nn}|$, we have

$$\left|\left|\begin{matrix}(n|m_\alpha)\\(n|m_{\alpha'})\end{matrix}\right|\right|_k \equiv \left|\left|\begin{matrix}(n|m_\alpha|k_1)\\(n|m_{\alpha'}|k_1)\end{matrix}\right|\left|\begin{matrix}(n|m_\alpha|k_2)\\(n|m_{\alpha'}|k_2)\end{matrix}\right| \cdots \left|\begin{matrix}(n|m_\alpha|k_\lambda)\\(n|m_{\alpha'}|k_\lambda)\end{matrix}\right|\right|$$
$$= \left|\begin{matrix}(n|m_\alpha)\\(n|m_{\alpha'})\end{matrix}\right|^{(m-1)_{k-1}}, \quad \text{where} \quad \lambda = (m)_k.$$

Operating on this by the Law of Extensible Minors we get

$$\left|\left|\begin{matrix}(\bar{n}|m_\alpha)(n|m_\alpha|k_1)\\(\bar{n}|m_{\alpha'})(n|m_{\alpha'}|k_1)\end{matrix}\right|\left|\begin{matrix}(\bar{n}|m_\alpha)(n|m_\alpha|k_2)\\(\bar{n}|m_{\alpha'})(n|m_{\alpha'}|k_2)\end{matrix}\right| \cdots \left|\begin{matrix}(\bar{n}|m_\alpha)(n|m_\alpha|k_\lambda)\\(\bar{n}|m_{\alpha'})(n|m_{\alpha'}|k_\lambda)\end{matrix}\right|\right|$$

(1) $$= |a_{nn}|^{(m-1)_{k-1}} \left|\begin{matrix}(\bar{n}|m_\alpha)\\(\bar{n}|m_{\alpha'})\end{matrix}\right|^{(m-1)_k}$$

Every element of the determinant on the left-hand side of this equation consists of the minor $|(\bar{n}\,|m_\alpha),\ (\bar{n}\,|m_\alpha)|$ bordered in all possible ways with k of the remaining rows and columns of the determinant A. We have therefore the following theorem:

If the elements of a determinant S (of order $\lambda=(m)_k$) consist of a minor M (of order $n-m$) of a determinant A (of order n) bordered in all possible ways with k of the remaining rows and columns not included in M, then S is equal to the $(m-1)_k$th power of M by the $(m-1)_{k-1}$th power of A. Thus

$$S = M^{(m-1)_k} A^{(m-1)_{k-1}}$$

The reader should observe the particular form which the left-hand side of (1) takes when the elements $\left.\begin{smallmatrix}(1r)\\(r1)\end{smallmatrix}\right\}=0$ for $r=1, 2, \cdots, p$. Thus when $n=6$, $p=3$ we have

$$\begin{vmatrix} 0 & (14) & (15) & (16) \\ (41) & \begin{vmatrix} 2\,3\,4 \\ 2\,3\,4 \end{vmatrix} & \begin{vmatrix} 2\,3\,4 \\ 2\,3\,5 \end{vmatrix} & \begin{vmatrix} 2\,3\,4 \\ 2\,3\,6 \end{vmatrix} \\ (51) & \begin{vmatrix} 2\,3\,5 \\ 2\,3\,4 \end{vmatrix} & \begin{vmatrix} 2\,3\,5 \\ 2\,3\,5 \end{vmatrix} & \begin{vmatrix} 2\,3\,5 \\ 2\,3\,6 \end{vmatrix} \\ (61) & \begin{vmatrix} 2\,3\,6 \\ 2\,3\,4 \end{vmatrix} & \begin{vmatrix} 2\,3\,6 \\ 2\,3\,5 \end{vmatrix} & \begin{vmatrix} 2\,3\,6 \\ 2\,3\,6 \end{vmatrix} \end{vmatrix} = A \begin{vmatrix} 2\,3 \\ 2\,3 \end{vmatrix}.$$

198. If S' denotes the determinant whose elements are M bordered in all possible ways with $(m-k)$ of the remaining rows and columns, and where the element of S' in the position (ij) is M bordered by the rows and columns not taken for the element in the position (ij) of S, then

$$S' = M^{(m-1)_{k-1}} \cdot A^{(m-1)_k}.$$

Therefore

$$S \cdot S' = M^{(m)_k} \cdot A^{(m)_k}.$$

199. In a manner similar to that used in §185 we may express a minor of S in terms of the complementary minor in S' times a power of A and of M. Thus

$$\left\| \begin{vmatrix} (\bar{n}\,|\,m_\alpha\,)(n\,|\,m_\alpha\,|\,k_1) \\ (\bar{n}\,|\,m_{\alpha'})(n\,|\,m_{\alpha'}\,|\,k_1) \end{vmatrix} \begin{vmatrix} (\bar{n}\,|\,m_\alpha\,)(n\,|\,m_\alpha\,|\,k_2) \\ (\bar{n}\,|\,m_{\alpha'})(n\,|\,m_{\alpha'}\,|\,k_2) \end{vmatrix} \cdots \begin{vmatrix} (\bar{n}\,|\,m_\alpha)(n\,|\,m_\alpha\,|\,k_h) \\ (\bar{n}\,|\,m_{\alpha'})(n\,|\,m_{\alpha'}\,|\,k_h) \end{vmatrix} \right\|$$

$$= A^{h-(m-1)k} \cdot M^{h-(m-1)k-1} \left\| \begin{matrix} (\bar{n} \mid m_{\alpha})(n \mid \bar{m}_{\alpha} \mid k_{h+1}) \\ (\bar{n} \mid m_{\alpha'})(n \mid \bar{m}_{\alpha'} \mid k_{h+1}) \end{matrix} \right| \cdots \left| \begin{matrix} (\bar{n} \mid m_{\alpha})(n \mid \bar{m}_{\alpha} \mid k_{\lambda}) \\ (\bar{n} \mid m_{\alpha'})(n \mid \bar{m}_{\alpha'} \mid k_{\lambda}) \end{matrix} \right\|$$

or

$$M_h = A^{h-(m-1)k} \cdot M^{h-(m-1)k-1} \cdot N'_{\lambda-h}$$

where M_h is the minor of order h of S and $N'_{\lambda-h}$ is its complementary in S'.

This may also be written

$$\frac{M_h}{N'_{\lambda-h}} = \frac{(A \cdot M)^h}{A^{(m-1)k} \cdot M^{(m-1)k-1}} .$$

200. Let

$$\Delta \equiv \begin{vmatrix} (a_{mm}) & (b_{mr}) \\ (c_{rm}) & (d_{rr}) \end{vmatrix},$$

where (a_{mm}) represents a square of m rows and m columns of a's; (b_{mr}) represents a rectangle of m rows and r columns of b's; (c_{rm}) represents a rectangle of r rows and m columns of c's; and (d_{rr}) represents a square of r rows and r columns of d's. Then forming the determinant S as in §197 using $(a_{mm}) \equiv |a_{mm}| \equiv A$ as the minor bordered k at a time, we have

(1) $$S = A^{(r-1)k} \cdot \Delta^{(r-1)k-1},$$

and

(2) $$M_h = A^{h-(r-1)k-1} \cdot \Delta^{h-(r-1)k} \cdot N'_{\rho-h}.$$

where $\rho = (r)_k$.

For the case where $k=1$, it is readily seen that M_h is an example of the theorem §197 as applied to a minor of Δ of order $m+h$, $(m+h \not> r)$, which we shall denote by Δ_{m+h}. It follows therefore that

(3) $$M_h = A^{h-1} \cdot \Delta_{m+h}.$$

If in Δ the d's are all zeros and $r > m$ then $\Delta = 0$. Three cases arise when

1. $k > m$.

Then every element of S is equal to 0.

2. $k = m$.

Then $S=0$, though every element is not necessarily zero, but breaks up into two factors one of which is common to every element in the same row. Taking out these common factors the remaining determinant has all its columns alike and therefore $M_h=0$ for $h \gtreqless 2$.

3. $k<m$.

The special interest here is when $h>m$ and $k=1$, in which case Δ_{m+h} is a minor of Δ with a square of zeros, h on one side, in the lower right-hand corner and is therefore zero. Then (3) shows that $M_h=0$; that is, every minor of Δ of order greater than m vanishes.

An important case of this is where $h=m+1$.

In the case where $k=1$, $r>m$ and the d's are all zero, we have

$$\begin{aligned} S &= A^{r-1}\cdot\Delta \\ &= 0. \end{aligned}$$

If we denote by D the determinant formed by subtracting from the element of S in the position (i, j), (which we may temporarily denote by A_{ij}), the product $d_{ij}A\,(i, j=1, 2, \cdots, n)$, then D is zero identically. For the element $A_{ij}-d_{ij}A$ of D does not contain d_{ij} and is what the element A_{ij} reduces to when d_{ij} is put equal to zero. That is, D is what S reduces to when the d's are put equal to zero and therefore $D=0$. The determinant D contains $m(m+2r)$ arbitrary quantitites.

If in Δ, $r=m$ and the d's are all zeros then Δ breaks up into two factors, thus $\Delta=(-1)^m\,|b_{mm}|\,|c_{mm}|=(-1)^m B\cdot C$ and the results containing Δ would be changed accordingly. Thus the ratio of the minors in §199 would be $(-1)^{my}M^xB^yC^y$ where $x=h-(m-1)_{k-1}$ and $y=h-(m-1)_k$.

201. Let

$$\Delta \equiv \begin{vmatrix} A_{n-\lambda} & \cdot & \\ & \cdot & \\ \cdots & \cdots & \cdots \\ & \cdot & \\ & \cdot & A_\lambda \end{vmatrix}$$

represent any determinant of order n and let $\Delta_{(m)}$ and $\Delta_{(n-m)}$ be the mth and $(n-m)$th compounds respectively, that is, they are adjugates of each other.

Form the mth compound of the elements A_λ and call it M. Then since M is a minor of $\Delta_{(m)}$ of order $(\lambda)_m$ its complementary which we will call N will be of order $\{(n)_m-(\lambda)_m\}$. Let the corresponding minors in $\Delta_{(n-m)}$ be M' and N' respectively.

We have then

(1) $$N = M' \cdot \Delta^{(n)_m-(\lambda)_m-(n-1)_m}.$$

But M' is a case of Sylvester's theorem where $A_{n-\lambda}$ is bordered in all possible ways with $(\lambda-m)$ of the remaining rows and columns. We have therefore

(2) $$M' = \Delta^{(\lambda-1)\cdot m} A_{n-\lambda}^{(\lambda-1)_{m-1}}$$

and consequently

(3) $$N = \Delta^{(n-1)_{m-1}-(\lambda-1)_{m-1}} \cdot A_{n-\lambda}^{(\lambda-1)_{m-1}}$$

This theorem, due to D'Ovidio, may be stated as follows:

If every possible m-line minor of Δ be taken whose rows are not all included in the last λ rows and whose columns are not all included in the last λ columns, the determinant of these minors is equal to

$$\Delta^{(n-1)_{m-1}-(\lambda-1)_{m-1}} \cdot A_{n-\lambda}^{(\lambda-1)_{m-1}}$$

For N' we have

(4) $$\begin{aligned} N' &= M \cdot \Delta^{(n)_m-(\lambda)_m-(n-1)_{m-1}} \\ &= A_\lambda^{(\lambda-1)_{m-1}} \cdot \Delta^{(n-1)_m-(\lambda)_m} \end{aligned}$$

since M is the mth compound of A_λ.

If we had formed the $(n-m)$th compound of $A_{n-\lambda}$ and denoted it by M_1, its complementary in $\Delta_{(n-m)}$ by N_1 and the corresponding, minors in $\Delta_{(m)}$ by M_1' and N_1', then

(5) $$N_1 = \Delta^{(n-1)_m-(n-\lambda-1)_{m-\lambda}} \cdot A_\lambda^{(n-\lambda-1)_{m-\lambda}}$$

202. *Bazin-Reiss-Picquet Theorem: If every set of q columns of $A=|a_{1n}|$ be replaced by every set of q columns of $B=|b_{1n}|$ and if the determinant of the square array of determinants thus obtained be denoted by Δ, then*

(1) $$\Delta = A^{(n-1)_q} B^{(n-1)_{q-1}}.$$

The truth of this theorem, which is due to Bazin and later to Reiss and Picquet, is apparent on observing that Δ is the product of the $(n-q)$th compound of A by the qth compound of B, since each element is the product of two columns one from each of these compounds.

If in the statement of this theorem the two determinants A and B be interchanged and Δ' represent the resulting determinant, then

$$(2) \qquad \Delta' = A^{(n-1)_q - 1} B^{(n-1)_q}$$

and

$$\Delta \cdot \Delta' = A^{(n)_q} B^{(n)_q}.$$

203. If $|\alpha_{\mu\mu}|$ and $|\beta_{\mu\mu}|$ represent the mth compounds of A and B respectively and if $|t_{\rho\rho}|$ and $|u_{\rho\rho}|$, where $\rho = (\mu)_q$, be the determinants formed from $|\alpha_{\mu\mu}|$ and $|\beta_{\mu\mu}|$ by the theorem of the last article, then

$$|t_{\rho\rho}| = |\alpha_{\mu\mu}|^{(\mu-1)_q} |\beta_{\mu\mu}|^{(\mu-1)_q - 1},$$

$$|u_{\rho\rho}| = |\alpha_{\mu\mu}|^{(\mu-1)_q - 1} \cdot |\beta_{\mu\mu}|^{(\mu-1)_q}$$

and of course the right-hand sides may be expressed in terms of A and B.

204. Let the elements of the determinants of §202 be represented by d's and d''s so that

$$\Delta \equiv |d_{\delta\delta}|$$

$$\Delta' \equiv |d'_{\delta\delta}|, \quad \text{where } \delta = (n)_q,$$

and let M_h denote a minor of Δ of order h. If we raise M_h to the order δ in the usual way by filling out the rows, adding a diagonal of ones under the added columns and making all other elements zeros and multiplied by Δ' we get

$$M_h \cdot \Delta' = (AB)^h N'_{\delta - h}$$

(by §135) where

$$N'_{\delta-h} = |d_{h+1,h+1} \cdots d_{\delta\delta}|,$$

Then since

$$\Delta' = A^{(n-1)_q - 1} B^{(n-1)_q},$$

we have

$$M_h = N'_{\delta-h} A^{h-(n-1)_q - 1} \cdot B^{h-(n-1)_q}$$

or

$$\frac{M_h}{N'_{\delta-h}} = \frac{(A \cdot B)^h}{A^{(n-1)_q - 1} \cdot B^{(n-1)_q}}.$$

If instead of A and B of the nth order we had taken them of the $(n+m)$th order and formed a determinant as before by supplanting every q columns of the first n columns of A by every q columns of the first n columns of B we would have a result which would be the extensional of theorem (1) of §202.

Let E represent the determinant of order $(n+m)$ with its first n columns the first n columns of B and the last m columns those of A. Then

$$\Delta = A^{(n-1)_q} \cdot E^{(n-1)_{q-1}}.$$

It is readily seen that we may in this write n for $n+m$ and h for n and have

(2) $$\Delta = A^{(h-1)_q} \cdot E^{(h-1)_{q-1}}.$$

If $q=1$, this becomes

(3) $$\Delta = A^{(h-1)} \cdot E.$$

When $q=1$ it is readily seen that any minor of Δ of order k, M_k say, is an example of the theorem where the E now (say E_k) has k columns of b's and $(n-k)$ columns of a's.

Therefore

(4) $$M_k = A^{(k-1)} \cdot E_k$$

which shows that every minor of Δ of order $\geqq 2$ contains A as a factor.

If $k=n-1$, then

(5) $$M_{k-1} = A^{n-2} \cdot E_{n-1}$$

which shows that every first minor of Δ contains A^{n-2} as a factor.

205. *If in the case of each row of a determinant the square root of the sum of the squares of the elements be taken, the product of the said square roots is greater than the determinant.*

Take $A = |a_1 b_2 c_3 d_4|$ and put

$$a_r^2 + b_r^2 + c_r^2 + d_r^2 \equiv s_r^2, \qquad A_r^2 + B_r^2 + C_r^2 + D_r^2 \equiv S_r^2$$

where A_r, B_r, etc. are the cofactors of a_r, b_r, etc. in A. Then

$$(a_r/s_r - A_r/S_r)^2 + (b_r/s_r - B_r/S_r)^2 + (c_r/s_r - C_r/S_r)^2 + (d_r/s_r - D_r/S_r)^2,$$

is a positive quantity, equal to

$$\frac{a_r^2+b_r^2+c_r^2+d_r^2}{s_r^2}-2\frac{a_rA_r+b_rB_r+c_rC_r+d_rD_r}{s_rS_r}+\frac{A_r^2+B_r^2+C_r^2+D_r^2}{S_r^2}$$

$$=1-2\frac{A}{s_rS_r}+1=2\left(1-\frac{A}{s_rS_r}\right)$$

thus

$$s_rS_r>A\,.$$

From this we see that

$$s_1\cdot s_2\cdot s_3\cdot s_4\cdot S_1\cdot S_2\cdot S_3\cdot S_4>A^4>A\cdot\left|\,A_1\,B_2\,C_3\,D_4\right|$$

Therefore $s_1\cdot s_2\cdot s_3\cdot s_4>A$, since to suppose otherwise would lead to a contradiction.

206. Let $x_1, x_2, \cdots, x_\lambda$ be the λ combinations of any n numbers (denoted by X) taken l at a time; and let $l\cdot\lambda$ integers (denoted by A) be broken up into λ sets of l in each, and let these sets be represented by $\alpha_1, \alpha_2, \cdots, \alpha_\lambda$.

Let the compound determinant $D_{[\lambda,l]}$ of order λ, with elements of order l be formed from a rectangular array of n rows and $l\cdot\lambda$ columns of elements. Thus

$$D_{[\lambda,l]}\equiv\left|\begin{matrix}\left|\begin{matrix}x_1\\ \alpha_1\end{matrix}\right| & \left|\begin{matrix}x_2\\ \alpha_1\end{matrix}\right| & \cdots & \left|\begin{matrix}x_\lambda\\ \alpha_1\end{matrix}\right|\\ \left|\begin{matrix}x_1\\ \alpha_2\end{matrix}\right| & & & \\ \cdot & & & \\ \cdot & & & \\ \cdot & & & \\ \left|\begin{matrix}x_1\\ \alpha_\lambda\end{matrix}\right| & & & \end{matrix}\right|$$

where writing but the first row and first column sufficiently indicates the law of formation, and where $\left|\begin{smallmatrix}x_i\\ \alpha_j\end{smallmatrix}\right|$ denotes the minor of order l formed by the elements of the array common to the ith selection of rows and the jth selection of columns. Let

$$
D'_{[\lambda,n-l]} \equiv \begin{vmatrix} \begin{vmatrix} x_1' \\ \beta_1 \end{vmatrix} & \begin{vmatrix} x_2' \\ \beta_1 \end{vmatrix} & \cdots & \begin{vmatrix} x_\lambda' \\ \beta_1 \end{vmatrix} \\ \begin{vmatrix} x_1' \\ \beta_2 \end{vmatrix} & & & \\ \cdot & & & \\ \cdot & & & \\ \cdot & & & \\ \begin{vmatrix} x_1' \\ \beta_\lambda \end{vmatrix} & & & \end{vmatrix}
$$

where x_i' is the complementary of the combination x_i with respect to X; and where $\beta_1, \beta_2, \cdots$, are the combinations of any set of n numbers (denoted by B) taken $(n-l)$ at a time.

Then the product

$$
D_{[\lambda,l]} \cdot D'_{[\lambda,n-l]} = \begin{vmatrix} \begin{vmatrix} X \\ \alpha_1\,\beta_1 \end{vmatrix} & \begin{vmatrix} X \\ \alpha_1\,\beta_2 \end{vmatrix} & \cdots & \begin{vmatrix} X \\ \alpha_1\,\beta_\lambda \end{vmatrix} \\ \begin{vmatrix} X \\ \alpha_2\,\beta_1 \end{vmatrix} & & & \\ \cdot & & & \\ \cdot & & & \\ \cdot & & & \\ \begin{vmatrix} X \\ \alpha_\lambda\,\beta_1 \end{vmatrix} & & & \end{vmatrix} = \Delta_{[\lambda,n]} \text{ say}.
$$

But $D'_{[\lambda,n-l]}$ being the $(n-l)$th compound of $\begin{vmatrix} X \\ B \end{vmatrix}$ is equal to $\begin{vmatrix} X \\ B \end{vmatrix}^{(n-1)_l}$

Therefore*

$$
(1) \qquad D_{[\lambda,l]} \begin{vmatrix} X \\ B \end{vmatrix}^{(n-1)_l} = \Delta_{[\lambda,n]}
$$

Instead of the multiplier $D'_{[\lambda,n-l]}$ let us use $D''_{[\lambda,n-l]}$, obtained from $D_{[\lambda,n-l]}$ by replacing the n numbers B by the n numbers C, where C contains h of those of B and $(n-h)$ of those of A; and where the $(n-h)$ numbers of A are those in $\alpha_{k+1}, \alpha_{k+2}, \cdots, \alpha_\lambda$. Let $\gamma_{k+1}, \gamma_{k+2}, \cdots, \gamma_\lambda$ be the complementaries of $\alpha_{k+1}, \alpha_{k+2}, \cdots, \alpha_\lambda$, respectively with respect to C.

* The β's here might be all different just as the α's are and give a more general result of which (1) is a special case.

Then the product

$$D_{[\lambda,l]}D''_{[\lambda,n-l]} = \begin{vmatrix} \begin{vmatrix} X \\ \alpha_1\gamma_1 \end{vmatrix} & \begin{vmatrix} X \\ \alpha_1\gamma_2 \end{vmatrix} & \cdots & \begin{vmatrix} X \\ \alpha_1\gamma_k \end{vmatrix} \\ \begin{vmatrix} X \\ \alpha_2\gamma_1 \end{vmatrix} & & & \\ \cdot & & & \\ \cdot & & & \\ \cdot & & & \\ \begin{vmatrix} X \\ \alpha_k\gamma_1 \end{vmatrix} & & & \end{vmatrix} \times \begin{vmatrix} X \\ C \end{vmatrix}^{\lambda-k}$$

$$= \Delta_{[k,n]} \begin{vmatrix} X \\ C \end{vmatrix}^{\lambda-k}.$$

Dividing out the common factor on both sides we have

$$D_{[\lambda,l]}\cdot \begin{vmatrix} X \\ C \end{vmatrix}^{(n-1)l-\lambda+k} = \Delta_{[k,n]} \tag{2}$$

Let $\rho+h=\lambda$, where $\rho=(r)_l$, and let $\alpha_1, \alpha_2, \cdots, \alpha_\lambda$ be $\alpha_1, \alpha_2, \cdots, \alpha_h$, $r_1, r_2, \cdots, r_\rho$ where $r_1, r_2, \cdots, r_\rho$ are the combinations of some r numbers taken l at a time, and denote the result by $D_{[\rho+h,l]}$. If now we use for $\beta_1, \beta_2, \cdots, \beta_\lambda$ the combinations $\gamma_1, \gamma_2, \cdots, \gamma_h$, r_1', r_2', $\cdots$, r_ρ', where these are the combinations $(n-l)$ at a time of the numbers C', consisting of the r numbers in $r_1, r_2, \cdots, r_\rho$ and $(n-r)$ others; where r_i' is the complementary of r_i; and denote the result by $D'''_{[\lambda,n-l]}$ then we have

$$D_{[\rho+h,l]}\cdot D'''_{[\lambda,n-l]} = \begin{vmatrix} \begin{vmatrix} X \\ \alpha_1\gamma_1 \end{vmatrix} & \begin{vmatrix} X \\ \alpha_1\gamma_2 \end{vmatrix} & \cdots & \begin{vmatrix} X \\ \alpha_1\gamma_h \end{vmatrix} \\ \begin{vmatrix} X \\ \alpha_2\gamma_1 \end{vmatrix} & & & \\ \cdot & & & \\ \cdot & & & \\ \cdot & & & \\ \begin{vmatrix} X \\ \alpha_h\gamma_1 \end{vmatrix} & & & \end{vmatrix} \times \begin{vmatrix} X \\ C' \end{vmatrix}^{\rho}$$

or

$$(3) \qquad D_{[\rho+h,l]}\cdot\begin{vmatrix} X \\ C' \end{vmatrix}^{(n-1)_l-\rho} = \Delta_{[h,n]}.$$

Various other assumptions for the numbers A and B may be made and corresponding results obtained, some of which will cause the Δ to break up into two or more factors.

Combining (1) and (2) we have

$$(4) \qquad \Delta_{[\lambda,n]}\begin{vmatrix} X \\ C \end{vmatrix}^{(n-1)_l-\lambda+k} = \Delta_{[k,n]}\begin{vmatrix} X \\ B \end{vmatrix}^{(n-1)_l}$$

This may be looked upon as a condensation formula for $\Delta_{[\lambda,n]}$.

If we put $l=n-1$, $k=n-1$, and therefore $\lambda=n$ and $h=1$, then (1), (2) and (4) become

$$(1a) \qquad D_{[n,n-1]}\begin{vmatrix} X \\ B \end{vmatrix} = \Delta_{[n,n]}$$

$$(2a) \qquad D_{[n,n-1]} = \Delta_{[n-1,n]}.$$

Since in this case the only trace of the multiplier is that embodied in $\Delta_{[n-1,n]}$ it is obvious that we may use any one of n different multipliers and obtain n different, but equivalent, Δ's on the right.

$$(4a) \qquad \Delta_{[n,n]} = \Delta_{[n-1,n]}\begin{vmatrix} X \\ B \end{vmatrix}.$$

In the special case of (3) where $l=n-1$, $r=n-1$ and therefore $\rho=1$ and $h=n-1$ we get the same as (2a).

As an illustration of (2a) we have the condition that the intersection of the planes

$$a_1x + b_1y + c_1z = 0,$$
$$a_2x + b_2y + c_2z = 0;$$

the intersection of

$$a_3x + b_3y + c_3z = 0,$$
$$a_4x + b_4y + c_4z = 0;$$

and the intersection of

$$a_5x + b_5y + c_5z = 0,$$
$$a_6x + b_6y + c_6z = 0,$$

will be in the same plane* is that

$$\begin{vmatrix} a_1 & a_2 & a_3 & a_4 & \cdot & \cdot \\ b_1 & b_2 & b_3 & b_4 & \cdot & \cdot \\ c_1 & c_2 & c_3 & c_4 & \cdot & \cdot \\ \cdot & \cdot & a_3 & a_4 & a_5 & a_6 \\ \cdot & \cdot & b_3 & b_4 & b_5 & b_6 \\ \cdot & \cdot & c_3 & c_4 & c_5 & c_6 \end{vmatrix} = 0$$

This determinant is readily seen to be the same as

$$\begin{Vmatrix} \begin{vmatrix} 1\,2\,3 \\ 1\,2\,3 \end{vmatrix} & \begin{vmatrix} 1\,2\,3 \\ 3\,5\,6 \end{vmatrix} \\ \begin{vmatrix} 1\,2\,3 \\ 1\,2\,4 \end{vmatrix} & \begin{vmatrix} 1\,2\,3 \\ 4\,5\,6 \end{vmatrix} \end{Vmatrix}$$

and this by (2a) is

$$-\begin{Vmatrix} \begin{vmatrix} 1\,2 \\ 1\,2 \end{vmatrix} & \begin{vmatrix} 2\,3 \\ 1\,2 \end{vmatrix} & \begin{vmatrix} 1\,3 \\ 1\,2 \end{vmatrix} \\ \begin{vmatrix} 1\,2 \\ 3\,4 \end{vmatrix} & \begin{vmatrix} 2\,3 \\ 3\,4 \end{vmatrix} & \begin{vmatrix} 1\,3 \\ 3\,4 \end{vmatrix} \\ \begin{vmatrix} 1\,2 \\ 5\,6 \end{vmatrix} & \begin{vmatrix} 2\,3 \\ 5\,6 \end{vmatrix} & \begin{vmatrix} 1\,3 \\ 5\,6 \end{vmatrix} \end{Vmatrix}.$$

207. It is worthy of note that there is a third form which is the equivalent of $D_{[n,n-1]}$ and $\Delta_{[n-1,n]}$, namely

$$A_{[n,n-1]} \equiv \begin{vmatrix} (A_1) & (0) & \cdots & (0) \\ (0) & (A_2) & \cdots & (0) \\ \cdot & \cdot & \cdots & \cdot \\ (0) & (0) & \cdots & (A_n) \\ (1) & (1) & \cdots & (1) \end{vmatrix}$$

where (A_i) denotes an array of $(n-1)$ rows and n columns; (1) denotes a determinant of order n having every element along the principal diagonal unity and all other elements zeros; (0) denotes a rectangle of $(n-1)$ rows and n columns of zeros.

The truth of the equality between $A_{[n,n-1]}$ and $D_{[n,n-1]}$ is seen by expanding $A_{[n,n-1]}$ by Albegiani's theorem and observing that the terms are just those of $D_{[n,n-1]}$.

* Cayley: Camb. Math. Jour. Vol. IV pp. 18-20.

Some interesting relations may be obtained from the equality of these three forms by giving some of the elements special values. Thus if $n=4$,

$$A_{[4,3]} = \begin{vmatrix} (1\,2\,3) & & & \\ & (4\,5\,6) & & \\ & & (7\,8\,9) & \\ & & & (\tau\,\sigma\,\rho) \\ (1) & (1) & (1) & (1) \end{vmatrix},$$

where $(\alpha\beta\gamma)$ stands for the array

$$\begin{matrix} a_{\alpha 1} & a_{\alpha 2} & a_{\alpha 3} & a_{\alpha 4} \\ a_{\beta 1} & a_{\beta 2} & a_{\beta 3} & a_{\beta 4} \\ a_{\gamma 1} & a_{\gamma 2} & a_{\gamma 3} & a_{\gamma 4} \end{matrix}$$

and the remaining elements are zeros.

This may readily be transformed into

$$A_{[4,3]} = \begin{vmatrix} (1\,2\,3) & & \\ & (4\,5\,6) & \\ & & (7\,8\,9) \\ (\tau\,\sigma\,\rho) & (\tau\,\sigma\,\rho) & (\tau\,\sigma\,\rho) \end{vmatrix} = D_{[4,3]} = \Delta_{[3,4]}.$$

If in this we put $a_{74}=a_{84}=a_{94}=0$ there results the relation, after taking out the common factor

$$\begin{vmatrix} a_{71} & a_{72} & a_{73} \\ a_{81} & a_{82} & a_{83} \\ a_{91} & a_{92} & a_{93} \end{vmatrix},$$

$$\begin{vmatrix} a_{11} & a_{12} & a_{13} & a_{14} & & & & & \\ a_{21} & a_{22} & a_{23} & a_{24} & & & & & \\ a_{31} & a_{32} & a_{33} & a_{34} & & & & & \\ & & & & a_{41} & a_{42} & a_{43} & a_{44} & 0 \\ & & & & a_{51} & a_{52} & a_{53} & a_{54} & 0 \\ & & & & a_{61} & a_{62} & a_{63} & a_{64} & 0 \\ a_{\tau 1} & a_{\tau 2} & a_{\tau 3} & a_{\tau 4} & a_{\tau 1} & a_{\tau 2} & a_{\tau 3} & a_{\tau 4} & a_{\tau 4} \\ a_{\sigma 1} & a_{\sigma 2} & a_{\sigma 3} & a_{\sigma 4} & a_{\sigma 1} & a_{\sigma 2} & a_{\sigma 3} & a_{\sigma 4} & a_{\sigma 4} \\ a_{\rho 1} & a_{\rho 2} & a_{\rho 3} & a_{\rho 4} & a_{\rho 1} & a_{\rho 2} & a_{\rho 3} & a_{\rho 4} & a_{\rho 4} \end{vmatrix}$$

$$= \left\| \begin{vmatrix} 1\,2\,3 \\ 1\,2\,4 \end{vmatrix} \begin{vmatrix} 4\,5\,6 \\ 1\,3\,4 \end{vmatrix} \begin{vmatrix} \tau\,\sigma\,\rho \\ 2\,3\,4 \end{vmatrix} \right\|$$

$$= \begin{vmatrix} \begin{vmatrix} 3\,4\,5\,6 \\ 1\,2\,3\,4 \end{vmatrix} & \begin{vmatrix} 2\,4\,5\,6 \\ 1\,2\,3\,4 \end{vmatrix} & \begin{vmatrix} 1\,4\,5\,6 \\ 1\,2\,3\,4 \end{vmatrix} \\ \begin{vmatrix} 3 \\ 4 \end{vmatrix} & \begin{vmatrix} 2 \\ 4 \end{vmatrix} & \begin{vmatrix} 1 \\ 4 \end{vmatrix} \\ \begin{vmatrix} 3\,\tau\,\sigma\,\rho \\ 1\,2\,3\,4 \end{vmatrix} & \begin{vmatrix} 2\,\tau\,\sigma\,\rho \\ 1\,2\,3\,4 \end{vmatrix} & \begin{vmatrix} 1\,\tau\,\sigma\,\rho \\ 1\,2\,3\,4 \end{vmatrix} \end{vmatrix}$$

208. If in $\Delta_{[\lambda,n]}$ α_i and β_i have no numbers in common, and if α_i and β_j have numbers in common for $\left\{\begin{smallmatrix} i=1,2,\cdots\lambda-1 \\ j=1,2,\cdots\lambda \end{smallmatrix}\right\} j>i$, then $\Delta_{[\lambda,n]}$ breaks up into factors and is equal to the product of the elements along the principal diagonal since all the elements on one side of it are zero. Thus (1) §206 becomes

$$(1b) \qquad D_{[\lambda,l]}\cdot \begin{vmatrix} X \\ B \end{vmatrix}^{(n-1)_l} = \begin{vmatrix} X \\ \alpha_1\,\beta_1 \end{vmatrix} \cdot \begin{vmatrix} X \\ \alpha_2\,\beta_2 \end{vmatrix} \cdots \begin{vmatrix} X \\ \alpha_\lambda\,\beta_\lambda \end{vmatrix}$$

(2) becomes

$$(2b) \quad D_{[\lambda,l]}\cdot \begin{vmatrix} X \\ C \end{vmatrix}^{(n-1)_l-\lambda+k} = \begin{vmatrix} X \\ \alpha_1\,\gamma_1 \end{vmatrix} \cdot \begin{vmatrix} X \\ \alpha_2\,\gamma_2 \end{vmatrix} \cdots \begin{vmatrix} X \\ \alpha_k\,\gamma_k \end{vmatrix}$$

(3) becomes

$$(3b) \quad D'_{[\lambda,l]}\cdot \begin{vmatrix} X \\ C' \end{vmatrix}^{(n-1)_l-\rho} = \begin{vmatrix} X \\ \alpha_1\,\gamma_1 \end{vmatrix} \cdot \begin{vmatrix} X \\ \alpha_2\,\gamma_2 \end{vmatrix} \cdots \begin{vmatrix} X \\ \alpha_h\,\gamma_h \end{vmatrix}.$$

If in addition, for some values of i, α_i and β_i have numbers in common then the right-hand side of (1b), (2b) and (3b) are zero.

209. In illustration of the foregoing let us take $n=5$, $l=2$, $h=1$, and therefore $\lambda=10$ and $k=8$. Let $\alpha_1, \alpha_2, \cdots, \alpha_{10}$ be the combinations $x_1x_2, x_3x_4, \cdots, x_{19}x_{20}$ and let X be the numbers 1, 2, 3, 4, 5. If we use for B, $x_{16}, x_{17}, x_{18}, x_{19}, x_{20}$, then (2) §206 becomes

$$D_{[10,2]}\cdot \begin{vmatrix} 1 & 2 & 3 & 4 & 5 \\ x_{16} & x_{17} & x_{18} & x_{19} & x_{20} \end{vmatrix}^4 = \Delta_{[8,5]} \equiv \Big\| \, |x_1\,x_2\,x_{16}\,x_{17}\,x_{19}|$$

$$(2'') \quad \begin{matrix} |x_3\;x_4\;x_{16}\,x_{17}\,x_{20}| & |x_5\;x_6\;x_{16}\,x_{18}\,x_{19}| & |x_7\;x_8\;x_{16}\,x_{18}\,x_{20}| \\ |x_9\;x_{10}\,x_{17}\,x_{18}\,x_{19}| & |x_{11}\,x_{12}\,x_{17}\,x_{18}\,x_{20}| & |x_{13}\,x_{14}\,x_{17}\,x_{19}\,x_{20}| \\ & |x_{15}\,x_{16}\,x_{18}\,x_{19}\,x_{20}| \, \Big\| & \end{matrix}$$

where the row numbers are dropped since they are all alike.

It may be noted here that on account of x_{16} in the column numbers of the third last line of $D_{[10,2]}$, there are but four elements in the last line of $\Delta_{[8,5]}$ which are not zero.

If we take for $\alpha_1, \alpha_2, \cdots, \alpha_{10}$ the combinations x_1x_2, x_3x_4, x_5x_6, x_7x_8, $\alpha\beta$, $\alpha\gamma$, $\alpha\delta$, $\beta\gamma$, $\beta\delta$, $\gamma\delta$, denoting the determinant in this case by $D'_{[10,2]}$, and using $\alpha, \beta, \gamma, \delta, \epsilon$ for B, then (3) §206 becomes

$$D'_{[10,2]} = \Big|\, |x_1\, x_2\, \beta\, \gamma\, \delta|\, |x_3\, x_4\, \alpha\, \gamma\, \delta|\, |x_5\, x_6\, \alpha\, \beta\, \delta|\, |x_7\, x_8\, \alpha\, \beta\, \gamma| \,\Big|$$

or

$$(3')\quad D'_{[10,2]} = \Delta_{[4,5]}.$$

If we take for $\alpha_1, \alpha_2, \cdots, \alpha_{10}$ the combinations $\alpha\beta$, $\alpha\gamma$, $\alpha\delta$, $\alpha\epsilon$, x_1x_2, x_3x_4, x_5x_6, x_7x_8, x_9x_{10}, $x_{11}x_{12}$, and denote the determinant in this case by $D''_{[10,2]}$ and use $\alpha, \beta, \gamma, \delta, \epsilon$ for B, then (3) §206 becomes

$$D''_{[10,2]}\cdot\begin{vmatrix} 1 & 2 & 3 & 4 & 5 \\ \alpha & \beta & \gamma & \delta & \epsilon \end{vmatrix}^2 = \begin{vmatrix} |x_1\; x_2\; \alpha\,\beta\,\gamma| & |x_1\, x_2\, \alpha\,\beta\,\delta| & \cdots & |x_1\, x_2\, \alpha\,\delta\,\epsilon| \\ |x_3\; x_4\; \alpha\,\beta\,\gamma| & & & \\ \cdot & & & \\ \cdot & & & \\ \cdot & & & \\ |x_{11}\, x_{12}\, \alpha\,\beta\,\gamma| & & & \end{vmatrix}$$

$$(3'')\qquad = \Delta_{[6,5]}.$$

If instead of using the third compound of $\begin{vmatrix} 12345 \\ \alpha\beta\gamma\delta\epsilon \end{vmatrix}$ as our multiplier we had used the third compound of $\begin{vmatrix} 1\,2\,3\,4\,5 \\ \alpha x_1 x_2 x_3 x_4 \end{vmatrix}$ we would get

$$(3'')\qquad D''_{[10,2]}\begin{vmatrix} 1 & 2 & 3 & 4 & 5 \\ \alpha & x_1 & x_2 & x_3 & x_4 \end{vmatrix}^4$$

$$= \begin{vmatrix} |\alpha\,\beta\, x_1\, x_2\, x_3| & |\alpha\,\beta\, x_1\, x_2\, x_4| & |\alpha\,\beta\, x_1\, x_3\, x_4| & |\alpha\,\beta\, x_2\, x_3\, x_4| \\ |\alpha\,\gamma\, x_1\, x_2\, x_3| & & & \\ |\alpha\,\delta\, x_1\, x_2\, x_3| & & & \\ |\alpha\,\epsilon\, x_1\, x_2\, x_3| & & & \end{vmatrix} \times$$

$$\begin{vmatrix} |x_5\; x_6\; \alpha\; x_1\; x_3| & |x_5\, x_6\, \alpha x_1\, x_4| & |x_5\, x_6\, \alpha x_2\, x_3| & |x_5\, x_6\, \alpha x_2\, x_4| \\ |x_7\; x_8\; \alpha\; x_1\; x_3| & & & \\ |x_9\; x_{10}\, \alpha\; x_1\; x_3| & & & \\ |x_{11}\, x_{12}\, \alpha\; x_1\; x_3| & & & \end{vmatrix}$$

$$= \Delta'_{[4,5]} \times \Delta_{[4,5]} \quad \text{say,}$$

But

$$\Delta'_{[4,5]} = \begin{vmatrix} 1 & 2 & 3 & 4 & 5 \\ \alpha & \beta & \gamma & \delta & \epsilon \end{vmatrix} \cdot \begin{vmatrix} 1 & 2 & 3 & 4 & 5 \\ \alpha & x_1 & x_2 & x_3 & x_4 \end{vmatrix}^3$$

since it is an example of the "Extensional" of the theorem of §202. We have, therefore,

$$(3''') \qquad D''_{[10,2]} \cdot \begin{vmatrix} 1 & 2 & 3 & 4 & 5 \\ \alpha & x_1 & x_2 & x_3 & x_4 \end{vmatrix} = \begin{vmatrix} 1 & 2 & 3 & 4 & 5 \\ \alpha & \beta & \gamma & \delta & \epsilon \end{vmatrix} \Delta_{[4,5]}.$$

From (3'') and (3''') it follows that

$$(4') \qquad \Delta_{[6,5]} \cdot \begin{vmatrix} 1 & 2 & 3 & 4 & 5 \\ \alpha & x_1 & x_2 & x_3 & x_4 \end{vmatrix} = \begin{vmatrix} 1 & 2 & 3 & 4 & 5 \\ \alpha & \beta & \gamma & \delta & \epsilon \end{vmatrix}^3 \Delta_{[4,5]}$$

If for $\alpha_1, \alpha_2, \cdots, \alpha_{10}$ we take $pq, r\beta, s\gamma, t\delta, \alpha\beta, \alpha\gamma, \alpha\delta, \beta\gamma, \beta\delta, \gamma\delta$ and denote the determinant by $D'''_{[10,2]}$, then using as multiplier the third compound of $\begin{vmatrix} 12345 \\ \alpha\beta\gamma\delta\epsilon \end{vmatrix}$, we get

$$(3'b) \qquad D'''_{[10,2]} = | p\, q\, \beta\, \gamma\, \delta | \, | r\, \alpha\, \beta\, \gamma\, \delta | \cdot | s\, \alpha\, \beta\, \gamma\, \delta | \cdot | t\, \alpha\, \beta\, \gamma\, \delta |$$

Here it is to be observed that p, q, r, s, t may be any numbers whatever as long as

(1) $\qquad p$ or $q \neq \beta$ or γ or δ

(2) $\qquad r$ or s or $t \neq \alpha$ or β or γ or δ.

If suitable values are given the column numbers the Δ's on the right of (2''), (3'), (3'') may be made to break up into factors. Thus

$$x_1 = x_3 = x_{18}\,; \quad x_2 = x_6 = x_9 = x_{20}\,; \quad x_4 = x_8 = x_{11} = x_{19}$$
$$x_5 = x_7 = x_{17}\,; \quad x_{13} = x_{16}$$

gives

$$(3''b) \qquad \left\| \begin{vmatrix} 1 & 2 \\ x_{18} & x_{20} \end{vmatrix} \begin{vmatrix} 1 & 3 \\ x_{18} & x_{19} \end{vmatrix} \begin{vmatrix} 1 & 4 \\ x_{17} & x_{20} \end{vmatrix} \begin{vmatrix} 1 & 5 \\ x_{17} & x_{19} \end{vmatrix} \begin{vmatrix} 2 & 3 \\ x_{20} & x_{10} \end{vmatrix} \begin{vmatrix} 2 & 4 \\ x_{19} & x_{12} \end{vmatrix} \begin{vmatrix} 2 & 5 \\ x_{16} & x_{14} \end{vmatrix} \begin{vmatrix} 3 & 4 \\ x_{15} & x_{16} \end{vmatrix} \begin{vmatrix} 3 & 5 \\ x_{17} & x_{18} \end{vmatrix} \begin{vmatrix} 4 & 5 \\ x_{19} & x_{20} \end{vmatrix} \right\|$$

$$= | x_{10}\, x_{17}\, x_{18}\, x_{19}\, x_{20} | \cdot | x_{12}\, x_{17}\, x_{18}\, x_{19}\, x_{20} |$$
$$\times | x_{14}\, x_{16}\, x_{17}\, x_{19}\, x_{20} | \cdot | x_{15}\, x_{16}\, x_{18}\, x_{19}\, x_{20} |$$

or as it may be written

$$\left\| \begin{vmatrix} 1 & 2 \\ \beta & \delta \end{vmatrix} \begin{vmatrix} 1 & 3 \\ \beta & \gamma \end{vmatrix} \begin{vmatrix} 1 & 4 \\ \alpha & \delta \end{vmatrix} \begin{vmatrix} 1 & 5 \\ \alpha & \gamma \end{vmatrix} \begin{vmatrix} 2 & 3 \\ s & \delta \end{vmatrix} \begin{vmatrix} 2 & 4 \\ r & \gamma \end{vmatrix} \begin{vmatrix} 2 & 5 \\ t & q \end{vmatrix} \begin{vmatrix} 3 & 4 \\ p & q \end{vmatrix} \begin{vmatrix} 3 & 5 \\ \alpha & \beta \end{vmatrix} \begin{vmatrix} 4 & 5 \\ \gamma & \delta \end{vmatrix} \right\|$$

$$(3''b) \qquad = \left| s\,\alpha\,\beta\,\gamma\,\delta \right| \cdot \left| r\,\alpha\,\beta\,\gamma\,\delta \right| \cdot \left| t\,q\,\alpha\,\gamma\,\delta \right| \cdot \left| p\,q\,\beta\,\gamma\,\delta \right|$$

In (3′) the substitution $x_1=\alpha$, $x_3=\beta$, $x_5=\gamma$, $x_7=\delta$, reduces it to the form (3′b)

$$\left\| \begin{vmatrix} 1 & 2 \\ \alpha & \beta \end{vmatrix} \begin{vmatrix} 1 & 3 \\ \alpha & \gamma \end{vmatrix} \begin{vmatrix} 1 & 4 \\ \alpha & \delta \end{vmatrix} \begin{vmatrix} 1 & 5 \\ \alpha & \epsilon \end{vmatrix} \begin{vmatrix} 2 & 3 \\ x_1 & x_2 \end{vmatrix} \begin{vmatrix} 2 & 4 \\ x_2 & x_4 \end{vmatrix} \begin{vmatrix} 2 & 5 \\ x_2 & x_8 \end{vmatrix} \begin{vmatrix} 3 & 4 \\ x_3 & x_4 \end{vmatrix} \begin{vmatrix} 3 & 5 \\ x_4 & x_{10} \end{vmatrix} \begin{vmatrix} 4 & 5 \\ x_{11} & x_{12} \end{vmatrix} \right\|$$

$$\times \begin{vmatrix} 1 & 2 & 3 & 4 & 5 \\ \alpha & x_1 & x_2 & x_3 & x_4 \end{vmatrix} = \left| \alpha\,\beta\,\gamma\,\delta\,\epsilon \right| \left| x_2\,x_4\,\alpha x_1\,x_3 \right| \left| x_2\,x_8\,\alpha x_1\,x_4 \right|$$

$$\times \left| x_4\,x_{10}\,\alpha x_2\,x_3 \right| \left| x_{11}\,x_{12}\,\alpha x_2\,x_4 \right|$$

If in (3′b) p or $q=\beta$, or γ, or δ then the right-hand side vanishes and therefore the left-hand side is zero. It may be noted that if r or s or $t=\alpha$, or β, or γ, or δ, then two columns of $D'''_{[10,2]}$ would be identical. If in (3″b) $x_{15}=x_{18}$ or x_{19} or x_{20} then the right-hand side vanishes, etc.

210. If the elements in the first column of $D_{[\lambda,l]}$ is the sum of two minors of order l then the elements in the first column of $\Delta_{[k,n]}$ will be the sum of two minors of order n. Thus if

$$D'_{[10,2]} \equiv \left\| \begin{vmatrix} 1 & 2 \\ x_1 & x_2 \end{vmatrix} + \begin{vmatrix} 1 & 2 \\ x_1' & x_2' \end{vmatrix} \begin{vmatrix} 1 & 3 \\ x_3 & x_4 \end{vmatrix} \begin{vmatrix} 1 & 4 \\ x_5 & x_6 \end{vmatrix} \begin{vmatrix} 1 & 5 \\ x_7 & x_8 \end{vmatrix} \begin{vmatrix} 2 & 3 \\ \alpha & \beta \end{vmatrix} \begin{vmatrix} 2 & 4 \\ \alpha & \gamma \end{vmatrix} \begin{vmatrix} 2 & 5 \\ \alpha & \delta \end{vmatrix} \right.$$

$$\left. \begin{vmatrix} 3 & 4 \\ \beta & \gamma \end{vmatrix} \begin{vmatrix} 3 & 5 \\ \beta & \delta \end{vmatrix} \begin{vmatrix} 4 & 5 \\ \gamma & \delta \end{vmatrix} \right\|$$

it is equal to

$$\left\| \left| x_1\,x_2\,\beta\,\gamma\,\delta \right| + \left| x_1'\,x_2'\,\beta\,\gamma\,\delta \right| \left| x_3\,x_4\,\alpha\,\gamma\,\delta \right| \left| x_5\,x_6\,\alpha\,\beta\,\delta \right| \left| x_7\,x_8\,\alpha\,\beta\,\gamma \right| \right\|$$

This follows readily as in (3′).

In general if

$$S_k = \sum_{i=1}^{h_k} \begin{vmatrix} (l_k^i) \\ (n \mid m_\alpha \mid l_k) \end{vmatrix}$$

and $\Delta = \left| S_1\,S_2 \cdots S_\lambda \right|$, then $\Delta\cdot\Delta' = \Delta'' = \left| S_1'\,S_2' \cdots S_\lambda' \right|$ where (l_k^i) is a set of l numbers and

$$S_k' = \sum_{i=1}^{h_k} \begin{vmatrix} (l_k^i)(n \mid \bar{m}_\beta \mid l_k) \\ (n \mid m_\alpha) \end{vmatrix}.$$

EXAMPLE.

$$\begin{vmatrix} \begin{vmatrix}123\\123\end{vmatrix}\begin{vmatrix}123\\126\end{vmatrix}-\begin{vmatrix}123\\135\end{vmatrix}+\begin{vmatrix}123\\234\end{vmatrix}\begin{vmatrix}123\\156\end{vmatrix}-\begin{vmatrix}123\\246\end{vmatrix}+\begin{vmatrix}123\\345\end{vmatrix}\begin{vmatrix}123\\456\end{vmatrix} \\ \begin{vmatrix}124\\123\end{vmatrix}\begin{vmatrix}124\\126\end{vmatrix}-\begin{vmatrix}124\\135\end{vmatrix}+\begin{vmatrix}124\\234\end{vmatrix}\begin{vmatrix}124\\156\end{vmatrix}-\begin{vmatrix}124\\246\end{vmatrix}+\begin{vmatrix}124\\345\end{vmatrix}\begin{vmatrix}124\\456\end{vmatrix} \\ \begin{vmatrix}134\\123\end{vmatrix}\begin{vmatrix}234\\126\end{vmatrix}-\begin{vmatrix}134\\135\end{vmatrix}+\begin{vmatrix}134\\234\end{vmatrix}\begin{vmatrix}134\\156\end{vmatrix}-\begin{vmatrix}134\\246\end{vmatrix}+\begin{vmatrix}134\\345\end{vmatrix}\begin{vmatrix}134\\456\end{vmatrix} \\ \begin{vmatrix}234\\123\end{vmatrix}\begin{vmatrix}234\\126\end{vmatrix}-\begin{vmatrix}234\\135\end{vmatrix}+\begin{vmatrix}234\\234\end{vmatrix}\begin{vmatrix}234\\156\end{vmatrix}-\begin{vmatrix}234\\246\end{vmatrix}+\begin{vmatrix}234\\345\end{vmatrix}\begin{vmatrix}234\\456\end{vmatrix} \end{vmatrix}$$

$$= \begin{vmatrix} |1234|\,|1345| - |1246|\,|1456| \\ |1235|\,|2345| - |1256|\,|2456| \\ |1236|\,|2346| - |1356|\,|3456| \end{vmatrix}.$$

211. In §185 we have seen how any minor of the compound of a determinant Δ may be expressed in terms of the complementary of corresponding minor of the adjugate and a power of Δ. These minors may be one of two types: (1) those fully resolvable, (2) those irresolvable. Thus if the elements of the minor, M say, consists of a minor of Δ bordered in all possible ways with other rows and columns, then §197 gives its value in terms of Δ and minors of Δ. If the elements of M are formed from two minors of Δ as in §202 its value is there given in terms of these two minors. The minor M may be the compound of some minor of Δ and is then expressible in terms of that minor. Again the minor M may be an "Extensional" such as

$$M = \begin{vmatrix} \begin{vmatrix}1\,2\,3\\1\,2\,4\end{vmatrix} & \begin{vmatrix}1\,2\,3\\1\,3\,4\end{vmatrix} & \begin{vmatrix}1\,2\,3\\1\,3\,5\end{vmatrix} \\ \begin{vmatrix}1\,2\,4\\1\,2\,4\end{vmatrix} & \begin{vmatrix}1\,2\,4\\1\,3\,4\end{vmatrix} & \begin{vmatrix}1\,2\,4\\1\,3\,5\end{vmatrix} \\ \begin{vmatrix}1\,3\,4\\1\,2\,4\end{vmatrix} & \begin{vmatrix}1\,3\,4\\1\,3\,4\end{vmatrix} & \begin{vmatrix}1\,3\,4\\1\,3\,5\end{vmatrix} \end{vmatrix} = a_{11}\begin{vmatrix}1\,2\,3\,4\\1\,2\,3\,4\end{vmatrix}\cdot\begin{vmatrix}1\,2\,3\,4\\1\,2\,3\,5\end{vmatrix}.$$

212. In further illustration we might find all possible types of minors of order 3 of the second compound of a determinant Δ of order 5. Starting with the irreducible minor

$$M \equiv \begin{vmatrix} \begin{vmatrix}1\,2\\1\,2\end{vmatrix} & \begin{vmatrix}1\,3\\1\,3\end{vmatrix} & \begin{vmatrix}2\,4\\2\,4\end{vmatrix} \end{vmatrix},$$

which may be found to be equal to

$$\begin{vmatrix} 1\,2\,3 \\ 1\,2\,3 \end{vmatrix} \cdot \begin{vmatrix} 1\,2\,4 \\ 1\,2\,4 \end{vmatrix} + \Delta \cdot \left(a_{11}\, a_{22} - \begin{vmatrix} 1\,2 \\ 1\,2 \end{vmatrix} \right), \tag{1}$$

we see that the numbers 3 and 4 play a unique part and may be interchanged giving another minor with this same value. There are $(4)_2 = 6$ such pairs of numbers which may play this unique part. Hence we may have 12 different sets of 3 pairs of numbers for rows of our irresolvable minor, and since there are as many for the columns we have $12 \times 12 = 144$ irresolvable minors.

If we change the column number 4 into a 3 in M and its equal we get

$$\left\| \begin{vmatrix} 1\,2 \\ 1\,2 \end{vmatrix} \begin{vmatrix} 1\,3 \\ 1\,3 \end{vmatrix} \begin{vmatrix} 2\,4 \\ 2\,3 \end{vmatrix} \right\| = \begin{vmatrix} 1\,2\,3 \\ 1\,2\,3 \end{vmatrix} \begin{vmatrix} 1\,2\,4 \\ 1\,2\,3 \end{vmatrix}. \tag{2}$$

We could have changed the 3 into a 4 just as well and hence have $6 \times 2 = 12$ of this type with the same row or column numbers. For each set of row numbers we can have 4 sets of column numbers and interchanging rows and columns we get $12 \times 4 \times 2 = 96$ of this type.

If now we change the 4 into a 3 in both rows and columns at the same time we get

$$\left\| \begin{vmatrix} 1\,2 \\ 1\,2 \end{vmatrix} \begin{vmatrix} 1\,3 \\ 1\,3 \end{vmatrix} \begin{vmatrix} 2\,3 \\ 2\,3 \end{vmatrix} \right\| = \begin{vmatrix} 1\,2\,3 \\ 1\,2\,3 \end{vmatrix}^2. \tag{3}$$

Of this type there are obviously 16.

The minor

$$\left\| \begin{vmatrix} 1\,2 \\ 1\,2 \end{vmatrix} \begin{vmatrix} 1\,3 \\ 1\,3 \end{vmatrix} \begin{vmatrix} 1\,4 \\ 1\,4 \end{vmatrix} \right\| = a_{11}^2 \begin{vmatrix} 1234 \\ 1234 \end{vmatrix} \tag{4}$$

and there are obviously 16 of this type.

The minor

$$\left\| \begin{vmatrix} 1\,2 \\ 1\,2 \end{vmatrix} \begin{vmatrix} 1\,3 \\ 1\,3 \end{vmatrix} \begin{vmatrix} 2\,4 \\ 1\,4 \end{vmatrix} \right\| = a_{11}a_{21} \begin{vmatrix} 1234 \\ 1234 \end{vmatrix} \tag{5}$$

as may be seen on multiplying by $\left\| \begin{smallmatrix} 134 \\ 134 \end{smallmatrix} \right\| \begin{smallmatrix} 124 \\ 124 \end{smallmatrix} \left\| \begin{smallmatrix} 123 \\ 123 \end{smallmatrix} \right\|$. As before we may interchange 3 and 4 in the rows and get 12 changes in the row numbers. In the columns we get 4 changes and interchanging rows and columns we get 96 of this type.

The minor

$$\left\| \begin{vmatrix}1&2\\1&2\end{vmatrix} \begin{vmatrix}1&3\\1&3\end{vmatrix} \begin{vmatrix}2&3\\1&4\end{vmatrix} \right\| = 0$$

as may be seen on making the 4 in the rows of (5) a 3. Of this type there are $4\times4\times2=32$.

213. The determinant M, obtained on writing $(n|\bar{m}_\alpha|l_\gamma|k_\delta)$ $(\bar{n}|m_\alpha|l-k_\epsilon)$ for $(n|m_\alpha|l_\gamma)$ in

$$\left\| \begin{vmatrix}(n|m_\alpha|l_1)\\(n|m_\beta|l_1)\end{vmatrix} \begin{vmatrix}(n|m_\alpha|l_2)\\(n|m_\beta|l_2)\end{vmatrix} \cdots \begin{vmatrix}(n|m_\alpha|l_\gamma)\\(n|m_\beta|l_\beta)\end{vmatrix} \right\|$$

vanishes. For if we multiply M by

$$\left\| \begin{vmatrix}(n|\bar{m}_\alpha|l_1)\\(n|\bar{m}_\beta|l_1)\end{vmatrix} \begin{vmatrix}(n|\bar{m}_\alpha|l_2)\\(n|\bar{m}_\beta|l_2)\end{vmatrix} \cdots \begin{vmatrix}(n|\bar{m}_\alpha|l_\lambda)\\(n|\bar{m}_\beta|l_\lambda)\end{vmatrix} \right\|$$

every constituent in the γth column will be zero since the upper line of every suffix of M contains some number in common with the combination $(n|\bar{m}_\alpha|l_\gamma)$. The product therefore vanishes, and since the multiplier is in general different from zero, M must be zero.

Thus, let $(n|m_\alpha)\equiv12345$ and $l=3$, $k=1$ and let $(n|m_\alpha|l_6)$ which is 145 be replaced by 367 in the row numbers of

$$\left\| \begin{vmatrix}1\,2\,3\\1\,2\,3\end{vmatrix} \begin{vmatrix}1\,2\,4\\1\,2\,4\end{vmatrix} \cdots \begin{vmatrix}3\,4\,5\\3\,4\,5\end{vmatrix} \right\|$$

and call the result M. Then M multiplied by

$$\left\| \begin{vmatrix}4\,5\\4\,5\end{vmatrix} \begin{vmatrix}3\,5\\3\,5\end{vmatrix} \cdots \begin{vmatrix}1\,2\\1\,2\end{vmatrix} \right\|$$

gives a result having all zeros in the 6th column.

We could of course make other substitutions at the same time.

EXERCISE. Do any of the minors of M in the preceding example vanish? If so what ones?

214. If

$$\Delta \equiv |\, a_1\, b_2\, c_3\, d_4 \,|$$

has a nullity three, that is if Δ vanishes and all minors of order two and three also vanish, then

$$\Delta' \equiv \begin{vmatrix} a_1 & a_2 & a_3 & a_4 & x_1 & y_1 & z_1 \\ b_1 & b_2 & b_3 & b_4 & x_2 & y_2 & z_2 \\ c_1 & c_2 & c_3 & c_4 & x_3 & y_3 & z_3 \\ d_1 & d_2 & d_3 & d_4 & x_4 & y_4 & z_4 \\ \alpha_1 & \alpha_2 & \alpha_3 & \alpha_4 & \xi_1 & \xi_2 & \xi_3 \\ \beta_1 & \beta_2 & \beta_3 & \beta_4 & \eta_1 & \eta_2 & \eta_3 \\ \gamma_1 & \gamma_2 & \gamma_3 & \gamma_4 & \zeta_1 & \zeta_2 & \zeta_3 \end{vmatrix}$$

$$(1) \qquad = \frac{1}{a_1}\begin{vmatrix} a_1 & x_1 & y_1 & z_1 \\ b_1 & x_2 & y_2 & z_2 \\ c_1 & x_3 & y_3 & z_3 \\ d_1 & x_4 & y_4 & z_4 \end{vmatrix}\begin{vmatrix} a_1 & a_2 & a_3 & a_4 \\ \alpha_1 & \alpha_2 & \alpha_3 & \alpha_4 \\ \beta_1 & \beta_2 & \beta_3 & \beta_4 \\ \gamma_1 & \gamma_2 & \gamma_3 & \gamma_4 \end{vmatrix}.$$

Because of the nullity of Δ it follows that Δ' is independent of the elements of the minor

$$\begin{vmatrix} \xi_1 & \xi_2 & \xi_3 \\ \eta_1 & \eta_2 & \eta_3 \\ \zeta_1 & \zeta_2 & \zeta_3 \end{vmatrix}$$

and therefore they may be replaced by zeros. If now we take Δ' thus modified and consider the minor complementary to a_1, which we may denote by A_1, it is obvious that

$$(2) \qquad A_1 = \begin{vmatrix} \alpha_2 & \alpha_3 & \alpha_4 \\ \beta_2 & \beta_3 & \beta_4 \\ \gamma_2 & \gamma_3 & \gamma_4 \end{vmatrix}\begin{vmatrix} x_2 & y_2 & z_2 \\ x_3 & y_3 & z_3 \\ x_4 & y_4 & z_4 \end{vmatrix}$$

If now we operate on (2) by the Law of Extensible Minors using the first row and column of Δ' we arrive at the relation (1).

The general theorem which may be proved in the same way is:

If an n-line determinant of nullity p have affixed to it a p-line border the product of the resulting determinant by any $(n-p)$-line minor of the original is expressible as the product of two n-line determinants.

215. If P and Q are any two determinants of order n and $\bar{Q}$ is the conjugate of Q, then

$$(QP\bar{Q})^{(m)}_{\nu s} = \sum r[Q^{(m)}_{\nu r}\{P^{(m)}_{r1}Q^{(m)}_{s1} + P^{(m)}_{r2}Q^{(m)}_{s2} + \cdots + P^{(m)}_{r\mu}Q^{(m)}_{s\mu}\}]$$

where $\mu = (n)_m$ and where $A^{(m)}{}_{rs}$ denotes the element in the position (r, s) in the mth compound of A.

This is obtained by first expressing the m-line minor of $Q(P\bar{Q})$ in terms of m-line minors of P and $P\bar{Q}$; then expressing the m-line minors of $P\bar{Q}$ in terms of m-line minors of P and Q.

If we put $\nu = s$ we get the expression for the m-line coaxial minors of $QP\bar{Q}$ and thence for the sum of all such minors the expression

$$\sum r \sum s[Q_{sr}^{(m)}\{P_{r1}^{(m)}Q_{s1}^{(m)} + \cdots + P_{r\mu}^{(m)}Q_{s\mu}^{(m)}\}]$$
$$= \sum r[P_{r1}^{(m)}\cdot M_{r1}^{(m)} + \cdots + P_{r\mu}^{(m)}\cdot M_{r\mu}^{(m)}]$$

where M is the determinant which equals Q^2.

216. The expression

$$|a_1b_2c_3e_4|\,|a_1b_2c_3d_4f_5| - |a_1b_2c_3d_4|\,|a_1b_2c_3e_4f_5| + |a_1b_2c_3f_4|\,|a_1b_2c_3e_4d_5|$$

is seen to the the extensional of

$$e_4|d_4f_5| - d_4|e_4f_5| + f_4|e_4d_5| = 0.$$

Dividing by $|a_1b_2c_3e_4|\,|a_1b_2c_3e_4d_5|$ we get

$$\frac{|a_1b_2c_3d_4f_5|}{|a_1b_2c_3e_4d_5|} - \frac{|a_1b_2c_3d_4|}{|a_1b_2c_3e_4|}\cdot\frac{|a_1b_2c_3e_4f_5|}{|a_1b_2c_3e_4d_5|} + \frac{|a_1b_2c_3f_4|}{|a_1b_2c_3e_4|} = 0.$$

Similarly

$$\frac{|a_1b_2c_3f_4|}{|a_1b_2e_3c_4|} - \frac{|a_1b_2c_3|}{|a_1b_2e_3|}\cdot\frac{|a_1b_2e_3f_4|}{|a_1b_2e_3c_4|} - \frac{|a_1b_2f_3|}{|a_1b_2e_3|} = 0$$

$$\frac{|a_1b_2f_3|}{|a_1e_2b_3|} - \frac{|a_1b_2|}{|a_1e_2|}\cdot\frac{|a_1e_2f_3|}{|a_1e_2b_3|} + \frac{|a_1f_2|}{|a_1e_2|} = 0$$

$$\frac{|a_1f_2|}{|e_1a_2|} - \frac{a_1}{e_1}\cdot\frac{|e_1f_2|}{|e_1a_2|} + \frac{f_1}{e_1} = 0.$$

Exercises. Show that

$$\frac{|a_3b_2|}{|a_1b_2|} = \frac{a_3}{a_1} + \frac{|a_3b_1|}{|a_1b_2|}\cdot\frac{a_2}{a_1}$$

$$\frac{|a_4b_2c_3|}{|a_1b_2c_3|} = \frac{a_4}{a_1} + \frac{|a_4b_1|}{|a_1b_2|}\cdot\frac{a_2}{a_1} + \frac{|a_4b_1c_2|}{|a_1b_2c_3|}\cdot\frac{|a_2b_3|}{|a_1b_2|}$$

$$\frac{|a_5b_2c_3d_4|}{|a_1b_2c_3d_4|} = \frac{a_5}{a_1} + \frac{|a_5b_1|}{|a_1b_2|}\cdot\frac{a_2}{a_1} - \frac{|a_5b_1c_2|}{|a_1b_2c_3|}\cdot\frac{|a_2b_3|}{|a_1b_2|} + \frac{|a_5b_1c_2d_3|}{|a_1b_2c_3d_4|}\cdot\frac{|a_2b_3c_4|}{|a_1b_2c_3|}$$

etc. (Hansen)

also show that

$$\frac{|a_1 b_2 c_3 d_4|}{|b_2 c_3 d_4|} + \frac{|a_2 c_3 d_4|}{|b_2 c_3 d_4|} \cdot \frac{|b_1 c_3 d_4|}{|c_3 d_4|} + \frac{|a_3 d_4|}{|c_3 d_4|} \cdot \frac{|c_1 d_4|}{d_4} + \frac{a_4 d_1}{d_4} = a_1$$

(Kronecker)

EXERCISES. SET XIII

1. If

$$\Delta \equiv |a_1 b_2 c_3 d_4| \text{ and } D \equiv |A_1 B_2 C_3 D_4|$$

is its adjugate show that

$$\begin{vmatrix} A_1A_2 & A_2A_3 & A_3A_4 & A_4A_1 \\ B_1B_2 & B_2B_3 & B_3B_4 & B_4B_1 \\ C_1C_2 & C_2C_3 & C_3C_4 & C_4C_1 \\ D_1D_2 & D_2D_3 & D_3D_4 & D_4D_1 \end{vmatrix} = \Delta^2 \begin{vmatrix} a_2A_2 & a_4A_2 & a_4A_4 & a_2A_4 \\ b_2B_2 & b_4B_2 & b_4B_4 & b_2B_4 \\ c_2C_2 & c_4C_2 & c_4C_4 & c_2C_4 \\ d_2B_2 & d_4D_2 & d_4D_4 & d_2D_4 \end{vmatrix}$$

$$= -\Delta^4 \begin{vmatrix} a_1a_2 & a_2a_3 & a_3a_4 & a_4a_1 \\ b_1b_2 & b_2b_3 & b_3b_4 & b_4b_1 \\ c_1c_2 & c_2c_3 & c_3c_4 & c_4c_1 \\ d_1d_2 & d_2d_3 & d_3d_4 & d_4d_1 \end{vmatrix}.$$

2. $$\begin{vmatrix} a_2A_2 & a_2A_4 & a_4A_4 & a_4A_2 \\ b_2B_2 & b_2B_4 & b_4B_4 & b_4B_2 \\ c_2C_2 & c_2C_4 & c_4C_4 & c_4C_2 \\ d_2D_2 & d_2D_4 & d_4D_4 & d_4D_2 \end{vmatrix} = \begin{vmatrix} a_1A_1 & a_1A_3 & a_3A_1 & a_3A_3 \\ b_1B_1 & b_1B_3 & b_3B_1 & b_3B_3 \\ c_1C_1 & c_1C_3 & c_3C_1 & c_3C_3 \\ d_1D_1 & d_1D_3 & d_3D_1 & d_3D_3 \end{vmatrix}.$$

3. $$\begin{aligned} &|A_2A_3A_4 \quad B_1B_3B_4 \quad C_1C_2C_4 \quad D_1D_2D_3| \\ = \Delta^2 &|a_2A_3A_4 \quad b_1B_3B_4 \quad c_4C_1C_2 \quad d_3D_1D_2| \end{aligned}$$

4. $$\begin{aligned} &|a_2a_3a_4 \quad b_1b_3b_4 \quad c_1c_2c_4 \quad d_1d_2d_3|\Delta^2 \\ = &|A_2a_3a_4 \quad B_1b_3b_4 \quad C_4c_1c_2 \quad D_3d_1d_2| \end{aligned}$$

5. If we put

$$X_{h,k} \text{ for } |y_h z_k|,$$
$$Y_{h,k} \text{ for } |z_h x_k|,$$
$$Z_{h,k} \text{ for } |x_h y_k|,$$

and $|Y_{hk}Z_{mn}|$ for $Y_{hk}Z_{mn}-Y_{mn}Z_{hk}$, then show that

$$\Delta \equiv \begin{vmatrix} |Y_{12}Z_{45}| & |Z_{12}X_{45}| & |X_{12}Y_{45}| \\ |Y_{34}Z_{61}| & |Z_{34}X_{61}| & |X_{34}Y_{61}| \\ |Y_{56}Z_{23}| & |Z_{56}X_{23}| & |X_{56}Y_{23}| \end{vmatrix}$$

$$= |x_1\,y_2\,z_6| \cdot |x_1\,y_3\,z_4| \cdot |x_2\,y_3\,z_5| \cdot |x_4\,y_5\,z_6|$$
$$- |x_1\,y_2\,z_3| \cdot |z_1\,y_4\,z_6| \cdot |x_2\,y_5\,z_6| \cdot |z_3\,y_4\,z_5|$$

Interpret the result geometrically.

(Hunyady)

6. If $|X_1Y_2Z_3|$, $|X_4Y_5Z_6|$ are the adjugates of $|x_1y_2z_3|$, $|x_4y_5z_6|$ respectively, show that

$$\big|\,|Y_1Z_4|\,|Z_2X_5|\,|X_3Y_6|\,\big|$$
$$= |x_1y_2z_3| \cdot |x_4\,y_5\,z_6| \cdot \big|\,|y_1\,z_4|\,|z_2\,x_5|\,|x_3\,y_6|\,\big|$$

Interpret the result geometrically.

(Hunyady)

7. Show that

$$\begin{vmatrix} |x_2\,y_3\,z_5|\,|x_2\,y_3\,z_6| & |x_2\,y_3\,z_6|\,|x_2\,y_3\,z_4| & |x_2\,y_3\,z_4|\,|x_2\,y_3\,z_5| \\ |x_3\,y_1\,z_5|\,|x_3\,y_1\,z_6| & |x_3\,y_1\,z_6|\,|x_3\,y_1\,z_4| & |x_3\,y_1\,z_4|\,|x_3\,y_1\,z_5| \\ |x_1\,y_2\,z_5|\,|x_1\,y_2\,z_6| & |x_1\,y_2\,z_6|\,|x_1\,y_2\,z_4| & |x_1\,y_2\,z_4|\,|x_1\,y_2\,z_5| \end{vmatrix}$$
$$= |x_1\,y_2\,z_3|^2\,|x_1^2\;y_2^2\;z_3^2\;y_4z_4\;z_5x_5\;x_6y_6|$$

(Pasch)

8. If P represents the second determinant on the right in problem 7, and if

$$Q \equiv \begin{vmatrix} x_1x_2 & y_1y_2 & z_1z_2 & y_1z_2-y_2z_1 & z_1x_2-z_2x_1 & x_1y_2-x_2y_1 \\ x_2x_3 & y_2y_3 & z_2z_3 & y_2z_3-y_3z_2 & z_2x_3-z_3x_2 & x_2y_3-x_3y_2 \\ \cdots & \cdots & \cdots & \cdots & \cdots & \cdots \\ x_6x_1 & y_6y_1 & z_6z_1 & y_6z_1-y_1z_6 & z_6x_1-z_1x_6 & x_6y_1-x_1y_6 \end{vmatrix}$$

then prove the relation

$$\Delta = -P = Q,$$

where Δ is the determinant so designated in problem 5.

CHAPTER VII

Rectangular Arrays

217. *If there be two sets of elements both consisting of m rows of n elements (n being greater than m) and a determinant be formed from them in the way in which the product of two determinants is formed, multiplying row by row,* then this determinant is equal to the sum of every product whose first factor is a determinant obtained by taking m columns of the first set of elements, and whose other factor is the determinant obtained by taking the corresponding m columns of the second set.*

Let the two sets of elements be

$$\left\{\begin{matrix} a_{11} & a_{12} & \cdots & a_{1m} & \cdots & a_{1n} \\ a_{21} & a_{22} & \cdots & a_{2m} & \cdots & a_{2n} \\ \cdots & \cdots & \cdots & \cdots & \cdots & \cdots \\ a_{m1} & a_{m2} & \cdots & a_{mm} & \cdots & a_{mn}, \end{matrix}\right.$$

denoted by (a_{mn}) and

$$\left\{\begin{matrix} b_{11} & b_{12} & \cdots & b_{1m} & \cdots & b_{1n} \\ b_{21} & b_{22} & \cdots & b_{2m} & \cdots & b_{2n} \\ \cdots & \cdots & \cdots & \cdots & \cdots & \cdots \\ b_{m1} & b_{m2} & \cdots & b_{mm} & \cdots & b_{mn}, \end{matrix}\right.$$

denoted by (b_{mn}) so that the determinant referred to, C say, is

$$\begin{vmatrix} a_{11}b_{11}+\cdots+a_{1n}b_{1n} & a_{11}b_{21}+\cdots+a_{1n}b_{2n} & \cdots & a_{11}b_{m1}+\cdots+a_{1n}b_{mn} \\ a_{21}b_{11}+\cdots+a_{2n}b_{1n} & a_{21}b_{21}+\cdots+a_{2n}b_{2n} & \cdots & a_{21}b_{m1}+\cdots+a_{2n}b_{mn} \\ \cdots & \cdots & \cdots & \cdots \\ a_{m1}b_{11}+\cdots+a_{mn}b_{1n} & a_{m1}b_{21}+\cdots+a_{mn}b_{2n} & \cdots & a_{m1}b_{m1}+\cdots+a_{mn}b_{mn} \end{vmatrix}.$$

Fixing upon any m terms of the first element of C, let us delete the other $n-m$ terms and the corresponding $n-m$ terms in all the other elements, and call the resulting determinant δ_1: again fixing upon any other m terms of the first element of C, let us proceed in the same way, and call the resulting determinant δ_2: and so on. For example, if it be the last $n-m$ terms which are deleted in obtaining δ_1,

* Either of the other ways of multiplying indicated in §154 would be the same as multiplying by determinants having rows of zeros. Thus the product $(a_{mn})(b_{mn})$ column-by-column gives the same result as the product of the two determinants $|a_{1n}|\,|b_{1n}|$, where in each of the last $n-m$ rows are zeros.

$$\delta_1 = \begin{vmatrix} a_{11}b_{11} + \cdots + a_{1m}b_{1m} & a_{11}b_{21} + \cdots + a_{1m}b_{2m} & \cdots & a_{11}b_{m1} + \cdots + a_{1m}b_{mm} \\ a_{21}b_{11} + \cdots + a_{2m}b_{1m} & a_{21}b_{21} + \cdots + a_{2m}b_{2m} & \cdots & a_{21}b_{m1} + \cdots + a_{2m}b_{mm} \\ \cdots\cdots & \cdots\cdots & \cdots & \cdots\cdots \\ a_{m1}b_{11} + \cdots + a_{mm}b_{1m} & a_{m1}b_{21} + \cdots + a_{mm}b_{2m} & \cdots & a_{m1}b_{m1} + \cdots + a_{mm}b_{mm} \end{vmatrix}$$

The number of such determinants is evidently the number of combinations of n things taken m at a time, that is,

$$\frac{n(n-1)(n-2)\cdots(n-m+1)}{m!}.$$

Returning to C we see that as each element consists of n terms and there are m columns, C may be partitioned into n^m determinants, the columns of each of which are got by taking from each of the columns of C a set of terms in the same vertical line. Of these determinants, however, those which do not vanish are in number only the number of permutations of n things taken m at a time, so that C is equal to the sum of $n(n-1)\cdots(n-m+1)$ determinants with monomial elements.

Now each of these monomial-element determinants is part of one of the determinants of the δ series, namely, that one in which its own columns occur as parts of columns. For example, $|(a_{11}b_{11})(a_{22}b_{22})\cdots(a_{mm}b_{mm})|$ is a part of δ_1, being one of the monomial-element determinants into which δ_1 can be partitioned. In other words, the monomial-element determinant which is formed by taking from the first column of C the set of xth terms, from the second column of C the set of yth terms, and so on, is part of that member of the δ series which is obtained by retaining the xth, yth, terms of each element of C and deleting the rest. Hence C is a part of $\delta_1+\delta_2+\cdots$. But when each member of the δ series is partitioned like C into monomial-element determinants, the number of those which do not vanish is (§175) $m!$ for each case and therefore for the whole series is

$$m!\frac{n(n-1)\cdots(n-m+1)}{m!} \quad \text{or} \quad n(n-1)\cdots(n-m+1),$$

that is to say, exactly the number of non-evanescent monomial-element determinants into which C can be partitioned.

Therefore $$C = \delta_1 + \delta_2 + \cdots.$$

Now (§153) $\delta_1, \delta_2, \cdots$, are each the product of two determinants, namely, that member of the δ series which is obtained by retaining the xth, yth, $\cdots$, terms of each element of C and deleting the rest is the product of the two determinants whose columns are the xth, yth, $\cdots$, columns of the first and second given set of elements respectively. Thus the theorem is established.

218. This theorem may also be proven simply as follows: Let the product of (a_{mn}) and (b_{mn}) be $|c_{mm}| \equiv C$ where

$$c_{ij} = a_{j1}b_{1i} + a_{j2}b_{2i} + \cdots + a_{jn}b_{ni}.$$

But

$$|c_{mm}| = \begin{vmatrix} c_{11} & c_{12} & \cdots & c_{1m} & b_{11} & b_{21} & \cdots & b_{n1} \\ c_{21} & c_{22} & \cdots & c_{2m} & b_{12} & b_{22} & \cdots & b_{n2} \\ \cdots & \cdots & \cdots & \cdots & \cdots & \cdots & \cdots & \cdots \\ c_{m1} & c_{m2} & \cdots & c_{mm} & b_{1m} & b_{2m} & \cdots & b_{nm} \\ 0 & 0 & \cdots & 0 & 1 & 0 & \cdots & 0 \\ \cdot & \cdot & \cdots & & \cdot & \cdot & \cdots & \cdot \\ 0 & 0 & \cdots & 0 & 0 & 0 & \cdots & 1 \end{vmatrix}$$

Multiplying the last n columns by $a_{i1}, \cdots, a_{in}$ respectively and subtracting the sum from the ith column we have*

$$C = (-1)^m \begin{vmatrix} 0 & \cdots & 0 & b_{11} & \cdots & b_{n1} \\ 0 & \cdots & 0 & b_{12} & \cdots & b_{n2} \\ \cdots & \cdots & \cdots & \cdots & \cdots & \cdots \\ 0 & \cdots & 0 & b_{1m} & \cdots & b_{nm} \\ a_{11} & \cdots & a_{m1} & 1 & \cdots & 0 \\ \cdots & \cdots & \cdots & \cdots & \cdots & \cdots \\ a_{1n} & \cdots & a_{mn} & 0 & \cdots & 1 \end{vmatrix}.$$

Expanding C in terms of minors of order m taken from the first m columns and their complementaries we have a series of products like α_i and β_i, where α_i and β_i are corresponding determinants from the matrices (a_{mn}) and (b_{mn}) respectively. The sign if every product is obviously positive for since α_i and β_i are corresponding determinants, whatever sign that is attached to one would be to the other and there would be $(-1)^m$ due to bringing α_i up into the left-hand corner, which together with the $(-1)^m$ outside gives $(-1)^{2m} = +1$.

* It is worthy of note to observe that the product of two arrays may be written as a determinant of order $(m+n)$.

219. In §217 if the condition $n>m$ was not fulfilled we could add columns of zeros to each of the two arrays to make the number of rows and columns the same. The resulting product determinant of the mth order would then be obviously zero since it would be the product of two determinants, both of which would be zero.

220. *If a rectangular array has two rows identical then its square will have two rows identical and therefore vanish. The arrays may be looked upon as entities having the value zero or otherwise.*

EXAMPLE. Formed from the two sets of elements

$$\begin{matrix} a & b & c \\ a' & b' & c', \end{matrix} \qquad \begin{matrix} x & y & z \\ x' & y' & z', \end{matrix}$$

the determinant

$$\begin{vmatrix} ax+by+cz & ax'+by'+cz' \\ a'x+b'y+c'z & a'x'+b'y'+c'z' \end{vmatrix}$$

$$= \begin{vmatrix} ax+by & ax'+by' \\ a'x+b'y & a'x'+b'y' \end{vmatrix} + \begin{vmatrix} ax+cz & ax'+cz' \\ a'x+c'z & a'x'+c'z' \end{vmatrix} + \begin{vmatrix} by+cz & by'+cz' \\ b'y+c'z & b'y'+c'z' \end{vmatrix}$$

$$= \begin{vmatrix} a & b \\ a' & b' \end{vmatrix}\begin{vmatrix} x & y \\ x' & y' \end{vmatrix} + \begin{vmatrix} a & c \\ a' & c' \end{vmatrix}\begin{vmatrix} x & z \\ x' & z' \end{vmatrix} + \begin{vmatrix} b & c \\ b' & c' \end{vmatrix}\begin{vmatrix} y & z \\ y' & z' \end{vmatrix}.$$

Although here the determinant with trinomial elements is expressed as the sum of three determinants with binomial elements, it must be noticed that this is not possible in the case of every determinant with trinomial elements. The general theorem in fact is

$$\begin{vmatrix} a_1+b_1+c_1 & x_1+y_1+z_1 \\ a_2+b_2+c_2 & x_2+y_2+z_2 \end{vmatrix} = \begin{vmatrix} a_1+b_1 & x_1+y_1 \\ a_2+b_2 & x_2+y_2 \end{vmatrix}$$

$$+ \begin{vmatrix} a_1+c_1 & x_1+z_1 \\ a_2+c_2 & x_2+z_2 \end{vmatrix}$$

$$+ \begin{vmatrix} b_1+c_1 & y_1+z_1 \\ b_2+c_2 & y_2+z_2 \end{vmatrix} - \begin{vmatrix} a_1 & x_1 \\ a_2 & x_2 \end{vmatrix} - \begin{vmatrix} b_1 & y_1 \\ b_2 & y_2 \end{vmatrix} - \begin{vmatrix} c_1 & y_1 \\ c_2 & z_2 \end{vmatrix}.$$

221. The theorem of §217 may be written

$$(1) \qquad C = \sum_{i=1}^{i=n_m} A_{(n|m_1),(n|m_i)} B_{(n|m_1),(n|m_i)}.$$

If the two sets of elements are the same then

$$(2) \qquad C = \sum_{i=1}^{i=n_m} A^2_{(n|m_1),(n|m_i)}.$$

EXAMPLE:

$$\begin{vmatrix} a^2+b^2+c^2 & aa'+bb'+cc' \\ aa'+bb'+cc' & a'^2+b'^2+c'^2 \end{vmatrix} = \begin{vmatrix} a & b \\ a' & b' \end{vmatrix}^2 + \begin{vmatrix} a & c \\ a' & c' \end{vmatrix}^2 + \begin{vmatrix} b & c \\ b' & c' \end{vmatrix}^2.$$

222. If $m=n$ the theorem of §217 reduces to that of §155. It is also worthy of note that the minors of the determinant product C of §153 have the form of the determinant C dealt with in §217. As a consequence of which we have the theorem:

Any minor of a product determinant is expressed as a sum of products of minors of the two factors, thus

$$(1) \quad C_{(n|m\alpha),(n|m\beta)} = \sum_{i=1}^{i=\lambda} A_{(n|m\alpha),(n|m_i)} B_{(n|m\beta),(n|m_i)} (\lambda = (n)_m)$$

where $C_{(n|m_\alpha),(n|m\beta)}$ denotes the minor of $|c_{ik}|$ of order m formed from the αth selection of m rows and the βth selection of m columns.

This shows that not only is the mth compound of the product of two determinants equal to the product of the mth compounds of the two factors but that they are equal element for element.

If we sum both sides of (1) with respect to α and β we get

$$\sum_{\beta=1}^{\lambda} \sum_{\alpha=1}^{\lambda} C_{(n|m\alpha),(n|m\beta)} = \sum_{\beta=1}^{\lambda} \sum_{\alpha=1}^{\lambda} \sum_{i=1}^{\lambda} A_{(n|m\alpha),(n|m_i)} B_{(n|m\beta),(n|m_i)}$$

$$= \sum_{i=1}^{\lambda} \left[\sum_{\alpha=1}^{\lambda} A_{(n|m\alpha),(n|m_i)} \sum_{\beta=1}^{\lambda} B_{(n|m\beta),(n|m_i)} \right]$$

or as it may as well be written

$$(2) \quad \sum_{\alpha=1}^{\lambda} \sum_{\beta=1}^{\lambda} C_{(n|m\alpha),(n|m\beta)} = \sum_{\beta=1}^{\lambda} \left[\sum_{\alpha=1}^{\lambda} A_{(n|m\alpha),(n|m\beta)} \sum_{\alpha=1}^{\lambda} B_{(n|m\alpha),(n|m\beta)} \right]$$

If $m=1$, then this reduces to the theorem that *if there be two determinants, and the sum of the elements of one first row be multiplied by the sum of the elements of the other first row, the sum of the elements of one second row by the sum of the elements of the other second row, and so on, then the sum of these products is equal to the sum of the elements of the product of the two determinants.*

223. In the case of coaxial minors we have

$$(3)\qquad C_{(n|m_j),(n|m_j)} = \sum_{i=1}^{\lambda} A_{(n|m_j),(n|m_i)} B_{(n|m_j),(n|m_i)}.$$

Taking the sum of all the coaxial minors of order m we have

$$(4)\qquad \sum_{j=1}^{\lambda} C_{(n|m_j),(n|m_j)} = \sum_{j=1}^{\lambda} \sum_{i=1}^{\lambda} A_{(n|m_j|,(n|m_i)} B_{(n|m_j),(n|m_i)}.$$

That is, *the sum of all the coaxial minors of order m in C is equal to the sum of the product of all minors of order m of A by the corresponding minors of B.*

It follows from this that the sum of the coaxial minors of order m in the product of A and B is the same however the product be formed, row by row, column by column, $A \cdot B$, $B \cdot A$, etc. and hence the theorem:

The sum of the m-line coaxial minors of the determinant which is the product of $|a_{1n}|$, $|b_{1n}|$, $|c_{1n}|$, $\cdots$, *is not altered when the factors are cyclically permuted.*

224. If $|a_{1n}| \equiv |b_{1n}|$, then §222 becomes

$$(1')\qquad C_{(n|m\alpha),(n|m\beta)} = \sum_{i=1}^{\lambda} A_{(n|m\alpha),(n|m_i)} A_{(n|m\beta),(n|m_i)}$$

and consequently

$$\sum_{\beta=1}^{\lambda} \sum_{\alpha=1}^{\lambda} C_{(n|m\alpha),(n|m\beta)} = \sum_{\alpha=1}^{\lambda} \left[\sum_{\beta=1}^{\lambda} A_{(n|m\beta),(n|m\alpha)} \right]^2,$$

(3) §223 becomes

$$(3')\qquad C_{(n|m_j),(n|m_j)} = \sum_{i=1}^{\lambda} A^2_{(n|m_j),(n|m_i)}$$

and (4) becomes

$$(4')\qquad \sum_{j=1}^{\lambda} C_{(n|m_j),(n|m_j)} = \sum_{j=1}^{\lambda} \sum_{i=1}^{\lambda} A^2_{(n|m_j),(n|m_i)}$$

In (1′) we have an expansion for any minor of the square of a determinant in terms of minors of the same order of the determinant itself. For a coaxial minor this expansion takes the form given in (3′). In (4′) we have a similar expression for the sum of the coaxial minors of the square of a determinant.

If $C_{(n|m_j),(n|m_j)}=0$, then $A_{(n|m_j),(n|m_i)}=0$ (for $i=1, 2, \cdots, \lambda$), by (3′).

If $\sum_{j=1}^{\lambda} C_{(n|m_j),(n|m_j)}=0$, then $A_{(n|m_j),(n|m_i)}=0$ for $\{i,j=1, 2, \cdots,\lambda\}$. That is, *if the sum of the coaxial minors of order m in the square of a determinant A is zero, then every minor of order m of A is zero.*

225. If we multiply the two arrays (a_{35}) and (b_{35}) we have

$$(a_{35})(b_{35}) = \begin{vmatrix} c_{11} & c_{12} & c_{13} \\ c_{21} & c_{22} & c_{23} \\ c_{31} & c_{32} & c_{33} \end{vmatrix} \equiv C$$

where

$$c_{ij} = a_{i1}b_{j1} + a_{i2}b_{j2} + a_{i3}b_{j3} + a_{i4}b_{j4} + a_{i5}b_{j5} = c'_{ij} + a_{i4}b_{j4} + a_{i5}b_{i5},$$

say,

then

$$C \equiv \begin{vmatrix} c'_{11}+a_{14}b_{14}+a_{15}b_{15} & c'_{12}+a_{14}b_{24}+a_{15}b_{25} & c'_{13}+a_{14}b_{34}+a_{15}b_{35} \\ c'_{21}+a_{24}b_{14}+a_{25}b_{15} & c'_{22}+a_{24}b_{24}+a_{25}b_{25} & c'_{23}+a_{24}b_{34}+a_{25}b_{35} \\ c'_{31}+a_{34}b_{14}+a_{35}b_{15} & c'_{32}+a_{34}b_{24}+a_{35}b_{25} & c'_{33}+a_{34}b_{34}+a_{35}b_{35} \end{vmatrix}$$

$$= \begin{vmatrix} c'_{11} & c'_{12} & c'_{13} & a_{14} & a_{15} \\ c'_{11} & c'_{22} & c'_{23} & a_{24} & a_{25} \\ c'_{31} & c'_{32} & c'_{33} & a_{34} & a_{35} \\ b_{14} & b_{24} & b_{34} & -1 & 0 \\ b_{15} & b_{25} & b_{35} & 0 & -1 \end{vmatrix} = \begin{vmatrix} c'_{11} & c'_{12} & c'_{13} \\ c'_{21} & c'_{22} & c'_{23} \\ c'_{31} & c'_{32} & c'_{33} \end{vmatrix}$$

$$- \begin{vmatrix} c'_{11} & c'_{12} & c'_{13} & a_{14} \\ c'_{21} & c'_{22} & c'_{23} & a_{24} \\ c'_{31} & c'_{32} & c'_{33} & a_{34} \\ b_{14} & b_{24} & b_{34} & 0 \end{vmatrix} - \begin{vmatrix} c'_{11} & c'_{12} & c'_{13} & a_{15} \\ c'_{21} & c'_{22} & c'_{23} & a_{25} \\ c'_{31} & c'_{32} & c'_{33} & a_{35} \\ b_{15} & b_{25} & b_{35} & 0 \end{vmatrix}$$

$$+ \begin{vmatrix} c'_{11} & c'_{12} & c'_{13} & a_{14} & a_{15} \\ c'_{21} & c'_{22} & c'_{23} & a_{24} & a_{25} \\ c'_{31} & c'_{32} & c'_{33} & a_{34} & a_{35} \\ b_{14} & b_{24} & b_{34} & 0 & 0 \\ b_{15} & b_{25} & b_{35} & 0 & 0 \end{vmatrix}$$

In this way we see that the product of two arrays may be written as the sum of single determinants.

226. We have, by §217.

$$\begin{Vmatrix} h_1a_{11} & h_2a_{12} & \cdots & h_na_{1n} \\ h_1a_{21} & h_2a_{22} & \cdots & h_na_{2n} \\ \cdots & \cdots & \cdots & \cdots \\ h_1a_{m1} & h_2a_{m2} & \cdots & h_na_{mn} \end{Vmatrix} \cdot \begin{Vmatrix} b_{11} & b_{12} & \cdots & b_{1n} \\ b_{21} & b_{22} & \cdots & b_{2n} \\ \cdots & \cdots & \cdots & \cdots \\ b_{m1} & b_{m2} & \cdots & b_{mn} \end{Vmatrix}$$

$$= \begin{vmatrix} h_1a_{11}\,b_{11}+h_2a_{12}\,b_{12}+\cdots+h_na_{1n}\,b_{1n} & \cdots & h_1a_{11}\,b_{m1}+h_2a_{12}\,b_{m2}+\cdots+h_na_{1n}\,b_{mn} \\ \cdots & \cdots & \cdots \\ h_1a_{m1}b_{11}+h_2a_{m2}b_{12}+\cdots+h_na_{mn}b_{1n} & \cdots & h_1a_{m1}\,b_{m1}+h_2a_{m2}\,b_{m2}+\cdots+h_na_{mn}b_{mn} \end{vmatrix}$$

$$= h_1h_2\cdots h_m \begin{vmatrix} a_{11} & a_{12} & \cdots & a_{1m} \\ a_{21} & a_{22} & \cdots & a_{2m} \\ \cdots & \cdots & \cdots & \cdots \\ a_{m1} & a_{m2} & \cdots & a_{mm} \end{vmatrix} \cdot \begin{vmatrix} b_{11} & b_{12} & \cdots & b_{1m} \\ b_{21} & b_{22} & \cdots & b_{2m} \\ \cdots & \cdots & \cdots & \cdots \\ b_{m1} & b_{m2} & \cdots & b_{mm} \end{vmatrix}$$

$$+ h_1h_2\cdots h_{m-1}h_{m+1} \begin{vmatrix} a_{11} & \cdots & a_{1,m-1} & a_{1,m+1} \\ a_{21} & \cdots & a_{2,m-1} & a_{2,m+1} \\ \cdots & \cdots & \cdots & \cdots \\ a_{m1} & \cdots & a_{m,m-1} & a_{m,m+1} \end{vmatrix}$$

$$\times \begin{vmatrix} b_{11} & \cdots & b_{1,m-1} & b_{1,m+1} \\ b_{21} & \cdots & b_{2,m-1} & b_{2,m+1} \\ \cdots & \cdots & \cdots & \cdots \\ b_{m1} & \cdots & b_{m,m-1} & b_{m,m+1} \end{vmatrix}$$

$+$ etc.

If the b's are the same as the a's this becomes

$$(2)\qquad \begin{Vmatrix} h_1a_{11} & h_2a_{12} & \cdots & h_na_{1n} \\ h_1a_{21} & h_2a_{22} & \cdots & h_na_{2n} \\ \cdots & \cdots & \cdots & \cdots \\ h_1a_{m1} & h_2a_{m2} & \cdots & h_na_{mn} \end{Vmatrix} \cdot \begin{Vmatrix} a_{11} & a_{12} & \cdots & a_{1n} \\ a_{21} & a_{22} & \cdots & a_{2n} \\ \cdots & \cdots & \cdots & \cdots \\ a_{m1} & a_{m2} & \cdots & a_{mn} \end{Vmatrix}$$

$$= \begin{vmatrix} \sum h_i a_{1i}^2 & \cdots & \sum h_i a_{1i} a_{mi} \\ \cdots & \cdots & \cdots \\ \sum h_i a_{mi} a_{1i} & \cdots & \sum h_i a_{mi}^2 \end{vmatrix}$$

$$= h_1 h_2 \cdots h_m \begin{vmatrix} a_{11} & a_{12} \cdots & a_{1m} \\ \cdots & \cdots & \cdots \\ a_{m1} & a_{m2} \cdots & a_{mm} \end{vmatrix}^2 + \text{etc.}$$

If $h_1 = h_2 = \cdots = h_n = 1$, we have

$$(3) \quad (a_{mn}) \cdot (b_{mn}) = \sum \begin{vmatrix} a_{11} & a_{12} \cdots & a_{1m} \\ \cdots & \cdots & \cdots \\ a_{m1} & a_{m2} \cdots & a_{mm} \end{vmatrix} \cdot \begin{vmatrix} b_{11} & b_{12} \cdots & b_{1m} \\ \cdots & \cdots & \cdots \\ b_{m1} & b_{m2} \cdots & b_{mm} \end{vmatrix}$$

$$= \begin{vmatrix} \sum a_{1i} b_{1i} & \cdots & \sum a_{1i} b_{mi} \\ \cdots & \cdots & \cdots \\ \sum a_{mi} b_{1i} & \cdots & \sum a_{mi} b_{mi} \end{vmatrix}$$

If the b's are the same as the a's (3) becomes

$$(4) \quad (a_{mn})^2 = \sum \left| a_{1m} \right|^2 = \begin{vmatrix} \sum a_{1i}^2 & \sum a_{1i} a_{2i} \cdots & \sum a_{1i} a_{mi} \\ \cdots & \cdots & \cdots \\ \sum a_{mi} a_{1i} & \sum a_{mi} a_{2i} \cdots & \sum a_{mi}^2 \end{vmatrix}.$$

227. If now

$$\begin{Vmatrix} h_1 a_{11} & h_2 a_{12} \cdots & h_n a_{1n} \\ h_1 a_{21} & h_2 a_{22} \cdots & h_n a_{2n} \\ \cdots & \cdots & \cdots \\ h_1 a_{m1} & h_2 a_{m2} \cdots & h_n a_{mn} \end{Vmatrix} \cdot \begin{Vmatrix} b_{11} & b_{12} \cdots & b_{1n} \\ b_{21} & b_{22} \cdots & b_{2n} \\ \cdots & \cdots & \cdots \\ b_{m1} & b_{m2} \cdots & b_{mn} \end{Vmatrix}$$

$$= \begin{vmatrix} c_{11} & c_{12} \cdots & c_{1m} \\ c_{21} & c_{22} \cdots & c_{2m} \\ \cdots & \cdots & \cdots \\ c_{m1} & c_{m2} \cdots & c_{mm} \end{vmatrix} \equiv \Delta \text{ say},$$

then

$$\begin{Vmatrix} x_1 & h_1 a_{11} & h_2 a_{12} \cdots & h_n a_{1n} \\ x_2 & h_1 a_{21} & h_2 a_{22} \cdots & h_n a_{2n} \\ \cdots & \cdots & \cdots & \cdots \\ x_m & h_1 a_{m1} & h_2 a_{m2} \cdots & h_n a_{mn} \end{Vmatrix} \begin{Vmatrix} x_1 & b_{11} & b_{12} \cdots & b_{1n} \\ x_2 & b_{21} & b_{22} \cdots & b_{2n} \\ \cdots & \cdots & \cdots & \cdots \\ x_m & b_{m1} & b_{m2} \cdots & b_{mn} \end{Vmatrix} \equiv \Delta'$$

$$= \begin{vmatrix} x_1^2+c_{11} & x_1x_2+c_{12} & \cdots & x_1x_m+c_{1m} \\ x_1x_2+c_{21} & x_2^2+c_{22} & \cdots & x_2x_m+c_{2m} \\ \cdots & \cdots & \cdots & \cdots \\ x_1x_m+c_{m1} & x_2x_m+c_{m2} & \cdots & x_m^2+c_{mm} \end{vmatrix}$$

$$= \begin{vmatrix} 1 & 0 & \cdots & 0 \\ x_1 & x_1^2+c_{11} & \cdots & x_1x_m+c_{1m} \\ x_2 & x_1x_2+c_{21} & \cdots & x_2x_m+c_{2m} \\ \cdots & \cdots & \cdots & \cdots \\ x_m & x_1x_m+c_{m1} & \cdots & x_m^2+c_{mm} \end{vmatrix}$$

$$= \begin{vmatrix} 1 & -x_1 & -x_2 & \cdots & -x_m \\ x_1 & c_{11} & c_{12} & \cdots & c_{1m} \\ x_2 & c_{21} & c_{22} & \cdots & c_{2m} \\ \cdots & \cdots & \cdots & \cdots & \cdots \\ x_m & c_{m1} & c_{m2} & \cdots & c_{mm} \end{vmatrix} = \Delta - \begin{vmatrix} 0 & x_1 & \cdots & x_m \\ x_1 & c_{11} & \cdots & c_{1m} \\ \cdots & \cdots & \cdots & \cdots \\ x_m & c_{m1} & \cdots & c_{mm} \end{vmatrix}.$$

But $\Delta'-\Delta$ are just those terms in the expansion of Δ', each factor of which has $x_1x_2 \cdots x_m$ for its first column.
Therefore

$$\begin{vmatrix} 0 & x_1 & \cdots & x_m \\ x_1 & c_{11} & \cdots & c_{1m} \\ x_2 & c_{21} & \cdots & c_{2m} \\ \cdots & \cdots & \cdots & \cdots \\ x_m & c_{m1} & \cdots & c_{mm} \end{vmatrix}$$

$$= - \sum h_\alpha h_\beta \cdots h_\delta \begin{vmatrix} x_1 & a_{1\alpha} & a_{1\beta} & \cdots & a_{1\delta} \\ x_2 & a_{2\alpha} & a_{2\beta} & \cdots & a_{2\delta} \\ \cdots & \cdots & \cdots & \cdots & \cdots \\ x_m & a_{m\alpha} & a_{m\beta} & \cdots & a_{m\delta} \end{vmatrix} \begin{vmatrix} x_1 & b_{1\alpha} & b_{1\beta} & \cdots & b_{1\delta} \\ x_2 & b_{2\alpha} & b_{2\beta} & \cdots & b_{2\delta} \\ \cdots & \cdots & \cdots & \cdots & \cdots \\ x_m & b_{m\alpha} & b_{m\beta} & \cdots & b_{m\delta} \end{vmatrix}$$

where the sigma means that we are to take all possible combinations $(\alpha\beta \cdots \delta)$, of the numbers $1 \cdots n$, $(m-1)$ at a time.

If the two sets of elements are the same then

$$\begin{vmatrix} 0 & x_1 & x_2 & \cdots & x_m \\ x_1 & c_{11} & c_{12} & \cdots & c_{1m} \\ \cdots & \cdots & \cdots & \cdots & \cdots \\ x_m & c_{m1} & c_{m2} & \cdots & c_{mm} \end{vmatrix} = - \sum h_\alpha h_\beta \cdots h_\delta \begin{vmatrix} x_1 & a_{1\alpha} & \cdots & a_{1\delta} \\ x_2 & a_{2\alpha} & \cdots & a_{2\delta} \\ \cdots & \cdots & \cdots & \cdots \\ x_m & a_{m\alpha} & \cdots & a_{m\delta} \end{vmatrix}^2.$$

228. By multiplying the arrays we have

$$\left\|\begin{matrix} 1 & a_{11} & a_{12} \cdots a_{1n} \\ 1 & a_{21} & a_{22} \cdots a_{2n} \\ \cdots & \cdots & \cdots \\ 1 & a_{m1} & a_{m2} \cdots a_{mn} \end{matrix}\right\| \cdot \left\|\begin{matrix} 1 & b_{11} & b_{12} \cdots b_{1n} \\ 1 & b_{21} & b_{22} \cdots b_{2n} \\ \cdots & \cdots & \cdots \\ 1 & b_{m1} & b_{m2} \cdots b_{mn} \end{matrix}\right\|$$

$$- (a_{mn})(b_{mn}) = \begin{vmatrix} 1+\sum a_{11}b_{11} & 1+\sum a_{11}b_{21} & \cdots & 1+\sum a_{11}b_{m1} \\ 1+\sum a_{21}b_{11} & 1+\sum a_{21}b_{21} & \cdots & 1+\sum a_{21}b_{m1} \\ \cdots & \cdots & \cdots & \cdots \\ 1+\sum a_{m1}b_{11} & 1+\sum a_{m1}b_{21} & \cdots & 1+\sum a_{m1}b_{m1} \end{vmatrix}_m$$

$$(1) \qquad - \begin{vmatrix} \sum a_{11}b_{11} & \sum a_{11}b_{21} & \cdots & \sum a_{11}b_{m1} \\ \sum a_{21}b_{11} & \sum a_{21}b_{21} & \cdots & \sum a_{21}b_{m1} \\ \cdots & \cdots & \cdots & \cdots \\ \sum a_{m1}b_{11} & \sum a_{m1}b_{21} & \cdots & \sum a_{m1}b_{m1} \end{vmatrix}_m$$

which difference is readily seen to be equal to

$$(2) \qquad - \begin{vmatrix} \cdot & 1 & 1 & \cdots & 1 \\ 1 & \sum a_{11}b_{11} & \sum a_{11}b_{21} & \cdots & \sum a_{11}b_{m1} \\ \cdots & \cdots & \cdots & \cdots & \cdots \\ 1 & \sum a_{m1}b_{11} & \sum a_{m1}b_{21} & \cdots & \sum a_{m1}b_{m1} \end{vmatrix}_{m+1} \equiv D \text{ say.}$$

If in this we multiply the last m rows by -2, divide the first column by -2 and then add $a^2{}_{ki}$ times the first row to the $(k+1)$st $(k=1,2, \cdots, m)$ and finally add $b^2{}_{1k}$ times the first column to the $(k+1)$st column $(k=1, 2, \cdots, m)$, it is seen that

$$(3) \quad D = \begin{vmatrix} & 1 & 1 & \cdots & 1 \\ 1 & \sum(a_{11}-b_{11})^2 & \sum(a_{11}-b_{21})^2 & \cdots & \sum(a_{11}-b_{m1})^2 \\ 1 & \sum(a_{21}-b_{11})^2 & \sum(a_{21}-b_{21})^2 & \cdots & \sum(a_{21}-b_{m1})^2 \\ \cdots & \cdots & \cdots & \cdots & \cdots \\ 1 & \sum(a_{m1}-b_{11})^2 & \sum(a_{m1}-b_{21})^2 & \cdots & \sum(a_{m1}-b_{m1})^2 \end{vmatrix}_{m+1} .$$

If $n=m-1$, then the second term on the left of (1) vanishes and the first term is the product of two determinants.

If $n<m-1$ both terms on the left are zero.

If $m=5$ and $n=3$ then (3) gives Cayley's relation existing between the distances of five points in space.

Exercises. Set XIV

1. Show that

$$\begin{vmatrix} \cdot & g & h & i \\ g & mx^2+m_1x_1^2+\cdots & mxy+m_1x_1y_1+\cdots & mxz+m_1x_1z_1+\cdots \\ h & mxy+m_1x_1y_1+\cdots & my^2+m_1y_1^2+\cdots & myz+m_1y_1z_1+\cdots \\ i & mxz+m_1x_1z_1+\cdots & myz+m_1y_1z_1+\cdots & mz^2+m_1z_1^2+\cdots \end{vmatrix}$$

$$= mm_1\begin{vmatrix} g & x & x_1 \\ h & y & y_1 \\ i & z & z_1 \end{vmatrix}^2 + mm_2\begin{vmatrix} g & x & x_2 \\ h & y & y_2 \\ i & z & z_2 \end{vmatrix}^2 + \cdots + m_1m_2\begin{vmatrix} g & x_1 & x_2 \\ h & y_1 & y_2 \\ i & z_1 & z_2 \end{vmatrix}^2 + \cdots$$

2. If $u_k = A_kx+B_ky+C_kz=0$ $(k=1, 2, \cdots, 6)$ represent six planes, show that the condition that $u_1, u_2; u_3, u_4; u_5, u_6$ lie in the same plane is

$$\begin{vmatrix} A_1 & A_2 & A_3 & A_4 & \cdot & \cdot \\ B_1 & B_2 & B_3 & B_4 & \cdot & \cdot \\ C_1 & C_2 & C_3 & C_4 & \cdot & \cdot \\ \cdot & \cdot & A_3 & A_4 & A_5 & A_6 \\ \cdot & \cdot & B_3 & B_4 & B_5 & B_6 \\ \cdot & \cdot & C_3 & C_4 & C_5 & C_6 \end{vmatrix} = 0.$$

3. Show that

$$\begin{vmatrix} |y_1z_2| & |y_4z_5| & |y_2z_3| & |y_5z_6| & \cdot & \cdot \\ |z_1x_2| & |z_4x_5| & |z_2x_3| & |z_5x_6| & \cdot & \cdot \\ |x_1y_2| & |x_4y_5| & |x_2y_3| & |x_5y_6| & \cdot & \cdot \\ \cdot & \cdot & |y_2z_3| & |y_5z_6| & |y_3z_4| & |y_6z_1| \\ \cdot & \cdot & |z_2x_3| & |z_5x_6| & |z_3x_4| & |z_6x_1| \\ \cdot & \cdot & |x_2y_3| & |x_5y_6| & |x_3y_4| & |x_6y_1| \end{vmatrix}$$

$$= \begin{vmatrix} |x_1y_2z_3| & \cdot & \cdot & |x_4y_2z_3| \\ |x_1y_4z_5| & |x_2y_4z_5| & \cdot & \cdot \\ |x_1y_5z_6| & |x_2y_5z_6| & |x_3y_5z_6| & |x_4y_5z_6| \\ \cdot & \cdot & |x_3y_6z_1| & |x_4y_6z_1| \end{vmatrix}.$$

229. Let $A \equiv (a_{mn})$, $B \equiv (b_{mn})$ and $C \equiv |c_{mm}| = A \cdot B$, then,

$$\text{(A)} \qquad C\frac{\partial C}{\partial c_{rs}} = \frac{\partial C}{\partial a_{r1}}\frac{\partial C}{\partial b_{s1}} + \frac{\partial C}{\partial a_{r2}}\frac{\partial C}{\partial b_{s2}} + \cdots + \frac{\partial C}{\partial a_{rm}}\frac{\partial C}{\partial b_{sm}}$$

We know that for any determinant

$$\text{(1)} \qquad c_{r1}\frac{\partial C}{\partial c_{s1}} + c_{r2}\frac{\partial C}{\partial c_{s2}} + \cdots + c_{rn}\frac{\partial C}{\partial c_{sn}} = \begin{matrix} C \\ 0 \end{matrix}\begin{cases} r = s \\ r \neq s \end{cases}$$

and that for the given determinant C

$$\text{(2)} \qquad \frac{\partial C}{\partial a_{r1}} = b_{11}\frac{\partial C}{\partial c_{r1}} + b_{21}\frac{\partial C}{\partial c_{r2}} + \cdots + b_{m1}\frac{\partial C}{\partial c_{rm}} \cdot$$

If now we use (2) to express the first factor of each term on the right of (A) we have

$$\begin{aligned} C\frac{\partial C}{\partial c_{rs}} = {} & \frac{\partial C}{\partial b_{s1}}\left(b_{11}\frac{\partial C}{\partial c_{r1}} + b_{21}\frac{\partial C}{\partial c_{r2}} + \cdots + b_{m1}\frac{\partial C}{\partial c_{rm}}\right) \\ & + \frac{\partial C}{\partial b_{s2}}\left(b_{12}\frac{\partial C}{\partial c_{r1}} + b_{22}\frac{\partial C}{\partial c_{r2}} + \cdots + b_{m2}\frac{\partial C}{\partial c_{rm}}\right) \\ & + \cdots\cdots\cdots\cdots\cdots\cdots\cdots\cdots \\ & + \frac{\partial C}{\partial b_{sn}}\left(b_{1n}\frac{\partial C}{\partial c_{r1}} + b_{2n}\frac{\partial C}{\partial c_{r2}} + \cdots + b_{mn}\frac{\partial C}{\partial c_{rm}}\right) \end{aligned}$$

Adding columnwise, and by (1) we see that the sum of each column is zero except one which is equal to $C\,\partial C/\partial c_{rs}$ and hence the truth of the statement.

The significance of this theorem is perhaps better seen if we write it in another form. Thus for $n=4$, $m=3$, $r=2$, and $s=3$ if we write

$$\left\|\begin{matrix} a_1 & a_2 & a_3 & a_4 \\ b_1 & b_2 & b_3 & b_4 \\ c_1 & c_2 & c_3 & c_4 \end{matrix}\right\| \left\|\begin{matrix} l_1 & l_2 & l_3 & l_4 \\ m_1 & m_2 & m_3 & m_4 \\ n_1 & n_2 & n_3 & n_4 \end{matrix}\right\| = \begin{vmatrix} \sum al & \sum am & \sum an \\ \sum bl & \sum bm & \sum bn \\ \sum cl & \sum cm & \sum cn \end{vmatrix}$$

then the theorem is

$$-\begin{vmatrix} \sum al & \sum am & \sum an \\ \sum bl & \sum bm & \sum bn \\ \sum cl & \sum cm & \sum cn \end{vmatrix} \cdot \begin{vmatrix} \sum al & \sum am \\ \sum cl & \sum cm \end{vmatrix}$$

$$= \begin{vmatrix} \sum al & \sum am & \sum an \\ l_1 & m_1 & n_1 \\ \sum cl & \sum cm & \sum cn \end{vmatrix} \cdot \begin{vmatrix} \sum al & \sum am & a_1 \\ \sum bl & \sum bm & b_1 \\ \sum cl & \sum cm & c_1 \end{vmatrix}$$

$$+ \begin{vmatrix} \sum al & \sum am & \sum an \\ l_2 & m_2 & n_2 \\ \sum cl & \sum cm & \sum cn \end{vmatrix} \cdot \begin{vmatrix} \sum al & \sum am & a_2 \\ \sum bl & \sum bm & b_2 \\ \sum cl & \sum cm & c_2 \end{vmatrix}$$

$$+ \begin{vmatrix} \sum al & \sum am & \sum an \\ l_3 & m_3 & n_3 \\ \sum cl & \sum cm & \sum cn \end{vmatrix} \cdot \begin{vmatrix} \sum al & \sum am & a_3 \\ \sum bl & \sum bm & b_3 \\ \sum cl & \sum cm & c_3 \end{vmatrix}$$

$$+ \begin{vmatrix} \sum al & \sum am & \sum an \\ l_4 & m_4 & n_4 \\ \sum cl & \sum cm & \sum cn \end{vmatrix} \cdot \begin{vmatrix} \sum al & \sum am & a_4 \\ \sum bl & \sum bm & b_4 \\ \sum cl & \sum cm & c_4 \end{vmatrix}.$$

230. In a precisely similar manner to that of §217 we may form the product of two rectangular arrays or *matrices* of n columns each, the first containing μ and the second ν rows. Thus from

$$\begin{Vmatrix} a_{11} & a_{12} \cdots a_{1n} \\ a_{21} & a_{22} \cdots a_{2n} \\ \cdots & \cdots \\ a_{\mu 1} & a_{\mu 2} \cdots a_{\mu n} \end{Vmatrix} \quad \text{and} \quad \begin{Vmatrix} b_{11} & b_{12} \cdots b_{1n} \\ b_{21} & b_{22} \cdots b_{2n} \\ \cdots & \cdots \\ b_{\nu 1} & b_{\nu 2} \cdots b_{\nu n} \end{Vmatrix}$$

we can form a third matrix

$$\begin{Vmatrix} c_{11} & c_{12} \cdots c_{1\nu} \\ c_{21} & c_{22} \cdots c_{2\nu} \\ \cdots & \cdots \\ c_{\mu 1} & c_{\mu 2} \cdots c_{\mu\nu} \end{Vmatrix}$$

containing μ rows and ν columns, the $\mu \cdot \nu$ elements of which are of the form

$$c_{ij} = a_{i1}b_{j1} + a_{i2}b_{j2} + \cdots + a_{in}b_{jn}.$$

$$i = 1, 2, \cdots, \mu; \quad j = 1, 2, \cdots, \nu.$$

If now $\mu+\nu=n$, and the $\mu\cdot\nu$ elements $c_{ij}=0$, then the first two are said to be *corresponding matrices.* We may also form from the first two matrices a determinant Δ of the nth order whose rows are the $\mu+\nu=n$ rows of the two matrices, thus:

$$\Delta \equiv \begin{vmatrix} a_{11} & a_{12} & \cdots & a_{1n} \\ a_{21} & a_{22} & \cdots & a_{2n} \\ \cdot & \cdot & \cdot & \cdot \\ a_{\mu 1} & a_{\mu 2} & \cdots & a_{\mu n} \\ b_{11} & b_{12} & \cdots & b_{1n} \\ b_{21} & b_{22} & \cdots & b_{2n} \\ \cdot & \cdot & \cdot & \cdot \\ b_{\nu 1} & b_{\nu 2} & \cdots & b_{\nu n} \end{vmatrix}.$$

Any minor of order μ formed from the first μ rows of this determinant together with its complementary are called *corresponding or complementary minors of the two corresponding matrices.*

231. *Complementary minors of corresponding matrices are proportional.*

Expanding the determinant Δ of the preceding article by Laplace's theorem and writing the expansion so that all terms are positive, which means that one of the two series of minors is signed, and we shall suppose it is the one involving the a's, we have

$$\Delta = \sum_{i=1}^{(n)_\mu} \mathcal{A}_{(n|\mu_1),(n|\mu_i)} B_{(n|\nu_1),(\bar{n}|\mu_i)}.$$

Squaring both sides of this equation we have

$$\Delta^2 = \left\{ \sum_{i=1}^{(n)_\mu} \mathcal{A}_{(n|\mu_1),(n|\mu_i)} B_{(n|\nu_1),(\bar{n}|\mu_i)} \right\}^2.$$

Let us now multiply the determinant Δ by itself, making use of the relations

$$c_{ij} = a_{i1}b_{j1} + \cdots + a_{in}b_{jn} = 0;$$

we have

$$\begin{vmatrix} \sum a_{1i}^2 & \sum a_{2i}a_{1i} \cdots & \sum a_{\mu i}a_{1i} & 0 & \cdots & 0 \\ \sum a_{1i}a_{2i} & \sum a_{2i}^2 \quad \cdots & \sum a_{\mu i}a_{2i} & 0 & \cdots & 0 \\ \sum a_{1i}a_{3i} & \sum a_{2i}a_{3i} \cdots & \sum a_{\mu i}a_{3i} & 0 & \cdots & 0 \\ \cdots & \cdots & \cdots & \cdots & \cdots & \cdots \\ \sum a_{1i}a_{\mu i} & \sum a_{2i}a_{\mu i} \cdots & \sum a_{\mu i}^2 & 0 & \cdots & 0 \\ 0 & 0 \quad \cdots & 0 & \sum b_{1i}^2 & \cdots & \sum b_{\nu i}b_{1i} \\ \cdots & \cdots & \cdots & \cdots & \cdots & \cdots \\ 0 & 0 \quad \cdots & 0 & \sum b_{1i}b_{\nu i} & \cdots & \sum b_{\nu i}^2 \end{vmatrix}$$

$$= \begin{Vmatrix} a_{11} & a_{12} \cdots & a_{1n} \\ \cdots & \cdots & \cdots \\ a_{\mu 1} & a_{\mu 2} \cdots & a_{\mu n} \end{Vmatrix}^2 \begin{Vmatrix} b_{11} & b_{12} \cdots & b_{1n} \\ \cdots & \cdots & \cdots \\ b_{\nu 1} & b_{\nu 2} \cdots & b_{\nu n} \end{Vmatrix}^2$$

$$= \sum_1^{(n)\mu} i \mathcal{A}^2_{(n|\mu_1),(n|\mu_i)} \cdot \sum_1^{(n)\nu} j B^2_{(n|\nu_1),(n|\nu_j)},$$

Equating these two forms for Δ^2 we have

$$\left\{ \sum_1^{(n)\mu} i \mathcal{A}_{(n|\mu_1),(n|\mu_i)} B_{(n|\nu_1),(\bar{n}|\mu_i)} \right\}^2 = \sum_1^{(n)\mu} i \mathcal{A}^2_{(n|\mu_1),(n|\mu_i)} \sum_1^{(n)\nu} j B^2_{(n|\nu_1),(n|\nu_j)}$$

Whence

$$\sum \left\{ \mathcal{A}_{(n|\mu_1),(n|\mu_i)} B_{(n|\nu_1),(\bar{n}|\mu_j)} - \mathcal{A}_{(n|\mu_1),(n|\mu_j)} B_{(n|\nu_1),(\bar{n}|\mu_i)} \right\}^2 = 0.$$

$$(i,j = 1,2, \cdots, n) \quad i \neq j.$$

Therefore

$$\frac{\mathcal{A}_{(n|\mu_1),(n|\mu_i)}}{B_{(n|\nu_1),(\bar{n}|\mu_i)}} = \frac{\mathcal{A}_{(n|\mu_1),(n|\mu_j)}}{B_{(n|\nu_1)\,(\bar{n}|\mu_j)}}.$$

232. *The sum of the squares of all the μ-line minors formed from the 1st μ rows multiplied by the sum of the squares of the ν-line minors of the last ν rows is equal to the square of Δ.*

The proof of this theorem may be seen on multiplying Δ by itself.

EXAMPLE. If we have the set of equations

$$a_1x + a_2y + a_3z + a_4w = 0$$
$$b_1x + b_2y + b_3z + b_4w = 0$$
$$c_1x + c_2y + c_3z + c_4w = 0$$

then the two arrays

$$\begin{matrix} a_1 & a_2 & a_3 & a_4, & x & y & z & w \\ b_1 & b_2 & b_3 & b_4 \\ c_1 & c_2 & c_3 & c_4 \end{matrix}$$

satisfy our conditions and the theorem gives

$$\frac{x}{|a_2\, b_3\, c_4|} = \frac{-y}{|a_1\, b_3\, c_4|} = \frac{z}{|a_1\, b_2\, c_4|} = \frac{-w}{|a_1\, b_2\, c_3|}$$

233. If

$$\left\| \begin{matrix} 11 & 12 \cdots 1n \\ 21 & 22 \cdots 2n \\ \cdots & \cdots \\ r1 & r2 \cdots rn \end{matrix} \right\| = 0$$

which means that all determinants of order r vanish, then

$$\left\| \begin{matrix} 11 & 12 \cdots 1n \\ 21 & 22 \cdots 2n \\ \cdots & \cdots \\ r1 & r2 \cdots rn \end{matrix} \right\| \cdot \left| \omega_{11}\omega_{22} \cdots \omega_{nn} \right| = 0$$

and

$$\left\| \begin{matrix} 11 & 12 \cdots 1n \\ 21 & 22 \cdots 2n \\ \cdots & \cdots \\ r1 & r2 \cdots rn \end{matrix} \right\| \cdot \left| \omega_{11}\omega_{22} \cdots \omega_{rr} \right| = 0$$

where $r<n$ and multiplication in the first instance is row-wise and in the second column-wise.

The truth of these are seen on observing that the resulting determinants in the product arrays are the corresponding determinants of the original multiplied by the ω determinant in each case and therefore are zero.

Conversely if

$$(a_{rn}) \cdot \Delta = 0$$

where Δ is a determinant of the nth or rth order, then

$$(a_{rn})=0, \text{ provided } \Delta\neq 0.$$

EXAMPLE. The product column-by-column

$$\left\|\begin{matrix} 11 & 12\cdots 1n \\ 21 & 22\cdots 2n \\ 31 & 32\cdots 3n \end{matrix}\right\| \cdot \left\|\begin{matrix} 1 & 0 & 0 \\ 0 & \lambda_1 & \lambda_2 \\ 0 & \mu_1 & \mu_2 \end{matrix}\right\|$$

$$= \begin{vmatrix} 11 & 12 & \cdots & 1n \\ \lambda_1 21+\mu_1 31 & \lambda_1 22+\mu_1 32 & \cdots & \lambda_1 2n+\mu_1 3n \\ \lambda_2 21+\mu_2 31 & \lambda_2 22+\mu_2 32 & \cdots & \lambda_2 2n+\mu_2 3n \end{vmatrix} = 0,$$

if

$$\left\|\begin{matrix} 11 & 12\cdots 1n \\ 21 & 22\cdots 2n \\ 31 & 32\cdots 3n \end{matrix}\right\| = 0\,;$$

similarly for r instead of 3.

234. *If in a given matrix a certain m-rowed determinant is not zero, and all the $(m+h)$-rowed determinants of which this m-rowed determinant is a minor are zero, then all the $(m+h)$-rowed determinants of the matrix are zero.*

Let*

$$M \equiv \begin{vmatrix} (n\,|\,m_\alpha) \\ (n\,|\,m_\alpha) \end{vmatrix} \neq 0 \text{ and } \begin{vmatrix} (n\,|\,m_\alpha)(\bar{n}\,|\,m_\alpha\,|\,h_\beta) \\ (n\,|\,m_\alpha)(\bar{n}\,|\,m_\alpha\,|\,h_\gamma) \end{vmatrix} = 0,$$

for $\left\{{\beta \atop \gamma}\right.=1, 2, \cdots, (n-m)_h$.

On account of identical columns we obviously have

$$\begin{vmatrix} (n\,|\,m_\alpha)(\bar{n}\,|\,m_\alpha\,|\,h+k_\delta) \\ (n\,|\,m_\alpha)(n\,|\,m_\alpha\,|\,k_i)(\bar{n}\,|\,m_\alpha\,|\,h_\gamma) \end{vmatrix} \equiv 0.$$

Expanding in terms of minors of order k and their complementaries we have

$$\sum_{j=1}^{(m)_k} \begin{vmatrix} (n\,|\,m_\alpha\,|\,k_j) \\ (n\,|\,m_\alpha\,|\,k_i) \end{vmatrix} \cdot \begin{vmatrix} (n\,|\,\bar{m}_\alpha\,|\,k_j)(\bar{n}\,|\,m_\alpha\,|\,h+k_\delta) \\ (n\,|\,m_\alpha)(\bar{n}\,|\,m_\alpha\,|\,h_\gamma) \end{vmatrix} = 0,$$

for $i=1, 2, \cdots, (m)_k$, and $k=0, 1, \cdots, m$.

* Obviously no generality is lost in taking the m-rowed determinant coaxial

The determinant of the coefficients† in this set of equations for each value of k is the kth compound of M and is therefore not zero. Giving k successively the values from 1 to m we have

$$\begin{vmatrix} (n \mid \bar{m}_\alpha \mid k_j)(\bar{n} \mid m_\alpha \mid h + k_\delta) \\ (n \mid m_\alpha)(\bar{n} \mid m_\alpha \mid h_\gamma) \end{vmatrix} = 0 \tag{1}$$

for all values of j, δ, γ, k.

For the values $1, 2, \cdots, m$ of k we have the result that *all minors of the matrix which have m columns and $(m-1), (m-2), \cdots, 0$, rows in common with M vanish.*

Therefore

$$\begin{vmatrix} (n \mid m + h_\epsilon) \\ (n \mid m_\alpha)(\bar{n} \mid m_\alpha \mid h_\gamma) \end{vmatrix} = 0, \tag{2}$$

for $\epsilon = 1, 2, \cdots, (n)_{m+h}$.

LEMMA. If

$$\begin{vmatrix} (n \mid m + h_\epsilon) \\ (n \mid m_\alpha)(\bar{n} \mid m_\alpha \mid h_\gamma) \end{vmatrix} = 0,$$

for all values of ϵ, γ, then

$$M' \equiv \begin{vmatrix} (n \mid m + h + k_\eta) \\ (n \mid m_\alpha)(\bar{n} \mid m_\alpha \mid h + k_\delta) \end{vmatrix} = 0. \tag{3}$$

That is, if all minors of order $(m+h)$ formed from the m columns $(n \mid m_\alpha)$ and h others vanish, then all minors of order $(m+h+k)$ formed from the same m columns and $(h+k)$ others will vanish.

For

$$M' = \sum_{i=1}^{(m+h)_k} \begin{vmatrix} (n \mid m + h + k_\eta \mid k_i) \\ (\bar{n} \mid m_\alpha \mid h + k_\delta \mid k_\rho) \end{vmatrix} \cdot \begin{vmatrix} (n \mid \overline{m + h + k_\eta} \mid k_i) \\ (n \mid m_\alpha)(\bar{n} \mid m_\alpha \mid \overline{h + k_\delta} \mid k_\rho) \end{vmatrix} = 0,$$

since the minors of order $(m+h)$ in every term are zero by hypothesis.

Since by (3) $M' = 0$, it is obvious that

$$\begin{vmatrix} (n \mid m_\alpha \mid k_j)(n \mid m + h_\epsilon) \\ (n \mid m_\alpha)(\bar{n} \mid m_\alpha \mid h + k_\delta) \end{vmatrix} = 0,$$

and we have

$$\sum_{i=1}^{(m)k} \begin{vmatrix} (n \mid m_\alpha \mid k_j) \\ (n \mid m_\alpha \mid k_i) \end{vmatrix} \cdot \begin{vmatrix} (n \mid m + h_\epsilon) \\ (n \mid \bar{m}_\alpha \mid k_i)(\bar{n} \mid m_\alpha \mid h + k_\delta) \end{vmatrix} = 0$$

for $k = 1, 2, \cdots, m$ and $j = 1, 2, \cdots, (m)_k$.

† By coefficient here is meant the first factor in each term.

The determinant of the coefficients in this set of equations is the kth compound of M and is therefore not zero.

It follows that

$$(4)\qquad \begin{vmatrix} (n \mid m + h_\epsilon) \\ (n \mid \bar{m}_\alpha \mid k_i)(\bar{n} \mid m_\alpha \mid h + k_\delta) \end{vmatrix} = 0$$

for all values of ϵ, i, δ, k.

Giving k successively the values $1, 2, \cdots, m$ we have our theorem proved, or stated symbolically it is:

If

$$\begin{vmatrix} (n \mid m_\alpha) \\ (n \mid m_\alpha) \end{vmatrix} \neq 0, \text{ and } \begin{vmatrix} (n \mid m_\alpha)(\bar{n} \mid m_\alpha \mid h_\beta) \\ (n \mid m_\alpha)(\bar{n} \mid m_\alpha \mid h_\gamma) \end{vmatrix} = 0,$$

for

$$\beta, \gamma = 1, 2, \cdots, (n - m)_h,$$

then

$$\begin{vmatrix} (n \mid \bar{m}_\alpha \mid g_i)(\bar{n} \mid m_\alpha \mid h + g_\delta) \\ (n \mid \bar{m}_\alpha \mid k_j)(\bar{n} \mid m_\alpha \mid h + k_\epsilon) \end{vmatrix} = 0,$$

for

$$g, k = 0, 1, \cdots, m;\ i, j = 1, 2, \cdots, (m)_k;\ \epsilon, \delta = 1, 2, \cdots, (n - m)_{h+k}.$$

or

$$\begin{vmatrix} (n \mid m + h_i) \\ (n \mid m + h_j) \end{vmatrix} = 0$$

for all values of i and j.

In this case the matrix is said to be of *rank m*. That is a matrix is said to be of rank m if it contains at least one determinant of order m which does not vanish while all determinants of order greater than m which the matrix contains do vanish. This non-vanishing minor has been called the *critical minor.*

235. From the foregoing it is seen that:

If a matrix of m rows and n columns ($m \gtreqless n$) *have at least one minor M of order r which is not zero, and if all the* $(n-r)_h \cdot (m-r)_h$ *determinants of order* $(r+h)$ *which contain M as a minor are zero, then all determinants of order* $(r+h)$ *contained in the matrix are zero.*

As a special case when $m=n$, we have the theorem:

If a determinant of order n is of rank r and if all the $\{(n-r)_h\}^2$ minors of order $(r+h)$ which contain the non-vanishing determinant as a minor vanish, then all the $\{(n)_{r+h}\}^2$ minors of order $(r+h)$ will be zero.

If $m=n-1$, $r=1$, $h=n-2$, then let the array be the first $n-1$ rows of $A\equiv|a_{1n}|$ and let the element $a_{1n}\neq 0$. Then from

$$(n1)[n1]+(n2)[n2]+\cdots+(nn)[nn]=A$$

we see that if $[n1]$, $[n2]$, $\cdots$, $[n,n-1]$ vanish then (nn) $[nn]$ must vanish and therefore A vanishes. That is, *if $[r2]$, $[r3]$, $\cdots$, $[rn]$ vanish and if $(s1)\neq 0$ for any value of $s\neq r$, then $[r1]=0$ and $A=0$.*

236. The required minimum of $(n-r)_h\times(m-r)_h$ vanishing minors or order $(r+h)$ in order to insure the vanishing of all minors of the matrix of order $(r+h)$ need not all have the same non-vanishing minor of order r. Thus when $n=7$, $m=5$, $r=3$, $h=2$ the matrix is

$$\begin{Vmatrix} 11 & 12\cdots 17\\ 21 & 22\cdots 27\\ \cdots & \cdots\cdots\\ 51 & 52\cdots 57\end{Vmatrix}$$

and let the given $(7-3)_2=6$ vanishing determinants be the 1st, 2nd, 3rd, 4th, 5th, 18th in the following list of $(7)_2=21$:

$$(1)\begin{vmatrix}1\,2\,3\,4\,5\\1\,2\,3\,4\,5\end{vmatrix},\quad (6)\begin{vmatrix}1\,2\,3\,4\,5\\1\,2\,3\,6\,7\end{vmatrix},\quad (11)\begin{vmatrix}1\,2\,3\,4\,5\\1\,3\,4\,5\,6\end{vmatrix},\quad (16)\begin{vmatrix}1\,2\,3\,4\,5\\2\,3\,4\,5\,6\end{vmatrix},$$

$$(2)\begin{vmatrix}1\,2\,3\,4\,5\\1\,2\,3\,4\,6\end{vmatrix},\quad (7)\begin{vmatrix}1\,2\,3\,4\,5\\1\,2\,4\,5\,6\end{vmatrix},\quad (12)\begin{vmatrix}1\,2\,3\,4\,5\\1\,3\,4\,5\,7\end{vmatrix},\quad (17)\begin{vmatrix}1\,2\,3\,4\,5\\2\,3\,4\,5\,7\end{vmatrix},$$

$$(3)\begin{vmatrix}1\,2\,3\,4\,5\\1\,2\,3\,4\,7\end{vmatrix},\quad (8)\begin{vmatrix}1\,2\,3\,4\,5\\1\,2\,4\,5\,7\end{vmatrix},\quad (13)\begin{vmatrix}1\,2\,3\,4\,5\\1\,3\,4\,6\,7\end{vmatrix},\quad (18)\begin{vmatrix}1\,2\,3\,4\,5\\2\,3\,4\,6\,7\end{vmatrix},$$

$$(4)\begin{vmatrix}1\,2\,3\,4\,5\\1\,2\,3\,5\,6\end{vmatrix},\quad (9)\begin{vmatrix}1\,2\,3\,4\,5\\1\,2\,4\,6\,7\end{vmatrix},\quad (14)\begin{vmatrix}1\,2\,3\,4\,5\\1\,3\,5\,6\,7\end{vmatrix},\quad (19)\begin{vmatrix}1\,2\,3\,4\,5\\2\,3\,5\,6\,7\end{vmatrix},$$

$$(5)\begin{vmatrix}1\,2\,3\,4\,5\\1\,2\,3\,5\,7\end{vmatrix},\quad (10)\begin{vmatrix}1\,2\,3\,4\,5\\1\,2\,5\,6\,7\end{vmatrix},\quad (15)\begin{vmatrix}1\,2\,3\,4\,5\\1\,4\,5\,6\,7\end{vmatrix},\quad (20)\begin{vmatrix}1\,2\,3\,4\,5\\2\,4\,5\,6\,7\end{vmatrix},$$

$$(21)\begin{vmatrix}1\,2\,3\,4\,5\\3\,4\,5\,6\,7\end{vmatrix}.$$

Expanding the determinant $\begin{vmatrix} 112345 \\ 123456 \end{vmatrix}$ which is identically zero, having two rows alike, we get

$$\begin{matrix}1\\1\end{matrix}\begin{vmatrix}1\,2\,3\,4\,5\\2\,3\,4\,5\,6\end{vmatrix}-\begin{matrix}1\\2\end{matrix}\begin{vmatrix}1\,2\,3\,4\,5\\1\,3\,4\,5\,6\end{vmatrix}+\begin{matrix}1\\3\end{matrix}\begin{vmatrix}1\,2\,3\,4\,5\\1\,2\,4\,5\,6\end{vmatrix}=0,$$

the other terms vanishing because (1), (2), (4), are given zero. Similarly

$$\begin{matrix}2\\1\end{matrix}\begin{vmatrix}1\,2\,3\,4\,5\\2\,3\,4\,5\,6\end{vmatrix}-\begin{matrix}2\\2\end{matrix}\begin{vmatrix}1\,2\,3\,4\,5\\1\,3\,4\,5\,6\end{vmatrix}+\begin{matrix}2\\3\end{matrix}\begin{vmatrix}1\,2\,3\,4\,5\\1\,2\,4\,5\,6\end{vmatrix}=0$$

and

$$\begin{matrix}3\\1\end{matrix}\begin{vmatrix}1\,2\,3\,4\,5\\2\,3\,4\,5\,6\end{vmatrix}-\begin{matrix}3\\2\end{matrix}\begin{vmatrix}1\,2\,3\,4\,5\\1\,3\,4\,5\,6\end{vmatrix}+\begin{matrix}3\\3\end{matrix}\begin{vmatrix}1\,2\,3\,4\,5\\1\,2\,4\,5\,6\end{vmatrix}=0.$$

Then if $\begin{vmatrix} 123 \\ 123 \end{vmatrix} \neq 0$ it follows that

$$\begin{vmatrix}1\,2\,3\,4\,5\\2\,3\,4\,5\,6\end{vmatrix}=\begin{vmatrix}1\,2\,3\,4\,5\\1\,3\,4\,5\,6\end{vmatrix}=\begin{vmatrix}1\,2\,3\,4\,5\\1\,2\,4\,5\,6\end{vmatrix}=0,$$

or $(7)=(11)=(16)=0$.

Expanding the determinant $\begin{vmatrix} 112345 \\ 123457 \end{vmatrix} \equiv 0$ we have

$$\begin{matrix}1\\1\end{matrix}\begin{vmatrix}1\,2\,3\,4\,5\\2\,3\,4\,5\,7\end{vmatrix}-\begin{matrix}1\\2\end{matrix}\begin{vmatrix}1\,2\,3\,4\,5\\1\,3\,4\,5\,7\end{vmatrix}+\begin{matrix}1\\3\end{matrix}\begin{vmatrix}1\,2\,3\,4\,5\\1\,2\,4\,5\,7\end{vmatrix}=0,$$

the other terms vanishing because (1), (3), (5) are given zero. Similarly

$$\begin{matrix}2\\1\end{matrix}\begin{vmatrix}1\,2\,3\,4\,5\\2\,3\,4\,5\,7\end{vmatrix}-\begin{matrix}2\\2\end{matrix}\begin{vmatrix}1\,2\,3\,4\,5\\1\,3\,4\,5\,7\end{vmatrix}+\begin{matrix}2\\3\end{matrix}\begin{vmatrix}1\,2\,3\,4\,5\\1\,2\,4\,5\,7\end{vmatrix}=0$$

$$\begin{matrix}3\\1\end{matrix}\begin{vmatrix}1\,2\,3\,4\,5\\2\,3\,4\,5\,7\end{vmatrix}-\begin{matrix}3\\2\end{matrix}\begin{vmatrix}1\,2\,3\,4\,5\\1\,3\,4\,5\,7\end{vmatrix}+\begin{matrix}3\\3\end{matrix}\begin{vmatrix}1\,2\,3\,4\,5\\1\,2\,4\,5\,7\end{vmatrix}=0.$$

Therefore

$$\begin{vmatrix}1\,2\,3\,4\,5\\2\,3\,4\,5\,7\end{vmatrix}=\begin{vmatrix}1\,2\,3\,4\,5\\1\,3\,4\,5\,7\end{vmatrix}=\begin{vmatrix}1\,2\,3\,4\,5\\1\,2\,4\,5\,7\end{vmatrix}=0$$

or $(8)=(12)=(17)=0$. Proceeding in a similar way and collecting the results in one table we have

Given as zero	There follows	
$(1)=(2)=(4)=0$	$(7)=(11)=(16)=0$	if $\left\|\begin{smallmatrix}123\\123\end{smallmatrix}\right\|\neq 0$
$(1)=(3)=(5)=0$	$(8)=(12)=(17)=0$	if $\left\|\begin{smallmatrix}123\\123\end{smallmatrix}\right\|\neq 0$
$(2)=(3)=(18)=0$	$(6)=(9)=(13)=0$	if $\left\|\begin{smallmatrix}234\\234\end{smallmatrix}\right\|\neq 0$
$(4)=(5)=(6)=0$	$(10)=(14)=(19)=0$	if $\left\|\begin{smallmatrix}123\\123\end{smallmatrix}\right\|\neq 0$
$(16)=(17)=(18)=0$	$(19)=(20)=(21)=0$	if $\left\|\begin{smallmatrix}234\\234\end{smallmatrix}\right\|\neq 0$
$(11)=(12)=(21)=0$	$(13)=(14)=(15)=0$	if $\left\|\begin{smallmatrix}345\\345\end{smallmatrix}\right\|\neq 0$

Instead of assuming that the non-vanishing determinants of the third order were coaxial we could have taken any determinant from the 1st, 2nd and 3rd columns, any from the 2nd, 3rd and 4th columns, and any from the 3rd, 4th and 5th columns of the matrix.

Other sets of six vanishing determinants of order five may be assumed with the result that all determinants of order five of the matrix vanish.

237. *If in an m-by-n ($m<n$) array all the m-line minors which include a fixed group of less then m columns vanish then either the array is wholly evanescent or the array composed of the fixed group of columns is evanescent.* For by §235 if there is a determinant from the fixed group of columns that is not zero the given array is wholly evanescent.

238. *If in an n-by-($n+1$) array $n-m$ of the n-line minors vanish and one does not vanish, the array common to the said $n-m$ minors is evanescent.*

Without loss of generality we may suppose the n-line minor which does not vanish to be the one consisting of the first n columns or, $|a_{1n}|$. Let the minors of order $(m+1)$ formed from the $(m+1)$ columns common to the vanishing minors be represented by $\alpha_1, \alpha_2, \cdots, \alpha_\lambda$ where $\lambda=(n)_{m+1}$, let $\beta=|\beta_{1\lambda}|$ be the $(n-m-1)$th compound of $|a_{1n}|$, and let α_r be formed from the rows not used in forming β_{rs}. Then as a consequence of $(n-m)$-line minors which are given zero and those which are zero because of identical columns we have

$$\beta_{1r}\alpha_1+\beta_{2r}\alpha_2+\cdots+\beta_{\lambda r}\alpha_\lambda=0, \qquad (r=1,2,\cdots,\lambda).$$

In this set of linear homogeoeous equations the determinant of the coefficients being a power of β is not zero. Therefore

$$\alpha_r=0 \quad (r=1,2,\cdots,\lambda).$$

Another way of stating this theorem is the following:

If $|a_{0m}|$ be a linear function of the first $m+1$ indeterminants

$$a_{00},\ a_{01},\ \cdots,\ a_{0m}$$

only, and the cofactor of a_{00} in it be not zero, then all the $(m+1)$-line determinants of the array

$$\left\|\begin{matrix} a_{10} & a_{11} & \cdots & a_{1m} \\ a_{20} & a_{21} & \cdots & a_{2m} \\ \cdots & \cdots & \cdots & \cdots \\ a_{n0} & a_{n1} & \cdots & a_{nm} \end{matrix}\right\|$$

must vanish.

If $m=0$, the foregoing becomes:

If an n-by-$(n+1)$ array be such that all its n-line determinants vanish except one, then the column common to those that vanish must consist of zeros.

239. *If in an n-by-$(n+1)$ array one minor of order n is not zero then at least one other is not zero unless the array has a column of zeros.*

Let

$$\begin{matrix} a_{11} & \cdots & a_{1n} & a_{1,n+1} \\ \cdots & \cdots & \cdots & \cdots \\ a_{n1} & \cdots & a_{nn} & a_{n,n+1} \end{matrix}$$

be the array, and let $|a_{1n}|=A\neq 0$. Form a zero determinant of order $n+1$ by repeating the kth row and expand in terms of the elements of this repeated row. This gives

$$a_{k1}\mathcal{A}_{n+1,1}+a_{k2}\mathcal{A}_{n+1,2}+\cdots+a_{kn}\mathcal{A}_{n+1,n}+a_{k,n+1}\mathcal{A}_{n+1,n+1}=0$$

and since $A_{n+1,n+1}$ is not zero one other minor of order n must not be zero unless $a_{k,n+1}=0$ for $k=1, 2, \cdots, n$.

240. *The determinants of the mth order of an m-by-n array involve $(n-m)m$ quantities that are absolutely independent and therefore they must be connected by $(n)_m-(n-m)m-1$ relations from which all others are deducible.*

Let the array be

$$\left\|\begin{matrix} a_{11} & a_{12} & \cdots & a_{1n} \\ a_{21} & a_{22} & \cdots & a_{2n} \\ \cdots & \cdots & \cdots & \cdots \\ a_{m1} & a_{m2} & \cdots & a_{mn} \end{matrix}\right\|$$

In this subject each element to the linear substitution

$$a_{rs}=\mu_{1r}b_{1s}+\mu_{2r}b_{2s}+\cdots+\mu_{mr}b_{ms}$$

obtaining

$$\begin{Vmatrix} b_{11} & b_{12} & \cdots & b_{1n} \\ b_{21} & b_{22} & \cdots & b_{2n} \\ \cdots & \cdots & \cdots & \cdots \\ b_{m1} & b_{m2} & \cdots & b_{mn} \end{Vmatrix} \cdot \begin{vmatrix} \mu_{11} & \mu_{12} & \mu_{1m} \\ \mu_{21} & \mu_{22} & \mu_{2m} \\ & & \\ \mu_{m1} & \mu_{m2} & \mu_{mm} \end{vmatrix}.$$

For every homogeneous equation connecting determinants of the a-array there exists a perfectly similar equation connecting the elements of the b-array, no matter what the elements of the modulus of substitution $|\mu_{1m}|$ may be. These m^2 elements of the modulus may be so chosen as to give definite values to m^2 of the b's and so leave only $m(n-m)$ independent elements for the $(n)_m-1$ ratios of the determinants of the array to be dependent upon. It follows that between such ratios there exists at most $(n)_m-1-(n-m)m$ relations.

These relations may be formed by using §202. Thus if the array is

$$\begin{Vmatrix} a_1 & b_1 & c_1 & f_1 & g_1 & h_1 \\ a_2 & b_2 & c_2 & f_2 & g_2 & h_2 \\ a_3 & b_3 & c_3 & f_3 & g_3 & h_3 \end{Vmatrix}$$

and denoting

$$\begin{vmatrix} a_1 & b_1 & c_1 \\ a_2 & b_2 & c_2 \\ a_3 & b_3 & c_3 \end{vmatrix} \text{ by } A, \begin{vmatrix} f_1 & g_1 & h_1 \\ f_2 & g_2 & h_2 \\ f_3 & g_3 & h_3 \end{vmatrix} \text{ by } B, \text{ and } \begin{vmatrix} a_1 & f_1 & g_1 \\ a_2 & f_2 & g_2 \\ a_3 & f_3 & g_3 \end{vmatrix} \text{ by } E$$

we have by the theorem, making $q=1$,

$$\begin{vmatrix} A & 0 & 0 \\ |f_1 b_2 c_3| & |a_1 f_2 c_3| & |a_1 b_2 f_3| \\ |g_1 b_2 c_3| & |a_1 g_2 c_3| & |a_1 b_2 g_3| \end{vmatrix} = A^2 \cdot E, \text{ or}$$

$$|a_1 b_2 c_3|\,|a_1 f_2 g_3| = |a_1 f_2 c_3|\,|a_1 b_2 g_3| - |a_1 g_2 c_3|\,|a_1 b_2 f_3|$$

Of this type there are $3 \cdot 3 = 9$ relations. Next using A and B for our two determinants, and leaving $q=1$, we have the tenth relation:

$$\begin{vmatrix} |f_1 b_2 c_3| & |a_1 f_2 c_3| & |a_1 b_2 f_3| \\ |g_1 b_2 c_3| & |a_1 g_2 c_3| & |a_1 b_2 g_3| \\ |h_1 b_2 c_3| & |a_1 h_2 c_3| & |a_1 b_2 h_3| \end{vmatrix} = A^2 \cdot B.$$

These relations may also be simply formed by multiplying columnwise any one of the minors of the array by the adjugate of the first of them and noting that every product of two columns in such a multiplication is equal to one of the minors of the array. Thus the array being

$$\left\|\begin{array}{cccccc} a_1 & b_1 & c_1 & f_1 & g_1 & h_1 \\ a_2 & b_2 & c_2 & f_2 & g_2 & h_2 \\ a_3 & b_3 & c_3 & f_3 & g_3 & h_3 \end{array}\right\|$$

there are $(6)_3$ multiplications to be performed. The first is nugatory the result being

$$|a_1 b_2 c_3|\,|A_1 B_2 C_3| = |a_1 b_2 c_3|^3;$$

so also are the nine, $3(6-3)$, in which the multiplicand has two columns in common with $|a_1 b_2 c_3|$, for example,

$$|a_1 b_2 f_3|\,|A_1 B_2 C_3| = \left|\begin{array}{ccc} |a_1 b_2 c_3| & \cdot & \cdot \\ \cdot & |a_1 b_2 c_3| & \cdot \\ |f_1 b_2 c_3| & |a_1 f_2 c_3| & |a_1 b_2 f_3| \end{array}\right|.$$

Then come those in which the multiplicand has only one column in common with $|a_1 b_2 c_3|$, an example being

$$|a_1 f_2 g_3|\,|A_1 B_2 C_3| = \left|\begin{array}{ccc} |a_1 b_2 c_3| & \cdot & \cdot \\ |f_1 b_2 c_3| & |a_1 f_2 c_3| & |a_1 b_2 f_3| \\ |g_1 b_2 c_3| & |a_1 g_2 c_3| & |a_1 b_2 g_3| \end{array}\right|$$

which on removing the factor $|a_1 b_2 c_3|$, becomes

$$|a_1 b_2 c_3|\,|a_1 f_2 g_3| = |a_1 b_2 f_3|\,|a_1 c_2 g_3| - |a_1 b_2 g_3|\,|a_1 c_2 f_3|$$

and lastly there is

$$|f_1 g_2 h_3|\,|A_1 B_2 C_3| = \left|\begin{array}{ccc} |f_1 b_2 c_3| & |a_1 f_2 c_3| & |a_1 b_2 f_3| \\ |g_1 b_2 c_3| & |a_1 g_2 c_3| & |a_1 b_2 g_3| \\ |h_1 b_2 c_3| & |a_1 h_2 c_3| & |a_1 b_2 h_3| \end{array}\right|$$

in which as many as eleven of the determinants of the array are involved.

The number of relations thus obtained is

$$(6)_3 - 1 - 3(6-3) = 10.$$

241. Let $A \equiv |a_{1n}|$, $B \equiv |b_{1n}|$, and $A \cdot B = C \equiv |c_{1n}|$. Let $A^{(\lambda)}_{(n|m\alpha)}$ represent the array of the αth selection of m rows of the determinant $|a^{(\lambda)}_{1n}|$, let $B^{(\mu)}_{(n|m_{\alpha'})}$ represent the array of the α'th selection of m rows of the determinant $|b^{(\mu)}_{1n}|$ and let $C^{(\lambda,\mu)}_{(n|m\alpha),(n|m_{\alpha'})}$ be the determinant of the mth order formed by taking their product, thus

$$A^{(\lambda)}_{(n|m\alpha)} \cdot B^{(\mu)}_{(n|m\alpha')} = C^{(\lambda,\mu)}_{(n|m\alpha),(n|m\alpha')}$$

which by the theorem of §217 gives

$$(1)\ A^{(\lambda)}_{(n|m\alpha)} \cdot B^{(\mu)}_{(n|m\alpha')} = C^{(\lambda,\mu)}_{(n|m\alpha),(n|m\alpha')} = \sum_{1}^{(n)m} i A^{(\lambda)}_{(n|m\alpha),(n|m_i)} B^{(\mu)}_{(n|m\alpha'),(n|m_i)}$$

and summing with respect to λ and μ we have

$$(2)\quad \begin{aligned} &\sum_{1}^{(n)m} \left[\sum_{1}^{r} \lambda A^{(\lambda)}_{(n|m\alpha),(n|m_i)} \sum_{1}^{s} \mu B^{(\mu)}_{(n|m\alpha'),(n|m_i)} \right] \\ &= \sum_{1}^{r} \lambda \sum_{1}^{s} \mu C^{(\lambda,\mu)}_{(n|m\alpha),(n|m\alpha')} \end{aligned}$$

where r and s are any two integers.

The truth of this relation is easily seen on observing that the left-hand side may be written

$$\sum_{1}^{r} \lambda \sum_{1}^{r} \mu \left[\sum_{1}^{(n)m} i A^{(\lambda)}_{(n|m\alpha),(n|m_i)} B^{(\mu)}_{(n|m\alpha'),(n|m_i)} \right]$$

which is equal to the right-hand side by (1).

If in (2) we make $r = s = (n)_m$ and use for $A^{(\lambda)}_{(n|m\alpha)}$ and $B^{(\mu)}_{(n|m_{\alpha'})}$ the $(n)_m$ selections m at a time of the rows of A and B respectively, we get

$$(3)\quad \begin{aligned} &\sum_{1}^{(n)m} j \left[\sum_{1}^{(n)m} i A_{(n|m_i),(n|m_j)} \sum_{1}^{(n)m} i A_{(n|m_i|,(n|m_j)} \right] \\ &= \sum_{1}^{(n)m} i \cdot \sum_{1}^{(n)m} j C_{(n|m_i),(n|m_j)} \end{aligned}$$

the result of §222 (2).

Expanding $C^{(\lambda,\mu)}_{(n|m\alpha),(n|m_{\alpha'})}$ by Laplace's theorem in terms of minors of order k and $m-k$, we have from (1)

$$(4)\quad \begin{aligned} A^{(\lambda)}_{(n|m\alpha)} B^{(\mu)}_{(n|m\alpha')} &= \sum_{1}^{(n)m} i A^{(\lambda)}_{(n|m\alpha),(n|m_i)} B^{(\mu)}_{(n|m\alpha'),(n|m_i)} = C^{(\lambda,\mu)}_{(n|m\alpha),(n|m\alpha')} \\ &= \sum_{1}^{(m)k} i (-1)^{\nu_1+\nu_2} P \cdot Q \end{aligned}$$

where

$$P = \sum_{1}^{(m)k} j\, A^{(\lambda)}_{(n|m\alpha|k\beta),(n|k_i)} B^{(\mu)}_{(n|m\alpha'|k_i),(n|k_j)}$$

and

$$Q = \sum_{1}^{(n)m-k} j\, A^{(\lambda)}_{(n|\bar{m}\alpha|k\beta),(n|m-k_j)} B^{(\mu)}_{(n|\bar{m}\alpha'|k_i),(n|m-k_j)}$$

where ν_1 = number of inversions in $(n|m_\alpha|k_\beta)(n|\bar{m}_\alpha|k_\beta)$, ν_2 = number of inversions in $(n|m_{\alpha'}|k_i)(n|\bar{m}_{\alpha'}|k_i)$.

Taking the sum of all possible expansions in terms of minors of order k and $m-k$, we have

$$(m)_k A^{(\lambda)}_{(n|m\alpha)} B^{(\mu)}_{(n|m\alpha'|} = (m)_k \sum_{1}^{(n)m} i\, A^{(\lambda)}_{(n|m\alpha),(n|m_i)} B^{(\mu)}_{(n|m\alpha'),(n|m_i)}$$

$$(5) \qquad = (m)_k C^{(\lambda,\mu)}_{(n|m\alpha),(n|m\alpha')} = \sum_{1}^{(m)k} \beta \sum_{1}^{(m)k} i (-1)^{\nu_1+\nu_2} P \cdot Q$$

Taking the sum with alternately positive and negative signs, we have

$$(6) \qquad \sum_{1}^{(m)k} \beta (-1)^{\beta+1} \sum_{1}^{(m)k} i (-1)^{\nu_1+\nu_2} P \cdot Q = A^{(\lambda)}_{(n|m\alpha)} B^{(\mu)}_{(n|m\alpha')},$$

or zero according as $(m)_k$ is odd or even.

From (2) and (3) we have

$$\sum_{1}^{(n)m} j \left[\sum_{1}^{r} \lambda A^{(\lambda)}_{(n|m\alpha),(n|mj)} \sum_{1}^{s} \mu B^{(\mu)}_{(n|m\alpha'),(n|mj)} \right] = \sum_{1}^{r} \lambda \sum_{1}^{s} \mu C^{(\lambda,\mu)}_{(n|m\alpha),(n|m\alpha')}$$

$$(7) \qquad = \sum_{1}^{r} \lambda \sum_{1}^{s} \mu \sum_{1}^{(m)k} i (-1)^{\nu_1+\nu_2} P \cdot Q.$$

If we take the sum of all possible expansions in terms of minors of order k and $m-k$ we get results corresponding to those of (5) and (6).

242. Let $(l)_k = \epsilon$, $(n)_{m+k} = \delta$, $(\delta > \epsilon)$, and let $|\alpha_{\delta\delta}|$, $|\beta_{\delta\delta}|$, $|\gamma_{\delta\delta}|$ represent the $(m+k)$th compounds of A, B, C, (§241) respectively. If then we represent the δ-by-ϵ array

$$\left\| \begin{matrix} A_{(n|m\alpha)(\bar{n}|m\alpha|l\beta|k_1),(n|m+k_1)} & \cdots & A_{(n|m\alpha)(\bar{n}|m\alpha|l\beta|k_1),(n|m+k\delta)} \\ \cdots & \cdots & \cdots \\ A_{(n|m\alpha)(\bar{n}|m\alpha|l\beta|k\epsilon),(n|m+k_1)} & \cdots & A_{(n|m\alpha)(\bar{n}|m\alpha|l\beta|k\epsilon),(n|m+k\delta)} \end{matrix} \right\|$$

by $(\alpha_{\epsilon\delta})$, and the array

$$\left\| \begin{matrix} B_{(n|m\alpha')(\bar{n}|m\alpha'|l\beta'|k_1),(n|m+k_1)} & \cdots & B_{(n|m\alpha')(\bar{n}|m\alpha'|l\beta'|k_1),(n|m+k\delta)} \\ \cdots & \cdots & \cdots \\ B_{(n|m\alpha')(\bar{n}|m\alpha'|l\beta'|k\epsilon),(n|m+k_1)} & \cdots & B_{(n|m\alpha')(\bar{n}|m\alpha'|l\beta'|k\epsilon),(n|m+k\delta)} \end{matrix} \right\|$$

by $(\beta\epsilon\delta)$ we have by (1) §241

$$(1') \qquad (\alpha_{\epsilon\delta})(\beta\epsilon\delta) = \sum_{1}^{(\delta)\epsilon} i \left| \alpha_{\epsilon\epsilon_i} \right| \cdot \left| \beta_{\epsilon\epsilon_i} \right|$$

where $|\alpha_{\epsilon\epsilon_i}|$ and $|\beta\epsilon\epsilon_i|$ denote the determinants formed by taking the ith selection of ϵ columns of $(\alpha\epsilon\delta)$ and $(\beta\epsilon\delta)$ respectively.

The arrays on the left of (1′) are ϵ rows of the $(m+k)$th compound of A and B respectively. The determinants on the right are minors of these compounds. The product of the two arrays is a determinant of order ϵ which may be represented by $|\gamma_{\epsilon\epsilon}|$, every element of which is the sum of products formed by multiplying each minor from a selection of $(m+k)$ rows of A by the corresponding minor from a selection of $(m+k)$ rows of B. Each element, therefore, is by (1) §241 the product of two arrays (one from A and one from B) and consequently is a determinant of order $(m+k)$. It is a minor of the determinant C and we have

$$|\gamma_{\epsilon\epsilon}| = \left| \begin{matrix} C_{(n|m\alpha)(\bar{n}|m\alpha|l\beta|k_1),(n|m\alpha')(\bar{n}|m\alpha'|l\beta'|k_1)} & \cdots & C_{(n|m\alpha')(\bar{n}|m\alpha|l\beta|k_1),(n|m\alpha')(\bar{n}|m\alpha'|l\beta'|k\epsilon)} \\ \cdots & \cdots & \cdots \\ C_{(n|m\alpha)(\bar{n}|m\alpha|l\beta|k\epsilon),(n|m\alpha')(\bar{n}|m\alpha'|l\beta'|k_1)} & \cdots & C_{(n|m\alpha)(\bar{n}|m\alpha|l\beta|k\epsilon),(n|m\alpha')(\bar{n}|m\alpha'|l\beta'|k\epsilon)} \end{matrix} \right|$$

It will be observed that the elements of this determinant are made up by bordering the minor $C_{(n|m\alpha),(n|m\alpha')}$ in all possible ways with k of the remaining rows and columns of the minor $C_{(n|m\alpha)(\bar{n}|m\alpha|l\beta),(n|m\alpha')(\bar{n}|m\alpha'|l\beta')}$. It follows, therefore, by §197 that

$$|\gamma_{\epsilon\epsilon}| = \left[C_{(n|m\alpha),(n|m\alpha')}\right]^{(l-1)k}\left[C_{(n|m\alpha)(\bar{n}|m\alpha|l\beta),(n|m\alpha')(\bar{n}|m\alpha'|l\beta')}\right]^{(l-1)k-1}$$

which by using (1) gives

$$(2) \quad (\alpha_{\epsilon\delta})(\beta_{\epsilon\delta}) = \left[\sum_{1}^{(n)m} i A_{(n|m\alpha),(n|m_i)} B_{(n|m\alpha'),(n|m_i)} \right]^{(l-1)k}$$

$$\times \left[\sum_{1}^{(n)m+l} i A_{(n|m\alpha)(\bar{n}|m\alpha|l\beta),(n|m+l_i)}\, B_{(n|m\alpha')(\bar{n}|m\alpha'|l\beta'),(n|m+l_i)} \right]^{(l-1)k-1}$$

From this and (2) §241 we have

$$(3) \qquad \sum_{1}^{(\delta)\epsilon} i \left[\sum_{1}^{r} \lambda \left| \alpha_{\epsilon\rho\epsilon_i}^{(\lambda)} \right| \sum_{1}^{s} \mu \left| \beta_{\epsilon\rho\epsilon_i}^{(\mu)} \right| \right] = \sum_{1}^{r} \lambda \sum_{1}^{s} \mu \left| \gamma_{\epsilon\rho\epsilon\rho'}^{(\lambda,\mu)} \right|$$

$$= \sum_{1}^{r} \lambda \sum_{1}^{s} \left[\sum_{1}^{(n)m} i A_{(n|m\alpha),(n|m_i)}^{(\lambda)} B_{(n|m\alpha'),(n|m_i)}^{(\mu)} \right]^{(l-1)k}$$

$$\times \left[\sum_{1}^{(n)m+l} i \; A_{(n|m\alpha)(\bar{n}|m\alpha|l\beta),(n|m+l_i)}^{(\lambda)}, B_{(n|m\alpha')(\bar{n}|m\alpha'|l\beta'),(n|m+l_i)}^{(\mu)} \right]^{(l-1)k-1}$$

243. There are various modifications and special cases of the foregoing results which might be obtained and we shall mention a few of them.

If in (1) §241 we take $B \equiv A$, and drop λ and μ it becomes

$$(1'') \qquad C_{(n|m\alpha),(n|m\alpha')} = \sum_{1}^{(n)m} i A_{(n|m\alpha),(n|m_i)} A_{(n|m\alpha'),(n|m_i)}$$

and if in addition $\alpha' \equiv \alpha$

$$(1''') \qquad C_{(n|m\alpha),(n|m\alpha)} = \sum_{1}^{(n)m} i A^{2}_{(n|m\alpha),(n|m_i)}.$$

If in (1) §241, $\alpha' \equiv \alpha$, and we sum with respect to α, we get

$$(1^{\mathrm{IV}}) \qquad \sum_{1}^{(n)m} \alpha C_{(n|m\alpha),(n|m\alpha)} = \sum_{1}^{(n)m} \alpha \sum_{1}^{(n)m} i A_{(n|m\alpha),(n|m_i)} B_{(n|m\alpha),(n|m_i)}.$$

We could express $B_{(n|m\alpha),(n|m_i)}$ in terms of the complementary of the corresponding minor of its adjugate and then write the double sum as a single sum of determinants of order n which have some columns a's and some columns from the adjugate of B. Thus

$$|c_{11}\; c_{22}| + |c_{11}\; c_{33}| + |c_{22}\; c_{33}| = \begin{vmatrix} b_{11} & b_{12} & A_{13} \\ b_{21} & b_{22} & A_{23} \\ b_{31} & b_{32} & A_{33} \end{vmatrix} + \begin{vmatrix} b_{11} & A_{12} & b_{13} \\ b_{21} & A_{22} & b_{23} \\ b_{31} & A_{32} & b_{33} \end{vmatrix} + \begin{vmatrix} A_{11} & b_{12} & b_{13} \\ A_{21} & b_{22} & b_{23} \\ A_{31} & b_{32} & b_{33} \end{vmatrix}.$$

If $B \equiv A$ then,

$$(1^{\mathrm{V}}) \qquad \sum_{1}^{(n)m} \alpha C_{(n|m\alpha),(n|m\alpha)} = \sum_{1}^{(n)m} \alpha \sum_{1}^{(n)m} i A^{2}_{(n|m\alpha),(n|m_i)}.$$

In (1″) we have an expansion for any minor of the square of a determinant in terms of minors of the same order of the determinant itself. For the principal or coaxial minors this takes the form given in (1‴).

In (1^{IV}) we have a similar expression for the sum of the principal minors of the product of two determinants expressed in terms of products of minors of the same order from the two determinants.

In (1^{V}) we have a similar expression for the sum of the principal minors of the square of a determinant.

If

$$\sum_{1}^{(n)m} \alpha\, C_{(n\,|\,m\alpha),\,(n\,|\,m\alpha)} = 0,$$

(1^{V}) shows that every minor of order m of A is zero, and (1″) shows that every minor of order m of A^2 is zero. Hence we have the theorem given at the end of §224.

244. In the relation $C=A\cdot B$, either A or B, or both may be considered the product of two or more factors, thus if $A=A'\cdot A''$, then from (1) §241

$$C_{(n\,|\,m\alpha),\,(n\,|\,m\alpha')} = \sum_{1}^{(n)m} i\left[\sum_{1}^{(n)m} j\, A'_{(n\,|\,m\alpha),\,(n\,|\,m_j)} A''_{(n\,|\,m_i),\,(n\,|\,m_j)}\right] B_{(n\,|\,m\alpha'),\,(n\,|\,m_i)}$$

If $B=B'\cdot B''$ also, then

$$C_{(n\,|\,m\alpha),\,(n\,|\,m\alpha')} = \sum_{1}^{(n)m} i[PQ]. \tag{1}$$

where

$$P = \sum_{1}^{(n)\,m} j\, A'_{(n|m\alpha),\,(n|m_j)}\, A''_{(n|m_i),(n|m_j)}$$

and

$$Q = \sum_{1}^{(n)\,m} k\, A'_{(n|m\alpha),\,(n|mk)}\, B''_{(n|m_i),(n|m_k)}$$

In this way we may get an expression for the minor of a determinant which is the product of any number of given determinants in terms of minors of the same order of the factors.

The relation (3) §241 is an expression for the sum of all the elements in the mth compound of the product of two determinants A and B in terms of the sum of elements from corresponding columns of the mth compounds of A and B.

It becomes when $B \equiv A$

$$(3') \quad \sum_{1}^{(n)m} j \left[\sum_{1}^{(n)m} i A_{(n|m_i),(n|m_j)} \right]^2 = \sum_{1}^{(n)m} i \sum_{1}^{(n)m} j C_{(n|m_i),(n|m_j)}.$$

Subtracting (1^{IV}) §243 from (3) §241 we have

$$(2) \quad \sum_{1}^{(n)m} i \sum_{1}^{(n)m} j C_{(n|m_i),(n|m_j)}$$

$$= \sum_{1}^{(n)m} k \left[\sum_{1}^{(n)m} i A_{(n|m_i),(n|m_k)} \sum_{1}^{(n)m} j B_{(n|m_j),(n|m_k)} \right]$$

where $i \neq j$.

This is an expression for the sum of all elements, except those along the principal diagonal, of the mth compound of the product of two determinants A and B.

If $B \equiv A$, $k=1$, and we drop λ and μ, then (4) §241 becomes

$$(4') \quad \sum_{1}^{(n)m} i A_{(n|m\alpha),(n|m_i)} A_{(n|m\alpha'|(n|m_i)} = C_{(n|m\alpha),(n|m\alpha')}$$

$$= \sum_{1}^{m} i(-1)^{\beta+i} \left[\sum_{1}^{n} j\, a_{\beta j} a_{ij} \sum_{1}^{(n)m-1} j A_{(n|\overline{m\alpha}|1\beta),(n|m-1_j)} A_{(n|\overline{m\alpha'}|1_i),(n|m-1_j)} \right]$$

which, if $\alpha' = \alpha$ and $\beta = m$, becomes

$$(4'') \quad \sum_{1}^{(n)m} i A^2_{(n|m\alpha),(n|m_i)} = C_{(n|m\alpha),(n|m\alpha)}$$

$$= \sum_{1}^{m} i(-1)^{m+i} \left[\sum_{1}^{n} j a_{mj} a_{ij} \sum_{1}^{(n)m-1} j A_{(n|\overline{m\alpha}|1m),(n|m-1_j)} A_{(n|\overline{m\alpha}|1_i),(n|m-1_j)} \right]$$

EXERCISE. By the use of (4″) prove the theorem.

In a matrix of m rows and n columns if at least one minor of order $(m-1)$ *formed from* $(m-1)$ *of the rows (say the βth selection) does not vanish and if every minor of the mth order, which has for its columns the βth selection of* $(m-1)$ *and one other vanishes, then every minor of order m will vanish.*

By a similar method prove the general theorem of §234.

245. If $l = n - m$, then (2) §242 becomes

$$(2') \quad (\alpha_{\epsilon\delta})(\beta_{\epsilon\delta}) = |\gamma_{\epsilon\epsilon}|$$

$$= \left[\sum_{1}^{(n)m} i A_{(n|m\alpha),(n|m_i)} B_{(n|m\alpha'),(n|m_i)} \right]^{(n-m-1)k} [A \cdot B]^{(n-m-1)k-1}.$$

If $B\equiv A$, $m=1$, $l=2$, $k=1$, and if $(n|1_\alpha)(\bar{n}|1_\alpha|2_\beta)$ and $(n|1_{\alpha'})(\bar{n}|1_{\alpha'}|2_{\beta'})$ represent the same numbers as $(n|3_\gamma)$, then (2) becomes

$$(2'') \qquad (\alpha_{\epsilon\delta})(\beta_{\epsilon\delta}) = |\gamma_{\epsilon\epsilon}| = \sum_1^n i a_{\alpha i} a_{\alpha' i} \sum_1^{(n)_3} i A^2_{(n|3\gamma),(n|3_i)}.$$

If $B\equiv A$, $\alpha'=\alpha$, $\beta'=\beta$, then it becomes

$$(2''') \quad (\alpha_{\epsilon\delta})^2 = |\gamma_{\epsilon\epsilon}| = \left[\sum_1^{(n)m} i A^2_{(n|m\alpha),(n|m_i)}\right]^{(l-1)k} \cdot \left[\sum_1^{(n)m+l} i A^2_{(n|m\alpha)(\bar{n}|m\alpha|l\beta),(n|m+l_i)}\right]^{(l-1)k-1}$$

This shows that if

$$|\gamma_{\epsilon\epsilon}| = 0,$$

then

$$A_{(n|m\alpha)(\bar{n}|m\alpha|l\beta),(n|m+l_i)} = 0$$

for $(i=1, 2, \cdots, (n)_{m+l})$.

It is to be observed that if the first of the two factors on the right-hand side of (2‴) vanishes, the second will also vanish.

If in (3) §242 $r=s=(n-m)_l$, and we use instead of $(\alpha^{\lambda}_{\epsilon\delta})$ and $(\beta^{\mu}_{\epsilon\delta})$ the $(n-m)_l$ selections ϵ at a time, of those rows of the $(m+k)$th compounds of A and B, all the elements of which have in common the row numbers $(n|m_\alpha)$ and $(n|m_\alpha')$ respectively, it becomes

$$(3') \quad \begin{aligned} \sum_1^{(\delta)\epsilon} i[\,|\alpha_{\epsilon\rho\epsilon_i}| \cdot |\beta_{\epsilon\rho'\epsilon_i}|\,] &= \sum_1^{s}\rho \sum_1^{s}\rho' |\gamma_{\epsilon\rho\epsilon\rho'}| \\ &= \sum_1^{s}\beta \sum_1^{s}\beta' \left[\sum_1^{(n)m} i A_{(n|m\alpha),(n|m_i)} B_{(n|m\alpha'),(n|m_i)}\right]^{(l-1)k} \\ &\cdot \left[\sum_1^{(n)_{m+l}} i A_{(n|m\alpha)(\bar{n}|m\alpha|l\beta),(n|m+l_i)} B_{(n|m\alpha')(\bar{n}|m\alpha'|l\beta'),(n|m+l_i)}\right]^{(l-1)_{k-1}} \end{aligned}$$

If $B\equiv A$ and $\alpha'=\alpha$ then this becomes

$$(3'') \quad \begin{aligned} \sum_1^{(\delta)\epsilon} i[\,|\alpha_{\epsilon\rho\epsilon_i}|\,]^2 &= \sum_1^{s}\rho \sum_1^{s}\rho' |\gamma_{\epsilon\rho\epsilon\rho'}| \\ &= \sum_1^{s}\beta \sum_1^{s}\beta' \left[\sum_1^{(n)m} i A^2_{(n|m\alpha),(n|m_i)}\right]^{(l-1)k} \\ &\cdot \left[\sum_1^{(n)m+l} i A_{(n|m\alpha)(\bar{n}|m\alpha|l\beta),(n|m+l_i)} A_{(n|m\alpha)(\bar{n}|m\alpha|l\beta'),(n|m+l_i)}\right]^{(l-1)k-1} \end{aligned}$$

If in (3) §242 $l=(n-m)$, $r=s=(n)_m$, and we used for $(\alpha_{\epsilon\delta}^{\lambda})$ and $(\beta_{\epsilon\delta}^{\mu})$ the $(n)_m$ selections ϵ at a time of those rows of the $(m+k)$th compounds of A and B, the elements of each selection of which having in common the same selection of row numbers, it becomes

$$(3''') \quad \sum_{1}^{(\delta)\epsilon} i\left[\sum_{1}^{(n)m}\rho \left|\alpha_{\epsilon\rho\epsilon_i}\right| \sum_{1}^{(n)m}\rho' \left|\beta_{\epsilon\rho'\epsilon_i}\right|\right] = \sum_{1}^{(n)m}\rho \sum_{1}^{(n)m}\rho' \left|\gamma_{\epsilon\rho\epsilon\rho'}\right|$$

$$= \sum_{1}^{(n)m}\alpha \sum_{1}^{(n)m}\alpha' \left[\sum_{1}^{(n)m} i\, A_{(n|m\alpha),(n|m_i)} B_{(n|m\alpha'),(n|m_i)}\right]^{(n-m-1)k} \left[A\cdot B\right]^{(n-m-1)k-1}.$$

246. Given the determinant $A\equiv|a_{nn}|$, $B\equiv|b_{nn}|$ and $C\equiv|c_{nn}|$ where $A\cdot B=C$. Denote the rth compounds of A, B, C by $|A|_r\equiv|\alpha_{\rho\rho}|$, $|B|_r\equiv|\beta_{\rho\rho}|$, and $|C|_r\equiv|\gamma_{\rho\rho}|$ respectively where $\rho=(n)_r$.

Denote the arrays consisting of the first m rows of A and of B by (a_{mn}) and (b_{mn}) respectively. Let $|c_{mm}|$ denote the minor of C which is the product of (a_{mn}) and (b_{mn}). Let $(\alpha_{\mu\rho})$ and $(\beta_{\mu\rho})$ be the arrays consisting of those rows of $|A|_r$ and $|B|_r$ which involve only the first m rows of A and B, $(\mu=(m)_r)$. Let $|\gamma_{\mu\mu}|$ be the minor of $|C|_r$ which is the product of $(\alpha_{\mu\rho})$ and $(\beta_{\mu\rho})$. Since $|\gamma_{\mu\mu}|$ is the rth compound of $|c_{mm}|$, we have

$$|\gamma_{\mu\mu}| = (\alpha_{\mu\rho})\,(\beta_{\mu\rho}) = |c_{mm}|^{(m-1)r-1}$$
$$= \left\{(a_{mn})\,(b_{mn})\right\}^{(m-1)r-1}$$

If $m=n$, then this becomes

$$|\gamma_{\rho\rho}| = \left\{|a_{nn}|\,|b_{nn}|\right\}^{(n-1)r-1}$$

or

$$|C|_r = |A|_r\cdot|B|_r.$$

That is the rth compound of the product of two determinants is equal to the product of the rth compounds of these two determinants.

If we write the product of A and B in the form

$$\begin{vmatrix} |a_{nn}| & |0_{nn}| \\ |0_{nn}| & |b_{nn}| \end{vmatrix} = C,$$

where $|0_{nn}|$ stands for a square of zeros,
and form the rth compound of each side, keeping the non-zero elements on the left in two squares by themselves, we have

$$\begin{vmatrix} |\alpha_{\rho\rho}| & |0_{\rho\rho}| \\ |0_{\rho\rho}| & |\beta_{\rho\rho}| \end{vmatrix} = |C|_r$$

or $|A|_r|B|_r=|C|_r$ as before.

247. Let $A=|a_{1n}|$, $B=|b_{1n}|$ and let A_{n-r} stand for a minor of order $n-r$ of A, A'_{n-r} stand for a minor of order $n-r$ of the adjugate of A; similarly for B.

If now

$$P = \left| \lambda a_{11} + \mu b_{11} \quad \cdots \quad \lambda a_{nn} + \mu b_{nn} \right| ,$$

and

$$Q = \left| \lambda \frac{B_{11}}{B} + \mu \frac{A_{11}}{A} \quad \cdots \quad \lambda \frac{B_{nn}}{B} + \mu \frac{A_{nn}}{A} \right| ,$$

then

$$P = A \cdot B \cdot Q$$

For the coefficient of $\lambda^{n-r} \cdot \mu^r$ in Q is

$$\frac{1}{B^{n-r}} \frac{1}{A^r} \sum (B'_{n-r} \text{ comp } A'_{n-r}) ,$$

where $(B'_{n-r} \text{ comp } A'_{n-r})$ represents the product of a minor of the adjugate of B of order $n-r$ times the complementary of the corresponding minor in the adjugate of A. Substituting for these minors their values in terms of minors of A and B we get for the coefficient of $\lambda^{n-r} \cdot \mu^r$

$$\frac{1}{B^{n-r} A^r} \sum (B^{n-r-1} \text{ comp } B_{n-r} \cdot A^{r-1} \cdot A_{n-r}), \quad \text{or}$$

$$\frac{1}{A \cdot B} \sum (A_{n-r} \text{ comp } B_{n-r}) = \frac{1}{A \cdot B} \cdot \text{coeff. of } \lambda^{n-r} \cdot \mu^r \text{ in } P.$$

The coefficient of $\lambda^{n-r} \cdot \mu^r$ in P is therefore $A \cdot B$ times the coefficient of $\lambda^{n-r} \cdot \mu^r$ in Q, and $P = A \cdot B \cdot Q$.

248. *If two determinants be reciprocals their kth compounds are also reciprocals.*

Let the two determinants be $A \equiv |a_{1n}|$ and $B \equiv |b_{1n}|$, then

$$|a_{1n}| \; |b_{1n}| = \begin{vmatrix} 1 & \cdot & \cdot & \cdot & \cdot \\ \cdot & 1 & \cdot & \cdot & \cdot \\ \cdot & \cdot & 1 & \cdot & \cdot \\ \cdot & \cdot & \cdot & \cdot & \cdot \\ \cdot & \cdot & \cdot & \cdot & 1 \end{vmatrix} .$$

Let the kth compounds of A and B be represented by

$$A' \equiv \begin{vmatrix} \alpha_{11} & \alpha_{12} \cdots & \alpha_{1\lambda} \\ \cdots & \cdots & \cdots \\ \alpha_{\lambda 1} & \alpha_{\lambda 2} \cdots & \alpha_{\lambda\lambda} \end{vmatrix} \text{ and } B' \equiv \begin{vmatrix} \beta_{11} & \beta_{12} \cdots & \beta_{1\lambda} \\ \cdots & \cdots & \cdots \\ \beta_{\lambda 1} & \beta_{\lambda 2} \cdots & \beta_{\lambda\lambda} \end{vmatrix}$$

where $\lambda = (n)_k$.

Then, since

$$\alpha_{11}\beta_{11} + \alpha_{12}\beta_{12} + \cdots + \alpha_{1\lambda}\beta_{1\lambda}$$

$$= \begin{Vmatrix} a_{11} & a_{12} \cdots & a_{1n} \\ \cdots & \cdots & \cdots \\ a_{k1} & a_{k2} \cdots & a_{kn} \end{Vmatrix} \cdot \begin{Vmatrix} b_{11} & b_{21} \cdots & b_{n1} \\ \cdots & \cdots & \cdots \\ b_{1k} & b_{2k} \cdots & b_{nk} \end{Vmatrix} = \begin{vmatrix} 1 & \cdot & \cdot & \cdot \\ \cdot & 1 & \cdot & \cdot \\ \cdot & \cdot & \cdot & \cdot \\ \cdot & \cdot & \cdot & 1 \end{vmatrix}_k = 1,$$

and similarly for the other elements of the product. It follows that

$$A' \cdot B' = \begin{vmatrix} 1 & \cdot & \cdot & \cdot & \cdot \\ \cdot & 1 & \cdot & \cdot & \cdot \\ \cdot & \cdot & \cdot & \cdot & \cdot \end{vmatrix}_\lambda = 1$$

or A' and B' are reciprocals.

249. *If two n-line determinants be such that their product equals*

$$\begin{vmatrix} s & \cdot & \cdot & \cdot & \cdot \\ \cdot & s & \cdot & \cdot & \cdot \\ \cdot & \cdot & s & \cdot & \cdot \\ \cdot & \cdot & \cdot & \cdot & \cdot \end{vmatrix}_n$$

then any m-line minor of either is equal to s^m *multiplied by the cofactor of the corresponding minor of the other and divided by that other.*

To prove this we have simply to multiply the m-line minor raised to the order n in the usual way by the other determinant, whence the result appears.

250. *If an array of* $n-1$ *rows and n columns be such that the sum of the elements in every one of the rows vanishes, the principal minor determinants of the array, when taken alternately positive and negative, are equal to one another.*

Let the array be

$$\begin{matrix} a_{11} & a_{12} & \cdots a_{1n} \\ a_{21} & a_{22} & \cdots a_{2n} \\ \cdots & \cdots & \cdots \\ a_{n-1,1} & a_{n-1,2} & \cdots a_{n-1,n} \end{matrix}$$

and let M_r denote the determinant formed from it by deleting the rth column.

The determinant

$$\Delta \equiv \begin{vmatrix} a_{11} + a_{1r} & a_{12} & \cdots a_{1,r-1} & a_{1,r+1} & \cdots a_{1n} \\ a_{21} + a_{2r} & a_{22} & \cdots a_{2,r-1} & a_{2,r+1} & \cdots a_{2n} \\ \cdots & \cdots & \cdots & \cdots & \cdots \\ a_{n-1,1} + a_{n-1,r} & a_{n-1,2} & \cdots a_{n-1,r-1} & a_{n-1,r+1} & \cdots a_{n-1,n} \end{vmatrix}$$

is zero, because the sum of the elements in each row is zero. Therefore, since

$$\Delta = (-1)^{r-2} M_1 + M_r,$$

we have $M_r = (-1)^{r-1} M_1$ for $(r = 1, 2, \cdots, n)$.

It follows from this that *if the sum of the elements in every row but one is zero then the cofactors of the elements of the excepted row are identical.* It also follows that *any determinant which has the sum of the elements in every row equal to zero has in the case of every row the same cofactor for every element of the row.* A symmetrical determinant which has this property would obviously have the cofactors of all the elements identical.

251. Let $\sum |x_1 y_2 z_3|$ represent the sum of the minors formed from the three sets of n quantitites

$$\begin{matrix} x_1 & x_2 & x_3 \cdots x_n \\ y_1 & y_2 & y_3 \cdots y_n \\ z_1 & z_2 & z_3 \cdots z_n \end{matrix}$$

then

$$(1) \quad \sum |x_1\, y_2\, z_3| = \sum x \sum |y_1\, z_2| + \sum y \sum |z_1\, x_2| + \sum z \sum |x_1\, y_2| .$$

If we expand all the determinants on the left having x_1, y_1, z_1 as their first column we get as their sum

$$\begin{aligned} &x_1(|y_2 z_3| + |y_2 z_4| + \cdots + |y_{n-1} z_n|) \\ &+ y_1(|z_2 x_3| + |z_2 x_4| + \cdots + |z_{n-1} x_n|) \\ &+ z_1(|x_2 y_3| + |x_2 y_4| + \cdots + |x_{n-1} y_n|). \end{aligned}$$

Then since

$$\begin{aligned} &x_1(|y_1 z_2| + |y_1 z_3| + \cdots + |y_1 z_n|) \\ &+ y_1(|z_1 x_2| + |z_1 x_3| + \cdots + |z_1 x_n|) \\ &+ z_1(|x_1 y_2| + |x_1 y_3| + \cdots + |x_1 y_n|) = 0, \end{aligned}$$

as may be seen by adding the columns, we may write the sum as follows:

$$\begin{aligned} &x_1(|y_1 z_2| + \cdots + |y_1 z_n| + |y_2 z_3| + \cdots + |y_{n-1} z_n|) \\ &+ y_1(|z_1 x_2| + \cdots + |z_1 x_n| + |z_2 x_3| + \cdots + |z_{n-1} x_n|) \\ &+ z_1(|x_1 y_2| + \cdots + |x_1 y_n| + |x_2 y_3| + \cdots + |x_{n-1} y_n|). \end{aligned}$$

Similarly for those terms on the left having x_2, y_2, z_2 as one column, x_3, y_3, z_3 as one column, and so on. If we add all these together we get

$$\sum x \sum |y_1 z_2| + \sum y \sum |z_1 x_2| + \sum z \sum |x_1 y_2|,$$

but each determinant on the left has been taken three times, twice with the positive sign and once with the negative sign, so that

$$\sum |x_1 y_2 z_3| = \sum x \sum |y_1 z_2| + \sum y \sum |z_1 x_2| + \sum z \sum |x_1 y_2|.$$

EXERCISE. Show that

$$\begin{aligned} &\sum w_1 \sum |x_1 y_2 z_3| - \sum x_1 \sum |y_1 z_2 w_3| \\ &\qquad + \sum y_1 \sum |z_1 w_2 x_3| - \sum z_1 \sum |w_1 x_2 y_3| = 0. \end{aligned}$$

252. From the equations

$$\begin{aligned} |x_1 y_2 z_3| &= x_1 |y_2 z_3| + y_1 |z_2 x_3| + z_1 |x_2 y_3| \\ &= x_2 |y_3 z_1| + y_2 |z_3 x_1| + z_2 |x_3 y_1| \\ &= x_3 |y_1 z_2| + y_3 |z_1 x_2| + z_3 |x_1 y_2| \\ |x_1 y_2 z_4| &= \text{etc.} \end{aligned}$$

by squaring and adding we get

$$\begin{aligned} (1)\quad 3 \sum |x_1 y_2 z_3|^2 = &\sum x^2 \sum |y_1 z_2|^2 + \sum y^2 \sum |z_1 x_2|^2 \\ &+ \sum z^2 \sum |x_1 y_2|^2 + 2 \sum yz \sum (|z_1 x_2| \cdot |x_1 y_2|) \\ &+ 2 \sum zx \sum (|x_1 y_2|\ |y_1 z_2|) \\ &+ 2 \sum xy \sum (|y_1 z_2|\ |z_1 x_2|) \end{aligned}$$

$$= 3 \left\| \begin{matrix} x_1 & \cdots & x_n \\ y_1 & \cdots & y_n \\ z_1 & \cdots & z_n \end{matrix} \right\|^2 .$$

Substituting for $\sum |y_1 z_2|^2$, $\sum |z_1 x_2|^2$, etc. their values given by the multiplication theorem we get

$$(2) \quad \sum | x_1\, y_2\, z_3 |^2 = \sum x^2 \sum y^2 \sum z^2 + 2 \sum xy \sum yz \sum zx$$
$$- \sum x^2 (\sum yz)^2 - \sum y^2 (\sum zx)^2 - \sum z^2 (\sum xy)^2 .$$

This, however, may be gotten directly by the multiplication theorem. In fact the simplest way to get (1) is to multiply both sides of (2) by 3 and then reverse the process of the multiplication theorem. Thus

$$\begin{aligned} 3 \sum | x_1\, y_2\, z_3 |^2 = \; & \sum x^2 [\sum y^2 \sum z^2 - (\sum yz)^2] \\ + \; & \sum y^2 [\sum x^2 \sum z^2 - (\sum xz)^2] \\ + \; & \sum z^2 [\sum x^2 \sum y^2 - (\sum xy)^2] \\ + \; & 2 \sum xy (\sum yz \sum xz - \sum xy \sum z^2) \\ + \; & 2 \sum yz (\sum zx \sum xy - \sum yz \sum x^2) \\ + \; & 2 \sum zx (\sum xy \sum yz - \sum xz \sum y^2) \end{aligned}$$

which by using the reverse of the multiplication theorem gives (1).

Writing (2) in the form

$$\begin{aligned} \sum | x_1\, y_2\, z_3 |^2 = \; & \sum y^2 [\sum z^2 \sum x^2 - (\sum zx)^2] \\ + \; & \sum z^2 [\sum x^2 \sum y^2 - (\sum xy)^2] \\ - \; & \sum x^2 [\sum y^2 \sum z^2 - (\sum yz)^2] \\ + \; & 2 \sum yz [\sum zx \sum xy - \sum yz \sum x^2] \end{aligned}$$

we get

$$\sum | x_1\, y_2\, z_3 |^2 = \sum y^2 \sum | z_1\, x_2 |^2 + \sum z^2 \sum | x_1\, y_2 |^2$$
$$- \sum x^2 \sum | y_1\, z_2 |^2 + 2 \sum yz \sum | x_1\, y_2 | \, | z_1\, x_2 | ,$$

similarly

$$= \sum z^2 \sum | x_1\, y_2 |^2 + \sum x^2 \sum | y_1\, z_2 |^2$$
$$- \sum y^2 \sum | z_1\, x_2 |^2 + 2 \sum zx \sum | y_1\, z_2 | \, | x_1\, y_2 |$$
$$= \sum x^2 \sum | y_1\, z_2 |^2 + \sum y^2 \sum | z_1\, x_2 |^2$$
$$(3) \qquad - \sum z^2 \sum | x_1\, y_2 |^2 + 2 \sum xy \sum | z_1\, x_2 | \, | y_1\, z_2 |$$

These results might be obtained from (4″) §244 on making therein $m = 3$.

By adding equations (3) in pairs we get

$$(4)\quad \begin{aligned}\sum |x_1\, y_2\, z_3|^2 &= \sum x^2 \sum |y_1\, z_2|^2 + \sum zx \sum |y_1\, z_2|\,|x_1\, y_2| \\ &\quad + \sum xy \sum |z_1\, x_2|\,|y_1\, z_2| = \sum y^2 \sum |z_1\, x_2|^2 \\ &\quad + \sum xy \sum |z_1 x_2|\,|y_1 z_2| + \sum yz \sum |x_1 y_2|\,|z_1 x_2| \\ &= \sum z^2 \sum |x_1\, y_2|^2 + \sum yz \sum |x_1\, y_2|\,|z_1\, x_2| \\ &\quad + \sum zx \sum |y_1\, z_2|\,|x_1\, y_2|.\end{aligned}$$

From the three rows of n quantities

$$\text{(a)}\qquad \begin{cases} x_1 & x_2 \cdots x_n \\ y_1 & y_2 \cdots y_n \\ z_1 & z_2 \cdots z_n \end{cases}$$

form three other rows of $\frac{1}{2}n(n-1)$ quantities

$$\text{(b)}\qquad \begin{cases} |y_1\, z_2|, & |y_1\, z_3|, \cdots |y_{n-1}\, z_n| \\ |z_1\, x_2|, & |z_1\, x_3|, \cdots |z_{n-1}\, x_n| \\ |x_1\, y_2|, & |x_1\, y_3|, \cdots |x_{n-1}\, y_n| \end{cases}$$

and from these in turn the following:

$$\text{(c)}\qquad \begin{cases} \begin{vmatrix} |z_1\, x_2| & |z_1\, x_3| \\ |x_1\, y_2| & |x_1\, y_3| \end{vmatrix}, \cdots \begin{vmatrix} |z_{n-2}\, x_n| & |z_{n-1}\, x_n| \\ |x_{n-2}\, y_n| & |x_{n-1}\, y_n| \end{vmatrix} \\ \begin{vmatrix} |x_1\, y_2| & |x_1\, y_3| \\ |y_1\, z_2| & |y_1\, z_3| \end{vmatrix}, \cdots \begin{vmatrix} |x_{n-2}\, y_n| & |x_{n-1}\, y_n| \\ |y_{n-2}\, z_n| & |y_{n-1}\, z_n| \end{vmatrix} \\ \begin{vmatrix} |y_1\, z_2| & |y_1\, z_3| \\ |z_1\, x_2| & |z_1\, x_3| \end{vmatrix}, \cdots \begin{vmatrix} |y_{n-2}\, z_n| & |y_{n-1}\, z_n| \\ |z_{n-2}\, x_n| & |z_{n-1}\, x_n| \end{vmatrix} \end{cases}$$

For the sake of convenience let us denote the quantities in (b) by

$$\text{(b)}\qquad \begin{cases} x_1', & x_2', \cdots x_\lambda' \\ y_1', & y_2', \cdots y_\lambda' \quad \text{where} \quad \lambda = (n)_2 = \frac{1}{2}n(n-1) \\ z_1', & z_2', \cdots z_\lambda' \end{cases}$$

and the quantities (c) by

$$\text{(c)}\qquad \begin{cases} x_1'', & x_2'', \cdots x_\mu'' \quad \text{where } \mu = \frac{1}{2}\lambda(\lambda-1) \\ y_1'', & y_2'', \cdots y_\mu'' \quad\quad = \frac{1}{2}\left\{\frac{1}{2}n(n-1)\right\}\left\{\frac{1}{2}n(n-1)-1\right\} \\ z_1'', & z_2', \cdots z_\mu'' \quad\quad = \frac{1}{8}(n+1)n(n-1)(n-2). \end{cases}$$

If we multiply the array consisting of the first two rows of (b) by itself we get

$$\begin{aligned}\sum \| y_1\, z_2 \| \, | z_1\, x_3 | \|^2 = {} & \sum | y_1\, z_2 |^2 \sum | z_1\, x_2 |^2 \\ & - \left(\sum | y_1\, z_2 | \, | z_1\, x_2 | \right)^2 = \Big\{ \sum y_1^2 \sum z_1^2 \\ & - \left(\sum y_1\, z_1 \right)^2 \Big\} \Big\{ \sum z_1^2 \sum x_1^2 - \left(\sum x_1\, z_1 \right)^2 \Big\} \\ & - \Big\{ \sum y_1\, z_1 \sum x_1\, z_1 - \sum y_1\, x_1 \sum z_1^2 \Big\}^2 \\ = {} & \sum z_1^2 \Big[\sum x_1^2 \sum y_1^2 \sum z_1^2 \\ & + 2 \sum y_1 z_1 \sum y_1 x_1 \sum z_1 x_1 - \sum y_1^2 \left(\sum x_1 z_1 \right)^2 \\ & - \sum z_1^2 \left(\sum x_1 y_1 \right)^2 - \sum x_1^2 \left(\sum y_1 z_1 \right)^2 \Big] \\ = {} & \sum z_1^2 \sum | x_1\, y_2\, z_3 |^2 \end{aligned}$$

or

$$\sum z_1''^2 = \sum z_1^2 \sum | x_1\, y_2\, z_3 |^2$$

similarly

$$\sum y_1''^2 = \sum y_1^2 \sum | x_1\, y_2\, z_3 |^2$$

(5)
$$\begin{aligned} \sum x_1''^2 &= \sum x_1^2 \sum | x_1\, y_2\, z_3 |^2 \\ \sum x_1'' y_1' &= \sum x_1 y_1 \sum | x_1\, y_2\, z_3 |^2 \\ \sum x_1'' z_1'' &= \sum x_1 z_1 \sum | x_1\, y_2\, z_3 |^2 \\ \sum y_1'' z_1'' &= \sum y_1 z_1 \sum | x_1\, y_2\, z_3 |^2. \end{aligned}$$

From (b) we have

$$\begin{aligned}\sum | x_1'\; y_2'\; z_3' |^2 = {} & \sum x_1'^2 \sum y_1'^2 \sum z_1'^2 + 2 \sum x_1' y_1' \sum x_1' z_1' \sum y_1' z_1' \\ & - \sum x_1'^2 \left(\sum y_1' z_1' \right)^2 - \sum y_1'^2 \left(\sum x_1' z_1' \right)^2 - \sum z_1'^2 \left(\sum x_1 y_1 \right)^2 \end{aligned}$$

Substituting on the right for

$$\begin{aligned} \sum x_1'^2 &= \sum y_1^2 \sum z_1^2 - \left(\sum y_1 z_1 \right)^2 \\ \sum y_1'^2 &= \sum z_1^2 \sum x_1^2 - \left(\sum x_1 z_1 \right)^2 \end{aligned}$$

etc. we get

(6)
$$\sum | x_1'\; y_2'\; z_3' |^2 = \Big\{ \sum | x_1\, y_2\, z_3 |^2 \Big\}^2$$

This result might have been obtained from (2) §242 on putting therein $m=0$, $k=2$, $l=3$, and $B \equiv A$.

Instead of 3 rows of n quantities we might take m rows of n quantities and form other sets of m rows by taking as elements the determinants of order $(m-1)$ formed from these and so on. For these we would find results corresponding to those of (1) · · · (6).

EXERCISE. Given an m-by-n array, show that all relations connecting the determinants of the mth order formed from this array are deducible from quadratic relations by means of multiplications and divisions.

253. If $A=|a_{1n}|$, and A_{nn} is its primary minor complementary to a_{nn}, and if S and s are the sums of the primary coaxial minors of A and A_{nn} respectively then

$$\begin{vmatrix} A_{nn} & s \\ A & S \end{vmatrix} = \begin{Vmatrix} a_{11} & a_{12} & \cdots & a_{1n} \\ \cdots & \cdots & \cdots & \cdots \\ a_{n-1,1} & a_{n-1,2} & \cdots & a_{n-1,n} \end{Vmatrix} \begin{Vmatrix} a_{11} & a_{21} & \cdots & a_{n1} \\ \cdots & \cdots & \cdots & \cdots \\ a_{1,n-1} & a_{2,n-1} & \cdots & a_{n,n-1} \end{Vmatrix}.$$

Thus for $n=4$ we have

$$\begin{vmatrix} A_{44} & |a_{11}\,a_{22}| + |a_{11}\,a_{33}| + |a_{22}\,a_{33}| \\ A & A_{11}+A_{22}+A_{33}+A_{44} \end{vmatrix}$$

$$= \begin{Vmatrix} a_{11} & a_{12} & \cdots & a_{14} \\ \cdots & \cdots & \cdots & \cdots \\ a_{31} & a_{32} & \cdots & a_{34} \end{Vmatrix} \begin{Vmatrix} a_{11} & a_{21} & \cdots & a_{41} \\ \cdots & \cdots & \cdots & \cdots \\ a_{13} & a_{23} & \cdots & a_{43} \end{Vmatrix}.$$

The truth of which is seen on observing that the right-hand member equals

$$A_{44}^2 + A_{34}A_{43} + A_{24}A_{42} + A_{14}A_{41}$$

and that

$$A[\,|a_{11}\,a_{22}| + |a_{11}\,a_{33}| + |a_{22}\,a_{33}|\,]$$
$$= |A_{33}\,A_{44}| + |A_{22}\,A_{44}| + |A_{11}\,A_{44}|$$

The truth of the general theorem follows in the same way.

254. The product of the arrays of the first 4 rows of the determinants $A\equiv|a_{15}|$ and $B\equiv|b_{15}|$ gives

$$\begin{Vmatrix} a_{11} & a_{12} & a_{13} & a_{14} & a_{15} \\ a_{21} & a_{22} & a_{23} & a_{24} & a_{25} \\ a_{31} & a_{32} & a_{33} & a_{34} & a_{35} \\ a_{41} & a_{42} & a_{43} & a_{44} & a_{45} \end{Vmatrix} \begin{Vmatrix} b_{11} & b_{12} & b_{13} & b_{14} & b_{15} \\ b_{21} & b_{22} & b_{23} & b_{24} & b_{25} \\ b_{31} & b_{32} & b_{33} & b_{34} & b_{35} \\ b_{41} & b_{42} & b_{43} & b_{44} & b_{45} \end{Vmatrix}$$
$$= A_{55}B_{55} + A_{54}B_{54} + A_{53}B_{53} + A_{52}B_{52} + A_{51}B_{51}$$

also

$$= \begin{vmatrix} c_{11}+a_{15}b_{15} & c_{12}+a_{15}b_{25} & c_{13}+a_{15}b_{35} & c_{14}+a_{15}b_{45} \\ c_{21}+a_{25}b_{15} & c_{22}+a_{25}b_{25} & c_{23}+a_{25}b_{35} & c_{24}+a_{25}b_{45} \\ c_{31}+a_{35}b_{15} & c_{32}+a_{35}b_{25} & c_{33}+a_{35}b_{35} & c_{34}+a_{35}b_{45} \\ c_{41}+a_{45}b_{15} & c_{42}+a_{45}b_{25} & c_{43}+a_{45}b_{35} & c_{44}+a_{45}b_{45} \end{vmatrix} \equiv C,$$

say, where $c_{rs}=a_{r1}b_{s1}+a_{r2}b_{s2}+a_{r3}b_{s3}+a_{r4}b_{s4}$. But

$$C = \begin{vmatrix} c_{11} & c_{12} & c_{13} & c_{14} & -a_{15} \\ c_{21} & c_{22} & c_{23} & c_{24} & -a_{25} \\ c_{31} & c_{32} & c_{33} & c_{34} & -a_{35} \\ c_{41} & c_{42} & c_{43} & c_{44} & -a_{45} \\ b_{15} & b_{25} & b_{35} & b_{45} & 1 \end{vmatrix}$$

$$= - \begin{vmatrix} c_{11} & c_{12} & c_{13} & c_{14} & a_{15} \\ c_{21} & c_{22} & c_{23} & c_{24} & a_{25} \\ c_{31} & c_{32} & c_{33} & c_{34} & a_{35} \\ c_{41} & c_{42} & c_{43} & c_{14} & a_{45} \\ b_{15} & b_{25} & b_{35} & b_{45} & 0 \end{vmatrix} + A_{55}B_{55}$$

or

$$-\Delta + A_{55}B_{55}, \quad \text{say},$$

Therefore

$$-\Delta = A_{54}B_{54} + A_{53}B_{53} + A_{52}B_{52} + A_{51}B_{51}.$$

This may be looked upon as stating that *the product of two determinants A and B each of order n bordered is equal to the sum of the products $A_k \cdot B_k$ where A_k is the determinant formed by taking the elements in the bordering column with the kth selection of $(n-1)$ of the columns of A and B_k is the determinant formed by taking for one column the elements of the bordering row together with the kth selection of columns of B.*

Give the result when $A \equiv B$.

255. The determinant

$$\Delta \equiv \begin{vmatrix} a_1 & a_2 & a_3 & a_4 & \cdot \\ b_1 & b_2 & b_3 & b_4 & \cdot \\ c_1 & c_2 & c_3 & c_4 & \cdot \\ d_1 & d_2 & d_3 & d_4 & m_1d_1+m_2d_2+m_3d_3+m_4d_4 \\ e_1 & e_2 & e_3 & e_4 & m_1e_1+m_2e_2+m_3e_3+m_4e_4 \end{vmatrix}$$

is obviously the equivalent of

$$\begin{vmatrix} a_1 & a_2 & a_3 & a_4 & -\sum ma \\ b_1 & b_2 & b_3 & b_4 & -\sum mb \\ c_1 & c_2 & c_3 & c_4 & -\sum mc \\ d_1 & d_2 & d_3 & d_4 & \cdot \\ e_1 & e_2 & e_3 & e_4 & \cdot \end{vmatrix}$$

Expanding Δ in this last form in terms of minors of order two and their complementaries formed from the elements of the last two rows we get:

$$\Delta = -\begin{vmatrix} d_1 & d_2 \\ e_1 & e_2 \end{vmatrix} \begin{vmatrix} a_3 & a_4 & \sum ma \\ b_3 & b_4 & \sum mb \\ c_3 & c_4 & \sum mc \end{vmatrix} + \begin{vmatrix} d_1 & d_3 \\ e_1 & e_3 \end{vmatrix} \begin{vmatrix} a_2 & a_4 & \sum ma \\ b_2 & b_4 & \sum mb \\ c_2 & c_4 & \sum mc \end{vmatrix} - \cdots$$

$$= |d_1 e_2| \cdot |D_1 m_2| + |d_1 e_3|\,|D_1 m_3| + \cdots + |d_3 e_4|\,|D_3 m_4|$$

or

$$\left\| \begin{matrix} d_1 & d_2 & d_3 & d_4 \\ e_1 & e_2 & e_3 & e_4 \end{matrix} \right\| \cdot \left\| \begin{matrix} D_1 & D_2 & D_3 & D_4 \\ m_1 & m_2 & m_3 & m_4 \end{matrix} \right\|$$

where D_r is the cofactor of d_r in $|a_1 b_2 c_3 d_4|$.

In precisely a similar way it is seen that the determinant

$$\begin{vmatrix} a_1 & a_2 & a_3 & a_4 & \cdot & \cdot \\ b_1 & b_2 & b_3 & b_4 & \cdot & \cdot \\ c_1 & c_2 & c_3 & c_4 & \cdot & \cdot \\ d_1 & d_2 & d_3 & d_4 & \sum md & \sum nd \\ e_1 & e_2 & e_3 & e_4 & \sum me & \sum ne \\ f_1 & f_2 & f_3 & f_4 & \sum mf & \sum nf \end{vmatrix}$$

$$= |d_1 e_2 f_3|\,|D_1 m_2 n_3| + |d_1 e_2 f_4|\,|D_1 m_2 n_4| + |d_1 e_3 f_4|\,|D_1 m_3 n_4| + |d_2 e_3 f_4|\,|D_2 m_3 n_4|$$

or

$$\left\| \begin{matrix} d_1 & d_2 & d_3 & d_4 \\ e_1 & e_2 & e_3 & e_4 \\ f_1 & f_2 & f_3 & f_4 \end{matrix} \right\| \left\| \begin{matrix} D_1 & D_2 & D_3 & D_4 \\ m_1 & m_2 & m_3 & m_4 \\ n_1 & n_2 & n_3 & n_4 \end{matrix} \right\|$$

EXERCISES.

1. Given

$$k\begin{Vmatrix} b_1 & b_2 & b_3 \\ c_1 & c_2 & c_3 \end{Vmatrix} = \begin{Vmatrix} x_1 & x_2 & x_3 \\ y_1 & y_2 & y_3 \end{Vmatrix}$$

where all the letters denote integers and suppose the highest common factor of $|b_1c_2|$, $|b_1c_3|$, $|b_2c_3|$ is 1, then four integers α, β, γ, δ may be found such that

$$\begin{vmatrix} \alpha & \beta \\ \gamma & \delta \end{vmatrix} \cdot \begin{Vmatrix} b_1 & b_2 & b_3 \\ c_1 & c_2 & c_3 \end{Vmatrix} = \begin{Vmatrix} x_1 & x_2 & x_3 \\ y_1 & y_2 & y_3 \end{Vmatrix}$$

2. If $|a_1b_2c_3| = 0$, then

$$\frac{|a_1\, b_2\, x_3|}{|a_1\, b_2|} = \frac{|a_1\, c_2\, x_3|}{|a_1\, c_2|} = \frac{|b_1\, c_2\, x_3|}{|b_1\, c_2|}.$$

3. The product

$$\begin{Vmatrix} \alpha_1 & \beta_1 & \gamma_1 & \delta_1 \\ \alpha_2 & \beta_2 & \gamma_2 & \delta_2 \\ \cdot & \cdot & \cdot & \cdot \\ \alpha_n & \beta_n & \gamma_n & \delta_n \end{Vmatrix}$$

$$\begin{Vmatrix} \cos\phi & \cos(s+1)\phi & \cdots & \cos(s+n-1)\phi \\ s\cos s\phi & (s+1)\cos(s+1)\phi & \cdots & (s+n-1)\cos(s+n-1)\phi \\ \sin s\phi & \sin(s+1)\phi & \cdots & \sin(s+n-1)\phi \\ s\sin s\phi & (s+1)\sin(s+1)\phi & \cdots & (s+n-1)\sin(s+n-1)\phi \end{Vmatrix}$$

is obviously zero for $n > 4$. Show that when $n = 4$ the product is independent of s.

CHAPTER VIII

Elimination

Linear Dependence

256. The m sets of n constants each,

$$a_{i1}, a_{i2}, \cdots, a_{in} \qquad (i = 1, 2, \cdots, m)$$

are said to be *linearly dependent* if there exists m constants $b_1, b_2, \cdots, b_m$ not all zero such that

$$b_1 a_{1j} + b_2 a_{2j} + \cdots + b_m a_{mj} = 0 \qquad (j = 1, 2, \cdots, n).$$

If there exists no such set of constants as to make the sum zero then the sets are said to be *linearly independent.*

The a's may be any constants, or they may represent polynomials.

As consequences of this definition we have

(1) If m sets of constants are linearly dependent it is always possible to express some one set as a linear function of the rest.

This follows because the b's are not all zero.

(2) If a smaller number than m of the sets are linearly dependent then the whole m sets are linearly dependent.

For if $k(k<m)$ sets are linearly dependent we may take zeros for the remaining constant multipliers.

(3) If any one of the m sets consists of zeros then the m sets are linearly dependent.

For we may multiply this set by $\alpha(\alpha \neq 0)$ and the others by zeros.

257. Consider the matrix

$$\begin{Vmatrix} a_{11} & a_{12} & \cdots & a_{1n} \\ a_{21} & a_{22} & \cdots & a_{2n} \\ \cdots & \cdots & \cdots & \cdots \\ a_{m1} & a_{m2} & \cdots & a_{mn} \end{Vmatrix}$$

of rank p, and suppose

$$(a) \quad m \leqq n.$$

If the m sets of constants are linearly dependent we can express one of the rows as a linear function of the others and the elements of this row may be reduced to zeros. From this it follows that all the determinants of order m formed from this matrix are zero.

Suppose now the determinants of order m of this matrix are all zero, and we may without loss of generality suppose that the minor of order p which does not vanish is the one in the upper left-hand corner or $|a_{1p}|$. Let $\mathcal{A}_{i\ p+1}$ be the cofactor of $a_{i\ p+1}$ in $|a_{1\ p+1}|$. Then

$$a_{1j}\mathcal{A}_{1,p+1} + a_{2j}\mathcal{A}_{2,p+1} + \cdots + a_{p+1,j}\mathcal{A}_{p+1,p+1} = 0$$
$$(j = p+1,\ p+2, \cdots, n)$$

as it is the expansion of a determinant of order $p+1$ which is therefore zero. The relation is also true for $j=1, 2, \cdots, p$ for then it is the expansion of a determinant with identical columns. From this we see that the first $p+1$ rows are linearly dependent and hence by §256 all are.

$$(b) \quad m>n=n+h \quad \text{say,}$$

In this case we add h zeros to each row and the matrix has but one determinant, which is zero because it has at least one column of zeros and by (a) the m sets of n constants are linearly dependent.

As consequences of the foregoing we have

I. The necessary and sufficient condition for the linear dependence of the m sets of n constants in the matrix

$$\begin{Vmatrix} a_{11} & a_{12} & \cdots & a_{1n} \\ a_{21} & a_{22} & \cdots & a_{2n} \\ \cdot & \cdot & \cdot & \cdot \\ a_{m1} & a_{m2} & \cdots & a_{mn} \end{Vmatrix}$$

is that it be evanescent.

II. Whenever $m>n$, the m sets of n constants are always linealy dependent.

258. If the sets of n constants are the coefficients in m polynomials in n independent variables then it is apparent that the necessary and sufficient condition that the polynomials are linearly dependent is that the m sets of coefficients are linearly dependent.

Linear Equations

I. Non Homogeneous

259. Let us consider the m equations in n variables

$$(1) \qquad a_{i1}x_1 + a_{i2}x_2 + \cdots + a_{in}x_n = y_i \qquad (i = 1, 2 \cdots m)$$

and let the matrix

$$\begin{Vmatrix} a_{11} & a_{12} & \cdots & a_{1n} \\ a_{21} & a_{22} & \cdots & a_{2n} \\ \cdots & \cdots & \cdots & \cdots \\ a_{m1} & a_{m2} & \cdots & a_{mn} \end{Vmatrix}$$

which is called the matrix of the set or system, be denoted by (A), and let the matrix

$$\begin{Vmatrix} a_{11} & a_{12} & \cdots & a_{1n} & y_1 \\ a_{21} & a_{22} & \cdots & a_{2n} & y_2 \\ \cdots & \cdots & \cdots & \cdots & \cdots \\ a_{m1} & a_{m2} & \cdots & a_{mn} & y_m \end{Vmatrix}$$

which is called the *augmented* matrix, be denoted by (B). Let the rank of (A) be p and without loss of generality we may suppose that $|a_{1p}| \equiv A \neq 0$.

It is obvious that the rank of (B) must be either equal to or greater than that of (A).

First suppose it greater and equal to $q(q>p)$. Then the matrix (B) has at least one minor of order q that does not vanish and which contains the column of y's for if not it would be a minor of (A). Suppose this minor is contained in the first q rows of (B), and suppose the left-hand sides of equations (1) be represented by $e_1, e_2, \cdots, e_m$. Then the ith equation could be written

$$e_i = y_i \quad \text{or} \quad e_i - y_i = 0. \tag{2}$$

Since the rank of (A) is less than q it follows that $e_1, e_2, \cdots, e_q$ are linearly dependent and therefore

$$b_1e_1 + b_2e_2 + \cdots + b_qe_q = 0,$$

where the b's are constants and not all zero.

From (2) we see that

$$b_1e_1 + b_2e_2 + \cdots + b_qe_q = b_1y_1 + b_2y_2 + \cdots + b_qy_q$$

or

$$b_1(e_1 - y_1) + b_2(e_2 - y_2) + \cdots + b_q(e_q - y_q) = 0, \tag{3}$$

but if the rank of (B) is q this cannot be zero. The equations are then inconsistent, for if they were consistent there would be some values of $x_1, x_2, \cdots, x_n$ which would make $e_i - y_i$ vanish for all values of i and therefore they would satisfy (3), but we have seen that this cannot be zero if the rank of (B) is greater than p.

Second suppose (B) has the same rank as (A) that is both have the rank p. Let us now consider the first p equations of (1) together with the rth and since the rank is p we have

$$b_1(e_1 - y_1) + b_2(e_2 - y_2) + \cdots + b_p(e_p - y_p) + b_r(e_r - y_r) = 0,$$

where the b's are constants, not all zero. But $e_1 - y_1, e_2 - y_2, \cdots, e_p - y_p$ are linearly independent and therefore b_r cannot be zero. It follows that whatever satisfies the first p equations will satisfy the rth. That is, the rth is a consequence of the first p.

We have then the theorem. *If the matrices (A) and (B) have the same rank p then a solution of the p equations whose matrix contains the non-vanishing minor of order p will be a solution of all of them.*

260. Next let us start with the consistency of the equations (1) §259 and let Δ_r denote the determinant

$$\begin{vmatrix} a_{11} \cdots a_{1p} & y_1 \\ a_{21} \cdots a_{2p} & y_2 \\ \cdots\cdots & \cdots \\ a_{p1} \cdots a_{pp} & y_p \\ a_{r1} \cdots a_{rp} & y_r \end{vmatrix}$$

For $r \overline{\overline{<}} p$ it is obvious that Δ_r is zero.

Let us write equations (1) §259 in the form

$$(4) \qquad a_{i1}x_1 + \cdots + a_{ip}x_p = y_i - a_{i,p+1}x_{p+1} - \cdots \quad a_{in}x_n$$
$$= y_i', \quad \text{say}.$$

If now we replace $y_1, y_2, \cdots, y_p, y_r$ in Δ_r by $y_1', y_2', \cdots, y_p', y_r'$ it will remain unchanged in value. For since the elements in the last column of the resulting determinant, Δ_r' say, are the sum of $n-p+1$ terms we can break it up into Δ_r and $n-p$ other determinants which after taking out the constant factor x are minors of (A) of order $p+1$ and are therefore zero. That is $\Delta_r' = \Delta_r$.

If we multiply the first p and the rth of equations (4) by the cofactors of the elements in the last column of Δ_r, that is, by $\mathcal{A}_{1\ p+1}$, $\mathcal{A}_{2\ p+1}, \cdots, \mathcal{A}_{p\ p+1}$, A and then add the results we get on the left zero, and on the right $\Delta_r + (n-p)$ terms each of which contains as a factor a minor of (A) of order $p+1$ and is therefore zero. Therefore

$\Delta_r = 0$ for $r = 1, 2, \cdots, m$, and therefore by §234 every minor of (B) of order $p+1$ is zero and the rank of (B) is the same as that of (A).

It follows therefore that: *The necessary and sufficient condition that a set of m linear equations in n variables be consistent is that the matrix of the set is of the same rank as the augmented matrix.*

261. Concerning the matrix (B) we should observe that the same linear homogeneous relation connects the elements of every one of its columns. For since the rank is p such a relation connects the elements of the columns in $p+1$ of the rows and therefore by §256 such a relation connects all of them.

That is

$$a_{r1} = \alpha_1 a_{11} + \alpha_2 a_{21} + \cdots + \alpha_p \alpha_{p1},$$
$$a_{r2} = \alpha_1 a_{12} + \alpha_2 a_{22} - \cdots + \alpha_p a_{p2},$$
$$\cdots\cdots\cdots\cdots\cdots\cdots\cdots\cdots$$
$$a_{rp} = \alpha_1 a_{1p} + \alpha_2 a_{2p} + \cdots + \alpha_p a_{pp},$$
$$y_r' = \alpha_1 y_1' + \alpha_2 y_2' + \cdots + \alpha_p y_p',$$

and these are true whatever $x_{p+1}, x_{p+2}, \cdots, x_n$ may be, so on substituting for y_i' we get

$$y_r = \alpha_1 y_1 + \alpha_2 y_2 + \cdots + \alpha_p y_p$$
$$a_{r,p+1} = \alpha_1 a_{1,p+1} + \alpha_2 a_{2,p+1} + \cdots + \alpha_p a_{p,p+1}$$
$$\cdots\cdots\cdots\cdots\cdots\cdots\cdots\cdots\cdots\cdots$$
$$a_{rn} = \alpha_1 a_{1n} + \alpha_2 a_{2n} + \cdots + \alpha_p a_{pn}.$$

From which we see that if we multiply the first p equations of (1) §259 by $\alpha_1, \alpha_2, \alpha_3, \cdots, \alpha_p$ respectively, and add, what we get will be exactly the rth equation. That is the rth equation is a consequence of the first p equations. Giving r any of the values $p+1, p+2, \cdots n$ we see that

If $\Delta_r = 0$ for all values of r, the system of m equations reduces to the first p, or the last $m-p$ equations are then only in a linear relation to the first p.

In the special case where $m = n$ and $p = n-1$ so that $|a_{1n}| \equiv A = 0$, it follows that if A_k, which is the result of substituting for the elements in the kth column of A the elements $y_1', y_2', \cdots, y_n'$, is zero for one value of k it is zero for all values provided the minors of order $n-1$ are not zero.

For suppose $A_1=0$, then multiplying the equations by $\mathcal{A}_{11}$, $\mathcal{A}_{21}, \cdots, \mathcal{A}_{n1}$ respectively and adding we have

$$y_1'\mathcal{A}_{11} + y_2'\mathcal{A}_{21} + \cdots + y_n'\mathcal{A}_{n1} = 0$$

and similarly $y_1'\mathcal{A}_{1k}+y_2'\mathcal{A}_{2k}+ \cdots +y_n'\mathcal{A}_{nk}=A_k$ $(k=2, 3, \cdots, n)$

The vanishing of A gives us

$$\frac{\mathcal{A}_{11}}{\mathcal{A}_{1s}} = \frac{\mathcal{A}_{21}}{\mathcal{A}_{2s}} = \cdots = \frac{\mathcal{A}_{n1}}{\mathcal{A}_{ns}},$$

provided $\mathcal{A}_{rs}\neq 0$. From this we see that with the above proviso $A_k=0$ for all values of k.

262. *If we have n linear equations determining n unknowns and $p=n$ then any given linear function of the same unknowns is readily expressible in terms of the coefficients.*

Let the equations be

$$a_{i1}x_1 + a_{i2}x_2 + \cdots + a_{i1}x_n = y_i \qquad (i = 1, 2, \cdots, n),$$

where $|a_{1n}|\equiv A\neq 0$, and let the given function be

$$c_1x_1 + c_2x_2 + \cdots + c_nx_n = v.$$

These $n+1$ equations require (if the x's are to have values other than zero) that

$$\begin{vmatrix} a_{11} & a_{12} & \cdots & a_{1n} & y_1 \\ \cdots & \cdots & \cdots & \cdots & \cdots \\ a_{n1} & a_{n2} & \cdots & a_{nn} & y_n \\ c_1 & c_2 & \cdots & c_n & v \end{vmatrix} = 0$$

or

$$A\cdot v = \begin{vmatrix} a_{11} & \cdots & a_{1n} & y_1 \\ \cdots & \cdots & \cdots & \cdots \\ c_1 & \cdots & c_n & 0 \end{vmatrix}$$

which gives v in terms of the coefficients.

263. If we multiply the first p equations of (1) §259 by the cofactors of the elements in the kth column of A respectively and add we get

$$x_k\cdot A = A_k.$$

Since $A\neq 0$ we have

(1) $$x_k = \frac{A_k}{1}.$$

Two cases arise

(a) $p=n$.

This is the case of n non-homogeneous equations in n variables, the determinant of whose coefficients is not zero. The right-hand side of (1) is constant and x_k has but one value for each value of k. The set has one and but one solution. It was treated in §80. The case where $m=n+1$ and $p=n$ was treated in §81.

(b) $p<n$.

In this case the right-hand side of (1) is a linear function in $x_{p+1}, x_{p+2}, \cdots, x_n$ and if we assign values to these, then $x_1, x_2, \cdots, x_p$ have finite and definite values. The set has in this case an $(n-p)$-fold infinity of solutions according to the values given to $x_{p+1}, x_{p+2}, \cdots, x_n$.

It may readily be seen that every solution is involved in (1).

To summarize we have the following

Hypothesis	Conclusions
I Rank of (A) less than rank of (B)	Equations inconsistent
II Rank of (A) the same as rank of (B)	Equations consistent (a) $p<n$ $(n-p)$-fold infinity of solutions. (b) $p=n$ one and only one solution.
III Equations consistent	Rank of (B) same as rank of (A).

II. Homogeneous

264. If in equations (1) §259 the y's are all zero we have m linear homogeneous equations in n variables:

$$(7)\qquad a_{i1}x_1 + a_{i2}x_2 + \cdots + a_{in}x_n = 0 \qquad (i = 1,2,\cdots,m).$$

The two matrices differ by a column of zeros and therefore they have the same rank.

It follows therefore that every set of m linear homogeneous equations has one or more solutions.

First: Suppose the rank of (A) is $p<n$.

Then we may assign arbitrary values to $n-p$ of the variables (those whose coefficients do not contain the non-vanishing minor of order p) and the other p will be definitely determined. There will then be a $(n-p)$-fold infinity of solutions depending upon the values assigned to the $n-p$ variables.

The case where $p=n-1$ and $m=n-1$ deserves special attention as being the case of $n-1$ linear homogeneous equations in n variables whose matrix is of rank $n-1$. Suppose $|a_{1,n-1}|\equiv A\neq 0$ then giving x_n a definite value we have the case of $n-1$ linear non-homogeneous equations in $n-1$ variables which has a definite solution for every value assigned to x_n.

Proceeding as before we have

$$x_k = \frac{A_k'}{A}$$

where A_k' is now the result of replacing the kth column of A by the negatives of $a_{1n}x_n, a_{2n}x_n, \cdots, a_{nn}x_n$. We may write therefore

$$x_k = (-1)^{n-k}\frac{c\cdot A_k}{A},$$

where c is the arbitrary value assigned to x_n, and A_k is the result of deleting the kth column of (A). The values of x_k are then proportional to A_k $(k=1, 2, \cdots, n)$.

Second: Suppose $p=n$.

In this case there will be but one set of values for $x_1, x_2, \cdots, x_n$, namely, zeros which will satisfy the equations. As a consequence we have:

The necessary and sufficient condition that a set of m linear homogeneous equations in n variables have a solution other than zeros for $x_1, x_2, \cdots, x_n$ is that the rank of the matrix of the set is less than n.

If m is greater than n, then since $p<m$ there is always a solution other than zeros.

If m is equal to n then A must be zero if there be a solution other than all zeros.

If one of the x's has a value different from zero then it is apparent that at least one other must have a non-zero value.

265. The condition that the two equations

$$a_1x + a_2y = 0$$

$$b_1x + b_2y = 0$$

should be consistent is that $|a_1b_2|=0$. If now we take the square of each and the product of the two and eliminate x^2, xy, y^2 we would get

$$\begin{vmatrix} a_1^2 & 2a_1a_2 & a_2^2 \\ a_1b_1 & a_1b_2 + a_2b_1 & a_2b_2 \\ b_1^2 & 2b_1b_2 & b_2^2 \end{vmatrix}.$$

This resultant must then be equal to $|a_1b_2|^3$.

Exercises. Set XV

1. Show that

$$\begin{vmatrix} a_1b_1 & a_1b_2 + a_2b_1 & a_2b_2 \\ a_1c_1 & a_1c_2 + a_2c_1 & a_2c_2 \\ b_1c_1 & b_1c_2 + b_2c_1 & b_2c_2 \end{vmatrix} = |\, a_1\, b_2 \,| \cdot |\, a_1\, c_2 \,| \cdot |\, b_1\, c_2 \,| \,.$$

2. What determinant of order four is equal to $|a_1b_2|^6$?

3. Find the values of $x_1, x_2, \cdots, x_n$ from the equations

$$\begin{aligned} (\alpha_1 - x_1) + \quad \alpha_2x_2 \quad + \alpha_3x_3 + \cdots + \alpha_nx_n &= 0 \\ \alpha_1x_1 + (\alpha_2 - x_2) + \alpha_3x_3 + \cdots + \alpha_nx_n &= 0 \\ \cdots\cdots\cdots\cdots\cdots\cdots\cdots\cdots\cdots & \\ \alpha_1x_1 + \quad \alpha_2x_2 \quad + \alpha_3x_3 + \cdots + (\alpha_n - x_n) &= 0. \end{aligned}$$

Noting the symmetry of each for α_r and x_r $(r=1, 2, \cdots, n)$ find the values of the α's in terms of the x's.

4. Solve for $x_1, x_2, \cdots, x_n$ in determinant form the set of equations

$$\begin{aligned} \frac{x_1}{b_1 - \beta_1} + \frac{x_2}{b_1 - \beta_2} + \cdots + \frac{x_n}{b_1 - \beta_n} &= 1 \\ \frac{x_1}{b_2 - \beta_1} + \frac{x_2}{b_2 - \beta_2} + \cdots + \frac{x_n}{b_2 - \beta_n} &= 1 \\ \cdots\cdots\cdots\cdots\cdots\cdots\cdots\cdots & \\ \frac{x_1}{b_n - \beta_1} + \frac{x_2}{b_n - \beta_2} + \cdots + \frac{x_n}{b_n - \beta_n} &= 1 \end{aligned}$$

and show that

$$x_r = -\frac{(\beta_r - b_1)(\beta_r - b_2)\cdots(\beta_r - b_n)}{(\beta_r - \beta_1)\cdots(\beta_r - \beta_{r-1})(\beta_r - \beta_{r+1})\cdots(\beta_r - \beta_n)} \qquad (r = 1,2, \cdots, n).$$

5. Show that the condition that z and w are zero in the four equations

$$a_kx + b_ky + c_kz + d_kw = e_k \quad (k = 1,2,3,4)$$

is

$$\left\| \begin{matrix} a_1 & a_2 & a_3 & a_4 \\ b_1 & b_2 & b_3 & b_4 \\ e_1 & e_2 & e_3 & e_4 \end{matrix} \right\| = 0.$$

6. Show that the eliminant of

$$
\begin{aligned}
(a_1x + a_2y)^3 &= 0 \\
(a_1x + a_2y)^2(b_1x + b_2y) &= 0 \\
(a_1x + a_2y)(b_1x + b_2y)^2 &= 0 \\
(b_1x + b_2y)^3 &= 0
\end{aligned}
$$

is $|a_1b_2|^6$.

7. Show that if

$$
\begin{aligned}
a + bx + cx^2 &= 0 \\
\alpha + \beta x + \gamma x^2 + \delta x^3 &= 0
\end{aligned}
$$

have a common root it is given by

$$
\begin{vmatrix} a + bx & c & \cdot \\ \alpha + \beta x & \gamma & \delta \\ ax & b & c \end{vmatrix} = 0.
$$

8. Show that if

$$
\begin{aligned}
a + bx + cx^2 + dx^3 &= 0 \\
\alpha + \beta x + \gamma x^2 + \delta x^3 + \epsilon x^4 &= 0
\end{aligned}
$$

have two common roots they are given by

$$
\begin{vmatrix} a + bx + cx^2 & d & \cdot \\ \alpha + \beta x + \gamma x^2 & \delta & \epsilon \\ ax + bx^2 & c & d \end{vmatrix} = 0, \quad \text{or} \quad \begin{vmatrix} b + cx + dx^2 & \alpha & \cdot \\ a + bx + cx^2 & \cdot & d \\ \beta + \gamma x + \delta x^2 & \alpha & \epsilon \end{vmatrix} = 0,
$$

$$
\text{or} \quad \begin{vmatrix} c + dx & a & b \\ b + cx + dx^2 & \cdot & a \\ \gamma + \delta x + \epsilon x^2 & \alpha & \beta \end{vmatrix} = 0.
$$

With the help of determinants solve the following sets of equations:

9.
$$
\begin{aligned}
4x + 7y + 3z - 2w &= 9 \\
2x - y - 4z + 3w &= 13 \\
3x + 2y - 7z - 4w &= 2 \\
5x - 3y + z + 5w &= 13
\end{aligned}
$$

10.
$$
\begin{aligned}
3x + 2y + 4z - w &= 13 \\
5x + y - z + 2w &= 9 \\
2x + 3y - 7z + 3w &= 14 \\
4x - 4y + 3z - 5w &= 4
\end{aligned}
$$

11.
$$\begin{aligned} v + w - y &= a \\ w + x - z &= b \\ x + y - v &= c \\ y + z - w &= d \\ z + v - x &= e \end{aligned}$$

12.
$$\begin{aligned} v + w + x - y &= a \\ w + x + y - z &= b \\ x + y + z - v &= c \\ y + z + v - w &= d \\ z + v + w - x &= e \end{aligned}$$

13.
$$\begin{aligned} w + x + y + z &= 1 \\ aw + bx + cy + dz &= e \\ a^2w + b^2x + c^2y + d^2z &= e^2 \\ a^3w + b^3x + c^3y + d^3z &= e^3 \end{aligned}$$

14.
$$\begin{aligned} w + x + y + z &= 1 \\ w + ax + by + cz &= 0 \\ w + a^2x + b^2y + c^2z &= 0 \\ w + a^3x + b^3y + c^3z &= 0 \end{aligned}$$

15.
$$\begin{aligned} v = 2w - 2x + y + 3z &= a \\ w - 2x - 2y + z + 3v &= b \\ x - 2y - 2z + v + 3w &= c \\ y - 2z - 2v + w + 3x &= d \\ z - 2v - 2w + x + 3y &= e. \end{aligned}$$

16. What relation must exist between a, b, c, d if the equations

$$\begin{aligned} ax + by + cz + d &= 0 \\ bx + ay + dz + c &= 0 \\ ax + cy + bz + d &= 0 \\ cx + ay + dz + b &= 0 \end{aligned}$$

be simultaneously true?

17. If the equations

$$\begin{aligned} a_1x^3 + b_1x^2 + c_1x + d_1 &= 0 \\ a_1x^4 + b_1x^3 + c_1x^2 + d_1x &= 0 \\ b_2x^2 + c_2x + d_2 &= 0 \\ b_2x^3 + c_2x^2 + d_2x &= 0 \\ b_2x^4 + c_2x^3 + d_2x^2 &= 0 \end{aligned}$$

be simultaneously true (which evidently will be the case if the first and third be simultaneously true, that is, have a common root), find the relation which must exist between $a_1, b_1, c_1, d_1, b_2, c_2, d_2$.

Similarly find the resultant in the case of each of the following pairs of equations:

18.
$$\begin{aligned} a_1x^2 + b_1x + c_1 &= 0 \\ a_2x^2 + b_2x + c_2 &= 0 \end{aligned}$$

19.
$$\begin{aligned} a_1x^4 + b_1x^3 + c_1x^2 + d_1x + e_1 &= 0 \\ c_2x^2 + d_2x + e_2 &= 0 \end{aligned}$$

20.
$$\begin{aligned} a_1x^3 + b_1x^2 + c_1x + d_1 &= 0 \\ a_2x^3 + b_2x^2 + c_2x + d_2 &= 0 \end{aligned}$$

21.
$$\begin{aligned} a_1x^4 + b_1x^3 + c_1x^2 + d_1x + e_1 &= 0 \\ b_2x^3 + c_2x^2 + d_2x + e_2 &= 0 \end{aligned}$$

22.
$$\begin{aligned} a_1x^4 + c_1x^2 + d_1x + e_1 &= 0 \\ a_2x^4 + b_2x^3 + d_2x + e_2 &= 0 \end{aligned}$$

23.
$$\begin{aligned} a_1x^6 + b_1x^2 + c_1 &= 0 \\ d_1x^4 + e_1x^2 + f_1 &= 0. \end{aligned}$$

Equations of Higher Degree

266. If we are given the two equations

$$\begin{aligned} a_1x^3 + a_2x^2 + a_3x + a_4 &= 0 \\ b_1x^4 + b_2x^3 + b_3x^2 + b\ x + b_5 &= 0 \end{aligned} \tag{1}$$

we may by multiplying the first successively by x, x^2 x^3 and the second successively by x, x^2, obtain the seven equations

$$\begin{aligned} a_1x^3 + a_2x^2 + a_3x + a_4 &= 0 \\ a_1x^4 + a_2x^3 + a_3x^2 + a_4x + 0 &= 0 \\ a_1x^5 + a_2x^4 + a_3x^3 + a_4x^2 + 0 + 0 &= 0 \\ a_1x^6 + a_2x^5 + a_3x^4 + a_4x^3 + 0 + 0 + 0 &= 0. \\ b_1x^4 + b_2x^3 + b_3x^2 + b_4x + b_5 &= 0 \\ b_1x^5 + b_2x^4 + b_3x^3 + b_4x^2 + b_5x + 0 &= 0 \\ b_1x^6 + b_2x^5 + b_3x^4 + b_4x^3 + b_5x^2 + 0 + 0 &= 0 \end{aligned} \tag{2}$$

These seven equations are non-homogeneous in the six quantities x^6, x^5, x^4, x^3, x^2, x considered as independent unknowns and therefore the eliminant is

$$\Delta \equiv \begin{vmatrix} a_1 & a_2 & a_3 & a_4 & \cdot & \cdot & \cdot \\ \cdot & a_1 & a_2 & a_3 & a_4 & \cdot & \cdot \\ \cdot & \cdot & a_1 & a_2 & a_3 & a_4 & \cdot \\ \cdot & \cdot & \cdot & a_1 & a_2 & a_3 & a_4 \\ b_1 & b_2 & b_3 & b_4 & b_5 & \cdot & \cdot \\ \cdot & b_1 & b_2 & b_3 & b_4 & b_5 & \cdot \\ \cdot & \cdot & b_1 & b_2 & b_3 & b_4 & b_5 \end{vmatrix}.$$

This method of eliminating is known as *Sylvester's Dialytic* method and the determinants of this type are called *bigradients.*

If $\Delta = 0$, then the two equations (1) have a common solution.

In general if we have two equations of the nth and mth degrees respectively

$$\begin{aligned} a_1x^n + a_2x^{n-1} + \cdots + a_{n+1} &= 0 \\ b_1x^m + b_2x^{m-1} + \cdots + b_{m+1} &= 0 \end{aligned} \tag{3}$$

We can in a precisely similar way get their eliminant as a determinant of the $(n+m)$ order. Thus

$$\Delta \equiv \begin{vmatrix} a_1 & a_2 \cdots a_{n+1} & \cdot & \cdots \\ \cdot & a_1 \cdots a_n & a_{n+1} & \cdots \\ \cdots & \cdots & \cdots & \cdots \\ b_1 & b_2 \cdots b_{m+1} & \cdot & \cdots \\ \cdot & b_1 \cdots b_m & b_{m+1} & \cdots \\ \cdots & \cdots & \cdots & \cdots \end{vmatrix}$$

where there are m lines of a's and n lines of b's.

267. If we take the three equations

$$\begin{aligned} a_1x^3 + a_2x^2 + a_3x + a_4 &= 0 \\ a_1x^4 + a_2x^3 + a_3x^2 + a_4x + 0 &= 0 \\ b_1x^4 + b_2x^3 + b_3x^2 + b_4x + b_5 &= 0 \end{aligned}$$

and multiply them by

$$\begin{vmatrix} a_1 & a_2 \\ b_1 & b_2 \end{vmatrix}, \qquad -\begin{vmatrix} \cdot & a_1 \\ b_1 & b_2 \end{vmatrix}, \qquad \begin{vmatrix} \cdot & a_1 \\ a_1 & a_2 \end{vmatrix}$$

respectively and add, there results

$$\begin{vmatrix} \cdot & a_1 & a_2 \\ a_1 & a_2 & a_3 \\ b_1 & b_2 & b_3 \end{vmatrix} x^2 + \begin{vmatrix} \cdot & a_1 & a_3 \\ a_1 & a_2 & a_4 \\ b_1 & b_2 & b_4 \end{vmatrix} x + \begin{vmatrix} \cdot & a_1 & a_4 \\ a_1 & a_2 & \cdot \\ b_1 & b_2 & b_5 \end{vmatrix} = 0$$

which is called the "*prime derivative*" *of the second order.*

We might have arrived at the same result by considering the three equations as non-homogeneous linear in x^4 and x^3 considered as independent unknowns and writing their resultant

$$\begin{vmatrix} \cdot & a_1 & a_2x^2 + a_3x + a_4 \\ a_1 & a_2 & a_3x^2 + a_4x + 0 \\ b_1 & b_2 & b_3x^2 + b_4x + b_5 \end{vmatrix} = 0$$

which is the same as the quadratic relation given.

If we take the five equations

$$\begin{aligned} a_1x^3 + a_2x^2 + a_3x + a_4 &= 0 \\ a_1x^4 + a_2x^3 + a_3x^2 + a_4x &= 0 \\ a_1x^5 + a_2x^4 + a_3x^3 + a_4x^2 &= 0 \\ b_1x^4 + b_2x^3 + b_3x^2 + b_4x + b_5 &= 0 \\ b_1x^5 + b_2x^4 + b_3x^3 + b_4x^2 + b_5x &= 0 \end{aligned}$$

and consider them as 5 linear non-homogeneous equations in x^5, x^4, x^3, x^2 considered as independent unknowns, the resultant is

$$\begin{vmatrix} \cdot & \cdot & a_1 & a_2 & a_3x + a_4 \\ \cdot & a_1 & a_2 & a_3 & a_4x + 0 \\ a_1 & a_2 & a_3 & a_4 & 0 + 0 \\ \cdot & b_1 & b_2 & b_3 & b_4x + b_5 \\ b_1 & b_2 & b_3 & b_4 & b_5x + 0 \end{vmatrix}$$

and this is

$$\begin{vmatrix} \cdot & \cdot & a_1 & a_2 & a_3 \\ \cdot & a_1 & a_2 & a_3 & a_4 \\ a_1 & a_2 & a_3 & a_4 & \cdot \\ \cdot & b_1 & b_2 & b_3 & b_4 \\ b_1 & b_2 & b_3 & b_4 & b_5 \end{vmatrix} x + \begin{vmatrix} \cdot & \cdot & a_1 & a_2 & a_4 \\ \cdot & a_1 & a_2 & a_3 & \cdot \\ a_1 & a_2 & a_3 & a_4 & \cdot \\ \cdot & b_1 & b_2 & b_3 & b_4 \\ b_1 & b_2 & b_3 & b_4 & \cdot \end{vmatrix}$$

which is the "*prime derivative*" *of the first order.*

In general if we wish to form the prime derivative of order r for equations (3) §266 we would take the $m-r$ equations formed by myltiplying the first successively by $x^0, x^1, x^2, \cdots, x^{m-r-1}$ with the $n-r$ equations formed by multiplying the second successively by $x^0, x^1, x^2, \cdots, x^{n-r-1}$ and eliminating $x^{r+1}, x^{r+2}, \cdots, x^{n+m-r-1}$ considered as independent quantities.

268. If any two equations

$$f(x, y) = 0,$$
$$\phi(x, y) = 0$$

have a common solution that solution must satisfy the equation $f(x, y) - \lambda\phi(x, y) = 0$ whatever the value of λ may be. That is the equation

$$f(x, y)\phi(x', y') - f(x', y')\phi(x, y) = 0$$

will be satisfied by the common solution whatever the values of x' and y'.

After dividing by $xy' - x'y$, which is obviously a factor, we may equate to zero the coefficients of the various powers of x' and y' and from these equations eliminate the powers of x and y considered as independent variables. The result is *Bezout's* form of the eliminant. Thus the eliminant of the two equations

$$a_1x^4 + a_2x^3 + a_3x^2 + a_4x + a_5 = 0$$
$$b_1x^4 + b_2x^3 + b_3x^2 + b_4x + b_5 = 0$$

is

$$B = \begin{vmatrix} |a_1 b_2| & |a_1 b_3| & |a_1 b_4| & |a_1 b_5| \\ |a_1 b_3| & |a_1 b_4| + |a_2 b_3| & |a_1 b_5| + |a_2 b_4| & |a_2 b_5| \\ |a_1 b_4| & |a_1 b_5| + |a_2 b_4| & |a_2 b_5| + |a_3 b_4| & |a_3 b_5| \\ |a_1 b_5| & |a_2 b_5| & |a_3 b_5| & |a_4 b_5| \end{vmatrix}.$$

It appears that every term in this eliminant must contain a_5 or b_5 since every term must contain an element from the last row and also one from the last column.

It appears also that the terms which contain a_5 or b_5 only in the first degree are $|a_4b_5|$ multiplied by the eliminant of the equations formed by putting a_5 and b_5 equal to zero, which is what the complementary minor of $|a_4b_5|$ in the determinant reduces to when $a_5 = b_5 = 0$.

EXERCISE. Show that the eliminant of the equations

$$\left\| \begin{matrix} a_1y - l_1x & a_2y - l_2x & a_3y - l_3x & a_4y - l_4x \\ b_1y - m_1x & b_2y - m_2x & b_3y - m_3x & b_4y - m_4x \\ c_1y - n_1x & c_2y - n_2x & c_3y - n_3x & c_4y - n_4x \end{matrix} \right\| = 0$$

is

$$\left| \begin{matrix} |a_1\, b_2\, c_3\, l_4| & |a_1\, b_2\, n_3\, l_4| + |a_1\, m_2\, c_3\, l_4| & |a_1\, m_2\, n_3\, l_4| \\ |a_1\, b_2\, c_3\, m_4| & |a_1\, b_2\, n_3\, m_4| + |l_1\, b_2\, c_3\, m_4| & |l_1\, b_2\, n_3\, m_4| \\ |a_1\, b_2\, c_3\, n_4| & |a_1\, m_2\, c_3\, n_4| + |l_1\, b_2\, c_3\, n_4| & |l_1\, m_2\, c_3\, n_4| \end{matrix} \right|$$

269. If we write Sylvester's eliminant of the equations

$$a_0x^4 + a_1x^3 + a_2x^2 + a_3x + a_4 = 0$$
$$b_0x^4 + b_1x^3 + b_2x^2 + b_3x + b_4 = 0$$

in the form

$$S \equiv \left| \begin{matrix} a_0 & a_1 & a_2 & a_3 & a_4 & \cdot & \cdot & \cdot \\ \cdot & a_0 & a_1 & a_2 & a_3 & a_4 & \cdot & \cdot \\ \cdot & \cdot & a_0 & a_1 & a_2 & a_3 & a_4 & \cdot \\ \cdot & \cdot & \cdot & a_0 & a_1 & a_2 & a_3 & a_4 \\ \cdot & \cdot & \cdot & b_0 & b_1 & b_2 & b_3 & b_4 \\ \cdot & \cdot & b_0 & b_1 & b_2 & b_3 & b_4 & \cdot \\ \cdot & b_0 & b_1 & b_2 & b_3 & b_4 & \cdot & \cdot \\ b_0 & b_1 & b_2 & b_3 & b_4 & \cdot & \cdot & \cdot \end{matrix} \right|$$

and denote the minor formed by the elements in the 2nd, 3rd, 7th rows and the 3rd, 5th, 7th columns by $\left|\begin{smallmatrix} 237 \\ 357 \end{smallmatrix}\right|$ so that

$$\left| \begin{matrix} 2\ 3\ 7 \\ 3\ 5\ 7 \end{matrix} \right| \equiv \left| \begin{matrix} a_1 & a_3 & \cdot \\ a_0 & a_2 & a_4 \\ b_1 & b_3 & \cdot \end{matrix} \right| .$$

Then the adjugate of Bezout's form of the eliminant of these same two equations is

$$\left| \begin{matrix} u_1 & u_2 & u_3 & u_4 \\ u_2 & u_3 & u_4 & u_5 \\ u_3 & u_4 & u_5 & u_6 \\ u_4 & u_5 & u_6 & u_7 \end{matrix} \right|$$

where

$$u_1 = \begin{vmatrix} 2\,3\,4\,5\,6\,7 \\ 3\,4\,5\,6\,7\,8 \end{vmatrix}, \quad u_2 = \begin{vmatrix} 2\,3\,4\,5\,6\,7 \\ 2\,4\,5\,6\,7\,8 \end{vmatrix} \cdots$$

the row-numbers of u_r being 234567 and the column-numbers being obtained by deleting the integer $r+1$ from 2345678. The u's are minors of S.

To prove that any one of the u's is equal to the complementary of the element in the corresponding position in Bezout's form B, we multiply the determinant

$$\begin{vmatrix} 1 & \cdot & \cdot & \cdot & \cdot & -b_0 \\ \cdot & 1 & \cdot & \cdot & -b_0 & -b_1 \\ \cdot & \cdot & 1 & -b_0 & -b_1 & -b_2 \\ \cdot & \cdot & \cdot & a_0 & a_1 & a_2 \\ \cdot & \cdot & \cdot & \cdot & a_0 & a_1 \\ \cdot & \cdot & \cdot & \cdot & \cdot & a_0 \end{vmatrix}$$

column-wise by the 6-by-7 array

$$\left\Vert \begin{matrix} 2\,3\,4\,5\,6\,7 \\ 2\,3\,4\,5\,6\,7\,8 \end{matrix} \right\Vert$$

and obtain

$$a_0^3 \left\Vert \begin{matrix} 2\,3\,4\,5\,6\,7 \\ 2\,3\,4\,5\,6\,7\,8 \end{matrix} \right\Vert$$

$$= \left\Vert \begin{matrix} a_0 & a_1 & a_2 & a_3 & a_4 & & \cdot \\ \cdot & a_0 & a_1 & a_2 & a_3 & a_4 & \cdot \\ \cdot & \cdot & a_0 & a_1 & a_2 & a_3 & a_4 \\ \cdot & \cdot & \cdot & |a_0\, b_1| & |a_0\, b_2| & |a_0\, b_3| & |a_0\, b_4| \\ \cdot & \cdot & \cdot & |a_0\, b_2| & |a_0\, b_3| + |a_1\, b_2| & |a_0\, b_4| + |a_1\, b_3| & |a_1\, b_4| \\ \cdot & \cdot & \cdot & |a_0\, b_3| & |a_0\, b_4| + |a_1\, b_3| & |a_1\, b_4| + |a_2\, b_3| & |a_2\, b_4| \end{matrix} \right\Vert$$

The seven determinants of this array are readily seen to be equivalents of the seven u's.

This may also be proved as follows

$$-u_1^2 = -\begin{vmatrix} 2\,3\,4\,5\,6\,7 \\ 3\,4\,5\,6\,7\,8 \end{vmatrix}^2 = -\begin{vmatrix} 2\,3\,4\,5\,6\,7 \\ 3\,4\,5\,6\,7\,8 \end{vmatrix} \begin{vmatrix} 5\,6\,7\,2\,3\,4 \\ 6\,7\,8\,3\,4\,5 \end{vmatrix}$$

$$= \begin{vmatrix} 2\,3\,4\,\bar{5}\,\bar{6}\,\bar{7} \\ 3\,4\,5\,6\,7\,8 \end{vmatrix} \begin{vmatrix} 5\,6\,7\,2\,3\,4 \\ 6\,7\,8\,3\,4\,5 \end{vmatrix}$$

where the bar above a row-number indicates that the signs of the elements in the row are to be changed. The result of multiplying this last product column-wise gives

$$-\left| \begin{array}{ccc} |a_0 b_3| + |a_1 b_2| & |a_0 b_4| + |a_1 b_3| & |a_1 b_4| \\ |a_0 b_4| + |a_1 b_3| & |a_1 b_4| + |a_2 b_3| & |a_2 b_4| \\ |a_1 b_4| & |a_2 b_4| & |a_3 b_4| \end{array} \right|^2$$

This suggests that Bezout's eliminant is variously expressible as the product of n columns of Sylvester's by a transformation of other n columns. Thus when n is 3 we have the following equivalents of Bezout's eliminant.

$$\left\| \begin{array}{l} 1\ 2\ 3\ 4\ 5\ 6 \\ 1\ 2\ 3 \end{array} \right\| \cdot \left\| \begin{array}{l} 4\ 5\ 6\ \bar{1}\ \bar{2}\ \bar{3} \\ 4\ 5\ 6 \end{array} \right\|, \quad \left\| \begin{array}{l} 1\ 2\ 3\ 4\ 5\ 6 \\ 4\ 5\ 6 \end{array} \right\| \cdot \left\| \begin{array}{l} \bar{4}\ \bar{5}\ \bar{6}\ 1\ 2\ 3 \\ 1\ 2\ 3 \end{array} \right\|,$$

$$\left\| \begin{array}{l} 1\ 2\ 3\ 4\ 5\ 6 \\ 2\ 3\ 4 \end{array} \right\| \cdot \left\| \begin{array}{l} \bar{6}\ 4\ 5\ 1\ \bar{3}\ \bar{2} \\ 1\ 5\ 6 \end{array} \right\|,$$

$$\left\| \begin{array}{l} 1\ 2\ 3\ 4\ 5\ 6 \\ 1\ 5\ 6 \end{array} \right\| \cdot \left\| \begin{array}{l} 6\ 4\ \bar{5}\ \bar{2}\ 3\ \bar{1} \\ 2\ 3\ 4 \end{array} \right\|, \quad \left\| \begin{array}{l} 1\ 2\ 3\ 4\ 5\ 6 \\ 3\ 4\ 5 \end{array} \right\| \cdot \left\| \begin{array}{l} \bar{5}\ \bar{6}\ 4\ \bar{3}\ 1\ 2 \\ 1\ 2\ 6 \end{array} \right\|,$$

$$\left\| \begin{array}{l} 1\ 2\ 3\ 4\ 5\ 6 \\ 1\ 2\ 6 \end{array} \right\| \cdot \left\| \begin{array}{l} 5\ \bar{6}\ \bar{4}\ \bar{3}\ 1\ 2 \\ 3\ 4\ 5 \end{array} \right\|.$$

270. In a similar way we can find the primary minors of B. Thus

$$\left\| \begin{array}{ccc} a_1 & a_2 & a_3 \\ a_0 & a_1 & a_2 \\ \cdot & a_0 & a_1 \\ \cdot & -b_0 & -b_1 \\ -b_0 & -b_1 & -b_2 \\ -b_1 & -b_2 & -b_3 \end{array} \right\| \left\| \begin{array}{ccc} b_2 & b_3 & b_4 \\ b_3 & b_4 & \cdot \\ b_4 & \cdot & \cdot \\ a_4 & \cdot & \cdot \\ a_3 & a_4 & \cdot \\ a_2 & a_3 & a_4 \end{array} \right\|$$

$$= \left| \begin{array}{ccc} |a_0 b_3| + |a_1 b_2| & |a_0 b_4| + |a_1 b_3| & |a_1 b_4| \\ |a_0 b_4| + |a_1 b_3| & |a_1 b_4| + |a_2 b_3| & |a_2 b_4| \\ |a_1 b_4| & |a_2 b_4| & |a_3 b_4| \end{array} \right|$$

which is u_1 or

$$\left\Vert\begin{array}{l}2\ 3\ 4\ 5\ 6\ 7\\3\ 4\ 5\end{array}\right\Vert\cdot\left\Vert\begin{array}{l}5\ 6\ 7\ 2\ 3\ 4\\6\ 7\ 8\end{array}\right\Vert = u_1.$$

The theorem resulting from this suggestion is as follows: *The adjugate of Bezout's eliminant of*

$$a_0x^4 + a_1x^3 + a_2x^2 + a_3x + a_4 = 0$$
$$b_0x^4 + b_1x^3 + b_2x^2 + b_3x + b_4 = 0$$

is

$$\begin{vmatrix} \nu_1 & \nu_2 & \nu_3 & \nu_4 \\ \nu_2 & \nu_3 & \nu_4 & \nu_5 \\ \nu_3 & \nu_4 & \nu_5 & \nu_6 \\ \nu_4 & \nu_5 & \nu_6 & \nu_7 \end{vmatrix}$$

where

$$\nu_1 = \left|\begin{array}{l}2\ 3\ 4\ \bar{5}\ \bar{6}\ \bar{7}\\3\ 4\ 5\end{array}\right|\cdot\left|\begin{array}{l}5\ 6\ 7\ 2\ 3\ 4\\6\ 7\ 8\end{array}\right|$$

$$\nu_2 = \left|\begin{array}{l}2\ 3\ 4\ \bar{5}\ \bar{6}\ \bar{7}\\2\ 4\ 5\end{array}\right|\cdot\left|\begin{array}{l}5\ 6\ 7\ 2\ 3\ 4\\6\ 7\ 8\end{array}\right|$$

. . . .

271. If in §268 $\phi(x)$ is the derivative of $f(x)$ then the resultant is the *discriminant*, the vanishing of which is the condition for equal roots in $f(x)=0$. In general the discriminant of a quantic in k variables is the eliminant of the k differentials with respect to each of the variables.

From the definition of a discriminant it is readily seen that the discriminant of

$$\begin{vmatrix} a_1x + a_2y + a_3z + a_4w & b_1x + b_2y + b_3z + b_4w \\ c_1x + c_2y + c_3z + c_4w & d_1x + d_2y + d_3z + d_4w \end{vmatrix}$$

is the product of

$$\begin{vmatrix} a_1 & b_1 & c_1 & d_1 \\ a_2 & b_2 & c_2 & d_2 \\ a_3 & b_3 & c_3 & d_3 \\ a_4 & b_4 & c_4 & d_4 \end{vmatrix} \cdot \begin{vmatrix} d_1 & -c_1 & -b_1 & a_1 \\ d_2 & -c_2 & -b_2 & a_2 \\ d_3 & -c_3 & -b_3 & a_3 \\ d_4 & -c_4 & -b_4 & a_4 \end{vmatrix}.$$

or $|a_1b_2c_3d_4|^2$.

272. Let

$$f(x) = a_1x^n + a_2x^{n-1} + \cdots + a_nx + a_{n+1},$$
$$= a_1(x - \alpha_1)(x - \alpha_2) \cdots (x - \alpha_n),$$

then

$$f'(x) = na_1x^{n-1} + (n-1)a_2x^{n-2} + \cdots.$$

The resultant of $f(x)=0$ and $f'(x)=0$ is

$$R \equiv \begin{vmatrix} a_1 & a_2 & \cdots a_n & a_{n+1} \cdots \\ \cdot & a_1 & \cdots a_{n-1} & a_n \quad \cdots \\ \cdots & \cdots & \cdots & \cdots \\ na_1 & (n-1)a_2 & \cdots a_n & \cdot \quad \cdots \\ \cdot & na_1 & \cdots 2a_{n-1} & a_n \quad \cdots \\ \cdots & \cdots & \cdots & \cdots \end{vmatrix}$$

of order $2n-1$. It is evident that $R=a_1R'$, where R' is a determinant of order $2n-2$.

We know that

$$f'(\alpha_1) = a_1(\alpha_1 - \alpha_2)(\alpha_1 - \alpha_3) \cdots (\alpha_1 - \alpha_n)$$
$$f'(\alpha_2) = (\alpha_2 - \alpha_1)a_1(\alpha_2 - \alpha_3) \cdots (\alpha_2 - \alpha_n)$$
$$\cdots\cdots\cdots\cdots$$
$$f'(\alpha_n) = (\alpha_n - \alpha_1)(\alpha_n - \alpha_2)(\alpha_n - \alpha_3) \cdots a_1$$

and therefore

$$f'(\alpha_1)\cdot f'(\alpha_2) \cdots f'(\alpha_n) = (-1)^{n(n-1)/2} a_1^n \zeta(\alpha_1, \alpha_2, \cdots \alpha_n)$$

where $\zeta(\alpha_1, \alpha_2, \cdots, \alpha_n)$ is the product of the squared differences of all the roots.

It is also known that

$$R = a_1^{n-1}f'(\alpha_1)\cdot f'(\alpha_2) \cdots f'(\alpha_n)$$

or

$$R' = a_1^{n-2}f'(\alpha_1)\cdot f'(\alpha_2) \cdots f'(\alpha_n)$$

and therefore

$$\frac{1}{a_1}\cdot R = R' = (-1)^{n(n-1)/2}a_1^{2(n-1)}\zeta(\alpha_1, \alpha_2, \cdots, \alpha_n)$$

which shows that if $R=0$, there must be two roots equal, provided $a_1 \neq 0$.

273. Suppose now we have three equations each of the nth degree.

$$f(x,y,z) = 0$$
$$\phi(x,y,z) = 0.$$
$$\psi(x,y,z) = 0$$

Theoretically, at least, it is possible to express the eliminant of these as a determinant.

Multiply each by the $\frac{1}{2}n(n-1)$ quantities $(x^{n-2},\ x^{n-3}y, \cdots)$ occuring in a polynomial of degree $(n-2)$ in three variables.

This gives us $\frac{3}{2}n(n-1)$ equations of the $(2n-2)$ degree and each having $\frac{1}{2}2n(2n-1)$ or $n(2n-1)$ terms. To solve dialytically we require, therefore, $n(2n-1)-\frac{3}{2}n(n-1)=\frac{1}{2}n(n+1)$ additional equations in these same powers of the variables. These are formed as follows: we write each of the three given equations in the form

$$Px^{\alpha} + Qy^{\beta} + Rz^{\gamma}$$

such that $\alpha+\beta+\gamma=n+2$ and form the determinant $|PQ'R''|$ which is of the $(2n-2)$ degree, and since the number of integral solutions of $\alpha+\beta+\gamma=n+2$ is $\frac{1}{2}n(n+1)$ we have the exact number of additional equations desired.

When the equations are not of the same degree this method will not give the eliminant free from extraneous factors.

If our three functions are quadratic, that is if

$$u_1 \equiv a_1x^2 + b_1y^2 + c_1z^2 + 2f_1zy + 2g_1zx + 2h_1xy \equiv \begin{array}{ccc|c} x & y & z & \\ \hline a_1 & h_1 & g_1 & x \\ h_1 & b_1 & f_1 & y \\ g_1 & f_1 & c_1 & z \end{array}$$

$$u_2 \equiv \begin{array}{ccc|c} x & y & z & \\ \hline a_2 & h_2 & g_2 & x \\ h_2 & b_2 & f_2 & y \\ g_2 & f_2 & c_2 & z \end{array} \qquad u_3 \equiv \begin{array}{ccc|c} x & y & z & \\ \hline a_3 & h_3 & g_3 & x \\ h_3 & b_3 & f_3 & y \\ g_3 & f_3 & c_3 & z \end{array}$$

We may find the eliminant of these as follows:

Form $|b_2c_3|u_1 - |b_1c_3|u_2 + |b_1c_2|u_3$ or $|u_1b_2c_3|$ which is

(a) $|a_1\, b_2\, c_3|\, x^2 + 2\,|f_1\, b_2\, c_3|\, yz + 2\,|g_1\, b_2\, c_3|\, zx + 2\,|h_1\, b_2\, c_3|\, xy$

and $|f_2c_3|u_1 - |f_1c_3|u_2 + |f_1c_2|u_3$ or $|u_1f_2c_3|$ which is

$$(b)\quad |a_1 f_2 c_3|\, x^2 + |b_1 f_2 c_3|\, y^2 + 2|g_1 f_2 c_3|\, zx + 2|h_1 f_2 c_3|\, xy.$$

From (a) and (b) we get

$$\frac{(a)\cdot y + (b)\cdot 2z}{x} = 0\cdot x^2 + 2|h_1 b_2 c_3|\, y^2 + 4|g_1 f_2 c_3|\, z^2$$

$$(c)\qquad + \{2|g_1 b_2 c_3| + 4|h_1 f_2 c_3|\}zy + 2|a_1 f_2 c_3|\, zx + |a_1 b_2 c_3|\, xy.$$

Similarly using u_2, u_3, u_1 as we just have u_1, u_2, u_3 we get

$$(d)\quad 4|h_1 g_2 a_3|\, x^2 + 0\cdot y^2 + 2|f_1 c_2 a_3|\, z^2 + |a_1 b_2 c_3|\, yz + \{2|h_1 c_2 a_3| + 4|f_1 g_2 a_3|\}zx + 2|b_1 g_2 a_3|\, xy$$

and again using u_3, u_1, u_2 we get

$$(e)\quad 2|g_1 a_2 b_3|\, x^2 + 4|f_1 h_2 b_3|\, y^2 + 0\cdot z^2 + 2|c_1 h_2 b_3|\, yz + |a_1 b_2 c_3|\, zx + \{2|f_1 a_2 b_3| + 4|g_1 h_2 b_3|\}xy.$$

The equations $(c)=(d)=(e)=0$, are not mere multiples of $u_1=u_2=u_3=0$ and therefore the six may be used to eliminate dialytically the six quantities $x^2, y^2, z^2, xy, xz, yz$ considered as independent variables.

The eliminant is

$$\begin{vmatrix} a_1 & b_1 & c_1 & 2f_1 & 2g_1 & 2h_1 \\ a_2 & b_2 & c_2 & 2f_2 & 2g_2 & 2h_2 \\ a_3 & b_3 & c_3 & 2f_3 & 2g_3 & 2h_3 \\ \cdot & 2[4]-4[3] & 2[8]+4[8'] & 2[6] & [0] \\ 4[1] & \cdot & 2[6] & [0] & 2[9]+4[9'] & 2[4] \\ 2[4] & 4[2] & \cdot & 2[5] & [0] & 2[7]+4[7'] \end{vmatrix}$$

where [0], [1] etc. are found in the table.

$$|a_1 b_2 c_3| = [0]$$

$$|f_1 g_2 h_3| = [0']$$

$$|a_1 g_2 h_3|,\ |b_1 h_2 f_3|,\ |c_1 f_2 g_3| = [1],\ [2],\ [3]$$

$$\begin{cases} |a_1 b_2 g_3|, & |b_1 c_2 h_3|, & |c_1 a_2 f_3| = [4], [5], [6] \\ |a_1 f_2 h_3|, & |b_1 g_2 f_3|, & |c_1 h_2 g_3| = [4'], [5'], [6'] \end{cases}$$

$$\begin{cases} |a_1 b_2 f_3|, & |b_1 c_2 g_3|, & |c_1 a_2 h_3| = [7], [8], [9] \\ |b_1 g_2 h_3|, & |c_1 h_2 f_3|, & |a_1 f_2 g_3| = [7'], [8'], [9'] \end{cases}$$

$$|a_1 b_2 h_3|, \quad |b_1 c_2 f_3|, \quad |c_1 a_2 g_3| = [10], [11], [12].$$

Another way of getting the three auxiliary equations $(c)=(d)=(e)=0$ is as follows:

Writing $u_1=u_2=u_3=0$ in the form

$$(a_1x + 2g_1z + 2h_1y)x + (b_1y + 2f_1z)y + c_1z^2 = 0,$$
$$(a_2x + 2g_2z + 2h_2y)x + (b_2y + 2f_2z)y + c_2z^2 = 0,$$
$$(a_3x + 2g_3z + 2h_3y)x + (b_3y + 2f_3z)y + c_3z^2 = 0,$$

and considering x, y, z^2, the variables the eliminant is

$$\begin{vmatrix} a_1x + 2g_1z + 2h_1y & b_1y + 2f_1z & c_1 \\ a_2x + 2g_2z + 2h_2y & b_2y + 2f_2z & c_2 \\ a_3x + 2g_3z + 2h_3y & b_3y + 2f_3z & c_3 \end{vmatrix}$$

which is (c), and interchanging cyclically in the three cycles a, b, c; f, g, h; and x, y, z we get the other two.

This suggests still another set of auxiliary equations.

We may write $u_1=u_2=u_3=0$ in the form

$$(a_1x + 2g_1z)x + (b_1y + 2f_1z + 2h_1x)y + c_1z^2 = 0$$
$$(a_2x + 2g_2z)x + (b_2y + 2f_2z + 2h_2x)y + c_2z^2 = 0$$
$$(a_3x + 2g_3z)x + (b_3y + 2f_3z + 2h_3x)y + c_3z^2 = 0$$

which on eliminating x, y, z^2 gives

$$\begin{vmatrix} a_1x + 2g_1z & b_1y + 2f_1z + 2h_1x & c_1 \\ a_2x + 2g_2z & b_2y + 2f_2z + 2h_2x & c_2 \\ a_3x + 2g_3z & b_3y + 2f_3z + 2h_3x & c_3 \end{vmatrix}$$

or

$$(f)\quad 2|a_1 h_2 c_3|\,x^2 + 4|g_1 f_2 c_3|\,z^2 + 2|g_1 b_2 c_3|\,zy$$
$$+ 2|a_1 f_2 c_3|\,xz - 4|c_1 h_2 g_3|\,xz + |a_1 b_2 c_3|\,xy$$

and two others formed by cyclical interchange. They are

$$\begin{aligned}&-4|b_1h_2f_3|y^2+2|b_1c_2g_3|z^2+\{2|b_1c_2h_3|-4|b_1g_2f_3|\}yz\\ (g)\qquad &\qquad +|a_1b_2c_3|zx+2|a_1b_2f_3|xy,\end{aligned}$$

$$\begin{aligned}&-4|a_1g_2h_3|x^2+2|a_1b_2f_3|y^2+|a_1b_2c_3|yz+2|c_1a_2h_3|xz\\ (h)\qquad &\qquad +\{2|a_1b_2g_3|-4|a_1f_2h_3|\}xy\end{aligned}$$

and the eliminant is

$$\begin{vmatrix} a_1 & b_1 & c_1 & 2f_1 & 2g_1 & 2h_1 \\ a_2 & b_2 & c_2 & 2f_2 & 2g_2 & 2h_2 \\ a_3 & b_3 & c_3 & 2f_3 & 2g_3 & 2h_3 \\ & -4[2] & 2[8] & 2[5]-4[5'] & [0] & 2[7] \\ 2[9] & & -4[3] & 2[8] & 2[6]-4[6'] & [0] \\ -4[1] & 2[7] & & [0] & 2[9] & 2[4]-4[4'] \end{vmatrix}$$

The cyclical operations that lead from (c) to (d) to (e) and from (f) to (h) to (g) are indicated by

$$\begin{pmatrix}1\,2\,3\\2\,3\,1\end{pmatrix}\quad\begin{pmatrix}f\,g\,h\\g\,h\,f\end{pmatrix}\quad\begin{pmatrix}x\,y\,z\\y\,z\,x\end{pmatrix}$$

274. It is also possible to find the eliminant as determinants of the fifth and third orders. Thus for the fifth order we take $-|u_1c_2a_3|$ or

$$0\cdot x^2+|a_1b_2c_3|y^2+0\cdot z^2-2|a_1c_2f_3|yz$$
$$\qquad -2|a_1c_2g_3|xz-2|a_1c_2h_3|xy,$$

$|u_1a_2b_3|$ or

$$0\cdot x^2+0\cdot y^2+|a_1b_2c_3|z^2+2|a_1b_2f_3|yz$$
$$\qquad +2|a_1b_2g_3|xz+2|a_1b_2h_3|xy,$$

$$\frac{1}{x}|u_1\quad b_2y+f_2z\quad c_3|$$

or

$$0\cdot x^2+2|b_1c_2h_3|y^2-4|c_1f_2g_3|z^2-\{2|c_1b_2g_3|+4|c_1f_2h_3|\}yz$$
$$\qquad -2|a_1c_2f_3|xz+|a_1b_2c_3|xy,$$

$$\frac{1}{x}\,|\,u_1 \;\; b_2 \;\; f_3y + c_3z\,|$$

or

$$-\,4\,|\,f_1\,b_2\,h_3\,|\,y^2 + 2\,|\,b_1\,c_2\,g_3\,|\,z^2 + \{2\,|\,b_1\,c_2\,h_3\,| + 4\,|\,b_1\,f_2\,g_3\,|\,\}\,yz$$
$$+\,|\,a_1\,b_2\,c_3\,|\,zx + 2\,|\,a_1\,b_2\,f_3\,|\,xy,$$

$$|\,u_1\,g_2\,h_3\,| + \frac{1}{y}\,|\,u_1 \;\; c_2z + g_2x \;\; a_3\,|$$

or

$$4\,|\,b_1\,g_2\,h_3\,|\,y^2 + \{2\,|\,a_1\,c_2\,f_3\,| - 4\,|\,c_1\,g_2\,h_3\,|\,\}\,z^2 + \{\,|\,a_1\,b_2\,c\,|_3$$
$$+\,8\,|\,f_1\,g_2\,h_3\,|\,\}\,yz - \{2\,|\,a_1\,c_2\,h_3\,| - 4\,|\,a_1\,f_2\,g_3\,|\,\}\,xz$$
$$+\,2\,|\,a_1\,b_2\,g_3\,|\,xy,$$

from which we may eliminate the five quantities y^2, z^2, yz, xz, xy dialytically and obtain

$$\begin{vmatrix} [0] & \cdot & -2[6] & -2[12] & -2[9] \\ \cdot & [0] & 2[7] & 2[4] & 2[10] \\ 2[5] & -4[3] & 2[8]+4[8'] & +2[6] & [0] \\ 4[2] & 2[8] & 2[5]-4[5'] & [0] & 2[7] \\ 4[7'] & +2[6]+4[6'] & [0]-8[0'] & +2[9]+4[9'] & 2[4] \end{vmatrix}.$$

The extraneous factor here is seen to be $|a_1b_2c_3|$.

EXERCISE. Show that the eliminant of

$$a_1x^2 + b_1y^2 + c_1z^2 + f_1yz + g_1zx = 0$$
$$a_2x^2 + b_2y^2 + c_2z^2 + f_2yz + g_2zx = 0$$
$$a_3x^2 + b_3y^2 + c_3z^2 + f_3yz + g_3zx = 0$$

is

$$\begin{vmatrix} \cdot & |a_1\,b_2\,g_3| & |a_1\,b_2\,f_3| & |a_1\,b_2\,c_3| \\ |a_1\,b_2\,g_3| & |a_1\,f_2\,g_3| & |a_1\,b_2\,c_3| & |a_1\,f_2\,c_3| \\ |a_1\,b_2\,f_3| & |a_1\,b_2\,c_3| & |b_1\,f_2\,g_3| & |b_1\,c_2\,g_3| \\ |a_1\,b_2\,c_3| & |a_1\,f_2\,c_3| & |b_1\,c_2\,g_3| & |f_1\,c_2\,g_3| \end{vmatrix}.$$

275. To get the eliminant as a determinant of the third order we may use

$$u_1 = 0,$$
$$u_2 = 0,$$
$$u_3 = 0,$$
$$\frac{1}{x}\,| \; u_1 \quad b_2y + f_2z \quad c_3 \; | = 0$$

to eliminate x^2, y^2, z^2 leaving an equation in xy, xz, yz from which by cyclical substitution we obtain two others and from the three eliminate xy, xz, yz giving a determinant of order three.

Various forms may be obtained by eliminating any three of the six quantities first and the remaining three second. Thus we might have eliminated xy, xz, yz first and x^2, y^2, z^2 second, or x^2, xy, y^2 and z^2, xz, yz second, etc.

These forms being of the 18th degree in the coefficients must each contain an extraneous factor of the sixth degree.*

Exercises. Set XVI

1. Denoting the eliminants of $u_1 = u_2 = u_3 = 0$ formed by eliminating x^2, y^2, z^2; xy, yz, zx; y^2, z^2, yz; x^2, z^2, xz; x^2, y^2, xy, by $\gamma_1, \gamma_2, \gamma_3, \gamma_4, \gamma_5$ respectively, express in the form of a determinant the extraneous factor in each case.

2. Show that the eliminant of the equations

$$ay^2 - 2hxy + bx^2 = 0$$
$$bz^2 - 2gyz + cy^2 = 0$$
$$cx^2 - 2gzx + az^2 = 0$$

$$= 2\begin{vmatrix} a & h & g \\ h & b & f \\ g & f & c \end{vmatrix}^2.$$

Deduce from these equations a set in x, y, z having for eliminant

$$\begin{vmatrix} A & H & G \\ H & B & F \\ G & F & C \end{vmatrix}$$

* For a fuller discussion of this problem see Muir: Trans. Royal Soc. Edinburgh, Vol. 39 pp. 667–684; Vol. 40 pp. 23–38; Vol. 41 pp. 387–397.

where A, H, etc. are the cofactors of a, h, etc. in the eliminant of the original set.

Show that if the minors complementary to a, b, c, in

$$\begin{vmatrix} a & h & g & x \\ h & b & f & y \\ g & f & c & z \\ x & y & z & \cdot \end{vmatrix}$$

vanish then all primary minors will vanish.

3. Show that

$$-\tfrac{1}{2}\begin{vmatrix} \cdot & C & B & -2A' & \cdot & \cdot \\ C & \cdot & A & \cdot & -2B' & \cdot \\ B & A & \cdot & \cdot & \cdot & -2C' \\ A' & \cdot & \cdot & A & -C' & -B' \\ \cdot & B' & \cdot & -C' & B & -A' \\ \cdot & \cdot & C' & -B' & -A' & C \end{vmatrix} =$$

$$\begin{vmatrix} B & A' & C' & BB'+2C'A' \\ A' & C & B' & CC'+2A'B' \\ C' & B' & A & AA'+2B'C' \\ BB'+2C'A' & CC'+2A'B' & AA'+2B'C' & ABC+8A'B'C' \end{vmatrix}$$

$$= \begin{vmatrix} B & A' & C' \\ A' & C & B' \\ C' & B' & A \end{vmatrix}^2 .$$

4. Find in determinant form the result of eliminating θ from the equations

$$x = (a^2 \cos^2 \theta + b^2 \sin^2 \theta)^{1/2} + p \cos \theta - q \sin \theta$$
$$y = (a^2 \sin^2 \theta + b^2 \cos^2 \theta)^{1/2} + p \sin \theta - q \cos \theta .$$

5. Find the eliminant of the equations

$$a_1x^2 + f_1yz + g_1xz + h_1xy = 0$$
$$b_2y^2 + f_2yz + g_2xz + h_2xy = 0$$
$$c_3z^2 + f_3yz + g_3xz + h_3xy = 0$$

first as a special case of §273 and second by using as auxiliary equations

$$\begin{vmatrix} a_1x + g_1z & f_1z + h_1x \\ g_2z + h_2y & b_2y + f_2z \end{vmatrix} = 0$$

and two others similarly formed.

Compare the results.

276. If we proceed from the case of three variables to that where there are four the problem presents new difficulties.

Consider the four equations in four unknowns

(1) $$Bx^2 - Dxy + Ay^2 = 0$$

(2) $$Cy^2 - Eyz + Bz^2 = 0$$

(3) $$Lz^2 - Kzw + Cw^2 = 0$$

(4) $$Aw^2 - Gwx + Lx^2 = 0.$$

The difficulty will be to find sufficient independent auxiliary equations If we take as our secondary variables

$$\frac{Cx^2 + Az^2}{xz} \cdot \frac{Ly^2 + Bw^2}{yw}, \quad \frac{Cx^2 + Az^2}{xz}, \quad \frac{Ly^2 + Bw^2}{yw}$$

or writing them for shortness sake

$$\phi\theta, \ \phi, \ \theta$$

we see that the cyclical substitutions

$$\begin{pmatrix} A & B & C & L \\ B & C & L & A \end{pmatrix}, \quad \begin{pmatrix} x & y & z & w \\ y & z & w & x \end{pmatrix}$$

change ϕ into θ, and θ into ϕ.

If we solve for D, E, K, G from our original equa tions, the values found will be seen to satisfy the equation

$$D\cdot\phi\theta - 2BG\phi - 2AE\cdot\theta + (4ABK + DEG - D^2K) = 0$$

From this by cyclical substitution we get three others which are

$$E\cdot\theta\phi - 2CD\cdot\theta - 2BK\phi + (4BGC + EKD - E^2G) = 0$$

$$K\cdot\phi\cdot\theta - 2LE\phi - 2CG\cdot\theta + (4CDL + KGE - K^2D) = 0$$

$$G\cdot\theta\phi - 2AK\cdot\theta - 2LD\phi + (4LEA + GDK - G^2E) = 0$$

Eliminating from these four $\phi\theta, \phi, \theta$ we get

$$\begin{vmatrix} D & BG & AE & 4AKB + D(EG - DK) \\ E & BK & CD & 4BGC + E(DK - EG) \\ K & LE & CG & 4CDL + K(EG - DK) \\ G & LD & AK & 4LEA + G(DK - EG) \end{vmatrix} = 0,$$

or

$$\begin{vmatrix} D & BG & AE & K(2AB - D^2) \\ E & BK & CD & G(2BC - E^2) \\ K & LE & CG & D(2CL - K^2) \\ G & LD & AK & E(2LA - G^2) \end{vmatrix} = 0$$

as the eliminant sought.*

EXERCISE. Show that the eliminant of

$$f(x) = 0,$$

$$f\left(\frac{1}{x}\right) = 0,$$

where $f(x) = ax^6 + bx^5 + cx^4 + dx^3 + ex^2 + fx + g$, is equal to

$$(a + b + c + d + e + f + g)(a - b + c - d + e - f + g)$$

$$\times \begin{vmatrix} a & b & c & d & e - g \\ \cdot & a & b & c - g & d - f \\ \cdot & \cdot & a - g & b - f & c - e \\ \cdot & -g & -f & a - e & b - d \\ -g & -f & -e & -d & a - c \end{vmatrix}^2$$

277. If our equations are lineo-linear such as

$$(a_1x_1 + a_2x_2 + a_3x_3)y_1 + (a_4x_1 + a_5x_2 + a_6x_3)y_2 = 0$$
$$(b_1x_1 + b_2x_2 + b_3x_3)y_1 + (b_4x_1 + b_5x_2 + b_6x_3)y_2 = 0$$
$$(c_1x_1 + c_2x_2 + c_3x_3)y_1 + (c_4x_1 + c_5x_2 + c_6x_3)y_2 = 0$$
$$(d_1x_1 + d_2x_2 + d_3x_3)y_1 + (d_4x_1 + d_5x_2 + d_6x_3)y_2 = 0$$

* For a fuller treatment of this problem see Muir: Proc. Royal Soc. Edinburgh: Vol. 21 pp. 328–341.

or as they may be conveniently written in matrix notation

$$\left(\begin{vmatrix} a_1 & a_2 & a_3 & a_4 & a_5 & a_6 \\ b_1 & b_2 & b_3 & b_4 & b_5 & b_6 \\ c_1 & c_2 & c_3 & c_4 & c_5 & c_6 \\ d_1 & d_2 & d_3 & d_4 & d_5 & d_6 \end{vmatrix} \between x_1,\ x_2,\ x_3 \between y_1,\ y_2\right) = 0$$

the eliminant may be formed by writing the equations in the form

$$a_1x_1y_1 + a_4x_1y_2 + (a_2y_1 + a_5y_2)x_2 + (a_3y_1 + a_6y_2)x_3 = 0$$

and eliminate y_1x_1, y_2x_1, x_2, x_3 from the four equations thus written.

The eliminant is

$$\begin{vmatrix} a_1 & a_4 & a_2y_1 + a_5y_2 & a_3y_1 + a_6y_2 \\ b_1 & b_4 & b_2y_1 + b_5y_2 & b_3y_1 + b_6y_2 \\ c_1 & c_4 & c_2y_1 + c_5y_2 & c_3y_1 + c_6y_2 \\ d_1 & d_4 & d_2y_1 + d_5y_2 & d_3y_1 + d_6y_2 \end{vmatrix} = 0, \text{ or}$$

$$|1\,2\,3\,4|\, y_1^2 + \{|1\,3\,4\,5| - |1\,2\,4\,6|\}y_1y_2 + |1\,4\,5\,6|\, y_2^2 = 0$$

where $|\alpha\beta\gamma\theta|$ is the determinant formed by using the αth, βth, γth, θth columns of the 4-by-6 array of the coefficients.

Similarly by eliminating x_1, x_2y_1, x_2y_2, x_3, we get

$$-|1\,2\,3\,5|\, y_1^2 + \{|1\,2\,5\,6| - |2\,3\,4\,5|\}y_1y_2 - |2\,4\,5\,6|\, y_2^2 = 0$$

and eliminating x_1, x_2, x_3y_1, x_3y_2, we get

$$|1\,2\,3\,6|\, y_1^2 + \{|2\,3\,4\,6| - |1\,3\,5\,6|\}y_1y_2 + |2\,4\,5\,6|\, y_2^2 = 0.$$

From these three equations we may eliminate y_1^2, y_1y_2, y_2^2 and get

$$\begin{Vmatrix} |1\,2\,3\,4| & |1\,3\,4\,5| - |1\,2\,4\,6| & |1\,4\,5\,6| \\ |1\,2\,3\,5| & |2\,3\,4\,5| - |1\,2\,5\,6| & |2\,4\,5\,6| \\ |1\,2\,3\,6| & |2\,3\,4\,6| - |1\,3\,5\,6| & |3\,4\,5\,6| \end{Vmatrix}$$

This type of compound determinant has already appeared in §210 and is one Sylvester called *double* determinant. Its equivalent as a determinant of the fourth order is there given.

It is readily seen that this process is perfectly general in its application so that if we wish the eliminant of the equations

$$\left(\begin{vmatrix} a_1 & a_2 \cdots a_{mn} \\ b_1 & b_2 \cdots b_{mn} \\ \cdots & \cdots \\ \multicolumn{2}{c}{(m + n - 1) \text{ rows}} \end{vmatrix} \between x_1, x_2, \cdots, x_m \between y_1, y_2, \cdots, y_n\right) = 0$$

we eliminate the x's in such a way as to obtain a set of

$$\frac{(m+n-2)!}{(m-1)!(n-1)!}$$

equations linear and homogeneous in the quantities $y_1^{m-1}, y_1^{m-2}y_2, \cdots, y_n^{m-1}$ considered as independent variables. We have then just the proper number of equations to eliminate these quantities also.

If we had eliminated the y's first and then the x's the result would have been the same except the rows of one are the columns of the other.

If one observes the symmetries involved it will be possible to obtain all the other equations after having that resulting from the first elimination.

278. If we are given equations in the form

$$\frac{a_1x+b_1y+c_1z}{\alpha_1x+\beta_1y+\gamma_1z}=\frac{a_2x+b_2y+c_2z}{\alpha_2x+\beta_2y+\gamma_2z}=\frac{a_3x+b_3y+c_3z}{\alpha_3x+\beta_3y+\gamma_3z}=\frac{1}{r}$$

we can deduce an equation containing only r and the coefficients of x, y, z. To this end consider the determinant

$$D \equiv \begin{vmatrix} a_1 & b_1 & c_1 & \alpha_1 & \beta_1 & \gamma_1 \\ a_2 & b_2 & c_2 & \alpha_2 & \beta_2 & \gamma_2 \\ a_3 & b_3 & c_3 & \alpha_3 & \beta_3 & \gamma_3 \\ \xi_1 & \eta_1 & \zeta_1 & r\xi_1 & r\eta_1 & r\zeta_1 \\ \xi_2 & \eta_2 & \zeta_2 & r\xi_2 & r\eta_2 & r\zeta_2 \\ \xi_3 & \eta_3 & \zeta_3 & r\xi_3 & r\eta_3 & r\zeta_3 \end{vmatrix}$$

where the ξ's, η's, ζ's are any quantities whatever.

Performing the operations $x\,\text{col}_1+y\,\text{col}_2+z\,\text{col}_3$, $x\,\text{col}_4+y\,\text{col}_5+z\,\text{col}_6$ we obtain a determinant equal to $x^2\cdot D$ having the fourth column r times the first and therefore vanishes. D is therefore zero for all values of the ξ's, η's, ζ's. But D is obviously equal to

$$\begin{vmatrix} a_1 & b_1 & c_1 & \alpha_1-ra_1 & \beta_1-rb_1 & \gamma_1-rc_1 \\ a_2 & b_2 & c_2 & \alpha_2-ra_2 & \beta_2-rb_2 & \gamma_2-rc_2 \\ a_3 & b_3 & c_3 & \alpha_3-ra_3 & \beta_3-rb_3 & \gamma_3-rc_3 \\ \xi_1 & \eta_1 & \zeta_1 & \cdot & \cdot & \cdot \\ \xi_2 & \eta_2 & \zeta_2 & \cdot & \cdot & \cdot \\ \xi_3 & \eta_3 & \zeta_3 & \cdot & \cdot & \cdot \end{vmatrix}$$

that is

$$-\left|\xi_1\,\eta_2\,\zeta_3\right|\cdot\left|\alpha_1 - ra_1\ \ \beta_2 - rb_2\ \ \gamma_3 - rc_3\right|$$

and therefore

$$\left|\alpha_1 - ra_1\ \ \beta_2 - rb_2\ \ \gamma_3 - rc_3\right| = 0$$

—a cubic equation for the determination of r in terms of the original coefficients of x, y, z.

279. If the number of equivalent fractions be one more than the number of unknowns, that is if $1/r$ be also equal to

$$\frac{a_4x + b_4y + c_4z}{\alpha_4x + \beta_4y + \gamma_4z}$$

we could get, by leaving out in succession the 4th, 3rd, 2nd, 1st fractions, the four equations

$$\begin{aligned}
\left|\alpha_1 - ra_1\ \ \beta_2 - rb_2\ \ \gamma_3 - rc_3\right| &= 0,\\
\left|\alpha_4 - ra_4\ \ \beta_1 - rb_1\ \ \gamma_2 - rc_2\right| &= 0,\\
\left|\alpha_3 - ra_3\ \ \beta_4 - rb_4\ \ \gamma_1 - rc_1\right| &= 0,\\
\left|\alpha_2 - ra_2\ \ \beta_3 - rb_3\ \ \gamma_4 - rc_4\right| &= 0,
\end{aligned}$$

from which we can eliminate r, r^2, r^3 and get as the eliminant

$$\Delta_{(4,3)} \equiv \begin{vmatrix}
\left|\alpha_1\beta_2\gamma_3\right| & \sum\left|a_1\beta_2\gamma_3\right| & \sum\left|\alpha_1 b_2 c_3\right| & \left|a_1 b_2 c_3\right| \\
\left|\alpha_2\beta_3\gamma_4\right| & \sum\left|a_2\beta_3\gamma_4\right| & \sum\left|\alpha_2 b_3 c_4\right| & \left|a_2 b_3 c_4\right| \\
\left|\alpha_3\beta_4\gamma_1\right| & \sum\left|a_3\beta_4\gamma_1\right| & \sum\left|\alpha_3 b_4 c_1\right| & \left|a_3 b_4 c_1\right| \\
\left|\alpha_4\beta_1\gamma_2\right| & \sum\left|a_4\beta_1\gamma_2\right| & \sum\left|\alpha_4 b_1 c_2\right| & \left|a_4 b_1 c_2\right|
\end{vmatrix}$$

where

$$\begin{aligned}
\sum\left|a_1\beta_2\gamma_3\right| &= \left|a_1\beta_2\gamma_3\right| + \left|\alpha_1 b_2\gamma_3\right| + \left|\alpha_1\beta_2 c_3\right|,\\
\sum\left|\alpha_1 b_2 c_3\right| &= \left|\alpha_1 b_2 c_3\right| + \left|a_1\beta_2 c_3\right| + \left|a_1 b_2\gamma_3\right|
\end{aligned}$$

etc.

The general theorem is: The eliminant of the set of equations

$$(1)\qquad \frac{a_1x_1 + b_1x_2 + \cdots + l_1x_n}{\alpha_1x_1 + \beta_1x_2 + \cdots + \lambda_1x_n} = \cdots$$

$$= \frac{a_{n+1}x_1 + b_{n+1}x_2 + \cdots + l_{n+1}x_n}{\alpha_{n+1}x_1 + \beta_{n+1}x_2 + \cdots + \lambda_{n+1}x_n}$$

is

$$\begin{vmatrix} D_1 & \sum D_1' & \sum D_1'' & \cdot & \cdot & \cdot \\ D_2 & \sum D_2' & \sum D_2'' & \cdot & \cdot & \cdot \\ \cdots & \cdots & \cdots & \cdots & \cdot & \cdot \\ D_{n+1} & \sum D_{n+1}' & \sum D_{n+1}'' & \cdot & \cdot & \cdot \end{vmatrix}$$

where

$$D_1 \equiv |\, a_1\, b_2 \cdots l_n \,|\,,\quad D_2 \equiv |\, a_2\, b_3 \cdots l_{n+1} \,|\,,\ \cdots,$$
$$D_{n+1} \equiv |\, a_{n+1}\, b_1 \cdots l_{n-1} \,|$$

and where D_r' indicates that any one of the letters in D_r has been replaced by the corresponding letter of the other alphabet, D_r'' that any two letters have been similarly treated, and so on.

280. In connection with the problem of the elimination of x, y, z when four equivalent fractions are given, let us now consider the determinant

$$\begin{vmatrix} a_1 & b_1 & c_1 & \alpha_1 & \beta_1 & \gamma_1 \\ a_2 & b_2 & c_2 & \alpha_2 & \beta_2 & \gamma_2 \\ a_3 & b_3 & c_3 & \alpha_3 & \beta_3 & \gamma_3 \\ a_4 & b_4 & c_4 & \alpha_4 & \beta_4 & \gamma_4 \\ \xi_1 & \eta_1 & \zeta_1 & r\xi_1 & r\eta_1 & r\zeta_1 \\ \xi_2 & \eta_2 & \zeta_2 & r\xi_2 & r\eta_2 & r\zeta_2 \end{vmatrix}$$

which as before is readily seen to be zero and also equal to

$$\begin{vmatrix} a_1 & b_1 & c_1 & \alpha_1 - ra_1 & \beta_1 - rb_1 & \gamma_1 - rc_1 \\ a_2 & b_2 & c_2 & \alpha_2 - ra_2 & \beta_2 - rb_2 & \gamma_2 - rc_2 \\ a_3 & b_3 & c_3 & \alpha_3 - ra_3 & \beta_3 - rb_3 & \gamma_3 - rc_3 \\ a_4 & b_4 & c_4 & \alpha_4 - ra_4 & \beta_4 - rb_4 & \gamma_4 - rc_4 \\ \xi_1 & \eta_1 & \zeta_1 & \cdot & \cdot & \cdot \\ \xi_2 & \eta_2 & \zeta_2 & \cdot & \cdot & \cdot \end{vmatrix}$$

so that the coefficients of $|\xi_1\eta_2|$, $|\xi_1\zeta_2|$, $|\eta_1\zeta_2|$ must be zero which gives us the following three equations:

$$|\, a_1\, \alpha_2\, \beta_3\, \gamma_4 \,| - \{\, |\, a_1\, \alpha_2\, b_3\, \gamma_4 \,| + |\, a_1\, \alpha_2\, \beta_3\, c_4 \,| \,\}r + |\, a_1\, \alpha_2\, b_3\, c_4 \,|\, r^2 = 0$$
$$|\, b_1\, \alpha_2\, \beta_3\, \gamma_4 \,| - \{\, |\, b_1\, \alpha_2\, \beta_3\, c_4 \,| + |\, b_1\, a_2\, \beta_3\, \gamma_4 \,| \,\}r + |\, b_1\, a_2\, \beta_3\, c_4 \,|\, r^2 = 0$$
$$|\, c_1\, \alpha_2\, \beta_3\, \gamma_4 \,| - \{\, |\, c_1\, a_2\, \beta_3\, \gamma_4 \,| + |\, c_1\, \alpha_2\, b_3\, \gamma_4 \,| \,\}r + |\, c_1\, a_2\, b_3\, \gamma_4 \,|\, r^2 = 0$$

from which we get as the eliminant

$$\Delta_{(3,4)} \equiv \begin{vmatrix} |a_1\,\alpha_2\,\beta_3\,\gamma_4| & |a_1\,\alpha_2\,\beta_3\,\gamma_4| + |a_1\,\alpha_2\,\beta_3\,c_4| & |a_1\,\alpha_2\,b_3\,c_4| \\ |b_1\,\alpha_2\,\beta_3\,\gamma_4| & |b_1\,\alpha_2\,\beta_3\,c_4| + |b_1\,a_2\,\beta_3\,\gamma_4| & |b_1\,a_2\,\beta_3\,c_4| \\ |c_1\,\alpha_2\,\beta_3\,\gamma_4| & |c_1\,a_2\,\beta_3\,\gamma_4| + |c_1\,\alpha_2\,b_3\,\gamma_4| & |c_1\,a_2\,b_3\,\gamma_4| \end{vmatrix} = 0$$

which from §210 we know is equal to $\Delta_{(4,3)}$ the eliminant as formed before.

In general we have for the eliminant of the equations (1) §279 the determinant

$$\begin{vmatrix} D_1 & \sum D_1' & \sum D_1'' & \cdot & \cdot & \cdot \\ D_2 & \sum D_2' & \sum D_2'' & \cdot & \cdot & \cdot \\ \cdot & \cdot & \cdot & \cdot & \cdot & \cdot \\ D_n & \sum D_n' & \sum D_n'' & \cdot & \cdot & \cdot \end{vmatrix}$$

where now

$$D_1 = |\,a_1\,b_2\,c_3 \cdots l_n\,\alpha_{n+1}\,|\,, \qquad D_2 = |\,a_1\,b_2\,c_3 \cdots l_n\,\beta_{n+1}\,| \cdots$$

and where D_r' indicates that any one of the italic letters of D_r, except the rth has been replaced by the corresponding Greek letter, D_r'' that any two letters except the rth have been similarly treated, and so on.

281. If we have two equations of the nth degree given in the form

$$(A) \qquad \sum_1^{n+1} s \frac{A_s}{x - r_s} = 0$$

$$(B) \qquad \sum_1^{n+1} s \frac{B_s}{x - r_s} = 0$$

and we increase the roots of each equation by r_1 we have

$$\frac{A_1}{x} + \frac{A_2}{x - \rho_2} + \frac{A_3}{x - \rho_3} + \cdots = 0$$

$$\frac{B_1}{x} + \frac{B_2}{x - \rho_2} + \frac{B_3}{x - \rho_3} + \cdots = 0$$

where $\rho_s = r_s - r_1$.

Clearing these of fractions and writing the results as an increasing power series in x we have

$$\rho_2\rho_3 \cdots \rho_{n+1}A_1 - x[A_1\beta + A_2\rho_3\rho_4 \cdots \rho_{n+1} + A_3\rho_2\rho_4 \cdots \rho_{n+1} + \cdots] + \cdots = 0$$

$$\rho_2\rho_3 \cdots \rho_{n+1}B_1 - x[B_1\beta + B_2\rho_3\rho_4 \cdots \rho_{n+1} + B_3\rho_2\rho_4 \cdots \rho_{n+1} + \cdots] + \cdots = 0$$

where β is the sum of the products of $\rho_2, \rho_3, \cdots, \rho_{n+1}$ taken $n-1$ at a time.

From the observations of §268 it is apparent that every term of the eliminant must contain one or the other of the independent terms of the two equations. It also appears that the terms which contain the independent terms only in the first degree are

$$\rho_2{}^2\rho_3{}^2 \cdots \rho_{n+1}^2 \sum_2^{n+1} s \,|\, A_1 B_s \,| \cdot 1/\rho_s$$

or

$$(r_2 - r_1)^2 \cdots (r_{n+1} - r_1)^2 \sum_2^{n+1} s \,|\, A_1 B_s \,| \cdot 1/(r_s - r_1)$$

times the eliminant of the equations formed by putting the independent terms equal to zero and dividing by x.

Every term in the eliminant must contain either A_s or B_s for every value of s, for when $A_s = 0$, $B_s = 0$, the two equations (A) and (B) have a common root $x = r_s$.

Also no term in the eliminant can contain A_s^2, A_sB_s, B_s^2 for all but one values of s, for then the degree of the eliminant would be $2n+1$ while it is of degree n in the A's and B's respectively.

It follows therefore from the symmetry in respect to A_s, B_s, r_s that the eliminant is determined when the terms containing A_1 or B_1 to the first power only are known.

Denoting the eliminant of (A) and (B) by $E(r_1, r_2, \cdots, r_{n+1})$ and the discriminant of

$$\prod_1^{n-1}(x - r_s) = 0 \quad \text{by} \quad D(r_1, r_2 \cdots r_{n+1})$$

we see that

$$E(r_1, r_2) = (r_1 - r_2) \,|\, A_1 \, B_2 \,| = D(r_1, r_2)(1, 2)$$

where (1, 2) denotes $|A_1B_2| \div (r_1 - r_2)$, and in general $(s, t) = (t, s) = |A_sB_t| \div (r_s - r_t)$. The terms in $E(r_1, r_2, r_3)$ which contain A_1 or B_1 to the first power, are

$$(r_2 - r_1)^2(r_3 - r_1)^2\{(1,2) + (1,3)\}(r_2 - r_3)^2(2,3)$$

or

$$(r_2 - r_1)^2(r_3 - r_1)^2\{(1,2) + (1,3)\}D(r_2, r_3)(2,3)$$

and we get

$$E(r_1, r_2, r_3) = D(r_1, r_2, r_3)\{(1,2)(1,3) + (2,1)(2,3) + (3,1)(3,2)\}$$

or as it may be written

$$D(r_1, r_2, r_3)\begin{vmatrix} \sum(1,2) & -(1,2) \\ -(2,1) & \sum(2,1) \end{vmatrix}$$

where

$$\sum(1,2) = (1,2) + (1,3),$$
$$\sum(2,1) = (2,1) + (2,3)$$

and in general

$$\sum(t,s) = (t,1) + (t,2) + \cdots + (t,t-1) + (t,t+1) + \cdots + (t,n+1),$$

the term (t, t) being omitted.

From this the terms in $E(r_1, r_2, r_3, r_4)$ of the first power in A_1 or B_1, will be

$$D(r_1, r_2, r_3, r_4)\{(1,2) + (1,3) + (1,4)\}\begin{vmatrix} \sum(2,3) & -(2,3) \\ -(3,2) & \sum(3,4) \end{vmatrix}$$

and the complete expansion is

$$D(r_1, r_2, r_3, r_4)\begin{vmatrix} \sum(1,s) & -(1,2) & -(1,3) \\ -(2,1) & \sum(2,s) & -(2,3) \\ -(3,1) & -(3,2) & \sum(3,s) \end{vmatrix}$$

where $\sum(1, s) = (1, 2) + (1, 3) + (1, 4)$ etc.

In general

$$E(r_1, r_2, \cdots, r_{n+1})$$
$$= D(r_1, r_2, \cdots, r_{n+1})\begin{vmatrix} \sum(1,s) & -(1,2) \cdots & -(1,n) \\ -(2,1) & \sum(2,s) \cdots & -(2,n) \\ \cdots & \cdots & \cdots \\ -(n,1) & -(n,2) \cdots & \sum(n,s) \end{vmatrix}$$

That this is true in general may be seen on observing that the law of formation shows that if true for $E(r_2, r_3 \cdots r_{n+1})$ it must be true for the terms which contain the number 1 only once and since these terms are $\{(1, 2)+(1, 3) \cdots +(1, n+1)\}E(r_2, r_3, \cdots, r_{n+1})$ we see by symmetry that the law applies to all terms of $E(r_1, r_2, \cdots, r_{n+1})$. We have seen that it applies to $E(r_1, r_2, r_3)$ and $E(r_1, r_2, r_3, r_4)$ and hence it applies generally.*

EXERCISE 1. If the quadratic $ax^2+by^2+cz^2+2lxy+2mxz+2nxy$ be increased by $(\alpha x+\beta y+\gamma z)^2$ then the discriminant is increased by

$$\begin{vmatrix} 0 & \alpha & \beta & \gamma \\ \alpha & a & n & m \\ \beta & n & b & l \\ \gamma & l & m & c \end{vmatrix}$$

2. Prove that if in the array

$$\begin{matrix} c_{00} & c_{01} \cdots c_{0,n-1} & f_{0n} & f_{0,n+1} \cdots \\ c_{10} & c_{11} \cdots c_{1,n-1} & f_{1n} & f_{1,n+1} \cdots \\ \cdots & \cdots & \cdots & \cdots \\ c_{n0} & c_{n1} \cdots c_{n,n-1} & f_{nn} & f_{n,n+1} \cdots \end{matrix}$$

where each column after the nth is an aggregate of multiples of the first n columns, there be at least one m-line minor of the c-array that does not vanish, while the minors of higher order all vanish, then each row of the whole array is a linear function of the rows from which the said non-vanishing minor is formed: further, if there be in the c-array an m-line minor that vanishes, its rows in their full extent are connected by a linear relation.

3. Show that any quadric whose discriminant is a zero-axial skew determinant vanishes identically.

* The form of the eliminant here given is due to Brochardt and was given by him in Crelle's Journal Bd. LVII pp. III. The proof followed was given by L. A. Dixon in Proceedings Lond. Math. Soc. Series 2, Vol. 6 pp. 468–478, which should be consulted for a fuller treatment. For the case of three quantics in two variables his papers in 2nd series vol. 7 pp. 49–69 and 473–492 should be consulted.

CHAPTER IX

Bipartites

282. In chapter II, §28, bipartites or *cumulants* as they are sometimes called were defined in connection with the lineo-linear function in 3 variables. In general if a row of n elements be taken, and closely following this array, but separated by a bar from it, we write n rows of n elements each; and closely following either outside *column* of this square array, but separated by a bar from it, we write n *columns* of n elements each; and closely following an outside *row* of this second square array, but separated by a bar from it, we write n *rows* of n elements each; and so on, passing from the rows or columns of one array to the columns or rows of the next, and ending not with a square array, but, as we began, with a single line of elements, we have the matrix representation of a *bipartite function.*

For example, when $n=3$ and the number of square arrays is 4, the representation is

$$\begin{array}{ccc|ccc|ccc}
 & & & h_3 & k_3 & l_3 & r_1 & r_2 & r_3\\
 & & & h_2 & k_2 & l_2 & n_1 & n_2 & n_3\\
a_1 & a_2 & a_3 & h_1 & k_1 & l_1 & m_1 & m_2 & m_3\\
\hline
b_1 & c_1 & d_1 & e_1 & e_2 & e_3 & x_1 & y_1 & z_1\\
b_2 & c_2 & d_2 & f_1 & f_2 & f_3 & & & \\
b_3 & c_3 & d_3 & g_1 & g_2 & g_3 & & &
\end{array}$$

or

$$\begin{array}{ccc|ccc|ccc}
a_1 & a_2 & a_3 & & & & & & \\
\hline
b_1 & c_1 & d_1 & e_1 & e_2 & e_3 & & & \\
b_2 & c_2 & d_2 & f_1 & f_2 & f_3 & & & \\
b_3 & c_3 & d_3 & g_1 & g_2 & g_3 & & & \\
\hline
 & & & h_1 & k_1 & l_1 & m_1 & m_2 & m_3\\
 & & & h_2 & k_2 & l_2 & n_1 & n_2 & n_3\\
 & & & h_3 & k_3 & l_3 & r_1 & r_2 & r_3\\
\hline
 & & & & & & x_1 & y_1 & z_1
\end{array}$$

283. The ordinary algebraical expression of the function is obtained from the matrix representation by forming every possible term containing as a factor one, and only one, element from each array, subject to the condition that the element to be taken from any one array must be in the same row or column with the element taken from the preceding array, and in the same column or row with the element taken from the following array; and then connecting, by means of plus signs, the terms thus formed.

284. If the number of elements in a row or column be n, the bipartite is said to be of the nth order: if the number of arrays, square or not, be m, it is evidently of the mth degree; and combining these we may speak of such a bipartite as being of the deg-order (m, n).

285. *The number of terms in the final expansion of a bipartite of deg-order* (m, n) *is* n^{m-1}.

For the deg-order $(2, n)$ the number is evidently n, that is, n^{2-1}: for the deg-order $(3, n)$ there must be one term, and one only, for every element in the square array, and therefore in all n^2 terms, that is, n^{3-1}; and if the number of terms in a bipartite of deg-order (p, n) be n^{p-1}, it is readily made evident that the number in the bipartite of deg-order $(p+1, n)$ is n^p: hence the statement is established.

286. *Each element of any one of the square arrays of a bipartite of deg-order* (m, n) *occurs* $n^{m-1} \div n^2$, *that is,* n^{m-3} *times in the final expansion; and each element of either of the other arrays occurs* $n^{m-1} \div n$, *that is,* n^{m-2} *times.*

For, one of the former, and only one, must occur in each term, and there are n^2 of them; and one of the latter, and only one, must occur in each term, and there are n of them.

The elements of the square arrays may therefore be called *secondary* elements, and the others *primary.*

The two lines of primary elements may be distinguished as *initial* and *final.* Strictly speaking, however, either is at the end; for the definition shows that the order of writing the arrays may be reversed without affecting the final expansion.

287. Also, it may be remarked, the law of formation of the terms would give the same result if the initial row of any bipartite were made into a *column*, and at the same time all the other rows and columns altered accordingly.

288. *If any two rows or two columns of a square array be interchanged, and, at the same time, the two collinear rows or columns in one of the adjacent arrays, the bipartite is in substance unaltered.*

Thus

$$\begin{array}{cc|cc|c} & & m & n & q \\ a & b & k & l & p \\ \hline c & d & g & i & \\ e & f & h & j & \end{array} = \begin{array}{cc|cc|c} & & m & n & q \\ a & b & k & l & p \\ \hline e & f & h & j & \\ c & d & g & i & \end{array}$$

$$= \begin{array}{cc|cc|c} & & k & l & p \\ a & b & m & n & q \\ \hline c & d & g & i & \\ e & f & h & j & \end{array}, \quad \cdots$$

289. *A bipartite is multiplied by any quantity if each of the elements of any one of its arrays be multiplied by that quantity.*

290. *A bipartite having every element of one of its square arrays a sum of p terms may be expressed as the sum of p bipartites, the first of which is got from the original by deleting all the terms of each of the p-termed elements except the first term, the second by deleting all the terms of each of the p-termed elements except the second term, and so on.*

291. *The cofactor of any one of the primary elements of a bipartite of deg-order (m, n) is expressible as a bipartite of deg-order $(m-1, n)$, which is obtained from the original bipartite by deleting, first, the line to which the said primary element belongs, and then the elements of the adjacent square array which are not collinear with the said primary element.*

Thus in

$$\begin{array}{ccc|ccc} a_1 & a_2 & a_3 & h_1 & k_1 & l_1 \\ \hline b_1 & c_1 & d_1 & e_1 & e_2 & e_3 \\ b_2 & c_2 & d_2 & f_1 & f_2 & f_3 \\ b_3 & c_3 & d_3 & g_1 & g_2 & g_3 \end{array}$$

the cofactor of a_2 is

$$\begin{array}{c|ccc} & h_1 & k_1 & l_1 \\ \hline c_1 & e_2 & e_2 & e_3 \\ c_2 & f_1 & f_2 & f_3 \\ c_3 & g_1 & g_2 & g_3 \end{array};$$

and the cofactor of h_1 is

$$\begin{array}{ccc|c} a_1 & a_2 & a_3 & \\ \hline b_1 & c_1 & d_1 & e_1 \\ b_2 & c_2 & d_2 & f_1 \\ b_3 & c_3 & d_3 & g_3 \end{array}$$

292. *A bipartite of deg-order* (m, n) *is thus expressible as a sum of* n *products of two factors each, the first factors being elements taken either all from the initial line or all from the final line, and the second factors being bipartites of deg-order* $(m-1, n)$.

Thus

$$\begin{array}{ccc|ccc} a_1 & a_2 & a_3 & h_1 & k_1 & l_1 \\ \hline b_1 & c_1 & d_1 & e_1 & e_2 & e_3 \\ b_2 & c_2 & d_2 & f_1 & f_2 & f_3 \\ b_3 & c_3 & d_3 & g_1 & g_2 & g_3 \end{array}$$

$$= a_1 \cdot \begin{array}{c|ccc} & h_1 & k_1 & l_1 \\ \hline b_1 & e_1 & e_2 & e_3 \\ b_2 & f_1 & f_2 & f_3 \\ b_3 & g_1 & g_2 & g_3 \end{array} + a_2 \cdot \begin{array}{c|ccc} & h_1 & k_1 & l_1 \\ \hline c_1 & e_1 & e_2 & e_3 \\ c_2 & f_1 & f_2 & f_3 \\ c_3 & g_1 & g_2 & g_3 \end{array} + a_3 \cdot \begin{array}{c|ccc} & h_1 & k_1 & l_1 \\ \hline d_1 & e_1 & e_2 & e_3 \\ d_2 & f_1 & f_2 & f_3 \\ d_3 & g_1 & g_2 & g_3 \end{array}$$

This recurrent law of formation of a bipartite might of course have been adopted as the definition.

293. *The cofactor of any one of the secondary elements belonging to the pth array of a bipartite of deg-order* (m, n) *is expressible as the product of two bipartites, one of deg-order* $(p-1, n)$ *and the other of deg-order* $(m-p, n)$, *the first being got from the first* $p-1$ *arrays by deleting from the* $(p-1)$*th array all the elements not collinear with the element in question, and the second being got from the last* $n-p$ *arrays in the same way.*

Thus in

$$\begin{array}{ccc|ccc|ccc|c} & & & h_3 & k_3 & l_3 & p_1 & p_2 & p_3 & \\ & & & h_2 & k_2 & l_2 & n_1 & n_2 & n_3 & \\ a_1 & a_2 & a_3 & h_1 & k_1 & l_1 & m_1 & m_2 & m_3 & \\ \hline b_1 & c_1 & d_1 & e_1 & e_2 & e_3 & r_1 & s_1 & u_1 & x_1 \\ b_2 & c_2 & d_2 & f_1 & f_2 & f_3 & r_2 & s_2 & u_2 & y_2 \\ b_3 & c_3 & d_3 & g_1 & g_2 & g_3 & r_3 & s_3 & u_3 & z_3 \end{array}$$

the cofactor of k_1 is

$$\begin{array}{ccc|c} a_1 & a_2 & a_3 & \\ \hline b_1 & c_1 & d_1 & e_2 \\ b_2 & c_2 & d_2 & f_2 \\ b_3 & c_3 & d_3 & g_2 \end{array} \cdot \begin{array}{ccc|c} m_1 & m_2 & m_3 & \\ \hline r_1 & s_1 & u_1 & x_1 \\ r_2 & s_2 & u_2 & y_1 \\ r_3 & s_3 & u_3 & z_1 \end{array},$$

and the cofactor of f_3 is

$$\begin{array}{ccc} a_1 & a_2 & a_3 \\ \hline b_2 & c_2 & d_2 \end{array} \cdot \begin{array}{c|ccc|c} l_3 & p_1 & p_2 & p_3 & \\ l_2 & n_1 & n_2 & n_3 & \\ l_1 & m_1 & m_2 & m_3 & \\ \hline & r_1 & s_1 & u_1 & x_1 \\ & r_2 & s_2 & u_2 & y_1 \\ & r_3 & s_3 & u_3 & z_1 \end{array},$$

or

$$\begin{array}{ccc} a_1 & a_2 & a_3 \\ \hline b_2 & c_2 & d_2 \end{array} \cdot \begin{array}{ccc|ccc} l_1 & l_2 & l_3 & x_1 & y_1 & z_1 \\ \hline m_1 & n_1 & p_1 & r_1 & r_2 & r_3 \\ m_2 & n_2 & p_2 & s_1 & s_2 & s_3 \\ m_3 & n_3 & p_3 & u_1 & u_2 & u_3 \end{array}.$$

294. *A bipartite of deg-order (m, n) is thus expressible as a sum of n^2 products of three factors each, the first factors being elements all taken from any one of the square arrays, the pth say, the second factors being minor bipartites of deg-order $(p-1, n)$, and the third factors being minors of deg-order $(m-p, n)$.*

Thus

$$\begin{array}{cc|cc|cc|cc} & & f_2 & g_2 & k_1 & k_2 & & \\ a_1 & a_2 & f_1 & g_1 & h_1 & h_2 & p_1 & q_1 \\ \hline b_1 & c_1 & d_1 & d_2 & i_1 & j_1 & m_1 & m_2 \\ b_2 & c_2 & e_1 & e_2 & i_2 & j_2 & n_1 & n_2 \end{array},$$

if we decide on taking the elements of its fourth array, is equal to

$$f_1 \cdot \begin{array}{cc|c} a_1 & a_2 & \\ \hline b_1 & c_1 & d_1 \\ b_2 & c_2 & e_1 \end{array} \cdot \begin{array}{cc|cc} h_1 & h_2 & p_1 & q_1 \\ \hline i_1 & j_1 & m_1 & m_2 \\ i_2 & j_2 & m_1 & m_2 \end{array} + g_1 \cdot \begin{array}{cc|c} a_1 & a_2 & \\ \hline b_1 & c_1 & d_2 \\ b_2 & c_2 & e_2 \end{array} \cdot \begin{array}{cc|cc} h_1 & h_2 & p_1 & q_1 \\ \hline i_1 & j_1 & m_1 & m_2 \\ i_2 & j_2 & n_1 & n_2 \end{array}$$

$$(\alpha)\qquad + f_2\cdot\begin{array}{cc|c}a_1 & a_2 & \\ \hline b_1 & c_1 & d_1\\ b_2 & c_2 & e_1\end{array}\cdot\begin{array}{cc|cc}k_1 & k_2 & p_1 & q_1\\ \hline i_1 & j_1 & m_1 & m_2\\ i_2 & j_2 & n_1 & n_2\end{array} + g_2\cdot\begin{array}{cc|c}a_1 & a_2 & \\ \hline b_1 & c_1 & d_2\\ b_2 & c_2 & e_2\end{array}\cdot\begin{array}{cc|cc}k_1 & k_2 & p_1 & q_1\\ \hline i_1 & j_1 & m_1 & m_2\\ i_2 & j_2 & n_1 & n_2\end{array}$$

or, if we decide on taking the elements of its third array, is equal to

$$(\beta)\qquad \begin{array}{l} d_1\cdot\begin{array}{cc}a_1 & a_2\\ \hline b_1 & c_1\end{array}\cdot\begin{array}{c|cc|cc}f_2 & k_1 & k_2 & & \\ f_1 & h_1 & h_2 & p_1 & q_1\\ \hline & i_1 & j_1 & m_1 & m_2\\ & i_2 & j_2 & n_1 & n_2\end{array} + e_1\cdot\begin{array}{cc}a_1 & a_2\\ \hline b_2 & c_2\end{array}\cdot\begin{array}{c|cc|cc}f_2 & k_1 & k_2 & & \\ f_1 & h_1 & h_2 & p_1 & a_1\\ \hline & i_1 & j_1 & m_1 & m_2\\ & i_2 & j_2 & n_1 & n_2\end{array}\\ + d_2\cdot\begin{array}{cc}a_1 & a_2\\ \hline b_1 & c_1\end{array}\cdot\begin{array}{c|cc|cc}g_2 & k_1 & k_2 & & \\ g_1 & h_1 & h_2 & p_1 & q_1\\ \hline & i_1 & j_1 & m_1 & m_2\\ & i_2 & j_2 & n_1 & n_2\end{array} + e_2\cdot\begin{array}{cc}a_1 & a_2\\ \hline b_2 & c_2\end{array}\cdot\begin{array}{c|cc|cc}g_2 & k_1 & k_2 & & \\ g_1 & h_1 & h_2 & p_1 & q_1\\ \hline & i_1 & j_1 & m_1 & m_2\\ & i_2 & j_2 & n_1 & n_2\end{array}.\end{array}$$

295. Since a bipartite function is linear with respect to the elements of any one of its arrays, the cofactor of any of the elements (which has been shown above to be expressible as a minor bipartite or as a product of minors) is expressible also as the first differential coefficient of the function with respect to the element in question.

Hence, B denoting the bipartite whose initial line is $a_1, a_2, a_3, \cdots, a_n$, the theorem of §292 may be alternatively stated in symbols thus:

$$B = \sum a_r \frac{\partial B}{\partial a_r} \qquad (r = 1, 2, \cdots, n).$$

and the elements of any square array of B being the elements of the determinant $|a_{1n}|$, the theorem of §294 is

$$B = \sum a_{rs} \frac{\partial B}{\partial a_{rs}} \qquad (r, s = 1, 2, \cdots, n)$$

296. *A bipartite of deg-order (m, n) is expressible as the sum of n products of two factors each, namely, a minor bipartite of any degree less than m, say of the degree p, and a minor of the degree $m-p$, the former being obtained from the first p arrays by deleting all the lines of the pth array except one, and the latter being obtained from the last $m-p$ arrays by deleting all the lines of the $(m-p)$th array except the line collinear with that formerly undeleted in the pth array from the beginning.*

This theorem is deduced from the theorem of §294 by combining those terms of the development there obtained which have a common factor. Thus, taking the first development of

$$\begin{array}{cc|cc|cc|cc} & & f_2 & g_2 & k_1 & k_2 & & \\ a_1 & a_2 & f_1 & g_1 & h_1 & h_2 & p_1 & q_1 \\ \hline b_1 & c_1 & d_1 & d_2 & i_1 & j_1 & m_1 & m_2 \\ b_2 & c_2 & e_1 & e_2 & i_2 & j_2 & n_1 & n_2 \end{array}$$

given as an example in §294, namely:

$$f_1 \cdot \begin{array}{cc|c} a_1 & a_2 & \\ \hline b_1 & c_1 & d_1 \\ b_2 & c_2 & e_1 \end{array} \cdot \begin{array}{cc|cc} h_1 & h_2 & p_1 & q_1 \\ \hline i_1 & j_1 & m_1 & m_2 \\ i_2 & j_2 & n_2 & n_2 \end{array} + g_1 \cdot \begin{array}{cc|c} a_1 & a_2 & \\ \hline b_1 & c_1 & d_2 \\ b_2 & c_2 & e_2 \end{array} \cdot \begin{array}{cc|cc} h_1 & h_2 & p_1 & q_1 \\ \hline i_1 & j_1 & m_1 & m_2 \\ i_2 & j_2 & n_1 & n_2 \end{array}$$

$$+ f_2 \cdot \begin{array}{cc|c} a_1 & a_2 & \\ \hline b_1 & c_1 & d_1 \\ b_2 & c_2 & e_1 \end{array} \cdot \begin{array}{cc|cc} k_1 & k_2 & p_1 & q_1 \\ \hline i_1 & j_1 & m_1 & m_2 \\ i_2 & j_2 & n_1 & n_2 \end{array} + g_2 \cdot \begin{array}{cc|c} a_1 & a_2 & \\ \hline b_1 & c_1 & d_2 \\ b_2 & c_2 & e_2 \end{array} \cdot \begin{array}{cc|cc} k_1 & k_2 & p_1 & q_1 \\ \hline i_1 & j_1 & m_1 & m_2 \\ i_2 & j_2 & n_1 & n_2 \end{array},$$

we observe that the first two terms have the common factor

$$\begin{array}{cc|cc} h_1 & h_2 & p_1 & q_1 \\ \hline i_1 & j_1 & m_1 & m_2 \\ i_2 & j_2 & n_1 & n_2 \end{array}$$

the full cofactor being

$$f_1 \cdot \begin{array}{cc|c} a_1 & a_2 & \\ \hline b_1 & c_1 & d_1 \\ b_2 & c_2 & e_1 \end{array} + g_1 \cdot \begin{array}{cc|c} a_1 & a_2 & \\ \hline b_1 & c_1 & d_2 \\ b_2 & c_2 & e_2 \end{array}$$

which, we know from §292, is equal to

$$\begin{array}{cc|cc} a_1 & a_2 & f_1 & g_1 \\ \hline b_1 & c_1 & d_1 & d_2 \\ b_2 & c_2 & e_1 & e_2 \end{array}.$$

Similarly, the cofactor of the factor common to the last two terms is seen to be

$$\begin{array}{cc|cc} a_1 & a_2 & f_2 & g_2 \\ \hline b_1 & c_1 & d_1 & d_2 \\ b_2 & c_2 & e_1 & e_2 \end{array}.$$

Hence we have as an example of the present theorem:

$$
(\alpha)\quad
\begin{array}{cc|cc|cc|cc}
 & & f_2 & g_2 & k_1 & k_2 & & \\
a_1 & a_2 & f_1 & g_1 & h_1 & h_2 & p_1 & q_1 \\ \hline
b_1 & c_1 & d_1 & d_2 & i_1 & j_1 & m_1 & m_2 \\
b_2 & c_2 & e_1 & e_2 & i_2 & j_2 & n_1 & n_2
\end{array}
=
\begin{array}{cc|cc|cc|cc}
a_1 & a_2 & f_1 & g_1\cdot & h_1 & h_2 & p_1 & q_1 \\ \hline
b_1 & c_1 & d_1 & d_2 & i_1 & j_1 & m_1 & m_2 \\
b_2 & c_2 & e_1 & e_2 & i_2 & j_2 & n_1 & n_2
\end{array}
+
\begin{array}{cc|cc|cc|cc}
a_1 & a_2 & f_2 & g_2\cdot & k_1 & k_2 & p_1 & q_1 \\ \hline
b_1 & c_1 & d_1 & d_2 & i_1 & j_1 & m_1 & m_2 \\
b_2 & c_2 & e_1 & e_2 & i_2 & j_2 & n_1 & n_2
\end{array}\,.
$$

Had we combined the first and third terms of the same development, and then the second and fourth, we should have obtained the example

$$
\begin{array}{cc|cc|cc|cc}
 & & f_2 & g_2 & k_1 & k_2 & & \\
a_1 & a_2 & f_1 & g_1 & h_1 & h_2 & p_1 & q_1 \\ \hline
b_1 & c_1 & d_1 & d_2 & i_1 & j_1 & m_1 & m_2 \\
b_2 & c_2 & e_1 & e_2 & i_2 & j_2 & n_1 & n_2
\end{array}
=
\begin{array}{cc|c}
a_1 & a_2 & \\ \hline
b_1 & c_1 & d_1 \\
b_2 & c_2 & e_1
\end{array}
\cdot
\begin{array}{c|cc|cc}
f_2 & k_1 & k_2 & & \\
f_1 & h_1 & h_2 & p_1 & q_1 \\ \hline
 & i_1 & j_1 & m_1 & m_2 \\
 & i_2 & j_2 & n_1 & n_2
\end{array}
+
\begin{array}{cc|c}
a_1 & a_2 & \\ \hline
b_1 & c_1 & d_2 \\
b_2 & c_2 & e_2
\end{array}
\cdot
\begin{array}{c|cc|cc}
g_2 & k_1 & k_2 & & \\
g_1 & h_1 & h_2 & p_1 & p_1 \\ \hline
 & i_1 & j_1 & m_1 & m_2 \\
 & i_2 & j_2 & n_1 & n_2
\end{array}
$$

Again, by combining the first and third terms of the *second* development in §294, we should have the case where the one factor is of the second degree, and the other of the sixth, namely:

$$
\begin{array}{cc|cc|cc|cc}
 & & f_2 & g_2 & k_1 & k_2 & & \\
a_1 & a_2 & f_1 & g_1 & h_1 & h_2 & p_1 & q_1 \\ \hline
b_1 & c_1 & d_1 & d_2 & i_1 & j_1 & m_1 & m_2 \\
b_2 & c_2 & e_1 & e_2 & i_2 & j_2 & n_1 & n_2
\end{array}
=
\begin{array}{cc}
a_1 & a_2 \\ \hline
b_1 & c_1
\end{array}
\cdot
\begin{array}{cc|cc|cc}
f_2 & g_2 & k_1 & k_2 & & \\
f_1 & g_1 & h_1 & h_2 & p_1 & q_1 \\ \hline
d_1 & d_2 & i_1 & j_1 & m_1 & m_2 \\
 & & i_2 & j_2 & n_1 & n_2
\end{array}
+
\begin{array}{cc}
a_1 & a_2 \\ \hline
b_2 & c_2
\end{array}
\cdot
\begin{array}{cc|cc|cc}
f_2 & g_2 & k_1 & k_2 & & \\
f_1 & g_1 & h_1 & h_2 & p_1 & q_1 \\ \hline
e_1 & e_2 & i_1 & j_1 & m_1 & m_2 \\
 & & i_2 & j_2 & n_1 & n_2
\end{array}\,.
$$

The case where the one factor is of the first degree and the other of the 7th falls under the theorem of §292 which may thus be looked on as a particular case of the present theorem.

297. Two minors such as those of each term of the development in the preceding paragraph—that is to say, minors which, when multiplied, give terms that are all terms of the parent bipartite—may be called *complementary minors.*

298. *A bipartite of deg-order* (m, n) *is expressible as a sum of* n^2 *products of three factors each, the first being an element of the initial line, the second an element of the final line, and the third the minor bipartite of deg-order* $(m-2, n)$, *which is obtained from the original by deleting the initial and final lines, and those lines of the first and last square arrays which are not collinear with one of the said pair of elements.*

Thus

$$\begin{array}{cc|cc} a_1 & a_2 & f_1 & g_1 \\ \hline b_1 & c_1 & d_1 & d_2 \\ b_2 & c_2 & e_1 & e_2 \end{array}$$

$$= a_1 f_1 \cdot \begin{array}{cc} b_1 & b_2 \\ \hline d_1 & e_1 \end{array} + a_1 g_1 \cdot \begin{array}{cc} b_1 & b_2 \\ \hline d_2 & e_2 \end{array} + a_2 f_1 \cdot \begin{array}{cc} c_1 & c_2 \\ \hline d_1 & e_1 \end{array} + a_2 g_1 \cdot \begin{array}{cc} c_1 & c_2 \\ \hline d_2 & e_2 \end{array} .$$

This of course is but the result of a double application of the theorem of §292.

299. A bipartite function may be expressed as a bipartite of lower degree, in which elements occur that are themselves bipartites, and to which, on that account, the name *compound bipartite* may be given.

300. The theorem of §296 is the case of this where the compound bipartite is of the *second* degree. Thus the identity (α) there given may be written also in the form

$$\begin{array}{cc|cc|cc|cc} & & f_2 & g_2 & k_1 & k_2 & & \\ a_1 & a_2 & f_1 & g_1 & h_1 & h_2 & p_1 & q_1 \\ \hline b_1 & c_1 & d_1 & d_2 & i_1 & j_1 & m_1 & m_2 \\ b_2 & c_2 & e_1 & e_2 & i_1 & j_2 & n_1 & n_2 \end{array} = \frac{\begin{array}{cc|cc} a_1 & a_2 & f_1 & g_1 \\ \hline b_1 & c_1 & d_1 & d_2 \\ b_2 & c_2 & e_1 & e_2 \end{array} , \quad \begin{array}{cc|cc} a_1 & a_2 & f_2 & g_2 \\ \hline b_1 & c_1 & d_1 & d_2 \\ b_2 & c_2 & e_1 & e_2 \end{array}}{\begin{array}{cc|cc} h_1 & h_2 & p_1 & q_1 \\ \hline i_1 & j_1 & m_1 & m_2 \\ i_2 & j_2 & n_1 & n_2 \end{array} , \quad \begin{array}{cc|cc} k_1 & k_2 & p_1 & q_1 \\ \hline i_1 & j_1 & m_1 & m_2 \\ i_2 & j_2 & n_1 & n_2 \end{array}} ;$$

and so of the others.

The theorem of §294 is the case where the compound bipartite is of the third degree, and its square array is a square array of the original bipartite. Thus the identity (α) there given may be written also in the form

$$\begin{array}{cc|cc|cc|cc} & & f_2 & g_2 & k_1 & k_2 & & \\ a_1 & a_1 & f_1 & g_1 & h_1 & h_2 & p_1 & q_1 \\ \hline b_1 & c_1 & d_1 & d_2 & i_1 & j_1 & m_1 & m_2 \\ b_2 & c_2 & e_1 & e_2 & i_2 & j_2 & n_1 & n_2 \end{array} = \begin{array}{cc|c} P_3 & Q_3 & \\ \hline f_2 & g_2 & R_4 \\ f_1 & g_1 & S_4 \end{array}$$

where

$$P_3 = \begin{array}{cc|c} a_1 & a_2 & \\ \hline b_1 & c_1 & d_1 \\ b_2 & c_2 & e_1 \end{array}, \qquad Q_3 = \begin{array}{cc|c} a_1 & a_2 & \\ \hline b_1 & c_1 & d_2 \\ b_2 & c_2 & e_2 \end{array},$$

$$R_4 = \begin{array}{cc|cc} k_1 & k_2 & p_1 & q_1 \\ \hline i_1 & j_1 & m_1 & m_2 \\ i_2 & j_2 & n_1 & n_2 \end{array}, \qquad S_4 = \begin{array}{cc|cc} h_1 & h_2 & p_1 & q_1 \\ \hline i_1 & j_1 & m_1 & m_2 \\ i_2 & j_2 & n_1 & n_2 \end{array}.$$

301. The theorem of §298 is the rather important case where the compound bipartite is again of the third degree, but having for its initial and final lines the initial and final lines of the original bipartite.

Thus

$$\begin{array}{cc|cc|cc|cc} & & f_2 & g_2 & k_1 & k_2 & & \\ a_1 & a_2 & f_1 & g_1 & h_1 & h_2 & p_1 & q_1 \\ \hline b_1 & c_1 & d_1 & d_2 & i_1 & j_1 & m_1 & m_2 \\ b_2 & c_2 & e_1 & e_2 & i_2 & j_2 & n_1 & n_2 \end{array} = \begin{array}{cc|c} a_1 & a_2 & \\ \hline U_6 & X_6 & p_1 \\ V_6 & Y_6 & q_1 \end{array}$$

where

$$U_6 = \begin{array}{c|cc|cc|c} & f_2 & g_2 & k_1 & k_2 & \\ & f_1 & g_1 & h_1 & h_2 & \\ \cline{2-5} b_1 & d_1 & d_2 & i_1 & j_1 & m_1 \\ b_2 & e_1 & e_2 & i_2 & j_2 & n_1 \end{array}, \qquad X_6 = \begin{array}{c|cc|cc|c} & f_2 & g_2 & k_1 & k_2 & \\ & f_1 & g_1 & h_1 & h_2 & \\ \cline{2-5} c_1 & d_1 & d_2 & i_1 & j_1 & m_1 \\ c_2 & e_1 & e_2 & i_2 & j_2 & n_1 \end{array},$$

$$V_6 = \begin{array}{c|cc|cc|c} & f_2 & g_2 & k_1 & k_2 & \\ & f_1 & g_1 & h_1 & h_2 & \\ \cline{2-5} b_1 & d_1 & d_2 & i_1 & j_1 & m_2 \\ b_2 & e_1 & e_2 & i_2 & j_2 & n_2 \end{array}, \qquad Y_6 = \begin{array}{c|cc|cc|c} & f_2 & g_2 & k_1 & k_2 & \\ & f_1 & g_1 & h_1 & h_2 & \\ \cline{2-5} c_1 & d_1 & d_2 & i_1 & j_1 & m_2 \\ c_2 & e_1 & e_2 & i_2 & j_2 & n_2 \end{array}.$$

302. There are, however, other modes of expressing a bipartite of a degree higher than the third as a compound bipartite of the third degree. These we obtain by making use of both §294 and §296.

For example, by §296 (or §294) we have

$$(\alpha)\quad \begin{array}{cc|cc|cc|cc} & & f_2 & g_2 & k_1 & k_2 & & \\ a_1 & a_2 & f_1 & g_1 & h_1 & h_2 & p_1 & q_1 \\ \hline b_1 & c_1 & d_1 & d_2 & i_1 & j_1 & m_1 & m_2 \\ b_2 & c_2 & e_1 & e_2 & i_2 & j_2 & n_1 & n_2 \end{array} = a_1 \begin{array}{c|cc|cc|cc} & f_2 & g_2 & k_1 & k_2 & & \\ & f_1 & g_1 & h_1 & h_2 & p_1 & q_1 \\ \hline b_1 & d_1 & d_2 & i_1 & j_1 & m_1 & m_2 \\ b_2 & e_1 & e_2 & i_2 & j_2 & n_1 & n_2 \end{array} + a_2 \cdot \begin{array}{c|cc|cc|cc} & f_2 & g_2 & k_1 & k_2 & & \\ & f_1 & g_1 & h_1 & h_2 & p_1 & q_1 \\ \hline c_1 & d_1 & d_2 & i_1 & j_1 & m_1 & m_2 \\ c_2 & e_1 & e_2 & i_2 & j_2 & n_1 & n_2 \end{array}$$

and by §294 the cofactor of a_1 is equal to

$$(\beta_1)\quad \begin{array}{c|cc|c} & f_2 & g_2 & k_1 \\ & f_1 & g_1 & h_1 \\ \hline b_1 & d_1 & d_2 & \\ b_2 & e_1 & e_2 & \end{array} \cdot \begin{array}{c|cc} & p_1 & q_1 \\ \hline i_1 & m_1 & m_2 \\ i_2 & n_1 & n_2 \end{array} + \begin{array}{c|cc|c} & f_2 & g_2 & k_2 \\ & f_1 & g_1 & h_2 \\ \hline b_1 & d_1 & d_2 & \\ b_2 & e_1 & e_2 & \end{array} \cdot \begin{array}{c|cc} & p_1 & q_1 \\ \hline j_1 & m_1 & m_2 \\ j_2 & n_1 & n_2 \end{array}$$

and the cofactor of a_2 is equal to

$$(\beta_2)\quad \begin{array}{c|cc|c} & f_2 & g_2 & k_1 \\ & f_1 & g_1 & h_1 \\ \hline c_1 & d_1 & d_2 & \\ c_2 & e_1 & e_2 & \end{array} \cdot \begin{array}{c|cc} & p_1 & q_2 \\ \hline i_1 & m_1 & m_2 \\ i_2 & n_1 & n_2 \end{array} + \begin{array}{c|cc|c} & f_2 & g_2 & k_2 \\ & f_1 & g_1 & h_2 \\ \hline c_1 & d_1 & d_2 & \\ c_2 & e_1 & e_2 & \end{array} \cdot \begin{array}{c|cc} & p_1 & q_1 \\ \hline j_1 & m_1 & m_2 \\ j_2 & n_1 & n_2 \end{array}$$

Hence

$$\begin{array}{cc|cc|cc|cc} & & f_2 & g_2 & k_1 & k_2 & & \\ a_1 & a_2 & f_1 & g_1 & h_1 & h_2 & p_1 & q_1 \\ \hline b_1 & c_1 & d_1 & d_2 & i_1 & j_1 & m_1 & m_2 \\ b_2 & c_2 & e_1 & e_2 & i_2 & j_2 & n_1 & n_2 \end{array} = \begin{array}{cc|c} a_1 & a_2 & \\ \hline P_4 & M_4 & R_3 \\ Q_4 & N_4 & S_3 \end{array}$$

where

$$P_4 = \begin{array}{c|cc|c} & f_2 & g_2 & k_1 \\ & f_1 & g_1 & h_1 \\ \hline b_1 & d_1 & d_2 & \\ b_2 & e_1 & e_2 & \end{array}, \qquad M_4 = \begin{array}{c|cc|c} & f_2 & g_2 & k_1 \\ & f_1 & g_1 & h_1 \\ \hline c_1 & d_1 & d_2 & \\ c_2 & e_1 & e_2 & \end{array}, \qquad R_3 = \begin{array}{c|cc} & p_1 & q_1 \\ \hline i_1 & m_1 & m_2 \\ i_2 & n_1 & n_2 \end{array}$$

$$Q_4 = \begin{array}{c|cc|c} & f_2 & g_2 & k_2 \\ & f_1 & g_1 & h_2 \\ \hline b_1 & d_1 & d_2 & \\ b_2 & e_1 & e_2 & \end{array}, \qquad N_4 = \begin{array}{c|cc|c} & f_2 & g_2 & k_2 \\ & f_1 & g_1 & h_2 \\ \hline c_1 & d_1 & d_2 & \\ c_2 & e_1 & e_2 & \end{array}, \qquad S_3 = \begin{array}{c|cc} & p_1 & q_1 \\ \hline j_1 & m_1 & m_2 \\ j_2 & n_1 & n_2 \end{array}$$

It will be observed that, in using §294 the second time here, it is necessary to do so in such a way that each term of (β_2) shall have a factor common to the corresponding term of (β_1).

303. When we note that in using the theorem of §296 in the preceding paragraph, several other identities might have been got in place of (α), and that in using the theorem of §294 we might have chosen several other pairs of identities in place of (β_1), (β_2), it is clear that we have not by any means exhausted the possible ways of expressing the given bipartite as a compound bipartite of the third degree.

There is little difficulty in seeing what the theorem is which includes these different forms in the same way as theorems of §296 and §294 include all the forms of §§300, 301 but the formulating of it would be troublesome.

304. The expression of a bipartite of a degree higher than the fourth as a compound bipartite of the fourth degree, and generally the expression of a bipartite of a degree higher than the mth as a compound bipartite of the mth degree, can be obtained in a manner similar to that of §302 by repeated use of the theorems of §§294, 296.

305. Of compound bipartites, some interest attaches to that special form, each of whose elements is the cofactor of the corresponding element in another bipartite, and which, in reference to an analogue in the theory of determinants, we may term the bipartite *adjugate* to the parent bipartite.

306. The elements of a determinant which is the product of m determinants of the nth order are bipartites of deg-order (m, n).

307. A continuant may be expressed as a bipartite. Thus

$$\begin{array}{cc|cc} p & 1 & & \\ \hline q & b & r & 1 \\ 1 & & c & \\ \hline & & s & d \end{array} = \begin{vmatrix} p & b & \cdot & \cdot \\ -1 & q & c & \cdot \\ \cdot & -1 & r & d \\ \cdot & \cdot & -1 & s \end{vmatrix}$$

308. If the product $|a_{1n}| \cdot |b_{1n}|$ with -1 affixed to each of its diagonal elements be denoted by $|P_{1n}|$, and the corresponding

result, when the order of the factors is reversed, by $|Q_{1n}|$, then each of the P's is expressible as a linear homogeneous function of the Q's, with coefficients involving only elements of $|a_{1n}|$. Thus if $|a_1 b_2 c_3|$ and $|\lambda_1 \mu_2 \nu_3|$ are the initiating determinants, then

$$P_{11} = \frac{\begin{array}{ccc} a_1 & a_2 & a_3 \end{array}}{\begin{array}{ccc|c} Q_{11} & Q_{21} & Q_{31} & A_1 \\ Q_{12} & Q_{22} & Q_{32} & A_2 \\ Q_{13} & Q_{23} & Q_{33} & A_3 \end{array}} \div |a_1 b_2 c_3|, \quad P_{12} = \frac{\begin{array}{ccc} a_1 & a_2 & a_3 \end{array}}{\begin{array}{ccc|c} Q_{11} & Q_{21} & Q_{31} & B_1 \\ Q_{12} & Q_{22} & Q_{32} & B_2 \\ Q_{13} & Q_{23} & Q_{33} & B_3 \end{array}} \div |a_1 b_2 c_3|$$

etc.

This is obviously an example of Caylean matrices and is included in the simple identity

$$uv - 1 = u(uv - 1)u^{-1},$$

where u and v are any matrices of the same order.

309. It is easily seen that

$$\frac{\begin{array}{cccc} a_1 & a_2 & a_3 & a_4 \end{array}}{\begin{array}{cccc|c} a_1 & b_1 & c_1 & d_1 & a_1 \\ a_2 & b_2 & c_2 & d_2 & b_1 \\ a_3 & b_3 & c_3 & d_3 & c_1 \\ a_4 & b_4 & c_4 & d_4 & d_1 \end{array}} - \frac{a_1, a_2, a_3, a_4}{a_1, b_1, c_1, d_1} \sum a_1$$

$$= a_1\{ |a_1 b_2| + |a_1 c_3| + |a_1 d_4| \}$$

$$+ \begin{vmatrix} 0 & a_2 & a_3 \\ b_1 & b_2 & b_3 \\ c_1 & c_2 & c_3 \end{vmatrix} + \begin{vmatrix} 0 & a_2 & a_4 \\ b_1 & b_2 & b_4 \\ d_1 & d_2 & d_4 \end{vmatrix}$$

$$+ \begin{vmatrix} 0 & a_3 & a_4 \\ c_1 & c_3 & c_4 \\ d_1 & d_3 & d_4 \end{vmatrix}$$

or adding $a_1 \sum |a_1 b_2|$ to both sides that

$$\frac{\begin{array}{cccc} a_1 & a_2 & a_3 & a_4 \end{array}}{\begin{array}{c|c} & a_1 \\ \Delta & b_1 \\ & c_1 \\ & d_1 \end{array}} - \frac{a_1, a_2, a_3, a_4}{a_1, b_1, c_1, d_1} \sum a_1 + a_1 \sum |a_1 b_2|$$

$$(1) \qquad = |\, a_1\, b_2\, c_3 \,| + |\, a_1\, b_2\, d_4 \,| + |\, a_1\, c_3\, d_4 \,|$$

where $\sum a_1 = a_1 + b_2 + c_3 + d_4$ and $\Delta \equiv |a_1 b_2 c_3 d_4|$. From this using a_2, b_2, c_2, d_2 in place of a_1, b_1, c_1, d_1 respectively we get

$$(2) \quad \begin{array}{c|c} a_1 \quad a_2 \quad a_3 \quad a_4 & \\ \hline \Delta & \begin{matrix} a_2 \\ b_2 \\ c_2 \\ d_2 \end{matrix} \end{array} - \frac{a_1, a_2, a_3, a_4}{a_2, b_2, c_2, d_2} \sum a_1 + a_2 \sum |\, a_1\, b_2 \,| = |\, a_2\, c_3\, d_4 \,|$$

where

$$\sum |\, a_1 b_2 \,| = |\, a_1 b_2 \,| + |\, a_1 c_3 \,| + |\, a_1 d_4 \,| + |\, b_2 c_3 \,| + |\, b_2 d_4 \,| + |\, c_3 d_4 \,| .$$

Similarly

$$(3) \quad \begin{array}{c|c} a_1 \quad a_2 \quad a_3 \quad a_4 & \\ \hline \Delta & \begin{matrix} a_3 \\ b_3 \\ c_3 \\ d_3 \end{matrix} \end{array} - \frac{a_1, a_2, a_3, a_4}{a_3, b_3, c_3, d_3} \sum a_1 + a_3 \sum |a_1 b_2| = - |a_1 b_3 d_4| ,$$

and

$$(4) \quad \begin{array}{c|c} a_1 \quad a_2 \quad a_3 \quad a_4 & \\ \hline \Delta & \begin{matrix} a_4 \\ b_4 \\ c_4 \\ d_4 \end{matrix} \end{array} - \frac{a_1, a_2, a_3, a_4}{a_4, b_4, c_4, d_4} \sum a_1 + a_4 \sum |\, a_1\, b_2 \,| = |\, a_2\, b_3\, c_4 \,| .$$

From these four we readily get

$$\begin{array}{c|c} a_1 \quad a_2 \quad a_3 \quad a_4 & a_1 \quad b_1 \quad c_1 \quad d_1 \\ \hline \Delta & \Delta \end{array} - \begin{array}{c|c} a_1 \quad a_2 \quad a_3 \quad a_4 & \\ \hline \Delta & \begin{matrix} a_1 \\ b_1 \\ c_1 \\ d_1 \end{matrix} \end{array} \sum a_1$$

$$+ \frac{a_1, a_2, a_3, a_4}{a_1, b_1, c_1, d_1} \sum |\, a_1\, b_2 \,| - a_1 \{ |\, a_1\, b_2\, c_3 \,| + |\, a_1 b_2 d_4 \,| + |\, a_1 c_3 d_4 \,|$$

$$(5) \quad + |\, b_2\, c_3\, d_4 \,| \} = - |\, a_1\, b_2\, c_3\, d_4 \,| .$$

From (5) using a_2, b_2, c_2, d_2 instead of a_1, b_1, c_1, d_1 respectively we get

$$\begin{array}{c|c} a_1\ a_2\ a_3\ a_4 & a_2\ b_2\ c_2\ d_2 \\ \hline \Delta & \Delta \end{array} - \frac{a_1\ a_2\ a_3\ a_4}{\Delta}\left|\begin{matrix} a_2\\ b_2\\ c_2\\ d_2\end{matrix}\right. \sum a_1$$

$$(6) \qquad + \frac{a_1, a_2, a_3, a_4}{a_2, b_2, c_2, d_2}\sum|\,a_1\,b_2\,| - a_2\sum|\,a_1\,b_2\,c_3\,| = 0.$$

Similarly using a_3, b_3, c_3, d_3 and a_4, b_4, c_4, d_4 or as we may write

$$\begin{array}{c|c} a_1\ a_2\ a_3\ a_4 & a_r\ b_r\ c_r\ d_r \\ \hline \Delta & \Delta \end{array} - \frac{a_1\ a_2\ a_3\ a_4}{\Delta}\left|\begin{matrix} a_r\\ b_r\\ c_r\\ d_r\end{matrix}\right. \sum a_1$$

(7)

$$+ \frac{a_1, a_2, a_3, a_4}{a_r, b_r, c_r, d_r}\sum|\,a_1\,b_2\,| - a_r\sum|\,a_1\,b_2\,c^{\varepsilon}\,| = -\begin{vmatrix} a_r & a_2 & a_3 & a_4\\ b_r & b_2 & b_3 & b_4\\ c_r & c_2 & c_3 & c_4\\ d_r & d_2 & d_3 & d_4\end{vmatrix}.$$

This connects bipartites and determinants of order four, but the law of formation is seen to be general. Thus for order six we have

$$\frac{a_1\ a_2\ a_3\ a_4\ a_5\ a_6\,|\,a_1\ b_1\ c_1\ d_1\ e_1\ f_1}{\Delta^4}$$

$$- \frac{a_1\ a_2\ a_3\ a_4\ a_5\ a_6\,|\,a_1\ b_1\ c_1\ d_1\ e_1\ f_1}{\Delta^3}\sum a_1$$

$$+ \frac{a_1\ a_2\ a_3\ a_4\ a_5\ a_6\,|\,a_1\ b_1\ c_1\ d_1\ e_1\ f_1}{\Delta^2}\sum|\,a_1\,b_2\,|$$

$$- \frac{a_1\ a_2\ a_3\ a_4\ a_5\ a_6\,|\,a_1\ b_1\ c_1\ d_1\ e_1\ f_1}{\Delta}\sum|\,a_1\,b_2\,c_3\,|$$

$$+ \frac{a_1\ a_2\ a_3\ a_4\ a_5\ a_6\,|\,a_1\ b_1\ c_1\ d_1\ e_1\ f_1}{\Delta^0}\sum|\,a_1\,b_2\,c_3\,d_4\,|$$

$$(8) \qquad - a_1\sum|\,a_1\,b_2\,c_3\,d_4\,e_5\,| + |\,a_1\,b_2\,c_3\,d_4\,e_5\,f_6\,| = 0.$$

310. If now we denote by $\Delta \equiv |A - x|$ the determinant of the nth order formed by subtracting x from the principal diagonal elements of the determinant $A \equiv |a_{1n}|$, and by Δ_{11} the complementary minor of the element in the position (11) of Δ then we may express the quotient $\Delta_{11} \div \Delta$ as a power series in x where the coefficients are bipartite functions.

The quotient is obviously of the form

$$- x^{-1} - A_1 x^{-2} - A_2 x^{-3} - A_3 x^{-4} - \cdots$$

and, for convenience, taking $n = 4$, the divisor is

$$x^4 - x^3 \sum a_{11} + x^2 \sum |a_{11}\, a_{22}| - x \sum |a_{11}\, a_{22}\, a_{33}| + |a_{11}\, a_{22}\, a_{33}\, a_{44}|$$
$$\text{or } x^4 - x^3 \sum a_{11} + x^2 \sum A_{12,\,12} - x \sum A_{123,\,123} + A\,.$$

If now we multiply both sides by this divisor we get

$$\Delta_{11} = \begin{array}{r|l|l|l|l} -x^3 - A_1 & x^2 - A_2 & x - A_2 & x^0 - A_4 & x^{-1} - \text{etc.} \\ -\sum a_{11} & +A_1 \sum a_{11} & +A_2 \sum a_{11} & +A_3 \sum a_{11} & \\ & -\sum A_{12,12} & -A_1 \sum A_{12,12} & -A_2 \sum A_{12,12} & \\ & & +\sum A_{123,123} & +A_1 \sum A_{123,123} & \\ & & & -A & \end{array}$$

Equating like powers of x we get

$$- A_1 + \sum a_{11} = a_{22} + a_{33} + a_{44}$$
$$- A_2 + A_1 \sum a_{11} - \sum A_{12,12} = - |a_{22}\, a_{33}|$$
$$- |a_{22}\, a_{44}| - |a_{33}\, a_{44}|$$
$$- A_3 + A_2 \sum a_{11} - A_1 \sum A_{12,12} + \sum A_{123,123} = |a_{22}\, a_{33}\, a_{44}|$$
$$- A_4 + A_3 \sum a_{11} - A_2 \sum A_{12,12} + A_1 \sum A_{123,123} - A = 0$$

. . . .

and from these equations we get, making use of relations (1), (5), (8) etc. §309,

$$A_1 = a_{11}$$

$$A_2 = \frac{a_{11},\ a_{12},\ a_{13},\ a_{14}}{a_{11},\ a_{21},\ a_{31},\ a_{41}}$$

$$A_3 = \begin{array}{c|c} a_{11}\ a_{12}\ a_{13}\ a_{14} & \\ \hline & a_{11} \\ & a_{21} \\ A & a_{31} \\ & a_{41} \end{array}$$

$$A_4 = \frac{a_{11}\ a_{12}\ a_{13}\ a_{14}}{A} \left| \frac{a_{11}\ a_{21}\ a_{31}\ a_{41}}{A} \right. \qquad \text{etc.}$$

We have then

$$\frac{\Delta_{11}}{\Delta} = -\ x^{-1} - a_1x^{-2} - \frac{a_{11},\ a_{12},\ a_{13},\ a_{14}}{a_{11},\ a_{21},\ a_{31},\ a_{41}}x^{-3}$$

$$- \frac{a_{11}\ a_{12}\ a_{13}\ a_{14}\ |\ a_{11}\ a_{21}\ a_{31}\ a_{41}}{A}x^{-4}$$

$$- \frac{a_{11}\ a_{12}\ a_{13}\ a_{14}\ |\ a_{11}\ a_{21}\ a_{31}\ a_{41}}{A^2}x^{-5} - \text{etc.}$$

The student who wishes to pursue this subject further should consult Muir's paper *On Bipartite Functions*, Trans. Royal Soc. Edinb. xxv p. 461+.

Many propositions similar to those for determinants are developed. Thus propositions corresponding to those of §§174, 175, 153, etc. are there found.

Exercises. Set XVII

1. Show that

$$\frac{\Delta}{\Delta_{11}} = -\ x + a_1 + \frac{a_{12},\ a_{13},\ a_{14}}{a_{21},\ a_{31},\ a_{41}}x^{-1} - \frac{a_{12}\ a_{13}\ a_{14}\ |\ a_{21}\ a_{31}\ a_{41}}{A_{11}}x^{-2}$$

$$+ \frac{a_{12}\ a_{13}\ a_{14}\ |\ a_{21}\ a_{31}\ a_{41}}{A^2{}_{11}}x^{-3} - \text{etc}$$

where A_{11} is the minor complementary to a_{11} in A.

2. If A, F, E, etc. are the complementary minors of a, f, e, etc. in the determinant

$$\Delta \equiv \begin{vmatrix} a & f & e \\ f & b & d \\ e & d & c \end{vmatrix}$$

and if

$$X = y_1z_2 - z_1y_2,\ \ Y = z_1x_2 - z_2x_1,\ \ Z = x_1y_2 - y_1x_2$$

show that

$$\begin{vmatrix} \begin{array}{c|c} x_1\ y_1\ z_1 & \\ \hline \Delta & \begin{array}{c} x_1 \\ y_1 \\ z_1 \end{array} \end{array} & \begin{array}{c|c} x_2\ y_2\ z_2 & \\ \hline \Delta & \begin{array}{c} x_1 \\ y_1 \\ z_1 \end{array} \end{array} \\ \begin{array}{c|c} x_1\ y_1\ z_1 & \\ \hline \Delta & \begin{array}{c} x_2 \\ y_2 \\ z_2 \end{array} \end{array} & \begin{array}{c|c} x_2\ y_2\ z_2 & \\ \hline \Delta & \begin{array}{c} x_2 \\ y_2 \\ z_2 \end{array} \end{array} \end{vmatrix} = \begin{array}{ccc|c} X & Y & Z & \\ \hline A & F & E & X \\ F & B & D & Y \\ E & D & C & Z \end{array}$$

3. Show that

$$\frac{a_1 a_2 \cdots a_6 \mid a_r b_r \cdots f_r}{\Delta^4} - \frac{a_1 a_2 \cdots a_6 \mid a_r b_r \cdots f_r}{\Delta^3} \sum a_1$$

$$+ \frac{a_1 a_2 \cdots a_6 \mid a_r b_r \cdots f_r}{\Delta^2} \sum \left| a_1 b_2 \right| - \cdots$$

$$- a_r \sum \left| a_1 b_2 c_3 d_4 e_5 \right| + \left| a_1 b_2 c_3 d_4 e_5 f_6 \right| = 0.$$

CHAPTER X

Aggregates

311. Aggregate is a term which might be used to indicate a relation between determinants or minors of determinants such as those given in §150 and others. In what follows, however, we shall confine our attention to relations involving minors and sums of minors of given determinants.

312. If A is a determinant of order $2n$, the sum of the minors* of order n

$$\sum_{i_1=1}^{(2n-k)_n} (-1)^{\nu} \begin{vmatrix} (2n \mid k_\alpha)(\overline{2n} \mid k_\alpha \mid n-k_{i_1}) \\ (\overline{2n} \mid \bar{k}_\alpha \mid n-k_{i_1}) \end{vmatrix}$$

which has k rows the same throughout may obviously be written in the form

$$\frac{1}{(n)_k} \sum_{i_1=1}^{(2n-k)_k} \sum_{i_2=1}^{(2n-2k)_{n-k}} (-1)^{\mu} \begin{vmatrix} (2n \mid k_\alpha)(\overline{2n} \mid \bar{k}_\alpha \mid k_{i_1} \mid n-k_{i_2}) \\ (\overline{2n} \mid k_\alpha \mid k_{i_1})(\overline{2n} \mid \bar{k}_\alpha \mid \bar{k}_{i_1} \mid n-k_{i_2}) \end{vmatrix}$$

which on expanding each term by Laplace's theorem in terms of minors formed from the αth selection of k rows with their complementaries, and finally collecting all the terms containing

$$\begin{vmatrix} (2n \mid k_\alpha) \\ (2n \mid k_\alpha \mid k_{i_1}) \end{vmatrix}$$

as a factor, takes the form

$$\sum_{i_1=1}^{(2n-k)_k} (-1)^{\lambda} \begin{vmatrix} (2n \mid k_\alpha) \\ (\overline{2n} \mid k_\alpha \mid k_{i_1}) \end{vmatrix} \sum_{i_2=1}^{(2n-2k)_{n-k}} \begin{vmatrix} (\overline{2n} \mid \bar{k}_\alpha \mid k_{i_1} \mid n-k_{i_2}) \\ (\overline{2n} \mid \bar{k}_\alpha \mid \bar{k}_{i_1} \mid n-k_{i_2}) \end{vmatrix},$$

On examining the inner sum of this form we easily see that any given selection of h $(1 \leqq h \leqq n-k)$ of the $2n-2k$ numbers in the combination $(\overline{2n} \mid \bar{k}_\alpha \mid k_{i_1})$ is constant as rows running through $(2n-2k-h)_{n-k}$ of the terms and constant as columns through as many more terms. This leads to the following theorem:

$$\sum_{i_1=1}^{(2n-k)_n} (-1)^{\nu_1} \begin{vmatrix} (2n \mid k_\alpha)(\overline{2n} \mid k_\alpha \mid n-k_{i_1}) \\ (\overline{2n} \mid \bar{k}_\alpha \mid n-k_{i_1}) \end{vmatrix}$$
$$= \sum_{i_1=1}^{(2n-k)_k} (-1)^{\nu_2} \begin{vmatrix} (2n \mid k_\alpha) \\ (\overline{2n} \mid k_\alpha \mid k_{i_1}) \end{vmatrix} \frac{1}{2(n-k)_h} \sum_{\beta=1}^{(2n-2k)_h} S_\beta,$$

* These would be minors along the secondary diagonal of the nth compound.

where S_β *denotes the sum of the two aggregates*

$$\sum_{i_2=1}^{(2n-2k-h)_{n-k}} (-1)^{\nu_3} \left| \begin{matrix} (\overline{2n} \mid \bar{k}_\alpha \mid k_{i_1} \mid h_\beta \between \overline{2n} \mid \bar{k}_\alpha \mid \bar{k}_{i_1} \mid h_\beta \mid n-k-h_{i_2}) \\ (\overline{2n} \mid \bar{k}_\alpha \mid \bar{k}_{i_1} \mid h_\beta \mid n-k-h_{i_2}) \end{matrix} \right|,$$

$$\sum_{i_2=1}^{(2n-2k-h)_{n-k}} (-1)^{\nu_4} \left| \begin{matrix} (\overline{2n} \mid \bar{k}_\alpha \mid \bar{k}_{i_1} \mid h_\beta \mid n-k-h_{i_2}) \\ (\overline{2n} \mid \bar{k}_\alpha \mid k_{i_1} \mid h_\beta \between \overline{2n} \mid \bar{k}_\alpha \mid \bar{k}_{i_1} \mid h_\beta \mid n-k-h_{i_2}) \end{matrix} \right|$$

in which, if

$$(2n \mid k_\alpha),\ (2n \mid k_\alpha \mid k_{i_1} \mid h_\beta),\ (\overline{2n} \mid \bar{k}_\alpha \mid \bar{k}_{i_1} \mid h_\beta \mid n-k-h_{i_2}),$$
$$(\overline{2n} \mid k_\alpha \mid k_{i_1}),\quad (\overline{2n} \mid \bar{k}_\alpha \mid \bar{k}_{i_1} \mid h_\beta \mid n-k-h_{i_2})$$

are denoted by (a), (b), (c), (d), (e) *respectively, the integers* ν_2, ν_3, ν_4 *denote the number of inversions in* $(a \between b \between c \between d \between e)$, $(b \between c \between e)$, $(e \between b \between c)$ respectively.

313. As an illustration of this theorem take $n, k, h = 4, 2, 1$ and we have

$$\begin{vmatrix} 1\,2\,3\,4 \\ 5\,6\,7\,8 \end{vmatrix} - \begin{vmatrix} 1\,2\,3\,5 \\ 4\,6\,7\,8 \end{vmatrix} + \begin{vmatrix} 1\,2\,3\,6 \\ 4\,5\,7\,8 \end{vmatrix} - \begin{vmatrix} 1\,2\,3\,7 \\ 4\,5\,6\,8 \end{vmatrix} + \begin{vmatrix} 1\,2\,3\,8 \\ 4\,5\,6\,7 \end{vmatrix}$$
$$+ \begin{vmatrix} 1\,2\,4\,5 \\ 3\,6\,7\,8 \end{vmatrix} - \begin{vmatrix} 1\,2\,4\,6 \\ 3\,5\,7\,8 \end{vmatrix} + \begin{vmatrix} 1\,2\,4\,7 \\ 3\,5\,6\,8 \end{vmatrix} - \begin{vmatrix} 1\,2\,4\,8 \\ 3\,5\,6\,7 \end{vmatrix} + \begin{vmatrix} 1\,2\,5\,6 \\ 3\,4\,7\,8 \end{vmatrix}$$
$$- \begin{vmatrix} 1\,2\,5\,7 \\ 3\,4\,6\,8 \end{vmatrix} + \begin{vmatrix} 1\,2\,5\,8 \\ 3\,4\,6\,7 \end{vmatrix} + \begin{vmatrix} 1\,2\,6\,7 \\ 3\,4\,5\,8 \end{vmatrix} - \begin{vmatrix} 1\,2\,6\,8 \\ 3\,4\,5\,7 \end{vmatrix} + \begin{vmatrix} 1\,2\,7\,8 \\ 3\,4\,5\,6 \end{vmatrix}$$
$$= \begin{vmatrix} 1 & 2 \\ 3 & 4 \end{vmatrix} \left\{ \begin{vmatrix} 5 & 6 \\ 7 & 8 \end{vmatrix} - \begin{vmatrix} 5 & 7 \\ 6 & 8 \end{vmatrix} + \begin{vmatrix} 5 & 8 \\ 6 & 7 \end{vmatrix} + \begin{vmatrix} 6 & 7 \\ 5 & 8 \end{vmatrix} - \begin{vmatrix} 6 & 8 \\ 5 & 7 \end{vmatrix} + \begin{vmatrix} 7 & 8 \\ 5 & 6 \end{vmatrix} \right\}$$
$$- \begin{vmatrix} 1 & 2 \\ 3 & 5 \end{vmatrix} \left\{ \begin{vmatrix} 4 & 6 \\ 7 & 8 \end{vmatrix} - \begin{vmatrix} 4 & 7 \\ 6 & 8 \end{vmatrix} + \begin{vmatrix} 4 & 8 \\ 6 & 7 \end{vmatrix} + \begin{vmatrix} 6 & 7 \\ 4 & 8 \end{vmatrix} - \begin{vmatrix} 6 & 8 \\ 4 & 7 \end{vmatrix} + \begin{vmatrix} 7 & 8 \\ 4 & 6 \end{vmatrix} \right\}$$
$$+ \cdots$$
$$+ \begin{vmatrix} 1 & 2 \\ 7 & 8 \end{vmatrix} \left\{ \begin{vmatrix} 3 & 4 \\ 5 & 6 \end{vmatrix} - \begin{vmatrix} 3 & 5 \\ 4 & 6 \end{vmatrix} + \begin{vmatrix} 3 & 6 \\ 4 & 5 \end{vmatrix} + \begin{vmatrix} 4 & 5 \\ 3 & 6 \end{vmatrix} - \begin{vmatrix} 4 & 6 \\ 3 & 5 \end{vmatrix} + \begin{vmatrix} 5 & 6 \\ 3 & 4 \end{vmatrix} \right\}$$

or as we may write it

$$\Sigma \begin{vmatrix} 1\ 2\ \overline{3\ 4} \\ \underline{5\ 6\ 7\ 8} \end{vmatrix} = \begin{vmatrix} 1 & 2 \\ 3 & 4 \end{vmatrix} \left\{ \Sigma \begin{vmatrix} 5 & \bar{6} \\ \underline{7} & \underline{8} \end{vmatrix} + \Sigma \begin{vmatrix} \overline{7} & \overline{8} \\ 5 & \underline{6} \end{vmatrix} \right\} - \cdots$$

where the sigma refers to the numbers 345678 taken two at a time etc. An important special case of the theorem is when $k=n-1$ and $h=1$.

314. The conjugate of every minor in the first aggregate of S_β is found in the second with the same or opposite sign according as $n-k$ is even or odd. This is readily seen on observing that in bringing $(b)(c)(e)$ to the form $(e)(b)(c)$ the $n-k$ numbers in (e) are carried over the $n-k$ numbers in (c) and (b) and therefore the exponent of the sign factor for the two differ, mod 2, by $(n-k)^2$ or $n-k$.

315. Since each aggregate of S_β is an expression involving minors of order $n-k$ the same in form as those on the left-hand side of the equation, we may apply the theorem to them. If $n-k$ is not odd we may select h, so that after applying the theorem a second time $(n-k-h)$ is odd and shall have within each of the resulting S_β's conjugate minors with opposite signs.

316. If now we take the sum

$$(n-k)_g \sum_{i_1=1}^{(2n-k)_n} (-1)^{\nu_1} \left| \begin{matrix} (2n \mid k_\alpha)(\overline{2n} \mid k_\alpha \mid n-k_{i_1}) \\ (\overline{2n} \mid \bar{k}_\alpha \mid n-k_{i_1}) \end{matrix} \right|$$

and, instead of expanding each minor in terms of minors formed from the αth selection of k rows as in §312, we expand in terms of minors formed from the βth selection of h of the αth selection of k rows and g others and finally collect like terms we get

$$(n-k)_g \sum_{i_1=1}^{(2n-k)_n} (-1)^{\nu_1} \left| \begin{matrix} (2n \mid k_\alpha)(\overline{2n} \mid k_\alpha \mid n-k_{i_1}) \\ \overline{2n} \mid \bar{k}_\alpha \mid n-k_{i_1}) \end{matrix} \right|$$

$$= \sum_{i_1=1}^{(2n-k)_{k+g}} (-1)^{\nu_2} P_{i_1},$$

where P_{i_1} denotes the product of the two aggregates

$$\sum_{i_2=1}^{(h+2g)_g} (-1)^{\nu_3} \left| \begin{matrix} (2n \mid k_\alpha \mid h_\beta)(\overline{2n} \mid k_\alpha \mid h+2g_{i_1} \mid g_{i_2}) \\ (\overline{2n} \mid k_\alpha \mid h+2g_{i_1} \mid g_{i_3}) \end{matrix} \right|$$

$$\sum_{i_3=1}^{(2n-2g-h-k)_{n-h-g}} (-1)^{\nu_4} \left| \begin{matrix} (2n \mid \bar{k}_\alpha \mid h_\beta)(\overline{2n} \mid \bar{k}_\alpha \mid h+2g_{i_1} \mid n-k-g_{i_3}) \\ (\overline{2n} \mid \bar{k}_\alpha \mid h+2g_{i_1} \mid n-k-g_{i_3}) \end{matrix} \right|.$$

If $h=k$ and $g=1$, then this theorem reduces to that of §312. In illustration take the example where $n, k, g, h=3, 1, 1, 1$ and we have

$$2\cdot\sum\begin{vmatrix}1\ \overline{2\ 3}\\ \underline{4\ 5\ 6}\end{vmatrix}=\left(\begin{matrix}2 & & 3\\ & - & \\ 3 & & 2\end{matrix}\right)\sum\begin{vmatrix}1\ \overline{4}\\ \underline{5\ 6}\end{vmatrix}-\left(\begin{matrix}2 & & 4\\ & - & \\ 4 & & 2\end{matrix}\right)\sum\begin{vmatrix}1\ \overline{3}\\ \underline{5\ 6}\end{vmatrix}$$

$$+\left(\begin{matrix}2 & & 5\\ & - & \\ 5 & & 2\end{matrix}\right)\sum\begin{vmatrix}1\ \overline{3}\\ \underline{4\ 6}\end{vmatrix}-\left(\begin{matrix}2 & & 6\\ & - & \\ 6 & & 2\end{matrix}\right)\sum\begin{vmatrix}1\ \overline{3}\\ \underline{4\ 5}\end{vmatrix}$$

$$+\left(\begin{matrix}3 & & 4\\ & - & \\ 4 & & 3\end{matrix}\right)\sum\begin{vmatrix}1\ \overline{2}\\ \underline{5\ 6}\end{vmatrix}-\left(\begin{matrix}3 & & 5\\ & - & \\ 5 & & 3\end{matrix}\right)\sum\begin{vmatrix}1\ \overline{2}\\ \underline{4\ 6}\end{vmatrix}$$

$$+\left(\begin{matrix}3 & & 6\\ & - & \\ 6 & & 3\end{matrix}\right)\sum\begin{vmatrix}1\ \overline{2}\\ \underline{4\ 5}\end{vmatrix}+\left(\begin{matrix}4 & & 5\\ & - & \\ 5 & & 4\end{matrix}\right)\sum\begin{vmatrix}1\ \overline{2}\\ \underline{3\ 6}\end{vmatrix}$$

$$-\left(\begin{matrix}4 & & 6\\ & - & \\ 6 & & 4\end{matrix}\right)\sum\begin{vmatrix}1\ \overline{2}\\ \underline{3\ 5}\end{vmatrix}+\left(\begin{matrix}5 & & 6\\ & - & \\ 6 & & 5\end{matrix}\right)\sum\begin{vmatrix}1\ \overline{2}\\ \underline{3\ 4}\end{vmatrix}.$$

It is to be observed here that these theorems take on different forms if some of the numbers in $(2n\,|k_\alpha)$ and $(\overline{2n}\,|k_\alpha)$ are alike. Thus in the example just given suppose 6 becomes 1, then we would have

$$\sum\begin{vmatrix}1\ \overline{2\ 3}\\ 1\ 4\ \underline{5}\end{vmatrix}=\left(\begin{matrix}2 & & 3\\ & - & \\ 3 & & 2\end{matrix}\right)\sum\begin{vmatrix}1\ \overline{4}\\ 1\ \underline{5}\end{vmatrix}-\left(\begin{matrix}2 & & 4\\ & - & \\ 4 & & 2\end{matrix}\right)\sum\begin{vmatrix}1\ \overline{3}\\ 1\ \underline{5}\end{vmatrix}+\text{etc.}$$

etc. If we take $n, k, h, g=5, 4, 2, 0$ and then in the theorem make $9, 10=1, 2$, respectively we would get

$$\sum\begin{vmatrix}\overline{1\ 2\ 3\ 4\ 5}\\ 7\ 8\ 1\ 2\ \underline{6}\end{vmatrix}=\begin{vmatrix}1 & 2\\ 7 & 8\end{vmatrix}\sum\begin{vmatrix}\overline{3\ 4\ 5}\\ 1\ 2\ \underline{6}\end{vmatrix}-\begin{vmatrix}1 & 3\\ 7 & 8\end{vmatrix}\sum\begin{vmatrix}\overline{2\ 4\ 5}\\ 1\ 2\ \underline{6}\end{vmatrix}$$

$$+\cdots+\begin{vmatrix}5 & 6\\ 7 & 8\end{vmatrix}\sum\begin{vmatrix}1\ 2\ \overline{3}\\ 1\ 2\ \underline{4}\end{vmatrix}.$$

317. Expand the twenty terms of $\sum\begin{vmatrix}123\\ 456\end{vmatrix}$ in terms of ${}^{1\,1\,1\,1\,1\,2\,3\,4\,5\,6}_{2,3,4,5,6,1,1,1,1,1}$ and their complementaries and collect terms containing ${}^{1}_{\alpha}$ and ${}^{\alpha}_{1}$ $(\alpha=2, 3, 4, 5, 6)$ and it will be found that they have the same coefficient so that

$$\sum\begin{vmatrix}\overline{1\ 2\ 3}\\ \underline{4\ 5\ 6}\end{vmatrix}=\left(\begin{matrix}1 & & 2\\ & - & \\ 2 & & 1\end{matrix}\right)\sum\begin{vmatrix}\overline{3\ 4}\\ \underline{5\ 6}\end{vmatrix}-\left(\begin{matrix}1 & & 3\\ & - & \\ 3 & & 1\end{matrix}\right)\sum\begin{vmatrix}\overline{2\ 4}\\ \underline{5\ 6}\end{vmatrix}$$

$$+\left(\begin{matrix}1 & - & 4\\ 4 & & 1\end{matrix}\right)\sum\left|\begin{matrix}\overline{2\ 3}\\ \underline{5\ 6}\end{matrix}\right| - \left(\begin{matrix}1 & - & 5\\ 5 & & 1\end{matrix}\right)\sum\left|\begin{matrix}\overline{2\ 3}\\ \underline{4\ 6}\end{matrix}\right|$$

$$+\left(\begin{matrix}1 & - & 6\\ 6 & & 1\end{matrix}\right)\sum\left|\begin{matrix}\overline{2\ 3}\\ \underline{4\ 5}\end{matrix}\right|$$

or dropping the dashes since all numbers after the sigma vary we have

$$\sum\left|\begin{matrix}1 & 2 & 3\\ 4 & 5 & 6\end{matrix}\right| = \sum\left|\begin{matrix}1\\ 2\end{matrix}\right|\sum\left|\begin{matrix}3 & 4\\ 5 & 6\end{matrix}\right| - \sum\left|\begin{matrix}1\\ 3\end{matrix}\right|\sum\left|\begin{matrix}2 & 4\\ 5 & 6\end{matrix}\right| + \sum\left|\begin{matrix}1\\ 4\end{matrix}\right|\sum\left|\begin{matrix}2 & 3\\ 5 & 6\end{matrix}\right|$$

$$- \sum\left|\begin{matrix}1\\ 5\end{matrix}\right|\sum\left|\begin{matrix}2 & 3\\ 4 & 6\end{matrix}\right| + \sum\left|\begin{matrix}1\\ 6\end{matrix}\right|\sum\left|\begin{matrix}2 & 3\\ 4 & 5\end{matrix}\right|$$

and since we could just as well have expanded in terms of $\begin{smallmatrix}2\\ \alpha\end{smallmatrix}$ and $\begin{smallmatrix}\alpha\\ 2\end{smallmatrix}$ or $\begin{smallmatrix}3\\ \alpha\end{smallmatrix}$ and $\begin{smallmatrix}\alpha\\ 3\end{smallmatrix}$ etc. We have

$$\sum\left|\begin{matrix}1 & 2 & 3\\ 4 & 5 & 6\end{matrix}\right| = -\sum\left|\begin{matrix}2\\ 1\end{matrix}\right|\sum\left|\begin{matrix}3 & 4\\ 5 & 6\end{matrix}\right| + \sum\left|\begin{matrix}2\\ 3\end{matrix}\right|\sum\left|\begin{matrix}1 & 4\\ 5 & 6\end{matrix}\right| - \sum\left|\begin{matrix}2\\ 4\end{matrix}\right|\sum\left|\begin{matrix}1 & 3\\ 5 & 6\end{matrix}\right|$$

$$+ \sum\left|\begin{matrix}2\\ 5\end{matrix}\right|\sum\left|\begin{matrix}1 & 3\\ 4 & 6\end{matrix}\right| - \sum\left|\begin{matrix}2\\ 6\end{matrix}\right|\sum\left|\begin{matrix}1 & 3\\ 4 & 5\end{matrix}\right| = \text{etc.}$$

The general theorem is

$$\sum i(-1)^{\nu_1}\left\{\left|\begin{matrix}(2n\,|\,1_\alpha)\\ (\overline{2n}\,|\,1_\alpha\,|\,1_i)\end{matrix}\right| - \left|\begin{matrix}(\overline{2n}\,|\,1_\alpha\,|\,1_i)\\ (2n\,|\,1_\alpha)\end{matrix}\right|\right\}\sum j\left|\begin{matrix}(\overline{2n}\,|\,\bar{1}_\alpha\,|\,1_i\,|\,n-1_j)\\ (\overline{2n}\,|\,\bar{1}_\alpha\,|\,\bar{1}_i\,|\,n-1_j)\end{matrix}\right|$$

$$= \sum i(-1)^{\nu_2}\left\{\left|\begin{matrix}(2n\,|\,1_\beta)\\ (\overline{2n}\,|\,1_\beta\,|\,1_i)\end{matrix}\right| - \left|\begin{matrix}(\overline{2n}\,|\,1_\beta\,|\,1_i)\\ (2n\,|\,1_\beta)\end{matrix}\right|\right\}\sum j\left|\begin{matrix}(\overline{2n}\,|\,\bar{1}_\beta\,|\,1_i\,|\,n-1_j)\\ (\overline{2n}\,|\,\bar{1}_\beta\,|\,\bar{1}_i\,|\,n-1_j)\end{matrix}\right|$$

318. If (m) denotes any combination of the $2n$ numbers 1, 2, $\cdots$, $2n$, then $(\overline{m})$ shall denote the combination formed by replacing every number in (m) by its defect from $2n+1$.

With this notation we have the theorem:

$$\sum_{i_1=1}^{(n)k}\left|\begin{matrix}(\overline{2n\,|\,n_\alpha\,|\,k_{i_1}})(2n\,|\,\bar{n}_\alpha\,|\,k_{i_1})\\ (2n\,|\,n_\alpha)\end{matrix}\right| - \sum_{i_1=1}^{(n)k}\left|\begin{matrix}(2n\,|\,n_\alpha)\\ (\overline{2n\,|\,n_\alpha\,|\,k_{i_1}})(2n\,|\,\bar{n}_\alpha\,|\,k_{i_1}\end{matrix}\right|$$

$$= \sum_{i_1=1}^{(n)k}\ \sum_{i_2=1}^{(n)k}(-1)^{\nu}\left|\begin{matrix}(2n\,|\,\bar{n}_\alpha\,|\,k_{i_1})\\ (2n\,|\,\bar{n}_\alpha\,|\,k_{i_2})\end{matrix}\right|\left\{\left|\begin{matrix}(\overline{2n\,|\,n_\alpha\,|\,k_{i_1}})\\ (2n\,|\,n_\alpha\,|\,k_{i_2})\end{matrix}\right| - \left|\begin{matrix}(2n\,|\,n_\alpha\,|\,k_{i_1})\\ (\overline{2n\,|\,|\,n_\alpha\,k_{i_2}})\end{matrix}\right|\right\}$$

where ν *= the sum of the inversions in* $(2n\,|\,\bar{n}_\alpha\,|\,k_{i_1})(2n\,|\,n_\alpha\,|\,k_{i_1})$ *and* $(2n\,|\,\bar{n}_\alpha\,|\,k_{i_2})(2n\,|\,n\,|\,k_{i_2})$.

The proof consists in expanding by Laplace's theorem each minor contained in the first sum on the left in terms of minors formed from the $n-k$ rows having for indices the numbers in $(2n\,|\bar{n}_\alpha\,|k_{i_1})$, with their complementaries; and each minor in the second sum on the left in terms of minors from the $(n-k)$ columns having the same numbers for column indices, with their complementaries, and finally collecting the terms in pairs one from each sum.

319. *If any two determinants A and B of the nth order be taken, and from these two sets of determinants be formed, namely, first, a set of* $(n)_r$ *determinants each of which is in r rows identical with A and in the remaining rows identical with B; and, second, a set of the same number of determinants each of which is in r columns identical with A and in the remaining columns with B, then the sum of the first set is equal to the sum of the second set.*

Let $\Delta\binom{ab}{\alpha\beta}$ denote the determinant of the nth order whose αth selection of a rows is the αth selection of a rows of A and whose remaining rows are the βth selection of b rows from B, where $a+b=n$ and where the αth and βth selections are complementary selections of the numbers $1, 2, \cdots, n$. The a rows from A and the b rows from B will therefore occupy the same positions in $\Delta\binom{ab}{\alpha\beta}$ as they do in A and B respectively. Let $\Delta\binom{\alpha\beta}{ab}$ denote the determinant formed similarly from the columns. Then the theorem stated symbolically is

$$\Delta\begin{pmatrix} a & b \\ \alpha & \beta \end{pmatrix} = \Delta\begin{pmatrix} \alpha & \beta \\ a & b \end{pmatrix},$$

where the number of determinants on each side is $n!/a!b!$.

Expanding the determinants involved by Laplace's theorem in terms of minors of order a formed from the elements from A and their complementaries formed from the elements from B we see that the coefficient of any minor involving the elements from A on the one side is equal to the coefficient of the same minor on the other side, and hence the truth of the theorem.

320. Let $a+b+c+\cdots+k=n$, and let the αth, βth, $\cdots$, γth selections of $a, b, c, \cdots, k$ respectively of the n numbers $1, 2, \cdots, n$ be a set of complementary selections. Let $\Delta\binom{abc\,\cdots k}{\alpha\beta\gamma\,\cdots \kappa}$ be the determinant formed by taking for its αth selection of a rows the αth selection of a rows from A, for its βth selection of b rows, the βth selection of b rows from B, and so on, so that the rows taken from the determinants $A, B, C, \cdots, K$ occupy the same positions in $\Delta\binom{abc\,\cdots k}{\alpha\beta\gamma\,\cdots \kappa}$ that they did in the determinant from which they were

taken. Let $\Delta\binom{\alpha\beta\gamma \cdots \kappa}{abc \cdots k}$ be the determinant formed similarly from the columns. Then we have the more general theorem:

$$\Delta\begin{pmatrix} abc & \cdots & k \\ \alpha\beta\gamma & \cdots & \kappa \end{pmatrix} = \Delta\begin{pmatrix} \alpha\beta\gamma & \cdots & \kappa \\ abc & \cdots & k \end{pmatrix}$$

where the number of determinants on each side is $n!/a!b! \cdots k!$.

Before taking up the proof of this theorem let us consider an example where $n=4$, $a=2$, $b=c=1$, and therefore the number of determinants formed in each set is $4!/2!1!1!=12$. We have

$$\begin{aligned} & R(a_{11}\, a_{22}\, b_{33}\, c_{44}) + R(a_{11}\, a_{22}\, c_{33}\, b_{44}) + R(a_{11}\, b_{22}\, a_{33}\, c_{44}) \\ & + R(a_{11}\, c_{22}\, a_{33}\, b_{44}) + R(a_{11}\, b_{22}\, c_{33}\, a_{44}) + R(a_{11}\, c_{22}\, b_{33}\, a_{44}) \\ & + R(b_{11}\, a_{22}\, a_{33}\, c_{44}) + R(c_{11}\, a_{22}\, a_{33}\, b_{44}) + R(b_{11}\, a_{22}\, c_{33}\, a_{44}) \\ & + R(c_{11}\, a_{22}\, b_{33}\, a_{44}) + R(b_{11}\, c_{22}\, a_{33}\, a_{44}) + R(c_{11}\, b_{22}\, a_{33}\, a_{44}) \\ = {} & C(a_{11}\, a_{22}\, b_{33}\, c_{44}) + C(a_{11}\, a_{22}\, c_{33}\, b_{44}) + \cdots + C(c_{11}\, b_{22}\, a_{33}\, a_{44}) \end{aligned}$$

where for convenience $R(a_{11}\, a_{22}\, b_{33}\, c_{44})$ stands for

$$\begin{vmatrix} a_{11} & a_{12} & a_{13} & a_{14} \\ a_{21} & a_{22} & a_{23} & a_{24} \\ b_{31} & b_{32} & b_{33} & b_{34} \\ c_{41} & c_{42} & c_{43} & c_{44} \end{vmatrix}$$

and $C(a_{11}\, a_{22}\, b_{33}\, c_{44})$ stands for

$$\begin{vmatrix} a_{11} & a_{12} & b_{13} & c_{14} \\ a_{21} & a_{22} & b_{23} & c_{24} \\ a_{31} & a_{32} & b_{33} & c_{34} \\ a_{41} & a_{42} & b_{43} & c_{44} \end{vmatrix}.$$

The truth of this relation is seen on expanding each determinant involved by Laplace's theorem, in terms of minors of the second order containing the a's and their complementaries. Thus it will be seen that the coefficient of any minor of the a's on the one side is the same as the coefficients of the same minor on the other side.

Thus taking the minor $\begin{vmatrix} a_{11} a_{12} \\ a_{21} a_{22} \end{vmatrix}$ it is readily seen that its coefficient on the left is

$$\begin{vmatrix} b_{33} & c_{34} \\ b_{43} & c_{44} \end{vmatrix} + \begin{vmatrix} c_{33} & b_{34} \\ c_{43} & b_{44} \end{vmatrix}$$

and on the right is

$$\begin{vmatrix} b_{33} & b_{34} \\ c_{43} & c_{44} \end{vmatrix} + \begin{vmatrix} c_{33} & c_{34} \\ b_{43} & b_{44} \end{vmatrix}$$

and the two are equal and this is seen to be an example of the theorem of the preceding article for the two determinants

$$\begin{vmatrix} b_{33} & b_{34} \\ b_{43} & b_{44} \end{vmatrix}, \quad \begin{vmatrix} c_{33} & c_{34} \\ c_{43} & c_{44} \end{vmatrix}.$$

If we had expanded in terms of the b's and their complementaries and equated the coefficients of any b such as b_{11} we would have

$$\begin{vmatrix} a_{22} & a_{23} & c_{24} \\ a_{32} & a_{33} & c_{34} \\ a_{42} & a_{43} & c_{44} \end{vmatrix} + \begin{vmatrix} a_{22} & c_{23} & a_{24} \\ a_{32} & c_{33} & a_{34} \\ a_{42} & c_{43} & a_{44} \end{vmatrix} + \begin{vmatrix} c_{22} & a_{23} & a_{24} \\ c_{32} & a_{33} & a_{34} \\ c_{42} & a_{43} & a_{44} \end{vmatrix}$$

$$= \begin{vmatrix} a_{22} & a_{23} & a_{24} \\ a_{32} & a_{33} & a_{34} \\ c_{42} & c_{43} & c_{44} \end{vmatrix} + \begin{vmatrix} a_{22} & a_{23} & a_{24} \\ c_{32} & c_{33} & c_{34} \\ a_{42} & a_{43} & a_{44} \end{vmatrix} + \begin{vmatrix} c_{22} & c_{23} & c_{24} \\ a_{32} & a_{33} & a_{34} \\ a_{42} & a_{43} & a_{44} \end{vmatrix}$$

which is another example of the theorem of the last article for the two determinants $|a_{22}a_{33}a_{44}|$ and $|c_{22}c_{33}c_{44}|$. For the proof of the general theorem it is sufficient to observe that if we take three determinants A, B, C and expand in terms of minors formed from the a's and their complementaries, then the coefficient of any minor of the a's on the one side is equal to the coefficient of the same minor on the other side by §319. Having thus established the theorem for three determinants it is extended in a similar manner to four, then to five, and so on up to k.

We might select for our k determinants any k mutually exclusive minors of order n in a determinant of order nk. It is apparent that unless the selection of rows and columns are mutually exclusive some of the terms will vanish having identical rows or columns.

CHAPTER XI

ALTERNANTS

321. When the elements of the first row of a determinant are all functions of one variable, the elements of the second row the like functions of a second variable, and so on, the determinant is called an *Alternant*: for example,

$$\begin{vmatrix} \sin x & \cos x & 1 \\ \sin y & \cos y & 1 \\ \sin z & \cos z & 1 \end{vmatrix}$$

or a more extended definition, any determinant which is an alternating function is called an alternant: for example,

$$\begin{vmatrix} 1 & a & a^2 & a(b^4c^4 + b^4d^4 + c^4d^4) \\ 1 & b & b^2 & b(c^4d^4 + c^4a^4 + d^4a^4) \\ 1 & c & c^2 & c(d^4a^4 + d^4b^4 + a^4b^4) \\ 1 & d & d^2 & d(a^4b^4 + a^4c^4 + b^4c^4) \end{vmatrix}$$

322. Every alternant of the nth order is evidently a function of n variables. To interchange two of these would be the same as to interchange two of the rows or columns of the determinant, and therefore would have the effect of merely changing the sign of the function; and as a function having this property is known as an *alternating* function, the origin of the name alternant is apparent.

323. *Every alternant with rational integral elements contains as a factor the difference-product of its variables.*

Let the variables be $x_1, x_2, \cdots, x_n$.

By substituting for x_n any other of the variables, we should cause the determinant to vanish, hence it follows that

$$x_n - x_{n-1},\ x_n - x_{n-2},\ \cdots,\ x_n - x_2,\ x_n - x_1$$

are factors of the determinant. Similarly

$$x_{n-1} - x_{n-2},\ \cdots,\ x_{n-1} - x_2,\ x_{n-1} - x_1$$

are factors, and in like manner

$$x_{n-2} - x_{n-3},\ \cdots,\ x_{n-2} - x_2,\ x_{n-2} - x_1$$

$$\cdot \quad \cdot \quad \cdot \quad \cdot \quad \cdot \quad \cdot \quad \cdot \quad \cdot \quad \cdot$$

$$x_3 - x_2,\ x_3 - x_1.$$

Hence, if from every variable there be subtracted every variable preceding it, and the differences thus got be multiplied together, the result, known as the *difference-product* of the variables, is a factor of the determinant.

Sylvester, who uses $\zeta(x_1, x_2, x_3, \cdots)$ or $\zeta(x_1x_2x_3 \cdots)$ for the second power of the difference-product of $x_1, x_2, x_3, \cdots$, denotes the difference-product itself by $\zeta^{1/2}(x_1, x_2, x_3, \cdots)$ or $\zeta^{1/2}(x_1x_2x_3 \cdots)$.

324. *A difference-product is itself expressible as an alternant, namely:*

$$\zeta^{1/2}(x_1x_2x_3 \cdots x_n) = \begin{vmatrix} 1 & x_1 & x_1^2 & \cdots & x_1^{n-1} \\ 1 & x_2 & x_2^2 & \cdots & x_2^{n-1} \\ 1 & x_3 & x_3^2 & \cdots & x_3^{n-1} \\ . & . & . & . & . \\ 1 & x_n & x_n^2 & \cdots & x_n^{n-1} \end{vmatrix}$$

For the alternant on the right has $\zeta^{1/2}(x_1x_2x_3 \cdots x_n)$ for a factor and the cofactor is readily seen, as in the particular case given, Example 2 §60, to be unity.

325. *In order that the alternant*

$$\begin{vmatrix} 1 & a_1 & a_1^2 & \cdots & \phi(a_1\,;\, a_2a_3 \cdots a_n) \\ 1 & a_2 & a_2^2 & \cdots & \phi(a_2\,;\, a_1a_3 \cdots a_n) \\ . & . & . & . & . \end{vmatrix} = k\cdot\zeta^{1/2}(a_1a_2 \cdots a_n),$$

where ϕ is a function of the a's following the semicolon, it is necessary and sufficient that ϕ be (1) *symmetric with respect to $a_2, a_3, a_4, \cdots, a_n$ and* (2) *of the $(n-1)$ degree.*

For, the alternant vanishes if any two of the variables are equal, and therefore is divisible by $\zeta^{1/2}(a_1a_2 \cdots a_n)$; also the degree of the alternant $\frac{1}{2}n(n-1)$ is the same as that of $\zeta^{1/2}$, so that the cofactor is constant.

To obtain the value of k, we denote $\phi(a_1; a_2, \cdots, a_n)$ by A; and since the appearance in A of a factor of the form a_1^m, as $a_1^m\phi(a_2a_3a_4 \cdots)$ does not affect the value of k corresponding to $\phi(a_2a_3 \cdots)$, it will only be necessary to consider functions of the latter form.

For instance suppose $A \equiv \phi(a_2a_3a_4) = \sum a_2^2\, a_3$ and the alternant is

$$\Delta \equiv \begin{vmatrix} 1 & a_1 & a_1^2 & \sum a_4^2\, a_2 \\ 1 & a_2 & a_2^2 & \sum a_1^2\, a_3 \\ 1 & a_3 & a_3^2 & \sum a_2^2\, a_4 \\ 1 & a_4 & a_4^2 & \sum a_3^2\, a_1 \end{vmatrix}.$$

We may write

$$\sum a_4^2 a_2 = \sum a_1^2 a_2 - a_1 \sum a_2^2 - a_1^2 \sum a_2$$
$$= \sum a_1^2 a_2 - a_1 \sum a_1^2 - a_1 \sum a_1 + 2a_1^3 .$$

Then the following operations

$$\text{col}_4 - \sum a_1^2 a_2 \cdot \text{col}_1 + \sum a_1^2 \cdot \text{col}_2 + \sum a_1 \cdot \text{col}_3$$

reduces Δ to the form

$$\begin{vmatrix} 1 & a_1 & a_1^2 & 2a_1^3 \\ 1 & a_2 & a_2^2 & 2a_2^3 \\ 1 & a_3 & a_3^2 & 2a_3^3 \\ 1 & a_4 & a_4^2 & 2a_4^3 \end{vmatrix} = 2\zeta^{1/2}(a_1\, a_2\, a_3\, a_4) .$$

326. In general A consists of the sum of a series of terms of the form $\sum a_2^{p_2} a_3^{p_3} \cdots a_n^{p_n}$, where we will suppose that α of the p's have the same value x, β of the remaining p's have the same value y, and so on, then from symmetric functions we know that A can be put into a form so as to be reduced by operations similar to those in §325 leaving a single term which is

$$a_1^{n-1}(-1)^{\alpha+\beta+\gamma+\cdots}\frac{(\alpha+\beta+\gamma+\cdots)!}{\alpha!\,\beta!\,\gamma!\cdots}$$

and therefore

$$k = (-1)^{\alpha+\beta+\gamma+\cdots}\frac{(\alpha+\beta+\gamma+\cdots)!}{\alpha!\,\beta!\,\gamma!\cdots}$$

Therefore to determine k we have to find the number of permutations of any number of things taken all together of which α are alike, β are alike and γ are alike, and so on. The sign of k is positive when the number of things is even and negative when odd.

Example. Let

$$A = \sum b^m c^n d^n e^p f^p h^p$$

which

$$= S - a^m \sum b^n c^n d^p e^p f^p - a^n \sum b^m c^n d^p e^p f^p - a^p \sum b^m c^n d^n e^p f^p ,$$

where S denotes the corresponding complete function involving a, then

$$k = 0 - \frac{5!}{2!3!} - \frac{5!}{3!} - \frac{5!}{2!2!} = \frac{5!}{2!3!}(1 + 2! + 3) = \frac{6!}{2!3!} \cdot$$

327. *In order that the alternant*

$$\begin{vmatrix} 1 & \phi_1(a\,;bc\cdots) & \phi_2(a\,;bc\cdots) & \phi_{n-1}(a\,;bc\cdots) \\ \cdot & \cdot & \cdot & \cdot \\ \cdot & \cdot & \cdot & \cdot \end{vmatrix}$$

$$= k\zeta^{1/2}(a\,b\,c\cdots 1)$$

the functions must be (1) *all symmetric with respect to b, c, d,* $\cdots$ *and* (2) *of the degrees 1, 2, 3,* $\cdots$, $n-1$ *respectively.*

To determine k for alternants of this form, let us consider

$$\begin{vmatrix} 1 & \sum b & \sum bc & \sum b^2c & \sum b^3c \\ \cdot & \cdot & \cdot & \cdot & \cdot \\ \cdot & \cdot & \cdot & \cdot & \cdot \end{vmatrix}$$

Since $\sum b = \sum a - a$, we can replace $\sum b$ by a on multiplying the alternant by -1. Also, since $\sum bc = \sum ab - a\sum a + a^2$, we can replace $\sum bc$ by a^2. And since $\sum b^2c = \sum a^2b - a^2\sum a - a\sum a^2 + 2a^3$, we can replace $\sum b^2c$ by a^3 on multiplying by 2. Thus the alternant is equal to $(-1)(+1)(2)(2)\zeta^{1/2}$, that is $k=-4$.

Now these multipliers are respectively the values of k corresponding to the functions taken in order. Hence we see that the value of k in the case of alternants of the above general form is equal to the product of the values of k corresponding to the several functions, as given in §326.

328. From the above theorems we conclude that *the most general form of alternants (involving integral functions) which are constant multiples of the difference-product of the variables is*

$$\begin{vmatrix} 1 & \phi_1(a\,b\,c\cdots) & \phi_2(a\,b\,c\cdots) & \phi_3(a\,b\,c\cdots)\cdots \\ \cdot & \cdot & \cdot & \cdot \\ \cdot & \cdot & \cdot & \cdot \end{vmatrix}$$

where

$$\phi_1(a\,b\,c\cdots) = m_1a + n_1\sum b$$

$$\phi_2(a\,b\,c\cdots) = m_2a^2 + n_2a\sum b + p_2\sum b^2 + q_2\sum bc$$

$$\phi_3(a\,b\,c\cdots) = m_3a^3 + n_3a^2\sum b + p_3a\sum b^2 + q_3a\sum bc + r_3\sum b^3 + s_3\sum b^3c + t_3\sum bcd$$

. . . .

and that the value of the constant multiplier is the product of

$$m_1 - n_1$$
$$m_2 - n_2 - p_2 + q_2$$
$$m_3 - n_3 - p_3 + q_3 - r_3 + 2s_3 - t_3$$
$$\cdot \quad \cdot \quad \cdot \quad \cdot \quad \cdot \quad \cdot \quad \cdot \quad \cdot \quad \cdot$$

329. *The difference-product of* $a_1, a_2, a_3, \cdots, a_{2n}$ *is equal to*

$$\sum (a_2 - a_1)(a_4 - a_3)(a_6 - a_5) \cdots (a_{2n} - a_{2n-1}) \times \overset{+}{\big|}(a_1a_2)^0(a_3a_4)^2 \cdots (a_{2n-1}a_n)^{2n-2}\overset{+}{\big|}$$

where $(a_2-a_1)(a_4-a_3) \cdots (a_{2n}-a_{2n-1})$ *is in magnitude and sign a term of*

$$\left| \begin{array}{llll} a_2 - a_1 & a_3 - a_1 & \cdots & a_{2n} - a_1 \\ & a_3 - a_2 & \cdots & a_{2n} - a_2 \\ & & \cdots & \cdot \quad \cdot \\ & & & a_{2n} - a_{2n-1} \end{array} \right|$$

and where each of the binary products $a_1a_2, a_3a_4, \cdots, a_{2n-1}a_{2n}$ *is formed from the original elements occurring in one of the linear factors immediately preceding.*

From §§100, 103 we have, on putting $n=2$

$$|a_1^0 a_2^1 \cdots a_{2n}^{2n-1}| = \overset{+}{|}\,|a_1^0 a_2^1|\,|a_3^2 a_4^3|\,|a_5^4 a_6^5| \cdots |a_{2n-1}^{2n-2} a_{2n}^{2n-1}|\,\overset{+}{|}$$
$$- \overset{+}{|}\,|a_1^0 a_2^1|\,|a_3^2 a_5^3|\,|a_4^4 a_6^5| \cdots |a_{2n-1}^{2n-2} a_{2n}^{2n-2}|\,\overset{+}{|} + \cdots$$

or

$$= \sum \overset{+}{|}\,|a_h^0 a_k^1|\,|a_p^2 a_q^3|\,|a_r^4 a_s^5| \cdots |a_y^{2n-2} a_z^{2n-1}|\,\overset{+}{|}$$

where $hk, pq, rs, \cdots, yz$ is one of the ways of separating the $2n$ suffixes $1, 2, 3, \cdots, 2n$ into n sets of 2 each, and where the sign of summation implies that all other such separations are to be taken, it being understood that the sign preceding any permanent is to be made the same as the sign of that particular term of the alternant which is brought into prominence by the notation employed in specifying the permanent. Since, however, in specifying the typical permanent the particular term of the alternant which makes its appearance is $a_h{}^0 a_k{}^1 a_p{}^2 a_q{}^3 \cdots a_y{}^{2n-2} a_z{}^{2n-1}$ the sign of which is $(-1)^\nu$

where ν is the number of inverted pairs in $h\,k\,p\,q \cdots y\,z$, a better way of writing this deduction from the general theorem is

$$
\left| a_1^0 a_2^1 \cdots a_{2n}^{2n-1} \right|
$$

$$
= \sum(-1)^\nu \begin{vmatrix} |a_h^0 a_k^1| & |a_h^2 a_k^3| & |a_h^4 a_k^5| & \cdots & |a_h^{2n-2} a_k^{2n-1}| \\ |a_p^0 a_q^1| & |a_p^2 a_q^3| & |a_p^4 a_q^5| & \cdots & |a_p^{2n-2} a_q^{2n-1}| \\ |a_r^0 a_s^1| & |a_r^2 a_s^3| & |a_r^4 a_s^5| & \cdots & |a_r^{2n-2} a_s^{2n-1}| \\ \cdot & \cdot & \cdot & \cdots & \cdot \\ |a_y^0 a_z^1| & |a_y^2 a_z^3| & |a_y^4 a_z^5| & \cdots & |a_y^{2n-2} a_z^{2n-1}| \end{vmatrix}
$$

It is then apparent that on the right the differences $a_k - a_h$, $a_q - a_p$, $a_r - a_s$, $\cdots$, $a_z - a_y$ are factors of the 1st, 2nd, 3rd, $\cdots$, nth rows respectively, and that if we remove them we shall have

$$
\left| a_1^0 a_2^1 \cdots a_{2n}^{2n-1} \right| = \sum(-1)^\nu (a_k - a_h)(a_q - a_p)(a_s - a_r) \cdots (a_z - a_y)
$$

$$
\times \begin{vmatrix} (a_h a_k)^0 & (a_h a_k)^2 & (a_h a_k)^4 & \cdots & (a_h a_k)^{2n-2} \\ (a_p a_q)^0 & (a_p a_q)^2 & (a_p a_q)^4 & \cdots & (a_p a_q)^{2n-2} \\ (a_r a_s)^0 & (a_r a_s)^2 & (a_r a_s)^4 & \cdots & (a_r a_s)^{2n-2} \\ \cdot & \cdot & \cdot & \cdots & \cdot \\ (a_y a_z)^0 & (a_y a_z)^2 & (a_y a_z)^4 & \cdots & (a_y a_z)^{2n-2} \end{vmatrix} .
$$

330. *If the expanded form of the difference-product of the n letters $a, b, c, \cdots, l$ be multiplied by $abc \cdots l$, and in the result every index to a letter be made a suffix to the same letter, the expression obtained is the determinant* $|a_1 b_2 c_3 \cdots l_n|$.

This is self-evident on writing the difference-product in its determinant form.

331. *The quotient of an alternant by the difference-product of its variables is a symmetric function of the variables.*

On the interchange of any pair of the variables both dividend and divisor change sign. Consequently when such an interchange is made the quotient remains unaltered, and therefore is by definition a symmetric function of its variables.

Example.

$$
\begin{vmatrix} 1 & a & a^3 \\ 1 & b & b^3 \\ 1 & c & c^3 \end{vmatrix} = \begin{vmatrix} 1 & a & a^2 \\ 1 & b & b^2 \\ 1 & c & c^2 \end{vmatrix} (a + b + c) = \zeta^{1/2}(abc) \sum a .
$$

332. An alternant of the nth order being known when we know in order its n functions and n variables, we may suitably use for it a notation in which only these $2n$ things are specified. Thus

$$\begin{vmatrix} \phi_1(x) & \phi_2(x) & \phi_3(x) \\ \phi_1(y) & \phi_2(y) & \phi_3(y) \\ \phi_1(z) & \phi_2(z) & \phi_3(z) \end{vmatrix} \equiv A(\phi_1(x)\ \phi_2(y)\ \phi_3(z)).$$

The alternants requiring first to be considered are those in which the functions are *powers*, such as

$$\begin{vmatrix} a^m & a^n & a^p \\ b^m & b^n & b^p \\ c^m & c^n & c^p \end{vmatrix} \equiv A(a^m b^n c^p)$$

or when possible by $A(m, n, p)$. This type of alternant is known as a *simple alternant.*

The form of alternant, for which the symmetric quotient referred to in §331 is most readily obtained, and which may be viewed as the *second simplest form*, is that in which all the powers except the last are in order the same as those of the difference-product alternant.

333. *The quotient of the alternant of the second simplest form by the corresponding difference-product is expressible as the sum of a series of fractions, namely:*

$$\frac{A(\alpha_1^0 \alpha_2^1 \cdots \alpha_{n-1}^{n-2} \alpha_n^r)}{A(\alpha_1^0 \alpha_2^1 \cdots \alpha_{n-1}^{n-2} \alpha_n^{n-1})} = \frac{\alpha_n^r}{(\alpha_n - \alpha_{n-1})(\alpha_n - \alpha_{n-2}) \cdots (\alpha_n - \alpha_1)}$$
$$= \frac{\alpha_{n-1}^r}{(\alpha_{n-1} - \alpha_n)(\alpha_{n-1} - \alpha_{n-2}) \cdots (\alpha_{n-1} - \alpha_1)}$$
$$+ \cdots\cdots\cdots\cdots$$
$$+ \frac{\alpha_1^r}{(\alpha_1 - \alpha_2)(\alpha_1 - \alpha_3) \cdots (\alpha_1 - \alpha_n)}$$

Expressing

$$\begin{vmatrix} 1 & \alpha_1^1 & \alpha_1^2 & \cdots & \alpha_1^{n-2} & \alpha_1^r \\ 1 & \alpha_2^1 & \alpha_2^2 & \cdots & \alpha_2^{n-2} & \alpha_2^r \\ \cdot & \cdot & \cdot & \cdots & \cdot & \cdot \\ 1 & \alpha_{n-1}^1 & \alpha_{n-1}^2 & \cdots & \alpha_{n-1}^{n-2} & \alpha_{n-1}^r \\ 1 & \alpha_n^1 & \alpha_n^2 & \cdots & \alpha_n^{n-2} & \alpha_n^r \end{vmatrix}$$

in terms of the elements of the last column and their complementaries which are all difference-products, we have

$$\alpha_n^r\zeta^{1/2}(\alpha_1\alpha_2\cdots\alpha_{n-1}) - \alpha_{n-1}^r\zeta^{1/2}(\alpha_1\alpha_2\cdots\alpha_{n-2}\alpha_n) + \alpha_{n-2}^r\zeta^{1/2}(\alpha_1\alpha_2\cdots\alpha_{n-3}\alpha_{n-1}\alpha_n) - \cdots + (-1)^{n-1}\alpha_1^r\zeta^{1/2}(\alpha_2\alpha_3\cdots\alpha_{n-1}\alpha_n).$$

Hence dividing by the difference-product of $a_1, a_2, \cdots, a_{n-1}, a_n$, and legitimately altering the signs so as to have in the second denominator a_{n-1} in every case the minuend, in the third denominator a_{n-2}, and so on, there results the identity as stated.

334. *The quotient of an alternant of the second simplest form by the corresponding difference-product is expressible by means of two simpler like quotients, namely:*

$$\frac{A(\alpha_1^0\alpha_2^1\cdots\alpha_{n-1}^{n-2}\alpha_n^r)}{\zeta^{1/2}(\alpha_1\alpha_2\cdots\alpha_{n-1}\alpha_n)} = \alpha_n\frac{A(\alpha_1^0\alpha_2^1\cdots\alpha_{n-1}^{n-2}\alpha_n^{r-1})}{\zeta^{1/2}(\alpha_1\alpha_2\cdots\alpha_{n-1}\alpha_n)} + \frac{A(\alpha_1^0\alpha_2^1\cdots\alpha_{n-2}^{n-3}\alpha_{n-1}^{r-1})}{\zeta^{1/2}(\alpha_1\alpha_2\cdots\alpha_{n-2}\alpha_{n-1})}.$$

From §28 we have $A(a_1^0a_2^1\cdots a_{n-1}^{n-2}a_n^r) - a_nA(a_1^0a_2^1\cdots a_{n-1}^{n-2}a_n^{r-1})$

$$= \begin{vmatrix} 1 & \alpha_1^1 & \alpha_1^2 & \cdots & \alpha_1^{n-2} & \alpha_1^r - \alpha_n\alpha_1^{r-1} \\ 1 & \alpha_2^1 & \alpha_2^2 & \cdots & \alpha_2^{n-2} & \alpha_2^r - \alpha_n\alpha_2^{r-1} \\ \cdot & \cdot & \cdot & \cdots & \cdot & \cdot \\ 1 & \alpha_{n-1}^1 & \alpha_{n-1}^2 & \cdots & \alpha_{n-1}^{n-2} & \alpha_{n-1}^r - \alpha_n\alpha_{n-1}^{r-1} \\ 1 & \alpha_n^1 & \alpha_n^2 & \cdots & \alpha_n^{n-2} & 0 \end{vmatrix},$$

and (§333) the right-hand member of this divided by $\zeta^{1/2}(a_1a_2\cdots a_n)$

$$= \left\{ \begin{aligned} & \frac{0}{(0-\alpha_{n-1})(0-\alpha_{n-2})\cdots(0-\alpha_1)} \\ &+ \frac{\alpha_{n-1}^r - \alpha_n\alpha_{n-1}^{r-1}}{(\alpha_{n-1}-\alpha_n)(\alpha_{n-1}-\alpha_{n-2})\cdots(\alpha_{n-1}-\alpha_1)} \\ &+ \cdots\cdots\cdots \\ &+ \frac{\alpha_1^r - \alpha_n\alpha_1^{r-1}}{(\alpha_1-\alpha_2)(\alpha_1-\alpha_3)\cdots(\alpha_1-\alpha_n)} \end{aligned} \right.$$

$$= \begin{cases} 0 \\ + \dfrac{\alpha_{n-1}^{r-1}}{(\alpha_{n-1} - \alpha_{n-2}) \cdots (\alpha_{n-1} - \alpha_1)} \\ + \cdots\cdots\cdots\cdots\cdots \\ + \dfrac{\alpha_1^{r-1}}{(\alpha_1 - \alpha_2)(\alpha_1 - \alpha_3) \cdots (\alpha_1 - \alpha_{n-1})} \end{cases}$$

$$= \frac{A(\alpha_1^0 \alpha_2^1 \cdots \alpha_{n-2}^{n-3} \alpha_{n-1}^{r-1})}{\zeta^{1/2}(\alpha_1 \alpha_2 \cdots \alpha_{n-2} \alpha_{n-1})} \qquad (\S 333);$$

whence the required identity.

EXAMPLE:

$$\frac{\begin{vmatrix} 1 & a & a^4 \\ 1 & b & b^4 \\ 1 & c & c^4 \end{vmatrix}}{\begin{vmatrix} 1 & a & a^2 \\ 1 & b & b^2 \\ 1 & c & c^2 \end{vmatrix}} = c \frac{\begin{vmatrix} 1 & a & a^3 \\ 1 & b & b^3 \\ 1 & c & c^3 \end{vmatrix}}{\begin{vmatrix} 1 & a & a^2 \\ 1 & b & b^2 \\ 1 & c & c^2 \end{vmatrix}} + \frac{\begin{vmatrix} 1 & a^3 \\ 1 & b^3 \end{vmatrix}}{\begin{vmatrix} 1 & a \\ 1 & b \end{vmatrix}},$$

$$= c(a + b + c) + \frac{b^3 - a^3}{b - a}, \qquad (\S 331, \text{Ex.})$$

$$= a^2 + b^2 + c^2 + ab + bc + ca = \sum a^2 + \sum ab.$$

Similarly with the help of this result we may find the like expansion of

$$\frac{A(a^0 b^1 c^5)}{\zeta^{1/2}(abc)},$$

thence that of

$$\frac{A(a^0 b^1 c^6)}{\zeta^{1/2}(abc)},$$

and so on.

335. *The quotient of an alternant of the second simplest form by the corresponding difference-product is equal to the sum of all the terms which can be formed by multiplying together such positive integral powers of the variables that the sum of their indices may equal the excess of the last index of the one alternant over the last index of the other.*

Let the quotient referred to be

$$\frac{A(\alpha_1^0\alpha_2^1 \cdots \alpha_{n-1}^{n-2}\alpha_n^r)}{\zeta^{1/2}(\alpha_1\alpha_2 \cdots \alpha_{n-1}\alpha_n)}\,.$$

Since the dividend consists of terms like its principal diagonal term

$$\alpha_1^0\alpha_2^1\alpha_3^2 \cdots \alpha_{n-1}^{n-2}\alpha_n^r,$$

and the divisor, similarly, of terms like

$$\alpha_1^0\alpha_2^1\alpha_3^2 \cdots \alpha_{n-1}^{n-2}\alpha_n^{n-1},$$

we see at once that the quotient must consist of terms of the kind specified in the statement of the theorem, and of no others. It thus only remains to be shown that all possible terms of this kind occur, and that, unlike the terms of dividend and divisor, all are positive.

As an instance of such a term, namely, one the sum of whose indices is $r-n+1$, let us take

$$\alpha_n^2\alpha_{n-1}^1 \cdots \alpha_3^5\alpha_2^3\alpha_1^4\,.$$

By repeated use of the theorem of §334 we have

$$\frac{A(\alpha_1^0 \cdots \alpha_{n-1}^{n-2}\alpha_n^r)}{\zeta^{1/2}(\alpha_1 \cdots \alpha_{n-1}\alpha_n)} = \frac{A(\alpha_1^0 \cdots \alpha_{n-2}^{n-3}\alpha_{n-1}^{r-1})}{\zeta^{1/2}(\alpha_1 \cdots \alpha_{n-1})} + \alpha_n\frac{A(\alpha_1^0 \cdots \alpha_{n-2}^{n-3}\alpha_{n-1}^{r-2})}{\zeta^{1/2}(\alpha_1 \cdots \alpha_{n-1})}$$

$$+\,\alpha_n^2\frac{A(\alpha_1^0 \cdots \alpha_{n-2}^{n-3}\alpha_{n-1}^{r-3})}{\zeta^{1/2}(\alpha_1 \cdots \alpha_{n-1})} + \cdots + \alpha_n^{r-n}\frac{A(\alpha_1^0 \cdots \alpha_{n-1}^{n-1})}{\zeta^{1/2}(\alpha_1 \cdots \alpha_{n-1})} + \alpha_n^{r-n+1},$$

the quotient we are concerned with being thus separated into $r-n+2$ groups of terms, namely, those independent of α_n, those containing α_n^1, those containing α_n^2, and so on. Now the third, if any, of these groups must contain the term in question. Taking it, therefore, and developing the cofactor of α_n^2 according to ascending powers of α_{n-1} in the same way, we see that the cofactor of $\alpha_n^2\alpha_{n-1}^1$ is

$$\frac{A(\alpha_1^0 \cdots \alpha_{n-3}^{n-4}\alpha_{n-2}^{r-5})}{\zeta^{1/2}(\alpha_1 \cdots \alpha_{n-3}\alpha_{n-2})}\,.$$

Expanding this coefficient in like manner, and so on, we shall at length come to the coefficient of $\alpha_n^2\alpha_{n-1}^1 \cdots \alpha_3^5$, which, as it must be of the seventh degree and must contain only α_2 and α_1, must be

$$\frac{A(\alpha_1^0\alpha_2^8)}{\zeta^{1/2}(\alpha_1\alpha_2)}\,.$$

This, however,

$$= \frac{\alpha_2{}^8 - \alpha_1}{\alpha_2 - \alpha_1}$$
$$= \alpha_2{}^7 + \cdots + \alpha_2^3\alpha_1^4 + \cdots + \alpha_1{}^7.$$

Thus $+\alpha_n{}^2\alpha_{n-1}^1 \cdots \alpha_3{}^5\alpha_2{}^3\alpha_1{}^4$ is a term of the expansion, and evidently the same may be proved in regard to any other term of the kind.

The quotient specified in this theorem is known as the *aleph* or "*complete symmetric function of* $a_1, a_2, a_3, \cdots, a_n$ *of the degree* $r-n+1$," and is denoted by $(a_1, a_2, a_3, \cdots, a_n)^{r-n+1}$, or, when there is no doubt as to the variables, by $(\ )^{r-n+1}$ or H_{r-n+1}. The theorem thus simply is

$$\frac{A(\alpha_1^0\alpha_2^1 \cdots \overset{n-2}{\alpha_{n-1}}\overset{r}{\alpha_n})}{A(\alpha_1^0\alpha_2^1 \cdots \overset{n-2}{\alpha_{n-1}}\overset{n-1}{\alpha_n})} \text{ or } \frac{A(\alpha_1^0 \cdots \overset{n-2}{\alpha_{n-1}}\overset{r}{\alpha_n})}{\zeta^{1/2}(\alpha_1 \cdots \alpha_{n-1}\alpha_n)}$$
$$= (\alpha_1, \alpha_2, \alpha_3, \cdots, \alpha_n)^{r-n+1}.$$

EXAMPLES:

$$\begin{vmatrix} 1 & a & a^5 \\ 1 & b & b^5 \\ 1 & c & c^5 \end{vmatrix} \div \begin{vmatrix} 1 & a & a^2 \\ 1 & b & b^2 \\ 1 & c & c^2 \end{vmatrix} = (a, b, c)^{5-2},$$
$$= \sum a^3 + \sum a^2b + \sum abc,$$
$$= (a^3 + b^3 + c^3) + (a^2b + a^2c + b^2a + b^2c + c^2a + c^2b) + abc.$$

and

$$\frac{A(a^0b^1c^2d^7)}{A(a^0b^1c^2d^3)} = (a, b, c, d)^{7-3},$$
$$= \sum a^4 + \sum a^3b + \sum a^2b^2 + \sum a^2bc + \sum abcd,$$
$$= (a^4 + b^4 + c^4 + d^4) + \cdots.$$

336. From the preceding it follows that every known theorem regarding complete symmetric functions becomes a theorem in alternants. Of these known theorems the simplest flow readily from the law of development

(1) $\quad (\alpha_1\alpha_2 \cdots \alpha_n)^s = \alpha_n(\alpha_1\alpha_2 \cdots \alpha_n)^{s-1} + (\alpha_1\alpha_2 \cdots \alpha_{n-1})^s$

established in §334 in regard to the functions in their alternant form.

Thus, developing the first term of the right-hand member of (1) by means of (1) itself, we have

$$(2)\ (\alpha_1 \cdots \alpha_n)^s = \alpha_n^s + \alpha_n^{s-1}(\alpha_1 \cdots \alpha_{n-1})^1 + \alpha_n^{s-2}(\alpha_1 \cdots \alpha_{n-1})^2 + \cdots + (\alpha_1 \cdots \alpha_{n-1})^s,$$

as has already been incidentally seen in §335. By development of the *second* term of the right-hand member in the same way, there results

$$(3)\ (\alpha_1 \cdots \alpha_n)^s = \alpha_n(\alpha_1 \cdots \alpha_n)^{s-1} + \alpha_{n-1}(\alpha_1 \cdots \alpha_{n-1})^{s-1} + \alpha_{n-2}(\alpha_1 \cdots \alpha_{n-2})^{s-1} + \cdots + \alpha_1^s.$$

By developing *both* terms, and making a third use of (1) to combine two of the terms resulting, we have

$$(\alpha_1 \cdots \alpha_n)^s = (\alpha_1 \cdots \alpha_{n-2})^s + (\alpha_{n-1}\alpha_n)^1(\alpha_1 \cdots \alpha_{n-1})^{s-1} + \alpha_n^2(\alpha_1 \cdots \alpha_n)^{s-2},$$

and thence

$$(\alpha_1 \cdots \alpha_n)^s = (\alpha_1 \cdots \alpha_{n-3})^s + (\alpha_{n-2}\alpha_{n-1}\alpha_n)^1(\alpha_1 \cdots \alpha_{n-2})^{s-1} + (\alpha_{n-1}\alpha_n)^2(\alpha_1 \cdots \alpha_{n-1})^{s-2} + \alpha_n^3(\alpha_1 \cdots \alpha_n)^{s-3},$$

and, finally,

$$(4)\ (\alpha_1 \cdots \alpha_n)^s = \alpha_1^s + (\alpha_2 \cdots \alpha_n)^1(\alpha_1\alpha_2)^{s-1} + (\alpha_3 \cdots \alpha_n)^2(\alpha_1\alpha_2\alpha_3)^{s-2} + \cdots + \alpha_n^{n-1}(\alpha_1 \cdots \alpha_n)^{s-n+1}.$$

Again, returning to the first term of the right-hand member of (1), and altering it by means of (1) in another way, we have

$$(\alpha_1 \cdots \alpha_n)^s = \alpha_n\{(\alpha_1 \cdots \alpha_{n+1})^{s-1} - \alpha_{n+1}(\alpha_1 \cdots \alpha_{n+1})^{s-2}\} + (\alpha_1 \cdots \alpha_{n-1})^s;$$

Similarly

$$(\alpha_1 \cdots \alpha_{n-1}\alpha_{n+1})^s = \alpha_{n+1}\{(\alpha_1 \cdots \alpha_{n+1})^{s-1} (-)\ \alpha_n(\alpha_1 \cdots \alpha_{n+1})^{s-2}\} + (\alpha_1 \cdots \alpha_{n-1})^s;$$

and therefore by subtraction and division

$$(5)\ \frac{(\alpha_1 \cdots \alpha_{n-1}\alpha_n)^s - (\alpha_1 \cdots \alpha_{n-1}\alpha_{n+1})^s}{\alpha_n - \alpha_{n+1}} = \alpha_1 \cdots \alpha_n\alpha_{n+1})^{s-1}.$$

337. *The quotient of an alternant in which the elements are the same as those of the difference-product with the ith power omitted by the corresponding difference-product is expressible as a sum of products of the variables.*

That is

$$\frac{\begin{vmatrix} a_1^n & \cdots & a_1^{i+1} & a_1^{i-1} & \cdots & 1 \\ a_2^n & \cdots & a_2^{i+1} & a_2^{i-1} & \cdots & 1 \\ \cdot & \cdot & \cdot & \cdot & \cdot & \cdot \\ a_n^n & \cdots & a_n^{i+1} & a_n^{i-1} & \cdots & 1 \end{vmatrix}}{\zeta^{1/2}(a_1 \cdots a_n)} = p_{n-i}$$

where p_{n-i} is the sum of the products $n-i$ at a time, without repetitions, of the quantities $a_1 \cdots a_n$.

Consider the obvious identity

$$\begin{vmatrix} x^n & x^{n-1} & \cdots & x & 1 \\ a_1^n & a_1^{n-1} & \cdots & a_1 & 1 \\ \cdot & \cdot & \cdot & \cdot & \cdot \\ a_n^n & a_n^{n-1} & \cdots & a_n & 1 \end{vmatrix} = \zeta^{1/2}(a_1a_2 \cdots a_n)f(x)$$

where

$$f(x) = (x - a_1)(x - a_2) \cdots (x - a_n)$$

or

$$= x^n - p_1x^{n-1} + p_2x^{n-2} - \cdots + (-1)^{n-i}p_{n-i}x^i + \cdots.$$

By equating coefficients of x^i on both sides and dividing by $\zeta^{1/2}(a_1 \cdots a_n)$ we obtain the theorem.

338. An extension of the above theorem may be given as follows: *the determinant*

$$\begin{vmatrix} x_1^{n+r-1} & x_1^{n+r-2} & \cdots & x_1 & 1 \\ \cdot & \cdot & \cdot & \cdot & \cdot \\ x_r^{n+r-1} & x_r^{n+r-2} & \cdots & x_r & 1 \\ a_1^{n+r-1} & a_1^{n+r-2} & \cdots & a_1 & 1 \\ \cdot & \cdot & \cdot & \cdot & \cdot \\ a_n^{n+r-1} & a_n^{n+r-2} & \cdots & a_n & 1 \end{vmatrix}$$

may be expressed as the product of three factors: first $\zeta^{1/2}(x_1 \cdots x_r)$, *second* $\zeta^{1/2}(a_1 \cdots a_n)$ *and third, the product of all such quantities as*

$$f(x_i) = (x_i - a_1)(x_i - a_2) \cdots (x_i - a_3)$$
$$= x_i^n - p_1 x_i^{n-1} - \cdots + (-1)^{n-k} p_{n-k} x_i^k + \cdots .$$

Its value is therefore

$$\begin{vmatrix} x_1^{r-1} & x_1^{r-2} \cdots x_1 & 1 \\ \cdot & \cdots & \cdot \\ x_r^{r-1} & x_r^{r-2} \cdots x_r & 1 \end{vmatrix} \zeta^{1/2}(a_1 \cdots a_n) f(x_1) \cdots f(x_r).$$

If we multiply the ith row by $f(x_i)$, and then equate coefficients of $x_1^\alpha x_2^\beta x_3^\gamma \cdots$, we get the theorem:

If $D_{\alpha\beta\gamma} \cdots$ *is the determinant of order n formed by suppressing the columns containing the* α*th,* β*th,* γ*th,* $\cdots$ *powers in the array*

$$\begin{matrix} a_1^{n+r-1} & a_1^{n+r-2} \cdots a_1 & 1 \\ \cdot & \cdots & \cdot \\ a_n^{n+r-1} & a_n^{n+r-2} \cdots a_n & 1 \end{matrix} ,$$

then

$$D_{\alpha\beta\gamma\cdots} = \begin{vmatrix} H_{n-\alpha+r-1} & H_{n-\alpha+r-2} \cdots H_{n-\alpha} \\ H_{n-\alpha+r-1} & H_{n-\alpha+r-2} \cdots H_{n-\alpha} \\ \cdots & \cdots & \cdots \end{vmatrix} \zeta^{1/2}(a_1 \cdots a_n)$$

where H_i is the sum of the products i at a time of $a_1 \cdots a_n$. Whenever i is negative or greater than n, $p_i=0$, $p_0=1$.

339. *The quotient of any simple alternant by the corresponding difference-product is expressible as a determinant whose elements are complete symmetric functions of the variables, namely:*

$$\frac{A(\alpha_1^p \alpha^q \cdots \alpha_n^z)}{\zeta^{1/2}(\alpha_1\alpha_2 \cdots \alpha_n)}$$

$$= \begin{vmatrix} (\alpha_1 \cdots \alpha_n)^p & (\alpha_1 \cdots \alpha_n)^q & \cdots (\alpha_1 \cdots \alpha_n)^z \\ (\alpha_1 \cdots \alpha_n)^{p-1} & (\alpha_1 \cdots \alpha_n)^{q-1} & \cdots (\alpha_1 \cdots \alpha_n)^{z-1} \\ \cdots & \cdots & \cdots \\ (\alpha_1 \cdots \alpha_n)^{p-n+1} & (\alpha_1 \cdots \alpha_n)^{q-n+1} & \cdots (\alpha_1 \cdots \alpha_n)^{z-n+1} \end{vmatrix}.$$

Subtracting each element of the first row of the given alternant from the corresponding element of all the following rows, we see that the factors a_2-a_1, a_3-a_1, $\cdots$, a_n-a_1 may be taken out, and that this being done the resulting determinant is

$$\begin{vmatrix} \alpha_1^p & \alpha_1^q & \cdots & \alpha_1^z \\ (\alpha_1\alpha_2)^{p-1} & (\alpha_1\alpha_2)^{q-1} & \cdots & (\alpha_1\alpha_2)^{z-1} \\ \cdot & \cdot & \cdot & \cdot \\ (\alpha_1\alpha_n)^{p-1} & (\alpha_1\alpha_n)^{q-1} & \cdots & (\alpha_1\alpha_n)^{z-1} \end{vmatrix}.$$

Treating this in the same way, the elements of the second row being now the subtrahends, we can (§336 (5)) remove the factors a_3-a_2, a_4-a_2, $\cdots$, (a_n-a_2); and continuing the process we find finally

$$\frac{A(\alpha_1^p\alpha_2^q\cdots\alpha_n^z)}{\zeta^{1/2}(\alpha_1\alpha_2\cdots\alpha_n)}$$

$$= \begin{vmatrix} \alpha_1^p & \alpha_1^q & \cdots & \alpha_1^z \\ (\alpha_1\alpha_2)^{p-1} & (\alpha_1\alpha_2)^{q-1} & \cdots & (\alpha_1\alpha_2)^{z-1} \\ (\alpha_1\alpha_2\alpha_3)^{p-2} & (\alpha_1\alpha_2\alpha_3)^{q-2} & \cdots & (\alpha_1\alpha_2\alpha_3)^{z-2} \\ \cdot & \cdot & \cdot & \cdot \\ (\alpha_1\cdots\alpha_n)^{p-n+1} & (\alpha_1\cdots\alpha_n)^{q-n+1} & \cdots & (\alpha_1\cdots\alpha_n)^{z-n+1} \end{vmatrix}.$$

Multiplying columnwise the right-hand member of this by unity in the form

$$\begin{vmatrix} 1 & 0 & 0 & \cdots 0 \\ (\alpha_2\cdots\alpha_n)^1 & 1 & 0 & \cdots 0 \\ (\alpha_3\cdots\alpha_n)^2 & (\alpha_3\cdots\alpha_n)^1 & 1 & \cdots 0 \\ \cdot & \cdot & \cdot & \cdot \\ \alpha_n^{n-1} & \alpha_n^{n-2} & \alpha_n^{n-3} & \cdots 1 \end{vmatrix},$$

and using (4) of §336 we obtain the result required.

If p be other than 0 it is better to remove the factor $a_1\,a_2^p\cdots a_n^p$ before applying the theorem.

EXAMPLE:

$$\begin{vmatrix} 1 & a^3 & a^4 \\ 1 & b^3 & b^4 \\ 1 & c^3 & c^4 \end{vmatrix} = \zeta^{1/2}(abc) \begin{vmatrix} (a,b,c)^0 & (a,b,c)^3 & (a,b,c)^4 \\ 0 & (a,b,c,)^2 & (a,b,c)^3 \\ 0 & (a,b,c)^1 & (a,b,c)^2 \end{vmatrix}$$

$$= \zeta^{1/2}(abc)\begin{vmatrix} \sum a^2 + \sum ab & \sum a^3 + \sum a^2b + \sum abc \\ \sum a & \sum a^2 + \sum ab \end{vmatrix}$$

$$= \zeta^{1/2}(abc)\begin{vmatrix} -\sum ab & -\sum a^2b - 2\sum abc \\ \sum a & \sum a^2 + \sum ab \end{vmatrix}$$

$$= \zeta^{1/2}(abc)\begin{vmatrix} -\sum ab & \sum abc \\ \sum a & -\sum ab \end{vmatrix}$$

$$= \zeta^{1/2}(abc) \times \left(\sum a^2b + \sum abc \right).$$

340. A particular case of the above can be obtained by substituting $p=n$, $q=n+1$, $r=n+2$, $\cdots$, so that

$$\begin{vmatrix} H_n & H_{n+1} & H_{n+2} & \cdots \\ H_{n-1} & H_n & H_{n+1} & \cdots \\ H_{n-2} & H_{n-1} & H_n & \cdots \\ \cdot & \cdot & \cdot & \cdot \end{vmatrix} = a_1^n a_2^n a_3^n \cdots .$$

Also from the above theorem we obtain the relation

$$\begin{vmatrix} H_{\alpha+\alpha'} & H_{\alpha+\beta'} & H_{\alpha+\gamma'} & \cdots \\ H_{\beta+\alpha'} & H_{\beta+\beta'} & H_{\beta+\gamma'} & \cdots \\ H_{\gamma+\alpha'} & H_{\gamma+\beta'} & H_{\gamma+\gamma'} & \cdots \\ \cdot & \cdot & \cdot & \cdot \end{vmatrix} = \frac{(-1)^{r(r-1)/2}(abc\cdots)^{r-1}}{\zeta(abc\cdots)}$$

$$\times \begin{vmatrix} a^\alpha & a^\beta & \cdots \\ b^\alpha & b^\beta & \cdots \\ c^\alpha & c^\beta & \cdots \\ \cdot & \cdot & \cdot \end{vmatrix} \cdot \begin{vmatrix} a^{\alpha'} & a^{\beta'} & \cdots \\ b^{\alpha'} & b^{\beta'} & \cdots \\ c^{\alpha'} & c^{\beta'} & \cdots \\ \cdot & \cdot & \cdot \end{vmatrix}$$

which is due to Jacobi.

341. From equation (1) §336 we have

$$(bcd\cdots)^s = (abcd\cdots)^s - a(abcd\cdots)^{s-1}$$

or using the notation with the H's

$$H_s' = \begin{vmatrix} 1 & a \\ H_{s-1} & H_s \end{vmatrix} \quad \text{or } H_{p+1}' = \begin{vmatrix} 1 & a \\ H_p & H_{p+1} \end{vmatrix}$$

and taking the next case

$$\begin{vmatrix} H'_{p+1} & H'_{p+2} \\ H'_{q+1} & H'_{q+2} \end{vmatrix} = \begin{vmatrix} H_{p+1} - aH_p & H_{p+2} - aH_{p+1} \\ H_{q+1} - aH_q & H_{q+2} - aH_{q+1} \end{vmatrix}$$

$$= \begin{vmatrix} H_{p+1} & H_{p+2} \\ H_{q+1} & H_{q+2} \end{vmatrix} - a \begin{vmatrix} H_p & H_{p+2} \\ H_q & H_{q+2} \end{vmatrix}$$

$$+ a^2 \begin{vmatrix} H_p & H_{p+1} \\ H_q & H_{q+1} \end{vmatrix}$$

$$= \begin{vmatrix} 1 & a & a^2 \\ H_p & H_{p+1} & H_{p+2} \\ H_q & H_{q+1} & H_{q+2} \end{vmatrix}$$

The general theorem is

$$\begin{vmatrix} H'_{p+1} & H'_{p+2} & H'_{p+3} & \cdots & H'_{p+m-2} \\ H'_{q+1} & H'_{q+2} & H'_{q+3} & \cdots & H'_{q-m-2} \\ \cdot & \cdot & \cdot & \cdots & \cdot \\ H'_{z+1} & H'_{z+2} & H'_{z+3} & \cdots & H'_{z+m-2} \end{vmatrix}$$

$$= \begin{vmatrix} 1 & a & a^2 & \cdots & a^{m-2} \\ H_p & H_{p+1} & H_{p+2} & \cdots & H_{p+m-3} \\ \cdot & \cdot & \cdot & \cdots & \cdot \\ H_z & H_{z+1} & H_{z+2} & \cdots & H_{z+m-2} \end{vmatrix}$$

the number of letters $p, q, \cdots, z$ being $m-2$.

342. *If the complementary minors of all the elements of the last column of an alternant be taken, their product is divisible by some power of* $\zeta^{1/2}(abc \cdots)$.

Denote by P_m the product of complementary minors of the elements of the last column.

(a) Let us consider the alternants of the form

$$| a^0 b^1 c^2 \cdots k^{m-2} l^n |$$

the number of letters being m. Since the factor $a-b$ occurs in the cofactors of the elements $c^m, d^m, \cdots, l^m$ that is, $m-2$ times, and similarly for the other factors in the difference-product of $a, b, c, \cdots, l$, it follows that

$$P_m = [\zeta^{1/2}(a,b,c, \cdots l)]^{m-2}.$$

(b) Let us consider the alternant of the form

$$\left| a^0 b^1 c^2 \cdots h^{m-3} k^z l^n \right|$$

where the last two indices are not consecutive. From §340 and §341 the minor

$$\begin{aligned} \left| b^0 c^1 d^2 \cdots k^{m-3} l^z \right| &= \zeta^{1/2}(bcd \cdots l) H'_{z-m+2} \\ &= \zeta^{1/2}(bcd \cdots l) \begin{vmatrix} 1 & a \\ H_{z-m+1} & H_{z-m+2} \end{vmatrix}, \end{aligned}$$

with $m-1$ like results for the other minors. We have P_m in $|a^0 b^1 c^2 \cdots h^{m-3} k^z l^n| = P_m$ in $|a^0 b^1 c^2 \cdots k^{m-2} l^n|$ multiplied by the product of H'_{z-m+2} and $m-1$ like expressions with $b, c, d, \cdots, l$ respectively in the first row instead of a. Therefore

$$P_m = \zeta^{(m-2)/2}(abc \cdots l)\Pi \begin{vmatrix} 1 & a \\ H_{z-m+1} & H_{z-m+2} \end{vmatrix}$$

where Π consists of m factors from a to l inclusive.

(c) Let us consider the alternant

$$\left| a^0 b^1 c^2 \cdots z^{m-4} h^y k^z l^n \right|$$

where the last three indices are not consecutive. From §340 and §341 the minor

$$\begin{aligned} &\left| b^0 c^1 d^2 \cdots h^{m-4} k^y l^z \right| \\ &\quad = \zeta^{1/2}(bcd \cdots l) \begin{vmatrix} H'_{z-m+2} & H'_{z-m+3} \\ H_{y-m+2} & H'_{y-m+3} \end{vmatrix} \\ &\quad = \zeta^{1/2}(bcd \cdots l) \begin{vmatrix} 1 & a & a^2 \\ H_{z-m+1} & H_{z-m+2} & H_{z-m+3} \\ H_{y-m+1} & H_{y-m+2} & H_{y-m+3} \end{vmatrix} \end{aligned}$$

with $m-1$ like results for the other minors. Therefore

$$P_m = \zeta^{(m-2)/2}(abc \cdots l)\Pi \begin{vmatrix} 1 & a & a^2 \\ H_{z-m+1} & H_{z-m+2} & H_{z-m+3} \\ H_{y-m+1} & H_{y-m+2} & H_{y-m+3} \end{vmatrix}.$$

(d) Lastly let us consider the most general type of simple alternant, that is

$$\left| a^0 b^p c^q \cdots g^x h^v k^z l^n \right|.$$

It follows from the above that the minor

$$\mid b^0c^pd^q \cdots h^xk^yl^z) = \zeta^{1/2}(bcd \cdots l) \begin{vmatrix} H'_{z-m+2} & H'_{z-m+3} & \cdots & H'_{z-1} \\ H'_{y-m+2} & H'_{y-m+3} & \cdots & H'_{y-1} \\ \cdot & \cdot & \cdots & \cdot \\ H'_{p-m+2} & H'_{p-m+3} & \cdots & H'_{p-1} \end{vmatrix}$$

with $m-1$ like results for the other minors, so that

P_m *in* $| a^0b^pc^q \cdots h^yk^zl^n |$

$$= \zeta^{(m-2)/2}(abc \cdots l)\Pi \begin{vmatrix} 1 & a & \cdots & a^{m-2} \\ H_{z-m+1} & H_{z-m+2} & \cdots & H_{z-1} \\ H_{y-m+1} & H_{y-m+2} & \cdots & H_{y-1} \\ \cdot & \cdot & \cdots & \cdot \\ H_{p-m+1} & H_{p-m+2} & \cdots & H_{p-1} \end{vmatrix},$$

Π consisting of m factors from a to l inclusive.

343. Consider the well known theorem of a determinant and its adjugate

$$\begin{vmatrix} A^\alpha & A_\beta & A^\gamma & \cdots \\ B^\alpha & B^\beta & B^\gamma & \cdots \\ C^\alpha & C^\beta & C^\gamma & \cdots \\ \cdot & \cdot & \cdot & \cdots \end{vmatrix}_r = \begin{vmatrix} a^\alpha & a^\beta & a^\gamma & \cdots \\ b^\alpha & b^\beta & b^\gamma & \cdots \\ c^\alpha & c^\beta & c^\gamma & \cdots \\ \cdot & \cdot & \cdot & \cdots \end{vmatrix}^{r-1} \equiv \Delta^{r-1}$$

where A^k is the complementary minor of a^k in Δ, and multiply both sides of the equation by

$$\begin{vmatrix} a^\lambda & a^\mu & a^\nu & \cdots \\ b^\lambda & b^\mu & b^\nu & \cdots \\ c^\lambda & c^\mu & c^\nu & \cdots \\ \cdot & \cdot & \cdot & \cdots \end{vmatrix}_r$$

and denoting by $\Delta_{\lambda,\mu}$ the determinant obtained by substituting λ for μ in Δ we get

$$\begin{vmatrix} \Delta_{\lambda,\alpha} & \Delta_{\lambda,\beta} & \Delta_{\lambda,\gamma} & \cdots \\ \Delta_{\mu,\alpha} & \Delta_{\mu,\beta} & \Delta_{\mu,\gamma} & \cdots \\ \Delta_{\nu,\alpha} & \Delta_{\nu,\beta} & \Delta_{\nu,\gamma} & \cdots \\ \cdot & \cdot & \cdot & \cdots \end{vmatrix}$$

$$= \begin{vmatrix} a^\lambda & a^\mu & a^\nu \cdots \\ b^\gamma & b^\mu & b^\nu \cdots \\ c^\gamma & c^\mu & c^\nu \cdots \\ \cdots & \cdots & \cdots \end{vmatrix} \cdot \begin{vmatrix} a^\alpha & a^\beta & a^\gamma \cdots \\ b^\alpha & b^\beta & b^\gamma \cdots \\ c^\alpha & c^\beta & c^\gamma \cdots \\ \cdots & \cdots & \cdots \end{vmatrix}^{r-1}$$

Multiplying this by the similar result obtained by interchanging $\alpha, \beta, \gamma, \cdots$ with $\lambda, \mu, \nu, \cdots$ respectively, we get

$$\begin{vmatrix} \Delta_{\lambda,\alpha} & \Delta_{\lambda,\beta} & \Delta_{\lambda,\gamma} \cdots \\ \Delta_{\mu,\alpha} & \Delta_{\mu,\beta} & \Delta_{\mu,\gamma} \cdots \\ \Delta_{\nu,\alpha} & \Delta_{\nu,\beta} & \Delta_{\nu,\gamma} \cdots \\ \cdots & \cdots & \cdots \end{vmatrix} \cdot \begin{vmatrix} \Delta_{\alpha,\lambda} & \Delta_{\alpha,\mu} & \Delta_{\alpha,\nu} \cdots \\ \Delta_{\beta,\lambda} & \Delta_{\beta,\mu} & \Delta_{\beta,\nu} \cdots \\ \Delta_{\gamma,\lambda} & \Delta_{\gamma,\mu} & \Delta_{\gamma,\nu} \cdots \\ \cdots & \cdots & \cdots \end{vmatrix}$$

$$= \begin{vmatrix} a^\alpha & a^\beta & a^\gamma \cdots \\ b^\alpha & b^\beta & b^\gamma \cdots \\ c^\alpha & c^\beta & c^\gamma \cdots \\ \cdots & \cdots & \cdots \end{vmatrix}^r \begin{vmatrix} a^\lambda & a^\mu & a^\nu \cdots \\ b^\lambda & b^\mu & b^\nu \cdots \\ c^\lambda & c^\mu & c^\nu \cdots \\ \cdots & \cdots & \cdots \end{vmatrix}^r$$

$$= \frac{\zeta^r(abc\cdots)}{(-1)^{r(r-1)/2}(abc\cdots)^{r(r-1)}} \begin{vmatrix} H_{\alpha+\lambda} & H_{\alpha+\mu} & H_{\alpha+\nu} \cdots \\ H_{\beta+\lambda} & H_{\beta+\mu} & H_{\beta+\nu} \cdots \\ H_{\gamma+\lambda} & H_{\gamma+\mu} & H_{\gamma+\nu} \cdots \\ \cdots & \cdots & \cdots \end{vmatrix}^r$$

344. If we use the theorem of §202 in connection with the two alternants $A(x^f y^g z^h w^k)$ and $A(x^0 y^1 z^2 w^3)$ we get

$$A(x^0y^1z^2w^3)^3 A(x^f y^g z^h w^k)$$
$$= \big|\; |x^f y^1 z^2 w^3| \;\; |x^0 y^g z^2 w^3| \;\; |x^0 y^1 z^h w^3| \;\; |x^0 y^1 z^2 w^k| \;\big|$$

or dividing both sides by $A(x^0y^1z^2w^3)^4$ and using $A(fghk)$ for $A(x^f y^g z^h w^k)$ we get

$$A(fghk) = |\, A(f123)\;\; A(0g23)\;\; A(01h3)\;\; A(012k)\,|$$

or putting $f=0$, as we obviously may,

$$A(0ghk) = |\, A(0g23)\;\; A(01h3)\;\; A(012k)\,| \,.$$

345. An alternant in which the elements are polynomial may often be expressed in terms of a simple alternant. This is evident when we consider that the converse is true, namely, that the mere transformation of a simple alternant or the multiplication of it by a non-alter-

nant expression may lead to an alternant with polynomial elements. For example, we obtain an alternant with polynomial elements if we multiply $A(\alpha_1{}^p\alpha_2{}^q \cdots \alpha_n{}^z)$ row-wise by $|\alpha_{1n}|$; the multiplication column wise gives of course the same alternating function, but the result is not an alternant in form.

EXAMPLE.

$$\begin{vmatrix} \lambda_1 a^3 + \mu_1 a^2 + \nu_1 & \lambda_2 a^2 + \mu_2 & 1 \\ \lambda_1 b^3 + \mu_1 b^2 + \nu_1 & \lambda_2 b^2 + \mu_2 & 1 \\ \lambda_1 c^3 + \mu_1 c^2 + \nu_1 & \lambda_2 c^2 + \mu_2 & 1 \end{vmatrix} = \lambda_2 \begin{vmatrix} \lambda_1 a^3 & a^2 & 1 \\ \lambda_1 b^3 & b^2 & 1 \\ \lambda_1 c^3 & c^2 & 1 \end{vmatrix}$$

$$= -\lambda_1\lambda_2\zeta^{1/2}(abc) \begin{vmatrix} (\ )^0 & (\ )^2 & (\ )^3 \\ 0 & (\ )^1 & (\ )^2 \\ 0 & (\ )^0 & (\ )^1 \end{vmatrix}$$

$$= -\lambda_1\lambda_2\zeta^{1/2}(abc) \begin{vmatrix} \sum a & \sum a^2 + \sum ab \\ 1 & \sum a \end{vmatrix},$$

$$= -\lambda_1\lambda_2\zeta^{1/2}(abc) \times \sum ab.$$

For another example let us consider the determinant

$$\begin{vmatrix} f_1(x_1) & f_2(x_1) & \cdots & f_n(x_1) \\ f_1(x_2) & f_2(x_2) & \cdots & f_n(x_2) \\ \cdot & \cdot & \cdot & \cdot \\ f_1(x_n) & f_2(x_n) & \cdots & f_n(x_n) \end{vmatrix},$$

where $f_i(x) = a_{1i}x^{n-1} + a_{2i}x^{n-2} + \cdots + a_{ni}$, the value of which is obtained by multiplying the two determinants

$$\begin{vmatrix} a_{11} & \cdots & a_{1n} \\ \cdot & \cdot & \cdot \\ a_{n1} & \cdots & a_{nn} \end{vmatrix} \quad \text{and} \quad \begin{vmatrix} x_1{}^{n-1} & x_1{}^{n-2} & \cdots & 1 \\ \cdot & \cdot & \cdot & \cdot \\ x_n{}^{n-1} & x_n{}^{n-2} & \cdots & 1 \end{vmatrix}$$

the product being equal to $|a_{1n}|\zeta^{1/2}(x_1x_2 \cdots x_n)$.

346. *In an alternant with rational integral elements the co-factor of the difference-product is expressible as a determinant whose elements are* (1) *the coefficients in the elements of the alternant,* (2) *those symmetric functions of the variables which are linear with respect to each variable, namely:*

$$\frac{A\{f_1(\alpha_1)f_2(\alpha_2)\cdots f_n(\alpha_n)\}}{\zeta^{1/2}(\alpha_1\alpha_2\cdots\alpha_n)}$$

$$= \begin{vmatrix} c_{01} & c_{11} & c_{21} & \cdots c_{n1} & c_{n+1,1} & \cdots c_{r-1,1} & c_{r1} \\ c_{02} & c_{12} & c_{22} & \cdots c_{n2} & c_{n+1,2} & \cdots c_{r-1,2} & c_{r2} \\ \cdot & \cdot & \cdot & \cdot & \cdot & \cdot & \cdot \\ c_{0n} & c_{1n} & c_{2n} & \cdots c_{nn} & c_{n+1,n} & \cdots c_{r-1,n} & c_{rn} \\ C_n & C_{n-1} & C_{n-2} & \cdots C_0 & 0 & \cdots 0 & 0 \\ 0 & C_n & C_{n-1} & \cdots C_1 & C_0 & \cdots 0 & 0 \\ 0 & 0 & C_n & \cdots C_2 & C_1 & \cdots 0 & 0 \\ \cdot & \cdot & \cdot & \cdot & \cdot & \cdot & \cdot \\ 0 & 0 & 0 & \cdots C_{r-n} & C_{r-n-1} & \cdots C_1 & C_0 \end{vmatrix},$$

where

$$f_s(x) = c_{0s} + c_{1s}x + c_{2s}x^2 + \cdots + c_{rs}x^r,$$

and

$$C_0 = 1, \quad C_1 = -\sum\alpha_1, \quad C_2 = \sum\alpha_1\alpha_2, \quad C_3 = -\sum\alpha_1\alpha_2\alpha_3, \quad \cdots.$$

Denoting the right-hand member by Δ, and multiplying it row-wise by $\zeta^{1/2}(\alpha_1\cdots\alpha_n\omega_0\omega_1\cdots\omega_{r-n})$, we have

$$\begin{vmatrix} f_1(\alpha_1) & f_2(\alpha_1) & \cdots f_n(\alpha_1) & \phi(\alpha_1) & \alpha_1\phi(\alpha_1) & \cdots \alpha_1^{r-n}\phi(\alpha_1) \\ f_1(\alpha_2) & f_2(\alpha_2) & \cdots f_n(\alpha_2) & \phi(\alpha_2) & \alpha_2\phi(\alpha_2) & \cdots \alpha_2^{r-n}\phi(\alpha_2) \\ \cdot & \cdot & \cdot & \cdot & \cdot & \cdot \\ f_1(\alpha_n) & f_2(\alpha_n) & \cdots f_n(\alpha_n) & \phi(\alpha_n) & \alpha_n\phi(\alpha_n) & \cdots \alpha_n^{r-n}\phi(\alpha_n) \\ f_1(\omega_0) & f_2(\omega_0) & \cdots f_n(\omega_0) & \phi(\omega_0) & \omega_0\phi(\omega_0) & \cdots \omega_0^{r-n}\phi(\omega_0) \\ f_1(\omega_1) & f_2(\omega_1) & \cdots f_n(\omega_1) & \phi(\omega_1) & \omega_1\phi(\omega_1) & \cdots \omega_1^{r-n}\phi(\omega_1) \\ \cdot & \cdot & \cdot & \cdot & \cdot & \cdot \\ f_1(\omega_{r-n}) & f_2(\omega_{r-n}) & \cdots f_n(\omega_{r-n}) & \phi(\omega_{r-n}) & \omega_{r-n}\phi(\omega_{r-n}) & \cdots \omega_{r-n}^{r-n}\phi(\omega_{r-n}) \end{vmatrix},$$

where for shortness we put

$$x^n + C_1x^{n-1} + \cdots + C_n \text{ or } (x-\alpha_1)(x-\alpha_2)\cdots(x-\alpha_n) = \phi(x).$$

In this result, however, $\phi(\alpha_1) = 0 = \phi(\alpha_2) = \cdots = \phi(\alpha_n)$, hence (107) it may be retransformed into a product, namely:

$$\begin{vmatrix} f_1(\alpha_1) & f_2(\alpha_1) & \cdots f_n(\alpha_1) \\ f_1(\alpha_2) & f_2(\alpha_2) & \cdots f_n(\alpha_2) \\ & & \\ f_1(\alpha_n) & f_2(\alpha_n) & \cdots f_n(\alpha_n) \end{vmatrix} \times \begin{vmatrix} \phi(\omega_0) & \omega_0\phi(\omega_0) & \cdots \omega_0^{r-n}\phi(\omega_0) \\ \phi(\omega_1) & \omega_1\phi(\omega_1) & \cdots \omega_1^{r-n}\phi(\omega_1) \\ & & \\ \phi(\omega_{r-n}) & \omega_{r-n}\phi(\omega_{r-n}) & \cdots \omega_{r-n}^{r-n}\phi(\omega_{r-n}) \end{vmatrix}$$

and therefore (§§39, 324) into

$$A\{f_1(\alpha_1)f_2(\alpha_2)\cdots f_n(\alpha_n)\}\times\phi(\omega_0)\phi(\omega_1)\cdots\phi(\omega_{r-n})$$
$$\times\,\zeta^{1/2}(\omega_0,\omega_1\cdots\omega_{r-n}).$$

Putting now the original from which this came, namely:

$$\Delta\zeta^{1/2}(\alpha_1\cdots\alpha_n\omega_0\omega_1\cdots\omega_{r-n})$$

into the form

$$\Delta\zeta^{1/2}(\alpha_1\cdots\alpha_n)\zeta^{1/2}(\omega_0\omega_1\cdots\omega_{r-n})\phi(\omega_0)\phi(\omega_1)\cdots\phi(\omega_{r-n}),$$

and removing the common factors, we have

$$A\{f_1(\alpha_1)f_2(\alpha_2)\cdots f_n(\alpha_n)\}=\zeta^{1/2}(\alpha_1,\alpha_2,\cdots,\alpha_n)\times\Delta,$$

as was to be proved.

EXAMPLE. Taking

$$f_1(x)=\nu_1+\mu_1x^2+\lambda_1x^3,$$
$$f_2(x)=\mu_2+\lambda_2x^2,$$
$$f_3(x)=1;$$

then

$$\frac{A\{f_1(a)f_2(b)f_3(c)}{\zeta^{1/2}(a,b,c)}=\begin{vmatrix}\nu_1 & 0 & \mu_1 & \lambda_1\\ \mu_2 & 0 & \lambda_2 & 0\\ 1 & 0 & 0 & 0\\ C_3 & C_2 & C_1 & C_0\end{vmatrix}=-\lambda_1\lambda_2C_2=-\lambda_1\lambda_2\sum ab,$$

as we have already found (§345, Ex.).

347. The foregoing theorem, though evidently much more general than that of §339, does not include the latter as a particular case. There the result is obtained in terms of *complete* symmetric functions of the variables, here in terms of *single* symmetric functions. Applying the general theorem to the case considered in §339, namely, where

$$f_1(x)=x^p,\quad f_2(x)=x^q,\quad\cdots,\quad f_n(x)=x^z,$$

we see that in Δ the elements of the first n rows are all 0 except the element in the first row and $(p+1)$th column, the element in the second row and $(q+1)$th column, and so on up to and including the element in the zth row and $(z+1)$th column, the excepted elements being 1. Consequently Δ may be reduced to a determinant Δ' of the $(z+1-n)$th order with the sign-factor

$$(-1)^{p+q+\cdots+z+n(n-1)/2},$$

the columns thrown out from Δ being the $(p+1)$th, $(q+1)$th, etc., or, what is the same thing, the columns ending with C_{z-p}, C_{z-q}, etc. Hence the last row of Δ' will contain all the C's except these, and as in any of the other rows the suffix of any C is less by unity than that of the C below i t, Δ' is thus fully determined.

EXAMPLE. Taking the alternant $A(a^0b^3c^4)$, used in exemplifying §339, we see that the omitted indices are 2, 1, and that the excesses of the highest index over these are 2, 3, etc. Thus the required quotient

$$= (-1)^{3+4+3(3-1)/2} \begin{vmatrix} C_2 & C_1 \\ C_3 & C_2 \end{vmatrix}$$

$$= \begin{vmatrix} \sum ab & -\sum a \\ -\sum abc & \sum ab \end{vmatrix},$$

$$= \sum a^2b^2 + \sum a^2bc.$$

The final result is thus obtained more simply than before. The advantage, however, is not always on the same side. The determinant in the first case is of the nth order, in the second case it is of the $(z+1-n)$th order, and of course either order may be higher than the other.

348. *The product of a simple alternant and a single* symmetric function of its variables is expressible as a sum of simple alternants, whose indices are got by arranging the variables in every term of the symmetric function in the same order and adding the indices of each term to the indices of the original alternant, the first to the first, the second to the second, and so on.*

Let $A(a^pb^qc^h\cdots)$ be the alternant and $\sum a^\mu b^\nu c^\sigma\cdots$ the symmetric function, the number of the variables $a, b, c, \cdots$ being n.

From the definitions of an alternating and a symmetric function it is at once clear that their product is an alternating function. Consequently, since $a^pb^qc^h\cdots$ is here a term of the one factor, $a^\mu b^\nu c^\sigma\cdots$ a term of the other, and therefore $a^{p+\mu}b^{q+\nu}c^{h+\sigma}\cdots$ a term of the product, there must occur in the product all the other terms of this type; that is to say, the alternant $A(a^{p+\mu}b^{q+\nu}c^{h+\sigma}\cdots)$ is part of the product. Taking thus in succession all the $n!$ terms of $\sum a^\mu b^\nu c^\sigma$

* As any symmetric function is a sum of single symmetric functions we can multiply by any symmetric function.

$\cdots$ we have part of the product proved to be the sum of $n!$ alternants. But $n!$ alternants of the nth order have $(n!)^2$ terms; and the product cannot contain more than $(n!)^2$ terms, for the number in each of the two factors is $n!$; therefore the sum of the $n!$ alternants is equal to the product.

EXAMPLE.

$$
\begin{aligned}
A(a^0b^1c^2) \times \sum a^4b &= A(a^0b^1c^2)\{a^4b^1c^0 + a^4b^0c^1 + a^1b^4c^0 + a^0b^4c^1 \\
&\qquad + a^1b^0c^4 + a^0b^1c^4\}, \\
&= A(012)\{(410) + (401) + (140) + (041) \\
&\qquad + (104) + (014)\}, \quad \text{say:} \\
&= A(422) + A(413) + A(152) + A(053) \\
&\qquad + A(116) + A(026), \\
&= A(134) - A(125) - A(035) + A(026).
\end{aligned}
$$

349. *If the alternant* $|a^0b^1c^2 \cdots k^{n-2}l^{n-1}|$ *be multiplied by any symmetric function of* $a, b, c, d, \cdots$ *of the* t*th degree,* t *being not greater than* n, *one term of the product is got from the multiplicand by increasing each of its last* t *indices by* 1, *and the coefficient of this term is the same symmetric function of the roots of the equation* $x^n - x^{n-1} + x^{n-2} - \cdots + (-1)^n \cdot 1 = 0$.

The symmetric function may be expressed in terms of $\sum a, \sum ab$ $\sum abc, \cdots$, the expression being of the form

$$
C_1(\sum a)^{\alpha_1}(\sum ab)^{\beta_1}(\sum abc)^{\gamma_1} \cdots
$$
$$
+ C_2(\sum a)^{\alpha_2}(\sum ab)^{\beta_2}(\sum abc)^{\gamma_2} \cdots + \cdots
$$

where $\alpha_1 + 2\beta_1 + 3\gamma_1 + \cdots = \alpha_2 + 2\beta_2 + 3\gamma_2 + \cdots = \cdots = t$.

Now the multiplication of $|a^0b^1c^2 \cdots k^{n-2}l^{n-1}|$ by $\sum a$ raises the index of l by 1, and in the multiplication of this result by $\sum a$ one term of the product will be got by raising the index of k by 1, and so on; consequently, the multiplication by $(\sum a)^{\alpha_1}$ will give rise to one term having each of its last indices increased by 1. Similarly in the multiplication by $\sum ab$, which then follows, there must arise a term got from the multiplicand by increasing the (α_1+1)th and (α_1+2)th indices from the end by 1 each, and in multiplying this result by $\sum ab$ a term must arise which is got from the multiplicand by increasing the (α_1+3)th and (α_1+4)th indices from the end by 1 each, and so forth through the remaining multiplications, the final result necessarily containing a term having the last $\alpha_1 + 2\beta_1 + 3\gamma_1 + \cdots$

(that is, t) of its indices increased by 1, and having for coefficient C_1. For the same reason a like term must occur with the coefficient C_2 and so on; so that the aggregate of its coefficients will be $C_1+C_2+\cdots$. Now this is exactly what the expression

$$C_1(\textstyle\sum a)^{\alpha_1}(\sum ab)^{\beta_1}(\sum abc)^{\gamma_1}\cdots$$
$$+\,C_2(\textstyle\sum a)(\sum ab)(\sum abc)\cdots+\cdots$$

becomes when we put $\sum a=\sum ab=\sum abc=\cdots=1$ and this value each of these functions would have if instead of $a, b, c, \cdots$ we take the roots of the equation

$$x^n - x^{n-1} + x^{n-2} - \cdots + (-1)^n\cdot 1 = 0.$$

350. From this law of multiplication we naturally proceed to the corresponding division, namely, to another way of finding the quotient of a simple alternant by the corresponding difference-product.

EXAMPLE.

$$A(a^0b^3c^4) \div A(a^0b^1c^2) = \textstyle\sum a^2b^2 + \sum a^2bc \qquad \text{(as before)}.$$

Process of Division.

$$\begin{array}{l|l|l}
A(012) & A(034) & (022)+(112). \\
 & A(034) - A(124) & \\
 & \qquad\qquad A(124) & \\
 & \qquad\qquad A(124) &
\end{array}$$

Here, subtracting the indices in the divisor from those in the dividend, we have 0, 2, 2, which gives the first term of the quotient: then we multiply, and proceed generally in the usual way of division.

351. A determinant may evidently be an alternant with respect to two sets of variables, the interchange of any two of the one set being equivalent to an interchange of rows, and of any two of the other to an interchange of columns. Such a *double* alternant is of the form

$$\begin{vmatrix} F(\alpha_1\beta_1) & F(\alpha_1\beta_2) & \cdots \\ F(\alpha_2\beta_1) & F(\alpha_2\beta_2) & \cdots \\ \cdot & \cdot & \cdot \end{vmatrix} \quad \text{or} \quad |F(\alpha_1\beta_1), F(\alpha_2\beta_2)\cdots F(\alpha_n\beta_n)|$$

and is divisible when the elements are integral and rational by $\zeta^{1/2}(\alpha_1\alpha_2\cdots\alpha_n)\zeta^{1/2}(\beta_1\beta_2\cdots\beta_n)$. An example is readily obtained as in §345 by multiplying row-wise the three determinants $|c_{0,n-1}|$ $\zeta^{1/2}(\alpha_1 \cdots \alpha_n)\zeta^{1/2}(\beta_1 \cdots \beta_n)$.

EXAMPLE.

$$\begin{vmatrix} c_{00}+c_{10}\alpha_1+c_{01}\beta_1+c_{11}\alpha_1\beta_1 & c_{00}+c_{10}\alpha_1+c_{01}\beta_2+c_{11}\alpha_1\beta_2 \\ c_{00}+c_{10}\alpha_2+c_{01}\beta_1+c_{11}\alpha_2\beta_1 & c_{00}+c_{10}\alpha_2+c_{01}\beta_2+c_{11}\alpha_2\beta_2 \end{vmatrix}$$

$$= \begin{vmatrix} c_{00} & c_{10} \\ c_{01} & c_{11} \end{vmatrix} \begin{vmatrix} 1 & \alpha_1 \\ 1 & \alpha_2 \end{vmatrix} \begin{vmatrix} 1 & \beta_1 \\ 1 & \beta_2 \end{vmatrix},$$

or

$$\begin{vmatrix} \sum c_{\kappa\lambda}\alpha_1{}^\kappa\beta_1{}^\lambda & \sum c_{\kappa\lambda}\alpha_1{}^\kappa\beta_2{}^\lambda \\ \sum c_{\kappa\lambda}\alpha_2{}^\kappa\beta_1{}^\lambda & \sum c_{\kappa\lambda}\alpha_2{}^\kappa\beta_2{}^\lambda \end{vmatrix}_{\kappa=0,1}^{\lambda=0,1} = |c_{01}|\,\zeta^{1/2}(\alpha_1\alpha_2)\zeta^{1/2}(\beta_1\beta_2).$$

352. *If in a double alternant the function be rational and integral with respect to both variables, the co-factor of the two difference-products is expressible as a determinant whose elements are the coefficients in the elements of the alternant, those symmetric functions of the first set of variables which are linear in respect to each of them, and the same symmetric functions of the second set, namely*:

$$\frac{|F(\alpha_1\beta_1) \cdots F(\alpha_n\beta_n)|}{\zeta^{1/2}(\alpha_1 \cdots \alpha_n)\zeta^{1/2}(\beta_1 \cdots \beta_n)}$$

$$= (-1)^{r+n+1} \begin{vmatrix} c_{00} & c_{10} & \cdots & c_{n0} & c_{n+1,0} & \cdots & c_{r0} & C_n' & 0 & 0 & \cdots & 0 \\ c_{01} & c_{11} & \cdots & c_{n1} & c_{n+1,1} & \cdots & c_{r1} & C_{n-1}' & C_n' & 0 & \cdots & 0 \\ \cdot & \cdot & \cdot & \cdot & \cdot & \cdot & \cdot & \cdot & \cdot & \cdot & \cdot & \cdot \\ c_{0n} & c_{1n} & \cdots & c_{nn} & c_{n+1,n} & \cdots & c_{rn} & C_0' & C_1' & C_2' & \cdots & 0 \\ \cdot & \cdot & \cdot & \cdot & \cdot & \cdot & \cdot & \cdot & \cdot & \cdot & \cdot & \cdot \\ c_{0r} & c_{1r} & \cdots & c_{nr} & c_{n+1,r} & \cdots & c_{rr} & 0 & 0 & 0 & \cdots & C_0' \\ C_n & C_{-1} & \cdots & C_0 & 0 & \cdots & 0 & 0 & 0 & 0 & \cdots & 0 \\ 0 & C_n & \cdots & C_1 & C_0 & \cdots & 0 & 0 & 0 & 0 & \cdots & 0 \\ 0 & 0 & \cdots & C_2 & C_1 & \cdots & 0 & 0 & 0 & 0 & \cdots & 0 \\ \cdot & \cdot & \cdot & \cdot & \cdot & \cdot & \cdot & \cdot & \cdot & \cdot & \cdot & \cdot \\ 0 & 0 & \cdots & 0 & C_{r-n+1} & \cdots & C_0 & 0 & 0 & 0 & \cdots & 0 \end{vmatrix},$$

where

$$F(xy) = \sum(c_{\kappa\lambda}x^\kappa y^\lambda) \quad (\kappa = 0,1,\cdots,n \quad \lambda = 0,1,\cdots,n)$$

$$C_0 = 1, \quad C_1 = -\sum\alpha_1, \quad C_2 = \sum\alpha_1\alpha_2, \cdots$$

and

$$C_0' = 1, \quad C_1' = -\sum\beta_1, \quad C_2' = \sum\beta_1\beta_2, \cdots$$

The proof of this is exactly similar to that in §346, the starting point being the multiplication of the right-hand member by the *two* determinants

$$\zeta^{1/2}(\alpha_1\alpha_2 \cdots \alpha_n\omega_0\omega_1 \cdots \omega_{r-n}), \; \zeta^{1/2}(\beta_1\beta_2 \cdots \beta_\nu\pi_0\pi_1 \cdots \pi_{r-n})$$

in succession.

353. Let us consider the double alternant

$$\begin{vmatrix} (x_1 - \alpha_1)^{-1} & (x_1 - \alpha_2)^{-1} \cdots & (x_1 - \alpha_n)^{-1} \\ (x_2 - \alpha_1)^{-1} & (x_2 - \alpha_2)^{-1} \cdots & (x_2 - \alpha_n)^{-1} \\ \cdot & \cdot & \cdot \\ (x_n - \alpha_1)^{-1} & (x_n - \alpha_2)^{-1} \cdots & (x_n - \alpha_n)^{-1} \end{vmatrix} \equiv D_{n;-1}, \text{ say*},$$

Multiplying the ith row by

$$f(x_i) = u_i = (x_i - \alpha_i)(x_i - \alpha_2) \cdots (x_i - \alpha_n)$$

we get

$$u_1u_2 \cdots u_nD_{n;-1} = \left|\frac{u_i}{x_i - \alpha_k}\right|.$$

But by §351 the double alternant on the right is divisible by

$$\zeta^{1/2}(x_1x_2 \cdots x_n)\zeta^{1/2}(\alpha_1\alpha_2 \cdots \alpha_n)$$

and the cofactor is numerical only, as both thc double alternant and the product of the difference-product have the same order.

To determine the value of the cofactor we put $x_1, x_2, \cdots, x_n$ equal to $\alpha_1, \alpha_2, \cdots, \alpha_n$ respectively; then all the elements vanish except those in the leading diagonal and if $x_i = \alpha_i$ we have

$$\frac{u_i}{x_i - \alpha_i} = (-1)^{i-1}(\alpha_1 - \alpha_i)(\alpha_2 - \alpha_i) \cdots (\alpha_{i-1} - \alpha_i)(\alpha_i - \alpha_{i+1}) \cdots (\alpha_i - \alpha_n)$$

and

$$\left|\frac{u_i}{x_i - \alpha_i}\right| = (-1)^{n(n-1)/2}\zeta(\alpha_1\alpha_2 \cdots \alpha_n)$$

* The general notation being $D_{n;p} = |(\alpha_1+\beta_1)^p(\alpha_2+\beta_2)^p \cdots (\alpha_n+\beta_n)^p|$.

and therefore

$$D_{n;-1} = \frac{(-1)^{n(n-1)/2}\zeta^{1/2}(y_1x_2 \cdots x_n)\zeta^{1/2}(\alpha_1\alpha_2 \cdots \alpha_n)}{u_1u_2 \cdots, u_n}.$$

354. Let us now consider the quotient resulting from dividing a double alternant of the form

$$\left| (\alpha_1 + \beta_1)^p(\alpha_2 + \beta_2)^p \cdots (\alpha_n + \beta_n)^p \right| \equiv D_{n;p}$$

by the difference-product of the α's and the difference-product of the β's.

(a) Let us take the case of $D_{n;n}$.

If we multiply row-by-row we have, starting with $n=3$,

$$\begin{vmatrix} \alpha_1^3 & 3\alpha_1^2 & 3\alpha_1 & 1 \\ \alpha_2^3 & 3\alpha_2^2 & 3\alpha_2 & 1 \\ \alpha_3^3 & 3\alpha_3^2 & 3\alpha_2 & 1 \\ -\beta_1\beta_2\beta_3 & \sum\beta_1\beta_2 & -\sum\beta_1 & 1 \end{vmatrix} \cdot \begin{vmatrix} 1 & \beta_1 & \beta_1^2 & \beta_1^3 \\ 1 & \beta_2 & \beta_2^2 & \beta_2^3 \\ 1 & \beta_3 & \beta_3^2 & \beta_3^3 \\ 1 & x & x^2 & x^3 \end{vmatrix} =$$

$$\begin{vmatrix} (\alpha_1+\beta_1)^3 & (\alpha_1+\beta_2)^3 & (\alpha_1+\beta_3)^3 & (\alpha_1+x)^3 \\ (\alpha_2+\beta_1)^3 & (\alpha_2+\beta_2)^3 & (\alpha_2+\beta_3)^3 & (\alpha_2+x)^3 \\ (\alpha_3+\beta_1)^3 & (\alpha_3+\beta_2)^3 & (\alpha_3+\beta_3)^3 & (\alpha_3+x)^3 \\ \cdot & \cdot & \cdot & (x-\beta_1)(x-\beta_2)(x-\beta_3) \end{vmatrix}$$

and dividing both sides of this by $(x-\beta_1)(x-\beta_2)(x-\beta_3)$ there results

$$D_{3;3} = \begin{vmatrix} \alpha_1^3 & 3\alpha_1^2 & 3\alpha_1 & 1 \\ \alpha_2^3 & 3\alpha_2^2 & 3\alpha_2 & 1 \\ \alpha_3^3 & 3\alpha_3^2 & 3\alpha_3 & 1 \\ -\beta_1\beta_2\beta_3 & \sum\beta_1\beta_2 & -\sum\beta_1 & 1 \end{vmatrix} \zeta^{1/2}(\beta_1\beta_2\beta_3)$$

Removing the factors $(\alpha_3-\alpha_2)$, $(\alpha_3-\alpha_1)$, $(\alpha_2-\alpha_1)$, we have

$$D_{3;3} = \begin{vmatrix} \alpha_1^3 & 3\alpha_1^2 & 3\alpha_1 & 1 \\ \alpha_2^2+\alpha_2\alpha_1+\alpha_1^2 & 3(\alpha_1+\alpha_2) & 3 & \cdot \\ \alpha_3+\alpha_2+\alpha_1 & 3 & \cdot & \cdot \\ -\beta_1\beta_2\beta_3 & \sum\beta_1\beta_2 & -\sum\beta_1 & 1 \end{vmatrix}$$
$$\times\, \zeta^{1/2}(\alpha_1\alpha_2\alpha_3)\zeta^{1/2}(\beta_1\beta_2\beta_3),$$

and performing the operations $\text{row}_1 - \alpha_1\text{row}_2 + \alpha_1\alpha_2\text{row}_3$; $\text{row}_2 - (\alpha_1 + \alpha_2)\text{row}_3$, upon this we get for the final result

$$\frac{D_{3;3}}{\zeta^{1/2}(\alpha_1\alpha_2\alpha_3)\zeta^{1/2}(\beta_1\beta_2\beta_3)} = \begin{vmatrix} \alpha_1\alpha_2\alpha_3 & \cdot & \cdot & 1 \\ -\sum\alpha_1\alpha_2 & \cdot & 3 & \cdot \\ \sum\alpha_1 & 3 & \cdot & \cdot \\ -\beta_1\beta_2\beta_3 & \sum\beta_1\beta_2 & -\sum\beta_1 & 1 \end{vmatrix}$$

$$= \begin{vmatrix} \alpha_1\alpha_2\alpha_3 + \beta_1\beta_2\beta_3 & \sum\beta_1\beta_2 & \sum\beta_1 \\ \sum\alpha_1\alpha_2 & \cdot & -3 \\ \sum\alpha_1 & -3 & \cdot \end{vmatrix}$$

Proceeding in exactly similar manner we obtain

$$\frac{D_{4;4}}{\zeta^{1/2}(\alpha_1\alpha_2\alpha_3\alpha_4)\zeta^{1/2}(\beta_1\beta_2\beta_3\beta_4)}$$

$$= \begin{vmatrix} \alpha_1\alpha_2\alpha_3\alpha_4 + \beta_1\beta_2\beta_3\beta_4 & \sum\beta_1\beta_2\beta_3 & \sum\beta_1\beta_2 & \sum\beta_1 \\ \sum\alpha_1\alpha_2\alpha_3 & \cdot & \cdot & -4 \\ \sum\alpha_1\alpha_2 & \cdot & -6 & \cdot \\ \sum\alpha_1 & -4 & \cdot & \cdot \end{vmatrix}$$

and so on, in general.*

The quotient obtained is invariant to the interchange of any two α's or any two β's, also to the simultaneous interchange of every α with the corresponding β.

(b) Let us take the case of $D_{n;n+1}$.

By the multiplication theorem we obtain

$$\begin{vmatrix} \alpha_1^3 & 3\alpha_1^2 & 3\alpha_1 & 1 \\ \alpha_2^3 & 3\alpha_2^2 & 3\alpha_2 & 1 \\ \beta_1\beta_2 & -\sum\beta_1 & 1 & \cdot \\ \cdot & \beta_1\beta_2 & -\sum\beta_1 & 1 \end{vmatrix} \begin{vmatrix} 1 & \beta_1 & \beta_1^2 & \beta_1^3 \\ 1 & \beta_2 & \beta_2^2 & \beta_2^3 \\ 1 & x & x^2 & x^3 \\ 1 & y & y^2 & y^3 \end{vmatrix} =$$

$$\begin{vmatrix} (\alpha_1 + \beta_1)^2 & (\alpha_1 + \beta_2)^2 & (\alpha_1 + x_2)^2 & (\alpha_1 + y)^3 \\ (\alpha_2 + \beta_1)^3 & (\alpha_2 + \beta_2)^3 & (\alpha_2 + x)^3 & (\alpha_2 + y)^3 \\ \cdot & \cdot & (x - \beta_1)(x - \beta_2) & (y - \beta_1)(y - \beta_2) \\ \cdot & \cdot & x(x - \beta_1)(x - \beta_2) & y(y - \beta_1)(y - \beta_2) \end{vmatrix}$$

* Compare result with ex. 39. Set XVIII

and dividing both sides by

$$(y - x)(y - \beta_2)(y - \beta_1)(x - \beta_2)(x - \beta_1)$$

we obtain

$$D_{2;3} = \begin{vmatrix} \alpha_1^3 & 3\alpha_1^2 & 3\alpha_1 & 1 \\ \alpha_2^3 & 3\alpha_2^2 & 3\alpha_2 & 1 \\ \beta_1\beta_2 & -\sum\beta_1 & 1 & \cdot \\ \cdot & \beta_1\beta_2 & -\sum\beta_1 & 1 \end{vmatrix} \cdot \zeta^{1/2}(\beta_1\beta_2)$$

Taking the first factor of the right side, removing $\alpha_2-\alpha_1$ and simplifying, we have

$$(\alpha_2 - \alpha_1)\begin{vmatrix} -\alpha_1\alpha_2\sum\alpha_1 & -3\alpha_1\alpha_2 & \cdot & 1 \\ \alpha_2^2 + \alpha_2\alpha_1 + \alpha_1^2 & 3\sum\alpha_1 & 3 & \cdot \\ \beta_1\beta_2 & -\sum\beta_1 & 1 & \cdot \\ \cdot & \beta_1\beta_2 & -\sum\beta_1 & 1 \end{vmatrix}$$

and by performing on this the operations $\text{col}_4\cdot\alpha_1\alpha_2$, $\text{row}_4\div\alpha_1\alpha_2$, $\text{row}_2+\sum\alpha_1\cdot\text{row}_1$ it takes the form

$$\begin{vmatrix} -\sum\alpha_1 & -3 & \cdot & \cdot \\ -\alpha_1\alpha_2 & \cdot & 3 & \sum\alpha_1 \\ \beta_1\beta_2 & -\sum\beta_1 & 1 & \cdot \\ \cdot & \beta_1\beta_2 & -\sum\beta_1 & \alpha_1\alpha_2 \end{vmatrix}(\alpha_2 - \alpha_1).$$

Thus we have

$$\frac{D_{2;3}}{\zeta^{1/2}(\alpha_1\alpha_2)\zeta^{1/2}(\beta_1\beta_2)} = -\begin{vmatrix} \beta_1\beta_2 & -\sum\beta_1 & 1 & \cdot \\ -\sum\alpha_1 & -3 & \cdot & 1 \\ \alpha_1\alpha_2 & \cdot & -3 & -\sum\alpha_1 \\ \cdot & \beta_1\beta_2 & -\sum\beta_1 & \alpha_1\alpha_2 \end{vmatrix}$$

The quotient* is invariant as stated in case (a).

355. If D_{ik} is the complementary minor of $(x_i-\alpha_k)^{-1}$ in $D_{n;-1}$ then D_{ik} is equal to the expression obtained by omitting x_i and α_k on the right, and multiplying by $(-1)^{i+k}$. That is

$$D_{ik} = (-1)^{i+k}$$
$$\times\frac{(-1)^{n(n-1)/2}\zeta^{1/2}(x_1\cdots x_{i-1}x_{i+1}\cdots x_n)\zeta^{1/2}(\alpha_1\cdots\alpha_{k-1}\alpha_{k+1}\cdots\alpha_n)}{w_1w_2\cdots w_{n-1}}$$

* For further extension see ex. 47, 48, 49.

where

$$w_1 w_2 \cdots w_{n-1}$$

$$= \frac{u_1}{x_1 - \alpha_k} \frac{u_2}{x_2 - \alpha_k} \cdots \frac{u_{i-1}}{x_{i-1} - \alpha_k} \frac{u_{i-1}}{x_{i+1} - \alpha_k} \cdots \frac{u_n}{x_n - \alpha_k}.$$

Now if we write

$$g(y) = (y - x_1)(y - x_2) \cdots (y - x_n),$$

$$f(x) = (x - \alpha_1)(x - \alpha_2) \cdots (x - \alpha_n),$$

then

$$\zeta^{1/2}(x_1 \cdots x_{i-1}x_{i+1} \cdots x_n)\zeta^{1/2}(\alpha_1 \cdots \alpha_{k-1}\alpha_{k+1} \cdots \alpha_n)$$

$$= \frac{(-1)^{i+k}\zeta^{1/2}(x_1 \cdots x_n)\zeta^{1/2}(\alpha_1 \cdots \alpha_n)}{g'(x_i)f'(\alpha_k)}$$

and

$$(x_1 - \alpha_k)(x_2 - \alpha_k) \cdots (x_{i-1} - \alpha_k)(x_{i+1} - \alpha_k) \cdots (x_n - \alpha_k)$$

$$= (-1)^n \frac{g(\alpha_k)}{x_i - \alpha_k}$$

hence

$$\frac{D_{ik}}{D_{n;-1}} = (-)^n \frac{f(x_i)g(\alpha_k)}{f'(\alpha_k)g'(x_1)} \cdot \frac{1}{x_i - \alpha_k}.$$

356. When the elements of an alternant are fractions with the variables occurring in the denominators, it will generally be found suitable to clear of fractions and then apply one of the theorems already given. When the elements are transcendental functions in form of sin, cos, $\cdots$, sinh, cosh, $\cdots$, the expression in their exponential form are substituted. In alternants of this kind the product of the sines of the halved differences of the variables often makes its appearance as a factor.

Exercises. Set XVIII

1. Prove that

$$\begin{vmatrix} 1 & x_2 + x_3 & x_2x_3 \\ 1 & x_3 + x_1 & x_1x_3 \\ 1 & x_1 + x_2 & x_1x_2 \end{vmatrix} = -\zeta^{1/2}(x_1x_2x_3)$$

2. Prove that if $F_k(x)=x^k+a_{1,k}x^{k-1}+a_{2,k}x^{k-2}+\cdots+a_{k,k}$

$$\begin{vmatrix} 1 & 1 & \cdots & 1 \\ F_1(x_1) & F_1(x_2) & \cdots & F_1(x_n) \\ F_2(x_1) & F_2(x_2) & \cdots & F_2(x_n) \\ \cdots & \cdots & \cdots & \cdots \\ F_{n-1}(x_1) & F_{n-1}(x_2) & \cdots & F_{n-1}(x_n) \end{vmatrix} = \zeta^{1/2}(x_1, x_2, \cdots, x_n).$$

3. Prove that if $F_k(x_1, x_r, \cdots, x_r)=(x_1, x_2, \cdots, x_r)_k+a_kF_{k-1}(x_1, x_2, \cdots, x_r)$ then

$$\begin{vmatrix} 1 & 1 & \cdots & 1 \\ F_1(x_1 \cdots x_i x_{i+1}) & F_1(x_1 \cdots x_i x_{i+2}) & \cdots & F_1(x_1 \cdots x_i x_n) \\ F_2(x_1 \cdots x_i x_{i+1}) & F_2(x_1 \cdots x_i x_{i+2}) & \cdots & F_2(x_1 \cdots x_i x_n) \\ \cdots & \cdots & \cdots & \cdots \\ F_{n-i-1}(x_1 \cdots x_i x_{i+1}) & F_{n-i-1}(x_1 \cdots x_i x_{i+2}) & \cdots & F_{n-i-1}(x_1 \cdots x_i x_n) \end{vmatrix}$$
$$= \zeta^{1/2}(x_{i+1}, x_{i+2}, \cdots, x_n).$$

4. Prove:

$$\begin{vmatrix} 1 & x_1 & x_1^2 & \cdots & x_1^{n-2} & x_2x_3 \cdots x_n \\ 1 & x_2 & x_2^2 & \cdots & x_2^{n-2} & x_1x_3 \cdots x_n \\ \cdots & \cdots & \cdots & \cdots & \cdots & \cdots \\ 1 & x_n & x_n^2 & \cdots & x_n^{n-2} & x_1x_2 \cdots x_{n-1} \end{vmatrix} = (-1)^{n-1}\zeta^{1/2}(x_1x_2 \cdots x_n).$$

5. Prove:

$$\begin{vmatrix} 1 & x_1 & x_1^2 & \cdots & (x_2+x_3+\cdots+x_n)^{n-1} \\ 1 & x_2 & x_2^2 & \cdots & (x_1+x_3+\cdots+x_n)^{n-1} \\ \cdots & \cdots & \cdots & \cdots & \cdots \\ 1 & x_n & x_n^2 & \cdots & (x_1+x_2+\cdots+x_{n-1})^{n-1} \end{vmatrix}$$
$$=(-1)^{n-1}\zeta^{1/2}(x_1x_2 \cdots x_n)$$

6. Prove:

$$\begin{vmatrix} 1 & a \cdots a^{n-2} & \dfrac{a^{n-1}(bcd \cdots l)_r^{n-1}}{\pi(cde \cdots 1)_r} \\ 1 & b \cdots b^{n-2} & \dfrac{b^{n-1}(acd \cdots l)_r^{n-1}}{\pi(cde \cdots l)_r} \\ \cdots & \cdots & \cdots \end{vmatrix} = \zeta^{1/2}(abc \cdots l)$$

and

$$\begin{vmatrix} 1 & a & \cdots & a^{n-2} & \dfrac{(bcd\cdots 1)_{r+1}^{n-1}}{\pi(cde\cdots 1)_r} \\ 1 & b & \cdots & b^{n-2} & \dfrac{(acd\cdots 1)_{r+1}^{n-1}}{\pi(cde\cdots 1)_r} \\ \cdot & \cdot & \cdot & \cdot & \cdot \end{vmatrix} = \pm\,\zeta^{1/2}(abc\cdots l)$$

where $r=0, 1, 2, \cdots, n-2$; and π consists of the $n-1$ factors of $(cd\cdots l)_r(bde\cdots)_r\cdots(bc\cdots k)_r$.

Find the expansions for

7. (a) $|a^0b^3c^5| \div |a^0b^1c^2|$. (c) $|a^0b^4c^5| \div |a^0b^1c^2|$.
   (b) $|a^0b^2c^6| \div |a^0b^1c^2|$. (d) $|a^0b^1c^5d^6| \div |a^0b^1c^2d^3|$.

where $|a^mb^nc^p|$ is written for $A(a^mb^nc^p)$.

8. Express $|a^0b^mc^nd^p|\sum a$ as a sum of alternants.

Prove the following identities, and generalize them in such a way as to leave the right-hand members unaltered in form:

9. $|a^0b^1c^3d^4| \div |a^0b^1c^2d^3| = \sum ab,$

10. $|a^0b^1c^3d^5| \div |a^0b^1c^2d^3| = \sum a^2b + 2\sum abc.$

11. $|a^0b^2c^3d^5| \div |a^0b^1c^2d^3| = \sum a^2bc + 3\sum abcd.$

12. $|a^0b^1c^4d^5| \div |a^0b^1c^2d^3| = \sum a^2b^2 + \sum a^2bc + 2\sum abcd.$

13. $|a^0b^1c^3d^6| \div |a^0b^1c^2d^3| = \sum a^3b + \sum a^2b^2 + 2\sum a^2bc + 3\sum abcd.$

14. Prove that

$$|a^0b^mc^n| - |a^0b^mc^{n-1}|\sum a + a^0b^m|c^{n-2}|\sum ab - |a^0b^mc^{n-3}|\,abc = 0$$

15. Prove that

$$|a^0b^1c^2d^7| \div |a^0b^1c^2d^3| = \begin{vmatrix} \sum a & -\sum a^2 & \sum a^3 & -\sum a^4 \\ 1 & \sum a & -\sum a^2 & \sum a^3 \\ 0 & 2 & \sum a & -\sum a^2 \\ 0 & 0 & 3 & \sum a \end{vmatrix}.$$

16. Express $\{3|a^0b^1c^2d^5| - 5|a^0b^1c^3d^4|\} \div |a^0b^1c^2d^3|$ as a sum of second powers.

17. Prove that

$$\begin{vmatrix} \cos\frac{1}{2}(a-b) & \cos\frac{1}{2}(b-c) & \cos\frac{1}{2}(c-a) \\ \cos\frac{1}{2}(a+b) & \cos\frac{1}{2}(b+c) & \cos\frac{1}{2}(c+a) \\ \sin\frac{1}{2}(a+b) & \sin\frac{1}{2}(b+c) & \sin\frac{1}{2}(c+a) \end{vmatrix}$$
$$= 2\sin\tfrac{1}{2}(a-b)\sin\tfrac{1}{2}(b-c)\sin\tfrac{1}{2}(c-a).$$

18. Show that $\{\,|a^0b^1c^2d^6|+2\,|a^0b^1c^3d^5|+|a^0b^2c^3d^4|\,\}\div|a^0b^1c^2d^3|$ is expressible as a third power.

19. Prove

$$\begin{vmatrix} H_{\alpha+\alpha'} & H_{\alpha+\beta'} & H_{\alpha+\gamma'} & \cdots \\ H_{\beta+\alpha'} & H_{\beta+\beta'} & H_{\beta+\gamma'} & \cdots \\ H_{\gamma+\alpha'} & H_{\gamma+\beta'} & H_{\gamma+\gamma'} & \cdots \\ \cdots & \cdots & \cdots & \cdots \end{vmatrix} = \frac{(-1)^{r(r-1)/2}(abc\cdots)^{r-1}}{\zeta(abc\cdots)}$$
$$\times \begin{vmatrix} S_{\alpha+\alpha'} & S_{\alpha+\beta'} & S_{\alpha+\gamma'} & \cdots \\ S_{\beta+\alpha'} & S_{\beta+\beta'} & S_{\beta+\gamma'} & \cdots \\ S_{\gamma+\alpha'} & S_{\gamma+\beta'} & S_{\gamma+\gamma'} & \cdots \\ \cdots & \cdots & \cdots & \cdots \end{vmatrix}$$

where $S_p = a^p+b^p+c^p+\cdots$. (See §340.)

20. Prove

$$|\,a^0b^1c^rd^{r+1}e^sf^{s+1}\,| = \sum \zeta^{1/2}(a,b)\zeta^{1/2}(c,d)\zeta^{1/2}(e,f)\cdot\begin{vmatrix} 1 & a^rb^r & a^sb^s \\ 1 & c^rd^r & c^sd^s \\ 1 & e^rf^r & e^sf^s \end{vmatrix}$$

where $\zeta^{1/2}(a,b)\zeta^{1/2}(c,d)\zeta^{1/2}(e,f)$ is any of the 15 terms of the Pfaffian

$$\begin{vmatrix} b-a & c-a & d-a & e-a & f-a \\ & c-b & d-b & e-b & f-b \\ & & d-c & e-c & f-c \\ & & & e-d & f-d \\ & & & & f-e \end{vmatrix}$$

21. Prove

$$\frac{\zeta^{1/2}(abcdefg)\zeta^{1/2}(efg)}{\zeta^{1/2}(abcd)} = (-1)^3 f'(e)f'(f)f'(g)$$

where $f'(e) = (e-a)(e-b)(e-c)(e-d)(e-f)(e-g)$.

22. Prove the identities

$$\{ | a^0b^2c^4 | + | a^0b^1c^5 | \} | a^0b^1c^2 | = | a^0b^1c^3 | \; | a^0b^1c^4 | ,$$

$$\{ | a^0b^3c^5 | + | a^0b^2c^6 | + | a^0b^1c^7 | \} | a^0b^1c^2 | = | a^0b^1c^4 | \; | a^0b^1c^5 | ;$$

and find a general identity including them.

23. Prove that if $f_r(a) = a_r + B_r a_{r-1} + \cdots + Z_r$

$$\begin{vmatrix} 1 & f_1(a) & \cdots & f_{n-1}(a) \\ 1 & f_1(b) & \cdots & f_{n-1}(b) \\ \cdot & \cdot & \cdot & \cdot \\ 1 & f_1(l) & \cdots & f_{n-1}(l) \end{vmatrix} = \zeta^{1/2}(ab \cdots l).$$

If the coefficient of a^r in $f_r(a)$ were A_r, what would the right-hand member require to be?

24. Show that the coefficient of x^{n+r} in the development

$$\frac{(s_1x - 1)^r}{(1 - ax)(1 - bx)(1 - cx)},$$

where $s_1 = a+b+c$ is

$$\frac{\begin{vmatrix} a^{n+2}(b+c)^r & a & 1 \\ b^{n+2}(c+a)^r & b & 1 \\ c^{n+2}(a+b)^r & c & 1 \end{vmatrix}}{\zeta^{1/2}(abc)}.$$

25. Show that the coefficient of x^{n+2r} in the development of

$$\frac{1 + (bc + ab + ca)x^2}{(1 - a^2x^2)(1 - b^2x^2)(1 - c^2x^2)}(s_2x^2 - 1)^r,$$

where $s_2 = a^2+b^2+c^2$, is

$$\frac{\begin{vmatrix} a^{n+2}(b^2+c^2)^r & a & 1 \\ b^{n+2}(c^2+a^2)^r & b & 1 \\ c^{n+2}(a^2+b^2)^r & c & 1 \end{vmatrix}}{\zeta^{1/2}(abc)}$$

26. Prove that

$$\left(\frac{\partial}{\partial a} + \frac{\partial}{\partial b} + \frac{\partial}{\partial c} + \frac{\partial}{\partial d}\right) | a^m b^n c^p d^q | = m | a^{m-1} b^n c^p d^q |$$

$$+ n | a^m b^{n-1} c^p d^q | + p | a^m b^n c^{p-1} d^q | + q | a^m b^n c^p d^{q-1} | .$$

For what values of m, n, p, q does this vanish, supposing $m < n < p < q$?

27. Find the co-factor of $\zeta^{1/2}(x_1 \cdots x_n)\zeta^{1/2}(y_1 \cdots y_n)$ in

$$\begin{vmatrix} (x_1-y_1)^{n-1} & (x_1-y_2)^{n-1} & \cdots & (x_1-y_n)^{n-1} \\ (x_2-y_1)^{n-1} & (x_2-y_2)^{n-1} & \cdots & (x_2-y_n)^{n-1} \\ \cdot & \cdot & \cdot & \cdot \\ (x_n-y_1)^{n-1} & (x_n-y_2)^{n-1} & \cdots & (x_n-y_n)^{n-1} \end{vmatrix}$$

28. Prove that

$$\left(\frac{\partial}{\partial a}+\frac{\partial}{\partial b}+\frac{\partial}{\partial c}+\frac{\partial}{\partial d}\right)|\, a^0b^1c^3d^4\,|$$
$$=\left(a^2\frac{\partial}{\partial a}+b^2\frac{\partial}{\partial b}+c^2\frac{\partial}{\partial c}+d^2\frac{\partial}{\partial d}\right)|\, a^0b^1c^2d^3\,| \;.$$

29. Prove that

$$|\cos n\alpha_0,\ \cos(n-1)\alpha_1,\ \cdots,\ \cos(0)\alpha_n|$$
$$\div\,|\cos^n\alpha_0,\ \cos^{n-1}\alpha_1,\ \cdots,\ \cos^0\alpha_n| = 2^{n(n-1)/2}.$$

30. For what value of ω does

$$\left(\frac{\partial}{\partial a}+\frac{\partial}{\partial b}+\frac{\partial}{\partial c}+\frac{\partial}{\partial d}\right)^{\omega}|\, a^mb^nc^pd^q\,|$$

vanish?

31. Prove that

$$|\sin(n+1)\alpha_0,\ \sin n\,\alpha_1\ \cdots,\ \sin\alpha_n| \div |\cos^n\alpha_0,\ \cos^{n-1}\alpha_1,\ \cdots,\ \cos^0\alpha_n|$$
$$= 2^{n(n-1)/2}\sin\alpha_0\sin\alpha_1\cdots\sin\alpha_n\,.$$

32. Prove that

$$\begin{vmatrix} |a^mb^1c^2d^3| & |a^nb^1c^2d^3| & |a^pb^1c^2d^3| & |a^qb^1c^2d^3| \\ |a^0b^mc^2d^3| & |a^0b^nc^2d^3| & |a^0b^pc^2d^3| & |a^0b^qc^2d^3| \\ |a^0b^1c^md^3| & |a^0b^1c^nd^3| & |a^0b^1c^pd^3| & |a^0b^1c^qd^3| \\ |a^0b^1c^2d^m| & |a^0b^1c^2d^n| & |a^0b^1c^2d^p| & |a^0b^1c^2d^q| \end{vmatrix}$$
$$= |\, a^0b^1c^2d^3\,|^3\, |\, a^mb^nc^pd^q\,|\;.$$

33. Find the co-factor of $\zeta^{1/2}(x_1 \cdots x_n)\zeta^{1/2}(y_1 \cdots y_n)$ in

$$\begin{vmatrix} (x_1-y_1)^{-1} & (x_1-y_2)^{-1} & \cdots & (x_1-y_n)^{-1} \\ (x_2-y_1)^{-1} & (x_2-y_2)^{-1} & \cdots & (x_2-y_n)^{-1} \\ \cdot & \cdot & \cdot & \cdot \\ (x_n-y_1)^{-1} & (x_n-y_2)^{-1} & \cdots & (x_n-y_n)^{-1} \end{vmatrix}$$

34. Find the co-factor of the product of the sines of the halved differences of $x_1, x_2, \cdots, x_n$, in

$$\begin{vmatrix} 1 & \sin x_1 & \cos x_1 & \sin 2x_1 & \cos 2x_1 & \cdots \\ 1 & \sin x_2 & \cos x_2 & \sin 2x_2 & \cos 2x_2 & \cdots \\ \cdot & \cdot & \cdot & \cdot & \cdot & \cdot \\ 1 & \sin x_{2n+1} & \cos x_{2n+1} & \sin 2x_{2n+1} & \cos 2x_{2n+1} & \cdots \end{vmatrix}$$

35. Prove that

$$\begin{vmatrix} \sin \alpha_1 & \sin \alpha_2 & \cdots & \sin \alpha_{2n} \\ \cos \alpha_1 & \cos \alpha_2 & \cdots & \cos \alpha_{2n} \\ \sin 2\alpha_1 & \sin 2\alpha_2 & \cdots & \sin 2\alpha_{2n} \\ \cos 2\alpha_1 & \cos 2\alpha_2 & \cdots & \cos 2\alpha_{2n} \\ \cdot & \cdot & \cdot & \cdot \\ \sin n\alpha_1 & \sin n\alpha_2 & \cdots & \sin n\alpha_{2n} \\ \cos n\alpha_1 & \cos n\alpha_2 & \cdots & \cos n\alpha_{2n} \end{vmatrix} = 2^{2n^2-2n+1}\pi \sin \tfrac{1}{2}(\alpha_i - \alpha_k) S$$

where $i > k$ and $S = \sum \cos \frac{1}{2}(\alpha_1 + \alpha_2 + \cdots + \alpha_n - \alpha_{n+1} - \cdots - \alpha_{2n})$.

36. If $s = a_1 + a_2 + \cdots + a_n$, $A_i = s - a_i$, prove

$$\begin{vmatrix} x - A_1 & a_2 & \cdots & a_n \\ a_1 & x - A_2 & \cdots & a_n \\ \cdot & \cdot & \cdot & \cdot \\ a_1 & a_2 & \cdots & x - A_n \end{vmatrix} = x(x - s)^{n-1}$$

37. Prove:

$$\begin{vmatrix} x - a_1 & A_2 & \cdots & A_n \\ A_1 & x - a_2 & \cdots & A_n \\ \cdot & \cdot & \cdot & \cdot \\ A_1 & A_2 & \cdots & x - a_n \end{vmatrix} = \{x + (n - 2)s\}(x - s)^{n-1}$$

38. Prove:

$$\begin{vmatrix} (x_1 - y_1)^{n-1} & (x_1 - y_2)^{n-1} & \cdots & (x_1 - y_n)^{n-1} \\ (x_2 - y_1)^{n-1} & (x_2 - y_2)^{n-1} & \cdots & (x_2 - y_n)^{n-1} \\ \cdot & \cdot & \cdot & \cdot \\ (x_n - y_1)^{n-1} & (x_n - y_2)^{n-1} & \cdots & (x_n - y_n)^{n-1} \end{vmatrix}$$

$$= {}_{n-1}C_1 \, {}_{n-1}C_2 \cdots {}_{n-1}C_{n-1} \zeta^{1/2}(x_1 \cdots x_n) \zeta^{1/2}(y_1 \cdots y_n).$$

39. Prove:

$$\begin{vmatrix} (x_1 - y_1)^n & \cdots & (x_1 - y_n)^n \\ (x_2 - y_1)^n & \cdots & (x_2 - y_n)^n \\ \cdot & \cdot & \cdot \\ (x_n - y_1)^n & \cdots & (x_n - y_n)^n \end{vmatrix}$$

$$= \frac{{}_nC_1\,{}_nC_2 \cdots {}_nC_r}{n!}\zeta^{1/2}(x_1 \cdots x_n)\zeta^{1/2}(y_1 \cdots y_n)$$

$$\times \overset{+}{\vphantom{x}}\begin{vmatrix} x_1 - y_1 & \cdots & x_1 - y_n \\ x_2 - y_1 & \cdots & x_2 - y_n \\ \cdot & \cdot & \cdot \\ x_n - y_1 & \cdots & x_n - y_n \end{vmatrix}^{+}.$$

40. Given

$$\frac{y_1}{x_1 - \alpha_1} + \frac{y_2}{x_1 - \alpha_2} + \cdots + \frac{y_n}{x_1 - \alpha_n} = u_1$$

$$\cdots\cdots\cdots\cdots$$

$$\frac{y_1}{x_n - \alpha_1} + \frac{y_2}{x_n - \alpha_2} + \cdots + \frac{y_n}{x_n - \alpha_n} = u_n,$$

prove that

$$y_k = -\frac{g(\alpha_k)}{f'(\alpha_k)}\left\{\frac{f(x_1)}{g'(x_1)}\frac{u_1}{x_1 - \alpha_k} + \cdots + \frac{f(x_n)}{g'(x_n)}\frac{u_n}{x_n - \alpha_k}\right\}$$

where the notation is as in §355.

41. Prove that

$$\begin{vmatrix} (x_1 - y_1)^{-2} & (x_1 - y_2)^{-2} & \cdots & (x_1 - y_n)^{-2} \\ \cdot & \cdot & \cdot & \cdot \\ (x_n - y_1)^{-2} & (x_n - y_2)^{-2} & \cdots & (x_n - y_n)^{-2} \end{vmatrix}$$

$$= \begin{vmatrix} (x_1 - y_1)^{-1} & \cdots & (x_1 - y_n)^{-1} \\ \cdot & \cdot & \cdot \\ (x_n - y_1)^{-1} & \cdots & (x_n - y_n)^{-1} \end{vmatrix} \cdot \overset{+}{\vphantom{x}}\begin{vmatrix} (x_1 - y_1)^{-1} & \cdots & (x_1 - y_n)^{-1} \\ \cdot & \cdot & \cdot \\ (x_n - y_1)^{-1} & \cdots & (x_n - y_n)^{-1} \end{vmatrix}^{+}.$$

42. Prove that the determinant of order $2n$

$$\begin{vmatrix} (x_1 - y_1)^{-1} & (x_1 - y_1)^{-2} & \cdots & (x_n - y_1)^{-1} & (x_n - y_1)^{-2} \\ (x_1 - y_2)^{-1} & (x_1 - y_2)^{-2} & \cdots & (x_n - y_2)^{-1} & (x_n - y_2)^{-2} \\ \cdot & \cdot & \cdot & \cdot & \cdot \\ (x_1 - y_{2n})^{-1} & (x_1 - y_{2n})^{-2} & \cdots & (x_n - y_{2n})^{-1} & (x_n - y_{2n})^{-2} \end{vmatrix}$$

$$= (-1)^n \frac{\{\zeta^{1/2}(y_1 y_2 \cdots y_n)\}^4 \zeta^{1/2}(x_1 x_2 \cdots x_n)}{[\phi(y_1)\phi(y_2) \cdots \phi(y_n)]^2}.$$

where

$$\phi(x) = (x - x_1)(x - x_2) \cdots (x - x_n).$$

State and prove the general theorem, that is where the determinant is of order rn.

43. If

$$\Delta = \begin{vmatrix} (x_1 - \alpha_1)^{-1} & (x_1 - \alpha_2)^{-1} & \cdots & (x_1 - \alpha_n)^{-1} \\ (x_2 - \alpha_1)^{-1} & (x_2 - \alpha_2)^{-1} & \cdots & (x_2 - \alpha_n)^{-1} \\ \cdot & \cdot & \cdot & \cdot \\ (x_n - \alpha_1)^{-1} & (x_n - \alpha_2)^{-1} & \cdots & (x_n - \alpha_n)^{-1} \end{vmatrix}$$

prove

$$\begin{vmatrix} (x_1 - \alpha_1)^{-1} & (x_1 - \alpha_2)^{-1} & \cdots & (x_1 - \alpha_n)^{-1} & 1 \\ \cdot & \cdot & \cdot & \cdot & \cdot \\ (x_n - \alpha_1)^{-1} & (x_n - \alpha_2)^{-1} & \cdots & (x_n - \alpha_n)^{-1} & 1 \\ 1 & 1 & \cdots & 1 & \cdot \end{vmatrix}$$

$$= \Delta \cdot (\alpha_1 + \alpha_2 + \cdots + \alpha_n - x_1 - x_2 - \cdots - x_n)$$

44. Prove that

$$\begin{vmatrix} (x_1 - \alpha_1)^{-1} & (x_1 - \alpha_2)^{-1} & \cdots & (x_1 - \alpha_n)^{-1} & \dfrac{x_1{}^r}{f(x_1)} \\ (x_2 - \alpha_1)^{-1} & (x_2 - \alpha_2)^{-1} & \cdots & (x_2 - \alpha_n)^{-1} & \dfrac{x_2{}^r}{f(x_2)} \\ \cdot & \cdot & \cdot & \cdot & \cdot \\ (x_n - \alpha_1)^{-1} & (x_n - \alpha_2)^{-1} & \cdots & (x_n - \alpha_n)^{-1} & \dfrac{x_n{}^r}{f(x_n)} \\ 1 & 1 & \cdots & 1 & \end{vmatrix} = -\Delta H_{r-n+1}$$

where Δ is the same as in ex. 43 and $f(x) = (x - \alpha_1)(x - \alpha_2) \cdots (x - \alpha_n)$.

45. Prove:

$$\begin{vmatrix} 1 & \cdots & 1 & 0 & \cdots & 0 \\ x_1 & \cdots & x_n & 1 & \cdots & 1 \\ x_1^2 & \cdots & x_n^2 & 2x_1 & \cdots & 2x_n \\ \cdot & \cdot & \cdot & \cdot & \cdot & \cdot \\ x_1^{2n-1} & \cdots & x_n^{2n-1} & (2n-1)x_1^{2n-2} & \cdots & (2n-1)x_n^{2n-2} \end{vmatrix} = (-1)^{n(n-1)/2}\left\{\zeta^{1/2}(x_1 \cdots x_n)\right\}^4$$

46. Obtain the quotient

$$\frac{D_{5;5}}{\zeta^{1/2}(\alpha_1\alpha_2 \cdots \alpha_5)\zeta^{1/2}(\beta_1\beta_2 \cdots \beta_5)}.$$

47. Obtain the quotient

$$\frac{D_{3;4}}{\zeta^{1/2}(\alpha_1\alpha_2 \cdots \alpha_4)\zeta^{1/2}(\beta_1\beta_2 \cdots \beta_4)}.$$

48. Prove

$$\frac{D_{2;4}}{\zeta^{1/2}(\alpha_1 \cdots \alpha_4)\zeta^{1/2}(\beta_1 \cdots \beta_4)}$$

$$= \begin{vmatrix} -\sum\alpha_1 & 1 & \cdot & -4 & \cdot & \cdot \\ \alpha_1\alpha_2 & -\sum\alpha_1 & 1 & \cdot & -6 & \cdot \\ \cdot & \alpha_1\alpha_2 & -\sum\alpha_1 & \cdot & \cdot & -4 \\ \beta_1\beta_2 & \cdot & \cdot & -\sum\beta_1 & 1 & \cdot \\ \cdot & \cdot & \cdot & \beta_1\beta_2 & -\sum\beta_1 & 1 \\ \cdot & \cdot & \alpha_1\alpha_3 & \cdot & \beta_1\beta_2 & -\sum\beta_1 \end{vmatrix}$$

49. Prove:

$$\begin{vmatrix} \dfrac{1}{f(x)} & \dfrac{x}{f(x)} & \dfrac{x^2}{f(x)} & \cdots \\ D^{(1)}\dfrac{1}{f(x)} & D^{(1)}\dfrac{x}{f(x)} & D^{(1)}\dfrac{x^2}{f(x)} & \cdots \\ D^{(2)}\dfrac{1}{f(x)} & D^{(2)}\dfrac{x}{f(x)} & D^{(2)}\dfrac{x^2}{f(x)} & \cdots \\ \cdot & \cdot & \cdot & \cdot \end{vmatrix}_r = \frac{(r-1)!(r-2)!\cdots 2!\,1}{\{f(x)\}^r}$$

where $D^{(k)}$ denotes the kth derivative with respect to x.

50. Prove that

$$\begin{vmatrix} \dfrac{(r+\alpha+\alpha'-1)!}{(\alpha+\alpha')!} & \dfrac{(r+\alpha+\beta'-1)!}{(\alpha+\beta')!} & \cdots \\ \dfrac{(r+\beta+\alpha'-1)!}{(\beta+\alpha')!} & \dfrac{(r+\beta+\beta'-1)!}{(\beta+\beta')!} & \cdots \\ \cdots & \cdots & \cdots \end{vmatrix}$$

$$= (-1)^{r(r-1)/2}\left[\frac{(r-1)!}{(r-1)!}\right]^r \zeta^{1/2}(\alpha,\beta,\gamma,\cdots)\zeta^{1/2}(\alpha',\beta',\gamma',\cdots).$$

51. Denoting by $(n)_r$ the coefficient of x^n in $(1+x)^n$, find the cofactor of $\zeta^{1/2}(ab\cdots c)$ in

$$\begin{vmatrix} 1 & (a)_1 & \cdots & (a)_{n-1} \\ 1 & (b)_1 & \cdots & (b)_{n-1} \\ \cdots & \cdots & \cdots & \cdots \\ 1 & (e)_1 & \cdots & (e)_{n-1} \end{vmatrix}$$

52. The determinant whose matrix is the sum of the matrices of $|a^0b^1c^2\cdots|_n$ and $|a^0b^{-1}c^{-2}\cdots|_n$ is equal to

$$2|(a+a^{-1})^0(b+b^{-1})^1(c+c^{-1})^2\cdots|_n$$

53. Given $(m+1)n$ linear independent polynomials $P_0(x)$, $P_1(x),\cdots$, of degree less than $(m+1)n$ and denoting their derivatives by P_0', $P_0''\cdots$, show that the determinant of order $(m+1)n$:

$$\begin{vmatrix} P_0(x_1) & P_0'(x_1)\cdots P_0^{(m)}(x_1) & P_0(x_2)\cdots P_0^{(m)}(x_2)\cdots P_0(x_n)\cdots P_0^{(m)}(x_n) \\ P_1(x_1) & P_1(x_1)\cdots P_1^{(m)}(x_1) & P_1(x_2)\cdots P_1^{(m)}(x_2)\cdots P_1(x_n)\cdots P_1^{(m)}(x_n) \\ \cdots & \cdots & \cdots \\ P_r(x_1) & P_r(x_1)\cdots P_r^{(m)}(x_1) & P_r(x_2)\cdots P_r^{(m)}(x_2)\cdots P_r(x_n)\cdots P_r^{(m)}(x_n) \\ \cdots & \cdots & \cdots \end{vmatrix}$$

$$= c[\zeta^{1/2}(x_1\cdots x_n)]^{(m+1)^2}.$$

54. If $f(x)=(x+\alpha_1)(x+\alpha_2)(x+\alpha_r)\cdots(x+\alpha_r)$ and D^λ denotes the λth derivative with respect to x, show that

$$\begin{vmatrix} D^{\alpha}\dfrac{1}{f(x)} & D^{\alpha}\dfrac{x}{f(x)} & D^{\alpha}\dfrac{x^2}{f(x)} & \cdots \\ D^{\beta}\dfrac{1}{f(x)} & D^{\beta}\dfrac{x}{f(x)} & D^{\beta}\dfrac{x^2}{f(x)} & \cdots \\ D^{\gamma}\dfrac{1}{f(x)} & D^{\gamma}\dfrac{x}{f(x)} & D^{\gamma}\dfrac{x^2}{f(x)} & \cdots \\ \cdots & \cdots & \cdots & \cdots \end{vmatrix}$$

$$= \frac{(-1)^{r(r-1)/1.2+\alpha+\beta+\cdots}\,\alpha!\beta!\gamma!\cdots f(x)^{\frac{r(r-3)}{1\cdot 2}}}{\zeta^{1/2}(\alpha_1\alpha_2\cdots)}$$

$$\times \left| (x+\alpha_1)^{-\alpha} \;\; (x+\alpha_2)^{-\beta} \;\; (x+\alpha_3)^{-\gamma} \cdots \right|$$

CHAPTER XII

SYMMETRIC DETERMINANTS

357. As we have seen before (§ 26) we may have symmetry with respect to a line or a point, and have therefore two general kinds of symmetric determinants to consider—axisymmetric and centrosymmetric.

SECTION I. CENTROSYMMETRIC DETERMINANTS

358. In a centrosymmetric determinant, the rth row reversed forms in every case the rth row from the end, that is to say, the determinant is the same when read backwards as when read forwards.

359. Two constituents are said to be conjugate with respect to the center of a determinant when they lie on a line through the center and equally distant from it. A determinant is centrosymmetric therefore, when every constituent is equal to its conjugate with respect to the center.

360. Let $\alpha \equiv (\alpha_1\alpha_2 \cdots \alpha_m)$ be a combination, m at a time, of the numbers $1, 2, 3, \cdots, 2m$, such that $\alpha_h+\alpha_k \neq 2m+1$ for all values of h and k from 1 to m. There are 2^m such combinations, for they may evidently be formed by writing the numbers $1, 2, 3, \ldots, 2m$ in m pairs, the sum of each pair being equal to $2m+1$, and taking one number from each pair.

Let $\beta \equiv (\beta_1\beta_2 \cdots \beta_m)$ be the complementary combination of α, then β is also the *reflex-combination* of α, that is β_k is, for each value of k from 1 to m, the defect from $2m+1$ for some one of the numbers in α. For by the hypothesis the defect of α_k from $2m+1$ is not found in α and therefore must be in β. It follows therefore that of the $(2m)_m$ combinations of the numbers $1, 2, \cdots, 2m$ taken m at a time there are 2^m, the complementary and reflex of each of which are alike.

361. Two minors of a determinant may be called *reflex* of each other where the rows and columns of one are the reflex combinations of the rows and columns respectively of the others. Thus $\left|{1246 \atop 1357}\right|$ and $\left|{3578 \atop 2468}\right|$ are reflex minors of a determinant of order 8.

Two minors are said to be *trans-reflex* of each other when the row numbers of the two are the same and the column numbers of the two are reflex combinations.

Two minors are said to be *sub-reflex* when the column numbers of the two are the same and the row numbers are reflex combinations.

362. *Every centrosymmetric determinant* Δ *of even order* $2m$ *is expressible as the product of two determinants each of order* m.

For if we perform the following operations:

$$r_1 + r_{2m},\ r_2 + r_{2m-1},\ \cdots,\ r_m + r_{m+1}$$

and

$$c_1 - c_{2m},\ c_2 - c_{2m-1},\ \cdots,\ c_m - c_{m+1}$$

the resulting determinant will have a square of m^2 zeros in the upper left-hand corner and therefore breaks up into the product of two determinants with binomial elements which we will call D and D'.

363. We may write

$$D = \left| a_{rs} + a_{rt} \right|,$$
$$D' = \left| a_{rs} - a_{rt} \right|,$$

where

$$\left\{ \begin{matrix} r \\ s \end{matrix} = 1,2,\ \cdots,m, \right\}$$

and

$$(t = 2m, 2m-1,\ \cdots,\ m+1).$$

The determinant D may be broken up into the sum of 2^m determinants with monomial elements concerning which it may be observed that:

(1) *For every determinant*

$$D_\alpha \equiv \left| \begin{matrix} 1\ 2 & \cdots & m \\ \alpha_1\alpha_2 & \cdots & \alpha_m \end{matrix} \right|,$$

there is another

$$D_\beta \equiv \left| \begin{matrix} 1\ 2 & \cdots & m \\ \beta_1\beta_2 & \cdots & \beta_m \end{matrix} \right|$$

where

$$\alpha_k + \beta_k = 2m + 1 \quad (k = 1,2,\ \cdots,\ m)$$

That is D_α and D_β are trans-reflex minors.

(2) *The signs of D_α and D_β, when the columns are arranged in their natural order, are the same or opposite according as $\frac{1}{2}m(m-1)$ is even or odd.*

For if there are g_k numbers following α_k smaller than α_k, there are g_k numbers following β_k larger than β_k. Therefore g_k is the number of inversions due to the position of α_k in D_α, and $m-k-g_k$ is the number of inversions due to the position of β_k in D_β. The sign factor, therefore, for D_α when the column numbers are written in their natural order is $(-1)^{g_1+g_2+\cdots+g_m}$ and that for D_β under similar conditions is $(-1)^{m^2-m(m+1)/2-(g_1+g_2+\cdots+g_m)}$ or $(-1)^{m(m-1)/2-(g_1+g_2+\cdots+g_m)}$ Since these exponents differ from $\frac{1}{2}m(m-1)$ by an even number the truth of the theorem appears.

(3) *In the series of 2^m determinants with monomial elements into the sum of which D may be expressed, there are as many positive as negative terms.*

Considering two consecutive cases, say when $m=k$ and $m=k+1$, we see that for every term of the series

$$\begin{vmatrix} 1\ 2 & \cdots & k \\ \alpha_1\alpha_2 & \cdots & \alpha_k \end{vmatrix} = M,$$

when $m=k$, there are two terms

$$\begin{vmatrix} 1 & 2 & 3 & \cdots & k+1 \\ 1 & \alpha_1+1 & \alpha_2+1 & \cdots & \alpha_k+1 \end{vmatrix} \equiv M',$$

and

$$\begin{vmatrix} 1 & 2 & 3 & \cdots & k+1 \\ 2k+2 & \alpha_1+1 & \alpha_2+1 & \cdots & \alpha_k+1 \end{vmatrix} \equiv M'',$$

when $m=k+1$.

The term M' will obviously have the same sign as M and M'' will have the same or opposite sign according as k is even or odd. It follows, therefore that if there are as many positive as negative terms when $m=k$ there will be as many of each when $m=k+1$, and since it is true when $m=2$ and $m=3$, it is true in general.

364. In the case of D' it is obvious from the method of formation that the same 2^m determinants occur as in D, and the signs of the various terms will be the same as in D except that whenever there is an odd number of columns with negative elements the sign will be changed. If k be the number of such columns taken to form D_k the sign factor of D_k will be multiplied by $(-1)^k$, and since there are $(m)_k$ such determinants the number of changes of sign on account of the negative elements would therefore be $(m)_1+(m)_3+\cdots+(m)_{2k+1}+\cdots=2^{m-1}$, which is just half of the whole number of terms.

365. From the foregoing it appears that *D is the sum of two sets of determinants of order m, and that D' is the difference of the same two sets of determinants, and therefore Δ, which is equal to their product may be expressed as the difference of two squares.*

366. If Δ is of odd order $2m+1$, then the $(m+1)$st row will be the same read backwards as forwards, and the $(m+1)$st column read from bottom to top as from top to bottom.

Performing the operations

$$r_1+r_{2m+1},\quad r_2+r_{2m},\quad \cdots,\quad r_m+r_{m+2}$$

and

$$c_1-c_{2m+1},\quad c_2-c_{2m},\quad \cdots,\quad c_m-c_{m+2}$$

will result in a determinant with a rectangle of $(m+1)$ rows and m columns of zeros in the upper right-hand corner and therefore breaks up into the product of two determinants, one of the mth and the other of the $(m+1)$th order.

367. *Any n-line determinant having the array of its last $(n-1)$ rows centrosymmetric is expressible as the product of two determinants.*

Thus when n is even we have

$$\begin{vmatrix} u & v & w & x & y & z \\ a_1 & a_2 & a_3 & a_4 & a_5 & a_6 \\ b_1 & b_2 & b_3 & b_4 & b_5 & b_6 \\ c_1 & c_2 & c_3 & c_3 & c_2 & c_1 \\ b_6 & b_5 & b_4 & b_3 & b_2 & b_1 \\ a_6 & a_5 & a_4 & a_3 & a_2 & a_1 \end{vmatrix}$$

$$= \begin{vmatrix} c_1 & c_2 & c_3 \\ b_6+b_1 & b_5+b_2 & b_4+b_3 \\ a_6+a_1 & a_5+a_2 & a_4+a_3 \end{vmatrix} \begin{vmatrix} u-z & v-y & w-x \\ b_6-b_1 & b_5-b_2 & b_4-b_3 \\ a_6-a_1 & a_5-a_2 & a_4-a_3 \end{vmatrix}$$

and when n is odd we have

$$\begin{vmatrix} u & v & w & x & y \\ a_1 & a_2 & a_3 & a_4 & a_5 \\ b_1 & b_2 & b_3 & b_4 & b_5 \\ b_5 & b_4 & b_3 & b_2 & b_1 \\ a_5 & a_4 & a_3 & a_2 & a_1 \end{vmatrix}$$

$$= \begin{vmatrix} b_5-b_1 & b_4-b_2 \\ a_5-a_1 & a_4-a_2 \end{vmatrix} \cdot \begin{vmatrix} w & v+x & u+y \\ a_3 & a_4+a_2 & a_5+a_1 \\ b_3 & b_4+b_2 & b_5+b_1 \end{vmatrix}.$$

368. A determinant is said to be *skew-centrosymmetric* when every constituent is the negative of its conjugate with respect to the center. One of odd order would therefore have its center element zero.

369. *Every skew-centrosymmetric determinant of even order 2m is expressible as the difference of two squares.*

For if we perform the operations

$$r_1 + r_{2m},\; r_2 + r_{2m-1},\; \cdots,\; r_m + r_{m+1},$$

and

$$c_1 + c_{2m},\; c_2 + c_{2m-1},\; \cdots,\; c_m + c_{m+1},$$

the result will be seen to break up into two determinants D and D' with binomial elements. Here as in the case of centrosymmetric determinants if the element in the rth row and sth column of D is $x+y$, then the element in the same position of D' (or $-D'$ if m is odd) is $x-y$, and hence the theorem follows as in §365.

370. *Every skew-centrosymmetric determinant of order* $2m+1$ *is zero.*

For performing the operations

$$r_1 + r_{2m+1},\; r_2 + r_{2m},\; \cdots,\; r_m + r_{m+2}$$

$$c_1 + c_{2m+1},\; c_2 + c_{2m},\; \cdots,\; c_m + c_{m+2}$$

the result will be a determinant having a square of $(m+1)^2$ zeros and therefore vanishes.

If the central element be not zero then the determinant is equal to the product of this central element and two m-line determinants.

Section II. Axisymmetric Determinants

371. In the case of an axisymmetric determinant; (1) conjugate lines are alike, (2) coaxial minors are axisymmetric, (3) conjugate minors are equal, (4) all compounds of the original are axisymmetric.

372. Of the general theorems whose forms are modified by the existence in the determinant of axisymmetry, the only one we note here is that of §110 which becomes

$$\Delta = a_{rr}A_{rr} - \sum a_{ri}^2 A_{ri,ri} - \sum(-1)^{i+j} a_{ri}a_{jr}A_{rj,ri}$$

where

$$\left(\begin{matrix} i \\ j \end{matrix} = 1, 2, \cdots r-1, r+1, \cdots, n\right), \quad i \neq j \neq r.$$

But since $a_{ri}a_{jr}$ is the same as $a_{rj}a_{ir}$ and $A_{rj,ri}$ is the same as $A_{ri,rj}$ it follows that the terms in the last sum are double.

373. From the law of multiplication it is apparent that *the square of any determinant is axisymmetric.*

374. *Any axisymmetric determinant myltiplied by the square of any determinant is expressible as an axisymmetric determinant.*

Let $A \equiv |a_{1n}|$, where $a_{rs} = a_{sr}$, $B \equiv |b_{1n}|$, and let $B \cdot A = C \equiv |c_{1n}|$ in which

$$\begin{aligned} c_{rs} &= b_{r1}a_{s1} + b_{r2}a_{s2} + \cdots \\ &= b_{r1}a_{1s} + b_{r2}a_{2s} + \cdots , \end{aligned}$$

since $a_{rs} = a_{sr}$.
Then

$$B \cdot C = D = | d_{1n} | ,$$

in which

$$\begin{aligned} d_{rs} &= b_{r1}c_{s1} + b_{r2}c_{s2} + \cdots \\ &= b_{s1}c_{r1} + b_{s2}c_{r2} + \cdots . \\ &= d_{sr} . \end{aligned}$$

Therefore $B^2 \cdot A$ is an axisymmetric determinant.

In this theorem it is obvious we may substitute any even power for the square.

375. If in §374 we let B be identical with A, then we have the theorem that *the cube of any axisymmetric determinant is expressible as an axisymmetric determinant.*

376. *Any power of an axisymmetric determinant is expressible as an axisymmetric determinant.*

This follows from §§374, 375.

377. *Any even power of any determinant is expressible as an axisymmetric determinant.*

This follows from §§373, 376.

378. *Any power of any determinant of the second order is expressible as an axisymmetric determinant.*

This follows from §376 and from the fact that any determinant of the second order is expressible as an axisymmetric determinant. Thus

$$\begin{vmatrix} a_{11} & a_{12} \\ a_{21} & a_{22} \end{vmatrix} = \begin{vmatrix} a_{11} & (a_{12}a_{21})^{1/2} \\ (a_{21}a_{12})^{1/2} & a_{22} \end{vmatrix} .$$

379. *The product of the kth power of an axisymmetric determinant by the second power of any determinant is expressible as an axisymmetric determinant.*

This follows from §§314, 376.

380. *If R represents a rectangular array then it is apparent that* R^2 *is an axisymmetric determinant and hence any even power of R is axisymmetric.*

381. The sum of any number of squares may be written as an axisymmetric determinant.*

Thus

$$a_1^2 + a_2^2 + a_3^2 + \cdots = -\begin{vmatrix} 0 & a_1 & a_2 & a_3 & \cdots \\ a_1 & 1 & 0 & 0 & \cdots \\ a_2 & 0 & 1 & 0 & \cdots \\ a_3 & 0 & 0 & 1 & \cdots \\ \cdot & \cdot & \cdot & \cdot & \cdots \end{vmatrix}.$$

382. In the expansion of a determinant there are, as we have seen, $n!$ terms, but when $a_{rs} = a_{sr}$ they are not all distinct. For the numerical coefficient of any term we have the following:

If $\nu_3, \nu_4, \cdots$, *be the number of ternary, quaternary,* $\cdots$ *index-cycles in any term of an axisymmetric determinant the coefficient of the term where condensation takes place is* $2^{\nu_3+\nu_4+\cdots}$.

For when we have a term with index-cycles higher than binary, we may, by reversing the order of indices in one of the said cycles, obtain another term of the development, and this will be equal to the former.

383. For A axisymmetric we have

$$A_{11}A_{22} - A_{12}^2 = A \cdot A_{12,12}$$

and if $A_{11} = 0$ we see that A and $A_{12,12}$ must have opposite signs.

Using this principle we reach the theorem: *In the series* A, A_{11}, $A_{12,12}$, $A_{123,123}$, $\cdots$, 1 *if any term (excluding the first and last) is zero then those adjoining terms have unlike signs.*

384. If in the foregoing we consider A_{11} as our original determinant and A as formed from it by bordering symmetrically and observing

* Certain general homogenous polynomials can be expressed as determinants with linear elements and the student who is interested should refer to a paper by Dickson: Trans. Am. Math. Soc., Vol. 22, no. 2, pp. 167; cf. Stouffer, ibid., vol. 26, no. 3, pp. 356.

that A_{12} is a linear homogeneous function of the bordering elements we have the following theorem:

If an axisymmetric determinant whose value is zero, be bordered symmetrically the product of the determinant so formed and its leading second minor is equal to the negative of the square of a linear homogeneous function of the bordering elements.

385. *If $A=0$ and if all coaxial minors of A of order $(n-m)$ vanish and if the sums of all coaxial minors of order greater than $(n-m)$ vanish, then all minors of order $(n-m)$ vanish.*

Let α_{ij} be the element in the ith row and jth column of the $(n-m)$th compound of A, let $A_{(n-m)2,ss}$ denote any coaxial minor of the second order of this $(n-m)$th compound, and let

$$A_{(n-m)2,ss} = \alpha_{ii}\alpha_{jj} - \alpha_{ij}\alpha_{ji}.$$

By hypothesis $\alpha_{ii}=0$ for all values of i and $\alpha_{ij}=\alpha_{ji}$ and therefore

$$A_{(n-m)2,ss} = - \alpha_{ij}^2$$

and

$$\sum sA_{(n-m)2,ss} = - \sum(\alpha_{ij})^2$$

But by §191 the left hand member may be expressed in terms of sums of minors of A of order higher than $(n-m)$ and these by hypothesis are zero, and we have $\sum(\alpha_{ij})^2=0$ and therefore $\alpha_{ij}=0$ for all values of i and j.

386. *If*

$$\begin{vmatrix} 1\,2\,3\,4 \cdots r \\ 1\,2\,3\,4 \cdots r \end{vmatrix} \neq 0$$

and if

$$\begin{vmatrix} 1\,2\,3 \cdots r\,\alpha \\ 1\,2\,3 \cdots r\,\alpha \end{vmatrix} = 0 = \begin{vmatrix} 1\,2\,3 \cdots r\,\alpha\,\beta \\ 1\,2\,3 \cdots r\,\alpha\,\beta \end{vmatrix}$$

for all values of α and β, then all minors of order $r+1$ vanish.

For

$$\begin{vmatrix} 1\,2 \cdots r\,\alpha \\ 1\,2 \cdots r\,\alpha \end{vmatrix} \begin{vmatrix} 1\,2 \cdots r\,\beta \\ 1\,2 \cdots r\,\beta \end{vmatrix} - \begin{vmatrix} 1\,2 \cdots r\,\alpha \\ 1\,2 \cdots r\,\beta \end{vmatrix}^2$$

$$= \begin{vmatrix} 1\,2 \cdots r \\ 1\,2 \cdots r \end{vmatrix} \begin{vmatrix} 1\,2 \cdots r\,\alpha\,\beta \\ 1\,2 \cdots r\,\alpha\,\beta \end{vmatrix}$$

or

$$0 \quad - \begin{vmatrix} 1\,2 \cdots r\,\alpha \\ 1\,2 \cdots r\,\beta \end{vmatrix}^2 = 0$$

for all values of α and β and hence by §234 all minors of order $r+1$ vanish.

It is also readily seen that if all coaxial minors of order r and also those of order $(r+1)$ vanish then all non-coaxial minors of the rth and higher order vanish.

387. *From the general relation* $A_{rp}A_{sq}-A_{ps}A_{rq}=A \cdot A_{rs,pq}$ *we see that if* $A=0$ *and* $A_{rp}=0$ *then either* A_{sp} *for all values of* s *or* A_{sq} *for all values of* q *must vanish.* That is all elements in the same row or column of the adjugate as A_{sp} must be zero.

If $p=r$, then since the adjugate is symmetrical it follows that all elements in both the rth row and rth column of the adjugate must vanish. If also $q=s$ then this relation becomes

$$A_{rr}A_{ss} - A_{rs}^2 = A \cdot A_{rs,rs}$$

and we see *that if* $A_{rs} \neq 0$ *and*

(1) *If* $A_{rs,rs}=0$, *then* A_{rr} *and* A_{ss} *have the same sign.*

(2) *If* $A=0$, *then all coaxial minors of order* $n-1$ *have the same sign.*

388. When $A=0$ we have

$$A_{r1}:A_{r2}:A_{r3}: \cdots = (A_{11})^{1/2}:(A_{22})^{1/2}:(A_{33})^{1/2}: \cdots$$

the signs for the radicals being undetermined.

Again by Laplace's theorem

$$(r1)[r1]+(r2)[r2]+ \cdots +(rn)[rn]=A=0$$

where $[rk]$ is the cofactor of the element (rk) in A. Therefore

$$\{(r1)[11]^{1/2}+(r2)[22]^{1/2}+ \cdots +(rn)[nn]^{1/2}\}[rr]^{1/2}=0$$

and since $[rr]$ is not in general zero, we have

$$(r1)[11]^{1/2}+(r2)[22]^{1/2}+ \cdots +(rn)[nn]^{1/2}=0$$

389. *If in an axisymmetric determinant the sum of the elements in every row is zero then all the primary minors are numerically equal.*

For convenience take $n=4$, and let the determinant be

$$\Delta \equiv \begin{vmatrix} a_{11} & a_{12} & a_{13} & a_{14} \\ a_{21} & a_{22} & a_{23} & a_{24} \\ a_{31} & a_{32} & a_{33} & a_{34} \\ a_{41} & a_{42} & a_{43} & a_{44} \end{vmatrix}.$$

Since the sum of the elements in each row is zero it follows that Δ vanishes. For if we add all the columns to the first it becomes a column of zeros.

We have given

$$\left.\begin{array}{l} a_{11} + a_{12} + a_{13} + a_{14} = 0 \\ a_{12} + a_{22} + a_{23} + a_{24} = 0 \\ a_{13} + a_{23} + a_{33} + a_{34} = 0 \\ a_{14} + a_{24} + a_{34} + a_{44} = 0 \end{array}\right\} \quad (1)$$

and expanding we get

$$\left.\begin{array}{l} a_{11}A_{11}^{1/2} + a_{12}A_{22}^{1/2} + a_{13}A_{33}^{1/2} + a_{14}A_{44}^{1/2} = 0 \\ a_{12}A_{11}^{1/2} + a_{22}A_{22}^{1/2} + a_{23}A_{33}^{1/2} + a_{24}A_{44}^{1/2} = 0 \\ a_{13}A_{11}^{1/2} + a_{23}A_{22}^{1/2} + a_{33}A_{33}^{1/2} + a_{34}A_{44}^{1/2} = 0 \\ a_{14}A_{11}^{1/2} + a_{24}A_{22}^{1/2} + a_{34}A_{33}^{1/2} + a_{44}A_{44}^{1/2} = 0 \end{array}\right\} \quad (2)$$

provided A_{11}, A_{22}, A_{33}, A_{44} are not zero. From these we get using (1)

$$\begin{array}{l} a_{12}(A_{11}^{1/2} - A_{22}^{1/2}) + a_{13}(A_{11}^{1/2} - A_{33}^{1/2}) + a_{14}(A_{11}^{1/2} - A_{44}^{1/2}) = 0 \\ a_{22}(A_{11}^{1/2} - A_{22}^{1/2}) + a_{23}(A_{11}^{1/2} - A_{33}^{1/2}) + a_{24}(A_{11}^{1/2} - A_{44}^{1/2}) = 0 \\ a_{23}(A_{11}^{1/2} - A_{22}^{1/2}) - a_{33}(A_{11}^{1/2} + A_{33}^{1/2}) + a_{44}(A_{11}^{1/2} - A_{44}^{1/2}) = 0 \end{array}$$

But the determinant of these (A_{14}) is not in general zero; therefore

$$A_{11}^{1/2} = A_{22}^{1/2} = A_{33}^{1/2} = A_{44}^{1/2}$$

and therefore by the last article all primary minors are numerically equal.

The same line of reasoning holds for determinants of any order.

390. Let

$$\left.\begin{array}{l} c_1a_{11} + c_2a_{12} + c_3a_{13} + c_4a_{14} = 0 \\ c_1a_{21} + c_2a_{22} + c_3a_{23} + c_4a_{24} = 0 \\ \cdots\cdots\cdots\cdots\cdots \end{array}\right\} \quad (1)$$

and we have

$$\left.\begin{array}{ll} a_{11}A_{11}^{1/2} + a_{12}A_{22}^{1/2} + a_{13}A_{33}^{1/2} + a_{14}A_{44}^{1/2} = 0 & \text{if} \quad A_{11} \neq 0 \\ a_{21}A_{11}^{1/2} + a_{22}A_{22}^{1/2} + a_{23}A_{33}^{1/2} + a_{24}A_{44}^{1/2} = 0 & \text{if} \quad A_{22} \neq 0 \\ a_{31}A_{11}^{1/2} + a_{32}A_{22}^{1/2} + a_{33}A_{33}^{1/2} + a_{34}A_{44}^{1/2} = 0 & \text{if} \quad A_{33} \neq 0 \\ a_{41}A_{11}^{1/2} + a_{42}A_{22}^{1/2} + a_{43}A_{33}^{1/2} + a_{44}A_{44}^{1/2} = 0 & \text{if} \quad A_{44} \neq 0 \end{array}\right\} \quad (2)$$

From (2) and (1) we have

$$\left.\begin{aligned}
a_{12}(c_1A_{22}^{1/2} - c_2A_{11}^{1/2}) + a_{13}(c_1A_{33}^{1/2} - c_3A_{11}^{1/2}) + a_{14}(c_1A_{44}^{1/2} - c_4A_{11}^{1/2}) &= 0\\
a_{22}(c_1A_{22}^{1/2} - c_2A_{11}^{1/2}) + a_{23}(c_1A_{33}^{1/2} - c_3A_{11}^{1/2}) + a_{24}(c_1A_{44}^{1/2} - c_4A_{11}^{1/2}) &= 0\\
a_{32}(c_1A_{22}^{1/2} - c_2A_{11}^{1/2}) + a_{33}(c_1A_{33}^{1/2} - c_3A_{11}^{1/2}) + a_{34}(c_1A_{44}^{1/2} - c_4A_{11}^{1/2}) &= 0\\
a_{42}(c_1A_{22}^{1/2} - c_2A_{11}^{1/2}) + a_{43}(c_1A_{33}^{1/2} - c_3A_{11}^{1/2}) + a_{44}(c_1A_{44}^{1/2} - c_4A_{11}^{1/2}) &= 0
\end{aligned}\right\} \quad (3)$$

If $A_{11} \neq 0$ then

$$\left.\begin{aligned}
c_1A_{22}^{1/2} &= c_2A_{11}^{1/2}\\
c_1A_{33}^{1/2} &= c_3A_{11}^{1/2}\\
c_1A_{44}^{1/2} &= c_4A_{11}^{1/2}
\end{aligned}\right\} \quad \text{or} \quad A_{11}^{1/2} : A_{22}^{1/2} : A_{33}^{1/2} : A_{44}^{1/2} = c_1 : c_2 : c_3 : c_4.$$

If one of the coaxial minors, say A_{44}, vanishes then equations (2) become

$$\left.\begin{aligned}
a_{11}A_{11}^{1/2} + a_{12}A_{22}^{1/2} + a_{13}A_{33}^{1/2} &= 0, \quad \text{if} \quad A_{11} \neq 0\\
a_{21}A_{11}^{1/2} + a_{22}A_{22}^{1/2} + a_{23}A_{33}^{1/2} &= 0, \quad \text{if} \quad A_{22} \neq 0\\
a_{31}A_{11}^{1/2} + a_{32}A_{22}^{1/2} + a_{33}A_{33}^{1/2} &= 0, \quad \text{if} \quad A_{33} \neq 0
\end{aligned}\right\} \quad (4)$$

Equations (3) become

$$\left.\begin{aligned}
a_{12}(c_1A_{22}^{1/2} - c_2A_{11}^{1/2}) + a_{13}(c_1A_{33}^{1/2} - c_3A_{11}^{1/2}) + a_{14}c_4A_{11}^{1/2} &= 0\\
a_{22}(c_1A_{22}^{1/2} - c_2A_{11}^{1/2}) + a_{23}(c_1A_{33}^{1/2} - c_3A_{11}^{1/2}) + a_{24}c_4A_{11}^{1/2} &= 0\\
a_{32}(c_1A_{22}^{1/2} - c_2A_{11}^{1/2}) + a_{33}(c_1A_{33}^{1/2} - c_3A_{11}^{1/2}) + a_{34}c_4A_{11}^{1/2} &= 0
\end{aligned}\right\} \quad (5)$$

Multiply these equations by c_1, c_2, c_3 respectively and add and we get

$$a_{42}(c_1A_{22}^{1/2} - c_2A_{11}^{1/2}) + a_{43}(c_1A_{33}^{1/2} - c_3A_{11}^{1/2}) + a_{44}c_4A_{11}^{1/2} = 0. \quad (6)$$

Since A_{44} is zero it follows that $A_{14} = A_{24} = A_{34} = 0$ and we cannot infer anything from equations (5).

If we take any two of (5) with (6) then if A_{11}, say, is not zero it follows that

$$c_1A_{22}^{1/2} - A_{11}^{1/2}c_2 = c_1A_{33}^{1/2} - c_3A_{11}^{1/2} = c_4A_{11}^{1/2} = 0$$

which requires that either c_4 or A_{11} vanish

If $c_4 = 0$ then $A_{11}^{1/2} : A_{22}^{1/2} : A_{33}^{1/2} = c_1 : c_2 : c_3$.

If $A_{11}^{1/2} = 0$ then $A_{22}^{1/2} = A_{33}^{1/2} = 0$, also, and therefore all primary minors vanish.

However, if $c_4=0$ we would not have (6), but taking any two of (5) we get

$$A_{11}^{1/2}:A_{22}^{1/2}:A_{33}^{1/2} = c_1:c_2:c_3,$$

unless all minors of order two from

$$\begin{Vmatrix} a_{12} & a_{22} & a_{32} \\ a_{13} & a_{23} & a_{33} \end{Vmatrix}$$

vanish, which in general is not the case.

We have therefore

$$A_{11}^{1/2}:A_{22}^{1/2}:A_{33}^{1/2} = c_1:c_2:c_3 \quad \text{if} \quad A_{44} = 0 \quad \text{and} \quad A_{11} \neq 0.$$

If $A_{44}=A_{11}=0$ then $A_{22}=A_{33}=0$ unless

$$\begin{vmatrix} a_{22} & a_{23} \\ a_{32} & a_{33} \end{vmatrix} = 0,$$

which is in general not the case.

391. *If in an axisymmetric determinant the signed primary minors are all equal then the sum of the elements in any row is* 0. For

$$A_{rr}A_{ss} - A_{rs}^{2} = A \cdot A_{rs,rs}$$

or $0=A \cdot A_{rs,rs}$, and therefore A is zero since in general $A_{rs,rs}\neq 0$. The theorem is then readily seen on expanding in terms of the elements of a row and then dividing by the common factor.

EXAMPLE: The determinant

$$A \equiv \begin{vmatrix} 2 & -1 & -1 \\ -1 & 2 & -1 \\ -1 & -1 & 2 \end{vmatrix}$$

which equals zero, has all its primary minors numerically equal or in other words has all its signed primary minors equal, and the sum of the elements in each row zero.

We observe that $a_{11}A_{11}^{1/2}+a_{12}A_{22}^{1/2}+a_{13}A_{33}^{1/2}$ and not $a_{11}A_{11}^{1/2}-a_{12}A_{22}^{1/2}+a_{13}A_{33}^{1/2}$ equals zero.

392. *If in an axisymmetric determinant the coaxial minor*

$$M \equiv \begin{vmatrix} (n \mid m_\alpha) \\ (n \mid m_\alpha) \end{vmatrix}$$

of order m is not zero and all coaxial minors of orders m+h, and m+h+1, which contain M as a minor are zero, then all minors of order m+h which contain M are zero.

That is if

$$M \equiv \begin{vmatrix} (n \mid m_\alpha) \\ (n \mid m_\alpha) \end{vmatrix} \neq 0, \quad \begin{vmatrix} (n \mid m_\alpha)(\bar{n} \mid m_\alpha \mid h_\beta) \\ (n \mid m_\alpha)(\bar{n} \mid m_\alpha \mid h_\beta) \end{vmatrix} = 0$$

and

$$D \equiv \begin{vmatrix} (n \mid m_\alpha)(\bar{n} \mid m_\alpha \mid h + 1_\gamma) \\ (n \mid m_\alpha)(\bar{n} \mid m_\alpha \mid h + 1_\gamma) \end{vmatrix} = 0$$

for all values of α, β and γ then

$$\begin{vmatrix} (n \mid m_\alpha)(\bar{n} \mid m_\alpha \mid h_i) \\ (n \mid m_\alpha)(\bar{n} \mid m_\alpha \mid h_j) \end{vmatrix} = 0$$

for all values of i and j. For

$$\begin{vmatrix} (n \mid m_\alpha)(\bar{n} \mid m_\alpha \mid h_i) \\ (n \mid m_\alpha)(\bar{n} \mid m_\alpha \mid h_i) \end{vmatrix} \cdot \begin{vmatrix} (n \mid m_\alpha)(\bar{n} \mid m_\alpha \mid h_j) \\ (n \mid m_\alpha)(\bar{n} \mid m_\alpha \mid h_j) \end{vmatrix} - \begin{vmatrix} (n \mid m_\alpha)(\bar{n} \mid m_\alpha \mid h_i) \\ (n \mid m_\alpha)(\bar{n} \mid m_\alpha \mid h_j) \end{vmatrix}^2$$

being a minor of the second order of the adjugate of D vanishes and since each factor of the first term is zero it follows that

$$\begin{vmatrix} (n \mid m_\alpha)(\bar{n} \mid m_\alpha \mid h_i) \\ (n \mid m_\alpha)(\bar{n} \mid m_\alpha \mid h_j) \end{vmatrix} = 0.$$

From this and the theorem of §234 *it follows that every minor of order m+h is zero.*

Exercise: Show that for the determinant

$$\begin{vmatrix} \cdot & 1 & 1 & 1 & \cdots \\ 1 & \cdot & d_{12} & d_{13} & \cdots \\ 1 & d_{12} & \cdot & d_{23} & \cdots \\ 1 & d_{13} & d_{23} & \cdot & \cdots \\ \cdot & \cdot & \cdot & \cdot & \cdot \end{vmatrix} = 0$$

we have

$$[22]^{1/2} + [33]^{1/2} + \cdots + [nn]^{1/2} = 0.$$

393. *If* $A = |a_{1n}|$ *be axisymmetric and positive and if the terms of the series of coaxial minors* a_{11}, $|a_{11}a_{22}|$, $|a_{11}a_{22}a_{33}|$, $\cdots$, *are all positive, then all coaxial minors are positive.*

This may be proven by induction. Given A of order three our hypothesis is that $A>0$, $|a_{11}a_{22}|>0$, $a_{11}>0$ and we are required to prove that $a_{22}>0$, $a_{33}>0$, $A_{11}>0$, $A_{22}>0$. Since $a_{11}a_{22}-a_{12}^2>0$ it follows that $a_{22}>0$. Since $A_{11}A_{33}-A_{13}^2=a_{22}A>0$, it follows that $A_{11}>0$. Since $A_{22}A_{33}-A_{23}^2=a_{11}A>0$, it follows that $A_{22}>0$. Since $A_{22}=(a_{11}a_{33}-a_{13}^2)>0$, it folows that $a_{33}>0$. The theorem is therefore true for determinants of the third order.

For the fourth order we are given $A>0$, $A_{44}>0$, $A_{34,34}>0$, $a_{11}>0$ and since the last three conditions are just those for the case of the minor A_{44} of order three we have $a_{22}>0$, $a_{33}>0$, $A_{24,24}>0$, $A_{14,14}>0$, and since

$$A_{rr}A_{44} - A_{r4}^2 = A \cdot A_{r4,r4} > 0 \text{ for } r = 1,2,3$$

we see that $A_{rr}>0$ for $r=1, 2, 3$ and all the conditions are satisfied for all four coaxial minors A_{11}, A_{22}, A_{33}, A_{44} and consequently $a_{44}>0$, $A_{12,12}>0$, $A_{13,13}>0$, $A_{23,23}>0$, and the theorem is true for determinants of order four.

Let us now assume that the theorem is true for a determinant A of order $(k-1)$; then all coaxial minors of A of all orders from 1 to $(k-1)$ are positive. Border A symmetrically with a kth row and kth column. Then the theorem will be true for the resulting determinant A' of order k and to show this it will be sufficient to show that it is true for every principal coaxial minor of A', since by hypothesis A' is positive.

We have

$$[kk][rr] - [kr]^2 = A'\begin{bmatrix} kr \\ kr \end{bmatrix}.$$

But $\left[\begin{smallmatrix} kr \\ kr \end{smallmatrix}\right]$ being a coaxial minor of A is positive and $[kk]$ is positive. Therefore $[rr]>0$, for $r=1, 2, \cdots, k-1$. The proposition is therefore true for the order k.

394. Again if $A=0$, then

$$A_{rr}A_{ss} - A_{rs}^2 = 0$$

or

$$A_{rs} = (A_{rr}A_{ss})^{1/2}.$$

The sum of the squares of the primary minors of A is $\sum A_{rr}^2+2\sum A_{rs}^2$ or $\sum A_{rr}^2+2\sum A_{rr}A_{ss}=\{\sum A_{rr}\}^2$. *That is the sum of the squares of the primary minors of an axisymmetric determinant which vanishes is equal to the square of the sum of the principal coaxial minors.*

395. By Laplace's theorem we have

$$A = (11)[11] + (12)[12] + \cdots + (1n)[1n],$$

where $[1r]$ is the cofactor of $(1r)$ in A.

But

$$A_{11}A_{rr} - A_{1r}^2 = A\begin{bmatrix}1r\\1r\end{bmatrix}$$
$$= 0,$$

if A or $\left[\begin{smallmatrix}1r\\1r\end{smallmatrix}\right]$ vanish.

In either case we may write

$$A = \{(11)[11]^{1/2} + (12)[22]^{1/2} + \cdots + (1n)[nn]^{1/2}\}[11]^{1/2}.$$

From which we see that:

If any principal coaxial minor $[ss]$ of an axisymmetric determinant A vanishes and if all the principal coaxial minors of $[ss]$ vanish then the determinant A vanishes.

396. Expanding A by Cauchy's theorem we have

$$A = (11)[11] - \sum(1i)^2\begin{bmatrix}1i\\1i\end{bmatrix} - (-1)^{i+j}2\sum a_{1i}a_{j1}\begin{bmatrix}1j\\1j\end{bmatrix}$$

$$\left\{\begin{matrix}i\\j\end{matrix} = 2, 3, \cdots, n\right\}, \quad i \neq j$$

But

$$\begin{bmatrix}1i\\1i\end{bmatrix}\begin{bmatrix}1j\\1j\end{bmatrix} - \begin{bmatrix}1i\\1j\end{bmatrix}^2 = [11]\begin{bmatrix}1ij\\1ij\end{bmatrix}$$
$$= 0,$$

if either $[11]$ or $\left[\begin{smallmatrix}1ij\\1ij\end{smallmatrix}\right]$ vanish. In either case

$$\begin{bmatrix}1i\\1j\end{bmatrix} = \left\{\begin{bmatrix}1i\\1i\end{bmatrix}\begin{bmatrix}1j\\1j\end{bmatrix}\right\}^{1/2}$$

and

$$A = (11)[11] - \left\{(12)\begin{bmatrix}12\\12\end{bmatrix}^{1/2} - (13)\begin{bmatrix}13\\13\end{bmatrix}^{1/2} + \cdots\right\}^2$$

$$= -\left\{(12)\begin{bmatrix}12\\12\end{bmatrix}^{1/2} - (13)\begin{bmatrix}13\\13\end{bmatrix}^{1/2} + \cdots\right\}^2,$$

if $[11]$ or $(11) = 0$.

This is the result obtained in §384. We get this same result, however, if instead of $[11]$ being zero we have $(11)=0$ and $\left[\begin{smallmatrix}1ij\\1ij\end{smallmatrix}\right]=0$. That is:

If one of the coaxial elements (rr) of A vanishes and if all the secondary coaxial minors of $[rr]$ vanish then A is the negative of the square of a linear homogeneous function of the elements in the rth row.

If $(1r)=0$ for $r=1, 2, \cdots, (n-2)$ it will be apparent that A will be a perfect square if the single minor $\left[\begin{smallmatrix}1n-1n\\1n-1n\end{smallmatrix}\right]$ vanishes.

397. Using Cauchy's theorem to expand $\left[\begin{smallmatrix}12\\12\end{smallmatrix}\right]$ we have

$$\begin{bmatrix}12\\12\end{bmatrix} = (33)\begin{bmatrix}123\\123\end{bmatrix} - \sum(34)^2\begin{bmatrix}1234\\1234\end{bmatrix} - 2\sum(34)(35)\begin{bmatrix}1234\\1235\end{bmatrix}$$

$$= -\left\{(34)\begin{bmatrix}1234\\1234\end{bmatrix}^{1/2} - (35)\begin{bmatrix}1235\\1235\end{bmatrix}^{1/2} + \cdots\right\}^2,$$

$$\text{if } \begin{bmatrix}123\\123\end{bmatrix} = 0.$$

Similarly

$$\begin{bmatrix}13\\13\end{bmatrix} = -\left\{(24)\begin{bmatrix}1234\\1234\end{bmatrix}^{1/2} - (25)\begin{bmatrix}1235\\1235\end{bmatrix}^{1/2} + \cdots\right\}^2$$

$$\text{if } \begin{bmatrix}123\\123\end{bmatrix} = 0 \text{ etc.}$$

Substituting these values in the expansion found for A in §396 we get

$$\begin{aligned} A = \Bigg[& (12)\left\{(34)\begin{bmatrix}1234\\1234\end{bmatrix}^{1/2} - (35)\begin{bmatrix}1235\\1235\end{bmatrix}^{1/2} + \cdots\right\} \\ & - (13)\left\{(24)\begin{bmatrix}1234\\1234\end{bmatrix}^{1/2} - (25)\begin{bmatrix}1235\\1235\end{bmatrix}^{1/2} + \cdots\right\} \\ & + (14)\left\{(25)\begin{bmatrix}1245\\1245\end{bmatrix}^{1/2} - (35)\begin{bmatrix}1345\\1345\end{bmatrix}^{1/2} + \cdots\right\} \\ & - (15)\left\{(24)\begin{bmatrix}1245\\1245\end{bmatrix}^{1/2} - (34)\begin{bmatrix}1345\\1345\end{bmatrix}^{1/2} + \cdots\right\} \\ & + \cdots \Bigg]^2 \end{aligned}$$

$$= \Bigg[\left\{\begin{vmatrix}14\\23\end{vmatrix}\begin{bmatrix}1234\\1234\end{bmatrix}^{1/2} - \begin{vmatrix}15\\23\end{vmatrix}\begin{bmatrix}1235\\1235\end{bmatrix}^{1/2} + \begin{vmatrix}16\\23\end{vmatrix}\begin{bmatrix}1236\\1236\end{bmatrix}^{1/2} - \cdots\right\}$$

$$+\left\{\begin{vmatrix}12\\45\end{vmatrix}\begin{bmatrix}1245\\1245\end{bmatrix}^{1/2} - \begin{vmatrix}13\\45\end{vmatrix}\begin{bmatrix}1345\\1345\end{bmatrix}^{1/2} + \begin{vmatrix}16\\45\end{vmatrix}\begin{bmatrix}1456\\1456\end{bmatrix}^{1/2} - \cdots\right\}$$

$$+ \cdots \Bigg]^2$$

$$= \left[\sum\alpha \sum k(-1)^{\nu}\begin{vmatrix}1 & \alpha\\ 2k & 2k+1\end{vmatrix}\cdot\begin{bmatrix}1 & 2k & 2k+1\,\alpha\\ 1 & 2k & 2k+1\,\alpha\end{bmatrix}^{1/2}\right]^2$$

when n is odd and where

$$k = 1,2,3,\cdots,$$

α takes all values from 1 to n except $2k$ and $2k+1$, and $\nu=1+2k+2k+1+\alpha=2(2k+1)+\alpha=\alpha$.

It would appear that under proper conditions this process might be continued.

It should be observed that the necessary conditions that A can be expressed as here given is that

$$(a) \qquad (11) = 0, \text{ and } \begin{bmatrix}1\,ij\\1\,ij\end{bmatrix} = 0,$$

or

$$(b) \qquad [11] = 0, \text{ and } \begin{bmatrix}1\,ij\\1\,ij\end{bmatrix} \text{ or } \begin{bmatrix}1\,ijrs\\1\,ijrs\end{bmatrix} = 0.$$

398. It is readily seen that if $\begin{bmatrix}1k\\1k\end{bmatrix}=0$, then

$$[kk] = -\left\{(13)\begin{bmatrix}13k\\13k\end{bmatrix}^{1/2} - (14)\begin{bmatrix}14k\\14k\end{bmatrix}^{1/2} + \cdots\right\}^2$$

and making this substitution in

$$(11)[11]^{1/2} - (12)[22]^{1/2} + (13)[33]^{1/2} - (14)[44]^{1/2} + \cdots - (-1)^{n-1}(1n)[nn]^{1/2}$$

it is readily seen that this factor of the norm of

$$(11)[11]^{1/2} + (12)[22]^{1/2} + \cdots + (1n)[nn]^{1/2}$$

reduces to $(11)[11]^{1/2}$ and from §395 we see that the norm is divisible by A.

That is, under the given condition

$$(11)A = \{(11)[11]^{1/2} + (12)[22]^{1/2} + \cdots + (1n)[nn]^{1/2}\}$$
$$\times\{(11)[11]^{1/2} - (12)[22]^{1/2} + \cdots + (-1)^{n-1}(1n)[nn]^{1/2}\}$$

Hence the theorem:

If the minors $\left[\begin{smallmatrix}1i\\1i\end{smallmatrix}\right]$ *of an axisymmetric determinant* A *vanish, then the norm contains* A *as a factor and vanishes if either* (11) *or* $[11]=0$.

399. *The axisymmetric determinant which has the differences* $a_1-a_2, a_1-a_3, \cdots, a_1-a_n$ *in the 1st, nth, 2nd,* $(n-1)$*th,* $\cdots$*, places of the principal diagonal, the differences* $a_2-a_3, a_2-a_4, \cdots, a_2-a_n$ *similarly disposed in the adjacent minor diagonal, the differences* $a_3-a_4, a_3-a_5, \cdots, a_3-a_n$ *similarly disposed in the next diagonal and so on, is resolvable into linear factors, being equal to the product of* $(n-1)$ *different expressions of the form*

$$a_1 + (\alpha + \alpha^{2n-2})a_2 + (\alpha^2 + \alpha^{2n-3})a_3 + \cdots + (\alpha^{n-1} + \alpha^n)a_n$$

where α *is a* $(2n-1)$*th root of unity.*

This may be established by performing the operation

$$\mathrm{col}_1 + (1 + \theta_1)\,\mathrm{col}_2 + (1 + \theta_1 + \theta_2)\,\mathrm{col}_3 + \cdots$$

where θ_r stands for $\alpha^r+\alpha^{2n-r-1}$.

For example when $n=4$ and α is an imaginary 7th root of unity, if we perform on the determinant

$$\begin{vmatrix} a_1 - a_2 & a_2 - a_3 & a_3 - a_4 \\ a_2 - a_3 & a_1 - a_4 & a_2 - a_4 \\ a_3 - a_4 & a_2 - a_4 & a_1 - a_3 \end{vmatrix}$$

the operation

$$\mathrm{col}_1 + (1 + \alpha + \alpha^6)\,\mathrm{col}_2 + (1 + \alpha + \alpha^6 + \alpha^2 + \alpha^5)\,\mathrm{col}_3,$$

the elements of the first column become

$$P,\ (1 + \alpha + \alpha^6)P,\ (1 + \alpha + \alpha^6 + \alpha^2 + \alpha^5)P,$$

where

$$P = a_1 + (\alpha + \alpha^6)a_2 + (\alpha^2 + \alpha^5)a_3 + (\alpha^3 + \alpha^4)a_4$$

showing that P is a factor. Similarly

$$a_1 + (\alpha^2 + \alpha^5)a_2 + (\alpha^3 + \alpha^4)a_3 + (\alpha + \alpha^6)a_4$$

and

$$a_1 + (\alpha^3 + \alpha^4)a_2 + (\alpha + \alpha^6)a_3 + (\alpha^2 + \alpha^5)a_4$$

are also factors.

400. Let $A = |a_{1n}|$, where $a_{rs} = a_{sr}$, and let $S = A_{11} + A_{22} + \cdots + A_{nn}$. Then

$$\left.\begin{aligned} A\{\mathcal{A}_{12,12} + \mathcal{A}_{13,13} + \cdots + \mathcal{A}_{1n,1n}\} \qquad & \\ = (\mathcal{A}_{11}\mathcal{A}_{22} - \mathcal{A}_{12}^2) + \cdots + (\mathcal{A}_{11}\mathcal{A}_{nn} - \mathcal{A}_{1n}^2) & \\ = -(\mathcal{A}_{11} - S)\mathcal{A}_{11} - \mathcal{A}_{12}^2 - \mathcal{A}_{13}^2 - \cdots - \mathcal{A}_{1n}^2, & \\ A\{\mathcal{A}_{12,12} + \mathcal{A}_{23,23} + \cdots + \mathcal{A}_{2n,2n}\} \qquad & \\ = -\mathcal{A}_{12}^2 - (\mathcal{A}_{22} - S)\mathcal{A}_{22} - \mathcal{A}_{23}^2 - \cdots - \mathcal{A}_{2n}^2 & \end{aligned}\right\} \quad (1)$$

etc. Denote

$$\mathcal{A}_{12,12} + \mathcal{A}_{13,13} + \cdots + \mathcal{A}_{1n,1n} \text{ by } M_{11}$$

$$\mathcal{A}_{13,23} + \cdots + \mathcal{A}_{1n,2n} \text{ by } M_{12} \qquad \text{etc.}$$

Then

$$\begin{aligned} a_{11}M_{11} + a_{12}M_{12} + \cdots + a_{1n}M_{1n} & \\ = a_{11}(\mathcal{A}_{12,12} + \mathcal{A}_{13,13} + \cdots + \mathcal{A}_{1n,1n}) & \\ + a_{12}(+\mathcal{A}_{13,23} + \cdots + \mathcal{A}_{1n,2n}) & \\ + a_{13}\ (\mathcal{A}_{12,32} + \cdots + \mathcal{A}_{1n,3n}) & \\ + \text{etc.} & \\ = \mathcal{A}_{22} + \cdots + \mathcal{A}_{nn} = S - \mathcal{A}_{11}. & \end{aligned}$$

Similarly

$$a_{21}M_{11} + a_{22}M_{12} + \cdots + a_{2n}M_{1n} = -\mathcal{A}_{12}$$

$$a_{31}M_{11} + a_{32}M_{12} + \cdots + a_{3n}M_{1n} = -\mathcal{A}_{13} \qquad \text{etc.}$$

Solving for the M's we have

$$-M_{11}\mathcal{A} = (\mathcal{A}_{11} - S)\mathcal{A}_{11} + \mathcal{A}_{12}^2 + \cdots + \mathcal{A}_{1n}^2$$

$$-M_{12}\mathcal{A} = (\mathcal{A}_{11} - S)\mathcal{A}_{12} + \mathcal{A}_{12}\mathcal{A}_{22} + \cdots + \mathcal{A}_{1n}\mathcal{A}_{2n}$$

. . . .

$$-M_{rs}\mathcal{A} = \mathcal{A}_{r1}\mathcal{A}_{s1} + \mathcal{A}_{r2}\mathcal{A}_{s2} + \cdots + (\mathcal{A}_{rr} - S)\mathcal{A}_{sr} + \cdots$$

From these relations we have

$$\begin{vmatrix} \mathcal{A}_{11}-S & \mathcal{A}_{12} & \cdots \mathcal{A}_{1n} \\ \mathcal{A}_{21} & \mathcal{A}_{22}-S & \cdots \mathcal{A}_{2n} \\ \cdot & \cdot & \cdot \\ \mathcal{A}_{n1} & \mathcal{A}_{n2} & \cdots \mathcal{A}_{nn}-S \end{vmatrix} \cdot \begin{vmatrix} \mathcal{A}_{11} & \mathcal{A}_{12} \cdots \mathcal{A}_{1n} \\ \mathcal{A}_{21} & \mathcal{A}_{22} \cdots \mathcal{A}_{2n} \\ \cdot & \cdot \\ \mathcal{A}_{n1} & \mathcal{A}_{n2} \cdots \mathcal{A}_{nn} \end{vmatrix}$$

$$= (-1)^n \cdot A^n \begin{vmatrix} M_{11} & M_{12} \cdots M_{1n} \\ M_{21} & M_{22} \cdots M_{2n} \\ \cdot & \cdot \\ M_{n1} & M_{n2} \cdots M_{nn} \end{vmatrix}.$$

But the second factor on the left is equal to A^{n-1} and therefore

$$\begin{vmatrix} M_{11} & M_{12} \cdots M_{1n} \\ M_{21} & M_{22} \cdots M_{2n} \\ \cdot & \cdot \\ M_{n1} & M_{n2} \cdots M_{nn} \end{vmatrix} = \frac{(-1)^n}{A} \begin{vmatrix} \mathcal{A}_{11}-S & \mathcal{A}_{12} & \cdots \mathcal{A}_{1n} \\ \mathcal{A}_{21} & \mathcal{A}_{22}-S & \cdots \mathcal{A}_{2n} \\ \cdot & \cdot & \cdot \\ \mathcal{A}_{n1} & \mathcal{A}_{n2} & \cdots \mathcal{A}_{nn}-S \end{vmatrix}$$

The determinant on the right may be expanded with the result

$$A\{A^{n-2} - A^{n-3}S_1S_{n-1} + A^{n-4}S_2 S_{n-1}^2 - \cdots - (-1)^n S_{n-2}S_{n-1}^{n-2}$$

where S_r = the sum of the coaxial minors of A of order r. Therefore

$$|M_{1n}| = S_{n-2}S_{n-1}^{n-2} - A \cdot S_{n-3} S_{n-1}^{n-3} + \cdots + (-1)^n A^{n-2}.$$

If $S_{n-1} \equiv S = 0$ then $|M_{1n}| = (-1)^n A^{n-2} = 0$, if $A=0$. But when $A=0$ and $S=0$, (1) shows that $A_{rs}=0$ for all values of r and s.

401. The norm contains 2^{n-1} factors and if we take as the first two

$$\{(11)[11]^{1/2} + (12)[22]^{1/2} + \cdots + (1n)nn]^{1/2}\}$$
$$\{(11)[11]^{1/2} - (12)[22]^{1/2} + \cdots + (-1)^{n-1}(1n)[nn]^{1/2}\}$$

it is readily seen that all the others may be taken in pairs and each pair derived from this pair by changing the sign of one or more of the elements (12), (13), $\cdots$, $(1n)$.

Thus in the case where $n=5$ we have, writing but the signs of the terms,

1. $(+ + + + +)(+ - + - +)$
2. $(+ - + + +)(+ + + - +)$

3. $(+ + - + +)(+ - - - +)$
4. $(+ + + + -)(+ - + - -)$
5. $(+ - - + +)(+ + - - +)$
6. $(+ - + + -)(+ + + - -)$
7. $(+ + - + -)(+ - - - -)$
8. $(+ + - - -)(+ - - + -)$

Here product 2 may be obtained from 1 by changing the sign of the element (12), product 3 may be obtained from 1 by changing the sign of the element (13), etc.

In §398 we saw that product

1. was equal to $A\cdot(11)$ and now we see that product
2. is equal to $A_1(11)$
3. is equal to $A_2(11)$

. . . .

8. is equal to $A_7(11)$,

where A_1 is derived from A by changing the signs of (12) and (21), A_2 is derived from A by changing the signs of (13) and (31), etc., A_7 is derived from A by changing the signs of (13), (14), (15) and (31), (41), (51). The norm is therefore equal to

$$A\cdot A_1A_2\cdots A_7(11)^8, \text{ if } \begin{bmatrix}1\,i\\ 1\,i\end{bmatrix} = 0, \qquad (i = 2,3,\cdots,n).$$

402. It should be observed that if (11) as well as $\left[\begin{smallmatrix}1i\\1i\end{smallmatrix}\right]$ vanishes then the eight products of pairs of factors of the norm reduce to half that number. If it is (12) instead of (11) that vanishes then there would be but eight distinct factors in the norm but they would be paired differently.

If $(12)=(13)=0$, then the number of factors in the norm is halved again, so that if r is the number of elements (12), (13), $\cdots$, which vanish then the number of factors in the norm is 2^{n-r-1}.

403. If $A=|a_{1n}|$ where $a_{rs}=a_{sr}$ and $a_{1r}=0$ for $r=2, 3, \cdots, n-3$ then by Cauchy's theorem

$$A = (11)[11] - (1\,n-1)^2\begin{bmatrix}1 & n-1\\ 1 & n-1\end{bmatrix}$$

$$- (1n)^2\begin{bmatrix}1n\\ 1n\end{bmatrix} + 2(1\,n-1)(1n)\begin{bmatrix}1 & n-1\\ 1 & n\end{bmatrix}$$

and

$$A_1 = (11)[11] - (1\,n-1)^2 \begin{bmatrix} 1\,n-1 \\ 1\,n-1 \end{bmatrix} - (1n)^2 \begin{bmatrix} 1n \\ 1n \end{bmatrix} - 2(1\,n-1)(1n) \begin{bmatrix} 1\,n-1 \\ 1\quad n \end{bmatrix}$$

where A_1 is derived from A by changing the sign of $(1n)$.

The product

$$A \cdot A_1 = \left\{ (11)[11] - (1\,n-1)^2 \begin{bmatrix} 1\,n-1 \\ 1\,n-1 \end{bmatrix} - (1n)^2 \begin{bmatrix} 1n \\ 1n \end{bmatrix} \right\}^2 - 4(1\,n-1)^2(1n)^2 \begin{bmatrix} 1\,n-1 \\ 1\quad n \end{bmatrix}^2$$

$$= - \begin{vmatrix} P & Q \\ Q & -2(1\,n-1)^2(1n)^2 \end{vmatrix} = \Delta, \text{ say},$$

in which

$$[11] \begin{bmatrix} 1\,n-1\,n \\ 1\,n-1\,n \end{bmatrix} - \begin{bmatrix} 1\,n-1 \\ 1\,n-1 \end{bmatrix} \begin{bmatrix} 1n \\ 1n \end{bmatrix}$$

has been put for

$$- \begin{bmatrix} 1\,n-1 \\ 1\quad n \end{bmatrix}^2.$$

and where

$$P = 2[11] \begin{bmatrix} 1\,n-1\,n \\ 1\,n-1\,n \end{bmatrix} - 2 \begin{bmatrix} 1\,n-1 \\ 1\,n-1 \end{bmatrix} \begin{bmatrix} 1n \\ 1n \end{bmatrix}$$

$$Q = (11)[11] - (1\,n-1)^2 \begin{bmatrix} 1\,n-1 \\ 1\,n-1 \end{bmatrix} - (1n)^2 \begin{bmatrix} 1n \\ 1n \end{bmatrix}.$$

But we may write

$$\Delta = \begin{vmatrix} P & Q & 0 & 0 \\ Q & -2(1\,n-1)^2(1n)^2 & 0 & 0 \\ \begin{bmatrix} 1n \\ 1n \end{bmatrix} & 0 & 0 & 1 \\ \begin{bmatrix} 1\,n-1 \\ 1\,n-1 \end{bmatrix} & 0 & 1 & 0 \end{vmatrix}$$

and the operations

$$r_1 + \begin{bmatrix} 1\,n-1 \\ 1\,n-1 \end{bmatrix} r_3 + \begin{bmatrix} 1n \\ 1n \end{bmatrix} r_4,\ r_2 + (1\,n-1)^2 r_4 + (1n)^2 r_3,$$

$$c_2 + (1n-1)^2 c_4 + (1n)^2 c_3$$

in succession give

$$\Delta = \begin{vmatrix} 2[11]\begin{bmatrix} 1\,n-1\,n \\ 1\,n-1\,n \end{bmatrix} & (11)[11] & \begin{bmatrix} 1n \\ 1n \end{bmatrix} & \begin{bmatrix} 1\,n-1 \\ 1\,n-1 \end{bmatrix} \\ (11)[11] & 0 & (1\,n-1)^2 & (1n)^2 \\ \begin{bmatrix} 1n \\ 1n \end{bmatrix} & (1\,n-1)^2 & 0 & 1 \\ \begin{bmatrix} 1\,n-1 \\ 1\,n-1 \end{bmatrix} & (1n)^2 & 1 & 0 \end{vmatrix}$$

$$= \begin{vmatrix} 2\begin{bmatrix} 1\,n-1\,n \\ 1\,n-1\,n \end{bmatrix} & (11) & \begin{bmatrix} 1n \\ 1n \end{bmatrix} & \begin{bmatrix} 1\,n-1 \\ 1\,n-1 \end{bmatrix} \\ (11) & 0 & (1\,n-1)^2 & (1n)^2 \\ \begin{bmatrix} 1n \\ 1n \end{bmatrix} & (1\,n-1)^2 & 0 & [11] \\ \begin{bmatrix} 1\,n-1 \\ 1\,n-1 \end{bmatrix} & (1n)^2 & [11] & 0 \end{vmatrix}$$

by dividing the first and second rows by $[11]$ and multiplying the third and fourth columns by the same.

The norm is the product

$$\begin{aligned} &\{(11)[11]^{1/2} + (1\,n-1)[n-1\,n-1]^{1/2} \\ &+ (1n)[nn]^{1/2}\}\{(11)[11]^{1/2} \\ &- (1\,n-1)[n-1\,n-1]^{1/2} + (1n)[nn]^{1/2}\} \\ \times &\{(11)[11]^{1/2} - (1\,n-1)[n-1\,n-1]^{1/2} \\ &- (1n)[nn]^{1/2}\}\{(11)[11]^{1/2} \\ &+ (1\,n-1)[n-1\,n-1]^{1/2} - (1n)[nn]^{1/2}\} \end{aligned}$$

which is equal to

$$\Big[\{(11)[11]^{1/2}+(1n)[nn]^{1/2}\}^2-(1\,n-1)^2[n-1\;n-1]\Big]\Big[\{(11)[11]^{1/2}$$

$$-(1n)[nn]^{1/2}\}^2-(1\,n-1)^2[n-1\;n-1]\Big]=(11)\Big\{(11)[11]$$

$$-(1\,n-1)^2\begin{bmatrix}1\;n-1\\1\;n-1\end{bmatrix}+(1n)^2\begin{bmatrix}1n\\1n\end{bmatrix}+2(1n)[11]^{1/2}[nn]^{1/2}\Big\}$$

$$\times(11)\Big\{(11)[11]$$

$$-(1\,n-1)^2\begin{bmatrix}1\;n-1\\1\;n-1\end{bmatrix}+(1n)^2\begin{bmatrix}1n\\1n\end{bmatrix}-2(1n)[11]^{1/2}[nn]^{1/2}\Big\}$$

$$=(11)^2\Big[\Big\{(11)[11]$$

$$-(1\,n-1)^2\begin{bmatrix}1\;n-1\\1\;n-1\end{bmatrix}+(1n)^2\begin{bmatrix}1n\\1n\end{bmatrix}^2\Big\}-4(1n)^2[11][nn]\Big]$$

and this may readily be shown equal to $(11)^2A\cdot A_1$. That is the norm is divisible by A and the quotient is $(11)^2A_1$.

If $(1\;n-1)$ is zero then the norm is equal to $(11)A$.

The norm of $x^{1/2}+y^{1/2}+z^{1/2}$ is readily seen to be

$$\begin{vmatrix}0 & x^{1/2} & y^{1/2} & z^{1/2}\\ x^{1/2} & 0 & z^{1/2} & y^{1/2}\\ y^{1/2} & z^{1/2} & 0 & x^{1/2}\\ z^{1/2} & y^{1/2} & x^{1/2} & 0\end{vmatrix}\quad\text{or}\quad\begin{vmatrix}0 & 1 & 1 & 1\\ 1 & 0 & z & y\\ 1 & z & 0 & x\\ 1 & y & x & 0\end{vmatrix}$$

404. If $A=|a_{1n}|$, where $a_{rs}=a_{sr}$ and $(1r)=0$ for $r=1, 2, \cdots, n-3$, then the norm is the product

$$\{(1\,n-2)[n-2\;n-2]^{1/2}+(1\,n-2)[n-1\;n-1]^{1/2}+(1n)[nn]^{1/2}\}$$
$$\times\{(1\,n-2)[n-2\;n-2]^{1/2}-(1\,n-2)[n-1\;n-1]^{1/2}+(1n)[nn]^{1/2}\}$$
$$\times\{(1\,n-2)[n-2\;n-2]^{1/2}-(1\,n-2)[n-1\;n-1]^{1/2}-(1n)[nn]^{1/2}\}$$
$$\times\{(1\,n-2)[n-2\;n-2]^{1/2}+(1\,n-2)[n-1\;n-1]^{1/2}-(1n)[nn]^{1/2}\}$$

$$=\begin{vmatrix}0 & 1 & 1 & 1\\ 1 & 0 & (1n)^2[nn] & (1\,n-1)^2[n-1\;n-1]\\ 1 & (1n)^2[nn] & 0 & (1\,n-2)^2[n-2\;n-2]\\ 1 & (1\,n-1)^2[n-1\;n-1] & (1\,n-2)^2[n-2\;n-2] & 0\end{vmatrix}=N,\text{ say}.$$

By §403 easy transformations give

$$N = \begin{vmatrix} 0 & (1\,n-2)^2 & (1\,n-1)^2 & (1n)^2 \\ (1\,n-2)^2 & 0 & [nn] & [n-1\,n-1] \\ (1\,n-1)^2 & [nn] & 0 & [n-2\,n-2] \\ (1n)^2 & [n-1\,n-1] & [n-2\,n-2] & 0 \end{vmatrix}$$

Substituting in this for

$$[n-2\,n-2] = -(1\,n-1)^2 \begin{bmatrix} 1\,n-2\,n-1 \\ 1\,n-2\,n-1 \end{bmatrix}$$

$$-(1n)^2 \begin{bmatrix} 1\,n-2\,n \\ 1\,n-2\,n \end{bmatrix} + 2(1\,n-1)(1n) \begin{bmatrix} 1\,n-2\,n-1 \\ 1\,n-2\,n-1 \end{bmatrix}$$

and similarly for $[n-1\,n-1]$ and $[nn]$, we get, after performing the operations

$$r_2 + \begin{bmatrix} 1\,n-1\,n \\ 1\,n-1\,n \end{bmatrix} r_1, \quad r_3 + \begin{bmatrix} 1\,n-2\,n \\ 1\,n-2\,n \end{bmatrix} r_1,$$

$$r_4 + \begin{bmatrix} 1\,n-2\,n-1 \\ 1\,n-2\,n-1 \end{bmatrix} r_1$$

then

$$c_2 + \begin{bmatrix} 1\,n-1\,n \\ 1\,n-1\,n \end{bmatrix} c_1, \quad c_3 + \begin{bmatrix} 1\,n-2\,n \\ 1\,n-2\,n \end{bmatrix} c_1,$$

$$c_4 + \begin{bmatrix} 1\,n-2\,n-1 \\ 1\,n-2\,n-1 \end{bmatrix} c_1$$

$$N = 4(1\,n-2)^2(1\,n-1)^2(1n)^2 \times$$

$$\begin{vmatrix} 0 & (1\,n-2) & (1\,n-1) & (1n) \\ (1\,n-2) & \begin{bmatrix} 1\,n-1\,n \\ 1\,n-1\,n \end{bmatrix} & \begin{bmatrix} 1\,n\,n-2 \\ 1\,n\,n-1 \end{bmatrix} & \begin{bmatrix} 1\,n-1\,n-2 \\ 1\,n-1\;\;n \end{bmatrix} \\ (1\,n-1) & \begin{bmatrix} 1\,n\,n-1 \\ 1\,n\,n-2 \end{bmatrix} & \begin{bmatrix} 1\,n-2\,n \\ 1\,n-2\,n \end{bmatrix} & \begin{bmatrix} 1\,n-2\,n-1 \\ 1\,n-2\;\;n \end{bmatrix} \\ (1n) & \begin{bmatrix} 1\,n-1\,n-2 \\ 1\,n-1\;\;n \end{bmatrix} & \begin{bmatrix} 1\,n-2\,n-1 \\ 1\,n-2\;\;n \end{bmatrix} & \begin{bmatrix} 1\,n-2\,n-1 \\ 1\,n-2\,n-1 \end{bmatrix} \end{vmatrix}$$

$$= 4(1\,n-2)^2(1\,n-1)^2(1n)^2 \begin{bmatrix} 1\,n-2\,n-1\,n \\ 1\,n-2\,n-1\,n \end{bmatrix} \cdot A$$

by Sylvester's theorem.

The norm will evidently vanish when any one of these factors vanishes.

Exercises: Set XIX

1. Show that

$$\begin{vmatrix} \cdot & x & y & z \\ x & \cdot & z & y \\ y & z & \cdot & x \\ z & y & x & \cdot \end{vmatrix} = (x+y+z)(x+y-z)(x-y+z)(-x+y+z).$$

Give the general equation and show that if $x, y, z, \cdots,$ are quadratic radicals the product of the linear factors is rational. (Briochi).

2. If

$$A \equiv \begin{vmatrix} 1 & 1 & 1 & \cdots & 1 \\ 1 & 1+a_1 & 1 & \cdots & 1 \\ 1 & 1 & 1+a_2 & \cdots & 1 \\ \cdot & \cdot & \cdot & \cdot & \cdot \\ & & & \cdots & 1+a_n \end{vmatrix}_{n+1},$$

$$\text{and } A_{11} \equiv \begin{vmatrix} 1+a_1 & 1 & \cdots & 1 \\ 1 & 1+a_2 & \cdots & 1 \\ \cdot & \cdot & \cdot & \cdot \\ 1 & 1 & \cdots & 1+a_n \end{vmatrix}.$$

then show that $A = a_1 a_2 \cdots a_n$, and

$$A_{11} \equiv A\left(1 + \frac{1}{a_1} + \frac{1}{a_2} + \cdots + \frac{1}{a_n}\right).$$

3. Show that

$$\begin{vmatrix} \cdot & 1 & 1 & 1 & \cdots \\ 1 & \cdot & a+b & a+c & \cdots \\ 1 & b+a & \cdot & b+c & \cdots \\ 1 & c+a & c+b & \cdot & \cdots \end{vmatrix} = abc \cdots \left(\frac{1}{a} + \frac{1}{b} + \frac{1}{c} + \cdots\right)$$

4. Show that

$$\begin{vmatrix} 1 & \cdot & \cdot & a_1 & a_2 & a_3 \\ \cdot & 1 & \cdot & b_1 & b_2 & b_3 \\ \cdot & \cdot & 1 & c_1 & c_2 & c_3 \\ a_1 & b_1 & c_1 & 1 & \cdot & \cdot \\ a_2 & b_2 & c_2 & \cdot & 1 & \cdot \\ a_3 & b_3 & c_3 & \cdot & \cdot & 1 \end{vmatrix}$$

$$= 1 - \sum a_1^2 + \sum |a_1 b_2|^2 - |a_1 b_2 c_3|^2$$

and in general

$$\begin{vmatrix} (1) & \Delta \\ \Delta' & (1) \end{vmatrix} = 1 - \sum M_1^2 + \sum M_2^2 - \sum M_3^2 + \cdots$$

where Δ' is the conjugate of Δ and $\sum M_r^2$ denotes the sum of the squares of all minors of Δ of order r.

405. If we denote the determinant

$$\begin{vmatrix} \sigma_1 - (11) & -(12) \cdots & -(1n) \\ -(21) & \sigma_2 - (22) \cdots & -(2n) \\ \cdots & \cdots & \cdots \\ -(n1) & -(n2) \cdots & \sigma_n - (nn) \end{vmatrix} \text{ by } (0,1,2,\cdots,n)$$

where $(rs)=(sr)$ and $\sigma_r=(r0)+(r1)+\cdots+(rn)$ then the co-factor of (01) is readily seen to be $(\overline{0+1}, 2, 3, \cdots, n)$ because it occurs only in the element $\sigma_1-(11)$. Again the co-factor of (01)(02) is $(\overline{0+1+2}, 3, \cdots, n)$ and finally the co-factor of $(01)(02)\cdots(0n)$ is 1.

If we add all the other rows to the first and then all the other columns to the first the determinant is not altered in value but it will be observed that these operations have the effect of interchanging the two numbers 0 and 1. Similarly it may be seen that interchanging any two umbrae does not alter the value of the determinant. There is no term of the determinant free from the umbra 0, for if we place all the elements $(0r)$, $(r=1, 2, \cdots, n)$ equal to zero the resulting determinant is seen to be zero. It follows therefore that

$$\begin{aligned}(0, 1, 2, \cdots, n) = & \sum(01)(1,2,\cdots,n) \\ & + \sum(01)(02)(\overline{1+2}, \cdots, n) \\ & + \cdots\end{aligned}$$

$$+ \sum(01)(02)\cdots(0r)(\overline{1+2+\cdots+r}, \\ r+1, \cdots, n)$$

$$+ \cdots$$

$$+ (01)(02)\cdots(0n)$$

Exercises: Set XX

1. Show that

(*a*)
$$\begin{vmatrix} a+b+c+d & a-b-c+d & a-b+c-d \\ a-b-c+d & a+b+c+d & a+b-c-d \\ a-b+c-d & a+b-c-d & a+b+c+d \end{vmatrix}$$
$$= 16(abc + abd + acd + bcd)$$

(*b*)
$$\begin{vmatrix} a & d & c \\ d & a & b \\ c & b & a \end{vmatrix}$$
$$= \tfrac{1}{4}[(a+b+c+d)(a+b-c-d)(a-b+c-d) + \cdots]$$

2. Show that

$$\Delta \equiv \begin{vmatrix} \cdot & a_1 + a_2 & a_1 + a_3 & \cdots \\ a_1 + a_2 & \cdot & a_2 + a_3 & \cdots \\ a_1 + a_3 & a_3 + a_2 & \cdot & \cdots \\ \cdots & \cdots & \cdots & \cdots \end{vmatrix}_n$$

$$= (-1)^{n-1}2^n a_1 a_2 \cdots a_n \cdot \left[n - 1 + \sum \frac{a_r - a_s}{4a_r a_s}\right]$$

where r and s are duads from 1 to n.

Also obtain the value of Δ in the form

$$(-2)^{n-2}\{(n-1)(n-4)a_1a_2\cdots a_n - \sum a_1^2 a_2 \cdots a_{n-1}\}$$

and note the result when $n = 4$.

3. In the determinant

$$\begin{vmatrix} 2a_1a_2 & a_1b_2 + a_2b_1 & a_1c_2 + a_2c_1 & \cdots \\ a_1b_2 + a_2b_1 & 2b_1b_2 & b_1c_2 + b_2c_1 & \cdots \\ a_1c_2 + a_2c_1 & b_1c_2 + b_2c_1 & 2c_1c_2 & \cdots \\ \cdots & \cdots & \cdots & \cdots \end{vmatrix}_n$$

show that (a) all coaxial minors of order greater than 2 are zero and (b) the sum of all coaxial minors of order two is

$$\left\| \begin{matrix} a_1 & b_1 & c_1 & \cdots \\ a_2 & b_2 & c_2 & \cdots \end{matrix} \right\|^2 = -\left(\sum a_1^2 \sum a_2^2 - S_n^2 \right)$$

where $S_n = a_1a_2 + b_1b_2 + \cdots$.

4. If Δ denote the determinant of the last example with S_n subtracted from each diagonal element, show that

$$\Delta = (-1)^n S_n^{n-2} \sum a_1^2 \sum a_2^2 .$$

Show that the result when all the a's become equal to a and all the b's equal to b is $(-1)^{n-1}(\sum a^2)^n$.

Give the results when all the letters but two are made zero; when all but three; etc.

Zero-Axial Skew Determinants

406. As we have seen (§26) a determinant having conjugate elements equal but opposite in sign (i.e. $a_{rs} = -a_{sr}$) is called a *skew* determinant; and if in addition $a_{rr} = 0$ it is called a *zero-axial skew* determinant. Zero-axial skew determinants have sometimes been called *skew-symmetric.*

The word *skew* as here used is meant to be contrasted with *symmetric*, every kind of symmetric determinant being matched by its corresponding skew determinant. But just as the term, symmetric, is often used in the narrower sense of 'axi-symmetric', so 'skew' is almost universally taken to mean 'skew with respect to the principal diagonal.'

407. In the case of zero-axial skew determinants it readily appears that (1) coaxial minors are zero-axial skew, (2) conjugate minors are equal or differ only in sign, according as they are of even or of odd order, (3) the adjugate determinant is skew if of even order and axi-symmetric if of odd order.

408. *A zero-axial skew determinant of odd order vanishes.*

For changing the signs of all the elements amounts to interchanging rows and columns which (§37) does not alter the determinant, yet the determinant being of odd order is changed in sign; therefore it is zero.

The adjugate would also be zero.

409. *The adjugate of a zero-axial skew determinant of even order is zero-axial.* This follows at once from §408.

410. For any zero-axial skew determinant A of order n we have

$$A_{rr}A_{ss} - A_{rs}A_{sr} = A \cdot A_{rs,rs}. \tag{1}$$

First, suppose n is even. Then A_{rr} and $A_{ss}=0$ (§408) and $A_{rs}=-A_{sr}$ (§407). Therefore

$$A_{rs}^2 = A \cdot A_{rs,rs}, \quad \text{or} \quad A = \frac{A_{rs}^2}{A_{rs,rs}} \tag{2}$$

and

$$A_{rs} = (A \cdot A_{rs,rs}).^{1/2} \tag{3}$$

The relation (3) shows that if A is zero then A_{rs} is zero for all r and s. *That is if a zero-axial skew of even order vanishes so do all its principal minors.*

From (3) and $A^2_{rsp,rst}=A_{rs,rs}A_{rstp,rstp}$ we see that A_{rs} and $A_{rsp,rst}$ have the common factor $A^{1/2}_{rs,rs}$.

Expanding A by Laplace's theorem we have

$$\begin{aligned} A &= (-1)^r\{a_{r1}A_{r1} - a_{r2}A_{r2} + \cdots\} \\ &= (-1)^r\{a_{r1}A^{1/2}\cdot A^{1/2}_{1r,1r} - a_{r2}A^{1/2}\cdot A^{1/2}_{2r,2r} + \cdots\} \\ &= (-1)^r A^{1/2}\{a_{r1}A^{1/2}_{1r,1r} - a_{r2}A^{1/2}_{2r,2r} + \cdots\} \end{aligned}$$

or

$$A^{1/2} = (-1)^r\{a_{r1}A^{1/2}_{1r,1r} - a_{r2}A^{1/2}_{2r,2r} + \cdots\}. \tag{4}$$

Second, suppose n is odd. In this case $A=0$ and (1) takes the form

$$A_{rs}^2 = A_{rr}A_{ss}$$

or

$$A_{rs} = (A_{rr}A_{ss})^{1/2}. \tag{5}$$

Giving s the values $1, 2, \cdots, n$ and substituting for $A_{r1}, A_{r2} \cdots,$ in the expansion of A we get

$$\begin{aligned} A &= (-1)^r\{a_{r1}A^{1/2}_{rr}A^{1/2}_{11} - a_{r2}A^{1/2}_{rr}A^{1/2}_{22} + \cdots\} \\ &= (-1)^r\{a_{r1}A^{1/2}_{11} - a_{r2}A^{1/2}_{22} + \cdots\}A^{1/2}_{rr} \end{aligned}$$

or

$$a_{r1}A^{1/2}_{11} - a_{r2}A^{1/2}_{22} + \cdots = 0, \tag{6}$$

since in general A_{rr} is not zero.

From (5)

$$(7) \qquad A_{r1}:A_{r2}:A_{r3}:\cdots = A_{11}^{1/2}:A_{22}^{1/2}:A_{33}^{1/2}:\cdots$$

which shows that the ratios are independent of r.

Relation (5) shows *that every principal coaxial minor is a factor of the adjugate of A.*

411. From (2) §410 we have the theorem:

A zero-axial skew determinant of even order is the second power of a rational function of the elements.

For (2) shows that it is true for A of even order n if it is true for $A_{rs,rs}$ of even order $n-2$, but it is obviously true for $n=2$, and therefore true in general.

412. Looking at the coaxial minors of order two of A we see that their product

$$a_{12}^2 a_{34}^2 a_{56}^2 \cdots a_{2n-1,2n}^2$$

is a term of the determinant and that therefore one square root of A contains the term $+a_{12}a_{34}\cdots a_{2n-1,2n}$ and the other $-a_{12}a_{34}\cdots a_{2n-1,2n}$.

The square root of a zero-axial skew determinant of order $2n$ which contains as a positive term the product of the elements in the places (1, 2), (3, 4), $\cdots$, $(2n-1, 2n)$ is called a *Pfaffian* function of the whole of the elements lying on the same side of the zero diagonal of the elements mentioned.

The usual notation for a Pfaffian is $'\!\mid a_{1\ 2n}\mid$ or $ff(a_{1\ 2n})$. Another notation is to use

$$\left|\begin{array}{ccc} a_2 & a_3 & a_4 \\ & b_3 & b_4 \\ & & c_4 \end{array}\right|$$

for $(a_2c_4 - a_3b_4 + b_3a_4)$. A Pfaffian which is of the nth degree in its element is said to be of the nth order.

413. A determinant having the complementary minor of one of its corner elements a zero-axial skew determinant, is called a *bordered* zero-axial skew determinant. The word is similarly applied in connection with other special determinant forms.

The first row of a Pfaffian of the nth order contains $2n-1$ elements: the line through the 1st column and 2nd row contains the same number: so also do the lines through the 2nd column and 3rd row, through the 3rd column and 4th row, and so on to the last column.

These $2n$ lines each containing $2n-1$ elements may be called the *frame-lines* of the Pfaffian and numbered 1st, 2nd, 3rd, etc. in order. Evidently every element of the Pfaffian belongs to *two* frame-lines, and is fully specified as to position when the numbers of these are given. In the Pfaffian $^{\|} a_{1,2n} |$, the suffixes of each element indicate the number of the frame-lines to which it belongs, the smaller number being always written first.

414. The terms *minor* and *adjugate* and the subsidiary terms connected with them are used in regard to Pfaffians just as in regard to determinants. Thus, if the frame-line of any element be deleted, the Pfaffian whose elements are in order the elements left is called a *first* or *primary* minor of the original Pfaffian and the *complementary* (minor) of the said element. The notation, also, which corresponds to this nomenclature may be made quite analogous for the two functions:

415. *A bordered zero-axial skew determinant is expressible as the product of two Pfaffians.*

As we have seen from §413 A_{rs} is a bordered zero-axial skew determinant and the relations

$$A_{rs} = A_{rr}^{1/2} A_{ss}^{1/2} \qquad \text{(when } n \text{ is odd)}$$
$$A_{rs} = A^{1/2} A_{rs,rs}^{1/2} \qquad \text{(when } n \text{ is even)}$$

where the two factors on the right are Pfaffians, proves the theorem.

From this it appears that *every principal minor of a zero-axial skew determinant is equal to the product of two Pfaffians.*

416. For a determinant of order $2n$ the relation (4), §410 gives an expression for a Pfaffian of order n in terms of Pfaffians of order $n-2$.

417. *Any determinant of the 2nth order is expressible as a Pfaffian of the nth order, viz.* $|a_{1,2n}| = {}^{\|} P_{1,2n} |$, where

$$P_{rs} = (a_{r1}, a_{r2}, \cdots, a_{r,2n})(a_{s2}, -a_{s1}, a_{s4}, -a_{s3}, \cdots, a_{s,2n}, -a_{s,2n-1})$$

Writing $|a_{1,2n}|$ in the form

$$\begin{vmatrix} a_{12} & -a_{11} & a_{14} & -a_{13} & \cdots & a_{1,2n} & -a_{1,2n-1} \\ a_{22} & -a_{21} & a_{24} & -a_{23} & \cdots & a_{2,2n} & -a_{2,2n-1} \\ \cdot & \cdot & \cdot & \cdot & \cdot & \cdot & \cdot \\ a_{2n,2} & -a_{2n,1} & a_{2n,4} & -a_{2n,3} & \cdots & a_{2n,2n} & -a_{2n,2n-1} \end{vmatrix}$$

and multiplying the two forms together row-wise we have

$$|a_{1,2n}|^2 = |P_{1,2n}|.$$

But it is evident that $P_{rs} = -P_{sr}$ and that $P_{rr} = 0$; hence, on extracting the square root, the theorem follows.

Treating the rows of $|a_{1,2n}|$ in the same way as the columns have been treated, and multiplying the result column-wise by $|a_{1,2n}|$, we obtain a different but equivalent Pfaffian.

418. *Any determinant may be expressed as a Pfaffian of the same order.*

Starting with the determinant Δ of order $2n$

$$\Delta \equiv \begin{vmatrix} \frac{1}{2}(a_{11}-a_{11}) & \frac{1}{2}(a_{12}-a_{21}) & \cdots & \frac{1}{2}(a_{12}+a_{21}) & \frac{1}{2}(a_{11}+a_{11}) \\ \frac{1}{2}(a_{21}-a_{12}) & \frac{1}{2}(a_{22}-a_{22}) & \cdots & \frac{1}{2}(a_{22}+a_{22}) & \frac{1}{2}(a_{21}+a_{12}) \\ \cdot & \cdot & \cdot & \cdot & \cdot \\ -\frac{1}{2}(a_{21}+a_{12}) & -\frac{1}{2}(a_{22}+a_{22}) & \cdots & -\frac{1}{2}(a_{22}-a_{22}) & -\frac{1}{2}(a_{21}-a_{12}) \\ -\frac{1}{2}(a_{11}+a_{11}) & -\frac{1}{2}(a_{12}+a_{21}) & \cdots & -\frac{1}{2}(a_{12}-a_{21}) & -\frac{1}{2}(a_{11}-a_{11}) \end{vmatrix},$$

which is zero-axial and skew centrosymmetric, its value is readily seen by §362 to be $|a_{1n}|^2$; hence $|a_{1n}| = \Delta^{1/2} = P$, where P is the Pfaffian function which is the square root of Δ. Since $|a_{1n}|$ is any determinant the truth of the theorem appears.

419. If we denote by B the coefficient of a_{rs} in the Pfaffian function P then B does not contain a_{rs} and is not therefore affected by changing the suffixes rs, to sr, but this changes the sign of a_{rs}, and it is thus seen that some of the terms of P', which we use to denote what P becomes on making the change, differ in sign from the corresponding terms of P. But since the change of rs into sr means the interchange of the signs of a row and column we have $P^2 = P'^2$ which shows that the signs of all the terms of P' differ from those of P. We have therefore the theorem that *interchanging two suffixes changes the sign of the Pfaffian.*

420. There are a number of properties of Pfaffians quite analogous to those of determinants. Thus the theorems of §§93, 174, 175, 179 may be transferred into the theory of Pfaffians without alteration. Others need some variation of statement, such as: Of the full number of terms of a Pfaffian there is one more positive than negative. Similarly for others.

421. It is obvious that a zero-axial skew determinant of even order

$$A = \begin{vmatrix} 0 & a_{12} & a_{13} & \cdots \\ -a_{12} & 0 & a_{23} & \cdots \\ -a_{13} & -a_{23} & 0 & \cdots \\ \cdot & \cdot & \cdot & \cdot \end{vmatrix}_{2n}$$

is not altered by being bordered thus

$$\begin{vmatrix} 1 & 1 & 1 & 1 & \cdots \\ -x & 0 & a_{12} & a_{13} & \cdots \\ -x & -a_{12} & 0 & a_{23} & \cdots \\ -x & -a_{13} & -a_{23} & 0 & \cdots \end{vmatrix}_{2n+1} \equiv \Delta, \text{ say},$$

for Δ is equal to A plus x times a zero-axial skew of odd order which vanishes.

If in Δ we add x times the first row to each of the others it reduces to

$$\begin{vmatrix} x & a_{12}+x & a_{13}+x & a_{14}+x & \cdots \\ -a_{12}+x & x & a_{23}+x & a_{24}+x & \cdots \\ -a_{13}+x & -a_{23}+x & x & a_{34}+x & \cdots \\ -a_{14}+x & -a_{24}+x & -a_{34}+x & x & \cdots \\ \cdots & \cdots & \cdots & \cdots & \cdots \end{vmatrix}_{2n} \equiv D, \text{ say}.$$

We have then $A=\Delta=D$, and the theorem:

A zero-axial skew determinant of even order is not altered by adding the same number x to each element.

The x may of course be positive or negative.

It follows from this theorem *that any determinant Δ of even order in which $a_{ii}=x$ and $a_{ij}+a_{ji}=2x$ may be written as a zero-axial skew of the same order.*

Thus we may write

$$\Delta \equiv \begin{vmatrix} x & a_{12} & a_{13} & a_{14} & \cdots \\ 2x-a_{12} & x & a_{23} & a_{24} & \cdots \\ 2x-a_{13} & 2x-a_{23} & x & a_{34} & \cdots \\ 2x-a_{14} & 2x-a_{24} & 2x-a_{34} & x & \cdots \\ \cdots & \cdots & \cdots & \cdots & \cdots \end{vmatrix}_{2n}$$

$$= \begin{vmatrix} 1 & 0 & 0 & 0 & \cdots \\ 1 & x & a_{12} & a_{13} & \cdots \\ 1 & 2x-a_{12} & x & a_{23} & \cdots \\ 1 & 2x-a_{13} & 2x-a_{23} & x & \cdots \\ \cdots & \cdots & \cdots & \cdots & \cdots \end{vmatrix}_{2n+1}$$

$$= \begin{vmatrix} 0 & a_{12} - x & a_{13} - x & \cdots \\ x - a_{12} & 0 & a_{23} - x & \cdots \\ x - a_{13} & x - a_{23} & 0 & \cdots \\ \cdot & \cdot & \cdot & \cdot \end{vmatrix}_{2n}$$

422. *If the same number x be added to every element of a zero-axial skew determinant of odd order the result is x times a perfect square.*

For if

$$\Delta \equiv \begin{vmatrix} x & a_{12} + x & a_{13} + x & \cdots \\ -a_{12} + x & x & a_{23} + x & \cdots \\ -a_{13} + x & -a_{23} + x & x & \cdots \\ \cdot & \cdot & \cdot & \cdot \end{vmatrix}_{2n+1}$$

then

$$\Delta = \begin{vmatrix} 1 & 0 & 0 & 0 & \cdots \\ -x & x & a_{12} + x & a_{13} + x & \cdots \\ -x & -a_{12} + x & x & a_{23} + x & \cdots \\ -x & -a_{13} + x & -a_{23} + x & x & \cdots \\ \cdot & \cdot & \cdot & \cdot & \cdot \end{vmatrix}_{2n+2}$$

and if on this we perform the operations c_2+c_1, c_3+c_1, $\cdots$, we have

$$\Delta = \begin{vmatrix} 1 & 1 & 1 & 1 & \cdots \\ -x & 0 & a_{12} & a_{13} & \cdots \\ -x & -a_{12} & 0 & a_{23} & \cdots \\ -x & -a_{13} & -a_{23} & 0 & \cdots \\ \cdot & \cdot & \cdot & \cdot & \cdot \end{vmatrix}_{2n+2}$$

$$= \begin{vmatrix} 0 & a_{12} & a_{13} & \cdots \\ -a_{12} & 0 & a_{23} & \cdots \\ -a_{13} & -a_{23} & 0 & \cdots \\ \cdot & \cdot & \cdot & \cdot \end{vmatrix}_{2n+1} + x \begin{vmatrix} 0 & 1 & 1 & 1 & \cdots \\ -1 & 0 & a_{12} & a_{13} & \cdots \\ -1 & -a_{12} & 0 & a_{23} & \cdots \\ -1 & -a_{13} & -a_{23} & 0 & \cdots \\ \cdot & \cdot & \cdot & \cdot & \cdot \end{vmatrix}_{2n+2}$$

$=0+x$ times a zero-axial skew of even order which is a perfect square. Hence the theorem.

423. *If we border a zero-axial skew of odd order by adding a column with elements all* 1's *and a row with elements all* -1's *with* 0 *in the intersection of the two, such a determinant may be condensed into a zero-axial skew of order two less. Thus if*

$$A \equiv \begin{vmatrix} 0 & a_{12} & a_{13} & a_{14} & a_{15} & 1 \\ -a_{12} & 0 & a_{23} & a_{24} & a_{25} & 1 \\ -a_{13} & -a_{23} & 0 & a_{34} & a_{35} & 1 \\ -a_{14} & -a_{24} & -a_{34} & 0 & a_{45} & 1 \\ -a_{15} & -a_{25} & -a_{35} & -a_{45} & 0 & 1 \\ -1 & -1 & -1 & -1 & -1 & 0 \end{vmatrix}$$

and we perform the operations

$$c_1 + a_{15}c_6, \quad c_2 + a_{25}c_6, \quad c_3 + a_{35}c_6, \quad c_4 + a_{45}c_6$$

followed by

$$r_1 + a_{15}r_6, \quad r_2 + a_{25}r_6, \quad r_3 + a_{35}r_6, \quad r_4 + a_{45}r_6$$

we get

$$A = \begin{vmatrix} 0 & a_{12}-a_{15}+a_{25} & a_{13}-a_{15}+a_{35} & a_{14}-a_{15}+a_{45} \\ -a_{12}+a_{15}-a_{25} & 0 & a_{23}-a_{25}+a_{35} & a_{24}-a_{25}+a_{45} \\ -a_{13}+a_{15}-a_{35} & -a_{23}+a_{25}-a_{35} & 0 & a_{34}-a_{35}+a_{45} \\ -a_{14}+a_{15}-a_{45} & -a_{24}+a_{25}-a_{45} & -a_{34}+a_{35}-a_{45} & 0 \end{vmatrix}$$

If we put $a_{15}=a_{25}=a_{35}=a_{45}=x$, then this relation becomes

$$A \equiv \begin{vmatrix} 0 & a_{12} & a_{13} & a_{14} & x & 1 \\ -a_{12} & 0 & a_{23} & a_{24} & x & 1 \\ -a_{13} & -a_{23} & 0 & a_{34} & x & 1 \\ -a_{14} & -a_{24} & -a_{34} & 0 & x & 1 \\ -x & -x & -x & -x & 0 & 1 \\ -1 & -1 & -1 & -1 & 1 & 0 \end{vmatrix} = \begin{vmatrix} 0 & a_{12} & a_{13} & a_{14} \\ -a_{12} & 0 & a_{23} & a_{24} \\ -a_{13} & -a_{23} & 0 & a_{34} \\ -a_{14} & -a_{24} & -a_{34} & 0 \end{vmatrix}$$

In other words A is independent of x and is equal to $A_{56,56}$.

EXERCISES:

1. Show that

$$\begin{vmatrix} x & ax_1 & bx_2 & abx_3 \\ -x_1 & x & -bx_3 & bx_2 \\ -x_2 & ax_3 & x & -ax_1 \\ -x_3 & -x_2 & x_1 & x \end{vmatrix} = (x^2 + ax_1^2 + bx_2^2 + abx_3^2)^2$$

2. Show that the number of distinct terms in a zero-axial skew determinant of order $2n$ is $1\cdot 3\cdot 5 \cdots (2n-1)V_n$ where V_n is determined from $V_n = (2n-1)V_{n-1} - (n-1)V_{n-2}$, and $V_0 = V_1 = 1$.

3. Show that V_n of problem 2 is equal to

$$\begin{vmatrix} 1 & 1 & \cdot & \cdots \\ 1 & 3 & 2 & \cdots \\ \cdot & 1 & 5 & 3\cdots \\ \cdot & \cdot & 1 & 7\cdots \\ \cdots & \cdots & \cdots & \cdots \end{vmatrix}.$$

424. If we expand an even-ordered zero-axial skew determinant Δ by Cauchy's theorem we get

$$\Delta = (11)[11] + \sum r(1r)^2 \begin{bmatrix} 1r \\ 1r \end{bmatrix} + 2\sum(-1)^{i+j} a_{1j} a_{i1} \begin{bmatrix} 1i \\ 1j \end{bmatrix}, \quad i \neq j$$

and since

$$\begin{bmatrix} 1i \\ 1i \end{bmatrix}\begin{bmatrix} 1j \\ 1j \end{bmatrix} - \begin{bmatrix} 1i \\ 1j \end{bmatrix}\begin{bmatrix} 1j \\ 1i \end{bmatrix} = [11]\begin{bmatrix} 1ij \\ 1ij \end{bmatrix} = 0,$$

both $[11]$ and $\left[{}^{1ij}_{1ij}\right]$ being zero-axial skew of odd order and therefore vanish, we have

$$\Delta = \left\{0 + (12)\begin{bmatrix} 12 \\ 12 \end{bmatrix}^{1/2} - (13)\begin{bmatrix} 13 \\ 13 \end{bmatrix}^{1/2} + \cdots + (1n)\begin{bmatrix} 1n \\ 1n \end{bmatrix}^{1/2}\right\}^2$$

$$= \left\{(12)\begin{bmatrix} 12 \\ 12 \end{bmatrix}^{1/2} + 0 + (23)\begin{bmatrix} 23 \\ 23 \end{bmatrix}^{1/2} - \cdots - (2n)\begin{bmatrix} 2n \\ 2n \end{bmatrix}^{1/2}\right\}^2$$

$$= \left\{-(13)\begin{bmatrix} 13 \\ 13 \end{bmatrix}^{1/2} + (23)\begin{bmatrix} 23 \\ 23 \end{bmatrix}^{1/2} + 0 + \cdots + (3n)\begin{bmatrix} 3n \\ 3n \end{bmatrix}^{1/2}\right\}^2$$

etc.

or denoting the Pfaffian by ff we have

$$ff = \left\{0 + (12)\begin{bmatrix}12\\12\end{bmatrix}^{1/2} - (13)\begin{bmatrix}13\\13\end{bmatrix}^{1/2} + \cdots + (1n)\begin{bmatrix}1n\\1n\end{bmatrix}^{1/2}\right\}$$

$$= \left\{(12)\begin{bmatrix}12\\12\end{bmatrix}^{1/2} + 0 + (23)\begin{bmatrix}23\\23\end{bmatrix}^{1/2} - \cdots - (2n)\begin{bmatrix}2n\\2n\end{bmatrix}^{1/2}\right\}$$

etc.

If we take the signs alternately positive and negative and add these equations we get

$$2\left\{- (13)\begin{bmatrix}13\\13\end{bmatrix}^{1/2} - (15)\begin{bmatrix}15\\15\end{bmatrix}^{1/2} - \cdots - (1\ n-1)\begin{bmatrix}1\ n-1\\1\ n-1\end{bmatrix}^{1/2}\right.$$

$$+ (24)\begin{bmatrix}24\\24\end{bmatrix}^{1/2} + (26)\begin{bmatrix}26\\26\end{bmatrix}^{1/2} + \cdots + (2n)\begin{bmatrix}2n\\2n\end{bmatrix}^{1/2}$$

$$- (35)\begin{bmatrix}35\\35\end{bmatrix}^{1/2} - (37)\begin{bmatrix}37\\37\end{bmatrix}^{1/2} - \cdots - (3\ n-1)\begin{bmatrix}3\ n-1\\3\ n-1\end{bmatrix}^{1/2}$$

$$+ \cdots\cdots\cdots\cdots\cdots\cdots$$

$$\left. + (n-2,\ n)\begin{bmatrix}n-2\ n\\n-2\ n\end{bmatrix}^{1/2}\right\} = 0.$$

Thus for $n=6$ we have

$$- (13)\begin{bmatrix}13\\13\end{bmatrix}^{1/2} - (15)\begin{bmatrix}15\\15\end{bmatrix}^{1/2} + (24)\begin{bmatrix}24\\24\end{bmatrix}^{1/2} + (26)\begin{bmatrix}26\\26\end{bmatrix}^{1/2}$$

$$- (35)\begin{bmatrix}35\\35\end{bmatrix}^{1/2} + (46)\begin{bmatrix}46\\46\end{bmatrix}^{1/2} = 0$$

or

$$(13)\begin{bmatrix}13\\13\end{bmatrix}^{1/2} + (15)\begin{bmatrix}15\\15\end{bmatrix}^{1/2} + (35)\begin{bmatrix}35\\35\end{bmatrix}^{1/2} = (24)\begin{bmatrix}24\\24\end{bmatrix}^{1/2}$$

$$+ (26)\begin{bmatrix}26\\26\end{bmatrix}^{1/2} + (46)\begin{bmatrix}46\\46\end{bmatrix}^{1/2}$$

For $n=6$, if we had taken the 1st, 5th, 6th positive and the 2nd, 3rd, 4th negative and added we would get

$$- (15)\begin{bmatrix}15\\15\end{bmatrix}^{1/2} + (16)\begin{bmatrix}16\\16\end{bmatrix}^{1/2} + (56)\begin{bmatrix}56\\56\end{bmatrix}^{1/2} = (23)\begin{bmatrix}23\\23\end{bmatrix}^{1/2}$$

$$- (24)\begin{bmatrix}24\\24\end{bmatrix}^{1/2} + (34)\begin{bmatrix}34\\34\end{bmatrix}^{1/2}.$$

Similarly other forms may be obtained.

425. From

$$\Delta = (12)\begin{bmatrix}12\\12\end{bmatrix}^{1/2} - (13)\begin{bmatrix}13\\13\end{bmatrix}^{1/2} + (14)\begin{bmatrix}14\\14\end{bmatrix}^{1/2} - (15)\begin{bmatrix}15\\15\end{bmatrix}^{1/2} + (16)\begin{bmatrix}16\\16\end{bmatrix}^{1/2}$$

we readily get

$$-(15)\begin{bmatrix}15\\15\end{bmatrix}^{1/2} + (16)\begin{bmatrix}16\\16\end{bmatrix}^{1/2} + (56)\begin{bmatrix}56\\56\end{bmatrix}^{1/2} = \Delta + (56)\begin{bmatrix}56\\56\end{bmatrix}^{1/2} - (12)\begin{bmatrix}12\\12\end{bmatrix}^{1/2} + (13)\begin{bmatrix}13\\13\end{bmatrix}^{1/2} - (14)\begin{bmatrix}14\\14\end{bmatrix}^{1/2}.$$

From expanding $\begin{vmatrix}234\\156\end{vmatrix}$ by Laplace's theorem and adding and subtracting the term (56) $\begin{bmatrix}56\\56\end{bmatrix}^{1/2}$ we get

$$-(12)\begin{bmatrix}12\\12\end{bmatrix}^{1/2} + (13)\begin{bmatrix}13\\13\end{bmatrix}^{1/2} - (14)\begin{bmatrix}14\\14\end{bmatrix}^{1/2} + (56)\begin{bmatrix}56\\56\end{bmatrix}^{1/2} = +(56)\begin{bmatrix}56\\56\end{bmatrix}^{1/2} - \begin{vmatrix}2\,3\,4\\1\,5\,6\end{vmatrix}$$

Therefore

$$-(15)\begin{bmatrix}15\\15\end{bmatrix}^{1/2} + (16)\begin{bmatrix}16\\16\end{bmatrix}^{1/2} + (56)\begin{bmatrix}56\\56\end{bmatrix}^{1/2} = \Delta - \begin{vmatrix}2\,3\,4\\1\,5\,6\end{vmatrix},$$

and also

$$(23)\begin{bmatrix}23\\23\end{bmatrix}^{1/2} - (24)\begin{bmatrix}24\\24\end{bmatrix}^{1/2} + (34)\begin{bmatrix}34\\34\end{bmatrix}^{1/2} = \Delta - \begin{vmatrix}2\,3\,4\\1\,5\,6\end{vmatrix}.$$

426. If in §424 we subtract the sum of the first k equations from the sum of the remaining $n-k$ we get

$$(n-2k)ff = 2\left\{(k+1, k+2)\begin{bmatrix}k+1 & k+2\\k+1 & k+2\end{bmatrix}^{1/2} - \cdots + (n-1, n)\begin{bmatrix}n-1 & n\\n-1 & n\end{bmatrix}^{1/2}\right\}.$$

If $2k=n$ then the left-hand side is zero.

427. If $A \equiv |a_{1n}|$ represents an even-ordered zero-axial skew determinant, and $|A_{1n}|$ represents its adjugate which as we have

seen is zero-axial skew, then $FF=(ff)^{n-1}$, where FF represents the Pfaffian function of $|A_{1n}|$.

428. Let $A\equiv|a_{1n}|$ be any determinant of order n and form from it another determinant Δ by subtracting from every element its conjugate. This gives

$$\Delta\equiv\begin{vmatrix} 0 & a_{12}-a_{21} & a_{13}-a_{31} & \cdots \\ a_{21}-a_{12} & 0 & a_{23}-a_{32} & \cdots \\ a_{31}-a_{13} & a_{32}-a_{23} & 0 & \cdots \\ \cdot & \cdot & \cdot & \cdot \end{vmatrix}_n$$

which is zero-axial skew.

Expanding we have, supposing n to be even,

$$\begin{aligned}\Delta &= -(a_{12}-a_{21})\Delta_{12}+(a_{13}-a_{31})\Delta_{13}-\cdots\\ &= -(a_{12}-a_{21})\Delta^{1/2}\Delta^{1/2}_{12,12}+(a_{13}-a_{31})\Delta^{1/2}\Delta^{1/2}_{13,13}-\cdots\end{aligned}$$

or

$$\Delta^{1/2}=-(a_{12}-a_{21})\Delta^{1/2}_{12,12}+(a_{13}-a_{31})\Delta_{13,13}-\cdots$$

a result corresponding to (4) §410.

It is obvious that Δ will vanish if $a_{ri}=a_{ir}$ $(i=1,2,\cdots,n)$ for any value of r. That is Δ will vanish if A has the elements of the rth row equal to their conjugates in the rth column. It will also be apparent that Δ is zero if A has any coaxial minor of order $\geqq m+1$ $(n=2m)$ axisymmetric.

429. The determinant Δ §428 having binomial elements may be written as the sum of 2^n determinants with monomial elements, so that if A' denotes the determinant, formed by interchanging the rows and columns of A and changing sign of every element, we have

$$\Delta=\sum|A_\alpha\cdot A'_\beta|=\begin{vmatrix} a_{12}-a_{21} & a_{13}-a_{31} & \cdots \\ & a_{23}-a_{32} & \cdots \\ & & \cdots \end{vmatrix}^2$$

where $(\alpha,\beta=0,1,\cdots,n)$ and $\alpha+\beta=n$, and where $|A_\alpha\cdot A'_\beta|$ denotes the determinant with monomial elements formed by taking α columns of A and β columns of A'.

430. By §313 we may write

$$\sum\begin{vmatrix}156\\234\end{vmatrix}=\sum\begin{vmatrix}1\\2\end{vmatrix}\sum\begin{vmatrix}56\\34\end{vmatrix}-\sum\begin{vmatrix}1\\3\end{vmatrix}\sum\begin{vmatrix}56\\24\end{vmatrix}+\sum\begin{vmatrix}1\\4\end{vmatrix}\sum\begin{vmatrix}56\\23\end{vmatrix}$$

$$-\sum\begin{vmatrix}1\\5\end{vmatrix}\sum\begin{vmatrix}46\\23\end{vmatrix}+\sum\begin{vmatrix}1\\6\end{vmatrix}\sum\begin{vmatrix}45\\23\end{vmatrix}=(a_{12}-a_{21})\sum\begin{vmatrix}56\\34\end{vmatrix}$$

$$-(a_{13}-a_{31})\sum\begin{vmatrix}56\\24\end{vmatrix}+(a_{14}-a_{41})\sum\begin{vmatrix}56\\23\end{vmatrix}-(a_{15}-a_{51})\sum\begin{vmatrix}46\\23\end{vmatrix}$$

$$+(a_{16}-a_{61})\sum\begin{vmatrix}45\\23\end{vmatrix}$$

and it may be easily verified that

$$\sum\begin{vmatrix}56\\34\end{vmatrix}=\Delta^{1/2}_{12,12},\quad \sum\begin{vmatrix}56\\24\end{vmatrix}=\Delta^{1/2}_{13,13},\quad \sum\begin{vmatrix}56\\23\end{vmatrix}=\Delta^{1/2}_{14,14},$$

etc. We have, therefore,

$$\Delta_6^{1/2}=-\sum\begin{vmatrix}1\ 5\ 6\\2\ 3\ 4\end{vmatrix},$$

where the subscript 6 denotes the order of Δ. Again

$$\sum\begin{vmatrix}1\ 6\ 7\ 8\\2\ 3\ 4\ 5\end{vmatrix}=\sum\begin{vmatrix}1\\2\end{vmatrix}\sum\begin{vmatrix}6\ 7\ 8\\3\ 4\ 5\end{vmatrix}-\sum\begin{vmatrix}1\\3\end{vmatrix}\sum\begin{vmatrix}6\ 7\ 8\\2\ 4\ 5\end{vmatrix}+\cdots$$

$$+\sum\begin{vmatrix}1\\8\end{vmatrix}\sum\begin{vmatrix}5\ 6\ 7\\2\ 3\ 4\end{vmatrix}=(a_{12}-a_{21})\Delta^{1/2}_{12,12}-(a_{13}-a_{31})\Delta^{1/2}_{13,13}+\cdots$$

$$+(a_{18}-a_{81})\Delta^{1/2}_{18,18},$$

by the preceding case. Therefore

$$\Delta_8^{1/2}=-\sum\begin{vmatrix}1\ 6\ 7\ 8\\2\ 3\ 4\ 5\end{vmatrix},$$

and in general

$$\Delta_{2m}^{1/2}=-\sum\begin{vmatrix}1 & m+2 & m+3 & \cdots & 2m\\ 2 & 3 & 4 & \cdots & m+1\end{vmatrix}$$

$$=-\sum\begin{vmatrix}1 & 2 & \cdots & m\\ m+1 & m+2 & \cdots & 2m\end{vmatrix}.$$

This sum will vanish under the same conditions as in §428.

431. If instead of subtracting every element of A from its conjugate to form Δ we subtract only those in the minor A_{11}, and replace the elements in the first column by the negatives of their conjugates, thus

$$\Delta' \equiv \begin{vmatrix} 0 & a_{12} & a_{13} & a_{14} & \cdots \\ -a_{12} & 0 & a_{23}-a_{32} & a_{24}-a_{42} & \cdots \\ -a_{13} & a_{32}-a_{23} & 0 & a_{34}-a_{43} & \cdots \\ -a_{14} & a_{42}-a_{24} & a_{43}-a_{34} & 0 & \cdots \\ \cdot & \cdot & \cdot & \cdot & \cdot \end{vmatrix}_{n=2m}$$

$$= -a_{12}\Delta'_{12} + a_{13}\Delta'_{13} - \cdots$$

$$= -a_{12}\Delta'^{1/2}\Delta'^{1/2}_{12,12} + a_{13}\Delta'^{1/2}\Delta'^{1/2}_{13,13} - \cdots$$

or

$$\Delta'^{1/2} = -a_{12}\Delta'^{1/2}_{12,12} + a_{13}\Delta'^{1/2}_{13,13} - \cdots - a_{1n}\Delta'^{1/2}_{1n,1n}.$$

This again is a Pfaffian and vanishes if $a_{ri} = a_{ir}$ $(i = 2, 3, \cdots, n)$ or if any coaxial minor of A_{11} of order $\overline{\gtrless} m+1$ is axisymmetric. Again we may write

$$\sum \begin{vmatrix} 1\overline{56} \\ \underline{234} \end{vmatrix} = a_{12}\sum \begin{vmatrix} 56 \\ 34 \end{vmatrix} - a_{13}\sum \begin{vmatrix} 56 \\ 24 \end{vmatrix} + \cdots + a_{16}\sum \begin{vmatrix} 45 \\ 23 \end{vmatrix}$$

$$= a_{12}\Delta'^{1/2}_{12,12} - a_{13}\Delta'^{1/2}_{13,13} + \cdots + a_{16}\Delta'^{1/2}_{16,16}$$

and therefore

$$\Delta_6'^{1/2} = -\sum \begin{vmatrix} \dot{1}\overline{56} \\ \underline{234} \end{vmatrix}$$

or as it is sometimes written

$$-\sum \begin{vmatrix} \dot{1}56 \\ 234 \end{vmatrix}$$

and in general

$$\Delta_{2m}'^{1/2} = -\sum \begin{vmatrix} \dot{1} & 2 & \cdots & m \\ m+1 & m+2 & \cdots & 2m \end{vmatrix}$$

More generally still we have

$$\Delta_{2m}^{(k)^{1/2}} = -\sum \begin{vmatrix} \dot{1} & \dot{2} & \cdots & \dot{k} & \cdots & m \\ m+1 & \cdots & \cdots & \cdots & \cdots & 2m \end{vmatrix},$$

where now the first k rows and columns have monomial elements and are the negative of their conjugates in A.

The conditions for vanishing in this case are
(1) if $a_{ri}=a_{ir}$ $(i=k+1,\ k+2\ \cdots\ n)$
or (2) if any coaxial minor of $A_{1\ldots k,\,1\ldots k}$ of order $m+1$ or greater is axisymmetric.

432. If Δ is a zero-axial skew of odd order then

$$(1)\qquad (11)[11]\ -(12)[12]\ +(13)[13]-\cdots + (1n)[1n] = \Delta = 0$$

or

$$(2)\qquad \{(11)[11]^{1/2}-(12)[22]^{1/2}+(13)[33]^{1/2}-\cdots + (1n)[nn]^{1/2}\}[11]^{1/2} = 0$$

or

$$(3)\qquad -(12)[22]^{1/2}+(13)[33]^{1/2}-\cdots + (1n)[nn]^{1/2} = 0$$

since $(11)=0$ and in general $[11]\neq 0$.

If $[rr]=0$ for any $n-2$ of the values $r=2,3,\cdots,n$ then it must be true for the remaining value, since the coefficient of the remaining term is in general not zero. In other words *if one, say* $[11]\neq 0$, *and if* $n-2$ *of the other principal coaxial minors vanish then* $[rr]=0$ *for all values of* r *except* 1.

Equation (3) may also be seen to be true by substituting for $[22]$, $[33]$, $\cdots$, their values by Cauchy's expansions where it will be seen that the terms cancel in pairs.

433. *If in a zero-axial skew determinant* Δ, *a coaxial minor of order* $2r$ *is not zero, and if every coaxial minor of order* $2r+2$ *formed from it by adding two additional rows and two additional columns does vanish, than all minors of order* $2r+1$, *which contain the given even-ordered minor are zero.* That is if

$$\begin{vmatrix} 1\cdots 2r \\ 1\cdots 2r\end{vmatrix} \neq 0,\ \text{and}\ \begin{vmatrix} 1\,2\cdots 2r\,\alpha\beta \\ 1\,2\cdots 2r\,\alpha\beta\end{vmatrix} = 0\left\{\begin{matrix}\alpha \\ \beta\end{matrix} = 2r+1,\cdots,n\right\},$$

then

$$\begin{vmatrix} 1\cdots 2r\alpha \\ 1\cdots 2r\beta\end{vmatrix} = 0.$$

For

$$\begin{vmatrix} 1\cdots 2r\alpha \\ 1\cdots 2r\alpha\end{vmatrix}\cdot\begin{vmatrix} 1\cdots 2r\beta \\ 1\cdots 2r\beta\end{vmatrix} + \begin{vmatrix} 1\cdots 2r\alpha \\ 1\cdots 2r\beta\end{vmatrix}^2 = \begin{vmatrix} 1\cdots 2r \\ 1\cdots 2r\end{vmatrix}\cdot\begin{vmatrix} 1\cdots 2r\alpha\beta \\ 1\cdots 2r\alpha\beta\end{vmatrix}$$

or

$$0 + \begin{vmatrix} 1 \cdots 2r\alpha \\ 1 \cdots 2r\beta \end{vmatrix}^2 = 0$$

and therefore

$$\begin{vmatrix} 1 \cdots 2r\alpha \\ 1 \cdots 2r\beta \end{vmatrix} = 0$$

This being true it follows from the general theorem on rank that all minors of order $2r+1$ vanish.

Since all minors of order $2r+1$ vanish all minors of higher order will vanish and therefore among them all minors of order $(2r+2)$.

434. *If a zero-axial skew determinant of order n vanishes and if the sum of all the coaxial minors of orders $n-1, n-2, \cdots, n-m$ vanish then all minors of order $n-m$ vanish.*

For since the sum of the coaxial minors of order r is zero and since it is either a sum of individual zeros (when r is odd) or a sum of squares (when r is even) it follows that every coaxial minor of order r is zero.

Again

$$\Delta_{(n-m)2ii} = \Delta_{(n-m)ii}\Delta_{(n-m)jj} - \Delta_{(n-m)ij}\Delta_{(n-m)ji}$$

where $\Delta_{(n-m)}$ is a minor of Δ of order $n-m$ and $\Delta_{(n-m)2ss}$ is a coaxial minor of order 2 of the $(n-m)$th compound of Δ.

But

$$\Delta_{(n-m)rr} = 0$$

and

$$\Delta_{(n-m)ij} = (-1)^{n-m}\Delta_{(n-m)ji}.$$

Therefore

$$\Delta_{(n-m)2ii} = (-1)^{n-m}\Delta^2_{(n-m)ij}$$

and

$$\sum\Delta_{(n-m)2ii} = (-1)^{n-m}\sum\Delta^2_{(n-m)ij}$$

But by hypothesis and §191

$$\sum\Delta_{(n-m)2ii} = 0,$$

and therefore

$$\sum\Delta^2_{(n-m)ij} = 0,$$

and $\Delta_{(n-m)ij}=0$ for all i, j.

435. If $\alpha_{rs} \neq 0$ and represents the element in the position (rs) of the mth compound of Δ and if all minors of Δ of order $m+1$ are zero, then

$$\alpha_{rr}\alpha_{ss} - \alpha_{rs}\alpha_{sr} = 0$$

or

$$\alpha_{rr}\alpha_{ss} + \alpha_{rs}^2 = 0.$$

From this we see that if $\alpha_{rs} \neq 0$, then α_{rr} and α_{ss} are not zero, and also that m must be even since all coaxial minors of odd order vanish.

If in §221 we make A an even-ordered zero-axial skew determinant, and B identical with it we have $-\Delta = \sum \alpha_r^2$, where α_r is a principal minor of A', and A' is formed by bordering A symmetrically, that is α_r is $A'_{n+1,r}$.

But

$$A'_{n+1,n+1} \cdot A'_{rr} - A'^2_{n+1,r} = A' \cdot A'_{r\ n+1,r\ n+1}$$
$$= 0,$$

since $A'_{r\ n+1,r\ n+1}$, is a zero-axial skew of odd order.

Substituting for α_r we get

$$(1) \qquad -\Delta = \sum rA'^2_{n+1,r} = \sum rA \cdot A'_{rr}, \quad \text{since} \quad A'_{n+1,n+1} = A$$
$$= A \sum rA'_{rr}.$$

From which it is seen that Δ contains A as a factor.

If the original determinant is of odd order, then

$$A'_{n+1,n+1}A'_{rr} - A'^2_{n+1,r} = A'A'_{r\ n+1,r\ n+1}$$

or

$$0 \qquad - A'^2_{n+1,r} = A'A'_{r\ n+1,r\ n+1}$$

and

$$(2) \qquad \Delta = A' \cdot \sum A'_{r\ n+1,r\ n+1}$$

in which case Δ contains A' as a factor.

Since the terms under the sigma on the right of (1) and (2) are even-ordered zero-axial skew determinants they may therefore be expressed as the squares of Pfaffians.

436. Let $A \equiv |a_{1n}|$ be any determinant and let $B \equiv |b_{1n}|$ be a zero-axial skew determinant. Let Δ_r be the determinant formed by replacing the rth column of B by the rth column of A. Thus

$$\Delta_1 = \begin{vmatrix} a_{11} & b_{12} & b_{13} \cdots b_{1n} \\ a_{21} & 0 & b_{23} \cdots b_{2n} \\ \cdots & \cdots & \cdots \\ a_{n1} & -b_{2n} & -b_{3n} \cdots 0 \end{vmatrix}$$

But by §415 when n is odd this is equal to

$$\begin{vmatrix} 0 & a_{11} & a_{12} \cdots a_{1n} \\ -a_{11} & 0 & b_{12} \cdots b_{2n} \\ \cdots & \cdots & \cdots \\ -a_{1n} & -b_{2n} & -b_{3n} \cdots 0 \end{vmatrix}^{1/2} \cdot \begin{vmatrix} 0 & b_{23} \cdots b_{2n} \\ -b_{23} & 0 \cdots b_{3n} \\ \cdots & \cdots \\ -b_{2n} & -b_{3n} \cdots 0 \end{vmatrix}^{1/2}$$

$$= \left\{ a_{11}B_{11}^{1/2} - a_{12}B_{22}^{1/2} + a_{13}B_{33}^{1/2} - \cdots \right\} B_{11}^{1/2}$$

since the two determinants are zero-axial skew of even order and therefore perfect squares.

Therefore

$$\sum_1^n r\Delta_r = \begin{array}{c|c} B_{11}^{1/2} \quad -B_{22}^{1/2} \cdots B_{nn}^{1/2} & \\ \hline \begin{matrix} a_{11} & a_{12} & \cdots a_{1n} \\ a_{21} & a_{22} & \cdots a_{2n} \\ \cdots & \cdots & \cdots \\ a_{n1} & a_{n2} & \cdots a_{nn} \end{matrix} & \begin{matrix} B_{11}^{1/2} \\ -B_{22}^{1/2} \\ \\ B_{nn}^{1/2} \end{matrix} \end{array}$$

$=0$ if A is zero-axial skew.

437. In the case when n is even we have

$$B_{1r} = B^{1/2} \cdot B_{1r,1r}^{1/2} \quad \text{since} \quad B_{rr} = 0,$$
$$\Delta_1 = \left\{ a_{11} \cdot 0 + a_{12}B_{12,12}^{1/2} - a_{13}B_{13,13}^{1/2} + \cdots + a_{1n}B_{1n,1n}^{1/2} \right\} B^{1/2}$$

and

$$\sum_1^n r\Delta_r = B^{1/2} \sum (a_{ij} - a_{ji}) B_{ij,ij}^{1/2}$$
$$= 0$$

when A is axisymmetric.

In this case both B and $B_{ij,ij}$ are even-ordered zero-axial skew and therefore perfect squares and their square roots may be taken.

Instead of replacing columns of B by the corresponding columns of A we might have replaced rows instead with the same result.

438. If Δ_r denote the determinant formed by replacing the rth column of A by the rth column of B, then whether n be even or odd we have

$$\sum_1^n r\Delta_r = \sum_1^{n-1} s \sum_1^n r b_{sr}(A_{rs} - A_{sr})$$

$$= 0,$$

when A is axisymmetric.

EXERCISE: Show that

$$\begin{vmatrix} b^2+c^2 & -ab-\gamma & -ac+\beta \\ -ba+\gamma & c^2+a^2 & -bc-\alpha \\ -ca-\beta & -cb+\alpha & a^2+b^2 \end{vmatrix} = \begin{Vmatrix} a & b & c \\ \alpha & \beta & \gamma \end{Vmatrix}^2$$

$$= \begin{array}{ccc|c} \alpha & \beta & \gamma & \\ \hline b^2+c^2 & -ab & -ac & \alpha \\ -ba & c^2+a^2 & -bc & \beta \\ -ca & -cb & a^2+b^2 & \gamma \end{array} = \begin{array}{ccc|c} a & b & c & \\ \hline \beta^2+\gamma^2 & -\alpha\beta & -\alpha\gamma & a \\ -\beta\alpha & \alpha^2+\gamma^2 & -\beta\gamma & b \\ -\gamma\alpha & -\gamma\beta & \alpha^2+\beta^2 & c \end{array}$$

SKEW DETERMINANTS

439. A skew determinant which is not zero-axial may, by Cayley's theorem, be expressed in terms of the diagonal elements and zero-axial skew determinants.

Thus

$$\begin{vmatrix} x_1 & a_{12} & a_{13} & a_{14} \\ -a_{12} & x_2 & a_{23} & a_{24} \\ -a_{13} & -a_{23} & x_3 & a_{34} \\ -a_{14} & -a_{24} & -a_{34} & x_4 \end{vmatrix} = \begin{vmatrix} 0 & a_{12} & a_{13} & a_{14} \\ -a_{12} & 0 & a_{23} & a_{24} \\ -a_{13} & -a_{23} & 0 & a_{34} \\ -a_{14} & -a_{24} & -a_{34} & 0 \end{vmatrix}$$

$$+ x_1x_2\begin{vmatrix} 0 & a_{34} \\ -a_{34} & 0 \end{vmatrix} + x_1x_3\begin{vmatrix} 0 & a_{24} \\ -a_{24} & 0 \end{vmatrix} + x_1x_4\begin{vmatrix} 0 & a_{23} \\ -a_{23} & 0 \end{vmatrix}$$

$$+ x_2x_3\begin{vmatrix} 0 & a_{14} \\ -a_{14} & 0 \end{vmatrix} + x_2x_4\begin{vmatrix} 0 & a_{13} \\ -a_{13} & 0 \end{vmatrix} + x_3x_4\begin{vmatrix} 0 & a_{12} \\ -a_{12} & 0 \end{vmatrix}$$

$$+ x_1x_2x_3x_4,$$

the terms having an odd number of diagonal elements obviously vanishing.

If $x_1 = x_2 = x_3 = x_4 = x$ then we have a power series in x with the squares of Pfaffians as coefficients.

EXAMPLE: From any three quantities l, m, n, form nine others the elements of $|\alpha_1\beta_2\gamma_3|$, so that making

$$x = \alpha_1 X + \alpha_2 Y + \alpha_3 Z$$
$$y = \beta_1 X + \beta_2 Y + \beta_3 Z$$
$$z = \gamma_1 X + \gamma_2 Y + \gamma_3 Z$$

we may have $x^2+y^2+z^2=X^2+Y^2+Z^2$.

Whatever ξ, η, ζ may be, if we put

$$(1) \qquad \begin{cases} x = \xi + l\eta + m\zeta \\ y = -l\xi + \eta + \eta\zeta \\ z = -m\xi - n\eta + \zeta \end{cases}$$

and

$$(2) \qquad \begin{cases} X = \xi - l\eta - m\zeta \\ Y = l\xi + \eta - n\zeta \\ Z = m\xi + n\eta + \zeta \end{cases}$$

we insure that $x^2+y^2+z^2=X^2+Y^2+Z^2$; and it then only remains to determine x, y, z in terms of X, Y, Z. This of course may be done by solving for ξ, η, ζ in (2) and substituting in (1); but the following method is neater. Denoting the skew determinant

$$\begin{vmatrix} 1 & l & m \\ -l & 1 & n \\ -m & -n & 1 \end{vmatrix}$$

by Δ we have from (2)

$$(3) \qquad \begin{cases} \xi\Delta = X\Delta_{11} - Y\Delta_{21} + Z\Delta_{31} \\ \eta\Delta = -X\Delta_{12} + Y\Delta_{22} - Z\Delta_{32} \\ \zeta\Delta = Z\Delta_{13} - Y\Delta_{23} + Z\Delta_{33} \end{cases}$$

But from (1) and (2) by addition there results

$$x + X = 2\xi, \quad y + Y = 2\eta, \quad z + Z = 2\zeta$$

so that substituting these values of $2\xi, 2\eta, 2\zeta$ in (3) we have

$$\begin{aligned} x &= (2\Delta_{11} - \Delta)X - 2\Delta_{21}Y + 2\Delta_{31}Z \\ y &= 2\Delta_{12}X + (2\Delta_{22} - \Delta)Y - 2\Delta_{32}Z \\ z &= 2\Delta_{13}X - 2\Delta_{23}Y + (2\Delta_{33} - \Delta)Z \end{aligned}$$

Hence the required values of $\alpha_1, \alpha_2, \alpha_3, \cdots$, are $2\Delta_{11}-\Delta$, $-2\Delta_{21}$, $2\Delta_{31}, \cdots$.

EXERCISE: Show that

1. $\sum \begin{vmatrix} \overline{1\ 2\ 3} \\ \underline{4\ 5\ 6} \end{vmatrix} = -8 \begin{Vmatrix} 1\ 2\ 3\ 4\ 5 \\ 2\ 3\ 4\ 5\ 6 \end{vmatrix}$ when $\begin{vmatrix} 1\ 2\ 3\ 4\ 5\ 6 \\ 1\ 2\ 3\ 4\ 5\ 6 \end{vmatrix}$ is skew

2. $\sum \begin{vmatrix} 1\ \overline{2\ 3} \\ \underline{4\ 5\ 6} \end{vmatrix} = -4 \begin{Vmatrix} 1\ 2\ 3\ 4\ 5 \\ 2\ 3\ 4\ 5\ 6 \end{vmatrix}$ " $\begin{vmatrix} 2\ 3\ 4\ 5\ 6 \\ 2\ 3\ 4\ 5\ 6 \end{vmatrix}$ " "

3. $\sum \begin{vmatrix} 1\ \overline{2\ 3} \\ 4\ \underline{5\ 6} \end{vmatrix} = -2 \begin{Vmatrix} 1\ 2\ 3\ 4\ 5 \\ 2\ 3\ 4\ 5\ 6 \end{vmatrix}$ " $\begin{vmatrix} 2\ 3\ 5\ 6 \\ 2\ 3\ 5\ 6 \end{vmatrix}$ " "

4. Show that any $2n$-line Pfaffian with k-termed elements is equal to the sum of k^n Pfaffians with monomial elements.

440. *If* $\Delta \equiv |a_{1n}|$ *be a skew determinant having its diagonal elements each equal to z, then the sum of the products of corresponding elements in any two rows or the corresponding two columns of the adjugate determinant are the same and contains* Δ *as a factor and the determinant of the* n^2 *cofactors equals* Δ^{n-2}.

For

$$a_{r1}[s1] + \cdots + a_{rr}[sr] + \cdots + a_{rn}[sn] \begin{array}{l} = \Delta, \text{ when } r = s \\ = 0, \quad \text{"} \quad r \neq s \end{array}$$

Subtracting $2z[sr]$ from both sides and changing signs we have

$$a_{1r}[s1] + \cdots + a_{rr}[sr] + \cdots + a_{nr}[sn] = -\Delta + 2z[sr], \text{when } r = s$$
$$= 2z[sr], \text{ when } r \neq s.$$

Also we have

$$a_{1r}[1s] + a_{2r}[2s] + \cdots + a_{rr}[rs] + \cdots + a_{nr}[ns] = \Delta, \text{ when } r = s$$
$$= 0 \text{ when } r \neq s.$$

Therefore whatever the value of s we have by addition.

$$(1) \qquad a_{1r}\{[s1] + [1s]\} + \cdots + a_{nr}\{[sn] + [ns]\} = 2z[sr]$$

Writing for $\{[sr]+[rs]\} \div 2z$, ω_{sr} and giving r the values $1, 2, \cdots n$ we have

$$\begin{array}{l} a_{11}\omega_{s1} + a_{21}\omega_{s2} + \cdots + a_{n1}\omega_{sn} = [s1] \\ a_{12}\omega_{s1} + a_{22}\omega_{s2} + \cdots + a_{n2}\omega_{sn} = [s2] \\ \cdots\cdots\cdots\cdots\cdots\cdots \\ a_{1n}\omega_{s1} + a_{2n}\omega_{s2} + \cdots + a_{nn}\omega_{sn} = [sn] \end{array}$$

Solving we have

$$\omega_{sr} = [s1][r1] + [s2][r2] + \cdots + [sn][rn] \div \Delta$$

or

$$[s1][r1] + \cdots + [sn][rn] = \omega_{sr}\cdot\Delta = \Delta\cdot\{[sr] + [rs]\} \div 2z$$

which shows that *the product of any two rows and by symmetry of the corresponding two columns of the adjugate of* Δ *are alike and is divisible by* Δ.

Using (1) to find the square of the adjugate of Δ we get

$$\Delta^{2n-2} = \Delta^n \left| \omega_{1n} \right|$$

Hence $\left|\omega_{1n}\right| = \Delta^{n-2}$ as was to be proved.

441. The $[rs]$ of §440 may be expressed as a power series in z. Thus

$$[rs] = \alpha_0 + \alpha_1 z + \alpha_2 z^2 + \cdots$$

but if n is even then

$$[sr] = -\alpha_0 + \alpha_1 z - \alpha_2 z^2 + \cdots$$

so that

$$[rs] + [sr] = 2z\{\alpha_1 + \alpha_3 z + \cdots\}$$

which shows that the sum of any primary minor and its conjugate is divisible by $2z$.

The primary minors $[sr]$ are readily seen to be bordered skew determinants with univarial diagonals. These last may be expanded by Cayley's theorem in terms of the diagonal elements and zero-axial skew determinants. It is possible therefore to express any primary minor as a power series in z with coefficients expressed in terms of products of Pfaffians. Muir has shown* this to be

$$A_{pq} = z^{n-2}[pq] - z^{n-3}\sum[p\alpha][q\alpha] + z^{n-4}\sum[pq\alpha\beta][\alpha\beta] - z^{n-5}\sum[p\alpha\beta\gamma][q\alpha\beta\gamma] + \cdots$$

where now $[p\ 123]$ stands for the Pfaffian

$$\begin{vmatrix} a_{p1} & a_{p2} & a_{p3} \\ & a_{12} & a_{13} \\ & & a_{23} \end{vmatrix}$$

442. If r_p denote the pth row of Δ, a skew determinant with univarial diagonal, and if R_p denote the pth row of the adjugate of Δ, then it is readily seen that

* Proc. Roy. Soc. Edin., vol. 29, 1908–1909.

$$R_pR_q\cdot\Delta = \begin{vmatrix} r_1r_1 & r_1r_2 & \cdots & r_1r_n \\ r_2r_1 & r_2r_2 & \cdots & r_2r_n \\ \cdot & \cdot & \cdot & \cdot \\ r_nr_1 & r_nr_2 & \cdots & r_nr_n \end{vmatrix} = (-1)^{p+q}\Delta(\Delta_{(q}\Delta_{(p})$$

where $\Delta_{(q}$ denotes the array got from Δ by deleting the qth row.

But from (2) §440

$$R_pR_q = \frac{1}{2}\{[pq] + [qp]\}\frac{\Delta}{z}$$

and hence

$$\Delta_{(p}\Delta_{(q} = (-1)^{p+q}\frac{1}{2}\{[pq] + [qp]\}\frac{\Delta}{z}\cdot$$

That is, *the product of any $n-1$ rows of Δ by the same or any other $n-1$ rows contains Δ as a factor.*

EXERCISE 1. If $A \equiv |a_{1n}|$, be a skew determinant with univarial diagonal, show that

$$\begin{vmatrix} r_1r_1 - z^2 & r_1r_2 & \cdots \\ r_2r_1 & r_2r_2 - z^2 & \cdots \\ \cdot & \cdot & \cdot \end{vmatrix}_n = \begin{cases} 0 & \text{if } n \text{ is odd} \\ F^4 & \text{if } n \text{ is even} \end{cases}$$

where z is the common diagonal element and F is the Pfaffian of the elements on the right of the diagonal.

2. Show that

$$\begin{vmatrix} A_{11} - \dfrac{A}{z} & A_{12} & \cdots & A_{1n} \\ A_{21} & A_{22} - \dfrac{A}{z} & \cdots & A_{2n} \\ \cdot & \cdot & \cdot & \cdot \end{vmatrix}_n = \begin{cases} 0 & \text{when } n \text{ is odd} \\ \dfrac{A^{n-1}F^2}{(-z)^n} & \text{when } n \text{ is even} \end{cases}$$

3.

$$\begin{vmatrix} A_{11} - \dfrac{A}{z} & \dfrac{1}{2}(A_{12} + A_{21}) & \cdots \\ \dfrac{1}{2}(A_{21} + A_{12}) & A_{22} - \dfrac{A}{z} & \cdots \\ \cdot & \cdot & \cdot \end{vmatrix}_n = \begin{cases} 0 & \text{when } n \text{ is odd} \\ \dfrac{A^{n-2}F^4}{(-z)^n} & \text{when } n \text{ is even} \end{cases}$$

For further treatment of this type of determinant consult Muir l.c.

443. Let D_1 represent the skew determinant of the sixth order with $x_1, x_2, x_3, \cdots x_5, 0$ as the elements in the positions (11), (22), $\cdots$ (66), then if we multiply the 2nd, 3rd, 4th, 5th, columns of D_1 by a_{16} and then multiply the result by x_1 in the form

$$\begin{vmatrix} 1 & \cdot & \cdot & \cdot & \cdot & \cdot \\ \cdot & \cdot & \cdot & \cdot & \cdot & x_1 \\ -a_{26} & 1 & \cdot & \cdot & \cdot & -a_{12} \\ -a_{36} & \cdot & 1 & \cdot & \cdot & -a_{13} \\ -a_{46} & \cdot & \cdot & 1 & \cdot & -a_{14} \\ -a_{56} & \cdot & \cdot & \cdot & 1 & -a_{15} \end{vmatrix}$$

and we have

$$D_1 x_1 a^4_{16} = \begin{vmatrix} x_1 & x_1a_{16} & -x_1a_{26} & -x_1a_{36} & -x_1a_{46} & -x_1a_{56} \\ -a_{12} & x_1a_{26} & x_2a_{16} & \left\|\begin{matrix}123\\236\end{matrix}\right| & \left\|\begin{matrix}124\\246\end{matrix}\right| & \left\|\begin{matrix}125\\256\end{matrix}\right| \\ -a_{13} & x_1a_{36} & -\left\|\begin{matrix}123\\236\end{matrix}\right| & x_3a_{16} & \left\|\begin{matrix}134\\346\end{matrix}\right| & \left\|\begin{matrix}135\\356\end{matrix}\right| \\ -a_{14} & x_1a_{46} & -\left\|\begin{matrix}124\\246\end{matrix}\right| & -\left\|\begin{matrix}134\\346\end{matrix}\right| & x_4a_{16} & \left\|\begin{matrix}145\\456\end{matrix}\right| \\ -a_{15} & x_1a_{56} & -\left\|\begin{matrix}125\\256\end{matrix}\right| & -\left\|\begin{matrix}135\\356\end{matrix}\right| & -\left\|\begin{matrix}145\\456\end{matrix}\right| & x_5a_{16} \\ -a_{16} & \cdot & \cdot & \cdot & \cdot & \cdot \end{vmatrix}$$

where $\left\|\begin{smallmatrix}123\\236\end{smallmatrix}\right|$ is the Pfaffian $a_{12}a_{36}-a_{13}a_{26}+a_{16}a_{23}$ etc. or

$$D_1 x_1 a^3_{16} = \begin{vmatrix} x_1a_{16} & x_1a_{26} & x_1a_{36} & x_1a_{46} & x_1a_{56} \\ -x_1a_{26} & x_2a_{16} & \left\|\begin{matrix}123\\236\end{matrix}\right| & \left\|\begin{matrix}124\\246\end{matrix}\right| & \left\|\begin{matrix}125\\256\end{matrix}\right| \\ -x_1a_{36} & -\left\|\begin{matrix}123\\236\end{matrix}\right| & x_3a_{16} & \left\|\begin{matrix}134\\346\end{matrix}\right| & \left\|\begin{matrix}135\\356\end{matrix}\right| \\ -x_1a_{46} & -\left\|\begin{matrix}124\\246\end{matrix}\right| & -\left\|\begin{matrix}134\\346\end{matrix}\right| & x_4a_{16} & \left\|\begin{matrix}145\\456\end{matrix}\right| \\ -x_1a_{56} & -\left\|\begin{matrix}125\\256\end{matrix}\right| & -\left\|\begin{matrix}135\\356\end{matrix}\right| & -\left\|\begin{matrix}145\\456\end{matrix}\right| & x_5a_{16} \end{vmatrix}$$

If we let D_2 be the determinant D_1 with $x_5=0$ then the right hand side of (1) is of the same form as the D_1 with which we started and therefore

$$(D_2x_1a^3_{16})\cdot a_{16}x_1(a_{56}x_1)^3 \div (a_{56}x_1) = D_2x_1^4a_{16}^4a_{56}^2$$

$$= \begin{vmatrix} a_{16}a_{56}x_1 & a_{16}x_1\left\|\begin{matrix}125\\256\end{matrix}\right| & a_{16}x_1\left\|\begin{matrix}125\\356\end{matrix}\right| & a_{16}x_1\left\|\begin{matrix}125\\456\end{matrix}\right| \\ -\,a_{16}x_1\left\|\begin{matrix}125\\256\end{matrix}\right| & a_{16}a_{56}x_1x_2 & P & Q \\ -\,a_{16}x_1\left\|\begin{matrix}125\\356\end{matrix}\right| & -\,P & a_{16}a_{56}x_1x_3 & R \\ -\,a_{16}x_1\left\|\begin{matrix}125\\456\end{matrix}\right| & -\,Q & -\,R & a_{16}a_{56}x_1x_4 \end{vmatrix}$$

where

$$P = \begin{vmatrix} a_{26}x_1 & a_{36}x_1 & a_{56}x_1 \\ & \left\|\begin{matrix}125\\236\end{matrix}\right| & \left\|\begin{matrix}125\\256\end{matrix}\right| \\ & & \left\|\begin{matrix}135\\356\end{matrix}\right| \end{vmatrix}$$

$$Q = \begin{vmatrix} a_{26}x_1 & a_{46}x_1 & a_{56}x_1 \\ & \left|\begin{matrix}124\\246\end{matrix}\right| & \left\|\begin{matrix}125\\256\end{matrix}\right| \\ & & \left\|\begin{matrix}145\\456\end{matrix}\right| \end{vmatrix}$$

$$R = \begin{vmatrix} a_{36}x_1 & a_{46}x_1 & a_{56}x_1 \\ & \left\|\begin{matrix}134\\345\end{matrix}\right| & \left\|\begin{matrix}135\\356\end{matrix}\right| \\ & & \left\|\begin{matrix}145\\456\end{matrix}\right| \end{vmatrix}$$

But it is readily seen that

$$P = x_1a_{16}\left\|\begin{matrix}235\\356\end{matrix}\right|, \quad Q = x_1a_{16}\left\|\begin{matrix}245\\456\end{matrix}\right|, \quad R = x_1a_{16}\left\|\begin{matrix}345\\456\end{matrix}\right|$$

and substituting and simplifying gives

$$(2)\quad a^2_{56}\cdot D_2 = \begin{vmatrix} a_{56}x_1 & \left\|\begin{matrix}125\\256\end{matrix}\right| & \left\|\begin{matrix}135\\356\end{matrix}\right| & \left\|\begin{matrix}145\\456\end{matrix}\right| \\ -\left\|\begin{matrix}125\\256\end{matrix}\right| & a_{56}x_2 & \left\|\begin{matrix}235\\356\end{matrix}\right| & \left\|\begin{matrix}245\\456\end{matrix}\right| \\ -\left\|\begin{matrix}135\\356\end{matrix}\right| & -\left\|\begin{matrix}235\\356\end{matrix}\right| & a_{56}x_3 & \left\|\begin{matrix}345\\456\end{matrix}\right| \\ -\left\|\begin{matrix}145\\456\end{matrix}\right| & -\left\|\begin{matrix}245\\456\end{matrix}\right| & -\left\|\begin{matrix}345\\456\end{matrix}\right| & a_{56}x_4 \end{vmatrix}$$

Similar procedure leads to

$$(3)\quad x_1\left\|\begin{matrix}145\\456\end{matrix}\right| D_3 = \begin{vmatrix} x_1\left\|\begin{matrix}145\\456\end{matrix}\right| & x_1\left\|\begin{matrix}245\\456\end{matrix}\right| & x_1\left\|\begin{matrix}345\\456\end{matrix}\right| \\ -x_1\left\|\begin{matrix}245\\456\end{matrix}\right| & -x_1\left\|\begin{matrix}145\\456\end{matrix}\right| & \left\|\begin{matrix}12345\\23456\end{matrix}\right| \\ -x_1\left\|\begin{matrix}345\\456\end{matrix}\right| & -\left\|\begin{matrix}12345\\23456\end{matrix}\right| & x_3\left\|\begin{matrix}145\\456\end{matrix}\right| \end{vmatrix}$$

and

$$D_4 = \begin{vmatrix} x_1\left\|\begin{matrix}345\\456\end{matrix}\right| & \left\|\begin{matrix}12345\\23456\end{matrix}\right| \\ -\left\|\begin{matrix}12345\\23456\end{matrix}\right| & x_2\left\|\begin{matrix}345\\456\end{matrix}\right| \end{vmatrix}.$$

EXERCISES. Set XXI

1. Put $x_1=x_2=x_3=x_4=0$ in (2) §443 and give the equality resulting from taking the square root of both sides.
2. Show that

$$\left|\begin{matrix} a_1+a_2 & a_1+a_3 & \cdots & a_1+a_{2n} \\ & a_2+a_3 & \cdots & a_2+a_{2n} \\ & & \cdots\cdots & \cdots \\ & & & a_{2n-1}+a_{2n} \end{matrix}\right|$$

$$= 2^{n-1}(a_1a_3\cdots a_{2n-1}+a_2a_4\cdots a_{2n}).$$

3. Show that

$$\begin{vmatrix} 1 & 1 & 1 \cdots 1 \\ -1 & \cdot & 1 \cdots 1 \\ -1 & -1 & \cdot \quad \cdots 1 \\ \cdot & \cdot\ \cdot\ \cdot & \cdot\ \cdot\ \cdot\ \cdot \\ -1 & -1 & -1 \cdots \cdot \end{vmatrix} = 1$$

4. Show that the Pfaffian

$$\left| \begin{matrix} x & -1 & \cdot & \cdot & \cdots \\ & x & -1 & \cdot & \cdots \\ & & x & -1 & \cdots \\ & & & \cdot\ \cdot & \cdot\ \cdot\ \cdot \end{matrix} \right|_{2n} = \begin{vmatrix} x & 1 & \cdot & \cdots \\ 1 & x & 1 & \cdots \\ \cdot & 1 & x & 1 \cdots \\ \cdot\ \cdot & \cdot\ \cdot & \cdot\ \cdot & \cdot\ \cdot\ \cdot \end{vmatrix}_{n}$$

$$= x^n - c_{n-1,1}x^{n-2} + c_{n-2,2}x^{n-4} - \cdots .$$

5. Show that

$$\begin{vmatrix} a_1 & a_2 & a_3 & a_4 \\ b_1 & b_2 & b_3 & b_4 \\ \cdot & c_2 & c_3 & c_4 \\ \cdot & d_2 & d_3 & d_4 \end{vmatrix} = \left| \begin{matrix} |a_1b_2| & |a_1b_3| & |a_1b_4| \\ & |c_2d_3| & |c_2d_4| \\ & & |c_3d_4| \end{matrix} \right|$$

6. Show that

$$m\begin{vmatrix} 1 & \alpha_1 & \alpha_2 \\ -\alpha_1 & 1 & \beta_1 \\ -\alpha_2 & -\beta_1 & 1 \end{vmatrix} + \begin{array}{ccc|c} h_1 & h_2 & h_3 & \\ \hline 1 & \alpha_1 & \alpha_2 & k_1 \\ -\alpha_1 & 1 & \beta_1 & k_2 \\ -\alpha_2 & -\beta_1 & 1 & k_3 \end{array}$$

$$= m + (h_1k_1 + h_2k_2 + h_3k_3) + \sum \alpha_1 \left| \begin{matrix} m & h_1 & h_2 \\ & k_1 & k_2 \\ & & \alpha_1 \end{matrix} \right|$$

Representing the left hand side by $mD_3 + B_3$ show that

$$mD_5 + B_5 = m + \sum h_1k_1 + \sum\left(\alpha_1 \left| \begin{matrix} m & h_1 & h_2 \\ & k_1 & k_2 \\ & & 1 \end{matrix} \right| \right)$$

$$+ \sum\left(m \left| \begin{matrix} \alpha_1\alpha_2 & \alpha_3 \\ \beta_1 & \beta_2 \\ & \gamma_1 \end{matrix} \right|^2 \right)$$

7. Express the square of any even ordered determinant as a zero-axial skew of the same order.

Persymmetric Determinants

444. A determinant such that each line perpendicular to the principal diagonal has all its elements alike is called a *persymmetric* determinant. In the persymmetric determinant of the nth order

$$\begin{vmatrix} a_1 & a_2 & a_3 & \cdots & a_n \\ a_2 & a_3 & a_4 & \cdots & a_{n-1} \\ a_3 & a_4 & a_5 & \cdots & a_{n-1} \\ \cdot & \cdot & \cdot & \cdot & \cdot \\ a_n & a_{n-1} & a_{n-2} & \cdots & a_{2n-1} \end{vmatrix}$$

there are evidently at most $2n-1$ distinct elements, viz. those of the principal diagonal and one adjacent minor diagonal. It may thus be shortly denoted by

$$P(a_1 a_2 \cdots a_{2n-1}).$$

It is apparent from the nature of a persymmetric determinant that, using double suffix notation, any minor is not altered if we add (or subtract) the same integer to each row-number provided we subtract (or add) it from the column numbers.

445. *The persymmetric determinant of* $a_1, a_2, \cdots, a_{2n-1}$, *is equal to the persymmetric determinant of* a_1, ma_1+a_2, $m^2a_1+2ma_2+a_3$, $m^3a_1+3m^2a_2+3ma_3+a_4$, etc.

This follows from multiplying the determinant row-wise by

$$\begin{vmatrix} 1 & 0 & 0 & 0 & 0 \cdots \\ m & 1 & 0 & 0 & 0 \cdots \\ m^2 & 2m & 1 & 0 & 0 \cdots \\ m^3 & 3m^2 & 3m & 1 & 0 \cdots \\ \cdot & \cdot & \cdot & \cdot & \cdot \end{vmatrix}$$

and repeating the operation upon the product.

Indicating row multiplication as practised in the multiplication of determinants by $\between$, —for example, writing $(a, b, c \between \alpha, \beta, \gamma)$ for $a\alpha+b\beta+c\gamma$,—the elements of the new persymmetric determinant here found are very conveniently denoted by a_1, $(a_1, a_2 \between m, 1)$, $(a_1, a_2, a_3 \between m, 1)^2$, $(a_1, a_2, a_3, a_4 \between m, 1)$,^{3} etc.

When $m = -1$, these elements are the first terms of the successive difference-series of $a_1, a_2, \cdots, a_{2n-1}$, so that as a special case we have the theorem—*The persymmetric determinant of* $a_1 \cdots a_{2n-1}$, *is not altered by substituting for its elements the* 0*th*, 1*st*, 2*nd*, $\cdots$, $(2n-2)$*th differences of the first of them.*

446. If we start with the equation

$$(1) \qquad a_0x^n + a_1x^{n-1} + \cdots + a_n = 0$$

whose roots are $x_1, x_2 \cdots x_n$, and substitute for x the roots one after another and add the results we get

$$(2) \qquad a_0s_n + a_1s_{n-1} + \cdots + a_ns_0 = 0.$$

Next multiply (1) by $x, x^2, \cdots, x^{n-1}$ and substitute in each case the roots as before and add and we get

$$\begin{aligned} a_0s_{n+1} &+ a_1s_n &&+ \cdots + a_ns_1 = 0 \\ a_{0n}s_{n+2} &+ a_1s_{n+1} &&+ \cdots + a_ns_n = 0 \\ &\cdots\cdots\cdots\cdots \\ a_0s_{2n-1} &+ a_1s_{2n-2} &&+ \cdots + a_ns_{n-1} = 0. \end{aligned}$$

Eliminating from these and (1) the a's we have

$$(3) \qquad \Delta \equiv \begin{vmatrix} s_0 & s_1 \cdots s_n \\ s_1 & s_2 \cdots s_{n+1} \\ \cdot & \cdots\cdots \\ s_{n-1} & s_n \cdots s_{2n-1} \\ 1 & x \cdots x^n \end{vmatrix} = 0$$

or as it may be expressed

$$\begin{vmatrix} s_1 - xs_0 & s_2 - xs_1 \cdots & s_n - xs_{n-1} \\ s_2 - xs_1 & s_3 - xs_2 \cdots & s_{n+1} - xs_n \\ \cdots & \cdots & \cdots \\ s_n - xs_{n-1} & s_{n+1} - xs_n \cdots & s_{2n-1} - xs_{2n-2} \end{vmatrix} = 0.$$

If we denote this last form by V_n and the coaxial minor formed by erasing the last r rows and the last r columns by V_{n-r}, then V_n, $V_{n-1}, V_{n-2} \cdots V_1$ possess the characteristic property of Sturm's remainders. That is if V_k vanish then V_{k-1} and V_{k+1} have opposite signs by §383.

If in (3) we multiply the columns by $a_n, a_{n-1}, \cdots, a_0$ and add the results to the last we get

$$\Delta = (a_0x^n + a_1x^{n-1} + \cdots + a_n)\begin{vmatrix} s_0 & s_1 \cdots s_{n-1} \\ s_1 & s_2 \cdots s_n \\ \cdot & \cdot \quad \cdot \quad \cdot \\ s_{n-1} & s_n \cdots s_{2n-2} \end{vmatrix}.$$

447. If we square

$$\begin{vmatrix} 1 & x_1 & x_1^2 & x_1^3 \cdots x_1^{n-1} \\ 1 & x_2 & x_2^2 & \cdot \quad \cdots x_2^{n-1} \\ \cdot & \cdot & \cdot & \cdot \quad \cdot \quad \cdot \\ 1 & x_n & x_n^2 & \cdot \quad \cdots x_n^{n-1} \end{vmatrix} \equiv \zeta^{1/2}(x_1x_2 \cdots x_n)$$

multiplying column-by-column we have

$$\zeta(x_1x_2 \cdots x_n) = \begin{vmatrix} s_0 & s_1 \cdots s_{n-1} \\ s_1 & s_2 \cdots s_n \\ \cdot & \cdot \quad \cdot \quad \cdot \\ s_{n-1} & s_n \cdots s_{2n-2} \end{vmatrix}.$$

If the number of distinct roots are less than n it is obvious that this determinant vanishes.

448. The determinant

$$\Delta \equiv \begin{vmatrix} s_{0+r} & s_{1+r} & \cdots s_{n+r} \\ s_{1+r} & s_{2+r} & \cdots s_{n+1+r} \\ \cdot & \cdot & \cdot \quad \cdot \quad \cdot \\ s_{n+r} & s_{n+1+r} & \cdots s_{2n+r} \end{vmatrix}$$

is the product of the alternants

$$\left| \overset{0}{x_1}\overset{1}{x_2} \cdots \overset{n-1}{x_n} \right| \text{ and } \left| \overset{r}{x_1}\overset{r+1}{x_2} \cdots \overset{r+n-1}{x_n} \right|$$

But $x_1^rx_2^r \cdots x_n^r\zeta(x_1x_2 \cdots x_n)$ is also the product of these two alternants and therefore

$$\Delta = x_1^rx_2^r \cdots x_n^r\zeta(x_1x_2 \cdots x_n).$$

It is again obvious that Δ vanishes when the roots are not all distinct.

When the number of rows of Δ exceeds the number of distinct roots of $x^n+a_1x^{n-1}+a_2x^{n-2}+\cdots+a_n=0$ then it is obvious that

$$\Delta \equiv \begin{vmatrix} s_{0+r} & s_{1+r} & s_{2+r} & \cdots \\ s_{1+r} & s_{2+r} & s_{3+r} & \cdots \\ s_{2+r} & s_{3+r} & s_{4+r} & \cdots \\ \cdot & \cdot & \cdot & \cdot \end{vmatrix} = 0$$

For

$$\Delta = \begin{Vmatrix} 1 & 1 & 1 & \cdots \\ x_1 & x_2 & x_3 & \cdots \\ x_1^2 & x_2^2 & x_3^2 & \cdots \\ \cdot & \cdot & \cdot & \cdot \end{Vmatrix} \cdot \begin{Vmatrix} x_1^r & x_2^r & x_3^r & \cdots \\ x_1^{r+1} & x_2^{r+1} & x_3^{r+1} & \cdots \\ x_1^{r+2} & x_2^{r+2} & x_3^{r+2} & \cdots \\ \cdot & \cdot & \cdot & \cdot \end{Vmatrix}$$

where $x_1, x_2, x_3 \cdots$ are the roots. Thus if the number of rows in Δ is greater than the number of roots we must attach columns of zeros to these two arrays to make the columns the same in number as the rows.

When the number of rows of Δ is just equal to the number of distinct roots, then Δ is equal to a multiple of the square of the difference product of the said roots.

449. Starting with the array

$$\begin{Vmatrix} x_1^{m-1} & x_1^{m-2} & \cdots & x_1 & 1 \\ x_2^{m-1} & x_2^{m-2} & \cdots & x_2 & 1 \\ \cdot & \cdot & \cdot & \cdot & \cdot \\ x_n^{m-1} & x_n^{m-2} & \cdots & x_n & 1 \end{Vmatrix}, \quad n > m$$

and squaring, multiplying column-by-column, we have a determinant Δ of order m whose elements

$$c_{ik} = x_1^{i-1}x_1^{k-1} + x_2^{i-1}x_2^{k-1} \cdots + x_n^{i-1}x_n^{k-1} = s_{i+k-2}$$

This determinant Δ is equal to the sum of the squares of all the $(n)_m$ determinants of order m formed by taking the rows m at a time.

We have therefore

$$\sum\{\zeta(x_px_q\cdots)\} = \begin{vmatrix} s_0 & s_1 & \cdots & s_{m-1} \\ s_1 & s_2 & \cdots & s_m \\ \cdot & \cdot & \cdot & \cdot \\ s_{m-1} & s_m & \cdots & s_{2m-2} \end{vmatrix}$$

where $x_p, x_q, \cdots$ are any m of the n quantities $x_1, x_2, \cdots, x_n$.

450. Every minor of order n formed from the array

$$\begin{matrix} T_1 & T_2 & \cdots T_{n+1} \\ T_2 & T_3 & \cdots T_{n+2} \\ \cdot & \cdot & \cdot \\ T_n & T_{n+1} & \cdots T_{2n} \end{matrix}$$

is zero if

$$T_i = \overset{r-i}{a_1}\overset{s+i}{b_1} + \cdots + \overset{r-i}{a_{n-1}}\overset{s+i}{b_{n-1}}.$$

This is seen on observing that every minor is the product of two zeros. Thus

$$T_i = (\overset{r-i}{a_1}\overset{r-i}{a_2} \cdots \overset{r-i}{a_{n-1}}0)(\overset{s+i}{b_1}\overset{s+i}{b_2} \cdots \overset{s+i}{b_{n-1}}0).$$

451. If we multiply the two alternating functions $|\alpha^0\beta^1\gamma^2\delta^3x^4|$ and $|\alpha^0\beta^1\gamma^2\delta^3y^4|$ written in the forms

$$\begin{vmatrix} 1 & \cdot & \cdot & \cdot & \cdot & \cdot \\ \cdot & 1 & 1 & 1 & 1 & 1 \\ \cdot & \alpha & \beta & \gamma & \delta & x \\ \cdot & \alpha^2 & \beta^2 & \gamma^2 & \delta^2 & x^2 \\ \cdot & \alpha^3 & \beta^3 & \gamma^3 & \delta^3 & x^3 \\ \cdot & \alpha^4 & \beta^4 & \gamma^4 & \delta^4 & x^4 \end{vmatrix}, \quad - \begin{vmatrix} \cdot & \cdot & \cdot & \cdot & \cdot & 1 \\ 1 & 1 & 1 & 1 & 1 & \cdot \\ y & \alpha & \beta & \gamma & \delta & \cdot \\ y^2 & \alpha^2 & \beta^2 & \gamma^2 & \delta^2 & \cdot \\ y^3 & \alpha^3 & \beta^3 & \gamma^3 & \delta^3 & \cdot \\ y^4 & \alpha^4 & \beta^4 & \gamma^4 & \delta^4 & \cdot \end{vmatrix},$$

respectively, we get

$$- \begin{vmatrix} \cdot & 1 & y & y^2 & y^3 & y^4 \\ 1 & s_0 & s_1 & s_2 & s_3 & s_4 \\ x & s_1 & s_2 & s_3 & s_4 & s_5 \\ x^2 & s_2 & s_3 & s_4 & s_5 & s_6 \\ x^3 & s_3 & s_4 & s_5 & s_6 & s_7 \\ x^4 & s_4 & s_5 & s_6 & s_7 & s_8 \end{vmatrix}.$$

But the two alternants may be written

$$|\alpha^0\beta^1\gamma^2\delta^3| \{x^4 - \sum\alpha x^3 + \sum\alpha\beta x^2 - \sum\alpha\beta\gamma x + \alpha\beta\gamma\delta\},$$
$$|\alpha^0\beta^1\gamma^2\delta^3| \{y^4 - \sum\alpha y^3 + \sum\alpha\beta y^2 - \sum\alpha\beta\gamma y + \alpha\beta\gamma\delta\}$$

and if we denote the adjugate of the cofactor of the zero element in this product determinant by $|S_{00}S_{11}S_{22}S_{33}S_{44}|$, and 1, $\sum\alpha$, $\sum\alpha\beta$, $\sum\alpha\beta\gamma$, $\alpha\beta\gamma\delta$ by c_0, c_1, c_2, c_3, c_4 respectively the relation takes the form

$$\begin{array}{ccccc|c} 1 & y & y^2 & y^3 & y^4 & \\ \hline S_{00} & S_{01} & S_{02} & S_{03} & S_{04} & 1 \\ S_{10} & S_{11} & S_{12} & S_{13} & S_{14} & x \\ S_{20} & S_{21} & S_{22} & S_{23} & S_{24} & x^2 \\ S_{30} & S_{31} & S_{32} & S_{33} & S_{34} & x^3 \\ S_{40} & S_{41} & S_{42} & S_{43} & S_{44} & x^4 \end{array}$$

$$= |\alpha^0\beta^1\gamma^2\delta^3|^2 \begin{array}{ccccc|c} y^4 & v^3 & y^2 & y & 1 & \\ \hline c_0^2 & -c_0c_1 & c_0c_2 & -c_0c_3 & c_0c_4 & x^4 \\ -c_1c_0 & c_1 & -c_1c_2 & c_1c_3 & -c_1c_4 & x^3 \\ c_2c_0 & -c_2c_1 & c_2^2 & -c_2c_3 & c_2c_4 & x^2 \\ -c_3c_0 & c_3c_1 & -c_3c_2 & c_3^2 & -c_3c_4 & x \\ c_4c_0 & -c_4c_1 & c_4c_2 & -c_4c_3 & c_4^2 & 1 \end{array}$$

and by equating coefficients of x^ry^s

$$S_{rs} = |\alpha^0\beta^1\gamma^2\delta^3|^2(-1)^{r+s}c_{n-r}c_{n-s}.$$

In other words we have the value of any primary minor of

$$\begin{vmatrix} s_0 & s_1 & \cdots & s_n \\ s_1 & s_2 & \cdots & s_{n-1} \\ \cdot & \cdot & \cdot & \cdot \\ s_n & s_{n-1} & \cdots & s_{2n} \end{vmatrix}$$

which vanishes having more than n rows.

452. If

$$a_0x^n + a_1x^{n-1} + \cdots + a_n \text{ or } a_0(x - x_1)(x - x_2)\cdots(x - x_n) \equiv f(x),$$
$$b_0x^m + b_1x^{m-1} + \cdots + b_m \equiv \phi(x), \quad (m < n),$$

then let

$$S_r = \frac{\overset{r}{x_1}\phi(x_1)}{f'(x_1)} + \frac{\overset{r}{x_2}\phi(x_2)}{f'(x_2)} + \cdots + \frac{\overset{r}{x_n}\phi(x_n)}{f'(x_n)}$$
$$= \overset{r}{x_1}A_1^2 + \overset{r}{x_2}A_2^2 + \cdots + \overset{r}{x_n}A_n^2,$$

where

$$A_k = \left\{ \frac{\phi(x_k)}{f'(x_k)} \right\}^{1/2}.$$

Squaring in two ways the determinant

$$\begin{vmatrix} A_1 & A_2 & \cdots A_n \\ A_1x_1 & A_2x_2 & \cdots A_nx_n \\ \cdot & \cdot & \cdot \\ A_1x_1^{n-1} & A_2x_2^{n-1} & \cdots A_nx_n^{n-1} \end{vmatrix}$$

we have on equating the results

$$\begin{vmatrix} S_0 & S_1 \cdots S_{n-1} \\ S_1 & S_2 \cdots S_n \\ \cdot & \cdot \\ S_{n-1} & S_n \cdots S_{2n-1} \end{vmatrix} = A_1^2A_2^2 \cdots A_n^2 \begin{vmatrix} s_0 & s_1 \cdots s_{n-1} \\ s_1 & s_2 \cdots s_n \\ \cdot & \cdot \\ s_{n-1} & s_n \cdots s_{2n-2} \end{vmatrix}$$

which, on putting $(-1)^{(n-1)/2}f'(x_1)f'(x_2) \cdots f'(x_n)$ for the determinant on the right and substituting for the A's becomes

$$\begin{vmatrix} S_0 & S_1 \cdots S_{n-1} \\ S_1 & S_2 \cdots S_n \\ \cdot & \cdot \\ S_{n-1} & S_n \cdots S_{2n-2} \end{vmatrix} = (-1)^{n(n-1)/2}\phi(x_1)\phi(x_2) \cdots \phi(x_n).$$

The determinant $P(s_0, s_1, \cdots, s_{2n-2})$ is very important in dealing with questions concerning the character, number of different roots, their multiplicities, etc. of an equation of the nth degree. (See Baur Math. Annalen lii p. 103+).

453. Sylvester has shown* that the problem to transform the binary $(2n+1)$ic into the sum of $(n+1)$ powers depends upon the solution of

$$D_{n+1} \equiv \begin{vmatrix} \lambda^{n+1} & \lambda^n & \cdots \lambda^0 \\ a_{n+1} & a_n & \cdots a_0 \\ a_{n+2} & a_{n+1} & \cdots a_1 \\ \cdot & \cdot & \cdot \\ a_{2n+1} & a_{2n} & \cdots a_n \end{vmatrix} = 0,$$

* Philos. Magazine II, pp. 391–410.

or its equivalent

$$\begin{vmatrix} a_{n+1} - a_n\lambda & a_n - a_{n-1}\lambda & \cdots & a_1 - a_0\lambda \\ a_{n+2} - a_{n+1}\lambda & a_{n+1} - a_n\lambda & \cdots & a_2 - a_1\lambda \\ \cdot & \cdot & \cdot & \cdot \\ a_{2n+1} - a_{2n}\lambda & a_{2n} - a_{2n-1}\lambda & \cdots & a_{n+1} - a_n\lambda \end{vmatrix} = 0,$$

which is a persymmetric determinant of order $n+1$.

In the case of a binary $2n-ic$ the problem depends upon the vanishing of a determinant of order $n+1$ persymmetric except for its center element.

EXAMPLES: Derive the recurrence formula

$$H_n^2 D_{n+1} + (H_n H'_{n+1} - H_{n+1} H'_n - H_n H_{n+1} x) D_n + H_{n+1}^2 D_{n-1} = 0$$

where

$$H_{n+1} \equiv \begin{vmatrix} a_n & a_{n-1} & \cdots & a_0 \\ a_{n+1} & a_n & \cdots & a_1 \\ \cdot & \cdot & \cdot & \cdot \\ a_{2n} & a_{2n-1} & \cdots & a_n \end{vmatrix} \quad \text{and} \quad H'_{n+1} \equiv \begin{vmatrix} a_{n+1} & a_{n-1} & \cdots & a_0 \\ a_{n+2} & a_n & \cdots & a_1 \\ \cdot & \cdot & \cdot & \cdot \\ a_{2n+1} & a_{2n-1} & \cdots & a_{2n} \end{vmatrix}$$

2. If we write the binary quartic $U \equiv (a_0 a_1 \cdots a_4 \between x, y)^4$ in the form

$$\begin{array}{ccc|c} x^2 & 2xy & y^2 & \\ \hline a_0 & a_1 & a_2 & x^2 \\ a_1 & a_2 & a_3 & 2xy, \\ a_2 & a_3 & a_4 & y^2 \end{array}$$

and the binary quintic $(a_0 a_1 \cdots a_5 \between x\ y)^5$ or

$$(a_0 x + a_1 y\ \ a_1 x + a_2 y \cdots a_4 x + a_5 y \between xy)^4$$

in the form

$$\begin{array}{ccc|c} x^2 & 2xy & y^2 & \\ \hline a_0 x + a_1 y & a_1 x + a_2 y & a_2 x + a_3 y & x^2 \\ a_1 x + a_2 y & a_2 x + a_3 y & a_3 x + a_4 y & 2xy \\ a_2 x + a_3 y & a_3 x + a_4 y & a_4 x + a_5 y & y^2 \end{array}$$

show that the condition that

$$\frac{\partial^2 u}{\partial x^2} = 0, \quad \frac{\partial^2 u}{\partial x \partial y} = 0 \quad \text{and} \quad \frac{\partial^2 u}{\partial y^2} = 0$$

is that

$$\begin{vmatrix} a_0 & a_1 & a_2 \\ a_1 & a_2 & a_3 \\ a_2 & a_3 & a_4 \end{vmatrix} = 0, \text{ etc.}$$

454. To find a function u of the form $N(x)/M(x)$, which shall have the values $u_1, u_2, \cdots, u_{n+m+1}$ when x has the values $x_1, x_2, \cdots, x_{n+m+1}$, it being understood that $N(x)$ and $M(x)$ are functions of x of the degrees n and m respectively.

The method can be illustrated for the case

$$u = \frac{\beta_0 + \beta_1 x}{\alpha_0 + \alpha_1 x + \alpha_2 x^2} \cdot$$

Starting with

$$\begin{vmatrix} 1 & x_1 & x_1^2 & x_1^p(\beta_0 + \beta_1 x_1) \\ 1 & x_2 & x_2^2 & x_2^p(\beta_0 + \beta_1 x_2) \\ 1 & x_3 & x_3^2 & x_3^p(\beta_0 + \beta_1 x_3) \\ 1 & x_4 & x_4^2 & x_4^p(\beta_0 + \beta_1 x_4) \end{vmatrix}$$

which is equal to zero when $p=0$ or 1, and therefore

$$\begin{vmatrix} 1 & x_1 & x_1^2 & u_1 x_1^p(\alpha_0 + \alpha_1 x_1 + \alpha_2 x_1^2) \\ 1 & x_2 & x_2^2 & u_2 x_2^p(\alpha_0 + \alpha_1 x_2 + \alpha_2 x_2^2) \\ 1 & x_3 & x_3^2 & u_3 x_3^p(\alpha_0 + \alpha_1 x_3 + \alpha_2 x_3^2) \\ 1 & x_4 & x_4^2 & u_4 x_4^p(\alpha_0 + \alpha_1 x_4 + \alpha_2 x_4^2) \end{vmatrix} = 0$$

when $p=0$ or 1.

This may be written

$$(1) \quad \begin{aligned} &\begin{vmatrix} 1 & x_1 & x_1^2 & x_1^p u_1 \\ 1 & x_2 & x_2^2 & x_2^p u_2 \\ 1 & x_3 & x_3^2 & x_3^p u_3 \\ 1 & x_4 & x_4^2 & x_4^p u_4 \end{vmatrix} \alpha_0 + \begin{vmatrix} 1 & x_1 & x_1^2 & u_1 x_1^{p+1} \\ 1 & x_2 & x_2^2 & u_2 x_2^{p+1} \\ 1 & x_0 & x_3^2 & u_3 x_3^{p+1} \\ 1 & x_4 & x_4^2 & u_4 x_4{}^{\ +1} \end{vmatrix} \alpha_1 \\ &\qquad + \begin{vmatrix} 1 & x_1 & x_1^2 & u_1 x_1^{p+2} \\ 1 & x_2 & x_2^2 & u_2 x_2^{p+2} \\ 1 & x_3 & x_3^2 & u_3 x_3^{p+2} \\ 1 & x_4 & x_4^2 & u_4 x_4^{p+2} \end{vmatrix} \alpha_2 = 0 \end{aligned}$$

From the two equations involved here ($p=0$ and 1), and the equation $\alpha_0+\alpha_1 x+\alpha_2 x^2 = M(x)$ we get

$$M(x) = \begin{vmatrix} 1 & x & x^2 \\ V_0 & V_1 & V_2 \\ V_1 & V_2 & V_3 \end{vmatrix},$$

where

$$V_p = \begin{vmatrix} 1 & x_1 & x_1^2 & x_1^p u_1 \\ 1 & x_2 & x_2^2 & x_2^p u_2 \\ 1 & x_3 & x_3^2 & x_3^p u_3 \\ 1 & x_4 & x_4^2 & x_4^p u_4 \end{vmatrix},$$

or

$$M(x) = \begin{vmatrix} W_0 & W_1 \\ W_1 & W_2 \end{vmatrix}$$

where $W_p = V_{p+1} - xV_p$.

Again

$$\frac{N(x)}{f(x)} \left| x_1^0 x_2^1 x_3^2 x_4^3 \right| = \begin{vmatrix} 1 & x_1 & x_1^2 & \dfrac{N(x_1)}{x - x_1} \\ \cdot & \cdot & \cdot & \cdot \\ 1 & x_4 & x_4^2 & \dfrac{N(x_4)}{x - x_4} \end{vmatrix}$$

$$= \begin{vmatrix} 1 & x_1 & x_1^2 & \dfrac{u_1(\alpha_0 + \alpha_1 x_1 + \alpha_2 x_1^2)}{x - x_1} \\ \cdot & \cdot & \cdot & \cdot \\ 1 & x_1 & x_1^2 & \dfrac{u_4(\alpha_0 + \alpha_1 x_4 + \alpha_2 x_4^2)}{x - x_4} \end{vmatrix}$$

where $f(x) = (x - x_1)(x - x_2)(x - x_3)(x - x_4)$.

This again may be written

$$\begin{vmatrix} 1 & x_1 & x_1^2 & \dfrac{u_1}{x - x_1} \\ \cdot & \cdot & \cdot & \cdot \\ 1 & x_4 & x_4^2 & \dfrac{u_4}{x - x_4} \end{vmatrix} \alpha_0 + \begin{vmatrix} 1 & x_1 & x_1^2 & \dfrac{u_1 x_1}{x - x_1} \\ \cdot & \cdot & \cdot & \cdot \\ 1 & x_4 & x_4^2 & \dfrac{u_4 x_4}{x - x_4} \end{vmatrix} \alpha_1$$

(2)

$$+ \begin{vmatrix} 1 & x_1 & x_1^2 & \dfrac{u_1 x_2^2}{x - x_1} \\ \cdot & \cdot & \cdot & \cdot \\ 1 & x_4 & x_4^2 & \dfrac{u_4 x_4^2}{x - x_4} \end{vmatrix} \alpha_2 .$$

This equation together with equations (1) gives

$$\frac{N(x)}{f(x)}\left| x_1^0 x_2^1 x_3^2\, x_4^p u_4 \right| = \begin{vmatrix} \rho_0 & \rho_1 & \rho_2 \\ V_0 & V_1 & V_2 \\ V_1 & V_2 & V_3 \end{vmatrix} = \begin{vmatrix} \rho_0 & \rho_1 & \rho_2 \\ \rho_1 & \rho_2 & \rho_3 \\ \rho_2 & \rho_3 & \rho_4 \end{vmatrix},$$

where $x\rho_p - V_p = \rho_{p+1}$. Therefore

$$u = \frac{N(x)}{M(x)} = f(x)\begin{vmatrix} \rho_0 & \rho_1 & \rho_2 \\ \rho_1 & \rho_2 & \rho_3 \\ \rho_2 & \rho_3 & \rho_4 \end{vmatrix} \div \left| x_1^0 x_2^1 x_3^2\, x_4^3 \right| \cdot \begin{vmatrix} W_0 & W_1 \\ W_1 & W_2 \end{vmatrix}.$$

455. If for positive integral values of s and r we have

$$A_{r+s} = a_1 A_{r+s-1} + a_2 A_{r+s-2} + \cdots + a_s A_r$$

then in the determinant

$$\begin{vmatrix} A_r & A_{r+1} \cdots A_{r+s-1} \\ A_{r+1} & A_{r+2} \cdots A_{r+s} \\ \cdots & \cdots \\ A_{r+s-1} & A_{r+s} \cdots A_{r+2s-2} \end{vmatrix}$$

which we may represent by $\Delta_{r,s}$, the last column may be changed into $a_s A_{r-1},\ a_s A_r, \cdots,\ a_s A_{r+s-2}$.

Therefore

$$\Delta_{r,s} = (-1)^{s-1} a_s \Delta_{r-1,s},$$

and hence

$$\Delta_{r,s} = (-1)^{r(s-1)} a_s^r \Delta_{0,s}$$

which shows that

$$\frac{\Delta_{r,s}}{(-1)^{r(s-1)} a_s^r} (= \Delta_{0,s})$$

is independent of r.

456. If

$$P_n \equiv \begin{vmatrix} \frac{1}{a} & \frac{1}{a+d} & \cdots & \frac{1}{a+(n-1)d} \\ \frac{1}{a+d} & \frac{1}{a+2d} & \cdots & \frac{1}{a+nd} \\ \cdots & \cdots & \cdots & \cdots \\ \frac{1}{a+(n-1)d} & \frac{1}{a+nd} & \cdots & \frac{1}{a+2(n-1)d} \end{vmatrix}$$

and if $\Pi(a+kd,\ d)$ stand for the product of all the denominators in the $(k+1)$st row or column, then it is easily seen that

$$P_n = \frac{1}{\Pi\{a+(n-1)d, d\}} \begin{vmatrix} \dfrac{a+(n-1)d}{a} & \dfrac{a+(n-1)d}{a+d} & \cdots 1 \\ \dfrac{a+nd}{a+d} & \dfrac{a+nd}{a+2d} & \cdots 1 \\ \cdots & \cdots & \cdots \\ \dfrac{a+2(n-1)d}{a+(n-1)d} & \dfrac{a+2(n-1)d}{a+nd} & \cdots 1 \end{vmatrix}$$

Next perform the operations $c_k - c_n$ for $k=1, 2, \cdots, n-1$ and we have after taking out the common factors

$$P_n = \frac{(n-1)!d^{n-1}}{\Pi\{a+(n-1)d, d\}} \begin{vmatrix} \dfrac{1}{a} & \dfrac{1}{a+d} & \cdots & \dfrac{1}{a+(n-2)d} & 1 \\ \dfrac{1}{a+d} & \dfrac{1}{a+2d} & \cdots & \dfrac{1}{a+(n-1)d} & 1 \\ \cdots & \cdots & \cdots & \cdots & \cdots \\ \dfrac{1}{a+(n-1)d} & \dfrac{1}{a+nd} & \cdots & \dfrac{1}{a+(2n-3)d} & 1 \end{vmatrix}$$

$$= \frac{(n-1)!d^{n-1}\{a+2(n-1)d\}}{[\Pi\{a+(n-1)d, d\}]^2}$$

$$\times \begin{vmatrix} \dfrac{a+(n-1)d}{a} & \dfrac{a+nd}{a+d} & \cdots & \dfrac{a+(2n-3)d}{a+(n-2)d} & 1 \\ \dfrac{a+(n-1)d}{a+d} & \dfrac{a+nd}{a+d} & \cdots & \dfrac{a+(2n-3)d}{a+(n-1)d} & 1 \\ \cdots & \cdots & \cdots & \cdots & \cdots \\ 1 & 1 & & 1 & 1 \end{vmatrix}$$

which on performing upon the determinant the following operations $r_k - r_n$ for $k=1, 2, \cdots, (n-1)$ we have

$$P_n = \frac{\{(n-1)!d^{n-1}\}^2\{a+2(n-1)d\}}{[\Pi\{a+(n-1)d, d\}]^2} \cdot P_{n-1}.$$

Using this as a reduction formula we get

$$P_n = \frac{\{(n-1)!d^{n-1}\}^2\{(n-2)!d^{n-2}\}^2\cdots d^2}{\prod(a,d)\prod(a+d,d)\cdots\prod\{a+(n-1)d,d\}}$$

EXAMPLE: Show that the eliminant of the set of equations

$$\begin{aligned} S_0 &= p_1 + p_2 \\ S_1 &= p_1\lambda_1 + p_2\lambda_2 \\ S_2 &= p_1\lambda_1^2 + p_2\lambda_2^2 \\ S_3 &= p_1\lambda_1^3 + p_2\lambda_2^3 \\ S_4 &= p_1\lambda_1^4 + p_2\lambda_2^4 \end{aligned} \quad \text{is} \quad \begin{vmatrix} S_0 & S_1 & S_2 \\ S_1 & S_2 & S_3 \\ S_2 & S_3 & S_4 \end{vmatrix} = 0$$

and show that every 3 consecutive S's are connected by a linear relation. (Schoute.)

457. If

$$\Delta \equiv \begin{vmatrix} 0 & 12 & 13 & 14 & 15 \\ -12 & 0 & 23 & 24 & 25 \\ -13 & -23 & 0 & 34 & 35 \\ -14 & -24 & -34 & 0 & 45 \\ -15 & -25 & -35 & -45 & 0 \end{vmatrix}$$

and

$$D \equiv \begin{vmatrix} 12 & 13 & 14 & 15 \\ 13 & 14+23 & 15+24 & 25 \\ 14 & 15+24 & 25+34 & 35 \\ 15 & 25 & 35 & 45 \end{vmatrix}$$

then it is readily seen that the adjugate of D will be persymmetric if $\Delta_{11} = \Delta_{22} = \Delta_{33} = \Delta_{44} = \Delta_{55} = 0$, where Δ_{ii} is a principal coaxial minor* of Δ.

If Δ is of even order then the corresponding D will have its adjugate persymmetric if Δ and all its coaxial second minors (that is minors of order 2 less than Δ) equal to zero.

If rs stands for $|a_r b_s|$ then these conditions are fulfilled and the adjugate is persymmetric.

Other similar results may be obtained from these by using the Law of Complementary Minors and the Law of Extensible Minors.

* It may be shown that if certain of the coaxial minors vanish the rest will vanish as a consequence. Thus if Δ_{rr} vanish for 3 of the values of r it will vanish for all 5 values.

The adjugate of the persymmetric determinant

$$\Delta \equiv \begin{vmatrix} a & b & c & d \\ b & c & d & e \\ c & d & e & f \\ d & e & f & g \end{vmatrix}$$

will be persymmetric if the principal minors of

$$\begin{matrix} a & b & c & d & e \\ b & c & d & e & f \\ c & d & e & f & g \end{matrix}$$

vanish or in general: *The persymmetric determinant of the nth order will have its adjugate also persymmetric if the principal minors of the corresponding persymmetric array of $n-1$ rows and $n+1$ columns all vanish.*

458. If $\overline{rs}$ denote the determinant got from the persymmetric m-by-$(m+2)$ array

$$\begin{matrix} a_1 & a_2 \cdots a_6 \\ a_2 & a_3 \cdots a_7 \\ a_3 & a_4 \cdots a_8 \\ a_4 & a_5 \cdots a_9 \end{matrix}$$

by deleting the rth and sth columns then the adjugate of the $(m+1)$-line persymmetric determinant $P(a_1, a_2, \cdots, a_9)$ is

$$\begin{vmatrix} \overline{12} & \overline{13} & \overline{14} & \overline{15} & \overline{16} \\ \overline{13} & \overline{14}+\overline{23} & \overline{15}+\overline{24} & \overline{16}+\overline{25} & \overline{26} \\ \overline{14} & \overline{15}+\overline{24} & \overline{16}+\overline{25}+\overline{34} & \overline{26}+\overline{35} & \overline{36} \\ \overline{15} & \overline{16}+\overline{25} & \overline{26}+\overline{35} & \overline{36}+\overline{45} & \overline{46} \\ \overline{16} & \overline{16} & \overline{36} & \overline{46} & \overline{56} \end{vmatrix}$$

The proof consists in observing that

$$\begin{vmatrix} a_1 & a_2 & a_4 & a_5 \\ a_2 & a_3 & a_5 & a_6 \\ a_4 & a_5 & a_7 & a_8 \\ a_5 & a_6 & a_8 & a_9 \end{vmatrix} = \overline{16} + \overline{25} + \overline{34}.$$

EXERCISE 1. If $\alpha, \beta, \gamma, \delta$ are column numbers in order of magnitude of the above array then show that

$$\overline{\alpha\beta}\cdot\overline{\gamma\delta} - \overline{\alpha\gamma}\cdot\overline{\beta\delta} + \overline{\alpha\delta}\cdot\overline{\beta\gamma} = 0.$$

2 Show that the adjugate of the adjugate of a persymmetric determinant is persymmetric.

459. *If P_5 stands for*

$$\begin{vmatrix} a_1x - a_2 & a_2x - a_3 & a_3x - a_4 & a_4x - a_5 & a_5x - a_6 \\ a_2x - a_3 & a_3x - a_4 & a_4x - a_5 & a_5x - a_6 & a_6x - a_7 \\ \cdot & \cdot & \cdot & \cdot & \cdot \\ a_5x - a_6 & a_6x - a_7 & a_7x - a_8 & a_8x - a_9 & a_9x - a_{10} \end{vmatrix}$$

and $\binom{rst\cdots}{uvw\cdots}$ denotes the determinant formed from the rth, sth, tth, $\cdots$, rows and the uth; vth, wth, $\cdots$ columns of the array

$$\begin{matrix} a_1 & a_2 \cdots a_6 \\ a_2 & a_3 \cdots a_7 \\ \cdot & \cdot \quad \cdot \\ a_5 & a_6 \cdots a_{10} \end{matrix}$$

then

$$\binom{1234}{1234}^2 P_5 + \binom{12345}{12345}^2 P_3 = \left\{\binom{1235}{1234}\binom{12345}{12345} - \binom{1234}{1234}\binom{12346}{12345} + \binom{1234}{1234}\binom{12345}{12345}.x\right\}P_4$$

a formula due to Jacobi.

For, starting with the identity $|e_5f_6| = e_5f_6 - e_6f_5$ and extending it by $a_1b_2c_3d_4$, we have

(1) $$|a_1b_2c_3d_4e_5f_6|\,|a_1b_2c_3d_4| = |a_1b_2c_3d_4e_5|\,|a_1b_2c_3d_4f_6| - |a_1b_2c_3d_4e_6|\,|a_1b_2c_3d_4f_5|$$

similarly taking the identity

$$d_4|d_5e_6| - d_5|d_4e_6| + d_6|d_4e_5| = 0$$

and extending it by $a_1b_2c_3$ we have

(2) $$|a_1b_2c_3d_4|\,|a_1b_2c_3d_5e_6| - |a_1b_2c_3d_5|\,|a_1b_2c_3d_4e_6| + |a_1b_2c_3d_6|\,|a_1b_2c_3d_4e_5|$$

Multiplying (1) by $|a_1b_2c_3d_4|$ and (2) by $|a_1b_2c_3d_4e_5|$ and adding the results we have

$$|a_1b_2c_3d_4e_5f_6|\,|a_1b_2c_3d_4|^2 + |a_1b_2c_3d_6|\,|a_1b_2c_3d_4e_5|^2$$

$$(3) = \{|a_1b_2c_3d_4e_5|\,|a_1b_2c_3d_5| - |a_1b_2c_3d_4f_5|\,|a_1b_2c_3d_4|\}\,|a_1b_2c_3d_4e_6|$$
$$+ |a_1b_2c_3d_4|\,|a_1b_2c_3d_4e_5|\,\{|a_1b_2c_3d_4f_5| - |a_1b_2c_3d_5e_6|\}$$

If in (3) we make

$$|a_1b_2c_3d_4e_5f_6| \equiv \begin{vmatrix} a_1 & a_2 & a_3 & a_4 & a_5 & 1 \\ a_2 & a_3 & a_4 & a_5 & a_6 & x \\ a_3 & a_4 & a_5 & a_6 & a_7 & x^2 \\ a_4 & a_5 & a_6 & a_7 & a_8 & x^3 \\ a_5 & a_6 & a_7 & a_8 & a_9 & x^4 \\ a_6 & a_7 & a_8 & a_9 & a_{10} & x^5 \end{vmatrix} \equiv F_5$$

and represent the persymmetric determinant of order 6

$$P(a_1a_2a_3a_4a_5a_6a_7a_8a_9a_{10}a_{11})$$

by P, then (3) becomes

$$(4)\quad F_5P^2_{56,56} + F_3P^2_{66} = (P_{46,46}P_{66} - P_{56,56}P_{56} + P_{56,56}P_{66}x)F_4.$$

Since F_k is equivalent to P_k this is seen to be Jacobi's formula.

EXERCISE. If

$$P_n = \begin{vmatrix} a_1x - a_2 & a_2x - a_3 & \cdots & a_nx - a_{n+1} \\ a_2x - a_3 & a_3x - a_4 & \cdots & a_{n+1}x - a_{n+2} \\ \cdot & \cdot & \cdot & \cdot \\ a_nx - a_{n+1} & a_{n+1}x - a_{n+2} & \cdots & a_{2n-1}x - a_{2n} \end{vmatrix}$$

show that

$$P_n - \left\{\frac{A\beta - B\alpha}{\alpha^2} + \frac{A}{\alpha}x\right\}P_{n-1} + \frac{A^2}{\alpha^2}P_{n-2} = 0$$

where A, $-B$, and α, $-\beta$ are the coefficients of the two highest powers of x in P_n and P_{n-1} respectively.

460. The persymmetric determinant

$$P(a_6, a_5, a_4, a_3, a_2, a_1, a_1, a_2, a_3, a_4, a_5)$$

may be written as follows:

$$P \equiv \begin{vmatrix} 0 & a_1 - a_2 & a_2 - a_3 & a_3 - a_4 & a_4 - a_5 & a_5 \\ a_2 - a_1 & 0 & a_1 - a_2 & a_2 - a_3 & a_3 - a_4 & a_4 \\ a_3 - a_2 & a_2 - a_1 & 0 & a_1 - a_2 & a_2 - a_3 & a_3 \\ a_4 - a_3 & a_3 - a_2 & a_2 - a_1 & 0 & a_1 - a_2 & a_2 \\ a_5 - a_4 & a_4 - a_3 & a_3 - a_2 & a_2 - a_1 & 0 & a_1 \\ a_6 - a_5 & a_5 - a_4 & a_4 - a_3 & a_3 - a_2 & a_2 - a_1 & a_1 \end{vmatrix}$$

which, if we border so as to make a skew-symmetric determinant Δ of order 7, the P will be the minor Δ_{76}, and $\Delta_{76} = \Delta_{66}^{\frac{1}{2}}\Delta_{77}^{\frac{1}{2}}$. Writing this out would give

$$P = \begin{vmatrix} a_2 - a_1 & a_3 - a_2 & a_4 - a_3 & a_5 - a_4 & a_6 - a_5 \\ & a_2 - a_1 & a_3 - a_2 & a_4 - a_3 & a_5 - a_4 \\ & & a_2 - a_1 & a_3 - a_2 & a_4 - a_3 \\ & & & a_2 - a_1 & a_3 - a_2 \\ & & & & a_2 - a_1 \end{vmatrix}$$

$$\cdot \begin{vmatrix} a_2 - a_1 & a_3 - a_2 & a_4 - a_3 & a_5 - a_4 & a_5 \\ & a_2 - a_1 & a_3 - a_2 & a_4 - a_3 & a_4 \\ & & a_2 - a_1 & a_3 - a_2 & a_3 \\ & & & a_2 - a_1 & a_2 \\ & & & & a_1 \end{vmatrix}$$

Equating coefficients of a_6 on both sides gives the result for the next lower order. $P(a_5, a_4, a_3, a_2, a_1, a_2, a_3, a_4, a_5)$ is seen to be centro-symmetric and therefore breaks up into two factors.

461. That the persymmetric determinant $P(1_0, 3_1, 5_2, 7_3, \cdots)$ where r_s stands for $r!/s!(r-s)!$ has a value unity may be seen by multiplying by unity in the form

$$\begin{vmatrix} 1 & \cdot & \cdot & \cdot & \cdot & \cdots \\ -3 & 1 & \cdot & \cdot & \cdot & \cdots \\ 5 & -5 & 1 & \cdot & \cdot & \cdots \\ -7 & 14 & -7 & 1 & \cdot & \cdots \\ 9 & -30 & 27 & -9 & 1 & \cdots \\ \cdot & \cdot & \cdot & \cdot & \cdot & \cdots \end{vmatrix}$$

the kth row of the multiplier being

$$(-1)^{k-1}(2k-1)\left\{1,\ -\frac{1}{3}(k)_2,\ \frac{1}{5}(k+1)_4,\ \cdots,\ (-1)^{k-1}\frac{1}{2k-1}(2k-2)_{2k-2}\right\}.$$

The property of combinatory numbers which underlies the multiplication is

$$(2h-1)_{h-1} - \frac{1}{3}(2h+1)_h\cdot k_2 + \frac{1}{5}(2h+3)_{h+1}(k+1)_4 - \frac{1}{7}(2k+5)_{h+2}(k+2)_6 + \cdots = (-1)^{k-1}\frac{1}{2k-1}(2h-1)_{h-k}$$

The persymmetric determinant

$$P\left\{\frac{1_0}{1},\ \frac{3_1}{3},\ \frac{5_2}{5},\ \frac{7_3}{7}\ \cdots\right\}$$

is seen to have the value unity by multiplying it by unity in the form

$$\begin{vmatrix} 1 & & & & \\ -1 & 1 & & & \\ 1 & -3 & 1 & & \\ -1 & 6 & -5 & 1 & \\ 1 & -10 & 15 & -7 & 1 \\ \cdot & \cdot & \cdot & \cdot & \cdot \end{vmatrix}$$

of which the kth row is

$$(-1)^{k-1}\{1,\ -(k)_2, (k+1)_4, (k+2)_6,\ \cdots,\ (-1)^{k-1}(2k-2)_{2k-2}\}$$

and the property of combinatory numbers underlying the multiplication being

$$\frac{1}{2h-1}(2h-1)_{h-1} - \frac{1}{2h+1}(2h+1)_h k_2 + \frac{1}{2h+3}(2h+3)_{h+2}(k+1)_4 + \cdots = (-1)^{k-1}\frac{2k-1}{2h-1}(2h-1)_{h-k}$$

462. Taking any n columns (say all but the $(k+1)$st) of the array

$$\begin{array}{cccccc} 1 & \frac{1}{2} & \frac{1}{3} & \frac{1}{4} & \cdots & \frac{1}{n} & \frac{1}{n+1} \\ \frac{1}{2} & \frac{1}{3} & \frac{1}{4} & \frac{1}{5} & \cdots & \frac{1}{n+1} & \frac{1}{n+2} \\ \cdot & \cdot & \cdot & \cdot & \cdot & \cdot & \cdot \\ \frac{1}{n} & \frac{1}{n+1} & \frac{1}{n+2} & \frac{1}{n+3} & \cdots & \frac{1}{2n+1} & \frac{1}{2n} \end{array}$$

and observing that the determinant D of them is a case of the double alternant

$$\begin{vmatrix} \frac{1}{\alpha_1 - \beta_1} & \frac{1}{\alpha_1 - \beta_2} & \cdots & \frac{1}{\alpha_1 - \beta_n} \\ \frac{1}{\alpha_2 - \beta_1} & \frac{1}{\alpha_2 - \beta_2} & \cdots & \frac{1}{\alpha_2 - \beta_n} \\ \cdot & \cdot & \cdot & \cdot \\ \frac{1}{\alpha_n - \beta_1} & \frac{1}{\alpha_n - \beta_2} & \cdots & \frac{1}{\alpha_n - \beta_n} \end{vmatrix} \equiv \Delta$$

where the α's have the values $n+2, n+3, \cdots, 2n+1$ and the β's the values $n+1, n, \cdots, n-k+2, n-k, n-k-1, \cdots, 1$ and where therefore

$$\Delta - (-1)^{n(n-1)/2} \zeta^{1/2}(\alpha_1 \cdots \alpha_n) \zeta^{1/2}(\beta_1 \cdots \beta_n)$$
$$\div \prod_{r=1}^{r=n} (\alpha_r - \beta_1)(\alpha_r - \beta_2) \cdots (\alpha_r - \beta_n).$$

Hence the value of D is

$$D = \frac{[1!2!3! \cdots (n-1)!]^3 n!(n+k)!}{(k!)^2(n-k)!(n+1)!(n+2)! \cdots (2n)!}$$

If in this $k=n$ then the value is

$$\frac{[1!2! \cdots (n-1)!]^3}{n!(n+1)!(n+2)! \cdots (2n-1)!}$$

463. The determinant

$$\begin{vmatrix} 1 & \alpha_1 & \alpha_1\alpha_2 & \alpha_1\alpha_2\alpha_3 & \cdots & \alpha_1\alpha_2\cdots\alpha_{n-1} \\ 1 & \alpha_2 & \alpha_2\alpha_3 & \alpha_2\alpha_3\alpha_4 & \cdots & \alpha_2\alpha_3\cdots\alpha_n \\ 1 & \alpha_3 & \alpha_3\alpha_4 & \alpha_3\alpha_4\alpha_5 & \cdots & \alpha_3\alpha_4\cdots\alpha_{n-1} \\ \cdot & \cdot & \cdot & \cdot & \cdot & \cdot \\ 1 & \alpha_n & \alpha_n\alpha_{n-1} & \alpha_n\alpha_{n-1}\alpha_{n+2} & \cdots & \alpha_n\alpha_{n+1}\cdots\alpha_{2n-2} \end{vmatrix} \equiv P' \text{ say}$$

may be made persymmetric by multiplying the rows by 1, α_1, $\alpha_1\alpha_2$, $\alpha_1\alpha_2\alpha_3, \cdots, \alpha_1\alpha_2\cdots\alpha_{n-1}$. Consequently

$$\alpha_1^{n-1}\alpha_2^{n-2}\cdots\alpha_{n-1}\cdot P' = P(1,\alpha_1,\alpha_1\alpha_2,\cdots,\alpha_1\alpha_2\cdots\alpha_{2n-2}).$$

Exercise. If

$$\alpha_k = \frac{k}{x+k\nu}(k=1,2,\cdots)$$

show that

$$P' = 1!2!\cdots(n-1)!\frac{(x,\nu)_1(x,\nu)_2\cdots(x,\nu)_{n-1}}{(x+\nu,\nu)_{n-1}(x+2\nu,\nu)_{n-1}\cdots(x+n\nu,\nu)_{n-1}}.$$

where $(x,\nu)_r = x(\alpha+\nu)(x+2\nu)\cdots(x+\overline{r-1}\nu)$.

464. It is easily seen that

$$P(a_0,a_1,\cdots,a_{2n-2}) = P(\Delta^{(0)},\Delta^{(1)},\Delta^{(2)},\cdots,\Delta^{(2n-2)})$$

where Δ^r is the rth difference-series of the a's.

Let $a_k=(x+k-\alpha_1)(x+k-\alpha_2)\cdots(x+k-\alpha_\mu)$ then the constituents of the μth difference-series are each equal to $\mu!$, and the value of the determinant is $(-1)^{n(n-1)/2}\{(n-1)!\}^n$ or 0 according as $\mu=$ or $<n-1$. In either case it is independent of the α's.

If the a's form an arithmetic series of order lower than $n-1$ then the determinant is zero. (See §62.)

If the a's form a geometrical series then the determinant vanishes.

Exercises. What is the result if

(1) $$a_k = s_k = \alpha_1^k + \alpha_2^k + \cdots + \alpha_n^k\,?$$

(2) $$a_k = u_k = a_0 + a_1k + a_2k^2 + \cdots + a_nk^n?$$

2. Show that

$$P(u_1,u_2\cdots u_{2r+1}) = (-1)^{n(n+1)/2}(a_nn!)^{n+1} \text{ when } r=n$$
$$= 0 \text{ when } r>n.$$

465. *If* c_0, c_1, c_2, $\cdots$, *be defined by the identity*

$$\frac{b_0x^{n-1} + b_1x^{n-2} + \cdots + b_{n-1}}{a_0x^n + a_1x^{n-1} + \cdots + a_n} = c_0x^{-1} + c_1x^{-2} + \cdots$$

then the persymmetric determinant

$$P(c_0, c_1, \cdots, c_{2m}) = \frac{(-1)^{m(m-1)/2}}{a_0^{2m+1}} \begin{vmatrix} a_0 & a_1 & a_2 \cdots & a_{2m} \\ & a_0 & a_1 \cdots & a_{2m-1} \\ & & a_0 \cdots & a_{2m-2} \\ \cdot & \cdot & \cdot & \cdot \\ \cdot & \cdot & \cdots & a_{m+1} \\ b_0 & b_1 & b_2 \cdots & b_{2m} \\ & b_0 & b_1 \cdots & b_{2m-1} \\ \cdot & \cdot & \cdot & \cdot \\ \cdot & \cdot & \cdots & b_m \end{vmatrix}$$

For proof express the b's in terms of the c's thus

$$b_0 = a_0c_0, \quad b_1 = a_0c_2 + a_1c_0, \quad b_2 = a_0c_2 + a_1c_1 + a_2c_0, \cdots$$

and then simplify.

Exercises. Set XXII.

1. Show that the eliminant of

$$\begin{vmatrix} 1 & x & x^2 & x^3 \\ a & b & c & d \\ b & c & d & e \\ e & d & e & f \end{vmatrix} = 0, \quad \begin{vmatrix} 1 & x & x^2 \\ a & b & c \\ b & c & d \end{vmatrix} = 0 \text{ is a power of } \begin{vmatrix} a & b & c \\ b & c & d \\ c & d & e \end{vmatrix}.$$

2. If we denote $(a_r, a_{r+1}, \cdots, \between x, y)^s$ by (r, s) and using the relation $(r, s) - x(r, s-1) = y(r+1, s-1)$ show that

$$\begin{vmatrix} (0, i-e) & (0, i-e+1) \cdots & (0, i) \\ (0, i-e+1) & (0, i-e+2) \cdots & (0, i+1) \\ \cdot & \cdot & \cdot \\ (0, i) & (0, i+1) & \cdots (0, i+e) \end{vmatrix}$$

$$= \begin{vmatrix} (0, i-e) & (1, i-e) & \cdots & (e, i-e) \\ (1, i-e) & (2, i-e) & \cdots & (e+i, i-e) \\ \cdot & \cdot & \cdot & \cdot \\ (e, i-e) & (e+i, i-e) & \cdots & (2e, i-e) \end{vmatrix} \cdot y^{e(e+1)}.$$

3. Show that

$$\begin{vmatrix} c_1 & a & a & \cdots & a \\ b & c_2 & a & \cdots & a \\ b & b & c_3 & \cdots & a \\ \cdot & \cdot & \cdot & \cdot & \cdot \\ b & b & b & \cdots & c_n \end{vmatrix} = \frac{a\phi(b) - b\phi(a)}{a - b},$$

where

$$\phi(x) = (c_1 - x)(c_2 - x) \cdots (c_n - x)$$

Note the special case when the c's are all equal.

4. Show that if the simple symmetric functions of $a, b, c, \cdots$, be denoted by $\sum_1, \sum_2, \cdots$, then any determinant

$$\begin{vmatrix} \sum_r & \sum_s & \sum_t & \cdots \\ \sum_{r'} & \sum_{s'} & \sum_{t'} & \cdots \\ \sum_{r''} & \sum_{s''} & \sum_{t''} & \cdots \\ \cdot & \cdot & \cdot & \cdot \end{vmatrix}$$

in which the suffixes diminish from right to left and from top to bottom is necessarily positive, so long as $a, b, c, \cdots$ are positive. (Brunn.)

5. Show that

$$\begin{vmatrix} 1 & \frac{1}{2} & \frac{1}{3} & \cdots & \frac{1}{n+1} \\ \frac{1}{2} & \frac{1}{3} & \frac{1}{4} & \cdots & \frac{1}{n+2} \\ \cdot & \cdot & \cdot & \cdot & \cdot \\ \frac{1}{n} & \frac{1}{n+1} & \frac{1}{n+2} & \cdots & \frac{1}{2n} \\ 1 & x & x^2 & \cdots & x^n \end{vmatrix} = 0$$

is equivalent to

$$\frac{\partial^n x^n (x-1)^n}{\partial x^n} = 0.$$

6. If $P_m(x)$ stands for $P(\phi_0, \phi_1, \cdots, \phi_{2m-2})$, where $\phi_k(x) = 1^k + 2^k + \cdots + x^k$, so that $\phi_k(x) - \phi_k(x-1) = x^k$ and $\phi_0(x) = x$ then show that

$$P_m(x) = A \cdot x^m (x^2 - 1^2)^{m-1} (x^2 - 2^2)^{m-2} \cdots (x^2 - \overline{m-1}^2),$$

where

$$A = \frac{\{1!2! \cdots (m-1)!\}^4}{1!2! \cdots (2m-1)!},$$

which when $m=x+1$ vanishes.

If instead of the ϕ's we use Bernoulli numbers B_1, B_2, $\cdots$, then

$$P(B_1, B_2, \cdots, B_{2m-1}) = \frac{1}{2^{2m}(m!)^3(2m)!} \frac{\{2!4!6! \cdots (2m)!\}^6}{2!4!6! \cdots (4m)!}.$$

7. If $a_0b_n+a_1b_{n-1}+\cdots+a_nb_0=0$ except when $n=0$ then

$$b_0^n P(a_0, a_1, \cdots, a_{2n}) = (-1)^n a_0^{n-1} P(b_{2n}, b_{2n-1}, \cdots, b_2).$$

8. If

$$\Delta_n = \begin{vmatrix} c & c+b & c+b & c+b & \cdots \\ a+c & c & c+b & c+b & \cdots \\ a+c & a+c & c & c+b & \cdots \\ a+c & a+c & a+c & c & \cdots \\ \cdot & \cdot & \cdot & \cdot & \cdot \end{vmatrix}_n$$

show that

$$\Delta_n + (a+b)\Delta_{n-1} + ab\Delta_{n-2} = 0$$

and hence

$$\Delta_n = (-1)^n\{(a+c)b^n - (b+c)a^n\} \div (a-b)$$

Also show that

$$\Delta_n = (\Delta_n)_{c=0} - cD_{n+1},$$

where

$$D_{n+1} = \begin{vmatrix} 0 & 1 & 1 & 1 & \cdots \\ 1 & 0 & b & b & \cdots \\ 1 & a & 0 & b & \cdots \\ 1 & a & a & 0 & \cdots \\ \cdot & \cdot & \cdot & \cdot & \cdot \end{vmatrix}_{n+1},$$

which is the negative of the sum of the signed primary minors of $(\Delta_n)_{c=0}$.

Hence

$$D_{n+1} = \frac{(\Delta_n)_{c=0} - \Delta_n}{c} = (-1)^n \frac{a^n - b^n}{a - b}.$$

When a and b are positive and n is odd the sum is positive.

Circulants

466. A determinant such that any row is got from the preceding row by passing the last element over the others to the first place is called a *circulant*. The circulant

$$\begin{vmatrix} a_1 & a_2 & \cdots & a_n \\ a_n & a_1 & \cdots & a_{n-1} \\ \cdot & \cdot & \cdot & \cdot \\ a_2 & a_3 & \cdots & a_1 \end{vmatrix}$$

whose first row is $a_1, a_2, \cdots, a_n$ may be denoted by $C(a_1, a_2, \cdots, a_n)$.

It is frequently more convenient to define circulant as a determinant such that any row is got from the preceding row by passing the first element over the others to the last place. The circulant

$$\begin{vmatrix} a_1 & a_2 & \cdots & a_n \\ a_2 & a_3 & \cdots & a_1 \\ \cdot & \cdot & \cdot & \cdot \\ a_n & a_1 & \cdots & a_{n-1} \end{vmatrix}$$

may be denoted by $C'(a_1, a_2, \cdots, a_n)$.

By transposition of rows it appears that

$$C'(a_1, a_2, \cdots, a_n) = (-1)^{(n-1)(n-2)/2} C(a_1, a_2, \cdots, a_n).$$

467. The determinant got by changing the signs of all the elements on one side of the principal diagonal of a circulant C is called a *skew circulant*. For it the functional symbol SC may be used.

A circulant is evidently a persymmetric determinant with but n distinct elements. Thus

$$C'(a_1, a_2, \cdots, a_n) = P(a_1, a_2, \cdots, a_n, a_1, a_2, \cdots, a_{n-1})$$

It is also readily seen that

$$C(a_1, a_2, \cdots, a_n) = (-1)^{n-1} C(a_2, a_3, \cdots, a_n, a_1),$$

and

$$C(a_1, a_2, \cdots, a_n) = C(a_1, a_n, a_{n-1}, \cdots, a_2), \qquad \text{etc.}$$

Circulants being persymmetric determinants the laws given for this type apply of course to circulants.

468. *A circulant contains as a factor the sum of the elements of one of its rows, the cofactor being expressible as a persymmetric determinant of the next lower order*, viz.

$$C'(a_1, \cdots, a_n) = (a_1 + a_2 + \cdots + a_n)$$
$$\cdot P(a_1 - a_2, a_2 - a_3, \cdots, a_n - a_1, a_1 - a_2, \cdots, a_{n-3} - a_{n-2})$$

Adding to each element of the last column of the circulant C' §466 the corresponding element of all the other columns, and then removing the factor $a_1 + a_2 + \cdots + a_n$ we have

$$\begin{vmatrix} a_1 & a_2 & a_3 \cdots a_{n-1} & 1 \\ a_2 & a_3 & a_4 \cdots a_n & 1 \\ a_3 & a_4 & a_5 \cdots a_1 & 1 \\ \cdot & \cdot & \cdot \cdots \cdot & \cdot \\ a_n & a_1 & a_2 \cdots a_{n-2} & 1 \end{vmatrix}$$

which passes into the required persymmetric determinant if each element of the first $n-1$ rows be diminished by the element immediately below it.

469. Using the operations $c_n - rc_{n-1}, c_{n-1} - rc_{n-2}, \cdots$, it is readily seen that

$$C(1, r, r^2, \cdots, r^{n-1}) = (1 - r^n)^{n-1}$$

and hence $C(a, ar, ar^2, \cdots, ar^{n-1}) = a^n(1-r^n)^{n-1}$.

EXERCISES: 1. Show that $C(a_1, a_2d, a_3d^2, \cdots, a_nd^{n-1})$ is a rational function of d^n.

2. If

$$\phi_r = \frac{x^r}{r!} - \frac{x^{n-r}}{(n-r)!} - \frac{x^{2n-r}}{(2n-r)!} - \cdots (r = 0, 1, 2, \cdots, \overline{n-1})$$

and therefore the ϕ's are solutions of $d^nu/dx^n = u$. Show that

$$\phi_0 + \phi_1 + \cdots + \phi_{n-1} = \epsilon^x$$
$$\phi_0 + \alpha_1\phi_1 + \cdots + \alpha_1^{n-1}\phi_{n-1} = \epsilon^{\alpha_1 x}$$
$$\cdots\cdots\cdots\cdots\cdots$$
$$\phi_0 + \alpha_{n-1}\phi_1 + \cdots \alpha_{n-1}^{n-1}\phi_{n-1} = \epsilon^{\alpha_{n-1} x}$$

and $C(\phi_0, \phi_1, \cdots, \phi_{n-1})=1$ where $\alpha_1, \alpha_2, \cdots, \alpha_{n-1}$ are the imaginary nth roots of unity.

3. Similarly if

$$\psi_r = x^r - \frac{x^{n+r}}{(n+r)!} + \cdots (r = 0,1, \cdots, \overline{n-1})$$

show that

$$SC(\psi_0,\psi_1, \cdots, \psi_{n-1}) = 1.$$

470. If in the circulant $C(a, a+d, a+2d, \cdots, a+\overline{n-1}d)$ we add all the rows to the first it will appear that $\{na+\frac{1}{2}n(n-1)d\}$ is a factor. Removing this and performing the operations c_r-c_n $(r=1, 2, \cdots, \overline{n-1})$ it will be seen that d is a factor of each column. Taking this out and reducing the order by one we get a determinant whose value is easily seen to be $(-1)^{n-1}(n)^{n-2}$.

Therefore $C=(-1)^{n-1}(nd)^{n-1}\{a+\frac{1}{2}(n-1)d\}$.

If $a=1=d$ this gives

$$C(1,2, \cdots, n) = (-1)^{n-1}\tfrac{1}{2}n^{n-1}(n+1).$$

Using the same method it may be seen that

$$C(a,b,b,b, \cdots, b)_n = (a + \overline{n-1}b)(a-b)^{n-1}.$$

Hence

$$C(-1,1,1, \cdots, 1)_n = (n-2)(-2)^{n-1},$$

and

$$C(n,-1,-1, \cdots, -1)_n = (n+1)^{n-1}.$$

471. If we multiply the circulant $C'(a_1, a_2, \cdots, a_n)$ column-wise by the alternant

$$\begin{vmatrix} 1 & 1 & \cdots & 1 \\ \alpha_1 & \alpha_2 & \cdots & \alpha_n \\ \alpha_1^2 & \alpha_2^2 & \cdots & \alpha_n^2 \\ \cdot & \cdot & \cdot & \cdot \\ \alpha_1^{n-1} & \alpha_2^{n-1} & & \alpha_n^{n-1} \end{vmatrix} \equiv \Delta$$

where α_r is an imaginary nth root of unity, we get, using θ_r to represent $a_1+a_2\alpha_r+a_3\alpha_r^2+\cdots+a_n\alpha_r^{n-1}$, $C'(a_1,a_2, \cdots, a_n)\cdot\Delta = \Delta'$

where

$$\Delta' \equiv \begin{vmatrix} \theta_1 & \theta_2 \cdots & \theta_n \\ \alpha_1^{n-1}\theta_1 & \alpha_2^{n-1}\theta_2 \cdots & \alpha_n^{n-1}\theta_n \\ \cdot & \cdot \quad \cdot & \cdot \\ \alpha_1\theta_1 & \alpha_2\theta_2 \cdots & \alpha_n\theta_n \end{vmatrix} = \theta_1\theta_2 \cdots \theta_n(-1)^{(n-1)(n-2)/2}\cdot\Delta.$$

Therefore $C'(a_1, a_2, \cdots, a_n) = (-1)^{(n-1)(n-2)/2}\theta_1\theta_2 \cdots \theta_n$.

It follows from this and §468 that

$$P(a_1 - a_2, \cdots, a_{n-3} - a_{n-2}) = (-1)^{(n-1)(n-2)/2}\theta_1\theta_2 \cdots \theta_{n-1}$$

if $\theta_n = \sum a = s$ say.

If in the circulant $C(a_1, a_2, \cdots, a_n)$, where n is even $=2m$, we perform the operations $c_1 - c_2 + c_3 - c_4 \cdots - c_n$ it will appear that $s' = a_1 - a_2 + a_3 - a_4 + \cdots - a_n$ is also a factor.

These factors may be taken in pairs whose product is a real quadratic expression. Thus the product

$$(a_1 + a_2\alpha + a_3\alpha^2 + \cdots + a_n\alpha^{n-1})(a_1 + a_2\alpha^{n-1} + \cdots + a_n\alpha)$$

is real since $\alpha^k + \alpha^{n-k}(k = 1, 2, \cdots, n-1)$ is real. *That is every circulant can be factored into real linear and quadratic factors.*

Another proof of this identity is obtained by considering the two members as different forms of the resultant of

$$a_1 + a_2x + a_3x^2 + \cdots + a_nx^{n-1} = 0 \quad \text{and} \quad 1 - x^n = 0.$$

Similarly a skew circulant of order n can be expressed as the product of n factors $\theta_r = a_1 + a_2\alpha_r + a_3\alpha_r^2 + \cdots + a_n\alpha_r^{n-1}$, where now α_r is an nth root of -1.

EXAMPLE: Show that $C(\theta_1, \theta_2, \cdots, \theta_n) = n^n a_1a_2 \cdots a_n$. Add the n equations $a_1 + a_2\alpha^k + a_3\alpha^{2k} + \cdots + a_n\alpha^{(n-1)k} = \theta_k\ (k = 1, 2, \cdots n)$ and we get $\theta_1 + \theta_2 + \cdots + \theta_n = na_1$. Then add after multiplying by $\alpha^{n-1}, \alpha^{n-2}, \cdots, 1$ and get $\theta_n + \alpha\theta_{n-1} + \cdots + \theta_1\alpha^{n-1} = na_2$ etc.

472. *The product of two circulants of the same order is expressible as a circulant.*

This is a direct result of the multiplication theorem.

473. *A circulant of the 2nth order is expressible as the product of two determinants of the nth order, a circulant and a skew circulant*, viz.

$$C(a_1, a_2 \cdots a_{2n}) = C_1(a_1 + a_{n+1}, a_2 + a_{n+2}, \cdots, a_n + a_{2n})$$
$$\times SC_2(a_1 - a_{n+1}, a_2 - a_{n+2}, \cdots, a_n - a_{2n}).$$

Reversing the order of the last n rows and then reversing the order of the last n columns the given circulant becomes centrosymmetric and thus (§362) the theorem follows.

Since in C_1 and SC_2 corresponding elements are a_k+a_{n+k} and a_k-a_{n+k} we see that C_1 is the sum of two sets of terms and SC_2 is the difference of the same two sets of terms, that is if $C_1=X+Y$, $SC_2=X-Y$ and therefore $C=C_1\cdot SC_2=X^2-Y^2$. Similarly

$$SC = X^2 + Y^2.$$

If in C we put $a_{n+1}=a_{n+2}= \cdots =a_{2n}=0$ then

$$C(a_1,a_2, \cdots , a_n,0,0, \cdots , 0)_{2n}$$
$$= C_1(a_1,a_2, \cdots , a_n)SC_1(a_1,a_2, \cdots , a_n).$$

474. *A circulant of the 2nth order is expressible as a circulant of the nth order, viz*

$$C(a_1, \cdots , a_{2n}) = C(x_1,x_2, \cdots , x_n),$$

where

$$x_r = (a_1, - a_2,a_3, \cdots , - a_{2n})(a_{2r-1},a_{2r-2}, \cdots , a_{2r+1},a_{2r}).$$

From $C(a_1, \cdots , a_{2n})$ a determinant equal to

$$(- 1)^{n(n+1)/2}C(a_1, \cdots , a_{2n})$$

is got by placing first the odd-numbered rows in order and then the even-numbered rows in order and altering the signs of the even-numbered columns; another equal to $(-1)^{(n-1)/2}C(a_1, \cdots , a_{2n})$ is got from this by deleting the negative signs, reversing the order of the rows and then reversing the order of the elements in each row. Multiplying these determinants together and expressing the result as the product of two of its minors, we have

$$(- 1)^nC^2(a_1, \cdots , a_{2n}) = (- 1)^nC^2(x_1, \cdots , x_n)$$

and hence the theorem required.

475. *A skew circulant of odd order is expressible as a circulant,* viz.

$$SC(a_1, \cdots , a_{2n+1}) = C(a_1,- a_2,a_3,- a_4, \cdots , a_{2n+1}).$$

This is at once obtained by changing the signs of all the elements in the even-numbered rows and then in the even-numbered columns.

476. *The circulant of order n, $C(a, a, \cdots , b, b, \cdots ,)_n$, where the number of a's is p and the number of b's is q ($p+q=n$) has for its value $(pa+qb)(a-b)^{n-1}$ when p is prime to q,* 0 *when p is not prime to q.*

That $pa+qb$ is a factor is readily seen on adding all the other columns to the first. The imaginary factors are

$$(a - b)(1 + \alpha^k + \alpha^{2k} + \cdots + \alpha^{pk})(k = 1, 2, \cdots, n - 1)$$

and their product is

$$(a - b)^{n-1} \prod_{k=1}^{n-1} (1 + \alpha^k + \alpha^{2k} + \cdots + \alpha^{pk}).$$

But

$$\prod_{k=1}^{n-1} (1 + \alpha^k + \alpha^{2k} + \cdots + \alpha^{pk})$$

is equal to 1 when p is prime to q, and 0 when p is not prime to q; hence the theorem.

477. The product of $a_1+\alpha a_2+\alpha^2 a_3+ \cdots +\alpha^{n-1}a_n$ and $b_1+\alpha b_2 +\alpha^2 b_3+ \cdots +\alpha^{n-1}b_n$ is readily seen to be

$$\begin{array}{ccccc|c}
1 & \alpha & \alpha^2 & \cdots & \alpha^{n-1} & \\
\hline
b_1 & b_2 & b_3 & \cdots & b_n & a_1 \\
b_n & b_1 & b_2 & \cdots & b_{n-1} & a_2 \\
b_{n-1} & b_n & b_1 & \cdots & b_{n-2} & a_3 \\
\cdot & \cdot & \cdot & \cdots & \cdot & \cdot \\
b_2 & b_3 & b_4 & \cdots & b_1 & a_n
\end{array}$$

similarly for three or more factors. Thus when $n=4$ we have for the product of

$$(a_1 + a_2\alpha + a_3\alpha^2 + a_4\alpha^3)(b_1 + b_2\alpha + b_3\alpha^2 + b_4\alpha^3)(c_1 + c_2\alpha + c_3\alpha^2 + c_4\alpha^3)$$

$$\begin{array}{cccc|cccc}
a_1 & a_2 & a_3 & a_4 & & & & \\
\hline
b_1 & b_4 & b_3 & b_2 & c_1 & c_2 & c_3 & c_4 \\
b_2 & b_1 & b_4 & b_3 & c_4 & c_1 & c_2 & c_3 \\
b_3 & b_2 & b_1 & b_4 & c_3 & c_4 & c_1 & c_2 \\
b_4 & b_3 & b_2 & b_1 & c_2 & c_3 & c_4 & c_1 \\
\hline
 & & & & 1 & \alpha & \alpha^2 & \alpha^3
\end{array}$$

Exercise: Show that

$$P \equiv \tfrac{1}{3} \sum \begin{vmatrix} a_1 + a_2\alpha + a_3\alpha^2 & b_1 + b_2\alpha + b_3\alpha^2 & c_1 + c_2\alpha + c_3\alpha^2 \\ d_1 + d_2\alpha + d_3\alpha^2 & e_1 + e_2\alpha + e_3\alpha^2 & f_1 + f_2\alpha + f_3\alpha^2 \\ g_1 + g_2\alpha + g_3\alpha^2 & h_1 + h_2\alpha + h_3\alpha^2 & k_1 + k_2\alpha + k_3\alpha^2 \end{vmatrix}$$

$$
= \begin{array}{ccc|c} a_1 & a_2 & a_3 & \\ \hline e_1 & e_3 & e_2 & k_1 \\ e_2 & e_1 & e_3 & k_3 \\ e_3 & e_2 & e_1 & k_2 \end{array}
+ \begin{array}{ccc|c} b_1 & b_2 & b_3 & \\ \hline f_1 & f_3 & f_2 & g_1 \\ f_2 & f_1 & f_3 & g_3 \\ f_3 & f_2 & f_1 & g_2 \end{array}
+ \begin{array}{ccc|c} c_1 & c_2 & c_3 & \\ \hline d_1 & d_3 & d_2 & h_1 \\ d_2 & d_1 & d_3 & h_3 \\ d_3 & d_2 & d_1 & h_2 \end{array}
$$

$$
- \begin{array}{ccc|c} a_1 & a_2 & a_3 & \\ \hline f_1 & f_3 & f_2 & h_1 \\ f_2 & f_1 & f_3 & h_3 \\ f_3 & f_2 & f_1 & k_2 \end{array}
- \begin{array}{ccc|c} b_1 & b_2 & b_3 & \\ \hline d_1 & d_3 & d_2 & k_1 \\ d_2 & d_1 & d_3 & k_3 \\ d_3 & d_2 & d_1 & k_2 \end{array}
- \begin{array}{ccc|c} c_1 & c_2 & c_3 & \\ \hline e_1 & e_3 & e_2 & g_1 \\ e_2 & e_1 & e_3 & g_3 \\ e_3 & e_2 & e_1 & g_2 \end{array}
$$

where the $\sum$ is used to indicate that α is to be replaced by α^2, α^3, α^4 and the results added.

478. If we multiply $C'(a_1, a_2, \cdots, a_n)$ by

$$
\begin{vmatrix} 1 & \alpha & \alpha^2 \cdots \alpha^{n-1} \\ 0 & 1 & 0 \cdots 0 \\ \cdot & \cdot & \cdot \; \cdot \; \cdot \; \cdot \\ 0 & 0 & 0 \cdots 1 \end{vmatrix}
$$

where α is an nth root of unity, and take out the common factor $a_1+a_2\alpha+ \cdots +a_n\alpha^{n-1}$ there is left a determinant whose expansion is seen to be $A_1+\alpha^{n-1}A_2+\alpha^{n-2}A_3+ \cdots +\alpha A_n$, where A_k is the signed complementary minor of the elements in the first row and kth column of C'. It follows that

$$
C' = (a_1 + \alpha a_2 + \cdots + \alpha^{n-1}a_n)(A_1 + A_2\alpha^{n-1} + \cdots + \alpha A_n),
$$

which taking $\alpha=1$ gives $C'=\sum a_1\sum A_1$. From this and §471 we have

$$
\sum A_1 = (-1)^{(n-1)(n-2)/2}\theta_1\theta_2 \cdots \theta_{n-1}
$$

which by §468 shows that $\sum A_1$ may be expressed as a persymmetric determinant.

When n is even $\alpha=-1$ is also a root showing that in this case $s'=a_1-a_2+a_3- \cdots -a_n$ is also a factor of C'.

479. If α_1 and α_2 are two of the nth roots of unity and we multiply C' by

$$
\begin{vmatrix} 1 & \alpha_1 & \alpha_1 \cdots \alpha_1^{n-1} \\ 1 & \alpha_2 & \alpha^2{}_2 \cdots \alpha_2^{n-1} \\ 0 & 0 & 1 \cdots 0 \\ \cdot & \cdot & \cdot \; \cdot \; \cdot \; \cdot \\ 0 & 0 & 0 \cdots 1 \end{vmatrix}.
$$

we get on taking out the common factors and expanding

$$(\alpha_2 - \alpha_1)C' = (a_1 + a_2\alpha_1 + \cdots + a_n\alpha_1^{n-1})(a_1 + a_2\alpha_2 + \cdots + a_n\alpha_2^{n-1})$$

$$\times \begin{vmatrix} 1 & 1 & a_3 & a_4 \cdots a_n \\ \alpha_1 & \alpha_2 & a_4 & a_5 \cdots a_1 \\ \alpha_1^2 & \alpha_2^2 & a_5 & a_6 \cdots a_2 \\ \cdot & \cdot & \cdot & \cdot \quad \cdot \\ \alpha_1^{n-1} & \alpha_2^{n-1} & a_2 & a_3 \cdots a_{n-1} \end{vmatrix}$$

Hence

$$C' = \theta_1\theta_2\left\{\begin{bmatrix}12\\12\end{bmatrix} - (\alpha_1 + \alpha_2)\begin{bmatrix}13\\12\end{bmatrix} + (\alpha^2{}_1 + \alpha_1\alpha_2 + \alpha^2{}_2)\begin{bmatrix}14\\12\end{bmatrix}\right.$$

$$+ \cdots - (-1)^r\frac{\alpha_1^{r-1} - \alpha_2^{r-1}}{\alpha_1 - \alpha_2}\begin{bmatrix}1\,r\\12\end{bmatrix}$$

$$\left. + \cdots - (-1)^{r+s+3}\alpha_1^{r-1}\alpha_2^{r-1}\frac{\alpha_1^{s-r} - \alpha_2^{s-r}}{\alpha_1 - \alpha_2}\begin{bmatrix}r\,s\\12\end{bmatrix} + \cdots\right\}$$

where $\left[{ij \atop 12}\right]$ is the minor formed by deleting the ith and jth rows and the 1st and 2nd columns.

This may be extended so as to take out any number of the linear factors but the remaining factor becomes more and more complicated if we expand.

When n is even $(=2m)$ and α_1 and α_2 are taken equal to 1 and -1 respectively we have

$$-2C' = s\cdot s'\begin{vmatrix} 1 & 1 & a_3 & a_4 \cdots a_n \\ 1 & -1 & a_4 & a_5 \cdots a_1 \\ 1 & 1 & a_5 & a_6 \cdots a_2 \\ \cdot & \cdot & \cdot & \cdot \quad \cdot \\ 1 & -1 & a_2 & a_3 \cdots a_{n-1} \end{vmatrix}$$

or

$$C' = s\cdot s'\left\{-\sum(-1)^r\begin{bmatrix}r\,s\\12\end{bmatrix}\right\} \quad \text{where} \quad \begin{cases} r = 1,2,\cdots,n-1, \\ s = 2,3,\cdots,n, \\ r < s, \text{ and } r+s \text{ is odd.}\end{cases}$$

It should be observed here that the cofactor of $s\cdot s'$ on the right may be reduced to a determinant of order $2m-2$ by performing the operations $r_{2m}-r_{2m-2},\ r_{2m-1}-r_{2m-3},\ r_{2m-2}-r_{2m-4},\ \cdots,\ r_3-r_1$. Thus if $m=4$ we have

$$-2C'(a_1,a_2,\cdots,a_8)$$

$$= s\cdot s'(-2)\begin{vmatrix} a_5-a_3 & a_6-a_4 & a_7-a_5 & a_8-a_6 & a_1-a_7 & a_2-a_8 \\ a_6-a_4 & a_7-a_5 & a_8-a_6 & a_1-a_7 & a_2-a_8 & a_3-a_1 \\ a_7-a_5 & a_8-a_6 & a_1-a_7 & a_2-a_8 & a_3-a_1 & a_4-a_2 \\ a_8-a_6 & a_1-a_7 & a_2-a_8 & a_3-a_1 & a_4-a_2 & a_5-a_3 \\ a_1-a_7 & a_2-a_8 & a_3-a_1 & a_4-a_2 & a_5-a_3 & a_6-a_4 \\ a_2-a_8 & a_3-a_1 & a_4-a_2 & a_5-a_3 & a_6-a_4 & a_7-a_5 \end{vmatrix}$$

or

$$C_1'(a_1,a_2,\cdots,a_8) = s\cdot s'P(a_5-a_3,\ a_6-a_4,\ a_7-a_5,\ a_8-a_6,$$
$$a_1-a_7,\ a_2-a_8,\ a_3-a_1,\ a_4-a_2,\ a_5-a_3,\ a_6-a_4,\ a_7-a_5).$$

In general (n even)

$$C'(a_1,a_2,\cdots,a_n) = s\cdot s'P(a_5-a_3,\ a_6-a_4,\ \cdots,\ a_{n-1}-a_{n-3},$$
$$a_n-a_{n-2},\ a_1-a_{n-1},\ a_2-a_n,\ a_3-a_1,$$
$$a_4-a_2,\ a_5-a_3,\ a_6-a_4,\ \cdots,\ a_{n-1}-a_{n-3}).$$

480. Take C' of odd order $2n+1$ and for convenience let $n=4$. The factors of the circulant other than s are

$$a_1+a_2\alpha^k+a_3\alpha^{2k}+\cdots+a_9\alpha^{8k}\equiv\theta_k(k=1,2,\cdots,8).$$

The product of θ_1 and θ_8 which we shall denote by $\theta_{1,8}$ is

$$\overset{0}{\sum}a^2{}_1+\overset{0}{\sum}a_1a_2(\alpha+\alpha^8)+\overset{0}{\sum}a_1a_3(\alpha^2+\alpha^7)$$
$$+\overset{0}{\sum}a_1a_4(\alpha^3+\alpha^6)+\overset{0}{\sum}a_1a_5(\alpha^4+a^5)$$

or (I)

$$S_1+S_2(\alpha+\alpha^8)+S_3(\alpha^2+\alpha^7)+S_4(\alpha^3+\alpha^6)+S_5(\alpha^4+\alpha^5)=\theta_{1,8}$$

where the cyclic sums are denoted as follows:

$$\overset{0}{\sum}a^2{}_1\equiv a^2{}_1+a^2{}_2+\cdots+a^2{}_9\equiv S_1$$

$$\overset{0}{\sum}a_1a_2\equiv a_1a_2+a_2a_3+\cdots+a_9a_1=S_2 \qquad \text{etc.}$$

Similarly

$$\text{(II)}\quad S_1 + S_2(\alpha^2+\alpha^7) + S_3(\alpha^4+\alpha^5) + S_4(\alpha^3+\alpha^6) + S_5(\alpha+\alpha^8) = \theta_{2,7}$$

$$\text{(III)}\quad S_1 + S_2(\alpha^3+\alpha^6) + S_3(\alpha^3+\alpha^6) + S_4(2) + S_5(\alpha^3+\alpha^6) = \theta_{3,6}$$

$$\text{(IV)}\quad S_1 + S_2(\alpha^4+\alpha^5) + S_3(\alpha+\alpha^8) + S_4(\alpha^3+\alpha^6) + S_5(\alpha^2+\alpha^7) = \theta_{4,5}$$

and $\theta_{1,8}$, $\theta_{2,7}$, $\theta_{3,6}$, $\theta_{4,5}$ are all real.

These four relations may be written

$$(S_1-S_2)+(S_2-S_3)(1+\alpha+\alpha^8)+(S_2-S_4)(1+\alpha+\alpha^8+\alpha^2+\alpha^7)$$
$$+(S_4-S_5)(1+\alpha+\alpha^8+\alpha^2+\alpha^7+\alpha^3+\alpha^6)=\theta_{1,8}$$

$$(S_2-S_3)(1+\alpha+\alpha^8+\alpha^2+\alpha^7)+(S_1-S_4)-(S_2-S_5)(1+\alpha+\alpha^8)$$
$$+(S_3-S_5)(1+\alpha+\alpha^8+\alpha^2+\alpha^7+\alpha^4+\alpha^5)=\theta_{2,7}$$

$$-(S_3-S_4)(1+\alpha^3+\alpha^6+\alpha^6+\alpha^9+\alpha^9)-(S_2-S_5)(1+\alpha^3+\alpha^6)+(S_1-S_5)$$
$$+(S_2-S_4)(1+\alpha^3+\alpha^6+\alpha^3+\alpha^6)=\theta_{3,6}$$

$$-(S_4-S_5)(1+\alpha+\alpha^8+\alpha^2+\alpha^7)+(S_3-S_5)(1+\alpha+\alpha^8)$$
$$-(S_2-S_4)(1+\alpha+\alpha^8+\alpha^2+\alpha^7+\alpha^3+\alpha^6)+(S_1-S_3)=\theta_{4,5}$$

EXERCISE: Show that

$$C' = s\begin{vmatrix} S_1-S_2 & S_2-S_3 & S_3-S_4 & S_4-S_5 \\ S_2-S_3 & S_1-S_4 & S_2-S_5 & S_3-S_5 \\ S_3-S_4 & S_2-S_5 & S_1-S_5 & S_2-S_4 \\ S_4-S_5 & S_3-S_5 & S_2-S_4 & S_1-S_3 \end{vmatrix}$$

and therefore the cofactor of s in C' of order $2n+1$ may be expressed as an axisymmetric determinant of order n.

Show that this is also true for the cofactor of $s\cdot s'$ in C' of order $2n+2$.

481. From §478 we have

$$\begin{aligned} C' &= \alpha_1(a_1+\alpha_1a_2+\cdots+\alpha_1^{n-1}a_n)(A_1\alpha_1^{n-1}+A_2\alpha_1^{n-2}+\cdots+A_n) \\ &= \alpha_2(a_1+\alpha_2a_2+\cdots+\alpha_2^{n-1}a_n)(A_1\alpha_2^{n-1}+A_2\alpha_2^{n-2}+\cdots+A_n) \\ &\quad \cdots\cdots\cdots\cdots\cdots\cdots \\ &= \alpha_n(a_1+\alpha_na_2+\cdots+\alpha_n^{n-1}a_n)(A_1\alpha_n^{n-1}+A_2\alpha_n^{n-2}+\cdots+A_n) \end{aligned}$$

Writing

$$A_1\alpha_k^{n-1}+A_2\alpha_k^{n-2}+\cdots+A_n=\phi_k(k=1,2,\cdots,n)$$

and multiplying these equations in order by $\alpha_1{}^h, \alpha_2{}^h, \cdots, \alpha_n{}^h$ and adding we get

$$\phi_1\alpha^h{}_1 + \phi_2\alpha^h{}_2 + \cdots + \phi_n\alpha^h{}_n = nA_n(h = 1,2, \cdots, n).$$

From this we see that

$$C'(\phi_1,\phi_2, \cdots, \phi_n) = n^n A_1 A_2 \cdots A_n.$$

482. *If the elements of the first row of a circulant of the nth order be multiplied by* $\alpha^n, \alpha^{n-1}, \cdots$ *respectively, the elements of the second row by* $\alpha^{n-1}, \alpha^{n-2}, \cdots, \alpha, \alpha^n$, *respectively, the elements of the third row by* $\alpha^{n-2}, \alpha^{n-3}, \cdots, \alpha^n, \alpha^{n-1}$ *respectively, and so on, the circulant is unaltered in value.*

This is seen by multiplying all the elements of the second row by α, all the elements of the third row by α^2, and so on. The elements of the 1st, 2nd, $\cdots$, nth column will have $\alpha^n, \alpha^{n-1}, \cdots$, respectively as factors. As these are just the factors introduced the truth of the theorem is seen.

483. *In every term of* $C'(a_1, a_2, \cdots, a_n)$ *the sum of the suffixes is divisible by* n.

The suffix of any element (i, j) is $i+j-1$ or $i+j-1-n$ according as $i+j-1 \not> n$ or $i+j-1>n$. From this it follows that in the case of any term $(1, r)(2, x)(3, t) \cdots$ the sum of the suffixes differs by a multiple of n from $1+r+2+s+3+t+ \cdots +n=n(n+1)-n=n^2$. Therefore the sum of the suffixes in any term is equal to n^2-kn where k is the number of elements from below the secondary diagonal.

484. It is apparent that in a circulant the complementary minor of any element found in one position can differ at most in sign from the complementary minor of the same element in any other position.

From the formation of a circulant C' it is readily seen that any minor $\begin{vmatrix} r & s & t & \cdots \\ u & v & w & \cdots \end{vmatrix}$ is equal to $\begin{vmatrix} r+a & s+a & t+a & \cdots \\ u-a & v-a & w-a & \cdots \end{vmatrix}$ with the proviso that when any row number becomes greater than n, the order of the circulant, it must be reduced by n, and when any column number becomes zero or negative it must be increased by n.

From this it follows that when $n=2m$ we have

$$A_{kk} = A_{1,2k-1} = A_{m+k,m+k} \ (k = 1,2, \cdots, m).$$

That is primary minors complementary to elements in the odd places of the first row have primary minors complementary to elements along the principal diagonal equal to them, but those complementary to elements in the even places have not. It follows that

$$\sum A_{11} = 2 \sum A_{1,2k-1}.$$

In the case of circulants of odd order every element in the first row has a signed complementary minor equal to the complementary minor of some element along the principal diagonal.

If $[p, q]_r$ denotes the complementary minor of the element in the pth row and qth column of the odd-ordered circulant C' after the rth column of the circulant has been replaced by units, then these relations are seen to be as follows.

1. When $1+r$ is even

$$[1,r]_s = [\tfrac{1}{2}(1+r),\tfrac{1}{2}(1+r)]_{s-(r-1)/2} \qquad \text{if } s > \tfrac{1}{2}(r-1),$$
$$= [\tfrac{1}{2}(1+r),\tfrac{1}{2}(1+r)]_{n+s-(r-1)/2} \qquad \text{if } s \lesseqqgtr \tfrac{1}{2}(r-1).$$

2. When $1+r$ is odd

$$[1,r]_s = -[\tfrac{1}{2}(n+r+1),\tfrac{1}{2}(n+r+1)]_{s-(n+r-1)/2}$$
$$\text{if } s > \tfrac{1}{2}(n+r-1),$$
$$= -[\tfrac{1}{2}(n+r+1),\tfrac{1}{2}(n+r+1)]_{s-(n-r+1)/2}$$
$$\text{if } s \lesseqqgtr \tfrac{1}{2}(n+r-1).$$

We also have

1. When $r+s$ is even

$$[r,r]_s = -[\tfrac{1}{2}(r+s),\tfrac{1}{2}(r+s)]_{n+(3r-s)/2} \qquad \text{if } 3r \lesseqqgtr s,$$
$$= -[\tfrac{1}{2}(r+s),\tfrac{1}{2}(r+s)]_{(3r-s)/2} \qquad \text{if } 3r > s;$$

2. When $r+s$ is odd

$$[r,r]_s = -[\tfrac{1}{2}(n+r+s),\tfrac{1}{2}(n+r+s)]_{n+(3r-n-s)/2} \qquad \text{if } 3r \leqq s+n$$
$$= -[\tfrac{1}{2}(n+r+s),\tfrac{1}{2}(n+r+s)]_{(3r-n-s)/2} \qquad \text{if } 3r > s+n.$$

485. If we multiply the n factors of a circulant $C'(a_0, a_1, \cdots, a_{n-1})$ together the terms are of the form $Aa_0^{l_0}a_1^{l_1}\cdots a_{n-1}^{l_{n-1}}$ where the l's are positive integers whose sum is n. Any a entering into this, say a_k, must be accompanied by one of the α's raised to the power k and therefore the form of A is known. A is a symmetric function of the roots of the equation $x^n-1=0$, and can be calculated.

For example the coefficient of $a_0^2a_1a_3$ in $C'(a_0, a_1, a_2, a_3)$ is

$$\sum\alpha_1^0\alpha_2^0\alpha_3^1\alpha_4^3\text{, i.e., } \alpha_3^1\alpha_4^3+\alpha_3^3\alpha_4^1+\alpha_2^1\alpha_4^3+\alpha_2^1\alpha_3^3+\alpha_2^3\alpha_4^1$$
$$+\alpha_2^3\alpha_3^1+\alpha_1^1\alpha_4^3+\alpha_1^1\alpha_3^3+\alpha_1^1\alpha_2^3+\alpha_1^3\alpha_4^1+\alpha_1^3\alpha_3^1+\alpha_1^3\alpha_2^1$$
$$\text{i.e., } (\alpha_1+\alpha_2+\alpha_3+\alpha_4)(\alpha_1^3+\alpha_2^3+\alpha_3^3+\alpha_4^3)$$
$$-(\alpha_1^4+\alpha_2^4+\alpha_3^4+\alpha_4^4)\text{, i.e., } 0\cdot 0-4.$$

From §483 we see that we have only to calculate those A's where the sum of the suffixes of the a's is zero or divisible by n.

In such expansions other labor saving devices may be used as follows:

(1) Other terms may be obtained by cyclical permutations of the suffixes. (§468.)

(2) If $A a_0^{l_0} a_1^{l_1} \cdots a_{n-1}^{l_{n-1}}$ is a term of the expansion there is also the term $A a_0^{l_0} a_m^{l_1} a_n^{l_2} \cdots a_t^{l_{n-1}}$, where $m, n, \cdots, t$ are numbers less than n, and such that $m, 2m, 3m, \cdots, (n-1)m \equiv m, n, \cdots, t \pmod{n}$.

This is also based on §468.

486. *If the signed primary minors of a circulant be equal and non-zero, the sum of the elements must vanish.*

This follows as in §391.

If the sum of the elements in every row of a circulant vanish then the signed primary minors are all equal.

This follows as in §389.

From this and §471 it appears that the unique minor is

$$A_1 = \frac{1}{n}(-1)^{n(n-1)/2}\theta_1\theta_2 \cdots \theta_{n-1}$$

487. Since the sum of the elements of every row of a circulant $C'(a_1, a_2, \cdots, a_n)$ is the same we see by §122 that

$$A_r - A_s = (-1)^{r+s-1} \sum a \cdot Q$$

where Q is got from C' by deleting the first row and the rth and sth columns and inserting a column of units in the first place.

488. If we subtract x from the principal diagonal elements of a circulant $C'(a_1, a_2, \cdots, a_n)$ the resulting determinant may be written

$$|C' - x| = -(x - s)\prod_{k=1}^{m}\{x^2 - (a_1 + \alpha_2 a^k + a_3\alpha^{2k} + \cdots + a_n\alpha^{(n-1)k}) \times (a_1 + a_2\alpha^{n-k} + \cdots + a_n\alpha^{n-(n-1)k})\}$$

or

$$= -(x - s)(x^2 - \theta_{1,n-1})(x^2 - \theta_{2,n-2}) \cdots (x^2 - \theta_{m,m+1})$$

or

$$= -x^n + sx^{n-1} + x^{n-2}\sum\theta_{1,n} - \cdots + (-1)^{k-1}x^{n-2k}$$

(1)
$$\sum\theta_{1,n-1}\theta_{2,n-2} \cdots \theta_{k,n-k} - (-1)^{k-1}sx^{n-2k-1}$$

$$\sum\theta_{1,n-1}\theta_{2,n-2} \cdots \theta_{k,n-k} + \cdots$$

$$+ (-1)^m s\theta_{1,n-1}\theta_{2,n-2} \cdots \theta_{m,n-m},$$

when $n=2m+1$, and

$$|C'-x| = (x-s)(x-s')\prod_{k=1}^{m-2}\{x^2-(a_1+a_2\alpha^k+\cdots+a_n\alpha^{(n-1)k})$$
$$\times(a_1+a_2\alpha^{n-k}+\cdots+a_n\alpha^{n-(n-1)k})$$

or

$$=(x-s)(x-s')(x^2-\theta_{1,n-2})(x^2-\theta_{2,n-2})\cdots(x^2-\theta_{m-2,m+2})$$

or

$$\begin{aligned}(2)\qquad &= x^{2m}-x^{2m-1}(s+s')+x^{2m-2}(ss'-\textstyle\sum\theta_{1,n-1})+\cdots\\ &\quad+(-1)^{k-1}x^{2m-2k}(ss'\textstyle\sum\theta_{1,n-1}\theta_{2,n-2}\cdots\theta_{k-1,n-k+1}\\ &\qquad-\textstyle\sum\theta_{1,n-1}\theta_{2,n-2}\cdots\theta_{k,n-k})\\ &\quad+(-1)^{k-1}x^{2m-2k-1}(s+s')\textstyle\sum\theta_{1,n-1}\theta_{2,n-2}\cdots\theta_{k,n-k}+\cdots\\ &\quad+(-1)^{m-1}ss'\theta_{1,n-1}\theta_{2,n-2}\cdots\theta_{m-1,n-m+1},\end{aligned}$$

when $n=2m$.

Expanding $|C'-x|$ in terms of powers of x and the sums of coaxial minors we have

$$(1')\quad |C'-x| = -x^n+x^{n-1}\textstyle\sum_1-x^{n-2}\sum_2+\cdots-x\sum_{n-1}+C',\qquad (n\text{ odd})$$

$$(2')\quad |C'-x| = x^{2m}-x^{2m-1}\textstyle\sum_1+x^{2m-2}\sum_2-\cdots-x\sum_{n-1}+C',\qquad (n\text{ even})$$

where $\sum_k$ represents the sum of the coaxial minors of order k.

Equating coefficients of like powers of x in corresponding expansions we have

I. When $n=2m+1$

$$(A)\qquad \begin{aligned}\textstyle\sum\theta_{1,n-1}\cdots\theta_{k,n-k} &= (-1)^k\textstyle\sum_{2k},\\ s\cdot\textstyle\sum\theta_{1,n-1}\cdots\theta_{k,n-k} &= (-1)^k\textstyle\sum_{2k+1}\end{aligned}$$

so that

$$(A')\qquad s=\frac{\sum_{2k+1}}{\sum_{2k}},$$

a constant ratio for all values of k from 1 to m.

When $k=m$,

$$s=\frac{C'}{\sum A_{11}}$$

and therefore $\sum A_{11}=\sum A_1$ (§478).

II. When $n=2m$.

$$(B)\ (s+s')\sum\theta_{1,n-1}\cdots\theta_{k-1,n-k+1}=(-1)^{k-1}\sum{}_{2k-1},$$

$$ss'\sum\theta_{1,n-1}\cdots\theta_{k-1,n-k+1}-\sum\theta_{1,n-1}\cdots\theta_{k,n-k}=(-1)^{k-1}\sum{}_{2k},$$

$$(s+s')\sum\theta_{1,n-1}\cdots\theta_{k,n-k}=(-1)^{k}\sum{}_{2k+1}.$$

That is

$$(-1)^{k-1}ss'\frac{\sum_{2k-1}}{s+s'}-(-1)^{k}\frac{\sum_{2k+1}}{s+s'}=(-1)^{k-1}\sum{}_{2k}$$

$$(B')\qquad ss'=\frac{(s+s')\sum_{2k}-\sum_{2k+1}}{\sum_{2k-1}},$$

a ratio constant for all values of k from 1 to $m-1$.

If $s=0$, then (B') gives

$$s'=\frac{\sum_{2k+1}}{\sum_{2k}}=\frac{\sum A_{11}}{\sum\begin{bmatrix}12\\12\end{bmatrix}},\quad\text{when}\quad k=m-1.$$

But when $s=0$ by §486 the signed primary minors are all equal and hence

$$A_1=\frac{s'\sum\begin{bmatrix}12\\12\end{bmatrix}}{2m}$$

is the value of this unique minor. But

$$C'=s\cdot\sum A_1=snA_1=(-1)^{n(n-1)/2}\theta_1\theta_2\cdots\theta_{n-1}\theta_n$$

and

$$nA_1=(-1)^{n(n-1)/2}\theta_1\theta_2\cdots\theta_{n-1},\ \text{if}\ \theta_n=s.$$

Hence

$$\sum\begin{bmatrix}12\\12\end{bmatrix}=(-1)^{n(n-1)/2}\theta_1\theta_2\cdots\theta_{n-2},\ \text{if}\ \theta_{n-1}=s'.$$

Similarly if $s'=0$, then

$$s=\frac{\sum_{2k+1}}{\sum_{2k}}=\frac{\sum A_{11}}{\sum\begin{bmatrix}12\\12\end{bmatrix}},\quad\text{when}\quad k=m-1.$$

If $s+s'=0$, then $s^2=s'^2=\sum_{2k+1}/\sum_{2k-1}$ by (B') and we see that the sums of coaxial minors of odd orders have the same signs. If in addition $s=0$ then $\sum_{2k+1}=0$ for all values of k. Equation (2′) then contains only even powers of x and since the values of x^2 are all real the signs of $\sum_2, \sum_4, \cdots$, must be alternately negative and positive, a property which is also true when $n=2m+1$ and $s=0$.

489. If A_r is the signed complementary minor of a_r in $C(a_1, a_2, a_3, a_4)$ we have

$$\begin{vmatrix} A_1 & A_2 & A_3 \\ A_4 & A_1 & A_2 \\ A_3 & A_4 & A_1 \end{vmatrix} = a_1 \begin{vmatrix} a_1 & a_2 & a_3 & a_4 \\ a_4 & a_1 & a_2 & a_3 \\ a_3 & a_4 & a_1 & a_2 \\ a_2 & a_3 & a_4 & a_1 \end{vmatrix}^2$$

Therefore

$$a_1 = \begin{vmatrix} A_1 & A_2 & A_3 \\ A_4 & A_1 & A_2 \\ A_3 & A_4 & A_1 \end{vmatrix} \div \begin{vmatrix} A_1 & A_2 & A_3 & A_4 \\ A_4 & A_1 & A_2 & A_3 \\ A_3 & A_4 & A_1 & A_2 \\ A_2 & A_3 & A_4 & A_1 \end{vmatrix}^{2/3}.$$

EXERCISE: Show that this is the same as solving

$$\begin{vmatrix} a_1 & a_2 & a_3 \\ a_4 & a_1 & a_2 \\ a_3 & a_4 & a_1 \end{vmatrix} = x^3, \qquad \begin{vmatrix} a_2 & a_3 & a_4 \\ a_1 & a_2 & a_3 \\ a_4 & a_1 & a_2 \end{vmatrix} = y^3, \qquad \begin{vmatrix} a_3 & a_4 & a_1 \\ a_2 & a_3 & a_4 \\ a_1 & a_2 & a_3 \end{vmatrix} = z^3,$$

$$\begin{vmatrix} a_4 & a_1 & a_2 \\ a_3 & a_4 & a_1 \\ a_2 & a_3 & a_4 \end{vmatrix} = u^3.$$

490. If $C(a_1, a_2, \cdots, a_n)$ be represented by C, and if $\partial C/\partial t$ denote the derivative of C with respect to any variable t of which the a's are functions, and if C_r denote the determinant got from C by replacing each element of the rth column by the derivative of that element, then $\partial C/\partial t=C_1+C_2+\cdots+C_n$, but in a circulant $C_r=C_h$ as may be seen by shifting rows and columns.

Therefore

$$\frac{\partial C}{\partial t} = nC_k \quad (k = 1, 2, \cdots, n).$$

From $\sum a \cdot \sum A = C$ we see that

$$\sum A + \sum a \frac{\partial \sum A}{\partial a_h} = \frac{\partial C}{\partial a_h} = n \cdot A_h.$$

Also

$$\sum A + \sum a \frac{\partial \sum A}{\partial a_k} = n \cdot A_k,$$

and therefore

$$\sum a \left(\frac{\partial \sum A}{\partial a_h} - \frac{\partial \sum A}{\partial a_k} \right) = n(A_h - A_k),$$

and

$$\left(\frac{\partial \sum A}{\partial a_h} - \frac{\partial \sum A}{\partial a_k} \right) = (-1)^{k+k-1} n \cdot Q \qquad (\S 487)$$

It follows that $A_h - A_k$ contains $\sum a$ as a factor.

If $\sum a = 0$, then $A_h = A_k$ (see §486) and $\sum A/n = A_h$ showing that $\sum A$ is divisible by n and the elements are integers.

491. If the nth roots of unity are $1, \alpha, \alpha^2, \cdots, \alpha^{n-1}$ and we denote the product of all the factors of the circulant $C(a_1, a_2, \cdots, a_n)$ except $\sum a$ (or s) and $a_1\alpha^n + a_2\alpha^{n-1} + \cdots + a_n$ or θ_2 by $b_1 + b_2\alpha + b_3\alpha^2 + \cdots + b_n\alpha^{n-1}$, or ϕ, then

$$\begin{aligned} \frac{C}{s} &= \theta_2 \cdot \phi = f(\alpha) \text{ say,} \\ &= a_1b_1 + a_2b_2 + \cdots + a_nb_n \\ &\quad + (a_1b_2 + a_2b_3 + \cdots + a_nb_1)\alpha \\ &\quad \cdots\cdots\cdots\cdots \\ &\quad + (a_1b_n + a_2b_1 + \cdots + a_nk_{n-1})\alpha^{n-1} \end{aligned}$$

Therefore

$$\begin{aligned} (n-1)\frac{C}{s} &= f(\alpha) + f(\alpha^2) + \cdots + f(\alpha^{n-1}) \\ &= n(a_1b_1 + a_2b_2 + \cdots + a_nb_n) \\ &\quad - (a_1 + a_2 + \cdots + a_n)(b_1 + b_2 + \cdots b_n) \\ &= n \cdot S - s \sum b_1 \text{ say.} \end{aligned}$$

When $\alpha = 1$ the value of ϕ or $\sum b_1$ is $(\sum a_1)^{n-2}$ or s^{n-2}. Therefore

$$(n-1)\frac{C}{s} = nS - s^{n-1}$$

or

$$C = \frac{nsS - s^n}{n - 1}$$

and

$$\frac{1}{n} \frac{\partial C}{\partial a_k} = \frac{nS - s^{n-1}}{n - 1} + \frac{s}{n - 1} \frac{\partial S}{\partial a_k}$$

$$= A_k \text{ by §490}$$

and

$$\sum \frac{\partial C}{\partial a_k} = n \sum A_k.$$

Therefore

$$A_h - A_k = \frac{s}{n - 1}\left(\frac{\partial S}{\partial a_h} - \frac{\partial S}{\partial a_k}\right)$$

$$= \frac{s}{n}\left(\frac{\partial \sum A}{\partial a_h} - \frac{\partial \sum A}{\partial a_k}\right) \qquad (§490)$$

492.

$$\sum \frac{\partial C}{\partial a_k} = \sum \frac{\partial}{\partial a_k}\left(\sum a \sum A\right)$$

$$= \sum A \cdot \sum \frac{\partial \sum a}{\partial a_k} + \sum a \sum \frac{\partial \sum A}{\partial a_k}$$

$$= n \sum A + \sum a \sum \frac{\partial \sum A}{\partial a_k}.$$

Therefore $\sum \partial \sum A / \partial a_k = 0$ (§491).

Differentiating $\sum a \sum A$ a second time with respect to a_k we have

$$\frac{\partial^2 C}{\partial a_k^2} = \sum a \frac{\partial^2 \sum A}{\partial a_k^2} + 2 \frac{\partial \sum A}{\partial a_k}.$$

Therefore

$$\sum \frac{\partial^2 C}{\partial a_k^2} = \sum a \sum \frac{\partial^2 \sum A}{\partial a_k^2}.$$

Again since

$$\frac{\partial A_k}{\partial a_k} = \frac{1}{n} \frac{\partial^2 C}{\partial a_k^2},$$

$$\sum \frac{\partial A_k}{\partial a_k} = \frac{1}{n} \sum \frac{\partial^2 C}{\partial a_k^2} = \frac{1}{n} \sum a\left(\sum \frac{\partial^2 \sum A_k}{\partial a_k^2}\right).$$

493. Neglecting for convenience the sign factor and taking

$$C = \theta_1\theta_2 \cdots \theta_n$$

we have

$$\frac{\partial C}{\partial a_1} = \frac{C}{\theta_1}\frac{\partial \theta_1}{\partial a_1} + \frac{C}{\theta_2}\frac{\partial \theta_2}{\partial a_1} + \cdots$$

$$\frac{\partial C}{\partial a_2} = \frac{C}{\theta_1}\frac{\partial \theta_1}{\partial a_2} + \frac{C}{\theta_2}\frac{\partial \theta_2}{\partial a_2} + \cdots$$

$$\cdots\cdots\cdots\cdots$$

and

$$\left(a_k\frac{\partial}{\partial a_1} + a_{k+1}\frac{\partial}{\partial a_2} + \cdots + a_{k-2}\frac{\partial}{\partial a_{n-1}} + a_{k-1}\frac{\partial}{\partial a_n}\right)C$$
$$= C(1 + \alpha + \alpha^2 + \cdots + \alpha^{n-1}) = 0.$$

That is

$$\sum_{h=1}^{n} a_{h+k}\frac{\partial C}{\partial a_h} = 0,$$

if $k \neq 0$ and $a_{h+k} = a + \epsilon$ when $h+k = n+\epsilon$, $\epsilon > 0$. When $k=0$, then by Euler's theorem we have

$$\sum_{h=1}^{n} a_h\frac{\partial C}{\partial a_h} = nC.$$

Also

$$\left(\sum_{h=1}^{n} a_h\frac{\partial}{\partial a_h}\right)A_k = (n-1)A_k$$

and

$$\left(\sum_{h=1}^{n} a_h\frac{\partial}{\partial a_h}\right)\sum A_k = (n-1)\sum A_k.$$

494. If we represent as in §480 the imaginary factors of the circulant C by $\theta_1, \theta_2, \cdots, \theta_{2m}$, their products $\theta_h\,\theta_{2m+1-h}$ $(h = 1, 2, \ldots)$ by $\theta_{h,\,2m+1-h}$ and

$$\sum_{1}^{(m)k}\theta_{1,2m}\cdots\theta_{k,2m-k+1} \text{ by } \sum_{j=1}^{(m)k}\theta_j^{(k)}$$

then it is readily seen that

$$(A) \qquad \frac{\partial}{\partial a_i}\sum_{j=1}^{(m)k}\theta_j^{(k)} = \left(\sum_{j=1}^{(m)k-1}\theta_j^{(k-1)}\right)\left(\sum_{h=1}^{m-k+1} I_{h1}\right)$$

where I_{h_1} represents $(\theta_{h_1}\alpha^{n-h_1(i-1)}+\theta_{2m+1-h_1}\alpha^{h_1(i-1)})$ and where for each $\theta_j^{(k)}$ in $\sum\theta_j^{(k-1)}$, the cofactor $\sum I_{h_1}$ contains no θ which is found in $\theta_j^{(k-1)}$.

Again

$$(B)\ \frac{\partial^2}{\partial a_i^2}\sum_{j=1}^{(m)k}\theta_j^{(k)} = 2(m-k+1)\sum_{j=1}^{(m)k-1}\theta_j^{(k-1)} + \left(\sum_{h=1}^{m-k+1} I_{h_1}\right)\frac{\partial}{\partial a_i}\sum_{j=1}^{(m-)k-1}\theta_j^{(k-1)}$$

and in general

$$(C)\ \frac{\partial^r}{\partial a^r_i}\sum_{j=1}^{(m)}\theta_j^{(k)} = 2(r-1)(m-k+1)\frac{\partial^{r-2}}{\partial a_i^{r-2}}\sum_{j=1}^{(m)k-1}\theta_j^{(k-1)} + \left(\sum_{h=1}^{m-k+1} I_{h_1}\right)\frac{\partial^{r-1}}{\partial a_i^{r-1}}\left(\sum_{j=1}^{(m)k-1}\theta_j^{(k-1)}\right)$$

Taking the sum with respect to i in each case we have

$$(A')\ \sum_{i=1}^{n}\frac{\partial}{\partial a_i}\sum_{j=1}^{(m)k}\theta_j^{(k)} = 0$$

$$(B')\ \sum_{i=1}^{n}\frac{\partial^2}{\partial a_i^2}\sum_{j=1}^{(m)k}\theta_j^{(k)} = 2\cdot n\cdot(m-k+1)\sum_{j=1}^{(m)k-1}\theta_j^{(k-1)}$$

$$(C')\ \sum_{i=1}^{n}\frac{\partial^r}{\partial a_i^r}\sum_{j=1}^{(m)k}\theta_j^{(k)} = 2(r-1)(m-k+1)\sum_{i=1}^{n}\frac{\partial^{r-2}}{\partial a_i^{r-2}}\sum_{j=1}^{(m)k-1}\theta_j^{(k-1)},$$

Since

$$\sum_{i=1}^{n}\left(\sum_{h_1=1}^{m-k+1} I_{h_1}\right).$$

vanishes.

495. From (C') we have when $r=2p$

$$(1)\ \sum_{i=1}^{n}\frac{\partial^{2p}}{\partial a_i^{2p}}\sum_{j=1}^{(m)k}\theta_j^{(k)} = 2(2p-1)(m-k+1)\sum_{i=1}^{n}\frac{\partial^{2p-2}}{\partial a_i^{2p-2}}\left(\sum_{j=1}^{(m)k-1}\theta_j^{(k-1)}\right)$$

$$= 2^p n(m-k+1)(m-k+2)\cdots (m-k+p)(2p-1)(2p-3)\cdots 3\cdot 1\sum_{j=1}^{(m)k-p}\theta_j^{(k-p)}$$

and when $r=2p+1$

$$(2)\qquad \sum_{i=1}^{n}\frac{\partial^{2p+1}}{\partial a_i^{2p+1}}\sum_{j=1}^{(m)k}\theta_j^{(k)} = 0.$$

I. For circulants of odd order, $n=2m+1$, we have from (1) and §488

$$(3)\quad \sum_{i=1}^{n} \frac{\partial^{2p}}{\partial a_i^{2p}} \sum\nolimits_{2k} = -2(2p-1)(m-k+1) \sum_{i=1}^{n} \frac{\partial^{2p-2}}{\partial a_i^{2p-2}} \sum\nolimits_{2k-2}$$
$$= (-)^p 2^p n(2p-1)(2p-3)\cdots 3\cdot 1$$
$$\times (m-k+1)\cdots(m-k+p) \sum\nolimits_{2k-2p},$$

which for $p=1$ and $k=m$ becomes

$$(4)\qquad \sum_{i=1}^{n} \frac{\partial^2}{\partial a_i^2} \sum A_{11} = -2n \sum\nolimits_{2m-2}$$
$$= \frac{-2n \sum_{2m-1}}{s} \quad \text{by } (A')\ \S 488.$$

From (2) and §488 we have

$$(5)\qquad \sum_{i=1}^{n} \frac{\partial^{2p+1}}{\partial a_i^{2p+1}} \sum\nolimits_{2k} = 0,$$

which for $p=0$ and $k=m$ is

$$(6)\qquad \sum_{i=1}^{n} \frac{\partial}{\partial a_i} \sum A_{11} = 0.$$

Using the relation $s\cdot\sum_{2k}=\sum_{2k-1}$ we have

$$(7)\qquad \frac{\partial}{\partial a_i} \sum\nolimits_{2k+1} = \sum\nolimits_{2k} + s\cdot\frac{\partial}{\partial a_i} \sum\nolimits_{2k},$$

$$(8)\qquad \frac{\partial^2}{\partial a_1^2} \sum\nolimits_{2k+1} = 2\frac{\partial}{\partial a_i} \sum\nolimits_{2k} + s\cdot\frac{\partial^2}{\partial a_i^2} \sum\nolimits_{2k}$$

and in general

$$(9)\qquad \frac{\partial^r}{\partial a_i^r} \sum\nolimits_{2k+1} = r\frac{\partial^{r-1}}{\partial a_i^{r-1}} \sum\nolimits_{2k} + s\cdot\frac{\partial^r}{\partial a_i^r} \sum\nolimits_{2k}.$$

If $r=2p$ this becomes

$$\frac{\partial^{2p}}{\partial a_i^{2p}} \sum\nolimits_{2k+1} = 2p\frac{\partial^{2p-1}}{\partial a_i^{2p-1}} \sum\nolimits_{2k} + s\cdot\frac{\partial^{2p}}{\partial a_i^{2p}} \sum\nolimits_{2k}$$

and

$$(10)\quad \sum_{i=1}^{n} \frac{\partial^{2p}}{\partial a_i^{2p}} \sum_{2k+1} = s\cdot \sum_{i=1}^{n} \frac{\partial^{2p}}{\partial a_i^{2p}} \sum_{2k} \qquad by\ (5)$$

$$= (-1)^p 2^p n(2p-1)(2p-3)\cdots 3\cdot 1$$

$$\times (m-k+1)\cdots(m-k+p)s \sum_{2k-2p} \text{ by } (3)$$

This for $p=1$ and $k=m$ is

$$(11)\qquad \sum_{i=1}^{n} \frac{\partial^2 C}{\partial a_i^2} = -2ns \sum_{n-3}$$

therefore by §492

$$(12)\qquad \sum_{i=1}^{n} \frac{\partial A_i}{\partial a_i} = 2s \sum_{n-3}.$$

If $r=2p+1$ we have

$$(13)\quad \sum_{i=1}^{n} \frac{\partial^{2p+1}}{\partial a_i^{2p+1}} \sum_{2k+1} = (2p+1) \sum_{i=1}^{n} \frac{\partial^{2p}}{\partial a_i^{2p}} \sum_{2k} \text{ by } (5)$$

$$= (-1)^p 2^p n(2p+1)(2p-1)\cdots 3\cdot 1$$

$$\times (m-k+1)\cdots(m-k+p) \sum_{2k-2p}$$

which for $p=0$ and $k=m$ gives

$$(14)\qquad \sum_{i=1}^{n} \frac{\partial C}{\partial a_i} = n \sum_{2m} = n \sum A_1.$$

From (7) we readily see that

$$(15)\qquad \frac{\partial}{\partial a_i} \sum_{2k+1} - \frac{\partial}{\partial a_j} \sum_{2k+1} = s\cdot\left(\frac{\partial}{\partial a_i} \sum_{2k} - \frac{\partial}{\partial a_j} \sum_{2k}\right)$$

which when $k=m$ gives

$$(16)\quad n(A_i - A_j) = s\left(\frac{\partial}{\partial a_i} \sum A_{11} - \frac{\partial}{\partial a_j} \sum A_{11}\right) \text{ by } (14)$$

II. For circulants of even order, $n=2m+2$ we have from §488. (B)

$$(17)\quad (-1)^k \frac{\partial}{\partial a_i} \sum_{2k+1}$$

$$= \left\{1 + (-1)^{i-1}\right\} \sum_{j=1}^{(m)k} \theta_j^{(k)} + (s+s') \frac{\partial}{\partial a_i} \sum_{j=1}^{(m)k} \theta_j^{(k)},$$

$$(18)\quad (-1)^k \frac{\partial^2}{\partial a_i^2} \sum_{2k+1}$$

$$= 2\{1 + (-1)^{i-1}\} \frac{\partial}{\partial a_i} \sum_{j=1}^{(m)k} \theta_j^{(k)} + (s + s') \frac{\partial^2}{\partial a_i^2} \sum_{j=1}^{(m)k} \theta_j^{(k)},$$

and in general

$$(19)\quad (-1)^k \frac{\partial^r}{\partial a_i^r} \sum_{2k+1}$$

$$= r\{1 + (-1)^{i-1}\} \frac{\partial^{r-1}}{\partial a_i^{r-1}} \sum_{i=1}^{(m)k} \theta_j^{(k)} + (s + s') \frac{\partial^r}{\partial a_i^r} \sum_{i=1}^{(m)k} \theta_j^{(k)}.$$

When $r = 2p$ we have from (19)

$$(20)\quad \sum_{i=1}^{n} \frac{\partial^{2p}}{\partial a_i^{2p}} \sum_{2k+1} = (-1)^k (s + s') \sum_{i=1}^{n} \frac{\partial^{2p}}{\partial a_i^{2p}} \sum_{i=1}^{(m)k} \theta_j^{(k)} \text{ by (2) §495}$$

$$= (-1)^p 2^p n (2p - 1)(2p - 3) \cdots 3 \cdot 1$$

$$\times (m - k + 1) \cdots (m - k + p) \sum_{2k-2p+1}$$

which when $p = 1$ and $k = m$, is

$$(21)\quad \sum_{i=1}^{n} \frac{\partial^2}{\partial a_i^2} \sum A_{11} = -2n \sum_{2m-1} = -2n \sum_{n-3}.$$

When $r = 2p + 1$ we have from (19)

$$(22)\quad \sum_{i=1}^{n} \frac{\partial^{2p+1}}{\partial a_i^{2p+1}} \sum_{2k+1} = (-1)^k (2p + 1) \sum_{i=1}^{n} \{1 + (-1)^{i-1}\}$$

$$\frac{\partial^{2p}}{\partial a_i^{2p}} \sum_{j=1}^{(m)k} \theta_j^{(k)}, \text{ by (2)}$$

$$= (-1)^{k+p} 2^p n (2p + 1)(2p - 1) \cdots 3 \cdot 1$$

$$\times (m - k + 1) \cdots (m - k + p) \frac{\sum_{2k-2p+1}}{\sum a_{11}}$$

which when $p = 0$ and $k = m$ is

$$(23)\quad \sum_{i=1}^{n} \frac{\partial}{\partial a_i} \sum A_{11} = n \frac{\sum A_{11}}{\sum a_{11}}.$$

Again we have from §488 (B)

$$(24)\quad (-1)^{k-1}\frac{\partial}{\partial a_i}\sum_{2k}$$

$$= \left\{s' + (-1)^{i-1}s\right\}\sum_{j=1}^{(m)k-1}\theta_j^{(k-1)} + s\cdot s'\frac{\partial}{\partial a_i}\sum_{j=1}^{(m)k-1}\theta_j^{(k-1)} - \frac{\partial}{\partial a_i}\sum_{j=1}^{(m)k}\theta_j^{(k}$$

$$(25)\quad (-1)^{k-1}\frac{\partial^2}{\partial a_i^2}\sum_{2k}$$

$$= 2\left\{s' + (-1)^{i-1}s\right\}\frac{\partial}{\partial a_i}\sum_{j=1}^{(m)k-1}\theta_j^{(k-1)} + s\cdot s'\frac{\partial^2}{\partial a_i^2}\sum_{j=1}^{(m)k-1}\theta_j^{(k-1)}$$

$$-\frac{\partial^2}{\partial a_i^2}\sum_{j=1}^{(m)k}\theta_j^{(k)} + 2(-1)^{i-1}\sum_{j=1}^{(m)k-1}\theta_j^{(k-1)}$$

and in general

$$(26)\quad (-1)^{k-1}\frac{\partial^r}{\partial a_i^r}\sum_{2k}$$

$$= r\left\{s' + (-1)^{i-1}s\right\}\frac{\partial^{r-1}}{\partial a_i^{r-1}}\sum_{j=1}^{(m)k-1}\theta_j^{(k-1)} + s\cdot s'\frac{\partial^r}{\partial a_i^r}\sum_{j=1}^{(m)k-1}\theta_j^{(k-1)}$$

$$-\frac{\partial^r}{\partial a_i^r}\sum_{j=1}^{(m)k}\theta_j^{(k)} + r(r-1)(-1)^{i-1}\frac{\partial^{r-2}}{\partial a_i^{r-2}}\sum_{j=1}^{(m)k-1}\theta_j^{(k-1)}$$

When $r=2p$ we have from (26)

$$(27)\quad (-1)^{k-1}\sum_{i=1}^{n}\frac{\partial^{2p}}{\partial a_i^{2p}}\sum_{2k}$$

$$= 2p\sum_{i=1}^{n}\left\{s' + (-1)^{i-1}s\right\}\frac{\partial^{2p-1}}{\partial a_i^{2p-1}}\sum_{j=1}^{(m)k-1}\theta_j^{(k-1)}$$

$$+\sum_{i=1}^{n}s\cdot s'\frac{\partial^{2p}}{\partial a_i^{2p}}\sum_{j=1}^{(m)k-1}\theta_j^{(k-1)} - \sum_{i=1}^{n}\frac{\partial^{2p}}{\partial a_i^{2p}}\sum_{j=1}^{(m)k}\theta_j^{(k)}$$

$$+ 2p(2p-1)\sum_{i=1}^{n}(-1)^{i-1}\frac{\partial^{2p-2}}{\partial a_i^{2p-2}}\sum_{j=1}^{(m)k-1}\theta_j^{(k-1)}$$

$$= (-1)^{k-p}2^p n(2p-1)(2p-3)\cdots 3\cdot 1$$

$$\times (m-k+2)\cdots(m-k+p)\left\{\frac{\sum_{2k-2p+1}}{\sum a_{11}}\right.$$

$$\left.-(m-k+p+1)\sum_{2k-2p}\right\}$$

(the first and fourth terms on the right vanishing) which, when $p=1$ and $k=m+1$, is

$$(28)\qquad \sum_{i=1}^{n} \frac{\partial^2 C}{\partial a_i^2} = 2n\left(\frac{\sum_{2m+1} - \sum a_{11} \sum_{2m}}{\sum a_{11}}\right)$$

$$= -2n\frac{s\cdot s'}{\sum a_{11}} \sum_{2m-1} = -2n\frac{s\cdot s'}{s+s'} \sum_{n-3}$$

But

$$\sum_{i=1}^{n} \frac{\partial^2 C}{\partial a_i^2} = n\sum_{i=1}^{n} \frac{\partial}{\partial a_i} A_i$$

therefore

$$(29)\qquad \sum_{i=1}^{n} \frac{\partial A_i}{\partial a_i} = -2\frac{s\cdot s'}{s+s'} \sum_{n-1}$$

When $r=2p+1$ we have from (26)

$$(30)\quad (-1)^{k-1}\sum_{i=1}^{n} \frac{\partial^{2p+1}}{\partial a_i^{2p+1}} \sum_{2k}$$

$$= (2p+1)\sum_{i=1}^{n}\{s'+(-1)^{i-1}s\}\frac{\partial^{2p}}{\partial a_i^{2p}} \sum_{j=1}^{(m)k-1}\theta_j^{(k-1)}$$

$$+ \sum_{i=1}^{n} s\cdot s'\frac{\partial^{2p+1}}{\partial a_i^{2p+1}} \sum_{i=1}^{(m)k-1}\theta_j^{(k-1)} - \sum_{i=1}^{n} \frac{\partial^{2p+1}}{\partial a_i^{2p+1}} \sum_{j}^{(m)k}\theta_j^{(k)}$$

$$+ (2p+1)2p\sum_{i=1}^{n}(-1)^i\frac{\partial^{2p-1}}{\partial a_i^{2p-1}} \sum_{j=1}^{(m)k}\theta_j^{(k)}$$

$$= (-1)^{k-p-1}2^p ns'(2p+1)(2p-1)\cdots 3\cdot 1$$

$$\times (m-k+2)\cdots(m-k+p+1)\frac{\sum_{2k-2p-1}}{\sum a_{11}}$$

(the second, third and fourth terms vanishing) which when $p=0$ and $k=1$, is

$$(31)\qquad \sum_{i=1}^{n} \frac{\partial}{\partial a_i} \sum_2 = n\cdot s'.$$

From $\sum A_1 = \sum A_{2k-1} + \sum A_{2k}$ we have

$$(32)\qquad \sum_1^n i\frac{\partial}{\partial a_i} \sum A_{2k-1} = -\sum_1^n i\frac{\partial}{\partial a_i} \sum A_{2k}$$

$$= \frac{n}{2}\,\frac{\sum A_{11}}{\sum a_{11}}\,.$$

If we denote

$$a_k\frac{\partial}{\partial a_1}+a_{k+1}\frac{\partial}{\partial a_2}+\cdots+a_n\frac{\partial}{\partial a_{n-k+1}}+a_1\frac{\partial}{\partial a_{n-k+2}}+\cdots+a_{k-1}\frac{\partial}{\partial a_n}$$

by $\sum a_k\partial/\partial a_1$ then

$$(33)\qquad \left(\sum a_k\frac{\partial}{\partial a_1}\right)A_h=-A_{h-k+1}\quad(\text{or } A_{n+h-k+1} \text{ if } k>h)$$

$$\begin{Bmatrix} h=1,2,\cdots,n \\ k=2,3,\cdots,n\end{Bmatrix}.$$

The truth of this is seen on observing that $(\sum a_k\partial/\partial a_1)A_h$ is the sum of n determinants which are obtained by increasing the subscripts of the elements in the first, second, and so on, columns of A_h by $k-1$. Of these all vanish, having identical columns, except the $(h-k+1)$th (or $\overline{n+h-k+1}$th if $h<k$) and it is $-A_{h-k+1}$ (or $-A_{n+h-k+1}$).

When $h=k$ then

$$(34)\qquad \left(\sum a_h\frac{\partial}{\partial a_1}\right)A_h=-A_1\quad(h=2,3,\cdots,n)$$

and

$$\left(\sum a_h\frac{\partial}{\partial a_1}\right)\sum A_1=-\sum A_1.$$

Exercises. Set XXIII

Prove the following relations

$$(1)\qquad \left(a_1\frac{\partial^2}{\partial a_k\partial a_1}+a_2\frac{\partial^2}{\partial a_k\partial a_2}+\cdots\right)C'=n(n-1)A_k$$

$$(2)\qquad \left(a_k\frac{\partial^2}{\partial a_k\partial a_1}+a_{k+1}\frac{\partial^2}{\partial a_k\partial a_2}+\cdots\right)C'=-nA_1$$

$$(3)\quad \sum a_1\left(a_1\frac{\partial^2}{\partial a_1{}^2}+a_2\frac{\partial^2}{\partial a_1\partial a_2}+\cdots\right)\sum A_1=(n-2)\left(nA_1-\sum A_1\right)$$

$$(4)\quad \left(\sum a_1\frac{\partial^2}{\partial a_1^2}\right)C'=\sum a_1\left(\sum a_1\frac{\partial^2}{\partial a_1^2}\right)\sum A_1+2(n-1)\sum A_1$$

$$(5)\quad \sum a_1\left(a_k\frac{\partial^2}{\partial a_k\partial a_1}+a_{k+1}\frac{\partial^2}{\partial a_k\partial a_2}+\cdots\right)\sum A_1$$
$$= 2\sum A_1 - n(A_1 + A_k)$$

$$(6)\quad \frac{\partial}{\partial a_k}\sum A_1 = \left(\sum_1^n i\frac{\partial}{\partial a_i}\right)A_k$$

$$(7)\quad \sum_1^n i\frac{\partial^2}{\partial a_k\partial a_i}\sum A_1 = 0.$$

496. Writing the circulant in the form

$$C = a_1A_1 + a_2A_2 + \cdots + a_nA_n$$

we have

$$\frac{\partial C}{\partial a_k} = a_1\frac{\partial A_1}{\partial a_k} + a_2\frac{\partial A_2}{\partial a_k} + \cdots + a_n\frac{\partial A_n}{\partial a_k}$$

or

$$\sum_1^n ia_i\frac{\partial A_i}{\partial a_k} = (n-1)A_k.\quad (k = 1,2,\cdots,n).$$

It should be observed here that the determinant of this set of equations is the Jacobian,* whose value is $(-1)^{n(n-1)/2}(n-1)C^{n-2}$. Solving for a_1 gives

$$a_1 = (n-1)\begin{vmatrix} A_1 & \frac{\partial A_2}{\partial a_1} & \frac{\partial A_3}{\partial a_1} & \cdots \\ A_2 & \frac{\partial A_2}{\partial a_2} & \frac{\partial A_3}{\partial a_2} & \cdots \\ \cdot & \cdot & \cdot & \cdot \end{vmatrix} \div (-1)^{n(n-1)/2}(n-1)C^{n-2},$$

or

$$\begin{vmatrix} A_1 & \frac{\partial A_2}{\partial a_1} & \frac{\partial A_3}{\partial a_1} & \cdots \\ A_2 & \frac{\partial A_2}{\partial a_2} & \frac{\partial A_3}{\partial a_1} & \cdots \\ \cdot & \cdot & \cdot & \cdot \end{vmatrix} = (-1)^{n(n-1)/2}a_1C^{n-2}.$$

* vide §694.

EXERCISES. SET XXIV

Show that

(1) $$\sum_1^n ia_i \frac{\partial^2}{\partial a_k^2} A_i = (n-2)A_k$$

(2) $$n\sum_1^n ia_i \frac{\partial^2}{\partial a_k^2} A_i = (n-2)\frac{\partial^2 C}{\partial a_k^2}$$

(3) $$\sum_1^n ia_i \ \sum_1^n k\frac{\partial}{\partial a_k} A_i = (n-1)\sum A_1$$

(4) $$\sum_1^n ia_i \ \sum_1^n k\frac{\partial}{\partial a_k} A_1 = \left(\sum a_1 \frac{\partial}{\partial a_1}\right)\sum A_1 .$$

497. If $\lambda = \log C(a_1, a_2, \cdots, a_n)$ then

$$C(a_1, a_2, \cdots, a_n) C\left(\frac{\partial\lambda}{\partial a_1} \frac{\partial\lambda}{\partial a_2}, \cdots, \frac{\partial\lambda}{\partial a_n}\right) = n^n.$$

To prove this we start with

$$C = \sum a_1 \pi(a_1 + \alpha_1 a_2 + \alpha_1^2 a_3 + \cdots + \alpha_1^{n-1} a_n),$$

then

$$\frac{\partial\lambda}{\partial a_i} = \frac{1}{\sum a_1} + \sum \frac{\alpha_1^{n-1}}{a_1 + \alpha_1 a_2 + \cdots + \alpha_1^{n-1} a_n}$$

and

$$\frac{\partial\lambda}{\partial a_1} + \frac{\partial\lambda}{\partial a_2} + \cdots + \frac{\partial\lambda}{\partial a_n} = \frac{n}{\sum a_1}$$

$$\frac{\partial\lambda}{\partial a_1} + \alpha_1\frac{\partial\lambda}{\partial a_2} + \cdots + \alpha_1^{n-1}\frac{\partial\lambda}{\partial a_n} = \frac{n}{a_1 + \omega_1 a_2 + \cdots + \omega_1^{n-1} a}$$

. . . .

$$\frac{\partial\lambda}{\partial a_1} + \alpha_{n-1}\frac{\partial\lambda}{\partial a_2} + \cdots + \alpha_{n-1}^{n-1}\frac{\partial\lambda}{\partial a_n} = \frac{n}{a_1 + \omega_{n-1} a_2 + \cdots + \omega_{n-1}^{n-1} a_n}$$

where $\omega_k \alpha_k = 1$.

It follows that

$$C\left(\frac{\partial\lambda}{\partial a_1}, \frac{\partial\lambda}{\partial a_2}, \cdots, \frac{\partial\lambda}{\partial a_n}\right) = \frac{n^n}{C(a_1, a_2, \cdots, a_n)}$$

or

$$C(a_1,a_2,\cdots,a_n)C\left(\frac{\partial\lambda}{\partial a_1},\cdots,\frac{\partial\lambda}{\partial a_n}\right)=n^n.$$

EXERCISES: Show that

(1) $C(1^2,2^2,\cdots,n^2)=(-1)^{n-1}\dfrac{(n+1)(2n+1)}{1\cdot 2}n^{n-2}\{(n+2)^n-n^n\}$

(2) $C(1,n-1,\frac{1}{2}(n-1)(n-2),\cdots,1)=\begin{cases}2^{n-1}\text{ when } n \text{ is odd.}\\ 0 \quad\text{ when } n \text{ is even.}\end{cases}$

498. If we take the circulant $C(t^{n-1}, t^{n-2}n_1, t^{n-3}n_2, \cdots, n_{n-1})$ where $n_k={}_nC_k$ and the elements are the terms of the expansions of $\{(t+1)^n-1\}\div t$, then C will contain only even powers of t and if in C we put τ for t^2 the roots of the equation $C=0$ are the squared differences of the roots of $t^n=1$.

Let $\alpha_1, \alpha_2, \cdots, \alpha_n$ be the nth roots of unity, then

$$(t-\alpha_1)(t-\alpha_2)\cdots(t-\alpha_n)\equiv t^n-1$$

and also

$$\{t-(\alpha_1-\alpha_r)\}\{t-(\alpha_2-\alpha_r)\}\cdots\{t-(\alpha_n-\alpha_r)\}\equiv(t+\alpha_r)^n-1,$$

the rth factor on the left being t. Giving r the values $1, 2, \cdots, n$ and taking the product we have

$$\pi\{t^2-(\alpha_r-\alpha_s)^2\}=\frac{(t+\alpha_1)^n-1}{t}\cdot\frac{(t+\alpha_2)^n-1}{t}\cdots\frac{(t+\alpha_n)^n-1}{t}$$

since the factors on the left combine in pairs.

But

$$\frac{(t+\alpha_r)^n-1}{t}=t^{n-1}+n_1t^{n-2}\alpha_r+n_2t^{n-3}\alpha_r^2+\cdots+n_{n-1}\alpha_r^{n-1}$$

which shows that $(t+\alpha_r)^2-1/t$ is a factor of C and therefore

$$C=\pi\{t^2-(\alpha_r-\alpha_s)^2\}$$

and hence the theorem.

The circulant C is the eliminant of the equation $x^n-1=0$ and the $n-1$ equations obtained on multiplying

$$t^{n-1}+n_1t^{n-2}x+n_2t^{n-3}x^2+\cdots+n_{n-1}x^{n-1}=0$$

by $x, x^2, \cdots, x^{n-1}$ successively.

499. The circulant $C(a_1, a_2, \cdots, a_n)$ is equal to the sum of the coefficients of the equation whose roots are the nth powers of the roots of the equation $a_1x^{n-1}+a_2x^{n-2}+\cdots+a_n=0$.

Thus when $n=4$ we have

$$\begin{aligned} C(a_1,a_2,a_3,a_4) &= \pi(a_1+a_2\alpha_1+a_3\alpha_1^2+a_4\alpha_1^3) \\ &= a_1(1-\alpha_1\beta_1)(1-\alpha_1\beta_2)(1-\alpha_1\beta_3) \\ &\quad \times a_1(1-\alpha_2\beta_1)(1-\alpha_2\beta_2)(1-\alpha_2\beta_3) \\ &\quad \times a_1(1-\alpha_3\beta_1)(1-\alpha_3\beta_2)(1-\alpha_3\beta_3) \\ &\quad \times a_1(1-\alpha_4\beta_1)(1-\alpha_4\beta_2)(1-\alpha_4\beta_3) \end{aligned}$$

where β_1, β_2, b_3 are the roots of

(1) $$a_1x^3+a_2x^2+a_3x+a_4=0.$$

Regrouping this product we have

$$C = a_1^4(1-\beta_1^4)(1-\beta_2^4)(1-\beta_3^4)$$

from which the theorem is apparent.

If in equation (1) we put $y^{1/4}$ for x and multiply the result by $y^{1/4}$, $y^{2/4}$, $y^{3/4}$ successively we get

$$\begin{aligned} a_1y^{3/4}+a_2y^{2/4}+a_3y^{1/4}+a_4 &= 0 \\ a_1y+a_2y^{3/4}+a_3y^{2/4}+a_4y^{1/4} &= 0 \\ a_1yy^{1/4}+a_2y+a_3y^{3/4}+a_4y^{2/4} &= 0 \\ a_1yy^{2/4}+a_2yy^{1/4}+a_3y+a_4y^{3/4} &= 0 \end{aligned}$$

from which on eliminating $y^{1/4}$, $y^{2/4}$, $y^{3/4}$ we get

(2) $$\begin{vmatrix} a_1 & a_2 & a_3 & a_4 \\ a_2 & a_3 & a_4 & a_1y \\ a_3 & a_4 & a_1y & a_2y \\ a_4 & a_1y & a_2y & a_3y \end{vmatrix} = 0$$

which is the equation having for its roots the 4th powers of the roots of equation (1). If in (2) we put $y=1$ then the left-hand side is the sum of the coefficients in the equation and gives the previous proposition.

If in (1) we put $y^{1/3}$ for x and multiply by $y^{1/3}$, $y^{2/3}$ successively we get

$$a_1y + a_2y^{2/3} + a_3y^{1/3} + a_4 = 0$$
$$a_1yy^{1/3} + a_2y + a_3y^{2/3} + a_4y^{1/3} = 0$$
$$a_1yy^{2/3} + a_2yy^{1/3} + a_3y + a_4y^{2/3} = 0$$

which on eliminating $y^{1/3}$, $y^{2/3}$ gives

$$\begin{vmatrix} a_2 & a_3 & a_1y + a_4 \\ a_3 & a_1y + a_4 & a_2y \\ a_1y + a_4 & a_2y & a_3y \end{vmatrix} = 0$$

which is the equation whose roots are the cubes of the roots of (1). Here again the sum of the coefficients is the circulant $C(a_2, a_3, a_1+a_4)$.

500. Any circulant of the $(r\cdot s)$th order is expressible as a circulant of the sth order and each of the elements of the latter is an aggregate of r-line minors, r^{s-1} in number, of the former.

To illustrate this proof let us take the case of $n=3\cdot 5=15$. Let $\gamma_1, \gamma_2, \gamma_3$ be the third and $\epsilon_1, \epsilon_2, \epsilon_3, \epsilon_4, \epsilon_5$ be the fifth roots of unity, then $\gamma_1\epsilon_1, \gamma_2\epsilon_2, \gamma_3\epsilon_3, \gamma_1\epsilon_4, \cdots, \gamma_3\epsilon_5$ are the fifteenth roots of unity since they are the third roots of each of the five fifth roots.

One factor of the circulant is then

$$a_1 + a_2(\gamma_1\epsilon_1^{1/3}) + a_3(\gamma_1\epsilon_1^{1/3})^2 + \cdots + a_{15}(\gamma_1\epsilon_1^{1/3})^{14}.$$

But this may be written

$$(a_1 + a_4\epsilon_1 + a_7\epsilon_1^2 + a_{10}\epsilon_1^3 + a_{13}\epsilon_1^4) + (a_2 + a_5\epsilon_1 + \cdots + a_{14}\epsilon_1^4)\gamma_1\epsilon_1^{1/3}$$
$$+ (a_3 + a_6\epsilon_1 + \cdots + a_{15}\epsilon_1^4)\gamma_1^2\epsilon_1^{2/3}.$$

There are two other factors that differ from this in having γ_2, γ_3 in place of γ_1 so that the product of the three is the circulant

$$\begin{vmatrix} a_1+a_4\epsilon_1+\cdots a_{13}\epsilon^4 & (a_2+a_5\epsilon_1+\cdots)\epsilon_1^{1/3} & (a_3+a_6\epsilon_1+\cdots)\epsilon_1^{2/3} \\ (a_3+a_6\epsilon_1+\cdots)\epsilon_1^{2/3} & (a_1+a_4\epsilon_1+\cdots) & (a_2+a_5\epsilon_1+\cdots)\epsilon_1^{1/3} \\ (a_2+a_5\epsilon_1+\cdots)\epsilon_1^{1/3} & (a_3+a_6\epsilon_1+\cdots) & (a_1+a_4\epsilon_1+\cdots) \end{vmatrix}$$
$$\equiv C_1 \text{ say}.$$

Multiplying the rows in order by $\epsilon_1^{\frac{1}{3}}$, $\epsilon_1^{\frac{2}{3}}$, ϵ and dividing the columns in order by the same we have

$$C_1 = \begin{vmatrix} a_1+a_4\epsilon_1+\cdots+a_{13}\epsilon_1^4 & a_2+a_5\epsilon_1+\cdots+a_{14}\epsilon_1^4 & a_3+a_6\epsilon_1+\cdots+a_{15}\epsilon_1^4 \\ a_{15}+a_3\epsilon_1+\cdots+a_{12}\epsilon_1^4 & a_1+a_4\epsilon_1+\cdots+a_{13}\epsilon_1^4 & a_2+a_5\epsilon_1+\cdots+a_{14}\epsilon_1^4 \\ a_{14}+a_2\epsilon_1+\cdots+a_{11}\epsilon_1^4 & a_{15}+a_3\epsilon_1+\cdots+a_{12}\epsilon_1^4 & a_1+a_4\epsilon_1+\cdots+a_{13}\epsilon_1^4 \end{vmatrix}.$$

This may be written as the sum of 5^3 determinants with monomial elements giving

$$C_1 = P + Q\epsilon_1 + R\epsilon_1^2 + S\epsilon_1^3 + T\epsilon_1^4$$

where each of the coefficients $P, Q, \cdots, T$ is the sum of 25 determinants. Taking the factors of C in sets of three as here we have

$$\begin{aligned} C = \quad & (P + Q\epsilon_1 + R\epsilon_1^2 + S\epsilon_1^3 + T\epsilon_1^4) \\ & \times (P + Q\epsilon_2 + R\epsilon_2^2 + S\epsilon_2^3 + T\epsilon_2^4) \\ & \quad \cdots\cdots\cdots\cdots\cdots\cdots \\ & \times (P + Q\epsilon_5 + R\epsilon_5^2 + S\epsilon_5^3 + T\epsilon_5^4) = C_1C_2C_3C_4C_5 \end{aligned}$$

so that

$$C(a_1, a_2, \cdots, a_{15}) = C(P, Q, R, S, T).$$

We might have taken the first factor in the form

$$(a_1 + a_6\gamma_1 + a_{11}\gamma_1^2) + (a_2 + a_7\gamma_1 + a_{12}\gamma_1^2)\epsilon_1\gamma^{1/5} + \cdots \\ + (a_5 + a_{10}\gamma_1 + a_{15}\gamma_1^2)\epsilon_1^4\gamma_1^{4/5}$$

and obtained

$$C = C_1'C_2'C_3' = C(X, Y, Z) \text{ where } C_1' = X + Y\gamma_1 + Z\gamma_1^2 \text{ etc.}$$

It is readily seen that a circulant factor of C which is free from the roots of unity is

$$C(a_1 + a_4 + \cdots a_{13},\ a_2 + a_5 + \cdots + a_{14},\ a_3 + a_6 + \cdots + a_{15}).$$

In like manner we have another

$$C(a_1 + a_6 + a_{11}, a_2 + a_7 + a_{12}, \cdots, a_5 + a_{10} + a_{15}).$$

501. *For every rational factor of $x^n - 1$ there is a rational factor of the circulant which is the eliminant of*

$$\left.\begin{aligned} a_1x^{n-1} + a_2x^{n-2} + \cdots + a_n &= 0 \\ x^n - 1 &= 0 \end{aligned}\right\}$$

For example we get from

$$ax^4 + bx^3 + cx^2 + dx + e = 0$$

and

$$\frac{x^5 - 1}{x - 1} = x^4 + x^3 + x^2 + x + 1 = 0$$

$$(b - a)x^3 + (c - a)x^2 + (d - a)x + (e - a) = 0.$$

Similarly

$$(c - b)x^3 + (d - b)x^2 + (e - b)x + (a - b) = 0,$$
$$(d - c)x^3 + (e - c)x^2 + (a - c)x + (b - c) = 0,$$
$$(e - d)x^3 + (a - d)x^2 + (b - d)x + (c - d) = 0,$$

and hence

$$\begin{vmatrix} b - a & c - a & d - a & e - a \\ c - b & d - b & e - b & a - b \\ d - c & e - c & a - c & b - c \\ e - d & a - d & b - d & c - d \end{vmatrix}$$

is the factor required.

Examples: (1). Find the factor of $C(a_1, a_2, \cdots, a_8)$ corresponding to $(x^8-1)\div(x^2-1)$ that is, the cofactor of ss'.

(2). Find the factor of $C(a_1, a_2, \cdots, a_6)$ corresponding to $(x^6-1)\div(x^2+x+1)$.

502. The determinant whose every element is the sum of the corresponding elements of the circulants

$$C(a_1, a_2, \cdots a_n), \qquad C'(b_1, b_2, \cdots b_n),$$

is divisible by

$$(a_1 + a_2\alpha + a_3\alpha^2 + \cdots + a_n\alpha^{n-1})(a_1 + a_2\alpha^{-1} + \cdots + a_n\alpha^{-n-1})$$
$$- (b_1 + b_2\alpha + b_3\alpha^2 + \cdots + b_n\alpha^{n-1})(b_1 + b_2\alpha^{-1} + \cdots + b_n\alpha^{-n-1}).$$

For convenience let us take $n=5$. The determinant in question is the eliminant of the set of equations

$$(a_1 + b_1)x_1 + (a_2 + b_2)x_2 + (a_3 + b_3)x_3 + (a_4 + b_4)x_4 + (a_5 + b_5)x_5 = 0$$
$$(a_5 + b_2)x_1 + (a_1 + b_3)x_2 + (a_2 + b_4)x_3 + (a_3 + b_5)x_4 + (a_4 + b_1)x_5 = 0$$
$$\cdots\cdots\cdots\cdots\cdots\cdots\cdots\cdots$$
$$(a_2 + b_5)x_1 + (a_3 + b_1)x_2 + (a_4 + b_2)x_3 + (a_5 + b_3)x_4 + (a_1 + b_4)x_5 = 0$$

Multiplying these by $\alpha^0, \alpha^1, \alpha^2, \alpha^3, \alpha^4$ respectively and adding we have

$$(a_1 + a_2\alpha^{-1} + a_3\alpha^{-2} + \cdots + a_5\alpha^{-4})(x_1 + x_2\alpha + x_3\alpha^2 + \cdots + x_5\alpha^4)$$
$$+ (b_1 + b_2\alpha + b_3\alpha^2 + \cdots + b_5\alpha^4)(x_1 + x_5\alpha + x_4\alpha^2 + \cdots + x_2\alpha^4) = 0$$

or

$$\frac{a_1 + a_2\alpha^{-1} + a_3\alpha^{-2} + \cdots + a_5\alpha^{-4}}{b_1 + b_2\alpha + b_3\alpha^2 + \cdots + b_5\alpha^4} = -\frac{x_1 + x_5\alpha + \cdots + x_2\alpha^4}{x_1 + x_2\alpha + \cdots + x_5\alpha^4}$$

If in these we substitute α^{-1} for α^1 and multiply the two results we get

$$\frac{(a_1 + a_2\alpha^{-1} + \cdots + a_5\alpha^{-4})(a_1 + a_2\alpha + \cdots + a_5\alpha^4)}{(b_1 + b_2\alpha + \cdots + b_5\alpha^4)(b_1 + b_2\alpha^{-1} + \cdots + b_5\alpha^{-4})} = 1$$

Since this is a particular eliminant of the set of equations it must be a factor of the determinant which is the general eliminant, and hence the theorem.

If n is odd there is the linear factor $a_1+a_2+\cdots+a_n+b_1+b_2+\cdots+b_n$ and if n is even there is in addition the factor

$$a_1 - a_2 + a_3 \cdots - a_n + b_1 - b_2 + \cdots - b_n.$$

If $b_s = a_s$ the quadratic factor becomes zero.

If we denote the determinant by Δ_5 and the quadratic factor by $F_aG_a - F_bG_b$, then the operations $\text{col}_1 + \alpha\ \text{col}_2 + \alpha^2\ \text{col}_3 + \cdots$ gives

$$\begin{matrix} F_a + F_b \\ \alpha F_a + \alpha^4 F_b \\ \alpha^2 F_a + \alpha^3 F_b \\ \alpha^3 F_a + \alpha^2 F_b \\ \alpha^4 F_a + \alpha F_b \end{matrix}$$

for the first column, and there results

$$\Delta_5 = F_a \begin{vmatrix} 1 & a_2 + b_2 & \cdots & a_5 + b_5 \\ \alpha & a_1 + b_3 & \cdots & a_4 + b_1 \\ \alpha^2 & a_5 + b_4 & \cdots & a_3 + b_2 \\ \alpha^3 & a_4 + b_5 & \cdots & a_2 + b_3 \\ \alpha^4 & a_3 + b_1 & \cdots & a_1 + b_4 \end{vmatrix} + F_b \begin{vmatrix} 1 & a_2 + b_2 & \cdots & a_5 + b_5 \\ \alpha^4 & a_1 + b_3 & \cdots & a_4 + b_1 \\ \alpha^3 & a_5 + b_4 & \cdots & a_3 + b_2 \\ \alpha^2 & a_4 + b_5 & \cdots & a_2 + b_3 \\ \alpha & a_3 + b_1 & \cdots & a_1 + b_4 \end{vmatrix}$$

or for shortness say $\Delta_5 = F_aU + F_bV$.

If in this we write α^{-1} for α then F_a, F_b become G_a, G_b respectively, U and V are interchanged and Δ_5 is unaltered.

We thus have the companion equality $\Delta_5 = G_bU + G_aV$ and from the pair we derive

$$(1) \qquad (F_a - G_b)\Delta_5 = (F_aG_a - F_bG_b)V$$

from which it follows that $F_aG_a - F_bG_b$ is an exact divisor of Δ_5 and that $F_a - G_b$ is an exact divisor of V. Denoting $F_aG_a - F_bG_b$ by Q_1 it is easily seen that Q_2 obtained from Q_1 by replacing α by α^2 is also a factor of Δ_5. The linear factor is obviously $\sum a + \sum b$.

We may now write (1) in the form

$$(2)\qquad (F_a - G_b)Q_1Q_2(\textstyle\sum a + \sum b) = Q_1V$$

or

$$V = (F_a - G_b)(\textstyle\sum a + \sum b)Q_2.$$

Similarly $U = (G_a - F_b)(\sum a + \sum b)Q_2$.

If we put $b_n = -x$, and $b_1 = b_2 = \cdots = b_{n-1} = 0$, then the determinant becomes

$$\begin{vmatrix} a_1 & a_2 \cdots & a_{n-1} & a_n - x \\ a_n & a_1 \cdots a_{n-2} - x & & a_{n-1} \\ \cdot & \cdot\ \cdot\ \cdot\ \cdot & \cdot\ \cdot\ \cdot & \cdot \\ a_2 - x & a_3 \cdots & a_n & a_1 \end{vmatrix}$$

and the factor becomes

$$-\{x^2 - (a_1 + \alpha a_2 + \cdots + \alpha^{n-1}a_n)(a_1 + \alpha^{-1}a_2 + \cdots + \alpha^{-n-1}a\ )\}$$

or

$$-\{x^2 - (a_1 + \alpha a_2 + \cdots + \alpha^{n-1}a_n)(a_1 + \alpha^{n-1}a_2 + \cdots + \alpha a_n).\}$$

Similarly by taking all the imaginary roots in pairs we get all the factors.

These quadratic factors may be written

$$\{x^2 - (A + Bi)(A - Bi)\} \text{ or } \{x^2 - (A^2 + B^2)\}$$

showing the reality of the roots.

The linear factors are $s + x$ and $s' + x$ (if n is even).

503. *The circulant C of order $r \cdot s$ where*

$$a_h = a_{kr+h}\begin{cases} h = 2,3, \cdots, r \\ k = 1,2, \cdots, \overline{s-1} \end{cases}$$

has for its value

$$(1)\ \frac{C^r(a_1, a_{1+r}, \cdots, a_{1+\overline{s-1}\cdot r})C(a_1 + a_{1+r} + \cdots a_{1+\overline{s-1}\cdot r}, sa_2, sa_3 \cdots sa_r)}{(a_1 + a_{1+r} + \cdots + a_{1+\overline{s-1}\cdot r})^r}$$

The theorem and method of proof may be seen from the case when $n = 3\cdot 4$, in which case

$$a_2 = a_5 = a_8 = a_{11}$$

$$a_3 = a_6 = a_9 = a_{12}.$$

The twelve factors are, on account of these relations between the a's and $1+\alpha^3+\alpha^6+\alpha^9=0$, seen to be

1. $a_1 + a_4\alpha^3 + a_7\alpha^6 + a_{10}\alpha^9$
2. $a_1 + a_4\alpha^9 + a_7\alpha^6 + a_{10}\alpha^3$
3. $a_1 + a_4\alpha^6 + a_7\alpha + a_{10}\alpha^6$
4. $(a_1 + a_4 + a_7 + a_{10}) + 4a_2\alpha^8 + 4a_3\alpha^4$
5. $(a_1 + a_4\alpha^9 + a_7\alpha^6 + a_{10}\alpha^3)$
6. $(a_1 + a_4\alpha^3 + a_7\alpha^6 + a_{10}\alpha^9)$
7. $(a_1 + a_4 + a_7 + a_{10}) + 4a_2\alpha^4 + 4a_3\alpha^8$
8. $(a_1 + a_4\alpha^6 + a_7 + a_{10}\alpha^6)$
9. $(a_1 + a_4\alpha^3 + a_7\alpha^6 + a_{10}\alpha^9)$
10. $(a_1 + a_4\alpha^9 + a_7\alpha^6 + a_{10}\alpha^3)$
11. $(a_1 + a_4\alpha^6 + a_7 + a_{10}\alpha^6)$
12. $(a_1 + a_4 + a_7 + a_{10}) + 4a_2 + 4a_3.$

The product of the first, third and fifth is

$$\frac{C(a_1, a_4, a_7, a_{10})}{a_1+a_4+a_7+a_{10}}.$$

The product of the second, sixth and eleventh as well as the eighth, ninth, tenth give the same result.

The product of the fourth, seventh and twelfth gives

$$C(a_1 + a_4 + a_7 + a_{10}, 4a_2, 4a_3).$$

Therefore

$$C = \frac{C^3(a_1, a_4, a_7, a_{10})C(a_1 + a_4 + a_7 + a_{10}, 4a_2, 4a_3)}{(a_1 + a_2 + a_7 + a_{10})^3}.$$

In (1) it should be observed that if $a_1 = a_{hr+1}$ $(h=1, 2, \cdots, \overline{s-1})$ then $C=0$.

In the case $a_2=a_3= \cdots =a_r$ then (1) becomes

$$(2)\quad C=\frac{C^r(a_1, a_{1+r}, \cdots, a_{1+\overline{s-1}\cdot r})(a_1+a_{1+r}+\cdots+a_{1+\overline{s-1}\cdot r}-sa_2)^{r-1}}{(a_1+a_{1+r}+\cdots+a_{1+\overline{s-1}\cdot r})^r}$$
$$\times\ (a_1 + a_{1+r} + \cdots + a_{1+s-1r} + \overline{r-1}\cdot sa_2)$$

which when $a_2=0$ becomes $C=C^r(a_1, a_{1+r}, \cdots, a_{1+\overline{s-1}.r})$. In case $s=2$ then (2) becomes

$$C = \frac{C^r(a_1,a_{1+r})(a_1 + a_{1+r} - 2a_2)^{r-1}(a_1 + a_{1+r} + 2\cdot\overline{r-1}\cdot a_2)}{(a_1 + a_{1+r})^r}$$

$$(3) \qquad = (a_1 - a_{1+r})^r(a_1 + a_{1+r} - 2a_2)^{r-1}(a_1 + a_{1+r} + 2\cdot\overline{r-1}\cdot a_2)$$

which when $a_2=0$ becomes $C=(a_1^2-a_{1+r}^2)^r$.

EXERCISES

(1) Take $r=4$ and $s=3$ and show that

$$C = \frac{C^4(a_1,a_5,a_9)C(a_1 + a_5 + a_9, 3a_2, 3a_3, 3a_4)}{(a_1 + a_5 + a_9)^4}\cdot$$

(2) Take $r=2$, $s=6$ and show that

$$C = \frac{C^2(a_1,a_3,a_5,a_7,a_9,a_{11})C(a_1 + a_3 + a_5 + a_7 + a_9 + a_{11}, 6a_2)}{(a_1 + a_3 + a_5 + a_7 + a_9 + a_{11})^2}\cdot$$

(3) Take $r=6$, $s=2$ and show that

$$C = \frac{C^6(a_1,a_7)C(a_1 + a_7, 2a_2, 2a_3, 2a_4, 2a_5, 2a_6)}{(a_1 + a_7)^6}$$

$$= (a_1- a_7)^6C(a_1 + a_7, 2a_2, 2a_3, 2a_4, 2a_5, 2a_6).$$

504. *The circulant C of order 2n where*

$$\left\{\begin{matrix} a_{2k} = a_{2k+2} \\ a_{2k-1} = a_{2k+1} \end{matrix}\right\} \quad (k = 2,3, \cdots, \overline{n-1})$$

has for its value

(a) *when n is even*

$$(4) \quad C=\frac{\{(a_1-a_{n+1})^n+(a_2-a_4)^n\}\{(a_1-2a_3+a_{n+1})^n-(a_2-a_4)^n\}\{(a_1+\overline{n-2}a_3+a_{n+1})^2-(a_2+\overline{n-1}a_4)^2\}}{\{(a_1-2a_3+a_{n+1})^2-(a_2-a_4)^2\}}$$

except that $a_{n+1} \neq a_3$

(b) *when n is odd*

$$(5) \quad C = \{(a_1 - a_3 - a_4 + a_{n+1})^n + (a_2 - a_4)^n\}\{(a_1 - a_3 + a_4 - a_{n+1})^n - (a_2 - a_4)^n\} \times \frac{\{(a_1 + \overline{n-1}\cdot a_3)^2 - (a_2 + \overline{n-2}\cdot a_4 + a_{n+1})^2\}}{(a_1 - a_3)^2 - (a_2 - 2a_4 + a_{n+1})^2}$$

except that $a_{n+1} \neq a_4$.

The proof of these may be made in a similar way to that in the last article.

If we put $a_{n+1}=a_3$ in (4) and $=a_4$ in (5) then both become

$$C = \frac{(a_1 - a_3)^{2n} - (a_2 - a_4)^{2n}}{(a_1 - a_3)^2 - (a_2 - a_4)^2}\left\{(a_1 + \overline{n-1}a_3)^2 - (a_2 + \overline{n-1}a_4)^2\right\}$$

which becomes when $a_2=a_4$

$$C = (a_1 - a_3)^{2n-2}\left\{(a_1 + \overline{n-1}a_3)^2 - (na_2)^2\right\}$$

and this when $a_1=a_3$ gives

$$C(a_1, a_2, a_1, a_2, \cdots a_1, a_2)_{2n} = 0\ (n > 1)$$

505. *The circulant of order* $2n+1$, *where*

$$a_1 = a_3 = \cdots = a_{2n+1}$$

$$a_2 = a_4 = \cdots = a_{2n}$$

has for value

$$C = (a_1 - a_2)^{2n}(\overline{n-1}a_1 + na_2).$$

The proof follows as in the other cases.

From this and §476 we see that *a circulant of odd order* $2n+1$ *having* n *a's and* $(n-1)$ *b's for elements has the same value whether the a's and b's occur alternately or all the b's follow all the a's.*

EXERCISE: If λ is an odd prime, α an imaginary root of the equation $\alpha^\lambda=1$, and γ a primitive root of the congruence $\gamma^{\lambda-1}\equiv 1 \pmod{\lambda}$ then show that the determinant

$$\begin{vmatrix} \alpha & \alpha^{\gamma} & \alpha^{\gamma 2} & \cdots & \alpha^{\gamma\lambda-2} \\ \alpha^{\gamma} & \alpha^{\gamma 2} & \alpha^{\gamma 3} & \cdots & \alpha \\ \cdot & \cdot & \cdot & \cdots & \cdot \\ \alpha^{\gamma\lambda-2} & \alpha & \alpha^{\gamma} & \cdots & \alpha^{\gamma\lambda-3} \end{vmatrix} = \lambda^{(\lambda-2)/2}.$$

506. If in the circulant $C(a_1, a_2, \cdots, a_7)$ we put $a_7=a_1$, $a_6=a_2$, $a_5=a_3$ it becomes

$$C(a_1, a_2, a_3, a_4, a_3, a_2, a_1) = (2a_1 + 2a_2 + 2a_3 + a_4)$$

$$\times \begin{vmatrix} 0 & 1-2 & 2-3 & 3-4 & 4-3 & 3-2 \\ 2-1 & 0 & 1-2 & 2-3 & 3-4 & 4-3 \\ 3-2 & 2-1 & 0 & 1-2 & 2-3 & 3-4 \\ 4-3 & 3-2 & 2-1 & 0 & 1-2 & 2-3 \\ 3-4 & 4-3 & 3-2 & 2-1 & 0 & 1-2 \\ 2-3 & 3-4 & 4-3 & 3-2 & 2-1 & 0 \end{vmatrix} = s \cdot F, \text{ say,}$$

where F is zero-axial skew and therefore the square of a Pfaffian.

If C is of order 8 and $a_8=a_2$, $a_7=a_3$, $a_6=a_4$ then

$$C(a_1,a_2,a_3,a_4,a_5,a_4,a_3,a_2)$$

$$= s\cdot s'\begin{vmatrix} 5-3 & 0 & 3-5 & 2-4 & 1-3 & 0 \\ 0 & 3-5 & 2-4 & 1-3 & 0 & 3-1 \\ 3-5 & 2-4 & 1-3 & 0 & 3-1 & 4-2 \\ 2-4 & 1-3 & 0 & 3-1 & 4-2 & 5-3 \\ 1-3 & 0 & 3-1 & 4-2 & 5-3 & 0 \\ 0 & 3-1 & 4-2 & 5-3 & 0 & 3-5 \end{vmatrix} = s\cdot s'G,$$

where G is zero-axial skew and therefore a perfect square.

Exercises. Set XXV

(1) If $C(x_1, x_2, \cdots, x_n)$ and $C(y_1, y_2, \cdots, y_n)$ be two circulants such that

$$\frac{X_1}{y_1}=\frac{X_2}{y_2}=\cdots=\frac{X_n}{y_n}$$

show that

$$\frac{x_1}{Y_1}=\frac{x_2}{Y_2}=\cdots=\frac{x_n}{Y_n}$$

where X_r and Y_r are primary minors.

(2) If $x_1-a_1=x_2-a_2=\cdots=x_n-a_n=y$ and $C(x, x_2, \cdots, x_n)=0$ then $C(y+a_1, y+a_2, \cdots, y+a_n)=0$ and

$$y = C(a_1,a_2, \cdots, a_n) \div \begin{vmatrix} 0 & 1 & 1 & 1 \cdots 1 \\ 1 & a_1 & a_2 & a_3 \cdots a_n \\ 1 & a_2 & a_3 & a_4 \cdots a_1 \\ \cdot & \cdot & \cdot & \cdots \cdot \\ 1 & a_n & a_1 & a_2 \cdots a_{n-1} \end{vmatrix}$$

(3) If $f(x)$ be a rational integral function of x, or a convergent infinite series in ascending powers of x, and f_r be a derived series whose terms are in order the rth, $(r-\lambda)$th, $(r-2\lambda)$th $\cdots$ terms of f, then the circulant $C(f_0, f_1, f_2 \cdots f_{\lambda-1})=f(x)f(\alpha_1 x)f(\alpha_2 x) \cdots f(\alpha_{\lambda-1}x)$ where $\alpha_1, \alpha_2, \cdots$, are the λth roots of 1.

(4) If

$$f = \epsilon^x + \epsilon^{\alpha x} + \epsilon^{\alpha^2 x} + \epsilon^{\alpha^3 x} + \epsilon^{\alpha^4 x},$$

α is a primitive fifth root of 1, and $f', f'', \cdots$ denotes the first second etc. derivatives of f, show that

$$C(f, f', f'', f''', f^{\text{IV}}) = C(5,0,0,0,0) = 5^5.$$

(5) If $C \equiv C(x_1, x_2, \cdots, x_n)$ and $P = \epsilon^C$ then show that

$$\frac{\partial^r P}{\partial x_1^r} + \frac{\partial^r P}{\partial x_2^r} + \cdots + \frac{\partial^r P}{\partial x_n^r} = \frac{(-1)^{r-1} n(r-1)!}{(\sum x)^r}.$$

(6) If $C \equiv C(x, y, z)$ show that

$$\left(\frac{\partial^3}{\partial x^3} + \frac{\partial^3}{\partial y^3} + \frac{\partial^3}{\partial z^3}\right) C^n = 27 n^3 C^{n-1}.$$

507. If we square the array formed by adding a column of x's to the circulant $C(a_1, a_2, \cdots, a_n)$ we see by §227 that the result is

$$\left\{nx^2 + (\textstyle\sum a_1)^2\right\}\left(\frac{C}{\sum a_1}\right)^2.$$

508. Determinants of order $2n$ where the first n rows are the same as those of one circulant and the last n rows the same as the first n rows of another circulant have been called *semicirculants*.

The semicirculant

$$\Delta \equiv \begin{vmatrix} a & b & c & d & e & f & g & h \\ h & a & b & c & d & e & f & g \\ g & h & a & b & c & d & e & f \\ f & g & h & a & b & c & d & e \\ h & g & f & e & d & c & b & a \\ a & h & g & f & e & d & c & b \\ b & a & h & g & f & e & d & c \\ c & b & a & h & g & f & e & d \end{vmatrix}$$

where the first n rows are the same as those of the circulant $C(a, b, c, d, e, f, g, h)$ and the last n rows of the reverse circulant $C(h, g, f, e, d, c, b, a)$, obviously contains $\sum a$ as a factor. If we denote the sum of the elements in the even positions of the first line by ϵ and those in the odd positions by ω and perform the following operations

$$r_1 + r_3 + r_5 + \cdots, \qquad r_2 + r_4 + r_6 + \cdots,$$

we get

$$\Delta = \begin{vmatrix} \omega & \epsilon & c & d & e & f & g & h \\ \epsilon & \omega & b & c & d & e & f & g \\ \omega & \epsilon & a & b & c & d & e & f \\ \epsilon & \omega & h & a & b & c & d & e \\ \epsilon & \omega & f & e & d & c & b & a \\ \omega & \epsilon & g & f & e & d & c & b \\ \epsilon & \omega & h & g & f & e & d & c \\ \omega & \epsilon & a & h & g & f & e & d \end{vmatrix}$$

Performing the operations c_7+c_5, c_8+c_6, r_5-r_4, r_6-r_3, r_7-r_2, r_8-r_1, r_4-r_2, r_3-r_1 gives

$$\begin{vmatrix} \omega & \epsilon & c & d & e & f & g+e & h+f \\ \epsilon & \omega & b & c & d & e & f+d & g+e \\ 0 & 0 & a-c & b-d & c-e & d-f & c-c & d-h \\ 0 & 0 & h-b & a-c & b-d & c-e & b-f & c-g \\ 0 & 0 & f-h & e-a & d-b & 0 & 0 & 0 \\ 0 & 0 & g-a & f-b & e-c & 0 & 0 & 0 \\ 0 & 0 & h-b & g-e & f-d & 0 & 0 & 0 \\ 0 & 0 & a-c & h-d & g-e & 0 & 0 & 0 \end{vmatrix}$$

$$= (\omega+\epsilon)(\omega-\epsilon)\cdot 0.$$

This shows that $(\omega+\epsilon)$, $(\omega-\epsilon)$ are factors and that the remaining factor is zero.

If we write Δ thus

$$\begin{vmatrix} d & e & f & a & b & c & f & g & h \\ c & d & e & h & a & b & e & f & g \\ b & c & d & g & h & a & d & e & f \\ a & b & c & f & g & h & c & d & e \\ e & d & c & h & g & f & c & b & a \\ f & e & d & a & h & g & d & c & b \\ g & f & e & b & a & h & e & d & c \\ h & g & f & c & b & a & f & e & d \\ 0 & 0 & 0 & 0 & 0 & 0 & 1 & 0 & 0 \end{vmatrix}$$

which is accomplished by moving bodily the 4th, 5th, 6th columns and making them the 1st, 2nd, and 3rd and repeating the 6th column as the 7th. In this form it is seen that the first 8 rows are centrosymmetric. Operate as follows r_1-r_8, r_2-r_7, r_3-r_6, r_4-r_5 and we get

$$\begin{vmatrix} d-h & e-g & 0 & a-c & 0 & c-a & 0 & g-e & h-d \\ c-g & d-f & 0 & h-b & 0 & b-h & 0 & f-d & g-c \\ b-f & c-e & 0 & g-a & 0 & a-g & 0 & e-c & f-b \\ a-e & b-d & 0 & f-h & 0 & h-f & 0 & d-b & e-a \\ e & d & c & h & g & f & c & b & a \\ f & e & d & a & h & g & d & c & b \\ g & f & e & b & a & h & e & d & c \\ h & g & f & c & b & a & f & e & d \\ 0 & 0 & 0 & 0 & 0 & 0 & 1 & 0 & 0 \end{vmatrix}$$

$\text{Col}_1+\text{col}_9$, $\text{col}_2+\text{col}_8$, $\text{col}_4+\text{col}_6$ gives a rectangle of zeros m by $m+1$ and therefore the determinant is zero.

509. If we take n odd, equal to $2m+1$, say, so that the order of the determinant is $4m+2$, then

$$\Delta \equiv \begin{vmatrix} a & b & c & d & e & f & g & h & m & n \\ h & a & b & c & d & e & f & g & h & m \\ m & n & a & b & c & d & e & f & g & h \\ h & m & n & a & b & c & d & e & f & g \\ g & h & m & n & a & b & c & d & e & f \\ n & m & h & g & f & e & d & c & b & a \\ a & n & m & h & g & f & e & d & c & b \\ b & a & n & m & h & g & f & e & d & c \\ c & b & a & n & m & h & g & f & e & d \\ d & c & b & a & n & m & h & g & f & e \end{vmatrix}$$

where for convenience we have taken $m=2$. Shift the last three $(m+1)$ columns bodily to occupy the first $(m+1)$ places, the determinant is then seen to be centrosymmetric and in consequence breaks up into two factors each of order $2m+1$.

From the one determinant comes $\omega+\epsilon$ as a factor and from the other comes $\omega-\epsilon$; the cofactors in the two cases are alike.

Comparing these two cases we see that, when n is even $=2m$ so that the order of the determinant is $4m$, the determinant cofactor of $(\omega^2-\epsilon^2)$ is of order $4m-2$, has a rectangle of zeros $2m$ by $(2m-1)$ and is therefore zero.

In the case n is odd $=2m+1$, say, the determinant cofactor of $(\omega^2-\epsilon^2)$ is of order $4m$ and has a rectangle of zeros $2m$ by $2m$ and therefore breaks up into two equal factors of order $2m$. We are led therefore to the theorems

(1) *Every semicirculant of order* $4m$ *vanishes*

(2) *Every semicirculant of order* $4m+2$ *is equal to* $(\omega^2-\epsilon^2)$ *times the square of a determinant of order* $2m$. *The determinant of order* $2m$ *is the difference between the matrices of two persymmetric determinants.*

Exercises. Set XXVI

(1) Show that if we make the cyclical changes of the second n rows the opposite of those in the first, the determinant is still resolvable into factors.

(2) Show that the determinant formed by taking two rows of $C(a, b, c, d)$ and two rows of $C(d, c, b, a)$ is resolvable into factors or vanishes.

(3) Show that every determinant formed by taking any three consecutive rows of $C(a, b, c, d, e)$ and any two consecutive rows of $C(e, d, c, b, a)$ is resolvable into three factors, one linear and two quadratic.

(4) If $C(a_1, a_2, a_3, a_4, a_5)$ is symmetrical with respect to the secondary diagonal and if $C(b_1, b_2, b_3, b_4, b_5)$ is symmetrical with respect to the primary diagonal, and if M represents the determinant of the sum of their matrices and N the determinant of the differences of the same two matrices, then show that

$$M \cdot N = C(U, V, W, W, V)$$

where

$$U = \sum a_1^2 - \sum b_1^2$$

$$V = \sum^{0} a_1 a_2 - \sum^{0} b_1 b_2,$$

$$W = \sum^{0} a_1 a_3 - \sum^{0} b_1 b_3$$

where

$$\sum^{0} a_1 a_2 = a_1 a_2 + a_2 a_3 + a_3 a_4 + a_4 a_5 + a_5 a_1$$

and hence

$$\frac{M}{\sum a_1 + \sum b_1}\cdot\frac{N}{\sum a_1 - \sum b_1} = \frac{C(U,V,W,W,V)}{U + 2V + 2W}$$

and

$$\frac{M}{\sum a_1 + \sum b_1} = \{U + V(\epsilon + \epsilon^{-1}) + W(\epsilon^2 + \epsilon^{-2})\}\{U + V(\epsilon^2 + \epsilon^{-2}) + W(\epsilon - \epsilon^{-1})\}$$

where ϵ is an imaginary fifth root of unity.

Block Circulants

510. Determinants of the form

$$\begin{vmatrix} A & B \\ B & A \end{vmatrix}, \quad \begin{vmatrix} A & B & C \\ B & C & A \\ C & A & B \end{vmatrix}, \quad \begin{vmatrix} A & B & C & D \\ B & C & D & A \\ C & D & A & B \\ D & A & B & C \end{vmatrix}, \quad \cdots,$$

where A, B, $\cdots$, are square arrays of any order n, might well be called *block circulants*. The blocks circulate in these in the same manner as the elements in the regular circulants.

511. If $A \equiv |a_{1m}|$, $B \equiv |b_{1m}|$, etc. then it is readily seen by adding all the columns in the symbolic form, that the determinant of order $n \cdot m$

$$\Delta \equiv \begin{vmatrix} A & B & C \cdots \\ B & C & D \cdots \\ C & D & E \cdots \\ \cdot & \cdot & \cdot \end{vmatrix}$$

has as a factor the determinant of the matrix $\|A+B+C+\cdots\|$, which, of course, is the determinant having for its rth column the elements formed by adding the corresponding elements of the rth columns of A, B, C, $\cdots$.

The other factor is obviously

$$\begin{vmatrix} C-B & D-C & E-D \cdots \\ D-C & E-D & F-E \cdots \\ E-D & F-E & G-F \cdots \\ \cdot & \cdot & \cdot \end{vmatrix}$$

which is persymmetric in block notation.

We have, therefore, the theorem *that the block circulation of n general m-line arrays is expressible as the product of two determinants, one of which may be expressed in persymmetric form.*

512. If in the preceding section $n=2$ then

$$\Delta \equiv \begin{vmatrix} A & B \\ B & A \end{vmatrix} = \|A^2 - B^2\| = \|A + B\|\cdot\|A - B\|.$$

That is, *the block circulant of two m-line arrays is expressible as the product of two m-line determinants.*

513. In §511 we have seen that the determinant of the matrix $\|A+B+C+\cdots\|$ is a factor of Δ. Let us suppose now for convenience sake that $n=3$ and let us perform on

$$\Delta \equiv \begin{vmatrix} A & B & C \\ B & C & A \\ C & A & B \end{vmatrix}$$

the operation $c_1+\alpha c_2+\alpha^2 c_3$ where α is a cube root of unity and where $\alpha\cdot A$ means that each element of A is to be multiplied by α etc. Then we get

$$\Delta \equiv \begin{vmatrix} A+\alpha B+\alpha^2 C & B & C \\ B+\alpha C+\alpha^2 A & C & A \\ C+\alpha A+\alpha^2 B & A & B \end{vmatrix}$$

$$= \|A+\alpha B+\alpha^2 C\|\cdot\begin{vmatrix} 1 & B & C \\ \alpha^2 & C & A \\ \alpha & A & B \end{vmatrix}$$

$$= \|A+\alpha B+\alpha^2 C\|\cdot\begin{vmatrix} C-\alpha A & A-\alpha B \\ A-\alpha B & B-\alpha C \end{vmatrix}$$

showing that the determinant of the matrix $\|A+\alpha B+\alpha^2 C\|$ is also a factor. Similarly we find that the determinant of the matrix $\|A+\alpha^2 B+\alpha C\|$ is a factor. We have, therefore

$$\Delta = -\|A+B+C\|\cdot\|A+\alpha B+\alpha^2 C\|\cdot\|A+\alpha^2 B+\alpha C\|\cdot$$

If we expand the determinant of $\|A+\alpha B+\alpha^2 C\|$ as a sum of determinants with monomial elements we get $P+\alpha Q+\alpha^2 R$ where P represents all the terms independent of α, Q all those containing α as a factor, and R all those containing α^2 as a factor. We have therefore

$$\Delta = -(P+Q+R)(P+\alpha Q+\alpha^2 R)(P+\alpha^2 Q+R)$$

or

$$\Delta = C(P,Q,R).$$

514. *In general, the block circulant Δ formed from n m-line general determinants is equal to the product of n determinants of order m.* That is

$$\Delta = h\cdot \sum_{k=0}^{n-1} \left\| A + \alpha^k B + \alpha^{2k} C + \cdots + \alpha^{(n-1)k} L \right\|,$$

where $h=(-1)^{(n-1)(n-2)/2}$

The proof follows as in §513.

515. If $A, B, C, \cdots$ are circulants then the determinant of $\|A+\alpha^k B+\alpha^{2k} C+ \cdots +\alpha^{(n-1)k} L\|$ is a circulant and breaks up into linear factors. Take the case when $n=3$ and $m=3$ and multiply

$$\Delta \equiv \begin{vmatrix} A & B & C \\ B & C & A \\ C & A & B \end{vmatrix},$$

where $A=C(a_1, a_2, a_3)$ etc. by

$$\Delta_1 \equiv \begin{vmatrix} \beta & \beta & \beta \\ \beta & \alpha^2\beta & \alpha\beta \\ \beta & \alpha\beta & \alpha^2\beta \end{vmatrix},$$

where

$$\beta \equiv \begin{vmatrix} 1 & 1 & 1 \\ 1 & \alpha^2 & \alpha \\ 1 & \alpha & \alpha^2 \end{vmatrix},$$

α is an imaginary cube root of unity, and $\alpha\cdot\beta$ means that each row of β is to be multiplied by α, etc.

Denoting the product of the 1st row of Δ by the kth row of Δ_1 by s_k and taking the product of Δ and Δ_1 in the non-symbolic form we have

$$\Delta\cdot\Delta_1 = \begin{vmatrix} s_1 & s_1 & s_1 & s_1 & s_1 & s_1 & s_1 & s_1 & s_1 \\ s_2 & s_2\alpha & s_2\alpha^2 & s_2 & s_2\alpha & s_2 & s_2\alpha^2 & s_2 & s_2\alpha^2 \\ s_3 & s_3\alpha^2 & s_3\alpha & s_3 & s_3\alpha^2 & s_3\alpha & s_3 & s_3\alpha^2 & s_3\alpha \\ s_4 & s_4 & s_4 & s_4\alpha & s_4\alpha & s_4\alpha & s_4\alpha^2 & s_4\alpha^2 & s_4\alpha^2 \\ s_5 & s_5\alpha & s_5\alpha^2 & s_5\alpha & s_5\alpha^2 & s_5 & s_5\alpha^2 & s_5 & s_5\alpha \\ s_6 & s_6\alpha^2 & s_6\alpha & s_6\alpha & s_6 & s_6\alpha^2 & s_6\alpha^2 & s_6\alpha & s_6 \\ s_7 & s_7 & s_7 & s_7\alpha^2 & s_7\alpha^2 & s_7\alpha^2 & s_7\alpha & s_7\alpha & s_7\alpha \\ s_8 & s_8\alpha & s_8\alpha^2 & s_8\alpha^2 & s_8 & s_8\alpha & s_8\alpha & s_8\alpha^2 & s_8 \\ s_9 & s_9\alpha^2 & s_9\alpha & s_9\alpha^2 & s_9\alpha & s_9 & s_9\alpha & s_9 & s_9\alpha^2 \end{vmatrix}$$

$$= s_1\cdot s_2\cdot s_3 \cdots s_9\cdot\Delta_1$$

therefore

$$\Delta = s_1 \cdot s_2 \cdot s_3 \cdots s_9.$$

516. If $m=2$ and we take another determinant of order 6

$$\Delta_1 \equiv \begin{vmatrix} A' & B' & C' \\ B' & C' & A' \\ C' & A' & B' \end{vmatrix}$$

and use it with Δ to form a new block determinant

$$\Delta_2 = \begin{vmatrix} \Delta & \Delta_1 \\ \Delta_1 & \Delta \end{vmatrix}$$

then

$$\begin{aligned} \Delta_2 &= \|\Delta^2 - \Delta_1^2\| = \|\Delta + \Delta_1\| \cdot \|\Delta_1 - \Delta\| \\ &= \|A+A'+B+B'+C+C'\| \ \|A+A'+\alpha(B+B')+\alpha^2(C+C')\| \\ &\quad \cdot \|A+A'+\alpha^2(B+B')+\alpha(C+C')\| \times \|A-A'+B-B'+C-C'\| \\ &\quad \cdot \|A-A'+\alpha(B-B')+\alpha^2(C-C')\| \cdot \|A-A'+\alpha^2(B-B')+\alpha(C-C')\| \end{aligned}$$

If A, B, $\cdots$, are circulants then

$$\begin{aligned} \Delta_2 = &\{(a_1+a_1'+b_1+b_1'+c_1+c_1')^2-(a_2+a_2'+b_2+b_2'+c_2+c_2')^2\} \\ &\times \{(a_1+a_1'+\alpha b_1+\alpha b_1'+\alpha^2 c_1+\alpha^2 c_1')^2 \\ &\quad - (a_2+a_2'+\alpha b_2+\alpha b_2'+\alpha^2 c_2+\alpha^2 c_2')^2\} \quad \text{etc.} \end{aligned}$$

That is Δ_2 is the product of twelve linear factors.

517. It will appear by shifting rows and columns, that the complementary minor of any element in one position is the same as that of the same element in any other position.

As a consequence of this it is seen that *the adjugate of a block circulant is a block circulant of the same type as the original.*

518. *The product of two block circulants is homogenetic, that is, the product is of the same form as the factors.*

We have seen (§472) that this is true for ordinary circulants and the product in symbolic form shows that it is true for block circulants. Thus

$$\begin{vmatrix} A & B & C & \cdots \\ B & C & D & \cdots \\ \cdots & \cdots & \cdots & \cdots \end{vmatrix} \cdot \begin{vmatrix} A' & B' & C' & \cdots \\ B' & C' & D' & \cdots \\ \cdots & \cdots & \cdots & \cdots \end{vmatrix}$$

$$= \begin{vmatrix} AA'+BB'+CC'+\cdots & AB'+BC'+CD'+\cdots \\ BA'+CB'+DC'+\cdots & BB'+CC'+DD'+\cdots \\ \cdots & \cdots \end{vmatrix}$$

Here, as in §472, the symmetry is with respect to the secondary diagonal.

If the order of a block circulant is $2n$ then complementary minors of order n differ at most in sign.

519. If a block circulant is of order, 2^μ, and such that the elements $(r, s)=(s, r)=(r+2^{k-1}, s+2^{k-1})$ for every $k \gtreqless \mu$ and for $r, s \lesseqgtr 2^k$, then the form of such a determinant, known as Pucta's would be

$$\begin{vmatrix} a_1 & a_2 & a_3 & a_4 & a_5 & a_6 & a_7 & a_8 & \cdots \\ a_2 & a_1 & a_4 & a_3 & a_6 & a_5 & a_8 & a_7 & \cdots \\ a_3 & a_4 & a_1 & a_2 & a_7 & a_8 & a_5 & a_6 & \cdots \\ a_4 & a_3 & a_2 & a_1 & a_8 & a_7 & a_6 & a_5 & \cdots \\ a_5 & a_6 & a_7 & a_8 & a_1 & a_2 & a_3 & a_4 & \cdots \\ a_6 & a_5 & a_8 & a_7 & a_2 & a_1 & a_4 & a_3 & \cdots \\ a_7 & a_8 & a_5 & a_6 & a_3 & a_4 & a_1 & a_2 & \cdots \\ a_8 & a_7 & a_6 & a_5 & a_4 & a_3 & a_2 & a_1 & \cdots \\ \cdot & \cdot & \cdot & \cdot & \cdot & \cdot & \cdot & \cdot & \cdot \end{vmatrix}$$

It is obviously centrosymmetric and therefore breaks up into two determinants each of order $2^{\mu-1}$.

The simplest case is $(\mu=1)$

$$(1) \qquad \Delta_2 \equiv \begin{vmatrix} a_1 & a_2 \\ a_2 & a_1 \end{vmatrix} = a_1^2 - a_2^2 = (a_1+a_2)(a_1-a_2).$$

The next case is $(\mu=2)$

$$(2)\ \Delta_4 \equiv \begin{vmatrix} a_1 & a_2 & a_3 & a_4 \\ a_2 & a_1 & a_4 & a_3 \\ a_3 & a_4 & a_1 & a_2 \\ a_4 & a_3 & a_2 & a_1 \end{vmatrix} = \begin{vmatrix} a_3 - a_2 & a_4 - a_1 \\ a_4 - a_1 & a_3 - a_2 \end{vmatrix} \begin{vmatrix} a_3 + a_2 & a_4 + a_1 \\ a_4 + a_1 & a_3 + a_2 \end{vmatrix}$$

$$= (a_3 - a_2 + a_4 - a_1)(a_3 - a_2 + a_4 + a_1)$$

$$(a_3 + a_2 + a_4 + a_1)(a_3 + a_2 - a_4 - a_1), \text{ by (1)}$$

The next case of order 8 would break up into two centrosymmetric factors each of order 4 and therefore by (2) each into four linear factors and give for Δ_8 eight linear factors. So we see in general that $\Delta_{2\mu}$ breaks up into 2^μ linear factors.

520. *If we use* P *to represent a Pucta determinant of order* 2^n *and* P_p *to denote the Pucta determinant whose elements are the linear factors of* P, *then*

$$P_p = (2^n)^{2^n} a_1 a_2 \cdots a_{2^n}.$$

Thus, taking $n=2$ for convenience, and denoting

$$a_1+a_2+a_3+a_4,\quad a_1-a_2+a_3-a_4,\quad a_1+a_2-a_3-a_4,\quad a_1-a_2-a_3+a_4$$

by $\sigma_1, \sigma_2, \sigma_3, \sigma_4$

respectively we have

$$P_p \equiv \begin{vmatrix} \sigma_1 & \sigma_2 & \sigma_3 & \sigma_4 \\ \sigma_2 & \sigma_1 & \sigma_4 & \sigma_3 \\ \sigma_3 & \sigma_4 & \sigma_1 & \sigma_2 \\ \sigma_4 & \sigma_3 & \sigma_2 & \sigma_1 \end{vmatrix} = \begin{vmatrix} a_1 & a_2 & a_3 & a_4 \\ a_1 & -a_2 & a_3 & -a_4 \\ a_1 & a_2 & -a_3 & -a_4 \\ a_1 & -a_2 & -a_3 & a_4 \end{vmatrix} \cdot \begin{vmatrix} 1 & 1 & 1 & 1 \\ 1 & -1 & 1 & -1 \\ 1 & 1 & -1 & -1 \\ 1 & -1 & -1 & 1 \end{vmatrix}$$

$$= a_1a_2a_3a_4 \begin{vmatrix} 1 & 1 & 1 & 1 \\ 1 & -1 & 1 & -1 \\ 1 & 1 & -1 & -1 \\ 1 & -1 & -1 & 1 \end{vmatrix}^2 = a_1a_2a_3a_4 \begin{vmatrix} 4 & \cdot & \cdot & \cdot \\ \cdot & 4 & \cdot & \cdot \\ \cdot & \cdot & 4 & \cdot \\ \cdot & \cdot & \cdot & 4 \end{vmatrix}$$

$$= (2^2)^{2^2} a_1a_2a_3a_4.$$

521. *If to the array of a Pucta determinant* P *of order* 2^m *there be annexed an additional column having all its elements identical, the determinant which is the square of this extended array is wholly resolvable into linear factors.*

To show this we observe that the square is a Pucta determinant with quadratic elements, and by §507 this square is seen to be

$$P^2 + 2^m x^2 \left(\frac{P}{\sum a}\right)^2 \quad \text{or} \quad \{2^m x^2 + (\sum a)^2\}\left(\frac{P}{\sum a}\right)^2$$

where x is the common element in the added column.

The factors of P being linear the truth of the theorem appears.

522. Quite similar properties are true of determinants where the iterating blocks, $A, B, \cdots$, are skew circulants. Thus the determinant akin to Pucta's generated by iterating the array $\begin{vmatrix} a & b \\ -b & a \end{vmatrix}$ as Pucta's is generated by iterating the array $\begin{vmatrix} a & b \\ b & a \end{vmatrix}$, can be readily seen to have the following properties:

(1) Resolvability into linear factors.

(2) Having its adjugate of the same form as itself.

(3) They are homogenetic.

523. The determinant

$$D \equiv \begin{vmatrix} a_{11} \cdots a_{1n} & -b_{11} \cdots -b_{1n} \\ \cdot \quad \cdot \quad \cdot \quad \cdot & \cdot \quad \cdot \quad \cdot \quad \cdot \\ a_{n1} \cdots a_{nn} & -b_{n1} \cdots -b_{nn} \\ b_{11} \cdots b_{1n} & a_{11} \cdots \quad a_{1n} \\ \cdot \quad \cdot \quad \cdot \quad \cdot & \cdot \quad \cdot \quad \cdot \quad \cdot \\ b_{n1} \cdots b_{nn} & a_{n1} \cdots a_{nn} \end{vmatrix}$$

is the eliminant of a set of linear homogeneous complex equations in the $2n$ quantities $p_1, \cdots, p_n, q_1, \cdots, q_n$

$$(1) \quad \begin{cases} (a_{11} + b_{11}i)(p_1 + q_1 i) + \cdots + (a_{1n} + b_{1n}i)(p_n + q_n i) = 0 \\ \cdot \quad \cdot \quad \cdot \quad \cdot \quad \cdot \quad \cdot \quad \cdot \quad \cdot \quad \cdot \quad \cdot \\ (a_{n1} + b_{n1}i)(p_1 + q_1 i) + \cdots + (a_{nn} + b_{nn}i)(p_n + q_n i) = 0 \end{cases}$$

This determinant D is obviously unaltered in value (except the sign will be changed when n is odd) by rearranging the last n rows and the last n columns so that they will be in the reverse order, and then multiplying the elements of the first n rows and the last n columns by i. In this form it is seen to be centrosymmetric with all the b's positive and all the a's multiplied by i. It may therefore be expressed as the product of two factors D' and D'', say. Thus $D = D' \cdot D''$ where

$$D' \equiv \begin{vmatrix} b_{1n} + a_{1n}i \cdots b_{11} + a_{11}i \\ \cdot \quad \cdot \quad \cdot \quad \cdot \quad \cdot \quad \cdot \\ b_{nn} + a_{nn}i \cdots b_{n1} + a_{n1}i \end{vmatrix}, \text{ or } \left| b_{rs} + a_{rs}i \right| ;$$

and

$$D'' \equiv \begin{vmatrix} b_{1n} - a_{1n}i \cdots b_{11} - a_{11}i \\ \cdot \quad \cdot \quad \cdot \quad \cdot \quad \cdot \quad \cdot \\ b_{nn} - a_{nn}i \cdots b_{n1} - a_{n1}i \end{vmatrix}, \text{ or } \left| b_{rs} - a_{rs}i \right| .$$

Then

$$D = (X + iY)(X - iY) = X^2 + Y^2.$$

That is D may be expressed as the sum of two squares.

The matrix of the determinant D may be expressed symbolically by

$$\begin{Vmatrix} (A_n) & (B_n) \\ (-B_n) & (A_n) \end{Vmatrix}$$

If instead of equations (1) we had taken the following

$$(2)\quad \begin{cases} (a_{11}+b_{11}i)(p_1+q_1i)+\cdots+(a_{1n}+b_{1n}i)(p_n+q_ni)=0\\ -(a_{12}+b_{12}i)(p_1+q_1i)+\cdots+(a_{2n}+b_{2n}i)(p_n+q_ni)=0\\ \cdots\cdots\cdots\cdots\cdots\cdots\cdots\cdots\\ -(a_{1n}+b_{1n}i)(p_1+q_1i)-\cdots+(a_{1n}+b_{1n}i)(p_n+q_ni)=0 \end{cases}$$

then the matrix of the determinant of this set may be expressed symbolically by

$$\left\| \begin{matrix} (A_2) & (B_2) & (C_2)\cdots \\ (-B_2) & (A_2) & (D_2)\cdots \\ (-C_2) & (-D_2) & (A_2)\cdots \end{matrix} \right\|$$

Exercises. Set XXVII

1. Show that $C(abc, def, ghi, jkl, mno) = C(adgjm, behkn, cfilo)$
2.

$$\text{If } \Delta \equiv \begin{vmatrix} a & b & c & d \\ b & a & d & c \\ c & d & a & b \\ d & c & b & a \end{vmatrix}$$

and A, B, C, D are the elements of the adjugate, show that

$$A+B+C+D=(a+b-c-d)(a-b+c-d)(a-b-c+d)$$

$$=\frac{\Delta}{a+b+c+d},$$

$$A+B-C-D=\frac{\Delta}{a+b-c-d}, \quad \text{etc.}$$

3. If the linear factors of Δ in example 2 are p, q, r, s and those of the adjugate are P, Q, R, S, then

$$Aa+Bb+Cc+Dd=\Delta=pqrs,$$

$$Ab+Ba+Cd+Dc=0,$$

$$Ac+Bd+Ca+Db=0,$$

$$Ad+Bc+Cb+Da=0,$$

where A, B, C, D are the signed primary minors of Δ.

Show that

(a) $P = qrs$, $Q = prs$, $R = pqs$, $S = pqr$, and $p^3 = \dfrac{QRS}{p^2}$, etc.

(b) $$\begin{vmatrix} a^2 & b^2 & c^2 & d^2 \\ b^2 & a^2 & d^2 & c^2 \\ c^2 & d^2 & a^2 & b^2 \\ d^2 & c^2 & b^2 & a^2 \end{vmatrix}$$
$$= 2^{-5}(A^2 + B^2 + C^2 + D^2)(AB + CD)(AC + BD)(AD + BC)$$

(c) $$\begin{vmatrix} A^2 & B^2 & C^2 & D^2 \\ B^2 & A^2 & D^2 & C^2 \\ C^2 & D^2 & A^2 & B^2 \\ D^2 & C^2 & B^2 & A^2 \end{vmatrix}$$
$$= 2^{11}(a^2 + b^2 + c^2 + d^2)(ab + dc)(ac + bd)(ad + bc)$$

(d) $$\frac{|\, A^0\, B^1\, C^2\, D^3 \,|}{|\, a^0\; b^1\; c^2\; d^3 \,|} = -\,(a + b + c + d)^4 \cdot \Delta^2.$$

4. If in a determinant of n rows the successive rows proceed from the first by permutations which form an Abelian group of order n (including identity), the determinant is expressible as the product of n linear factors.

5. If

$$\Delta = \begin{vmatrix} A & B \\ B & C \end{vmatrix},$$

where $A = C(a, b, c)$, $B = C(d, e, f)$ then show that

$$\Delta = (a + b + c + d + e + f)(a + b + c - d - e - f)$$
$$\cdot \left\{ \frac{C(a,b,c)}{a + b + c} - \frac{C(d,e,f)}{d + e + f} \right\}^2$$

6. Show that the circulant of the arrays of $C(a_1, a_2, a_3)$, $C(a_4, a_5, a_6)$, $C(a_7, a_8, a_9)$

(a) is the same as the circulant of the arrays of $C(a_1, a_4, a_7)$, $C(a_2, a_5, a_8)$, $C(a_3, a_6, a_9)$.

(b) is divisible by $C(a_1 + a_2 + a_3,\ a_4 + a_5 + a_6,\ a_7 + a_8 + a_9)$.

7. Show that

$$\begin{vmatrix} a & b & c & d & e & f & g & h \\ -b & a & d & -c & f & -e & -h & g \\ -c & -d & a & b & g & h & -e & -f \\ -d & c & -b & a & h & -g & f & -e \\ -e & -f & -g & -h & a & b & c & d \\ -f & e & -h & g & -b & a & -d & c \\ -g & h & e & -f & -c & d & a & -b \\ -h & -g & f & e & -d & -c & b & a \end{vmatrix} = (\sum a^2)^4$$

Section III. Aggregates

(a) Centrosymmetry

524. Under the conditions of centrosymmetry the right-hand side of the theorem of §318 vanishes and we have

$$\sum_1^{(m)k} i \begin{vmatrix} \overline{(2m \mid m_\alpha \mid k_i)}(2m \mid \overline{m}_\alpha \mid k_i) \\ (2m \mid m_\alpha) \end{vmatrix} - \sum_1^{(m)k} i \begin{vmatrix} (2m \mid m_\alpha) \\ \overline{(2m \mid m_\alpha \mid k_i)}(2m \mid \overline{m}_\alpha \mid k_i) \end{vmatrix} = 0$$

It is to be observed here that we do not get a true vanishing aggregate for every selection of the $2m$ numbers m at a time. For if x and y be any two of the m numbers in any given selection, then whenever $x+y=2m+1$ some of the minors in the first sum have their conjugates in the second and cancel each other while others will vanish identically. Those that remain, if any, form an aggregate which is the extensional of a vanishing aggregate of minors of lower order. Thus for a determinant of order six if we take the selection 256 we get

$$\begin{vmatrix} 2\,5\,1 \\ 2\,5\,6 \end{vmatrix} + \begin{vmatrix} 2\,2\,6 \\ 2\,5\,6 \end{vmatrix} + \begin{vmatrix} 5\,5\,6 \\ 2\,5\,6 \end{vmatrix} - \begin{vmatrix} 2\,5\,6 \\ 2\,5\,1 \end{vmatrix} - \begin{vmatrix} 2\,5\,6 \\ 2\,2\,6 \end{vmatrix} - \begin{vmatrix} 2\,5\,6 \\ 5\,5\,6 \end{vmatrix},$$

in which the 1st and 4th are conjugate and cancel each other and the others vanish having identical rows or columns:

For a determinant of order eight taking the selection 1678 we get

$$\begin{vmatrix} 1\,6\,7\,1 \\ 1\,6\,7\,8 \end{vmatrix} + \begin{vmatrix} 1\,6\,2\,8 \\ 1\,6\,7\,8 \end{vmatrix} + \begin{vmatrix} 1\,3\,7\,8 \\ 1\,6\,7\,8 \end{vmatrix} + \begin{vmatrix} 8\,6\,7\,8 \\ 1\,6\,7\,8 \end{vmatrix}$$

$$- \begin{vmatrix} 1\,6\,7\,8 \\ 1\,6\,7\,1 \end{vmatrix} - \begin{vmatrix} 1\,6\,7\,8 \\ 1\,6\,2\,8 \end{vmatrix} - \begin{vmatrix} 1\,6\,7\,8 \\ 1\,3\,7\,8 \end{vmatrix} - \begin{vmatrix} 1\,6\,7\,8 \\ 8\,6\,7\,8 \end{vmatrix},$$

which reduces to the extensional of $\begin{vmatrix} 6 & 2 \\ 6 & 7 \end{vmatrix} + \begin{vmatrix} 3 & 7 \\ 6 & 7 \end{vmatrix} - \begin{vmatrix} 6 & 7 \\ 6 & 2 \end{vmatrix} - \begin{vmatrix} 6 & 7 \\ 3 & 7 \end{vmatrix}$. Neglecting these degenerate aggregates we have, if λ denotes the number of true vanishing aggregates of each type (that is for each of the $m-1$ values of k) for the determination of λ the following

$$\lambda = \frac{1}{2}\left\{C_{2m,m} - \sum_{1}^{m/2} r \frac{m(m-1)\cdots(m-2r+1)}{(r!)^2} 2^{m-2r}\right\}.$$

This we see by counting the number of combinations containing one, two, etc. pairs of numbers satisfying the equation $x+y=2m+1$ and subtracting the result from $C_{2m,m}$ and observing that for every relation as given there is another

$$\sum_{1}^{(m)_k} i \left| \frac{\overline{(\overline{2m} \mid m_\alpha \mid k_i)}(\overline{2m} \mid \bar{m}_\alpha \mid k_i)}{(\overline{2m} \mid m_\alpha)} \right| - \sum_{1}^{(m)_k} i \left| \frac{(\overline{2m} \mid m_\alpha)}{(\overline{2m} \mid m_\alpha \mid k_i)(\overline{2m} \mid \bar{m}_\alpha \mid k_i)} \right| = 0$$

which in virtue of the centrosymmetry is not different from the first.

On examining a set of λ true vanishing aggregates of any type k it will be seen that:

(a) Each aggregate contains $(m)_k$ terms together with the negative of their conjugates.

(b) Every minor $\begin{vmatrix} (a) \\ (b) \end{vmatrix}$ enters two aggregates in one with (b) as the selection of columns and in another with (a) (or $(\bar{a})$) as the selection of columns.

(c) There are $(m)_k \cdot \lambda$ independent minors $\frac{1}{2}(m)_k \cdot \lambda$ with their conjugates).

(d) For every set of $\frac{1}{2}\lambda$ aggregates there is a set of $\frac{1}{2}\lambda$ minors (one in each aggregate) whose conjugates are found in the remaining $\frac{1}{2}\lambda$ aggregates.

(e) The sum of the λ aggregates vanishes.

It follows from (a) and (b) that every minor enters one aggregate with the positive and one with the negative sign, and that any two aggregates have either two or no minors in common.

It follows from (d) that no fewer than λ aggregates can vanish. It follows that the number of linearly independent aggregates of each type is $\lambda-1$.

(b) Skew-Centrosymmetry

525. Since every minor of order m of a skew-centrosymmetric determinant Δ of order $2m$ is equal to the reflex when m is even and to the negative of the reflex when m is odd it is evident that the vanishing aggregate of type k for skew-symmetric determinants is

$$\sum_{1}^{(m)k} i \left| \begin{matrix} (\overline{2m \mid m_\alpha \mid k_i})(2m \mid \bar{m}_\alpha \mid k_i) \\ (2m \mid m_\alpha) \end{matrix} \right|$$

$$= (-1)^m \sum_{1}^{(m)k} i \left| \begin{matrix} (2m \mid m_\alpha) \\ (\overline{2m \mid m_\alpha \mid k_i})(2m \mid \bar{m}_\alpha \mid k_i) \end{matrix} \right|$$

(c) Axisymmetry

526. If in §312 we impose the condition that the original determinant has the coaxial minor $\left| \begin{smallmatrix} (\overline{2n} \mid k_\alpha) \\ (\overline{2n} \mid k_\alpha) \end{smallmatrix} \right|$ axisymmetric the right-hand side of the theorem will vanish. Thus for the example there given we would have the left-hand side, which we shall write $\sum \left| \begin{smallmatrix} 1\,2\,\overline{3\,4} \\ \underline{5\,6\,7\,8} \end{smallmatrix} \right|$, equal to zero.

527. If $k=n-1$, then the theorem of §526 becomes

$$\sum \left| \begin{matrix} 1 & 2 & 3 & \cdots & \bar{n} \\ \underline{n+1} & \underline{n+2} & \underline{n+3} & \underline{\cdots} & \underline{2n} \end{matrix} \right| = 0,$$

a result due to Kronecker. It should be observed that the theorem does not require that the whole determinant of order $2n$ be axisymmetric, but it is sufficient that the minor $\left| \begin{smallmatrix} 1\,2\,\cdots\,n+1 \\ 1\,2\,\cdots\,n+1 \end{smallmatrix} \right|$ be axisymmetric to have

$$\sum \left| \begin{matrix} \overline{1} & \overline{2} & \overline{\cdots} & \overline{n} \\ \underline{n+1} & n+2 & \cdots & 2n \end{matrix} \right| = 0.$$

528. It is apparent that

$$(1) \qquad \sum \left| \begin{matrix} 12\overline{34} \\ \underline{5678} \end{matrix} \right| = \sum \left| \begin{matrix} 123\bar{4} \\ \underline{5678} \end{matrix} \right| + \sum \left| \begin{matrix} 12\overline{45} \\ 3\underline{678} \end{matrix} \right|, \text{ where } \sum \left| \begin{matrix} 12\overline{45} \\ 3\underline{678} \end{matrix} \right|$$

has 1, 2 constant in the rows and 3 constant in the columns of every term. This is dependent upon the relation $(6)_2=(5)_2+(5)_1$. The general to which this leads depends upon

$$(m+k)_k = (m+1)_k + (k-1)_1(m+1)_{k-1}$$
$$+ (k-1)_2(m+1)_{k-2} + \cdots + (k-1)_{k-2}(m+1)_2 + (m+1)_1$$

and is that, *an axisymmetric determinant of order* $2m$

$$\sum \left| \begin{matrix} x_1 \; x_2 \cdots x_{m-k}\overline{y_1 y_2 \cdots y_k} \\ \underline{y_{k+1} y_{k+2} \cdots\cdots\cdots y_{m+k}} \end{matrix} \right|,$$

where the x*'s and the* y*'s together are the numbers* 1, 2, 3, $\cdots$, $2m$, *may be written as a sum of sigmas where under each sigma the* $(m-k)x$*'s and* $(k-1)$ *of the* y*'s remain constant while the remaining* $(m+1)y$*'s vary from term to term.*

Thus if we denote the numbers indicated by the x's and the selection of $(k-1)y$'s by $\alpha_1, \alpha_2, \cdots, \alpha_{m-1}$ and the remaining $(m+1)y$'s by $\beta_1, \beta_2, \cdots, \beta_{m+1}$ then we may write

$$\sum \left| \begin{matrix} x_1 \; x_2 \cdots x_{m-k}\overline{y_1 \cdots y_k} \\ \underline{y_{k+1} \cdots\cdots\cdots y_{m+k}} \end{matrix} \right|$$

$$= \sum_0^{k-1} h(k-1)_h \sum \beta \left| \begin{matrix} \alpha_1 \cdots \alpha_{m-k+h}\overline{\beta_1 \cdots \beta_{k-h}} \\ \alpha_{m-k+h+1} \; \alpha_{m-1} \; \underline{\beta_{k-h+1} \cdots {}_{m+1}} \end{matrix} \right|$$

529. By §526 we know that the term on the left of (1) §528 is zero, and by §527 we know that the first term on the right is zero and therefore $\sum \left|\begin{smallmatrix} 12\overline{45} \\ 3\underline{678} \end{smallmatrix}\right| = 0$.

This may be looked upon as the extensional of $\sum \left|\begin{smallmatrix} 2\overline{45} \\ \underline{678} \end{smallmatrix}\right| = 0$.

Similarly $\sum \left|\begin{smallmatrix} 13\overline{45} \\ 2\underline{678} \end{smallmatrix}\right| = 0$, so that we have

$$\sum \left| \begin{matrix} 123\overline{4} \\ \underline{5678} \end{matrix} \right| = \sum \left| \begin{matrix} 12\overline{45} \\ 3\underline{678} \end{matrix} \right| = \sum \left| \begin{matrix} 13\overline{45} \\ 2\underline{678} \end{matrix} \right| = \sum \left| \begin{matrix} 1\overline{678} \\ 23\underline{45} \end{matrix} \right| = 0$$

or

$$\sum \left| \begin{matrix} \alpha\beta\gamma\overline{\delta} \\ \underline{\epsilon\zeta\eta\theta} \end{matrix} \right| = \sum \left| \begin{matrix} \alpha\beta\overline{\delta\epsilon} \\ \gamma\underline{\zeta\eta\theta} \end{matrix} \right| + \sum \left| \begin{matrix} \alpha\gamma\overline{\delta\epsilon} \\ \beta\underline{\zeta\eta\theta} \end{matrix} \right| = \sum \left| \begin{matrix} \alpha\overline{\delta\epsilon\zeta} \\ \beta\gamma\underline{\eta\theta} \end{matrix} \right| = 0$$

In order to have the Law of Extensible minors applicable to any given identity it will be seen that the proof of this theorem requires

(a) that the extended determinant be of the same form as the original and contain the latter as a minor, and (b) the adjugate be a determinant of the same form as the original. These requirements are obviously satisfied in case of axisymmetry.

530. From the two relations

$$(1)\quad \sum\begin{vmatrix}\overline{1234}\\ \underline{5}678\end{vmatrix}=\begin{vmatrix}1234\\5678\end{vmatrix}-\begin{vmatrix}1235\\4678\end{vmatrix}+\begin{vmatrix}1245\\3678\end{vmatrix}-\begin{vmatrix}1345\\2678\end{vmatrix}+\begin{vmatrix}2345\\1678\end{vmatrix}=0$$

$$(2)\quad \sum\begin{vmatrix}123\overline{4}\\ \underline{5678}\end{vmatrix}=\begin{vmatrix}1234\\5678\end{vmatrix}-\begin{vmatrix}1235\\4678\end{vmatrix}+\begin{vmatrix}1236\\4578\end{vmatrix}-\begin{vmatrix}1237\\4568\end{vmatrix}+\begin{vmatrix}1238\\4567\end{vmatrix}=0$$

we have

$$\begin{vmatrix}1245\\3678\end{vmatrix}-\begin{vmatrix}1345\\2678\end{vmatrix}+\begin{vmatrix}2345\\1678\end{vmatrix}=\begin{vmatrix}1236\\4578\end{vmatrix}-\begin{vmatrix}1237\\4568\end{vmatrix}+\begin{vmatrix}1236\\4567\end{vmatrix},$$

or

$$\sum\begin{vmatrix}\overline{12}45\\ \underline{3}678\end{vmatrix}=\sum\begin{vmatrix}123\overline{6}\\ 45\underline{78}\end{vmatrix}$$

From (1) we have

$$\sum\begin{vmatrix}123\overline{4}\\ \underline{5}678\end{vmatrix}=-\sum\begin{vmatrix}\overline{12}45\\ \underline{3}678\end{vmatrix}$$

Therefore

$$\sum\begin{vmatrix}\overline{12}45\\ \underline{3}678\end{vmatrix}=-\sum\begin{vmatrix}123\overline{4}\\ \underline{5}678\end{vmatrix}=\sum\begin{vmatrix}123\overline{6}\\ 45\underline{78}\end{vmatrix}$$

or

$$\sum\begin{vmatrix}\overline{\alpha\beta}\gamma\delta\\ \underline{\epsilon}\zeta\eta\theta\end{vmatrix}=-\sum\begin{vmatrix}\alpha\beta\epsilon\overline{\gamma}\\ \underline{\delta}\zeta\eta\theta\end{vmatrix}=\sum\begin{vmatrix}\alpha\beta\epsilon\overline{\zeta}\\ \gamma\delta\underline{\eta\theta}\end{vmatrix}.$$

It is to be observed that the $2m$ numbers are divided into three groups, the first containing x, the second y, and the third z of the $2m$ numbers, where $(x+y+z)=2m$ and in the first sigma the x numbers, in the second the y numbers, and in the third the z numbers vary while the others remain invariant in each case.

531. If $s+t=s'+t'=m$, then the general statement* of the relations in the last article may be expressed as follows:

$$\sum i \begin{vmatrix} (\overline{2m} \mid t' + s_{\alpha_1})(2m \mid \overline{t' + s_{\alpha_1}} \mid \overline{s_{\alpha_2}} \mid t_i) \\ (2m \mid t' + s_{\alpha_1} \mid s_{\alpha_2})(2m \mid \overline{t' + s_{\alpha_1}} \mid s_{\alpha_2} \mid t_i) \end{vmatrix}$$

$$= (-1)^t \sum i \begin{vmatrix} (2m \mid \overline{t' + s_{\alpha_1}} \mid s_{\alpha_2})(\overline{2m} \mid \overline{t' + s_{\alpha_1}} \mid t_i) \\ (2m \mid t' + s_{\alpha_1} \mid s_{\alpha_2})(\overline{2m} \mid t' + s_{\alpha_1} \mid t_i) \end{vmatrix}$$

$$= (-1)^{s-s'} \sum i \begin{vmatrix} (\overline{2m} \mid t' + s_{\alpha_1})(2m \mid \mid t' + s_{\alpha_1} \mid s_{\alpha_2} \mid s_i') \\ (2m \mid \overline{t' + s_{\alpha_1}} \mid s_{\alpha_2})(2m \mid t' + s_{\alpha_1} \mid s_{\alpha_2} \mid s_i') \end{vmatrix}$$

where the groups of the $2m$ numbers involved are

$$(2m \mid \overline{t' + s_{\alpha_1}} \mid s_{\alpha_2}), \quad (\overline{2m} \mid t' + s_{\alpha_1}), \quad (2m \mid t' + s_{\alpha_1} \mid s_{\alpha_2}).$$

532. In §316 we see that when the determinant is axisymmetric the left-hand side is a sum of Kronecker expressions, and denoting it by $(n-k)_g K(n, k)$, we may write the theorem schematically

$$(1) \qquad (n-k)_g K(n,k) = \sum K(h+g, h) K(n-g-h, k-h).$$

If in this $g=1$, then $K(h+g, h)$ is a Kronecker expression, and if in addition $k=n-2$, then $K(n-g-h, k-h)$ is a Kronecker expression so that under these conditions the theorem expresses a sum of Kronecker expressions for minors of order n as a sum of products of Kronecker expressions for minors of order $h+g$ and $n-h-g$ respectively. It is apparent that, by repeated applications of the theorem and by a proper choice of g, we may reduce both factors of each term on the right of (1) to Kronecker expressions.

If all the $2n$ numbers involved are not distinct then some of the terms will disappear on account of having identical rows or columns, and the sum on the left becomes an extensional as also some of those on the right.

Thus in the example

$$\sum \begin{vmatrix} \overline{12345} \\ \underline{6}789\tau \end{vmatrix} = \begin{vmatrix} 12 \\ 78 \end{vmatrix} \sum \begin{vmatrix} \overline{345} \\ \underline{6}9\tau \end{vmatrix} - \begin{vmatrix} 13 \\ 78 \end{vmatrix} \sum \begin{vmatrix} \overline{245} \\ \underline{6}9\tau \end{vmatrix} + \cdots$$

$$- \begin{vmatrix} 46 \\ 78 \end{vmatrix} \sum \begin{vmatrix} \overline{123} \\ \underline{5}9\tau \end{vmatrix} + \begin{vmatrix} 56 \\ 78 \end{vmatrix} \sum \begin{vmatrix} \overline{123} \\ \underline{4}9\tau \end{vmatrix}$$

* Proc. Royal Soc. Edin., vol. xxv, pt. v, p. 716.

if we put 9, $\tau=1$, 2 respectively then

$$\sum\begin{vmatrix}12\overline{345}\\ \underline{6}7812\end{vmatrix}=\begin{vmatrix}12\\78\end{vmatrix}\sum\begin{vmatrix}\overline{345}\\ \underline{6}12\end{vmatrix}-\begin{vmatrix}13\\78\end{vmatrix}\sum\begin{vmatrix}\overline{245}\\ \underline{6}12\end{vmatrix}+\cdots$$
$$-\begin{vmatrix}46\\78\end{vmatrix}\sum\begin{vmatrix}12\overline{3}\\ \underline{5}12\end{vmatrix}+\begin{vmatrix}56\\78\end{vmatrix}\sum\begin{vmatrix}12\overline{3}\\ \underline{4}12\end{vmatrix}.$$

If the minor $\left|{123456 \atop 123456}\right|$ is axisymmetric then the sums involved here all vanish since they are either Kronecker relations or the extensions of Kronecker relations.

533. We may obviously write

$$(1)\qquad \begin{vmatrix}a_1 & a_2 & a_3\\ b_1 & b_2 & b_3\\ c_1 & c_2 & c_3\end{vmatrix}=\begin{vmatrix}\cdot & a_2 & a_3\\ \cdot & b_2 & b_3\\ c_1 & \cdot & \cdot\end{vmatrix}+\begin{vmatrix}a_1 & \cdot & a_3\\ b_1 & \cdot & b_3\\ \cdot & c_2 & \cdot\end{vmatrix}+\begin{vmatrix}a_1 & a_2 & \cdot\\ b_1 & b_2 & \cdot\\ \cdot & \cdot & c_3\end{vmatrix}$$

and fill the empty spaces in the columns of the three determinants on the right with any numbers A, B and still have a true equation. It is also possible to fill the empty places in the third rows on the right and still have a true relation. Thus

$$\begin{vmatrix}a_1 & a_2 & a_3\\ b_1 & b_2 & b_3\\ c_1 & c_2 & c_3\end{vmatrix}=\begin{vmatrix}A & a_2 & a_3\\ B & b_2 & b_3\\ c_1 & \alpha & \beta\end{vmatrix}+\begin{vmatrix}a_1 & A & a_3\\ b_1 & B & b_3\\ \alpha & c_2 & \gamma\end{vmatrix}+\begin{vmatrix}a_1 & a_2 & A\\ b_1 & b_2 & B\\ \beta & \gamma & c_3\end{vmatrix}.$$

For the total coefficient of A is

$$\begin{vmatrix}b_2 & b_3\\ \alpha & \beta\end{vmatrix}-\begin{vmatrix}b_1 & b_3\\ \alpha & \gamma\end{vmatrix}+\begin{vmatrix}b_1 & b_2\\ \beta & \gamma\end{vmatrix}$$

and this is an example of Kronecker's theorem made from the array $b_1b_2b_3$ and the axisymmetric determinant

$$\begin{vmatrix}0 & \alpha & \beta\\ \alpha & 0 & \gamma\\ \beta & \gamma & 0\end{vmatrix}$$

by taking with the array the rows of the determinant one after another and deleting the column containing the zero in each case. It is therefore zero as may be otherwise seen. The same thing is true of the total coefficient of B. The coefficient of α, β, γ are seen to be zero and therefore the relation is true independent of the numbers introduced.

Instead of expanding in terms of the elements of a single row and their complementaries we might expand in terms of minors of order k formed from k rows and their complementaries and fill in the vacant places with numbers in such a way as to still have a true relation. Thus expanding $|a_{55}|$ in terms of minors of order 2 and their complementaries of order 3 and filling in the empty places we have

$$
\begin{aligned}
|a_{55}| = {} & \begin{vmatrix} a_{11} & a_{12} & P_1 & P_2 & P_3 \\ a_{21} & a_{22} & Q_1 & Q_2 & Q_3 \\ x_4 & x_5 & a_{33} & a_{34} & a_{35} \\ x_4 & x_5 & a_{43} & a_{44} & a_{45} \\ x_4 & x_5 & a_{53} & a_{54} & a_{55} \end{vmatrix} + \begin{vmatrix} a_{11} & P_1 & a_{13} & P_4 & P_5 \\ a_{21} & Q_1 & a_{23} & Q_4 & Q_5 \\ x_3 & a_{32} & x_5 & a_{34} & a_{35} \\ x_3 & a_{42} & x_5 & a_{44} & a_{45} \\ x_3 & a_{52} & x_5 & a_{54} & a_{55} \end{vmatrix} \\
& + \begin{vmatrix} a_{11} & P_1 & P_4 & a_{14} & P_6 \\ a_{21} & Q_1 & Q_4 & a_{24} & Q_6 \\ x_2 & a_{32} & a_{33} & x_5 & a_{35} \\ x_2 & a_{42} & a_{43} & x_5 & a_{45} \\ x_2 & a_{52} & a_{53} & x_5 & a_{55} \end{vmatrix} + \begin{vmatrix} a_{11} & P_3 & P_5 & P_6 & a_{15} \\ a_{21} & Q_3 & Q_5 & Q_6 & a_{25} \\ x_1 & a_{32} & a_{33} & a_{34} & x_5 \\ x_1 & a_{42} & a_{43} & a_{44} & x_5 \\ x_1 & a_{52} & a_{53} & a_{54} & x_5 \end{vmatrix} \\
& + \begin{vmatrix} P_1 & a_{12} & a_{13} & P_7 & P_8 \\ Q_1 & a_{22} & a_{23} & Q_7 & Q_8 \\ a_{31} & x_3 & x_4 & a_{34} & a_{35} \\ a_{41} & x_3 & x_4 & a_{44} & a_{45} \\ a_{51} & x_3 & x_4 & a_{54} & a_{55} \end{vmatrix} \quad \begin{vmatrix} P_2 & a_{12} & P_7 & a_{14} & P_8 \\ Q_2 & a_{22} & Q_7 & a_{24} & Q_8 \\ a_{31} & x_2 & a_{33} & x_4 & a_{35} \\ a_{41} & x_2 & a_{43} & x_4 & a_{45} \\ a_{51} & x_2 & a_{53} & x_4 & a_{55} \end{vmatrix} \\
& + \begin{vmatrix} P_3 & a_{12} & P_8 & P_9 & a_{15} \\ Q_3 & a_{22} & Q_8 & Q_9 & a_{25} \\ a_{31} & x_1 & a_{33} & a_{34} & x_4 \\ a_{41} & x_1 & a_{43} & a_{44} & x_4 \\ a_{51} & x_1 & a_{53} & a_{54} & x_4 \end{vmatrix} + \begin{vmatrix} P_4 & P_7 & a_{13} & a_{14} & P_{10} \\ Q_4 & Q_7 & a_{23} & a_{24} & Q_{10} \\ a_{31} & a_{32} & x_2 & x_3 & a_{35} \\ a_{41} & a_{42} & x_2 & x_3 & a_{45} \\ a_{51} & a_{52} & x_2 & x_3 & a_{55} \end{vmatrix} \\
& + \begin{vmatrix} P_5 & P_8 & a_{13} & P_{10} & a_{15} \\ Q_5 & Q_8 & a_{23} & Q_{10} & a_{25} \\ a_{31} & a_{32} & x_1 & a_{34} & x_3 \\ a_{41} & a_{42} & x_1 & a_{44} & x_3 \\ a_{51} & a_{52} & x_1 & a_{54} & x_3 \end{vmatrix} + \begin{vmatrix} P_6 & P_9 & P_{10} & a_{14} & a_{15} \\ Q_6 & Q_9 & Q_{10} & a_{24} & a_{25} \\ a_{31} & a_{32} & a_{33} & x_1 & x_2 \\ a_{41} & a_{42} & a_{43} & x_1 & x_2 \\ a_{51} & a_{52} & a_{53} & x_1 & x_2 \end{vmatrix}
\end{aligned}
\tag{2}
$$

The truth of this is seen by showing that the total coefficient of any of the introduced numbers vanishes. Thus the coefficient of x_1 is seen to be

$$\begin{vmatrix} P_3 & P_5 & P_6 & a_{15} \\ Q_3 & Q_5 & Q_6 & a_{25} \\ \alpha_2 & \alpha_3 & \alpha_4 & 0 \\ \beta_3 & \beta_3 & \beta_4 & 0 \end{vmatrix} - \begin{vmatrix} P_3 & P_8 & P_9 & a_{15} \\ Q_3 & Q_8 & Q_9 & a_{25} \\ \alpha_1 & \alpha_3 & \alpha_4 & 0 \\ \beta_1 & \beta_3 & \beta_4 & 0 \end{vmatrix} + \begin{vmatrix} P_5 & P_8 & P_{10} & a_{15} \\ Q_5 & Q_8 & Q_{10} & a_{25} \\ \alpha_1 & \alpha_2 & \alpha_4 & 0 \\ \beta_1 & \beta_2 & \beta_4 & 0 \end{vmatrix} - \begin{vmatrix} P_6 & P_9 & P_{10} & a_{15} \\ Q_6 & Q_9 & Q_{10} & a_{25} \\ \alpha_1 & \alpha_2 & \alpha_3 & 0 \\ \beta_1 & \beta_2 & \beta_3 & 0 \end{vmatrix}$$

where

$$\alpha_1 = a_{31} - a_{41}, \quad \alpha_2 = a_{32} - a_{42}, \quad \alpha_3 = a_{33} - a_{43}, \quad \alpha_4 = a_{34} - a_{44}$$

$$\beta_1 = a_{41} - a_{51}, \quad \beta_2 = a_{42} - a_{52}, \quad \beta_3 = a_{43} - a_{53}, \quad \beta_4 = a_{44} - a_{54}.$$

The coefficient of a_{25} in this is

$$\begin{vmatrix} P_3 & P_5 & P_6 \\ \alpha_2 & \alpha_3 & \alpha_4 \\ \beta_2 & \beta_3 & \beta_4 \end{vmatrix} - \begin{vmatrix} P_3 & P_8 & P_7 \\ \alpha_1 & \alpha_3 & \alpha_4 \\ \beta_1 & \beta_3 & \beta_4 \end{vmatrix} + \begin{vmatrix} P_5 & P_8 & P_{10} \\ \alpha_1 & \alpha_2 & \alpha_4 \\ \beta_1 & \beta_2 & \beta_4 \end{vmatrix} - \begin{vmatrix} P_6 & P_9 & P_{10} \\ \alpha_1 & \alpha_2 & \alpha_3 \\ \beta_1 & \beta_2 & \beta_3 \end{vmatrix}$$

which is seen to be a case of Kronecker's theorem formed by combining with the array

$$\begin{matrix} \alpha_1 & \alpha_2 & \alpha_3 & \alpha_4 \\ \beta_1 & \beta_2 & \beta_3 & \beta_4 \end{matrix}$$

the rows of the zero-axial symmetric determinant

$$\begin{vmatrix} 0 & P_3 & P_5 & P_6 \\ P_3 & 0 & P_8 & P_9 \\ P_5 & P_8 & 0 & P_{10} \\ P_6 & P_9 & P_{10} & 0 \end{vmatrix}$$

one after another and striking out the column containing the zero in each case. The coefficient of a_{25} is therefore zero. Similarly it is seen that the coefficient of a_{15} is zero and consequently the coefficient of x_1 is zero. Similarly the coefficients of any of the x's is seen to be zero.

The total coefficient of P_1 is

$$\begin{vmatrix} a_{21} & a_{22} & Q_2 & Q_3 \\ x_4 & x_5 & a_{34} & a_{35} \\ x_4 & x_5 & a_{44} & a_{45} \\ x_4 & x_5 & a_{54} & a_{55} \end{vmatrix} - \begin{vmatrix} a_{21} & a_{23} & Q_4 & Q_5 \\ x_3 & x_5 & a_{34} & a_{35} \\ x_3 & x_5 & a_{44} & a_{45} \\ x_3 & x_5 & a_{54} & a_{55} \end{vmatrix} + \begin{vmatrix} a_{22} & a_{23} & Q_7 & Q_6 \\ x_3 & x_4 & a_{34} & a_{35} \\ x_3 & x_4 & a_{44} & a_{45} \\ x_3 & x_4 & a_{54} & a_{55} \end{vmatrix}$$

which is readily seen to be zero, and similarly for the coefficient of any other P. What is true of the coefficients of the P's is also true for the coefficients of the Q's. It follows therefore that the relation is not altered by inserting the $5+2\cdot 10=25$ new numbers.

534. To see how the new numbers are introduced it is best to write the right-hand side in diagrammatic form. Using $|a_{nn}|_{(n-k)\cdot k}$ to denote the expansion of the determinant of the nth order in terms of minors of order k and their complementaries of order $n-k$, the number of terms in $|a_{55}|_{3\cdot 2}$ would be $(5)_2=10$, and if we use **45** 123 to denote

$$\begin{vmatrix} a_{11} & a_{12} & P_1 & P_2 & P_3 \\ a_{21} & a_{22} & Q_1 & Q_2 & Q_3 \\ x_4 & x_5 & a_{33} & a_{34} & a_{35} \\ x_4 & x_5 & a_{43} & a_{44} & a_{45} \\ x_4 & x_5 & a_{53} & a_{54} & a_{55} \end{vmatrix}$$

where the numbers in heavy type are the subscripts of the x's and those in light type are the subscripts of the P's and Q's. The ten determinants on the right of (2) would then be expressed in the following diagram:

4	**5**	1	2	3
3	1	**5**	4	5
2	2	4	**5**	6
1	3	5	6	**5**
1	**3**	**4**	7	8
2	**2**	7	**4**	9
3	**1**	8	9	**4**
4	7	**2**	**3**	10
5	8	**1**	10	**3**
6	9	10	**1**	**2**

This not only gives the subscripts of the P's, Q's and x's, but locates them and thereby determines the a's for that particular term.

The right-hand side of $|a_{77}|_{4\cdot 3}$ is given in the diagram

15	**20**	**21**	1	2	3	4
14	**19**	1	**21**	5	6	7
13	**18**	2	5	**21**	8	9
12	**17**	3	6	8	**21**	10
11	**16**	4	7	9	10	**21**
10	1	**19**	**20**	11	12	13
9	2	**18**	11	**20**	14	15
8	3	**17**	12	14	**20**	16
7	4	**16**	13	15	16	**20**
6	5	11	**18**	**19**	17	18
5	6	12	**17**	17	**19**	19
4	7	13	**16**	18	19	**19**
3	8	14	17	**17**	**18**	20
2	9	15	18	**16**	20	**18**
1	10	16	19	20	**16**	**17**
1	**10**	**14**	**15**	21	22	23
2	**9**	**13**	21	**15**	24	25
3	**8**	**12**	22	24	**15**	26
4	**7**	**11**	23	25	26	**15**
5	**6**	21	**13**	**14**	27	28
6	**5**	22	**12**	27	**14**	29
7	**4**	23	**11**	28	29	**14**
8	**3**	24	27	**12**	**13**	30
9	**2**	25	28	**11**	30	**13**
10	**1**	26	29	30	**11**	**12**
11	21	**6**	**9**	**10**	31	32
12	22	**5**	**8**	31	**10**	33
13	23	**4**	**7**	32	33	**10**
14	24	**3**	31	**8**	**9**	34
15	25	**2**	32	**7**	34	**9**
16	26	**1**	33	34	**7**	**8**
17	27	31	**3**	**5**	**6**	35
18	28	32	**2**	**4**	35	**6**
19	29	33	**1**	35	**4**	**5**
20	30	34	35	**1**	**2**	**3**

The law of formation of these is perhaps apparent but the following observations for the general case may be made as a guide.

1) The number of determinants represented in the diagram is $(n)_k$.
2) The number of distinct x's introduced is $(n)_{k-1}$.
3) The number of distinct P's introduced is $(n)_{n-k-1}=(n)_{k+1}$.
4) The total number of new numbers introduced is $(n)_{k-1}+k(n)_{k+1}$.
5) The number of times x_i occurs is $k(n)_k/(n)_{k-1}=(n-k+1)$.
6) The number of times P_i occurs is $(n-k)(n)_k/(n)_{n-k-1}=k+1$.
7) The x's involved in the total coefficient of any P form by themselves a zero-axial symmetric determinant. Thus in $|a_{77}|_{4\cdot 3}$ the total coefficient of P_{35} involves

$$\begin{vmatrix} x_3 & x_5 & x_6 & 0 \\ x_2 & x_4 & 0 & x_6 \\ x_1 & 0 & x_4 & x_5 \\ 0 & x_1 & x_2 & x_3 \end{vmatrix}$$

and the coefficient of P_1 involves

$$\begin{vmatrix} x_{15} & x_{20} & x_{21} & 0 \\ x_{14} & x_{19} & 0 & x_{21} \\ x_{10} & 0 & x_{19} & x_{20} \\ 0 & x_{10} & x_{14} & x_{15} \end{vmatrix}.$$

Similarly for the coefficients of the other P's.

8) In the case of $|a_{nn}|_{(n-k)\cdot k}$, using d's to represent the suffixes of the x's, the x's involved in the total coefficient of P_1 are those (using subscripts only) of the determinant

$$\begin{vmatrix} d_{11} & d_{12} & \cdots & d_{1k} & 0 \\ d_{21} & d_{22} & \cdots & 0 & d_{2,k+1} \\ \cdot & \cdot & \cdots\cdot & & \cdot \\ \cdot & \cdot & \cdots\cdot & & \cdot \\ 0 & d_{k+1,2} & \cdots & \cdot & d_{k+1,k+1} \end{vmatrix}$$

where

$$d_{rs} = d_{k+2-s,k+2-r},$$

$$d_{1h} = (n-1)_{k-1} + (n-2)_{k-2} \cdots + (n-h)_{k-h}, \quad (h = 1,2, \cdots k),$$

$$\begin{aligned} d_{gh} &= d_{1h} - (n-k+g-2)_{g-2}, \\ &= 0 \text{ when } g+h = k+2. \end{aligned} \qquad \begin{Bmatrix} g = 1,2, \cdots k \\ g+h \leqq k+1 \end{Bmatrix}$$

9) Similarly the P's involved in the total coefficient of any x form by themselves a similar determinant. Thus the coefficient of x_{19} in $|a_{77}|_{4\cdot 3}$ involves the P's of the determinant

$$\begin{vmatrix} 0 & P_1 & P_5 & P_6 & P_7 \\ P_1 & 0 & P_{11} & P_{12} & P_{13} \\ P_5 & P_{11} & 0 & P_{17} & P_{18} \\ P_6 & P_{12} & P_{17} & 0 & P_{19} \\ P_7 & P_{13} & P_{18} & P_{19} & 0 \end{vmatrix}$$

10) In the case of $|a_{nn}|_{(n-k)\cdot k}$, using c's to represent the suffixes of the P's, the P's involved in the total coefficient of x_1 are those (using subscripts only) of the determinant

$$\begin{vmatrix} 0 & c_{12} & c_{13} & \cdots & c_{1,n-k+1} \\ c_{21} & 0 & c_{23} & \cdots & c_{2,n-k+1} \\ \cdot & \cdot & \cdot & \cdots & \cdot \\ \cdot & \cdot & \cdot & \cdots & \cdot \\ c_{n-k+1,1} & c_{n-k+1,2} & c_{n-k+1,3} & \cdots & 0 \end{vmatrix}$$

where

$$c_{rs} = c_{sr},$$

$$c_{rr} = c_{ss} = 0,$$

$$c_{1h} = (n-2)_{n-k-1} + (n-3)_{n-k-2} + \cdots + (n-h)_{n-k-h+1}, \quad (h = 2,3, \cdots, n-k+1),$$

$$c_{gh} = c_{g-1,h}(n-q)_{n-k-g} \quad (g = 2,3, \cdots, k).$$

Similarly for the coefficient of any x.

535. The general theorem, the proof of which would proceed on precisely the same lines, is that, if we expand the determinant $|a_{nn}|_{(n-k)\cdot k}$ as in (1) and then fill the vacant places by new numbers according to the plan indicated by the diagrams and the observations 1)–10) of §534, the relation remains true.

536. That the relation

$$\begin{vmatrix} a_{11} & a_{12} & a_{13} & a_{14} \\ a_{21} & a_{22} & a_{23} & a_{24} \\ a_{31} & a_{32} & a_{33} & a_{34} \\ a_{41} & a_{42} & a_{43} & a_{44} \end{vmatrix} = \begin{vmatrix} a_{11} & s & t & u \\ x & a_{22} & a_{23} & a_{24} \\ y & a_{32} & a_{33} & a_{34} \\ z & a_{42} & a_{43} & a_{44} \end{vmatrix} + \begin{vmatrix} s & a_{12} & r & q \\ a_{21} & x & a_{23} & a_{24} \\ a_{31} & y & a_{33} & a_{34} \\ a_{41} & z & a_{43} & a_{44} \end{vmatrix}$$

$$(1) \qquad + \begin{vmatrix} t & r & a_{13} & p \\ a_{21} & a_{22} & x & a_{24} \\ a_{31} & a_{32} & y & a_{34} \\ a_{41} & a_{42} & z & a_{44} \end{vmatrix} + \begin{vmatrix} u & q & p & a_{14} \\ a_{21} & a_{22} & a_{23} & x \\ a_{31} & a_{32} & a_{33} & y \\ a_{41} & a_{42} & a_{43} & z \end{vmatrix}.$$

is true as being an example of Kronecker's theorem may be seen on taking the sum of the determinants resulting from taking with the array

$$\begin{matrix} x & a_{21} & a_{22} & a_{23} & a_{24} \\ y & a_{31} & a_{32} & a_{33} & a_{34} \\ z & a_{41} & a_{42} & a_{43} & a_{44} \end{matrix}$$

the rows, one after another, of the determinant

$$\begin{vmatrix} 0 & a_{11} & a_{12} & a_{13} & a_{14} \\ a_{11} & 0 & s & t & u \\ a_{12} & s & 0 & p & q \\ a_{13} & t & p & 0 & r \\ a_{14} & u & q & r & 0 \end{vmatrix}$$

and striking out the column containing the zero in each case.

Again, expanding $|a_{44}|$ in terms of the elements of the first column and their complementaries and filling in similarly as in (1) we get

$$\begin{vmatrix} a_{11} & a_{12} & a_{13} & a_{14} \\ a_{21} & a_{22} & a_{23} & a_{24} \\ a_{31} & a_{32} & a_{33} & a_{34} \\ a_{41} & a_{42} & a_{43} & a_{44} \end{vmatrix} = \begin{vmatrix} a_{11} & x' & y' & z' \\ l & a_{22} & a_{23} & a_{24} \\ m & a_{32} & a_{33} & a_{34} \\ n & a_{42} & a_{43} & a_{44} \end{vmatrix} + \begin{vmatrix} l & a_{12} & a_{13} & a_{14} \\ a_{21} & x' & y' & z' \\ \gamma & a_{32} & a_{33} & a_{34} \\ \beta & a_{42} & a_{43} & a_{44} \end{vmatrix}$$

$$(2) \qquad + \begin{vmatrix} m & a_{12} & a_{13} & a_{14} \\ \gamma & a_{22} & a_{23} & a_{24} \\ a_{31} & x' & y' & z' \\ \alpha & a_{42} & a_{43} & a_{44} \end{vmatrix} + \begin{vmatrix} n & a_{12} & a_{13} & a_{14} \\ \beta & a_{22} & a_{23} & a_{24} \\ \alpha & a_{32} & a_{23} & a_{34} \\ a_{41} & x' & y' & z' \end{vmatrix}.$$

Equating the right-hand sides of (1) and (2) we get

$$\begin{vmatrix} a_{11} & s & t & u \\ x & a_{22} & a_{23} & a_{24} \\ y & a_{32} & a_{33} & a_{34} \\ z & a_{42} & a_{43} & a_{44} \end{vmatrix} + \begin{vmatrix} s & a_{12} & r & q \\ a_{21} & x & a_{23} & a_{24} \\ a_{31} & y & a_{33} & a_{34} \\ a_{41} & z & a_{43} & a_{44} \end{vmatrix} + \begin{vmatrix} t & r & a_{13} & p \\ a_{21} & a_{22} & x & a_{24} \\ a_{31} & a_{32} & y & a_{34} \\ a_{41} & a_{42} & z & a_{44} \end{vmatrix}$$

$$
(3) + \begin{vmatrix} u & q & p & a_{14} \\ a_{21} & a_{22} & a_{23} & x \\ a_{31} & a_{32} & a_{33} & y \\ a_{41} & a_{42} & a_{43} & z \end{vmatrix} = \begin{vmatrix} a_{11} & x' & y' & z' \\ l & a_{22} & a_{23} & a_{24} \\ m & a_{32} & a_{33} & a_{34} \\ n & a_{42} & a_{43} & a_{44} \end{vmatrix} + \begin{vmatrix} l & a_{12} & a_{13} & a_{14} \\ a_{21} & x' & y' & z' \\ \gamma & a_{32} & a_{33} & a_{34} \\ \beta & a_{42} & a_{43} & a_{44} \end{vmatrix}
$$

$$
+ \begin{vmatrix} m & a_{12} & a_{13} & a_{14} \\ \gamma & a_{22} & a_{23} & a_{24} \\ a_{31} & x' & y' & z' \\ \alpha & a_{42} & a_{43} & a_{44} \end{vmatrix} + \begin{vmatrix} n & a_{12} & a_{13} & a_{14} \\ \beta & a_{22} & a_{23} & a_{24} \\ \alpha & a_{32} & a_{33} & a_{34} \\ a_{41} & x' & y' & z' \end{vmatrix}
$$

a relation between eight determinants of order four containing 34 letters, which should be compared with Muir's similar relation* containing 31 letters.

If in (3) $x', y', z', = s, t, u$, and $l, m, n = x, y, z$ respectively, then the first determinant on the left cancels with the first on the right and we have

$$
(4) \quad \begin{vmatrix} s & a_{12} & r & q \\ a_{21} & x & a_{23} & a_{24} \\ a_{31} & y & a_{33} & a_{34} \\ a_{41} & z & a_{43} & a_{44} \end{vmatrix} + \begin{vmatrix} t & r & a_{13} & p \\ a_{21} & a_{22} & x & a_{24} \\ a_{31} & a_{32} & y & a_{34} \\ a_{41} & a_{42} & z & a_{44} \end{vmatrix} + \begin{vmatrix} u & q & p & a_{14} \\ a_{21} & a_{22} & a_{23} & x \\ a_{31} & a_{32} & a_{33} & y \\ a_{41} & a_{42} & a_{43} & z \end{vmatrix}
$$

$$
= \begin{vmatrix} x & a_{12} & a_{13} & a_{14} \\ a_{21} & s & t & u \\ \gamma & a_{32} & a_{33} & a_{34} \\ \alpha & a_{42} & a_{43} & a_{44} \end{vmatrix} + \begin{vmatrix} y & a_{12} & a_{13} & a_{14} \\ \gamma & a_{22} & a_{23} & a_{24} \\ a_{31} & s & t & u \\ \alpha & a_{42} & a_{43} & a_{44} \end{vmatrix} = \begin{vmatrix} z & a_{12} & a_{13} & a_{14} \\ \beta & a_{22} & a_{23} & a_{24} \\ \alpha & a_{32} & a_{33} & a_{34} \\ a_{41} & s & t & u \end{vmatrix}
$$

a relation involving six determinants and 27 letters which should be compared with that obtained from equating the two Kronecker relations

$$
\begin{vmatrix} 1\,2\,3\,\bar{4} \\ \underline{5\,6\,7\,8} \end{vmatrix} = 0 = \begin{vmatrix} \overline{1\,2\,3\,4} \\ \underline{5}\,6\,7\,8 \end{vmatrix}
$$

which involves six determinants and 28 letters.

* Muir: Cayley's Linear Relation between Minors of a Special Three-Row Array, Mess. Math., new ser., No. 482, June 1911.

If in (3) we put $x', y', z' = x, y, z$ and $l, m, n = s, t, u$ respectively and make $|a_{22}\, a_{33}\, a_{44}|$ symmetric then the first term on the left cancels with the first term on the right and we have

$$(5)\quad \begin{vmatrix} s & a_{12} & r & q \\ a_{21} & x & a_{23} & a_{24} \\ a_{31} & y & a_{33} & a_{34} \\ a_{41} & z & a_{43} & a_{44} \end{vmatrix} + \begin{vmatrix} t & r & a_{13} & p \\ a_{21} & a_{22} & x & a_{24} \\ a_{31} & a_{32} & y & a_{34} \\ a_{41} & a_{42} & z & a_{44} \end{vmatrix} + \begin{vmatrix} u & q & p & a_{14} \\ a_{21} & a_{22} & a_{23} & x \\ a_{31} & a_{32} & a_{33} & y \\ a_{41} & a_{42} & a_{43} & z \end{vmatrix}$$

$$= \begin{vmatrix} s & a_{12} & a_{13} & a_{14} \\ a_{21} & x & y & z \\ \gamma & a_{32} & a_{33} & a_{34} \\ \beta & a_{42} & a_{43} & a_{44} \end{vmatrix} + \begin{vmatrix} t & a_{12} & a_{13} & a_{14} \\ \gamma & a_{22} & a_{23} & a_{24} \\ a_{31} & x & y & z \\ \alpha & a_{42} & a_{43} & a_{44} \end{vmatrix} + \begin{vmatrix} u & a_{12} & a_{13} & a_{14} \\ \beta & a_{22} & a_{23} & a_{24} \\ \alpha & a_{32} & a_{33} & a_{34} \\ a_{41} & x & y & z \end{vmatrix}$$

which is a relation involving 6 determinants and 24 letters.

537. In general if we expand the determinant $|a_{nn}|$ in terms of minors of order k formed from the first k rows and their complementaries, filling in the zero places as in §534 and then expand in terms of minors of order k formed from the first k columns and their complementaries, filling in as before with different letters, we may equate these two expansions and have a relation between $2(n)_k$ determinants. If we introduce the same letters in the first term of the second expansion as were introduced in the first term of the first expansion except that they be placed in the conjugate positions, then these two will cancel provided $|a_{11} \cdots a_{kk}|$ and $|a_{k+1,k+1} \cdots a_{nn}|$ are both symmetric and we have a relation between $2(n)_k - 2$ determinants.

(d) Persymmetry

538. Since a persymmetric is a particular case of an axisymmetric determinant it follows that every type of vanishing aggregate for axisymmetric determinants is also a vanishing aggregate for persymmetric determinants.

Using the umbral notation we have

(1) *Given any identical relation between the minors of order m of a persymmetric determinant of order n, we may obtain another identity by increasing every row number by α and every column number by β.*

As this has the effect of increasing the subscripts of every element in each minor of the identity by $\alpha+\beta$, the truth of the theorem is apparent.

(2) *Given any identical relation between the minors of order m of a persymmetric determinant of order n, where k of the row numbers and h of the column numbers are invariant* $(k+h<n)$, *we may obtain another identity by increasing each of the other* $m-k$ *row numbers by* α *and each of the other* $m-h$ *column numbers by* β.

The original identity in this case is the extensional of another having either no column numbers or no row numbers invariant according as $h \lesseqgtr k$.

Let $k-h=g$ and starting with the unextended form we see that if we expand by LaPlace's theorem each minor of the identity in terms of minors of order g and their complementaries, formed from the g invariant rows, then since these complementaries are minors of a persymmetric determinant, it follows from §526 that the total coefficient of any minor of order g, formed from the g invariant rows vanishes, which establishes the theorem for the unextended form and hence for the extended form.

(3) *In* (2) *we might have increased the invariant numbers instead of the variant ones.*

For by (1) we might increase all the row-numbers by α and all the column-numbers by β and then decrease the variant row-numbers by α and the variant column numbers by β.

Since the truth of such identities does not depend upon the invariant rows or columns it appears we may increase part of these invariant numbers without increasing all of them and still have an identity.

In these theorems certain obvious restrictions rest upon α and β as for instance in (1) the largest row number plus $\alpha \ngtr n$. It is also apparent that α and β under these restrictions may have any values, positive, zero, or negative.

It is also apparent that we may add the same positive or negative integer to the row-numbers of any individual minor provided we subtract it from the column-numbers as this does not change the minor. Again we may write column-numbers for row-numbers and vice-versa.

539. If now we apply these theorems to vanishing aggregates for axisymmetric determinants we get results true for persymmetric determinants, but no longer true for axisymmetric determinants.

Thus starting with the Kronecker relation

$$\sum \begin{vmatrix} a & \bar{k} \\ \underline{h} & \underline{1} \end{vmatrix} \equiv \begin{vmatrix} a & k \\ h & 1 \end{vmatrix} - \begin{vmatrix} a & h \\ k & 1 \end{vmatrix} + \begin{vmatrix} a & 1 \\ k & h \end{vmatrix} = 0$$

and that obtained from it by replacing each minor by its complementary

$$\begin{bmatrix} a & k \\ h & 1 \end{bmatrix} - \begin{bmatrix} a & h \\ k & 1 \end{bmatrix} + \begin{bmatrix} a & 1 \\ k & h \end{bmatrix} = 0$$

and applying to these theorem (1) we have

$$\begin{vmatrix} a & k+\alpha \\ h+\beta & 1+\beta \end{vmatrix} - \begin{vmatrix} a & h+\alpha \\ k+\beta & 1+\beta \end{vmatrix} + \begin{vmatrix} a & 1+\alpha \\ k+\beta & h+\beta \end{vmatrix} = 0$$

and

$$\begin{bmatrix} a & k+\alpha \\ h+\beta & 1+\beta \end{bmatrix} - \begin{bmatrix} a & h+\alpha \\ k+\beta & 1+\beta \end{bmatrix} + \begin{bmatrix} a & 1+\alpha \\ k+\beta & h+\beta \end{bmatrix} = 0$$

Applying theorem (2) to

$$\sum \begin{vmatrix} 1\overline{23} \\ \underline{456} \end{vmatrix} \equiv \begin{vmatrix} 123 \\ 456 \end{vmatrix} - \begin{vmatrix} 125 \\ 436 \end{vmatrix} + \begin{vmatrix} 126 \\ 435 \end{vmatrix} + \begin{vmatrix} 135 \\ 426 \end{vmatrix} - \begin{vmatrix} 136 \\ 425 \end{vmatrix} + \begin{vmatrix} 156 \\ 423 \end{vmatrix} = 0$$

Making $\alpha=\beta=1$ we get

$$\begin{vmatrix} 134 \\ 467 \end{vmatrix} - \begin{vmatrix} 136 \\ 447 \end{vmatrix} + \begin{vmatrix} 137 \\ 446 \end{vmatrix} + \begin{vmatrix} 146 \\ 437 \end{vmatrix} - \begin{vmatrix} 147 \\ 436 \end{vmatrix} + \begin{vmatrix} 167 \\ 434 \end{vmatrix}$$

or

$$\begin{vmatrix} 134 \\ 467 \end{vmatrix} + \begin{vmatrix} 146 \\ 437 \end{vmatrix} - \begin{vmatrix} 147 \\ 436 \end{vmatrix} = 0, \text{ or } \sum \begin{vmatrix} 14\overline{3} \\ \underline{467} \end{vmatrix} = 0 = \sum \begin{vmatrix} 1\overline{3} \\ \underline{67} \end{vmatrix}.$$

If we apply theorem (2) with $k=0$, $\alpha=0$, $\beta=-s$ to the Kronecker relation

$$\sum \begin{vmatrix} \overline{b_1 \quad b_2 \quad b_3} \\ \underline{c_1} \quad d_1 \quad d_2 \end{vmatrix} \equiv \begin{vmatrix} b_1 & b_2 & b_3 \\ c_1 & d_1 & d_2 \end{vmatrix} - \begin{vmatrix} b_1 & b_2 & c_1 \\ b_3 & d_1 & d_2 \end{vmatrix}$$
$$+ \begin{vmatrix} b_1 & b_3 & c_1 \\ b_2 & d_1 & d_2 \end{vmatrix} - \begin{vmatrix} b_2 & b_3 & c_1 \\ b_1 & d_1 & d_2 \end{vmatrix} = 0$$

we have

$$\begin{vmatrix} b_1 & b_2 & b_3 \\ c_1-s & d_1 & d_2 \end{vmatrix} - \begin{vmatrix} b_1 & b_2 & c_1 \\ b_3-s & d_1 & d_2 \end{vmatrix} + \begin{vmatrix} b_1 & b_3 & c_1 \\ b_2-s & d_1 & d_2 \end{vmatrix} - \begin{vmatrix} b_2 & b_3 & c_1 \\ b_1-s & d_1 & d_2 \end{vmatrix} = 0$$

or replacing for convenience $c_1 - s$ by c_1

$$(1)\quad \begin{vmatrix} b_1 & b_2 & b_3 \\ c_1 & d_1 & d_2 \end{vmatrix} - \begin{vmatrix} b_1 & b_2 & c_1+s \\ b_3-s & d_1 & d_2 \end{vmatrix} + \begin{vmatrix} b_1 & b_3 & c_1+s \\ b_2-s & d_1 & d_2 \end{vmatrix} - \begin{vmatrix} b_2 & b_3 & c_1+s \\ b_1-s & d_1 & d_2 \end{vmatrix} = 0$$

Giving in this s successively the value s_1 and s_2 and subtracting the results we have

$$(2)\quad \begin{aligned} &\begin{vmatrix} b_1 & b_2 & c_1-s_1 \\ b_3-s_1 & d_1 & d_2 \end{vmatrix} - \begin{vmatrix} b_1 & b_3 & c_1+s_1 \\ b_2-s_1 & d_1 & d_2 \end{vmatrix} + \begin{vmatrix} b_2 & b_3 & c_1+s_1 \\ b_1-s_1 & d_1 & d_2 \end{vmatrix} \\ -&\begin{vmatrix} b_1 & b_2 & c_1+s_2 \\ b_3-s_2 & d_1 & d_2 \end{vmatrix} + \begin{vmatrix} b_1 & b_3 & c_1+s_2 \\ b_2-s_2 & d_1 & d_2 \end{vmatrix} - \begin{vmatrix} b_2 & b_3 & c_1+s_2 \\ b_1-s_2 & d_1 & d_2 \end{vmatrix} = 0 \end{aligned}$$

Remembering that elements are equal when the sum of the row- and column-numbers of one is the same as that of the other, we see that equation (2) contains 19 variables, six occuring four times, four occuring three times and nine occuring twice.

If either $c_1 + s_1 = b_1$ or b_2 or b_3, or if either d_1 or $d_2 = b_1 - s_2$ or $b_2 - s_2$ or $b_3 - s_2$, then one or more of these terms vanish. Thus if $b_2 - s_1 = b_1 - s_2 = d_2$, then the 2nd and 6th vanish, and the relation between the remaining four may be written as follows:

$$(3)\quad \begin{vmatrix} n & a & b \\ f & c & d \\ m & g & h \end{vmatrix} + \begin{vmatrix} b & c & d \\ n & e & f \\ l & g & h \end{vmatrix} - \begin{vmatrix} f & a & b \\ o & c & d \\ m & k & l \end{vmatrix} + \begin{vmatrix} d & a & b \\ o & e & f \\ h & k & l \end{vmatrix} = 0$$

where the elements in the positions b_1d_1, b_1d_2, b_2d_1, b_2d_2, b_3d_1, b_3d_2, $(c_1+s_1)d_1$, $(c_1+s_1)d_2$, $(c_1+s_2)d_1$, $(c_1+s_2)d_2$, $(c_1+s_1)(b_3-s_1)$, $b_1(b_3-s_1)$, $b_2(b_3-s_2)$ are a, b, c, d, e, f, g, h, k, l, m, n, o, respectively.

EXAMPLE: Show that

$$\begin{vmatrix} ka & kb & \omega & \theta_2 \\ b & c & d & b' \\ c & D & e & c' \\ \beta & \gamma & \delta & b'' \end{vmatrix} - \begin{vmatrix} kb & kD & ke & \theta_3 \\ a & c & d & a' \\ b & D & e & b' \\ \alpha & \gamma & \delta & a'' \end{vmatrix} + \begin{vmatrix} ka & kb & kd & \theta_1 \\ a & b & d & c' \\ b & c & e & d' \\ \alpha & \beta & \delta & c'' \end{vmatrix}$$

$$- \begin{vmatrix} \omega & kd & ke & \theta_4 \\ a & b & c & b' \\ b & c & D & c' \\ \alpha & \beta & \gamma & \delta \end{vmatrix} + \begin{vmatrix} \theta_2 & \theta_3 & \theta_1 & \theta_4 \\ a & b & c & d \\ b & c & d & e \\ \alpha & \beta & \gamma & \delta \end{vmatrix} = 0$$

(Muir, Metzler and Rice.)

(e) Circulants

540. Starting with the determinant

$$\begin{vmatrix} b & c+a & d+e & 1 \\ c & d+b & e+a & 1 \\ d & e+c & a+b & 1 \\ e & a+d & b+c & 1 \end{vmatrix}$$

which obviously vanishes since two columns may be made identical, we have for $C(abcde)$

$$(1) \quad \begin{vmatrix} b & c & d & 1 \\ c & d & e & 1 \\ d & e & a & 1 \\ e & a & b & 1 \end{vmatrix} + \begin{vmatrix} b & c & e & 1 \\ c & d & a & 1 \\ d & e & b & 1 \\ e & a & c & 1 \end{vmatrix} + \begin{vmatrix} b & a & d & 1 \\ c & b & e & 1 \\ d & c & a & 1 \\ e & d & b & 1 \end{vmatrix} + \begin{vmatrix} b & a & e & 1 \\ c & b & a & 1 \\ d & c & b & 1 \\ e & d & c & 1 \end{vmatrix} = 0$$

or $[1, 4]_5 - [1, 5]_3 + [1, 2]_4 - [1, 3]_2 = 0$, where $[p, q]_r$ denotes the complementary minor of the element in the pth row and qth column of the circulant after the rth column has been replaced by units.

Making use of the properties given in §484 equation (1) takes the form

$$[2,2]_1 + [3,3]_1 + [4,4]_1 + [5,5]_1 = 0.$$

If we had started with

$$\begin{vmatrix} b & c+e+a & d & 1 \\ c & d+a+b & e & 1 \\ d & e+b+c & a & 1 \\ e & a+c+d & b & 1 \end{vmatrix}$$

which is also equal to zero, we would have

$$(2) \qquad [1,4]_5 - [1,5]_2 + [1,2]_4 = 0.$$

Similarly

$$(3) \qquad [1,3]_5 + [1,5]_2 - [1,2]_3 = 0.$$

From (2) and (3) we get

$$[1,3]_5 - [1,2]_3 + [1,4]_5 + [1,2]_4 = 0$$

or

$$[3,3]_1 + [2,2]_1 + [5,5]_1 + [4,4]_1 = 0$$

which is (1). Again

$$[1,1]_2 + [3,3]_2 + [4,4]_2 + [5,5]_2 = 0,$$

and since

$$[5,5]_1 = -[3,3]_2,$$

we have

(4) $[1,1]_2 + [4,4]_2 + [5,5]_2 + [2,2]_1 + [3,3]_1 + [4,4]_1 = 0$

The general law for (2) is, in the case of circulants of odd order

(5) $(-1)^{r-1}[1,r]_s + (-1)^{s-1}[1,s]_t + (-1)^{t-1}[1,t]_r = 0.$

The general law for (1) is

(6) $(-1)^{r-1}[1,r]_s + (-1)^{s-1}[1,s]_t + (-1)^{t-1}[1,t]_u + (-1)^{u-1}[1,n]_r = 0$

but as has been seen (6) is made up of two of (5).

The method here used will give vanishing aggregates of any order. Thus from

$$\begin{vmatrix} 1 & a+b+c+d & e \\ 1 & b+c+d+e & a \\ 1 & c+d+e+a & b \end{vmatrix} = 0$$

we have, for the same circulant $C(abcde)$,

$$\begin{vmatrix} 1 & a & e \\ 1 & b & a \\ 1 & c & b \end{vmatrix} + \begin{vmatrix} 1 & b & e \\ 1 & c & a \\ 1 & d & b \end{vmatrix} + \begin{vmatrix} 1 & c & e \\ 1 & d & a \\ 1 & e & b \end{vmatrix} + \begin{vmatrix} 1 & d & e \\ 1 & e & a \\ 1 & a & b \end{vmatrix} = 0,$$

or

$$-\begin{bmatrix} 45 \\ 23 \end{bmatrix}_4 + \begin{bmatrix} 45 \\ 34 \end{bmatrix}_1 - \begin{bmatrix} 45 \\ 41 \end{bmatrix}_2 + \begin{bmatrix} 45 \\ 12 \end{bmatrix}_3 = 0$$

or

$$-\begin{bmatrix} 12 \\ 12 \end{bmatrix}_5 + \begin{bmatrix} 12 \\ 12 \end{bmatrix}_4 - \begin{bmatrix} 15 \\ 15 \end{bmatrix}_3 - \begin{bmatrix} 15 \\ 15 \end{bmatrix}_2 = 0$$

where $\begin{bmatrix} rs \\ tu \end{bmatrix}_v$ is the complementary of the minor of order two formed by the elements in the rth and sth rows, and the tth and uth columns, after the elements in the vth columns of the circulant have been replaced by units.

The method is not confined to circulants of odd order but may be applied to circulants of even order. Thus for the circulant $C(abcd)$ we have

$$\begin{vmatrix} 1 & a+b+c & d \\ 1 & b+c+d & a \\ 1 & c+d+a & b \end{vmatrix} = 0,$$

whence

$$\begin{vmatrix} 1 & a & c \\ 1 & b & a \\ 1 & c & b \end{vmatrix} + \begin{vmatrix} 1 & b & d \\ 1 & c & a \\ 1 & d & b \end{vmatrix} + \begin{vmatrix} 1 & c & d \\ 1 & d & a \\ 1 & a & b \end{vmatrix} = 0$$

or

$$-[1,1]_2 + [1,2]_4 + [1,4]_1 = 0$$

which is in accord with (5).

Again

$$\begin{vmatrix} 1 & a+c & b+d \\ 1 & b+d & c+a \\ 1 & c+a & d+b \end{vmatrix} = 0$$

or

$$[1,2]_3 - [1,3]_4 + [1,4]_1 - [1,1]_2 = 0,$$

which is in accord with (6).

It is evident we may remove from (5) and (6) the restriction that n is odd, and have them true for circulants of any order.

EXAMPLE: Show that

$$\begin{bmatrix} 1 & r \\ 1 & s \end{bmatrix} = \begin{bmatrix} 1 & r+1 \\ 1 & s-1 \end{bmatrix} + \begin{bmatrix} 1 & 2 \\ s-1 & r \end{bmatrix} \qquad \text{(Cazzaniga).}$$

CHAPTER XIII

CONTINUANTS

541. DEFINITION. A *continuant* is a determinant all of whose elements are zero except those in the main diagonal and in the two adjacent diagonal lines parallel to and on either side of the main diagonal.

We shall denote the continuant

$$\begin{vmatrix} a_1 & b_1 & & \cdot\ \cdot\ \cdot\ \cdot \\ c_1 & a_2 & b_2 & \cdot\ \cdot\ \cdot \\ \cdot & c_2 & a_3 & \cdot\ \cdot\ \cdot \\ \cdot & \cdot & \cdot & \cdot\ \cdot\ \cdot\ \cdot\ \cdot \end{vmatrix}$$

of order n by

$$K\begin{pmatrix} & b_1 & b_2 & \cdots & b_{n-1} & \\ a_1 & a_2 & a_3 & \cdots & & a_n \\ & c_1 & c_2 & \cdots & c_{n-1} & \end{pmatrix} \quad \text{or} \quad K(1,n).$$

The diagonal of b's will be called the *upper* diagonal, and the diagonal of c's the *lower* diagonal. Together the upper and lower diagonals are called the *minor* diagonals of the continuant.

542 *No term of the continuant can contain two consecutive b's or c's.*

For suppose, if possible, in a determinant of order n a term could contain b_r and b_{r+1}; then denoting by an x that the element in that place cannot be taken we would have

$$\begin{vmatrix} \cdot & & & & & & & \\ & \cdot & & & & & & \\ & & x & x & b_r & & & \\ & & & x & x & b_{r+1} & & \\ & & & & x & x & & \\ & & & & & x & & \\ & & & & & & \cdot & \\ & & & & & & & \cdot \end{vmatrix}$$

and as every term must contain an element from each row and column this term must contain b_{r+1} and b_{r+2} also, and continuing we find that the term must contain all the b's and we would have to take with them the zero from the first column and nth row as these are not represented in the term. Therefore no term can contain two consecutive b's. Similarly with respect to the c's.

543. *If a term contains b_r it also contains c_r.*

For by §542 we have

$$\begin{vmatrix} \cdot & & & & & & & \\ & \cdot & & & & & & \\ & & \cdot & x & & & & \\ & & x & x & b_r & & & \\ & & & \cdot & x & x & & \\ & & & & x & \cdot & & \\ & & & & & & \cdot & \\ & & & & & & & \cdot \end{vmatrix}$$

which shows that we must have an element from the $(r+1)$st row and the rth column, that is we must have c_r to go with b_r.

544. One term of the continuant is obviously $a_1a_2 \cdots a_n$.

No term of the continuant can be formed in which an odd number of consecutive a's are omitted.

For any term which contains a_{r-1}, but not a_r must contain b_r and c_r and therefore cannot contain a_{r+1} or b_{r+1} or c_{r+1}. If it does not contain a_{r+2} it must contain b_{r+2} and c_{r+2} and therefore cannot contain a_{r+3}. Hence the truth of the theorem.

545. *Other terms can be formed from $a_1a_2 \cdots a_n$ by replacing any pair of consecutive a's by the product of the b and c having the same suffix as the first a of the pair with a negative sign.* For example we may replace a_ra_{r+1} by $-b_rc_r$.

This is obvious from the definition and from the fact that to get b_r and c_r into the positions of a_r and a_{r+1} one interchange is necessary.

546. All terms of the continuant are obtained by using the process of the last article to the full extent.

It follows from §544 that every term in a continuant of odd order must contain a main diagonal element. In a continuant of even order there is one term which does not, namely

$$(-1)^{n/2}b_1c_1b_3c_3 \cdots$$

From this it is clear that a zero axial continuant of odd order is zero, and one of even order is equal to the single term above indicated.

547. It is evident that the general continuant $K(1, n)$ is equal to the sum of two A and B say, where A is the sum of all the terms of $K(1, n)$ which do not contain $b_r c_r$, while B is equal to the sum of all the terms of $K(1, n)$ which do contain $b_r c_r$. Thus

$$(1) \quad K(1,n) = K(1,r)K(r+1,n) - K(l,r-1)b_r c_r K(r+2,n)$$

We may use (1) to break up $K(1, r)$ and $K(1, r-1)$ and get

$$\begin{aligned}(2) \quad K(1,n) = \{&K(1,h)K(h+1,r) \\ &- K(1,h-1)b_h c_h K(h+2,r)\}K(r+1,n) \\ &- \{K(1,h)K(h+1,r-1) \\ &- K(1,h-1)b_h c_h K(h+2,r-1)\}b_r c_r K(r+2,n)\end{aligned}$$

Equation (1) gives us the expansion of a continuant in terms of minors of the rth order taken from the first r rows and their complementaries.

By actual expansion in terms of elements of the first row and their complementaries we have

$$(3) \qquad K(1,n) = a_1 K(2,n) - b_1 c_1 K(3,n).$$

Expansion in terms of elements of the last row gives

$$(4) \qquad K(1,n) = a_n K(1,n-1) - c_{n-1} b_{n-1} K(1,n-2)$$

and in terms of elements in the rth row

$$\begin{aligned}(5)\ K(1,n) = a_r K(1,r-1)K(r+1,n) - b_r c_r K(1,r-1)K(r+2,n) \\ - b_{r-1} c_{r-1} K(1,r-2)K(r+1,n)\end{aligned}$$

548. *If in $K(1, n)$ the element a_r be the sum of two numbers, thus $a_r = a_r' + a_r''$, the continuant may be expressed in the form*

$$K\begin{pmatrix} & b_1 \cdots & b_{r-1} & \\ a_1 & \cdots\ a_{r-1} & & a_r' \\ & c_1 \cdots & c_{r-1} & \end{pmatrix} K(r+1,n)$$

$$+ K(1,r-1)K\begin{pmatrix} & b_r \cdots b_{n-1} & \\ a_r'' & \cdots & a_n \\ & c_r \cdots c_{n-1} & \end{pmatrix}$$

For the non-zero elements of the rth column written from top to bottom are $b_{r-1}+0$, $a_r' + a_r''$, $0+c_r$ and therefore $K(1, n)$ may be written as the sum of two determinants, each of which breaks up into factors as above given.

Conversely the sum of two terms as above written may be combined into a single continuant.

As a consequence of this we have

$$K\begin{pmatrix} & b_1 \cdots & \\ a_1' + a_1'' & & a_2 \cdots \\ & c_1 \cdots & \end{pmatrix} = a_1' K\begin{pmatrix} & b_2 \cdots \\ a_2 \cdots & \\ & c_2 \cdots \end{pmatrix} + K\begin{pmatrix} & b_1 & b_2 \cdots \\ a_1'' & a_2 \cdots & \\ & c_1 & c_2 \cdots \end{pmatrix}$$

(1) or

$$= K\begin{pmatrix} & b_1 \cdots \\ a_1' \cdots & \\ & c_1 \cdots \end{pmatrix} + a_1'' K\begin{pmatrix} & b_2 \cdots \\ a_2 \cdots & \\ & c_2 \cdots \end{pmatrix}$$

and

(2)
$$K\begin{pmatrix} & b_1 \cdots b_{n-1} & \\ a_1 \cdots & & a_n' + a_n'' \\ & c_1 \cdots c_{n-1} & \end{pmatrix} = K\begin{pmatrix} & b_1 \cdots b_{n-1} & \\ a_1 \cdots & & a_n' \\ & c_1 \cdots c_{n-1} & \end{pmatrix}$$

$$+ a_n'' K\begin{pmatrix} & b_1 \cdots b_{n-2} & \\ a_1 \cdots & & a_{n-1} \\ & c_1 \cdots c_{n-2} & \end{pmatrix}$$

549. From §543 it is evident that any factor of an element of one of the minor diagonals may be transferred to the corresponding element of the other minor diagonal, and hence

$$K(1,n) = K\begin{pmatrix} & -b_1c_1 & & -b_2c_2 \cdots & -b_{n-1}c_{n-1} & \\ a_1 & & a_2 \cdots & & & a_n \\ & -1 & & -1 & \cdots -1 & \end{pmatrix}$$

which, if we write d_r for $-b_rc_r$, gives

$$K(1,n) = K\begin{pmatrix} & d_1 \cdots & d_{n-1} & \\ a_1 \cdots & & & a_n \\ & -1 \cdots & -1 & \end{pmatrix}$$

or

$$= K(1,n)_d \text{ say.}$$

If $a_1 = 1$, then obviously

$$K(1,n)_d = K\begin{pmatrix} & d_2 \cdots \\ d_1 + a_2 & a_3 \cdots \end{pmatrix}$$

550. If we divide the 2nd, 3rd, 4th, $\cdots$, rows by c_2, c_3, c_2c_4, c_3c_5, $c_2c_4c_6$, $\cdots$, respectively and then multiply the 3rd, 4th, $\cdots$, columns by the same we readily see that

$$K(1,n) = K\begin{pmatrix} 1 & 1 & 1 \cdots & 1 \\ a_1 & A_2 & A_3 \cdots & A_n \\ 1 & 1 & 1 \cdots & 1 \end{pmatrix},$$

where

$$A_r = a_r\frac{b_{r-1}b_{r-3}b_{r-5}\cdots}{b_rb_{r-2}b_{r-4}\cdots}.$$

551. *If the elements* b_{r-1}, a_r, b_r *have a common factor then this is a factor of the continuant.*

This is readily seen on observing that the factor of b_{r-1} may be transferred to c_{r-1}.

552. *If* a_1', a_2', a_3', $\cdots$, *be any given numbers then*

$$K(1,n)_d = \frac{a_1a_2\cdots a_n}{a_1'a_2'\cdots a_n'}$$

$$\times\begin{vmatrix} a_1' & \dfrac{a_1'a_2'}{a_1a_2}d_1 & & & \\ -1 & a_2' & \dfrac{a_2'a_3'}{a_2a_3}d_2 & & \\ \cdots & \cdots & \cdots & \cdots & \cdots \\ & & -1 & a_{n-1}' & \dfrac{a_{n-1}'a_n'}{a_{n-1}a_n}d_{n-1} \\ & & & -1 & a_n' \end{vmatrix}$$

The truth of this is seen on multiplying row r of $K(1, n)_d$ by a_r'/a_r, for all values of r and then transferring the factor from the lower to the upper diagonal.

As corollaries of this we have

$$(1)\quad K(1,n)_d = \frac{1}{h^n}\begin{vmatrix} ha_1 & h^2d_1 & & & \\ -1 & ha_2 & h^2d_2 & & \\ \cdots & \cdots & \cdots & \cdots & \cdots \\ & & -1 & ha_{n-1} & h^2d_{n-1} \\ & & & -1 & ha_n \end{vmatrix}$$

$$(2)\ K(1,n)_d = \frac{a_1a_2\cdots a_n}{\alpha^n}\begin{vmatrix} \alpha & \frac{\alpha^2}{a_1a_2}d_1 & & & \\ -1 & \alpha & \frac{\alpha^2}{a_2a_3}d_2 & & \\ \cdot & \cdot & \cdot & \cdot & \cdot \\ & & -1 & \alpha & \frac{\alpha^2}{a_{n-1}a_n}d_{n-1} \\ & & & -1 & \alpha \end{vmatrix}$$

Taking $\alpha^2 d_k/a_ka_{k+1}=\beta_k^2$ we see that

$$K(1,n)_d = A\cdot\begin{vmatrix} \alpha & \beta_1 & \cdots & \\ -\beta_1 & \alpha & \beta_2 & \cdots \\ & -\beta_2 & \alpha & \cdots \\ \cdot & \cdot & \cdot & \cdot \end{vmatrix}$$

that is a skew determinant.

553. If in (1) §547 we put $r=1$, it becomes

$$K(1,n) = a_1K(2,n) - K(1,0)b_1c_1K(3,n)$$

but this must be the same as (3) and we must therefore interpret $K(1,\,0)$ to be unity. Similarly if r is put equal to $n-1$ we must put $K(n+1,\,n)=1$.

With the understanding, then, that $K(p,\,p-1)=1$ for all values of p, we may assert that equation (1) §547 is true for all values of r between 0 and n.

In adopting this meaning for $K(p,\,p-1)$ we do not assume that every relation will hold true for values of the letters which cause such a form to appear, but we simplify the statement of such relations as do then hold true.

554. There is a more general relation from which (1) of §547 comes as a specia' case. It is

$$(1)\ K(1,n)K(h,r) - K(1,r)K(h,n) - b_{h-1}c_{h-1}b_hc_h\cdots b_rc_rK(1,h-2)K(r+2,n) = 0.$$

If in this we put $h=r+1$, then it reduces to (1) of §547.

That this relation is true is seen on observing that it is merely the result of applying the theorem of §148 to a continuant.

Special cases of (1) are, using the $K(1, n)_d$ form

$$(2)\qquad K(1,n)K(2,n-1) - K(1,n-1)K(2,n) + (-1)^{n-1}d_2d_3\cdots d_n = 0$$

$$(3)\qquad K(1,n)K(2,n-2) - K(1,n-2)K(2,n) + (-1)^{n}d_2d_3\cdots d_{n-1}a_n = 0.$$

If in (1) $n=2m-1$, $r=h=m$, then we have

$$(4)\qquad a_mK\begin{pmatrix} & b_1 & \cdots & b_{m-1} & b_{m-1} & & \cdots & b_1 & \\ a_1 & a_2 & \cdots & & a_m & & a_{m-1} & \cdots & a_1 \\ & c_1 & \cdots & c_{m-1} & c_{m-1} & & \cdots & c_1 & \end{pmatrix} = [K(1,m)]^2 - [b_{m-1}c_{m-1}K(1,n-2)]^2.$$

555. The element of the adjugate of $K(1, n)$ in the position (r, s) is readily seen to be

$$K(1,r-1)c_rc_{r+1}\cdots c_{s-1}K(s+1,n),\qquad \text{if } r<s,$$

$$K(1,r-1)K(r+1,n),\qquad \text{if } r=s,$$

$$K(1,s-1)b_sb_{s+1}\cdots b_{r-1}K(r+1,n),\qquad \text{if } r>s,$$

where $K(t+1, n)$ must be interpreted as unity when $t\geqq n-1$.

556. *If the signs of the elements in the main diagonal be changed, the continuant is multiplied by* $(-1)^n$.

The truth of this is seen on multiplying the n rows by -1 and then removing the factor -1 from the one minor diagonal to the other which leaves the elements in these two diagonals the same as originally but those in the main diagonal the negative of what they were.

557. From (4) §554 we see that for a centrosymmetric continuant of even order, $n=2m$, we have

$$(1)\qquad K\begin{pmatrix} & b_1 & \cdots & b_{m-1} & \theta & c_{m-1} & \cdots & c_1 & \\ a_1 & & \cdots & a_m & a_m & & \cdots & & a_1 \\ & c_1 & \cdots & c_{m-1} & \theta & b_{m-1} & \cdots & b_1 & \end{pmatrix} = [K(1,m)]^2 - [\theta K(1,m-1)]^2.$$

For one of odd order, $n=2m-1$, we have

$$(2)\qquad K\begin{pmatrix} & b_1 & \cdots & b_{m-1} & c_{m-1} & \cdots & c_1 & \\ a_1 & \cdots & & a_m & & \cdots & & a_1 \\ & c_1 & \cdots & c_{m-1} & b_{m-1} & \cdots & b_1 & \end{pmatrix} = K(1,m-1)[K(1,m) - b_{m-1}c_{m-1}K(1,m-2)].$$

If in (2) we replace a_m by $2a_m$, then it becomes

$$
\begin{aligned}
&K\begin{pmatrix} b_1 \cdots b_{m-1} & c_{m-1} \cdots c_1 \\ a_1 \quad \cdots \quad 2a_m & \cdots \quad a_1 \\ c_1 \cdots c_{m-1} & b_{m-1} \cdots b_1 \end{pmatrix} \\
&= K(1,m-1)[K(1,m) + a_m K(1,m-1) \\
&\qquad - b_{m-1}c_{m-1}K(1,m-2)] \qquad (3) \\
&= 2K(1,m)K(1,m-1).
\end{aligned}
$$

558. *A continuant of even order may be transformed by alternately multiplying and dividing the successive elements of the main diagonal by any given number; and a continuant of odd order may be so transformed if in addition the continuant be divided by this number.*

This is readily seen on performing the following operations: multiply row_1, row_3, $\cdots$, by k, then divide col_2, col_4, $\cdots$, by k. Thus

$$
\begin{vmatrix} a_1 & 1 & \cdot & \cdot \\ c_1 & a_2 & b_2 & \cdot \\ \cdot & c_2 & a_3 & b_3 \\ \cdot & \cdot & c_3 & a_4 \end{vmatrix} = \begin{vmatrix} ka_1 & b_1 & \cdot & \cdot \\ c_1 & \dfrac{a_2}{k} & b_2 & \cdot \\ \cdot & c_2 & ka_3 & b_3 \\ \cdot & \cdot & c_3 & \dfrac{a_4}{k} \end{vmatrix}
$$

$$
\begin{vmatrix} a_1 & b_1 & \cdot & \cdot & \cdot \\ c_1 & a_2 & b_2 & \cdot & \cdot \\ \cdot & c_2 & a_3 & b_3 & \cdot \\ \cdot & \cdot & c_3 & a_4 & b_4 \\ \cdot & \cdot & \cdot & c_4 & a_5 \end{vmatrix} = \frac{1}{k}\begin{vmatrix} ka_1 & b_1 & & & \\ c_1 & \dfrac{a_2}{k} & b_2 & & \\ & c_2 & ka_3 & b_3 & \\ & & c_3 & \dfrac{a_4}{k} & b_4 \\ & & & c_4 & ka_5 \end{vmatrix}
$$

The multiplications and divisions by k may stop before the last a is reached or they may not begin until the $(2h-1)$st row is reached and give corresponding results.

559. *The continuant* $K(1, n)_d$ *equals*

$$
\frac{d_{n-1}d_{n-3}d_{n-5}\cdots}{d'_{n-1}d'_{n-3}d'_{n-5}\cdots}
$$

$$
\times \begin{vmatrix}
a_1 & d_1' & & & & & \\
-1 & \dfrac{d_1'}{d_1}a_2 & d_2' & & & & \\
 & -1 & \dfrac{d_2'd_1}{d_2d_1'}a_3 & d_3' & & & \\
 & & -1 & \dfrac{d_3'd_2d_1'}{d_3d_2'd_1}a_4 & & & \\
 & & & & \ddots & & \\
 & & & & & -1 & \dfrac{d_{n-1}'d_{n-2}d_{n-3}'\cdots}{d_{n-1}d_{n-2}'d_{n-3}\cdots}a_n
\end{vmatrix}
$$

where $d_{n-1}', d_{n-3}', d_{n-3}', \cdots$, *etc. are any given numbers.*

To see the truth of this begin at a_n in $K(1, n)_d$ and multiply (divide) the columns by d_1/d_1' if n is odd (even), divide (multiply) a_{n-1} by the same, and so on upwards until we come to col_2 when division by d_1/d_1' will give d_1' in place of d_1 in the upper diagonal. Do likewise with respect to all the d's.

As corollaries of the foregoing we have

$$
(1) \qquad K(1,n)_d = \frac{1}{\delta^\nu}\begin{vmatrix}
a_1 & \delta d_1 & & \\
-1 & a_2\delta & d_2\delta & \\
 & -1 & a_3 & d_3\delta \\
\cdots & \cdots & \cdots & \cdots
\end{vmatrix}
$$

where ν is the greatest integer in $n/2$.

(2) $K(1,n)_d$

$$
= \frac{d_{n-1}d_{n-3}\cdots}{\lambda\lambda}\begin{vmatrix}
a_1 & \lambda & \cdot & \cdot & \cdot & \cdots \\
-1 & \lambda\dfrac{a_2}{d_1} & \lambda & \cdot & \cdot & \cdots \\
\cdot & -1 & a_3\dfrac{d_1}{d_2} & \lambda & \cdot & \cdots \\
\cdot & \cdot & -1 & \lambda a_4\dfrac{d_2}{d_1d_3} & \lambda & \cdots \\
\cdot & \cdot & \cdot & -1 & a_5\dfrac{d_1d_3}{d_2d_4} & \cdots \\
\cdots & \cdots & \cdots & \cdots & \cdots & \cdots
\end{vmatrix}
$$

560. *The whole number of terms in a continuant is*

$$\binom{n}{0}+\binom{n-1}{1}+\binom{n-2}{2}\cdots+\binom{n-r}{r}+\cdots$$

For a term formed by replacing r pairs of consecutive a's by the appropriate b's and c's may be regarded as composed of $(n-r)$ things, namely, $(n-2r)$ a's and r pairs of b's and c's. There are as many such terms, therefore, as there are arrangements of $n-r$ things of which $n-2r$ are of one kind and the remaining r of another kind, that is $\binom{n-r}{r}$.

EXERCISE: Show that the number of terms in a continuant is equal to

(a) $$\frac{1}{2^{n+1}(5)^{1/2}}[(1+(5)^{1/2})^{n+1}-(1-(5)^{1/2})^{n+1}]$$

(b) $$\frac{1}{2^n}\left[\binom{n+1}{1}+5\binom{n+1}{3}+5^2\binom{n+1}{5}+\cdots\right].$$

561. Since $K(1, n)=K(1, r)K(r+1, n)-b_rc_rK(1, r-1)K(r+2, n)$ we have

(1) $$\frac{\partial K}{\partial b_r}=-c_rK(1,r-1)K(r+2,n)$$

(2) $$\frac{\partial K}{\partial c_r}=-b_rK(1,r-1)K(r+2,n)$$

and therefore

(3) $$\frac{\partial K}{\partial b_r}\div\frac{\partial K}{\partial c_r}=\frac{c_r}{b_r}$$

or

$$b_r\cdot\frac{\partial K}{\partial b_r}=c_r\frac{\partial K}{\partial c_r}\,.$$

In particular

$$\frac{\partial K}{\partial b_1}=-c_1K(3,n),\qquad\frac{\partial K}{\partial b_{n-1}}=-c_{n-1}K(1,n-2)\,,$$

$$\frac{\partial K}{\partial c_1}=-bK(3,n),\qquad\frac{\partial K}{\partial c_{n-1}}=-b_{n-1}K(1,n-2)\,.$$

Again since

$$K(1,n) = \{K(1,r-2)b_{r-1} - K(1,r-1)a_r\}K(r+1,n) - b_rc_rK(1,r-1)K(r+2,n)$$

we have

(4) $$\frac{\partial K}{\partial a_r} = K(1,r-1)K(r+1,n)$$

which is the cofactor of a_r. Similarly

(5) $$\frac{\partial^2 K}{\partial a_r \partial a_s} = K(1,r-1)K(r+1,s-1)K(s+1,n), \quad (s > r)$$

and so on.

If $s=r+1$ we have

(6) $$\frac{\partial^2 K}{\partial a_r \partial a_{r+1}} = K(1,r-1)K(r+2,n),$$

which is the cofactor of a_ra_{r+1}.

Similarly for higher derivatives with respect to elements in the minor diagonals, except it is to be observed that derivatives with respect to consecutive elements in minor diagonals are zero.

From (1) and (6) we have

$$\frac{\partial^2 K}{\partial a_r \partial a_{r+1}} \div \frac{\partial K}{\partial b_r} = -\frac{1}{c_r}.$$

From (4) we have

$$\frac{1}{K}\frac{\partial K}{\partial a_r} = \frac{K(1,r-1)K(r+1,n)}{K}$$

or

$$\frac{\partial K}{\partial a_r}\log K = \frac{K(1,r-1)K(r+1,n)}{K}.$$

Again we have

$$\frac{\partial K}{\partial a_r} \div \frac{\partial K}{\partial b_r} = -\frac{1}{c_r}\frac{K(r+1,n)}{K(r+2,n)}$$

$$= -\frac{1}{c_r}\left[a_{r+1} + \frac{b_{r+1}}{a_{r+2} + \dfrac{b_{r+2}}{a_{r+3} + \cdots + \dfrac{{}_nb_{-1}}{a_n}}}\right].$$

562. If S_n represent the sum of n terms of the harmonical progression

$$\frac{1}{a} + \frac{1}{a+b} + \frac{1}{a+2b} + \cdots$$

we know from Euler's transformation that

$$S_n = \cfrac{1}{a - \cfrac{a^2}{2a+b - \cfrac{(a+b)^2}{2a+3b - \cfrac{(a+2b)^2}{2a+5b - \cdots - \cfrac{(a+\overline{n-2}b)^2}{2a+(2n-3)b}}}}}$$

and therefore

$$S_n = \frac{\begin{vmatrix} 2a+b & (a+b)^2 & \cdot & \\ 1 & 2a+3b & (a+2b)^2 & \\ \cdot & \cdot & \cdot & \cdots \\ & & & \ddots \end{vmatrix}}{\begin{vmatrix} a & a^2 & \cdot & \cdot \\ 1 & 2a+b & (a+b)^2 & \cdot \\ \cdot & 1 & 2a+3b & (a+2b)^2 \\ & & & \ddots \end{vmatrix}} = \frac{K_{n-1}}{H_n}, \text{ say.}$$

From this we get

$$\begin{aligned} S_n &= \frac{(2a+\overline{2n-3}\cdot b)K_{n-2} - (a+\overline{n-2}\cdot b)^2 K_{n-3}}{(2a+\overline{2n-3}\cdot b)H_{n-1} - (a+\overline{n-2}\cdot b)^2 K_{n-2}} \\ &= \frac{K_{n-2}}{H_{n-1}} + \frac{\{K_{n-2} - (a+\overline{n-2}\cdot b)K_{n-3}\}(a+\overline{n-2}\cdot b)}{H_n} \\ &= \frac{K_{n-2}}{H_{n-1}} + \frac{H_{n-2}(a+\overline{n-2}\cdot b)}{H_n} \\ &= S_{n-1} + \frac{1}{a-\overline{n-1}\cdot b} \end{aligned}$$

since $H_n = a(a+b)(a+2b)\cdots(a+\overline{n-1}\cdot b)$, and

$$K_{n-2} - (a+\overline{n-2}\cdot b)K_{n-3} = a(a+b)\cdots(a+\overline{n-3}\cdot b) = H_{n-2}$$

563. If we are given the difference equation

$$u_x = a_{x-1}u_{x-1} + b_{x-2}u_{x-2}$$

we may obtain an expression for u_x in terms of u_1 and u_0.

Thus taking for convenience $x=6$ we have the equations

$$\begin{aligned}
-u_6 + a_5u_5 + b_4u_4 &= 0\\
-u_5 + a_4u_4 + b_3u_3 &= 0\\
-u_4 + a_3u_3 + b_2u_2 &= 0\\
-u_3 + a_2u_2 + b_1u_1 &= 0\\
-u_2 + a_1u_1 + b_0u_0 &= 0
\end{aligned}$$

from which we have at once

$$u_6 = \begin{vmatrix} a_5 & b_4 & \cdot & \cdot & \cdot \\ -1 & a_4 & b_3 & \cdot & \cdot \\ \cdot & -1 & a_3 & b_2 & \cdot \\ \cdot & \cdot & -1 & a_2 & b_1u_1 \\ \cdot & \cdot & \cdot & -1 & a_1u_1 + b_0u_0 \end{vmatrix}$$

$$= K\begin{pmatrix} & b_4 & & b_3 & & b_2 & & b_1 & \\ a_5 & & a_4 & & a_3 & & a_2 & & a_1 \end{pmatrix}u_1 + K\begin{pmatrix} & b_4 & & b_3 & & b_2 & \\ a_5 & & a_4 & & a_3 & & a_2 \end{pmatrix}u_0.$$

EXERCISES:

1. (a) If $u_x = a_{x-1}u_{x-1} + b_{x-2}u_{x-2} + c_{x-2}$ find u_6.

(b) If $(x-2)u_x - x(x-2)u_{x-1} - xu_{x-2} = (-1)^{x-1}\cdot 4$ show that $u_6 = 80$. $(u_2=0,\ u_3=1)$.

2. If $\frac{1}{1-bx+acx^2} = \beta_0 + \beta_1x + \beta_2x^2 + \ldots$

show that

$$\beta_n = \begin{vmatrix} b & a & & & & \\ c & b & a & & & \\ & c & b & a & & \\ & & & \cdot & & \\ & & & & \cdot & \\ & & & & & \cdot \end{vmatrix}_n$$

564. *Any determinant can be expressed as the quotient of a continuant, whose elements are minors of the determinant, divided by a product of such minors.*

Thus if $A = |a_{1n}|$ we have

$$A = \frac{K}{A_{r_2n,12}A_{r_3n,23}\cdots A_{r_nn,(n-1)n}},$$

where K is a continuant whose non-zero element in the first line are

$$A_{n,2},\ A_{n,1}\ ;$$

kth line are

$$A_{r_kn,k(k+1)},\ A_{r_kn,(k-1)(k+1)},\ A_{r_kn,(k-1)k}\ ;$$

last line are

$$A_{r_n,n},\ A_{r_n,(n-1)}.$$

where $r_1, r_2, \cdots, r_n$ may each have any of the values $1, 2, \cdots, n-1$. That is if a continuant K be formed each of whose elements is that minor of A resulting from deleting (1) the nth row of A in the case of every element of K except the two elements in the last row (2) a row of A other than the nth, in the case of every element of K except the two elements in the first row, this deleted row of A being the same for all the elements in the same row of K, but not necessarily the same for elements in different rows of K, and (3) the column or columns of A whose column-number or column-numbers, in the case of any particular element, is or are the same as the column-number or column-numbers giving the position of the remaining elements in the same row of K; then A is equal to K divided by the product of $n-1$ minors of A of which the first $n-2$ are identical with the last $n-2$ elements in the upper diagonal of K and the last one results from deleting from the last main diagonal element of K the last row and column.

From the n equations

$$(1)\qquad \begin{cases} a_{11}x_1 + a_{12}x_2 + \cdots + a_{1n}x_n = 0 \\ a_{21}x_1 + a_{22}x_2 + \cdots + a_{2n}x_n = 0 \\ \cdots\cdots\cdots\cdots\cdots\cdots \\ a_{n1}x_1 + a_{n2}x_2 + \cdots + a_{nn}x_n = 1 \end{cases}$$

eliminate, first, all the variables but x_1 and x_2 from all but the nth equation, second all the variables but x_{k-1}, x_k and x_{k+1} from all but the nth and one other equation ($k=2, 3, \cdots, n-1$), third, all the variables but x_{n-1} and x_n from all but the hth equation ($h \neq n$). We have

$$(2)\quad \begin{cases} A_{n,2}x_1 + A_{n,1}x_2 & = 0 \\ A_{r_2n,2\,3}x_1 + A_{r_2n,1\,3}x_2 + A_{r_2n,1\,2}x_3 & = 0 \\ \cdots\cdots\cdots\cdots\cdots\cdots\cdots\cdots & \\ A_{r_{n-1}n,(n-1)n}x_{n-2} + A_{r_{n-1}n,(n-2)n}x_{n-1} + A_{r_{n-1}n,(n-2)(n-1)}x_n & = 0 \\ A_{r_n,n}x_{n-1} + A_{r_n,n-1}x_n & = A_{r_n n,n-1\,n} \end{cases}$$

Solving for x_1 from (2) we have

$$x_1 = \frac{(-1)^{n-1}A_{n,1}A_{r_2n,1\,2}A_{r_3n,2\,3}\cdots A_{r_nn,(n-1)n}}{K}.$$

But from (1)

$$x_1 = (-1)^{n-1}A_{n,1} \div A.$$

Therefore

$$A = \frac{K}{A_{r_2n,1\,2}\cdots A_{r_nn,(n-1)n}}.$$

Solving equations (2) for x_n we have

$$K \cdot x_n = \begin{vmatrix} A_{n,2} & A_{n,1} & & & \\ A_{r_2n,2\,3} & A_{r_2n,1\,3} & A_{r_2n,1\,2} & & \\ \cdots & \cdots & \cdots & \cdots & \cdots \\ & & & A_{r_n,n} & A_{r_nn,(n-1)n} \end{vmatrix}$$

$$= A_{r_nn,(n-1)n}K_{n,n}.$$

Solving for x_n from (1) we have

$$A \cdot x_n = A_{n,n}$$

whence

$$\frac{A \cdot A_{r_nn,(n-1)n}}{A_{n,n}} = \frac{K}{K_{n,n}} = A_{n,2} - \cfrac{A_{n,1}A_{r_2n,2\,3}}{A_{r_2n,1\,3} - \cfrac{A_{r_2n,1\,2}A_{r_3n,3\,4}}{A_{r_4n,2\,4}}}$$

Other relations may be obtained by applying the Law of Complementaries.

565. The theorem of the foregoing article may be generalized by giving to r_k all the possible $n-1$ values in eliminating all the variables but x_{k-1}, x_k, x_{k+1}, and adding together the $n-1$ equations thus obtained, after multiplying each by one of the $n-1$ arbitrary quantities $m_{k1}, m_{k2}, \cdots, m_{k,n-1}$. Thus

$$\begin{aligned} m_{k1}A_{1\;n,k(k+1)}x_{k-1} + m_{k1}A_{1\;n,(k-1)(k+1)}x_k + m_{k1}A_{1\;n,(k-1)k}\,x_{k+1} &= 0 \\ -m_{k2}A_{2\;n,k\;(k+1)}x_{k-1} - m_{k2}A_{2\;n,(k-1)(k+1)}x_k - m_{k2}A_{2\;n,(k-1)k}x_{k+1} &= 0 \\ \cdots\cdots\cdots\cdots\cdots\cdots\cdots\cdots\cdots\cdots\cdots\cdots\cdots\cdots & \end{aligned}$$

whence, using $\binom{\mathrm{AM}}{\mathrm{k\;k}}_{n,k+1}$ to denote the determinant formed by deleting the nth row and $k+1$ column of the determinant resulting on replacing the kth column of A by the kth column of

$$M \equiv \begin{vmatrix} m_{11} & m_{12} & \cdots & m_{1n} \\ \cdot & \cdot & \cdots & \cdot \\ m_{n1} & m_{n2} & \cdots & m_{nn} \end{vmatrix}$$

we have on summing

$$\begin{pmatrix} A & M \\ k & k \end{pmatrix}_{n,k+1} x_{k-1} + \begin{pmatrix} A & M \\ k-1 & k \end{pmatrix}_{n,k+1} x_k + \begin{pmatrix} A & M \\ k-1 & k \end{pmatrix}_{n,k} x_{k+1} = 0$$

Hence

$$A = \begin{vmatrix} A_{n,2} & A_{n,1} & & \\ \begin{pmatrix} A & M \\ 2 & 2 \end{pmatrix}_{n,3} & \begin{pmatrix} A & M \\ 1 & 2 \end{pmatrix}_{n,3} & \begin{pmatrix} A & M \\ 1 & 2 \end{pmatrix}_{n,2} & \\ \cdot & \begin{pmatrix} A & M \\ 3 & 3 \end{pmatrix}_{n,4} & \begin{pmatrix} A & M \\ 2 & 3 \end{pmatrix}_{n,4} & \begin{pmatrix} A & M \\ 2 & 3 \end{pmatrix}_{n,3} \\ \cdots & \cdots & \cdots & \cdots \end{vmatrix} \div \begin{pmatrix} A & M \\ 1 & 2 \end{pmatrix}_{n,2} \begin{pmatrix} A & M \\ 2 & 3 \end{pmatrix}_{n,3} \cdots.$$

566. *A continuant of even order, $n=2m$, having the main diagonal elements in the even-numbered rows equal, is expressible as a continuant of order m, the relation being:*

$$\Phi_{2m} = \begin{vmatrix} \theta_1 & d_1 & \cdot & \cdot & \cdot & & & & \\ -1 & \phi & d_2 & \cdot & \cdot & & & & \\ \cdot & -1 & \theta_2 & d_3 & \cdot & & & & \\ \cdot & \cdot & -1 & \phi & d_4 & & & & \\ \cdots & \cdots & \cdots & \cdots & \cdots & \cdots & \cdots & \cdots & \cdots \\ & & & & -1 & \phi & d_{2m-3} & \cdot \\ & & & & \cdot & -1 & \theta_m & d_{2m-1} \\ & & & & \cdot & \cdot & -1 & \phi \end{vmatrix}_{2m}$$

$$= \begin{vmatrix} \theta_1\phi + d_1 & d_1 & \cdot & & & \\ d_2 & d_2 + \theta_2\phi + d_3 & d_3 & & & \\ & d_4 & d_4 + \theta_3\phi + d_5 & d_5 & & \\ & & & \cdot & & \\ & & & & \cdot & \\ & & & & & \cdot \end{vmatrix} \equiv K_m, \text{ say},$$

where the non-zero elements in the $(m-k+1)$th row are

$$d_{2m-2k}, d_{2m-2k} + \theta_{m-k+1}\phi + d_{2m-2k+1}, d_{2m-2k+1}, d_{2m-2k+1}$$

the last row containing but the first two.

To prove this multiply Φ_{2m}, row-by-row, by ϕ^m in the form

$$\begin{vmatrix} \phi & 1 & \cdot & \cdot & \cdot & \cdots \\ \cdot & 1 & \cdot & \cdot & \cdot & \cdots \\ \cdot & -d_2 & \phi & 1 & & \cdots \\ \cdot & \cdot & \cdot & 1 & \cdot & \cdots \\ \cdot & \cdot & \cdot & -d_4 & \phi & 1 \cdots \\ & & & & & \cdot \\ & & & & & & \cdot \\ & & & & & & & \cdot \end{vmatrix}_n,$$

giving

$$\Phi_{2m}\,\phi^m = \begin{vmatrix} \phi\theta_1 + d_1 & d_1 & -d_1d_2 & \cdot & \cdot & \cdots \\ \cdot & \phi & \cdot & \cdot & \cdot & \cdots \\ -1 & -1 & d_2 + \phi\theta_2 + d_3 & d_3 & -d_3d_4 & \cdots \\ \cdot & \cdot & \cdot & \phi & \cdot & \cdot \\ \cdots & \cdots & \cdots & \cdots & \cdots & \cdots \\ \cdots & \cdots & \cdots & \cdots & \cdots & \cdots \end{vmatrix}_n$$

$$= \phi^m K_m$$

Hence the truth of the theorem.

Similarly

$$\Phi_{2m} \equiv \begin{vmatrix} \phi & d_1 & \cdot & \cdot & & & \\ -1 & \theta_1 & d_2 & \cdot & & & \\ \cdot & -1 & \phi & d_3 & & & \\ & & & \cdot & & & \\ & & & & \cdot & & \\ & & & & & \cdot & \end{vmatrix}_{2m}$$

(2)

$$= \begin{vmatrix} d_1 + \phi\theta_1 + d_2 & d_2 & \cdot & & \\ d_3 & d_3 + \phi\theta_2 + d_4 & d_4 & & \\ & & \cdot & & \\ & & & \cdot & \\ & & & & \cdot \end{vmatrix}_m$$

If in (1) we make the θ's all equal to ϕ then it becomes

$$\begin{vmatrix} \phi & d_1 & \cdot & \cdot & & & & & \\ -1 & \phi & d_2 & \cdot & & & & & \\ \cdot & 1 & \phi & d_3 & & & & & \\ & & & \cdot & & & & & \\ & & & & \cdot & & & & \\ & & & & & \cdot & & & \\ & & & & & -1 & \phi & d_{2m-1} \\ & & & & & \cdot & -1 & \phi \end{vmatrix}$$

(3)

$$= \begin{vmatrix} \phi^2 + d_1 & d_1 & \cdot & & & \\ d_2 & \phi^2 + d_2 + d_3 & d_3 & & & \\ & & \cdot & & & \\ & & & \cdot & & \\ & & & & \cdot & \\ & & & & d_{2m-2} & \phi^2 + d_{2m-2} + d_{2m-1} \end{vmatrix}_m$$

EXAMPLE: Show that

$$a_1\begin{vmatrix} a_1 & b_1 & & & \\ -1 & a_2 & b_2 & & \\ & -1 & a_3 & b_3 & \\ & & \cdot & \cdot & \cdot \end{vmatrix} = \begin{vmatrix} |a_1a_2| & b_1b_2 & \cdot & \cdot \\ a_4 & |a_2a_3a_4| & b_3b_4 & \cdot \\ \cdot & a_6a_2 & |a_4a_5a_6| & b_5b_6 \\ \cdot & \cdot & a_8a_4 & |a_6a_7a_8| \end{vmatrix}$$

$$\div\, a_2a_4a_6$$

567. If in (1) of §566 we put $d_{2m-1}=0$ then we have

$$(1)\quad \Phi_{2m-1}\equiv\begin{vmatrix} \theta_1 & d_1 & \cdot & \cdot & & & & \\ -1 & \phi & d_2 & \cdot & & & & \\ \cdot & -1 & \theta_2 & d_3 & & & & \\ & & & \cdot & & & & \\ & & & & \cdot & & & \\ & & & & & \cdot & & \\ & & & & & -1 & \phi & d_{2m-2} \\ & & & & & & -1 & \theta_m \end{vmatrix}_{2m-1}$$

$$=\frac{1}{\phi}K'_m$$

where K'_m differs from K_m simply in having $d_{2m-1}=0$.

That is a continuant of odd order $2m-1$ *having the main diagonal elements of even numbered rows equal, is expressible as a multiple of a continuant of order* m.

If we use Φ'_{2m-1} and K_m'' to represent the two sides of (1) when the θ's are all equal and if we perform on K_m'' the operations

$$r_k + r_{k+1},(k = 1,2, \cdots m);$$

and then diminish the second column by the first, the third by the new second, and so on, we get, on reducing the order by one

$$\Phi'_{2m-1}=\theta$$

$$\times\begin{vmatrix} \phi\theta + d_1 + d_2 & d_3 & & & \\ d_2 & \theta\phi + d_3 + d_4 & d_5 & & \\ & & \cdot & & \\ & & & \cdot & \\ & & & & \cdot \\ & & & d_{2m-4} & d_{2m-2} + d_{2m-3} + \theta\phi \end{vmatrix}_{m-1}$$

EXAMPLE:

$$\begin{vmatrix} A+b & b & \cdot & \cdot \\ c & A+c+d & d & \cdot \\ \cdot & e & A+e+f & f \\ \cdot & \cdot & g & g+A \end{vmatrix}$$

$$= A\begin{vmatrix} A+b+c & c & \cdot \\ d & A+d+e & e \\ \cdot & f & f+A+g \end{vmatrix}$$

Similarly putting $d_1 = d_{2m-2} = 0$ in (1) we have, on relettering the elements

$$(3)\ \Phi_{2m-3} \equiv \begin{vmatrix} \phi & d_1 & \cdot & \cdot & & & & \\ -1 & \theta_1 & d_2 & \cdot & & & & \\ \cdot & -1 & \phi & d_3 & & & & \\ & & & \cdot & & & & \\ & & & & \cdot & & & \\ & & & & & \cdot & & \\ & & & & & -1 & \theta_{m-2} & d_{2m-4} \\ & & & & & & -1 & \phi \end{vmatrix}_{2m-3}$$

$$= \phi\cdot\begin{vmatrix} \phi\theta_1 + d_1 + d_2 & d_2 & & \\ d_3 & d_3 + \theta_2\phi + d_4 & d_4 & \\ & & \cdot & \cdot \\ & & & \cdot \\ & & & \cdot \end{vmatrix}_{m-2}$$

where the non-zero elements in the $(m-k)$th row are $d_{2m-2k-1}$, $d_{2m-2k-1} + \theta_{m-k}\phi + d_{2m-2k}$, d_{2m-2k}.

EXERCISES. SET XXVIII

1. $$\begin{vmatrix} P & b & & & & \\ -1 & K & c & & & \\ & -1 & Q & d & & \\ & & -1 & K & e & \\ & & & -1 & R & f \\ & & & & -1 & K \end{vmatrix}_6$$

$$= \begin{vmatrix} PK+b & b & \\ c & c+QK+d & d \\ & e & e+RK+f \end{vmatrix}_3$$

2. $$\begin{vmatrix} P & b & & & \\ -1 & K & c & & \\ & -1 & Q & d & \\ & & -1 & K & e \\ & & & -1 & R \end{vmatrix}_5 = \frac{1}{K} \begin{vmatrix} PK+b & b & \\ c & c+QK+d & d \\ & e & e+RK \end{vmatrix}_3$$

3. $$\begin{vmatrix} K & b & & & & & \\ -1 & P & c & & & & \\ & -1 & K & d & & & \\ & & -1 & Q & e & & \\ & & & -1 & K & f & \\ & & & & -1 & R & g \\ & & & & & -1 & K \end{vmatrix}_7$$

$$= K \begin{vmatrix} b+PK+c & c & \\ d & d+QK+e & e \\ & f & f+RK+g \end{vmatrix}_3$$

4. $$\begin{vmatrix} K & b & & & & \\ -1 & P & c & & & \\ & -1 & K & d & & \\ & & -1 & Q & e & \\ & & & -1 & K & f \\ & & & & -1 & R \end{vmatrix}_6$$

$$= \begin{vmatrix} b+PK+c & c & \\ d & d+QK+e & e \\ & f & f+RK \end{vmatrix}_3$$

5. $$\begin{vmatrix} x & b & & & \\ -1 & x & c & & \\ & -1 & x & d & \\ & & -1 & x & e \\ & & & -1 & x \end{vmatrix}_5 = \frac{1}{x} \begin{vmatrix} x^2+b & b & \\ c & c+x^2+d & d \\ & e & e+x^2 \end{vmatrix}_3$$

6. $$\begin{vmatrix} x & b & & & & \\ -1 & x & c & & & \\ & -1 & x & d & & \\ & & -1 & x & e & \\ & & & -1 & x & f \\ & & & & -1 & x \end{vmatrix}_6 = \begin{vmatrix} x^2+b & b & \\ c & c+x^2+d & d \\ & e & e+x^2+f \end{vmatrix}_3$$

7. $$\begin{vmatrix} \cdot & b & & & & \\ -1 & 1 & c & & & \\ & -1 & \cdot & d & & \\ & & -1 & 1 & e & \\ & & & -1 & \cdot & f \\ & & & & -1 & 1 \end{vmatrix}_6 = \begin{vmatrix} b & b & \\ c & c+d & d \\ & e & e+f \end{vmatrix} = bdf$$

8. $$\begin{vmatrix} \cdot & b & & & \\ -1 & 1 & c & & \\ & -1 & \cdot & d & \\ & & -1 & 1 & e \\ & & & -1 & \cdot \end{vmatrix}_5 = \begin{vmatrix} b & b & \\ c & c+d & d \\ & e & e \end{vmatrix} = 0$$

9. $$\begin{vmatrix} 1 & b & & & & & \\ -1 & \cdot & c & & & & \\ & -1 & 1 & d & & & \\ & & -1 & \cdot & e & & \\ & & & -1 & 1 & f & \\ & & & & -1 & \cdot & g \\ & & & & & -1 & 1 \end{vmatrix}_7 = \begin{vmatrix} b+c & c & \\ d & d+e & e \\ & f & f+g \end{vmatrix}$$
$$= ceg + beg + bdg + bdf$$

568. *If Δ be a continuant wherein each main diagonal element, except the first and last is the sum of the minor diagonal elements in the same row, say*

$$\Delta \equiv \begin{vmatrix} d_1+d_2 & d_2 & & & \\ d_3 & d_3+d_4 & d_4 & & \\ & & \cdot & & \\ & & & \cdot & \\ & & d_{2m-3} & d_{2m-3}+d_{2m-2} & d_{2m-2} \\ & & & d_{2m-1} & d_{2m-1}+d_{2m} \end{vmatrix}_m$$

then

$$\Delta = \sum_{0}^{m} i d_1 d_3 \cdots d_{2i-1} d_{2i+2} d_{2i+4} \cdots d_{2m}. \tag{1}$$

To prove this put $\phi=1$ and $\theta_1=\theta_2=\cdots=0$ in (2) of §566 and we have

$$\Delta = \begin{vmatrix} 1 & d_1 & & & & \\ -1 & 0 & d_2 & & & \\ & -1 & 1 & d_3 & & \\ & & & \cdot & & \\ & & & & \cdot & \\ & & & & & \cdot \end{vmatrix}_{2m} = \begin{vmatrix} d_1+d_2 & d_2 & & & \\ d_3 & d_3+d_4 & d_4 & & \\ & & \cdot & & \\ & & & \cdot & \\ & & & & \cdot \end{vmatrix}_{m}$$

But on expanding Δ we see that the only non-zero terms are those having a 1 from the main diagonal and the remaining factors from the minor diagonals, and these are just the terms included in the sigma.

If in (1) we put $d_{2k-1}=-d_{2k+2}$ $(k=1, 2, \cdots, m)$ then we get

$$\begin{vmatrix} d_2-d_4 & d_2 & & & \\ -d_6 & d_4-d_6 & d_4 & & \\ & & \cdot & & \\ & & & \cdot & \\ & & & \cdot & \\ & & & -d_{2m+2} & d_{2m}-d_{2m+2} \end{vmatrix}_{m}$$

$$= d_4 d_6 \cdots d_{2m}(d_2 - d_4 + d_6 - d_8 + \cdots \pm d_{2m+2}). \tag{2}$$

The left-hand side of (2) might be written as a determinant of the $(2m+1)$st order; thus for $m=4$

$$\begin{vmatrix} d_2-d_4 & d_2 & & \\ -d_6 & d_4-d_6 & d_4 & \\ & -d_8 & d_6-d_8 & d_6 \\ & & -d_{10} & d_8-d_{10} \end{vmatrix}$$

$$= \begin{vmatrix} 1 & -d_4 & & & & & & & \\ -1 & \cdot & d_2 & & & & & & \\ & -1 & 1 & -d_6 & & & & & \\ & & -1 & \cdot & d_4 & & & & \\ & & & -1 & 1 & -d_8 & & & \\ & & & & -1 & \cdot & d_6 & & \\ & & & & & -1 & 1 & -d_{10} & \\ & & & & & & -1 & \cdot & d_8 \\ & & & & & & & -1 & 1 \end{vmatrix}_9 .$$

569. *Any continuant is expressible as a continuant of one less than twice the order, the relation being:*

$$K(1,m) = \begin{vmatrix} a_1 & b_1 & & & \\ c_1 & a_2 & b_2 & & \\ & & \cdot & & \\ & & & \cdot & \\ & & & & \cdot \end{vmatrix}_m$$

$$= \begin{vmatrix} a_1 - b_1 & b_1 & & & & & & \\ -1 & 1 & c_1 & & & & & \\ & -1 & a_2 - c_1 - b_2 & b_2 & & & & \\ & & -1 & 1 & c_2 & & & \\ & & & & \cdot & & & \\ & & & & & \cdot & & \\ & & & & & & \cdot & \\ & & & & & -1 & 1 & c_{m-1} \\ & & & & & & -1 & a_m - c_{m-1} \end{vmatrix}_{2m-1}$$

The truth of this is readily seen on putting $\phi = 1$,

$\theta_r = a_r - c_{r-1} - b_r (r = 1, 2, \cdots, m)$,

$$\left\{ \begin{matrix} c_0 = 0 \\ b_m = 0 \end{matrix} \right\}, \; d_{2r-1} = b_r, \; d_{2r} = c_r$$

in (1) of §567.

570. If on the continuant

$$\begin{vmatrix} \frac{b_1}{s_1}+A & \frac{b_1}{s_1} & & & \\ s_1c_1 & \frac{b_2}{s_2}+A+s_1c_1 & \frac{b_2}{s_2} & & \\ & s_2c_2 & \frac{b_3}{s_3}+A+s_2c_2 & \frac{b_3}{s_3} & \\ & & s_3c_3 & \frac{b_4}{s_4}+A\cdot\frac{s_3}{r_3}+s_3c_3 & \frac{b_4}{s_4} \\ & & & s_4c_4 & \frac{b_5}{s_5}+\frac{s_4}{r_4}A+s_4c_4 \end{vmatrix}$$

we perform the operations $\text{row}_1+\text{row}_2$, $\text{row}_2+\text{row}_3$, $\text{row}_3+s_3/r_3\ \text{row}_4$ followed by $\text{col}_2-\text{col}_1$, $\text{col}_3-\text{col}_2$, $\text{col}_4-\text{col}_3$, $\text{col}_5-\text{col}_4$ in order we get

$$\begin{vmatrix} \frac{b_1}{s_1}+A+s_1c_1 & \frac{b_2}{s_2} & & & \\ s_1c_1 & \frac{b_2}{s_2}+A+s_2c_2 & \frac{b_3}{s_3} & & \\ & s_2c_2 & \frac{b_3}{s_3}+A+r_3c_3 & \frac{b_4}{s_4}\cdot\frac{r_3}{s_3} & \\ & & r_3c_3 & \frac{b_4}{s_4}+\frac{s_3}{r_3}A & -\frac{s_3}{r_3}A \\ & & & s_4c_4 & \frac{b_5}{s_5}+\frac{s_4}{r_4}A \end{vmatrix}$$

EXAMPLES:

1. $$\begin{vmatrix} A+b_1 & b_1 & & & \\ c_1 & c_1+A+b_2 & b_2 & & \\ & c_2 & c_2+A+b_3 & b_3 & \\ & & c_3 & c_3+A+b_4 & b_4 \\ & & & c_4 & c_4+A+b_5 \end{vmatrix}$$

$$= \begin{vmatrix} b_1 + A + c_1 & b_2 & & & \\ c_1 & b_2 + A + c_2 & b_3 & & \\ & c_2 & b_3 + A + c_3 & b_4 & \\ & & c_3 & b_4 + A + c_4 & b_5 \\ & & & c_4 & b_5 + A \end{vmatrix}$$

2. Show that

$$K(a_1 + x \,{}^{w+a_1x} a_2 \,{}^{w+a_2x} \cdots {}^{w+a_{n-1}x} a_n) = K(a_1 \,{}^{w+a_2x} a_2 \,{}^{w+a_3x} \cdots {}^{w+a_nx} a_n + x).$$

That the continuant

$$\begin{vmatrix} a_1 + x & w + a_1x & & & & \\ -1 & a_2 & w + a_2x & & & \\ & & \cdot & & & \\ & & & \cdot & & \\ & & & & \cdot & \\ & & & -1 & a_{n-1} & w + a_{n-1}x \\ & & & & -1 & a_n \end{vmatrix}$$

is equal to the continuant

$$\begin{vmatrix} a_1 & w + a_2x & & & & \\ -1 & a_2 & w + a_3x & & & \\ & & \cdot & & & \\ & & & \cdot & & \\ & & & & \cdot & \\ & & & -1 & a_{n-1} & w + a_nx \\ & & & & -1 & a_n + x \end{vmatrix}$$

may be seen on performing the following operations:

$$r_1 + xr_2,\ r_2 + xr_3, \ldots$$
$$\mathrm{col}_2 - x\,\mathrm{col}_1,\ \mathrm{col}_3 - x \cdot \text{the new } \mathrm{col}_2, \cdots.$$

571. If in $K(1, n)_d$, $d_1 = d_2 = \cdots = d_{n-1} = 1$ the result is called a *simple continuant*, and is denoted by $K(a_1a_2 \cdots a_n)$, $K(1, n)_1$ or $K(1, n)$ when no ambiguity can arise. The terms of $K(a_1a_2 \cdots a_n)$ consist of $a_1a_2 \cdots a_n$ and all the terms formed from this by omitting one or more pairs of consecutive a's.

Thus

$$K(a_1a_2a_3) = a_1a_2a_3 + a_1 + a_3$$
$$K(a_1a_2a_3a_4) = a_1a_2a_3a_4 + a_1a_4 + a_3a_4 + 1.$$

For simple continuants some of the foregoing relations take the following forms:

(1) $K(1,n)_1 = K(1,r)_1K(r+1,n)_1 + K(1,r-1)_1K(r+2,n)_1$

(2) $K(1,n)_1 = \{K(1,h)_1K(h+1,r)_1 + K(1,h-1)_1K(h+2,r)\} \times K(r+1,n)_1 + \{K(1,h)_1K(h+1,r-1)_1 + K(1,h-1)_1K(h+2,r-1)_1\}K(r+2,n)_1$

(3) $K(1,n)_1 = a_1K(2,n)_1 + K(3,n)_1$

(4) $K(1,n)_1 = a_nK(1,n-1)_1 - K(1,n-2)_1$

(5) $K(1,n)_1 = a_rK(1,r-1)_1K(r+1,n)_1 + K(1,r-1)_1K(r+2,n)_1 + K(1,r-2)_1K(r+1,n)_1$

(6) $K(1,n)_1K(h,r)_1 = K(1,r)_1K(h,n)_1 - (-1)^{r-h}K(1,h-2)_1K(r+2,n)_1$

(7) $K(1,n)_1K(2,n-1)_1 - K(1,n-1)_1K(2,n)_1 + (-1)^{n-1} = 0$

(8) $K(1,n)_1K(2,n-2)_1 - K(1,n-2)_1K(2,n)_1 + (-1)^n = 0.$

572. From (3) §571 we have

$$\frac{K(1,n)_1}{K(2,n)_1} = a_1 + \frac{1}{\dfrac{K(2,n)_1}{K(3,n)_1}} = a_1 + \frac{1}{a_2 + \dfrac{K(3,n)_1}{K(4,n)_1}}$$

$$= a_1 + \frac{1}{a_2 +}\,\frac{1}{a_3 +}\,\frac{1}{a_4 +}\cdots$$

From (7) we have

$$\frac{K(1,n)_1}{K(2,n)_1} - \frac{K(1,n-1)_1}{K(2,n-1)_1} = \frac{(-1)^n}{K(2,n)_1K(2,n-1)_1}$$

and since we may write

$$\frac{K(1,n)_1}{K(2,n)_1} = \frac{K(1,1)_1}{K(2,1)_1} + \left\{\frac{K(1,2)_1}{K(2,2)_1} - \frac{K(1,1)_1}{K(2,1)_1}\right\} + \cdots + \left\{\frac{K(1,n)_1}{K(2,n)_1} - \frac{K(1,n-1)_1}{K(2,n-1)_1}\right\}$$

we have

$$\frac{K(1,n)_1}{K(2,n)_1} = a_1 - \frac{1}{K(2,1)_1K(2,2)_1} - \frac{1}{K(2,2)_1K(2,3)_1} + \cdots + \frac{(-1)^n}{K(2,n-1)_1K(2,n)_1}.$$

EXAMPLE: Show that

$$\frac{K\begin{pmatrix} & b_1 & b_2 & \cdots & b_{n-1} & \\ a_1 & a_2 & a_3 & \cdots & & a_n \end{pmatrix}}{K\begin{pmatrix} & b_2 & \cdots & b_{n-1} & \\ a_2 & a_3 & \cdots & & a_n \end{pmatrix}} = a_1 + \cfrac{b_1}{a_2 + \cfrac{b_2}{a_3 + \cdots + \cfrac{b_{n-1}}{a_n}}}$$

$$= \frac{d}{da_1} \log_e K.$$

573. *The two simple continuants* $K(1, n)_1$ and $K(2, n)_1$ *are prime to each other provided* $a_1, a_2, \cdots, a_n$ *are integers.*

For the necessary and sufficient condition that two integers c and d be relatively prime is that there exist two integers x and y such that $xc+yd=1$. But

$$K(1,n)_1K(2,n-1)_1 - K(1,n-1)_1K(2,n)_1 = (-1)^n$$

and therefore if the a's are integers or such as to make all four continuants integers it follows that $K(1, n)_1$ and $K(2, n)_1$ are prime to each other.

574. The continuant $K(1, n)$ will factor into

$$a_1\left(a_2 - \frac{b_2c_2}{a_1}\right)\left(a_3 - \frac{b_3c_3a_1}{K(1,2)}\right)\left(a_4 - \frac{b_4c_4K(1,2)}{K(1,3)}\right)\cdots$$

$$\left(a_n - \frac{b_nc_nK(1,n-2)}{K(1,n-1)}\right)$$

since this product is equal by (4) §547 to

$$a_1\cdot\frac{K(1,2)}{a_1}\cdot\frac{K(1,3)}{K(1,2)}\cdots\frac{K(1,n)}{K(1,n-1)}=K(1,n).$$

This shows that if

$$a_r=\frac{b_rc_rK(1,r-2)}{K(1,r-1)}$$

for any value of r, then $K(1, n)=0$.

575. The continuant

$$K\begin{pmatrix} & \delta_2 & & (\alpha_2+\delta_2)\delta_3 & & \cdots & & (\alpha_{n-1}+\delta_{n-1})\delta_n & \\ 1 & & \alpha_2 & & \alpha_3 & & \alpha_{n-1} & & \alpha_n \end{pmatrix}$$

factors into $(\alpha_2+\delta_2)(\alpha_3+\delta_3)\cdots(\alpha_n+\delta_n)$.

For by §549 the given continuant reduces to

$$K\begin{pmatrix} & (\alpha_2+\delta_2)\delta_3 & & \cdots & & (\alpha_{n-1}+\delta_{n-1})\delta_n & \\ \delta_2+\alpha_2 & & \alpha_3 & & \alpha_{n-1} & & \alpha_n \end{pmatrix}$$

$$=(\alpha_2+\delta_2)K\begin{pmatrix} & \delta_3 & & (\alpha_3+\delta_3)\delta_4 & & \cdots \\ 1 & & \alpha_3 & & \alpha_4 & \end{pmatrix}$$

$$=(\alpha_2+\delta_2)(\alpha_3+\delta_3)K\begin{pmatrix} & \delta_4 & & (\alpha_4+\delta_4)\delta_5 & & \cdots \\ 1 & & \alpha_4 & & \alpha_5 & \end{pmatrix}$$

$$=(\alpha_2+\delta_2)(\alpha_3+\delta_3)\cdots(\alpha_n+\delta_n).$$

576. *The continuant*

$$\Delta_n\equiv\begin{vmatrix} a & b & & & & \\ -(n-1)c & a-(b+c) & 2b & & & \\ & -(n-2)c & a-2(b+c) & 3b & & \\ & & & \cdot & & \\ & & & & \cdot & \\ & & & & \cdot & \\ & & & & -c & a-(n-1)(b+c) \end{vmatrix}_n$$

$\equiv\phi_n(a,b,c)$ say,

$$=(a-\overline{n-1}c)(a-\overline{n-2}c-b)(a-\overline{n-3}c-2b)\cdots(a-\overline{n-1}b).$$

That $(a-\overline{n-1}c)$ is a factor is readily seen on adding all the rows to the first. Then taking out the factor common to the first row and reducing the order of what remains by one we have

$$\Delta_n = (a - \overline{n-1}c)\Delta'_{n-1}$$

say, and if we add to each row of Δ'_{n-1} the sum of all the rows below it and then perform the operations

$$c_n - c_{n-1},\ c_{n-1} - c_{n-2},\ \cdots,\ c_2 - c_1$$

we get

$$\phi_n(a,b,c) = (a - \overline{n-1}c)\phi_{n-1}(a-b,b,c)$$

and hence the truth of the theorem.

Another, but longer way to get the remaining factors is to perform on Δ'_{n-1} the following operations:

$$\text{row}_1 + 2\,\text{row}_2 + 3\,\text{row}_3 + \cdots + (n-1)\,\text{row}_{n-1}$$

when it will appear that $\{a-(n-2)c-b\}$ is a factor. Taking it out and reducing the order of the cofactor by one we have

$$\Delta'_{n-1} = (a - \overline{n-2}c - b)\Delta'_{n-2} \text{ say.}$$

Performing on Δ'_{n-2} the operations $\text{row}_1+3\,\text{row}_2+6\,\text{row}_3+\cdots$ the next factor $a-(n-3)c-2b$ appears. The multipliers of the rows to bring out the successive factors are indicated by the triangular array of integers

$$\begin{matrix} 1 & 1 & 1 & 1 & 1 & \cdots \\ 1 & 2 & 3 & 4 & \cdot & \cdots \\ 1 & 3 & 6 & \cdot & \cdot & \cdots \\ 1 & 4 & \cdot & \cdot & \cdot & \cdots \\ 1 & \cdot & \cdot & \cdot & \cdot & \cdots \end{matrix}$$

577. *The foregoing leads to the theorem that the value of the continuant* Δ_n *is not altered by adding to its matrix the matrix of the continuant*

$$D_n \equiv \begin{vmatrix} (n-1)x & x & & & & \\ (1-n)x & (n-3)x & 2x & & & \\ & (2-n)x & (n-5)x & & & \\ & & (3-n)x & & & \\ & & & \cdot & & \\ & & & \cdot & & \\ & & & & -(n-3)x & (n-1)x \\ & & & & -x & -(n-1)x \end{vmatrix}.$$

For treating D_n in the same manner as we have Δ_n it readily appears that the elements in the first row in each reduction are zero. The operations then on the combined matrix will produce the same results as on the original.

578. Two special cases of the theorem of §576 are worthy of notice.

First, when $c=-(b-1)$ we have the result, where for convenience n is taken equal to 4,

$$\begin{vmatrix} a & b & & \\ 3(b-1) & a-1 & 2b & \\ & 2(b-1) & a-2 & 3b \\ & & (b-1) & a-3 \end{vmatrix}$$
$$= (a-b-1)(a-3b)(a+3b-3)(a+b-2).$$

Second, when $c=-b=1$, we have $(n=6)$.

$$\begin{vmatrix} a & 1 & & & & \\ 5 & a & 2 & & & \\ & 4 & a & 3 & & \\ & & 3 & a & 4 & \\ & & & 2 & a & 5 \\ & & & & 1 & a \end{vmatrix} = (a^2-1^2)(a^2-3^2)(a^2-5^2).$$

A centrosymmetric continuant of even order may be expressed as the product of two continuants of half that order.

Thus the one last written is equal to the product of

$$\begin{vmatrix} a & 1 & \\ 5 & a & 2 \\ & 4 & a+3 \end{vmatrix} \quad \text{and} \quad \begin{vmatrix} a & 1 & \\ 5 & a & 2 \\ & 4 & a-3 \end{vmatrix}$$

If the first is denoted by $f(a)$ the second would be $(-1)^{n/2}f(-a)$. That is the factors of $f(a)$ are $(a+1)(a-3)(a+5)$ and those of $(-1)^{n/2}f(-a)$ are $(a-1)(a+3)(a-5)$.

579. We might assume a set of multipliers and then determine the most general continuant resolvable by means of them. Thus taking the set of column-multipliers

$$\left\{\begin{array}{cccccccccc}
1 & 1 & 1 & 1 & 1 & 1 & 1 & 1 & \cdots \\
 & 1 & -1 & 2 & -2 & 3 & -3 & 4 & \cdots \\
 & & 1 & 1 & 3 & 3 & 6 & 6 & \cdots \\
 & & & 1 & -1 & 4 & -4 & 10 & \cdots \\
 & & & & 1 & 1 & 5 & 5 & \cdots \\
 & & & & & 1 & -1 & 6 & \cdots \\
 & & & & & & 1 & 1 & \cdots \\
 & & & & & & & 1 & \cdots
\end{array}\right.$$

it may be found that *the continuant of even order* $2m$

$$\left|\begin{array}{cccc}
A_1 & \beta_1 & \cdot & \cdots \\
\gamma_1 & A_2 & \beta_2 & \cdots \\
\cdot & \gamma_2 & A_3 & \cdots \\
\cdot & \cdot & \cdot & \cdots
\end{array}\right|_{2m}$$

is resolvable into linear factors if we have

$$A_1 = a + mc,\ A_{2\theta} = a + \frac{\theta}{2\theta - 1}(2b - s + c),$$

$$A_{2\theta+1} = a + \frac{\theta}{2\theta + 1}(2b - s + c)$$

$$\beta_1 = b, \qquad \beta_{2\theta} = (m - \theta)c,\ \beta_{2\theta+1} = \frac{1}{2\theta + 1}\{\theta s - b - 2\theta(\theta + 1)c\}$$

$$\gamma_1 = s - b,\ \gamma_{2\theta} = (m + \theta)c,\ \gamma_{2\theta+1} = \frac{1}{2\theta + 1}\{(\theta+1)s - b + 2\theta(\theta+1)c\},$$

the factors being

$$\begin{array}{ll}
\{a + b + mc\} & \{a + b - s - (m - 2)c\} \\
\{a + b + (m - 2)c\} & \{a + b - s - (m - 4)c\} \\
\{a + b + (m - 4)c\} & \{a + b - s - (m - 6)c\} \\
\cdots\cdots & \cdots\cdots \\
\{a + b + (2 - m)c\} & \{a + b - s + mc\}
\end{array}$$

That this is true may be readily verified, but the actual work will be left to the reader.

In the case of odd order $2m-1$ *we have the continuant*

$$\begin{vmatrix} A_1 & \beta_1 & \cdot & \cdots \\ \gamma_1 & A_2 & \beta_2 & \cdots \\ \cdot & \gamma_2 & A_3 & \cdots \\ \cdot & \cdot & \cdot & \cdots \end{vmatrix}_{2m-1}$$

resolvable into factors if

$$A_1 = a + b, \qquad A_{2\theta} = a + \frac{\theta(2m-1)}{2(2\theta-1)}(c-s),$$

$$A_{2\theta+1} = a + \frac{\theta(2m-1)}{2(2\theta+1)}(c-s),$$

$$\beta_1 = (m-1)c, \qquad \beta_{2\theta} = b - \frac{\theta}{2}(s+c),$$

$$\beta_{2\theta+1} = \frac{m-\theta-1}{2\theta+1}\{(c+\theta(s+c)\},$$

$$\gamma_1 = s-(m-1)c, \quad \gamma_{2\theta} = b + \frac{\theta}{2}(s+c),$$

$$\gamma_{2\theta+1} = \frac{m+\theta}{2\theta+1}\{(\theta+1)(s+c)-c\},$$

the factors being

$$\{a+b+(m-1)c\}\{a-b+(m-1)c+(s+c)\}$$
$$\{a+b+(m-1)c-(s+c)\}\{a-b+(m-1)c+2(s+c)\}$$
$$\{a+b+(m-1)c-2(s+c)\}\{a-b+(m-1)c+3(s+c)\}$$
$$\cdots\cdots\cdots\cdots\cdots\cdots$$
$$\{a+b+(m-1)c-(m-1)(s+c)\}.$$

Special cases of these worthy of note are, for even order (1) when $s=2b+c$, (2) $s=2c$, (3) $c=2$, $b=2m-1$, $s=2b+c=4m$; for odd order (1) $s=-c$, (2) $s=c$, (3) $c=0$.

580. *The continuant*

$$K_a \equiv \begin{vmatrix} A_1 & 2(n-1)\beta_1 & & & \\ n\gamma_1 & A_2 & (n-2)\beta_2 & & \\ & (n+1)\gamma_2 & A_3 & (n-3)\beta_3 & \\ & & (n+2)\gamma_3 & A_4 & \\ & & & & \ddots \end{vmatrix}_n,$$

where

$$A_m = a - 2(m-1)^2c + \frac{(m-1)^2}{(2m-3)(2m-1)}\{2m(m-2)+3n\}5d$$

$$\beta_m = b - (m-1)c + \frac{(m-1)(m-2)}{2(2m-1)}5d$$

$$\gamma_m = b + mc - \frac{(m+1)m}{2(2m-1)}\cdot 5d$$

is resolvable into linear factors

$$\begin{aligned} X\{&X - 2Y - 1\cdot nZ\} \\ \{&X - 4Y - 2(n-1)Z\} \\ \{&X - 6Y - 3(n-2)Z\} \\ &\cdots\cdots\cdots \\ \{&X - (2n-2)Y - (n-1)2Z\}, \quad \textit{where} \end{aligned}$$

$$X = a + 2(n-1)b, \quad Y = 2b - 3c, \quad Z = 4c - 5d.$$

If we represent the continuant K_a by $f_n(a, b, c)$, then it may be seen that

$$f_n(a,b,c) = \{a + 2(n-1)b\}f_{n-1}\left(a - 4b + 2c + 5d, b - 2c + \frac{5d}{2}, c\right)$$

and hence the factors.

To obtain the nil-factor continuant put $X = Y = Z = 0$ and solving for a, b, c, in terms of d we obtain the continuant

$$\begin{vmatrix} -\frac{2}{1}(n-1)e & 2(n-1)e & & & \\ -ne & \frac{2}{1\cdot 3}(n+1)e & (n-2)\frac{1}{3}e & & \\ & -(n+1)\frac{1}{3}e & \frac{2}{3\cdot 5}(n+7)e & (n-3)\frac{1}{5}e & \\ & & -(n+2)\frac{1}{5}e & \frac{2}{5\cdot 7}(n+17)e & \\ & & & & \ddots \end{vmatrix}$$

where $e = 15\,d/8$.

581. *Similarly the continuant*

$$K_b \equiv \begin{vmatrix} A_1 & (n-1)\beta_1 & & & \\ (n+2)\gamma_1 & A_2 & (n-2)\beta_2 & & \\ & (n+3)\gamma_2 & A_3 & (n-3)\beta_3 & \\ & & (n+4)\gamma_3 & A_4 & \\ & & & & \cdot \\ & & & & \quad \cdot \\ & & & & \quad\quad \cdot \end{vmatrix}_n$$

where

$$A_m = a + 2(m^2-1)c + \frac{m^2-1}{4m^2-1}(2m^2+2+5n)\cdot 7d,$$

$$\beta_m = b - (m-1)c + \frac{(m-1)(m-2)}{2(2m+1)}\cdot 7d,$$

$$\gamma_m = b + (m+2)c - \frac{(m+3)(m+2)}{2(2m+1)}\cdot 7d,$$

is resolvable into the factors

$$\begin{aligned} X&\{X - 2Y - (n+2)Z\} \\ &\{X - 2\cdot 2Y - 2(n+1)Z\} \\ &\{X - 2\cdot 3Y - 3(n)Z\} \\ &\{X - 2\cdot 4Y - 4(n-1)Z\} \\ &\cdots\cdots\cdots\cdots\cdots\cdots \\ &\{X - 2(n-1)Y - (n-1)5Z\}, \end{aligned}$$

where $X = a + 2(n-1)b$, $Y = 2b - 5c$, $Z = 4c - 7d$.

Here if K_b is represented by $f_n(a, b, c)$ then it may be shown that

$$f_n(a,b,c) = \{a + 2(n-1)b\} f_{n-1}(a - 4b - 2c + 3\cdot 7d, b - 2c + \tfrac{1}{2}\cdot 7d, c)$$

582. If in K_a, $Y = -Z$, that is, if $2b - 3c = 4c - 5d$ or $5d = 2b + c$, then the last factor is like the first, the second last like the second, and so on. If, therefore, n is even, K_a is a perfect square. Again if $Y = Z = 0$, that is, if $8b = 12c = 15d$, then $K_a = X^n$. Similarly for K_b. The special cases of K_a and K_b when $d = 0$ should be noted.

EXERCISE: Show that the set of column-multipliers* which will give the factors of K_a is

$$\begin{array}{cccccc} 1, & 1, & 1, & 1, & 1, & \cdots 1 \\ & 1, & 4, & 9, & 16, & \cdots \frac{r}{1} C_{r,1} \\ & & 1, & 6, & 20, & \cdots \frac{r}{2} C_{r+1,3} \\ & & & 1, & 8, & \cdots \frac{r}{3} C_{r+2,5} \\ & & & & \cdots & \cdots \end{array}$$

and of K_b is

$$\begin{array}{cccccc} 1, & 2, & 3, & 4, & 5, & \cdots C_{r,1} \\ & 1, & 4, & 10, & 20, & \cdots C_{r+2,3} \\ & & 1, & 6, & 21, & \cdots C_{r+4,5} \\ & & & 1, & 8, & \cdots C_{r+6,7} \\ & & & & \cdots & \cdots \end{array}$$

583. *The continuant*

$$K_c \equiv \begin{vmatrix} A_1 & \beta_1 & \cdot & \cdots \\ \gamma_1 & A_2 & \beta_2 & \cdots \\ \cdot & \gamma_2 & A_3 & \cdots \\ \cdots & \cdots & \cdots & \cdots \end{vmatrix}$$

where

$$\begin{aligned} A_m &= a + (m-1)(b-c) - (m-1)\{n - 2(m-1)\}\gamma, \\ \beta_m &= (n-m)\{b + (m-1)\gamma\}, \\ \gamma_m &= m\{c - (m-1)\gamma\}, \end{aligned}$$

has as factors

$$\{a + (n-1)b\}\{a + (n-2)b - c + 1\cdot(n-2)\gamma\}$$
$$\{a + (n-3)b - 2c + 2(n-3)\gamma\} \cdots \{a - (n-1)c\}.$$

* Vide Muir, Trans. Roy. Soc. Edin., vol. xli, Part II, pp. 343–358.

This may be seen on using the column multipliers

$$\begin{matrix} 1, & 1, & 1, & 1, & 1, & \cdots \\ & 1, & 2, & 3, & 4, & \cdots \\ & & 1, & 3, & 6, & \cdots \\ & & & & 1, & \cdots \\ & & & & & \cdots \end{matrix}$$

Also if $K_c = f_n(a, b, c)$, then

$$f_n(a,b,c) = \{a + (n-1)b\} f_{n-1}(a - b - c - n\gamma, b + 2\gamma, c).$$

EXERCISES. SET XXIX

(1) Compare the sum of the factors with the sum of the principal diagonal elements in K_a and K_b.

(2) If in such continuants as K_a and K_c we put $a=0$, then the continuant reduces to the product of the elements in the places (1, 2) and (2, 1) times a continuant of order one less, and that the product of the elements in these two places is a multiple of the product of the first and last factors. It follows, therefore, that this reduced determinant must equal a multiple of the product of the remaining factors. Write the determinant with its equivalent factors in each case. Also put the elements in one or both of the positions $(r\ r)$, $(r-1, r-1)$ equal to zero and note the results.

(3) Find the value of the fraction

$$\frac{K}{K_{11}} \text{ when } K = \begin{vmatrix} z - c_0 & c_0 & \cdot & \\ c_1 & z - c_1 - c_2 & c_2 & \\ \cdot & c_3 & z - c_3 - c_4 & \\ & & & \cdot \\ & & & \quad\cdot \\ & & & \qquad\cdot \end{vmatrix}$$

and K_{11} is the principal minor of K complementary to (1, 1) and where

$$c_0 = \frac{m(m+1)}{2}, \quad c_{k-1} = \frac{1}{4}(m - k + 1)(m + k)(k = 2,3, \cdots, m)$$

(4) Find the nil-factor continuants corresponding to the set of column-multipliers §579 for both even and odd order.

(5) Add the nil-factor matrix to the matrix of the corresponding continuant and solve for x so as to cause the element in the place (1, 2) to vanish and thereby reduce the order by one.

584. In §579 we have seen that given a set of multipliers we can form the most general continuant factorable by the use of these multipliers. On the other hand if a given continuant is factorable we can determine the set of multipliers which will produce the factors. For instance a set of multipliers might be formed* to bring out the factors of such continuants as

$$T_a \equiv \begin{vmatrix} a & \frac{(2n-1)\beta}{2n-1} & & \\ \frac{1(\beta+2n-2)}{3-2n} & a & \frac{(2n-2)(\beta+1)}{2n-3} & \\ & \frac{2(\beta+2n-3)}{5-2n} & a & \\ & & & \ddots \end{vmatrix}_{2n}$$

where the non-zero elements in the kth row are

$$\frac{(k-1)(\beta+2n-k)}{(2k-1)-2n}, \quad a, \quad \frac{(2n-k)(\beta+k-1)}{2n-(2k-1)},$$

$$T_b \equiv \begin{vmatrix} a & \frac{(2n-1)\beta}{2n-1} & & \\ \frac{a+(2n-2)\gamma}{3-2n} & b & \frac{(2n-2)(a+\gamma)}{2n-3} & \\ & \frac{2\{\beta+(2n-3)\delta\}}{5-2n} & a & \\ & & & \ddots \end{vmatrix}_{2n}$$

Where the non-zero elements in the $(2k-1)$th row are

$$\frac{(2k-2)\{\beta+(2n-2k+1)\delta\}}{(4k-3)-2n}, \quad a, \quad \frac{(2n-2k+1)\{\beta+(2k-2)\delta\}}{2n-(4k-3)};$$

and in the $(2k)$th row are

$$\frac{(2k-1)\{a+(2n-2k)\gamma\}}{(4k-1)-2n}, \quad b, \quad \frac{(2n-2k)\{a+(2k-1)\gamma\}}{2n-(4k-1)},$$

* Such a set is given in Proc. Roy. Soc. Edinb., vol. 34, Part III, p. 224 and is not reproduced here.

$$T_c \equiv \left| \begin{array}{cccc} a & \dfrac{(2n-1)\alpha(\beta+2n-2)}{2n-1} & & \\ \dfrac{\gamma(\delta+2n-2)}{3-2n} & b & \dfrac{(2n-2)\{\delta(\gamma+2n-3)+\gamma\}}{2n-3} & \\ & \dfrac{2\{\beta(\alpha+2n-3)+\alpha\}}{5-2n} & a & \\ & & & \ddots \end{array} \right|_{2n}$$

where the non-zero elements in the $(2k-1)$th row are

$$\frac{(2k-2)\{\beta(\alpha+2n-2k+1)+(2k-3)\alpha\}}{(4k-3)-2n}, \quad a, \quad \frac{(2n-2k+1)\{\alpha(\beta+2n-2k)+(2k-2)\beta\}}{2n-(4k-3)};$$

and in the $(2k)$th row are

$$\frac{(2k-1)\{\gamma(\delta+2n-2k)+(2k-2)\delta\}}{(4k-1)-2n}, \quad b, \quad \frac{(2n-2k)\{\delta(\gamma+2n-2k-1)+(2k-1)\gamma\}}{2n-(4k-1)}.$$

If in T_a we add to every odd-numbered column the sum of odd-numbered columns which follow it, add to every even-numbered column the sum of all the even-numbered columns which follow it, and then subtract from every row the second row above it, the determinant breaks up into $(a^2-\beta^2)$ and a continuant of order $(2n-2)$, which on interchanging the denominators of conjugate elements is of the same form as T_a with n one less and β two more. Thus if T_a is represented by $f_{2n}(a, \beta)$ we have

$$\begin{aligned} (1) \qquad T_a = f_{2n}(a,\beta) &= (a^2-\beta^2)f_{2n-2}(a,\beta+2) \\ &= (a^2-\beta^2)\{a^2-(\beta+2)^2\}\{a^2 \\ &\quad -(\beta+4)^2\}\cdots\{a^2-(\beta+2n-2)^2\}. \end{aligned}$$

Similarly we may find that

$$\begin{aligned} (2) \qquad T_b = f_{2n}(ab,\alpha\beta) &= (ab-\alpha\beta)f_{2n-2}(ab,\overline{\alpha+2\gamma}\cdot\overline{\beta+2\delta}) \\ &= (ab-\alpha\beta)\{ab-(\alpha+2\gamma)(\beta+2\delta)\}\cdots \\ &\quad \{ab-(\alpha+\overline{2n-2}\cdot\gamma)(\beta+\overline{2n-2}\cdot\gamma)\} \end{aligned}$$

and

$$(3)\quad T_c = f_{2n}(ab, \alpha(2n-2)+\alpha\beta, \delta(2n-2)+\delta\gamma)$$
$$= \{ab - (\alpha\beta + \alpha\overline{2n-2})(\delta\gamma + \delta\overline{2n-2})\}$$
$$\times f_{2n-2}\{(ab, \alpha(2n-4)+\alpha\beta+2\beta, \delta(2n-4)+\delta\gamma+2\gamma)\}$$
$$= \{ab - \alpha(\beta+2n-2)\delta(\gamma+2n-2)\}$$
$$\{ab - (\alpha\cdot\overline{\beta+2n-4}+2\beta)(\delta\cdot\overline{\gamma+2n-4}+2\gamma)\}$$
$$\cdots\{ab - \beta(\alpha+2n-2)\gamma(\delta+2n-2)\}.$$

585. Since T_a is centro-symmetric it may be broken up into the product of two continuants each of order n. If we represent one of these by $f_n(a, \beta)$, then it is readily seen that the other will be

$(-1)^n f_n(-a, \beta)$ and

$$f_n(a,\beta) = (a+\beta)f_{n-1}(a, \beta+2),\ \text{hence}$$
$$f_n(a,\beta) = (a+\beta)(a+\beta+2)\cdots(a+\beta+2n-2),$$
$$f_n(-a,\beta) = (-1)^n(a-\beta)(a-\beta-2)\cdots(a-\beta-2n+2).$$

If in T_b we put $\beta=\alpha$, $\delta=\gamma$ and $b=a$ it is centro-symmetric and breaks up into two factors which we will denote by $f_n(a, \alpha, \gamma)$ and $F_n(a, \alpha, \gamma)$ respectively, where $f_n(a, \alpha, \gamma)$ is the sum of two terms and $F_n(a, \alpha, \gamma)$ is the difference of the same two terms.

As before it is seen that $F_n(a, \alpha, \gamma) = (-1)^n f_n(-a, \alpha, \gamma)$ and it is left to the reader to find the factors of $f_n(a, \alpha, \gamma)$ and $F_n(a, \alpha, \gamma)$.

586. It is readily seen that if in T_b we put

(1) $\gamma=\delta=0$, then $T_b=(ab-\alpha\beta)^n=(a^2-\alpha^2)^n$, if $b=a$ and $\beta=\alpha$

(2) $b=a$, $\beta=\alpha$ and $\gamma=\delta=1$ it reduces to T_a

(3) $\gamma=-\alpha$ and $\delta=-\beta$ two factors become alike and

$$T_b = (ab-\alpha\beta)^2(ab-9\alpha\beta)\cdots(ab-\overline{2n-3}^2\alpha\beta).$$

(4) $\alpha=\gamma=1/\beta=1/\delta$, and $b=a$, then

$$T_b = (a^2-1^2)(a^2-3^2)\cdots(a^2-\overline{2n-1}^2).$$

Other forms of the continuant may be seen to yield these same factors. Thus if $\beta=\alpha=2n-1$, $\gamma=\delta=-2$ or if $\beta=\alpha=2n-1$, $\gamma=\delta=-1$

(5) $\beta=-\alpha=2n-1$, $\gamma=-\delta=2$, then

$$T_b = (a^2+1^2)(a^2+3^2)\cdots(a^2+\overline{2n-1}^2)$$

(6) $\alpha=\beta=0$, $\gamma=\delta=1$. and $b=a$, then

$$T_b = a^2(a^2 - 2^2)(a^2 - 4^2) \cdots (a^2 - \overline{2n-2}^2)$$

(7) $\alpha=\beta=1$, $\gamma=\delta=\frac{1}{2}$ and $b=a$, then

$$T_b = (a^2 - 1^2)(a^2 - 2^2) \cdots (a^2 - n^2)$$

(8) $\alpha=-\beta=1$, $\gamma=-\delta=\frac{1}{2}$ and $b=a$, then

$$T_b = (a^2 + 1^2)(a^2 + 2^2) \cdots (a^2 + n^2)$$

(9) $\beta=-2n+1$, $\gamma=-2$, $b=a$, $\alpha=2n-1$, $\delta=2$, then

$$T_b = (a^2 + 1^2)(a^2 + 3^2) \cdots (a^2 + \overline{2n-3}^2)(a^2 + \overline{2n-1}^2)$$

etc.

587. If in T_c we put $b=a$, $\delta=\alpha$, $\gamma=\beta$, then it is centrosymmetric and breaks up into two factors which we will denote by $f_n(a, \alpha\beta)$ and $F_n(a, \alpha\beta)$ respectively,

$$f_n(a, \alpha\beta) = \{a + \alpha(\beta + 2n - 2)\} f_{n-1}(a, \alpha\beta + 2\beta)$$
$$F_n(a, \alpha\beta) = \{a - \alpha(\beta + 2n - 2)\} F_{n-1}(a, \alpha\beta + 2\beta).$$

If in T_c we put

(1) $\delta=\alpha$, $\gamma=\beta$ and $b=a$ then

$$T_c = \{a^2 - \alpha^2(\beta + 2n - 2)^2\}\{a^2 - (\alpha\cdot\overline{\beta + 2n - 4} + 2\beta)^2\} \cdots \{a^2 - \beta^2(\alpha + 2n - 2)^2\}.$$

(2) $\delta=\gamma=\beta=\alpha$ and $b=a$, then

$$T_c = \{a^2 - \alpha^2(\alpha + 2n - 2)^2\}^n$$

(3) $\delta=\gamma=\beta=\alpha=b=a$, then

$$T_c = a^{2n}(2n + a - 1)^n(3 - a - 2n)^n$$
$$T_c = 2^{2n}\{n(1 - n)\}^n, \text{ if } a = 1.$$

or if $a+2n-2=x$, then

$$T_c = (1^2 - x^2)^n(x - 2n + 2)^n = 0, \text{ if } x = \pm 1 \text{ or } 2n - 2$$

(4) $\delta=\gamma=\beta=\alpha=1$, $b=a=x(2n-1)$, then

$$T_c = (2n - 1)^{2n}(x^2 - 1)^n$$
$$= 0, \text{ if } x = \pm 1.$$

588. If we denote by $f_{2n}(x, \beta, b)$ the continuant

$$T_d \equiv \left|\begin{array}{cccccccc} x & b & & & & & & \\ -\beta+\overline{n-1}\cdot\delta & x & (n-1)\delta & & & & & \\ & -d & x & b-d & & & & \\ & & -\beta+\overline{n-2}\cdot\delta & x & (n-2)\delta & & & \\ & & & -2d & x & & & \\ & & & & \cdot\quad\cdot\quad\cdot & & & \\ & & & & & -(n-1)d & x & b-\overline{n-1}d \\ & & & & & & -\beta & x \end{array}\right|_{2n}$$

where the non-zero elements in the $(2k)$th row are

$$-\beta+(n-k)\delta,\ x,\ (n-k)\delta\ ;$$

and in the $(2k+1)$th row are

$$-kd, \qquad x, \qquad b-kd,$$

then it is easily seen that

$$\begin{aligned} f_{2n}(x,\beta,b) &= (x^2+\beta b)f_{2n-2}(x,\beta-\delta,b-d) \\ &= \{x^2+\beta b\}\{x^2+(\beta-\delta)(b-d)\}f_{2n-4}(x,\beta-2\delta,b-2d) \\ &= \{x^2+\beta b\}\{x^2+(\beta-\delta)(b-d)\}\cdots \\ &\qquad \{x^2+(\beta-\overline{n-1}\delta)(b-\overline{n-1}d)\}. \end{aligned}$$

The operations which show this are

$$\mathrm{col}_r - \mathrm{col}_{r+2} + \mathrm{col}_{r+4} - \cdots\ (r = 1,2,\cdots,2n-2)$$

followed by

$$\mathrm{row}_r + \mathrm{row}_{r-2},\ (r = 2n, 2n-1, \cdots, 3).$$

It should be observed that if in the factors of T_b we put

$$a = b = x,\quad \alpha = \beta,\quad \beta = -b,\quad \gamma = -\frac{\delta}{2},\quad \delta = \frac{d}{2},$$

then they become those of T_d. The two continuants should be compared under these conditions.

589. If in $K(1, n)_d$, $a_r = 0$, and $d_r = d_{r+1} = \delta$ then

$$K(1,n)_d = \delta[K(1,r-1)K(r+2,n) + K(1,r-2)K(r+1,n)].$$

This follows at once from expansion in terms of elements of the rth row.

If $a_r = d_r = d_{r+1} = 0$, then $K(1, n)_d = 0$.

If in $K(1, n)_d$, $a_r = a_{r+1} = 0$ and $d_r = d_{r+1} = \delta$ then

$$K(1,n)_d = \delta[K(1,r-1)K(r+2,n) + K(1,r-2)d_{r+2}K(r+3,n)].$$

If in $K(1, n)_d$, $a_r = a_{r+1} = a_{r+2} = 0$, $d_r = d_{r+1} = \delta$, and $d_{r+2} = d_{r+3} = \gamma$, then

$$K(1,n)_d = \gamma\delta[K(1,r-1)K(r+4,n) + K(1,r-2)K(r+3,n)]$$

590. If

$$\begin{aligned} c_1 &= a_1\alpha + \delta, \\ c_2 &= a_2c_1 + d_2\alpha + \pi_1, \\ c_3 &= a_3c_2 + d_3c_1 + \pi_2, \\ &\cdots\cdots\cdots\cdots \\ c_r &= a_rc_{r-1} + d_rc_{r-2} + \pi_{r-1} \end{aligned}$$

then, solving for c_r we get, after increasing the order by one and simplifying:

$$c_r = \begin{vmatrix} \alpha & \delta & \pi_1 & \cdots & \pi_{r-1} \\ -1 & a_1 & d_2 & \cdots & \cdot \\ & -1 & a_2 & \cdots & \cdot \\ \cdot & \cdot & \cdot & \cdot & \cdot \\ & & & a_{r-1} & d_r \\ & & & -1 & a_r \end{vmatrix}_{r+1}$$

and hence

$$c_r = \alpha K(1,r)_d + \delta K(2,r)_d + \pi_1 K(3,r)_d + \cdots + \pi_{r-2}a_r + \pi_{r-1}.$$

591. If F_n is used to represent $\{[a]x + [b]y\}^n$, where

$$[a]^r = a(a-1)(a-2)\cdots(a-r+1),$$

then from the relation

$$\begin{aligned} \{[a]x + [b]y\}^{n+1} = \{(a-n)x + (b-n)y\}\{[a]x + [b]y\}^n \\ - nxy(a+b-n+1)\{[a]x + [b]y\}^{n-1} \end{aligned}$$

we get

$$\begin{aligned} &F_1 - (ax + by) &&= 0 \\ &F_2 - \{(a-1)x + (b-1)y\}F_1 + xy(a+b) &&= 0 \\ &F_3 - \{(a-2)x + (b-2)y\}F_2 + 2xy(a+b-1)F_1 &&= 0 \\ &F_4 - \{(a-3)x + (b-3)y\}F_3 + 3xy(a+b-2)F_2 &&= 0 \\ &\cdots\cdots\cdots\cdots\cdots\cdots \end{aligned}$$

and hence

$$\{[a]x + [b]y\}^n$$

$$= \begin{vmatrix} ax + by & xy & & \\ a + b & (a-1)x + (b-1)y & 2xy & \\ & a + b - 1 & (a-2)x + (b-2)y & \cdots \\ & & a + b - 2 & \cdots \\ \cdots & \cdots & \cdots & \cdots \end{vmatrix}_n$$

Exercise: Show that

$$\begin{vmatrix} 1 & (a-2)x - by & 1 & \cdot \\ 3y & (a-2)(x^2 + xy) & ax - (b-1)y & 2 \\ 3y^2 & \cdot & (a-1)(x^2 - xy) & (a+2)x - (b-2)y \\ y^3 & \cdot & \cdot & a(x^2 + xy) \end{vmatrix}$$

$$= \begin{vmatrix} (a-2)x^2 + by^2 & 1 & \cdot \\ (a-2)(x^4 - x^2y^2) & ax^2 + (b-1)y^2 & 2 \\ \cdot & (a-1)(x^4 - x^2y^2) & (a+2)x^2 + (b-2)y^2 \end{vmatrix}$$

$$= \{[a]x^2 + [b]y^2\}^3.$$

592. If A in §310 is a continuant then the quotient Δ_{11}/Δ is a continued fraction which is expressed as a series in descending powers of x.

Exercises: 1. Under the conditions that A is a continuant show that the coefficients $A_1, A_2, A_3, \cdots$, can be expressed in the form

$$\begin{vmatrix} S_1 - \sigma_1 & 1 & \\ S_2 - \sigma_2 & S_1 & 1 \\ \cdots & \cdots & \cdots \end{vmatrix}$$

where S_k denotes the sum of the coaxial minors of A of order k, and where σ_k denotes the same for A_{11}. (Muir.)

2. Show that A_k is equal to the (1, 1)th or leading element in the kth power of the continuant. (Whittaker.)

593. If S represents the continued fraction

$$\frac{n}{a_1 +} \frac{n + a_1}{a_2} + \frac{n + a_2}{a_3} + \cdots + \frac{n + a_{r-1}}{a_r},$$

that is

$$S = \frac{nK\begin{pmatrix} & n + a_2 & \cdots & n + a_{r-1} & \\ a_2 & & \cdots & & a_r \end{pmatrix}}{K\begin{pmatrix} & n + a_1 & n + a_2 & \cdots & n + a_{r-1} & \\ a_1 & & a_2 & \cdots & & a_r \end{pmatrix}}$$

then

$$S + 1 = \frac{K\begin{pmatrix} & n & n + a_1 & \cdots & n + a_{r-1} & \\ 1 & & a_1 & \cdots & & a_r \end{pmatrix}}{K\begin{pmatrix} & n + a_1 & n + a_2 & \cdots & n + a_{r-1} & \\ a_1 & & a_2 & \cdots & & a_r \end{pmatrix}}$$

$$= \frac{(n + a_1)K\begin{pmatrix} & n + a_2 & n + a_3 & \cdots & n + a_{r-1} & \\ a_2 + 1 & & a_3 & \cdots & & a_r \end{pmatrix}}{K\begin{pmatrix} & n + a_1 & \cdots & n + a_{r-1} & \\ a_1 & & \cdots & & a_r \end{pmatrix}}$$

Also

$$S + n = \frac{nK\begin{pmatrix} & n + a_1 & \cdots & n + a_{r-1} & \\ a_1 + 1 & & \cdots & & a_r \end{pmatrix}}{K\begin{pmatrix} & n + a_1 & \cdots & n + a_{r-1} & \\ a_1 & & \cdots & & a_r \end{pmatrix}}.$$

Therefore

$$\frac{n(S + 1)}{S + n} = \frac{(n + a_1)K\begin{pmatrix} & n + a_2 & \cdots & n + a_{r-1} & \\ a_2 + 1 & & & & a_r \end{pmatrix}}{K\begin{pmatrix} & n + a_1 & \cdots & n + a_{r-1} & \\ a_1 + 1 & & \cdots & & a_r \end{pmatrix}}$$

$$= \frac{(n+a_1)K\begin{pmatrix} n+a_3 & n+a_4 \cdots n+a_r & \\ a_2 & a_3 \quad\cdots & a_r+1 \end{pmatrix}}{K\begin{pmatrix} n+a_2 & n+a_3 \cdots n+a_r & \\ a_1 & a_2 \quad\cdots & a_r+1 \end{pmatrix}}$$

by ex. 2 §570

$= T$, where

$$T = \frac{n+a_1}{a_1} + \frac{n+a_2}{a_2} + \frac{n+a_3}{a_3} + \cdots + \frac{n+a_r}{a_r+1}.$$

Exercise: If

$$S = \frac{b_1c_0}{d+b_2-c_0} + \frac{b_2c_1}{d+b_3-c_1} + \cdots + \frac{b_rc_{r-1}}{d+b_{r+1}-c_{r-1}}$$

and

$$T = \frac{b_2c_0}{d+b_2-c_1} + \frac{b_3c_1}{d+b_3-c_2} + \cdots + \frac{b_{r+1}c_{r-1}}{d+b_{r+1}}$$

show that

$$\frac{(c_0-d)S+b_1c_0}{S+b_1} = T.$$

594. If we represent the continuant

$$\begin{vmatrix} a_0+a_1 & -\lambda_1a_1 & & \\ -\lambda_1a_1 & a_1+a_2 & -\lambda_2a_2 & \\ & -\lambda_2a_2 & a_2+a_3 & \\ & & & \ddots \end{vmatrix}_n, \quad \text{by } D_n$$

and if the a's are all positive and $\lambda_r^2 > 1$ for all values of r then D_n is necessarily positive.

For it is readily seen that

$$\begin{aligned} D_n &= (a_{n-1} + a_n)D_{n-1} - \lambda_{n-1}^2 a_{n-1}^2 D_{n-2} \\ &= a_n D_{n-1} + a_{n-1}(D_{n-1} - \lambda_{n-1}^2 a_{n-1} D_{n-2}) \end{aligned}$$

which shows that $D_n > a_n D_{n-1}$ if $D_{n-1} > a_{n-1}D_{n-2}$, but obviously $D_2 > a_2 D_1$, and therefore D_n is positive.

EXERCISES

1. If in D_n the λ's are all put equal to unity show that the resulting continuant is equal to

$$a_0 a_1 \cdots\cdots a_n \left(\frac{1}{a_0} + \frac{1}{a_1} + \cdots + \frac{1}{a_n}\right).$$

2. Show that the sum of the signed primary minors of D_n is

$$a_0 a_1 a_2 \cdots a_n(\sigma_1 + 4\sigma_2 + 9\sigma_3 + \cdots + n^2\sigma_n) = S \text{ say},$$

where

$$\sigma_k = \frac{1}{a_0 a_k} + \frac{1}{a_1 a_{k+1}} + \frac{1}{a_2 a_{k+2}} + \cdots + \frac{1}{a_{n-k} a_n}.$$

3. Show that

$$\begin{vmatrix} x + a_0 + a_1 & x - a_1 & & & \\ x - a_1 & x + a_1 + a_2 & x - a_2 & & \\ & x - a_2 & x + a_2 + a_3 & & \\ & & & \cdot & \\ & & & & \cdot \\ & & & & & \cdot \end{vmatrix}_n = D_n + xS.$$

EXERCISES. SET XXX

1. Evaluate the continuant

$$\begin{vmatrix} \theta & x & & & \\ 1 & \theta & 2(x-1) & & \\ & 1 & \theta & 3(x-2) & \\ & & & \cdot & \\ & & & & \cdot \\ & & & & & \cdot \end{vmatrix}_n$$

2. Show that

$$\begin{vmatrix} -r & a\lambda & & & & & \\ f\mu & -q & b\lambda & & & & \\ & e\mu & -p & c\lambda & & & \\ & & d\mu & \cdot & d\lambda & & \\ & & & c\mu & p & e\lambda & \\ & & & & b\mu & q & f\lambda \\ & & & & & a\mu & r \end{vmatrix} = 0$$

3. Show that

$$\begin{vmatrix} 1+x^2 & x & & & \\ x & 1+x^2 & x & & \\ & x & 1+x^2 & & \\ & & & \ddots & \\ & & & & \cdot \end{vmatrix}_n = 1 + x^2 + x^4 + \cdots + x^{2n}$$

4. Show that

$$(a^2+b)^{1/2} = \begin{vmatrix} |a| & b & & & \\ -1 & 2|a| & b & & \\ & -1 & 2|a| & & \\ & & & \ddots & \\ & & & & \cdot \end{vmatrix} \div \begin{vmatrix} 2|a| & b & & \\ -1 & 2|a| & b & \\ & & \ddots & \\ & & & \cdot \end{vmatrix}$$

5. If

$$A_{n+1} \equiv \begin{vmatrix} 2x & 1 & & & & \\ 2n & 2x & 1 & & & \\ & 2(n-1) & 2x & 1 & & \\ & & 2(n-2) & 2x & & \\ & & & & \ddots & \\ & & & & & \cdot \end{vmatrix}_{n+1},$$

show that the first derivative is $2(n+1)A_n$.

6. Show that the continuant

$$\begin{vmatrix} 1 & x & & & & & \\ a+n-1 & a & x & & & & \\ & 1(1-n) & a+1 & x & & & \\ & & a(a+n) & a+2 & x & & \\ & & & 2(2-n) & a+3 & x & \\ & & & & (a+1)(a+n+1) & a+4 & \\ & & & & & & \ddots \end{vmatrix}_{2n}$$

$$= a(a+1)(a+2)\cdots(a+2n-3)(a+2n-2)(1-x)^n$$

7. Show that the continuant

$$\begin{vmatrix} 1 & x & & \\ e_2 & g_2 & x & \\ & e_3 & g_3 & \\ & & & \ddots \end{vmatrix}_{2n}$$

$$= \{(1+y)(1+qy)\cdots(1+q^{2n-3}y)\}\{(1-x)(1-qx)\cdots(1-q^{n-1}x)\}$$

where

$$e_{2m} = (1+q^{m-2}y)(1+q^{m+n-2}y)q^{m-1},$$

$$e_{2m+1} = (q^m-1)(q^{n-m}-1)yq^{2m-2},$$

and

$$g_r = 1+q^{r-3}y.$$

8. If K_n denote the persymmetric continuant

$$\begin{vmatrix} c & a & & \\ b & c & a & \\ & b & c & a \\ \cdot & \cdot & \cdot & \cdot\ \cdot\ \cdot\ \cdot \end{vmatrix}_n$$

show that $K_n = cK_{n-1} - abK_{n-2}$ and

$$K_n = \frac{u^{n+1}-v^{n+1}}{(u-v)2^{n+1}},$$

where u and v are the roots of $x^2 - cx + ab = 0$.

9. From §112 it is known that the determinant

$$D_{n+1} \equiv \begin{vmatrix} \cdot & 1 & 1 & 1 \cdots \\ 1 & c & a & \cdots \\ 1 & b & c & a \cdots \\ \cdot & \cdot & \cdot & \cdot \end{vmatrix}_{n+1}$$

is the sum of the signed primary minors of K_n in the previous example. Show that

$$D_{n+1} + (a+b)D_n + abD_{n-1} = -nK_{n-1}$$

and hence

$$\begin{aligned} D_{n+1} = & -nK_{n-1} + (n-1)(a+b)K_{n-2} \\ & -(n-2)(a^2+ab+b^2)K_{n-3} + \cdots \\ & + (-1)^n(a^{n-1} + a^{n-2}b + \cdots + b^{n-1}). \end{aligned}$$

10. Show that

$$\cos n\theta = \begin{vmatrix} \cos\theta & 1 & & \\ 1 & 2\cos\theta & 1 & \\ & 1 & 2\cos\theta & 1 \\ \cdot & \cdot & \cdot & \cdot \end{vmatrix}_n$$

11. Show that

$$K\begin{pmatrix} & -b^2 & -b^2 \cdots & -b^2 & \\ a+b & & a & \cdots & a+b \end{pmatrix}_n$$

$$= (a+2b)K\begin{pmatrix} & -b^2 & -b^2 \cdots & -b^2 & \\ a & & a & \cdots & a \end{pmatrix}_{n-1}$$

12. Show that

$$D_r \equiv \begin{vmatrix} 1 & n-1 & & \\ -1 & 2 & n-1 & \\ & -1 & 2 & n-1 \\ \cdot & \cdot & \cdot & \cdot \end{vmatrix}_r = \frac{((n)^{1/2}+1)^r + (-1)^r((n)^{1/2}-1)^r}{2}$$

and

$$D'_r \equiv \begin{vmatrix} 2 & n-1 & & \\ -1 & 2 & n-1 & \\ & -1 & 2 & n-1 \\ \cdot & \cdot & \cdot & \cdot \end{vmatrix}_r = \frac{((n)^{1/2}+1)^{r+1} + (-1)^r((n)^{1/2}-1)^{r+1}}{2(n)^{1/2}}$$

and hence we have a convergent to $\sqrt{n}$ in the quotient of the two determinants.

CHAPTER XIV

ORTHOGONANTS

595. The determinant $A = |a_{1n}|$ is called an *orthogonant* when it is the determinant of an orthogonal substitution, that is, when the following relations hold:

$$a_{i1}^2 + a_{i2}^2 + \cdots + a_{in}^2 = 1$$

$$a_{i1}a_{j1} + a_{i2}a_{j2} + \cdots + a_{in}a_{jn} = 0$$

where

$$i, j = 1, 2, \cdots, n\} \quad i \neq j.$$

596. The square of an orthogonant is equal to unity as may be seen on multiplying it by itself. *It follows that the orthogonant is equal to* ± 1. An orthogonant having the value $+1$ is said to be *proper* and one which has the value -1 is said to be *improper*.

597. By definition we have the equations

$$(1) \qquad \left\{ \begin{array}{l} a_{11}a_{i1} + a_{12}a_{i2} + \cdots + a_{1n}a_{in} = 0 \\ \cdot \quad \cdot \quad \cdot \quad \cdot \quad \cdot \quad \cdot \quad \cdot \quad \cdot \quad \cdot \quad \cdot \quad \cdot \quad \cdot \\ a_{i1}a_{i1} + a_{i2}a_{i2} + \cdots + a_{in}a_{in} = 1 \\ \cdot \quad \cdot \quad \cdot \quad \cdot \quad \cdot \quad \cdot \quad \cdot \quad \cdot \quad \cdot \quad \cdot \quad \cdot \quad \cdot \\ a_{n1}a_{i1} + a_{n2}a_{i2} + \cdots + a_{nn}a_{in} = 0 \end{array} \right.$$

Denoting the algebraic complement of a_{ij} in A by $\mathcal{A}_{ij}$, and multiplying the foregoing equations by $\mathcal{A}_{1j}, \mathcal{A}_{2j}, \cdots, \mathcal{A}_{nj}$ respectively, and then adding the results we get

$$(2) \qquad A \cdot a_{ij} = \mathcal{A}_{ij}.$$

This shows that *every element in A is numerically equal to its complement.*

If we multiply both sides of (2) by a_{ik} we have

$$A \cdot a_{ij}a_{ik} = \mathcal{A}_{ij}a_{ik},$$

and giving i all values from 1 to n and adding the results we have

$$A(a_{1j}a_{1k} + \cdots + a_{nj}a_{nk}) = \mathcal{A}_{1j}a_{1k} + \cdots + \mathcal{A}_{nj}a_{nk}$$
$$= A \quad \text{or} \quad 0,$$

according as $j=k$ or $j\neq k$. From this it follows that

$$(3)\qquad \begin{aligned} &a_{1i}^2+a_{2i}^2+\cdots+a_{ni}^2=1, \quad \begin{pmatrix} i \\ k \end{pmatrix} = 1,2,\cdots,n \\ &a_{1i}a_{1j}+\cdots+a_{ni}a_{nj}=0, \quad i\neq j. \end{aligned}$$

598. *Every minor M of order m of A is equal to its algebraic complement, multiplied by the given determinant A.*

The truth of this may be seen on multiplying every element of M by A, and therefore M by A^m, then using (2) §597 we get a minor of order m of the adjugate of A, which is equal to the complement of M in A (which we may denote by N) multiplied by A^{m-1}. That is

$$M\cdot A^m = N\cdot A^{m-1},$$

or

$$M\cdot A \;= N,$$

or

$$M\cdot A^2 = M = N\cdot A.$$

If

$$\begin{cases} x_1 = a_{11}y_1+a_{12}y_2+\cdots+a_{1n}y_n \\ x_2 = a_{21}y_1+a_{22}y_2+\cdots+a_{2n}y_n \\ \cdots\cdots\cdots\cdots\cdots\cdots \\ x_n = a_{n1}y_1+a_{n2}y_2+\cdots+a_{nn}y_n \end{cases}$$

and $|a_{nn}|$ is an orthogonant then it follows from the foregoing that

$$\begin{cases} y_1 = a_{11}x_1+a_{21}x_2+\cdots+a_{n1}x_n \\ y_2 = a_{12}x_1+a_{22}x_2+\cdots+a_{n2}x_n \\ \cdots\cdots\cdots\cdots\cdots\cdots \\ y_n = a_{1n}x_1+a_{2n}x_2+\cdots+a_{nn}x_n \end{cases}$$

and

$$x_1^2+x_2^2+\cdots+x_n^2 = y_1^2+y_2^2+\cdots+y_n^2.$$

599. If we multiply both sides of

$$M_i = N_iA,$$

by M_i we get

$$M_i^2 = N_iM_iA$$

and summing with respect to i we have

$$\sum_1^\mu iM_i^2 = \sum_1^\mu iN_iM_iA = A^2 = 1 \quad (\mu = (n)_m).$$

If we multiply both sides by M', where M' is the minor corresponding to M but formed from a different set of rows, we get

$$\sum_1^\mu iM_iM_i' = \sum_1^\mu iM_i'N_i. \qquad = 0$$

It follows, therefore, *that the mth compound of an orthogonant A is itself an orthogonant.*

600. *The product of two orthogonants is an orthogonant.*

For if $A \equiv |a_{nn}|$ and $B \equiv |b_{nn}|$ are two orthogonants, and $A \cdot B = C \equiv |c_{nn}|$, where $c_{ij} = a_{i1}b_{j1} + \cdots + a_{in}b_{jn}$, then

$$\sum_1^n ic_{ij}^2 = b_{ji}^2 \sum_1^n ia_{i1}^2 + \cdots + b_{jn}^2 \sum_1^n ia^2{}_{in} + 2b_{j1}b_{j2} \sum_1^n ia_{i1}a_{i2} + \cdots$$

$$= \sum_1^n ib^2{}_{j1} = 1.$$

Similarly $\sum c_{ij}c_{hk} = 0$, where $i \neq h$ and $j \neq k$.

The determinant C is therefore orthogonal.

601. *If two n-line proper orthogonants be taken and the rows of one be multiplied by $h_1, h_2, \cdots, h_n$ respectively, and the rows of the other by $k_1, k_2, \cdots, k_n$ respectively, and a new determinant be formed, each of whose elements is the sum of the corresponding elements in the orthogonants thus modified, this determinant is such that it will remain unaltered on the interchange of the h's and k's.* That is if $A \equiv |a_{nn}|$, and $B \equiv |b_{nn}|$ are the two orthogonants, then

$$S = |h_na_{nn} + k_nb_{nn}| = |k_na_{nn} + h_nb_{nn}|$$

The product $|h_na_{nn} + k_nb_{nn}| \cdot |a_{nn}|$

$$= \begin{vmatrix} h_1 + k_1 \sum b_{11}a_{11} & k_1 \sum b_{11}a_{11} & \cdots & k_1 \sum b_{11}a_{n1} \\ k_2 \sum b_{21}a_{11} & h_2 + k_2 \sum b_{21}a_{21} & \cdots & k_2 \sum b_{21}a_{n1} \\ \cdot & \cdot & \cdot & \cdot \\ k_n \sum b_{n1}a_{11} & k_n \sum b_{n1}a_{21} & \cdots & h_n + k_n \sum b_{n1}a_{n1} \end{vmatrix} = P_1, \text{ say.}$$

The product $|k_n a_{nn}+h_n b_{nn}|\cdot|b_{nn}|$

$$=\begin{vmatrix} h_1+k_1\sum a_{11}b_{11} & k_1\sum a_{11}b_{21} & \cdots & k_1\sum a_{11}b_{n1} \\ k_2\sum a_{21}b_{11} & h_2+k_2\sum a_{21}b_{21} & \cdots & k_2\sum a_{21}b_{n1} \\ \cdots & \cdots & \cdots & \cdots \\ k_n\sum a_{n1}b_{11} & k_n\sum a_{n1}b_{21} & \cdots & h_n+k_n\sum a_{n1}b_{n1} \end{vmatrix}=P_2, \text{ say.}$$

The determinant P_1 may readily be transformed into P_2 by first dividing the rows by $k_1, k_2, \cdots, k_n$ respectively, and then multiplying the columns in order by the same numbers.

It follows therefore that

$$|h_n a_{nn}+k_n b_{nn}| = |k_n a_{nn}+h_n b_{nn}|$$

602. The determinant $S=|a_{nn}+b_{nn}|$ when multiplied by A or B gives

$$A\cdot S=\begin{vmatrix} \sum a_{11}b_{11}+1 & \sum a_{11}b_{21} & \cdots & \sum a_{11}b_{n1} \\ \sum a_{21}b_{11} & \sum a_{21}b_{21}+1 & \cdots & \sum a_{21}b_{n1} \\ \cdots & \cdots & \cdots & \cdots \\ \sum a_{n1}b_{11} & \sum a_{n1}b_{21} & \cdots & \sum a_{n1}b_{n1}+1 \end{vmatrix}=D, \text{ say,}$$

showing that the determinant of the sum of two orthogonal matrices is equal to the determinant of the sum of their product and unity. Since the product of two orthogonants is orthogonal, D is an orthogonal determinant with 1 added to the diagonal elements.

603. If in §601 we put all the h's equal to 1, and all the k's equal to -1, then when n is odd S vanishes giving the theorem: *If from each element of an odd-ordered orthogonant there be subtracted the corresponding element of another orthogonant of the same order, the determinant so formed vanishes.*

604. If A and B have opposite signs, that is, if one is proper and the other is improper, then

$$|h_n a_{nn}+k_n b_{nn}| = -|k_n a_{nn}+h_n b_{nn}|$$

and

(1) If $h_1=h_2=\cdots=h_n=1=k_1=k_2=\cdots=k_n$ then $S=0$.

(2) If $h_1=h_2=\cdots=h_n=1$, and $k_1=k_2=\cdots=k_n=-1$ then

$$\begin{aligned} S\equiv|a_{nn}-b_{nn}| &= -|-a_{nn}+b_{nn}| \\ &= -(-1)^n|a_{nn}-b_{nn}| \\ &= 0 \end{aligned}$$

when n is even.

If the determinant B is taken as positive or negative unity, thus

$$B \equiv \pm \begin{vmatrix} 1 & 0 & \cdots & 0 \\ 0 & 1 & \cdots & 0 \\ \cdot & \cdot & \cdot & \cdot \\ 0 & 0 & \cdots & 1 \end{vmatrix}$$

then $|a_{nn}+b_{nn}|$ becomes an orthogonant with unity added to or subtracted from the elements along the principal diagonal.

605. *The result of increasing the elements along the main diagonal of a proper orthogonant by* $x_1, x_2, \cdots, x_n$ *is the same as increasing them by* $x^{-1}, x^{-1}, \cdots, x^{-1}$, *as long as* $x_1x_2 \cdots x_n = 1$.

For if $A = |a_{nn}| = 1$ is an orthogonant then

$$\begin{vmatrix} a_{11}+x_1 & a_{12} & \cdots & a_{1n} \\ a_{21} & a_{22}+x_2 & \cdots & a_{2n} \\ \cdot & \cdot & \cdot & \cdot \\ a_{n1} & a_{n2} & \cdots & a_{nn}+x_n \end{vmatrix} A = \begin{vmatrix} a_{11}x_1+1 & a_{21}x_1 & \cdots & a_{n1}x_1 \\ a_{12}x_2 & a_{22}x_2+1 & \cdots & a_{n2}x_2 \\ \cdot & \cdot & \cdot & \cdot \\ a_{1n}x_n & a_{2n}x_n & \cdots & a_{nn}x_n+1 \end{vmatrix}$$

$$= x_1x_2 \cdots x_n \begin{vmatrix} a_{11}+x^{-1} & a_{12} & \cdots & a_{1n} \\ a_{21} & a_{22}+x_2^{-1} & \cdots & a_{2n} \\ \cdot & \cdot & \cdot & \cdot \\ a_{n1} & a_{n2} & \cdots & a_{nn}+x_n^{-1} \end{vmatrix}$$

which, since A and the product $x_1x_2 \cdots x_n$ are both unity, proves the theorem.

606. If $A \equiv |a_{nn}|$ be an orthogonant and

$$C \equiv \begin{vmatrix} a_{11}+1 & a_{12} & \cdots & a_{1n} \\ a_{21} & a_{22}+1 & \cdots & a_{2n} \\ \cdot & \cdot & \cdot & \cdot \\ a_{n1} & a_{n2} & \cdots & a_{nn}+1 \end{vmatrix}, \quad C' \equiv \begin{vmatrix} a_{11}-1 & a_{12} & \cdots & a_{1n} \\ a_{21} & a_{22}-1 & \cdots & a_{2n} \\ \cdot & \cdot & \cdot & \cdot \\ a_{n1} & a_{n2} & \cdots & a_{nn}-1 \end{vmatrix}$$

then, if we let C_{ij} be the minor of C corresponding to (complementary of) the element in the position (ij), by multiplication it is readily seen that:

$$(1) \quad \begin{cases} C \cdot A = C \\ \text{or } C(A-1) = 0 \end{cases} \qquad \begin{matrix} C'A = (-1)^n C' \\ \text{or} \quad C'\{A-(-1)^n\} = 0 \end{matrix}$$

$$(2)\quad\begin{cases} C_{\kappa\kappa}\cdot A = C - C_{\kappa\kappa} & C'_{\kappa\kappa}A = (-1)^{n-1}\{C' + C'_{\kappa\kappa}\} \\ \text{or } C_{\kappa\kappa}(A+1) = C & \text{or } C'_{\kappa\kappa}\{A - (-1)^{n-1}\} = (-1)^{n-1}C' \\ \text{or } \quad C_{\kappa\kappa} = \dfrac{C}{A+1} & \\ C_{\kappa\lambda}\cdot A = -C_{\lambda\kappa} & C'_{\kappa\lambda}\cdot A = (-1)^{n-1}C'_{\lambda\kappa} \end{cases}$$

$$(3)\quad\begin{cases} C_{\lambda\kappa,\lambda\kappa}A = C - C_{\lambda\lambda} - C_{\kappa\kappa} & C'_{\kappa\lambda,\kappa\lambda}\cdot A \\ \qquad + C_{\lambda\kappa,\lambda\kappa} & \quad = (-1)^{n-2}\{C' + C'_{\kappa\kappa} + C'_{\kappa\lambda,\kappa\lambda}\} \\ \text{or } C_{\lambda\kappa,\lambda\kappa}(A-1) = C - C_{\lambda\lambda} & \\ \qquad - C_{\kappa\kappa} & \text{or } C'_{\kappa\lambda,\kappa\lambda}\{A - (-1)^{n-2}\} \\ & \quad = (-1)^{n-2}(C' + C'_{\kappa\kappa} + C'_{\lambda\lambda}) \end{cases}$$

$$(4)\quad\begin{cases} C_{\kappa\lambda,\kappa\mu}\cdot A = -(C_{\mu\lambda} - C_{\kappa\mu,\kappa\lambda}) & C'_{\kappa\lambda,\kappa\mu}\cdot A = (-1)^{n-4}\{C'_{\kappa\lambda} + C'_{\kappa\mu,\kappa\lambda}\} \\ C_{\kappa\lambda,\mu\nu}\cdot A = C_{\mu\nu,\kappa\lambda} & C'_{\kappa\lambda,\mu\nu}\cdot A = (-1)^{n-4}C'_{\mu\nu,\kappa\lambda}. \end{cases}$$

In general

$$(5)\quad\begin{cases} C_{\alpha_1\alpha_2\cdots\alpha_m,\alpha_1\alpha_2\cdots\alpha_m}\cdot A = C - (C_{\alpha_1\alpha_1} + C_{\alpha_2\alpha_2} + \cdots + C_{\alpha_m\alpha_m}) \\ \quad + C_{\alpha_1\alpha_2,\alpha_1\alpha_2} + C_{\alpha_1\alpha_3,\alpha_1\alpha_3} + \cdots + C_{\alpha_{m-1}\alpha_m,\alpha_{m-1}\alpha_m} \cdots \\ \quad + (-1)^{m-1}(C_{\alpha_1\alpha_2\cdots\alpha_{m-1},\alpha_1\alpha_2\cdots\alpha_{m-1}} + \cdots \\ \quad + C_{\alpha_2\alpha_3\cdots\alpha_m,\alpha_2\alpha_3\cdots\alpha_m}) + (-1)^m C_{\alpha_1\alpha_2\cdots\alpha_m,\alpha_1\alpha_2\cdots\alpha_m} \\ \text{or } C_{\alpha_1\alpha_2\cdots\alpha_m,\alpha_1\alpha_2\cdots\alpha_m}\{A - (-1)^m\} = C - \sum C_{\alpha_1\alpha_1} \\ \quad + \sum C_{\alpha_1\alpha_2,\alpha_1\alpha_2} - \cdots + (-1)^{m-1}C_{\alpha_1\cdots\alpha_{m-1},\alpha_1\cdots\alpha_{m-1}} \\ C_{\alpha_1\alpha_2\cdots\alpha_m,\beta_1\beta_2\cdots\beta_m}\cdot A = (-1)^m C_{\beta_1\beta_2\cdots\beta_m,\alpha_1\alpha_2\cdots\alpha_m} \\ \quad + \text{terms containing minors* of order higher than } n-m \\ C'_{\alpha_1\cdots\alpha_m,\alpha_1\cdots\alpha_m}\{A - (-1)^{n-m}\} = (-1)^{n-m}\{C' + \sum C'_{\alpha_1\alpha_1} \\ \quad + \cdots + \sum C'_{\alpha_1\cdots\alpha_{m-1},\alpha_1\cdots\alpha_{m-1}}\} \\ C'_{\alpha_1\cdots\alpha_m,\beta_1\cdots\beta_m}\cdot A = (-1)^{n-m-2}C'_{\beta_1\cdots\beta_m,\alpha_1\cdots\alpha_m} \\ \quad + \text{terms containing minors* of order higher than } n-m\} \end{cases}$$

It may be observed that we may also write

$$C_{12\cdots m,12\cdots m}\cdot A = A + \sum A_{m+1,m+1} + \sum A_{m+1\, m+2,m+1\, m+2} + \cdots + \sum A_{m+1\cdots n-1,m+1\cdots n-1} + A_{m+1\cdots n,m+1\cdots n}.$$

* These terms do not exist when $\alpha_1\alpha_2\cdots\alpha_m$ and $\beta_1\beta_2\cdots\beta_m$ have no numbers in common.

Substituting in the right-hand side for $A_{m+1\ m+1}$ etc. by (2) §597 and, using inclusive notation instead of exclusive, we get after dividing through by A

$$C_{m+1\cdots n,m+1\cdots n} = 1 + \sum A_{m+1,m+1} + \sum A_{m+1\ m+2,m+1\ m+2} + \cdots + A_{m+1\cdots n,m+1\cdots n}.$$

607. From the foregoing results we see that

I. When $A=+1$.

(a) We have from (1), $C'=0$ when n is odd and hence the theorem: *If unity be subtracted from the main diagonal elements of a proper orthogonant of odd degree the resulting determinant vanishes*; a result which might have been obtained from §603 by making B the unit orthogonant.

(b) From (2) we have, $C_{\kappa\kappa}=\frac{1}{2}C$ and hence the theorem: *If unity be added to the principal diagonal elements of a proper orthogonant then the determinant thus formed has the complementary minors of its diagonal elements all equal to one-half itself.*

When n is even $C'_{\kappa\kappa}=-\frac{1}{2}C'$ and we have the theorem: *If unity be subtracted from the principal diagonal elements of an even ordered proper orthogonant the determinant thus formed has the complementary minors of its diagonal elements all equal to the negative half of itself.*

(c) We also have $C_{\kappa\lambda}=-C_{\lambda\kappa}$, which with

$$C_{\kappa\kappa}C_{\lambda\lambda} - C_{\kappa\lambda}C_{\lambda\kappa} = C\,C_{\kappa\lambda,\kappa\lambda}, \text{ and } C_{\kappa\kappa} = \tfrac{1}{2}C$$

gives

$$C_{\kappa\lambda}{}^2 = C\left(C_{\kappa\lambda,\kappa\lambda} - \frac{C}{4}\right).$$

Similarly

$$\pm C'_{\kappa\lambda}{}^2 = C'\left(C'_{\kappa\lambda,\kappa\lambda} - \frac{C'}{4}\right).$$

These show that: *If* $\left\{{C \atop C'}=0\right\}$ *then all minors of* $\left\{{C \atop C'}\right\}$ *of order* $n-1$ *are also zero.*

The adjugate of C is skew, and when n is even that of C′ is also.

(d) From (4) we see that, under the same conditions as in the last theorem, *every minor of C or C′ of order* $n-2$ *is equal to the positive or negative of its conjugate.*

(e) From (5) we see that: *When* $\left\{{m \atop n-m}\right\}$ *is odd, and all* coaxial minors of* $\left\{{C \atop C'}\right\}$ *of order higher than* $n-m$ *vanish then all coaxial minors of order* $n-m$ *vanish.*

* This could be replaced by the sums of all.

It also shows that: *If all minors of order higher than $n-m$ vanish then every minor of order $n-m$ is equal to the positive or negative of its conjugate.*

But, using α for $\alpha_1 \cdots \alpha_m$ and β for $\beta_1 \cdots \beta_m$,

$$C_{\alpha,\alpha} \cdot C_{\beta,\beta} - C_{\alpha,\beta} \cdot C_{\beta,\alpha} = 0$$

since it is equal (§191) to an expression involving minors of order higher than $n-m$.

It follows therefore that

$$C_{\alpha,\beta} = 0$$

Similarly

$$C'_{\alpha,\beta} = 0$$

and we have the theorem: (*A*) *If* $\left\{ {n \atop n-m} \right\}$ *is odd and all minors of* $\left\{ {C \atop C'} \right\}$ *of order higher than $n-m$ vanish then all minors of order $n-m$ vanish.*

II. When $A = -1$.

(f) In this case (1) shows that $C = 0$ and we have the theorem: *If unity be added to the principal diagonal elements of an improper orthogonant the resulting determinant is zero.*

It also follows that *all coaxial minors of C of order $n-1$ have the same signs and the adjugate is axisymmetric.*

(g) If n is even then $C' = 0$ and we have the theorem: *If unity be subtracted from the principal diagonal elements of an even ordered improper orthogonant the resulting determinant is zero.*

In this case *all coaxial minors of C' of order $n-1$ have the same signs and the adjugate is axisymmetric.*

(h) From (2) we see that when n is odd $C'_{\kappa\kappa} = -\frac{1}{2}C'$ and we have the theorem: *If unity be subtracted from the principal diagonal elements of an odd ordered improper orthogonant the determinant thus formed has the complementary minors of its diagonal elements all equal to the negative half of itself.*

The adjugate of C' in this case is skew.

(k) If $C' = 0$ and n is odd then it follows that every minor of C' of order $n-1$ is zero giving the theorem: *If unity be subtracted from the principal diagonal elements of an odd ordered improper orthogonant and if the resulting determinant is zero then all its minors of order $n-1$ are zero also.*

(l) From (3) we have

$$C_{\kappa\lambda,\kappa\lambda} = \tfrac{1}{2}(C_{\kappa\kappa} + C_{\lambda\lambda})$$

$$C'_{\kappa\lambda,\kappa\lambda} = -\tfrac{1}{2}(C'_{\kappa\kappa} + C'_{\lambda\lambda}) \quad \text{when } n \text{ is even;}$$

and we have the theorem: *If all coaxial minors of* $\left\{\begin{smallmatrix} C \\ C' \end{smallmatrix}\right\}$ *of order* $n-1$ *vanish when* $\left\{\begin{smallmatrix} n \text{ any number} \\ n \text{ even} \end{smallmatrix}\right\}$ *then all coaxial minors of order* $n-2$ *will vanish also.*

Or we may say that: *If* $\left\{\begin{smallmatrix} C \\ C' \end{smallmatrix} = 0\right\}$ *and all coaxial minors of order* $n-1$ *are also zero then* (α) *all minors of order* $n-1$ *vanish and* (β) *all coaxial minors of C of order* $n-2$ *vanish and all coaxial minors of* C' *of order* $n-2$ *vanish when n is even.* Under these conditions (4) shows that all minors of order $n-2$ are zero.

(m) From (5) we see that: *When* $\left\{\begin{smallmatrix} m \\ n-m \end{smallmatrix}\right\}$ *is even and all coaxial minors of* $\left\{\begin{smallmatrix} C \\ C' \end{smallmatrix}\right\}$ *of order greater than* $n-m$ *vanish then all coaxial minors of order* $n-m$ *vanish.*

(n) It also follows that: (*B*) *When* $\left\{\begin{smallmatrix} m \\ n-m \end{smallmatrix}\right\}$ *is even and all minors of* $\left\{\begin{smallmatrix} C \\ C' \end{smallmatrix}\right\}$ *of order greater than* $n-m$ *vanish then all minors of order* $n-m$ *vanish.*

608. Since it is true that if all minors of order h of any determinant vanish then all minors of higher order vanish as a consequence and since any odd power of an orthogonant is equal to the orthogonant itself we may state theorems (A) and (B) §607 as follows:

(A_1) *If* $+1$ *be added to each of the diagonal elements of an odd power of a proper orthogonant and the resulting determinant be such that its minors of order* $n-2k$ *are all zero, then all its minors of order* $n-2k-1$ *are zero also.*

(A_2) *If* -1 *be added to each of the diagonal elements of an odd power of a proper orthogonant and the resulting determinant be such that its minors of order* $2k$ *are all zero, then all its minors of order* $2k-1$ *are zero also.*

(B_1) *If* $+1$ *be added to each of the diagonal elements of an odd power of an improper orthogonant, and the resulting determinant be such that all minors of order* $n-2k+1$ *are zero, then all minors of order* $2n-2k$ *are zero also.*

609. These theorems may be recombined as follows:

(C_1) *If* $+1$ *be added to each of the diagonal elements of an orthogonant and the resulting determinant be such that all minors of order* $n-m$ *vanish, but not all minors of order* $n-m-1$, *then m is odd or even according as A is positive or negative.*

(C_2) *If* -1 *be added to each of the diagonal elements of an orthogonant and the resulting determinant be such that minors of order* $n-m$ *are all zero, but not all minors of order* $n-m-1$, *then* $(n-m)$ *is odd or even according as A is positive or negative.* In this case it must be noticed that n must be even.

610. The primary minors of S (§602) can be expressed as linear functions of the primary minors of D as may be seen on multiplying one of them by A, and we know (§607 (c)) that if $D=0$, then all the primary minors of D vanish also. It follows therefore that *if* $D=0$ *and consequently* $S=0$, *that all the primary minors of* S *also vanish.*

611. If

$$(1)\qquad \left.\begin{aligned} x_i &= \sum_1^n {}_j a_{ij} y_j \\ u_i &= \sum_1^n {}_j b_{ij} v_j \end{aligned}\right\},\quad (i = 1, 2, \cdots, n)$$

where $|a_{1n}|$ and $|b_{1n}|$ are orthogonants, then we know that

$$(2)\qquad \begin{cases} \sum x_i^2 = \sum y_i^2 \\ \sum u_i^2 = \sum v_i^2 \end{cases}$$

Let us consider the transformation

$$(3)\qquad \sum c_{ij} x_j u_i \equiv \sum G_i y_i v_i,\quad \binom{i}{j} = 1, 2, \cdots, n)$$

and for convenience let $n=3$.

From this identity (3) we have the nine (n^2) equations

$$(4)\qquad c_{ij} = G_1 a_{j1} b_{i1} + G_2 a_{j2} b_{i2} + G_3 a_{j3} b_{i3},\quad \binom{i}{j} = 1, 2, 3)$$

and from these nine we readily get the following two sets of nine each

$$(5)\qquad G_i a_{ji} = \sum {}_h b_{hi} c_{hj},$$

and

$$\{h, i, j, = 1, 2, 3\}$$

$$(6)\qquad G_i b_{ji} = \sum {}_h a_{ih} c_{jh}.$$

Solving (6) for a_{ij} we have

$$(7)\qquad \frac{Ca_{1j}}{G_j} = |\, b_{1j} c_{22} c_{33} \,|,\ \text{etc.}$$

and solving these for C_{ij} we get

$$(8)\qquad \frac{C_{ij}}{C} = \sum {}_h \frac{a_{jh} b_{ih}}{G_h}.$$

But

$$C_{11} \equiv \begin{vmatrix} c_{22} & c_{23} \\ c_{32} & c_{33} \end{vmatrix},$$

which by (4) is equal to

$$\begin{Vmatrix} G_1a_{21} & G_2a_{22} & G_3a_{23} \\ G_1a_{31} & G_2a_{32} & G_3a_{33} \end{Vmatrix} \cdot \begin{Vmatrix} b_{21} & b_{22} & b_{23} \\ b_{31} & b_{32} & b_{33} \end{Vmatrix}$$

$$= G_1G_2A_{13}B_{13} + G_1G_3A_{12}B_{12} + G_2G_3A_{11}B_{11}$$

$$= G_1G_2a_{13}b_{13} + G_1G_3a_{12}b_{12} + G_2G_3a_{11}b_{11}.$$

Comparing this with C_{11} from (8) we see that

$$G_1 \cdot G_2 \cdot G_3 = C$$

Comparing (4) and (8) we observe a dualism. Each equation in (8) being obtained from the corresponding equation in (4) by substituting C_{ij}/C for C_{ij} and $1/G_i$ for G_i.

This dualism extends to the transformation so that (3) is necessarily accompanied by

$$\sum C_{ij}x_ju_i = \sum \frac{1}{G_i} y_iv_i.$$

Exercises:

1. If

$$\begin{array}{cccc|c} x_1 & x_2 & \cdots & x_n & \\ \hline a_{11} & a_{12} & \cdots & a_{1n} & x_1 \\ a_{21} & a_{22} & \cdots & a_{2n} & x_2 \\ \cdot & \cdot & \cdots & \cdot & \vdots \\ a_{n1} & a_{n2} & \cdots & a_{nn} & x_n \end{array} = G_1y_1^2 + G_2y_2^2 + \cdots + G_ny_n^2$$

then

$$\begin{array}{cccc|c} x_1 & x_2 & \cdots & x_n & \\ \hline & & & & x_1 \\ & & & & x_2 \\ & & A^p & & \vdots \\ & & & & x_n \end{array} = G_1^p y_1^2 + G_2^p y_2^2 + \cdots + G_n^p y_n^2$$

where A^p is the pth power of A.

2. Given

$$\sum c_{ij}x_iy_j = \sum G_ky_k^2$$

Show that if $|a_{nn}|$ is the determinant of the orthogonal transformation which, with the G's, produces it, then the G's are the roots of the equation $|C-x|=0$, and

$$a_{jk} = \frac{(C_{ij}^{(k)})^{1/2}}{\left\{C_{11}^{(k)} + C_{22}^{(k)} + \cdots + C_{nn}^{(k)}\right\}^{1/2}}$$

where $C_{ii}^{(m)}$ represents what C_{ii} becomes on substituting $a_{11}-G_m$, $a_{22}-G_m, \cdots$, for $a_{11}, a_{22}, \cdots$, respectively.

612. Let us consider the following equations:

$$(1)\quad \begin{cases} b_{11}x_1 + b_{12}x_2 + \cdots + b_{1n}x_n = b_{11}y_1 + b_{21}y_2 + \cdots + b_{n1}y_n \\ b_{21}x_1 + b_{22}x_2 + \cdots + b_{2n}x_n = b_{12}y_1 + b_{22}y_2 + \cdots + b_{n2}y_n \\ \cdots\cdots\cdots\cdots\cdots\cdots\cdots\cdots\cdots\cdots \\ b_{n1}x_1 + b_{n2}x_2 + \cdots + b_{nn}x_n = b_{1n}y_1 + b_{2n}y_2 + \cdots + b_{nn}y_n \end{cases}$$

where the determinant of the b's, B say, is skew, that is, $b_{ij}=-b_{ji}$ and $b_{ii}=\omega$.

Solving (1) let us write

$$y_i = c_{i1}x_1 + c_{i2}x_2 + \cdots + c_{in}x_n$$

$$x_j = d_{j1}y_1 + d_{j2}y_2 + \cdots + d_{jn}y_n$$

where $B\cdot c_{ij}=B_{i1}b_{1j}+B_{i2}b_{2j}+\cdots+B_{in}b_{nj}$, and B_{ij} is the cofactor of b_{ij} in B.

Let

$$\epsilon = B_{i1}b_{j1} + B_{i2}b_{j2} + \cdots + B_{in}b_{jn},$$

then

$$B\cdot c_{ij} + \epsilon = 2B_{ij}\cdot\omega.$$

But $\epsilon=B$ or 0, according as $i=j$ or $i\neq j$. Therefore

$$c_{ij} = \frac{2B_{ij}\cdot\omega}{B}, \qquad c_{ii} = \frac{2B_{ii}\omega - B}{B}.$$

Similarly

$$d_{ji} = \frac{2B_{ij}\omega}{B}, \qquad d_{ii} = \frac{2B_{ii}\omega - B}{B},$$

and therefore $c_{ij}=d_{ji}$.

We may then write

$$y_i = c_{i1}x_1 + c_{i2}x_2 + \cdots + c_{in}x_n$$

$$x_i = c_{1i}y_1 + c_{2i}y_2 + \cdots + c_{ni}y_n$$

which on substituting from the second in the first gives on equating coefficients

$$c_{i1}^2 + c_{i2}^2 + \cdots + c_{in}^2 = 1,$$

$$c_{i1}c_{j1} + c_{i2}c_{j2} + \cdots + c_{in}c_{jn} = 0.$$

Substituting from the first in the second, we have

$$c_{1i}^2 + c_{2i}^2 + \cdots + c_{ni}^2 = 1,$$

$$c_{1i}c_{1j} + c_{2i}c_{2j} + \cdots + c_{ni}c_{nj} = 0.$$

Starting with the equations

$$B_{i1}b_{i1} + B_{i2}b_{i2} + \cdots + B_{in}b_{in} = B$$

$$B_{i1}b_{j1} + B_{i2}b_{j2} + \cdots + B_{in}b_{jn} = 0 \qquad (i \neq j)$$

and after multiplying by 2ω we may write

$$2\omega B_{i1}b_{i1} + \cdots + (2\omega B_{ii} - B)\omega + \cdots + 2\omega B_{in}b_{in} = B\cdot\omega$$

$$2\omega B_{i1}b_{j1} + \cdots + (2\omega B_{ii} - B)b_{ji} + \cdots + 2\omega B_{in}b_{jn} = B\cdot b_{ij}.$$

These show that the determinant of the c's, C say, is such that

$$C\cdot B = B, \text{ or } C = 1.$$

It follows therefore that C is an orthogonant with the value $+1$, and, when $\omega=1$, is known as *Cayley's*.

613. If in

$$C \equiv \begin{vmatrix} \dfrac{2\omega B_{11} - B}{B} & \dfrac{2\omega B_{12}}{B} & \cdots & \dfrac{2\omega B_{1n}}{B} \\ \dfrac{2\omega B_{21}}{B} & \dfrac{2\omega B_{22} - B}{B} & \cdots & \dfrac{2\omega B_{2n}}{B} \\ \cdots & \cdots & \cdots & \cdots \\ \dfrac{2\omega B_{n1}}{B} & \dfrac{2\omega B_{n2}}{B} & \cdots & \dfrac{2\omega B_{nn} - B}{B} \end{vmatrix}$$

we expand B in powers of ω we have when n is odd an expression for B containing only odd powers of ω, beginning with the nth and the last will be ω times a sum of squares since zero-axial skew determinants of odd order vanish and those of even order are perfect squares. Dividing the numerator and denominator of each element of C by ω and then if we put $\omega=0$, C becomes an axisymmetric orthogonant involving $\frac{1}{2}n(n-1)$ constants, the b's. When n is even, however, we cannot divide out ω from numerator and denominator of the elements unless $|b_{nn}|_0=0$, where $|b_{nn}|_0$ represents what B becomes on putting $\omega=0$. Thus if $n=4$, and

$$B \equiv \begin{vmatrix} \omega & a & b & c \\ -a & \omega & -h & g \\ -b & h & \omega & -f \\ -c & -g & f & \omega \end{vmatrix} = \omega^2(a^2+b^2+c^2+\omega^2+f^2+g^2+h^2) + (af+bg+ch)^2$$

$$= \omega^2\Delta + (af+bg+ch)^2$$

$$= \omega^2\{\Delta+\theta^2\}, \text{ say,}$$

then

$$C \equiv \begin{vmatrix} B_{fgh} & \dfrac{2\omega B_{12}}{B} & \dfrac{2\omega B_{13}}{B} & \dfrac{2\omega B_{14}}{B} \\ \dfrac{2\omega B_{21}}{B} & B_{bcf} & \dfrac{2\omega B_{23}}{B} & \dfrac{2\omega B_{24}}{B} \\ \dfrac{2\omega B_{31}}{B} & \dfrac{2\omega B_{32}}{B} & B_{acg} & \dfrac{2\omega B_{34}}{B} \\ \dfrac{2\omega B_{41}}{B} & \dfrac{2\omega B_{42}}{B} & \dfrac{2\omega B_{43}}{B} & B_{abh} \end{vmatrix}$$

where B_{fgh} represents

$$\frac{2\omega\{\omega^3+\omega(f^2+g^2+h^2)\}-B}{B}, \text{ etc.}$$

This becomes when $(af+bg+ch)=0$, and ω has been put equal to zero

$$\begin{vmatrix} \dfrac{\Delta_{fgh}}{\Delta} & \dfrac{2(bh-cg)}{\Delta} & \dfrac{2(cf-ah)}{\Delta} & \dfrac{2(ag-bf)}{\Delta} \\ \dfrac{2(bh-cg)}{\Delta} & \dfrac{\Delta_{bcf}}{\Delta} & \dfrac{2(fg-ab)}{\Delta} & \dfrac{2(fh-ac)}{\Delta} \\ \dfrac{2(cf-ah)}{\Delta} & \dfrac{2(fg-ab)}{\Delta} & \dfrac{\Delta_{acg}}{\Delta} & \dfrac{2(gh-bc)}{\Delta} \\ \dfrac{2(ag-bf)}{\Delta} & \dfrac{2(fh-ac)}{\Delta} & \dfrac{2(gh-bc)}{\Delta} & \dfrac{\Delta_{abh}}{\Delta} \end{vmatrix}$$

where $\Delta_{fgh}=f^2+g^2+h^2-a^2-b^2-c^2$, etc.

The determinant C may be factored thus

$$\begin{vmatrix} \frac{\omega-\theta}{(\Delta')^{1/2}} & +\frac{a+f}{(\Delta')^{1/2}} & +\frac{b+g}{(\Delta')^{1/2}} & +\frac{c+h}{(\Delta')^{1/2}} \\ -\frac{a+f}{(\Delta')^{1/2}} & \frac{\omega-\theta}{(\Delta')^{1/2}} & +\frac{c+h}{(\Delta')^{1/2}} & -\frac{b+g}{(\Delta')^{1/2}} \\ -\frac{b+g}{(\Delta')^{1/2}} & -\frac{c+h}{(\Delta')^{1/2}} & \frac{\omega-\theta}{(\Delta')^{1/2}} & +\frac{a+f}{(\Delta')^{1/2}} \\ -\frac{c+h}{(\Delta')^{1/2}} & +\frac{b+g}{(\Delta')^{1/2}} & -\frac{a+f}{(\Delta')^{1/2}} & \frac{\omega-\theta}{(\Delta')^{1/2}} \end{vmatrix}$$

$$\cdot\begin{vmatrix} \frac{\omega+\theta}{(\Delta')^{1/2}} & \frac{f-a}{(\Delta')^{1/2}} & \frac{g-b}{(\Delta')^{1/2}} & \frac{h-c}{(\Delta')^{1/2}} \\ \frac{a-f}{(\Delta')^{1/2}} & \frac{\omega+\theta}{(\Delta')^{1/2}} & \frac{c-h}{(\Delta')^{1/2}} & \frac{g-b}{(\Delta')^{1/2}} \\ \frac{b-g}{(\Delta')^{1/2}} & \frac{h-c}{(\Delta')^{1/2}} & \frac{\omega+\theta}{(\Delta')^{1/2}} & \frac{a-f}{(\Delta')^{1/2}} \\ \frac{c-h}{(\Delta')^{1/2}} & \frac{b-g}{(\Delta')^{1/2}} & \frac{f-a}{(\Delta')^{1/2}} & \frac{\omega+\theta}{(\Delta')^{1/2}} \end{vmatrix}$$

where $\Delta' = \Delta + \theta^2$, and it is readily seen that each of the factors is a proper orthogonant.

Since the only way to have C axi-symmetric is to have $\theta = 0$, it follows that Cayley's orthogonant cannot be axi-symmetric. It can, however, be made to approximate axi-symmetry.*

If $|c_{nn}|$ is an orthogonant with real elements then the product of $|c_{nn}|$ by itself after $(n-m)$ rows have been altered in sign, namely, $(-1)^{n-m}|c_{nn}|^2$ is an axi-symmetric orthogonant.

614. If unity be subtracted from the principal diagonal elements of C, then 2 is a factor of every line and therefore 2^n is a factor of the determinant thus affected which we may denote by C'. Multiplying C' by B we get

$$C'B = (-1)^n 2^n |b_{nn}|_0$$

or

$$C' = \frac{(-1)^n 2^n |b_{nn}|_0}{|b_{nn}|},$$

* Kronecker, Werke III (1), pp. 371–459.

where $|b_{nn}|_0$ is what $B=|b_{nn}|$ becomes on making all the diagonal elements zero. That is: *If the Cayleyan orthogonant obtained from the skew determinant $|b_{nn}|$ has unity subtracted from each of its diagonal elements, the resulting determinant is equal to*

$$(-1)^n 2^n \frac{|b_{nn}|_0}{|b_{nn}|}.$$

If n is odd then $|b_{nn}|_0=0$ being a skew-symmetric of odd order. The sign factor therefore may be dropped. The fact that C' vanishes when n is odd shows that every Cayleyan orthogonant of odd order has $+1$ for a latent root.

615. If in a determinant of the form of C §613 we multiply the rows by B and call the resulting determinant S, then S has been called by Muir the *superadjugate* of B. It is readily seen that

$$S\cdot B = \begin{vmatrix} b_{11}B & -b_{12}B & -b_{13}B \cdots \\ -b_{21}B & b_{22}B & -b_{23}B \cdots \\ -b_{31}B & -b_{32}B & b_{33}B \cdots \\ \cdots & \cdots & \cdots \end{vmatrix} = B^n\cdot\overline{B}, \text{ say,}$$

where $\overline{B}$ is what B becomes on changing the signs of all the non-diagonal elements.

It follows that

$$S = B^{n-1}\cdot\overline{B} = B^n$$

since B is a skew determinant and therefore $\overline{B}$ is B with rows and columns interchanged.

We have therefore the theorem that *the superadjugate of any skew determinant B is equal to B^n.*

This determinant S has many properties akin to those of skew determinants with univarial diagonal.

616. If we add unity to each of the diagonal elements of C §613, the result, C'' say, is readily seen to equal $(2\omega)^n/B$ and vanishes only when ω vanishes. It follows therefore *that Cayley's orthogonant cannot have -1 for a latent root and orthogonants having -1 for a latent root cannot be represented by Cayley's formula.*

If we expand B in descending powers of ω we have

$$C'' = \frac{2^n\omega^n}{\omega^n + \cdots + \omega\Delta' + \Delta}$$

where $\Delta = |b_{nn}|_0$ and Δ' is the sum of the coaxial minors of Δ of order $n-1$, etc.

If n is odd then $\Delta = 0$ and $\Delta' \neq 0$, so that C'' vanishes with ω.

If n is even then $\Delta \neq 0$ and C'' vanishes with ω.

It follows, therefore, that *when* $\omega = 0$, *all minors of* C'' *down to those of order two vanish and when n is even the elements are individually zero.*

617. The actual representation of an orthogonant in terms of the $\frac{1}{2}n(n-1)$ elements of a skew determinant with the elements $(ii) = 1$ may be found to be as follows:

For $n = 2$ and

$$B = \begin{vmatrix} 1 & \lambda \\ -\lambda & 1 \end{vmatrix}$$

$$C \equiv \begin{vmatrix} \dfrac{1-\lambda^2}{1+\lambda^2} & -\dfrac{2\lambda}{1+\lambda^2} \\[2ex] \dfrac{2\lambda}{1+\lambda^2} & \dfrac{1-\lambda^2}{1+\lambda^2} \end{vmatrix}$$

for $n = 3$ and

$$B = \begin{vmatrix} 1 & \lambda & -\mu \\ -\lambda & 1 & \nu \\ \mu & -\nu & 1 \end{vmatrix}$$

$$C \equiv \begin{vmatrix} \dfrac{1+\nu^2-\mu^2-\lambda^2}{B} & 2\dfrac{\lambda+\nu\mu}{B} & 2\dfrac{-\mu+\lambda\nu}{B} \\[2ex] 2\dfrac{-\lambda+\nu\mu}{B} & \dfrac{1+\mu^2- \quad -\lambda^2}{B} & 2\dfrac{\nu+\mu\lambda}{B} \\[2ex] 2\dfrac{\mu+\lambda\nu}{B} & 2\dfrac{-\nu+\mu\lambda}{B} & \dfrac{1-\nu^2+\lambda^2-\mu^2}{B} \end{vmatrix}$$

for $n = 4$ and

$$B = \begin{vmatrix} 1 & a & b & c \\ -a & 1 & -h & g \\ -b & h & 1 & -f \\ -c & -g & f & 1 \end{vmatrix}$$

C would be the same as in §613 with $\omega = 1$.

618. These representations fail for those orthogonants having both $+1$ and -1 for latent roots. It is readily seen that for a determinant of order three, if we take

$$B \equiv \begin{vmatrix} 1 & \lambda & \mu \\ -\lambda & 1 & 0 \\ -\mu & 0 & 1 \end{vmatrix} = 1 + \mu^2 + \lambda^2,$$

and form the orthogonant as Cayley's is formed we get

$$\begin{vmatrix} \dfrac{1-\lambda^2-\mu^2}{B} & \dfrac{2\lambda}{B} & \dfrac{2\mu}{B} \\ -\dfrac{2\lambda}{B} & \dfrac{1+\mu^2-\lambda^2}{B} & -\dfrac{2\lambda\mu}{B} \\ -\dfrac{2\mu}{B} & -\dfrac{2\lambda\mu}{B} & \dfrac{1+\lambda^2-\mu^2}{B} \end{vmatrix} \equiv \phi$$

Let

$$\phi_1 = \left(\phi \begin{vmatrix} -1 & 0 & 0 \\ 0 & 1 & 0 \\ 0 & 0 & 1 \end{vmatrix}\right)$$

then

$$|\phi_1 + 1| = \left(\frac{2}{B}\right)^3 \cdot \begin{vmatrix} \lambda^2+\mu^2 & -\lambda & -\mu \\ -\lambda & 1+\mu^2 & -\lambda\mu \\ -\mu & -\lambda\mu & 1+\lambda^2 \end{vmatrix} = 0$$

$$|\phi_1 - 1| = -\left(\frac{2}{B}\right)^3 \cdot \begin{vmatrix} 1 & \lambda & \mu \\ \lambda & \lambda^2 & \lambda\mu \\ \mu & \lambda\mu & \mu^2 \end{vmatrix} = 0$$

showing that ϕ_1 has both $+1$ and -1 for latent roots.

619. Cayley represented an orthogonant by $\psi = (1-\theta)/(1+\theta)$ where θ is a zero-axial skew determinant. If n is even then

$$|1+\theta| = |\theta - 1| = +|1-\theta| \text{ so that } |\psi| = +1.$$

If n is odd then

$$|1+\theta| = -|\theta - 1| = |1-\theta| \text{ and } |\psi| = +1.$$

The orthogonant ψ is necessarily proper and an orthogonant which has -1 as a latent root cannot be put in this form. Frobenius*, however, showed that we could find a function, θ_ρ, whose coefficients are rational functions of a parameter ρ of which one at least is infinite for $\rho=0$, such that

$$\phi = \underset{\rho=0}{Lt}\,(1+\theta_\rho)^{-1}(1-\theta_\rho)$$

and where -1 is a root of $|\phi - x| = 0$.

Any orthogonal substitution can be expressed as the product of a symmetric orthogonant and a Cayley orthogonant. That is

$$\phi = \phi_0(1+\theta)^{-1}(1-\theta),$$

where ϕ_0 is symmetric,

$$\begin{aligned}\phi^2 &= \phi_0^2\,[(1+\theta)^{-1}(1-\theta)]^2\\ &= [(1+\theta)^{-1}(1-\theta)]^2,\end{aligned}$$

since ϕ_0^2 is equal to 1.

It follows therefore that every orthogonal substitution which is the square of an orthogonal substitution can be expressed as the square of Cayley's expression.

The square of every improper orthogonal substitution is also the square of a proper orthogonal substitution. But every proper orthogonal substitution cannot be put equal to Cayley's expression.

Orthogonants may therefore be divided into two classes, first and second, according as they are or are not the square of an orthogonal substitution.

Cipolla has made a classification by taking for ϕ_0 a determinant whose every diagonal element is either $+1$ or -1 and every non-diagonal element is zero and basing the distinction on the number of negative units in the diagonal of the multiplier.

620. Besides Cayley's there are other specialized forms of orthogonants as has been shown**. Thus

* Crelle, vol. 84.

** Muir, Proc. R.S.E., Vol. 38, Part II, p. 146–153.

$$\begin{vmatrix} -\dfrac{a_1}{\sigma} & \dfrac{a_2}{\sigma} & \dfrac{a_3}{\sigma} & \dfrac{a_4}{\sigma} \\ -\dfrac{a_2}{\sigma} & -\dfrac{a_1}{\sigma} & \dfrac{a_4}{\sigma} & -\dfrac{a_3}{\sigma} \\ -\dfrac{a_3}{\sigma} & -\dfrac{a_4}{\sigma} & -\dfrac{a_1}{\sigma} & \dfrac{a_2}{\sigma} \\ -\dfrac{a_4}{\sigma} & \dfrac{a_3}{\sigma} & -\dfrac{a_2}{\sigma} & -\dfrac{a_1}{\sigma} \end{vmatrix}$$

where $\sigma^2 = a_1^2 + a_2^2 + a_3^2 + a_4^2$, is a skew orthogonant as may readily be seen.

In general:

1) *The determinant*

$$\begin{vmatrix} a_{11} - 1 & a_{12} & \cdots & a_{1n} \\ a_{21} & a_{22} - 1 & \cdots & a_{2n} \\ \cdot & \cdot & \cdot & \cdot \\ a_{n1} & a_{n2} & \cdots & a_{nn} - 1 \end{vmatrix}$$

will be an orthogonant, if the (r, s)*th element of* $|a_{nn}|$ *be taken equal to* $2b_rb_s/\sigma$, *where* $\sigma = b_1^2 + b_2^2 + \cdots + b_n^2$. Thus

$$\begin{vmatrix} \dfrac{2b_1^2}{\sigma} - 1 & \dfrac{2b_1b_2}{\sigma} & \cdots & \dfrac{2b_1b_n}{\sigma} \\ \dfrac{2b_2b_1}{\sigma} & \dfrac{2b_2^2}{\sigma} - 1 & \cdots & \dfrac{2b_2b_n}{\sigma} \\ \cdot & \cdot & \cdot & \cdot \\ \dfrac{2b_nb_1}{\sigma} & \dfrac{2b_nb_2}{\sigma} & \cdots & \dfrac{2b_n^2}{\sigma} - 1 \end{vmatrix} \equiv B, \text{ say.}$$

That this is an orthogonant may readily be verified. Concerning B three things may be noticed (1) It is axisymmetric, (2) It contains but n arbitrary parameters, (3) It has the value $+1$ when n is odd and -1 when n is even.

2) *The determinant*

$$\begin{vmatrix} a_{11} - 1 & a_{12} & \cdots & a_{1n} \\ a_{21} & a_{22} - 1 & \cdots & a_{2n} \\ \cdot & \cdot & \cdot & \cdot \\ a_{n1} & a_{n2} & \cdots & a_{nn} - 1 \end{vmatrix}$$

will be an orthogonant, if the square of $|a_{nn}|$ *be equal elementally to its duplicant.* That is if

$$|a_{nn}|^2 = \begin{vmatrix} a_{11}+a_{11} & a_{12}+a_{21} & \cdots & a_{1n}+a_{n1} \\ a_{21}+a_{12} & a_{22}+a_{22} & \cdots & a_{2n}+a_{n2} \\ \cdot & \cdot & \cdot & \cdot \\ a_{n1}+a_{1n} & a_{n2}+a_{2n} & \cdots & a_{nn}+a_{nn} \end{vmatrix}$$

The truth of this theorem follows immediately from the definition.

That the orthogonant B of 1) satisfies the condition of 2) is readily seen. It is also true of Cayley's form.

621. The matrix of the adjugate of a unit-axial skew determinant, Δ say, may be partitioned into n parts, the first containing the elements of Δ to the 0th degree, the second containing the elements to the 1st degree, the third containing the elements to the 2nd degree, and so on, the nth containing the elements to the $(n-1)$st degree. Muir has shown that the determinant of the third matrix containing the elements of the second degree, which we may denote by P, is such that, under certain conditions, its square is elementally equal to a multiple of its duplicant. The rule for the formation of P is as follows:

$$P \equiv \begin{vmatrix} S-r_1^2 & -r_1r_2 & \cdots & -r_1r_n \\ -r_1r_2 & S-r_2^2 & \cdots & -r_2r_n \\ -r_1r_3 & -r_2r_3 & \cdots & -r_3r_n \\ \cdot & \cdot & \cdot & \cdot \\ -r_1r_n & -r_2r_n & \cdots & S-r_n^2 \end{vmatrix}$$

where r_hr_k represents the product of the hth and kth rows of Δ_0, where Δ_0 is what Δ becomes when the units along the principal diagonal are replaced by zeros, and where S is the sum of the squares of the elements on one side of the principal diagonal.

The cases when n is odd and when n is even require separate consideration.

First let n be even, $=4$, say.

Let

$$\Delta \equiv \begin{vmatrix} 1 & a_{12} & a_{13} & a_{14} \\ -a_{12} & 1 & a_{23} & a_{24} \\ -a_{13} & -a_{23} & 1 & a_{34} \\ -a_{14} & -a_{24} & -a_{34} & 1 \end{vmatrix}$$

then

$$\Delta_0 \equiv \begin{vmatrix} 0 & a_{12} & a_{13} & a_{14} \\ -a_{12} & 0 & a_{23} & a_{24} \\ -a_{13} & -a_{23} & 0 & a_{34} \\ -a_{14} & -a_{24} & -a_{34} & 0 \end{vmatrix}$$

$$= (a_{12}a_{34} - a_{13}a_{24} + a_{14}a_{23})^2.$$

Forming P and denoting the (r, s)th element of P by p_{rs} we have

$$P^2 = \begin{vmatrix} p_{11}S - \Delta_0 & p_{12}S & p_{13}S & p_{14}S \\ p_{21}S & p_{22}S - \Delta_0 & p_{23}S & p_{24}S \\ p_{31}S & p_{32}S & p_{33}S - \Delta_0 & p_{34}S \\ p_{41}S & p_{42}S & p_{43}S & p_{44}S - \Delta_0 \end{vmatrix}$$

$$= S^4 \begin{vmatrix} p_{11} - \dfrac{\Delta_0}{S} & p_{12} & p_{13} & p_{14} \\ p_{21} & p_{22} - \dfrac{\Delta_0}{S} & p_{23} & p_{24} \\ p_{31} & p_{32} & p_{33} - \dfrac{\Delta_0}{S} & p_{34} \\ p_{41} & p_{42} & p_{43} & p_{44} - \dfrac{\Delta_0}{S} \end{vmatrix}$$

which when $\Delta_0 = 0$ may be put in the form

$$P^2 = \left(\frac{S}{2}\right)^4 \begin{vmatrix} 2p_{11} & 2p_{12} & 2p_{13} & 2p_{14} \\ 2p_{21} & 2p_{22} & 2p_{23} & 2p_{24} \\ 2p_{31} & 2p_{32} & 2p_{33} & 2p_{34} \\ 2p_{41} & 2p_{42} & 2p_{43} & 2p_{44} \end{vmatrix}$$

showing that the square of P is equal to a multiple of its duplicant and therefore

$$\begin{vmatrix} p_{11} - \tfrac{1}{2}S & p_{12} & p_{13} & p_{14} \\ p_{21} & p_{22} - \tfrac{1}{2}S & p_{23} & p_{24} \\ p_{31} & p_{32} & p_{33} - \tfrac{1}{2}S & p_{34} \\ p_{41} & p_{42} & p_{43} & p_{44} - \tfrac{1}{2}S \end{vmatrix}$$

is an orthogonant.

Next let n be odd, $=5$, say, and let

$$\Delta = \begin{vmatrix} 1 & a_{12} & a_{13} & a_{14} & a_{15} \\ -a_{12} & 1 & a_{23} & a_{24} & a_{25} \\ -a_{13} & -a_{23} & 1 & a_{34} & a_{35} \\ -a_{14} & -a_{24} & -a_{34} & 1 & a_{45} \\ -a_{15} & -a_{25} & -a_{35} & -a_{45} & 1 \end{vmatrix}$$

In this case, if we denote the coaxial pirmary minors of Δ_0 by A_{11}, A_{22}, A_{33}, A_{44}, A_{55}, and using $\sum^2{}_{\alpha\alpha}$ to represent the sum of the squares of all except $A_{\alpha\alpha}$ we have

$$P^2 = \begin{vmatrix} p_{11}S - (\sum^2_{11}) & p_{12}S - A_{11}A_{22} & \cdots & p_{15}S - A_{11}A_{55} \\ p_{21}S - A_{22}A_{11} & p_{22}S - (\sum^2_{22}) & \cdots & p_{25}S - A_{22}A_{55} \\ \cdots & \cdots & \cdots & \cdots \\ p_{51}S - A_{11}A_{55} & p_{52}S - A_{22}A_{55} & \cdots & p_{55}S - (\sum^2_{55}) \end{vmatrix}$$

which when $A_{11}=A_{22}=A_{33}=A_{44}=A_{55}=0$, shows P^2 to be a multiple of its duplicant and therefore .

$$\begin{vmatrix} p_{11} - \frac{1}{2}S & p_{12} & \cdots & p_{15} \\ p_{21} & p_{22} - \frac{1}{2}S & \cdots & p_{25} \\ \cdots & \cdots & \cdots & \cdots \\ p_{51} & p_{52} & \cdots & p_{55} - \frac{1}{2}S \end{vmatrix}$$

is an orthogonant.

622. We may extend the idea of orthogonant by having the sum of the squares of the elements in a row equal to σ instead of 1. We may then speak of it as an orthogonant to the base σ. The value of such an orthogonant would obviously be $\sigma^{n/2}$.

It is apparent that the product of two r-by-n arrays formed by taking two sets of r rows of an orthogonant is always zero, unless the arrays be identical, in which case the product is σ^r.

623. The following theorems follow readily for this type of orthogonant:

1) *If the hth row be replaced by the kth column the resulting determinant is equal to* $\sigma^{n/2-1}(\text{row}_h \cdot \text{col}_k)$.

2) *The determinant got by replacing the hth row by the kth column is equal to the determinant got by replacing the kth column by the hth row.*

3) *If any column be multiplied by each of the rows, the sum of the squares of the resulting products is* σ^2.

4) *The sum of the squares of the n determinants formed by replacing the hth column with all its rows in succession is equal to* σ^n.

5) *The cofactor of any k-line minor is equal to said minor times* $\sigma^{n/2-k}$.

Again

1′) *If the hth and kth rows be replaced by the pth and qth columns, the resulting determinant is equal to*

$$\sigma^{2/n-2}\left\|\begin{matrix}\text{row } h\\ \text{row } k\end{matrix}\right\|\cdot\left\|\begin{matrix}\text{col } p\\ \text{col } q\end{matrix}\right\|.$$

2′) *The determinant got by replacing the hth and kth rows by the pth and qth columns is equal to that got by replacing the pth and qth columns by the hth and kth rows.*

3′) *If any pair of columns be multiplied by each pair of rows, the sum of the squares of the resulting products is* σ^4.

4′) *The sum of the squares of the* $\frac{1}{2}n(n-1)$ *determinants formed by replacing a fixed pair of columns by every possible pair of rows is* σ^n.

It will be apparent that there is a corresponding set of theorems where the number of replaced rows etc. is any number m.

624. *If each row of any determinant be multiplied by the corresponding column the sum of the resulting products is* $Saxm_1^2 - 2Saxm_2$, *where* $Saxm_r$ *stands for the sum of the coaxial minors of order r.*

The sum of the products in question is evidently

$$\begin{aligned} a^2_{11} + 2a_{21}a_{12} + 2a_{31}a_{13} + \cdots &+ 2a_{n1}a_{1n}\\ + a^2_{22} + 2a_{32}a_{23} + \cdots &+ 2a_{n2}a_{2n}\\ + \cdots&\cdots\\ + a^2_{nn}& \end{aligned}$$

and the simultaneous addition and subtraction of

$$\begin{aligned} 2a_{11}a_{22} + 2a_{11}a_{33} + \cdots &+ 2a_{11}a_{nn}\\ + 2a_{22}a_{33} + \cdots &+ 2a_{22}a_{nn}\\ + \cdots&\cdots\\ &+ 2a_{n-1,n-1}a_{nn} \end{aligned}$$

gives

$$(a_{11} + a_{22} + \cdots + a_{nn})^2 - 2\left\{\left|\begin{matrix}12\\12\end{matrix}\right| + \left|\begin{matrix}13\\13\end{matrix}\right| + \cdots + \left|\begin{matrix}n-1\ n\\ n-1\ n\end{matrix}\right|\right\}$$

or

$$Saxm_1^2 - 2Saxm_2.$$

625. *In an n-line orthogonant of base σ the sum of the coaxial minors of order (n−r) is equal to the sum of the coaxial minors of order r times* $\sigma^{2/n-r}$, *or*

$$Saxm_{n-r} = \sigma^{n/2-r}Saxm_r.$$

This is seen to be an immediate consequence of the relation between the complementary minors of an orthogonant.

626. *If the rows and columns of an orthogonant to the base σ be denoted by*

$$r_1, r_2, \cdots, r_n$$

$$c_1, c_2, \cdots, c_n$$

and the rows and columns of its adjugate be denoted by

$$R_1, R_2, \cdots, R_n$$

$$C_1, C_2, \cdots, C_n$$

then

$$\sum R_1C_1 = \sigma^{n-2}(Saxm_1^2 - 2Saxm_2)$$

For

$$A_{rs} = \sigma^{n/2-1}a_{rs}$$

and

$$A_{rs}A_{sr} = \sigma^{n-2}a_{rs}a_{sr}$$

and therefore

$$\begin{aligned}\sum R_1C_1 &= \sigma^{n-2}\sum r_1c_1 \\ &= \sigma^{n-2}(Saxm_1^2 - 2Saxm_2) \quad \text{by} \quad \S 624.\end{aligned}$$

627. *The sum of the n determinants formable from an n-line orthogonant of base σ by deleting a column and inserting the corresponding row is equal to*

$$\sigma^{n/2-1}(Saxm_1^2 - 2Saxm_2)$$

By 1) §623 the sum in question is

$$\sigma^{n/2-1}(r_1c_1 + r_2c_2 + \cdots + r_nc_n)$$

which by §626 gives the desired result.

628. *The sum of the* $\frac{1}{2}n(n-1)$ *determinants formed from an n-line orthogonant with the base* σ *by deleting every pair of columns and inserting in their places the corresponding rows is equal to* $\sigma^{n/2-2}\{Saxm_2^2-2\sum$ (minors of order two of second compound)$\}$.

By 1′) §623 this sum is

$$\sigma^{n/2-2}\sum\left\|\begin{matrix} r_1 \\ r_2 \end{matrix}\right\|\cdot\left\|\begin{matrix} c_1 \\ c_2 \end{matrix}\right\|$$

$$= \sigma^{n/2-2}\sum\begin{vmatrix} a_{i1} & a_{i2} & \cdots & a_{in} \\ a_{j1} & a_{j2} & \cdots & a_{jn} \end{vmatrix}\cdot\begin{vmatrix} a_{1i} & a_{2i} & \cdots & a_{ni} \\ a_{1j} & a_{2j} & \cdots & a_{nj} \end{vmatrix}\left\{\begin{matrix} i \\ j \end{matrix} = 1,2,\cdots,n\right\}$$

$$= \sigma^{n/2-2}\left\{\begin{vmatrix} 12 \\ 12 \end{vmatrix}^2 + \begin{vmatrix} 12 \\ 13 \end{vmatrix}\cdot\begin{vmatrix} 13 \\ 12 \end{vmatrix} + \begin{vmatrix} 12 \\ 14 \end{vmatrix}\begin{vmatrix} 14 \\ 12 \end{vmatrix} + \cdots + \begin{vmatrix} 12 \\ 1n \end{vmatrix}\cdot\begin{vmatrix} 1n \\ 12 \end{vmatrix}\right.$$

$$+ \begin{vmatrix} 13 \\ 12 \end{vmatrix}\cdot\begin{vmatrix} 12 \\ 13 \end{vmatrix} + \begin{vmatrix} 13 \\ 13 \end{vmatrix}^2 + \begin{vmatrix} 13 \\ 14 \end{vmatrix}\cdot\begin{vmatrix} 14 \\ 13 \end{vmatrix} + \cdots + \begin{vmatrix} 13 \\ 1n \end{vmatrix}\cdot\begin{vmatrix} 1n \\ 13 \end{vmatrix}$$

$$+ \cdots\cdots\cdots\cdots\cdots\cdots\cdots$$

$$\left. + \begin{vmatrix} 1n \\ 12 \end{vmatrix}\cdot\begin{vmatrix} 12 \\ 1n \end{vmatrix} + \begin{vmatrix} 1n \\ 13 \end{vmatrix}\cdot\begin{vmatrix} 13 \\ 1n \end{vmatrix} + \begin{vmatrix} 1n \\ 14 \end{vmatrix}\cdot\begin{vmatrix} 14 \\ 1n \end{vmatrix} + \cdots + \begin{vmatrix} 1n \\ 1n \end{vmatrix}^2\right\}$$

$$= \sigma^{n/2-2}\left\{\sum\begin{vmatrix} 12 \\ 12 \end{vmatrix}^2 + 2\sum\begin{vmatrix} 12 \\ 13 \end{vmatrix}\cdot\begin{vmatrix} 13 \\ 12 \end{vmatrix}\right\}$$

$$= \sigma^{n/2-2}\left\{\sum\begin{vmatrix} 12 \\ 12 \end{vmatrix}^2 + 2\sum\begin{vmatrix} 12 \\ 12 \end{vmatrix}\cdot\begin{vmatrix} 13 \\ 13 \end{vmatrix}\right.$$

$$\left. - 2\sum\begin{vmatrix} 12 \\ 12 \end{vmatrix}\cdot\begin{vmatrix} 13 \\ 13 \end{vmatrix} + 2\sum\begin{vmatrix} 12 \\ 13 \end{vmatrix}\cdot\begin{vmatrix} 13 \\ 12 \end{vmatrix}\right\}$$

$$= \sigma^{n/2-2}\left[\left\{\sum\begin{vmatrix} 12 \\ 12 \end{vmatrix}\right\}^2 - 2\begin{vmatrix} \begin{vmatrix} 12 \\ 12 \end{vmatrix} & \begin{vmatrix} 12 \\ 13 \end{vmatrix} \\ \begin{vmatrix} 13 \\ 12 \end{vmatrix} & \begin{vmatrix} 13 \\ 13 \end{vmatrix} \end{vmatrix}\right]$$

$$= \sigma^{n/2-2}\left[\left\{\sum\begin{vmatrix} 12 \\ 12 \end{vmatrix}\right\}^2 - 2\left\{\sum a_{11}\sum\begin{vmatrix} 123 \\ 123 \end{vmatrix} - \sum\begin{vmatrix} 1234 \\ 1234 \end{vmatrix}\right\}\right]$$

When $n=4$ this becomes

$$Saxm_2^2 - 2Saxm_1^2 + 2\sigma^2.$$

629. *The sum of the three-line coaxial minors of the duplicant of a four-line orthogonant of base* σ *is equal to*

$$2Saxm_1\,(Saxm_2 - 2\sigma).$$

If the orthogonant be $|\alpha_1\beta_2\gamma_3\delta_4|$ the first of the coaxial minors in question

$$\begin{vmatrix} 2\alpha_1 & \alpha_2+\beta_1 & \alpha_3+\gamma_1 \\ \beta_1+\alpha_2 & 2\beta_2 & \beta_3+\gamma_2 \\ \gamma_1+\alpha_3 & \gamma_2+\beta_3 & 2\gamma_3 \end{vmatrix}$$

which is equal to

$$\begin{aligned}
&2|\alpha_1\beta_2\gamma_3| + 2\left\{\begin{vmatrix} \alpha_1 & \beta_1 & \alpha_3 \\ \alpha_2 & \beta_2 & \beta_3 \\ \alpha_3 & \beta_3 & \gamma_3 \end{vmatrix} + \begin{vmatrix} \alpha_1 & \beta_2 & \gamma_1 \\ \alpha_2 & \beta_2 & \gamma_2 \\ \alpha_3 & \gamma_2 & \gamma_3 \end{vmatrix} + \begin{vmatrix} \alpha_1 & \beta_1 & \gamma_1 \\ \beta_1 & \beta_2 & \gamma_2 \\ \gamma_1 & \beta_3 & \gamma_3 \end{vmatrix}\right\} \\
&= 2|\alpha_1\beta_2\gamma_3| + 2\{\alpha_1|\beta_2\gamma_3| - \alpha_2|\alpha_2\gamma_3| + \alpha_3|\alpha_2\beta_3| - \beta_1|\beta_1\gamma_3| \\
&\quad + \beta_2|\alpha_1\gamma_3| - \beta_3|\alpha_1\beta_3| + \gamma_1|\beta_1\gamma_2| - \gamma_2|\alpha_1\gamma_2| + \gamma_3|\alpha_1\beta_2|\} \\
&= 2|\alpha_1\beta_2\gamma_3| + 2\{\alpha_1(|\beta_2\gamma_3| + |\alpha_1\delta_4|) + \beta_2(|\alpha_1\gamma_3| + |\beta_2\delta_4|) \\
&\quad + \gamma_3(|\alpha_1\beta_2| + |\gamma_3\delta_4|)\} - 2\{\alpha_1|\alpha_1\delta_4| + \beta_2|\beta_2\delta_4| + \gamma_3|\gamma_3\delta_4|\} \\
&\quad - 2\{\alpha_2^2\gamma_3 + \alpha_3^2\beta_2 + \beta_1^2\gamma_3 + \beta_3^2\alpha_1 + \gamma_1^2\beta_2 + \gamma_2^2\alpha_1\} \\
&\quad + 2\{\alpha_2\gamma_2\alpha_3 + \alpha_3\alpha_2\beta_3 + \beta_1\gamma_1\beta_3 + \beta_1\beta_3\alpha_3 + \beta_1\gamma_1\gamma_2 + \gamma_1\gamma_2\alpha_2\} \\
&= 2|\alpha_1\beta_2\gamma_3| + 2\{\alpha_1(|\beta_2\gamma_3| + |\alpha_1\delta_4|) + \beta_2(|\alpha_1\gamma_3| + |\beta_2\delta_4|) \\
&\quad + \gamma_3(|\alpha_1\beta_2| + |\gamma_3\delta_4|)\} - 2\{\alpha_1(\alpha_4^2 + \beta_3^2 + \gamma_2^2) \\
&\quad + \beta_2(\alpha_3^2 + \beta_4^2 + \gamma_1^2) + \gamma_3(\alpha_2^2 + \beta_1^2 + \gamma_4^2) + \delta_4(\alpha_1^2 + \beta_2^2 + \gamma_3^2\} \\
&\quad + 2\{\alpha_4(\alpha_1\delta_1 + \alpha_4\delta_4) + \alpha_3\alpha_2\gamma_2 + \alpha_2\alpha_3\beta_3 + - + \beta_4(\beta_2\delta_2 + \beta_4\delta_4) \\
&\quad + \beta_3\beta_1\gamma_1 + - + \beta_1\alpha_3\beta_3 + \gamma_4(\gamma_3\delta_3 + \gamma_4\delta_4) + - + \gamma_2\beta_1\gamma_1 \\
&\quad + \gamma_1\alpha_2\gamma_2\} + 2\{\alpha_1\alpha_4^2 + \beta_2\beta_4^2 + \gamma_3\gamma_4^2 + \delta_4\delta_4^2\} \\
&\quad - 2\delta_4(\alpha_4^2 + \beta_4^2 + \gamma_4^2 + \delta_4^2).
\end{aligned}$$

The other minors are obtained from this by performing simultaneously the substitutions

$$\alpha,\beta,\gamma,\delta \to \beta,\gamma,\delta,\alpha$$
$$1,2,3,4 \to 2,3,4,1.$$

Making these substitutions and summing we have from:

the first terms, $+2Saxm_3 = 2\sigma Saxm_1$ (by §625)
the second terms, $+2Saxm_1Saxm_2$
the third terms, $-6\sigma Saxm_1$
the fourth terms, 0
the fifth terms, $+2\sigma Saxm_1$
the sixth terms, $-2\sigma Saxm_1$.

The sum of all is therefore

$$2Saxm_1(Saxm_2 - 2\sigma).$$

Concerning the fourth term we readily see that it divides itself into 3 parts as follows:

(1) $$\alpha_4(\alpha_1\delta_1 + \alpha_4\delta_4) + \beta_1\alpha_3\beta_3 + \gamma_2\beta_1\gamma_1$$

(2) $$\beta_4(\beta_2\delta_2 + \beta_4\delta_4) + \alpha_2\gamma_2\gamma_1 + \alpha_3\alpha_2\gamma_2$$

(3) $$\gamma_4(\gamma_3\delta_3 + \gamma_4\delta_4) + \alpha_2\alpha_3\beta_3 + \beta_3\beta_1\gamma_1.$$

Each of these parts vanishes separately when summed cyclically.

Since every term such as $\beta_1\alpha_3\beta_3$ after four substitutions returns to itself it is apparent that we may replace it by any of those terms into which it is transformed. We may therefore write

(1) $$\alpha_4(\alpha_1\delta_1 + \alpha_2\delta_2 + \alpha_3\delta_3 + \alpha_4\delta_4)$$

which shows that it is zero.

The same thing may be done with the third and other terms.

630. *The sum of the six determinants formable from a four-line orthogonant of base σ by deleting a pair of columns and inserting in their place the corresponding rows is equal to*

$$Saxm_2^2 - 2Saxm_1^2 + 2\sigma^2.$$

If the rows be denoted by 1234 and the columns by $1'2'3'4'$ then the sum in question is

$$|1'2'34| + |1'23'4| + |1'234'| + |12'3'4| + |12'34'| + |123'4'|$$

which by 1′) §623 is

$$\left\|\begin{matrix}\alpha_1 & \alpha_2 & \alpha_3 & \alpha_4\\ \beta_1 & \beta_2 & \beta_3 & \beta_4\end{matrix}\right\| \cdot \left\|\begin{matrix}\alpha_1 & \beta_1 & \gamma_1 & \delta_1\\ \alpha_2 & \beta_2 & \gamma_2 & \delta_2\end{matrix}\right\| + \cdots$$
$$+\left\|\begin{matrix}\gamma_1 & \gamma_2 & \gamma_3 & \gamma_4\\ \delta_1 & \delta_2 & \delta_3 & \delta_4\end{matrix}\right\| \cdot \left\|\begin{matrix}\alpha_3 & \beta_3 & \gamma_3 & \delta_3\\ \alpha_4 & \beta_4 & \gamma_4 & \delta_4\end{matrix}\right\|$$

$$= |\alpha_1\beta_2|^2 + |\alpha_1\beta_3|\cdot|\alpha_1\gamma_2| + \cdots + |\alpha_3\beta_4|\cdot|\gamma_1\delta_2|$$
$$+ |\alpha_1\gamma_2|\cdot|\alpha_1\beta_3| + |\alpha_1\gamma_3|^2 + \cdots + |\alpha_3\gamma_4|\cdot|\gamma_1\delta_3|$$
$$+ \cdots\cdots\cdots\cdots\cdots\cdots\cdots\cdots$$
$$+ |\gamma_1\delta_2|\cdot|\alpha_3\beta_4| + |\gamma_1\delta_3|\cdot|\alpha_3\gamma_4| + \cdots + |\gamma_3\delta_4|^2$$
$$= \sum|\alpha_1\beta_2|^2 + 2\sum\{|\alpha_1\beta_3|\cdot|\alpha_1\gamma_2|\}$$
$$= \{\sum|\alpha_1\beta_2|\}^2 - 2\sum\{|\alpha_1\beta_2|\cdot\alpha_1\gamma_3|\} + 2\{|\alpha_1\beta_3|\cdot|\alpha_1\gamma_2|\}$$
$$= \{\sum|\alpha_1\beta_2|\}^2 - 2\sum\begin{vmatrix} |\alpha_1\beta_2| & |\alpha_1\beta_3| \\ |\alpha_1\gamma_2| & |\alpha_1\gamma_3| \end{vmatrix}$$
$$= Saxm_2^2 - 2\sigma(\sigma - Saxm_1^2).$$

631. If we use $\{Dup\}_r$ to denote the duplicant of an r-line orthogonant with base σ, then

$$\begin{aligned}\{Dup\}_3 &= 2\sigma^{3/2} + 2\sigma^{1/2}(Saxm_1^2 - 2Saxm_2)\\ &= 2\sigma^{1/2}[\sigma + Saxm_1^2 - 2\sigma^{1/2}Saxm_1]\\ &= 2\sigma^{1/2}(\sigma^{1/2} - Saxm_1)^2.\end{aligned}$$

Again

$$\begin{aligned}\{Dup\}_4 &= 2\sigma^2 + 2\sigma(Saxm_1^2 - 2Saxm_2)\\ &\quad + (Saxm_2^2 - 2\sigma Saxm_1^2 + 2\sigma^2)\\ &= 4\sigma^2 - 4\,Saxm_2\sigma + Saxm_2^2\\ &= (2\sigma - Saxm_2)^2.\end{aligned}$$

632. Let $A = |a_{nn}|$, and $B = |b_{nn}|$, be n-line orthogonants of bases σ_1 and σ_2 respectively. Denote the columns of A by $1, 2, 3, \cdots, n$ and of B by $1', 2', \cdots, n'$. If we multiply the determinant C formed by the first $n-1$ columns of A and the pth column of B by A we get

$$C\cdot A = \begin{vmatrix} \sigma_1 & 0 & \cdots & 0 \\ 0 & \sigma_1 & \cdots & 0 \\ \cdot & \cdot & \cdot & \cdot \\ p'\cdot 1 & \cdot p'\cdot 2 & \cdot p & \cdot '\cdot n \end{vmatrix}$$

where $p'i$ denotes the product of the pth column of B by the ith column of A. Therefore

$$C\cdot\sigma_1^{n/2} = \sigma_1^{n-1}(p'\cdot n)$$

or

$$C = \sigma_1^{n/2-1}(p' \cdot n).$$

If instead of the last column of A we had replaced the qth then we would have had the theorem:

(1) *If the qth column of A be replaced by the pth column of B, the resulting determinant is equal to $\sigma^{n/2-1}(p' \cdot q)$.*

Similarly *if the pth column of B be replaced by the qth column of A, the resulting determinant is equal to $\sigma_2^{n/2-1}(q \cdot p')$.*

The ratio of the two determinants is therefore

$$(2) \qquad \left(\frac{\sigma_1}{\sigma_2}\right)^{n/2-1}$$

633. *If any column of A be multiplied by each column of B, the sum of the squares of the resulting products is $\sigma_1 \cdot \sigma_2$.*

Let the selected column of A be the pth then

$$p \cdot 1' = a_{1p}b_{11} + a_{2p}b_{21} + \cdots + a_{np}b_{n1}$$
$$p \cdot 2' = a_{1p}b_{12} + a_{2p}b_{22} + \cdots + a_{np}b_{n2}$$
$$\cdots\cdots\cdots\cdots\cdots\cdots$$
$$p \cdot n' = a_{1p}b_{1n} + a_{2p}b_{2n} + \cdots + a_{np}b_{nn}$$

Squaring and summing we get on the right-hand side

$$\begin{aligned} &\sum a^2{}_{1p}(b_{11}{}^2 + b_{12}{}^2 + \cdots + b_{1n}{}^2) \\ &+ 2\sum a_{1p}a_{2p}(b_{11}b_{21} + b_{12}b_{22} + \cdots + b_{1n}b_{2n}) \\ &= \sum a^2{}_{1p}\sigma_2 + 0 = \sigma_1\sigma_2. \end{aligned}$$

634. *The sum of the squares of the n determinants formed from A by interchanging its pth column with all of the columns of B in succession is equal to $\sigma_1{}^{n-1} \cdot \sigma_2$.*

For the sum in question is by (1) §632 equal to

$$\begin{aligned} &\sigma_1^{n-2}\{(1' \cdot p)^2 + (2' \cdot p)^2 + \cdots + (n' \cdot p)^2\} \\ &= \sigma_1{}^{n-2} \cdot \sigma_1 \cdot \sigma_2 \text{ by §633} \\ &= \sigma_1^{n-1} \cdot \sigma_2. \end{aligned}$$

If we replace the pth and qth columns of A by the hth and kth columns of B, the resulting determinant is equal to

$$\sigma_1{}^{n/2-2} \left\| \begin{matrix} c_p{}' \\ c_q{}' \end{matrix} \right\| \cdot \left\| \begin{matrix} c_h \\ c_k \end{matrix} \right\|.$$

This is proven in precisely the same manner as in §632 by multiplying the resulting determinant by A and dividing out by $\sigma_1{}^{n/2-1}$.

Interchanging A and B in this proposition gives a determinant whose value is

$$\sigma^{n/2-2}\begin{Vmatrix} c_h \\ c_k \end{Vmatrix}\cdot\begin{Vmatrix} c_p{}' \\ c_q{}' \end{Vmatrix}.$$

The ratio of the two is $(\sigma_1/\sigma_2)^{n/2-2}$.

635. *If any pair of columns of A be multiplied by each pair of columns of B the sum of the squares of the resulting products is equal* $\sigma_1^2\cdot\sigma_2^2$.

Let the selected columns of A be the pth and qth, then

$$\begin{Vmatrix} c_p \\ c_q \end{Vmatrix}\cdot\begin{Vmatrix} c_1{}' \\ c_2{}' \end{Vmatrix} = \sum | a_{1p}a_{2q} | \cdot | b_{11}b_{22} |$$

$$\begin{Vmatrix} c_p \\ c_q \end{Vmatrix}\cdot\begin{Vmatrix} c_1{}' \\ c_3{}' \end{Vmatrix} = \sum | a_{1p}a_{2q} | \cdot | b_{11}b_{23} |$$

. . . .

$$\begin{Vmatrix} c_p \\ c_q \end{Vmatrix}\cdot\begin{Vmatrix} c_{(n-1)}{}' \\ c_n{}' \end{Vmatrix} = \sum | a_{1p}a_{2q} | \cdot | b_{n-1,n-1}b_{nn} |$$

Squaring and adding we have on the right

$$\begin{aligned} &\sum | a_{1p}a_{2q} |^2\{ | b_{11}b_{22} |^2 + | b_{11}b_{23} |^2 + \cdots + | b_{n-1,n-1}b_{nn} |^2\} \\ &\quad + 2\sum | a_{1p}a_{2q} | \cdot | a_{1p}a_{3q} | \{ | b_{11}b_{22} | \cdot | b_{11}b_{32} | + \cdots \} \\ &\qquad = \sigma_2^2 \sum | a_{1p}a_{2q} |^2 + 0, \\ &\qquad = \sigma_2^2\cdot\sigma_1^2, \end{aligned}$$

since $\{ |b_{11}b_{22}|\cdot|b_{11}b_{32}| + \cdots \}$ is the product of

$$\begin{vmatrix} b_{11} & b_{12} & \cdots & b_{1n} \\ b_{21} & b_{22} & \cdots & b_{2n} \end{vmatrix}\cdot\begin{vmatrix} b_{11} & b_{12} & \cdots & b_{1n} \\ b_{31} & b_{32} & \cdots & b_{3n} \end{vmatrix} = \begin{vmatrix} \sigma_2 & 0 \\ 0 & 0 \end{vmatrix} = 0$$

636. *The sum of the squares of the* $\frac{1}{2}n(n-1)$ *determinants formed from A by replacing a fixed pair of its columns by every possible pair of columns of B is equal to*

$$\sigma_1^{n-2}\cdot\sigma_2^2 .$$

Let the fixed pair of columns be the hth and kth, then the sum in question is by §634 equal to

$$\sigma_1^{n-4}\left\{\begin{Vmatrix} c_1' \\ c_2' \end{Vmatrix}\cdot\begin{Vmatrix} c_h \\ c_k \end{Vmatrix}\right\}^2 + \sigma_1^{n-4}\left\{\begin{Vmatrix} c_1' \\ c_3' \end{Vmatrix}\cdot\begin{Vmatrix} c_h \\ c_k \end{Vmatrix}\right\}^2 + \cdots$$

$$= \sigma_1^{n-4}\sum\left\{\begin{Vmatrix} c_1' \\ c_2' \end{Vmatrix}\cdot\begin{Vmatrix} c_h \\ c_k \end{Vmatrix}\right\}^2$$

$$= \sigma_1^{n-4}\cdot\sigma_1^2\cdot\sigma_2^2 \quad by \quad \S 635$$

$$= \sigma_1^{n-2}\cdot\sigma_2^2 .$$

637. In general we may formulate corresponding theorems where the number of columns replaced or interchanged is r. This leads to the following theorem:

If there be two n-line orthogonants A and B whose bases are σ_1 *and* σ_2, *and from their 2n columns every possible set of n columns be taken, the sum of the squares of the determinants so formable is equal to* $(\sigma_1+\sigma_2)^n$.

The number of determinants in question is $(2n)_n$ and may be partitioned into

$$1^2 + n^2 + \left\{\frac{n(n-1)}{1\cdot 2}\right\}^2 + \cdots + n^2 + 1^2 .$$

The first 1^2 corresponding to A and the n^2 to those formed from A by replacing each column by a column from B, the $\{n(n-1)/1\cdot 2\}^2$ to those formed from A by replacing each pair of columns by a pair from B, etc.

But we know that

$$A^2 = \sigma_1^n ;$$

that the sum of the squares of those obtained from A by replacing a single column is

$$n\sigma_1^{n-1}\cdot\sigma_2 ;$$

that the sum of the squares of those obtained by replacing two columns is

$$\frac{n(n-1)}{1\cdot 2}\cdot\sigma_1^{n-2}\ \sigma_2^2$$

and so on. The total is therefore

$$\sigma_1^n + n\sigma_1^{n-1}\sigma_2 + \frac{n(n-1)}{1\cdot 2}\sigma_1^{n-2}\sigma_2^2 + \cdots = (\sigma_1+\sigma_2)^n .$$

638. *If* $|a_{nn}|$ *and* $|b_{nn}|$ *be such that the product of any two rows of the one is equal to the product of the corresponding two rows of the other, then* $|a_{nn}| = \pm |b_{nn}|$.

For $|a_{nn}|^2 = |b_{nn}|^2$ by hypothesis hence the theorem.

639. *If* $|a_{nn}|$ *be such that the sum of the squares of the elements in any line is equal to* α *and the sum of the products of corresponding elements in any two parallel lines is* β, *then the determinant is equal to* $\{(\alpha-\beta)^{n-1}(\alpha+\overline{n-1}\beta)\}^{1/2}$.

For squaring $|a_{nn}|$ we get a determinant with α's for elements along the principal diagonal and β's for all the other elements, which by the operations c_1-c_2, c_2-c_3, $\cdots$, $c_{n-1}-c_n$, taking out $(\alpha-\beta)$ from $n-1$ columns and then adding the sum of all the rows above it to the $(n-1)$st and $\{\alpha+(n-1)\beta\}$ will be the result.

640. Using the notation of matrices and being given

$$(A\between x_{i1}, x_{i2}, \cdots, x_{in}) = X_{i1}, X_{i2}, \cdots, X_{in},$$

where A stands for the matrix

$$\left(\left|\begin{matrix} a_{11} & a_{12} & \cdots & a_{1n} \\ a_{21} & a_{22} & \cdots & a_{2n} \\ \cdot & \cdot & \cdot & \cdot \\ a_{n1} & a_{n2} & \cdots & a_{nn} \end{matrix}\right|\right)$$

whose determinant is orthogonal, then it follows that

$$|x_{nn}|^2 = |X_{nn}|^2$$

and that if multiplication is row-by-row the determinant resulting from squaring are identical element for element.

641. $A \equiv |a_{nn}|$ *is an orthogonant which has all its elements functions of one and the same variable x and if* $B \equiv |b_{nn}|$, *where* b_{ij} *is the derivative of* a_{ij} *with respect to x, then B is equal to zero or the square of a Pfaffian according as n is odd or even.*

Starting with the relations

$$a_{r1}^2 + a_{r2}^2 + \cdots + a_{rn}^2 = 1,$$

$$a_{r1}a_{s1} + a_{r2}a_{s2} + \cdots + a_{rn}a_{sn} = 0,$$

we have by differentiation

$$a_{r1}b_{r1} + a_{r2}b_{r2} + \cdots + a_{rn}b_{rn} = 0$$

$$(a_{r1}b_{s1} + a_{r2}b_{s2} + \cdots + a_{rn}b_{sn}) + (a_{s1}b_{r1} + a_{s2}b_{r2} + \cdots + a_{sn}b_{rn}) = 0$$

which shows that if $|p_{nn}| \equiv P \equiv A \cdot B$, then $p_{rr}=0$ and $p_{rs}+p_{sr}=0$.

The determinant P or B is therefore skew-symmetric and hence the truth of the theorem.

EXERCISES:

1. Show that P^2 and B^2 are element-by-element identical.
2. If $A \equiv |a_{nn}|$ is an orthogonant and if

$$
\begin{aligned}
(A)(x_1, x_2, \cdots, x_n) &= X_1, X_2, \cdots, X_n\\
(A)(y_1, y_2, \cdots, y_n) &= Y_1, Y_2, \cdots, Y_n\\
&\cdots\cdots\cdots\\
(A)(w_1, w_2, \cdots, w_n) &= W_1, W_2, \cdots, W_n
\end{aligned}
$$

show that

$$
\begin{vmatrix} x_1 & x_2 & \cdots & x_n\\ y_1 & y_2 & \cdots & y_n\\ \cdot & \cdot & \cdot & \cdot\\ w_1 & w_2 & \cdots & w_n \end{vmatrix}^2 = \begin{vmatrix} X_1 & X_2 & \cdots & X_n\\ Y_1 & Y_2 & \cdots & Y_n\\ \cdot & \cdot & \cdot & \cdot\\ W_1 & W_2 & \cdots & W_n \end{vmatrix}^2 .
$$

642. *If $\phi_1, \phi_2, \cdots, \phi_n$ be functions of $x_1, x_2, \cdots, x_{n+r}$ the ϕ's will be mutually independent if*

$$
\left\Vert \begin{matrix} \dfrac{\partial\phi_1}{\partial x_1} & \dfrac{\partial\phi_1}{\partial x_2} & \cdots & \dfrac{\partial\phi_1}{\partial x_{n+r}}\\ \cdot & \cdot & \cdot & \cdot\\ \dfrac{\partial\phi_n}{\partial x_1} & \dfrac{\partial\phi_n}{\partial x_2} & \cdots & \dfrac{\partial\phi_n}{\partial x_{n+r}} \end{matrix} \right\Vert
$$

be not evanescent.

Take the case where $n=6$ and $r=3$, and let the equations be

$$
\begin{aligned}
w_{11}^2 + w_{12}^2 + w_{13}^2 &\quad 1\\
w_{11}w_{21} + w_{12}w_{22} + w_{13}w_{23} &= 0\\
w_{11}w_{31} + w_{12}w_{32} + w_{13}w_{33} &= 0\\
w_{21}^2 + w_{22}^2 + w_{23}^2 &= 1\\
w_{21}w_{31} + w_{22}w_{32} + w_{23}w_{33} &= 0\\
w_{31}^2 + w_{32}^2 + w_{33}^2 &= 1
\end{aligned}
$$

then

$$\begin{Vmatrix} 2w_{11} & 2w_{12} & 2w_{13} & 0 & 0 & 0 & 0 & 0 & 0 \\ w_{21} & w_{22} & w_{23} & w_{11} & w_{12} & w_{13} & 0 & 0 & 0 \\ w_{31} & w_{32} & w_{33} & 0 & 0 & 0 & w_{11} & w_{12} & w_{13} \\ 0 & 0 & 0 & 2w_{21} & 2w_{22} & 2w_{23} & 0 & 0 & 0 \\ 0 & 0 & 0 & w_{31} & w_{32} & w_{33} & w_{21} & w_{22} & w_{23} \\ 0 & 0 & 0 & 0 & 0 & 0 & 2w_{31} & 2w_{32} & 2w_{33} \end{Vmatrix}$$

when multiplied by itself gives

$$\begin{vmatrix} 4 & \cdot & \cdot & \cdot & \cdot & \cdot \\ \cdot & 2 & \cdot & \cdot & \cdot & \cdot \\ \cdot & \cdot & 2 & \cdot & \cdot & \cdot \\ \cdot & \cdot & \cdot & 4 & \cdot & \cdot \\ \cdot & \cdot & \cdot & \cdot & 2 & \cdot \\ \cdot & \cdot & \cdot & \cdot & \cdot & 4 \end{vmatrix} = 2^9.$$

It follows therefore that all minors of order 6 of the array are not zero.

643. If in the expression (§215)

$$\sum s(QP\bar{Q})^{(m)}_{ss} = \sum r \sum s[P^{(m)}_{r1} M^{(m)}_{r1} + \cdots + P^{(m)}_{r\mu} M^{(m)}_{r\mu}]$$

we consider Q to be an orthogonant, then

$$M^{(m)}_{rr} = 1 \quad \text{and} \quad M^{(m)}_{rs} = 0,$$

and therefore

$$\sum s(QP\bar{Q})^{(m)}_{ss} = \sum sP^{(m)}_{ss}.$$

That is, *if Q be an orthogonant and P any other determinant of the same order then the sum of the m-line coaxial minors of $QP\bar{Q}$ is the same as the sum of the m-line coaxial minors of P.*

644. *If two general determinants D_1 and D_2 be multiplied row-wise by one and the same orthogonant P, and the first product thus obtained be multiplied row-wise by the second the resulting determinant is equal to the product of D_1 and D_2.*

For

$$(D_1P')(D_2P')' = (D_1P')(PD_2') = D_1PP'D_2' = D_1D_2'$$

where D' stands for the conjugate of D.

645. A determinant whose elements are complex and which is such that its complex conjugate is its determinantal reciprocal is called a *hyperorthogonant.* Thus if the product row-by-row of

$$\begin{vmatrix} \alpha_1 & \alpha_2 & \alpha_3 \\ \beta_1 & \beta_2 & \beta_3 \\ \gamma_1 & \gamma_2 & \gamma_3 \end{vmatrix} \cdot \begin{vmatrix} \bar{\alpha}_1 & \bar{\alpha}_2 & \bar{\alpha}_3 \\ \bar{\beta}_1 & \bar{\beta}_2 & \bar{\beta}_3 \\ \bar{\gamma}_1 & \bar{\gamma}_2 & \bar{\gamma}_3 \end{vmatrix} = \begin{vmatrix} 1 & \cdot & \cdot \\ \cdot & 1 & \cdot \\ \cdot & \cdot & 1 \end{vmatrix}$$

where $\bar{\alpha}_i$ is the conjugate of α_i, then $|\alpha_1\beta_2\gamma_3|$ is a hyperorthogonant.

646. The following theorems concerning hyperorthogonants are readily proved.

1) *If H be a hyperorthogonat and $\overline{H}$ its complex conjugate, then the product of any element of H by $\overline{H}$ is equal to the cofactor of the corresponding element in $\overline{H}$.*

2) *The product of H and $\overline{H}$ row-wise gives the same result as if taken column-wise.*

3) *The product of any minor of H by $\overline{H}$ is equal to the cofactor of the corresponding minor in $\overline{H}$.*

4) *If $H\,{}^c_c\,A = B$, then $\overline{H}\,{}^r_c\,B = A$, where A is any determinant and $H\,{}^c_c\,A$ means multiplication col-by-col.*

5) *If (a_{kn}) stands for the array consisting of k rows of H and (α_{kn}) stands for the corresponding array of $\overline{H}$, then*

$$(a_{kn})(\alpha_{kn}) = 1.$$

6) *If P is any determinant then the sum of the diagonal elements of the product $HP\overline{H}$ is the same as the corresponding sum in P.*

Exercises. Set XXXI

1. If the elements of the determinant

$$\Delta \equiv \begin{vmatrix} 1 & x_{01} & x_{02} \cdots & x_{0n} \\ 1 & x_{11} & x_{12} \cdots & x_{1n} \\ \cdot & \cdot & \cdot & \cdot \\ 1 & x_{n1} & x_{n2} \cdots & x_{nn} \end{vmatrix}$$

be such that the square of each row is equal to 2, then show that for a maximum value of Δ we have

$$\Delta^2 = \begin{vmatrix} 2 & 1 - 1/n \cdots & 1 - 1/n \\ 1 - 1/n & 2 \cdots & 1 - 1/n \\ \cdot & \cdot & \cdot \\ 1 - 1/n & 1 - 1/n \cdots & 2 \end{vmatrix}$$

and hence

$$\Delta = \left\{ \frac{(n+1)^{n+1}}{n^n} \right\}^{1/2} \qquad \text{(Borsch)}$$

2. Show that the basic for transforming $\sum x\bar{x}$ into itself where $\bar{x}$ is the conjugate complex of x is obtained by adding to Cayley's basic matrix the product of $(-1)^{1/2}$ and an arbitrary axisymmetric matrix.

3. If the square of $|u_1v_2w_3|$ got by column-by-column multiplication be identical in elements with the square of $|x_1y_2z_3|$ got in the same way, then

$$\begin{vmatrix} |x_1\ v_2\ w_3| & |y_1\ v_2\ w_3| & |z_1\ v_2\ w_3| \\ |u_1\ x_2\ w_3| & |u_1\ y_2\ w_3| & |u_1\ z_2\ w_3| \\ |u_1\ v_2\ x_3| & |u_1\ v_2\ y_3| & |u_1\ v_2\ z_3| \end{vmatrix}$$

is an orthogonant whose basic constant is $|u_1v_2w_3|^2$.

4. If the determinant which is the square of

$$\begin{Vmatrix} u_1 & v_1 & w_1 \\ u_2 & v_2 & w_2 \end{Vmatrix}$$

be identical in elements with the determinant which is the square of

$$\begin{Vmatrix} x_1 & y_1 & z_1 \\ x_2 & y_2 & z_2 \end{Vmatrix},$$

then

$$\begin{vmatrix} & |y_1\ z_2| & |z_1\ x_2| & |x_1\ y_2| \\ |v_1\ w_2| & |u_1\ x_2| & |u_1\ y_2| & |u_1\ z_2| \\ |w_1\ u_2| & |v_1\ x_2| & |v_1\ y_2| & |v_1\ z_2| \\ |u_1\ v_2| & |w_1\ x_2| & |w_1\ y_2| & |w_1\ z_2| \end{vmatrix}$$

is an orthogonant whose constant base is

$$\begin{Vmatrix} u_1 & v_1 & w_1 \\ u_2 & v_2 & w_2 \end{Vmatrix}^2.$$

5. If $|\alpha_1\beta_2|$ is a positive unit orthogonant then

$$\begin{vmatrix} \alpha_1^2 & 2^{1/2}\alpha_1\alpha_2 & \alpha_2^2 \\ 2^{1/2}\alpha_1\beta_1 & \alpha_1\beta_2 + \alpha_2\beta_1 & 2^{1/2}\alpha_2\beta_2 \\ \beta_1^2 & 2^{1/2}\beta_1\beta_2 & \beta_2^2 \end{vmatrix}$$

is also a positive unit orthogonant.

6. If A be an m-line and B an n-line orthogonant then the Zehfuss determinant equal to A^nB^m (§168) is an orthogonant also.

CHAPTER XV

DETERMINANTAL EQUATIONS

647. If we subtract (or add) a variable x from each of the diagonal elements of a determinant of the nth order we have a polynomial in x of the nth degree, which equated to zero has n roots and is called a *determinantal equation*. If the determinant of the nth order is $|a_{nn}| \equiv A$, then $|A-x|=0$ may be used to represent the equation. The equation $|A-x|=0$, is sometimes called the *characteristic equation*, and $|A-x|$ is called the *latent function*. Let $g_1, g_2, \cdots, g_n$ represent the roots which are known as *latent roots*.

648. *If $A_{(m)}$ denotes the mth compound of A and $|A_{(m)}-x|=0$ the equation formed by subtracting x from each of the diagonal elements, then the roots of $|A_{(m)}-x|=0$ are the products of the g's taken m at a time.*

Let

$$\rho \cdot x_\alpha = a_{\alpha 1}x_1 + a_{\alpha 2}x_2 + \cdots + a_{\alpha n}x_n \quad (\alpha = 1, 2, \cdots, n)$$

be a system of n equations with arbitrary coefficients—the a's.

The condition that these equations are simultaneously true for values of the x's other than all zero is that

$$\begin{vmatrix} a_{11}-\rho & a_{12} & \cdots & a_{1n} \\ a_{21} & a_{22}-\rho & \cdots & a_{2n} \\ \cdot & \cdot & \cdot & \cdot \\ a_{n1} & a_{n2} & \cdots & a_{nn}-\rho \end{vmatrix} \equiv \phi(\rho) = 0.$$

But this is an equation of the nth degree in ρ and in general has n distinct roots. For each value of ρ there is a value system

$$x_{\beta 1}, x_{\beta 2}, \cdots, x_{\beta n}$$

so that we may write

$$\rho_\beta x_{\beta\alpha} = a_{\alpha 1}x_{\beta 1} + a_{\alpha 2}x_{\beta 2} + \cdots + a_{\alpha n}x_{\beta n}, \qquad (\beta = 1, 2, \cdots, n).$$

Let

$$\Delta \equiv \begin{vmatrix} x_{11} & x_{12} & \cdots & x_{1n} \\ x_{21} & x_{22} & \cdots & x_{2n} \\ \cdot & \cdot & \cdot & \cdot \\ x_{n1} & x_{n2} & \cdots & x_{nn} \end{vmatrix}$$

Of this determinant we may observe (1) It is not zero, for no two rows can be proportional since no two values of ρ are alike, (2) Not all the x's in any row are zero.

Let $X_{i1}, X_{i2}, \cdots, X_{i\mu}\{\mu=(n)_m\}$ be the minors of order m of the matrix

$$\begin{Vmatrix} x_{i_11} & x_{i_12} & \cdots & x_{i_1n} \\ x_{i_21} & x_{i_22} & \cdots & x_{i_2n} \\ \cdot & \cdot & \cdot & \cdot \\ x_{i_m1} & x_{i_m2} & \cdots & x_{i_mn} \end{Vmatrix}$$

so that

$$X_{ik} \equiv \begin{vmatrix} x_{i_1k_1} & x_{i_1k_2} & \cdots & x_{i_1k_m} \\ x_{i_2k_1} & x_{i_2k_2} & \cdots & x_{i_2k_m} \\ \cdot & \cdot & \cdot & \cdot \\ x_{i_mk_1} & x_{i_mk_2} & \cdots & x_{i_mk_m} \end{vmatrix}$$

Now $X_{ik}\neq 0$ for some values of i and k, otherwise Δ would be zero. Let the μ minors of order m of the matrix

$$\begin{Vmatrix} a_{k_11} & a_{k_12} & \cdots & a_{k_1n} \\ a_{k_21} & a_{k_22} & \cdots & a_{k_2n} \\ \cdot & \cdot & \cdot & \cdot \\ a_{k_m1} & a_{k_m2} & \cdots & a_{k_mn} \end{Vmatrix}$$

be represented by

$$A_{(m)k1}, A_{(m)k2}, \cdots, A_{(m)k\mu},$$

then

$$A_{(m)k1}X_{i1} + A_{(m)k2}X_{i2} + \cdots + A_{(m)k\mu}X_{i\mu}$$

$$= \begin{vmatrix} a_{k_11} & \cdots & a_{k_1n} \\ \cdot & \cdot & \cdot \\ a_{k_m1} & \cdots & a_{k_mn} \end{vmatrix} \cdot \begin{vmatrix} x_{i_11} & \cdots & x_{i_1n} \\ \cdot & \cdot & \cdot \\ x_{i_m1} & \cdots & x_{i_mn} \end{vmatrix}$$

$$= \begin{vmatrix} \sum a_{k_11}x_{i_11} & \sum a_{k_11}x_{i_21} & \cdots & \sum a_{k_11}x_{m1} \\ \cdot & \cdot & \cdot & \cdot \\ \sum a_{k_m1}x_{i_11} & \sum a_{k_m1}x_{i_21} & \cdots & \sum a_{k_m1}x_{i_m1} \end{vmatrix}$$

$$= \begin{vmatrix} \rho_{i_1}x_{i_1k_1} & \cdots & \rho_{i_m}x_{i_mk_1} \\ \cdot & \cdot & \cdot \\ \rho_{i_1}x_{i_1k_m} & \cdots & \rho_{i_m}x_{i_mk_m} \end{vmatrix}$$

$$= \rho_{i_1}\,\rho_{i_2} \cdots \rho_{i_m} X_{ik}$$

Now since

$$\rho_{i_1}\rho_{i_2}\cdots\rho_{i_m}X_{ik} = A_{(m)k1}X_{i1} + \cdots + A_{(m)k\mu}X_{i\mu}\ (k = 1,2,\cdots,\mu)$$

or

$$\rho' X_{ik} = A_{(m)k1}X_{i1} + \cdots + A_{(m)k\mu}X_{i\mu}$$

for all values of i and k from 1 to μ and not all the X's vanish, it follows that

$$\begin{vmatrix} A_{(m)11} - \rho' & A_{(m)12} \cdots A_{(m)1\mu} \\ \cdots & \cdots \\ A_{(m)\mu 1} & A_{(m)\mu 2} \cdots A_{(m)\mu\mu} - \rho' \end{vmatrix} = 0.$$

This equation in ρ' is precisely similar to that in ρ and the values of ρ' are the products of the ρ's taken m at a time.

If the original set of equations had multiple roots we would take our μ, X_{ik} from those equations corresponding to m distinct roots and proceed as before, the principal of continuity showing the theorem still true even though there are multiple roots.

649. Denoting the roots of $|A_{(m)} - x| = 0$, by $\alpha_1, \alpha_2, \cdots$, we have just seen that the α's are the products of the g's taken m at a time. From this relation and from the fact that symmetric functions of the g's are expressible in terms of the coefficients $p_1, p_2, \cdots, p_n$ in $|A - x| = 0$, that is in terms of the sums of coaxial minors of A; and also from the fact that the sums of coaxial minors of $A_{(m)}$ are expressible as symmetric functions of the α's and hence as symmetric functions of the g's, it follows that if we had symmetric function tables* giving the values of functions of the α's in terms of the p's we could write down the value of the sum of the coaxial minors of order k of the mth compound of A. That is we would have

$$\sum A_{(m)kii} = x_1 \sum A_{km} + x_2 \sum a_{ii} \sum A_{km-1} + \cdots \\ + x_h \sum A_{h-1} \sum A_{km-h+1} + \cdots$$

where $\sum_{(m)kii}$ is the sum of the coaxial minors of order k of the mth compound of A, and where $\sum A_r$ is the sum of the coaxial minors of order r of A. The coefficients $x_1, x_2, \cdots$, are to be found from the tables in question. Thus from the table which gives symmetric func-

* Such tables for weights up to sixteen were given in a paper "Some New Symmetric Function Tables" by W. H. Metzler—Trans. Royal Soc., Canada, 3rd series, vol. II, 1908–1909.

tions of the α's, which are the products of the g's two at a time, in terms of the p's

	9	81	72	71^2	63	621	54	531	52^2	4^21	432	3^3
3	3	−3	−3	3	6	−3	−3	−3	3	3	−3	1
21	−3	3	3	−3	−8	3	3	5	−3	−3	1	0
1^3	1	−1	−1	1	3	−1	−1	−2	1	1	0	0

we may write down at once

$$\begin{aligned}\sum A_{(3)3ii} = \sum A_9 - \sum a_{ii} \sum A_8 - \sum A_2 \sum A_7 + 3 \sum A_3 \sum A_6 \\ - \sum A_4 \sum A_5 - \sum a_{ii} \sum A_2 \sum A_6 - 2 \sum a_{ii} \sum A_3 \sum A_5 \\ + \left(\sum a_{ii}\right)^2 \sum A_7 + \left(\sum A_2\right)^2 \sum A_5 + \sum a_{ii}\left(\sum A_4\right)^2.\end{aligned}$$

From the expansion for $\sum A_{(m)kii}$ it should be observed that it will vanish when the sums of coaxial minors of A of order higher than m vanish. The case where $k=2$ should be noted. It is

$$\begin{aligned}\sum A_{(m)2ii} = \sum A_{m-1} \sum A_{m+1} - \sum A_{m-2} \sum A_{m+2} + \cdots \\ + (-1)^{m-2} \sum a_{ii} \sum A_{2m-1} + (-1)^{m-1} \sum A_{2m}.\end{aligned}$$

650. If from the set of equations

$$\begin{aligned}\lambda \cdot x_1 &= a_{11}x_1 + a_{12}x_2 + a_{13}x_3 \\ \lambda \cdot x_2 &= a_{21}x_1 + a_{22}x_2 + a_{23}x_3 \qquad (1) \\ \lambda \cdot x_3 &= a_{31}x_1 + a_{32}x_2 + a_{33}x_3\end{aligned}$$

we eliminate the x's we get

$$\begin{vmatrix} a_{11} - \lambda & a_{12} & a_{13} \\ a_{21} & a_{22} - \lambda & a_{23} \\ a_{31} & a_{32} & a_{33} - \lambda \end{vmatrix} = 0 \qquad (2)$$

Multiply the equations (1) by λ and substituting in the results for λx_1, λx_2, λx_3 their values from (1) we get

$$\begin{aligned}\lambda^2 \cdot x_1 &= x_1(a_{11}^2 + a_{12}a_{21} + a_{13}a_{31}) + x_2(a_{11}a_{12} + a_{12}a_{22} + a_{13}a_{32}) \\ &\quad + x_3(a_{11}a_{13} + a_{12}a_{23} + a_{13}a_{33}) \\ \lambda^2 \cdot x_2 &= x_1(a_{21}a_{11} + a_{22}a_{21} + a_{23}a_{31}) + x_2(a_{21}a_{12} + a_{22}^2 + a_{23}a_{32}) \\ &\quad + x_3(a_{21}a_{13} + a_{22}a_{23} + a_{23}a_{33}) \\ \lambda^2 \cdot x_3 &= x_1(a_{31}a_{11} + a_{32}a_{21} + a_{33}a_{31}) + x_2(a_{31}a_{12} + a_{32}a_{22} + a_{33}a_{32}) \\ &\quad + x_3(a_{31}a_{13} + a_{32}a_{23} + a_{33}^2)\end{aligned}$$

or

$$(3)\qquad \begin{aligned} \lambda^2 \cdot x_1 &= x_1 a_{11}^{(2)} + x_2 a_{12}^{(2)} + x_3 a_{13}^{(2)} \\ \lambda^2 \cdot x_2 &= x_1 a_{21}^{(2)} + x_2 a_{22}^{(2)} - x_3 a_{23}^{(2)} \\ \lambda^2 \cdot x_3 &= x_1 a_{31}^{(2)} + x_2 a_{32}^{(2)} + x_3 a_{33}^{(2)} \quad \text{say.} \end{aligned}$$

Eliminating as before we have

$$(4)\qquad \begin{vmatrix} a_{11}^{(2)} - \lambda^2 & a_{12}^{(2)} & a_{13}^{(2)} \\ a_{21}^{(2)} & a_{22}^{(2)} - \lambda^2 & a_{23}^{(2)} \\ a_{31}^{(2)} & a_{32}^{(2)} & a_{33}^{(2)} - \lambda^2 \end{vmatrix} = 0.$$

In (4) we have an equation of precisely the same form as (2), whose roots are the squares of the roots of (2). It is to be observed that the elements $a_{ij}^{(2)} = a_{i1}a_{1j} + a_{i2}a_{2j} + a_{i3}a_{3j}$, that is the determinant of the $a^{(2)}$'s is the square of the determinant of the a's where the multiplication is row by column. It is also to be observed that (4) may be written

$$|a - \lambda| \cdot |a + \lambda| = 0,$$

where $|a-\lambda|$ denotes the determinant of the a's with λ subtracted from the elements along the principal diagonal, and where multiplication of the two determinants $|a-\lambda|$ and $|a+\lambda|$ is row by column.

Multiplying equations (3) by λ and using (1) for the values λx_1, λx_2, λx_3 as before we get

$$(5)\qquad \begin{aligned} \lambda^3 \cdot x_1 &= a_{11}^{(3)} x_1 + a_{12}^{(3)} x_2 + a_{13}^{(3)} x_3 \\ \lambda^3 \cdot x_2 &= a_{21}^{(3)} x_1 + a_{22}^{(3)} x_2 + a_{23}^{(3)} x_3 \\ \lambda^3 \cdot x_3 &= a_{31}^{(3)} x_1 + a_{32}^{(3)} x_2 + a_{33}^{(3)} x_3 \end{aligned}$$

where

$$a_{ij}^{(3)} = a_{i1} a_{1j}^{(2)} + a_{i2} a_{2j}^{(2)} + a_{i3} a_{3j}^{(2)} .$$

The eliminant of (5) is

$$(6)\qquad \begin{vmatrix} a_{11}^{(3)} - \lambda^3 & a_{12}^{(3)} & a_{13}^{(3)} \\ a_{21}^{(3)} & a_{22}^{(3)} - \lambda^3 & a_{23}^{(3)} \\ a_{31}^{(3)} & a_{32}^{(3)} & a_{33}^{(3)} - \lambda^3 \end{vmatrix} = 0.$$

The roots of (6) are the cubes of the roots of (2). Continuing this process we get a set of equations whose eliminant is

$$\begin{vmatrix} \overset{(p)}{a_{11}} - \lambda^p & \overset{(p)}{a_{12}} & \overset{(p)}{a_{13}} \\ \overset{(p)}{a_{21}} & \overset{(p)}{a_{22}} - \lambda^p & \overset{(p)}{a_{23}} \\ \overset{(p)}{a_{31}} & \overset{(p)}{a_{32}} & \overset{(p)}{a_{33}} - \lambda^p \end{vmatrix} = 0$$

and whose roots are the pth powers of the roots of (2).

For convenience we have used three variables and three equations but the reasoning is obviously just as applicable to n equations in n variables.

If the determinant of the a's $|a_{nn}| \equiv A$ is axisymmetric then multiplication row-by-row is the same as row-by-column and we have the theorem.

If

$$\begin{vmatrix} [11] & [12] \cdots \\ [21] & [22] \cdots \\ \cdots & \cdots \end{vmatrix}$$

denotes the pth power of the axisymmetric determinant

$$\begin{vmatrix} 11 & 12 \cdots \\ 21 & 22 \cdots \\ \cdots & \cdots \end{vmatrix}$$

then

$$\begin{vmatrix} [11] - x & [12] & \cdots \\ [21] & [22] - x & \cdots \\ \cdots & \cdots & \cdots \end{vmatrix} = 0$$

has for its roots the pth powers of the roots of

$$\begin{vmatrix} 11 - x & 12 & \cdots \\ 21 & 22 - x & \cdots \\ \cdots & \cdots & \cdots \end{vmatrix} = 0$$

651. If the roots of the resultant of the original set of n equations are $g_1, g_2, \cdots, g_n$, when $a_{rs} = a_{sr}$, we have

$$g_1^p + g_2^p + \cdots + g_n^p = \overset{(p)}{a_{11}} + \overset{(p)}{a_{22}} + \cdots + \overset{(p)}{a_{nn}}$$

or

$$S_p = \sum a_{ii}^{(p)}$$

From the law of formation we have

$$a_{ij}^{(r+s)} = \sum_1^n k a_{ki}^{(r)} a_{kj}^{(s)}$$

or

$$a_{ij}^{(p)} = \sum_1^n k a_{ki}^{(r)} a_{ki}^{(p-r)}$$

and hence S_p can have any one of $p-1$ expressions got from

$$\sum_1^n k \sum_1^n k a_{ki}^{(r)} a_{ki}^{(p-r)}$$

by giving r the values $1, 2, \cdots, (p-1)$.

It has been seen (§447) that the determinant

$$\Delta \equiv \begin{vmatrix} s_0 & s_1 & \cdots & s_{n-1} \\ s_1 & s_2 & \cdots & s_n \\ s_2 & s_3 & \cdots & s_{n+1} \\ \cdot & \cdot & \cdot & \cdot \\ s_{n-1} & s_n & \cdots & s_{2n-2} \end{vmatrix}$$

is equal to the squared differences of the roots.

If now we agree to call $a^{(0)}{}_{ij}$ 1 or 0, according as i is the same or different from j, then it is seen that Δ is the square of the array

$$\begin{Vmatrix} a_{11}^{(0)} & a_{12}^{(0)} & \cdots & a_{1n}^{(0)} & a_{21}^{(0)} & a_{22}^{(0)} & \cdots & a_{2n}^{(0)} & \cdots & a_{nn}^{(0)} \\ a_{11}^{(1)} & a_{12}^{(1)} & \cdots & a_{1n}^{(1)} & a_{21}^{(1)} & a_{22}^{(1)} & \cdots & a_{2n}^{(1)} & \cdots & a_{nn}^{(1)} \\ \cdot & \cdot & \cdot & \cdot & \cdot & \cdot & \cdot & \cdot & \cdot & \cdot \\ a_{11}^{(n-1)} & a_{12}^{(n-1)} & \cdots & a_{1n}^{(n-1)} & a_{21}^{(n-1)} & a_{22}^{(n-1)} & \cdots & a_{2n}^{(n-1)} & \cdots & a_{nn}^{(n-1)} \end{Vmatrix}$$

and is therefore equal to a sum of squares and consequently positive. We thus have a proof of the reality of the g's, the roots of the equation (2), §650 which is known as *Lagrange's equation.*

652. Given two sets of linear homogeneous equations in n_1 and n_2 variables respectively

(1) $$\sum_1^{n_1} g a_{fg} x_g = \rho \cdot x_f, \quad (f = 1, 2, \cdots, n_1)$$

$$(2) \qquad \sum_{1}^{n_2} {}_i b_{hi} y_i = \sigma \cdot y_h \quad (h = 1, 2, \cdots, n_2)$$

and let us form from them another set by multiplying every equation of (1) by every equation of (2). Thus

$$\sum_{1}^{n_1} {}_g a_{fg} x_g \cdot \sum_{1}^{n_2} {}_i b_{hi} y_i = \rho \cdot \sigma \cdot x_f \cdot y_h$$

or

$$(A) \qquad \sum_{i=1\; g=1}^{i=n_2\; g=n_1} p_{fhgi} \cdot \omega_{gi} = \lambda \cdot \omega_{fh}, \quad \begin{Bmatrix} f = 1, 2, \cdots, n_1 \\ h = 1, 2, \cdots, n_2 \end{Bmatrix}$$

where

$$p_{\alpha\beta\gamma\delta} = a_{\alpha\gamma} \cdot b_{\beta\delta}, \quad \omega_{\alpha\beta} = x_\alpha \cdot y_\beta, \quad \text{and} \quad \lambda = \rho \cdot \sigma.$$

This set (A) of $n = n_1 \cdot n_2$ equations is linear and homogeneous in the n variables ω. It is of the same form as sets (1) and (2) and the determinant of it equated to zero is a determinantal equation in λ which has for its roots the products of the roots of the determinantal equations in ρ and σ formed from equating to zero the determinants of (1) and (2) respectively. If now we take another set of n_3 linear homogenous equations

$$(3) \qquad \sum_{1}^{n_3} {}_j c_{kj} z_j = \tau \cdot z_k \quad (k = 1, 2, \cdots, n_3)$$

and form from (1), (2) and (3) a new set by multiplying in all possible ways, taking one equation from each set. Then

$$(B) \qquad \sum_{1}^{n_1} {}_g a_{fg} \cdot x_g \cdot \sum_{1}^{n_2} {}_i b_{hi} \cdot y_i \cdot \sum_{1}^{n_3} {}_j c_{kj} z_j = \rho \cdot \sigma \cdot \tau \cdot x_f y_h \cdot z_k$$

or

$$\sum_{j=1\; i=1\; g=1}^{j=n_3\; i=n_2\; g=n_1} q_{fhkgij} \omega_{gij} = \lambda \cdot \omega_{fhk}.$$

where

$$q_{\alpha\beta\gamma\delta\epsilon\eta} = a_{\alpha\delta} \cdot b_{\beta\epsilon} \cdot c_{\gamma\eta}, \quad \omega_{\alpha\beta\gamma} = x_\alpha \cdot y_\beta \cdot z_\gamma, \quad \text{and} \quad \lambda = \rho \cdot \sigma \cdot \tau.$$

This set (B) of $n = n_1 \cdot n_2 \cdot n_3$ equations is linear and homogeneous in the n variables ω. It is of the same form as (1), (2), and (3) and the determinant of the set equated to zero is a determinantal equation in λ which has for its roots the products of the roots of the determinantal equations of (1), (2) and (3).

Continuing in this manner we may form a new set of equations (C) from k linear homogeneous sets (1), (2), $\cdots$, (k) in $n_1, n_2, \cdots, n_k$ variables respectively. This new set will contain $n = n_1 \cdot n_2 \cdots n_k$ equations in the same number of variables and the determinant of this set equated to zero is a determinantal equation in λ whose roots are the products of the roots of the determinantal equations of the sets (1), (2), $\cdots$, (k).

If we denote the determinants on the left-hand sides of (1), (2), $\cdots$, (k), by $A, B, \cdots, K$, respectively and that of the left-hand side of (C) by R, then R is the eliminant of the equations (C) **when** λ is zero and it is known that

(I) $$R = A^{n/n_1} \cdot B^{n/n_2} \cdots K^{n/nk} \quad \text{by} \quad \S 168$$

If $n_1 = n_2 = \cdots = n_k = m$, then

(II) $$R = A^{m^{k-1}} \cdot B^{m^{k-1}} \cdots K^{m^{k-1}}.$$

If further, all the sets become alike, except that the variables may or may not be made alike, then

(III) $$R = A^{k \cdot m^{k-1}}.$$

In this case when the variables are alike, the terms in any one equation of the set (C) are not all distinct and if we bring together all terms having the same variables there will remain as many distinct terms as there are terms in the expansion of the multinomial of m terms raised to the kth power namely, $(m+k-1)!/k!(m-1)!$. There will also be the same numbers of distinct equations which we shall denote by (D) and for the eliminant* in this case we have

(IV) $$R = A^{\nu}, \text{ where } \nu = (m + k - 1)_m.$$

If, when the sets all become alike, we use different variables for each set, then it appears that from the way the equations (D) are formed the roots of the λ-equation are the various terms without the multinomial coefficients, of $(g_1 + g_2 + \cdots + g_m)^k$, where $g_1, g_2, \cdots, g_m$ are the roots of the λ-equation of (A). For if we set the roots $g_1, g_2, \cdots, g_m$ down k times forming k groups we see that the values of λ are the products of the g's taken one and but one from each group. For example if the two sets (1) and (2) become alike,

* Muir, South African Assoc. Adv. Sci., vol. 1 (1903).

then the determinantal equation of (C) has for its roots the squares of the g's and their products two at a time, each repeated. The roots of the determinantal equation of (D) are the squares of the g's and their products, two at a time, without repetitions.

653. Another simple proof that the roots in this last case are the squares and unrepeated products of the g's is furnished by starting with $|C_{ii}-\lambda|$, where $|C_{ii}-\lambda|=0$ is the determinantal equation of (C), and by easy combinations of rows and columns break it up into the factors $|D_{ii}-\lambda|$ and $|A_{(2)ii}-\lambda|$, where $A_{(2)ij}$ is the element in the ith row and jth column of the second compound of the determinant A. The roots of $|A_{(2)ii}-\lambda|=0$ are the products of the g's two at a time (§648), and it follows, therefore, that the roots of $|D_{ii}-\lambda|$ are the squares and unrepeated products, two at a time, of the g's. Thus for the case $m=3$ and $k=2$, we have for the determinantal equation of (C)

$$\begin{vmatrix}
a_{11}^2-\lambda & a_{12}^2 & a_{13}^2 & a_{11}a_{12} & a_{11}a_{13} & a_{12}a_{13} & a_{11}a_{12} & a_{11}a_{13} & a_{12}a_{13} \\
a_{21}^2 & a_{22}^2-\lambda & a_{23}^2 & a_{21}a_{22} & a_{21}a_{23} & a_{22}a_{23} & a_{21}a_{22} & a_{21}a_{23} & a_{22}a_{23} \\
a_{31}^2 & a_{32}^2 & a_{33}^2-\lambda & a_{31}a_{32} & a_{31}a_{33} & a_{32}a_{33} & a_{31}a_{32} & a_{31}a_{33} & a_{32}a_{33} \\
a_{11}a_{21} & a_{12}a_{22} & a_{13}a_{23} & a_{11}a_{22}-\lambda & a_{11}a_{23} & a_{12}a_{23} & a_{12}a_{21} & a_{13}a_{21} & a_{13}a_{22} \\
a_{11}a_{31} & a_{12}a_{32} & a_{13}a_{33} & a_{11}a_{32} & a_{11}a_{33}-\lambda & a_{12}a_{33} & a_{12}a_{31} & a_{13}a_{31} & a_{13}a_{32} \\
a_{21}a_{31} & a_{22}a_{32} & a_{23}a_{33} & a_{21}a_{32} & a_{21}a_{33} & a_{22}a_{33}-\lambda & a_{22}a_{31} & a_{23}a_{31} & a_{23}a_{32} \\
a_{21}a_{11} & a_{22}a_{12} & a_{23}a_{13} & a_{21}a_{12} & a_{21}a_{13} & a_{23}a_{13} & a_{22}a_{11}-\lambda & a_{23}a_{11} & a_{23}a_{12} \\
a_{31}a_{11} & a_{32}a_{12} & a_{33}a_{13} & a_{31}a_{12} & a_{31}a_{13} & a_{32}a_{13} & a_{32}a_{11} & a_{33}a_{11}-\lambda & a_{33}a_{12} \\
a_{31}a_{21} & a_{32}a_{22} & a_{33}a_{23} & a_{31}a_{22} & a_{31}a_{23} & a_{32}a_{23} & a_{32}a_{21} & a_{33}a_{21} & a_{33}a_{22}-\lambda
\end{vmatrix}=0$$

which by performing the following operations

Col. 4 + col. 7 ; col. 5 + col. 8 ; col. 6 + col. 9 ;

then

Row 7 − row 4 ; row 8 − row 5 ; row 9 − row 6 ;

gives the result

$$|D_{ii}-\lambda|\cdot|A_{(2)ii}-\lambda|=0,$$

and since the roots of $|A_{(2)ii}-\lambda|=0$ are g_1g_2, g_1g_3, g_2g_3 it follows that the roots of $|D_{ii}-\lambda|=0$ are g_1^2, g_2^2, g_3^2, g_1g_2, g_1g_3, g_2g_3.

If we denote the square of A (row-by-column) by A^2 and if $|A^2_{ii}-\lambda|=0$ denotes the determinantal equation formed by subtracting λ from the elements along the principal diagonal of A^2, then since the roots of $|A^2_{ii}-\lambda|=0$, are g_1^2, g_2^2, g_3^2 (§650) it follows that

$$|D_{ii}-\lambda|=|A^2_{ii}-\lambda|\cdot|A_{ii}-\lambda|$$

654. If A stands for $|a_{nn}|$ and Δ for its recpirocal the equation whose roots are A times the reciprocals of the roots of the equation $|A-x|=0$ is

$$\begin{vmatrix} A_{11}-x & A_{12} & \cdots & A_{1n} \\ A_{21} & A_{22}-x & \cdots & A_{2n} \\ \cdot & \cdot & \cdot & \cdot \\ A_{n1} & A_{n2} & \cdots & A_{nn}-x \end{vmatrix} = 0 \quad \text{or} \quad |\Delta - x| = 0, \text{ say.}$$

Starting with

$$|A - x| \equiv \begin{vmatrix} a_{11}-x & a_{12} \cdots a_{1n} \\ a_{21} & a_{22}-x \cdots a_{2n} \\ \cdot & \cdot \\ a_{n1} & a_{n2} \cdots a_{nn}-x \end{vmatrix} = 0,$$

and substituting A/y for x and expanding as a power series in A/y we have

$$(-1)^n\left[\left(\frac{A}{y}\right)^n - \left(\frac{A}{y}\right)^{n-1}\sum a_{11} + \left(\frac{A}{y}\right)^{n-2}\sum A_{(2)ii} - \left(\frac{A}{y}\right)^{n-3}\sum A_{(3)ii} + \cdots\right] = 0$$

or

$$(-1)^n\frac{A}{y^n}\left[A^{n-1} - y\sum A_{11} + \cdots\right] = 0$$

or

$$(-1)^n\frac{A}{y^n}\left[\Delta - y\sum \Delta_{11} + y^2\sum \Delta_{(2)ii} - \cdots\right] = 0$$

which proves our theorem.

655. If we transform $\sum a_{rs}x_rx_s$ into $\sum A_ry_r^2$ by the orthogonal transformation

$$x_r = w_{r1}y_1 + w_{r2}y_2 + \cdots + w_{rn}y_n,$$

then to determine the A's we have

$$y_r = w_{1r}x_1 + w_{2r}x_2 + \cdots + w_{nr}x_n,$$

which gives

$$a_{rs} = A_1 w_{r1} w_{s1} + A_2 w_{r2} w_{s2} + \cdots + A_n w_{rn} w_{sn}.$$

Placing in this $s=1, 2, \cdots, n$ and using as multipliers

$$w_{1s}, w_{2s}, \cdots w_{ns},$$

$$a_{r1} w_{1s} + a_{r2} w_{2s} + \cdots + a_{rn} w_{ns} = A_s w_{rs}$$

whence putting $r=1, 2, \cdots, n$, we get

$$\begin{vmatrix} a_{11} - A_s & a_{12} & \cdots & a_{1n} \\ a_{21} & a_{22} - A_s & \cdots & a_{2n} \\ \cdot & \cdot & \cdot & \cdot \\ a_{n1} & a_{n2} & \cdots & a_{nn} - A_s \end{vmatrix} = 0.$$

656. Let $A \equiv |a_{nn}|$ in which $a_{rs} = a_{sr}$, then the product

$$|a + x| \cdot |a - x| = |q - x^2|,$$

where

$$q_{rs} = a_{r1} a_{1s} + a_{r2} a_{2s} + \cdots + a_{rn} a_{ns}$$

and therefore $|q_{nn}| = A^2$.

Expanding $|q - x^2|$ as a power series in x^2 we have

$$(-1)^n [(x^2)^n - c_1 (x^2)^{n-1} + c_2 (x^2)^{n-2} - \cdots$$

where $c_1, c_2, \cdots$, being the squares of rectangular arrays, are seen to be the sums of squares and consequently the values of x^2 in $|q - x^2| = 0$ are all positive, and therefore the values of x are not pure imaginary.

To show that the equation has not a root of the form $\alpha + \beta i$, put $a_{11} - \alpha = a'_{11}$, $a_{22} - \alpha = a'_{22}$, etc. Then

$$\begin{vmatrix} a'_{11} - x & a_{12} & \cdots & a_{1n} \\ a_{21} & a'_{22} - x & \cdots & a_{2n} \\ \cdot & \cdot & \cdot & \cdot \\ a_{n1} & a_{n2} & \cdots & a'_{nn} - x \end{vmatrix}$$

cannot have a root of the form βi *and therefore* $|A - x| = 0$ *cannot have any but real roots*, as already seen in §651.

Still another proof of this fact is obtained as follows:

Let $A \equiv |a_{nn}|$ be an axisymmetric determinant and let

$$\Delta_n \equiv \begin{vmatrix} a_{11}+x & a_{12} & \cdots & a_{1n} \\ a_{21} & a_{22}+x & \cdots & a_{2n} \\ \cdot & \cdot & \cdot & \cdot \\ a_{n1} & a_{n2} & \cdots & a_{nn}+x \end{vmatrix},$$

$$\Delta_{n-1} \equiv \begin{vmatrix} a_{11}+x & a_{12} & \cdots & a_{1,n-1} \\ a_{21} & a_{22}+x & \cdots & a_{2,n-1} \\ \cdot & \cdot & \cdot & \cdot \\ a_{n-1,1} & a_{n-1,2} & \cdots & a_{n-1,n-1}+x \end{vmatrix}$$

$$\cdots, \Delta_1 \equiv \left| a_{11}+x \right|, \Delta_0 \equiv 1.$$

be polynomials in x of degree $n, n-1, \cdots, 1, 0$ respectively and where Δ_{r-1} is obtained by deleting the last row and column of Δ_r. We have seen §383 that Δ_{n-r} and Δ_{n-r-2} have opposite signs when Δ_{n-r-1} vanishes.

When $+\infty$ is substituted in the series of polynomials in x the signs are all positive and when $-\infty$ is substituted the signs are alternately positive and negative (beginning with Δ_0). If x be regarded as increasing continuously, n changes of sign must be lost from $-\infty$ to $+\infty$. It follows that the vanishing of any function Δ_s(Δ_n and Δ_0 excluded) will cause Δ_{s+1} and Δ_{s-1} to have opposite signs. Δ_0 has the same sign throughout. It follows therefore that a change of sign can never be lost except when x passes through a real root of $\Delta_n=0$. This equation must therefore have n real roots in order to have that many changes of sign lost in x passing from $-\infty$ to $+\infty$.

It follows that every equation $\Delta_s=0$ *has all its roots real* since it is of the same form as $\Delta_n=0$.

The roots of $\Delta_{s-1}=0$ *lie between those of* $\Delta_s=0$, for in order that a change of sign be lost between Δ_s and Δ_{s-1} at the passage of each of two consecutive roots of Δ_s, the value of Δ_{s-1} must change sign between these two values of x.

From this it is seen that the roots of each equation of the series

$$\Delta_n = 0, \ \Delta_{n-1} = 0, \ \cdots \Delta_1 = 0$$

lie between the roots of the equation immediately preceding it.

If $\Delta_n=0$ has r roots equal to α, then
$\Delta_{n-1}=0$ has $r-1$ roots equal to α,
$\Delta_{n-2}=0$ has $r-2$ roots equal to α etc.

In the foregoing proof $\Delta_n = 0$ was supposed to have all its roots distinct. In case of equal roots we can give a small increment ϵ to one of the coefficients a_{rs} and as ϵ can be made as small as we please without the theorem ceasing to be valid, the validity would remain at the limit 0 for ϵ.

Since we might have deleted the rth row and rth column of Δ_n instead of the nth row and nth column what is true of the roots of Δ_{n-1} is also true of any of the principal coaxial minors. That is if $\Delta_n = 0$ has r roots equal to α, then α will be a root $(r-1)$ times of the equations $\Delta_{(n-1)ii} = 0$ $(i = 1, 2, \cdots n)$, that is of the equations obtained by equating each of the principal coaxial minors equal to zero. Similarly it will be a root $(r-2)$ times of the equations obtained by equating each of the coaxial minors of order $(n-2)$ equal to zero, and so on.

Again

$$\Delta_{(n-1)ii}\Delta_{(n-1)jj} - \Delta_{(n-1)ij}\Delta_{(n-1)ji} = \Delta_n\Delta_{(n-2)ij,ij}$$

and if $(x-\alpha)^r$ is a factor of Δ_n, then by what we have just seen $(x-\alpha)^{r-1}$ is a factor of $\Delta_{(n-1)ii}$ and $\Delta_{(n-1)jj}$, and $(x-\alpha)^{r-2}$ is a factor of $\Delta_{(n-2)ij,ij}$. It follows, therefore, that $(x-\alpha)^{r-1}$ is a factor of $\Delta_{(n-1)ij}$ and of all minors of order $(n-1)$. Similar reasoning will show that $(x-\alpha)^{r-2}$ is a factor of all minors of order $(n-2)$ and in general $(x-\alpha)^{r-k}$ will be a factor of all minors of order

$$(n-k) \qquad (k = 1, 2, \cdots r-1)$$

657. With the help of the foregoing the problem to determine the equation whose roots give the maxima and minima of the homogeneous function of the second degree

$$s \equiv A_{xx}x^2 + A_{yy}y^2 + A_{zz}z^2 + \cdots + 2A_{xy}xy + 2A_{xz}xz + \cdots,$$

where the coefficients are real and the variables limited by the condition

$$x^2 + y^2 + z^2 + \cdots = 1$$

is easily obtained.

For these values we have the equations

$$\frac{\dfrac{\partial s}{\partial x}}{x} = \frac{\dfrac{\partial s}{\partial y}}{y} = \frac{\dfrac{\partial s}{\partial z}}{z} = \cdots = \frac{x\dfrac{\partial s}{\partial x} + y\dfrac{\partial s}{\partial y} + \cdots}{x^2 + y^2 + z^2 + \cdots}$$

$$= 2s$$

that is

$$\frac{1}{2}\frac{\partial s}{\partial x} = sx, \quad \frac{1}{2}\frac{\partial s}{\partial y} = sy, \cdots$$

or

$$\begin{aligned} (A_{xx} - s)x + A_{xy}y + A_{xz}z + \cdots &= 0 \\ A_{yx}x + (A_{yy} - s)y + A_{yz}z + \cdots &= 0 \\ A_{zx}x + A_{zy}y + (A_{zz} - s)z + \cdots &= 0 \\ \cdots\cdots\cdots\cdots\cdots\cdots & \end{aligned}$$

from which

$$\begin{vmatrix} A_{xx} - s & A_{xy} & A_{xz} & \cdots \\ A_{yx} & A_{yy} - s & A_{yz} & \cdots \\ A_{zx} & A_{zy} & A_{zz} - s & \cdots \\ \cdots & \cdots & \cdots & \cdots \end{vmatrix} \equiv S = 0.$$

EXERCISE: Find the values of $x_r, y_r, z_r, \cdots$, corresponding to the root s_r of $S=0$.

658. If the roots of $S=0$, denoted by $s_1, s_2, \cdots, s_m$, are all distinct it follows that $x_rx_s+y_ry_s+\cdots=0$, where $x_r, y_r, \cdots$, are the values of the variables corresponding to the root s_r.

EXERCISE: If we have

$$\begin{aligned} x &= x_1\xi + x_2\eta + x_3\zeta + \cdots \\ y &= y_1\xi + y_2\eta + y_3\zeta + \cdots \\ z &= z_1\xi + z_2\eta + z_3\zeta + \cdots \end{aligned}$$

show that

(1) $$x^2 + y^2 + z^2 + \cdots = \xi^2 + \eta^2 + \zeta^2 + \cdots$$

(2) $$S = s_1\xi^2 + s_2\eta^2 + s_3\zeta^2 + \cdots.$$

659. *The necessary and sufficient condition that α is a root r times of $\Delta_n=0$ is that all minors of order $n-r+1$ vanish when x is put equal to α.* For if the minors of order $n-r+1$ vanish then the minors of all higher orders will vanish under the same conditions and $\sum\Delta_{(n-h)ii}=0$ $(h=1, 2, \cdots, (r-1))$ where $\Delta_{(n-h)ii}$ is a coaxial minor of Δ_n of order $n-h$. It follows that $(x-\alpha)^r$ must be a factor of Δ_n.

EXERCISE: If $\Delta \equiv |a_{1n}|$ where $a_{rs}=a_{sr}$ and where the elements are integral functions of x with real coefficients, then show that the series of functions $\Delta, \Delta_1, \Delta_2, \cdots, \Delta_n$, where Δ_h is the minor of Δ complementary to the product $a_{11}a_{22}a_{33}\cdots a_{hh}$, is a Sturm's series for the equation $\Delta=0$, under the conditions

(1) None of the functions vanish for all values of x.

(2) The sum

$$\sum a'_{rs}A_{1r}A_{1s}\left(\begin{matrix} r \\ s \end{matrix} = 1, 2, \cdots, n\right)$$

shall not vanish in the interval a to b (Hazzidakis).

660. Since a symmetric determinant remains symmetric when the kth row and kth columns are multiplied by the same quantity, we see that

$$\Delta \equiv \begin{vmatrix} a_{11}+h_1^2x & a_{12} & a_{13} & \cdots \\ a_{21} & a_{22}+h_2^2x & a_{23} & \cdots \\ a_{31} & a_{32} & a_{33}+h_3^2x & \cdots \\ \cdots & \cdots & \cdots & \cdots \end{vmatrix},$$

where the h's are real, is axisymmetric if $a_{rs}=a_{sr}$ and the roots of $\Delta=0$ are real.

Changing x into its reciprocal obviously does not affect the reality of the roots, so the roots of

$$\begin{vmatrix} a_{11}x+h_1^2 & a_{12}x & a_{13}x & \cdots \\ a_{21}x & a_{22}x+h_2^2 & a_{23}x & \cdots \\ a_{31}x & a_{32}x & a_{33}x+h_3^2 & \cdots \\ \cdots & \cdots & \cdots & \cdots \end{vmatrix} = 0$$

are also real. In particular the roots of the continuant

$$\begin{vmatrix} a_{11}x+h_1^2 & a_{12}x & \cdot & \cdots \\ a_{21}x & a_{22}x+h_2^2 & a_{23}x & \cdots \\ \cdot & a_{32}x & a_{33}x+h_3^2 & \cdots \\ \cdots & \cdots & \cdots & \cdots \end{vmatrix} = 0$$

are real.

Since we may replace a_{rs} and a_{sr} by $(a_{rs}a_{sr})^{1/2}$ without altering the continuant and thereby make it axisymmetric it follows that the roots are real when $a_{rs}\neq a_{sr}$ as long as they have the same sign.

From the foregoing it follows that the roots of

$$\begin{vmatrix} h_1^2 & \lambda a_{12} & \cdot & \cdots \\ a_{21} & h_2^2 & \lambda a_{23} & \cdots \\ \cdot & a_{32} & h_3^2 & \cdots \\ \cdot & \cdot & \cdot & \cdots \end{vmatrix} = 0$$

are

(1) Real and positive if $a_{r,r+1}$ and $a_{r+1,r}$ have the same signs

(2) Real and negative if $a_{r,r+1}$ and $a_{r+1,r}$ have opposite signs.

EXERCISES. 1. The determinant

$$\begin{vmatrix} h_1^2 & \cdot & xa_{13} & \cdot & \cdots \\ \cdot & h_2^2 & \cdot & xa_{24} & \cdots \\ a_{31} & \cdot & h_3^2 & \cdot & \cdots \\ \cdot & a_{42} & \cdot & h_4^2 & \cdots \\ \cdot & \cdot & \cdot & \cdot & \cdots \end{vmatrix}$$

which has a row of zeros lying between the main diagonal and each of the minor diagonals, is a continuant which has the same properties as that just considered.

2. If $A \equiv |a_{1n}|$ is axisymmetric and we multiply the rows in order by $\mu_1, \mu_2, \cdots, \mu_n$ and the columns in order by $\nu_1, \nu_2, \cdots, \nu_n$, where the μ's and ν's are any real positive magnitudes, then show that the latent roots of the resulting determinant are all real (Quintilli-Nicoletti).

661. *The roots of the equation*

$$\begin{vmatrix} x & a_{12} & a_{13} & \cdots \\ -a_{12} & x & a_{23} & \cdots \\ -a_{13} & -a_{23} & x & \cdots \\ \cdot & \cdot & \cdot & \cdots \end{vmatrix} = 0,$$

or $f(x) = 0$ *say, are all pure imaginary.*

Expanding as a power series in x we have

$$x^n + x^{n-1}(c_1) + x^{n-2}(c_2) + \cdots + x(c_{n-1}) + \Delta = 0,$$

where Δ is the determinant of the a's. The coefficients $c_1, c_3, \cdots$, are all zero being sums of zero-axial skew determinants of odd order, and $c_2, c_4, \cdots$, are all positive being the sums of zero-axial skew determinants of even orders, and therefore sums of squares.

If n is odd $\Delta=0$ and one root at least is zero, the others are seen to be pure imaginary, and if n is even all roots are pure imaginary. For suppose $f(x)=x^r f'(x)$, then $f'(x)$ in either case is the sum of a series of positive terms whether x is positive or negative.

The roots of the equation $f(x+\alpha)=0$ *are readily seen to be complex since* $x+\alpha$ must be of the form βi and therefore $x=-\alpha+\beta i$.

662. *The equation* $|A-x|=0$, *where* $A=|a_{1n}|$, *will have all its roots real, if in the case of every pair* μ, ν *of the* $(n-1)$ *indices* 2, 3, $\cdots$, n, *we have*

$$a_{1\mu}\cdot a_{\mu\nu}\cdot a_{\nu 1} = a_{\mu 1}\cdot a_{\nu\mu}\cdot a_{1\nu},$$

and $a_{1\mu}\cdot a_{\mu 1}$ *positive.*

Form the product

$$|A-x|\cdot|A+x| \equiv |Q-x^2|,$$

where

$$q_{rs} = a_{r1}a_{1s} + a_{r2}a_{2s} + \cdots + a_{rn}a_{ns},$$

and $Q=|q_{1n}|$ then the given relations lead to

I $$a_{\lambda\mu}\cdot a_{\mu\nu}\cdot a_{\nu\lambda} = a_{\mu\lambda}\cdot a_{\nu\mu}\cdot a_{\lambda\nu},$$

where λ, μ, ν are any three of the indices 1, 2, $\cdots$, n and

II $$\begin{cases} \dfrac{\dfrac{a_{rs}}{a_{sr}}}{\dfrac{a_{ps}}{a_{sp}}} = \text{constant for } s = 1, 2, \cdots, n \\[2ex] \dfrac{\dfrac{a_{rs}}{a_{sr}}}{\dfrac{a_{rp}}{a_{pr}}} = \text{constant for } r = 1, 2, \cdots, n \end{cases}$$

Multiplying

$$q_{rs} = a_{r1}a_{1s} + a_{r2}a_{2s} + \cdots,$$

and

$$q_{sr} = a_{s1}a_{1r} + a_{s2}a_{2r} + \cdots.$$

by a_{sr} and a_{rs} respectively we get

$$q_{rs}a_{sr} = a_{1s}a_{sr}a_{r1} + a_{2s}a_{sr}a_{r2} + \cdots$$

and

$$q_{sr} \cdot a_{rs} = a_{1r} a_{rs} a_{s1} + a_{2r} a_{rs} a_{s2} + \cdots .$$

From I we see that $q_{rs} \cdot a_{sr} = q_{sr} \cdot a_{rs}$ and this and II show that any minor of $|Q - x^2|$ is equal to the conjugate minor multiplied by a positive constant. It follows therefore that every coefficient of every power of x^2 in $|Q - x^2|$ is positive and the values of x^2 in $|Q - x^2| = 0$ are all positive and hence the values of x are all real.

663. *The equation* $|A - x| = 0$ or $f(x) = 0$ *will have all its roots pure imaginary if* λ, μ, ν *are any three of the indices* 1, 2, $\cdots$, n *and we have*

$$a_{\lambda\mu} \cdot a_{\mu\nu} \cdot a_{\nu\lambda} = - a_{\mu\lambda} \cdot a_{\nu\mu} \cdot a_{\mu\nu},$$

$a_{\mu\nu} \cdot a_{\nu\mu}$ *is negative and* $a_{\mu\mu} = 0$.

The proof of this proceeds in a precisely similar manner to that of the last article.

664. *Let*

$$(1) \quad \left\{ \begin{array}{l} a_{11}x_1 + a_{12}x_2 + \cdots + a_{1n}x_n + b_{11}y_1 + b_{12}y_2 \cdots + b_{1k}y_k = \lambda x_1 \\ \cdots\cdots\cdots\cdots\cdots\cdots\cdots\cdots\cdots \\ a_{n1}x_1 + a_{n2}x_2 + \cdots + a_{nn}x_n + b_{n1}y_1 + b_{n2}y_2 + \cdots + b_{nk}y_k = \lambda x_n \\ b_{11}x_1 + b_{21}x_2 + \cdots + b_{n1}x_n = 0 \\ \cdots\cdots\cdots\cdots\cdots\cdots\cdots\cdots\cdots \\ b_{1k}x_1 + b_{2k}x_2 + \cdots + b_{nk}x_n = 0 \end{array} \right\}$$

where $a_{rs} = a_{sr}$, *and the x's and y's may be complex;*

and let

$$\begin{vmatrix} a_{11} - \lambda & a_{12} & \cdots & a_{1n} & b_{11} \cdots b_{1k} \\ a_{21} & a_{22} - \lambda & \cdots & a_{2n} & b_{21} \cdots b_{2k} \\ \cdots & \cdots & \cdots & \cdots & \cdots \\ a_{n1} & a_{n2} & \cdots & a_{nn} - \lambda & b_{n1} \cdots b_{nk} \\ b_{11} & b_{21} & \cdots & b_{n1} & 0 \cdots 0 \\ \cdots & \cdots & \cdots & \cdots & \cdots \\ b_{1k} & b_{2k} & \cdots & b_{nk} & 0 \cdots 0 \end{vmatrix}$$

be denoted by $f(\lambda)$, *then* $f(\lambda) = 0$ *has all its roots real.*

If $\lambda_1 = \alpha + \beta i$ could be a root then $\lambda_2 = \alpha - \beta i$ must also be a root.

Let x'_r and y'_r be x_r and y_r when $-i$ is put for i and let

$$x_r = z_r + iz'_r, \text{ and } x'_r = z_r - iz'_r \qquad (r = 1, 2, \cdots, n);$$

and form a set of equations (2) by replacing $x_1, x_2, \cdots, x_n, y_1, y_2, \cdots, y_k$ in (1) by $x_1', x_2', \cdots, x_n', y_1', \cdots, y_k'$ and let λ_2 be the value of λ which goes with the primes to satisfy equations (2). Multiply equations (1) by $x_1', x_2', \cdots, x_n', y_1', \cdots, y_k'$ respectively and add using equations (2) to simplify the coefficients and we get

$$x_1(\lambda_2 x_1') + x_2(\lambda_2 x_2') + \cdots + x_n(\lambda_2 x_n') = \lambda_1(x_1x_1' + \cdots)$$

or

$$(\lambda_1 - \lambda_2)(x_1x_1' + x_2x_2' + \cdots) = 0,$$

But $\sum x_1x_1' = \sum z_1^2 + \sum z_1'^2$ and therefore is not zero unless the z's and z''s are all zero in which case the x's would be all zero and we suppose this not the case. It follows therefore that $\lambda_1 = \lambda_2$ or $\beta = 0$ and the roots of $f(\lambda) = 0$ are all real.

665. Let

$$(1)\quad \begin{cases} a_{11}x_1 + a_{12}x_2 + \cdots + a_{1n}x_n = k(b_{11}x_1 + b_{12}x_2 + \cdots + b_{1n}x_n) \\ a_{21}x_1 + a_{22}x_2 + \cdots + a_{2n}x_n = k(b_{21}x_1 + b_{22}x_2 + \cdots + b_{2n}x_n) \\ \cdots\cdots\cdots\cdots\cdots\cdots\cdots\cdots\cdots \\ a_{n1}x_1 + a_{n2}x_2 + \cdots + a_{nn}x_n = k(b_{n1}x_1 + b_{n2}x_2 + \cdots + b_{nn}x_n) \end{cases}$$

where $a_{rs} = a_{sr}$ and $b_{rs} = b_{sr}$, be n equations in n variables, the x's, the a's and b's being all real. The condition that they are consistent for values of the x's not all zero is

$$|A - kB| \equiv f(k) = 0.$$

If possible let $f(k) = 0$ have the imaginary root $k = \alpha + \beta i$ and if so it must have a second $k' = \alpha - \beta i$. If $x_1, x_2, \cdots, x_n$ go with k to satisfy equations (1), let $x_1', x_2', \cdots, x_n'$ (the results of replacing $+i$ by $-i$ in $x_1, x_2, \cdots, x_n$) go with k' to satisfy equations (2), which are obtained from (1) by substituting for $x_1, x_2, \cdots, x_n$ the accented x's.

Multiply equations (1) by $x_1', x_2', \cdots, x_n'$, respectively and adding using equations (2) to simplify the coefficients we get*

$$(3)\qquad (k' - k)\ \left[\sum b_{rr}x_rx_r' + \sum b_{rs}(x_rx_s' + x_sx_r')\right] = 0,$$

If

and

$$\left.\begin{aligned} x_r &= z_r + iz_r' \\ x_r' &= z_r - iz_r' \end{aligned}\right\} \qquad (r = 1, 2, \cdots n)$$

* If we had eliminated the b's instead of the a's the result would have been

$$(k' - k)\left[\sum a_{rr}x_rx_r' + 2a_{rs}(x_rx_s' + x_sx_r')\right] = 0$$

then

$$x_r x_r' = z_r^2 + z_r'^2, \text{ and } x_r x_s' + x_s x_r' = 2(z_r z_s + z_r' z_s')$$

and equation (3) becomes

$$(k' - k)[(\sum b_{rr} z_r^2 + 2 \sum b_{rs} z_r z_s) + (\sum b_{rr} z_r'^2 + 2 \sum b_{rs} z_r' z_s')] = 0,$$

which shows that *if* $\sum b_{rr} z_r^2 + 2\sum b_{rs} z_r z_s$ *is always of one sign and never vanishes for real values, not all zero, of the z's, then k' must equal k or* $\beta = 0$ *and equations* $f(x) = 0$ *has all its roots real.*

It is to be observed the roots are also real if the restrictions are placed upon the a's instead of the b's.

The theorem of §656 may be considered a special case of this where $b_{rs} = 0$ $(r \neq s)$, and $b_{rr} = 1$.

666. The restriction that $\sum b_{rr} z_r^2 + 2b_{rs} z_r z_s$, or $\sum a_{rr} z_r^2 + 2a_{rs} z_r z_s$ must never be zero for real values of the z's may be removed leaving the condition that it must always be of one sign when not zero.

To the quadratic form $\sum b_{rr} z_r^2 + 2b_{rs} z_r z_s$ add ϵ times a sufficient number of squares of linearly independent functions of the z's to insure that the quadratic form thus altered is not zero. Let the altered form be $\sum b'_{rr} z_r^2 + 2\sum b'_{rs} z_r z_s$ and if we take ϵ real and of the same sign as the given form, then by making ϵ small each accented b may be made to differ from the corresponding unaccented b by a small quantity. The result of replacing in the determinantal equation the unaccented by the accented b's is

$$f(k) + \epsilon\phi_1(k) + \epsilon^2\phi_2(k) + \cdots + \epsilon^n\phi_n(k) = 0,$$

where the coefficients in $\phi_r(k)$ are finite, and where none of these functions of k are of a degree greater than n. This altered equation has real roots by what has been proven, and it remains true, however small we make ϵ. Now if $f(k) = 0$ have an imaginary root we might take ϵ small enough to make the altered equation to have an imaginary root. Hence $f(k) = 0$ has only real roots. The only exception is when $f(k) = 0$ is an identity satisfied by all values of k. The final result is therefore, *that* $f(k) = 0$ *can have an imaginary root only when*

(1) *It is an identity satisfied by all values of k or*

(2) *Both the quadratic forms*

$$\sum a_{rr} z_r^2 + 2 \sum a_{rs} z_r z_s,$$
$$\sum b_{rr} z_r^2 + 2 \sum b_{rs} z_r z_s$$

can assume both signs for real values of the z's.

For real roots there may be linear relations between the left-hand or right-hand members of equations (1) as long as the equations are linearly independent.

667. The necessary and sufficient condition that the equations

$$
(1)\quad \begin{cases}
b_{12}x_2+b_{13}x_3+\cdots+b_{1n}x_n=-k(a_{11}x_1+a_{12}x_2+\cdots+a_{1n}x_n) \\
-b_{12}x_1+b_{23}x_3+\cdots+b_{2n}x_n=-k(a_{21}x_1+a_{22}x_2+\cdots+a_{2n}x_n) \\
\cdots\cdots\cdots\cdots\cdots\cdots\cdots\cdots\cdots\cdots \\
-b_{1n}x_1-b_{2n}x_2-b_{3n}x_3-\cdots=-k(a_{n1}x_1+a_{n2}x_2+\cdots+a_{nn}x_n),
\end{cases}
$$

where $a_{rs}=a_{sr}$, be consistent for values of the x's not all zero is that

$$
\begin{vmatrix}
ka_{11} & ka_{12}+b_{12} & \cdots & ka_{1n}+b_{1n} \\
ka_{21}-b_{12} & ka_{22} & \cdots & ka_{2n}+b_{2n} \\
\cdot & \cdot & \cdot & \cdot \\
ka_{n1}-b_{1n} & ka_{12}-b_{2n} & \cdots & ka_{nn}
\end{vmatrix} = 0,
$$

or $f(k)=0$, say.

Suppose there is a real root of $f(k)=0$, then equations (1) with this real value of k are satisfied by real values of $x_1, x_2, \cdots, x_n$.

Multiply equations (1) by these real values of $x_1, x_2, \cdots, x_n$, respectively and add and we get

$$k\{\sum a_{rr}x_r^2 + 2\sum a_{rs}x_rx_s\} = 0$$

If now the second factor on the left does not change sign and is not zero for every set of real x's, not all zero, it follows that $k=0$, and $f(k)=0$ has no non-vanishing real root.

Let $k=\alpha+\beta i$ be an imaginary root of $f(k)=0$, and let $x_1, x_2, \cdots, x_n$ go with it. The conjugate imaginary, $k'=\alpha-\beta i$ must also be a root and let $x_1', x_2', \cdots, x_n'$ (which are the results of replacing $+i$ by $-i$ in $x_1, x_2, \cdots, x_n$) go with it.

Multiplying equations (1) by $x_1', x_2', \cdots, x_n'$ respectively and adding we get*

$$(k+k')[\sum a_{rr}x_rx_r' + \sum a_{rs}(x_rx_s'+x_sx_r')] = 0,$$

which, when $x_r=z_r+iz'_r$ and $x'_r=z_r-iz'_r$ becomes

$$(k+k')[(\sum a_{rr}z_r^2+2\sum a_{rs}z_rz_s)+(\sum a_{rr}z_r'^2+2\sum a_{rs}z_r'z_s')] = 0.$$

* If the a's instead of the b's had been eliminated we would have

$$(k+k')[\Sigma b_{rs}(x_rx_s'-x_sx_r')]=0$$

If now $\sum a_{rr}z_r^2+2\sum a_{rs}z_rz_s$ does not change sign nor vanish for all real values of the z's, it follows that $k+k'=0$ or $\alpha=0$. The equation $f(k)=0$, is therefore of the form

$$k^{n-2m}(k^2+\alpha_1^2)(k^2+\alpha_2^2)\cdots(k^2+\alpha_m^2)=0.$$

It is to be noted here that the fact that there is no non-zero real root does not depend on the determinant of the a's being axisymmetric provided $\sum a_{rr}x_r^2+\sum(a_{rs}+a_{sr})x_rx_s$ does not change sign or vanish. Similar observations may be made in §665.

In a manner similar to that used in §666 we may remove the restriction that the quadratic form $\sum a_{rr}z_r^2+2\sum a_{rs}z_rz_s$ be incapable of vanishing.

It follows, therefore, that *when the equations* (1) *are linearly independent the roots of* $f(k)=0$ *are zero or pure imaginary as long as either* $\sum a_{rr}z_r^2+2\sum a_{rs}z_rz_s$ *does not change sign or* $\sum b_{rs}(z_rz_s'-z_sz_r')$ *does not vanish for all real values of the* z*'s.*

The theorem of §661 may be considered a special case of this when $a_{rs}=0$ $(r\neq s)$ and $a_{rr}=1$

668. If in the foregoing article we divide by k^n and put $k=1/k$ then it is seen that *there can be no real finite root of the equation*

$$\begin{vmatrix} a_{11} & a_{12}+kb_{12} & \cdots & a_{1n}+kb_{1n} \\ a_{12}-kb_{12} & a_{22} & \cdots & a_{2n}+kb_{2n} \\ \cdot & \cdot & \cdot & \cdot \\ a_{1n}-kb_{1n} & a_{2n}-kb_{2n} & \cdots & a_{nn} \end{vmatrix}=0\,;$$

and when $\sum a_{rr}z_r^2+2\sum a_{rs}z_rz_s$ *does not vanish or change sign for real values of the* z*'s not all zero it is of the form*

$$(\alpha_1^2k^2+1)(\alpha_2^2k^2+1)\cdots(\alpha_m^2k^2+1)=0$$

or

$$(k^2+\beta_1^2)(k^2+\beta_2^2)\cdots(k^2+\beta_m^2)=0,$$

where $2m\lesseqgtr n$ *and all roots are pure imaginary.*

The very form of the equation shows that for every root α there is root $-\alpha$.

669. Since subtracting (or adding) x from the principal diagonal elements of a centro symmetric determinant does not destroy the centrosymmetry it follows that the latent roots of

$$\begin{vmatrix} a & b & c & d \\ g & e & f & h \\ h & f & e & g \\ d & c & b & a \end{vmatrix}$$

are the latent roots of

$$\begin{vmatrix} a+d & b+c \\ g+h & e+f \end{vmatrix} \quad \text{and} \quad \begin{vmatrix} a-d & b-c \\ g-h & e-f \end{vmatrix}.$$

EXERCISES. 1. Show that the absolute value of any latent root of $|a_{1n}|$ is not greater than n times the greatest in absolute value of the elements a_{rs} (Hirsch).

2. Show that the real part of any latent root of $|a_{1n}|$ lies between the greatest and least roots of the determinant $\left|\frac{1}{2}(a_{rs}+a_{sr})\right|_n$; and the imaginary part is in absolute value not greater than $\sqrt{n(n-1)/2}$ times the greatest of the quantities $\pm(a_{rs}+a_{sr})$ (Benedixen).

670. *If* $A \equiv |a_{1n}|$ *be an orthogonant, then* $|A-x|=0$, *or* $f(x)=0$, *is a reciprocal equation, that is, for every root* α *there is a root* $1/\alpha$.

It is readily seen that the product

$$A \cdot f(x) = (-1)^n x^n f\left(\frac{1}{x}\right),$$

from which form the truth of the theorem appears.

This relation may be written in the form

$$f(x) = (-1)^n \frac{x^n}{A} \cdot f\left(\frac{1}{x}\right)$$

from which we see that

I. If $A=+1$, then $f(+1)=(-1)^n f(+1)$, and hence $f(+1)=0$ when $x=+1$ and n is odd.

It is seen therefore that $+1$ *is a root of* $f(x)$ *when* A *is a proper orthogonant of odd order.*

II. If $A=-1$, then

$$f(-1) = (-1)^{2n-1} f(-1) = -f(-1) = 0, \quad \text{when } x=-1;$$

$f(+1)=-f(+1)=0$, when $x=+1$, and n is even.

These show that *when* A *is improper* -1 *is a root of* $f(x)=0$, *and that* $+1$ *is a root when* n *is even.*

671. The product

$$\frac{f(x)\cdot f(-x)}{x^n} \text{ is } \begin{vmatrix} \frac{1}{x}-x & a_{12}-a_{21} & \cdots & a_{1n}-a_{n1} \\ a_{21}-a_{12} & \frac{1}{x}-x & \cdots & a_{2n}-a_{n2} \\ \cdot & \cdot & \cdot & \cdot \\ a_{n1}-a_{1n} & a_{n2}-a_{2n} & \cdots & \frac{1}{x}-x \end{vmatrix} \equiv F(Z) \text{ say}$$

where $Z=1/x-x$.

Expanding as a power series in Z we have

$$F(Z) = Z^n + C_1Z^{n-1} + C_2Z^{n-2} + \cdots + C_n,$$

but as we have seen $c_1, c_3, \cdots$, are all zero and $c_2, c_4, \cdots$, are sums of squares.

Therefore

$$\begin{aligned} F(Z) &\equiv Z^n + C_2Z^{n-2} + C_4Z^{n-4} + \cdots \\ &= Z(Z^{n-1} + C_2Z^{n-3} + C_4x^{n-5} + \cdots) \text{ when } n \text{ is odd and } A = \pm 1 \\ &= Z^2(Z^{n-2} + C_2Z^{n-4} + \cdots) \qquad \text{when } n \text{ is even and } A = -1. \end{aligned}$$

For the case when n is even and $A=-1$ we have seen that both $f(x)$ and $f(-x)$ were zero when $x=\pm 1$. It follows therefore, *that, when A is an even-ordered improper orthogonant, the zero-axial skew determinant formed by subtracting from each element of A its conjugate element is zero, or $|A-\breve{A}|=0$, where $\breve{A}$ is the conjugate of A.*

672. Multiplying $f(x)$ by itself we get

$$\{f(x)\}^2=\phi(y)\cdot 2^n(-1)^n x^n,$$

where

$$\phi(y) = \begin{vmatrix} a_{11}-y & \frac{a_{12}+a_{21}}{2} & \cdots & \frac{a_{1n}+a_{n1}}{2} \\ \frac{a_{12}+a_{21}}{2} & a_{22}-y & \cdots & \frac{a_{2n}+a_{u2}}{2} \\ \cdot & \cdot & \cdot & \cdot \\ \frac{a_{1n}+a_{n1}}{2} & \frac{a_{2n}+a_{n2}}{2} & \cdots & a_{nn}-y \end{vmatrix}, \text{ and } y = \frac{x^2+1}{2x}.$$

From this we see that y has the same value for every pair of reciprocal roots. Therefore for n even and $A=+1$, $\phi(y)$ is the square of a polynomial in y. That is $\phi(y)=(y-y_1)^2(y-y_2)^2\cdots(y-y_k)^2$, where $k=n/2$, and where y_r is the root of $\phi(y)=0$, corresponding to the rth pair of reciprocal roots of $f(x)=0$. *It follows, therefore, that for each of its double roots $\phi(y)$ and its principal coaxial minors vanish and consenquently all minors of order $n-1$ vanish.*

In the case of generalized orthogonants, that is, $\sum_s a^2_{rs}=m$, and $A^2=m^n$, the product of two reciprocal roots would be m instead of 1 and the real roots of $f(x)=0$ would be $\pm\sqrt{m}$ instead of ± 1. Similar changes throughout §§670, 671 would give the corresponding results for this type of orthogonant.

EXERCISES. 1. Show that if A is a Cayley's orthogonant then $|A-x|=0$, has its roots occurring in pairs of reciprocal imaginaries except when n is odd in which case there is a single real root 1.

2. If

$$A\equiv\begin{vmatrix} a & h & g \\ h & b & f \\ g & f & c \end{vmatrix}$$

is an orthogonant obtain the discriminant of $|A-x|=0$ as the sum of 7, 10, 13 and 15 squares.

673. If $A\equiv|a_{1n}|$ be a generalized orthogonant with the value $m^{n/2}$, let D_1 represent the determinant

$$\begin{vmatrix} a_{11}-\rho & a_{12} & \cdots & a_{1n} & u_1 & v_1 & w_1 & \cdots \\ a_{21} & a_{22}-\rho & \cdots & a_{2n} & u_2 & v_2 & w_2 & \cdots \\ \cdot & \cdot & \cdot & \cdot & \cdot & \cdot & \cdot & \cdot \\ a_{n1} & a_{n2} & \cdots & a_{nn}-\rho & u_n & v_n & w_n & \cdots \\ \bar{U}_1 & \bar{U}_2 & \cdots & \bar{U}_n & 0 & 0 & 0 & \cdots \\ \bar{V}_1 & \bar{V}_2 & \cdots & \bar{V}_n & 0 & 0 & 0 & \cdots \\ \bar{W}_1 & \bar{W}_2 & \cdots & \bar{W}_n & 0 & 0 & 0 & \cdots \\ \cdot & \cdot & \cdot & \cdot & \cdot & \cdot & \cdot & \cdot \end{vmatrix}$$

obtained by bordering $|A-\rho|\equiv\phi(\rho)$, say, by k additional rows and columns of arbitrary constants and a square of zeros k on a side to fill out the right-hand lower corner.

In the expansion of D_1 in terms of minors of order k from each of the borders the factors from $\phi(\rho)$ would be of order $n-k$. Forming the product of A and D_1 we have

$$A\cdot D_1=(-1)^{n-k}\rho^{n-k}\begin{vmatrix} a_{11}-\rho' & \cdots & a_{1n} & \sum a_{1i}\bar{U}_i & \sum a_{1i}\bar{V}_i & \cdots \\ \cdot & \cdot & \cdot & \cdot & \cdot & \cdot \\ a_{n1} & \cdots & a_{nn}-\rho' & \sum a_{ni}\bar{U}_i & \sum a_{ni}\bar{V}_i & \cdots \\ u_1 & \cdots & u_n & 0 & 0 & \cdots \\ v_1 & \cdots & v_n & 0 & 0 & \cdots \\ w_1 & \cdots & w_n & 0 & 0 & \cdots \\ \cdot & \cdot & \cdot & \cdot & \cdot & \cdot \end{vmatrix},$$

where $\rho'=m/\rho$.

If now we multiply the first column by $\bar{U}_1$ the second by $\bar{U}_2$, etc. and subtract the sum from the $(n+1)$st column, similarly for the $(n+2)$nd using $\bar{V}_1, \bar{V}_2, \cdots$, and so on, we get

$$A\cdot D_1=(-1)^{n-k}\rho^{n-k}\begin{vmatrix} a_{11}-\rho' & \cdots & a_{1n} & \rho'\bar{U}_1 & \rho'\bar{V}_1 & \cdots \\ \cdot & \cdot & \cdot & \cdot & \cdot & \cdot \\ a_{n1} & \cdots & a_{nn}-\rho' & \rho'\bar{U}_n & \rho'\bar{V}_n & \cdots \\ u_1 & \cdots & u_n & -\sum u\bar{U} & -\sum u\bar{V} & \cdots \\ v_1 & \cdots & v_n & -\sum v\bar{U} & -\sum v\bar{V} & \cdots \\ w_1 & \cdots & w_n & -\sum w\bar{U} & -\sum w\bar{V} & \cdots \\ \cdot & \cdot & \cdot & \cdot & \cdot & \cdot \end{vmatrix}$$

$$= (-1)^{n-k}\rho^{n-k}\rho'^{k}\begin{vmatrix} a_{11}-\rho' & \cdots & a_{1n} & \bar{U}_1 & \bar{V}_1 & \bar{W}_1 & \cdots \\ \cdot & \cdot & \cdot & \cdot & \cdot & \cdot & \cdot \\ a_{n1} & \cdots & a_{nn}-\rho' & \bar{U}_n & \bar{V}_n & \bar{W}_n & \cdots \\ u_1 & \cdots & u_n & 0 & 0 & 0 & \cdots \\ v_1 & \cdots & v_n & 0 & 0 & 0 & \cdots \\ w_1 & \cdots & w_n & 0 & 0 & 0 & \cdots \\ \cdot & \cdot & \cdot & \cdot & \cdot & \cdot & \cdot \end{vmatrix}$$

$$+ (-1)^{n-k}\rho^{n-k}(-1)^{k}\phi(\rho')(u)(\bar{U}),$$

where $(u)(\bar{U})$ stands for the product of the two arrays used in bordering.

We may write this relation thus

$$(1)\quad D_1\cdot A = (-1)^{n-k}\cdot\rho^{n-k}\cdot\rho'^{k}\cdot D_2 + (-1)^{n}\rho^{n-k}\phi(\rho')(u)(\bar{U}).$$

Where D_2 is D_1 with ρ' in the place of ρ and with (u) and $(\overline{U})$ interchanged.

When $k=0$, that is when (u) and $(\overline{U})$ do not exist, then (1) takes the form

$$D_1 \cdot A = (-1)^n \rho^n (\phi(\rho'))$$

or

$$\text{(2)} \qquad \phi(\rho) = \frac{(-1)^n \rho^n \phi\left(\dfrac{m}{\rho}\right)}{\pm (m^{1/2})^n} .$$

This shows the reciprocal nature of the roots of $\phi(\rho)=0$.

If D_1' is D_1 with $+\rho$ in the place of $-\rho$ we have in place of (1)

$$\text{(1')} \qquad D_1' \cdot A = \rho^{n-k} \cdot \rho'^k \cdot D_2' + (-1)^k \phi(-\rho')(u)(\overline{U}).$$

In the place of (2) we have

$$\text{(2')} \qquad \phi(-\rho) = \frac{\rho^n \cdot \phi\left(-\dfrac{m}{\rho}\right)}{\pm (m^{1/2})^n} .$$

If $A = -(\sqrt{m})^n$ then (2) shows that $\phi(-\sqrt{m})=0$ whether n is odd or even, $\phi(+\sqrt{m})=0$, when n is even.

If $A=+(\sqrt{m})^n$, then (2) gives $\phi(+\sqrt{m})=0$, when n is odd.

We see therefore that

(a) $\sqrt{m}$ is a root of $\phi(\rho)=0$, when n is odd and $A=+(\sqrt{m})^n$ also when n is even and $A=-(\sqrt{m})^n$

(b) $-\sqrt{m}$ is a root of $\phi(\rho)=0$, when n is odd or even and $A=-(\sqrt{m})^n$.

674. If the determinants of order k formed from (u) and $(\overline{U})$ be denoted by $\mu_1, \mu_2, \cdots, \mu_\lambda$ and $\nu_1, \nu_2, \cdots, \nu_\lambda$ respectively, where $\lambda=(n)_k$, then since (u) and $(\overline{U})$ are arbitrary the coefficient of $\mu_i \cdot \nu_i$ on the one side of (1) §673 must be the same as that on the other side.

We have therefore

$$\text{(1)} \quad \phi_{(n-k)ii}(\rho) \cdot A = (-1)^{n-k} \rho^{n-k} \cdot \rho'^k \cdot \phi_{(n-k)ii}(\rho') + (-1)^n \rho^{n-k} \phi(\rho'),$$

where $\phi_{(n-k)ii}(\rho)$ denotes a coaxial minor of $\phi(\rho)$ of order $n-k$.

This shows that all coaxial minors of order $n-k$ are equal when $\rho=\rho'$.

If α is a root of $\phi(\rho)=0$, then m/α is also a root and (1) becomes

$$\phi_{(n-k)ii}(\alpha) \cdot A = (-1)^{n-k} \alpha^{n-2k} m^k \phi_{(n-k)ii}\left(\frac{m}{\alpha}\right)$$

or

$$(2) \qquad \phi_{(n-k)ii}(\alpha) = \frac{(-1)^{n-k}\alpha^{n-2k}\phi_{(n-k)ii}\left(\dfrac{m}{\alpha}\right)}{\pm (m^{1/2})^{n-2k}} \cdot$$

This equation (2) shows the reciprocal nature of the roots of the coaxial minors of $\phi(\rho)$.

If $\rho = \rho'$, then (1) becomes

$$(3) \qquad \phi_{(n-k)ii}(\rho) = \frac{(-1)^n \rho^{n-k}\phi(\rho)}{A - (-1)^{n-k}\rho^n} \cdot$$

Similarly the coefficients of $\mu_i\nu_j$ on the two sides of (1) §673 must be the same giving us

$$(4) \quad \phi_{(n-k)ij}(\rho)\cdot A = (-1)^{n-k}\rho^{n-k}\cdot\rho'^{k}\cdot\phi_{(n-k)ji}(\rho') + (-1)^n\rho^{n-k}\phi(\rho'),$$

$$(5) \qquad \phi_{(n-k)ij}(\alpha) = \frac{(-1)^{n-k}\alpha^{n-2k}\phi_{(n-k)ji}\left(\dfrac{m}{\alpha}\right)}{\pm (m^{1/2})^{n-2k}} \cdot$$

The relation (5) shows the reciprocal nature of the roots of non-coaxial minors.

I. When $A = +(\sqrt{m})^n$, then

$$(6) \qquad \phi_{(n-k)ii}(m^{1/2}) = \frac{(-1)^n\phi(m^{1/2})}{(m^{1/2})^k\left\{1 - (-1)^{n-k}\right\}},$$

from (3),

$$= \frac{\phi(m^{1/2})}{2(m^{1/2})^k}$$

when n is even and k is odd,

$$= 0$$

when n is odd and k even since $\phi(\sqrt{m}) = 0$,

$$= \frac{0}{0}$$

when n is odd and k is odd,

$$= \infty$$

when n is even and k is even.

(7) $$\phi_{(n-k)ii}(-m^{1/2}) = \frac{(-1)^k\phi(-m^{1/2})}{(m^{1/2})^k\{1-(-1)^k\}}$$

$$= \frac{-\phi(-m^{1/2})}{2(m^{1/2})^k}$$

when k is odd,

$$= \infty$$

when k is even since $\phi(-\sqrt{m}) \neq 0$.

From (4) we have when n is odd

(8) $$\phi_{(n-k)ij}(m^{1/2}) = (-1)^{n-k}\phi_{(n-k)ji}(m^{1/2})$$

since $\phi(\sqrt{m}) = 0$,

$$= \phi_{(n-k)ji}(m^{1/2})$$

when k is odd,

$$= -\phi_{(n-k)ji}(m^{1/2})$$

when k is even.

(9) $$\phi_{(n-k)ij}(-m^{1/2}) = (-1)^k\phi_{(n-k)ji}(-m^{1/2}) + \frac{(-1)^k\phi(-m^{1/2})}{(m^{1/2})^k}$$

From (6) *and* (8) *we see that when* $A = (\sqrt{m})^n$ *the odd-ordered compounds of* $\phi(\sqrt{m})$ *are all zero-axial skew determinants and therefore vanish*

II. When $A = -(\sqrt{m})^n$ we have

(10) $$\phi_{(n-k)ii}(m^{1/2}) = -\frac{(-1)^n\phi(m^{1/2})}{(m^{1/2})^k\{1+(-1)^{n-k}\}}$$

$$= -\frac{\phi(m^{1/2})}{2(m^{1/2})^k}$$

when n is odd and k is odd,

$$= \infty$$

when n is odd and k is even,

$$= 0$$

when n is even and k is even,

$$= \frac{0}{0}$$

when n is even and k is odd.

(11) $$\phi_{(n-k)ii}(-m^{1/2}) = -\frac{(-1)^k\phi(-m^{1/2})}{(m^{1/2})^k\{1+(-1)^k\}}$$

$$= 0$$

when k is even,

$$= \frac{0}{0}$$

when k is odd.

(12) $$\phi_{(n-k)ij}(m^{1/2}) = -(-1)^{n-k}\phi_{(n-k)ji}(m^{1/2}) - \frac{(-1)^n\phi(m^{1/2})}{(m^{1/2})^k}$$

$$= -(-1)^{n-k}\phi_{(n-k)ji}(m^{1/2})$$

when n is even,

$$= -\phi_{(n-k)ji}(m^{1/2})$$

when n is even and k is even,

$$= \phi_{(n-k)ji}(m^{1/2})$$

when n is even and k is odd.

From (11) *and* (12) *we see that when A is an improper orthogonant of even order, all the even-ordered compounds of $\phi(-\sqrt{m})$ are zero-axial skew and therefore are perfect squares.*

675. Since

$$\phi_{(n-k)ii}(\rho)\phi_{(n-k)jj}(\rho) - \phi_{(n-k)ij}(\rho)\phi_{(n-k)ji}(\rho) = E,$$

where the terms of E contain minors of order higher than $n-k$, and therefore E vanishes where such minors vanish it follows that, *when $A=(\sqrt{m})^n$, n odd, and all minors of order greater than $n-k$ vanish, then $\phi_{(n-k)ij}(\sqrt{m})=0$ for all values of i and j.*

For

$$\phi_{(n-k)ii}(\rho)\phi_{(n-k)jj}(\rho) + \{\phi_{(n-k)ij}(\rho)\}^2 = 0$$

and $\phi_{(n-k)ii}(\sqrt{m})=0$ when n is odd and k is even and must also vanish when n is odd and k is odd, for then the first term is a perfect square. We have therefore the theorem: *when A is an odd-ordered orthogonant of value $(m^{1/2})^n$ and all minors of $\phi(\sqrt{m})$ of order greater than $n-k$ vanish, then all minors of order $n-k$ will vanish also.*

This is the same as saying that *if the orthogonant $A=(\sqrt{m})^n$ is of odd order and if all minors of order $2r+1$ and greater vanish then all minors of order $2r$ will vanish also.*

Again since $\phi_{(n-k)ii}(\sqrt{m})=0$ and $\phi_{(n-k)ij}(\sqrt{m})=-\phi_{(n-k)ji}(\sqrt{m})$, when $A=-(\sqrt{m})^n$ and both n and k even, we have the following theorem:

If all minors of $\phi(\sqrt{m})$ of order greater than $n-k$ vanish, then all minors of order $n-k$ will vanish also.

This is the same as saying that *if the orthogonant $A=-(\sqrt{m})^n$ is of even order and all minors of $\phi(\sqrt{m})$ of order $2r+1$ and greater vanish, then all minors of order $2r$ will vanish also.*

EXERCISE: If the c's are real and the members of the sequence

$$c_0, \quad \begin{vmatrix} c_0 & c_1 \\ c_1 & c_2 \end{vmatrix}, \quad P(c_0, c_1, \cdots, c_4), \cdots, P(c_0, c_1, \cdots, c_{2n+2})$$

are positive show that all the roots of the equation

$$c_0 + c_1 x + \cdots + c_{2n} x^{2n} = 0$$

and all but one of the roots of the equation

$$c_0 + c_1 x + \cdots + c_{2n+1} x^{2n+1} = 0$$

are imaginary.

CHAPTER XVI

JACOBIANS

676. If there be n functions all of the same n variables, the determinant which in every case has the element in its rth row and sth column equal to the differential coefficient of the rth function with respect to the sth variable is called the *Jacobian* of the set of functions with respect to the said variables. Thus the Jacobian of $y_1, y_2, \cdots, y_n$ with respect to $x_1, x_2, \cdots, x_n$ is

$$\left| \frac{\partial y_1}{\partial x_1} \frac{\partial y_2}{\partial x_2} \cdots \frac{\partial y_n}{\partial x_n} \right| \quad \text{or} \quad \left| a_{1n} \right| \text{ if } a_{ij} = \frac{\partial y_i}{\partial x_j} \cdot$$

It is usually denoted by $J(y_1 y_2 \cdots y_n)$ where there is no possibility of ambiguity in regard to the independent variables.

677. If the variables $x_1, x_2, \cdots, x_n$ simultaneously receive the increments $\Delta_r x_1, \Delta_r x_2, \cdots, \Delta_r x_n$ respectively and in consequence of this simultaneous change the functions $y_1, y_2, \cdots, y_n$ receive the increments $\Delta_r y_1, \Delta_r y_2, \cdots, \Delta_r y_n$ respectively, then, as in the case of a single function, the limiting value of*

$$\left| \Delta_1 y_1 \, \Delta_2 y_2 \cdots \Delta_n y_n \right| \div \left| \Delta_1 x_1 \, \Delta_2 x_2 \cdots \Delta_n x_n \right|$$

when the elements of the latter determinant are indefinitely diminished is equal to

$$\left| \frac{\partial y_1}{\partial x_1} \frac{\partial y_2}{\partial x_2} \cdots \frac{\partial y_n}{\partial x_n} \right| .$$

The product

$$\left| \frac{\partial y_1}{\partial x_1} \frac{\partial y_2}{\partial x_2} \cdots \frac{\partial y_n}{\partial x_n} \right| \cdot \left| \Delta_1 x_1 \, \Delta_2 x_2 \cdots \Delta_n x_n \right| = \left| Q_{1n} \right|$$

where

$$Q_{rs} = \frac{\partial y_r}{\partial x_1} \Delta_s x_1 + \frac{\partial y_r}{\partial x_2} \Delta_s x_2 + \cdots + \frac{\partial y_r}{\partial x_n} \Delta_s x_n = \Delta_s y_r$$

and

$$\left| Q_{1n} \right| = \left| \Delta_1 y_1 \Delta_2 y_2 \cdots \Delta_n y_n \right|$$

* That is when there is a definite limiting value; vide Peano, Giornaie di Mat., xxvii, pp. 226–228.

The limiting value of the quotient might have been taken as the definition of the Jacobian and the relations between Jacobians present close resemblance to the formulae of differential calculus. The notation

$$\frac{d(y_1y_2 \cdots y_n)}{d(x_1x_2 \cdots x_n)} \text{ or } \frac{\partial(y_1y_2 \cdots y_n)}{\partial(x_1x_2 \cdots x_n)}$$

for the Jacobian is valuable as exhibiting this relation.

678. *If $y_{m+1}, y_{m+2} \cdots y_n$ be constant with respect to $x_1, x_2, \cdots, x_m$, or $y_1, y_2, \cdots, y_m$ be constant with respect to $x_{m+1}, \cdots, x_n$ then*

$$\frac{d(y_1 \cdots y_my_{m+1} \cdots y_n)}{d(x_1 \cdots x_mx_{m+1} \cdots x_n)} = \frac{d(y_1 \cdots y_m)}{d(x_1 \cdots x_m)} \cdot \frac{d(y_{m+1} \cdots y_n)}{d(x_{m+1} \cdots x_n)}$$

and in particular

$$\frac{d(y_1 \cdots y_mx_{m+1} \cdots x_n)}{d(x_1 \cdots x_mx_{m+1} \cdots x_n)} = \frac{d(y_1 \cdots y_m)}{d(x_1 \cdots x_m)} \cdot$$

The truth of this is apparent from §108.

679. *If $y_1y_2 \cdots y_n$ be functions of $v_1v_2 \cdots v_n$ and $v_1v_2 \cdots v_n$ be functions of $x_1x_2 \cdots x_n$, then*

$$\frac{d(y_1y_2 \cdots y_n)}{d(x_1x_2 \cdots x_n)} = \frac{d(y_1y_2 \cdots y_n)}{d(v_1v_2 \cdots v_n)} \cdot \frac{d(v_1v_2 \cdots v_n)}{d(x_1x_2 \cdots x_n)}$$

Multiplying row-by-row the two Jacobians on the right we get a determinant $|X_{1n}|$ where

$$\begin{aligned} x_{rs} &= \frac{\partial y_r}{\partial v_1}\frac{\partial v_1}{\partial x_s} + \frac{\partial y_r}{\partial v_2}\frac{\partial v_2}{\partial x_s} + \cdots + \frac{\partial y_r}{\partial v_n}\frac{\partial v_n}{\partial x_s} \\ &= \frac{\partial y_r}{\partial x_s} \cdot \end{aligned}$$

therefore

$$|X_{1n}| = \left| \frac{\partial y_1}{\partial x_1} \frac{\partial y_2}{\partial x_2} \cdots \frac{\partial y_n}{\partial x_n} \right|$$

as was to be proved.

In like manner if $y_1y_2 \cdots y_m$ are functions of $v_1v_2 \cdots v_n$ and $v_1v_2 \cdots v_n$ are functions of $x_1x_2 \cdots x_m$ we see that

$$\frac{d(y_1y_2 \cdots y_m)}{d(x_1x_2 \cdots x_m)} = 0$$

if $m>n$,

$$= \sum \frac{d(y_1 \cdots y_m)}{d(v_\alpha v_\beta \cdots v_\gamma)} \frac{d(v_\alpha v_\beta \cdots v_\gamma)}{d(x_1 x_2 \cdots x_m)},$$

if $m<n$ *where* $\alpha, \beta, \cdots,$ *are a set of* m *numbers selected from* $1, 2, \cdots, n$.

680. *If* $y_1 y_2 \cdots y_n$ *be independent functions of* $x_1 x_2 \cdots x_n$, *the Jacobian of* $y_1 y_2 \cdots y_n$ *with respect to* $x_1 x_2 \cdots x_n$ *is the reciprocal of the Jacobian of* $x_1 x_2 \cdots x_n$ *with respect to* $y_1 y_2 \cdots y_n$.

This follows at once from the theorem of the preceding article. For $x_1 x_2 \cdots x_n$ are independent functions of $y_1 y_2 \cdots y_n$ and the product of the two Jacobians gives

$$\frac{d(x_1 x_2 \cdots x_n)}{d(x_1 x_2 \cdots x_n)} = 1.$$

681. *If* $y_1, y_2, \cdots, y_n$ *be given implicitly in terms of* $x_1, x_2, \cdots, x_n$ *in the form*

$$F_1(x_1 \cdots x_n y_1 \cdots y_n) = 0,$$
$$F_2(x_1 \cdots x_n y_1 \cdots y_n) = 0,$$
$$\cdots\cdots\cdots\cdots$$
$$F_n(x_1 \cdots x_n y_1 \cdots y_n) = 0$$

then

$$\frac{d(y_1 y_2 \cdots y_n)}{d(x_1 x_2 \cdots x_n)} = (-1)^n \frac{d(F_1 F_2 \cdots F_n)}{d(x_1 x_2 \cdots x_n)} \div \frac{d(F_1 F_2 \cdots F_n)}{d(y_1 y_2 \cdots y_n)}.$$

This is readily seen to be a direct result of the multiplication theorem, and

$$\frac{\partial F_r}{\partial y_1}\frac{\partial y_1}{\partial x_s} + \frac{\partial F_r}{\partial y_2}\frac{\partial y_2}{\partial x_s} + \cdots + \frac{\partial F_r}{\partial y_n}\frac{\partial y_n}{\partial x_s} = -\frac{\partial F_r}{\partial x_s}.$$

682. *The Jacobian of a set of* n *functions is always expressible as a product of* n *differential coefficients; namely*

$$\frac{d(y_1 y_2 \cdots y_n)}{d(x_1 x_2 \cdots x_n)} = \frac{\partial \phi_1}{\partial x_1}\frac{\partial \phi_2}{\partial x_2} \cdots \frac{\partial \phi_n}{\partial x_n},$$

where ϕ_r *is a function of* $y_1 y_2 \cdots y_{r-1} x_r \cdots x_n$.

Since we have y_1 given as a function of $x_1 x_2 \cdots x_n$, it follows that x_1 is a function of $y_1 x_2 \cdots x_n$, and therefore also that y_2 is a function of the latter set of variables. Again from this we have x_2 a function of $y_1y_2x_3 \cdots x_n$; therefore also y_3 is a function of $y_1y_2x_3 \cdots x_n$. Similarly it is evident that y_4 is a function of $y_1y_2y_3x_4 \cdots x_n$, and so on; so that we have

$$\begin{aligned} y_1 - \phi_1(x_1x_2 \cdots x_n) &= 0 \\ y_2 - \phi_2(y_1x_2 \cdots x_n) &= 0 \\ y_3 - \phi_3(y_1y_2x_3 \cdots x_n) &= 0 \\ \cdots\cdots&\cdots\cdots \\ y_n - \phi_n(y_1 \cdots y_{n-1}x_n) &= 0. \end{aligned}$$

Hence from §681 there results

$$\frac{d(y_1y_2 \cdots y_n)}{d(x_1x_2 \cdots x_n)} = (-1)^n \begin{vmatrix} -\frac{\partial\phi_1}{\partial x_1} & -\frac{\partial\phi_1}{\partial x_2} & \cdots & -\frac{\partial\phi_1}{\partial x_n} \\ 0 & -\frac{\partial\phi_2}{\partial x_2} & \cdots & -\frac{\partial\phi_2}{\partial x_n} \\ \cdots & \cdots & \cdots & \cdots \\ 0 & 0 & \cdots & -\frac{\partial\phi_n}{\partial x_n} \end{vmatrix}$$

$$\div \begin{vmatrix} 1 & 0 & 0 & \cdots & \\ -\frac{\partial\phi_2}{\partial y_1} & 1 & 0 & \cdots & \\ -\frac{\partial\phi_3}{y_1} & -\frac{\partial\phi_3}{\partial y_2} & 1 & \cdots & \\ \cdots & \cdots & \cdots & \cdots & \cdots \\ -\frac{\partial\phi_n}{\partial y_1} & -\frac{\partial\phi_n}{\partial y_2} & -\frac{\partial\phi_n}{\partial y_3} & \cdots & 1 \end{vmatrix} = \frac{\partial\phi_1}{\partial x_1}\frac{\partial\phi_2}{\partial x_2} \cdots \frac{\partial\phi_n}{\partial x_n}.$$

683. From §682 it follows that if the Jacobian vanishes one of the factors on the right is zero, say $\partial\phi_m/\partial x_m = 0$. Consequently ϕ_m does not involve x_m, that is y_m is a function of $y_1 \cdots y_{m-1}x_{m-1} \cdots x_n$. But from this it follows that x_{m+1} is a function of $y_1 \cdots y_mx_{m+2} \cdots x_n$, hence y_{m+1}, which equals $\phi_{m+1}(y_1 \cdots y_mx_{m+1} \cdots x_n)$ is a function of $y_1 \cdots y_mx_{m+2} \cdots x_n$. Similarly we may show that y_{m+2} is a function of $y_1 \cdots y_{m+1}x_{m+3} \cdots x_n$, and finally that y_n is a function of $y_1 \cdots y_{n-1}$ which is what was to be proved.

684. *If* $y_1 \cdots y_n$ *be independent functions of* $x_1 \cdots x_n$ *then*

$$\frac{d(y_1 \cdots y_n)}{d(x_1 \cdots x_n)} \cdot \frac{d(x_1 \cdots x_m)}{d(y_1 \cdots y_m)} = \frac{d(y_{m+1} \cdots y_n)}{d(x_{m+1} \cdots x_n)}.$$

For the left-hand member

$$= \frac{d(y_1 \cdots y_n)}{d(x_1 \cdots x_n)} \frac{d(x_1 \cdots x_m x_{m+1} \cdots x_n)}{d(y_1 \cdots y_m x_{m+1} \cdots x_n)} \qquad (\S 678)$$

$$= \frac{d(y_1 \cdots y_m y_{m+1} \cdots y_n)}{d(y_1 \cdots y_m x_{m+1} \cdots x_n)} \qquad (\S 679)$$

$$= \frac{d(y_{m+1} \cdots y_n)}{d(x_{m+1} \cdots x_n)}. \qquad (\S 678)$$

685. *If* $y_1 \cdots y_n$ *be functions of* $n+1$ *variables* $x_1, \cdots, x_{n+1}$, *then*

$$\frac{\partial}{\partial x_1} \frac{d(y_1 \cdots y_n)}{d(x_2 \cdots x_{n+1})} - \frac{\partial}{\partial x_2} \frac{d(y_1 \cdots y_n)}{d(x_1 x_3 \cdots x_{n+1})} + \cdots + (-1)^n \frac{\partial}{\partial x_{n+1}} \frac{d(y_1 \cdots y_n)}{d(x_1 \cdots x_n)} = 0.$$

For each term on the left is expressible as the sum of n determinants and the $n(n+1)$ determinants so obtained vanish in pairs in virtue of the identity

$$\frac{\partial^2 y_m}{\partial x_r \partial x_s} = \frac{\partial^2 y_m}{\partial x_s \partial x_r}.$$

686. *If* n *functions of* $x_1 x_2 \cdots x_n$ *be not independent, their Jacobian with respect to* $x_1 x_2 \cdots x_n$ *vanishes.*

Let the n functions be $y_1 y_2 \cdots y_n$, and the relation between them

$$f(y_1 y_2 \cdots y_n) = 0.$$

Differentiating with respect to the x's in succession we have

$$\frac{\partial f}{\partial y_1}\frac{\partial y_1}{\partial x_1} + \frac{\partial f}{\partial y_2}\frac{\partial y_2}{\partial x_1} + \cdots + \frac{\partial f}{\partial y_n}\frac{\partial y_n}{\partial x_1} = 0$$

$$\frac{\partial f}{\partial y_1}\frac{\partial y_1}{\partial x_2} + \frac{\partial f}{\partial y_2}\frac{\partial y_2}{\partial x_2} + \cdots + \frac{\partial f}{\partial y_n}\frac{\partial y_n}{\partial x_2} = 0$$

$$\cdots\cdots\cdots\cdots\cdots\cdots$$

$$\frac{\partial f}{\partial y_1}\frac{\partial y_1}{\partial x_n} + \frac{\partial f}{\partial y_2}\frac{\partial y_2}{\partial x_n} + \cdots + \frac{\partial f}{\partial y_n}\frac{\partial v_n}{\partial x_n} = 0$$

Eliminate

$$\frac{\partial f}{\partial y_1}, \frac{\partial f}{\partial y_2}, \ldots, \frac{\partial f}{\partial y_n}$$

and we have the Jacobian equal to zero.

687. *If the functions* $y_1 y_2 \cdots y_n$ *are fractions with the same denominator so that*

$$y_k = \frac{u_k}{v},$$

and

$$v^2\frac{dy_k}{dx_k} = v\frac{du_k}{dx_k} - u_k\frac{dv}{dx_k}$$

then

$$\frac{d(y_1 \cdots y_n)}{d(x_1 \cdots x_n)} = \frac{1}{v^{2n+1}}\begin{vmatrix} v & 0 & \cdots & 0 \\ u_1 & v\dfrac{du_1}{dx_1} - u_1\dfrac{dv}{dx_1} & \cdots & v\dfrac{du_1}{dx_n} - u_1\dfrac{dv}{dx_n} \\ \cdot & \cdot & \cdot & \cdot \\ u_n & v\dfrac{du_n}{dx_1} - u_n\dfrac{dv}{dx_1} & \cdots & v\dfrac{du_n}{dx_n} - u_n\dfrac{dv}{dx_n} \end{vmatrix}_{n+1}$$

which on simplifying gives

$$v^{2n+1}\frac{d(y_1 \cdots y_n)}{d(x_1 \cdots x_n)} = \begin{vmatrix} v & v\dfrac{dv}{dx_1} & \cdots & v\dfrac{dv}{dx_n} \\ u_1 & v\dfrac{du_1}{dx_1} & \cdots & v\dfrac{du_1}{dx_n} \\ \cdot & \cdot & \cdot & \cdot \\ u_n & v\dfrac{du_n}{dx_1} & \cdots & v\dfrac{du_n}{dx_n} \end{vmatrix}_{n+1}$$

and finally

$$\frac{d(y_1 \cdots y_n)}{d(x_1 \cdots x_n)} = \frac{1}{v^{n+1}}\begin{vmatrix} v & \dfrac{dv}{dx_1} & \cdots & \dfrac{dv}{dx_n} \\ u_1 & \dfrac{du_1}{dx_1} & \cdots & \dfrac{du_1}{dx_n} \\ \cdot & \cdot & \cdot & \cdot \\ u_n & \dfrac{du_n}{dx_1} & \cdots & \dfrac{du_n}{dx_n} \end{vmatrix}$$

The determinant on the right has been called the *pre-Jacobian* and is denoted by $K(v, u_1, u_2, \cdots, u_n)$ so that the relation may be written

$$J\left(\frac{u_1}{v}, \frac{u_2}{v}, \cdots, \frac{u_n}{v}\right) = \frac{1}{v^{n+1}} K(v, u_1, u_2, \cdots, u_n).$$

If J vanishes then K will also vanish.

688. *If the functions $y_1 y_2 \cdots y_n$ have a common factor so that*

$$y_k = v \cdot u_k$$

then it is readily seen that

$$\frac{d(y_1 \cdots y_n)}{d(x_1 \cdots x_n)} = 2v^n \frac{d(u_1 \cdots u_n)}{d(x_1 \cdots x_n)} - v^{n-1} K(v, u_1, \cdots, u_n)$$

or

$$J(vu_1, vu_2, \cdots, vu_n) = 2v^n J(u_1, u_2, \cdots, u_n) - v^{n-1} K(v . u_1, u_2, \cdots, u_n)$$

EXERCISE: Show that the theorems of §683 and §686 are true for pre-Jacobians.

689. *If*

$$\frac{d(y_1 \cdots y_n)}{d(x_1 \cdots x_n)} \equiv A, \quad \frac{d(x_1 \cdots x_n)}{d(y_1 \cdots y_n)} \equiv B,$$

and if A_{ik}, B_{ik} are the complementary minors of dy_i/dx_k and dx_i/dy_k in A and B respectively, then

$$A \frac{dx_i}{dy_k} = A_{ki}, \quad B \frac{dy_i}{dx_k} = B_{ki},$$

$$A \frac{d(x_1 \cdots x_m)}{d(y_1 \cdots y_m)} = \frac{d(y_{m+1} \cdots y_n)}{d(x_{m+1} \cdots x_n)},$$

and

$$B \frac{d(y_1 \cdots y_m)}{d(x_1 \cdots x_m)} = \frac{d(x_{m+1} \cdots x_n)}{d(y_{m+1} \cdots y_n)}.$$

For

$$\frac{\partial y_i}{\partial x_1} \frac{\partial x_1}{\partial y_k} + \frac{\partial y_i}{\partial x_2} \frac{\partial x_2}{\partial y_k} + \cdots + \frac{\partial y_i}{\partial x_n} \frac{\partial x_n}{\partial y_k} = 0 \text{ or } 1$$

according as $i \neq k$ or $i = k$.

Giving i the values 1, 2, $\cdots$, n and multiplying these successive equations by A_{1i}, A_{2i}, $\cdots$, A_{ni} and adding gives

$$A\frac{\partial x_i}{\partial y_k} = A_{ki}.$$

Similarly

$$B\frac{\partial y_i}{\partial x_k} = B_{ki}$$

From this it follows that

$$A_{ij}B_{rs} = A\cdot B\cdot\frac{\partial x_j}{\partial y_i}\,\frac{\partial y_s}{\partial x_r}$$

$$= \frac{\partial x_j}{\partial y_i}\,\frac{\partial y_s}{\partial x_r},$$

since $A\cdot B=1$.

Again

$$\begin{vmatrix} A_{11} & A_{12} \cdots A_{1m} \\ A_{21} & A_{22} \cdots A_{2m} \\ \cdot\ \cdot\ \cdot & \cdot\ \cdot\ \cdot\ \cdot \\ A_{m1} & A_{m2} \cdots A_{mm} \end{vmatrix} = A^{m-1}\frac{d(y_{m+1}\cdots y_n)}{d(x_{m+1}\cdots x_n)}$$

and substituting the value for A_{ki} just found we have

$$A\cdot\frac{d(x_1\cdots x_m)}{d(y_1\cdots y_m)} = \frac{d(y_{m+1}\cdots y_n)}{d(x_{m+1}\cdots x_n)}\,.$$

Similarly

$$B\cdot\frac{d(y_1\cdots y_m)}{d(x_1\cdots x_m)} = \frac{d(x_{m+1}\cdots x_n)}{d(y_{m+1}\cdots y_n)}\,.$$

690. If all the functions depend upon a single variable t, then we have

$$\frac{\partial A}{\partial t} = \sum\frac{\partial A}{\partial a_{ik}}\cdot\frac{\partial a_{ik}}{\partial t} = \sum A_{ik}\frac{\partial}{\partial t}\left(\frac{\partial y_i}{\partial x_k}\right)$$

$$= \sum A_{ik}\frac{\partial^2 y_i}{\partial t\partial x_k} = \sum A\frac{\partial x_k}{\partial y_i}\,\frac{\partial^2 y_i}{\partial t\partial x_k} \qquad (i,k = 1,2,\cdots,n)$$

$$= A\sum\left(\frac{\partial^2 y_i}{\partial t\partial x_1}\,\frac{\partial x_1}{\partial y_i} + \frac{\partial^2 y_i}{\partial t\partial x_2}\,\frac{\partial x_2}{\partial y_i} + \cdots\right)$$

$$= A\sum\frac{\partial}{\partial y_i}\left(\frac{\partial y_i}{\partial t}\right)$$

or

$$\frac{\partial}{\partial t}\log A = \sum \frac{\partial}{\partial y_i}\left(\frac{\partial y_i}{\partial t}\right)$$

Similarly

$$\frac{\partial B}{\partial t} = B \sum \frac{\partial}{\partial x_i}\left(\frac{\partial x_i}{\partial t}\right).$$

691. *If* A_{ik} *is the cofactor of* $\partial y_i/\partial x_k$ *in the Jacobian*

$$\frac{d(y_1 \cdots y_n)}{d(x_1 \cdots x_n)} \equiv A,$$

then

$$\frac{\partial A_{i1}}{\partial x_1} + \frac{\partial A_{i2}}{\partial x_2} + \cdots + \frac{\partial A_{in}}{\partial x_n} = 0$$

Starting with

$$\frac{\partial B}{\partial y_k} = B \cdot \sum \frac{\partial}{\partial x_i}\left(\frac{\partial x_i}{\partial y_k}\right)$$

and using the fact that A and B are reciprocals we have

$$\frac{\partial A}{\partial y_k} + A \sum \frac{\partial}{\partial x_i}\left(\frac{\partial x_i}{\partial y_k}\right) = 0.$$

But

$$\frac{\partial A}{\partial y_k} = \sum \frac{\partial A}{\partial x_i} \frac{\partial x_i}{\partial y_k}$$

and therefore

$$\sum \frac{\partial}{\partial x_i}\left(A \frac{\partial x_i}{\partial y_k}\right) = 0,$$

or

$$\sum \frac{\partial}{\partial x_i}(A_{ki}) = 0,$$

which is the theorem.

692. *If $f_1, f_2, \cdots, f_n$ are n-thics in $x_1, x_2, \cdots, x_n$ which vanish, then the Jacobian $d(f_1 \cdots f_n)/d(x_1 \cdots x_n) \equiv \phi$, and each of its differential coefficients will also vanish.*

For

$$\begin{vmatrix} x_1 & x_2 \cdots & x_n \\ 0 & 1 \cdots & 0 \\ \cdot & \cdot \cdots & \cdot \\ 0 & 0 \cdots & 1 \end{vmatrix} \cdot \begin{vmatrix} \dfrac{\partial f_1}{\partial x_1} & \dfrac{\partial f_1}{\partial x_2} \cdots & \dfrac{\partial f_1}{\partial x_n} \\ \cdot & \cdot \cdots & \cdot \\ \dfrac{\partial f_n}{\partial x_1} & \dfrac{\partial f_n}{\partial x_2} \cdots & \dfrac{\partial f_n}{\partial x_n} \end{vmatrix}$$

$$= x_1\phi = \begin{vmatrix} nf_1 & \dfrac{\partial f_1}{\partial x_2} \cdots & \dfrac{\partial f_1}{\partial x_n} \\ nf_2 & \dfrac{\partial f_2}{\partial x_2} \cdots & \dfrac{\partial f_2}{\partial x_n} \\ \cdot & \cdot \cdots & \cdot \\ nf_n & \dfrac{\partial f_n}{\partial x_2} \cdots & \dfrac{\partial f_n}{\partial x_n} \end{vmatrix}$$

which shows that values of the x's which cause the functions to vanish will also make ϕ vanish. Differentiating both sides with respect to x_1 we have on the left $x_1\partial\phi/\partial x_1+\phi$ and on the right the sum of n determinants, the first of which is $n\cdot\phi$ and all the others have $f_1, f_2, \cdots, f_n$ for the first column. It follows therefore that $\partial\phi/\partial x_1$ will also vanish when the x's are such as to make the f's to vanish.

Combining the two we have the theorem that *when n n-ary n-thics vanish, their Jacobian and each of its first differential-coefficients will vanish also.*

Exercises. Set XXXII

1. Show that the necessary and sufficient condition that n functions in $n+h$ variables shall be connected by an equation independent of said variables is that the Jacobian of the functions with respect to any n of the variables shall vanish. (Bertrand).

2. Show that the Jacobian of $a^2+b^2+c^2$, $ax+by+cz$, $x^2+y^2+z^2$, $bx-cy$, $cx-az$, $ay-bx$, with respect to a, b, c, x, y, z vanishes.

3. If $u_1, u_2, \cdots, u_n$ be homogeneous functions of $x_1x_2 \cdots x_n$, the first $n-1$ of them being of the pth degree and the last of the qth degree, then

$$x_r\frac{\partial J}{\partial x_s} = p\sum u_t\frac{\partial[tr]}{\partial x_s} + 0 + (q-p)\frac{\partial}{\partial x_s}(u_n[nr])$$

$$x_r\frac{\partial J}{\partial x_r} = p\sum u_t\frac{\partial[tr]}{\partial x_r} + (p-1)J + (q-p)\frac{\partial}{\partial x_r}(u_n[nr])$$

where J stands for the Jacobians of the u's with respect to the x's, and $|[11][22]\cdots[nn]|$ for the adjugate of the Jacobian.

4. The necessary and sufficient condition that the functions $u_1u_2\cdots u_n$ of the independent variables $x_1x_2\cdots x_n$ shall satisfy p relations independent of these variables is that the leading minor of the $(n-p)$th order in the n-by-m array

$$\begin{matrix}
\frac{\partial u_1}{\partial x_1} & \frac{\partial u_1}{\partial x_2} & \cdots & \frac{\partial u_1}{\partial x_{n-p}} & \frac{\partial u_1}{\partial x_{n-p+1}} & \cdots & \frac{\partial u_1}{\partial x_m} \\
\frac{\partial u_2}{\partial x_1} & \frac{\partial u_2}{\partial x_2} & \cdots & \frac{\partial u_2}{\partial x_{n-p}} & \frac{\partial u_2}{\partial x_{n-p+1}} & \cdots & \frac{\partial u_2}{\partial x_m} \\
\cdot & \cdot & \cdot & \cdot & \cdot & \cdot & \cdot \\
\frac{\partial u_{n-p}}{\partial x_1} & \frac{\partial u_{n-p}}{\partial x_2} & \cdots & \frac{\partial u_{n-p}}{\partial x_{n-p}} & \frac{\partial u_{n-p}}{\partial x_{n-p+1}} & \cdots & \frac{\partial u_{n-p}}{\partial x_m} \\
\cdot & \cdot & \cdot & \cdot & \cdot & \cdot & \cdot \\
\frac{\partial u_n}{\partial x_1} & \frac{\partial u_n}{\partial x_2} & \cdots & \frac{\partial u_n}{\partial x_{n-p}} & \frac{\partial u_n}{\partial x_{n-p+1}} & \cdots & \frac{\partial u_n}{\partial x_m}
\end{matrix}$$

shall not vanish, but all the minors of the $(n-p+1)$th order obtained by bordering the said leading minor shall vanish. (Trzaska.)

Note the special cases $(p=1,\ m=n+h)$, $(p=1,\ m=n)$.

5. If $f, f_1, f_2, \cdots, f_n$ be functions of $x, x_1, x_2, \cdots, x_n$ and by legitimate operations the functions $f_1, f_2, \cdots, f_n$ be introduced into the expression for f which thereby takes the form of ϕ, then

$$\frac{d(f, f_1, \cdots, f_n)}{d(x, x_1, \cdots, x_n)} = \frac{d(\phi, f_1, \cdots, f_n)}{d(x, x_1, \cdots, x_n)}$$

(Jacobi-Muir.)

693. *If* $\Delta \equiv |a_{1n}|$ *represents an axisymmetric determinant, then*

$$\frac{d(A_{11}A_{22}\cdots A_{nn}A_{n-1,n}\cdots A_{12})}{d(a_{11}a_{22}\cdots a_{nn}a_{n-1,n}\cdots a_{12})} = (1-n)\left(\frac{|A_{1n}|}{|a_{1n}|}\right)^{(n+1)/2}$$

where A_{ij} *is the signed complementary minor of* a_{ij} *in* Δ.

For convenience sake in writing take $n=4$, and

$$\Delta \equiv \begin{vmatrix} a_1 & a_2 & a_3 & a_4 \\ a_2 & b_2 & b_3 & b_4 \\ a_3 & b_3 & c_3 & c_4 \\ a_4 & b_4 & c_4 & d_4 \end{vmatrix}$$

then J or

$$\frac{d(A_1B_2C_3D_4C_4B_4B_3A_4A_3A_2)}{d(a_1b_2c_3d_4c_4b_4b_3a_4b_3a_2)}$$

$$= \begin{vmatrix} & c_3d_4 - c_4^2 & b_2d_4 - b_4^2 & \cdots & . \\ c_3d_4 - c_4^2 & & a_1d_4 - a_4^2 & \cdots & . \\ b_2d_4 - b_4^2 & a_1d_4 - a_4^2 & & \cdots & 2(a_4b_4 - a_2d_4) \\ \cdots & \cdots & \cdots & \cdots & \cdots \\ & & a_4b_4 - a_2d_4 & \cdots & c_4^2 - c_3d_4 \end{vmatrix} \equiv \Delta'$$

say.

Taking the product

$$\Delta' \cdot \begin{vmatrix} a_1^2 & a_2^2 & a_3^2 & \cdots & a_1a_2 \\ a_2^2 & b_2^2 & b_3^2 & \cdots & a_2b_2 \\ \cdots & \cdots & \cdots & \cdots & \cdots \\ 2a_1a_2 & 2a_2b_2 & 2a_3b_3 & \cdots & a_1b_2 + a_2b_1 \end{vmatrix}$$

we readily obtain

$$a_1A_1 \cdot b_2B_2 \cdot c_3C_3 \cdots 2a_3A_3 \cdot 2a_2A_2$$

$$\times \begin{vmatrix} 1 - \dfrac{\Delta}{a_1A_1} & 1 & \cdots & 1 \\ 1 & 1 - \dfrac{\Delta}{b_2B_2} & \cdots & 1 \\ \cdots & \cdots & \cdots & \cdots \\ 1 & 1 & \cdots & 1 - \dfrac{\Delta}{2a_2A_2} \end{vmatrix}$$

But the second factor on the left is equal to Δ^5 by §170 and the determinant on the right is equal to

$$\frac{\Delta^{10}}{a_1A_1\cdot b_2B_2\cdots 2a_2A_2}\left\{1-\frac{a_1A_1+b_2B_2+\cdots+2a_2A_2}{\Delta}\right\}$$

by §75 and therefore

$$J = \Delta^5(1-4).$$

The reasoning is true in general.

694. The circulant has n distinct elements and if we take the Jacobian of the signed primary minors of the circulant corresponding to these elements its value may readily be found by multiplying the Jacobian by the circulant. Thus

$$\frac{d(A_1A_2\cdots A_n)}{d(a_1a_2\cdots a_n)}\cdot\begin{vmatrix} a_1 & a_2\cdots & a_n \\ a_2 & a_3\cdots & a_1 \\ \cdot & \cdots & \cdot \\ a_n & a_1\cdots & a_{n-1}\end{vmatrix}$$

gives

$$\begin{vmatrix} (n-1)A_1 & -A_n & -A_{n-1}\cdots & -A_2 \\ (n-1)A_2 & -A_1 & -A_n\quad\cdots & -A_3 \\ \cdot & \cdot & \cdots & \cdot \\ (n-1)A_n & -A_{n-1} & -A_{n-2}\cdots & -A_1\end{vmatrix} \text{ by } \S 493$$

Therefore

$$J(A_1\cdots A_n)\cdot C(a_1\cdots a_n) = (-1)^{n(n-1)/2}(n-1)\{C(a_1\cdots a_n)\}^{n-1}$$

or

$$J(A_1,\cdots,A_n) = (-1)^{n(n-1)/2}(n-1)\{C(a_1\cdots a_n)\}^{n-2}$$

Exercises. Set XXXIII

1. The square of the Jacobian of $2k$ binary quantics each of degree not less than $2k$ is a homogeneous quadratic function of the quantics, the coefficients of the function being sums of products of related covariants. (Paige.)

2. The Jacobian of $2k+1$ binary quantics each of degree not less than $2k+1$ is a homogeneous linear function of the quantics, the coefficients of the function being sums of products of related covariants. (Paige.)

3. Show that

$$J(u_1v_1, u_2v_2, \cdots, u_nv_n)$$

$$= \begin{vmatrix} u_1 & \cdot & \cdots & \cdot & -v_1 & \cdot & \cdots & \cdot \\ \cdot & u_2 & \cdots & \cdot & \cdot & -v_2 & \cdots & \cdot \\ \cdot & \cdot & \cdots & \cdot & \cdot & \cdot & \cdots & \cdot \\ \cdot & \cdot & \cdots & u_n & \cdot & \cdot & \cdots & -v_n \\ \dfrac{\partial u_1}{\partial x_1} & \dfrac{\partial u_2}{\partial x_1} & \cdots & \dfrac{\partial u_n}{\partial x_1} & \dfrac{\partial v_1}{\partial x_1} & \dfrac{\partial v_2}{\partial x_1} & \cdots & \dfrac{\partial v_n}{\partial x_1} \\ \cdot & \cdot & \cdots & \cdot & \cdot & \cdot & \cdots & \cdot \\ \dfrac{\partial u_1}{\partial x_n} & \dfrac{\partial u_2}{\partial x_n} & \cdots & \dfrac{\partial u_n}{\partial x_n} & \dfrac{\partial v_1}{\partial x_n} & \dfrac{\partial v_2}{\partial x_n} & \cdots & \dfrac{\partial v_n}{\partial x_n} \end{vmatrix}$$

4. If

$$u_r = (a_{r0} + a_{r1}x_1 + \cdots + a_{rn}x_n) \div (a_{00} + a_{01}x_1 + \cdots + a_{0n}x_n), \qquad (r = 1, 2, \cdots, n)$$

show that the Jacobian of the u's is equal to

$$| a_{00}a_{11} \cdots a_{nn} | \div (a_{00} + a_{01}x_1 + \cdots + a_{0n}x_n)^{n+1}.$$

695. *If* u_i, $(i=1, 2, \cdots, \overline{n+1})$ *are homogeneous functions of the* m*th degree in* x_1, x_2, $\cdots$, x_n, *then*

$$(1) \qquad \begin{vmatrix} u_1 & \dfrac{\partial u_1}{\partial x_1} & \cdots & \dfrac{\partial u_1}{\partial x_n} \\ u_2 & \dfrac{\partial u_2}{\partial x_1} & \cdots & \dfrac{\partial u_2}{\partial x_n} \\ \cdot & \cdot & \cdots & \cdot \\ u_{n+1} & \dfrac{\partial u_{n+1}}{\partial x_1} & \cdots & \dfrac{\partial u_{n+1}}{\partial x_n} \end{vmatrix} = 0.$$

For by Euler's theorem $x_1 \operatorname{col}_2 + x_2 \operatorname{col}_3 + \cdots + x_n \operatorname{col}_{n-1} = m \cdot \operatorname{col}_1$.

If

$$v_i = (-1)^{i-1} J(u_1 \cdots u_{i-1}u_{i+1} \cdots u_{n+1}), \qquad (i = 1, 2, \cdots, \overline{n+1})$$

then (1) may be written

$$(2) \qquad u_1v_1 + u_2v_2 + \cdots + u_{n+1}v_{n+1} = 0.$$

Differentiating (2) with respect to the x's we have

$$(3)\quad \begin{cases} u_1\dfrac{\partial v_1}{\partial x_1}+u_2\dfrac{\partial v_2}{\partial x_1}+\cdots+u_{n+1}\dfrac{\partial v_{n+1}}{\partial x_1}=0 \\ u_1\dfrac{\partial v_1}{\partial x_2}+u_2\dfrac{\partial v_2}{\partial x_2}+\cdots+u_{n+1}\dfrac{\partial v_{n+1}}{\partial x_2}=0 \\ \cdots\cdots\cdots\cdots\cdots\cdots \\ u_1\dfrac{\partial v_1}{\partial x_n}+u_2\dfrac{\partial v_2}{\partial x_n}+\cdots+u_{n+1}\dfrac{\partial v_{n+1}}{\partial x_n}=0. \end{cases}$$

For the derivative with respect to x_1 is

$$\sum u_h\frac{\partial v_h}{\partial x_1}+\sum v_h\frac{\partial u_h}{\partial x_1}=0$$

and the second sum is zero because it is the determinant in (1) with one of the other columns put in the place of the first column.

In (3) we have n homogeneous equations in $n+1$ u's and therefore

$$\frac{u_i}{u_k}=\frac{w_i}{w_k}$$

where the w's are Jacobians formed from the v's as the v's are Jacobians formed from the u's.

The general theorem may be stated as follows: *If $u_1\cdots u_{n+1}$, be homogeneous integral functions of the mth degree in the n variables $x_1x_2\cdots x_n$; and $v_1\cdots v_{n+1}$ be the set of Jacobians formed from the u's; and $w_1\cdots w_{n+1}$ be the set of Jacobians similarly formed from the v's, then the w's differ from the u's by a common factor only, that is*

$$W_k=Mu_k,\qquad (k=1,2,\cdots,\overline{n+1})$$

696. A similar theorem to that in the last article, when stated for four functions, is: *If u_1, u_2, u_3, u_4 be homogeneous integral functions of the mth degree in x, y, and*

$$v_1=\begin{vmatrix} \dfrac{\partial^2u_2}{\partial x^2} & \dfrac{\partial^2u_2}{\partial x\partial y} & \dfrac{\partial^2u_2}{\partial y^2} \\ \dfrac{\partial^2u_3}{\partial x^2} & \dfrac{\partial^2u_3}{\partial x\partial y} & \dfrac{\partial^2u_3}{\partial y^2} \\ \dfrac{\partial^2u_4}{\partial x^2} & \dfrac{\partial^2u_4}{\partial x\partial y} & \dfrac{\partial^2u_4}{\partial y^2} \end{vmatrix}=R(u_2u_3u_4)\text{ say,}$$

$$v_2 = -R(u_1u_3u_4), \quad v_3 = R(u_1u_2u_4), \quad v_4 = -R(u_1u_2u_3),$$

then

$$u_1v_1 + u_2v_2 + u_3v_3 + u_4v_4 = 0.$$

For this last equation is the same as

$$\begin{vmatrix} u_1 & \frac{\partial^2 u_1}{\partial x^2} & \frac{\partial^2 u_1}{\partial x \partial y} & \frac{\partial^2 u_1}{\partial y^2} \\ u_2 & \frac{\partial^2 u_2}{\partial x^2} & \frac{\partial^2 u_2}{\partial x \partial y} & \frac{\partial^2 u_2}{\partial y^2} \\ u_3 & \frac{\partial^2 u_3}{\partial x^2} & \frac{\partial^2 u_3}{\partial x \partial y} & \frac{\partial^2 u_3}{\partial y^2} \\ u_4 & \frac{\partial^2 u_4}{\partial x^2} & \frac{\partial^2 u_4}{\partial x \partial y} & \frac{\partial^2 u_4}{\partial y^2} \end{vmatrix} = 0$$

as is seen on observing that

$$x^2 \operatorname{col}_2 + 2xy \operatorname{col}_3 + y^2 \operatorname{col}_4 = m(m-1) \operatorname{col}_1.$$

EXERCISES: Besides $\sum u_1v_1 = 0$ establish the following

$$\sum \frac{\partial u_1}{\partial x} \frac{\partial v_1}{\partial x} = 0$$

$$\sum \frac{\partial u_1}{\partial y} \frac{\partial v_1}{\partial y} = 0$$

$$\sum v_1 \frac{\partial u_1}{\partial x} = 0 \qquad \sum u_1 \frac{\partial v_1}{\partial x} = 0 \qquad \sum v_1 \frac{\partial u_1}{\partial y} = 0 \qquad \sum u_1 \frac{\partial v_1}{\partial y} = 0$$

$$\sum v_1 \frac{\partial^2 u_1}{\partial x^2} = 0 \qquad \sum u_1 \frac{\partial^2 v_1}{\partial y^2} = 0 \qquad \sum v_1 \frac{\partial^2 u_1}{\partial x \partial y} = 0 \qquad \sum u_1 \frac{\partial^2 v_1}{\partial x \partial y} = 0$$

$$\sum v_1 \frac{\partial^2 u_1}{\partial y^2} = 0 \qquad \sum u_1 \frac{\partial^2 v_1}{\partial y^2} = 0 \qquad \sum \frac{\partial v_1}{\partial y} \frac{\partial u_1}{\partial x} = 0 \qquad \sum \frac{\partial v_1}{\partial x} \frac{\partial u_1}{\partial y} = 0$$

697. *If the differential coefficients of $y_1y_2 \cdots y_n$ with respect to x be proportional to the differential coefficients with respect to z, the common ratio being R, and if the y's be also functions of $w_1, w_2, \cdots, w_n$, then*

$$\frac{\partial}{\partial x}\left\{\frac{d(y_1 \cdots y_n)}{d(w_1 \cdots w_n)}\right\} = \frac{\partial}{\partial z}\left\{R\frac{d(y_1 \cdots y_n)}{d(w_1 \cdots w_n)}\right\} - \frac{d(y_1 \cdots y_nR)}{d(w_1 \cdots w_nz)}.$$

To show this it will be sufficient to take $n=3$ and the reasoning used will be applicable in general.

The first term on the right is, on performing the differentiation indicated, equal to

$$\frac{d(y_1y_2y_3)}{d(w_1w_2w_3)}\cdot\frac{\partial R}{\partial z}+R\frac{\partial}{\partial z}\left\{\frac{d(y_1y_2y_3)}{d(w_1w_2w_3)}\right\}$$

and the second term on the right is readily seen to be equal to

$$-\left\{\frac{d(y_1y_2y_3)}{d(w_1w_2w_3)}\right\}\cdot\frac{\partial R}{\partial z}-\begin{vmatrix}\frac{\partial y_1}{\partial w_1} & \frac{\partial y_1}{\partial w_2} & \frac{\partial y_1}{\partial w_3} & \frac{\partial y_1}{\partial z}\\ \frac{\partial y_2}{\partial w_1} & \frac{\partial y_2}{\partial w_2} & \frac{\partial y_2}{\partial w_3} & \frac{\partial y_2}{\partial z}\\ \frac{\partial y_3}{\partial w_1} & \frac{\partial y_3}{\partial w_2} & \frac{\partial y_3}{\partial w_3} & \frac{\partial y_3}{\partial z}\\ \frac{\partial R}{\partial w_1} & \frac{\partial R}{\partial w_2} & \frac{\partial R}{\partial w_3} & \cdot\end{vmatrix}.$$

The right-hand side reduces therefore to

$$R\frac{\partial}{\partial z}\left\{\frac{d(v_1v_2v_3)}{d(w_1w_2w_3)}\right\}-\Delta \quad \text{say}.$$

The left-hand side is the sum of three determinants which we may denote by

$$\sum\begin{vmatrix}\frac{\partial}{\partial x}\frac{\partial y_1}{\partial w_1} & \frac{\partial y_1}{\partial w_2} & \frac{\partial y_1}{\partial w_3}\\ \frac{\partial}{\partial x}\frac{\partial y_2}{\partial w_1} & \frac{\partial y_2}{\partial w_2} & \frac{\partial y_2}{\partial w_3}\\ \frac{\partial}{\partial x}\frac{\partial y_3}{\partial w_1} & \frac{\partial y_3}{\partial w_2} & \frac{\partial y_3}{\partial w_3}\end{vmatrix}\equiv\sum D_x \quad \text{say}.$$

But

$$\frac{\partial}{\partial x}\frac{\partial y_r}{\partial w_s}=\frac{\partial}{\partial w_s}\frac{\partial y_r}{\partial x}$$
$$=\frac{\partial}{\partial w_s}\left\{R\frac{\partial y_r}{\partial z}\right\}$$
$$=R\frac{\partial^2 y_r}{\partial w_s\partial z}+\frac{\partial y_r}{\partial z}\frac{\partial R}{\partial w_s}$$

and substituting in the D's we get

$$\sum D_x = R\sum D_z + \sum \begin{vmatrix} \frac{\partial R}{\partial w_1} & \frac{\partial y_1}{\partial z} & \frac{\partial y_1}{\partial w_2} & \frac{\partial y_1}{\partial w_3} \\ \frac{\partial R}{\partial w_1} & \frac{\partial y_2}{\partial z} & \frac{\partial y_2}{\partial w_2} & \frac{\partial y_2}{\partial w_3} \\ \frac{\partial R}{\partial w_1} & \frac{\partial y_3}{\partial z} & \frac{\partial y_3}{\partial w_2} & \frac{\partial y_3}{\partial w_3} \end{vmatrix}.$$

The first term here is the same as the first term in the reduced right-hand side and the second here is seen to be condensable into the second on the right.

The theorem is thus established.

698. Jacobians occur in the problem of changing the variables in a multiple integral. Thus if in

$$\dot{I} = \int\int \cdots f(y_1 \cdots y_n)dy_1dy_2 \cdots dy_n$$

we wish to change to an integral with respect to $x_1x_2 \cdots x_n$, where the y's are functions of the x's, the result will be found to be

$$\dot{I} = \int\int \cdots f(x_1)\frac{d(y_1 \cdots y_n)}{d(x_1 \cdots x_n)}dx_1dx_2 \cdots dx_n,$$

where $f(x)$ is the result of substituting for the y's in terms of x's in $f(y_1 \cdots y_n)$.

699. *If* $y_1y_2 \cdots y_n$ *are functions of the variables* $x_1x_2 \cdots x_n$ *and if the variables are transformed by the substitution*

$$x_i = a_{i1}z_1 + a_{i2}z_2 + \cdots + a_{in}z_n \qquad (i = 1,2, \cdots, n)$$

and if $y_1', y_2' \cdots y_n'$ *be what* $y_1y_2 \cdots y_n$ *become in consequence, then*

$$\frac{d(y_1'y_2' \cdots y_n')}{d(z_1z_2 \cdots z_n)} = \frac{d(y_1y_2 \cdots y_n)}{d(x_1x_2 \cdots x_n}\left| a_{1n} \right|$$

where $|a_{1n}|$ *is the modulus of the substitution.*

For

$$\frac{\partial y_i}{\partial z_k} = \frac{\partial y_i}{\partial x_1}\frac{\partial x_1}{\partial z_k} + \frac{\partial y_i}{\partial x_2}\frac{\partial x_2}{\partial z_k} + \cdots + \frac{\partial y_i}{\partial x_n}\frac{\partial x_n}{\partial z_k}$$

$$= a_{ik}\frac{\partial y_i}{\partial x_1} + \cdots + a_{nk}\frac{\partial y_i}{\partial x_n}$$

and we see the truth of the theorem. Jacobians are thus seen to be covariants.

If the y's are linear functions thus

$$y_i = a_{i1}x_1 + a_{i2}x_2 + \cdots + a_{in}x_n$$

then obviously

$$\frac{d(y_1 \cdots y_n)}{d(x_1 \cdots x_n)} = | a_{1n} | .$$

EXERCISES. 1. If u, v, w, are functions of x, y, z, show that the problems of determining $\theta(x, y, z)$ so that $J(\theta u, \theta v, \theta w) = \theta^3 J(y, v, w)$ is the same as solving

$$P\frac{\partial\theta}{\partial x} + Q\frac{\partial\theta}{\partial y} + R\frac{\partial\theta}{\partial z} = 0.$$

2. If $$D(v; u_1, u_2 \cdots u_n) \equiv \begin{vmatrix} \cdot & \frac{\partial v}{\partial x_1} & \frac{\partial v}{\partial x_2} & \cdots & \frac{\partial v}{\partial x_n} \\ u_1 & \frac{\partial u_1}{\partial x_1} & \frac{\partial u_1}{\partial x_2} & \cdots & \frac{\partial u_1}{\partial x_n} \\ \cdot & \cdot & \cdot & \cdot & \cdot \\ u_n & \frac{\partial u_n}{\partial x_1} & \frac{\partial u_n}{\partial x_2} & \cdots & \frac{\partial u_n}{\partial x_n} \end{vmatrix}$$

Show that $D = K(v_1, u_1, \cdots, u_n) - uJ(u_1, \cdots, u_n)$ and generalize when the v's are m in number (Usai).

3. If u and v are binary quadrics and J_1 their Jacobian then if the Jacobian of u and J_1 is J_2, and of v and J_1 is J_2' and so on show that there are but five, namely, u, v, J_1, J_2, J_2' independent numbers of the series. (Hayashi.)

CHAPTER XVII

Hessians

700. The Jacobian of the first differential coefficients of a function of n variables is called the *Hessian* of the function; in symbols

$$H(u) = J\left(\frac{\partial u}{\partial x_1}, \frac{\partial u}{\partial x_2}, \cdots, \frac{\partial u}{\partial x_n}\right) = \left| \frac{\partial^2 u}{\partial x_1^2} \; \frac{\partial^2 u}{\partial x_2^2} \; \cdots \; \frac{\partial^2 u}{\partial x_n^2} \right|.$$

On account of the relation

$$\frac{\partial^2 u}{\partial x_i \partial x_j} = \frac{\partial^2 u}{\partial x_j \partial x_i}$$

it is apparent that the Hessian is a symmetric determinant.

It is also apparent that if the differential coefficients of u are not independent the Hessian vanishes.

701. *A Hessian is a covariant.* For since it is the Jacobian of

$$\frac{\partial u}{\partial x_1}, \frac{\partial u}{\partial x_2}, \cdots \frac{\partial u}{\partial x_n}$$

we have, if u' be the transformed function and

$$x_i = a_{i1}z_1 + a_{i2}z_2 + \cdots + a_{in}z_n \qquad (i = 1, 2, \cdots, n)$$

$$H(u') = \frac{d\left(\dfrac{\partial u'}{\partial z_1} \cdots \dfrac{\partial u'}{\partial z_n}\right)}{d(z_1 \cdots z_n)}$$

But

$$\frac{\partial}{\partial z_i} = a_{1i}\frac{\partial}{\partial x_1} + a_{2i}\frac{\partial}{\partial x_2} + \cdots,$$

and

$$\frac{\partial^2}{\partial x_i \partial z_i} = \frac{\partial^2}{\partial z_i \partial x_i}$$

therefore

$$H(u) \cdot |a_{1n}|^2 = \frac{d\left(\dfrac{\partial u}{\partial x_1} \cdots \dfrac{\partial u}{\partial x_n}\right)}{d(x_1 \cdots x_n)} |a_{1n}|^2$$

$$= \frac{d\left(\frac{\partial u}{\partial x_1} \cdots \frac{\partial u}{\partial x_n}\right)}{d(z_1 \cdots z_n)} \mid a_{1n} \mid$$

$$= \frac{d\left(\frac{\partial u'}{\partial z_1} \cdots \frac{\partial u'}{\partial z_n}\right)}{d(z_1 \cdots z_n)}$$

$$= H(u').$$

702. *If u is a function of $x_1 x_2 \cdots x_n$ and we use u_r to denote $\partial u/\partial x_r$, and u_{rs} to denote $\partial^2 y/\partial x_s \partial x_r$ and if $c_1 u_1 + c_2 u_2 + \cdots + c_n u_n = 0$ where the c's are constant, then the Hessian of u vanishes and u may be transformed into a function with one less variable.*

That $H(u)$ vanishes under the conditions given follows from §686 and from the substitution

$$x_k = a_{k1} y_1 + a_{k2} y_2 + \cdots + a_{k,n-1} y_{n-1} + c_k y_n.$$

Hence

$$\frac{du}{dy_n} = \frac{\partial u}{\partial x_1} \frac{\partial x_1}{\partial y_n} + \cdots + \frac{\partial u}{\partial x_n} \frac{\partial x_n}{\partial y_n}$$

$$= c_1 u_1 + c_2 u_2 + \cdots + c_n u_n = 0.$$

That is u does not contain y_n, but is a function of $y_1 y_2 \cdots y_{n-1}$ only.

It is similarly seen that if we have a second such equation with constant coefficients, namely

$$d_1 u_1 + d_2 u_2 + \cdots d_n u_n = 0$$

then n is transformable into a function with two fewer variables.

Exercise: Show that if it is possible to transform u by a linear substitution into a function with one less variable then $H(u)$ vanishes and if it is changed into a function with two less variables then $H(u)$ and all its primary minors vanish.

703. The theorem that if the Hessian of a function u vanish identically, then it is possible to transform u by a linear transformation into a function with one less variable has been shown to be true only for binary, ternary, and quaternary quantics.* (Gordon-Noether.)

* The n-ary quadric might be included.

Thus for u a binary quantic we have

$$x_1 u_{11} + x_2 u_{12} = (m-1)u_1$$
$$x_1 u_{21} + x_2 u_{22} = (m-1)u_2$$

and if $H(u)=0$, then

$$\frac{u_{11}}{u_{21}} = \frac{u_{12}}{u_{22}} = \frac{u_1}{u_2}$$

and consequently $c_1u_1+c_2u_2=0$.

704. From the equations

$$\begin{aligned} x_1u_1 + x_2u_2 + \cdots + x_nu_n &= mu \\ x_1u_{11} + x_2u_{21} + \cdots + x_nu_{n1} &= (m-1)u_1 \\ \cdots\cdots\cdots&\cdots\cdots \\ x_1u_{1n} + x_2u_{2n} + \cdots + x_nu_{nn} &= (m-1)u_n \end{aligned}$$

we get

$$\begin{aligned} -(m-1):x_1:x_2:\cdots:x_n &= H:v_1:v_2:\cdots:v_n, \\ &= v_1:v_{11}:v_{21}:\cdots:v_{n1}, \\ &\cdots\cdots\cdots \\ &= v_n:v_{1n}:v_{2n}:\cdots:v_{nn}, \end{aligned}$$

where H is the Hessian of u and the v's are the cofactors of the corresponding u's in the resultant of the set of equations.

From these we get

$$v_r = -\frac{x_r}{m-1}\cdot H$$

and

$$v_{sr} = -\frac{v_r x_s}{(m-1)} = \frac{x_r x_s}{(m-1)^2}\cdot H.$$

These show *that all the primary minors of the vanishing determinant*

$$\Delta \equiv \begin{vmatrix} \dfrac{m}{m-1}u & u_1 & u_2 & \cdots & u_n \\ u_1 & u_{11} & u_{12} & \cdots & u_{1n} \\ \cdot & \cdot & \cdot & \cdots & \cdot \\ u_n & u_{n1} & u_{n2} & \cdots & u_{nn} \end{vmatrix}$$

have the Hessian as a factor and consequently if the Hessian vanishes then all the primary minors vanish.

705. *If u be a homogeneous integral function of the mth degree in $r+1$ variables $x_0 x_1 \cdots x_r$, the Hessian of which with respect to these variables is H_{r+1} and with respect to the variables $x_1 x_2 \cdots x_r$ is H_r, then the determinant which is the result of bordering H_r by prefixing*

$$0, \frac{\partial u}{\partial x_1}, \cdots, \frac{\partial u}{\partial x_r}$$

as a first row and as a first column is equal to

$$-\frac{m}{m-1} u \cdot H_r + \frac{x_0^2}{(m-1)^2} H_{r+1}.$$

Since by Euler's theorem

$$\begin{aligned}(m-1)u_j &= x_0 u_{j0} + x_1 u_{j1} + \cdots + x_r u_{jr} \\ (m-1)u_i &= x_0 u_{0i} + x_1 u_{1i} + \cdots + x_r u_{ri}\end{aligned} \qquad \left(\begin{matrix} i \\ j \end{matrix} = 0,1,2, \cdots, r\right)$$

the bordered determinant is seen to be equal to

$$\frac{1}{m-1}\begin{vmatrix} x_0 u_0 - mu & u_1 & u_2 & \cdots & u_r \\ x_0 u_{10} & u_{11} & u_{12} & \cdots & u_{1r} \\ \cdot & \cdot & \cdot & \cdots & \cdot \\ x_0 u_{r0} & u_{r1} & u_{r2} & \cdots & u_{rr} \end{vmatrix}$$

which is equal to

$$-\frac{m}{m-1} u \cdot H_r + \frac{x_0}{m-1}\begin{vmatrix} u_0 & u_1 & \cdots & u_r \\ u_{10} & u_{11} & \cdots & u_{1r} \\ \cdot & \cdot & \cdots & \cdot \\ u_{r0} & u_{r1} & \cdots & u_{rr} \end{vmatrix}$$

$$= -\frac{m}{m-1} \cdot u \cdot H_r + \frac{x_0^2}{(m-1)^2} \cdot H_{r+1}$$

as required.

If $x_0 = 0$ then the bordered Hessian is equal to $-m/m-1 \cdot u \cdot H_r$.

706. From the equations

$$x_1 u_{1i} + x_2 u_{2i} + \cdots + x_n u_{ni} = (m-1)u_i \qquad (i = 1, 2, \cdots n)$$

we get on solving

$$(1) \qquad x_i : x_j = U_{ri} : U_{rj}$$

and

$$(2) \qquad x_i H = (m-1)\{U_{i1}u_1 + \cdots + U_{in}u_n\}$$

which on differentiating both sides of (2) with respect to x_k $(k \neq i)$ gives

$$x_i \frac{\partial H}{\partial x_k} = (m-1)\left\{u_1 \frac{\partial U_{i1}}{\partial x_k} + u_2 \frac{\partial U_{i2}}{\partial x_k} + \cdots + u_n \frac{\partial U_{in}}{\partial x_k}\right\}$$

since $U_{i1}u_{k1} + \cdots + U_{in}u_{kn} = 0$ when $k \neq i$. Differentiating again with respect to h, where $h \neq k \neq i$, we get

$$x_i \frac{\partial^2 H}{\partial x_h \partial x_k} = (m-1)\left\{u_1 \frac{\partial^2 U_{i1}}{\partial x_h \partial x_k} + \cdots + u_n \frac{\partial^2 U_{in}}{\partial x_h \partial x_k}\right\}$$
$$+ (m-1)\left\{u_{h1} \frac{\partial U_{i1}}{\partial x_k} + \cdots + u_{hn} \frac{\partial U_{in}}{\partial x_k}\right\}.$$

But if $h \neq i$ we have

$$u_{h1}U_{i1} + \cdots + u_{hn}U_{in} = 0$$

which by differentiation gives

$$u_{h1} \frac{\partial U_{i1}}{\partial x_k} + \cdots + u_{hn} \frac{\partial U_{in}}{\partial x_k}$$
$$+ U_{i1} \frac{\partial u_{h1}}{\partial x_k} + \cdots + U_{in} \frac{\partial u_{hn}}{\partial x_k} = 0$$

Consequently by substitution we have

$$(3) \qquad x_i \frac{\partial^2 H}{\partial x_h \partial x_k} = (m-1)\left\{u_1 \frac{\partial^2 U_{i1}}{\partial x_h \partial x_k} + \cdots + u_n \frac{\partial^2 U_{in}}{\partial x_h \partial x_k}\right\}$$
$$- (m-1)\left\{U_{i1} \frac{\partial u_{h1}}{\partial x_k} + \cdots + U_{in} \frac{\partial u_{hn}}{\partial x_k}\right\}.$$

From (1) we see that

$$\frac{U_{ij}}{U_{hk}} = \frac{x_i x_j}{x_h x_k} \quad \text{or} \quad U_{ij} = K \cdot x_i x_j$$

where K is constant. Making this substitution in (3) we get

$$x_i\frac{\partial^2 H}{\partial x_h \partial x_k} = -(m-1)K\left\{x_i x_1 \frac{\partial u_{kh}}{\partial x_1} + \cdots + x_i x_n \frac{\partial u_{kh}}{\partial x_n}\right\}$$

$$= -(m-1)Kx_i(m-2)u_{kh}$$

whence

$$\frac{\partial^2 H}{\partial x_h \partial x_k} = -(m-1)(m-2)K\cdot\frac{\partial^2 u}{\partial x_k \partial x_h}\cdot$$

We have then the theorem: *If the first differential-quotients of a homogeneous rational integral function all vanish, the elements of the Hessian of the function are proportional to the elements of the Hessian of the Hessian.*

Exercises. Set XXXIV

1. If u_{33} is a ternary cubic show that

$$H\{H(u_{33})\} = 3S^3 u_{33} - 2TH(u_{33})$$

where S and T are the two invariants of the ternary cubic.

2. If

$$u_{24} = ax^4 + 4bx^3y + 6cx^2y^2 + 4dxy^3 + ey^4$$

then

$$H\{H(u_{24})\} = 12^3\{432Iu_{24} - JH(u_{24})\}$$

where

$$I = ace + 2bcd - ad^2 - eb^2 - c^3,$$

$$J = ae + 3c^2 + 4bd.$$

3. Show that the adjugate of Δ §704 vanishes independently of the value of the Hessian.

4. If the ternary cubic u_{33} is such that $u_{33}/H(u_{33})$ is a constant, then show that $H(u_{33})$ is resolvable into linear factors.

5. Show that if an integral function of n variables be divisible by the mth power of a function of the same variables the Hessian of the former is divisible by the $n(m-1)$th power of the latter.

Also show that if a function of n variables be divisible by a function of μ of these variables, the Hessian of the former is divisible by the $(n-2\mu)$th power of the latter. (Casorati.)

6. If u be a binary n-thic show that if u have p roots each equal to α, the Hessian must have $2(p-1)$ roots equal to α.

7. If $\alpha_1\alpha_2 \cdots \alpha_n$ be the roots of a binary quantic, its Hessian is an arithmetical multiple of

$$- \sum(\alpha_1 - \alpha_2)^2(x - \alpha_3)^2(x - \alpha_4)^2 \cdots (x - \alpha_n)^2$$

and hence if m of the roots of the quantic be identical and all the rest different, the number of imaginary roots of the Hessian is $2(n-m-1)$.

8. If

$$u = \sum(x_r x_s)^2 \quad \left(\begin{matrix} r \\ s \end{matrix} = 1, 2, \cdots, n \quad r \neq s\right)$$

and

$$\sigma = \tfrac{1}{3} \sum x_r^2 \qquad (r = 1, 2, \cdots, n).$$

show that

$$H(u) = \sigma^n(\sigma - x_1^2)(\sigma - x_2^2) \cdots (\sigma - x_n^2)\left\{1 + \frac{2}{3} \sum \frac{x_1^2}{\sigma - x_1^2}\right\}.$$

9. If $u_1, u_2, \cdots, u_n$ be linear functions of $x_1x_2 \cdots x_n$ show that

$$H(\log u_1u_2 \cdots u_n) = (-1)^n \left\{\frac{d(u_1u_2 \cdots u_n)}{d(x_1x_2 \cdots x_n)}\right\}^2 \quad (u_1u_2 \cdots u_n)^{-2}$$

10. If

$$u_r = a_{r1}x_1 + a_{r2}x_2 + \cdots + a_{rk}x_k \qquad (r = 1, 2, \cdots, n)$$

then

$$H(u_1u_2 \cdots u_n) = (-1)^{k-1}(n-1)(u_1u_2 \cdots u_n)^{k-2} \sum \left\{ |a_{ik}| u_{k+1} \cdots u_n \right\}^2$$

11. If $\phi(u, x_1x_2 \cdots x_n) = 0$ then the Hessian of u with respect to the x's is

$$\frac{(-1)^{n-1}}{\phi_0^{n+2}} \begin{vmatrix} \phi_{00} & \phi_{01} & \cdots & \phi_{0n} & \phi_0 \\ \phi_{10} & \phi_{11} & \cdots & \phi_{1n} & \phi_1 \\ \cdot & \cdot & \cdot & \cdot & \cdot \\ \phi_{n0} & \phi_{n1} & \cdots & \phi_{nn} & \phi_n \\ \phi_0 & \phi_1 & \cdots & \phi_n & \cdot \end{vmatrix}$$

where the suffixes $0, 1, 2, \cdots, n$ denote differentiation with respect to $u, x_1 \cdots x_n$ respectively.

707. Besides the Hessian $H(u)$ bordered by the differential-quotients $u_1, u_2, \cdots$, we may have an additional border. Thus for

$$\Delta \equiv \begin{vmatrix} u_{11} & u_{12} & u_{13} & u_1 & \alpha_1 \\ u_{21} & u_{22} & u_{23} & u_2 & \alpha_2 \\ u_{31} & u_{32} & u_{33} & u_3 & \alpha_3 \\ u_1 & u_2 & u_3 & 0 & 0 \\ \alpha_1 & \alpha_2 & \alpha_3 & 0 & 0 \end{vmatrix}$$

if we perform the operations

$$(n-1)\,\mathrm{col}_4 - x_1\,\mathrm{col}_1 - x_2\,\mathrm{col}_2 - x_3\,\mathrm{col}_3,$$

followed by

$$(n-1)\,\mathrm{row}_4 - x_1\,\mathrm{row}_1 - x_2\,\mathrm{row}_2 - x_3\,\mathrm{row}_3,$$

we get

$$\Delta = -\frac{n}{n-1}u \begin{vmatrix} u_{11} & u_{12} & u_{13} & \alpha_1 \\ u_{21} & u_{22} & u_{23} & \alpha_2 \\ u_{31} & u_{32} & u_{33} & \alpha_3 \\ \alpha_1 & \alpha_2 & \alpha_3 & \end{vmatrix} - \frac{1}{(n-1)^2}\left(\sum \alpha_1 x_1\right)^2 H(u).$$

Similarly for further bordering.

CHAPTER XVIII

WRONSKIANS

708. If there be n functions of one and the same variable x, the determinant which has in every case the element in its rth row and sth column, the $(r-1)$st differential coefficient of the sth function may be called the *Wronskian* of the functions with respect to x. Thus the Wronskian of y_1, y_2, y_3 is

$$\begin{vmatrix} y_1 & y_2 & y_3 \\ \dfrac{dy_1}{dx} & \dfrac{dy_2}{dx} & \dfrac{dy_3}{dx} \\ \dfrac{d^2y_1}{dx^2} & \dfrac{d^2y_2}{dx^2} & \dfrac{d^2y_3}{dx^2} \end{vmatrix}$$

which we will denote by $W_x(y_1y_2y_3)$.

709. It is apparent that the only non-vanishing term in the differential coefficient of a Wronskian is the one got by differentiating each element of the last row. Thus if $y_{r(s)}$ denotes the sth differential coefficient of y_r, then

$$W_x(y_1y_2 \cdots y_n) \equiv | \, y_1y_{2(1)}y_{3(2)} \cdots y_{n(n-1)} \, |$$

and

$$\frac{d}{dx}W_x(y_1y_2 \cdots y_n) = | \, y_1y_{2(1)} \cdots y_{n-1(n-2)}y_{n(n)} \, |$$

710. *If* $y_1, y_2, \cdots, y_n$ *be functions of* x *and* x *be a function of* t, *then*

$$W_x(y_1y_2 \cdots y_n) = \left(\frac{dt}{dx}\right)^{n(n-1)/2} W_t(y_1y_2 \cdots y_n).$$

If we change the independent variables in each element of the left-hand member we readily obtain the result on the right by the application of §§39, 57.

711. By the use of §§39, 57 it is readily seen that

$$W(yy_1, yy_2, \cdots, yy_n) = y^nW(y_1y_2 \cdots y_n).$$

This may also be seen by using for y^n the determinant

$$\begin{vmatrix} y & \cdot & \cdot & \cdots \\ y' & y & \cdot & \cdots \\ y'' & 2y' & y & \cdots \\ \cdots & \cdots & \cdots & \cdots \\ y^{(n-1)} & (n-1)y^{(n-2)} & \frac{1}{2}(n-1)(n-2)y^{(n-3)} & \cdots\ y \end{vmatrix}$$

and multiplying row-by-column fashion.

If in the theorem we put $y=1/y_1$ and observe that

$$\frac{d}{dx}\frac{y_h}{y_1}=\frac{W(y_1y_h)}{y_1^2}$$

then it becomes

$$W(y_1\,y_2\cdots y_n)=\frac{1}{y_1^{n-2}}W\{W(y_1y_2),W(y_1y_3)\cdots W(y_1y_n)\}.$$

EXERCISE: Show that

$$W\left(\frac{y_1}{z},\frac{y_2}{z},\cdots,\frac{y_n}{z}\right)=\frac{1}{z^n}W(y_1y_2\cdots y_n).$$

712. *If* $y_1, y_2, \cdots, y_n$ *be functions of one and the same variable, then*

$$\begin{aligned} &W(y_1,y_2,\cdots,y_m,\cdots,y_n)\\ &=W\{W(y_1\cdots y_my_{m+1}),W(y_1\cdots y_m,y_{m+2}),\cdots,\\ &\quad W(y_1\cdots y_my_n)\}\div W(y_1\cdots y_m)^{n-m-1}. \end{aligned}$$

This is obviously an example of Sylvester's theorem §197 obtained by bordering the minor $|y_1y_{2(1)}\cdots y_{m(m-1)}|$ in all possible ways with one of the remaining rows and columns.

As a particular case we have when $m=n-2$

$$W(y_1y_2\cdots y_n)=W\{W(y_1\cdots y_{n-2}y_{n-1}),\\ W(y_1\cdots y_{n-2}y_n)\}\div W(y_1\cdots y_{n-2})$$

or as we may write it

$$W(y,y_1\cdots y_n)W(y_1\cdots y_{k-1}y_{k+1}\cdots y_n)\\ =W\{W(y_1\cdots y_n),W(y,y_1\cdots y_{k-1}y_{k+1}\cdots y_n)\}.$$

This is equivalent to

$$W(y, y_1, \cdots, y_n)W(y_1, \cdots, y_{k-1}, y_{k+1}, \cdots, y_n)$$
$$= W(y_1, \cdots, y_n)W'(y, y_1, \cdots, y_{k-1}, y_{k+1} \cdots y_n)$$
$$- W'(y_1, \cdots, y_n)W(y, y_1, \cdots, y_{k-1}, y_{k+1}, \cdots, y_n)$$

or

$$\frac{W(y, y_1, \cdots, y_n)}{W(y_1, \cdots, y_n)} \cdot \frac{W(y_1, \cdots, y_{k-1}, y_{k+1}, \cdots, y_n)}{W(y_1, \cdots, y_n)}$$
$$= \frac{\partial}{\partial x} \cdot \frac{W(y, y_1 \cdots y_{k-1} y_{k+1} \cdots y_n)}{W(y_1, \cdots, y_n)}.$$

713. *If a set of n functions of the same variable be connected by a linear relation with coefficients which are constant with respect to the variable the Wronskian of the functions vanishes.*

Let the relation be

$$c_1 y_1 + c_2 y_2 + \cdots + c_n y_n = 0.$$

Differentiating $(n-1)$ times we have in all n equations from which to eliminate the c's. The result of elimination gives

$$W(y_1, y_2, \cdots, y_n) = 0.$$

It is here assumed that each of the functions has a finite derivative of each of the first $(n-1)$ orders at every point of the interval over which x ranges.

714. *If the n functions y_1, y_2, $\cdots$, y_n of a real variable x each have, at every point of the interval I over which x ranges, finite derivatives of the first $n-1$ orders, it is a sufficient condition for their linear dependence that the Wronskian vanishes identically, provided that at least one set of $n-1$ of the functions can be so selected that their Wronskian and its first derivative do not vanish together at any point of the interval I.*

Let D_1, D_2, $\cdots$, D_n be the cofactors of the elements in the last row of the Wronskian and let E_1, E_2, $\cdots$, E_n be the cofactors of the elements of the second last row of W.

It is readily seen that

$$D_i' = -E_i \quad (i = 1, 2, \cdots, n)$$

and

$$D_1 y_1 + D_2 y_2 + \cdots + D_n y_n = 0. \tag{1}$$

Since by hypothesis $W=0$, and not all the D's vanish we have

$$\frac{E_1}{D_1} = \frac{E_2}{D_2} = \cdots = \frac{E_n}{D_n}$$

Therefore

$$\frac{D_1'}{D_1} = \frac{D_2'}{D_2} = \cdots = \frac{D_n'}{D_n} = \lambda = \frac{D_1' D_1 + D_2' D_2 + \cdots + D_n' D_n}{D_1^2 + D_2^2 + \cdots + D_n^2}.$$

If M represents the matrix of the first $n-1$ rows of W, then

$$D_1^2 + D_2^2 + \cdots + D_n^2 = M^2 \neq 0,$$

and

$$D_1' D_1 + D_2' D_2 + \cdots + D_n' D_n = M \cdot M'.$$

Hence

$$\frac{D_1'}{D_1} = \frac{D_2'}{D_2} = \cdots = \frac{D_n'}{D_n} = \frac{M'}{M}$$

or

$$\frac{MD_i' - M'D_i}{M^2} = 0 \quad (i = 1, 2, \cdots, n)$$

from which we get

$$\frac{d}{dx}\left(\frac{D_i}{M}\right) = 0 \quad \text{or} \quad \frac{D_i}{M} = c_i.$$

Substituting in (1) we get on dividing out the M

$$c_1 y_1 + c_2 y_2 + \cdots + c_n y_n = 0$$

where the c's cannot all be zero.

715. Various forms of the test for linear dependence of the y's have been given by Peano, Vivanti, Bocher, Curtiss and others.

Curtiss' statement is: If $y_1, y_2, \cdots, y_n$ be functions (non-analytic) of x which at every point of the interval I have finite derivatives of the first k orders ($k \geqq n-1$) while the $(n-1)$-line determinants of the $(n-1)$-by-$(k+1)$ Wronskian array do not all vanish at any point of I; and if $W(y_1, y_2, \cdots, y_n)$ be identically zero, then the y's are linearly dependent.

Martinotte's statement is that if the y's are developable by Taylor series the vanishing of the Wronskian is the necessary and sufficient condition that the y's be linearly dependent.

Pascal has shown that the condition $W=0$ is equivalent to the condition

$$y_1 \int W_{(1)}\partial x - y_2 \int W_{(2)}\partial x + \cdots + (-1)^{n-1} y_n \int W_{(n)}\partial x = 0,$$

where $W_{(r)}$ stands for the Wronskian of all the y's except y_r.

716. We have assumed in the foregoing theorems that x was real, but similar theorems are true when x is complex. Thus for the theorem of §714 we have: *The identical vanishing of the Wronskian of n analytic functions of a complex variable is a necessary and sufficient condition for their linear dependence.*

717. If we have the differential equation

$$y''' + ay'' + by' + cy = 0,$$

where the accents represent differentiations, and if p_1, p_2, p_3 are particular integrals, then

$$p_1''' + ap_1'' + bp_1' + cp_1 = 0,$$

$$p_2''' + ap_2'' + bp_2' + cp_2 = 0,$$

$$p_3''' + ap_3'' + bp_3' + cp_3 = 0,$$

from which and the given equation we may eliminate a, b, c and obtain the Wronskian

$$| y''' p_1'' p_2' p_3 | = 0 \text{ or } W(y, p_1, p_2, p_3) = 0$$

showing that in linear differential equations the Wronskian is the analogue of the alternant in ordinary algebraic equations.

718. *If the Wronskian, W say, of y_1, y_2, $\cdots$, y_n be a known function of the independent variable x, then any one of the y's, y_r say, can be expressed in terms of the others and W.*

For

$$W \equiv \begin{vmatrix} y_1 & y_2 & \cdots & y_n \\ y_{1(1)} & y_{2(1)} & \cdots & y_{n(1)} \\ \cdot & \cdot & \cdot & \cdot \\ y_{1(n-1)} & y_{2(n-1)} & \cdots & y_{n(n-1)} \end{vmatrix}$$

and if we denote the cofactor of $y_{i(k)}$ by $Y_{i(k)}$ and that of the minor

$$\begin{vmatrix} y_{i(k)} & y_{j(k)} \\ y_{i(k+1)} & y_{j(k+1)} \end{vmatrix} \text{ by } Y_{(k)i,(k+1)j},$$

then we have, observing that

$$\frac{d}{dx}Y_{i(k)} = Y_{i(k-1)},$$

$$\frac{d}{dx}\left(\frac{Y_{s(n-1)}}{Y_{r(n-1)}}\right) = \frac{-Y_{r(n-1)}Y_{s(n-2)} + Y_{s(n-1)}Y_{r(n-2)}}{Y^2_{r(n-1)}}$$

$$= \frac{W \cdot Y_{(n-1)s,(n-2)r}}{Y^2_{r(n-1)}}.$$

Integrating both sides with respect to x we have

$$Y_{s(n-1)} = Y_{r(n-1)} \int \frac{W \cdot Y_{(n-1)s,(n-2)r}}{Y^2_{r(n-1)}} dx \equiv I_s \text{ say},$$

where s may take any of the values from 1 to n except r. We have then but to substitute in the identity

$$y_1 Y_{1(n-1)} + y_2 Y_{2(n-1)} + \cdots + y_n Y_{n(n-1)} = 0$$

and get

$$y_1 I_1 + y_2 I_2 + \cdots + y_r + \cdots + y_n I_n = 0$$

which is the theorem.

719. The Wronskian

$$W \equiv \begin{vmatrix} u & u' & u'' \\ v & v' & v'' \\ w & w' & w'' \end{vmatrix} = \begin{vmatrix} u \cdot 1 & (u \cdot 1)' & (u \cdot 1)'' \\ \left(u \cdot \frac{v}{u}\right) & \left(u \cdot \frac{v}{u}\right)' & \left(u \cdot \frac{v}{u}\right)'' \\ \left(u \cdot \frac{w}{u}\right) & \left(u \cdot \frac{w}{u}\right)' & \left(u \cdot \frac{w}{u}\right)'' \end{vmatrix}$$

$$= u^3 \begin{vmatrix} 1 & (1)' & (1)'' \\ \frac{v}{u} & \left(\frac{v}{u}\right)' & \left(\frac{v}{u}\right)'' \\ \frac{w}{u} & \left(\frac{w}{u}\right)' & \left(\frac{w}{u}\right)'' \end{vmatrix} = u^3 \begin{vmatrix} \left(\frac{v}{u}\right)' & \left(\frac{v}{u}\right)'' \\ \left(\frac{w}{u}\right)' & \left(\frac{w}{u}\right)'' \end{vmatrix}.$$

Treating the resulting two line determinant in a similar fashion we get

$$W = u^3\left\{\left(\frac{v}{u}\right)'\right\}^2 \left\{\frac{\left(\frac{W}{u}\right)'}{\left(\frac{v}{u}\right)'}\right\}.$$

Similar procedure would lead to

$$W(f, f_1, f_2, \cdots, f_n) = f^{n+1} \cdot f_{10}{}^n \cdot f_{21}{}^{n-1} \cdots f_{n,n-1}$$

where

$$f_{\mu,k} = \frac{\partial}{\partial x}\left(\frac{f_{\mu,k-1}}{f_{k,k-1}}\right).$$

720. The determinant

$$\begin{vmatrix} y_1 & y_2 & \cdots y_n \\ y_{1(1)} & y_{2(1)} & \cdots y_{n(1)} \\ y_{1(2)} & y_{2(2)} & \cdots y_{n(2)} \\ \cdot & \cdot & \cdot \\ y_{1(n-2)} & y_{2(n-2)} & \cdots y_{n(n-2)} \\ y_{1(\lambda)} & y_{2(\lambda)} & \cdots y_{n(\lambda)} \end{vmatrix}$$

is obviously zero for $\lambda < n-1$ and is equal to $W(y_1 \cdots y_n)$ for $\lambda = n-1$. Developing in terms of elements in the last row we have

$$y_{1(\lambda)}z_1 + y_{2(\lambda)}z_2 + \cdots + y_{n(\lambda)}z_n$$
$$= \begin{Bmatrix} 0 \text{ if } \lambda < n-1 \\ 1 \text{ if } \lambda = n-1 \end{Bmatrix} \quad (\lambda = 0, 1, \cdots, \overline{n-1})$$

where

$$z_k = (-1)^{n+k}\frac{W(y_1 \cdots y_{k-1}, y_{k+1} \cdots y_n)}{W(y_1 \cdots y_n)},$$

and where $y_k(_0)$ is the same as y_k, that is, the first equation of the series is

$$y_1z_1 + y_2z_2 + \cdots + y_nz_n = 0.$$

Differentiating this we have

$$y_1z_{1(1)} + y_2z_{2(1)} + \cdots + y_nz_{n(1)} = 0.$$

Similarly

$$y_1 z_{1(2)} + y_2 z_{2(2)} + \cdots + y_n z_{n(2)} = 0,$$
$$y_1 z_{1(3)} + y_2 z_{2(3)} + \cdots + y_n z_{n(3)} = 0,$$
$$\cdots\cdots\cdots\cdots\cdots$$
$$y_1 z_{1(n-2)} + y_2 z_{2(n-2)} + \cdots + y_n z_{n(n-2)} = 0,$$
$$y_1 z_{1(n-1)} + y_2 z_{2(n-1)} + \cdots + y_n z_{n(n-1)} = (-1)^{n-1}.$$

From these we get

$$y_k = (-1)^{k-1} \frac{W(z_1 \cdots z_{k-1}, z_{k+1} \cdots z_n)}{W(z_1 \cdots z_n)}.$$

By use of the multiplication theorem we see that

$$W(y_1 \cdots y_n) \cdot W(z_{k+1} \cdots z_n) = W(y_1 \cdots y_k)$$

which when $k=0$ is

$$W(y_1 \cdots y_n) W(z_1 \cdots z_n) = 1.$$

The functions z have been termed the *adjunct functions* of the y's.

721. If $D_4 y$ denotes the persymmetric determinant

$$\begin{vmatrix} y & y_1 & y_2 & y_3 \\ y_1 & y_2 & y_3 & y_4 \\ y_2 & y_3 & y_4 & y_5 \\ y_3 & y_4 & y_5 & y_6 \end{vmatrix}$$

where y_r stands for $d^r y/dx^r$ then it is easy to see that

$$\frac{d}{dx}(D_3 y) = \begin{vmatrix} y & y_1 & v_2 \\ y_1 & y_2 & y_3 \\ y_3 & y_4 & y_5 \end{vmatrix}, \quad \frac{d^2}{dx^2}(D_3 y) = \begin{vmatrix} y & x_1 & y_3 \\ y_1 & y_2 & y_4 \\ y_3 & y_4 & y_6 \end{vmatrix}$$

and therefore

$$\begin{vmatrix} D_3 y & \dfrac{d}{dx}(D_3 y) \\ \dfrac{d}{dx}(D_3 y) & \dfrac{d^2}{dx^2}(D_3 y) \end{vmatrix}$$

that is $D_2(D_3 y)$ is a minor of the adjugate of $D_4 y$ and hence

$$D_2(D_3 y) = D_2 y D_4 y.$$

For the general case this is

$$D_2(D_m y) = D_{m-1}y \cdot D_{m+1}y$$

722. Extensions have been made where a more general operator than d/dx has been used. Thus if

$$\begin{aligned} \partial y &\text{ stands for } u_1\frac{\partial y}{\partial x_1} + u_2\frac{\partial y}{\partial x_2} + \cdots \\ \partial^2 y &\text{ stands for } \partial(\partial y) \\ \partial^3 y &\text{ stands for } \partial(\partial^2 y) \\ &\cdots\cdots\cdots\cdots \end{aligned}$$

and the u's are any magnitudes whatever. The determinant

$$\begin{vmatrix} y_1 & y_2 & \cdots & y_n \\ \partial y_1 & \partial y_2 & \cdots & \partial y_n \\ \cdots & \cdots & \cdots & \cdots \\ \partial^{n-1}y_1 & \partial^{n-1}y_2 & \cdots & \partial^{n-1}y_n \end{vmatrix}$$

has been considered by Pasch.

Rosanes considered the determinant

$$\begin{vmatrix} \dfrac{\partial^2 u_1}{\partial x^2} & \dfrac{\partial^2 u_1}{\partial x \partial y} & \dfrac{\partial^2 u_1}{\partial y^2} \\ \dfrac{\partial^2 u_2}{\partial x^2} & \dfrac{\partial^2 u_2}{\partial x \partial y} & \dfrac{\partial^2 u_2}{\partial y^2} \\ \dfrac{\partial^2 u_3}{\partial x^2} & \dfrac{\partial^2 u_3}{\partial x \partial y} & \dfrac{\partial^2 u_3}{\partial y^2} \end{vmatrix}$$

where the u's are binary quantics, and Pasch showed that

$$\begin{vmatrix} \dfrac{\partial^{n-1}u_1}{\partial x^{n-1}} & \dfrac{\partial^{n-1}u_1}{\partial x^{n-2}\partial y} & \cdots & \dfrac{\partial^{n-1}u_1}{\partial y^{n-1}} \\ \dfrac{\partial^{n-1}u_2}{\partial x^{n-1}} & \dfrac{\partial^{n-1}u_2}{\partial x^{n-2}\partial y} & \cdots & \dfrac{\partial^{n-1}u_2}{\partial y^{n-1}} \\ \cdots & \cdots & \cdots & \cdots \\ \dfrac{\partial^{n-1}u_n}{\partial x^{n-1}} & \dfrac{\partial^{n-1}u_n}{\partial x^{n-2}\partial y} & \cdots & \dfrac{\partial^{n-1}u_n}{\partial y^{n-1}} \end{vmatrix}$$

$$= \frac{m(m-1)^1(m-2)^2\cdots(m-n+1)^{n-1}}{y^{n(n-1)/2}} \cdot W(u_1, u_2, \cdots, u_n)$$

where the u's are binary m-thics.

In connection with these generalized determinants Bortolotti has shown that if $\phi_1, \phi_2, \cdots, \phi_n$ be analytic functions having a common domain of convergence, and 0 be a functional operation that is distributive and uniquely determinate, the necessary and sufficient condition for the functions being connected by a linear homogeneous relation with coefficients constant with respect to 0 is

$$| \phi_1 \, 0\phi_2 \, 0^2\phi_3 \cdots 0^{n-1}\phi_n | \equiv 0.$$

723. The determinant represented by $|\int_a^b f_r \bar{f}_s dx|$ or $G(a, b)$, where the element in the place (rs) is $\int_a^b f_r \bar{f}_s dx$, and where $\bar{f}$ is the conjugate of f has been called by some the *Gramian*.

If we denote the Wronskian of the f's by $W(x)$, then Meader has shown that

$$\frac{d^{n^2}}{db^{n^2}} G(a, b) = N \cdot W(a),$$

where

$$N = \frac{(n^2)!}{1^2 2^2 \cdots (n-1)^2 n^n (n+1)^{n-1} \cdots (2n-1)^1}.$$

Exercises. Set XXXV

1. Show that if $W(y_1 \cdots y_n) = 0$, then every minor of order n formed from the array

$$\begin{Vmatrix} y_1 & \frac{dy_1}{dx} & \frac{d^2y_1}{dx^2} & \cdots \\ y_2 & \frac{dy_2}{dx} & \frac{d^2y_2}{dx^2} & \cdots \\ \cdot & \cdot & \cdot & \cdot \\ y_n & \frac{dx_n}{dx} & \frac{d^2y_n}{dx^2} & \cdots \end{Vmatrix}$$

vanishes. (Pasch.)

2. When $x_1, x_2, \cdots, x_n$ have the common value x, show that

$$\frac{|\phi_1(x_1)\,\phi_2(x_2)\cdots\phi_n(x_n)|}{|\psi_1(x_1)\,\psi_2(x_2)\cdots\psi_n(x_n)|} = \frac{|\phi_1(x)\,\phi_2'(x)\cdots\phi_n^{(n-1)}(x)|}{|\psi_1(x)\,\psi_2'(x)\cdots\psi_n^{(n-1)}(x)|}$$

3. Show that the necessary and sufficient condition for $y_1 y_2 \cdots y_n$ being connected by a linear homogeneous relation with constant coefficients is

$$\begin{vmatrix} y_1 & y_2 & \cdots & y_n \\ \Delta y_1 & \Delta y_2 & \cdots & \Delta y_n \\ \cdot & \cdot & \cdot & \cdot \\ \Delta^{n-1}y_1 & \Delta^{n-1}y_2 & \cdots & \Delta^{n-1}y_n \end{vmatrix} = 0,$$

where the Δ's are difference operators.

4. If $y_1, y_2, \cdots, y_{n-1}$ be functions of x which at every point of an interval I have continuous derivatives of the first n orders, and if the Wronskian of $y_1, y_2, \cdots, y_n$ vanish identically then the Wronskian of $y_1, y_2, \cdots, y_{n-1}$ likewise vanishes identically. (Bocher) see Pasch Ex. 1.

5. Show that

$$\frac{d^n x}{dy^n} = \frac{W\left(\dfrac{d^2y}{dx^2}, \dfrac{d^2y^2}{dx^2} \cdots \dfrac{d^2y^{n-1}}{dx^2}\right)}{1!2!\cdots(n-1)!} \times \frac{1}{\left(\dfrac{dy}{dx}\right)^{n(n-1)/2}}$$

(Mina.)

CHAPTER XIX

Bordered Determinants

724. Bordered determinants may be of the type indicated in §200 where the determinant $|a_{1m}|$ may be looked upon as bordered by r rows and columns or it may be of the type where the d's are all zeros.

Of both types we have had numerous examples.

The theorem of §254 may be extended so as to have the product bordered by r rows and r columns. Thus if

$$(a_{3,6}) \equiv \left\| \begin{matrix} a_{11} & a_{12} \cdots a_{16} \\ a_{21} & a_{22} \cdots a_{26} \\ a_{31} & a_{32} \cdots a_{36} \end{matrix} \right\|$$

$$(b_{3,6}) \equiv \left\| \begin{matrix} b_{11} & b_{12} \cdots b_{16} \\ b_{21} & b_{22} \cdots b_{26} \\ b_{31} & b_{32} \cdots b_{36} \end{matrix} \right\|$$

and if

$$c_{11} = a_{11}b_{11} + a_{12}b_{12} + a_{13}b_{13} + a_{14}b_{14}$$

$$c_{12} = \text{etc.}$$

then

$$\begin{vmatrix} c_{11} & c_{12} & c_{13} & a_{15} & a_{16} \\ c_{21} & c_{22} & c_{23} & a_{25} & a_{26} \\ c_{31} & c_{32} & c_{33} & a_{35} & a_{36} \\ b_{15} & b_{25} & b_{35} & \cdot & \cdot \\ b_{16} & b_{26} & b_{36} & \cdot & \cdot \end{vmatrix} = \left\| \begin{matrix} a_{11} & a_{12} & a_{13} & a_{14} & a_{15} & a_{16} & \cdot & \cdot \\ a_{21} & a_{22} & a_{23} & a_{24} & a_{25} & a_{26} & \cdot & \cdot \\ a_{31} & a_{32} & a_{33} & a_{34} & a_{35} & a_{36} & \cdot & \\ \cdot & \cdot & \cdot & \cdot & \cdot & \cdot & 1 & \cdot \\ \cdot & \cdot & \cdot & \cdot & \cdot & \cdot & \cdot & 1 \end{matrix} \right\|$$

$$\times \left\| \begin{matrix} b_{11} & b_{12} & b_{13} & b_{14} & \cdot & \cdot & b_{15} & b_{16} \\ b_{21} & b_{22} & b_{23} & b_{24} & \cdot & \cdot & b_{25} & b_{26} \\ b_{31} & b_{32} & b_{33} & b_{34} & \cdot & \cdot & b_{35} & b_{36} \\ \cdot & \cdot & \cdot & \cdot & 1 & \cdot & \cdot & \cdot \\ \cdot & \cdot & \cdot & \cdot & \cdot & 1 & \cdot & \cdot \end{matrix} \right\| = -|a_{156}| \cdot |b_{156}|$$

$$+ |a_{256}| \quad |b_{256}| + |a_{356}| \cdot |b_{356}| + |a_{456}| \cdot |b_{456}|$$

where $|a_{156}|$ denotes the minor of $(a_{3,6})$ formed by taking the 1st, 5th, 6th columns, etc.

The same reasoning will apply in general.

725. Let $(a.\,\alpha, \beta)$ represent any determinant $|a_{1n}|$ bordered horizontally by $\alpha_1, \alpha_2, \cdots, \alpha_n$ and vertically by $\beta_1, \beta_2, \cdots, \beta_n$ and more generally let $(a.\,\alpha\beta, \gamma\delta)$ represent the determinant $|a_{1n}|$ bordered horizontally by $\alpha_1, \alpha_2, \cdots, \alpha_n$ and $\beta_1, \beta_2, \cdots, \beta_n$; and vertically by $\gamma_1, \gamma_2, \cdots, \gamma_n$ and $\delta_1, \delta_2, \cdots, \delta_n$. When no ambiguity can arise the 'a' part of this symbol may be omitted.

Let $(u.\,\alpha, \beta)$ represent the Hessian $H(u)$ bordered vertically by $\beta_1, \beta_2, \beta_3, \cdots, \beta_n$ and horizontally by $\alpha_1, \alpha_2, \cdots, \alpha_n$. Then using the theorem concerning the two-line minor of the adjugate of $(u.\,\alpha\gamma, \beta\delta)$ we have

$$(1) \qquad (u.\alpha,\beta)(u.\gamma,\delta) - (u.\alpha,\delta)(u.\gamma,\beta) = H(u)(u.\alpha\gamma,\beta\delta).$$

Inserting in this ξ and ϵ for β and γ respectively a similar result is obtained and from these two by eliminating $(u.\,\alpha, \delta)$ we get

$$(2) \quad (u.\alpha,\beta)(u.\gamma,\delta)(u.\epsilon,\xi) - (u.\gamma,\beta)(u.\epsilon,\delta)(u.\alpha,\xi)$$
$$= H(u)\left\{(u.\epsilon,\xi)(u.\alpha\gamma,\beta\delta) - (u.\gamma,\beta)(u.\alpha\epsilon,\xi\delta)\right\}.$$

If in this we make cyclical substitutions on α, γ, ϵ and β, δ, ξ simultaneously the left-hand side is not altered and therefore it follows that the three cofactors of $H(u)$ in the three results obtained must be equal and therefore

$$\begin{aligned} &(u.\epsilon,\xi)(u.\alpha\gamma,\beta\delta) - (u.\gamma,\beta)(u.\alpha\epsilon,\xi\delta) \\ (3) \qquad &= (u.\alpha,\beta)(u.\gamma\epsilon,\delta\xi) - (u.\epsilon,\delta)(u.\gamma\alpha,\beta\xi) \\ &= (u.\gamma,\delta)(u.\epsilon\alpha,\xi\beta) - (u.\alpha,\xi)(u.\epsilon\gamma,\delta\beta). \end{aligned}$$

If in (2) we make the cyclical substitution α, γ, ϵ alone then the three cofactors of $H(u)$ have a vanishing sum, since the sum of the other sides is zero. That is

$$\begin{aligned} &(u.\epsilon,\xi)(u.\alpha\gamma,\beta\delta) - (u.\gamma,\delta)(u.\alpha\epsilon,\xi\delta) \\ + &(u.\alpha,\xi)(u.\gamma\epsilon,\beta\delta) - (u.\epsilon,\delta)(u.\gamma\alpha,\xi\delta) \\ + &(u.\gamma,\xi)(u.\epsilon\alpha,\beta\delta) - (u.\alpha,\delta)(u.\epsilon\gamma,\xi\delta) = 0. \end{aligned}$$

Again if in (1) we perform the cyclic substitution on β, δ, ξ there is obtained another triad of equalities, and if on these we use the multipliers $(u.\,\epsilon, \xi)$, $(u.\,\epsilon, \beta)$, $(u.\,\epsilon, \delta)$ and thereafter add, there results

$$H(u)\{(u.\epsilon,\xi)(u.\alpha\gamma,\beta\delta)+(u.\epsilon,\beta)(u.\alpha\gamma,\delta\xi)+(u.\epsilon,\delta)(u.\alpha\gamma,\xi\beta)\}$$
$$=\begin{vmatrix}(u.\alpha,\beta) & (u.\alpha,\delta) & (u.\alpha,\xi)\\ (u.\gamma,\beta) & (u.\gamma,\delta) & (u.\gamma,\xi)\\ (u.\epsilon,\beta) & (u.\epsilon,\delta) & (u.\epsilon,\xi)\end{vmatrix}$$

These four results are important in connection with the generalization of Pascal's theorem.

726. If we multiply the determinant

$$V\equiv\begin{vmatrix}a_1 & a_2 & a_3 & a_4 & m_1 & n_1\\ b_1 & b_2 & b_3 & b_4 & m_2 & n_2\\ c_1 & c_2 & c_3 & c_4 & m_3 & n_3\\ d_1 & d_2 & d_3 & d_4 & m_4 & n_4\\ r_1 & r_2 & r_3 & r_4 & \cdot & \cdot\\ s_1 & s_2 & s_3 & s_4 & \cdot & \cdot\end{vmatrix} \text{ by } \begin{vmatrix}A_1 & A_2 & A_3 & A_4 & \cdot & \cdot\\ B_1 & B_2 & B_3 & B_4 & \cdot & \cdot\\ C_1 & C_2 & C_3 & C_4 & \cdot & \cdot\\ D_1 & D_2 & D_3 & D_4 & \cdot & \cdot\\ \cdot & \cdot & \cdot & \cdot & 1 & \cdot\\ \cdot & \cdot & \cdot & \cdot & \cdot & 1\end{vmatrix}$$

where A_1 is the complementary minor of a_1 in $\Delta\equiv|a_1b_2c_3d_4|$, etc. we have

$$V\cdot|a_1b_2c_3d_4|^3=\begin{vmatrix}\Delta & \cdot & \cdot & \cdot & m_1 & n_1\\ \cdot & \Delta & \cdot & \cdot & m_2 & n_2\\ \cdot & \cdot & \Delta & \cdot & m_3 & n_3\\ \cdot & \cdot & \cdot & \Delta & m_4 & n_4\\ \sum rA & \sum rB & \sum rC & \sum rD & \cdot & \cdot\\ \sum sA & \sum sB & \sum sC & \sum sD & \cdot & \cdot\end{vmatrix}.$$

Multiplying the 5th and 6th rows by Δ and then dividing Δ from the first four columns we get

$$V\cdot\Delta=\begin{vmatrix}1 & \cdot & \cdot & \cdot & m_1 & n_1\\ \cdot & 1 & \cdot & \cdot & m_2 & n_2\\ \cdot & \cdot & 1 & \cdot & m_3 & n_3\\ \cdot & \cdot & \cdot & 1 & m_4 & n_4\\ \sum rA & \sum rB & \sum rC & \sum rD & \cdot & \cdot\\ \sum sA & \sum sB & \sum sC & \sum sD & \cdot & \cdot\end{vmatrix}$$
$$=\begin{Vmatrix}m_1 & m_2 & m_3 & m_4\\ n_1 & n_2 & n_3 & n_4\end{Vmatrix}\begin{Vmatrix}\sum rA & \sum rB & \sum rC & \sum rD\\ \sum sA & \sum sB & \sum sC & \sum sD\end{Vmatrix}$$

$$= \left\Vert \frac{\begin{vmatrix} a_1 & a_2 & a_3 & a_4 & m_1 \\ b_1 & b_2 & b_3 & b_4 & m_2 \\ c_1 & c_2 & c_3 & c_4 & m_3 \\ d_1 & d_2 & d_3 & d_4 & m_4 \\ r_1 & r_2 & r_3 & r_4 & \end{vmatrix} \quad \begin{vmatrix} a_1 & a_2 & a_3 & a_4 & n_1 \\ b_1 & b_2 & b_3 & b_4 & n_2 \\ c_1 & c_2 & c_3 & c_4 & n_3 \\ d_1 & d_2 & d_3 & d_4 & n_4 \\ r_1 & r_2 & r_3 & r_4 & \end{vmatrix}}{\begin{vmatrix} a_1 & a_2 & a_3 & a_4 & m_1 \\ b_1 & b_2 & b_3 & b_4 & m_2 \\ c_1 & c_2 & c_3 & c_4 & m_3 \\ d_1 & d_2 & d_3 & d_4 & m_4 \\ s_1 & s_2 & s_3 & s_4 & \end{vmatrix} \quad \begin{vmatrix} a_1 & a_2 & a_3 & a_4 & n_1 \\ b_1 & b_2 & b_3 & b_4 & n_2 \\ c_1 & c_2 & c_3 & c_4 & n_3 \\ d_1 & d_2 & d_3 & d_4 & n_4 \\ s_1 & s_2 & s_3 & s_4 & \end{vmatrix}} \right\Vert$$

EXERCISE: If in the bordered determinant $(a.\ x, y)$ the x's be transformed by a linear substitution with modulus M, and the y's be similarly transformed by a substitution whose modulus is N show that

$$(a\,.\,x,\ y) = M \cdot A \cdot N, \text{ where } A \equiv |\, a_{1n} \,|$$

727. If in §705 we used a bilinear function for u, that is if

$$u \equiv \begin{array}{cccc|c} x_1 & x_2 & \cdots & x_n & \\ \hline a_{11} & a_{12} & \cdots & a_{1n} & y_1 \\ a_{21} & a_{22} & \cdots & a_{2n} & y_2 \\ \cdot & \cdot & \cdots & \cdot & \cdot \\ a_{n1} & a_{n2} & \cdots & a_{nn} & y_n \end{array}$$

and if now u_{ij} denotes the second derivative of u with respect to x_i and y_j and therefore $|u_{1n}| \equiv |a_{1n}|$; and if we use $(a.\ u_x, u_y)$ to denote the determinant formed by bordering the determinant $|a_{1n}|$ by

$$\frac{\partial u}{\partial x_1}, \frac{\partial u}{\partial x_2}, \ldots, \frac{\partial u}{\partial x_n} \text{ and } \frac{\partial u}{\partial y_1}, \frac{\partial u}{\partial y_2}, \ldots, \frac{\partial u}{\partial y_n}$$

then

$$(a\,.\,u_x, u_y) = -\,u\,|\,a_{1n}\,|\,.$$

The proof of this would follow as in the Hessian.

Similarly

$$(A\,.\,x, y) = -\,u\,|\,a_{1n}\,|^{\,n-2},$$

where $(A.\ x, y)$ denotes the adjugate of $|a_{1n}|$ bordered by the x's and y's. *That is the product of a bipartite of degree-order* $(3, n)$ *by the* $(n-2)$*th* *power of the determinant of its square array is expressible as a bordered determinant of the $(n+1)$th order.*

728. If the quaternary quadric

$$F \equiv \begin{array}{cccc|c} x_1 & x_2 & x_3 & x_4 & \\ \hline a & e & f & g & x_1 \\ e & b & h & k & x_2 \\ f & h & c & l & x_3 \\ g & k & l & d & x_4 \end{array}$$

vanishes, and if

$$\delta_1 = \frac{1}{2}\frac{\partial F}{\partial x_1}, \quad \delta_2 = \frac{1}{2}\frac{\partial F}{\partial x_2}, \quad \delta_3 = \frac{1}{2}\frac{\partial F}{\partial x_3}, \quad \delta_4 = \frac{1}{2}\frac{\partial F}{\partial x_4},$$

$$b_1x_1 + b_2x_2 + b_3x_3 + b_4x_4 = 0, \qquad c_1x_1 + c_2x_2 + c_3x_3 + c_4x_4 = 0,$$

and M_r represents the cofactor of m_r in $|m_1\delta_2b_3c_4|$, then, as an example, we see that

$$M_4^2 = -\begin{vmatrix} a & e & f & \delta_1 & b_1 & c_1 \\ e & b & h & \delta_2 & b_2 & c_2 \\ f & h & c & \delta_3 & b_3 & c_3 \\ \delta_1 & \delta_2 & \delta_3 & \cdot & \cdot & \cdot \\ b_1 & b_2 & b_3 & \cdot & \cdot & \cdot \\ c_1 & c_2 & c_3 & \cdot & \cdot & \cdot \end{vmatrix}$$

which after performing the operations

$$\text{col}_4 - x_1\,\text{col}_1 - x_2\,\text{col}_2 - x_3\,\text{col}_3,$$

$$\text{row}_4 - x_1\,\text{row}_1 - x_2\,\text{row}_2 - x_3\,\text{row}_3,$$

becomes

$$M_4^2 = -\begin{vmatrix} a & e & f & g & b_1 & c_1 \\ e & b & h & k & b_2 & c_2 \\ f & h & c & 1 & b_3 & c_3 \\ g & k & 1 & d & b_4 & c_4 \\ b_1 & b_2 & b_3 & b_4 & \cdot & \cdot \\ c_1 & c_2 & c_3 & c_4 & \cdot & \cdot \end{vmatrix} x_4^2 = -\,x_4^2\Delta, \text{ say.}$$

Similarly we may show that

$$M_3^2 = -\,x_3^2\Delta, \qquad M_2^2 = -\,x_2^2\Delta, \qquad M_1^2 = -\,x_1^2\Delta$$

and hence

$$|m_1\delta_2b_3c_4| = (m_1x_1 + m_2x_2 + m_3x_3 + m_4x_4)(-\Delta)^{1/2}.$$

The same reasoning applies in general.

Exercises: 1. Show that the two values of x/y that satisfy the equations

$$ax^2 + by^2 + cz^2 + 2fyz + 2gzx + 2hxy = 0$$
$$px + qy + rz = 0$$

are given by

$$\begin{vmatrix} A & H & x \\ H & B & y \\ x & y & \cdot \end{vmatrix} = 0$$

or are

$$\frac{H \pm (H^2 - AB)^{1/2}}{B},$$

where the adjugate of

$$\begin{vmatrix} a & h & g & p \\ h & b & f & q \\ g & f & c & r \\ p & q & r & \cdot \end{vmatrix} \quad \text{is} \quad \begin{vmatrix} A & H & G & P \\ H & B & F & Q \\ G & F & C & R \\ P & Q & R & \cdot \end{vmatrix}.$$

Similarly for y/z and z/x.

2. If $A = |a_{1n}|$ be axisymmetric and bordered with p rows and columns of n elements each then in order to have this bordered determinant identically zero A must have a nullity of $(p+1)$. (Darboux.)

3. Show that

$$\begin{vmatrix} a & h & g & \alpha_1 & \alpha_2 \\ h & b & f & \beta_1 & \beta_2 \\ g & f & c & \gamma_1 & \gamma_2 \\ \alpha_1 & \beta_1 & \gamma_1 & \cdot & \cdot \\ \alpha_2 & \beta_2 & \gamma_2 & \cdot & \cdot \end{vmatrix} = \begin{array}{ccc|c} |\beta_1\gamma_2| & |\gamma_1\alpha_2| & |\alpha_1\beta_2| & \\ \hline a & h & g & |\beta_1\gamma_2| \\ h & b & f & |\gamma_1\alpha_2| \\ g & f & c & |\alpha_1\beta_2| \end{array}$$

$$= \begin{vmatrix} \begin{vmatrix} a & \alpha_1 & \alpha_2 \\ h & \beta_1 & \beta_2 \\ g & \gamma_1 & \gamma_2 \end{vmatrix} & \begin{vmatrix} h & \alpha_1 & \alpha_2 \\ b & \beta_1 & \beta_2 \\ f & \gamma_1 & \gamma_2 \end{vmatrix} & \begin{vmatrix} g & \alpha_1 & \alpha_2 \\ f & \beta_1 & \beta_2 \\ c & \gamma_1 & \gamma_2 \end{vmatrix} \\ \alpha_1 & \beta_1 & \gamma_1 \\ \alpha_2 & \beta_2 & \gamma_2 \end{vmatrix}$$

CHAPTER XX

Determinants whose Elements are Combinatory Numbers

729. By using the formula

$$(x)_k - (x-1)_k = (x-1)_{k-1}$$

on the determinant

$$\begin{vmatrix} 1 & (c+m)_1 & (c+m+1)_2 & \cdots & (c+2m-1)_m \\ 1 & (c+m+1)_1 & (c+m+2)_2 & \cdots & (c+2m)_m \\ \cdot & \cdot & \cdot & \cdot & \cdot \\ 1 & (c+2m)_1 & (c+2m+1)_2 & \cdots & (c+3m-1)_m \end{vmatrix}_{m+1}$$

it is readily seen that its value is 1, and is, therefore, independent of c and m.

730. In a similar manner the determinant

$$\begin{vmatrix} (c+m)_m & (c+m+1)_m & \cdots & (c+2m)_m \\ (c+m+1)_m & (c+m+2)_m & \cdots & (c+2m+1)_m \\ \cdot & \cdot & \cdot & \cdot \\ (c+2m)_m & (c+2m+1)_m & \cdots & (c+3m)_m \end{vmatrix}$$

is seen to have the value $(-1)^{(m+1)/2}$.

It may be observed that this is an illustration of the theorem of §463.

731. The determinant

$$D \equiv \begin{vmatrix} (d)_1 & (d)_2 & \cdots & (d)_r \\ (2d)_1 & (2d)_2 & \cdots & (2d)_r \\ \cdot & \cdot & \cdot & \cdot \\ (rd)_1 & (rd)_2 & \cdots & (rd)_r \end{vmatrix}$$

is readily seen to be equal to

$$\frac{d \cdot 2d \cdot 3d \cdots rd}{2 \cdot 3 \cdots r} \begin{vmatrix} 1 & (d-1)_1 & \cdots & (d-1)_{r-1} \\ 1 & (2d-1)_1 & \cdots & (2d-1)_{r-1} \\ \cdot & \cdot & \cdot & \cdot \\ 1 & (rd-1)_1 & \cdots & (rd-1)_{r-1} \end{vmatrix}.$$

If now we perform on this last determinant the operations $\text{col}_k - \text{col}_{(k-1)}$, where $k=2, 3, \cdots, r$, we get

$$D = d^r \begin{vmatrix} 1 & (d)_1 & \cdots & (d)_{r-1} \\ 1 & (2d)_1 & \cdots & (2d)_{r-1} \\ \cdot & \cdot & \cdot & \cdot \\ 1 & (rd)_1 & \cdots & (rd)_{r-1} \end{vmatrix} = d^r D_{(r,d)}, \text{ say.}$$

If now we multiply $D_{(r,d)}$ by unity in the form

$$\begin{vmatrix} (-d)_0 & 0 & \cdots & 0 \\ (-d)_1 & (-d)_0 & \cdots & 0 \\ \cdot & \cdot & \cdot & \cdot \\ (-d)_{r-1} & (-d)_{r-2} & \cdots & (-d)_0 \end{vmatrix}$$

we get

$$D_{(r,d)} = d^{r-1} D_{(r-1,d)}$$

from which we get

$$\begin{aligned} D_{(r,d)} &= d^{r-1} d^{r-2} \cdots d D_{(1,d)} \\ &= d^{r(r-1)/2}. \end{aligned}$$

Therefore

$$D = d^{r(r+1)/2}.$$

732. By taking common factors from the rows and columns of the determinant

$$\Delta_{(m,p,r,d)} \equiv \begin{vmatrix} (m)_p & (m)_{p-1} & \cdots & (m)_{p-r} \\ (m+d)_p & (m+d)_{p-1} & \cdots & (m+d)_{p-r} \\ \cdot & \cdot & \cdot & \cdot \\ (m+rd)_p & (m+rd)_{p+1} & \cdots & (m+rd)_{p+r} \end{vmatrix}_{r+1}$$

we have

$$\Delta_{(m,p,r,d)} = \frac{m(m+d)\cdots(m+rd)}{p(p+1)\cdots(p+r)} \cdot \Delta_{(m-1,p-1,r,d)}$$

which used upon itself yields

$$\Delta_{(m,p,r,d)} = \frac{(m)_p (m+d)_p \cdots (m+rd)_p}{(p)_p (p+1)_p \cdots (p+r)_p} \Delta_{(m-p,0,r,d)}.$$

If now we multiply $\Delta_{(m-p,o,r,d)}$ by unity in the form

$$\begin{vmatrix} (p-m)_0 & 0 & \cdots & 0 \\ (p-m)_1 & (p-m)_0 & \cdots & 0 \\ \cdot & \cdot & \cdot & \cdot \\ (p-m)_r & (p-m)_{r-1} & \cdots & (p-m)_0 \end{vmatrix}$$

we get

$$\Delta_{(m-p,0,r,d)} = \begin{vmatrix} 1 & 0 & \cdots & 0 \\ 1 & (d)_1 & \cdots & (d)_r \\ 1 & (2d)_1 & \cdots & (2d)_r \\ \cdot & \cdot & \cdot & \cdot \\ 1 & (rd)_1 & \cdots & (rd)_r \end{vmatrix}$$

$$= d^{r(r+1)/2}.$$

Therefore

$$\Delta_{(m,p,r,d)} = \frac{(m)_p(m+d)_p\cdots(m+rd)_p}{(p)_p(p+1)_p\cdots(p+r)_p} d^{r(r+1)/2}.$$

When $p=0$ and $d=1$ the value of the determinant is seen to be 1, or

$$\Delta_{(m,0,r,1)} = 1.$$

733. If we treat the determinant

$$\Delta_{(m,r+1,p,d)} \equiv \begin{vmatrix} (m+rd-p+1)_{r+1} & \cdots & (m+rd-p+1)_{r+p} \\ (m+rd-p+2)_{r+1} & \cdots & (r+rd-p+2)_{r+p} \\ \cdot & \cdot & \cdot \\ (m+rd)_{r+1} & \cdots & (m+rd)_{r+p} \end{vmatrix}_p$$

in a similar way we get

$$\Delta_{(m,r+1,p,d)}$$

$$= \frac{(m+rd)\cdots(m+rd-p+1)}{(r+1)\cdots(r+p)} \frac{(m+rd-1)\cdots(m+rd-p)}{(r)\cdots(r+p-1)}\cdots$$

$$\frac{(m+rd-r+1)\cdots(m+rd-p-r+2)}{2\cdots(p+1)}$$

$$\frac{(m+rd-r)\cdots(m+rd-p-r+1)}{1\cdots p}\cdot\Delta_{(m-r-1,0,p,d)}$$

$$= \frac{(m+rd)_{r+1}(m+rd-1)_{r+1}\cdots(m+rd-p+1)_{r+1}}{(r+1)_{r+1}(r+2)_{r+1}\cdots(r+p)_{r+1}}\cdot\Delta_{(m-r-1,0,p,d)}$$

$$= \frac{(m+rd)_p(m+rd-1)_p\cdots(m+rd-r)_p}{(r+p)_p(r+p-1)_p\cdots(p)_p}\cdot\Delta_{(m-r-1,0,p,d)}$$

But

$$\Delta_{(m-r-1,0,p,d)} = 1.$$

Therefore

$$\Delta_{(m,r+1,p,d)} = \frac{(m+rd)_{r+1}\cdots(m+rd-p+1)_{r+1}}{(r+1)_{r+1}\cdots(r+p)_{r+1}}$$

$$= \frac{(m+rd)_p\cdots(m+rd-r)_p}{(r+p)_p\cdots(p)_p}$$

When $d=1$ this last result is the same as that given in §732 for $\Delta_{(m,p,r,d)}$ when $d=1$. We see therefore that

$$\Delta_{(m,p,r,1)} = \Delta_{(m,r+1,p,1)}.$$

By repeated use of the formula in §729 we readily see that the determinant

$$\Delta'_{(m,p,r,1)} \equiv \begin{vmatrix} (m)_p & (m)_{p+1} & \cdots & (m)_{p+r} \\ (m)_{p-1} & (m)_p & \cdots & (m)_{p+r-1} \\ \cdot & \cdot & \cdot & \cdot \\ (m)_{p-r} & (m)_{p-r+1} & \cdots & (m)_p \end{vmatrix}_{r+1}$$

is the equivalent of $\Delta_{(m,p,r,1)}$.

If $p=1$, then

$$\Delta'_{(m,1,r,1)} = \begin{vmatrix} (m)_1 & (m)_2 & \cdots & (m)_{r+1} \\ (m)_0 & (m)_1 & \cdots & (m)_r \\ \cdot & \cdot & \cdot & \cdot \\ 0 & 0 & \cdots & (m)_1 \end{vmatrix}_{r+1}$$

$$= (m+r)_{r+1}.$$

By repeated use of the formula of §729 we may also see that the determinant

$$\Delta''_{(m,p,r,1)} \equiv \begin{vmatrix} (m)_p & (m+1)_{p+1} & \cdots & (m+r)_{p+r} \\ (m+1)_p & (m+2)_{p+1} & \cdots & (m+r+1)_{p+r} \\ \cdot & \cdot & \cdot & \cdot \\ (m+r)_p & (m+r+1)_{p+1} & \cdots & (m+2r)_{p+r} \end{vmatrix}$$

is equivalent to $\Delta_{(m,p,r,1)}$ of §732.

Similarly the determinant

$$\Delta'''_{(m,p,r,1)} \equiv \begin{vmatrix} (m+r)_p & (m+r+1)_{p+1} & \cdots & (m+2r)_{p+r} \\ (m+r-1)_{p-1} & (m+r)_p & \cdots & (m+2r-1)_{p+r-1} \\ \cdots & \cdots & \cdots & \cdots \\ (m)_{p-r} & (m+1)_{p-r+1} & \cdots & (m+r)_p \end{vmatrix}$$

is also seen to be equivalent to $\Delta_{(m,p,r,1)}$ of §732.

734. By making use of the formula

$$(m)_{p+e} = \frac{(m)_p(m-p)_e}{(p+e)_e}$$

the determinant

$$D \equiv \begin{vmatrix} (m)_p & (m)_{p+e} & \cdots & (m)_{p+re} \\ (m+1)_p & (m+1)_{p+e} & \cdots & (m+1)_{p+re} \\ \cdots & \cdots & \cdots & \cdots \\ (m+r)_p & (m+r)_{p+e} & \cdots & (m+r)_{p+re} \end{vmatrix}_{r+1}$$

is seen to be equal to

$$\frac{(m)_p(m+1)_p \cdots (m+r)_p}{(p+e)_e(p+2e)_{2e} \cdots (p+re)_{re}}$$

$$\times \begin{vmatrix} 1 & (m-p)_e & (m-r)_{2e} & \cdots & (m-p)_{re} \\ 1 & (m-p+1)_e & (m-p+1)_{2e} & \cdots & (m-p+1)_{re} \\ \cdots & \cdots & \cdots & \cdots & \cdots \\ 1 & (m-p+r)_e & (m-p+r)_{2e} & \cdots & (m-p+r)_{re} \end{vmatrix}_{r+1}$$

$$= \frac{(m)_p \cdots (m+r)_p}{(p+e)_e \cdots (p+re)_{re}} \Delta_{(m-p,e,r+1)}, \text{ say.}$$

By subtracting rows we have

$$\Delta_{(m-p,e,r+1)} = \begin{vmatrix} (m-p)_{e-1} & \cdots & (m-p)_{re-1} \\ (m-p+1)_{e-1} & \cdots & (m-p+1)_{re-1} \\ \cdots & \cdots & \cdots \\ (m-p+r-1)_{e-1} & \cdots & (m-p+r-1)_{re-1} \end{vmatrix}_r$$

$$= \frac{(m-p)_{e-1} \cdots (m-p+r-1)_{e-1}}{(2e-1)_e \cdots (re-1)_{(r-1)e}}$$

$$\times\begin{vmatrix} 1 & (m-p-e+1)_e & \cdots & (m-p-e+1)_{(r-1)e} \\ 1 & (m-p-e+2)_e & \cdots & (m-r-e+2)_{(r-1)e} \\ \cdot & \cdot & \cdot & \cdot \\ 1 & (m-p-e+r)_e & \cdots & (m-p-e+r)_{(r-1)e} \end{vmatrix}$$

$$= \frac{(m-p)_{e-1}\cdots(m-p+r-1)_{e-1}}{(2e-1)_{e-1}\cdots(re-1)_{e-1}}\Delta_{(m-p-e+1,e,r)}$$

$$= \frac{(m-p)_{e-1}\cdots(m-p+r-1)_{e-1}}{(2e-1)_{e-1}\cdots(re-1)_{e-1}}$$

$$\frac{(m-p-e+1)_{e-1}\cdots(m-p-e+r-1)_{e-1}}{(2e-1)_{e-1}\cdots(\overline{r-1}\cdot e-1)_{e-1}}\cdots$$

$$\frac{(m-p-\overline{r-2}\cdot e+r-2)_{e-1}(m-p-\overline{r-2}\cdot e+r-1)_{e-1}}{(2e-1)_{e-1}}$$

$$\frac{(m-p-\overline{r-1}\cdot e+r-1)_{e-1}}{1}$$

$$= \frac{P(m-p,r-1)P(m-p-e+1,r-2)\cdots P(m-p-\overline{r-1}\cdot e,1)}{\{(2e-1)_{e-1}\}^{r-1}\{(3e-1)_{e-1}\}^{r-2}\cdots\{(re-1)_{e-1}\}^{1}}$$

where $P(m-p,\ r-1)=(m-p)_{e-1}\cdots(m-p+r-1)_{e-1}$, etc. For the value of D we have, therefore,

$$D = \frac{(m)_p(m+1)_p\cdots(m+r)_p}{(p+e)_e(p+2e)_{2e}\cdots(p+re)_{re}}$$
$$\frac{P(m-p,r-1)\cdots P(m-p-\overline{r-1}\cdot e,1)}{\{2e-1)_{e-1}\}^{r-1}\cdots\{(re-1)_{e-1}\}^{1}}.$$

If in D we put $p=1$, $e=2$, $r=m-1$ we get

$$(a)\quad \begin{vmatrix} (m)_1 & (m)_3 & (m)_5 & \cdots \\ (m)_0 & (m)_1 & (m)_3 & \cdots \\ 0 & (m)_0 & (m)_1 & \cdots \\ \cdot & \cdot & \cdot & \cdot \end{vmatrix}_m = 2^{m-1}\cdot 2^{m-2}\cdots 2^1 = 2^{m(m-1)/2},\ \text{since}$$

$$\frac{m(m+1)\cdots(2m-1)}{3\ 5\cdots(2m-1)} = \frac{(2m-1)!}{(m-1)!3\,5\cdots(2m-1)}$$
$$= \frac{2^{m-1}(m-1)!}{(m-1)!} = 2^{m-1},\ \text{etc.}$$

If in D we put $e=1$, we get

$$(b)\qquad \begin{vmatrix} (m)_p & (m)_{p+1} & \cdots & (m)_{p+r} \\ (m+1)_p & (m+1)_{p+1} & \cdots & (m+1)_{p+r} \\ \cdot & \cdot & \cdot & \cdot \\ (m+r)_p & (m+r)_{p+1} & \cdots & (m+r)_{p+r} \end{vmatrix}_{r+1}$$

$$= \frac{(m)_p(m+1)_p \cdots (m+r)_p}{(p+1)_1(p+2)_2 \cdots (p+r)_r}.$$

If in D we put $p=1$ and $e=1$ we get

$$(c)\qquad \begin{vmatrix} (m)_1 & (m)_2 & \cdots & (m)_{r+1} \\ (m+1)_1 & (m+1)_2 & \cdots & (m+1)_{r+1} \\ \cdot & \cdot & \cdot & \cdot \\ (m+r)_1 & (m+r)_2 & \cdots & (m+r)_{r+1} \end{vmatrix} = \frac{m(m+1)\cdots(m+r)}{(2)_1(3)_2\cdots(r+1)_r}$$

$$= (m+r)_{r+1}.$$

The two results (b) and (c) of this article might have been obtained from $\Delta_{(m,p,r,d)}$ of §732 by putting $d=1$ to get (b) and $d=1$ and $p=1$ to get (c).

The determinant

$$D' \equiv \begin{vmatrix} (m)_p & (m)_{p+e} & \cdots & (m)_{p+re} \\ (m)_{p-1} & (m)_{p+e-1} & \cdots & (m)_{p+re-1} \\ \cdot & \cdot & \cdot & \cdot \\ (m)_{p-r} & (m)_{p+e-r} & \cdots & (m)_{p+re-r} \end{vmatrix}$$

by the use of the formula of §729 may be seen to be equivalent to the determinant D of this article.

735. The determinant

$$\Delta \equiv \begin{vmatrix} (m)_0 & (m)_1 & \cdots & (m)_r \\ (m+1)_0 & (m+1)_1 & \cdots & (m+1)_r \\ \cdot & \cdot & \cdot & \cdot \\ (m+h)_0 & (m+h)_1 & \cdots & (m+h)_r \\ (m+h+k)_0 & (m+h+k)_1 & \cdots & (m+h+k)_r \\ \cdot & \cdot & \cdot & \cdot \\ (m+r+k-1)_0 & (m+r+k-1)_1 & \cdots & (m+r+k-1)_r \end{vmatrix}_{r+1}$$

where the base increases by 1 down the columns except at one point where it increases by k, if multiplied by unity in the form

$$\begin{vmatrix} (m)_0 & 0 & 0 & \cdots \\ -(m)_1 & (m)_0 & 0 & \cdots \\ (m+1)_2 & -(m)_1 & (m)_0 & \cdots \\ -(m+2)_3 & (m+1)_2 & -(m)_1 & \cdots \\ \cdots & \cdots & \cdots & \cdots \end{vmatrix}$$

gives a result which readily reduces to

$$\begin{vmatrix} (h+k)_{h+1} & \cdots & (h+k)_{h+k} & 0 & \cdots & \cdot \\ (h+k+1)_{h+1} & \cdots & (h+k+1)_{h+k} & (h+k+1)_{h+k+1} & \cdots & \cdot \\ \cdots & \cdots & \cdots & \cdots & \cdots & \cdots \\ (r+k-1)_{h+1} & \cdots & (r+k-1)_{h+k} & (r+k-1)_{h+k+1} & \cdots & (r+k-1)_r \end{vmatrix}_{r-h}$$

$$= \frac{(r+1)_{h+1} \cdots (r+k-1)_{h+1}}{(h+1)_{h+1} \cdots (h+k-1)_{h+1}} = 1 \qquad \text{(by §732)}$$

when $k=1$ (by §734 (b)).

736. The determinant

$$\Delta \equiv \begin{vmatrix} (m)_p & (m)_{p+1} & \cdots & (m)_{p+r} \\ (m+1)_p & (m+1)_{p+1} & \cdots & (m+1)_{p+r} \\ \cdots & \cdots & \cdots & \cdots \\ (m+h)_p & (m+h)_{p+1} & \cdots & (m+h)_{p+r} \\ (m+h+k)_p & (m+h+k)_{p+1} & \cdots & (m+h+k)_{p+r} \\ \cdots & \cdots & \cdots & \cdots \\ (m+k+r-1)_p & (m+k+r-1)_{p+1} & \cdots & (m+k+r-1)_{p+r} \end{vmatrix}_{r+1}$$

$$= \frac{(m)_p(m+1)_p \cdots (m+h)_p(m+h+k)_p \cdots (m+k+r-1)_p}{(p+1)_r(p+2)_r \cdots (p+r)_r} \cdot \Delta',$$

where

$$\Delta' = \left| (m-p)_0(m-p+1)_1 \cdots (m-p+h)_h \right.$$
$$\left. (m-p+h+k)_{h+1} \cdots (m-p+k+r)_r \right|_{r+1}$$

But this by §735 is equal to

$$\frac{(r+1)_{h+1} \cdots (r+k-1)_{h+1}}{(h+1)_{h+1} \cdots (h+k-1)_{h+1}}.$$

Therefore

$$\Delta = \frac{(m)_p(m+1)_p \cdots (m+h)_p(m+h+k)_p \cdots (m+r+k-1)_p}{(p+1)_1(p+2)_2 \cdots (p+r)_r}$$

$$\times \frac{(r+1)_{h+1} \cdots (r+k-1)_{h+1}}{(h+1)_{h+1} \cdots (h+k-1)_{h+1}} = (m+r)_{r+1},$$

when $p=1$ and $k=1$.

737. By using the formula of §729 the determinant

$$\begin{vmatrix} (m)_0 & \cdots & (m)_p & (m)_{p+q} & \cdots & (m)_{r+q-1} \\ (m+1)_0 & \cdots & (m+1)_p & (m+1)_{p+q} & \cdots & (m+1)_{r+q-1} \\ \cdot & \cdot & \cdot & \cdot & \cdot & \cdot \\ (m+p-1)_0 & \cdots & (m+p+1)_p & (m+p+1)_{p+q} & \cdots & (m+p+1)_{r+q-1} \\ \cdot & \cdot & \cdot & \cdot & \cdot & \cdot \\ (m+r)_0 & \cdots & (m+r)_p & (m+r)_{p+q} & \cdots & (m+r)_{r+q-1} \end{vmatrix}_{r+1}$$

may readily be reduced to the determinant

$$\begin{vmatrix} (m+1)_q & \cdots & (m+1)_{r+q-p-1} \\ \cdot & \cdot & \cdot \\ (m+r-p)_q & \cdots & (m+r-p)_{r+q-p-1} \end{vmatrix}_{r-p}$$

which by (b) §734 is

$$= \frac{(m+1)_q(m+2)_q \cdots (m+r-p)_q}{(q+1)_q(q+2)_q \cdots (r+q-p-1)_q}.$$

If $q=m$ then

$$\Delta = (m+r-p)_m$$

738. The determinant

$$D \equiv \begin{vmatrix} (\alpha_1)_0 & (\alpha_2)_0 & \cdots & (\alpha_m)_0 \\ (\alpha_1)_1 & (\alpha_2)_1 & \cdots & (\alpha_m)_1 \\ \cdot & \cdot & \cdot & \cdot \\ (\alpha_1)_m & (\alpha_2)_m & \cdots & (\alpha_m)_m \end{vmatrix}$$

after taking out the factors

$$\frac{1}{1!},\ \frac{1}{2!},\ \cdots,\ \frac{1}{m!},$$

is easily seen to be a simple alternant and equal to $\zeta^{1/2}(\alpha_1\alpha_2\cdots\alpha_m)$. Therefore

$$D = \frac{\zeta^{1/2}(\alpha_1\cdots\alpha_m)}{2!3!\cdots m!} = \frac{\zeta^{1/2}(\alpha_1\cdots\alpha_m)}{2^{m-1}3^{m-2}\cdots(m-1)^2m^1}.$$

739. The determinant

$$D \equiv \begin{vmatrix} (\alpha_1)_p & (\alpha_2)_p & \cdots (\alpha_{m+1})_p \\ (\alpha_1)_{p+1} & (\alpha_2)_{p+1} & \cdots (\alpha_{m+1})_{p+1} \\ \cdot & \cdot & \cdot \\ (\alpha_1)_{p+m} & (\alpha_2)_{p+m} & \cdots (\alpha_{m+1})_{p+m} \end{vmatrix}_{m+1}$$

on removing $(\alpha_1)_p, (\alpha_2)_p \cdots$ from the columns gives

$$D = \frac{(\alpha_1)_p(\alpha_2)_p\cdots(\alpha_{m+1})_p}{(p)_p(p+1)_p\cdots(p+m)_p}\left|(\alpha_1-p)_0\ (\alpha_2-p)_1\cdots(\alpha_{m+1}-p)_m\right|$$

$$= \frac{(\alpha_1)_p\cdots(\alpha_{m+1})_p}{(p)_p(p+1)_p\cdots(p+m)_p}\cdot\frac{\zeta^{1/2}(\alpha_1\alpha_2\cdots\alpha_{m+1})}{2^{m-1}3^{m-2}\cdots(m-1)^2m^1} \quad \text{(by §738)}$$

740. If in the determinant $|(m)_o\ (m+1)_1\cdots(m+r)_r|$ we substitute for the last column $x^n, (x+1)^n, \cdots, (x+r)^n$ and then expand in terms of the elements in the last column we get

$$(x+r)^n - (r)_{r-1}(x+r-1)^n + (r)_{r-2}(x+r-2)^n\cdots = S_n, \text{ say.}$$

The coefficient of $(x+r-q)^n$ in this expansion is given by §735 to be $(r)_{r-q}$. It is known from algebra that

$$S_r = r!,\ S_{r+1} = (\tfrac{1}{2}r + x)(r+1)! \text{ and } S_k = 0 \text{ for } k < r.$$

Exercise: If we substitute $x^n, (x+1)^n, \cdots, (x+r)^n$ for the last column of $|(m)_1\ (m+1)_2\cdots(m+r)_{r+1}|$ show that the result is

$$(-1)^r(x-m)^n \quad \text{when} \quad n \gtreqless r.$$

741. By making use of the formula

$$(a+b)_s = (a)_s + (a)_{s-1}(b)_1 + (a)_{s-2}(b)_2 + \cdots + (b)_s$$

it is readily seen that the determinant

$$\Delta \equiv \begin{vmatrix} (m+m')_r & (m+n')_r & \cdots & (m+p')_r & (m+q')_r \\ (n+m')_r & (n+n')_r & \cdots & (n+p')_r & (n+q')_r \\ \cdot & \cdot & \cdot & \cdot & \cdot \\ (p+m')_r & (p+n')_r & \cdots & (p+p')_r & (p+q')_r \\ (q+m')_r & (q+n')_r & \cdots & (q+p')_r & (q+q')_r \end{vmatrix}_{r+1}$$

is equal to the product

$$\begin{vmatrix} (m)_0 & (m)_1 & \cdots & (m)_r \\ (n)_0 & (n)_1 & \cdots & (n)_r \\ \cdot & \cdot & \cdot & \cdot \\ (p)_0 & (p)_1 & \cdots & (p)_r \\ (q)_0 & (q)_1 & \cdots & (q)_r \end{vmatrix} \begin{vmatrix} (m')_r & (m')_{r-1} & \cdots & (m')_0 \\ (n')_r & (n')_{r-1} & \cdots & (n')_0 \\ \cdot & \cdot & \cdot & \cdot \\ (p')_r & (p')_{r-1} & \cdots & (p')_0 \\ (q')_r & (q')_{r-1} & \cdots & (q')_0 \end{vmatrix}$$

which gives

$$\Delta = \frac{\zeta^{1/2}(m,n,\cdots,p,q)\zeta^{1/2}(m',n',\cdots,p',q')}{\{1^{r-1}2^{r-2}\cdots(r-1)^1\}^2}.$$

742. *The determinant formed by multiplying each element of*

$$\Delta_{(m,p,r,1)} \equiv \begin{vmatrix} (m)_p & (m)_{p+1} & \cdots & (m)_{p+r} \\ (m+1)_p & (m+1)_{p+1} & \cdots & (m+1)_{p+r} \\ \cdot & \cdot & \cdot & \cdot \\ (m+r)_p & (m+r)_{p+1} & \cdots & (m+r)_{p+r} \end{vmatrix}_{r+1}$$

by the corresponding element of the array

$$\begin{matrix} (m-p) & \alpha & \beta & \cdots & \rho \\ (m-p+1) & \alpha+1 & \beta+1 & \cdots & \rho+1 \\ \cdot & \cdot & \cdot & \cdot & \cdot \\ (m-p+r) & \alpha+r & \beta+r & \cdots & \rho+r \end{matrix}$$

is equal to the original determinant multiplied by

$$(m-p)(m-p+1)\cdots(m-p+r)$$

and is therefore independent of $\alpha, \beta, \cdots, \rho$.

To show the truth of this we remove

$$(m-p)(m)_p,\ (m-p+1)(m+1)_p,\ \cdots,(m-p+r)(m+r)_p$$

from the rows and

$$\frac{1}{1}, \frac{1}{p+1}, \frac{1}{(p+1)(p+2)}, \frac{1\cdot 2}{(p+1)(p+2)(p+3)}, \ldots, \frac{1\cdot 2\cdots(r-1)}{(p+1)\cdots(p+r)}$$

from the columns. What is left is

$$D \equiv \begin{vmatrix} 1 & \alpha & \beta(m-p-1)_1 & \gamma(m-p-1)_2 & \cdots & \rho(m-p-1)_{r-1} \\ 1 & \alpha+1 & (\beta+1)(m-p)_1 & (\gamma+1)(m-p)_2 & \cdots & (\rho+1)(m-p)_{r-1} \\ \cdot & \cdot & \cdot & \cdot & \cdot & \cdot \\ 1 & \alpha+r & (\beta+r)(m-p+r-1)_1 & (\gamma+r)(m-p+r-1)_2 & \cdots & (\rho+r)(m-p+r-1)_{r-1} \end{vmatrix}_{r+1}$$

Subtracting each row of D from the one following it and repeating this process on the result we get

$$D = \begin{vmatrix} 2 & \gamma+2(m-p)_1 & \cdots & \rho(m-p-1)_{r-3}+2(m-p)_{r-2} \\ 2 & \gamma+1+2(m-p+1)_1 & \cdots & (\rho+1)(m-p)_{r-3}+2(m-p+1)_{r-2} \\ \cdot & \cdot & \cdot & \cdot \\ 2 & (\gamma+r-2)+2(m-p+r-2)_1 & \cdots & (\rho+r-2)(m-p+r-3)_{r-3}+2(m-p+r-2)_{r-2} \end{vmatrix}_{r-1}$$

Taking out the factor 2 and repeating the process we get finally $r!$ as the value of D. It follows, therefore, that the theorem as stated is true.

743. *If each element of the determinant*

$$\begin{vmatrix} (m)_k & (m)_p & (m)_{p+1} & \cdots & (m)_{p+r-1} \\ (m+1)_k & (m+1)_p & (m+1)_{p+1} & \cdots & (m+1)_{p+r-1} \\ \cdot & \cdot & \cdot & \cdot & \cdot \\ (m+r)_k & (m+r)_p & (m+r)_{p+1} & \cdots & (m+r)_{p+r-1} \end{vmatrix}_{r+1}$$

where k has any value from p to $p+r-1$ inclusive, be multiplied by the corresponding element of the array

$$\begin{matrix} 1 & \alpha & \beta & \cdots & \rho \\ 1 & \alpha+1 & \beta+1 & \cdots & \rho+1 \\ \cdot & \cdot & \cdot & \cdot & \cdot \\ 1 & \alpha+r & \beta+r & \cdots & \rho+r \end{matrix}$$

the resulting determinant is equal to

$$(k-p+1)(k-p+2)\cdots r\frac{(m)_p(m+1)_p\cdots(m+r)_p}{(p)_p(p+1)_p\cdots(p+r-1)_p(k)_p}$$
$$(m-p-\alpha)(m-p-\beta-1)(m-p-\gamma-2)\cdots,$$

where the number of factors following the fraction is $k-p$.

The modified determinant is

$$D \equiv \begin{vmatrix} (m)_k & \alpha(m)_p & \beta(m)_{p+1} & \cdots & \rho(m)_{p+r-1} \\ (m+1)_k & (\alpha+1)(m+1)_p & (\beta+1)(m+1)_{p+1} & \cdots & (\rho+1)(m+1)_{p+r-1} \\ \cdot & \cdot & \cdot & \cdot & \cdot \\ (m+r)_k & (\alpha+r)(m+r)_p & (\beta+r)(m+r)_{p+1} & \cdots & (\rho+r)(m+r)_{p+r-1} \end{vmatrix}$$

and if we take from the rows the factors $(m)_p, (m+1)_p \cdots (m+r)_p$ and from the columns the factors

$$\frac{1}{(k)_p}, \frac{1}{(p+1)_p}, \frac{1}{(p+2)_p}, \cdots, \frac{1}{(p+r-1)_p},$$

what remains is

$$D' \equiv \begin{vmatrix} (m-p)_{k-p} & \alpha & \beta(m-p)_1 & \cdots & \rho(m-p)_{r-1} \\ (m-p+1)_{k-p} & \alpha+1 & (\beta+1)(m-p+1)_1 & \cdots & (\rho+1)(m-p+1)_{r-1} \\ \cdot & \cdot & \cdot & \cdot & \cdot \\ (m-p+k-p)_{k-p} & \alpha+k-p & (\beta+k-p)(m-p+k-p)_1 & \cdots & (\rho+k-p)(m-p+k-p)_{r-1} \\ \cdot & \cdot & \cdot & \cdot & \cdot \\ (m-p+r)_{k-p} & \alpha+r & (\beta+r)(m-p+r)_1 & \cdots & (\rho+r)(m-p+r)_{r-1} \end{vmatrix}_{r+1}$$

From the $(h+1)$st row of D' subtract the hth row first for $h=1, 2, \cdots, r$, then for $h=2, 3, \cdots, r$, and so on down to $h=r$ and we get

$$D' \equiv \begin{vmatrix} (m-p)_{k-p} & \alpha & \beta(m-p)_1 & \gamma(m-p)_2 & \cdots & \rho(m-p)_{r-1} \\ (m-p)_{p-k-1} & 1 & \beta+(m-p-1)_1 & \gamma(m-p)_1+(m-p+1)_2 & \cdots & \rho(m-p)_{r-2}+(m-p+1)_{r-2} \\ (m-p)_{k-p-2} & 0 & 2 & \gamma+2(m-p+1)_1 & \cdots & \rho(m-p)_{r-3}+2(m-p+1)_{r-2} \\ (m-p)_{k-p-3} & 0 & 0 & 3 & \cdots & \rho(m-p)_{r-4}+3(m-p+1)_{r-3} \\ \cdot & \cdot & \cdot & \cdot & \cdot & \cdot \\ (m-p)_0 & 0 & 0 & 0 & \cdots & \rho(m-p)_{r-k+p-1}+(k-p)(m-p+1)_{r-k+p} \\ \cdot & \cdot & \cdot & \cdot & \cdot & \cdot \\ 0 & 0 & 0 & 0 & 0 \cdots & r \end{vmatrix}_{r+1}$$

$$= (k-p+1)(k-p+2)\cdots r \times D'' \qquad \text{where}$$

$$D'' = \begin{vmatrix} (m-p)_{k-p} & \alpha & \beta(m-p)_1 & \cdots & \theta(m-p)_{k-p-1} \\ (m-p)_{k-p-1} & 1 & \beta+(m-p-1)_1 & \cdots & \theta(m-p)_{k-p-2}+(m-p+1)_{k-p-1} \\ (m-p)_{k-p-2} & 0 & 2 & \cdots & \theta(m-p)_{k-p-3}+2(m-p+1)_{k-p-2} \\ \cdot & \cdot & \cdot & \cdot & \cdot \\ (m-p)_1 & 0 & 0 & \cdots & \theta+(k-p-1)(m-p+1)_1 \\ (m-p)_0 & 0 & 0 & \cdots & k-p \end{vmatrix}_{k-p+1}$$

If in D'' we subtract $(k-p)$ times the first column from the last we get

$$D'' = (-1)^{k-p}$$

$$\times \begin{vmatrix} \alpha & \beta(m-p)_1 & \gamma(m-p)_2 & \cdots & -(m-p)_{k-p-1}(m-k-\theta+1) \\ 1 & \beta+(m-p+1)_1 & \gamma(m-p)_1+(m-p+1)_2 & \cdots & -(m-p)_{k-p-2}(m-k-\theta+1) \\ 0 & 2 & \gamma+2(m-p+1)_1 & \cdots & -(m-p)_{k-p-3}(m-k-\theta+1) \\ \cdot & \cdot & \cdot & \cdot & \cdot \\ 0 & 0 & 0 & \cdots & -(m-k-\theta+1) \end{vmatrix}_{k-p}$$

from which we may take the factor $-(m-k-\theta+1)$ leaving the determinant

$$\begin{vmatrix} \alpha & \beta(m-p)_1 & \cdots & \eta(m-p)_{k-p-2} & (m-p)_{k-p-1} \\ 1 & \beta+(m-p)_1 & \cdots & \eta(m-p)_{k-p-3}+(m-p+1)_{k-p-2} & (m-p)_{k-p-2} \\ 0 & 2 & \cdots & \eta(m-p)_{k-p-4}+2(m-p+1)_{k-p-3} & (m-p)_{k-p-3} \\ \cdot & \cdot & \cdot & \cdot & \cdot \\ 0 & 0 & \cdots & k-p-1 & 1 \end{vmatrix}_{k-p}$$

If in this we take $k-p-1$ times the last column from the second last we get on taking out the factor $-(m-k-\eta+2)$

$$\begin{vmatrix} \alpha & \beta(m-p)_1 & \cdots & (m-p)_{k-p-2} \\ 1 & \beta+(m-p)_1 & \cdots & (m-p)_{k-p-3} \\ 0 & 2 & \cdots & (m-p)_{k-p-4} \\ \cdot & \cdot & \cdot & \cdot \\ 0 & 0 & \cdots & 1 \end{vmatrix}_{k-p-1}$$

By continuing this process we find that

$$D'' = (m-k-\theta+1)(m-k-\eta+2)\cdots(m-p-\alpha),$$

and, therefore,

$$D = (m-p-\alpha)\cdots(m-k-\theta+1)\frac{(m)_p\cdots(m+r)_p}{(p)_p\cdots(p+r-1)_p}\,\frac{1}{(k)_p}$$

$$(k-p+1)\cdots r.$$

It may be observed that this result does not involve the quantities from θ to and including ρ. When $k=p$, then $D'=r!$.

744. *If each element of the determinant*

$$\begin{vmatrix} 1 & (m)_0 & (m)_1 & \cdots & (m)_{r-1} \\ 1 & (m+d)_0 & (m+d)_1 & \cdots & (m+d)_{r-1} \\ 1 & (m+2d)_0 & (m+2d)_1 & \cdots & (m+2d)_{r-1} \\ \cdot & \cdot & \cdot & \cdots & \cdot \\ 1 & (m+rd)_0 & (m+rd)_1 & \cdots & (m+rd)_{r-1} \end{vmatrix}_{r+1}$$

be multiplied by the corresponding element of the array

$$\begin{matrix} 1 & \alpha & \beta & \cdots & \rho \\ 1 & \alpha+d & \beta+d & \cdots & \rho+d \\ \cdot & \cdot & \cdot & \cdots & \cdot \\ 1 & \alpha+rd & \beta+rd & \cdots & \rho+rd \end{matrix}$$

the resulting determinant, D say, is equal to $r!d^{r(r+1)/2}$.

For if we subtract each row of D from the one following, the determinant reduces to the order r and d comes out as a factor. Repeating this process we get determinants of order one less each time from which $2d^2, 3d^3, \cdots, rd^r$ come out as factors, and

$$D = r!d^{r(r+1)/2}.$$

745. *If each element of the determinant*

$$\begin{vmatrix} (m)_k & (m)_0 & \cdots & (m)_{r-1} \\ (m+1)_k & (m+1)_0 & \cdots & (m+1)_{r-1} \\ \cdot & \cdot & \cdots & \cdot \\ (m+r)_k & (m+r)_0 & \cdots & (m+r)_{r-1} \end{vmatrix}$$

be multiplied by the corresponding element of the array

$$\begin{matrix} 1 & \alpha & \beta & \cdots & \rho \\ 1 & \alpha+d & \beta+d & \cdots & \rho+d \\ \cdot & \cdot & \cdot & \cdots & \cdot \\ 1 & \alpha+rd & \beta+rd & \cdots & \rho+rd \end{matrix}$$

the resulting determinant, D say, is equal to

$$(k+1)(k+2)\cdots r\cdot d^{r-k}(md-\alpha)(\overline{m-1}d-\beta)\cdots$$

where the number of factors following d^{r-k} is k.

By continued subtraction of each row from the one following we get

$$\begin{vmatrix} (m)_k & \alpha & \beta(m)_1 & \cdots & \theta(m)_{k-1} \\ (m)_{k-1} & d & \beta + d(m+1)_0 & \cdots & \theta(m)_{k-2} + d(m+1)_{k-1} \\ \cdot & \cdot & \cdot & \cdots & \cdot \\ 1 & 0 & 0 & \cdots & kd \end{vmatrix}_{k+1}$$

$\times (k+1)d(k+2)d \cdots rd.$

But the determinant factor here is, by the method used to find D'' §743 seen to be

$$(md - \alpha)(\overline{m-1}d - \beta) \cdots (\overline{m-k+1}d - \theta).$$

Therefore

$$D = (md-\alpha)(\overline{m-1}d - \beta) \cdots (\overline{m-k+1}d - \theta)(k+1)(k+2) \cdots rd^{r-k}$$

746. If from the determinant

$$\begin{vmatrix} (m)_k & \alpha(m)_p & \cdots & \theta(m)_{p+k-p-1} & \cdots & \rho(m)_{p+r-1} \\ (m+1)_k & (\alpha+d)(m+1)_p & \cdots & (\theta+d)(m+1)_{k-1} & \cdots & (\rho+d)(m+1)_{p+r-1} \\ \cdot & \cdot & \cdots & \cdot & \cdots & \cdot \\ (m+r)_k & (\alpha+rd)(m+r)_p & \cdots & (\theta+rd)(m+r)_{k-1} & \cdots & (\rho+rd)(m+r)_{p+r-1} \end{vmatrix}$$

$\equiv D$, we take the factors

$$\frac{(m)_p(m+1)_p \cdots (m+r)_p}{(p)_p(p+1)_p \cdots (p+r-1)_p(k)_p}$$

what remains is

$$D' \equiv \begin{vmatrix} (m-p)_{k-p} & \alpha & \cdots & \theta(m-p)_{k-p-1} & \cdots & \rho(m-p)_{r-1} \\ (m-p+1)_{k-p} & \alpha+d & \cdots & (\theta+d)(m-p+1)_{k-p-1} & \cdots & (\rho+d)(m-p+1)_{r-1} \\ \cdot & \cdot & \cdots & \cdot & \cdots & \cdot \\ (m-p+r)_{k-p} & \alpha+rd & \cdots & (\theta+rd)(m-p+r)_{k-p-1} & \cdots & (\rho+rd)(m-p+r)_{r-1} \end{vmatrix}_{r+1}.$$

By continued subtraction of each row from the one following as in §743 we get

$$D' = \begin{vmatrix} (m-p)_{k-p} & \alpha \cdots & \theta(m-p)_{k-p-1} & \cdots & \rho(m-p)_{r-1} \\ (m-p)_{k-p-1} & d \cdots & \theta(m-p)_{k-p-2} + d(m-p+1)_{k-p-1} & \cdots & \rho(m-p)_{r-2} + d(m-p+1)_{r-1} \\ \cdot & \cdot & \cdot & \cdots & \cdot \\ (m-p)_0 & 0 \cdots & (k-p)d & \cdots & \rho(m-p)_{r-k+p-1} + (k-p)(m-p+1)_{r-k+p} \\ \cdot & \cdot & \cdot & \cdots & \cdot \\ 0 & 0 \cdots & 0 & \cdots & rd \end{vmatrix}$$

$$= (k-p+1)(k-p+2) \cdots rd^{r-k+p}D'',$$

where

$$D'' = \begin{vmatrix} (m-p)_{k-p} & \alpha & \beta(m-p)_1 & \cdots & \theta(m-p)_{k-p-1} \\ (m-p)_{k-p-1} & d & \beta + d(m-p+1)_1 & \cdots & \theta(m-p)_{k-p-2}d + (m-p+1)_{k-p-1} \\ \cdots & \cdots & \cdots & \cdots & \cdots \\ 1 & 0 & 0 & \cdots & (k-p)d \end{vmatrix}$$

Subtracting $(k-p)d$ times the first column from the last in D'' we see that the factor $-(\overline{m-k+1}d-\theta)$ comes out, and repeating the process on what is left as for D'' in §743 we get

$$D'' = (\overline{m-k+1}d - \theta) \cdots (\overline{m-p-1}d - \beta)(\overline{m-p}d - \alpha).$$

Therefore

$$D = (\overline{m-k+1}d - \theta) \cdots (\overline{m-p-1}d - \beta)(\overline{m-p}d - \alpha)$$
$$\frac{(m)_p(m+1)_p \cdots (m+r)_p}{(p)_p(p+1)_p \cdots (p+r-1)_p(k)_p}(k-p+1)(k-p+2) \cdots rd^{r-k+p}.$$

The result given in §743 is a particular case of this when $d=1$.
(a) When $k=1$ and $p=1$, then

$$D = \frac{(m+r)!}{m!} \cdot d^r$$

(b) When $k=1$, $p=1$ and $d=2$, then

$$D = \frac{(m+r)!}{m!} 2^r$$

That is

$$\begin{vmatrix} (m)_1 & (m)_1 & (m)_2 & \cdots & (m)_r \\ (m+1)_1 & 3(m+1)_1 & 3(m+1)_2 & \cdots & 3(m+1)_r \\ (m+2)_1 & 5(m+2)_1 & 5(m+2)_2 & \cdots & 5(m+2)_r \\ \cdots & \cdots & \cdots & \cdots & \cdots \\ (m+r)_1 & (2r+1)(m+r)_1 & (2r+1)(m+r)_2 & \cdots & (2r+1)(m+r)_r \end{vmatrix}$$
$$= \frac{(m+r)!}{m!} 2^r.$$

Dividing both sides of this relation by

$$3 \cdot 5 \cdots (2r+1) \frac{(m+r)!}{m!}$$

we have

$$\begin{vmatrix} 1 & 1 & \frac{(m-1)_1}{2} & \cdots & \frac{(m-1)_{r-1}}{r} \\ \frac{1}{3} & 1 & \frac{m}{2} & \cdots & \frac{(m)_{r-1}}{r} \\ \frac{1}{5} & 1 & \frac{m+1}{2} & \cdots & \frac{(m+1)_{r-1}}{r} \\ \cdot & \cdot & \cdot & \cdot & \cdot \\ \frac{1}{2r+1} & 1 & \frac{m+r-1}{2} & \cdots & \frac{(m+r-1)_{r-1}}{r} \end{vmatrix} = \frac{2^r}{3 \cdot 5 \cdots (2r+1)}$$

or

$$\Delta \equiv \begin{vmatrix} 1 & 1 & (m-1)_1 & \cdots & (m-1)_{r-1} \\ \frac{1}{3} & 1 & (m)_1 & \cdots & (m)_{r-1} \\ \cdot & \cdot & \cdot & \cdot & \cdot \\ \frac{1}{2r+1} & 1 & (m+r-1)_1 & \cdots & (m+r-1)_{r-1} \end{vmatrix}$$

$$= \frac{2 \cdot 3 \cdots r}{3 \cdot 5 \cdots (2r+1)} 2^r = \frac{2 \cdot 4 \cdot 6 \cdots 2r}{3 \cdot 5 \cdots (2r+1)} \cdot$$

From this we get by Wallis' theorem

$$\lim_{r=\infty} (2r+1)\Delta^2 = \frac{\pi}{2} \cdot$$

747. *If we multiply the elements of the determinant*

$$\begin{vmatrix} (m)_0 & (m)_1 & \cdots & (m)_r \\ (m+1)_0 & (m+1)_1 & \cdots & (m+1)_r \\ \cdot & \cdot & \cdot & \cdot \\ (m+r)_0 & (m+r)_1 & \cdots & (m+r)_r \end{vmatrix}$$

by the corresponding elements of the array

$$\begin{matrix} 1 & 1 & \cdots & 1 \\ \alpha & \alpha+1 & \cdots & \alpha+r \\ \beta & \beta+1 & \cdots & \beta+r \\ \cdot & \cdot & \cdot & \cdot \\ \rho & \rho+1 & \cdots & \rho+r \end{matrix}$$

the value of the resulting determinant, D say, is

$$(m+\alpha+1)(m+\beta+2)\cdots(m+\rho+r).$$

In the determinant

$$D = \begin{vmatrix} 1 & (m)_1 & (m)_2 & \cdots & (m)_r \\ \alpha & (\alpha+1)(m+1)_1 & (\alpha+2)(m+1)_2 & \cdots & (\alpha+r)(m+1)_r \\ \cdot & \cdot & \cdot & \cdot & \cdot \\ \rho & (\rho+1)(m+r)_1 & (\rho+2)(m+r)_2 & \cdots & (\rho+r)(m+r)_r \end{vmatrix}$$

subtract α times the first row from the second, β times the first from the third and so on. The result is

$$\begin{vmatrix} 1 & (m)_1 & \cdots & (m)_r \\ 0 & m+\alpha+1 & \cdots & (m)_{r-1}(m+\alpha+1) \\ 0 & (m+\beta+2)+\beta & \cdots & (m+1)_{r-1}(m+\beta+2)+(m)_{r-1}\beta \\ 0 & (m+\gamma+3)+2\gamma & \cdots & (m+2)_{r-1}(m+\gamma+3)+(m+1)_{r-1}\gamma+(m)_{r-1}\gamma \\ \cdot & \cdot & \cdot & \cdot \\ 0 & (m+\rho+r)+(r-1)\rho & \cdots & (m+r-1)_{r-1}(m+\rho+r)+\{(m+r-2)_{r-1}+\cdots+(m)_{r-1}\}\rho \end{vmatrix}_{r+1}$$

If in this we reduce the order to r, take out the factor $(m+\alpha+1)$ and in the resulting form subtract β times the first row from the second, γ times the first from the third and so on, we get a result from which we may take $(m+\beta+2)$ as a factor. Continuing this process we get

$$D = (m+\alpha+1)(m+\beta+2)\cdots(m+\rho+r).D'$$

where

$$D' = \begin{vmatrix} 1 & (m)_1 & \cdots & (m)_{r-1} \\ 1 & (m+1)_1 & \cdots & (m+1)_{r-1} \\ \cdot & \cdot & \cdot & \cdot \\ 1 & (m+r)_1 & \cdots & (m+r)_{r-1} \end{vmatrix}_r$$

which has the value 1.

Therefore

$$D = (m + \alpha + 1)(m + \beta + 2) \cdots (m + \rho + r).$$

748. The determinant

$$\begin{vmatrix} (m)_0\alpha^m & (m)_1\alpha^{m-t} & \cdots & (m)_r\alpha^{m-rt} \\ (m+d)_0\alpha^{m+d} & (m+d)_1\alpha^{m+d-t} & \cdots & (m+d)_r\alpha^{m+d-rt} \\ (m+2d)_0\alpha^{m+2d} & (m+2d)_1\alpha^{m+2d-t} & \cdots & (m+2d)_r\alpha^{m+2d-rt} \\ \cdots & \cdots & \cdots & \cdots \\ (m+rd)_0\alpha^{m+rd} & (m+rd)_1\alpha^{m+rd-t} & \cdots & (m+rd)_r\alpha^{m+rd-rt} \end{vmatrix}$$

may be freed of the α's by multiplying the columns by $\alpha^0, \alpha^t, \cdots, \alpha^{rt}$ and then dividing the rows by $\alpha^m, \alpha^{m+d}, \cdots, \alpha^{m+rd}$ so that $\alpha^{\{2m+r(d-t)\}(r+1)/2}$ is a factor of D. The remaining factor is

$$\begin{vmatrix} (m)_0 & (m)_1 & \cdots & (m)_r \\ (m+d)_0 & (m+d)_1 & \cdots & (m+d)_r \\ \cdots & \cdots & \cdots & \cdots \\ (m+rd)_0 & (m+rd)_1 & \cdots & (m+rd)_r \end{vmatrix}$$

which by §732 is equal to $d^{r(r+1)/2}$.

Therefore

$$D = \alpha^{\{2m+r(d-t)\}r(r+1)/2} \cdot d^{r(r+1)/2}.$$

Exercises. Set XXXVI

1. Show that

$$\begin{vmatrix} (5)_2 & (5)_3 & (5)_4 \\ (6)_2 & (6)_3 & (6)_4 \\ (7)_2 & (7)_3 & (7)_4 \end{vmatrix}, \quad \begin{vmatrix} (5)_2 & (5)_3 & (5)_4 \\ (5)_1 & (5)_2 & (5)_3 \\ (5)_0 & (5)_1 & (5)_2 \end{vmatrix} \quad \text{and} \quad \begin{vmatrix} (5)_2 & (6)_3 & (7)_4 \\ (6)_2 & (7)_3 & (8)_4 \\ (7)_2 & (8)_3 & (9)_4 \end{vmatrix}$$

are equivalent forms and each equal to 175.

2. Show that

$$\begin{vmatrix} 1 & (p+1)_1 & (p+2)_2 & \cdots & (p+n-1)_{n-1} \\ 1 & (p+2)_1 & (p+3)_2 & \cdots & (p+n)_{n-1} \\ \cdots & \cdots & \cdots & \cdots & \cdots \\ 1 & (p+n)_1 & (p+n+1)_2 & \cdots & (p+2n-2)_{n-1} \end{vmatrix} = 1$$

3. Show that

$$\begin{vmatrix} 2_1 & 3_1 & 4_1 & \cdots & n_1 \\ 2_2 & 3_2 & 4_2 & \cdots & n_2 \\ \cdot & \cdot & \cdot & \cdot & \cdot \\ 2_{n-1} & 3_{n-1} & 4_{n-1} & \cdots & n_{n-1} \end{vmatrix} =$$

$$\begin{vmatrix} 2_1 & 3_1 & 4_1 & \cdots & n_1 \\ 3_2 & 4_2 & 5_2 & \cdots & (n+1)_2 \\ \cdot & \cdot & \cdot & \cdot & \cdot \\ n_{n-1} & (n+1)_{n-1} & (n+2)_{n-1} & \cdots & (2n-2)_{n-1} \end{vmatrix} = n$$

4. Show that

$$\begin{vmatrix} (m)_0\alpha^0 & (m+d)_0\alpha^d & \cdots & (m+zd)_0\alpha^{zd} \\ (m)_1\alpha^0 & (m+d)_1\alpha^d & \cdots & (m+zd)_1\alpha^{zd} \\ \cdot & \cdot & \cdot & \cdot \\ (m)_{p-1}\alpha^0 & (m+d)_{p-1}\alpha^d & \cdots & (m+zd)_{p-1}\alpha^{zd} \\ (m)_0\beta^0 & (m+d)_0\beta^d & \cdots & (m+zd)_0\beta^{zd} \\ \cdot & \cdot & \cdot & \cdot \\ (m)_{q-1}\beta^0 & (m+d)_{q-1}\beta^d & \cdots & (m-zd)_{q-1}\beta^{zd} \end{vmatrix},$$

where $z=p+q-1$, is equal to

$$d^{p(p-1)/2+q(q-1)/2}\cdot\alpha^{p(p-1)d/2}\beta^{q(q-1)d/2}(\beta^d-\alpha^d)^{pq}.$$

5. Show that

$$\begin{vmatrix} (n)_k & (n)_{k-1} & (n)_{k-2} & \cdots \\ (n)_{k+1} & (n)_k & (n)_{k-1} & \cdots \\ (n)_{k+2} & (n)_{k-1} & (n)_k & \cdots \\ \cdot & \cdot & \cdot & \cdot \end{vmatrix}_{m-k+1}$$

$$= \begin{vmatrix} (n+m-k)_{m-k+1} & (n+m-k-1)_{m-k} & \cdots \\ (n+m-k+1)_{m-k+2} & (n+m-k)_{m-k+1} & \cdots \\ (n+m-k+2)_{m-k+3} & (n+m-k+1)_{m-k+2} & \cdots \\ \cdot & \cdot & \cdot \end{vmatrix}_k$$

(Studnicka.)

CHAPTER XXI

RECURRENTS

749. It is readily seen that any binary quantic

$$a_0x^n + a_1x^{n-1}y + a_2x^{n-2}y^2 + \cdots + a_ny^n$$

can be written in the form of a determinant of the $(n+1)$st order,

$$\begin{vmatrix} a_0 & a_1 & a_2 & \cdots & a_n \\ y & x & 0 & \cdots & 0 \\ 0 & y & x & \cdots & 0 \\ \cdot & \cdot & \cdot & \cdot & \cdot \\ 0 & 0 & 0 & \cdots & x \end{vmatrix}.$$

750. The determinant

$$\Delta_{n+1} \equiv \begin{vmatrix} 1 & -1 & 0 & 0 & 0 \cdots \\ x & h & -1 & 0 & 0 \cdots \\ x^2 & hx & h & -1 & 0 \cdots \\ x^3 & hx^2 & hx & h & -1 \cdots \\ \cdot & \cdot & \cdot & \cdot & \cdot \end{vmatrix}_{n+1} = (x+h)^n$$

as is obvious from the *reduction* or *recurrent* formula

$$\Delta_{n+1} = (x+h)\begin{vmatrix} 1 & -1 & 0 & 0 \cdots \\ x & h & -1 & 0 \cdots \\ x^2 & hx & h & -1 \\ \cdot & \cdot & \cdot & \cdot \end{vmatrix}_n = (x+h)\cdot\Delta_n$$

751. The determinant

$$\Delta_{n+1} \equiv \begin{vmatrix} a_0 & a_1 & a_2 & \cdots & a_n \\ -a_{11} & a_{11}x - a_{12} & a_{12}x - a_{13} & \cdots & a_{1n}x \\ -a_{21} & a_{21}x - a_{22} & a_{22}x - a_{23} & \cdots & a_{2n}x \\ -a_{31} & a_{31}x - a_{32} & a_{32}x - a_{33} & \cdots & a_{3n}x \\ \cdot & \cdot & \cdot & \cdot & \cdot \end{vmatrix}_{n+1}$$

$$= |a_{nn}|(a_0x^n + a_1x^{n-1} + \cdots + a_n)$$

For performing the operations

$$c_{n+1} + xc_n + x^2c_{n-1} + \cdots + x^nc_1$$

we get the result

$$(-1)^n(a_0x^n + a_1x^{n-1} + \cdots + a_n)\cdot\begin{vmatrix} -a_{11} & a_{11}x - a_{12} & a_{12}x - a_{13}\cdots \\ -a_{21} & a_{21}x - a_{22} & a_{22}x - a_{33}\cdots \\ -a_{31} & a_{31}x - a_{32} & a_{32}x - a_{33}\cdots \\ \cdots & \cdots & \cdots \end{vmatrix}_n$$

$$= (a_0x^n + a_1x^{n-1} + \cdots + a_n)\cdot|a_{nn}|$$

as may be seen by adding x times each column to the one following it.

752. The determinant

$$\Delta_n \equiv \begin{vmatrix} a_1 & -1 & 0 & 0\cdots & \cdot \\ a_2 & b_2 & -1 & 0\cdots & \cdot \\ a_3 & 0 & b_3 & -1\cdots & \cdot \\ \cdots & \cdots & \cdots & \cdots & \cdots \\ a_{n-1} & 0 & 0 & 0\cdots b_{n-1} & -1 \\ a_n & 0 & 0 & 0\cdots 0 & b_n \end{vmatrix}$$

$$= a_1b_2b_3\cdots b_n + \begin{vmatrix} a_2 & -1 & 0\cdots \\ a_3 & b_3 & -1\cdots \\ \cdots & \cdots & \cdots \end{vmatrix}_{n-1}$$

$$= a_1b_2b_3\cdots b_n + \Delta_{n-1}$$

$$= a_1b_2\cdots b_n + a_2b_3\cdots b_n + a_3b_4\cdots b_n + \cdots + a_{n-1}b_n + a_n.$$

753. By continued expansion in terms of elements of the last row the determinant

$$\begin{vmatrix} n!a_0 & -n & 0 & 0 & \cdots & 0 & 0 \\ (n-1)!a_1 & x & -(n-1) & 0 & \cdots & 0 & 0 \\ (n-2)!a_2 & 0 & x & -(n-2) & \cdots & 0 & 0 \\ \cdots & \cdots & \cdots & \cdots & \cdots & \cdots & \cdots \\ 2!a_{n-2} & 0 & 0 & 0 & \cdots & -2 & 0 \\ a_{n-1} & 0 & 0 & 0 & \cdots & x & -1 \\ a_n & 0 & 0 & 0 & \cdots & 0 & x \end{vmatrix}_{n+1}$$

is seen to be equal to

$$= n!(a_0x^n + a_1x^{n-1} + \cdots + a_n).$$

754. Another way to write the equation

$$x^n + a_1x^{n-1} + a_2x^{n-2} + \cdots + a_n = 0$$

in determinant form is to take it with the equations

$$\begin{aligned} &S_1 + a_1 &&= 0 \\ &S_2 + a_1S_1 + 2a_2 &&= 0 \\ &S_3 + a_1S_2 + a_2S_1 + 3a_3 &&= 0 \\ &\cdots\cdots\cdots\cdots \end{aligned}$$

and eliminate the a's with the result

$$\begin{vmatrix} S_1 & 1 & 0 & 0 & \cdots 0 \\ S_2 & S_1 & 2 & 0 & \cdots 0 \\ S_3 & S_2 & S_1 & 3 & \cdots 0 \\ \cdots & \cdots & \cdots & \cdots & \cdots \\ S_n & S_{n-1} & S_{n-2} & S_{n-3} & \cdots n \\ x^n & x^{n-1} & x^{n-2} & x^{n-3} & \cdots 1 \end{vmatrix}_{n+1} = 0$$

755. Expanding the determinant

$$\Delta_{n-1} \equiv \begin{vmatrix} x+y & xy & 0 & 0 \cdots \\ 1 & x+y & xy & 0 \cdots \\ 0 & 1 & x+y & xy \cdots \\ \cdots & \cdots & \cdots & \cdots \end{vmatrix}_{n-1}$$

in terms of the elements of the first row gives

$$\begin{aligned} \Delta_{n-1} &= (x+y)\Delta_{n-2} - xy\Delta_{n-3} \\ &= \frac{x^3 - y^3}{x - y}\Delta_{n-3} - xy\frac{x^2 - y^2}{x - y}\Delta_{n-4} \\ &= \frac{x^4 - y^4}{x - y}\Delta_{n-4} - xy\frac{x^3 - y^3}{x - y}\Delta_{n-5} \\ &\cdots\cdots\cdots\cdots \\ &= \frac{x^n - y^n}{x - y}. \end{aligned}$$

756. If $(u/v)^r$ is used to express the rth differential-quotient of u/v in terms of the rth and lower differential-quotients of u and of v, then we can write

$$\left(\frac{u}{v}\right)' = \begin{vmatrix} v & v' \\ u & u' \end{vmatrix} \cdot v^{-2}, \quad \left(\frac{u}{v}\right)'' = -\begin{vmatrix} 0 & v & 2v' \\ v & v' & v'' \\ u & u' & u'' \end{vmatrix} \cdot v^{-3},$$

$$\left(\frac{u}{v}\right)''' = \begin{vmatrix} 0 & 0 & v & 3v' \\ 0 & v & 2v' & 3v'' \\ v & v' & v'' & v''' \\ u & u' & u'' & u''' \end{vmatrix} \cdot v^{-4},$$

$$\left(\frac{u}{v}\right)^{\mathrm{IV}} = -\begin{vmatrix} 0 & 0 & 0 & v & 4v' \\ 0 & 0 & v & 3v' & 6v'' \\ 0 & v & 2v' & 3v'' & 4v''' \\ v & v' & v'' & v''' & v^{\mathrm{IV}} \\ u & u' & u'' & u''' & u^{\mathrm{IV}} \end{vmatrix} \cdot v^{-5},$$

etc.

757. Starting with Newton's formulae

$$a_1 + S_1 = 0$$
$$2a_2 + a_1S_1 + S_2 = 0$$
$$3a_3 + a_2S_1 + a_3S_2 + S_3 = 0$$
$$\cdots\cdots\cdots\cdots\cdots$$

we have from the first r of the equations

$$S_r = (-1)^r \begin{vmatrix} a_1 & 1 & 0 & \cdots 0 \\ 2a_2 & a_1 & 1 & \cdots 0 \\ 3a_3 & a_2 & a_1 & \cdots 0 \\ \cdots & \cdots & \cdots & \cdots \\ ra_r & a_{r-1} & a_{r-2} & \cdots a_1 \end{vmatrix}_r$$

which by Waring's formula gives for $r = n$,

$$S_n = n \sum \epsilon \frac{(-1)^{n+\epsilon}(\epsilon - 1)!}{\epsilon_1!\epsilon_2! \cdots \epsilon_n!} \cdot a_1^{\epsilon_1} a_2^{\epsilon_2} \cdots a_n^{\epsilon_n},$$

where

$$\epsilon_1 + \epsilon_2 + \cdots + \epsilon_n = \epsilon,$$

$$\epsilon_1 + 2\epsilon_2 + 3\epsilon_3 + \cdots + n\epsilon_n = n;$$

and where the ϵ's are to have all positive integral values, zero included, which satisfy the second of these two equations.

EXERCISE: Show that the determinant expression for S_r gives the sum of the divisors of any integer r if

$$a_{p(p+1)/2} = (-1)^p(2p+1)$$

and the other a's vanish.

From the same set of Newton's equations we find on solving for a_r

$$a_r = \frac{(-1)^r}{r!}\begin{vmatrix} S_1 & 1 & 0 & \cdots \\ S_2 & S_1 & 1 & \cdots \\ S_3 & S_2 & S_1 & \cdots \\ \cdot & \cdot & \cdot & \cdot \end{vmatrix}_r$$

758. If

$$\phi(x) = c_0x^m + c_1x^{m-1} + \cdots + c_m,$$

$$f(x) = a_0x^n + a_1x^{n-1} + \cdots + a_n,$$

and

$$\frac{\phi(x)}{f(x)} = A_0x^{m-n} + A_1x^{m-n-1} + A_2x^{m-n-2} + \cdots$$

We have on equating coefficients on both sides of this equation

$$c_0 = a_0A_0$$

$$c_1 = A_0a_1 + A_1a_0$$

$$c_2 = A_0a_2 + A_1a_1 + A_2a_0$$

$$\cdots\cdots\cdots$$

From which, solving for A_r, we get

$$(a) \qquad A_r = (-1)^r\frac{1}{a_0{}^{r-1}}\begin{vmatrix} c_0 & a_0 & 0 & 0 & \cdots 0 \\ c_1 & a_1 & a_0 & 0 & \cdots 0 \\ c_2 & a_2 & a_1 & a_0 & \cdots 0 \\ \cdot & \cdot & \cdot & \cdot & \cdot \\ c_{r-1} & a_{r-1} & a_{r-2} & a_{r-3} & \cdots a_0 \\ c_r & a_r & a_{r-1} & a_{r-2} & \cdots a_1 \end{vmatrix}_{r+1}$$

We have therefore

$$\frac{\phi(x)}{f(x)}=\frac{c_0}{a_0}x^{m-n}-\frac{|c_0a_1|}{a_0^2}x^{m-n-1}+\frac{1}{a_0^3}\begin{vmatrix} c_0 & a_0 & 0 \\ c_1 & a_1 & a_0 \\ c_2 & a_2 & a_1 \end{vmatrix}x^{m-n-2}-\cdots$$

If we had written

$$\phi(x) = 1 + b_1x + b_2x^2 + \cdots,$$
$$f(x) = 1 + a_1x + a_2x^2 + \cdots,$$

and

$$\frac{\phi(x)}{f(x)} = 1 - A_1x + A_2x^2 - A_3x^3 + \cdots,$$

then we would find

$$\text{(b)} \qquad A_r = \begin{vmatrix} a_1 - b_1 & 1 & 0 & 0\cdots \\ a_2 - b_2 & a_1 & 1 & 0\cdots \\ a_3 - b_3 & a_2 & a_1 & 1\cdots \\ \cdots & \cdots & \cdots & \cdots \end{vmatrix}_r$$

If $b_1 = b_2 = \cdots = 0$ so that

$$(1 + a_1x + a_2x^2 + \cdots)^{-1} = 1 - A_1x + A_2x^2 - A_3x^3 + \cdots$$

then

$$\text{(c)} \qquad A_r = \begin{vmatrix} a_1 & 1 & 0 & 0\cdots \\ a_2 & a_1 & 1 & 0\cdots \\ a_3 & a_2 & a_1 & 1\cdots \\ \cdots & \cdots & \cdots & \cdots \end{vmatrix}_r$$

and A_r is the same function of the a's as a_r is of the A's.

If in addition

$$(1 + b_1x + b_2x^2 + \cdots)^{-1} = 1 - B_1x + B_2x^2 - B_3x^3 + \cdots$$

then

$$\begin{vmatrix} a_1 - b_1 & 1 & 0 & 0\cdots \\ a_2 - b_2 & a_1 & 1 & 0\cdots \\ a_3 - b_3 & a_2 & a_1 & 1\cdots \\ \cdots & \cdots & \cdots & \cdots \end{vmatrix}_r = (-1)^r \begin{vmatrix} B_1 - A_1 & 1 & 0 & 0\cdots \\ B_2 - A_2 & B_1 & 1 & 0\cdots \\ B_3 - A_3 & B_2 & B_1 & 1\cdots \\ \cdots & \cdots & \cdots & \cdots \end{vmatrix}_r$$

759. If we write

$$\frac{\phi(x)}{f(x)} = A_0x^{m-n} + A_1x^{m-n-1} + \cdots + \frac{b_0x^{n-1} + b_1x^{n-2} + \cdots}{a_0x^n + a_1x^{n-1} + \cdots}$$

and then multiply both sides by $f(x)$ we get, on equating coefficients of like powers of x, the necessary equations for determining the b's. They are

$$\frac{1}{a_0}\begin{vmatrix} c_0 & c_1 \\ a_0 & a_1 \end{vmatrix}, \quad \frac{1}{a_0}\begin{vmatrix} c_0 & c_2 \\ a_0 & a_2 \end{vmatrix}, \quad \frac{1}{a_0}\begin{vmatrix} c_0 & c_3 \\ a_0 & a_3 \end{vmatrix}, \cdots$$

when $m=n$,

$$\frac{1}{a_0^2}\begin{vmatrix} c_0 & c_1 & c_2 \\ 0 & a_0 & a_1 \\ a_0 & a_1 & a_2 \end{vmatrix}, \quad \frac{1}{a_0^2}\begin{vmatrix} c_0 & c_1 & c_3 \\ 0 & a_0 & a_2 \\ a_0 & a_1 & a_3 \end{vmatrix}, \cdots$$

when $m=n+1$ etc.

760. As an immediate consequence of the results of §756 we may write

$$\frac{\phi(x)}{f(x)} = \frac{\phi}{f} - \left(\frac{1}{f}\right)^2 \begin{vmatrix} \phi & f \\ \phi' & f' \end{vmatrix} x + \left(\frac{1}{f}\right)_3 \begin{vmatrix} \phi & f & \cdot \\ \phi' & f' & f \\ \phi'' & f'' & f' \end{vmatrix} \frac{x^2}{2!} - \cdots$$

where $\phi, \phi', \cdots$, denote the values of $\phi(x), \phi'(x) \cdots$ when $x=0$.

761. There have been developed thus far three general formulas for expressing the quotient of two functions as a power series in which the coefficients are determined. The first of these, due to Faure, is that given in §758; the second, due to Trudi, is that given in §759; the third, due to Hammond, is that given in §760.

By giving the functions $\phi(x)$ and $f(x)$ properly selected values an indefinite number of interesting relations may be obtained.

Using the theorem that if

$$f(x) = a_0(x - x_1)(x - x_2) \cdots (x - x_n)$$

then

$$\psi(x_1) + \psi(x_2) + \cdots + \psi(x_n) = \text{Coefficient of } x^{-1} \text{ in } \frac{f'(x)\psi(x)}{f(x)}$$

and using for $\psi(x)$, $x^r\phi(x)/f'(x)$, we have, making $m=n$ in the two functions $\phi(x)$ and $f(x)$ as given in §758

$$\frac{x_1{}^r\phi(x_1)}{f'(x_1)}+\frac{x_2{}^r\phi(x_2)}{f'(x_2)}+\cdots+\frac{x_n{}^r\phi(x_n)}{f'(x_n)}=\text{Coeffi. of } x^{-r-1} \text{ in } \frac{\phi(x)}{f(x)}$$

$$=A_{r+1}=(-1)^{r+1}\frac{1}{a_0{}^{r+2}}\begin{vmatrix} c_0 & a_0 & 0 & \cdots 0 \\ c_1 & a_1 & a_0 & \cdots 0 \\ \cdot & \cdot & \cdot & \cdot\cdot\cdot \\ c_r & a_r & a_{r-1} & \cdots a_0 \\ c_{r-1} & a_{r-1} & a_r & \cdots a_1 \end{vmatrix}_{r+2}$$

If we make $m=n-1$, $c_0=na_0$, $c_1=(n-1)a_1$, etc. so that $\phi(x)=f'(x)$, then

$$x_1{}^r+x_2{}^r+\cdots+x_n{}^r=S_r=(-1)^r\frac{1}{a_0{}^{r+1}}\begin{vmatrix} na_0 & a_0 & & \\ (n-1)a_1 & a_1 & a_0 & \\ \cdot & \cdot & \cdot & \cdot\cdot\cdot \\ (n-r)a_r & a_r & a_{r-1} & \cdots \end{vmatrix}_{r+1}$$

$$=(-1)^r\frac{1}{a_0{}^r}\begin{vmatrix} a_1 & a_0 & 0 & \cdots 0 \\ 2a_1 & a_1 & a_0 & \cdots 0 \\ \cdot & \cdot & \cdot & \cdot\cdot\cdot \\ ra_r & a_{r-1} & a_{r-2} & \cdots a_1 \end{vmatrix}$$

as in §757.

762. If in §738 we put $\phi(x)=x$ and $f(x)=\epsilon^x-1$, then

$$\frac{x}{\epsilon^x-1}=\frac{1}{1+\dfrac{x}{2!}+\dfrac{x^2}{3!}+\cdots}$$

$$=x^{-n+1}-\frac{1}{2!}x^{-n}+\begin{vmatrix} \frac{1}{2!} & 1 \\ \frac{1}{3!} & \frac{1}{2!} \end{vmatrix}x^{-n-1}-\begin{vmatrix} \frac{1}{2!} & 1 & 0 \\ \frac{1}{3!} & \frac{1}{2!} & 1 \\ \frac{1}{4!} & \frac{1}{3!} & \frac{1}{2!} \end{vmatrix}x^{-n-2}+\cdots$$

The nth Bernoulli's number is therefore given by

$$B_n = (-1)^{n-1}(2n)!\begin{vmatrix} \frac{1}{2!} & 1 & 0 & 0 \cdots \\ \frac{1}{3!} & \frac{1}{2!} & 1 & 0 \cdots \\ \frac{1}{4!} & \frac{1}{3!} & \frac{1}{2!} & 1 \cdots \\ \cdot & \cdot & \cdot & \cdot \end{vmatrix}_{2n}$$

which by easy transformation may be written in the form

$$B_n = (-1)^{n-1}\frac{1}{(2n+1)!}\begin{vmatrix} 1 & 2 & 0 & 0 & 0 \cdots \\ 1 & 3 & 3 & 0 & 0 \cdots \\ 1 & 4 & 6 & 4 & 0 \cdots \\ 1 & 5 & 10 & 10 & 5 \cdots \\ \cdot & \cdot & \cdot & \cdot & \cdot \end{vmatrix}_{2n}$$

With these we have the concomitant results of the determinants of odd order vanishing. Thus

$$\begin{vmatrix} \frac{1}{2!} & 1 & 0 & 0 \cdots \\ \frac{1}{3!} & \frac{1}{2!} & 1 & 0 \cdots \\ \frac{1}{4!} & \frac{1}{3!} & \frac{1}{2!} & 1 \cdots \\ \cdot & \cdot & \cdot & \cdot \end{vmatrix}_{2n-1} = 0 \quad \text{and} \quad \begin{vmatrix} 1 & 2 & 0 & 0 \cdots \\ 1 & 3 & 3 & 0 \cdots \\ 1 & 4 & 6 & 4 \cdots \\ \cdot & \cdot & \cdot & \cdot \end{vmatrix}_{2n-1} = 0$$

763. From the results of the last article we may therefore write

$$\frac{x}{e^x - 1} = 1 - \tfrac{1}{2}x + c_1x^2 + c_2x^4 + c_3x^6 + \cdots$$

$$x = (e^x - 1)(1 - \tfrac{1}{2}x + c_1x^2 + c_2x^4 + c_3x^6 + \cdots)$$

where

$$c_n = (-1)^{n+1}\frac{B_n}{(2n)!}.$$

Equating coefficients of odd powers of x on both sides we have

$$c_1 - \frac{1}{2\cdot 2!} + \frac{1}{3!} = 0$$

$$\frac{c_1}{3!} + c_2 - \frac{1}{2\cdot 4!} + \frac{1}{5!} = 0$$

$$\frac{c_1}{5!} + \frac{c_2}{3!} + c_3 - \frac{1}{2\cdot 6!} + \frac{1}{7!} = 0^{*}$$

$$\cdots\cdots\cdots\cdots$$

From these we get, on noting that

$$\frac{1}{2(2n)!} - \frac{1}{(2n+1)!} = \frac{2n-1}{2(2n+1)!}$$

$$B_n = \tfrac{1}{2}(2n)!\begin{vmatrix} \frac{1}{3!} & 1 & 0 & 0\cdots \\ \frac{3}{5!} & \frac{1}{3!} & 1 & 0\cdots \\ \frac{5}{7!} & \frac{1}{5!} & \frac{1}{3!} & 1\cdots \\ \cdots & \cdots & \cdots & \cdots \end{vmatrix}_n$$

giving B_n as a recurrent of order n. Equating coefficients of the even powers of x on both sides of our equation we have

$$\frac{c_1}{2!} - \frac{1}{2\cdot 3!} + \frac{1}{4!} = 0$$

$$\frac{c_1}{4!} + \frac{c_2}{2!} - \frac{1}{2\cdot 5!} + \frac{1}{6!} = 0$$

$$\cdots\cdots\cdots\cdots$$

from which we get

$$B_n = 2^n(2n)!\begin{vmatrix} \frac{1}{4!} & \frac{1}{2!} & 0 & 0\cdots \\ \frac{2}{6!} & \frac{1}{4!} & \frac{1}{2!} & 0\cdots \\ \frac{3}{8!} & \frac{1}{6!} & \frac{1}{4!} & \frac{1}{2!}\cdots \\ \cdots & \cdots & \cdots & \cdots \end{vmatrix}_n$$

764. If we put $\phi(x)=e^x-1$ and $f(x)=e^x+1$, then

$$\frac{\phi(x)}{f(x)}=\frac{e^x-1}{e^x+1}=\frac{e^x+e^{-x}-2}{e^x-e^{-x}}=\frac{\frac{1}{2}x+\frac{1}{4}\frac{x^3}{3!}+\frac{1}{6}\frac{x^5}{5!}+\cdots}{1+\frac{1}{3}\frac{x^2}{2!}+\frac{1}{5}\frac{x^4}{4!}+\cdots}$$

$$=(2^2-1)B_1x-\frac{(2^4-1)}{2}B_2\frac{x^3}{3!}+\frac{(2^6-1)}{3}B_3\frac{x^5}{5!}-\cdots$$

from which we may find

$$B_n=\frac{1}{(2^{2n}-1)2^n(n-1)!}\begin{vmatrix} 1 & 2 & 0 & 0\cdots \\ 1 & 4 & 4 & 0\cdots \\ 1 & 6 & 20 & 6\cdots \\ \cdot & \cdot & \cdot & \cdot\cdots \end{vmatrix}_n$$

765. Let $x_1, x_2, \cdots, x_n$ represent the roots of

$$a_0x^n+a_1x^{n-1}+a_2x^{n-2}+\cdots+a_n=0,$$

and let H_r represent the rth "Aleph" function, that is, the complete homogeneous function of degree r. It is known that H_r is the coefficient of z^r in the product

$$(1+x_1z+x_1^2z^2+\cdots)(1+x_2z+x_2^2z^2+\cdots)\cdots$$
$$(1+x_nz+x_n^2z^2+\cdots)$$

or

$$\frac{1}{(1-x_1z)(1-x_2z)\cdots(1-x_nz)}\equiv\frac{1}{\phi(z)}\quad\text{say}.$$

From this we have

$$\frac{1}{\phi(z)}=1+H_1z+H_2z^2+H_3z^3+\cdots$$

$$=\frac{a_0}{(a_0+a_1z+a_2z^2+\cdots+a_nz^n)}.$$

Therefore

$$a_0=(1+H_1z+H_2z^2+\cdots)(a_0+a_1z+a_2z^2+\cdots+a_nz^n)$$

from which we get

$$a_r + a_{r-1}H_1 + a_{r-2}H_2 + \cdots + a_0 H_r = 0, \quad (r = 1,2, \cdots, r)$$

and therefore

$$H_r = \frac{(-1)^r}{a_0^r}\begin{vmatrix} a_1 & a_0 & 0 & \cdots 0 \\ a_2 & a_1 & a_0 & \cdots 0 \\ \cdot & \cdot & \cdot & \cdot \\ a_r & a_{r-1} & a_{r-2} & \cdots a_1 \end{vmatrix}_r$$

Comparing this value of H_r when $a_0 = 1$ with S_r we have

$$-S_r = a_1 H_{r-1} + 2a_2 H_{r-2} + 3a_3 H_{r-3} + \cdots + ra_r.$$

766. The coefficient of z^r in

$$\frac{a_0}{a_0 + a_1 z + a_2 z^2 + \cdots + a_n z^n}$$

is by the multinomial theorem

$$\sum (-1)^e e! a_0^{-e} \cdot \frac{a_1^{e_1} a_2^{e_2} \cdots a_n^{e_n}}{e_1! e_2! \cdots e_n!}$$

where

$$e_1 + e_2 + \cdots + e_n = e,$$

and

$$e_1 + 2e_2 + \cdots + ne_n = n.$$

We have therefore

$$\begin{vmatrix} a_1 & a_0 & 0 & 0 & \cdots 0 \\ a_2 & a_1 & a_0 & 0 & \cdots 0 \\ a_3 & a_2 & a_1 & a_0 & \cdots 0 \\ \cdot & \cdot & \cdot & \cdot & \cdot \\ a_n & a_{n-1} & a_{n-2} & a_{n-3} & \cdots a_1 \end{vmatrix}_n = \sum (-1)^{n+e} e! a_0^{n-e} \frac{a_1^{e_1} a_2^{e_2} \cdots a_n^{e_n}}{e_1! e_2! \cdots e_n!}$$

EXERCISE: Show that

$$\begin{vmatrix} a_1 & 1 & & & \\ a_2 & a_1 & 2 & & \\ a_3 & a_2 & a_1 & 3 & \\ \cdot & \cdot & \cdot & \cdot & \cdot \\ a_n & a_{n-1} & a_{n-2} & a_{n-3} & \cdots a_1 \end{vmatrix}_n$$

$$= \sum (-1)^{n+e} \frac{a_1^{e_1} a_2^{e_2} \cdots a_n^{e_n}}{1^{e_1} 2^{e_2} \cdots n^{e_n} \cdot e_1! e_2! \cdots e_n!}$$

From the two forms given in §765 for $1/\phi(z)$ we get by differentiation

$$-\frac{\phi'(z)}{\phi(z)} = \frac{H_1 + 2H_2 z + 3H_3 z^2 + \cdots}{1 + H_1 z + H_2 z^2 + \cdots},$$

and also

$$\begin{aligned} &\frac{x_1}{1 - x_1 z} + \frac{x_2}{1 - x_2 z} + \cdots + \frac{x_n}{1 - x_n z} \\ &= x_1 + x_1^2 z + x_1^3 z^2 + \cdots \\ &\quad + x_2 + x_2^2 z + x_2^3 z^2 + \cdots \\ &\quad \cdot \cdot \cdot \cdot \cdot \cdot \cdot \cdot \cdot \cdot \\ &\quad + x_n + x_n^2 z + x_n^3 z^2 + \cdots \\ &= S_1 + S_2 z + S_3 z^2 + \cdots . \end{aligned}$$

From these two forms we have

$$\begin{aligned} H_1 &= S_1 \\ 2H_2 &= S_2 + H_1 S_1 \\ 3H_3 &= S_3 + H_1 S_2 + H_2 S_1 \\ &\cdot \cdot \cdot \cdot \cdot \cdot \cdot \cdot \cdot \cdot \\ rH_r &= S_r + H_1 S_{r-1} + H_2 S_{r-2} + \cdots + H_{r-1} S_1, \end{aligned}$$

and from these we get

$$H_r = \frac{1}{r!} \begin{vmatrix} S_1 & -1 & 0 & 0 & \cdots 0 \\ S_2 & S_1 & -2 & 0 & \cdots 0 \\ S_3 & S_2 & S_1 & -3 & \cdots 0 \\ \cdot & \cdot & \cdot & \cdot & \cdot \\ S_r & S_{r-1} & S_{r-2} & S_{r-3} & \cdots S_1 \end{vmatrix}_r$$

and

$$S_r = \begin{vmatrix} H_1 & 1 & 0 & 0 & \cdots 0 \\ 2H_2 & H_1 & 1 & 0 & \cdots 0 \\ 3H_3 & H_2 & H_1 & 1 & \cdots 0 \\ . & . & . & . & . \\ rH_r & H_{r-1} & H_{r-2} & H_{r-3} & \cdots H_1 \end{vmatrix}.$$

767. If

$$x^n + a_1x^{n-1} + \cdots + a_n \equiv (x - \alpha_1)(x - \alpha_2) \cdots (x - \alpha_n) = 0$$

and ϕ is any symmetric function of order r of the roots $\alpha_1, \alpha_2, \cdots, \alpha_n$, then

$$\frac{\partial\phi}{\partial s_r} = \frac{\partial\phi}{\partial a_1}\cdot\frac{\partial a_1}{\partial s_r} + \frac{\partial\phi}{\partial a_2}\frac{\partial a_2}{\partial s_r} + \cdots + \frac{\partial\phi}{\partial a_n}\frac{\partial a_n}{\partial s_r}.$$

By differentiating Newton's relations we have

$$\frac{\partial a_1}{\partial s_r} = -\frac{\partial s_1}{\partial s_r}$$

$$\frac{\partial a_2}{\partial s_r} = -a_1\frac{\partial s_1}{\partial s_r} - \frac{1}{2}\frac{\partial s_2}{\partial s_r}$$

$$\frac{\partial a_3}{\partial s_r} = -a_2\frac{\partial s_1}{\partial s_r} - \frac{1}{2}a_1\frac{\partial s_2}{\partial s_r} - \frac{1}{3}\frac{\partial s_3}{\partial s_r}$$

$$\cdots\cdots\cdots\cdots$$

and from these we get

$$\frac{\partial a_h}{\partial s_r} = -\frac{1}{r}a_{h-r}$$

$$= 0 \text{ for } h < r.$$

Substituting in the expression for $\partial\phi/\partial s_r$ we have

$$\frac{\partial\phi}{\partial a_r} + a_1\frac{\partial\phi}{\partial a_{r+1}} + \cdots + a_{n-r}\frac{\partial\phi}{\partial a_n} = -r\frac{\partial\phi}{\partial s_r}$$

Giving r the values from 1 to r in this relation and noting that $\partial\phi/\partial a_k = 0$ for $k > n$ since ϕ is isoberic in the a's of weight r, there results the following equations:

$$\frac{\partial\phi}{\partial a_1} + a_1\frac{\partial\phi}{\partial a_2} + \cdots + a_{r-1}\frac{\partial\phi}{\partial a_r} + \frac{\partial\phi}{\partial s_1} = 0$$

$$\frac{\partial\phi}{\partial a_2} + \cdots + a_{r-2}\frac{\partial\phi}{\partial a_r} + 2\frac{\partial\phi}{\partial s_2} = 0$$

$$\cdots\cdots\cdots\cdots$$

$$\frac{\partial\phi}{\partial a_r} + r\frac{\partial\phi}{\partial s_r} = 0,$$

which with

$$a_1\frac{\partial\phi}{\partial a_1} + 2a_2\frac{\partial\phi}{\partial a_2} + \cdots + ra_r\frac{\partial\phi}{\partial a_r} = r\cdot\phi$$

gives

$$r\cdot\phi = \begin{vmatrix} r\dfrac{\partial\phi}{\partial s_r} & 1 & 0 & \cdots 0 \\ (r-1)\dfrac{\partial\phi}{\partial s_{r-1}} & a_1 & 1 & \cdots 0 \\ \cdots & \cdots & \cdots & \cdots \\ \dfrac{\partial\phi}{\partial s_1} & a_{r-1} & a_{r-2} & \cdots 1 \\ 0 & ra_r & (r-1)a_{r-1} & \cdots a_1 \end{vmatrix}$$

768. If to the persymmetric recurrent

$$(-1)^v \begin{vmatrix} \dfrac{1}{1!} & 1 & \cdots & \\ \dfrac{1}{2!} & \dfrac{1}{1!} & \cdots & \\ \dfrac{1}{3!} & \dfrac{1}{2!} & \cdots & \\ \cdots & \cdots & \cdots & \cdots \\ \dfrac{1}{(n-p)!} & \dfrac{1}{(n-p-1)!} & \cdots & \dfrac{1}{1!} \end{vmatrix},$$

where

$$v = \tfrac{1}{2}(n-p)(n-p+1),$$

we prefix as a first row $1/p, 1, 0, 0, \cdots$, and as a first column

$$\frac{1}{p}, \frac{1}{1!(p-1)!}, \cdots, \frac{1}{(n-p)!n}$$

and denote the result by D_{n-p+1}, we have, after performing the operations, $\mathrm{col}_2 - p \cdot \mathrm{col}_1$

$$D_{n-p+1} = \frac{1}{p} \cdot D_{n-p}$$

$$= \frac{1}{p(p+1)\cdots n} = \frac{(p-1)!}{n!}$$

when $p=1$, this becomes

$$D_n = \frac{1}{n!}.$$

It is to be noticed that when $p=1$ D_n is a persymmetric recurrent exactly like the original.

769. If, starting with the persymmetric recurrent

$$D_n = \begin{vmatrix} (m)_1 & 1 & 0 & \cdots 0 \\ (m)_2 & (m)_1 & 1 & \cdots 0 \\ \cdot & \cdot & \cdot & \cdot \\ (m)_{n-1} & (m)_{n-2} & (m)_{n-3} & \cdots 1 \\ (m)_n & (m)_{n-1} & (m)_{n-2} & \cdots (m)_1 \end{vmatrix}$$

we form a new determinant, $D'_{(n,p)}$ say, by having for its last row the elements of the last row of D_n modified as follows: The first element $(m)_n$ is multiplied by $1/(n+1)(n+2) \cdots (n+p)$ and all the other elements are multiplied by $1/n(n+1) \cdots (n+p-1)$. Each of the other rows are formed from the one just below it by giving n the value one less.

If on $D'_{(n,p)}$ we perform the operations $\mathrm{col}_2 - (p+1)\,\mathrm{col}_1/m$ we get

$$D'_{(n,p)} = (m+p+1)\frac{(n-1)!}{(p+1)!}D'_{(n-1,p+1)}$$

$$= \frac{(m+p+1)}{\dfrac{(p+1)!}{(n-1)!}} \frac{(m+p+2)}{\dfrac{(p+2)!}{(n-2)!}} \cdots \frac{(m+p+n-1)}{\dfrac{(p+n-1)!}{1!}} \frac{m}{(p+n)!}$$

If in this value for $D'_{(n,p)}$ we put $p=0$ we get

$$D_n = (m + n - 1)_n$$

as we should §734.

770. Let

$$a(x) = 1 + a_1x + a_2x^2 + a_3x^3 + \cdots$$
$$b(x) = 1 + b_1x + b_2x^2 + b_3x^3 + \cdots$$

and let the generating function of the persymmetric recurrent

$$\Delta \equiv \begin{vmatrix} a_1 & 1 & & \\ a_2 & a_1 & 1 & \\ a_3 & a_2 & a_1 & 1 \\ \cdot & \cdot & \cdot & \cdot \end{vmatrix}_n$$

be denoted by $1/a(x)$.

Denote by A_n the determinant formed from Δ by replacing the a's in the even-numbered columns by corresponding b's, and by B_n the determinant formed from Δ by replacing the a's in the odd-numbered columns by the corresponding b's. Then to determine the generating functions of A_n and B_n we have by expansion

$$A_n - a_1B_{n-1} + a_2A_{n-2} - a_3B_{n-3} + \cdots = 0$$
$$B_n - b_1A_{n-1} + b_2B_{n-2} - b_3A_{n-3} + \cdots = 0.$$

Let

$$A(x) = 1 - B_1x + A_2x^2 - B_3x^3 + \cdots$$

and

$$B(x) = 1 - A_1x + B_2x^2 - A_3x^3 + \cdots,$$

then

$$a(x)A(x) + a(-x)A(-x) = 2,$$
$$b(x)A(x) - b(-x)A(-x) = 0,$$
$$a(x)B(x) - a(-x)B(-x) = 0,$$
$$b(x)B(x) + b(-x)B(-x) = 2,$$

from which we get

$$A(x) = \frac{2b(-x)}{a(x)b(-x) + a(-x)b(x)}$$

$$B(x) = \frac{2a(-x)}{a(x)b(-x) + a(-x)b(x)}$$

771. Writing the expression $(x-\alpha)(x-\beta)(x-\gamma)(x-\delta)$ in the form

$$\frac{|\,\alpha^0\beta^1\gamma^2\delta^3\mathrm{x}^4\,|}{|\,\alpha^0\beta^1\gamma^2\delta^3\,|},$$

and observing the coefficients of the powers of x, which are quotients of alternants we see that

$$\begin{vmatrix} 1 & 0 & 0 & 0 & 1 \\ H_1 & 1 & 0 & 0 & x \\ H_2 & H_1 & 1 & 0 & x^2 \\ H_3 & H_2 & H_1 & 1 & x^3 \\ H_4 & H_3 & H_2 & H_1 & x^4 \end{vmatrix} = x^4 - \sum\alpha x^3 + \sum\alpha\beta x^2 - \sum\alpha\beta\gamma x + \alpha\beta\gamma\delta$$

and equating coefficients of like powers of x we have expressions for $\sum\alpha, \sum\alpha\beta, \cdots$, in terms of Aleph functions. Thus

$$\sum\alpha\beta = \begin{vmatrix} 1 & 0 & 0 & 0 \\ H_1 & 1 & 0 & 0 \\ H_3 & H_2 & H_1 & 1 \\ H_4 & H_3 & H_2 & H_1 \end{vmatrix} \quad \text{etc.}$$

772. The determinant

$$\begin{vmatrix} z & 1 & \cdot & \cdot & \cdots & \cdot & \cdot \\ z^2 & 1 & 2 & \cdot & \cdots & \cdot & \cdot \\ z^3 & 1 & 3 & 3 & \cdots & \cdot & \cdot \\ z^4 & 1 & 4 & 6 & \cdots & \cdot & \cdot \\ \cdot & \cdot & \cdot & \cdot & \cdot & \cdot & \cdot \\ z^r & 1 & (r)_1 & (r)_2 & \cdots & (r)_{r-2} & (r)_{r-1} \\ z^{r+1} & 1 & (r+1)_1 & (r+1)_2 & \cdots & (r+1)_{r-2} & (r+1)_{r-1} \end{vmatrix} \equiv R(z)$$

$$= R(z+1) - (r+1)!z^r,$$

as may be seen on performing the following operations

$$\mathrm{col}_1 + \mathrm{col}_2 + z\cdot\mathrm{col}_3 + z^2\cdot\mathrm{col}_4 + \cdots + z^{r-1}\cdot\mathrm{col}_{r+1}.$$

From the recurrence formula

$$R(z) = R(z+1) - (r+1)!z^r$$

there follows

$$R(z+1) = R(z+2) - (r+1)!(z+1)^r$$
$$\cdots\cdots\cdots\cdots\cdots\cdots$$
$$R(z+n) = R(z+n+1) - (r+1)!(z+n)^r.$$

Adding and dividing out we have

$$R(z+n+1) = R(z) + (r+1)!\{z^r + (z+1)^r + \cdots + (z+n)^r\}$$

which on putting $z=0$ gives

(I) $$R(n+1) = (r+1)!(1^r + 2^r + \cdots + n^r)$$

If we had used the operations

$$\text{col}_1 + \text{col}_2 - z\cdot\text{col}_3 + z^2\text{col}_4 - \cdots$$

we would get

$$R(-z) = R(-z+1) + (-1)^r(r+1)!z^r$$

and finally

(II) $$(-1)^rR(-n) = (r+1)!(1^r + 2^r + \cdots + n^r)$$

From I and II we have

$$R(n+1) = (-1)^rR(-n)$$

773. If

$$(x+1)(x+2)\cdots(x+n) = x^n + k_{n,1}x^{n-1} + k_{n,2}x^{n-2} + \cdots$$

then from the known relations

$$\begin{aligned} k_{n,1} - (n+1)_2 &= 0 \\ 2k_{n,2} - (n)_2k_{n,1} - (n+1)_3 &= 0 \\ 3k_{n,3} - (n-1)_2k_{n,2} - (n)_3k_{n,1} - (n+1)_4 &= 0 \\ \cdots\cdots\cdots\cdots\cdots\cdots \end{aligned}$$

we have

$$k_{n,p} = \frac{1}{p!}\begin{vmatrix} (n+1)_2 & -1 & \cdot & \cdots & \cdot \\ (n+1)_3 & (n)_2 & -2 & \cdots & \cdot \\ (n+1)_4 & (n)_3 & (n-1)_2 & \cdots & \cdot \\ \cdots & \cdots & \cdots & \cdots & \cdots \\ (n+1)_{p+1} & (n)_p & (n-1)_{p-1} & \cdots & (n-p+2)_2 \end{vmatrix}_p$$

774. We have seen from §757 that

$$-\frac{1}{n}S_n = (-1)^{n-1}\frac{1}{n}\begin{vmatrix} a_1 & 1 & 0 & 0\cdots \\ 2a_2 & a_1 & 1 & 0\cdots \\ 3a_3 & a_2 & a_1 & 1\cdots \\ \cdot & \cdot & \cdot & \cdot \end{vmatrix}_n$$

$$= \sum(-1)^{e-1}(e-1)!\frac{a_1^{e_1}a_2^{e_2}\cdots a_n^{e_n}}{e_1!e_2!\cdots e_n!}$$

which is the coefficient of x^n in $\log(1+a_1x+a_2x^2+\cdots+a_nx^n)$. Denoting this coefficient by b_n we have

$$\exp(b_1x+b_2x^2+\cdots) = 1+a_1x+a_2x^2+\cdots$$

and therefore

$$a_n = \frac{1}{n!}\begin{vmatrix} b_1 & -1 & \cdot & \cdots \\ 2b_2 & b_1 & -2 & \cdots \\ 3b_3 & 2b_2 & b_1 & -3\cdots \\ \cdot & \cdot & \cdot & \cdot \end{vmatrix}_n = \sum\frac{b_1^{e_1}b_2^{e_2}\cdots b_n^{e_n}}{e_1!e_2!\cdots e_n!}$$

as may be seen on equating coefficients.

If we put n for $a_n(n=1, 2, \cdots)$, and notice that the function expanded in powers of x is then

$$\log\frac{(1+x^3)}{(1-x)(1-x^2)}$$

we find that

$$\begin{vmatrix} 1^2 & 1 & 0 & 0 & 0\cdots \\ 2^2 & 1 & 1 & 0 & 0\cdots \\ 3^2 & 2 & 1 & 1 & 0\cdots \\ 4^2 & 3 & 2 & 1 & 1\cdots \\ \cdot & \cdot & \cdot & \cdot & \cdot \end{vmatrix}_n = 0, 1, -3, 4, -3, 1$$

according as, to the modulus 6, $n=0, 1, 2, 3, 4, 5$.

If in a similar manner we put $n+1$ for a_n and notice that the function expanded is now $-2\log(1-x)$ or $\log 1/(1-x)^2$ we find

$$\begin{vmatrix} 1 & 1 & \cdot & \cdot & \cdots \\ (3)_1 & 2 & 1 & \cdot & \cdots \\ (4)_2 & 3 & 2 & 1 & \cdots \\ (5)_3 & 4 & 3 & 2 & 1 \cdots \\ \cdot & \cdot & \cdot & \cdot & \cdots \end{vmatrix}_n = (-1)^{n-1}.$$

Again if we put $(n+1)^2$ for a_n, the function now being

$$\log \frac{(1+x)}{(1-x)^3}$$

we find

$$\begin{vmatrix} 1^3 & 1^2 & & \\ 2^3 & 2^2 & 1^2 & \\ 3^3 & 3^2 & 2^2 & 1^2 \\ \cdot & \cdot & \cdot & \cdots \end{vmatrix}_n = \begin{cases} -4 \text{ when } n \text{ is even,} \\ 2 \text{ when } n \text{ is odd and } > 1. \end{cases}$$

775. If we multiply the determinant

$$\Delta_n \equiv \begin{vmatrix} (m)_1 & 1 & \cdot & \cdot & \cdots & \cdot \\ (m)_2 & (m-1)_1 & 1 & \cdot & \cdots & \cdot \\ (m)_3 & (m-1)_2 & (m-2)_1 & 1 & \cdots & \cdot \\ \cdot & \cdot & \cdot & \cdot & \cdots & \cdot \\ (m)_{n-1} & (m-1)_{n-2} & (m-2)_{n-3} & (m-3)_{n-4} & \cdots & 1 \\ (m)_n & (m-1)_{n-1} & (m-2)_{n-2} & (m-3)_{n-3} & \cdots & (m-n-1)_1 \end{vmatrix}_n$$

by unity in the form

$$\begin{vmatrix} 1 & -(m)_1 & (m)_2 & -(m)_3 & \cdots \\ \cdot & 1 & -(m-1)_1 & (m-1)_2 & \cdots \\ \cdot & \cdot & \cdot & \cdot & \cdots \end{vmatrix}_n$$

the result is

$$\begin{vmatrix} \cdot & 1 & \cdot \\ \cdot & \cdot & \cdot \\ & & \\ \cdot & \cdot & 1 \\ (-1)^{n-1}(m)_n & (-1)^{n-2}(m-1)_{n-1} \cdots & (m-n+1)_1 \end{vmatrix} = (m)_n$$

Therefore $\Delta_n = (m)_n$.

776. Other forms of B_n may be obtained as follows:
(a) Putting $\phi(x)/f(x) = x/\sin x$ we may find

$$B_n = \begin{vmatrix} 1 & 3 & 0 & 0 \cdots \\ 1 & 10 & 5 & 0 \cdots \\ 1 & 21 & 35 & 7 \cdots \\ \cdots & \cdots & \cdots & \cdots \end{vmatrix}_n \times \frac{2^{n-1} n!}{(2n+1)!(2^{2n-1}-1)}$$

(b) Putting $\phi(x)/f(x) = \sin x/\cos x$ we may find

$$B_n = \frac{2n}{2^{2n}(2^{2n}-1)} \begin{vmatrix} 1 & 1 & 0 & 0 \cdots \\ 1 & 3 & 1 & 0 \cdots \\ 1 & 5 & 10 & 1 \cdots \\ 1 & 7 & 35 & 21 \cdots \\ \cdots & \cdots & \cdots & \cdots \end{vmatrix}_n$$

(c) Additional forms are

$$B_n = \begin{vmatrix} 1 & 3_2 & 0 & 0 & 0 \cdots \\ 2 & 5_2 & 5_4 & 0 & 0 \cdots \\ 3 & 7_2 & 7_4 & 7_6 & 0 \cdots \\ 4 & 9_2 & 9_4 & 9_6 & 9_8 \cdots \\ \cdots & \cdots & \cdots & \cdots & \cdots \end{vmatrix}_n \frac{1}{3 \cdot 5 \cdot 7 \cdots (2n+1) 2^{2n-1}}$$

$$= \begin{vmatrix} 1 & 3_2 & 0 & 0 & 0 \cdots \\ 3 & 5_2 & 5_4 & 0 & 0 \cdots \\ 5 & 7_2 & 7_4 & 7_6 & 0 \cdots \\ 7 & 9_2 & 9_4 & 9_6 & 9_8 \cdots \\ \cdots & \cdots & \cdots & \cdots & \cdots \end{vmatrix}_n \frac{1}{3 \cdot 5 \cdots (2n+1) 2^{4n-3}}$$

$$= \begin{vmatrix} 1 & 3_1 & 0 & 0 & 0 \cdots \\ 1 & 5_1 & 5_3 & 0 & 0 \cdots \\ 1 & 7_1 & 7_3 & 7_5 & 0 \cdots \\ 1 & 9_1 & 9_3 & 9_5 & 9_7 \cdots \\ \cdots & \cdots & \cdots & \cdots & \cdots \end{vmatrix}_n \frac{2^{n-1}}{(2n-1)!}$$

EXERCISE: Show by multiplication and division that this is equal to

$$\begin{vmatrix} \frac{1}{3!} & \frac{1}{2!} & 0 & 0 & \cdots \\ \frac{1}{5!} & \frac{1}{4!} & \frac{1}{2!} & 0 & \cdots \\ \frac{1}{7!} & \frac{1}{6!} & \frac{1}{4!} & \frac{1}{2!} & \cdots \end{vmatrix}_n 2^{n-1}(2n-1).$$

777. If $(a_0+a_1x+a_2x^2+\cdots)^n \equiv (A_0+A_1x+A_2x^2+\cdots)^m$ and we write the series of equations obtained by equating coefficients of like powers of x, it may then be found that

$$A_r = \frac{a_0^{n/(m-r)}}{r!m^r}\begin{vmatrix} na_1 & -ma_0 & 0 & 0 & \cdots \\ 2na_2 & (n-m)a_1 & -2ma_0 & 0 & \cdots \\ 3na_3 & (2n-m)a_2 & (n-2m)a_1 & -3ma_0 & \cdots \\ 4na_4 & (3n-m)a_3 & (2n-2m)a_2 & (n-3m)a_1 & \cdots \\ \cdots & \cdots & \cdots & \cdots & \cdots \end{vmatrix}$$

EXERCISES. SET XXXVII

1. If

$$(a_0 + a_1x + a_2x^2 \cdots)^{-1} = \frac{1}{a_0} - \frac{g_1}{a_0^2}x + \frac{g_2}{2!a_0^3}x^2 - \cdots$$

show that

$$g_n = \begin{vmatrix} 2a_1 & a_0 & 0 & \cdots & \\ 4a_2 & 3a_1 & 2a_0 & \cdots & \\ 6a_3 & 5a_2 & 4a_1 & \cdots & \\ \cdots & \cdots & \cdots & \cdots & \cdots \\ (2n-2)a_{n-1} & (2n-3)a_{n-2} & (2n-4)a_{n-3} & \cdots & (n-1)a_0 \\ na_n & (n-1)a_{n-1} & (n-2)a_{n-2} & \cdots & a_1 \end{vmatrix}_n$$

Compare this result with that for A_n when $n=-1$ and $m=1$; also with A_n from §758 (c).

If we put $a_k=1/k!$, $(k=1, 2, \cdots, n)$ with $a_0=1$ then $g_n=1$. Compare the result thus obtained with A_n §758 (a).

2. If

$$(a_0 + a_1x + a_2x^2 + \cdots)^{1/2} = \sqrt{a_0} + \frac{H_1}{1!2^1a_0^{1-1/2}}x - \frac{H_2}{2!2^2a_0^{2-1/2}}x^2 + \cdots$$

show that

$$H_n = \begin{vmatrix} a_1 & 2a_0 & 0 & \cdots & \\ 2a_2 & 3a_1 & 4a_0 & \cdots & \\ 3a_3 & 4a_2 & 5a_1 & \cdots & \\ \cdots & \cdots & \cdots & \cdots & \cdots \\ (n-1)a_{n-1} & na_{n-2} & (n+1)a_{n-3} & \cdots & (2n-2)a_0 \\ na_n & (n-1)a_{n-1} & (n-2)a_{n-2} & \cdots & 1\cdot a_1 \end{vmatrix}_n$$

If in this we put $a_0 = 1$ and $a_k = 1/k!$ then

$$H_n = 1.$$

Using $\tan x/x$ for $\phi(x)/f(x)$ show that

$$B_n = \frac{(2n)!}{2^{2n-1}(2^{2n}-1)} \begin{vmatrix} \frac{1}{3!} & 1 & 1 & \cdots \\ \frac{2}{5!} & \frac{1}{2!} & 1 & \cdots \\ \frac{3}{7!} & \frac{1}{4!} & \frac{1}{2!} & \cdots \\ \cdots & \cdots & \cdots & \cdots \end{vmatrix}_{n-1}$$

3. Using $x/\tan x$ for $\phi(x)/f(x)$ show that

$$B_n = \frac{(2n)!}{2^{2n-1}} \begin{vmatrix} \frac{1}{3!} & 1 & 0 & \cdots \\ \frac{2}{5!} & \frac{1}{3!} & 1 & \cdots \\ \frac{3}{7!} & \frac{1}{5!} & \frac{1}{3!} & \cdots \\ \cdots & \cdots & \cdots & \cdots \end{vmatrix}_n$$

Also obtain this directly by multiplication and division from (c) §776.

4. Using $x/\log(1+x)$ for $\phi(x)/f(x)$ show that

$$\int_0^1 x(x-1)\cdots(x-n+1)dx = n!\begin{vmatrix} \frac{1}{2} & 1 & 0 & \cdots \\ \frac{1}{3} & \frac{1}{2} & 1 & \cdots \\ \frac{1}{4} & \frac{1}{3} & \frac{1}{2} & \cdots \\ \cdot & \cdot & \cdot & \cdots \end{vmatrix}_n$$

5. Using $1/(1-2hx+h^2)^{1/2}$ for $\phi(x)/f(x)$ show that the coefficient of h^n is

$$\Delta \equiv \begin{vmatrix} 1 & 1 & 0 & 0 & 0 & 0 & \cdots \\ 0 & x & 1 & 0 & 0 & 0 & \cdots \\ \frac{1}{2}(x^2-1) & x^2 & 2x & 1 & 0 & 0 & \cdots \\ 0 & x^3 & 3x^2 & 3x & 1 & 0 & \cdots \\ \frac{1}{2}\cdot\frac{3}{4}(x^2-1)^2 & x^4 & 4x^3 & 6x^2 & 4x & 1 & \cdots \\ \cdot & \cdot & \cdot & \cdot & \cdot & \cdot & \cdots \end{vmatrix}_{n+1}$$

Investigate the relation between Δ and the continuant

$$P_m(x) = \frac{1}{m!}\begin{vmatrix} x & 1 & & & & \\ 1 & 3x & 2 & & & \\ \cdot & 2 & 5x & 3 & & \\ \cdot & \cdot & \cdot & \cdot & \cdot & \cdot \\ & & & & m-1 & (2m-1)x \end{vmatrix}$$

778. If we put $\phi(x)/f(x) = 1/\cos x$ we know that

$$\frac{1}{\cos x} = \frac{1}{1 - \dfrac{x^2}{2!} + \dfrac{x^4}{4!} - \cdots} = 1 + \sum E_n \frac{x^{2n}}{(2n)!}$$

where E_n is the nth of Euler's numbers, then

$$\left(1 - \frac{x^2}{2!} + \frac{x^4}{4!} - \cdots\right)^{-1} = 1 + \sum E_n \frac{x^{2n}}{(2n)!}$$

From this we may find that

$$E_n = \begin{vmatrix} 1 & 1 & 0 & 0 & 0 & \cdots \\ 1 & 6 & 1 & 0 & 0 & \cdots \\ 1 & 15 & 15 & 1 & 0 & \cdots \\ 1 & 28 & 70 & 28 & 1 & \cdots \\ \cdot & \cdot & \cdot & \cdot & \cdot & \cdots \end{vmatrix}_n$$

779. From

$$\left(1 - \frac{x^2}{2!} + \frac{x^4}{4!} - \cdots\right)^{-n} = \left(1 + \frac{A_1 x}{2!} + \frac{A_2 x}{3!} + \cdots\right)^n$$

we may find that

$$(-1)^r \begin{vmatrix} \frac{n}{2!} & 1 & 0 & \cdots \\ \frac{2n}{4!} & \frac{n-1}{2!} & 2 & \cdots \\ & \frac{2n-1}{4!} & \frac{n-2}{2!} & \cdots \\ \cdots & \cdots & \cdots & \cdots \end{vmatrix}_r$$

$$= \begin{vmatrix} \frac{n}{2!}A_1 & -1 & 0 & \cdots \\ \frac{2n}{4!}A_2 & \frac{n-1}{2!}A_1 & -2 & \cdots \\ \frac{3n}{6!}A_3 & \frac{2n-1}{4!}A_2 & \frac{n-2}{2!}A_1 & \cdots \\ \cdots & \cdots & \cdots & \cdots \end{vmatrix}_r$$

EXERCISE: From the relation

$$\left(1 + \frac{x}{2!} + \frac{x^2}{3!} + \cdots\right)^{-n} = \left(1 - \frac{x}{2} + \frac{B_1}{2!}x^2 - \frac{B_2}{4!}x^4 + \cdots\right)^n$$

obtain the corresponding relation involving the B's.

780. From the known relation

$$\frac{2}{\exp(x) + \exp(-x)} = 1 + \sum(-1)^n E_n \frac{x^{2n}}{(2n)!}$$

by multiplying the left-hand side by $\{\exp(x)+\exp(-x)\}/2$ and the right-hand side by its equivalent

$$1 + \sum E_n \frac{x^{2n}}{(2n)!}$$

we get

$$1 = \left\{1 + \sum(-1)^n E_n \frac{x^{2n}}{(2n)!}\right\}\left\{1 + \sum E_n \frac{x^{2n}}{(2n)!}\right\}$$

where the E's are to be determined so as to make this an identity. Therefore

$$E_n - (2n)_2 E_{n-1} + (2n)_4 E_{n-2} - \cdots + (-1)^{n-1}(2n)_{2n-2}E_1 + (-1)^n = 0$$

from which we get

$$\begin{aligned} 1 &= E_1 \\ 1 &= (4)_2E_1 - E_2 \\ 1 &= (6)_2E_1 - (6)_4E_2 + E_3 \\ 1 &= (8)_2E_1 - (8)_4E_2 + (8)_6E_3 - E_4 \\ &\cdots\cdots\cdots\cdots\cdots \end{aligned}$$

Solving from these we have

$$E_n = \begin{vmatrix} 1 & 1 & 0 & 0 & 0 & \cdots & \\ 1 & (4)_2 & 1 & 0 & 0 & \cdots & \\ 1 & (6)_2 & (6)_4 & 1 & 0 & \cdots & \\ 1 & (8)_2 & (8)_4 & (8)_6 & 1 & \cdots & \\ \cdot & \cdot & \cdot & \cdot & \cdot & \cdot & \cdot \\ 1 & (2n)_2 & (2n)_4 & (2n)_6 & (2n)_8 & \cdots & (2n)_{2(n-1)} \end{vmatrix}_n$$

or what is easily seen as an equivalent

$$E_n = (2n)! \begin{vmatrix} \frac{1}{2} & 1 & 0 & 0 \cdots \\ \frac{1}{4!} & \frac{1}{2!} & 1 & 0 \cdots \\ \frac{1}{6!} & \frac{1}{4!} & \frac{1}{2!} & 1 \cdots \\ \cdot & \cdot & \cdot & \cdot \end{vmatrix}_n$$

781. Starting with the relation

$$(a + d)^r - a^r = (r)_1 a^{r-1}d + (r)_2 a^{r-2}d^2 + \cdots + (r)_r a^0 d^r,$$

giving a the values $a+d,\ a+2d,\ \cdots,\ a+nd$, and summing the results we get

$$(a+\overline{n+1}d)^r - a^r = (r)_1 S_{r-1} d - (r)_2 S_{r-2} d^2 + \cdots + (r)_r S_0 d^r$$

where

$$S_k = a^k + (a+d)^k + (a+2d)^k + \cdots + (a+nd)^k.$$

If in this we put $a_k = a + kd$, then

$$(a) \qquad a^r{}_{n+1} - a^r = (r)_1 S_{r-1} d + (r)_2 S_{r-2} d^2 + \cdots + (r)_r S_0 d^r$$
$$= (S+d)^r_{\downarrow} - S_r, \text{ say,}$$

and in this giving r the successive integral values from 1 to m we have

$$m!d\cdot S_{m-1} = (-1)^{m-1} \begin{vmatrix} (a_{n-1}+a)d^{m-1} & (1)_1 & \cdots & 0 & 0 \\ (a^2_{n+1}-a^2)d^{m-2} & (2)_2 & \cdots & 0 & 0 \\ \cdot & \cdot & \cdot & \cdot & \cdot \\ (a^{m-1}_{n+1}-a^{m-1})d & (m-1)_{m-1} & \cdots & (m-1)_2 & (m-1)_1 \\ (a^m_{n+1}-a^m) & (m)_m & \cdots & (m)_3 & (m)_2 \end{vmatrix}$$

If in (a) we put $d=1$, and $a=n$ we have

$$(n+1)^r - n^r = (r)_1 n^{r-1} + (r)_2 n^{r-2} + \cdots + (r)_r n$$

and in this giving n successively the values $1, 2, \cdots, n$ we get after adding

$$(b) \qquad (n+1)^r - 1 = (S+1)^r_{\downarrow} - S_r.$$

If, in (b), we replace n by $n - 1/2$ we find, after putting $n-1, 2, \cdots, n$ successively and adding, that

$$\begin{aligned}(n+\tfrac{1}{2})^r - (\tfrac{1}{2})^r = \quad & (r)_1[(n-\tfrac{1}{2})^{r-1} + (n-\tfrac{3}{2})^{r-1} + \cdots + (\tfrac{1}{2})^{r-1}] \\ & + (r)_2[(n-\tfrac{1}{2})^{r-2} + (n-\tfrac{3}{2})^{r-2} + \cdots \qquad] \\ & \cdots\cdots\cdots\cdots\cdots\cdots \\ & + (r)_r[(n-\tfrac{1}{2})^0 + (n-\tfrac{3}{2})^0 + \cdots \qquad]\end{aligned}$$

or

$$\begin{aligned}(c)\ (2n+1)^r - 1 = {} & (r)_1 2[(2n-1)^{r-1} + (2n-3)^{r-1} + \cdots + 1^{r-1}] \\ & + (r)_2 2^2[(2n-1)^{r-2} + (2n-3)^{r-2} + \cdots + 1^{r-2}] \\ & \cdots\cdots\cdots\cdots\cdots\cdots \\ & + (r)_r 2^r[(2n-1)^0 + (2n-3)^0 + \cdots + 1^0 \qquad]\end{aligned}$$

or

$$(2n-1)^{r-h}+(2n-3)^{r-h}+\cdots+(1)^{r-h}$$
$$=2^rS_{r-h}-2^{r-1}(r-h)_1S_{r-h-1}+2^{r-2}(r-h)_2S_{r-h-2}-\cdots$$

Expanding

$$\left(n-\frac{2k+1}{2}\right)^{r-h}\equiv\left(\overline{n-k}-\frac{1}{2}\right)^{r-h}=(n-k)^{r-h}$$
$$-(r-h)_1(n-k)^{r-h-1}\frac{1}{2}+(r-h)_2(n-k)^{r-h-2}\left(\frac{1}{2}\right)^2-\quad\cdots$$

giving k the values $k=0, 1, 2, \cdots, , \overline{n-1}$, and adding we have

$$(n-\tfrac{1}{2})^{r-h}+(n-\tfrac{3}{2})^{r-h}+\cdots+(\tfrac{1}{2})^{r-h}$$
$$=S_{r-h}-(r-h)_1S_{r-h-1}\cdot\tfrac{1}{2}+(r-h)_2S_{r-h-2}\cdot(\tfrac{1}{2})^2-\cdots$$

In this giving h the values $1, 2, \cdots, r$ and substituting in (c), we get, on observing that the coefficient of S_{r-i} is zero when i is even,

$$(2n+1)^r-1=(2S+1)^{r}_{\downarrow}-(2S-1)^{r}_{\downarrow}$$

Giving r in this last relation the values $1, 3, 5, \cdots$, we get

$$\begin{aligned}
(2n+1)-1&=2S_0\\
(2n+1)^3-1&=2S_0+2^3(3)_2S_2\\
(2n+1)^5-1&=2S_0+2^3(5)_2S_2+2^5(5)_4S_4\\
(2n+1)^7-1&=2S_0+2^3(7)_2S_2+2^5(7)_4S_4+2^7(7)_6S_6\\
&\cdots\cdots\cdots\cdots\cdots\cdots
\end{aligned}$$

where $S_0=1$.

From these relations we get

$$2^{2r+1}1\cdot3\cdots(2r+1)S_{2r}$$
$$=(-1)^{r(r+1)/2}\begin{vmatrix}
(2n+1)^1 & 1 & 0 & 0 & \cdots & 0\\
(2n+1)^3 & 1 & (3)_2 & 0 & \cdots & 0\\
(2n+1)^5 & 1 & (5)_2 & (5)_4 & \cdots & 0\\
\cdots & \cdots & \cdots & \cdots & \cdots & \cdots\\
(2n+1)^{2r-1} & 1 & (2r-1)_2 & (2r-1)_4 & \cdots & (2r-1)_{2r-2}\\
(2n+1)^{2r+1} & 1 & (2r+1)_2 & (2r+1)_4 & \cdots & (2r+1)_{2r-2}
\end{vmatrix}$$

EXERCISE: If $B_r = (-1)^{r-1}\beta_{2r}$, where B_r is the rth Bernoulli's number, show that

$$r(S_{r-1} - n^{r-1}) = (\beta + n)^{r}_{\downarrow} - \beta_r.$$

782. If $u \equiv f(x)$ and $v \equiv \phi(x)$, then

$$(a)\ f(x + h) = u + h\frac{\partial u}{\partial x} + \frac{h^2}{2!}\frac{\partial^2 u}{\partial x^2} + \cdots$$

$$(b)\ \phi(x + h) = v + h\frac{\partial v}{\partial x} + \frac{h^2}{2!}\frac{\partial^2 v}{\partial x^2} + \cdots$$

$$(c)\ f(x + h) = f\phi^{-1}[v + \{\phi(x + h) - v\}]$$

$$= u + \{\phi(x + h) - v\}\frac{\partial u}{\partial v} + \left\{\frac{\phi(x+h) - v}{2!}\right\}^2 \frac{\partial^2 u}{\partial v^2} + \cdots$$

Substituting for $\{\phi(x+h) - v\}$ from (b) in (c) and then equating coefficients of like powers of h in (a) and (c) we have

$$u_1 = v_1\frac{\partial u}{\partial v}$$

$$u_2 = v_2\frac{\partial u}{\partial v} + \frac{v_1^2}{2!}\frac{\partial^2 u}{\partial v^2}$$

$$u_3 = v_3\frac{\partial u}{\partial u} + \frac{2v_1v_2}{2!}\frac{\partial^2 u}{\partial v^2} + \frac{v_1^3}{3!}\frac{\partial^3 u}{\partial v^3}$$

$$u_4 = v_4\frac{\partial u}{\partial v} + \frac{2v_1v_3 + v_2^2}{2!}\frac{\partial^2 u}{\partial v^2} + \frac{3v_1v_2}{3!}\frac{\partial^3 u}{\partial v^3} + \frac{v_1^4}{4!}\frac{\partial^4 u}{\partial v^4}$$

. . . .

where

$$u_r = \frac{1}{r!}\frac{\partial^r u}{\partial x^r} \quad \text{and} \quad v_r = \frac{1}{r!}\frac{\partial^r v}{\partial x^r}.$$

From these equations we get

$$\frac{1}{n!}\frac{\partial^n u}{\partial v^n} = \; - 1)^n \frac{\begin{vmatrix} u_1 & v_1 & 0 & 0 & 0 & \cdots \\ u_2 & v_2 & v_1^2 & 0 & 0 & \cdots \\ u_3 & v_3 & 2v_1v_2 & v_1^3 & 0 & \cdots \\ u_4 & v_4 & 2v_1v_3 + v_2^2 & 3v_1^2v_2 & v_1^4 & \cdots \\ u_5 & v_5 & 2(v_1v_4 + v_2v_3) & 3(v_1^2v_3 + v_1v_2^2) & 4v_1^3v_2 & \cdots \\ \cdot & \cdot & \cdot & \cdot & \cdot & \cdot \end{vmatrix}}{v_1v_1^2 \cdots v_1^n}$$

or

$$\frac{1}{n!}\frac{\partial^n u}{\partial v^n} = (-1)^n \frac{\begin{vmatrix} u_1 & S_1^1 & 0 & \cdots 0 \\ u_2 & S_2^1 & S_2^2 & \cdots 0 \\ u_3 & S_3^1 & S_3^2 & \cdots 0 \\ \cdot & \cdot & \cdot & \cdot \\ u_{n-1} & S_{n-1}^1 & S_{n-1}^2 & \cdots S_{n-1}^{n-1} \\ u_n & S_n^1 & S_n^2 & \cdots S_n^{n-1} \end{vmatrix}}{v_1 v_1^2 v_1^3 \cdots v_1^n}$$

where $S_m{}^n$ is the sum of the terms of weight m in the expansion of $(v_0+v_1+v_2+\cdots)^n$ with the proviso that $S_m{}^p=0$ for $p>m$ and that $S_m{}^m=v_1{}^m$.

783. From the set of recurrence formulas

$$\begin{aligned} a_1P_1 + a_2Q_1 &= 0 \\ b_1P_2 + b_2P_1Q_1 + b_3Q_2 &= 0 \\ c_1P_3 + c_2P_2Q_1 + c_3P_1Q_2 + c_4Q_3 &= 0 \\ \cdots\cdots\cdots\cdots & \end{aligned}$$

we readily find that

$$P_3 = -\begin{vmatrix} a_2Q_1 & a_1 & 0 \\ b_3Q_2 & b_2Q_1 & b_1 \\ c_4Q_3 & c_3Q_2 & c_2Q_1 \end{vmatrix} \div a_1b_1c_1$$

$$Q_3 = -\begin{vmatrix} a_1P_1 & a_2 & 0 \\ b_1P_2 & b_2P_1 & b_3 \\ c_1P_3 & c_2P_2 & c_3P_1 \end{vmatrix} \div a_2b_3c_4$$

etc.

EXERCISE: Show that

$$\begin{vmatrix} (m+2)_1 & 1 & 0 & 0 & \cdots \\ (m+4)_2 & (m+4)_1 & 1 & 0 & \cdots \\ (m+6)_3 & (m+6)_2 & (m+6)_1 & 1 & \cdots \\ \cdot & \cdot & \cdot & \cdot & \cdot \\ (m+2n)_n & (m+2n)_{n-1} & (m+2n)_{n-2} & (m+2n)_{n-3} & \cdots (m+2n)_1 \end{vmatrix} = \frac{(m+2n)(m+n-1)!}{m!n!}.$$

784. By using the set of row multipliers

$$\begin{matrix} 1 & -1 & 1 & -1 & 1 \\ 1 & -2 & 3 & -4 & \\ 1 & -3 & 6 & & \\ 1 & -4 & & & \end{matrix}$$

it may be shown that the recurrent

$$R \equiv \begin{vmatrix} x & \beta & c-b+3d & & \\ 4(\gamma-3c-3d) & x+P & 2(\beta-b) & 3(c-b+4d) & \\ \cdot & 3(\gamma-2c-d) & x+Q & 3(\beta-2b-3d) & 6(c-b+5d) \\ \cdot & \cdot & 2(\gamma-c) & x+R & 4(\beta-3b+9d) \\ \cdot & \cdot & \cdot & \gamma & x+S \end{vmatrix}$$

$$=(x-4\gamma+12c+12d)(x+\beta-3\gamma+6c+3d)(x+2\beta-2\gamma-b$$
$$+3c+3d)(x+3\beta-\gamma-3b+3c+9d)(x+4\beta-6b+6c+18d),$$

where

$$P = \beta - \gamma + 6c + 9d,$$
$$Q = 2\beta - 2\gamma - b + 9c + 9d,$$
$$R = 3\beta - 3\gamma - 3b + 9c + 9d,$$
$$S = 4\beta - 4\gamma + 6c + 18d.$$

If in R we put $\gamma=\beta+6c$, $b=-3c$ and then $c=1$ we get a case worthy of special note.

If in R we put $c=b$ and $d=0$, then R becomes a factorable continuant.

785. Multiplying the elements of the last row of the determinant

$$R_n \equiv \begin{vmatrix} 1 & -2 & 0 & 0 & \cdots 0 \\ (3)_1 & 1 & -4 & 0 & \cdots 0 \\ (5)_2 & (3)_1 & 1 & -6 & \cdots 0 \\ \cdot & \cdot & \cdot & \cdot & \cdot \\ (2n-1)_{n-1} & (2n-3)_{n-2} & (2n-5)_{n-3} & (2n-7)_{n-4} & \cdots 1 \end{vmatrix}_n$$

by $(2n+1)$ we readily see on expanding in terms of the elements of the last column

$$(2n+1)R_n = (2n+1)R_{n-1} + 2(n-1)\Delta,$$

where

$$\Delta = \begin{vmatrix} 1 & -2 & \cdots & \\ (3)_1 & 1 & \cdots & \\ \cdot & \cdot & \cdot & \cdot \\ (2n-5)_{n-3} & (2n-7)_{n-4} & \cdots & -(2n-4) \\ (2n+1)(2n-1)_{n-1} & (2n+1)(2n-3)_{n-2} & \cdots & (2n+1)(3)_1 \end{vmatrix}_{n-1}$$

But $\Delta = (8n-1)R_{n-1}$ as may be seen on performing the operations

$$r_{n-1} - \tfrac{1}{2}(2)_1 r_{n-2} - \tfrac{1}{3}(4)_2 r_{n-3} - \tfrac{1}{4}(6)_3 r_{n-4} - \tfrac{1}{5}(8)_4 r_{n-5} - \tfrac{1}{6}(10)_5 r_{n-6} - \cdots .$$

Therefore

$$R_n(2n+1) = (4n-1)(4n-3)R_{n-1}$$

and

$$R_n = (2n+3)(2n+5)\cdots(4n-3)(4n-1).$$

Exercise: If S_n is the same determinant as R_n except that the signs of all the elements are positive then show that

$$(2n-1)S_n = -(4n-5)(4n-3)S_{n-1}$$

and hence

$$S_n = (-1)^{n-1}(2n+1)(2n+3)\cdots(4n-3).$$

786. The determinant

$$\begin{vmatrix} 1 & 0 & 0 & 0 & \cdots & & U_0 \\ 1 & (1)_1 & 0 & 0 & \cdots & & U_1 \\ 1 & (2)_1 & (2)_2 & 0 & \cdots & & U_2 \\ 1 & (3)_1 & (3)_2 & (3)_3 & \cdots & & U_3 \\ \cdot & \cdot & \cdot & \cdot & \cdot & \cdot & \cdot \\ 1 & (n-2)_1 & (n-2)_2 & (n-2)_3 & \cdots & (n-2)_{n-2} & U_{n-2} \\ 1 & (n-1)_1 & (n-1)_2 & (n-1)_3 & \cdots & (n-1)_{n-2} & U_{n-1} \end{vmatrix}_n$$

$$= U_{n-1} - (n-1)_1 U_{n-2} + (n-1)_2 U_{n-3} \cdots + (-1)^{n-2}(n-1)_{n-2}U_1 + (-1)^{n-1}U_0$$

as may be seen on performing the operations

$$r_n - (n-1)_1 r_{n-1} + (n-1)_2 r_{n-2} - (n-1)_3 r_{n-3} + \cdots ,$$

This will reduce all the elements in the last row to zero except the last which is the value of the determinant.

If in this determinant we put

$$U_{n-1} = x^{n-1}, \quad U_{n-2} = x^{n-2}h, \quad U_{n-3} = x^{n-3}h^2 \cdots$$

then its value is $(x-h)^{n-1}$.

Exercise: Show that the determinant

$$\begin{vmatrix} b^{n/a} & b^{1/b} & 0 & 0 & \cdots & 0 \\ (2b)^{n/a} & (2b)^{1/b} & (2b)^{2/b} & 0 & \cdots & 0 \\ (3b)^{n/a} & (3b)^{1/b} & (3b)^{2/b} & (3b)^{3/b} & \cdots & 0 \\ \cdot & \cdot & \cdot & \cdot & \cdot & \cdot \\ (nb)^{n/a} & (nb)^{1/b} & (nb)^{2/b} & (nb)^{3/b} & \cdots & (nb)^{(n-1)/b} \end{vmatrix}_n$$
$$= (-1)^{n-1} b^{n(n+1)/2} 1!2!3! \cdots n!,$$

where

$$b^{n/a} = b(b-a)(b-2a) \cdots (b-\overline{n-1}a).$$

CHAPTER XXII

Invariant Factors

787. As we have seen (§38) a general determinant Δ may be expressed as a linear function of the elements of any line, and if the elements are independent quantities it follows that the determinant cannot be factored.

For if Δ has factors they must be factors of the primary minors which are determinants of order $n-1$ and unless these have factors Δ can have no factors. But it is true for $n=1$ and therefore true in general that Δ is irreducible.

If, however, the elements are not independent but integers or functions of a single variable with integral coefficients, then Δ and all its minors may have factors and we shall suppose that every set of the minors of Δ may be expected to have a highest-common-factor. Thus the determinant

$$\Delta \equiv \begin{vmatrix} \lambda - \mu & 3\lambda & 4\lambda \\ 2\lambda & 2\lambda - \mu & 4\lambda \\ 2\lambda & 3\lambda & 3\lambda - \mu \end{vmatrix} = (\lambda + \mu)^2(8\lambda - \mu)$$

has the factors

$$1,\ \lambda + \mu,\ (\lambda + \mu)^2,\ (8\lambda - \mu),\ (\lambda + \mu)(8\lambda - \mu),\ (\lambda + \mu)^2(8\lambda - \mu).$$

It is also spoken of as having the linear factors $\lambda+\mu$, $\lambda+\mu$, $8\lambda-\mu$, but it is the group $\lambda+\mu$, $(\lambda+\mu)(8\lambda-\mu)$ with which we shall presently be concerned.

788. If D_s represents the highest common factor of the minors of Δ of order s, then if

$$(1) \qquad D_s/D_{s-1} \equiv E_s$$

the group in question is $E_n, E_{n-1}, \cdots, E_1$, where we assume $D_0=1$ and therefore $D_1=E_1$.

That D_s is divisible by D_{s-1} is seen on observing that if any minor of order s is expanded in terms of the elements of a line and their complementaries, these latter are minors of order $s-1$ and therefore all divisible by D_{s-1}.

Since D_n differs from $\Delta(\Delta\neq 0)$ by a constant factor at most, we have from (1)

$$D_s = E_s E_{s-1} \cdots E_1,$$

and

$$D_n = E_n E_{n-1} \cdots E_1$$
$$= \Delta,$$

except a possible constant factor.

The E's are all factors of Δ and have been variously called *elementary factors*, *critical factors*, and *invariant factors*. The D's have been called *elementary divisors*.

It has been shown* that D_s/D_{s-1} is the highest common factor of the quotients obtained by dividing each minor of order s by the highest common factor of its own first minors.

It has also been shown* that D_{s-1}/D_s is the highest common factor of the functions obtained by dividing each minor of order $s-1$ by the highest common factor of its first majors. The term *major* is used to indicate a determinant of which another determinant is a minor. Thus if M is a minor of A, then A is a major of M.

789. It is apparent that factors of D_s are also factors of Δ. Let p be a prime factor of D_s and let p^{l_s} be the highest power of p contained in D_s, then it follows from §788 (1) that

$$l_n \geqq l_{n-1} \geqq \cdots \geqq l_1$$

Let

$$(1) \qquad l_s - l_{s-1} = e_s$$

then with the assumption that $l_0 = 0$ we have the series $e_n, e_{n-1}, \cdots, e_1$ all of which are positive integers.

From (1) we have

$$e_1 + e_2 + \cdots + e_s = l_s.$$

From (1) §788 we have

$$E_s = D_s/D_{s-1} = D_s' p^{l_s}/D_{s-1}' p^{l_{s-1}} \text{ say},$$
$$= p^{e_s}(D_s'/D_{s-1}')$$
$$= \pi p^{e_s}$$

where the product is to include all the prime factors of Δ_s.

790. If there is at least one minor of Δ of order s which contains p^{l_s} but no higher power of p, then such a minor is called *regular*.

* H. J. S. Smith: Phil. Trans., vol. cli, pp. 293– ; Lond. Math. Soc., vol. iv, pp. 236–

791. Let

$$M \equiv \left| a_{ij} \right| \neq 0 \quad \begin{pmatrix} i = i_1, i_2, \cdots, i_r \\ j = j_1, j_2, \cdots, j_r \end{pmatrix}$$

be a minor of Δ of order r and let p^{λ_s} be the highest power of p which is contained in the minors of M of order s. Let N be a minor of M of order $r-2$ which contains $p^{\lambda_{r-2}}$ but no higher power of p, and let A, B, C, D be minors of M of order $r-1$ which satisfy the identity

$$MN \equiv AB - CD.$$

The left-hand side of this identity is divisible by $p^{\lambda_r+\lambda_{r-2}}$ but no higher power of p. The right-hand side is divisible by $p^{2\lambda_{r-1}}$ at least. It follows therefore that

$$\lambda_r + \lambda_{r-2} \geqq 2\lambda_{r-1}$$

or

(2) $$\lambda_r - \lambda_{r-1} \geqq \lambda_{r-1} - \lambda_{r-2}.$$

Let

$$E \equiv \left| a_{hk} \right| \quad \begin{pmatrix} h = h_1, h_2, \cdots, h_{r-1} \\ k = k_1, k_2, \cdots, k_{r-1} \end{pmatrix}$$

be any minor of Δ of order $r-1$ and let E_{ij} be E bordered by the row and column of M which contains a_{ij}. Then by Kronecker's theorem we have

$$\left| E_{ij} - a_{ij}E \right| = 0 \quad \begin{pmatrix} i = i_1, i_2, \cdots, i_r \\ j = j_1, j_2, \cdots, j_r \end{pmatrix}$$

identically.

Expanding in powers of E we have

(3) $$E^rM = E^{r-1}M_1 + E^{r-2}M_2 + \cdots + M_r$$

where M_k is a function homogeneous in the quantities E_{ij}, the coefficients being minors of M of order $(r-k)$.

If now p^l is the highest power of p contained in E, and $p^{l'}$ is the highest power of p contained in the highest common factor of all the determinants E_{ij}, then E^rM contains p to the power $rl+\lambda_r$ and $E^{r-k}M_k$ contains p to a power at least equal to $(r-k)l+kl'+\lambda_{r-k} \equiv \rho_k$ say.

Hence

(4) $$\rho_{k+1} - \rho_k = (l' - l) - (\lambda_{r-k} - \lambda_{r-k-1}).$$

Then from (2) we have

$$\rho_{k+1} - \rho_k \gtreqless (l' - l) - (\lambda_r - \lambda_{r-1}).$$

But $\rho_{k+1} < \rho_k$ for otherwise every term on the right-hand side of (3) would contain p to a power greater than $rl + \lambda_r (\equiv \rho_0)$ which is impossible. Therefore

$$l' - l \lesseqgtr \lambda_r - \lambda_{r-1},$$

or

(5) $$\lambda_r + l \gtreqless \lambda_{r-1} + l'$$

If E and M are regular with respect to p then $\lambda_r = l_r$ and $l = l_{r-1}$, so that

$$l_r + l_{r-1} \gtreqless \lambda_{r-1} + l'.$$

But $l' \gtreqless l_r$ and $\lambda_{r-1} \gtreqless l_{r-1}$ and therefore it must be that

$$\lambda_{r-1} = l_{r-1} \quad \text{and} \quad l' = l_r.$$

That is if $\lambda_r = l_r$, then $\lambda_{r-1} = l_{r-1}$ and if $l = l_{r-1}$, then

$$l' = l_r \quad (r = 1, 2, \cdots, r).$$

The first of these shows that

(I) $$e_r \gtreqless e_{r-1}.$$

That is, not only the first order of differences of the l's but also the second are not negative.

The second shows that

II. *Every minor of order r which is regular with respect to p has at least one first minor which is also regular.*

It also appaers that

III. *Every regular minor of order r−1 is a first minor of at least one regular minor of order r.*

If p^t represent the highest power of p contained in all those minors of Δ which are majors of E, then $l' \gtreqless t$ and consequently from (5) we have

(6) $$\lambda_r + l \gtreqless \lambda_{r-1} + t.$$

If B_r and B_{r-1} are any two minors of Δ of orders r and $r-1$ respectively, then (6) shows that the product $B_r B_{r-1}$ is divisible by the product of the highest common factor of all the minors of B_r which are of order $r-1$ by the highest common factor of all the majors which have B_{r-1} as a first minor.

792. Suppose now that Δ is of rank k then

$$D_n = D_{n-1} = \cdots = D_{k-1} = 0,$$

and

$$E_n = E_{n-1} = \cdots = E_{k-1} = 0.$$

Also

$$e_k \gtreqqless e_{k-1} \gtreqqless \cdots \gtreqqless e_1$$

and

$$e_1 + e_2 + \cdots + e_k = l_k,$$

where l_k is the index of the highest power of p found in D_k.

With the understanding that the minors of Δ of order s and r are not of higher order than k the theorems of the preceding article still hold true.

793. Let A and B be two determinants of order n whose elements belong to the same domain, and let $AB=C$. Let $A_{r-1,r-1}$ be a regular minor of order $r-1$ taken from the rows $i_1, i_2, \cdots, i_{r-1}$ and columns $j_1, j_2, \cdots, j_{r-1}$ of A, and let C_{rr} be a regular minor of C of order r taken from the rows $h_1, h_2, \cdots, h_r$ and columns $k_1, k_2, \cdots, k_r$, then form the determinant

$$\Delta \equiv \begin{vmatrix} (A_{r-1,r-1}) & (A_{r-1,r}) \\ (C_{r,r-1}) & (C_{rr}) \end{vmatrix}$$

where $(A_{r-1,r})$ represents the elements of A taken from the $i_1, i_2, \cdots, i_{r-1}$ rows and $k_1, k_2, \cdots, k_r$ columns and where $(C_{r,r-1})$ represents the elements of C taken from the $h_1, h_2, \cdots, h_r$ rows and $j_1, j_2, \cdots, j_{r-1}$ columns. If now we form a determinant Δ' whose elements are $A_{r-1,r-1}$ bordered in all possible ways with one of the remaining rows and columns of Δ according to Sylvester's theorem we have from theorem I §791

$$\lambda_r + l_{r-1} \gtreqqless \lambda_{r-1} + l'$$

where l_r and λ_r are the exponents of the highest power of the prime p contained in determinant factors of order r of A and C respectively, and l' is the highest power of p contained in the elements of Δ'. But $l' \gtreqq l_r$ since each of the elements of Δ' can be expressed as a linear homogeneous function of the minors of A of order r; and therefore

$$\lambda_r - \lambda_{r-1} \gtreqq l_r - l_{r-1}$$

which shows that the factor of A is contained into the corresponding factor of C. Similarly for the factors of B. We have then the theorem that: *Every invariant factor of C is divisible by the corresponding invariant factors of A and B.*

The special case where either A or B is equal to 1 deserves attention and it may be shown that if $B=1$ then A and C are *equivalent*, that is one may be transformed into the other by means of a unit determinant.

794. If the elements of Δ are all integers then Δ is an integer and in general has integral prime factors. In earlier paragraphs we have seen the effects of the following operations:

(1) The interchange of two rows
(2) The interchange of two columns
(3) The addition to any row of a constant times another
(4) The addition to any column of a constant times another.

We wish now to show that by means of these operations we may transform any determinant $A \equiv |a_{nn}|$ into one of the form

$$\begin{vmatrix} E_1 & \cdot & \cdot & \cdots & \cdot \\ \cdot & E_2 & \cdot & \cdots & \cdot \\ \cdot & \cdot & E_3 & \cdots & \cdot \\ \cdot & \cdot & \cdot & \cdots & \cdot \\ \cdot & \cdot & \cdot & \cdots & E_n \end{vmatrix}$$

in which the principal diagonal elements are the invariant factors of A and all the other elements are zeros.

It is apparent these operations do not change (save as to arrangement) the system of minors of any order, and consequently they leave unaltered the invariant factors E_s. It is also readily seen that there is a unit matrix of value $+1$ or -1 which when used as a premultiplier or a post-multiplier will accomplish the transformation obtained by these four operations.

In the determinant A whose elements are not all zero some one element is as small or smaller than any of the others and denoting this element by a we may bring it into the position (1, 1) by the operations (1) and (2). Then by the use of (4) we may reduce all the other elements in the first row to zero or to numbers smaller than a. Having done this we may by (2) rearrange the elements of this row so as to bring the smallest non-zero element (a' say) into the position (1, 1). Reducing the other elements as before and continuing the process we must ultimately reach a point where all the elements of the first row are zero except the first and it is smaller than any other which was previously there. If we next treat the elements in the first column in a similar fashion we come by a finite number of operations to a determinant, A', say, with all the elements in the first row and first column zeros except the first which we denote by e.

Let

$$A' \equiv \begin{vmatrix} e & \cdot & \cdot & \cdots & \cdot \\ \cdot & b_{11} & b_{12} & \cdots & b_{1,n-1} \\ \cdot & b_{21} & b_{22} & \cdots & b_{2,n-1} \\ \cdot & \cdot & \cdot & \cdot & \cdot \\ \cdot & b_{n-1,1} & b_{n-1,2} & \cdots & b_{n-1,n-1} \end{vmatrix}$$

If there is an element among the b's which is not a multiple of e then add the row containing it to the first and reduce as before. Repeating this process, which must come to an end, until every b is a multiple of e.

Put $e=e_1$ and $b_{ij}=e_1c_{ij}$ and we have

$$A' \equiv \begin{vmatrix} e_1 & \cdot & \cdots & \cdot \\ \cdot & e_1c_{11} & \cdots & e_1c_{1,n-1} \\ \cdot & \cdot & \cdot & \cdot \\ \cdot & e_1c_{n-1,1} & \cdots & e_1c_{n-1,n-1} \end{vmatrix}$$

Let $C = |c_{1,n-1}|$ and let us treat C in exactly the same way arriving at

$$A'' = \begin{vmatrix} e_1 & \cdot & \cdot & \cdots & \cdot \\ \cdot & e_1e_2 & \cdot & \cdots & \cdot \\ \cdot & \cdot & e_1e_2d_{11} & \cdots & e_1e_2d_{1,n-2} \\ \cdot & \cdot & \cdot & \cdot & \cdot \\ \cdot & \cdot & e_1e_2d_{n-2,1} & \cdots & e_1e_2d_{n-2,n-2} \end{vmatrix}$$

Continuing in this way we arrive at

$$A = \begin{vmatrix} E_1 & \cdot & \cdot & \cdot & \cdot & \cdot & \cdot \\ \cdot & E_2 & \cdot & \cdot & \cdot & \cdot & \cdot \\ \cdot & \cdot & E_3 & \cdot & \cdot & \cdot & \cdot \\ \cdot & \cdot & \cdot & \cdot & \cdot & \cdot & \cdot \\ \cdot & \cdot & \cdot & \cdot & \cdot & \cdot & E_n \end{vmatrix} \equiv E \text{ say,}$$

where $E_s = e_1 e_2 \cdots e_s$.

Since all the elements of A are not zero it follows that e_1 cannot be zero but it may happen that at some stage (the kth) of the process all of the elements of the minor, corresponding to C at the second stage, has all its elements zero. In such case we put

$$e_{k+1} = e_{k+2} = \cdots = e_n = 0.$$

795. From the preceding article we see that the matrix A may be reduced to the matrix E by the operation $E = M \cdot A \cdot N$ where M and N are unit matrices.

The sth invariant factor of A will coincide with the sth invariant factor of E, which is obviously E_s, and

$$D_s = E_1 E_2 \cdots E_s$$

or

$$D_s / D_{s-1} = E_s$$

It follows that if P and Q are two matrices having the same rank and the same set of invariant factors they are equivalent.

796. If A is rectangular instead of square as we have heretofore assumed we may make it square by the addition of rows or columns of zeros and then proceed as before.

CHAPTER XXIII

MULTILINEANTS

797. Determinants of infinite order, that is where the number of rows and the number of columns are infinite are called *Multilineants*.

Starting with the square array of order n

$$\Delta_n \equiv \begin{vmatrix} a_{11} & a_{12} \cdots a_{1n} \\ a_{21} & a_{22} \cdots a_{2n} \\ \cdot & \cdot\ \cdot\ \cdot\ \cdot\ \cdot \\ a_{n1} & a_{n2} \cdots a_{nn} \end{vmatrix}$$

we may conceive of it enlarging in all four directions. Thus

$$\Delta_{2m+1} \equiv \begin{vmatrix} \cdot & \cdot & \cdot & \cdot & \cdot \\ \cdots a_{-2-2} & a_{-2-1} & a_{-20} & a_{-21} & a_{-22} \cdots \\ \cdots a_{-1-2} & a_{-1-1} & a_{-10} & a_{-11} & a_{-12} \cdots \\ \cdots a_{0-2} & a_{0-1} & a_{00} & a_{01} & a_{02} \cdots \\ \cdots a_{1-2} & a_{1-1} & a_{10} & a_{11} & a_{12} \cdots \\ \cdots a_{2-2} & a_{2-1} & a_{20} & a_{21} & a_{22} \cdots \\ \cdot & \cdot & \cdot & \cdot & \cdot \end{vmatrix}_{2m+1}$$

where a_{rs} represents the elements, both r and s taking all values, (including 0) from $-m$ to $+m$.

The element a_{00} is called the *central element* and the elements $\cdots a_{-1-1}, a_{00}, a_{11}, \cdots$, are called the diagonal elements.

Δ_{2m+1} may or may not approach a definite finite limit as $m \doteq \infty$. Denote the case in which it converges to a definite limit by Δ.

Since we may associate with (r, s) another set of indices which assume independently the positive integral values $1, 2, \cdots$, it is apparent that we may take for Δ the determinant $|a_{ij}|$ $\left(\begin{smallmatrix} i \\ j \end{smallmatrix}=1, 2, \cdots\right)$

799. The multilineant Δ is said to be *normal* if the product of the diagonal elements is absolutely convergent, and if the sum of all the other elements is absolutely convergent. That is if πa_{ii} and

$$\sum_j \sum_k a_{jk} \quad \left\{\begin{matrix} i \\ j = 1, \cdots, \infty \\ k \end{matrix}\right\} j \neq k,$$

are absolutely convergent.

800. *A normal multilineant is convergent.*

If the multilineant Δ is normal then $\sum a_{ij}$ $(i \neq j)$ and πa_{ii} are both absolutely convergent. Let $a_{ii} = b_{ii} + 1$ and $a_{ij} = b_{ij}$. Then πb_{ii} is convergent and therefore $\sum b_{ii}$ is convergent. It follows therefore that $\sum b_{ij}$ is convergent for all values of i and j. But if $\sum |b_{ij}|$ is convergent we know that

$$\pi_n = \pi(1 + \sum | b_{ij} |), \quad \left(\begin{matrix} i \\ j \end{matrix} = 1, 2, \cdots, n\right)$$

is absolutely convergent as n increases without limit.

From the product π_n we can get all the terms of Δ_n by giving the coefficients of the terms in the product the values of $+1$, -1, or 0.

It follows therefore that $\Delta_n < \pi_n$.

If we make all the non-diagonal elements of the last p rows and columns of Δ_{n+p} equal to zero and put all the last p diagonal elements equal to 1 then Δ_{n+p} reduces to Δ_n and at the same time π_{n+p} reduces to π_n. We have then

$$| \Delta_{n+p} - \Delta_n | < \pi_{n+p} - \pi_n.$$

Hence

$$\lim_{n=\infty} \pi_{n+p} = \lim_{n=\infty} \pi_n$$

and

$$\lim_{n=\infty} \Delta_n = \Delta$$

is convergent.

801. Poincaré and Koch have shown that we may substitute for the elements in any finite number of lines of Δ quantities all less in absolute value than a certain positive number and still have the resulting determinant convergent. Thus if $x_1, x_2, x_3 \cdots$ where $|x_i| < k$ for all values of i, be substituted for the elements in any line the resulting determinant, Δ', say, is convergent. For if we denote by π' the product of the sums of the absolute values of all lines of Δ except the one replaced, then any term of Δ' will be a term of π multiplied by one of the quantities x, or by this quantity changed in sign and therefore $|\Delta_n| < k \cdot \pi_n$ and finally Δ' is convergent.

802. If in a normal multilineant Δ we replace the element in the position (i, j) by unity and all the other elements of the ith row and jth column by zeros we get a first minor of Δ which we shall denote by A_{ij} or $(-1)^{i+j} A_{ij}$.

Then we may write as in determinants of finite order

$$\Delta = a_{i1}\mathcal{A}_{i1} + a_{i2}\mathcal{A}_{i2} + \cdots = \sum_{j=1}^{\infty} a_{ij}\mathcal{A}_{ij}$$

and

$$\Delta = a_{1j}\mathcal{A}_{1j} + a_{2j}\mathcal{A}_{2j} + \cdots = \sum_{i=1}^{\infty} a_{ij}\mathcal{A}_{ij}$$

Let

$$\Delta_n \equiv |\, a_{nn} \,| = a_{i1}\mathcal{A}'_{i1} + a_{i2}\mathcal{A}'_{i2} \cdots = \sum_{j=1}^{n} a_{ij}\mathcal{A}'_{ij}$$

where $\mathcal{A}'_{ij}$ is a first minor of Δ_n, then

$$\Delta - \Delta_n = \sum_{j=1}^{n} a_{ij}(\mathcal{A}_{ij} - \mathcal{A}'_{ij}) + \sum_{j=n+1}^{\infty} a_{ij}\mathcal{A}_{ij}.$$

If now n increases indefinitely the first sum on the right approaches zero and so also does the second and therefore the expansion is proper.

We may extend the idea of a first minor and have one of any order and show in a similar manner that we may write the expansion as in Laplace's theorem.

803. It may also be shown that the following theorems concerning finite determinants are also true for multilineants: those of §§37, 45, 49, 57, 154, 174.

804. If $A = |a_{1\infty}|$, $B = |b_{1\infty}|$ and $b_{ij} = x_i a_{ij}/y_j$, where

$$\prod_{k=1}^{\infty} \frac{x_k}{y_k} = P$$

is absolutely convergent then if A is a normal multilineant so also will B be normal; but if B is normal it does not follow that A will be. A is then said to be *semi-normal* and the quantities x_i/y_i have been called the *reducent* of A.

Since $|b_{1n}| = P_n |a_{1n}|$, where P_n is the product of x_k/y_k for $k = 1, 2, \cdots n$, then when n is increased indefinitely we have

$$|\, b_{1\infty} | = P |\, a_{1\infty} | \quad \text{or} \quad |\, a_{1\infty} | = P^{-1} |\, b_{1\infty} | \,.$$

CHAPTER XXIV

Determinants of Higher Class

805. Cayley[1] in 1843 defined what is now called a full-sign determinant of higher class, the *class* being the number of directions in the matrix or the number of suffixes or *indices* needed to specify an element. He gave a law of multiplication based on the law for ordinary determinants, but he noted that it was inapplicable if both factors were of odd class.[2] He gave also a rule of decomposition of a determinant into a sum of determinants of lower class. In 1879 Scott[3] described another kind of multiplication. Determinants more general in resepct to signancy were introduced in 1918,[4] which admit of multiplication without restriction upon the class. The processes of Cayley and Scott as applied to these determinants will here be explained.

The cubic or 3-way matrix and determinant are first considered but only so far as to illustrate some of the topographical terms found useful in treating the general class and to afford simple examples of the above-mentioned processes. The 4-way matrix with its determinants of different signancies is then described, in which first appear some fundamental features of the theory. Formulas for decomposition and multiplication of p-way determinants close the chapter. The reader is referred to the literature for further developments and applications.[5]

[1] A. Cayley, On the theory of determinants (§2 of the paper), Trans. Cambridge Phil. Soc., vol. VIII (1843), p. 75; Coll. Math. Papers, vol. I, p. 63.

[2] An exception ignored by almost all writers prior to 1911, as noted by M. Lecat in his work, Abrégé de la Théorie des Déterminants à n Dimensions (Gand, Hoste, 1911), and in his work, Coup d'Oeil sur la Théorie des Déterminants Supérieurs dans son État Actuel (Bruxelles, Lamertin, 1927). In the latter work is continued Lecat's Bibliographie des Déterminants Supérieurs in the appendix of his Bibliographie de la Relativité (Bruxelles, Lamertin, 1924).

[3] R. F. Scott, On cubic determinants, etc., Proc. London Math. Soc., vol. XI (1879), p. 17.

[4] L. H. Rice, P-way determinants, Am. J. of Math., vol. XL (1918); p. 242.

[5] F. L. Hitchcock, Multiple invariants and generalized rank of a p-way matrix or tensor, J. of Math. and Phys. of the Mass. Inst. of Tech., vol. vii (1927), p. 39. L. H. Rice, Compounds of Cayley products of determinants of higher class, same Journal, vol. vi (1926), p. 33; File multiplication of ordered determinants, same Journal, vol. iv (1925), p. 200; A Taylor's expansion of a determinant, same Journal, vol. iv (1925), p. 62; A contribution to the generalization of a determinantal theorem of Frobenius,

806. The 3-way Matrix and Determinant. If n square ordinary or 2-way matrices of order n be placed one behind another they form a 3-way matrix M of order n:

$$(1)\qquad M = \left\|\begin{matrix} A_{\alpha 11} & A_{\alpha 12} \cdots A_{\alpha 1n} \\ A_{\alpha 21} & A_{\alpha 22} \cdots A_{\alpha 2n} \\ \cdots & \cdots \\ A_{\alpha n1} & A_{\alpha n2} \cdots A_{\alpha nn} \end{matrix}\right\|, \quad (\alpha = 1, 2, \cdots, n).$$

The n^3 elements $A_{\alpha\beta\gamma}$ not only lie in the n *α-layers*, but also lie in n *β-layers* (horizontal and seen edgewise in (1)), and again in n *γ-layers* (vertical and seen edgewise in (1)). The n^2 rows and n^2 columns of the α-layers are the *rows* and *columns* of M. Perpendicular to both rows and columns are n^2 *normals* (horizontal and seen endwise in (1)). Rows, columns, and normals are all called *files*. A layer is of *aspect* α or β or γ; a file is of *direction* α or β or γ.

A *transversal* of M is a set of n elements of which no two lie in the same layer of any aspect, formed into a product. The leading transversal is $A_{111}A_{222} \cdots A_{nnn}$; as viewed in (1) it is shaped like the leading term of a 2-way determinant, and is one of $n!$ transversals so shaped when thus viewed. There are $n!$ transversals shaped like each term of a 2-way determinant. Prefix to every transversal of each such set the sign of the corresponding term of a 2-way determinant. The sum of the signed transversals is a *determinant* of M.

This determinant A may be expressed in symbols. As usual let $\epsilon_{i^{(1)}i^{(2)}\cdots i^{(n)}}$ mean $+1$, -1, or 0, according as $i^{(1)}i^{(2)} \cdots i^{(n)}$ is an even (or positive) or an odd (or negative) permutation of $12 \cdots n$ or is not a permutation of $1\ 2 \cdots n$ (that is, has two or more i's equal in value).[6] Then

$$(2)\qquad A = \sum_{\beta,\gamma}^{(n!)^2} \epsilon_{\beta^{(1)}\cdots\beta^{(n)}}\epsilon_{\gamma^{(1)}\cdots\gamma^{(n)}} A_{\alpha^{(1)}\beta^{(1)}\gamma^{(1)}} \cdots A_{\alpha^{(n)}\beta^{(n)}\gamma^{(n)}}.$$

In this determinant α is *nonsignant* while β and γ are *signant*. There are two other determinants of M, in which respectively β and

same Journal, vol. iii (1924), p. 118; On the expression of the sum of any two determinants as a determinant of more dimensions, same Journal, vol i (1922), p. 160. These papers by Hitchcock and Rice, with others cited therein, are reprinted as Publications of the Massachusetts Institute of Technology, Series II, Nos. 17, 39, 59, 76, 86, 96, 100, 101, 102, 105, 113, 115, 123, 134.

[6] Another notation is $\pm i^{(1)}i^{(2)} \cdots i^{(n)}$.

γ are nonsignant instead of α; the three determinants may be denoted[7] by $|A^{+\pm\pm}_{\alpha\,\beta\,\gamma}|$, $|A^{\pm+\pm}_{\alpha\,\beta\,\gamma}|$, $|A^{\pm\pm+}_{\alpha\,\beta\,\gamma}|$. Of the second order they are:

$$(3) \qquad A_{111}A_{222} \,\overset{-}{\underset{+}{-}}\, A_{112}A_{221} \,\overset{-}{\underset{-}{+}}\, A_{121}A_{212} \,\overset{+}{\underset{-}{-}}\, A_{122}A_{221}.$$

Only one determinant of the same matrix will herein appear at the same time, and any convenient arrangement of indices will be used.

The *permanent* $|A^{+++}_{\alpha\,\beta\,\gamma}|$ of M is the sum of the transversals each with a positive sign.

An interchange of two α-layers of M will evidently leave the determinant A unchanged, while an interchange of two β-layers or of two γ-layers will change the sign of A.

As A is a homogeneous linear function of the elements of any layer, a factor of a layer is a factor of A. If all the elements of a layer are k-nomials, A is expressible as a sum of k determinants with monomials in that layer. If a layer consists of zeros, $A=0$.

If two β-layers or two γ-layers are equal or proportional, $A=0$, but not if two α-layers are equal or proportional; and if all the α-layers are equal, A is equal to the 2-way determinant of an α-layer, multiplied by $n!$.

Any multiple of a β-layer or of a γ-layer, but not of an α-layer, can be added to another such layer without changing the value of A.

807. Decomposition of 3-way Determinants and Permanents. The determinant A is equal to a sum of 2-way determinants, and is also equal to an algebraic sum of 2-way permanents.

A *perjunctive* set of files consists of n files of the same direction no two of which lie in the same layer of any aspect.

Take n perjunctive normals of A as the rows of a 2-way permanent. Its $n!$ terms are transversals of A to which must be prefixed the same sign to make them terms of A; prefix this sign to the permanent. The algebraic sum of all $(n!)$ such permanents is equal to A; it is a *decomposition* of A and the signed permanents are *components*:

$$(4) \qquad A = \sum^{n} \epsilon_{\gamma^{(1)}\ldots\gamma^{(n)}} \left|\begin{matrix} A_{11\gamma^{(1)}} & A_{21\gamma^{(1)}} & \cdots & A_{n1\gamma^{(1)}} \\ A_{12\gamma^{(2)}} & A_{22\gamma^{(2)}} & \cdots & A_{n2\gamma^{(2)}} \\ \cdot & \cdot & \cdots & \cdot \\ A_{1n\gamma^{(n)}} & A_{2n\gamma^{(n)}} & \cdots & A_{nn\gamma^{(n)}} \end{matrix}\right|^{+}_{+}.$$

[7] Another notation is the mark $\frown$ over a signant index and $\smile$ over a nonsignant index.

On the other hand, take n perjunctive columns of A as the columns of a 2-way determinant, letting them stand in the order in which they are seen in (1). Its $n!$ terms are terms of A. So also for rows. The results are:

$$
\text{(5)}\qquad
\begin{aligned}
A &= \sum_{\alpha}^{n!} \begin{vmatrix} A_{\alpha^{(1)}11} & A_{\alpha^{(2)}12} & \cdots & A_{\alpha^{(n)}1n} \\ A_{\alpha^{(1)}21} & A_{\alpha^{(2)}22} & \cdots & A_{\alpha^{(n)}2n} \\ \cdot & \cdot & \cdot & \cdot \\ A_{\alpha^{(1)}n1} & A_{\alpha^{(2)}n2} & \cdots & A_{\alpha^{(n)}nn} \end{vmatrix} \\
&= \sum_{\alpha}^{n!} \begin{vmatrix} A_{\alpha^{(1)}11} & A_{\alpha^{(1)}12} & \cdots & A_{\alpha^{(1)}1n} \\ A_{\alpha^{(2)}21} & A_{\alpha^{(2)}22} & \cdots & A_{\alpha^{(2)}2n} \\ \cdot & \cdot & \cdot & \cdot \\ A_{\alpha^{(n)}n1} & A_{\alpha^{(n)}n2} & \cdots & A_{\alpha^{(n)}nn} \end{vmatrix} .
\end{aligned}
$$

A 3-way *permanent* may of course be decomposed by rows, columns, or normals, into a sum of 2-way permanents.

These decompositions may be more briefly expressed. The *locant* of an element $A_{\alpha\beta\ldots}$ is $\alpha\beta\cdots$; the locant of a transversal is made up of the locants of its elements; the locant of a file $(A_{1\beta\gamma}, \cdots, A_{n\beta\gamma})$ is $\beta\gamma$; and so on. A symbol like $(X_y)_z$ means that the element whose locant was y has now the locant z. If λ and μ are two indices the notation $\lambda^{(\mu)}\mu$ means that the value of λ is a function of the value of μ, one value of λ corresponding to each value of μ. The formulas then are:

$$
\text{(6a, 6b)}\qquad \left| A^{+\pm\pm}_{\alpha\beta\gamma} \right| = \sum_{\beta}^{n!} \epsilon_\beta \epsilon_\gamma \left| (A_{\alpha\beta^{(\gamma)}\gamma})^{++}_{\alpha\gamma} \right| = \sum_{\alpha}^{n!} \left| (A_{\alpha^{(\beta)}\beta\gamma})^{\pm\pm}_{\beta\gamma} \right| .
$$

For Example,[8] if $n=2$,

$$
A = \begin{vmatrix} A_{1,11} & A_{1,22} \\ A_{2,11} & A_{2,22} \end{vmatrix}^{\dagger\,\dagger} - \begin{vmatrix} A_{1,21} & A_{2,21} \\ A_{1,12} & A_{2,12} \end{vmatrix}^{+\,+} = \begin{vmatrix} A_{11,1} & A_{11,2} \\ A_{22,1} & A_{22,2} \end{vmatrix} + \begin{vmatrix} A_{21,1} & A_{21,2} \\ A_{12,1} & A_{12,2} \end{vmatrix} .
$$

808. The Cayley Product of a 2-way and a 3-way Determinant. The 3-way determinant $A = \left| {}^{+\,\pm\,\pm}_{\alpha\,\beta\,\gamma} \right|$ and the 2-way determinant $B = \left| {}^{\pm\,\pm}_{\rho\,\sigma} \right|$ can be multiplied together like two 2-way determinants, row into column, the locant of the row and the locant of the column combining to form the locant of the element of the (3-way) product:

[8] In writing formulas, commas may be inserted in locants at pleasure; they do not change but only bring out the meaning.

$$(1)\qquad |A^{+\pm\pm}_{\alpha\beta\gamma}|\cdot|B^{\pm\pm}_{\rho\sigma}| = \left|\left(\sum_{i=1}^{n} A_{\alpha\beta i}B_{i\sigma}\right)^{+\pm\pm}_{\alpha\beta\sigma}\right|,$$

where each α-layer of the product is the matrix product of that α-layer of A by the matrix of B, and the matrix of the product, seen as in (1), is

$$(2)\qquad \left\|\begin{matrix} \sum_i A_{\alpha 1i}B_{i1} & \sum_i A_{\alpha 1i}B_{i2} \cdots & \sum_i A_{\alpha 1i}B_{in} \\ \sum_i A_{\alpha 2i}B_{i1} & \sum_i A_{\alpha 2i}B_{i2} \cdots & \sum_i A_{\alpha 2i}B_{in} \\ \cdots & \cdots & \cdots \\ \sum_i A_{\alpha ni}B_{i1} & \sum_i A_{\alpha ni}B_{i2} \cdots & \sum_i A_{\alpha ni}B_{in} \end{matrix}\right\| \quad (\alpha = 1, 2, \cdots, n).$$

For, in the decomposition by rows each component is the product of the corresponding component of A into B.

As with two 2-way determinants, four (equal) products of A and B can be formed. But *normals* may *not* be multiplied into rows or columns.

809. The Scott Product of 2-way Determinants and Permanents. The product of two 2-way determinants, or of a 2-way determinant and permanent, can be expressed as a 3-way determinant; and that of two 2-way permanents as a 3-way permanent. Each element of the product is the product of two elements one from each factor, the product element whose locant is $\alpha\beta\gamma$ being the product of the elements whose locants are $\alpha\beta$ and $\beta\gamma$. The signancy of the indices follows them into the product, so that α and γ have the same signancy in the product as in the factors, while β is nonsignant if nonsignant in both factors, signant if signant in only one factor, and nonsignant if signant in both factors (since $\epsilon_\beta^2 = 1$ for any arrangement of the values of β) The formulas may be written:

$$(1)\qquad AB = |A^{\pm\pm}_{\alpha\beta}|\cdot|B^{\pm\pm}_{\beta\gamma}| = |(A_{\alpha\beta}B_{\beta\gamma})^{\pm+\pm}_{\alpha\beta\gamma}|,$$

$$(2)\qquad AB = |A^{\pm\pm}_{\alpha\beta}|\cdot|B^{++}_{\beta\gamma}| = |(A_{\alpha\beta}B_{\beta\gamma})^{\pm\pm+}_{\alpha\beta\gamma}|,$$

$$(3)\qquad AB = |A^{++}_{\alpha\beta}|\cdot|B^{++}_{\beta\gamma}| = |(A_{\alpha\beta}B_{\beta\gamma})^{+++}_{\alpha\beta\gamma}|.$$

For, if a *range* be the set of n values of an index in the locant of a transversal, the locants of the transversals of the right member can be formed by writing against a β-range each possible α-range with every possible γ-range, which results in bringing every transversal of B against each transversal of A; and the prescribed signancy gives to each such product the product of the signs of the combined transversals, so that the right member is the product of A and B multiplied term into term.

In the case $n=2$ the matrix of the product is

$$\left\| \begin{matrix} A_{11}B_{11} & A_{11}B_{12} & A_{21}B_{11} & A_{21}B_{12} \\ A_{12}B_{21} & A_{12}B_{22} & A_{22}B_{21} & A_{22}B_{22} \end{matrix} \right\|$$

810. The 4-way Matrix. The 4-way matrix $\|A_{\alpha\beta\gamma\delta}\|$ of order n has n^3 elements. They lie in n 3-way α-layers of order n

$$\|A_{1\beta\gamma\delta}\|, \ \|A_{2\beta\gamma\delta}\|, \ \cdots, \ \|A_{n\beta\gamma\delta}\|,$$

and similarly in n 3-way β-layers, γ-layers, and δ-layers. Again, the elements lie in n^3 1-way α-running files of order n

$$\|A_{\alpha 111}\|, \ \|A_{\alpha 112}\|, \ \cdots, \ \|A_{\alpha nnn}\|,$$

and similarly in n^3 β-running, γ-running, and δ-running files.

The n^2 elements common to an α-layer $\|A_{\alpha^{(i)}\beta\gamma\delta}\|$ and a β-layer $\|A_{\alpha\beta^{(j)}\gamma\delta}\|$ form a *2-way couche* $\|A_{\alpha^{(i)}\beta^{(j)}\gamma\delta}\|$ of order n, of *aspect* $\alpha\beta$ and *direction* $\gamma\delta$; and there are n^2 such *parallel* 2-way couches. Similarly there are n^2 parallel 2-way couches of each of the aspects $\alpha\gamma$, $\alpha\delta$, $\beta\gamma$, $\beta\delta$, and $\gamma\delta$. Files are *1-way couches*, the α-running files being the $\beta\gamma\delta$-couches or couches of aspect $\beta\gamma\delta$, each consisting of the elements common to (or formed by the intersection of) a β-layer, a γ-layer, and a δ-layer. An $\alpha\beta$-couche and a $\gamma\delta$-couche intersect in a single element, and so do any two *contraspective* couches.

Two or more parallel couches are *conjunctive* if no two of them lie in the same layer of any aspect; n conjunctive couches are *perjunctive.* Conjunctive or perjunctive elements are such as form a part or the whole of a transversal.

811. 4-way Determinants and Permanents. There are $(n!)^3$ transversals in the matrix just described. Their sum is the *permanent* of the matrix. The matrix has a full-sign determinant and six other determinants. The *full-sign determinant* is the algebraic sum of the transversals each bearing the sign which is the product of the signs of the four ranges in that transversal:

$$(1) \qquad \left| A^{\pm\pm\pm\pm}_{\alpha\beta\gamma\delta} \right| = \sum_{\beta\gamma\delta}^{(n!)^{\delta}} \epsilon_\alpha \epsilon_\beta \epsilon_\gamma \epsilon_\delta A_{\alpha^{(1)}\beta^{(1)}\gamma^{(1)}\delta^{(1)}} \cdots A_{\alpha^{(n)}\beta^{(n)}\gamma^{(n)}\delta^{(n)}}.$$

In each of the other six determinants two indices are signant and two are nonsignant: they are

$$(2)\qquad \begin{array}{lll} \left| A^{\pm\pm++}_{\alpha\beta\gamma\delta} \right|, & \left| A^{\pm+\pm+}_{\alpha\beta\gamma\delta} \right|, & \left| A^{\pm++\pm}_{\alpha\beta\gamma\delta} \right|, \\ \left| A^{+\pm+\pm}_{\alpha\beta\gamma\delta} \right|, & \left| A^{+\pm+\pm}_{\alpha\beta\gamma\delta} \right|, & \left| A^{++\pm\pm}_{\alpha\beta\gamma\delta} \right|. \end{array}$$

The elementary properties stated in §806 with respect to signant and nonsignant directions hold here.

812. Decomposition of 4-way Determinants and Permanents. Take n perjunctive δ-running files of $\|A_{\alpha\beta\gamma\delta}\|$ as the rows of a 2-way matrix

$$(1)\qquad \left\| (A_{\alpha^{(\gamma)}\beta^{(\gamma)}\gamma\delta})_{\gamma\delta} \right\|.$$

There are $(n!)^2$ such 2-way matrices, and the $n!$ transversals of each are transversals of the 4-way matrix. These matrices give the following decompositions:

$$(2)\qquad \left| A^{\pm\pm\pm\pm}_{\alpha\beta\gamma\delta} \right| = \sum_{\alpha\beta} \epsilon_\alpha \epsilon_\beta \left| (A_{\alpha^{(\gamma)}\beta^{(\gamma)}\gamma\delta})^{\pm\pm}_{\gamma\delta} \right|,$$

$$(3)\qquad \left| A^{++\pm\pm}_{\alpha\beta\gamma\delta} \right| = \sum_{\alpha\beta} \left| (A_{\alpha^{(\gamma)}\beta^{(\gamma)}\gamma\delta})^{\pm\pm}_{\gamma\delta} \right|,$$

$$(4)\qquad \left| A^{\pm\pm++}_{\alpha\beta\gamma\delta} \right| = \sum_{\alpha\beta} \epsilon_\alpha \epsilon_\beta \left| (A_{\alpha^{(\gamma)}\beta^{(\gamma)}\gamma\delta})^{++}_{\gamma\delta} \right|,$$

$$(5)\qquad \left| A^{++++}_{\alpha\beta\gamma\delta} \right| = \sum_{\alpha\beta} \left| (A_{\alpha^{(\gamma)}\beta^{(\gamma)}\gamma\delta})^{++}_{\gamma\delta} \right|.$$

Formula (2) says that if all four indices are signant the determinant may be decomposed indifferently by files of any direction δ into an algebraic sum of 2-way determinants whose indices are δ and $\alpha\beta\gamma$, the sign of any term of any one of these determinants being the product of the sign of the δ range and (say) the γ range; the sign prefixed to the determinant being the product of the signs of the α and β ranges (which is the same in every term of that determinant). Thus ultimately each transversal receives the same sign that it has in the 4-way determinant.[9] For example, if $n=2$ the decomposition is

$$\begin{vmatrix} A_{111,1} & A_{111,2} \\ A_{222,1} & A_{222,2} \end{vmatrix} - \begin{vmatrix} A_{211,1} & A_{211,2} \\ A_{122,1} & A_{122,2} \end{vmatrix} - \begin{vmatrix} A_{121,1} & A_{121,2} \\ A_{212,1} & A_{212,2} \end{vmatrix} + \begin{vmatrix} A_{221,1} & A_{221,2} \\ A_{112,1} & A_{112,2} \end{vmatrix}.$$

Formulas (3) and (4) say that if only two indices are signant, files of a signant direction should be combined into 2-way determinants,

[9] And this is a ready test of the remaining formulas of this section.

the sign of a term being the product of the signs of the two signant ranges and each determinant having a positive sign; while files of a nonsignant direction should be combined into 2-way permanents each having the sign which is the product of the signs of the signant ranges.

Again, take n perjunctive $\alpha\beta$-couches of $\|A_{\alpha\beta\gamma\delta}\|$ as parallel layers of a 3-way matrix

$$(6) \qquad \|(A_{\alpha^{(\beta)}\beta\gamma\delta})_{\beta\gamma\delta}\|.$$

There are $n!$ such 3-way matrices, and the $(n!)^2$ transversals of each are transversals of the 4-way matrix. These matrices give the following decompositions:

$$(7) \qquad |A_{\alpha\beta\gamma\delta}^{\pm\pm\pm\pm}| = \sum_{\alpha} \epsilon_\alpha \epsilon_\beta |(A_{\alpha^{(\beta)}\beta\gamma\delta})_{\beta\gamma\delta}^{+\pm\pm}|,$$

$$(8) \qquad |A_{\alpha\beta\gamma\delta}^{++\pm\pm}| = \sum_{\alpha} |(A_{\alpha^{(\beta)}\beta\gamma\delta})_{\beta\gamma\delta}^{+\pm\pm}|,$$

$$(9) \qquad |A_{\alpha\beta\gamma\delta}^{+\pm\pm+}| = \sum_{\alpha} |(A_{\alpha^{(\beta)}\beta\gamma\delta})_{\beta\gamma\delta}^{\pm\pm+}|,$$

$$(10) \qquad |A_{\alpha\beta\gamma\delta}^{\pm\pm++}| = \sum_{\alpha} \epsilon_\alpha \epsilon_\beta |(A_{\alpha^{(\beta)}\beta\gamma\delta})_{\beta\gamma\delta}^{+++}|,$$

$$(11) \qquad |A_{\alpha\beta\gamma\delta}^{++++}| = \sum_{\alpha} |(A_{\alpha^{(\beta)}\beta\gamma\delta})_{\beta\gamma\delta}^{+++}|.$$

Crossed decomposition. By decomposing the right members of (7) to (10) by means of (6a) and (6b) §807, formulas (2) to (4) may be obtained, and in addition other formulas appear, based on 2-way matrices of the type

$$(12) \qquad \|(A_{\alpha^{(\beta)}\beta\gamma^{(\delta)}\delta})_{\beta\delta}\|,$$

$(n!)^2$ in number, whose transversals together are the transversals of the 4-way matrix. These *crossed decompositions* are:

$$(13) \qquad |A_{\alpha\beta\gamma\delta}^{\pm\pm\pm\pm}| = \sum_{\alpha\gamma} \epsilon_\alpha \epsilon_\beta \epsilon_\gamma \epsilon_\delta |(A_{\alpha^{(\beta)}\beta\gamma^{(\delta)}\delta})_{\beta\delta}^{++}|,$$

$$(14) \qquad |A_{\alpha\beta\gamma\delta}^{++\pm\pm}| = \sum_{\alpha\gamma} \epsilon_\gamma \epsilon_\delta |(A_{\alpha^{(\beta)}\beta\gamma^{(\delta)}\delta})_{\beta\delta}^{++}|,$$

$$(15) \qquad |A_{\alpha\beta\gamma\delta}^{+\pm+\pm}| = \sum_{\alpha\gamma} |(A_{\alpha^{(\beta)}\beta\gamma^{(\delta)}\delta})_{\beta\delta}^{\pm\pm}|.$$

Note that the matrix (12) is formed by the intersection of n perjunctive $\alpha\beta$-couches and n perjunctive $\gamma\delta$-couches, the elements of each row lying in the same $\alpha\beta$-couche and the elements of each column in the same $\gamma\delta$-couche.

813. The Cayley Product of Two 3-way Determinants. Two 3-way determinants A and B may be multiplied by files, files of a signant direction of A into files of a signant direction of B. Two signant indices are lost in the process. The signancy of the remaining indices follows them into the product, and the result is a 4-way determinant with two signant and two nonsignant indices:

$$(1) \qquad \left| A_{\alpha\beta\lambda}^{+\pm\pm} \right| \cdot \left| B_{\gamma\delta\mu}^{+\pm\pm} \right| = \left| \left(\sum_{i=1}^{n} A_{\alpha\beta i} B_{\gamma\delta i} \right)_{\alpha\beta\gamma\delta}^{+\pm+\pm} \right|$$

For, by (15) §812 the right member is equal to

$$\sum_{\alpha\gamma} \left| \left(\sum_i A_{\alpha^{(\beta)}\beta i} B_{\gamma^{(\delta)}\delta i} \right)_{\beta\delta}^{\pm\pm} \right| ;$$

and each of the $(n!)^2$ 2-way determinants in this sum is an ordinary product determinant, so that the sum is equal to

$$\sum_{\alpha} \left| (A_{\alpha^{(\beta)}\beta\lambda})_{\beta\lambda}^{\pm\pm} \right| \cdot \sum_{\gamma} \left| (B_{\gamma^{(\delta)}\delta\mu})_{\delta\mu}^{\pm\pm} \right| ,$$

which by (6*b*) §807 equals the left member of (1). In brief, a crossed decomposition of the right member is the product of decompositions of the factors in the left member.

814. The p-way Matrix and Its Determinants. The p-way matrix $\|A_{\alpha_1\alpha_2\cdots\alpha_p}\|$ of order n consists of n^p elements. It has n layers of n^{p-1} elements, of each aspect α_i $(i=1, 2, \cdots, p)$; it has n^{p-1} files of n elements, of each direction α_i; and in general it has n^{p-k} couches of n^k elements, of each aspect $(\alpha_{r_1}\alpha_{r_2} \cdots \alpha_{r_{p-k}})$ and direction

$$(a_{r_{p-k+1}} a_{r_{p-k+2}} \cdots a_{r_p}) ,$$

there being $\binom{p}{k}$ possible aspects and directions, for each value of k from 1 to $p-1$, files and layers being thus included as the extremes. The intersection of any number of layers of different directions forms a couche. The matrix has $(n!)^{p-1}$ transversals of n elements each.

A *determinant* of the matrix is obtained by making signant any even number of indices and taking all the transversals each with the sign which is the product of the signs of its signant ranges. A *full-sign* determinant has all its indices signant if p is even, and all but one if p is odd. The *permanent* of the matrix has all its indices nonsignant.

The elementary properties stated for 3-way determinants and permanents in §806 hold for p-way determinants and permanents.

815. Decomposition of p-way Determinants and Permanents Take n perjunctive $(\alpha_1 \cdots \alpha_r\beta_1 \cdots \beta_s)$-couches of the p-way matrix

(1) $$\left\| A_{\alpha_1 \cdots \alpha_r\beta_1 \cdots \beta_s\gamma_1 \cdots \gamma_t\delta_1 \cdots \delta_u} \right\|$$

of order n as parallel layers of a $(p-r-s+1)$-way matrix

(2) $$\left\| \left(A \frac{(\beta_s)}{\alpha_1 \cdots \alpha_r\beta_1 \cdots \beta_{s-1}\beta_s\gamma_1 \cdots \gamma_t\delta_1 \cdots \delta_u} \right)_{\beta_s\gamma_1 \cdots \gamma_t\delta_1 \cdots \delta_u} \right\|$$

There are $(n!)^{r+s-1}$ such matrices, and their transversals are together the transversals of (1). Let

(3) $$A = \left| A \frac{+}{\alpha_1 \cdots \alpha_r} \frac{\pm}{\beta_1 \cdots \beta_s} \frac{+}{\gamma_1 \cdots \gamma_t} \frac{\pm}{\delta_1 \cdots \delta_u} \right| .$$

Then if s and u are odd,

(4) $$A = \sum_{\alpha\beta} \epsilon_{\beta_1} \cdots \epsilon_{\beta_{s-1}} \left| \left(A \frac{(\beta_s)}{\alpha_1 \cdots \alpha_r\beta_1 \cdots \beta_{s-1}\beta_s\gamma_1 \cdots \gamma_t\delta_1 \cdots \delta_u} \right) \frac{\pm \quad +}{\beta_s\gamma_1 \cdots \gamma_t} \frac{\pm}{\delta_1 \cdots \delta_u} \right| ;$$

while if s and u are even,

(5) $$A = \sum_{\alpha\beta} \epsilon_{\beta_1} \cdots \epsilon_{\beta_s} \left| \left(A \frac{(\beta_s)}{\alpha_1 \cdots \alpha_r\beta_1 \cdots \beta_{s-1}\beta_s\gamma_1 \cdots \gamma_t\delta_1 \cdots \delta_u} \right) \frac{+}{\beta_s} \frac{+}{\gamma_1 \cdots \gamma_t} \frac{\pm}{\delta_1 \cdots \delta_u} \right| ,$$

with the alternative formula

(6) $$A = \sum_{\alpha\beta} \epsilon_{\beta_1} \cdots \epsilon_{\beta_s} \left| \left(A \frac{(\alpha_1)}{\alpha_1\alpha_2 \cdots \alpha_r\beta_1 \cdots \beta_s\gamma_1 \cdots \gamma_t\delta_1 \cdots \delta_u} \right) \frac{+}{\alpha_1} \frac{+}{\gamma_1 \cdots \gamma_t} \frac{\pm}{\delta_1 \cdots \delta_u} \right| ,$$

and the latter must be used if $s=0$.

For the permanent,

(7) $$\left| A \frac{+}{\alpha_1 \cdots \alpha_r\gamma_1 \cdots \gamma_t} \right| = \sum_{\alpha} \left| \left(A \frac{(\alpha_r)}{\alpha_1 \cdots \alpha_{r-1}\alpha_r\gamma_1 \cdots \gamma_t} \right) \frac{+}{\alpha_r\gamma_1 \cdots \gamma_t} \right| .$$

Complete decomposition into 2-way components whose rows are perjunctive files of (1) falls under (4) if the files are signant (no γ's, one δ), and under (5) and (6) if the files are nonsignant (one γ, no δ's).

Crossed decomposition.[10] The intersection of n perjunctive $(\alpha_1 \cdots \alpha_r\beta_1 \cdots \beta_s)$-couches and n perjunctive $(\gamma_1 \cdots \gamma_t\delta_1 \cdots \delta_u)$-couches (contraspective couches) is a 2-way matrix of order n,

$$(8) \qquad \left\| \left(A \frac{(\beta_s)}{\alpha_1 \cdots \alpha_r\beta_1 \cdots \beta_{s-1}\beta_s} \frac{(\delta_u)}{\gamma_1 \cdots \gamma_t\delta_1 \cdots \delta_{u-1}\delta_u} \right)_{\beta_s\delta_u} \right\|,$$

and the transversals of the $(n!)^{r+s+t+u-2}$ matrices of this type are together the transversals of (1). Hence if s and u are odd,

$$(9) \quad A = \sum_{\alpha\beta\gamma\delta} \epsilon_{\beta_1} \cdots \epsilon_{\beta_{s-1}} \epsilon_{\delta_1} \cdots \epsilon_{\delta_{u-1}}$$
$$\left| \left(A \frac{(\beta_s)}{\alpha_1 \cdots \alpha_r\beta_1 \cdots \beta_{s-1}\beta_s} \frac{(\delta_u)}{\gamma_1 \cdots \gamma_t\delta_1 \cdots \delta_{u-1}\delta_u} \right)^{\pm\,\pm}_{\beta_s\,\delta_u} \right|.$$

This formula may also be obtained by applying (4) to itself.

816. The Cayley Product of Any Two Determinants. A p-way determinant A and a p'-way determinant A'.

$$(1)\ A = \left| A \frac{+}{\nu_1 \cdots \nu_g} \frac{\pm}{\sigma_1 \cdots \sigma_h} \right|, \quad A' = \left| A' \frac{+}{\nu'_1 \cdots \nu'_{g'}} \frac{\pm}{\sigma'_1 \cdots \sigma'_{h'}} \right|$$

of order n, may be multiplied together, files of a signant direction into files of a signant direction, so as to form a $(p+p'-2)$-way determinant $\bar{A}$ of order n:

$$(2)\ AA' = \bar{A} = \left| \left(\sum_i A_{\nu_1 \cdots \nu_g\sigma_1 \cdots \sigma_{h-1}i} A'_{\nu'_1 \cdots \nu'_{g'}\sigma'_1 \cdots \sigma'_{h'-1}i} \right)^{+\,\pm}_{\nu\ \sigma} \right|,$$

where $\overset{+}{\nu}\ \overset{\pm}{\sigma}$ stands for

$$\frac{+}{\nu_1 \cdots \nu_g\nu'_1 \cdots \nu'_{g'}} \quad \frac{\pm}{\sigma_1 \cdots \sigma_{h-1}\sigma'_1 \cdots \sigma'_{h'-1}}$$

that is, the product of any two files becomes the element of $\bar{A}$ whose locant consists of the locants of the files, and the signancy of the indices in these locants follows them into $\bar{A}$.

PROOF. Crossed components of $\bar{A}$ are the Cayley products of 2-way components of A and A':

[10] Only enough is here given to serve in the next section.

$$\bar{A} = \sum_{\nu\sigma\nu'\sigma'} \epsilon_{\sigma_1} \cdots \epsilon_{\sigma_{h-2}} \epsilon_{\sigma_1'} \cdots \epsilon_{\sigma'_{h'-2}}$$

$$\left| \left(\sum_i A \frac{(\sigma_{h-1})}{\nu_1 \cdots \nu_g \sigma_1 \cdots \sigma_{h-2} \sigma_{h-1} i} \, A' \frac{(\sigma'_{h'-1})}{\nu_1' \cdots \nu'_{g'} \sigma' \cdots \sigma'_{h'-2} \sigma'_{h'-1}} \right) \frac{\pm \quad \pm}{\sigma_{h-1} \sigma'_{h'-1}} \right|$$

$$= \sum_{\nu\sigma} \epsilon_{\sigma_1} \cdots \epsilon_{\sigma_{h-2}} \left| \left(A \frac{(\sigma_{h-1})}{\nu_1 \cdots \nu_g \sigma_1 \cdots \sigma_{h-2} \sigma_{h-1} \sigma_h} \right) \frac{\pm \quad \pm}{\sigma_{h-1} \sigma_h} \right|$$

$$\times \sum_{\nu'\sigma'} \epsilon_{\sigma_1'} \cdots \epsilon_{\sigma'_{h'-2}} \left| \left(A \frac{(\sigma'_{h'-1})}{\nu_1' \cdots \nu'_{g'} \sigma'_1 \cdots \sigma'_{h'-2} \sigma'_{h'-1} \sigma'_{h'}} \right) \frac{\pm \quad \pm}{\sigma'_{h'-1} \sigma'_{h'}} \right|$$

817. Scott Products of Determinants and Permanents. (See §809). Given a p-way matrix and a q-way matrix of order n,

(1a,b) $$\| A_{\alpha_1 \cdots \alpha_p} \|, \qquad \| B_{\beta_1 \cdots \beta_q} \|,$$

let the $(p+q-1)$-way matrix of order n,

(2) $$\| (A_{\alpha_1 \cdots \alpha_{p-1} \lambda} B_{\lambda \beta_1 \cdots \beta_q})_{\alpha_1 \cdots \alpha_{p-1} \lambda \beta_2 \cdots \beta_q} \|,$$

be formed, each λ-layer being the outer product of the corresponding α_p-layer of A and the corresponding β_1-layer of B. Each transversal is the product of two transversals of the given matrices, and each possible product appears once and only once. Let

(3) $$A = \left| A \frac{(\pm)_\theta}{\alpha_1 \cdots \alpha_p} \right|$$

denote any determinant or the permanent of (1a), and

(4) $$B = \left| B \frac{(\pm)_\phi}{\beta_1 \cdots \beta_q} \right|$$

denote any determinant or the permanent of (1b), $(\pm)_\theta$ and $(\pm)_\phi$ standing for specific distributions of $\pm$ and $+$ signs over the indices. Let

(5) $$C = \left| (A_{\alpha_1 \cdots \alpha_{p-1} \lambda} B_{\lambda \beta_2 \cdots \beta_q}) \frac{(\pm)_\theta \qquad (+)_\phi}{\alpha_1 \cdots \alpha_{p-1} \lambda \beta_2 \cdots \beta_q} \right|,$$

the notation meaning that the same distributions of $\pm$ and $+$ signs as in (3) and (4) have been made, whereby the index λ has been doubly affected, and inversions of order in the λ-range are counted twice, once, or not at all, according as both, one, or neither of the indices α_p and β_1 are signant; and thus λ is to be regarded, respectively, as nonsignant, signant, or nonsignant. The reasoning of §809 shows that

(6) $$C = AB.$$

CHAPTER XXV

Less Common Forms

Determinants with Complex Elements

818. *Any determinant* Δ *whose conjugate elements are conjugate imaginaries as* $a_{rs}+b_{rs}i$, $a_{rs}-b_{rs}i$ *and whose diagonal elements are* a_{rr} *must be real.*

For if possible let Δ be of the form $A+Bi$. If now we substitute $-i$ for $+i$ on both sides of $\Delta=A+Bi$ we get $\Delta'=A-Bi$, but Δ' is Δ with the rows and columns interchanged and therefore unaltered in value. Consequently

$$A + Bi = A - Bi$$

or $B=0$ and Δ is real.

Exercises. 1. *The equation* $|\Delta-x|=0$ has all its roots real. (Hermite.)

2. The determinant

$$\begin{vmatrix} 1 & 1 & 1 \\ 1 & \alpha & \alpha^2 \\ 1 & \alpha^2 & \alpha \end{vmatrix} = (-\ 3^3)^{1/2}$$

where α is a primitive cube root of 1.

Also

$$\begin{vmatrix} 1 & 1 & 1 & 1 & 1 \\ 1 & \beta & \beta^2 & \beta^3 & \beta^4 \\ 1 & \beta^2 & \beta^4 & \beta^6 & \beta^8 \\ 1 & \beta^3 & \beta^6 & \beta^9 & \beta^{12} \\ 1 & \beta^4 & \beta^8 & \beta^{12} & \beta^{16} \end{vmatrix} \quad \text{or} \quad \begin{vmatrix} 1 & 1 & 1 & 1 & 1 \\ 1 & \beta & \beta^2 & \beta^3 & \beta^4 \\ 1 & \beta^2 & \beta^4 & \beta & \beta^3 \\ 1 & \beta^3 & \beta & \beta^4 & \beta^2 \\ 1 & \beta^4 & \beta^3 & \beta^2 & \beta \end{vmatrix} = (5^5)^{1/2}$$

where β is a primitive fifth root of 1.

3. The product of $|a_{1n}-b_{1n}i|\cdot|a_{1n}+b_{1n}i|$ is expressible as the sum of two squares.

819. If $|a_1b_2 \cdots|$ and $|a_1'\,b_2' \ \cdots|$ be two determinants such that corresponding elements are complex conjugates and

$$\begin{vmatrix} (aa') & (ab') & \cdots \\ (a'b) & (bb') & \cdots \\ \cdot & \cdot & \cdot\ \cdot\ \cdot \end{vmatrix}$$

be their product row-by-row then it is clear that

$$(1) \qquad \begin{vmatrix} (aa') & (ab') \\ (ba') & (bb') \end{vmatrix} \ngtr (aa')(bb')$$

and

$$\begin{vmatrix} (aa') & (ab') & (ac') \\ (ba') & (bb') & (bc') \\ (ca') & (cb') & (cc') \end{vmatrix} = \begin{Vmatrix} \begin{vmatrix} (aa') & (ab') \\ (ba') & (bb') \end{vmatrix} & \begin{vmatrix} (aa') & (ac') \\ (ba') & (bc') \end{vmatrix} \\ \begin{vmatrix} (aa') & (ab') \\ (ca') & (cb') \end{vmatrix} & \begin{vmatrix} (aa') & (ac') \\ (ca') & (cc') \end{vmatrix} \end{Vmatrix} \div (aa'),$$

which by (1)

$$\ngtr \begin{vmatrix} (aa') & (ab') \\ (ba') & (bb') \end{vmatrix} \cdot \begin{vmatrix} (aa') & (ac') \\ (ca') & (cc') \end{vmatrix} \div (aa'),$$

and again by (1) $\ngtr (aa')(bb')(cc')$, and so on for any order. This leads to the theorem: *The row-by-row product of two determinants whose corresponding elements are complex conjugates is not greater than its own principal diagonal term.*

It is apparent that we may replace row-by-row by column-by-column and still have the theorem true.

820. Denote

$$\begin{vmatrix} a_1 + a_1' i & a_2 + a_2' i & a_3 + a_3' i \\ b_1 + b_1' i & b_2 + b_2' i & b_3 + b_3' i \\ c_1 + c_1' i & c_2 + c_2' i & c_3 + c_3' i \end{vmatrix} \text{ by } \mu,$$

and let μ' denote what μ becomes on putting $-i$ for i. Let M and M' be the adjugate of μ and μ' respectively, then

$$\mu = (a_r + a_r' i)(\mathcal{A}_r + \mathcal{A}_r' i) + (b_r + b_r' i)(\mathcal{B}_r + \mathcal{B}_r' i) + (c_r + c_r' i)(\mathcal{C}_r + \mathcal{C}_r' i)$$

$$(1) \qquad = (\textstyle\sum a_r\mathcal{A}_r - \sum a_r'\mathcal{A}_r') + (\sum a_r\mathcal{A}_r' + \sum a_r'\mathcal{A}_r)i$$

$$(2) \qquad \mu\mu' = (\textstyle\sum a_r\mathcal{A}_r - \sum a_r'\mathcal{A}_r')^2 + (\sum a_r\mathcal{A}_r' + \sum a_r'\mathcal{A}_r)^2.$$

But by §819 we have

$$\mu\mu' \ngtr s_1^2 s_2^2 s_3^2, \quad \text{where} \quad s_r^2 = \textstyle\sum a_r^2 + \sum a_r'^2 \quad (r = 1,2,3)$$

and

$$(3)\qquad \left\| \begin{matrix} a_r + a_r' i & b_r + b_r' i & c_r + c_r' i \\ \mathcal{A}_r - \mathcal{A}_r' i & \mathcal{B}_r - \mathcal{B}_r' i & \mathcal{C}_r - \mathcal{C}_r' i \end{matrix} \right\| \cdot \left\| \begin{matrix} a_r - a_r' i & b_r - b_r' i & c_r - c_r' i \\ \mathcal{A}_r + \mathcal{A}_r' i & \mathcal{B}_r + \mathcal{B}_r' i & \mathcal{C}_r + \mathcal{C}_r' i \end{matrix} \right\| \overline{\gtrless}\ 0$$

because it is equal to the sum of three terms each of which is the product of a complex quantity by its conjugate. Hence

$$s_r^2 S_r^2 \overline{\gtrless} \left(\sum a_r \mathcal{A}_r - \sum a_r' \mathcal{A}_r' \right)^2 + \left(\sum a_r \mathcal{A}_r' + \sum a_r' \mathcal{A}_r \right)^2$$

$\overline{\gtrless}\, \mu\mu'$ by (2), where

$$S_r = \sum \mathcal{A}_r^2 + \sum \mathcal{A}_r'^2 .$$

Therefore

$$s_1^2 s_2^2 s_3^2 S_1^2 S_2^2 S_3^2 \overline{\gtrless} (\mu\mu')^3$$

and hence $s_1 s_2 s_3 \overline{\gtrless} |\mu|$.

The limit $s_1^2 s_2^2 s_3^2$ which $|\mu|^2$ cannot equal will actually be reached when the sum of the three terms which the left-hand side of (3) is equal to is zero and this can only happen when

$$(4)\qquad \begin{cases} \dfrac{a_1 + a_1' i}{\mathcal{A}_1 - \mathcal{A}_1' i} = \dfrac{b_1 + b_1' i}{\mathcal{B}_1 - \mathcal{B}_1' i} = \dfrac{c_1 + c_1' i}{\mathcal{C}_1 - \mathcal{C}_1' i} = \rho_1 \text{ say}, \\[2ex] \dfrac{a_2 + a_2' i}{\mathcal{A}_2 - \mathcal{A}_2' i} = \dfrac{b_2 + b_2' i}{\mathcal{B}_2 - \mathcal{B}_2' i} = \dfrac{c_2 + c_2' i}{\mathcal{C}_2 - \mathcal{C}_2' i} = \rho_2 \text{ say}, \\[2ex] \dfrac{a_3 + a_3' i}{\mathcal{A}_3 - \mathcal{A}_3' i} = \dfrac{b_3 + b_3' i}{\mathcal{B}_3 - \mathcal{B}_3' i} = \dfrac{c_3 + c_3' i}{\mathcal{C}_3 - \mathcal{C}_3' i} = \rho_3 \text{ say} \end{cases}$$

and hence

$$\mu \underset{c}{\overset{c}{\times}} \mu' = \begin{vmatrix} \rho_1\mu' & \cdot & \cdot \\ \cdot & \rho_2\mu' & \cdot \\ \cdot & \cdot & \rho_3\mu' \end{vmatrix}$$

Similarly if $s_a^2 = (a_1^2 + a_1'^2 + a_2^2 + a_2'^2 + a_3^2 + a_3'^2)$, etc. then the limit $s_a^2 s_b^2 s_c^2$ will be reached when the elements of each row of μ are proportional to the elements of the corresponding row of M' and

$$\mu \underset{r}{\overset{r}{\times}} \mu' = \begin{vmatrix} \rho_1\mu' & \cdot & \cdot \\ \cdot & \rho_2\mu' & \cdot \\ \cdot & \cdot & \rho_3\mu' \end{vmatrix} .$$

Both limits will be reached and will therefore coalesce when all the elements of μ are proportional to the corresponding elements of M' and then the same result will be obtained from row-by-row multiplication as by column-by-column multiplication. When this is true the determinant may be said to have a *maximum value* or be a *maximum determinant.* Such is evidently possible when μ is axisymmetric, axially skew or when the elements are equimodular.

821. It is seen that when the elements of μ are proportional to the corresponding elements of M' and are equimodular the product of any element of μ by the corresponding element of M is constant or in Sylvester's language is "*inversely orthogonal.*" Also if μ be "inversely orthogonal" and have equimodular elements, the elements must be proportional to those of M' and therefore by a preceding result μ must have its maximum value. The problem of finding inversely-orthogonal determinants is thus closely connected with the problem of finding determinants of maximum value.

Exercises. Set XXXVIII

1. Show that the alternant $|x^0y^1z^2w^3 \cdots|$ of the nth order will be inversely orthogonal if $x, y, z, \cdots$, are the roots of $\omega^n = a$.

2. Show that

$$\begin{vmatrix} 1 & 1 & 1 & 1 \\ 1 & 1 & -1 & -1 \\ 1 & -1 & d & -d \\ 1 & -1 & -d & d \end{vmatrix}$$

is inversely orthogonal whatever d may be.

3. If in addition to requiring the elements of μ to be unimodular we insist on them being real—in other words if the elements are $+1$ or -1, the problem is solvable only for certain orders of determinants.

4. The maximum determinant with complex elements is inversely orthogonal (*ant-orthogonant*) only when the elements are equimodular, and that a maximum determinant with real elements is always an orthogonant.

In the latter case, however, it has to be noted that if the only elements permissible be $+1$ and -1, the destinction between ant-orthogonant and orthogonant disappears. If a solution be obtained for order r it is easy to give a solution for order $2r$. Those of order 2^m are axisymmetric.

5. If M be an oblong array of complex elements, then MM' is not greater than the product of any two complementary coaxial minors of MM'. (Szász.)

6. If every coaxial minor of a Hermitean* determinant be positive, then the determinant is not greater than the product of any pair of its complementary coaxial minors (Szász).

7. Any positive axisymmetric determinant whose coaxial minors of every order are positive can never be greater than the product of its leading diagonal elements. (Nanson.)

8. If the moduli of the elements be at most 1, the modulus of the determinant is at most $n^{n/2}$. (Hadamard.)

9. The determinant of every positive definite Hermitean form is not greater than the product of any coaxial minor and its complementary. (Fischer.)

Cross or Anharmonic Ratio

822. The determinant

$$\Delta \equiv \begin{vmatrix} 1 & \alpha & \alpha' & \alpha\alpha' \\ 1 & \beta & \beta' & \beta\beta' \\ 1 & \gamma & \gamma' & \gamma\gamma' \\ 1 & \delta & \delta & \delta\delta' \end{vmatrix} = 0$$

states that the anharmonic ratio of the four points α, β, γ, δ is the same as that of the four points α', β', γ', δ', that is

$$\frac{(\gamma-\alpha)(\beta-\delta)}{(\alpha-\beta)(\gamma-\delta)} = \frac{(\gamma'-\alpha')(\beta'-\delta')}{(\alpha'-\beta')(\gamma'-\delta')}$$

and

$$\begin{aligned} \Delta = {} & (\gamma-\alpha)(\beta-\delta)(\alpha'-\beta')(\gamma'-\delta') \\ & - (\alpha-\beta)(\gamma-\delta)(\gamma'-\alpha')(\beta'-\delta'). \end{aligned}$$

If

$$\begin{aligned} P &= (\alpha-\beta)(\gamma-\delta) \\ Q &= (\alpha-\gamma)(\beta-\delta) \\ R &= (\alpha-\delta)(\beta-\gamma) \end{aligned}$$

then Δ may be written in the three equivalent forms

$$|PQ'|, \quad |QR'|, \quad -|RP'|.$$

* Determinants whose elements are complex.

UNISIGNANTS

823. The determinant

$$S \equiv \begin{vmatrix} a_1 + a_2 + a_3 & -a_2 & -a_3 \\ -b_1 & b_1 + b_2 + b_3 & -b_3 \\ -c_1 & -c_2 & c_1 + c_2 + c_3 \end{vmatrix}$$

$$= \begin{cases} a_1b_2c_3 + (c_1 + c_2)a_1b_2 + (a_2 + a_3)b_2c_3 + (b_1 + b_3)a_1c_3 \\ +a_1(b_1c_1+b_1c_2+c_1b_3)+b_2(a_2c_2+c_2a_3+c_1a_2)+c_3(a_3b_3+b_1a_3+a_2b_3) \end{cases}$$

in the expanded form has all its terms positive and on that account such determinants have been called *unisignants*. This type is due to Sylvester.

824. Another type of unisignant is the following:

$$M \equiv \begin{vmatrix} a + b_2 + b_3 + c_1 & a + b_3 & a + b_2 \\ a + b_3 & a + b_1 + b_3 + c_2 & a + b_1 \\ a + b_2 & a + b_1 & a + b_1 + b_2 + c_3 \end{vmatrix}$$

$$= \begin{cases} c_1c_2c_3 + (c_1c_2 + c_1c_3 + c_2c_3)a \\ \quad + (c_1 + c_2)(c_3 + a)b_3 + (c_1 + c_3)(c_2 + a)b_2 \\ \quad + (c_1 + a)(c_2 + c_3)b_1 + (c_1 + c_2 + c_3 + a)(b_1b_2 + b_1b_3 + b_2b_3) \\ \quad + 4b_1b_2b_3, \end{cases}$$

where we have for convenience taken the order to be 3. This type is due to Muir.

825. It will be observed that in either form there are in the positions (1, 1), (2, 2), (3, 3) $\cdots$ numbers which are found in these positions only. Thus in S, a_1, b_2, c_3, $\cdots$ and in M, c_1, c_2, c_3, $\cdots$ and it is apparent it will not alter the unisignancy of these types to add (or subtract as long as it is not numerically greater than the smallest of the quantities from which it is subtracted) a positive number.

It is also to be observed that to increase by the same positive quantity every a, b, c, $\cdots$ involved in either S or M does not affect the unisignancy. Doing this is equivalent to subtracting the same quantity from every element of S and then adding a multiple of that quantity to the principal diagonal elements. In the type M it is equivalent to adding the same positive quantity to each element and then in addition adding a multiple of that quantity to the principal diagonal elements.

826. It may be seen from §825 that the following properties are true:

I. Coaxial minors of unisignants are also unisignants.

II. The unisignancy is not affected by

(1) Adding the same positive quantity to each diagonal element in either S or M.

(2) Subtracting the same positive quantity from each non-diagonal element of S provided we add a proper multiple of this quantity to the diagonal elements.

(3) Adding the same positive quantity to each element of M.

(4) Subtracting the same positive quantity from each element of M.

827. The following result may readily be seen.

If we border S, horizontally and vertically by the elements 0, 1, 1, $\cdots$, then the resulting determinant has all the terms in its final development negative.

828. Boole gave* the following theorem: *If each element of an axisymmetric determinant be an aggregate of multiples of one and the same series of variables, the multipliers in the diagonal elements being all positive and if the multipliers of any one of the variables in any row be in order proportional to the multipliers of the same variable in any other row, then the final development of the determinant contains nothing but positive terms.* Thus for order 3 the determinant would be

$$\begin{vmatrix} \alpha_1 a + \alpha_2 b + \alpha_3 c + \cdots & \beta_1 a + \beta_2 b + \beta_3 c + \cdots & \gamma_1 a + \gamma_2 b + \gamma_3 c + \cdots \\ \beta_1 a + \beta_2 b + \beta_3 c + \cdots & \dfrac{\beta_1^2}{\alpha_1}a + \dfrac{\beta_2^2}{\alpha_2}b + \dfrac{\beta_3^2}{\alpha_3}c + \cdots & \dfrac{\beta_1\gamma_1}{\alpha_1}a + \dfrac{\beta_2\gamma_2}{\alpha_2}b + \dfrac{\beta_3\gamma_3}{\alpha_1}c + \cdots \\ \gamma_1 a + \gamma_2 b + \gamma_3 c + \cdots & \dfrac{\beta_1\gamma_1}{\alpha_1}a + \dfrac{\beta_2\gamma_2}{\alpha_2}b + \dfrac{\beta_3\gamma_3}{\alpha_3}c + \cdots & \dfrac{\gamma_1^2}{\alpha_1}a + \dfrac{\gamma_2^2}{\alpha_2}b + \dfrac{\gamma_3^2}{\alpha_3}c + \cdots \end{vmatrix}$$

This is equal to the product

$$\begin{Vmatrix} \alpha_1 a & \alpha_2 b & \alpha_3 c & \alpha_4 d \cdots \\ \beta_1 a & \beta_2 b & \beta_3 c & \beta_4 d \cdots \\ \gamma_1 a & \gamma_2 b & \gamma_3 c & \gamma_4 d \cdots \end{Vmatrix} \cdot \begin{Vmatrix} 1 & 1 & 1 & 1 \cdots \\ \dfrac{\beta_1}{\alpha_1} & \dfrac{\beta_2}{\alpha_2} & \dfrac{\beta_3}{\alpha_3} & \dfrac{\beta_4}{\alpha_4} \cdots \\ \dfrac{\gamma_1}{\alpha_1} & \dfrac{\gamma_2}{\alpha_2} & \dfrac{\gamma_3}{\alpha_3} & \dfrac{\gamma_6}{\alpha_4} \cdots \end{Vmatrix}$$

$$= abc\frac{|\alpha_1\beta_2\gamma_3|^2}{\alpha_1\alpha_2\alpha_3} + abd\frac{|\alpha_1\beta_2\gamma_4|^2}{\alpha_1\alpha_2\alpha_4} + \cdots$$

* Phil. Trans., clii, pp. 225.

which shows that all the terms are positive. It also shows that, as far as the multipliers are concerned all that is necessary is to limit the first ones to be positive.

EXERCISE: Show that

$$\begin{vmatrix} \cdot & a_2 & a_3 & a_4 \cdots \\ b_1 & \cdot & a_3 & a_4 \cdots \\ b_1 & b_2 & \cdot & a_4 \cdots \\ b_1 & b_2 & b_3 & \cdots\cdot \\ \cdot & \cdot & \cdot & \cdots\cdot \end{vmatrix} = (-1)^{n-1}(b_1 a_2 a_3 \cdots a_n + b_1 b_2 a_3 a_4 \cdots a_n + b_1 b_2 b_3 a_4 \cdots a_n + \cdots + b_1 b_2 \cdots b_{n-1} a_n).$$

ELEMENTS INTEGRALS

829. If

$$\phi_i(x) = (x-a_0)^{m_0}(x-a_1)^{m_1} \cdots (x-a_i)^{m_i}(a_{i+1}-x)^{m_{i+1}} \cdots (a_n-x)^{m_n}$$

where the m's are all less than 1 and $a_{n+1}=\infty$ then the determinant having

$$\int_{a_{s-1}}^{a_s} e^{-x}\frac{x^r dx}{\phi_{s-1}(x)}$$

for its (r, s)th element has been considered. Thus when the order is 2 we have

$$\begin{vmatrix} \int_{a_0}^{a_1} e^{-x}\frac{dx}{\phi_0(x)} & \int_{a_1}^{\infty} e^{-x}\frac{dx}{\phi_1(x)} \\ \int_{0}^{a_1} e^{-x}\frac{dx}{\phi_0(x)} & \int_{a_1}^{\infty} e^{-x}\frac{xdx}{\phi_1(x)} \end{vmatrix}$$

$$= \int_{a_0}^{a_1}\int_{a_1}^{\infty} \frac{e^{-x-x_1}(x_1 - x)dxdx_1}{(a - x_0)^{m_0}(a_1 - x)^{m_1}(x_1 - a_0)^{m_0}(x_1 - a_1)^{m_1}}$$

$$= \Gamma(1 - m_0)\Gamma(1 - m_1)(a_1 - a_0)^{1-m_0-m_1}e^{-a_0-a_1}.$$

INDEX

The numbers following titles are paragraph numbers.

CATALOG OF DOVER BOOKS

BOOKS EXPLAINING SCIENCE AND MATHEMATICS

THE COMMON SENSE OF THE EXACT SCIENCES, W. K. Clifford. Introduction by James Newman, edited by Karl Pearson. For 70 years this has been a guide to classical scientific and mathematical thought. Explains with unusual clarity basic concepts, such as extension of meaning of symbols, characteristics of surface boundaries, properties of plane figures, vectors, Cartesian method of determining position, etc. Long preface by Bertrand Russell. Bibliography of Clifford. Corrected, 130 diagrams redrawn. 249pp. 5⅜ x 8.
T61 Paperbound **$1.60**

SCIENCE THEORY AND MAN, Erwin Schrödinger. This is a complete and unabridged reissue of SCIENCE AND THE HUMAN TEMPERAMENT plus an additional essay: "What is an Elementary Particle?" Nobel Laureate Schrödinger discusses such topics as nature of scientific method, the nature of science, chance and determinism, science and society, conceptual models for physical entities, elementary particles and wave mechanics. Presentation is popular and may be followed by most people with little or no scientific training. "Fine practical preparation for a time when laws of nature, human institutions . . . are undergoing a critical examination without parallel," Waldemar Kaempffert, N. Y. TIMES. 192pp. 5⅜ x 8.
T428 Paperbound **$1.35**

PIONEERS OF SCIENCE, O. Lodge. Eminent scientist-expositor's authoritative, yet elementary survey of great scientific theories. Concentrating on individuals—Copernicus, Brahe, Kepler, Galileo, Descartes, Newton, Laplace, Herschel, Lord Kelvin, and other scientists—the author presents their discoveries in historical order adding biographical material on each man and full, specific explanations of their achievements. The clear and complete treatment of the post-Newtonian astronomers is a feature seldom found in other books on the subject. Index. 120 illustrations. xv + 404pp. 5⅜ x 8.
T716 Paperbound **$1.50**

THE EVOLUTION OF SCIENTIFIC THOUGHT FROM NEWTON TO EINSTEIN, A. d'Abro. Einstein's special and general theories of relativity, with their historical implications, are analyzed in non-technical terms. Excellent accounts of the contributions of Newton, Riemann, Weyl, Planck, Eddington, Maxwell, Lorentz and others are treated in terms of space and time, equations of electromagnetics, finiteness of the universe, methodology of science. 21 diagrams. 482pp. 5⅜ x 8.
T2 Paperound **$2.00**

THE RISE OF THE NEW PHYSICS, A. d'Abro. A half-million word exposition, formerly titled THE DECLINE OF MECHANISM, for readers not versed in higher mathematics. The only thorough explanation, in everyday language, of the central core of modern mathematical physical theory, treating both classical and modern theoretical physics, and presenting in terms almost anyone can understand the equivalent of 5 years of study of mathematical physics. Scientifically impeccable coverage of mathematical-physical thought from the Newtonian system up through the electronic theories of Dirac and Heisenberg and Fermi's statistics. Combines both history and exposition; provides a broad yet unified and detailed view, with constant comparison of classical and modern views on phenomena and theories. "A must for anyone doing serious study in the physical sciences," JOURNAL OF THE FRANKLIN INSTITUTE. "Extraordinary faculty . . . to explain ideas and theories of theoretical physics in the language of daily life," ISIS. First part of set covers philosophy of science, drawing upon the practice of Newton, Maxwell, Poincaré, Einstein, others, discussing modes of thought, experiment, interpretations of causality, etc. In the second part, 100 pages explain grammar and vocabulary of mathematics, with discussions of functions, groups, series, Fourier series, etc. The remainder is devoted to concrete, detailed coverage of both classical and quantum physics, explaining such topics as analytic mechanics, Hamilton's principle, wave theory of light, electromagnetic waves, groups of transformations, thermodynamics, phase rule, Brownian movement, kinetics, special relativity, Planck's original quantum theory, Bohr's atom, Zeeman effect, Broglie's wave mechanics, Heisenberg's uncertainty, Eigen-values, matrices, scores of other important topics. Discoveries and theories are covered for such men as Alembert, Born, Cantor, Debye, Euler, Foucault, Galois, Gauss, Hadamard, Kelvin, Kepler, Laplace, Maxwell, Pauli, Rayleigh, Volterra, Weyl, Young, more than 180 others. Indexed. 97 illustrations. ix + 982pp. 5⅜ x 8.
T3 Volume 1, Paperbound **$2.00**
T4 Volume 2, Paperbound **$2.00**

CONCERNING THE NATURE OF THINGS, Sir William Bragg. Christmas lectures delivered at the Royal Society by Nobel laureate. Why a spinning ball travels in a curved track; how uranium is transmuted to lead, etc. Partial contents: atoms, gases, liquids, crystals, metals, etc. No scientific background needed; wonderful for intelligent child. 32pp. of photos, 57 figures. xii + 232pp. 5⅜ x 8.
T31 Paperbound **$1.35**

THE UNIVERSE OF LIGHT, Sir William Bragg. No scientific training needed to read Nobel Prize winner's expansion of his Royal Institute Christmas Lectures. Insight into nature of light, methods and philosophy of science. Explains lenses, reflection, color, resonance, polarization, x-rays, the spectrum, Newton's work with prisms, Huygens' with polarization, Crookes' with cathode ray, etc. Leads into clear statement of 2 major historical theories of light, corpuscle and wave. Dozens of experiments you can do. 199 illus., including 2 full-page color plates. 293pp. 5⅜ x 8.
S538 Paperbound **$1.85**

BRIDGES AND THEIR BUILDERS, David Steinman and Sara Ruth Watson. Engineers, historians, everyone who has ever been fascinated by great spans will find this book an endless source of information and interest. Dr. Steinman, recipient of the Louis Levy medal, was one of the great bridge architects and engineers of all time, and his analysis of the great bridges of history is both authoritative and easily followed. Greek and Roman bridges, medieval bridges, Oriental bridges, modern works such as the Brooklyn Bridge and the Golden Gate Bridge, and many others are described in terms of history, constructional principles, artistry, and function. All in all this book is the most comprehensive and accurate semipopular history of bridges in print in English. New, greatly revised, enlarged edition. 23 photographs, 26 line drawings. Index. xvii + 401pp. 5⅜ x 8. T431 Paperbound **$2.00**

FADS AND FALLACIES IN THE NAME OF SCIENCE, Martin Gardner. Examines various cults, quack systems, frauds, delusions which at various times have masqueraded as science. Accounts of hollow-earth fanatics like Symmes; Velikovsky and wandering planets; Hoerbiger; Bellamy and the theory of multiple moons; Charles Fort; dowsing, pseudoscientific methods for finding water, ores, oil. Sections on naturopathy, iridiagnosis, zone therapy, food fads, etc. Analytical accounts of Wilhelm Reich and orgone sex energy; L. Ron Hubbard and Dianetics; A. Korzybski and General Semantics; many others. Brought up to date to include Bridey Murphy, others. Not just a collection of anecdotes, but a fair, reasoned appraisal of eccentric theory. Formerly titled IN THE NAME OF SCIENCE. Preface. Index. x + 384pp. 5⅜ x 8. T394 Paperbound **$1.50**

See also: **A PHILOSOPHICAL ESSAY ON PROBABILITIES, P. de Laplace; ON MATHEMATICS AND MATHEMATICIANS, R. E. Moritz; AN ELEMENTARY SURVEY OF CELESTIAL MECHANICS, Y. Ryabov; THE SKY AND ITS MYSTERIES, E. A. Beet; THE REALM OF THE NEBULAE, E. Hubble; OUT OF THE SKY, H. H. Nininger; SATELLITES AND SCIENTIFIC RESEARCH, D. King-Hele; HEREDITY AND YOUR LIFE, A. M. Winchester; INSECTS AND INSECT LIFE, S. W. Frost; PRINCIPLES OF STRATIGRAPHY, A. W. Grabau; TEACH YOURSELF SERIES.**

HISTORY OF SCIENCE AND MATHEMATICS

DIALOGUES CONCERNING TWO NEW SCIENCES, Galileo Galilei. This classic of experimental science, mechanics, engineering, is as enjoyable as it is important. A great historical document giving insights into one of the world's most original thinkers, it is based on 30 years' experimentation. It offers a lively exposition of dynamics, elasticity, sound, ballistics, strength of materials, the scientific method. "Superior to everything else of mine," Galileo. Trans. by H. Crew, A. Salvio. 126 diagrams. Index. xxi + 288pp. 5⅜ x 8. S99 Paperbound **$1.65**

A DIDEROT PICTORIAL ENCYCLOPEDIA OF TRADES AND INDUSTRY, Manufacturing and the Technical Arts in Plates Selected from "L'Encyclopédie ou Dictionnaire Raisonné des Sciences, des Arts, et des Métiers" of Denis Diderot. Edited with text by C. Gillispie. This first modern selection of plates from the high point of 18th century French engraving is a storehouse of valuable technological information to the historian of arts and science. Over 2000 illustrations on 485 full page plates, most of them original size, show the trades and industries of a fascinating era in such great detail that the processes and shops might very well be reconstructed from them. The plates teem with life, with men, women, and children performing all of the thousands of operations necessary to the trades before and during the early stages of the industrial revolution. Plates are in sequence, and show general operations, closeups of difficult operations, and details of complex machinery. Such important and interesting trades and industries are illustrated as sowing, harvesting, beekeeping, cheesemaking, operating windmills, milling flour, charcoal burning, tobacco processing, indigo, fishing, arts of war, salt extraction, mining, smelting, casting iron, steel, extracting mercury, zinc, sulphur, copper, etc., slating, tinning, silverplating, gilding, making gunpowder, cannons, bells, shoeing horses, tanning, papermaking, printing, dyeing, and more than 40 other categories. Professor Gillispie, of Princeton, supplies a full commentary on all the plates, identifying operations, tools, processes, etc. This material, presented in a lively and lucid fashion, is of great interest to the reader interested in history of science and technology. Heavy library cloth. 920pp. 9 x 12. T421 Two volume set **$18.50**

DE MAGNETE, William Gilbert. This classic work on magnetism founded a new science. Gilbert was the first to use the word "electricity", to recognize mass as distinct from weight, to discover the effect of heat on magnetic bodies; invent an electroscope, differentiate between static electricity and magnetism, conceive of the earth as a magnet. Written by the first great experimental scientist, this lively work is valuable not only as an historical landmark, but as the delightfully easy to follow record of a perpetually searching, ingenious mind. Translated by P. F. Mottelay. 25 page biographical memoir. 90 figures. lix + 368pp. 5⅜ x 8. S470 Paperbound **$2.00**

CHARLES BABBAGE AND HIS CALCULATING ENGINES, edited by P. Morrison and E. Morrison. Babbage, leading 19th century pioneer in mathematical machines and herald of modern operational research, was the true father of Harvard's relay computer Mark I. His Difference Engine and Analytical Engine were the first machines in the field. This volume contains a valuable introduction on his life and work; major excerpts from his autobiography, revealing his eccentric and unusual personality; and extensive selections from "Babbage's Calculating Engines," a compilation of hard-to-find journal articles by Babbage, the Countess of Lovelace, L. F. Menabrea, and Dionysius Lardner. 8 illustrations, Appendix of miscellaneous papers. Index. Bibliography. xxxviii + 400pp. 5⅜ x 8. T12 Paperbound **$2.00**

A HISTORY OF ASTRONOMY FROM THALES TO KEPLER, J. L. E. Dreyer. (Formerly A HISTORY OF PLANETARY SYSTEMS FROM THALES TO KEPLER.) This is the only work in English to give the complete history of man's cosmological views from prehistoric times to Kepler and Newton. Partial contents: Near Eastern astronomical systems, Early Greeks, Homocentric Spheres of Eudoxus, Epicycles, Ptolemaic system, medieval cosmology, Copernicus, Kepler, etc. Revised, foreword by W. H. Stahl. New bibliography. xvii + 430pp. 5⅜ x 8.
S79 Paperbound **$1.98**

A SHORT HISTORY OF ANATOMY AND PHYSIOLOGY FROM THE GREEKS TO HARVEY, Charles Singer. Corrected edition of THE EVOLUTION OF ANATOMY, classic work tracing evolution of anatomy and physiology from prescientific times through Greek & Roman periods, Dark Ages, Renaissance, to age of Harvey and beginning of modern concepts. Centered on individuals, movements, periods that definitely advanced anatomical knowledge: Plato, Diocles, Aristotle, Theophrastus, Herophilus, Erasistratus, the Alexandrians, Galen, Mondino, da Vinci, Linacre, Sylvius, others. Special section on Vesalius; Vesalian atlas of nudes, skeletons, muscle tabulae. Index of names, 20 plates. 270 extremely interesting illustrations of ancient, medieval, Renaissance, Oriental origin. xii + 209pp. 5⅜ x 8. T389 Paperbound **$1.75**

FROM MAGIC TO SCIENCE, Charles Singer. A great historian examines aspects of medical science from the Roman Empire through the Renaissance. Includes perhaps the best discussion of early herbals, and a penetrating physiological interpretation of "The Visions of Hildegarde of Bingen." Also examined are Arabian and Galenic influences; the Sphere of Pythagoras; Paracelsus; the reawakening of science under Leonardo da Vinci, Vesalius; the Lorica of Gildas the Briton; etc. Frequent quotations with translations. New Introduction by the author. New unabridged, corrected edition. 158 unusual illustrations from classical and medieval sources. Index. xxvii + 365pp. 5⅜ x 8. T390 Paperbound **$2.00**

HISTORY OF MATHEMATICS, D. E. Smith. Most comprehensive non-technical history of math in English. Discusses lives and works of over a thousand major and minor figures, with footnotes supplying technical information outside the book's scheme, and indicating disputed matters. Vol I: A chronological examination, from primitive concepts through Egypt, Babylonia, Greece, the Orient, Rome, the Middle Ages, the Renaissance, and up to 1900. Vol 2: The development of ideas in specific fields and problems, up through elementary calculus. Two volumes, total of 510 illustrations, 1355pp. 5⅜ x 8. Set boxed in attractive container.
T429, 430 Paperbound, the set **$5.00**

A SHORT ACCOUNT OF THE HISTORY OF MATHEMATICS, W. W. R. Ball. Most readable non-technical history of mathematics treats lives, discoveries of every important figure from Egyptian, Phoenician mathematicians to late 19th century. Discusses schools of Ionia, Pythagoras, Athens, Cyzicus, Alexandria, Byzantium, systems of numeration; primitive arithmetic; Middle Ages, Renaissance, including Arabs, Bacon, Regiomontanus, Tartaglia, Cardan, Stevinus, Galileo, Kepler; modern mathematics of Descartes, Pascal, Wallis, Huygens, Newton, Leibnitz, d'Alembert, Euler, Lambert, Laplace, Legendre, Gauss, Hermite, Weierstrass, scores more. Index. 25 figures. 546pp. 5⅜ x 8. S630 Paperbound **$2.00**

A SOURCE BOOK IN MATHEMATICS, D. E. Smith. Great discoveries in math, from Renaissance to end of 19th century, in English translation. Read announcements by Dedekind, Gauss, Delamain, Pascal, Fermat, Newton, Abel, Lobachevsky, Bolyai, Riemann, De Moivre, Legendre, Laplace, others of discoveries about imaginary numbers, number congruence, slide rule, equations, symbolism, cubic algebraic equations, non-Euclidean forms of geometry, calculus, function theory, quaternions, etc. Succinct selections from 125 different treatises, articles, most unavailable elsewhere in English. Each article preceded by biographical, historical introduction. Vol. I: Fields of Number, Algebra. Index. 32 illus. 338pp. 5⅜ x 8. Vol. II: Fields of Geometry, Probability, Calculus, Functions, Quaternions. 83 illus. 432pp. 5⅜ x 8.
Vol. 1: S552 Paperbound **$1.85**
Vol. 2: S553 Paperbound **$1.85**
2 vol. set, boxed **$3.50**

A HISTORY OF THE CALCULUS, AND ITS CONCEPTUAL DEVELOPMENT, Carl B. Boyer. Provides laymen and mathematicians a detailed history of the development of the calculus, from early beginning in antiquity to final elaboration as mathematical abstractions. Gives a sense of mathematics not as a technique, but as a habit of mind, in the progression of ideas of Zeno, Plato, Pythagoras, Eudoxus, Arabic and Scholastic mathematicians, Newton, Leibnitz, Taylor, Descartes, Euler, Lagrange, Cantor, Weierstrass, and others. This first comprehensive critical history of the calculus was originally titled "The Concepts of the Calculus." Foreword by R. Courant. Preface. 22 figures. 25-page bibliography. Index. v + 364pp. 5⅜ x 8.
S509 Paperbound **$2.00**

A CONCISE HISTORY OF MATHEMATICS, D. Struik. Lucid study of development of mathematical ideas, techniques from Ancient Near East, Greece, Islamic science, Middle Ages, Renaissance, modern times. Important mathematicians are described in detail. Treatment is not anecdotal, but analytical development of ideas. "Rich in content, thoughtful in interpretation," U.S. QUARTERLY BOOKLIST. Non-technical; no mathematical training needed. Index. 60 illustrations, including Egyptian papyri, Greek mss., portraits of 31 eminent mathematicians. Bibliography. 2nd edition. xix + 299pp. 5⅜ x 8. T255 Paperbound **$1.75**

See also: **NON-EUCLIDEAN GEOMETRY, R. Bonola; THEORY OF DETERMINANTS IN HISTORICAL ORDER OF DEVELOPMENT, T. Muir; HISTORY OF THE THEORY OF ELASTICITY AND STRENGTH OF MATERIALS, I. Todhunter and K. Pearson; A SHORT HISTORY OF ASTRONOMY, A. Berry; CLASSICS OF SCIENCE.**

PHILOSOPHY OF SCIENCE AND MATHEMATICS

FOUNDATIONS OF SCIENCE: THE PHILOSOPHY OF THEORY AND EXPERIMENT, N. R. Campbell. A critique of the most fundamental concepts of science in general and physics in particular. Examines why certain propositions are accepted without question, demarcates science from philosophy, clarifies the understanding of the tools of science. Part One analyzes the presuppositions of scientific thought: existence of the material world, nature of scientific laws, multiplication of probabilities, etc.: Part Two covers the nature of experiment and the application of mathematics: conditions for measurement, relations between numerical laws and theories, laws of error, etc. An appendix covers problems arising from relativity, force, motion, space, and time. A classic in its field. Index. xiii + 565pp. 5⅝ x 8⅜. S372 Paperbound **$2.95**

WHAT IS SCIENCE?, Norman Campbell. This excellent introduction explains scientific method, role of mathematics, types of scientific laws. Contents: 2 aspects of science, science & nature, laws of science, discovery of laws, explanation of laws, measurement & numerical laws, applications of science. 192pp. 5⅜ x 8. S43 Paperbound **$1.25**

THE VALUE OF SCIENCE, Henri Poincaré. Many of the most mature ideas of the "last scientific universalist" covered with charm and vigor for both the beginning student and the advanced worker. Discusses the nature of scientific truth, whether order is innate in the universe or imposed upon it by man, logical thought versus intuition (relating to math, through the works of Weierstrass, Lie, Klein, Riemann), time and space (relativity, psychological time, simultaneity), Hertz's concept of force, interrelationship of mathematical physics to pure math, values within disciplines of Maxwell, Carnot, Mayer, Newton, Lorentz, etc. Index. iii + 147pp. 5⅜ x 8. S469 Paperbound **$1.35**

SCIENCE AND METHOD, Henri Poincaré. Procedure of scientific discovery, methodology, experiment, idea-germination—the intellectual processes by which discoveries come into being. Most significant and most interesting aspects of development, application of ideas. Chapters cover selection of facts, chance, mathematical reasoning, mathematics, and logic; Whitehead, Russell, Cantor; the new mechanics, etc. 288pp. 5⅜ x 8. S222 Paperbound **$1.35**

SCIENCE AND HYPOTHESIS, Henri Poincaré. Creative psychology in science. How such concepts as number, magnitude, space, force, classical mechanics were developed, and how the modern scientist uses them in his thought. Hypothesis in physics, theories of modern physics. Introduction by Sir James Larmor. "Few mathematicians have had the breadth of vision of Poincaré, and none is his superior in the gift of clear exposition," E. T. Bell. Index. 272pp. 5⅜ x 8. S221 Paperbound **$1.35**

PHILOSOPHY AND THE PHYSICISTS, L. S. Stebbing. The philosophical aspects of modern science examined in terms of a lively critical attack on the ideas of Jeans and Eddington. Discusses the task of science, causality, determinism, probability, consciousness, the relation of the world of physics to that of everyday experience. Probes the philosophical significance of the Planck-Bohr concept of discontinuous energy levels, the inferences to be drawn from Heisenberg's Uncertainty Principle, the implications of "becoming" involved in the 2nd law of thermodynamics, and other problems posed by the discarding of Laplacean determinism. 285pp. 5⅜ x 8. T480 Paperbound **$1.65**

EXPERIMENT AND THEORY IN PHYSICS, Max Born. A Nobel laureate examines the nature and value of the counterclaims of experiment and theory in physics. Synthetic versus analytical scientific advances are analyzed in the work of Einstein, Bohr, Heisenberg, Planck, Eddington, Milne, and others by a fellow participant. 44pp. 5⅜ x 8. S308 Paperbound **60¢**

BIBLIOGRAPHIES

GUIDE TO THE LITERATURE OF MATHEMATICS AND PHYSICS, N. G. Parke III. Over 5000 entries included under approximately 120 major subject headings, of selected most important books, monographs, periodicals, articles in English, plus important works in German, French, Italian, Spanish, Russian (many recently available works). Covers every branch of physics, math, related engineering. Includes author, title, edition, publisher, place, date, number of volumes, number of pages. A 40-page introduction on the basic problems of research and study provides useful information on the organization and use of libraries, the psychology of learning, etc. This reference work will save you hours of time. 2nd revised edition. Indices of authors, subjects. 464pp. 5⅜ x 8. S447 Paperbound **$2.49**

THE STUDY OF THE HISTORY OF MATHEMATICS & THE STUDY OF THE HISTORY OF SCIENCE, George Sarton. Scientific method & philosophy in 2 scholarly fields. Defines duty of historian of math., provides especially useful bibliography with best available biographies of modern mathematicians, editions of their collected works, correspondence. Observes combination of history & science, will aid scholar in understanding science today. Bibliography includes best known treatises on historical methods. 200-item critically evaluated bibliography. Index. 10 illustrations. 2 volumes bound as one. 113pp. + 75pp. 5⅜ x 8. T240 Paperbound **$1.25**

MATHEMATICAL PUZZLES

AMUSEMENTS IN MATHEMATICS, Henry Ernest Dudeney. The foremost British originator of mathematical puzzles is always intriguing, witty, and paradoxical in this classic, one of the largest collections of mathematical amusements. More than 430 puzzles, problems, and paradoxes. Mazes and games, problems on number manipulation, unicursal and other route problems, puzzles on measuring, weighing, packing, age, kinship, chessboards, joiners', crossing river, plane figure dissection, and many others. Solutions. More than 450 illustrations. vii + 258pp. 5⅜ x 8. T473 Paperbound **$1.25**

THE CANTERBURY PUZZLES, Henry Ernest Dudeney. Chaucer's pilgrims set one another problems in story form. Also Adventures of the Puzzle Club, the Strange Escape of the King's Jester, the Monks of Riddlewell, the Squire's Christmas Puzzle Party, and others. All puzzles are original, based on dissecting plane figures, arithmetic, algebra, elementary calculus, and other branches of mathematics, and purely logical ingenuity. "The limit of ingenuity and intricacy . . ." The Observer. Over 110 puzzles. Full solutions. 150 illustrations. viii + 225pp. 5⅜ x 8. T474 Paperbound **$1.25**

SYMBOLIC LOGIC and THE GAME OF LOGIC, Lewis Carroll. "Symbolic Logic" is not concerned with modern symbolic logic, but is instead a collection of over 380 problems posed with charm and imagination, using the syllogism, and a fascinating diagrammatic method of drawing conclusions. In "The Game of Logic," Carroll's whimsical imagination devises a logical game played with 2 diagrams and counters (included) to manipulate hundreds of tricky syllogisms. The final section, "Hit or Miss" is a lagniappe of 101 additional puzzles in the delightful Carroll manner. Until this reprint edition, both of these books were rarities costing up to $15 each. Symbolic Logic: Index, xxxi + 199pp. The Game of Logic: 96pp. Two vols. bound as one. 5⅜ x 8. T492 Paperbound **$1.50**

PILLOW PROBLEMS and A TANGLED TALE, Lewis Carroll. One of the rarest of all Carroll's works, "Pillow Problems" contains 72 original math puzzles, all typically ingenious. Particularly fascinating are Carroll's answers which remain exactly as he thought them out, reflecting his actual mental processes. The problems in "A Tangled Tale" are in story form, originally appearing as a monthly magazine serial. Carroll not only gives the solutions, but uses answers sent in by readers to discuss wrong approaches and misleading paths, and grades them for insight. Both of these books were rarities until this edition, "Pillow Problems" costing up to $25, and "A Tangled Tale" $15. Pillow Problems: Preface and introduction by Lewis Carroll. xx + 109pp. A Tangled Tale: 6 illustrations. 152pp. Two vols. bound as one. 5⅜ x 8. T493 Paperbound **$1.50**

DIVERSIONS AND DIGRESSIONS OF LEWIS CARROLL. A major new treasure for Carroll fans! Rare privately published puzzles, mathematical amusements and recreations, games. Includes the fragmentary Part III of "Curiosa Mathematica." Also contains humorous and satirical pieces: "The New Belfry," "The Vision of the Three T's," and much more. New 32-page supplement of rare photographs taken by Carroll. Formerly titled "The Lewis Carroll Picture Book." Edited by S. Collingwood. x + 375pp. 5⅜ x 8. T732 Paperbound **$1.50**

THE BOOK OF MODERN PUZZLES, G. L. Kaufman. More than 150 word puzzles, logic puzzles. No warmed-over fare but all new material based on same appeals that make crosswords and deduction puzzles popular, but with different principles, techniques. Two-minute teasers, involved word-labyrinths, design and pattern puzzles, puzzles calling for logic and observation, puzzles testing ability to apply general knowledge to peculiar situations, many others. Answers to all problems. 116 illustrations. 192pp. 5⅜ x 8. T143 Paperbound **$1.00**

NEW WORD PUZZLES, Gerald L. Kaufman. Contains 100 brand new challenging puzzles based on words and their combinations, never published before in any form. Most are new types invented by the author—for beginners or experts. Chess word puzzles, addle letter anagrams, double word squares, double horizontals, alphagram puzzles, dual acrostigrams, linkogram lapwords—plus 8 other brand new types, all with solutions included. 196 figures. 100 brand new puzzles. vi + 122pp. 5⅜ x 8. T344 Paperbound **$1.00**

MATHEMATICAL RECREATIONS

MATHEMATICS, MAGIC AND MYSTERY, Martin Gardner. Card tricks, feats of mental mathematics, stage mind-reading, other "magic" explained as applications of probability, sets, theory of numbers, topology, various branches of mathematics. Creative examination of laws and their applications with scores of new tricks and insights. 115 sections discuss tricks wtih cards, dice, coins; geometrical vanishing tricks, dozens of others. No sleight of hand needed; mathematics guarantees success. 115 illustrations. xii + 174pp. 5⅜ x 8. T335 Paperbound **$1.00**

MATHEMATICAL EXCURSIONS, Helen A. Merrill. Fun, recreation, insights into elementary problem-solving. A mathematical expert guides you along by-paths not generally travelled in elementary math courses—how to divide by inspection, Russian peasant system of multiplication; memory systems for pi; building odd and even magic squares; dyadic systems; facts about 37; square roots by geometry; Tchebichev's machine; drawing five-sided figures; dozens more. Solutions to more difficult ones. 50 illustrations. 145pp. 5⅜ x 8. T350 Paperbound **$1.00**

CRYPTOGRAPHY, L. D. Smith. Excellent elementary introduction to enciphering, deciphering secret writing. Explains transposition, substitution ciphers; codes; solutions. Geometrical patterns, route transcription, columnar transposition, other methods. Mixed cipher systems; single-alphabet, polyalphabetical substitution; mechanical devices; Vigenere system, etc. Enciphering Japanese; explanation of Baconian Biliteral cipher; frequency tables. More than 150 problems provide practical application. Bibliography. Index. 164pp. 5⅜ x 8. T247 Paperbound **$1.00**

CRYPTANALYSIS, Helen F. Gaines. (Formerly ELEMENTARY CRYPTANALYSIS.) A standard elementary and intermediate text for serious students. It does not confine itself to old material, but contains much that is not generally known, except to experts. Concealment, Transposition, Substitution ciphers; Vigenere, Kasiski, Playfair, multafid, dozens of other techniques. Appendix with sequence charts, letter frequencies in English, 5 other languages, English word frequencies. Bibliography. 167 codes. New to this edition: solution to codes. vi + 230pp. 5⅜ x 8. T97 Paperbound **$1.95**

MAGIC SQUARES AND CUBES, W. S. Andrews. Only book-length treatment in English, a thorough non-technical description and analysis. Here are nasik, overlapping, pandiagonal, serrated squares; magic circles, cubes, spheres, rhombuses. Try your hand at 4-dimensional magical figures! Much unusual folklore and tradition included. High school algebra is sufficient. 754 diagrams and illustrations. viii + 419pp. 5⅜ x 8. T658 Paperbound **$1.85**

PAPER FOLDING FOR BEGINNERS, W. D. Murray and F. J. Rigney. A delightful introduction to the varied and entertaining Japanese art of origami (paper folding), with a full crystal-clear text that anticipates every difficulty; over 275 clearly labeled diagrams of all important stages in creation. You get results at each stage, since complex figures are logically developed from simpler ones. 43 different pieces are explained: place mats, drinking cups, bonbon boxes, sailboats, frogs, roosters, etc. 6 photographic plates. 279 diagrams. 95pp. 5⅝ x 8⅜. T713 Paperbound **$1.00**

CHESS, CHECKERS, GAMES, GO

A TREASURY OF CHESS LORE, edited by Fred Reinfeld. A delightful collection of anecdotes, short stories, aphorisms by and about the masters, poems, accounts of games and tournaments, photographs. Hundreds of humorous, pithy, satirical, wise, and historical episodes, comments, and word portraits. A fascinating "must" for chess players; revealing and perhaps seductive to those who wonder what their friends see in the game. 49 photographs (14 full page plates). 12 diagrams. xi + 306pp. 5⅜ x 8. T458 Paperbound **$1.75**

GO AND GO-MOKU: THE ORIENTAL BOARD GAMES, Edward Lasker. Best introduction to Go and its easier sister-game, Go-Moku—games new to Western world, but ancient in China, Japan. Extensively revised work by famed chess master Lasker, Go-player for over 50 years, stresses theory rather than brute memory, presents step-by-step explanation of strategy, gives examples of world championship matches, in game which has replaced chess as favorite of many physicists, mathematicians. 72 diagrams. xix + 215 pp. 5⅜ x 8.
T613 Paperbound **$1.45**

FICTION

FLATLAND, E. A. Abbott. A science-fiction classic of life in a 2-dimensional world that is also a first-rate introduction to such aspects of modern science as relativity and hyperspace. Political, moral, satirical, and humorous overtones have made FLATLAND fascinating reading for thousands. 7th edition. New introduction by Banesh Hoffmann. 16 illustrations. 128pp. 5⅜ x 8. T1 Paperbound **$1.00**

THE WONDERFUL WIZARD OF OZ, L. F. Baum. Only edition in print with all the original W. W. Denslow illustrations in full color—as much a part of "The Wizard" as Tenniel's drawings are of "Alice in Wonderland." "The Wizard" is still America's best-loved fairy tale, in which, as the author expresses it, "The wonderment and joy are retained and the heartaches and nightmares left out." Now today's young readers can enjoy every word and wonderful picture of the original book. New introduction by Martin Gardner. A Baum bibliography. 23 full-page color plates. viii + 268pp. 5⅜ x 8. T691 Paperbound **$1.45**

THE MARVELOUS LAND OF OZ, L. F. Baum. This is the equally enchanting sequel to the "Wizard," continuing the adventures of the Scarecrow and the Tin Woodman. The hero this time is a little boy named Tip, and all the delightful Oz magic is still present. This is the book with the Animated Saw-horse, the Woggle-Bug, and Jack Pumpkinhead. All the original John R. Neill illustrations, 16 in full color. 287pp. 5⅜ x 8. T692 Paperbound **$1.45**

FIVE GREAT DOG NOVELS, edited by Blanche Cirker. The complete original texts of five classic dog novels that have delighted and thrilled millions of children and adults throughout the world with stories of loyalty, adventure, and courage. Full texts of Jack London's "The Call of the Wild"; John Brown's "Rab and His Friends"; Alfred Ollivant's "Bob, Son of Battle"; Marshall Saunders' "Beautiful Joe"; and Ouida's "A Dog of Flanders." 21 illustrations from the original editions. 495pp. 5⅜ x 8. T777 Paperbound **$1.50**

3 ADVENTURE NOVELS by H. Rider Haggard. Complete texts of "She," "King Solomon's Mines," "Allan Quatermain." Qualities of discovery; desire for immortality; search for primitive, for what is unadorned by civilization, have kept these novels of African adventure exciting, alive to readers from R. L. Stevenson to George Orwell. 636pp. 5⅜ x 8.
T584 Paperbound **$2.00**

The Space Novels of Jules Verne

TO THE SUN? and OFF ON A COMET!, Jules Verne. Complete texts of two of the most imaginative flights into fancy in world literature display the high adventure that have kept Verne's novels read for nearly a century. Only unabridged edition of the best translation, by Edward Roth. Large, easily readable type. 50 illustrations selected from first editions. 462pp. 5⅜ x 8.
T634 Paperbound **$1.75**

FROM THE EARTH TO THE MOON and ALL AROUND THE MOON, Jules Verne. Complete editions of two of Verne's most successful novels, in finest Edward Roth translations, now available after many years out of print. Verne's visions of submarines, airplanes, television, rockets, interplanetary travel; of scientific and not-so-scientific beliefs; of peculiarities of Americans; all delight and engross us today as much as when they first appeared. Large, easily readable type. 42 illus. from first French edition. 476pp. 5⅜ x 8.
T633 Paperbound **$1.75**

THE CASTING AWAY OF MRS. LECKS AND MRS. ALESHINE, F. R. Stockton. A charming light novel by Frank Stockton, one of America's finest humorists (and author of "The Lady, or the Tiger?"). This book has made millions of Americans laugh at the reflection of themselves in two middle-aged American women involved in some of the strangest adventures on record. You will laugh, too, as they endure shipwreck, desert island, and blizzard with maddening tranquility. Also contains complete text of "The Dusantes," sequel to "The Casting Away." 49 original illustrations by F. D. Steele. vii + 142pp. 5⅜ x 8. T743 Paperbound **$1.00**

GESTA ROMANORUM, trans. by Charles Swan, ed. by Wynnard Hooper. 181 tales of Greeks, Romans, Britons, Biblical characters, comprise one of greatest medieval story collections, source plots for writers including Shakespeare, Chaucer, Gower, etc. Imaginative tales of wars, incest, thwarted love, magic, fantasy, allegory, humor, tell about kings, prostitutes, philosophers, fair damsels, knights, Noah, pirates, all walks and stations of life. Introduction. Notes. 500pp. 5⅜ x 8. T535 Paperbound **$1.85**

THREE PROPHETIC NOVELS BY H. G. WELLS, edited by E. F. Bleiler. Complete texts of "When the Sleeper Wakes" (1st book printing in 50 years), "A Story of the Days to Come," "The Time Machine" (1st complete printing in book form). Exciting adventures in the future are as enjoyable today as 50 years ago when first printed. Predict TV, movies, intercontinental airplanes, prefabricated houses, air-conditioned cities, etc. First important author to foresee problems of mind control, technological dictatorships. "Absolute best of imaginative fiction," N. Y. Times. Introduction. 335pp. 5⅜ x 8. T605 Paperbound **$1.45**

SEVEN SCIENCE FICTION NOVELS, H. G. Wells. Full unabridged texts of 7 science-fiction novels of the master. Ranging from biology, physics, chemistry, astronomy to sociology and other studies, Mr. Wells extrapolates whole worlds of strange and intriguing character. "One will have to go far to match this for entertainment, excitement, and sheer pleasure . . . ," NEW YORK TIMES. Contents: The Time Machine, The Island of Dr. Moreau, First Men in the Moon, The Invisible Man, The War of the Worlds, The Food of the Gods, In the Days of the Comet. 1015pp. 5⅜ x 8. T264 Clothbound **$3.95**

28 SCIENCE FICTION STORIES OF H. G. WELLS. Two full unabridged novels, MEN LIKE GODS and STAR BEGOTTEN, plus 26 short stories by the master science-fiction writer of all time. Stories of space, time, invention, exploration, future adventure—an indispensable part of the library of everyone interested in science and adventure. PARTIAL CONTENTS: Men Like Gods, The Country of the Blind, In the Abyss, The Crystal Egg, The Man Who Could Work Miracles, A Story of the Days to Come, The Valley of Spiders, and 21 more! 928pp. 5⅜ x 8. T265 Clothbound **$3.95**

DAVID HARUM, E. N. Westcott. This novel of one of the most lovable, humorous characters in American literature is a prime example of regional humor. It continues to delight people who like their humor dry, their characters quaint, and their plots ingenuous. First book edition to contain complete novel plus chapter found after author's death. Illustrations from first illustrated edition. 192pp. 5⅜ x 8. T580 Paperbound **$1.15**

HUMOR

THE WIT AND HUMOR OF OSCAR WILDE, ed. by Alvin Redman. Wilde at his most brilliant, in 1000 epigrams exposing weaknesses and hypocrisies of "civilized" society. Divided into 49 categories—sin, wealth, women, America, etc.—to aid writers, speakers. Includes excerpts from his trials, books, plays, criticism. Formerly "The Epigrams of Oscar Wilde." Introduction by Vyvyan Holland, Wilde's only living son. Introductory essay by editor. 260pp. 5⅜ x 8. T602 Paperbound **$1.00**

A NONSENSE ANTHOLOGY, collected by **Carolyn Wells.** 245 of the best nonsense verses ever written, including nonsense puns, absurd arguments, mock epics and sagas, nonsense ballads, odes, "sick" verses, dog-Latin verses, French nonsense verses, songs. By Edward Lear, Lewis Carroll, Gelett Burgess, W. S. Gilbert, Hilaire Belloc, Peter Newell, Oliver Herford, etc., 83 writers in all plus over four score anonymous nonsense verses. A special section of limericks, plus famous nonsense such as Carroll's "Jabberwocky" and Lear's "The Jumblies" and much excellent verse virtually impossible to locate elsewhere. For 50 years considered the best anthology available. Index of first lines specially prepared for this edition. Introduction by Carolyn Wells. 3 indexes: Title, Author, First lines. xxxiii + 279pp. 5⅜ x 8. T499 Paperbound **$1.25**

THE BAD CHILD'S BOOK OF BEASTS, MORE BEASTS FOR WORSE CHILDREN, and A MORAL ALPHABET, H. Belloc. Hardly an anthology of humorous verse has appeared in the last 50 years without at least a couple of these famous nonsense verses. But one must see the entire volumes—with all the delightful original illustrations by Sir Basil Blackwood—to appreciate fully Belloc's charming and witty verses that play so subacidly on the platitudes of life and morals that beset his day—and ours. A great humor classic. Three books in one. Total of 157pp. 5⅜ x 8. T749 Paperbound **$1.00**

THE DEVIL'S DICTIONARY, Ambrose Bierce. Sardonic and irreverent barbs puncturing the pomposities and absurdities of American politics, business, religion, literature, and arts, by the country's greatest satirist in the classic tradition. Epigrammatic as Shaw, piercing as Swift, American as Mark Twain, Will Rogers, and Fred Allen. Bierce will always remain the favorite of a small coterie of enthusiasts, and of writers and speakers whom he supplies with "some of the most gorgeous witticisms of the English language." (H. L. Mencken) Over 1000 entries in alphabetical order. 144pp. 5⅜ x 8. T487 Paperbound **$1.00**

THE PURPLE COW AND OTHER NONSENSE, Gelett Burgess. The best of Burgess's early nonsense, selected from the first edition of the "Burgess Nonsense Book." Contains many of his most unusual and highly original pieces: 37 nonsense quatrains, the Poems of Patagonia, Alphabet of Famous Goops, and the other hilarious (and rare) adult nonsense that places him in the forefront of American humorists. All pieces are accompanied by the original Burgess illustrations. 123 illustrations. xiii + 113pp. 5⅜ x 8.
T772 Paperbound **$1.00**

THE HUMOROUS VERSE OF LEWIS CARROLL. Almost every poem Carroll ever wrote, the largest collection ever published, including much never published elsewhere: 150 parodies, burlesques, riddles, ballads, acrostics, etc., with 130 original illustrations by Tenniel, Carroll, and others. "Addicts will be grateful . . . there is nothing for the faithful to do but sit down and fall to the banquet," N. Y. Times. Index to first lines. xiv + 446pp. 5 x 8.
T654 Paperbound **$1.85**

DIVERSIONS AND DIGRESSIONS OF LEWIS CARROLL. A major new treasure for Carroll fans! Rare privately published humor, fantasy, puzzles, and games by Carroll at his whimsical best, with a new vein of frank satire. Includes many new mathematical amusements and recreations, among them the fragmentary Part III of "Curiosa Mathematica." Contains "The Rectory Umbrella," "The New Belfry," "The Vision of the Three T's," and much more. New 32-page supplement of rare photographs taken by Carroll. x + 375pp. 5⅜ x 8.
T732 Paperbound **$1.50**

THE COMPLETE NONSENSE OF EDWARD LEAR. This is the only complete edition of this master of gentle madness available at a popular price. A BOOK OF NONSENSE, NONSENSE SONGS, MORE NONSENSE SONGS AND STORIES in their entirety with all the old favorites that have delighted children and adults for years. The Dong With A Luminous Nose, The Jumblies, The Owl and the Pussycat, and hundreds of other bits of wonderful nonsense. 214 limericks, 3 sets of Nonsense Botany, 5 Nonsense Alphabets. 546 drawings by Lear himself, and much more. 320pp. 5⅜ x 8. T167 Paperbound **$1.00**

PECK'S BAD BOY AND HIS PA, George W. Peck. The complete edition, containing both volumes, one of the most widely read of all American humor books. The endless ingenious pranks played by bad boy "Hennery" on his pa and the grocery man, the outraged pomposity of Pa, the perpetual ridiculing of middle class institutions, are as entertaining today as they were in 1883. No pale sophistications or subtleties, but rather humor vigorous, raw, earthy, imaginative, and, as folk humor often is, sadistic. This peculiarly fascinating book is also valuable to historians and students of American culture as a portrait of an age. 100 original illustrations by True Williams. Introduction by E. F. Bleiler. 347pp. 5⅜ x 8.
T497 Paperbound **$1.35**

FABLES IN SLANG & MORE FABLES IN SLANG, George Ade. 2 complete books of major American humorist in pungent colloquial tradition of Twain, Billings. 1st reprinting in over 30 years includes "The Two Mandolin Players and the Willing Performer," "The Base Ball Fan Who Took the Only Known Cure," "The Slim Girl Who Tried to Keep a Date that was Never Made," 42 other tales of eccentric, perverse, but always funny characters. "Touch of genius," H. L. Mencken. New introduction by E. F. Bleiler. 86 illus. 203pp. 5⅜ x 8.
T533 Paperbound **$1.00**

SINGULAR TRAVELS, CAMPAIGNS, AND ADVENTURES OF BARON MUNCHAUSEN, R. E. Raspe, with 90 illustrations by Gustave Doré. The first edition in over 150 years to reestablish the deeds of the Prince of Liars exactly as Raspe first recorded them in 1785—the genuine Baron Munchausen, one of the most popular personalities in English literature. Included also are the best of the many sequels, written by other hands. Introduction on Raspe by J. Carswell. Bibliography of early editions. xliv + 192pp. 5⅜ x 8. T698 Paperbound **$1.00**

HOW TO TELL THE BIRDS FROM THE FLOWERS, R. W. Wood. How not to confuse a carrot with a parrot, a grape with an ape, a puffin with nuffin. Delightful drawings, clever puns, absurd little poems point out farfetched resemblances in nature. The author was a leading physicist. Introduction by Margaret Wood White. 106 illus. 60pp. 5⅜ x 8.
T523 Paperbound **75¢**

MATHEMATICS, ELEMENTARY TO INTERMEDIATE

HOW TO CALCULATE QUICKLY, Henry Sticker. This handy volume offers a tried and true method for helping you in the basic mathematics of daily life—addition, subtraction, multiplication, division, fractions, etc. It is designed to awaken your "number sense" or the ability to see relationships between numbers as whole quantities. It is not a collection of tricks working only on special numbers, but a serious course of over 9,000 problems and their solutions, teaching special techniques not taught in schools: left-to-right multiplication, new fast ways of division, etc. 5 or 10 minutes daily use will double or triple your calculation speed. Excellent for the scientific worker who is at home in higher math, but is not satisfied with his speed and accuracy in lower mathematics. 256pp. 5 x 7¼. T295 Paperbound **$1.00**

TEACH YOURSELF books. For adult self-study, for refresher and supplementary study.

The most effective series of home study mathematics books on the market! With absolutely no outside help, they will teach you as much as any similar college or high-school course, or will helpfully supplement any such course. Each step leads directly to the next, each question is anticipated. Numerous lucid examples and carefully-wrought practice problems illustrate meanings. Not skimpy outlines, not surveys, not usual classroom texts, these 204- to 380-page books are packed with the finest instruction you'll find anywhere for adult self-study.

TEACH YOURSELF ALGEBRA, P. Abbott. Formulas, coordinates, factors, graphs of quadratic functions, quadratic equations, logarithms, ratio, irrational numbers, arithmetical, geometrical series, much more. 1241 problems, solutions. Tables. 52 illus. 307pp. 6⅞ x 4¼. Clothbound **$2.00**

TEACH YOURSELF GEOMETRY, P. Abbott. Solids, lines, points, surfaces, angle measurement, triangles, theorem of Pythagoras, polygons, loci, the circle, tangents, symmetry, solid geometry, prisms, pyramids, solids of revolution, etc. 343 problems, solutions. 268 illus. 334pp. 6⅞ x 4¼. Clothbound **$2.00**

TEACH YOURSELF TRIGONOMETRY, P. Abbott. Geometrical foundations, indices, logarithms, trigonometrical ratios, relations between sides, angles of triangle, circular measure, trig. ratios of angles of any magnitude, much more. Requires elementary algebra, geometry. 465 problems, solutions. Tables. 102 illus. 204pp. 6⅞ x 4¼. Clothbound **$2.00**

TEACH YOURSELF THE CALCULUS, P. Abbott. Variations in functions, differentiation, solids of revolution, series, elementary differential equations, areas by integral calculus, much more. Requires algebra, trigonometry. 970 problems, solutions. Tables. 89 illus. 380pp. 6⅞ x 4¼. Clothbound **$2.00**

TEACH YOURSELF THE SLIDE RULE, B. Snodgrass. Fractions, decimals, A-D scales, log-log scales, trigonometrical scales, indices, logarithms. Commercial, precision, electrical, dualistic, Brighton rules. 80 problems, solutions. 10 illus. 207pp. 6⅞ x 4¼. Clothbound **$2.00**

See also: **TEACH YOURSELF ELECTRICITY, C. W. Wilman; TEACH YOURSELF HEAT ENGINES, E. De Ville; TEACH YOURSELF MECHANICS, P. Abbott.**

* * *

HOW DO YOU USE A SLIDE RULE? by A. A. Merrill. Not a manual for mathematicians and engineers, but a lucid step-by-step explanation that presents the fundamental rules clearly enough to be understood by anyone who could benefit by the use of a slide rule in his work or business. This work concentrates on the 2 most important operations: multiplication and division. 10 easy lessons, each with a clear drawing, will save you countless hours in your banking, business, statistical, and other work. First publication. Index. 2 Appendixes. 10 illustrations. 78 problems, all with answers. vi + 36pp. 6⅛ x 9¼. T62 Paperbound **60¢**

THEORY OF OPERATION OF THE SLIDE RULE, J. P. Ellis. Not a skimpy "instruction manual", but an exhaustive treatment that will save you uncounted hours throughout your career. Supplies full understanding of every scale on the Log Log Duplex Decitrig type of slide rule. Shows the most time-saving methods, and provides practice useful in the widest variety of actual engineering situations. Each operation introduced in terms of underlying logarithmic theory. Summary of prerequisite math. First publication. Index. 198 figures. Over 450 problems with answers. Bibliography. 12 Appendices. ix + 289pp. 5⅜ x 8. S727 Paperbound **$1.50**

ARITHMETICAL EXCURSIONS: AN ENRICHMENT OF ELEMENTARY MATHEMATICS, H. Bowers and J. Bowers. For students who want unusual methods of arithmetic never taught in school; for adults who want to increase their number sense. Little known facts about the most simple numbers, arithmetical entertainments and puzzles, figurate numbers, number chains, mysteries and folklore of numbers, the "Hin-dog-abic" number system, etc. First publication. Index. 529 numbered problems and diversions, all with answers. Bibliography. 50 figures. xiv + 320pp. 5⅜ x 8. T770 Paperbound **$1.65**

APPLIED MATHEMATICS FOR RADIO AND COMMUNICATIONS ENGINEERS, C. E. Smith. No extraneous material here!—only the theories, equations, and operations essential and immediately useful for radio work. Can be used as refresher, as handbook of applications and tables, or as full home-study course. Ranges from simplest arithmetic through calculus, series, and wave forms, hyperbolic trigonometry, simultaneous equations in mesh circuits, etc. Supplies applications right along with each math topic discussed. 22 useful tables of functions, formulas, logs, etc. Index. 166 exercises, 140 examples, all with answers. 95 diagrams. Bibliography. x + 336pp. 5⅜ x 8. S141 Paperbound **$1.75**

HIGHER MATHEMATICS FOR STUDENTS OF CHEMISTRY AND PHYSICS, J. W. Mellor. Not abstract, but practical, building its problems out of familiar laboratory material, this covers differential calculus, coordinate, analytical geometry, functions, integral calculus, infinite series, numerical equations, differential equations, Fourier's theorem, probability, theory of errors, calculus of variations, determinants. "If the reader is not familiar with this book, it will repay him to examine it," CHEM. & ENGINEERING NEWS. 800 problems. 189 figures. Bibliography. xxi + 641pp. 5⅜ x 8. S193 Paperbound **$2.25**

TRIGONOMETRY REFRESHER FOR TECHNICAL MEN, A. Albert Klaf. 913 detailed questions and answers cover the most important aspects of plane and spherical trigonometry. They will help you to brush up or to clear up difficulties in special areas. The first portion of this book covers plane trigonometry, including angles, quadrants, trigonometrical functions, graphical representation, interpolation, equations, logarithms, solution of triangle, use of the slide rule and similar topics. 188 pages then discuss application of plane trigonometry to special problems in navigation, surveying, elasticity, architecture, and various fields of engineering. Small angles, periodic functions, vectors, polar coordinates, de Moivre's theorem are fully examined. The third section of the book then discusses spherical trigonometry and the solution of spherical triangles, with their applications to terrestrial and astronomical problems. Methods of saving time with numerical calculations, simplification of principal functions of angle, much practical information make this a most useful book. 913 questions answered. 1738 problems, answers to odd numbers. 494 figures. 24 pages of useful formulae, functions. Index. x + 629pp. 5⅜ x 8. T371 Paperbound **$2.00**

CALCULUS REFRESHER FOR TECHNICAL MEN, A. Albert Klaf. This book is unique in English as a refresher for engineers, technicians, students who either wish to brush up their calculus or to clear up uncertainties. It is not an ordinary text, but an examination of most important aspects of integral and differential calculus in terms of the 756 questions most likely to occur to the technical reader. The first part of this book covers simple differential calculus, with constants, variables, functions, increments, derivatives, differentiation, logarithms, curvature of curves, and similar topics. The second part covers fundamental ideas of integration, inspection, substitution, transformation, reduction, areas and volumes, mean value, successive and partial integration, double and triple integration. Practical aspects are stressed rather than theoretical. A 50-page section illustrates the application of calculus to specific problems of civil and nautical engineering, electricity, stress and strain, elasticity, industrial engineering, and similar fields.—756 questions answered. 566 problems, mostly answered. 36 pages of useful constants, formulae for ready reference. Index. v + 431pp. 5⅜ x 8. T370 Paperbound **$2.00**

TEXTBOOK OF ALGEBRA, G. Chrystal. One of the great mathematical textbooks, still about the best source for complete treatments of the topics of elementary algebra; a chief reference work for teachers and students of algebra in advanced high school and university courses, or for the mathematician working on problems of elementary algebra or looking for a background to more advanced topics. Ranges from basic laws and processes to extensive examination of such topics as limits, infinite series, general properties of integral numbers, and probability theory. Emphasis is on algebraic form, the foundation of analytical geometry and the key to modern developments in algebra. Prior course in algebra is desirable, but not absolutely necessary. Includes theory of quotients, distribution of products, arithmetical theory of surds, theory of interest, permutations and combinations, general expansion theorems, recurring fractions, and much, much more. Two volume set. Index in each volume. Over 1500 exercises, approximately half with answers. Total of xlviii + 1187pp. 5⅜ x 8.

S750 Vol I Paperbound **$2.35**
S751 Vol II Paperbound **$2.35**
The set **$4.70**

COLLEGE ALGEBRA, H. B. Fine. Standard college text that gives a systematic and deductive structure to algebra; comprehensive, connected, with emphasis on theory. Discusses the commutative, associative, and distributive laws of number in unusual detail, and goes on with undetermined coefficients, quadratic equations, progressions, logarithms, permutations, probability, power series, and much more. Still most valuable elementary-intermediate text on the science and structure of algebra. Index. 1560 problems, all with answers. x + 631pp. 5⅜ x 8. T211 Paperbound **$2.00**

THE CONTINUUM AND OTHER TYPES OF SERIAL ORDER, E. V. Huntington. This famous book gives a systematic elementary account of the modern theory of the continuum as a type of serial order. Based on the Cantor-Dedekind ordinal theory, which requires no technical knowledge of higher mathematics, it offers an easily followed analysis of ordered classes, discrete and dense series, continuous series, Cantor's transfinite numbers. 2nd edition. Index. viii + 82pp. 5⅜ x 8.

S129 Clothbound **$2.75**
S130 Paperbound **$1.00**

A TREATISE ON PLANE AND ADVANCED TRIGONOMETRY, E. W. Hobson. Extraordinarily wide coverage, going beyond usual college level trig, one of the few works covering advanced trig in full detail. By a great expositor with unerring anticipation and lucid clarification of potentially difficult points. Includes circular functions; expansion of functions of multiple angle; trig tables; relations between sides and angles of triangle; complex numbers; etc. Many problems solved completely. "The best work on the subject." Nature. Formerly entitled "A Treatise on Plane Trigonometry." 689 examples. 6 figures. xvi + 383pp. 5⅜ x 8.

S353 Paperbound **$1.95**

FAMOUS PROBLEMS OF ELEMENTARY GEOMETRY, Felix Klein. Expanded version of the 1894 Easter lectures at Göttingen. 3 problems of classical geometry, in an excellent mathematical treatment by a famous mathematician: squaring the circle, trisecting angle, doubling cube. Considered with full modern implications: transcendental numbers, pi, etc. Notes by R. Archibald. 16 figures. xi + 92pp. 5⅜ x 8. T348 Clothbound **$1.50**
T298 Paperbound **$1.00**

* * *

ELEMENTARY MATHEMATICS FROM AN ADVANCED STANDPOINT, Felix Klein.

This classic text is an outgrowth of Klein's famous integration and survey course at Göttingen. Using one field of mathematics to interpret, adjust, illuminate another, it covers basic topics in each area, illustrating its discussion with extensive analysis. It is especially valuable in considering areas of modern mathematics. "Makes the reader feel the inspiration of . . . a great mathematician, inspiring teacher . . . with deep insight into the foundations and interrelations," BULLETIN, AMERICAN MATHEMATICAL SOCIETY.

Vol. 1. ARITHMETIC, ALGEBRA, ANALYSIS. Introducing the concept of function immediately, it enlivens abstract discussion with graphical and geometrically perceptual methods. Partial contents: natural numbers, extension of the notion of number, special properties, complex numbers. Real equations with real unknowns, complex quantities. Logarithmic, exponential functions, goniometric functions, infinitesimal calculus. Transcendence of e and pi, theory of assemblages. Index. 125 figures. ix + 274pp . 5⅜ x 8. S150 Paperbound **$1.75**

Vol. 2. GEOMETRY. A comprehensive view which accompanies the space perception inherent in geometry with analytic formulas which facilitate precise formulation. Partial contents: Simplest geometric manifolds: line segment, Grassmann determinant principles, classification of configurations of space, derivative manifolds. Geometric transformations: affine transformations, projective, higher point transformations, theory of the imaginary. Systematic discussion of geometry and its foundations. Indexes. 141 illustrations. ix + 214pp. 5⅜ x 8.
S151 Paperbound **$1.75**

* * *

COORDINATE GEOMETRY, L. P. Eisenhart. Thorough, unified introduction. Unusual for advancing in dimension within each topic (treats together circle, sphere; polar coordinates, 3-dimensional coordinate systems; conic sections, quadric surfaces), affording exceptional insight into subject. Extensive use made of determinants, though no previous knowledge of them is assumed. Algebraic equations of 1st degree, 2 and 3 unknowns, carried further than usual in algebra courses. Over 500 exercises. Introduction. Appendix. Index. Bibliography. 43 illustrations. 310pp. 5⅜ x 8. S600 Paperbound **$1.65**

MONOGRAPHS ON TOPICS OF MODERN MATHEMATICS, edited by **J. W. A. Young.** Advanced mathematics for persons who haven't gone beyond or have forgotten high school algebra. 9 monographs on foundation of geometry, modern pure geometry, non-Euclidean geometry, fundamental propositions of algebra, algebraic equations, functions, calculus, theory of numbers, etc. Each monograph gives proofs of important results, and descriptions of leading methods, to provide wide coverage. New introduction by Prof. M. Kline, N. Y. University. 100 diagrams. xvi + 416pp. 6⅛ x 9¼. S289 Paperbound **$2.00**

MATHEMATICS, INTERMEDIATE TO ADVANCED

Geometry

THE FOUNDATIONS OF EUCLIDEAN GEOMETRY, H. G. Forder. The first rigorous account of Euclidean geometry, establishing propositions without recourse to empiricism, and without multiplying hypotheses. Corrects many traditional weaknesses of Euclidean proofs, and investigates the problems imposed on the axiom system by the discoveries of Bolya and Lobatchefsky. Some topics discussed are Classes and Relations; Axioms for Magnitudes; Congruence and Similarity; Algebra of Points; Hessenberg's Theorem; Continuity; Existence of Parallels; Reflections; Rotations; Isometries; etc. Invaluable for the light it throws on foundations of math. Lists: Axioms employed, Symbols, Constructions. 295pp. 5⅜ x 8.
S481 Paperbound **$2.00**

ADVANCED EUCLIDEAN GEOMETRY, R. A. Johnson. For years the standard textbook on advanced Euclidean geometry, requires only high school geometry and trigonometry. Explores in unusual detail and gives proofs of hundreds of relatively recent theorems and corollaries, many formerly available only in widely scattered journals. Covers tangent circles, the theorem of Miquel, symmedian point, pedal triangles and circles, the Brocard configuration, and much more. Formerly "Modern Geometry." Index. 107 diagrams. xiii + 319pp. 5⅜ x 8.
S669 Paperbound **$1.65**

NON-EUCLIDEAN GEOMETRY, Roberto Bonola. The standard coverage of non-Euclidean geometry. It examines from both a historical and mathematical point of view the geometries which have arisen from a study of Euclid's 5th postulate upon parallel lines. Also included are complete texts, translated, of Bolyai's THEORY OF ABSOLUTE SPACE, Lobachevsky's THEORY OF PARALLELS. 180 diagrams. 431pp. 5⅜ x 8. S27 Paperbound **$1.95**

ELEMENTS OF NON-EUCLIDEAN GEOMETRY, D. M. Y. Sommerville. Unique in proceeding step-by-step, in the manner of traditional geometry. Enables the student with only a good knowledge of high school algebra and geometry to grasp elementary hyperbolic, elliptic, analytic non-Euclidean geometries; space curvature and its philosophical implications; theory of radical axes; homothetic centres and systems of circles; parataxy and parallelism; absolute measure; Gauss' proof of the defect area theorem; geodesic representation; much more, all with exceptional clarity. 126 problems at chapter endings provide progressive practice and familiarity. 133 figures. Index. xvi + 274pp. 5⅜ x 8. S460 Paperbound **$1.50**

HIGHER GEOMETRY: AN INTRODUCTION TO ADVANCED METHODS IN ANALYTIC GEOMETRY, F. S. Woods. Exceptionally thorough study of concepts and methods of advanced algebraic geometry (as distinguished from differential geometry). Exhaustive treatment of 1-, 2-, 3-, and 4-dimensional coordinate systems, leading to n-dimensional geometry in an abstract sense. Covers projectivity, tetracyclical coordinates, contact transformation, pentaspherical coordinates, much more. Based on M.I.T. lectures, requires sound preparation in analytic geometry and some knowledge of determinants. Index. Over 350 exercises. References. 60 figures. x + 423pp. 5⅜ x 8. S737 Paperbound **$2.00**

ELEMENTS OF PROJECTIVE GEOMETRY, L. Cremona. Outstanding complete treatment of projective geometry by one of the foremost 19th century geometers. Detailed proofs of all fundamental principles, stress placed on the constructive aspects. Covers homology, law of duality, anharmonic ratios, theorems of Pascal and Brianchon, foci, polar reciprocal figures, etc. Only ordinary geometry necessary to understand this honored classic. Index. Over 150 fully worked out examples and problems. 252 diagrams. xx + 302pp. 5⅜ x 8. S668 Paperbound **$1.75**

A TREATISE ON THE DIFFERENTIAL GEOMETRY OF CURVES AND SURFACES, L. P. Eisenhart. Introductory treatise especially for the graduate student, for years a highly successful textbook. More detailed and concrete in approach than most more recent books. Covers space curves, osculating planes, moving axes, Gauss' method, the moving trihedral, geodesics, conformal representation, etc. Last section deals with deformation of surfaces, rectilinear congruences, cyclic systems, etc. Index. 683 problems. 30 diagrams. xii + 474pp. 5⅜ x 8. S667 Paperbound **$2.75**

A TREATISE ON ALGEBRAIC PLANE CURVES, J. L. Coolidge. Unabridged reprinting of one of few full coverages in English, offering detailed introduction to theory of algebraic plane curves and their relations to geometry and analysis. Treats topological properties, Riemann-Roch theorem, all aspects of wide variety of curves including real, covariant, polar, containing series of a given sort, elliptic, polygonal, rational, the pencil, two parameter nets, etc. This volume will enable the reader to appreciate the symbolic notation of Aronhold and Clebsch. Bibliography. Index. 17 illustrations. xxiv + 513pp. 5⅜ x 8. S543 Paperbound **$2.45**

AN INTRODUCTION TO THE GEOMETRY OF N DIMENSIONS, D. M. Y. Sommerville. An introduction presupposing no prior knowledge of the field, the only book in English devoted exclusively to higher dimensional geometry. Discusses fundamental ideas of incidence, parallelism, perpendicularity, angles between linear space; enumerative geometry; analytical geometry from projective and metric points of view; polytopes; elementary ideas in analysis situs; content of hyper-spacial figures. Bibliography. Index. 60 diagrams. 196pp. 5⅜ x 8. S494 Paperbound **$1.50**

GEOMETRY OF FOUR DIMENSIONS, H. P. Manning. Unique in English as a clear, concise introduction. Treatment is synthetic, and mostly Euclidean, although in hyperplanes and hyperspheres at infinity, non-Euclidean geometry is used. Historical introduction. Foundations of 4-dimensional geometry. Perpendicularity, simple angles. Angles of planes, higher order. Symmetry, order, motion, hyperpyramids, hypercones, hyperspheres; figures with parallel elements; volume, hypervolume in space; regular polyhedroids. Glossary. 70 figures. ix + 348pp. 5⅜ x 8. S182 Paperbound **$1.95**

ELEMENTARY CONCEPTS OF TOPOLOGY, P. Alexandroff. First English translation of the famous brief introduction to topology for the beginner or for the mathematician not undertaking extensive study. This unusually useful intuitive approach deals primarily with the concepts of complex, cycle, and homology, and is wholly consistent with current investigations. Ranges from basic concepts of set-theoretic topology to the concept of Betti groups. "Glowing example of harmony between intuition and thought," David Hilbert. Translated by A. E. Farley. Introduction by D. Hilbert. Index. 25 figures. 73pp. 5⅜ x 8. S747 Paperbound **$1.00**

THE WORKS OF ARCHIMEDES, edited by **T. L. Heath.** All the known works of the great Greek mathematician are contained in this one volume, including the recently discovered Method of Archimedes. Contains: On Sphere & Cylinder, Measurement of a Circle, Spirals, Conoids, Spheroids, etc. This is the definitive edition of the greatest mathematical intellect of the ancient world. 186-page study by Heath discusses Archimedes and the history of Greek mathematics. Bibliography. 563pp. 5⅜ x 8. S9 Paperbound **$2.00**

THE THIRTEEN BOOKS OF EUCLID'S ELEMENTS, edited by **Sir Thomas Heath.** Definitive edition of one of the very greatest classics of Western world. Complete English translation of Heiberg text, together with spurious Book XIV. Detailed 150-page introduction discussing aspects of Greek and Medieval mathematics. Euclid, texts, commentators, etc. Paralleling the text is an elaborate critical apparatus analyzing each definition, proposition, postulate, covering textual matters, mathematical analysis, commentators of all times, refutations, supports, extrapolations, etc. This is the FULL EUCLID. Unabridged reproduction of Cambridge U. 2nd edition. 3 volumes. Total of 995 figures, 1426pp. 5⅜ x 8.
S88,89,90, 3 volume set, paperbound **$6.00**

THE GEOMETRY OF RENE DESCARTES. With this book Descartes founded analytical geometry. Excellent Smith-Latham translation, plus original French text with Déscartes' own diagrams. Contains Problems the Construction of Which Requires Only Straight Lines and Circles; On the Nature of Curved Lines; On the Construction of Solid or Supersolid Problems. Notes. Diagrams. 258pp. 5⅜ x 8.
S68 Paperbound **$1.50**

See also: **FOUNDATIONS OF GEOMETRY, B. Russell; THE PHILOSOPHY OF SPACE AND TIME, H. Reichenbach; FAMOUS PROBLEMS OF ELEMENTARY GEOMETRY, F. Klein; MONOGRAPHS ON TOPICS OF MODERN MATHEMATICS, ed. by J. W. Young.**

Calculus and function theory, Fourier theory, real and complex functions, determinants

A COLLECTION OF MODERN MATHEMATICAL CLASSICS, edited by R. Bellman. 13 classic papers, complete in their original languages, by Hermite, Hardy and Littlewood, Tchebychef, Fejér, Fredholm, Fuchs, Hurwitz, Weyl, van der Pol, Birkhoff, Kellogg, von Neumann, and Hilbert. Each of these papers, collected here for the first time, triggered a burst of mathematical activity, providing useful new generalizations or stimulating fresh investigations. Topics discussed include classical analysis, periodic and almost periodic functions, analysis and number theory, integral equations, theory of approximation, non-linear differential equations, and functional analysis. Brief introductions and bibliographies to each paper. xii + 292pp. 6 x 9.
S730 Paperbound **$2.00**

MATHEMATICS OF MODERN ENGINEERING, E. G. Keller and R. E. Doherty. Written for the Advanced Course in Engineering of the General Electric Corporation, deals with the engineering use of determinants, tensors, the Heaviside operational calculus, dyadics, the calculus of variations, etc. Presents underlying principles fully, but purpose is to teach engineers to deal with modern engineering problems, and emphasis is on the perennial engineering attack of set-up and solve. Indexes. Over 185 figures and tables. Hundreds of exercises, problems, and worked-out examples. References. Two volume set. Total of xxxiii + 623pp. 5⅜ x 8.
S734 Vol I Paperbound **$1.65**
S735 Vol II Paperbound **$1.65**
The set **$3.30**

MATHEMATICAL METHODS FOR SCIENTISTS AND ENGINEERS, L. P. Smith. For scientists and engineers, as well as advanced math students. Full investigation of methods and practical description of conditions under which each should be used. Elements of real functions, differential and integral calculus, space geometry, theory of residues, vector and tensor analysis, series of Bessel functions, etc. Each method illustrated by completely-worked-out examples, mostly from scientific literature. 368 graded unsolved problems. 100 diagrams. x + 453pp. 5⅝ x 8⅜.
S220 Paperbound **$2.00**

THEORY OF FUNCTIONS AS APPLIED TO ENGINEERING PROBLEMS, edited by R. Rothe, F. Ollendorff, and K. Pohlhausen. A series of lectures given at the Berlin Institute of Technology that shows the specific applications of function theory in electrical and allied fields of engineering. Six lectures provide the elements of function theory in a simple and practical form, covering complex quantities and variables, integration in the complex plane, residue theorems, etc. Then 5 lectures show the exact uses of this powerful mathematical tool, with full discussions of problem methods. Index. Bibliography. 108 figures. x + 189pp. 5⅜ x 8.
S733 Paperbound **$1.35**

ADVANCED CALCULUS, E. B. Wilson. An unabridged reprinting of the work which continues to be recognized as one of the most comprehensive and useful texts in the field. It contains an immense amount of well-presented, fundamental material, including chapters on vector functions, ordinary differential equations, special functions, calculus of variations, etc., which are excellent introductions to these areas. For students with only one year of calculus, more than 1300 exercises cover both pure math and applications to engineering and physical problems. For engineers, physicists, etc., this work, with its 54 page introductory review, is the ideal reference and refresher. Index. ix + 566pp. 5⅜ x 8.
S504 Paperbound **$2.45**

ELLIPTIC INTEGRALS, H. Hancock. Invaluable in work involving differential equations containing cubics or quartics under the root sign, where elementary calculus methods are inadequate. Practical solutions to problems that occur in mathematics, engineering, physics: differential equations requiring integration of Lamé's, Briot's, or Bouquet's equations; determination of arc of ellipse, hyperbola, lemiscate; solutions of problems in elastica; motion of a projectile under resistance varying as the cube of the velocity; pendulums; many others. Exposition is in accordance with Legendre-Jacobi theory and includes rigorous discussion of Legendre transformations. 20 figures. 5 place table. Index. 104pp. 5⅛ x 8.
S484 Paperbound **$1.25**

FIVE VOLUME "THEORY OF FUNCTIONS' SET BY KONRAD KNOPP

This five-volume set, prepared by Konrad Knopp, provides a complete and readily followed account of theory of functions. Proofs are given concisely, yet without sacrifice of completeness or rigor. These volumes are used as texts by such universities as M.I.T., University of Chicago, N. Y. City College, and many others. "Excellent introduction . . . remarkably readable, concise, clear, rigorous," JOURNAL OF THE AMERICAN STATISTICAL ASSOCIATION.

ELEMENTS OF THE THEORY OF FUNCTIONS, Konrad Knopp. This book provides the student with background for further volumes in this set, or texts on a similar level. Partial contents: foundations, system of complex numbers and the Gaussian plane of numbers, Riemann sphere of numbers, mapping by linear functions, normal forms, the logarithm, the cyclometric functions and binomial series. "Not only for the young student, but also for the student who knows all about what is in it," MATHEMATICAL JOURNAL. Bibliography. Index. 140pp. 5⅜ x 8.
S154 Paperbound **$1.35**

THEORY OF FUNCTIONS, PART I, Konrad Knopp. With volume II, this book provides coverage of basic concepts and theorems. Partial contents: numbers and points, functions of a complex variable, integral of a continuous function, Cauchy's integral theorem, Cauchy's integral formulae, series with variable terms, expansion of analytic functions in power series, analytic continuation and complete definition of analytic functions, entire transcendental functions, Laurent expansion, types of singularities. Bibliography. Index. vii + 146pp. 5⅜ x 8.
S156 Paperbound **$1.35**

THEORY OF FUNCTIONS, PART II, Konrad Knopp. Application and further development of general theory, special topics. Single valued functions, entire, Weierstrass, Meromorphic functions. Riemann surfaces. Algebraic functions. Analytical configuration, Riemann surface. Bibliography. Index. x + 150pp. 5⅜ x 8.
S157 Paperbound **$1.35**

PROBLEM BOOK IN THE THEORY OF FUNCTIONS, VOLUME 1, Konrad Knopp. Problems in elementary theory, for use with Knopp's THEORY OF FUNCTIONS, or any other text, arranged according to increasing difficulty. Fundamental concepts, sequences of numbers and infinite series, complex variable, integral theorems, development in series, conformal mapping. 182 problems. Answers. viii + 126pp. 5⅜ x 8.
S158 Paperbound **$1.35**

PROBLEM BOOK IN THE THEORY OF FUNCTIONS, VOLUME 2, Konrad Knopp. Advanced theory of functions, to be used either with Knopp's THEORY OF FUNCTIONS, or any other comparable text. Singularities, entire & meromorphic functions, periodic, analytic, continuation, multiple-valued functions, Riemann surfaces, conformal mapping. Includes a section of additional elementary problems. "The difficult task of selecting from the immense material of the modern theory of functions the problems just within the reach of the beginner is here masterfully accomplished," AM. MATH. SOC. Answers. 138pp. 5⅜ x 8.
S159 Paperbound **$1.35**

* * *

LECTURES ON THE THEORY OF ELLIPTIC FUNCTIONS, H. Hancock. Reissue of the only book in English with so extensive a coverage, especially of Abel, Jacobi, Legendre, Weierstrasse, Hermite, Liouville, and Riemann. Unusual fullness of treatment, plus applications as well as theory, in discussing elliptic function (the universe of elliptic integrals originating in works of Abel and Jacobi), their existence, and ultimate meaning. Use is made of Riemann to provide the most general theory. 40 page table of formulas. 76 figures. xxiii + 498pp.
S483 Paperbound **$2.55**

THE THEORY AND FUNCTIONS OF A REAL VARIABLE AND THE THEORY OF FOURIER'S SERIES, E. W. Hobson. One of the best introductions to set theory and various aspects of functions and Fourier's series. Requires only a good background in calculus. Provides an exhaustive coverage of: metric and descriptive properties of sets of points; transfinite numbers and order types; functions of a real variable; the Riemann and Lebesgue integrals; sequences and series of numbers; power-series; functions representable by series sequences of continuous functions; trigonometrical series; representation of functions by Fourier's series; complete exposition (200pp.) on set theory; and much more. "The best possible guide," Nature. Vol. I: 88 detailed examples, 10 figures. Index. xv + 736pp. Vol. II: 117 detailed examples, 13 figures. Index. x + 780pp. 6⅛ x 9¼.
Vol. I: S387 Paperbound **$3.00**
Vol. II: S388 Paperbound **$3.00**

ALMOST PERIODIC FUNCTIONS, A. S. Besicovitch. This unique and important summary by a well-known mathematician covers in detail the two stages of development in Bohr's theory of almost periodic functions: (1) as a generalization of pure periodicity, with results and proofs; (2) the work done by Stepanoff, Wiener, Weyl, and Bohr in generalizing the theory. Bibliography. xi + 180pp. 5⅜ x 8.
S18 Paperbound **$1.75**

THE TAYLOR SERIES, AN INTRODUCTION TO THE THEORY OF FUNCTIONS OF A COMPLEX VARIABLE, P. Dienes. This book investigates the entire realm of analytic functions. Only ordinary calculus is needed, except in the last two chapters. Starting with an introduction to real variables and complex algebra, the properties of infinite series, elementary functions, complex differentiation and integration are carefully derived. Also biuniform mapping, a thorough two part discussion of representation and singularities of analytic functions, overconvergence and gap theorems, divergent series, Taylor series on its circle of convergence, divergence and singularities, etc. Unabridged, corrected reissue of first edition. Preface and index. 186 examples, many fully worked out. 67 figures. xii + 555pp. 5⅜ x 8.
S391 Paperbound **$2.75**

INTRODUCTION TO BESSEL FUNCTIONS, Frank Bowman. A rigorous self-contained exposition providing all necessary material during the development, which requires only some knowledge of calculus and acquaintance with differential equations. A balanced presentation including applications and practical use. Discusses Bessel Functions of Zero Order, of Any Real Order; Modified Bessel Functions of Zero Order; Definite Integrals; Asymptotic Expansions; Bessel's Solution to Kepler's Problem; Circular Membranes; much more. "Clear and straightforward . . . useful not only to students of physics and engineering, but to mathematical students in general," Nature. 226 problems. Short tables of Bessel functions. 27 figures. Index. x + 135pp. 5⅜ x 8. S462 Paperbound **$1.35**

MODERN THEORIES OF INTEGRATION, H. Kestelman. Connected and concrete coverage, with fully-worked-out proofs for every step. Ranges from elementary definitions through theory of aggregates, sets of points, Riemann and Lebesgue integration, and much more. This new revised and enlarged edition contains a new chapter on Riemann-Stieltjes integration, as well as a supplementary section of 186 exercises. Ideal for the mathematician, student, teacher, or self-studier. Index of Definitions and Symbols. General Index. Bibliography. x + 310pp. 5⅝ x 8⅜.
S572 Paperbound **$2.00**

A TREATISE ON THE THEORY OF DETERMINANTS, T. Muir. Unequalled as an exhaustive compilation of nearly all the known facts about determinants up to the early 1930's. Covers notation and general properties, row and column transformation, symmetry, compound determinants, adjugates, rectangular arrays and matrices, linear dependence, gradients, Jacobians, Hessians, Wronskians, and much more. Invaluable for libraries of industrial and research organizations as well as for student, teacher, and mathematician; very useful in the field of computing machines. Revised and enlarged by W. H. Metzler. Index. 485 problems and scores of numerical examples. iv + 766pp. 5⅜ x 8. S670 Paperbound **$2.95**

THEORY OF DETERMINANTS IN THE HISTORICAL ORDER OF DEVELOPMENT, Sir Thomas Muir. Unabridged reprinting of this complete study of 1,859 papers on determinant theory written between 1693 and 1900. Most important and original sections reproduced, valuable commentary on each. No other work is necessary for determinant research: all types are covered—each subdivision of the theory treated separately; all papers dealing with each type are covered; you are told exactly what each paper is about and how important its contribution is. Each result, theory, extension, or modification is assigned its own identifying numeral so that the full history may be more easily followed. Includes papers on determinants in general, determinants and linear equations, symmetric determinants, alternants, recurrents, determinants having invariant factors, and all other major types. "A model of what such histories ought to be," NATURE. "Mathematicians must ever be grateful to Sir Thomas for his monumental work," AMERICAN MATH MONTHLY. Four volumes bound as two. Indices. Bibliographies. Total of lxxxiv + 1977pp. 5⅜ x 8. S672-3 The set, Clothbound **$10.00**

A COURSE IN MATHEMATICAL ANALYSIS, Edouard Goursat. Trans. by E. R. Hedrick, O. Dunkel. Classic study of fundamental material thoroughly treated. Exceptionally lucid exposition of wide range of subject matter for student with 1 year of calculus. Vol. 1: Derivatives and Differentials, Definite Integrals, Expansion in Series, Applications to Geometry. Problems. Index. 52 illus. 556pp. Vol. 2, Part I: Functions of a Complex Variable, Conformal Representations, Doubly Periodic Functions, Natural Boundaries, etc. Problems. Index. 38 illus. 269pp. Vol. 2, Part 2: Differential Equations, Cauchy-Lipschitz Method, Non-linear Differential Equations, Simultaneous Equations, etc. Problems. Index. 308pp. 5⅜ x 8.
Vol. 1 S554 Paperbound **$2.25**
Vol. 2 part 1 S555 Paperbound **$1.65**
Vol. 2 part 2 S556 Paperbound **$1.65**
3 vol. set **$5.00**

INFINITE SEQUENCES AND SERIES, Konrad Knopp. First publication in any language! Excellent introduction to 2 topics of modern mathematics, designed to give the student background to penetrate farther by himself. Sequences & sets, real & complex numbers, etc. Functions of a real & complex variable. Sequences & series. Infinite series. Convergent power series. Expansion of elementary functions. Numerical evaluation of series. Bibliography. v + 186pp. 5⅜ x 8. S152 Clothbound **$3.50**
S153 Paperbound **$1.75**

TRIGONOMETRICAL SERIES, Antoni Zygmund. Unique in any language on modern advanced level. Contains carefully organized analyses of trigonometric, orthogonal, Fourier systems of functions, with clear adequate descriptions of summability of Fourier series, proximation theory, conjugate series, convergence, divergence of Fourier series. Especially valuable for Russian, Eastern European coverage. Bibliography. 329pp. 5⅜ x 8. S290 Paperbound **$1.50**

COLLECTED WORKS OF BERNHARD RIEMANN. This important source book is the first to contain the complete text of both 1892 Werke and the 1902 supplement, unabridged. It contains 31 monographs, 3 complete lecture courses, 15 miscellaneous papers, which have been of enormous importance in relativity, topology, theory of complex variables, and other areas of mathematics. Edited by R. Dedekind, H. Weber, M. Noether, W. Wirtinger. German text. English introduction by Hans Lewy. 690pp. 5⅜ x 8. S226 Paperbound **$2.85**

See also: **A HISTORY OF THE CALCULUS, C. B. Boyer; CALCULUS REFRESHER FOR TECHNICAL MEN, A. A. Klaf; MONOGRAPHS ON TOPICS OF MODERN MATHEMATICS, ed. by J. W. A. Young; THE CONTINUUM AND OTHER TYPES OF SERIAL ORDER, E. V. Huntington.**

Symbolic logic

AN INTRODUCTION TO SYMBOLIC LOGIC, Susanne K. Langer. Probably the clearest book ever written on symbolic logic for the philosopher, general scientist and layman. It will be particularly appreciated by those who have been rebuffed by other introductory works because of insufficient mathematical training. No special knowledge of mathematics is required. Starting with the simplest symbols and conventions, you are led to a remarkable grasp of the Boole-Schroeder and Russell-Whitehead systems clearly and quickly. PARTIAL CONTENTS: Study of forms, Essentials of logical structure, Generalization, Classes, The deductive system of classes, The algebra of logic, Abstraction of interpretation, Calculus of propositions, Assumptions of PRINCIPIA MATHEMATICA, Logistics, Logic of the syllogism, Proofs of theorems. "One of the clearest and simplest introductions to a subject which is very much alive. The style is easy, symbolism is introduced gradually, and the intelligent non-mathematician should have no difficulty in following the argument," MATHEMATICS GAZETTE. Revised, expanded second edition. Truth-value tables. 368pp. 5⅜ x 8. S164 Paperbound **$1.75**

THE ELEMENTS OF MATHEMATICAL LOGIC, Paul Rosenbloom. First publication in any language. This book is intended for readers who are mature mathematically, but have no previous training in symbolic logic. It does not limit itself to a single system, but covers the field as a whole. It is a development of lectures given at Lund University, Sweden, in 1948. Partial contents: Logic of classes, fundamental theorems, Boolean algebra, logic of propositions, logic of propositional functions, expressive languages, combinatory logics, development of mathematics within an object language, paradoxes, theorems of Post and Goedel, Church's theorem, and similar topics. iv + 214pp. 5⅜ x 8. S227 Paperbound **$1.45**

A SURVEY OF SYMBOLIC LOGIC: THE CLASSIC ALGEBRA OF LOGIC, C. I. Lewis. Classic survey of the field, comprehensive and thorough. Indicates content of major systems, alternative methods of procedure, and relation of these to the Boole-Schroeder algebra and to one another. Contains historical summary, as well as full proofs and applications of the classic, or Boole-Schroeder, algebra of logic. Discusses diagrams for the logical relations of classes, the two-valued algebra, propositional functions of two or more variables, etc. Chapters 5 and 6 of the original edition, which contained material not directly pertinent, have been omitted in this edition at the author's request. Appendix. Bibliography. Index. viii + 352pp. 5⅝ x 8⅜. S643 Paperbound **$2.00**

INTRODUCTION TO SYMBOLIC LOGIC AND ITS APPLICATIONS, R. Carnap. One of the clearest, most comprehensive, and rigorous introductions to modern symbolic logic by perhaps its greatest living master. Symbolic languages are analyzed and one constructed. Applications to math (symbolic representation of axiom systems for set theory, natural numbers, real numbers, topology, Dedekind and Cantor explanations of continuity), physics (the general analysis of concepts of determination, causality, space-time-topology, based on Einstein), biology (symbolic representation of an axiom system for basic concepts). "A masterpiece," Zentralblatt für Mathematik und ihre Grenzgebiete. Over 300 exercises. 5 figures. Bibliography. Index. xvi + 241pp. 5⅜ x 8. S453 Paperbound **$1.85**
Clothbound **$4.00**